Light and Heavy
Vehicle Technology

When theory and practice do not agree, you should examine the facts to see what is wrong with the theory
Charles F. Kettering, General Motors (1934)

Light and Heavy Vehicle Technology

Fourth edition

M.J. Nunney

CGIA, MSAE, MIMI

Amsterdam • Boston • Heidelberg • London • New York • Oxford
Paris • San Diego • San Francisco • Singapore • Sydney • Tokyo
Butterworth-Heinemann is an imprint of Elsevier

Butterworth-Heinemann is an imprint of Elsevier
Linacre House, Jordan Hill, Oxford OX2 8DP
30 Corporate Drive, Suite 400, Burlington, MA 01803

First published 1988
Reprinted 1991
Second edition 1992
Third edition 1998
Reprinted 1998, 2000, 2001, 2002, 2003, 2004, 2005
Fourth edition 2007

Notice
No responsibility is assumed by the publisher for any injury and/or damage to persons
or property as a matter of products, liability, negligence or otherwise, or from any
use or operation of any methods, products, instructions or ideas contained in
the material herein. Because of rapid advances in the medical sciences, in particular,
independent verification of diagnoses and drug dosages should be made.

British Library Cataloguing in Publication Data
A catalogue record for this book is available from the British Library

Library of Congress Cataloging-in-Publication Data
A catalog record for this book is available from the Library of Congress

ISBN-13: 978-0-7506-8037-0
ISBN-10: 0-7506-8037-7

For information on all Butterworth-Heinemann publications
visit our web site at http://books.elsevier.com

Typeset by Charon Tec Ltd (A Macmillan Company), Chennai, India
www.charontec.com
Printed and bound in the UK

06 07 08 09 10 10 9 8 7 6 5 4 3 2 1

Contents

Preface

The purpose of this new fourth edition of *Light and Heavy Vehicle Technology* remains one of providing readily accessible information, which bridges the gap between the purely basic and the more advanced treatments of the subject. By understanding the reasons behind the design, construction and operation of the many and varied components of modern motor vehicles, the technician should be better equipped to deal with their servicing and overhaul. Some references to past automotive practice have been retained, not only because a technician may still be required to test and repair older vehicles, but also to provide a convenient transition to later practice.

Two entirely new sections of the book provide a topical introduction to alternative power sources and fuels, and battery-electric, hybrid and fuel-cell vehicles. Also, the number of entries in the list of automotive technical abbreviations has now increased to over 200. Finally, as in previous editions of the book, the tradition of including brief historical notes on the development of modern automotive concepts has been continued.

M.J. Nunney

Acknowledgements

The illustrations have been chosen for their special relevance to the text and, apart from those originating from the author, grateful acknowledgement is due not only to the publishers for allowing the use of illustrations from certain of their technical books, but also to the following firms and organizations who so kindly supplied the remaining illustrations and much useful background information:

AE Piston Products Ltd, Alfa-Romeo (GB) Ltd, Alpha Automotive Productions Ltd, Automotive Products Ltd, Bainbridge Silencers Ltd, Bendix Ltd, BL-MG, Robert Bosch Ltd, John Bradshaw Vehicles, Bridgestone Tyre UK Ltd, British Rubber Manufacturers' Association Ltd, Brockhouse Transmissions Ltd, Brown Brothers Ltd, Burman and Sons Ltd, Castrol (UK) Ltd, Champion Sparking Plug Co., Chillcotts Ltd (Reinz/Mann), Citroen UK Ltd, The Colt Car Co. Ltd (Mitsubishi), Con-Vel Division of Dana Corp., Coopers Payen Ltd, Cummins Engine Co. Ltd, David Brown Gear Industries Ltd, Davies Magnet, Deutz Engines Ltd, Dunlop Automotive Division, Eaton Ltd, FAG Kugelfischer (Germany), Ferodo Ltd, Fiat Auto (UK) Ltd, Ford of Europe Inc., Frigoblock (UK) Ltd, L. Gardner & Sons Ltd, General Motors Ltd (AC Delco) (Powertrain Lansing), GKN (Hardy Spicer) (Kent Alloys) (Laycock) (Pistons) (Salisbury Transmissions) (SDF) (Tadchurch), Glacier Metal Co. Ltd, Gleason Works, USA, Goodyear Tyre & Rubber Co., Haldex Brake Products, Sweden, Hendrickson Norde, Honda (UK) Ltd, Holset Engineering Co. Ltd, Hope Technical Developments Ltd, Interlube (Tecalemit UK), Jacobs Europe, Jaguar Cars Ltd, Johnson Matthey (Catalytic Systems Division), Lada Cars, Lancia (Fiat), Laystall Engineering Co. Ltd, Lipe-Rollway Ltd, Lucas Ltd (CAV) (Diesel Systems) (Electrical) (Girling) (Kienzle), Mazda Cars (UK) Ltd, Mercedes-Benz (UK) Ltd, Midcyl Productions, Mintex Ltd, Nissan Motors (GB) Ltd, Nissan (UK) (Europe), NSK-RHP, Perkins Engines Ltd, Pirelli Ltd, Pressed Steel Fisher, Renault (UK) Ltd, Renold Automotive (France) (Power Transmissions Ltd), RHP Ltd, Rover Cars, Rubery Owen-Rockwell, SAAB (GB) Ltd, Sachs Automotive Components Ltd, Scania (Great Britain) Ltd, Schrader Automotive Products Division, Seddon Atkinson Vehicles Ltd and International Harvester, Self-Changing Gears Ltd, SKF (UK) Ltd and 'Ball Bearing Journal', Smallman Lubricants Ltd, Smiths Industries Ltd, Start Pilot Ltd, Subaru (UK) Ltd, SU Butec, Suzuki GB (Cars) Ltd, Telma Retarder Ltd, Alfred Teves GMBH, The Timken Co, Torotrak (Development) Ltd, Toyota (GB) Ltd, TRW Cam Gears Ltd, Steering Systems Ltd and Valves Ltd, Turner-Spicer, Valeo Clutches Ltd, Vandervell Ltd, VL Churchill Ltd, Volkswagen (GB) Ltd (Audi), Volvo Concessionaires Ltd and Volvo Trucks, Yamaha-Mitsui Machinery Sales (UK) Ltd, The Zenith Carburettor Co. Ltd, ZF Group, UK.

My thanks to the Rolls-Royce Heritage Trust for their permission to use the Dr Stanley Hooker quote on page 6 General note – the ever-increasing sophistication in the design and construction of modern passenger cars and commercial vehicles, makes it more than ever essential for service personnel to consult the vehicle manufacturer for up-to-date technical information and adjustments data in relation to a particular model, both in the interests of vehicle safety and customer satisfaction.

Automotive technical abbreviations

ABC	Active body control (Mercedes-Benz)
ABS	Anti-Blockier-System (German) – anti-lock braking system
ABDC	After bottom dead centre (engine timing)
AC	Alternating current
A/C	Air conditioning
ACL	Automatic chassis lubrication (commercial vehicles)
ACT	Air charge temperature
A/F	Air/fuel ratio
AIR	Air injection reactor (emission control)
ALB	Anti-lock brakes (Honda)
ARCS	Active roll control system (Citroen)
ASD	Automatic slip-control differential
ASF	Audi space frame (aluminium body construction)
ASR	Antriebs-Schlupf-Regelung (German) – anti-slip regulation or traction control
ATC	Automatic temperature control
ATDC	After top dead centre (engine timing)
ATF	Automatic transmission fluid
AWD	All-wheel drive (also 4WD)
AWS	All-wheel steering (also 4WS)
BAS	Brake assist system
BBDC	Before bottom dead centre (engine timing)
BDC	Bottom dead centre (engine timing)
BEV	Battery-electric vehicle
BHP	Brake horsepower
BMEP	Brake mean effective pressure
BOFT	Bearing oil film thickness
BSFC	Brake specific fuel consumption
BTDC	Before top dead centre (engine timing)
CAD	Computer aided design
CAFE	Corporate average fuel economy (American)
CAG	Computer aided gearshift (Scania)
CATS	Computer active technology suspension (Jaguar)
CB	Contact-breaker
CBE	Cab behind engine (commercial vehicles)
C_d	Coefficient of drag (vehicle aerodynamics)
CD	Capacity discharge (ignition system)
CFC	Chlorofluorocarbon (refrigerant)
CGI	Compact graphite iron
CI	Compression ignition (diesel engines)
CN	Cetane number (diesel fuel ignition rating)
CNG	Compressed natural gas (fuels)
CO	Carbon monoxide (emission control)
CO_2	Carbon dioxide (global warming)
COE	Cab over engine (commercial vehicles)
CP	Centre of pressure (vehicle aerodynamics)
CR	Compression ratio (engine)
CRS	Common rail system (diesel fuel injection)
CTX	Continuously variable transaxle (Ford)
CV	Constant velocity (universal joints)
CVT	Continuously variable transmission

C_w	Coefficient of drag (German) – vehicle aerodynamics
C_x	Coefficient of drag (French) – vehicle aerodynamics
DC	Direct current
DERV	Diesel engine road vehicles (fuel)
DI	Direct injection
DIS	Direct ignition system (no distributor)
DISI	Direct injection spark ignition
DOHC	Double overhead camshafts
DRP	Dynamic rear proportioning (brakes)
DSC	Dynamic stability control
DSG	Direct shift gearbox (Volkswagen group)
DWB	Double wishbone suspension
EBA	Emergency brake assist
EBFD	Electronic brake force distribution
EBS	Electronic braking system (air brakes)
ECI	Electronically controlled injection
ECM	Electronic control module
ECS	Evaporative control system (fuel system) Electronically controlled suspension
ECT	Engine coolant temperature
ECU	Electronic control unit
EDC	Electronic diesel control
EFI	Electronic fuel injection
EGR	Exhaust gas recirculation (emission control)
ELV	End-of-life vehicle (materials recycling)
EMS	Engine management system
EP	Extreme pressure (lubricants)
EPAS	Electrical power-assisted steering (NSK-RHP)
EPHS	Electrically powered hydraulic steering (TRW)
EPS	Electric power steering
ESP	Electronic stability programme
ETC	Electronic traction control
ETS	Enhanced traction system (General Motors)
EUI	Electronic unit injector (Lucas Diesel)
EVC	Exhaust valve closed (engine timing)
EVO	Exhaust valve open (engine timing)
FCEV	Fuel-cell electric vehicle
FHP	Friction horsepower
FWD	Front-wheel drive
GCW	Gross combination weight (articulated vehicles)
GCWR	Gross combined weight rating (vehicle and trailer)
GDI	Gasoline direct injection (Mitsubishi)
GRP	Glass reinforced plastics
GTW	Gross train weight (drawbar vehicles)
GV	Governor valve (automatic transmissions)
GVW	Gross vehicle weight (rigid vehicles)
GVWR	Gross vehicle weight rating
GWP	Greenhouse warming potential (refrigerants)
HC	Hydrocarbons (emission control)
HDC	Hill descent control (ABS system)
HEV	Hybrid-electric vehicle

HFC	Hydrofluorocarbon (refrigerant)
HGV	Heavy goods vehicle
HT	High tension
HUCR	Highest useful compression ratio
HVAC	Heating, ventilation and air conditioning
IFS	Independent front suspension
IHP	Indicated horsepower
INJ	Injection (timing mark)
IOE	Inlet over exhaust (obsolete valve layout)
IPM	Integrated power module (hybrid electric vehicles)
IRS	Independent rear suspension
IVC	Inlet valve closed (valve timing)
IVO	Inlet valve open (valve timing)
KPI	King-pin inclination (steering)
LCV	Light commercial vehicle
LGV	Large goods vehicle
LI	Load index (tyres)
LNG	Liquefied natural gas (fuels)
LPG	Liquid petroleum gas (fuels)
LS	Leading shoe (drum brakes)
LSD	Limited slip differential
MAF	Mass air flow (engines)
MAP	Manifold absolute pressure
MOFT	Minimum oil film thickness
MON	Motor octane number (more demanding ON test)
MPI	Multi-point injection
MPV	Multi-purpose vehicle (people carrier)
NO	Nitrogen oxides (emission control)
NOAT	Nitrite organic acid technology (coolants)
NVH	Noise, vibration and harshness (vehicle refinement testing)
OAT	Organic acid technology (coolants)
OBD	On-board diagnosis
OD	Overdrive
ODP	Ozone depletion potential (refrigerants)
OHC	Overhead camshaft
OHV	Overhead valves
ON	Octane number (petrol anti-knock rating)
PAS	Power-assisted steering
PBD	Polybutadiene (tyres)
PCM	Power train control module (engine and transmission)
PCV	Positive crankcase ventilation (emission control)
PEM	Polymer electrolyte membrane (fuel cells) (or proton exchange membrane)
PFI	Port fuel injection (petrol engines)
PM	Particulate matter (diesel emission control)
PR	Ply-rating (tyres)
PSV	Public service vehicle
PTFE	Polytetrafluoroethylene
PTO	Power take-off (commercial vehicles)
PVC	Polyvinyl chloride
PZEV	Partial zero emission vehicle
RC	Roll-centre (suspension geometry)
RON	Research octane number (less demanding ON test)
RTV	Room temperature vulcanizing (sealant)

RWD	Rear-wheel drive
SAMT	Semi-automated mechanical transmission (Eaton)
SBC	Stand-by-control (electronic transmission control ZF)
SBR	Styrene-butadiene rubber (tyres)
SCA	Supplemental coolant additives
SCR	Selective catalytic reduction (emission control)
SCS	Stop control system (Girling)
SEFI	Sequential electronically controlled fuel injection (Ford)
SFC	Specific fuel consumption
SFI	Sequential fuel injection
SG	Spheroidal graphite (high-strength cast iron)
SI	Spark ignition (petrol engines)
SLA	Short and long arm (American) – suspension linkage
SOHC	Single overhead camshaft
SPI	Single point injection (petrol engines)
SRS	Supplemental restraint system (airbags)
SUV	Sports utility vehicle
SV	Side valves (obsolete valve layout)
TAC	Thermostatic air cleaner
TBI	Throttle body injection (SPI)
TC	Twin carburettors
TCI	Transistorized coil ignition
TCM	Transmission control module
TCS	Transmission controlled spark (engine intervention system)
TDC	Top dead centre (engine timing)
TDI	Turbocharged direct injection (diesel engines)
TEL	Tetra ethyl lead (petrol anti-knock additive)
TML	Tetra methyl lead (as above)
TPS	Throttle position sensor
TS	Trailing shoe (drum brakes)
TV	Throttle valve (engine and automatic transmissions)
TVS	Thermal vacuum switch (exhaust gas recirculation)
TWC	Three-way catalyst (emission control)
TXV	Thermostatic expansion valve (refrigeration)
UJ	Universal joint
ULEV	Ultra-low emission vehicle
VCP	Variable cam phasing (valve timing)
VCU	Viscous coupling unit (transmission)
VDC	Vehicle dynamics control (Bosch)
VGT	Variable-geometry turbocharger
VI	Viscosity index (lubricants)
VIP	Vehicle intrustion protection (Toyota)
VIVT	Variable inlet valve timing
VKPI	Virtual king-pin inclination (steering)
VSC	Vehicle skid control
VTG	Variable turbine geometry (turbocharging)
VTT	Variable twin turbo (turbocharging)
VVT	Variable valve timing
VVTL	Variable valve timing and lift
WOT	Wide-open throttle
ZEV	Zero emission vehicle

1 The reciprocating piston petrol engine

1.1 MODERN REQUIREMENTS

General background

The motor vehicle engine is basically a device for converting the internal energy stored in its fuel into mechanical energy. It is classified as an internal combustion engine by virtue of this energy conversion taking place within the engine cylinders.

Since the term 'energy' implies the capacity to perform work, the engine is thus able to propel the vehicle along the road and, within limits, overcome unwanted opposition to its motion arising from rolling friction, gradient resistance and air drag. To facilitate this process the engine is combined with a transmission system, the functioning of which is discussed later.

The vast majority of car engines are of the reciprocating piston type and utilize spark ignition to initiate the combustion process in the cylinders. However, the compression ignition or diesel principle to initiate combustion is increasingly challenging the petrol engine for car applications, especially in Europe. Both petrol and diesel engines operate on the four-stroke principle in which the piston travels one complete stroke for each of the successive events of induction, compression, combustion and exhaust.

The late Laurence Pomeroy, a distinguished motoring historian, once summarized the early history of the motor car as follows: From 1885 to 1895 men struggled to make the car go. From 1896 to 1905 they contrived to make it go properly. Between 1907 and 1915 they succeeded in making it go beautifully! What then are the requirements for the engine of the modern passenger car as reflected in many decades of further development and, not least, in the light of present-day energy conservation and environmental pollution considerations? These requirements can now add up to quite a formidable list. As we pursue our studies into the whys and wherefores of engine construction and operation, it will become evident that although some of the requirements are complementary, others are not, and therefore (as in most engineering) some compromise has generally to be accepted in the final product.

Modern requirements

Optimum performance

With modern advances in engine design it is not particularly difficult to obtain sufficient power to give the car a high top speed, especially since the recent trend towards car bodies of lighter construction and more efficient aerodynamic shape. Today, however, a more important engine requirement than a further increase in top speed is an improved accelerating capability together with better flexibility in the low to middle speed range, or what is sometimes termed 'driveability'. A further performance requirement of a new engine design is that it must usually allow for possible future increases in cylinder size.

Good fuel economy

The overall aim of improving the fuel economy of cars is to minimize the amounts of crude oil used to provide petrol for their engines, because of constraints imposed by limited petroleum resources and rising costs. Fuel economy may also be made the subject of legislation, as it already is in America, where each manufacturer has to comply with corporate average fuel economy standards (or CAFE standards, as they are generally termed). For these reasons, further engine requirements are those of minimum weight so as to reduce total car weight; improved combustion efficiency, better to utilize the fuel; and reduced friction losses between the working parts.

Low pollution

Since the late 1960s increasingly stringent legislation has been applied to limit the levels of atmospheric pollutants emitted from car engines, especially the American FTP (Federal Test Procedure), the Japanese and later the European Community ECE/EEC test cycles, all of which differ in their requirements and are therefore not directly comparable. In Britain The Road Vehicles (Construction and Use) Regulations are also now such that there is a requirement for every motor vehicle to be so constructed that no avoidable smoke or visible vapour is emitted therefrom, and another that makes it an offence to use a vehicle which emits substances likely to cause damage to property or injury to persons. In general, legislation is concerned with carbon monoxide, which has toxic effects; unburned hydrocarbons, which contribute to atmospheric smog; and nitrogen oxides, which cause irritation to the eyes and lungs, and also combine with water to produce acid rain that destroys vegetation. To reduce these harmful emissions, not only is very careful control of the combustion process required in modern engine design, but also various sophisticated devices may have to be added for after-treatment of the exhaust gases. Of further concern to the environmentalist is the emission of carbon dioxide which, although non-toxic, is nevertheless an unwanted contributor to global warming. This has to the development of systems for deactivating half the number of cylinders on some large capacity V8 and V12 engines, to reduce fuel consumption and therefore the emission of carbon dioxide when full power is not required.

Minimum noise level

Noise is generally defined as unwanted sound. Reducing interior noise makes a car more attractive to the buyer. Reducing exterior noise to socially acceptable limits has been the subject of increasingly stringent legislation in the European Community and other countries since the early 1980s, and in

Britain is included in the Provision of the Motor Vehicles (Construction and Use) Regulations relating to noise. A similar function is performed in America by the EPA (Environmental Protection Agency) noise regulations. Since the engine is an obvious source of noise an important requirement is that its design and installation should minimize noise emission, not only that directly radiated from the engine itself to the exterior, but also that arising from vibrations transmitted through its mounting system to the car body interior.

Easy cold starting

An essential driver requirement of any engine, whether it be of past or present design, is that it should possess good cold starting behaviour and then continue to run without hesitation during the warming-up period. A present-day additional requirement is that the cold starting process should be accomplished with the least emission of polluting exhaust gases and detriment to fuel economy. To monitor the required enrichment of the air and fuel mixture for cold starting, increasingly sophisticated controls were applied first to carburettor automatic choke systems and then later to fuel injection cold start systems. These controls form part of what are now termed 'engine management systems'.

Economic servicing

An important owner requirement of a car is that its engine design should acknowledge the need to reduce servicing costs. This aim may be approached by minimizing the number of items that need periodic attention by a service engineer. For example, the use of hydraulic tappets eliminates altogether the need for adjustment of the valve clearances. It is also promoted by allowing ready access to those items of the engine involved in routine preventive maintenance, such as the drive-belt tensioner, spark plugs, and petrol and oil filters.

Acceptable durability

In order to reduce fuel consumption while still maintaining good car performance, it is now the trend to develop engines of smaller size with relatively higher power output. Furthermore, the installation of a turbo-charger permits an increase in power without imposing a corresponding increase in the size or weight of the engine itself. However, the greater heating effect on certain engine components may require changes to their material specifications and also the addition of an oil cooler. The components of modern engines have therefore tended to become more highly stressed, so that engine testing of ever-increasing severity by the manufacturers is now required to maintain durability in extremes of customer service.

Least weight

Another important design requirement of the modern petrol engine is that it should be made as light as possible. This is because a corresponding reduction in car weight can make significant improvements not only in fuel economy and acceleration capability, but also in general handling and ease of manoeuvring the car. Since reducing engine weight is not always consistent with maintaining durability, the need for adequate testing of the engine components is confirmed. Also special manufacturing techniques may have to be adopted to avoid damage to such items as castings with very thin walls.

Compact size

For the modern car, the manufacturer strives to provide the maximum interior space for the minimum possible exterior dimensions. Thus the trend is inevitably towards having the front wheels driven, with the power unit (engine and transmission) installed transversely between them; the conventional arrangement was to have a longitudinally mounted power unit from which the drive was taken to the rear wheels. It follows that the requirement now is for a more compact engine. This is because the engine length is controlled by the distance available between the steerable front wheels, less that required by the transaxle (combined gearbox and final drive); its width by the distance available between the radiator and the dash structure, less that required by the engine auxiliaries; and its height by the need for a low and sloping bonnet line, which contributes to an efficient aerodynamic body shape.

Economic manufacture

This is clearly a most important requirement for any new design of engine, since putting it into production demands a massive initial investment on the part of the car manufacturer. It is, of course, for this reason that the smaller specialist car manufacturer generally uses an existing engine from a volume producer. For economic manufacture a new design of engine should lend itself as far as possible to existing automatic production processes and require the minimum of special tooling. The cost of materials will be reduced in building a smaller engine, and the construction should be as simple as possible to minimize the number of parts to be assembled and thereby further reduce manufacturing costs. Similarly, to produce a range of large capacity V6 and V8 engines a modular design approach may be adopted, so that their major components can be produced on the same machinery.

Aesthetic appearance

In early years the under-bonnet appearance of high-grade cars of the 1920s and 1930s, such as Bugatti, Hispano-Suiza and Rolls-Royce, was much admired for the elegant proportions and beautiful finish of their engines. More recently manufacturers have recognized the customer appeal of a pleasing under-bonnet appearance. Not so much of the engine itself, which is usually buried deeper within the engine compartment, but of the neat arrangement and smooth contours of the modern comprehensive air intake system and its manifold runners that now lie above the engine.

1.2 ENGINE NOMENCLATURE

To understand the information given in an engine specification table, such as those included in a manufacturer's service manual or published in the motoring press, it is necessary to become familiar with some commonly used terms (Figure 1.1). The 'language' of the reciprocating piston engine is summarized in the following sections.

Top dead centre

The top dead centre (TDC) is of general application in engineering; it is any position of a hinged linkage in which three successive joints lie in a straight line. In the case of a motor

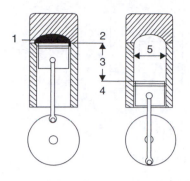

Figure 1.1 Engine nomenclature (*Yamaha*)

1 volume of combustion chamber
2 top dead centre (TDC)
3 stroke
4 bottom dead centre (BDC)
5 bore

Piston displacement: volume of gases displaced by the piston as it moves from BDC to TDC.

vehicle engine, top dead centre refers to the position of the crankshaft when the piston has reached its closest point to the cylinder head. This results in the main, big-end and small-end bearings lying in a straight line. A motor vehicle service engineer often needs to establish top dead centre for checking the ignition and valve timing of an engine.

Bottom dead centre

The bottom dead centre (BDC) is, of course, the opposite extreme of crankshaft rotation when the piston has reached its furthest point from the cylinder head.

Piston stroke

In a general engineering sense, the stroke is the movement of a reciprocating component from one end of its travel to the other. In the motor vehicle engine the piston stroke, therefore, is the distance travelled by the piston in its movement from BDC to TDC or, of course, vice versa, and is expressed in millimetres (mm).

Cylinder bore

In engineering practice the term *bore* may refer to a hole through a bushing or pipe, or to the cutting of a large-diameter cylindrical hole, or to an actual measurement of the inside diameter of a hollow cylinder. It is the last named with which we are concerned here, where the bore refers to the inside diameter of the engine cylinder expressed in millimetres (mm).

Piston displacement

This term refers to the volume of cylinder displaced or swept by a single stroke of the piston, and is also referred to as *swept volume*. It is expressed in cubic centimetres (cm^3) and may be simply calculated as follows:

$$V_h = \frac{\pi d^2 s}{4000} \, cm^3$$

where V_h is the piston displacement or swept volume (cm^3), d is the cylinder bore (mm) and s is the piston stroke (mm).

Engine capacity

Here we are referring to the total piston displacement or the swept volume of all cylinders. For example, if the swept volume of one cylinder of an engine is 375 cm^3 and the engine has four cylinders, then the engine capacity is 1500 cm^3 or 1.5 litres (1). This can be simply stated as:

$$V_H = V_h z$$

where V_H is the engine capacity (cm^3), V_h is the piston displacement (cm^3) and z is the number of cylinders.

Stroke/bore ratio

Reference is sometimes made to the stroke/bore ratio of an engine and although the term itself is self-explanatory, its significance deserves further explanation because up to the 1950s the length of the piston stroke almost invariably exceeded the cylinder diameter, whereas until recently the converse situation has usually applied. The reason for the change from so-called *under-square* to *over-square* cylinder proportions from the 1950s onwards, was that for taxation and insurance purposes engines had previously been rated for horse-power by an RAC formula dating from the early years of motoring, which bore little relationship to the actual power they developed. Unfortunately it also meant that to keep the rated horse-power low for taxation purposes, but the actual power high, the engine designer was restricted to small bore and long stroke cylinder proportions. However, the introduction in 1947 of a flat tax on all passenger cars regardless of their engine size, relieved designers of this artificial restraint on cylinder bore size and led to improvements in engine performance.

The advantages claimed at the time for engines with larger bores and shorter strokes included increased size of valves and ports for better engine breathing; lower piston speeds that not only reduced mechanical loses due to friction and therefore improved fuel consumption, but also increased life expectancy of the cylinder bores; greater rigidity for the crankshaft by virtue of a smaller crankthrow (Section 1.8); and last but not least a reduction in engine height for a lower bonnet line. In current practice engines tend to be neither under-square nor over-square but to have a stroke/bore ratio closer to unity, so as better to achieve a compact combustion chamber and reduce harmful exhaust gas emission (Section 3.1).

Mean effective pressure

This term is used because the gas pressure in the cylinder varies from a maximum at the beginning of the power stroke to a minimum near its end. From this value must, of course, be subtracted the mean or average pressures that occur on the non-productive exhaust, induction and compression strokes. Engine mean effective pressure can be expressed in kilo-newtons per square metre (kN/m^2).

Indicated and brake power

The most important factor about a motor vehicle engine is the rate at which it can do work or, in other words, the power it can develop. It is at this point that we must distinguish between the rate at which it might be expected to work (as calculated from the mean effective pressure in the cylinder, the piston displacement, the number of effective working

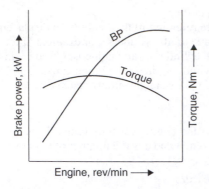

Figure 1.2 Engine power and torque curves

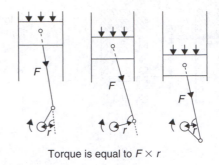

Torque is equal to $F \times r$

Figure 1.3 Engine torque or turning effort

strokes in a given time and the number of cylinders) and the rate at which it actually does work (as measured in practice when the engine is running against a braking device known as a dynamometer).

The significance of this is that the *brake power* delivered at the crankshaft is always less than the *indicated power*, owing to internal friction losses in the engine. A simple expression for calculating in kilowatts (kW) the indicated power of an engine is as follows:

$$P = psAEz$$

where P is the indicated power (kW), p is the mean effective pressure (kN/m^2), s is the piston stroke (m), A is the piston area (m^2), E is the number of effective working strokes per second and z is the number of cylinders. (Since in the four-stroke cycle engine there is only one power stroke for every two complete revolutions of the crankshaft, the number of effective working strokes per second will correspond to one-half the number of engine revolutions per second.)

As mentioned earlier, a dynamometer is used in an engine testing laboratory to measure the brake power (or effective power) of an engine, because it acts as a brake to balance the torque or turning effort at the crankshaft through a range of speeds. A graph of the engine *power curve* can be drawn by plotting brake power values against engine speeds (Figure 1.2). Various standardized test procedures may be adopted in engine testing, such as those established by the American Society of Automotive Engineers (SAE), the German Deutsche Institut für Normung (DIN) and the Italian Commissione tecnica di Unificazione nell' Automobile (CUNA). In an engine specification table only the maximum brake power and corresponding crankshaft speed are usually quoted. For example, the 5.9 litre V twelve-cylinder 48-valve engine used in the high-performance Aston Martin DB9 Volante car is claimed to develop 330 kW at 6000 rev/min.

Engine torque

Also usually included on an engine performance graph is a *torque curve*, which is obtained by plotting crankshaft torque, or turning effort, against engine speed (Figure 1.2). The engine torque, of course, is derived from combustion pressure acting upon the cross-sectional area of the piston, the resulting force from which applies a turning effort to the crankshaft through the connecting rod and crankthrow arrangements (Figure 1.3).

Engine torque, therefore, may be considered as the force of rotation acting about the crankshaft axis at any given instant.

Engine torque T may be expressed in newton metres (Nm), and generally reaches a peak value at some speed below that at which maximum power is developed; the reason for this is explained at a later stage. An engine that provides good pulling power is typically one in which maximum torque is developed at moderate engine speeds. For example, the 1.4 litre four-cylinder 16-valve K Series engine of advanced construction (Figure 1.30), as originally designed by the Rover Group, developed its peak power of 70 kW at 6250 rev/min and a maximum torque of 124 Nm at 4000 rev/min. At the other extreme a heavy-vehicle diesel engine, such as the 11 litre six-cylinder turbocharged Scania with a peak power of 280 kW at a governed maximum speed of 1900 rev/min, develops a maximum torque of 1660 Nm at 1300 rev/min, which really is pulling power!

1.3 OPERATING PRINCIPLES

The four-stroke petrol engine

As with the various repeating cycles of events in nature, so does the motor vehicle petrol engine need to operate on a constantly repeating cycle known as the four-stroke principle.

It would seem to be generally accepted that the first internal combustion engine to operate successfully on the four-stroke cycle was constructed in 1876 by Nicolaus August Otto (1829–91). This self-taught German engineer was to become one of the most brilliant researchers of his time and also a partner in the firm of Deutz near Cologne, which for many years was the largest manufacturer of internal combustion engines in the world.

Although the Otto engine ran on gas, which was then regarded as a convenient and reliable fuel to use, it nevertheless incorporated the essential ideas that led to the development in 1889 of the first successful liquid-fuelled motor vehicle engine. This was the twin-cylinder Daimler engine, patented and built by the German automotive pioneer Gottlieb Daimler (1834–1900) who, like Otto, had been connected with the Deutz firm. The Daimler engine was subsequently adopted by several other car manufacturers and, in most respects, it can be regarded as the true forerunner of the modern four-stroke petrol engine (Figure 1.4).

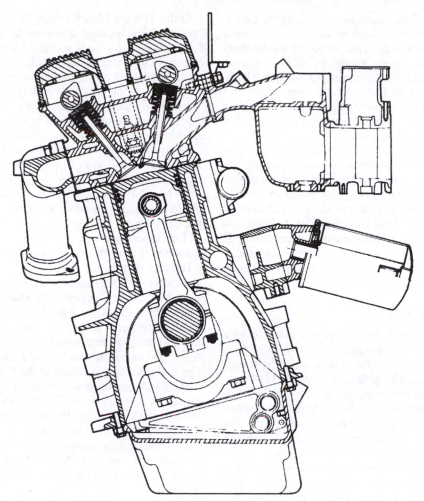

Figure 1.4 Cross-section of a modern four-stroke petrol engine (*Jaguar*)

In this type of engine the following sequence of events is continuously repeated all the time it is running (Figure 1.5):

1 The *induction* stroke, during which the combustible charge of air and fuel is taken into the combustion chamber and cylinder, as a result of the partial vacuum or depression created by the retreating piston.

2 The *compression* stroke, which serves to raise both the pressure and temperature of the combustible charge as it is compressed into the lesser volume of the combustion chamber by the advancing piston.

3 The *power* stroke, immediately preceding which the combustible charge is ignited by the sparking plug and during which the gases expand and perform useful work on the retreating piston.

4 The *exhaust* stroke, during which the products of combustion are purged from the cylinder and combustion chamber by the advancing piston, and discharged into the exhaust system.

It thus follows that one complete cycle of operations occupies two complete revolutions of the engine crankshaft. Since energy is necessarily required to perform the initial induction and compression strokes of the engine piston before firing occurs, an electrical starter motor is used for preliminary

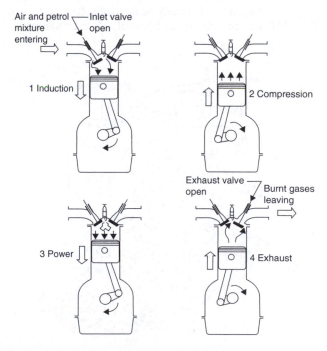

Figure 1.5 The four-stroke petrol engine cycle

cranking of the engine. Once the engine is running the energy required for performing subsequent induction, compression and exhaust strokes is derived from the crankshaft and flywheel system, by virtue of its kinetic energy of rotation. *Kinetic energy* is a term used to express the energy possessed by a body due to its mass and motion. The principle of an engine flywheel is therefore to act as a storage reservoir for rotational kinetic energy, so that it absorbs energy upon being speeded up, and delivers it when slowed down.

In the four-stroke cycle, the functions of admitting the combustible charge before its compression, and releasing the burnt gases after their expansion, are performed by the engine inlet and exhaust valves. The opening and closing of the inlet and exhaust valves are not, in actual practice, timed to coincide exactly with the beginning and ending of the induction and exhaust strokes; nor is the spark timed to occur exactly at the beginning of the power stroke. At a later stage the reasons for these departures in *valve and ignition timing* from the basic four-stroke operating cycle will be made clear.

The two-stroke petrol engine

As its designation implies, the two-stroke petrol engine (Figure 1.6) completes its working cycle in only two strokes of the piston, so that a combustible charge is ignited at each revolution of the crankshaft. Although in its simplest construction the two-stroke petrol engine needs no valves, the induction and exhaust process must be facilitated by a system of *scavenging* or forcible clearing of the cylinder gases. This may either take the form of a separate engine-driven pump, or utilize the motion of the engine piston itself in a sealed crankcase. The flow of gases entering and leaving the cylinder is controlled by the reciprocating movement of the engine piston, which thus acts as a slide valve in conjunction with ports cut in the cylinder wall. Although the two-stroke petrol engine was once favoured in Europe for some small inexpensive passenger cars, it generally became obsolescent because of the difficulty in reducing its harmful exhaust emissions.

At the 11th Sir Henry Royce Memorial Lecture in 1966, Cyril Lovesey recalled a remark by another distinguished Rolls-Royce aero engine specialist, Dr Stanley Hooker, who amusingly described a four-stroke engine as one with 'one stroke to produce power and three strokes to wear it out'. It is therefore perhaps not surprising that from time to time attempts are made to revive interest in the two-stroke petrol engine for automotive use, albeit in much more sophisticated forms of which an example will be later described.

It may be of interest to recall that the two-stroke and the four-stroke engine both originated in the late 1870s, so it might reasonably be assumed that both types of engine started out in life with an equal chance of success. The fact that the four-stroke engine became by far the more widely adopted type can probably be explained by its having a greater potential for further development. This is a criterion that can often be applied to rival ideas in all branches of engineering.

The first successful application of the two-stroke cycle of operation to an early gas engine is generally attributed to a Scottish mechanical engineer, Sir Dugald Clerk (1854–1932). It is for this reason, of course, that the two-stroke cycle is sometimes referred to as the *Clerk cycle*. Dugald Clerk, like several other pioneer researchers of the internal combustion engine, was later to achieve high academic distinction, culminating in his election as Fellow of the Royal Society in 1908.

The Clerk engine was scavenged by a separate pumping cylinder. A few early motor vehicle two-stroke petrol engines followed the same principle, but it later became established practice to utilize the underside of the piston in conjunction with a sealed crankchamber to form the scavenge pump. This idea was patented in 1889 by Joseph Day & Son of Bath, England and represented the simplest type of two-cycle engine.

In the two-stroke or Clerk cycle, as applied by Day, the following sequence of events is continuously repeated all the time the engine is running (Figure 1.7):

1 The *induction-compression* stroke. A fresh charge of air and fuel is taken into the crankchamber as a result of the depression created below the piston as it advances towards the cylinder head. At the same time, final compression of the charge transferred earlier in the stroke from the crankchamber to the cylinder takes place above the advancing piston.
2 The *power-exhaust* stroke. The combustible charge in the cylinder is ignited immediately preceding the power stroke,

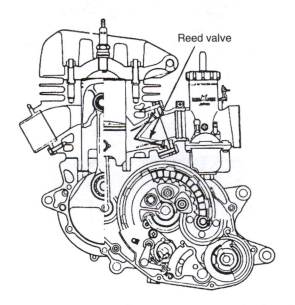

Figure 1.6   Cross-section of a two-stroke petrol engine (*Honda*)

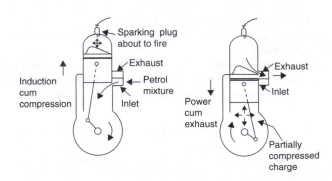

Figure 1.7   The two-stroke petrol engine cycle

during which the gases expand and perform useful work on the retreating piston. At the same time, the previously induced charge trapped beneath the retreating piston is partially compressed. Towards the end of the stroke, the exhaust gases are evacuated from the cylinder, a process that is facilitated by the scavenging action of the new charge transferred from the crankcase.

The uncovering and covering of the cylinder ports of the piston, or *port timing*, is determined by considerations similar to those affecting the valve timing of the four-stroke engine and will explained at a later stage.

Four-stroke versus two-stroke engines

The following generalizations may be made concerning the relative merits of two-stroke and four-stroke petrol engines in their basic form:

1 The two-stroke engine performs twice as many power strokes per cylinder per revolution. In theory at least, this might be expected to produce twice the performance of a four-stroke engine of equivalent size. Unfortunately, this is not realized in practice because of the difficulties encountered in effectively purging the exhaust gases from the cylinder and then filling it completely with a fresh combustible charge. The scavenging efficiency of the basic two-stroke petrol engine is therefore poor.

2 In performing twice as many power strokes per revolution, the two-stroke engine can deliver a smoother flow of power, but this may be less true at low engine speeds when irregular firing or 'four-stroking' can result from poor scavenging.

3 An obvious practical advantage of the basic two-stroke engine is the mechanical simplicity conferred by its valveless construction, which contributes to a more compact and lighter engine that should be less expensive to make.

4 Reduced maintenance requirements might reasonably be expected with the basic two-stroke engine by virtue of point 3. There is, however, the well-known tendency for carbon formation to have a blocking effect on the exhaust ports, which impairs engine performance by reducing scavenging efficiency.

5 The fuel consumption of the basic two-stroke engine is adversely affected by the poor cylinder scavenging, which allows part of the fresh charge of air and fuel to escape through the exhaust port before final compression of the charge takes place.

6 There is a greater danger of overheating and piston seizure with a two-stroke engine, which can set a limit on the maximum usable performance. It is more difficult to cool satisfactorily, because it does not have the benefit of the second revolution in the four-stroke cycle when no heat is being generated.

7 Lubrication of the two-stroke petrol engine is complicated by the need to introduce oil into the fuel supply to constitute what is generally termed a *petroil* mixture. The working parts of the engine are thus lubricated in aerosol fashion by oil in the air and fuel charge, and this tends to increase harmful exhaust emissions. It is for this reason that the basic two-stroke petrol engine is now obsolescent for cars.

Scavenging: further details

Frequent reference has been made to the inherently poor scavenging efficiency of the basic two-stroke petrol engine. The word 'basic' has been used deliberately and is intended to apply to the Day type of early two-stroke engine, which had a deflector-head piston to promote a cross-scavenging effect on the burnt charge leaving the cylinder (Figure 1.8a). This not entirely successful scheme persisted until the mid 1920s, when Dr E. Schnürle of Germany developed an alternative loop-scavenging system. In this the deflector on the piston head is omitted and two transfer ports with angled passages are disposed on either side of, instead of opposite, the exhaust port (Figure 1.8b). The loop-scavenge effect produced is such that before the two streams of fresh charge intermingle, they converge upon the cylinder wall at a point furthest away from the exhaust port, so there is less chance of escape.

The Day type of early two-stroke engine also used what would now be classified as a three-port system of scavenging. This system comprises inlet, transfer and exhaust ports, all in the cylinder wall, and necessarily imposes a restriction on the period during which a fresh charge of air and fuel may enter the crankcase.

To achieve more complete filling of the crankcase, the later two-port system of scavenging is now generally employed. In this system only the transfer and exhaust ports are in the cylinder wall, the inlet port being situated in the crankcase itself and controlled by either an automatic flexible reed valve (Figure 1.9a) or an engine-driven rotary disc valve (Figure 1.9b) which improve the torque and power characteristics respectively of an engine. The two-port system of scavenging thus allows the fresh charge to continue entering the crankcase

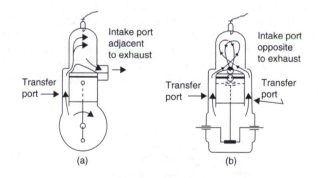

Figure 1.8 Early three-port scavenging systems: (a) cross-scavenging (b) loop-scavenging

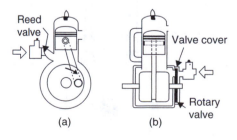

Figure 1.9 Later two-port scavenging systems: (a) reed valve (b) rotary valve (*Yamaha*)

during the whole, instead of part, of the induction-compression stroke, albeit with a little extra mechanical complication.

Further development of the two-stroke petrol engine

We take as an example the advanced concept of the Toyota S-2 (supercharged two-stroke) engine (Figure 1.10). The development objective of this engine is to deliver greater power more smoothly than a conventional one, and also to confer very high torque at low engine speeds. A Roots-type supercharger or blower (Section 9.2) is used to achieve positive scavenging of the exhaust gases. Unlike the two-stroke petrol engines so far described, the Toyota engine borrows many features from established four-stroke practice. It has a four-valve cylinder head with two valves for intake and two for exhaust, which do of course open and close twice as often in this two-stroke application. The engine is provided with electronically controlled ignition and fuel injection systems, with fuel being injected directly into the cylinders.

The importance of engine compression

Early internal combustion engines were very inefficient because they were provided with a combustible charge that was ignited at atmospheric pressure. However, it was recognized as early as 1838 by William Barnett in the UK that compression of the charge before combustion was advantageous. Nearly 25 years later, a French railway engineer with the splendid name of Alphonse Beau de Rochas was granted a patent in respect of several ideas that related to the practical and economical operation of the internal combustion engine. Among these ideas he stated a requirement for the maximum possible expansion of the cylinder gases during the power stroke, since the cooler they become the more of their energy is transformed into useful work on the retreating piston. To assist in this aim, there was a further requirement for the maximum possible pressure at the beginning of the expansion process, which the motor vehicle service engineer recognizes as the *compression pressure* of an engine. The first successful engine to utilize this principle was, as mentioned earlier, the Otto engine.

Within certain limitations, which will be better understood at a later stage, a high-compression engine is relatively more efficient than a low-compression one in terms of either improved fuel economy or greater power. This is simply because it better utilizes the internal energy received from its fuel or, in other words, it possesses a higher *thermal efficiency*. It is this particular feature of engine operation that explains the importance of checking compression pressures before attempting to tune an engine in service.

Compression ratio

The extent to which the air and fuel charge in a petrol engine is compressed, prior to the power stroke, is known as the *compression ratio* of an engine. It is calculated as the ratio of the total volume enclosed above the piston at BDC to the volume remaining above the piston at TDC.

The swept volume has already been explained in Section 1.2. A complementary term is *clearance volume*, which is that volume remaining above the piston when it reaches TDC. In some engines the combustion chamber is formed mainly in the piston head (Section 3.3), so that the clearance volume is concentrated within, rather than above, the piston. Hence, the compression ratio (usually abbreviated to CR) may also be expressed as the swept volume plus the clearance volume divided by the clearance volume (Figure 1.11), as follows:

$$\varepsilon = \frac{V_h + V_c}{V_c}$$

where ε is the compression ratio, V_h is the cylinder swept volume (cm^3), and V_c is the combustion space clearance volume (cm^3).

The calculated compression ratios for petrol engines are typically in the range 8:1 to 9.5:1. It should be appreciated, however, that in the petrol engine the calculated compression ratio is realized in practice only when the engine is running with a wide open throttle.

Furthermore, in the case of a two-stroke engine V_h is generally considered as the cylinder volume from the point of exhaust port closing to TDC, which is therefore less than the

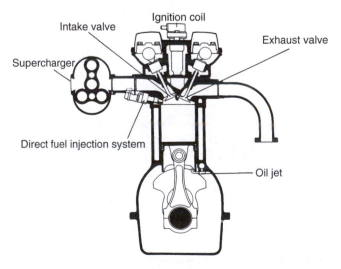

Figure 1.10   Toyota S2 supercharged two-stroke engine

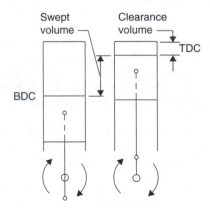

Figure 1.11   Engine compression ratio

swept volume of an equivalent four-stroke engine and gives a lower compression ratio.

1.4   BASIC STRUCTURE AND MECHANISM

It is convenient to introduce the various components of the petrol engine in the following groups:

Cylinder block and crankcase
Piston and rings
Connecting rod and bearings
Crankshaft assembly and bearings
Cylinder head and gasket
Valve train and timing drive
Engine support mountings.

Cylinder block and crankcase

In combination, the cylinder block and crankcase form the main structural component of the engine and perform several important functions, as follows:

1 Each cylinder must act not only as a pressure vessel in which the process of combustion can take place, but also as a guide and bearing surface for the piston sliding within it.
2 Since the engine cylinders have to be cooled effectively, the cylinder block must also form a jacket to contain the liquid coolant.
3 The crankcase provides an enclosure for the crankshaft and various other parts of the engine mechanism and must preserve accurate alignment of their supporting bearings under all operating conditions.
4 Pressure conduits in the form of either drilled or cast in ducts must also be incorporated in the crankcase to convey oil to the engine working parts.
5 The cylinder block is required to provide external mounting surfaces for various engine auxiliary units.

It has long since been established practice to combine the cylinder block and crankcase into a single unit, this generally being termed *monobloc* construction (Figure 1.12). The historical origins of this form of construction date back to the early 1920s, when there was a general trend towards simplification of the engine structure. Until that time the cylinder block and crankcase were produced as separate units, in cast iron and aluminium alloy respectively, and then bolted together.

In relation to modern engine design, the monobloc construction provides a very necessary rigid foundation for the engine and reduces manufacturing costs. It should be added, however, that this particular form of construction is not always the best compromise for heavy-duty diesel engines, which will be considered at a later stage.

Piston and rings

The main function of the piston itself is twofold:

1 It acts as a moving pressure transmitter, by means of which the force of combustion is impressed upon the crankshaft through the medium of the connecting rod and its bearings.
2 By supporting a gudgeon pin the piston provides a guiding function for the small end of the connecting rod.

The piston assumes a *trunk* form to present a sliding bearing surface against the cylinder wall, which thus reacts against the side thrust arising from the angular motion of the connecting rod. Since the piston is a major reciprocating part, it must of necessity be light in weight to minimize the inertia forces created by its changing motion – bearing in mind that the piston momentarily stops at each end of its stroke! Another, perhaps obvious, requirement is that the piston must be able to withstand the heat of combustion and should operate quietly in its cylinder, both during warm-up and at the normal running temperature of the engine.

To perform its sealing function efficiently, the upper part of the piston is encircled by flexible metal sealing rings known as the *piston rings*, of which there are typically three in number for petrol engines (Figure 1.13). In combination the piston rings perform several important functions, as follows:

1 The upper *compression rings* must maintain an effective seal against combustion gases leaking past the pistons into the crankcase.
2 These rings also provide a means by which surplus heat is transmitted from the piston to the cylinder wall and thence to the cooling jacket.
3 The lower *oil control ring* serves to control and effectively distribute the lubricating oil thrown on to the cylinder walls, consistent with maintaining good lubrication and an acceptable oil consumption.

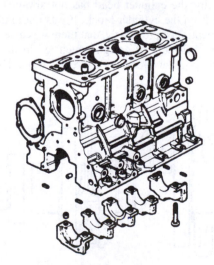

Figure 1.12   Cylinder block and crankcase

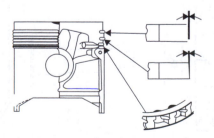

Figure 1.13   Piston and rings (*Toyota*)

Connecting rod and bearings

The function of the connecting rod and its bearings is to serve as a constraining link between the reciprocating piston and the rotating crankshaft. A fair analogy of this particular conversion of motion is to be found in the pedalling of a cycle, where the knee, calf and foot of the cyclist may be likened to the piston, connecting rod and crankpin of an engine.

The connecting rod is attached at what is termed its *big end* to the crankpin and at its *small end* to the gudgeon pin of the piston. Each connecting rod big-end bearing is divided into two half-liners, so as to make possible its assembly around the crankpin (Figure 1.14). The big-end bearing housing is therefore formed partly by the lower end of the connecting rod and partly by a detachable cap, the two halves being bolted together. No such complication arises in the case of the small-end bearing arrangements, this end of the connecting rod being formed as a continuous eye.

As a consequence of its reciprocating and partly rotating motion, the connecting rod is subjected to appreciable inertia forces. Its detail design must therefore be such as to ensure the maximum rigidity with the minimum weight.

Crankshaft assembly and bearings

The crankshaft represents the final link in the conversion of reciprocating motion at the piston to one of rotation at the flywheel. In the case of the multicylinder engine, the crankshaft has to control the relative motions of the pistons, whilst simultaneously receiving their power impulses.

Figure 1.14   Connecting rod and bearings

A one-piece construction is most commonly used for the motor vehicle crankshaft, which extends the whole length of the engine and must therefore possess considerable rigidity. The timing drive for the engine valve mechanism is taken from the front end of the crankshaft, as is the pulley and belt drive for the engine auxiliaries, such as the cooling fan and the alternator for the electrical system. Attached to the rear end of the crankshaft is the engine flywheel.

The crankshaft is supported radially in the crankcase by a series of bearings, known as the engine *main bearings*. Each main bearing is divided into two half-liners, similar to the big-end bearings, and again to allow assembly around the journals of the one-piece construction crankshaft (Figure 1.15).

Cylinder head and gasket

The functions of the cylinder head may be listed as follows:

1 It must provide a closure or chamber for the upper part of each cylinder, so that the gas pressure created by the combustion process is constrained to act against the piston.
2 Associated with function 1 is the need to incorporate a gas porting system with inlet and exhaust valves, as well as a platform upon which to mount their operating mechanism. Provision must also be made for a screwed boss to retain the sparking plug.
3 Similar to the cylinder block, the head must form a jacket that allows liquid coolant to circulate over the high-temperature metal surfaces.
4 It is required to contribute to the overall rigidity of the engine structure and maintain a uniform clamping pressure on its sealing gasket with the cylinder block.

The sealing gasket is generally known as the *cylinder head gasket* or simply 'head gasket'. In liquid-cooled engines the function of the cylinder head gasket is to seal the combustion chambers and coolant and oil passages at the joint faces of the cylinder block and head. The gasket is therefore specially shaped to conform to these openings, and is also provided with numerous holes through which pass either the studs or the set bolts for attaching the cylinder head to the block (Figure 1.16).

Historically, the cylinder head has not always been made detachable from the cylinder block. It was not until the early 1920s, as mentioned previously, that there was a general trend towards simplification of design which led to the abandonment of the cylinder block with integral head. This change

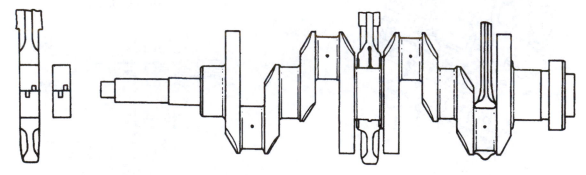

Figure 1.15   Crankshaft and bearings (*Alfa-Romeo*)

facilitated both the production and the servicing of the motor vehicle engine, although it did introduce the risk of incorrect tightening down of the cylinder head, which can cause at least joint trouble and at worst distortion of the cylinder bores.

Valve train and timing drive

The overall function of the valve train and timing drive is to provide first for the admission and then the retention of the combustible charge within the cylinder, and finally for the release of the burnt gases from the cylinder, all in synchronism with the motion of the pistons (Figure 1.17).

To perform this sequence of events in accordance with the requirements of the four-stroke cycle, a cam-and-follower mechanism driven at one-half crankshaft speed is used to operate the engine inlet and exhaust valves. There are several different methods of operating the valves from the cam-and-follower mechanism, but in all cases it is necessary for one or more camshafts to be driven from the front end of the crankshaft by what is termed the *timing drive*, and which will be examined in detail at a later stage.

It is, of course, the engine valves themselves that actually perform the functions of admitting the air and fuel charge before its compression, sealing it in the cylinder during compression and combustion, and then releasing the burnt gases

after their expansion. Either single or paired inlet and exhaust valves serve each cylinder, and these are mounted in the cylinder head combustion chambers. Valve springs are fitted to ensure that the motion of the valves and their operating mechanism follows faithfully that intended by the cams, and also to maintain adequate sealing pressure when the valves are closed.

Engine support mountings

The engine is subject to complex vibration effects, which may produce six different free motions or degrees of freedom and combinations of the same. These motions can generate what are termed 'bounce' and 'yaw' about a vertical axis, fore-and-aft movement and roll about a horizontal longitudinal axis, and sideways shake and pitch about a horizontal lateral axis. The three inertia axes intersect at the centre of gravity of the engine (Figure 1.18). For mounting the engine, a resilient anti-vibration system is therefore required, so that the vibratory forces are reduced to the relatively small spring forces transmitted by the support mountings themselves. Apart from supporting the static load of the engine unit and isolating the car structure from engine vibrations, the engine mounts also insulate the engine against vibrations of the car structure and mechanism. By thus minimizing unwanted movements of the engine under all running conditions, any contribution to car shake from this source is reduced.

The engine support mountings almost invariably feature rubber as their spring medium, since this material is highly resilient when loaded in shear (Figure 1.19a). If a rubber block is simply loaded in compression it inevitably bulges sideways (Figure 1.19b), because it is deformable as opposed to being compressible, so the effect is to increase the area of rubber under load and hence reduce its flexibility as a spring. In actual practice, a compromise is often sought by loading the rubber partly in compression and partly in shear by inclining the rubber mountings (Figure 1.19c). A widely used type of support mounting is the sandwich unit, which generally consists of either a rectangular or a circular block of rubber with metal attachment plates bonded to its upper and lower faces. For some applications, one or more metal interleaves may be incorporated in the mount (Figure 1.19d). These serve

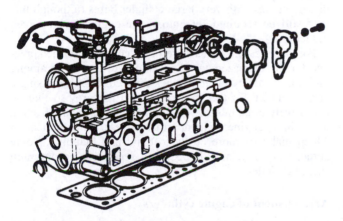

Figure 1.16   Cylinder head and gasket

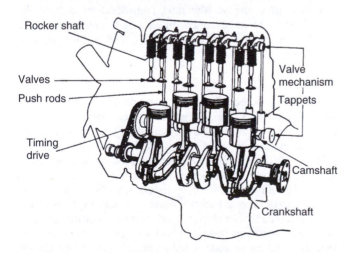

Figure 1.17   Valve train for a push-rod overhead valve engine

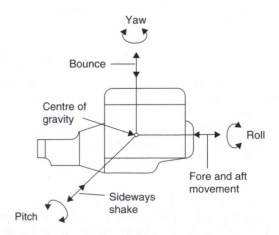

Figure 1.18   Vibratory motions of an engine

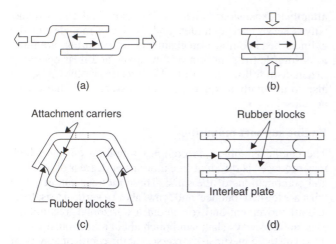

Figure 1.19 Engine support mountings: (a) rubber loaded in shear (b) rubber loaded in compression (c) rubber loaded in combined compression and shear (d) rubber sandwich with interleaf

to restrict sideways bulging of the rubber, thus further reducing its flexibility under compression loading, but leaving its shear loading characteristics unaltered, and they also assist in dissipating the heat of internal friction created by flexing of the rubber. Another advantage of using rubber, rather than steel, for the spring medium is that transmission of sound through the mountings is less, since there is no metal-to-metal path for the sound to travel along. The various types of resilient mounting system employed for longitudinally and transversely disposed engine and transmission units are described in Section 30.1.

1.5 CYLINDER AND CRANKTHROW ARRANGEMENTS

Cylinder number and displacement

Every new engine must be designed with a specific type of service in view, which then determines its general characteristics. Important among these for the car is smooth and efficient operation over a wide range of speeds and loads. Herein lies the explanation why no manufacturer lists a single-cylinder engine and few produce an engine with fewer than four cylinders. The following considerations are pertinent.

Smoothness
With a single-cylinder engine operating on the four-stroke cycle, it will be recalled that only one power impulse occurs for every two revolutions of the crankshaft. The fluctuations in crankshaft torque of a single-cylinder engine would, therefore, be quite unacceptable in motor vehicle operation. Hence the greater number of cylinders used, the shorter will be the interval between the power impulses and the smoother will be the flow of torque from the engine.

Mechanics
It will be evident from the explanation of the factors governing engine power given in Section 1.2 that the power output obtainable from a single-cylinder engine of realistic dimensions and running at a reasonable speed is unlikely to be

sufficient for motor vehicle, as distinct from motorcycle, requirements. This is because a practical limit is set on individual cylinder size by dynamic factors, namely the inertia forces created by accelerating and decelerating the reciprocating masses comprising the piston assembly and the upper portion of the connecting rod.

If an unusually large, and consequently heavy, piston were adopted in a single-cylinder engine intended for high-speed operation, the dynamic effects could be such as to increase the magnitude of the inertia forces to a level that at least would make engine imbalance unacceptable and at worst would prove mechanically destructive. This is partly because the inertia forces are proportional to the cube of the piston mass (e.g. doubling the piston mass will cause the inertia forces to become eight times as great), and partly because they also vary as the square of the engine speed (e.g. doubling the engine speed will cause them to become four times as great).

Temperature
Another problem arising from the use of an unduly large cylinder bore is that cooling of the piston and valves can be seriously impaired, and this may lead to their failure from thermal overstressing. If this is to be avoided it would be necessary to incorporate special features of design, such as oil cooling of the piston, which is often practised in marine diesel engines with very large cylinder bores (although this may still be required for automotive engines with forced induction).

For these reasons it has therefore become established practice for the displacement of an engine to be shared among multiple small cylinders, rather than confined to a single large cylinder. To this must be added the proviso that an excessively large number of cylinders increases the friction losses in an engine, quite apart from the extra complication which makes it more costly to build and maintain. Some authorities now claim that $330 \, cm^3$ represents an optimum size of cylinder.

Arrangement of engine cylinders

Once the displacement and number of cylinders have been decided in relation to the required performance characteristics of a new engine, the next consideration is how the cylinders are to be arranged. In cars they may be arranged in three different ways, each with its own advantages and disadvantages.

In-line cylinders
As would be expected, in this arrangement all the cylinders are mounted in a straight line along the crankcase, which confers a degree of mechanical simplicity. Such engines are now produced with any number of cylinders from two to six (Figure 1.20), with four cylinders continuing to represent a very widely used cylinder arrangement (Figures 1.30 and 3.10). The single bank of cylinders may be contained in either a vertical or an inclined plane. The latter type is sometimes referred to as a sloper or slant engine. For this particular mounting of an in-line cylinder engine, the advantages usually claimed include a reduction in overall installation height and improved accessibility for routine servicing.

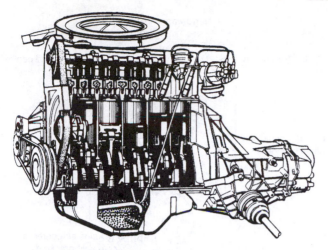

Figure 1.20 An interesting in-line five-cylinder engine (*Audi*)

Figure 1.21 A modern horizontally opposed four-cylinder engine (*Alfa-Romeo*)

With the exception of the in-line two-cylinder or parallel twin engine where both pistons move in unison to obtain even firing intervals, this arrangement of cylinders provides generally satisfactory balance in respect of the reciprocating parts, especially in six-cylinder versions. Apart from space requirements, a mechanical limitation is placed on the acceptable length of an in-line cylinder engine, because of the difficulty in controlling torsional vibrations of the crankshaft. This is a topic that will be discussed at a later stage.

Horizontally opposed cylinders
Horizontally opposed engines have their cylinders mounted on the crankcase in two opposite banks and are sometimes referred to as flat or boxer engines. They are typically produced in two-, four- and six-cylinder versions (Figure 1.21). The main advantages usually claimed for them include inherently good balance of the reciprocating parts, a low centre of gravity, which contributes to car stability, and a short engine structure. It is the latter feature that makes this arrangement of cylinders particularly suitable both for front-wheel-drive and rear-engined cars, since the engine can be mounted either ahead of or behind the driven wheels with the minimum of overhang. By virtue of its low overall height, the horizontally opposed engine can readily allow a sloping bonnet line in front-engined cars and also provide additional space for stowing luggage above it in rear-engined cars. Furthermore, it lends itself admirably to air cooling because with an in-line cylinder arrangement it is difficult to get the rear cylinders to run as cool as the front ones, unless the engine is installed transversely.

The disadvantages associated with horizontally opposed cylinders include the need for lengthy intake manifolds if a central carburettor is used, the duplication of coolant inlet and outlet connections in the case of liquid cooling, and much reduced accessibility for the cylinder heads and valve mechanism. Its greater width can also impose restrictions on the available steering movements of the wheels.

V-formation cylinders
With V engines the cylinders are mounted on the crankcase in two banks set at either a right angle or an acute angle to

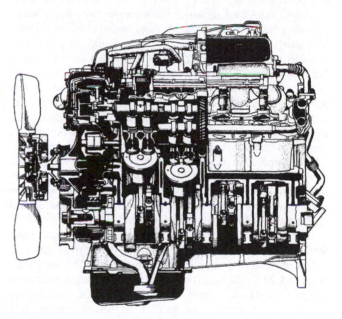

Figure 1.22 A high-grade V eight-cylinder engine (*Lexus/Toyota*)

each other. They may be produced in four-, six-, eight- and occasionally twelve-cylinder versions (Figure 1.22). Where a V cylinder arrangement has been adopted in preference to mounting the cylinders in line, it has usually been in the interests of providing a more compact and less heavy engine. In particular, the overall length of the engine can be appreciably reduced, so that both the structure and the crankshaft can be made more rigid. The former is thus better able to accept greater combustion loads and the latter is less prone to torsional vibrations. For typically large-displacement V engines, the inherently wider cylinder spacings ensure adequate size of coolant passages, both around the cylinder walls and in the hot exhaust valve regions in the cylinder heads (Figure 1.36).

On the debit side, the V cylinder arrangement generally presents a more difficult balancing problem and also demands a more elaborate intake manifold from a central carburettor or a single-point fuel injection system. In common with horizontally opposed cylinder layouts, V engines tend to be more costly to produce, since there are more surfaces to machine and some duplication of structural features.

However, the V six-cylinder engine has become widely adopted in modern practice for medium class cars with engine capacities, ranging between 2.4 and 3.2 litres. Once described by D.A. Martens of Chevrolet as 'a product of necessity', the V6 engine can provide improved performance and enhanced smoothness when transversely mounted in front-wheel-drive cars, where an in-line four-cylinder engine may also be available as a lower cost option.

W-formation cylinders

A more recent development has been the 'double-V' or 'W' cylinder engine, introduced in eight-, twelve- and sixteen-cylinder versions by the Volkswagen group. Basically the W engine, as it is now generally known, comprises two wide-angle cylinder banks, each of which incorporates two offset rows of cylinders inclined at a much narrower angle. In the early 1990s Volkswagen had introduced a single bank narrow-V six-cylinder engine, which externally resembled an in-line unit, but internally had two offset rows of three cylinders inclined at an angle of only 15°. It therefore followed that if two narrow-V six-cylinder banks could themselves be mounted in a wide-V formation of 72° and share a common crankshaft, a W12 engine then becomes a practicality. Several advantages may be gained from this unorthodox, but albeit complex, layout when compared to a conventional V12 engine. A W12 engine can offer a more compact design and indeed may require no more installation space than a conventional V8 engine. It can also be made less heavy, whilst possessing greater structural rigidity with reduced vibration levels.

A distinguishing feature of the narrow-V type cylinder bank arrangement is that the axes of the cylinder pairs intersect below, instead of coinciding with, the axis of the crankshaft (Figure 1.23). This departure from the operating geometry of a conventional wide-angle V-cylinder layout becomes necessary to accommodate the motion of the pistons as they approach their bottom dead centre positions. Similar consideration applies to the need for either separate, or split-type, staggered crankpins for the cylinder pairs.

Historically, the narrow-V cylinder layout had its origins in piston aero engine practice. It was conceived by the Lancia company in Italy at the beginning of World War I, the aim being to reduce the width of a conventional V-cylinder engine for aircraft installation. Smaller narrow-V four-cylinder engines were subsequently used for many years in the passenger cars made by this company, the Lancia Lambda model of the 1920s being a well-known classic example.

Crankthrow arrangements

The arrangement of the crankthrows in relation to the disposition of the engine cylinders and their number is determined by two sometimes conflicting considerations: acceptable engine balance, and equal firing intervals.

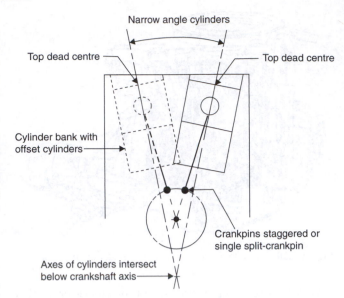

Figure 1.23 Basic arrangement of cylinder pair in narrow-V engine

Acceptable engine balance

Theoretically, a perfectly balanced engine is one which, when running and 'suspended in space' from its centre of gravity, would exhibit no vibratory movements whatsoever. In reality, of course, the reciprocating engine can never be perfectly balanced. Apart from any rotation imbalance, there are also the inevitable torque irregularities, although as stated previously these can be minimized by the use of more than one cylinder.

Multiple cylinders further allow a much better standard of general engine balance, provided that the choice of a particular arrangement and number of cylinders takes into account the presence of what are termed primary and secondary inertia forces. The effect of these forces, which act in the reciprocating sense, will now be explained in simple qualitative terms.

Primary inertia forces These arise from the force that must be applied to accelerate the piston over the first half of its stroke, and similarly from the force developed by the piston as it decelerates over the second half of its stroke (Figure 1.24a). When the piston is around the mid-stroke position it is then moving at the same speed as the crankpin and no inertia force is being generated.

For an engine to be acceptable in practice, the arrangement and number of its cylinders must be so contrived that the primary inertia forces generated in any particular cylinder are directly opposed by those of another cylinder. Where the primary inertia forces cancel one another out in this manner, as for example in an in-line or a horizontally opposed four-cylinder engine with the outer and inner pair of pistons moving in opposite directions, the engine is said to be in *primary balance* (Figure 1.24b).

Secondary inertia forces These are due to the angular variations that occur between the connecting rod and the cylinder axis as the piston performs each stroke. As a consequence of this departure from straight-line motion of the connecting rod, the piston is caused to move more rapidly over the outer

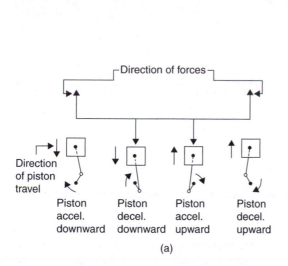

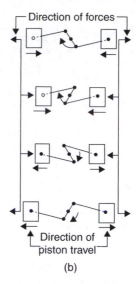

Figure 1.24 Primary inertia forces: (a) unbalanced (single-cylinder) (b) balanced (horizontally opposed cylinders)

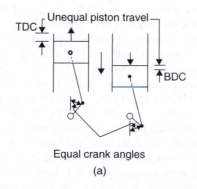

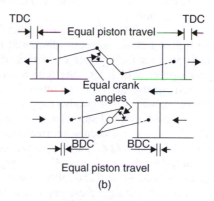

Figure 1.25 Secondary inertia forces: (a) unbalanced (single-cylinder) (b) balanced (horizontally opposed cylinders)

half of its stroke than it does over the inner half. That is, the piston travel at the two ends of the stroke differs for the same angular movements of the crankshaft (Figure 1.25a). The resulting inequality of piston accelerations and decelerations produces corresponding differences in the inertia forces generated. Where these differing inertia forces can be both matched and opposed in direction between one cylinder and another, as for example in a horizontally opposed four-cylinder engine with corresponding pistons in each bank moving over identical parts of their stroke (Figure 1.25b) the engine is said to be in *secondary balance*.

It is not always practicable for the cylinders to be arranged so that secondary balance can be obtained, but fortunately the vibration effects resulting from this type of imbalance are much less severe than those associated with primary imbalance and can usually be minimized by the flexible mounting system of the engine. This is confirmed by the long-established and popular in-line four-cylinder engine, which possesses primary balance but lacks secondary balance. However, the continuing search for greater refinement of running with this type of engine led, in the mid 1970s, to a revival of interest in the use of twin counterbalancing shafts for cancelling out these secondary inertia forces. The modern application of

this system of harmonic balancing will be described in Section 1.8.

Equal firing intervals

The arrangement of the crankthrows is also determined by the requirements for even firing intervals of the cylinders and for spacing the successive power impulses as far apart as possible along the crankshaft, so as to reduce torsional deflections or twisting effects. For any four-stroke engine the firing intervals must, if they are to be even, be equal to 720° divided by the number of cylinders.

For in-line four-cylinder engines the first and fourth crankthrows are therefore indexed on one side of the crankshaft and the second and third throws on the other side (Figure 1.26a). The *firing order* of these engines, numbering from the front, may then be either 1-3-4-2 or 1-2-4-3 at 180° intervals. Similarly, in the case of in-line six-cylinder engines, the crankthrows are spaced in pairs with an angle of 120° between them. Hence, the first and sixth crankthrows are paired, as are the second and fifth, and likewise the third and fourth (Figure 1.26b). The firing order may then be such that no two adjacent cylinders fire in succession; that is, either 1-5-3-6-2-4 or 1-4-2-6-3-5 at, of course, 120° intervals.

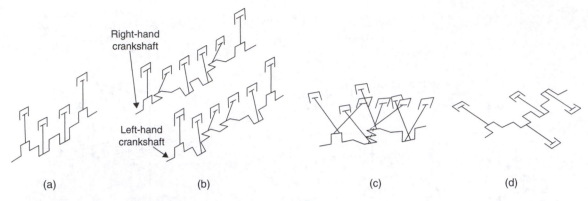

Figure 1.26 Cylinder and crankthrow arrangements: (a) in-line four-cylinder (b) in-line six-cylinder (c) eight-cylinder (d) horizontally opposed four-cylinder

For V and horizontally opposed engines (Figures 1.26c and d), the firing orders follow a sequence similar to those of in-line engines, but the crankthrows are so disposed that firing alternates between the cylinder banks as it proceeds to and fro along the crankshaft.

The V6 engine is nevertheless deserving of special comment. When the Italian Lancia company pioneered the V6 engine layout in 1950, they inclined the two banks of cylinders at 60° to each other, which together with a six-throw crankshaft conferred equally spaced firing intervals of 120°. However, this was not generally true of other early V6 engines, where their designers had adopted an included angle of 90° between the cylinder banks and used a shorter three-throw crankshaft with the three pairs of connecting rods each sharing common, double-length, crankpins. This arrangement resulted in firing intervals of 90° and 150° alternately instead of even 120° intervals, so although these engines were more economical on fuel than a V8 engine they were not particularly smooth running. To overcome the problem of uneven firing intervals with a 90° V6 engine, Buick in America retained a three-throw crankshaft but ingeniously replaced the common, double-length, crankpins by adjacent single crankpins that were staggered by 15° in opposite directions to produce a so-called 'split-pin' crankshaft. It may be of interest to recall that the reason for adopting a 90° included angle between the cylinder banks of early V6 engines was basically one of production economy. In the case of American V6's it was to utilize tooling that already existed for their (90°) V8 engines, while conversely for European V6's it was to create new tooling that could also be used to meet any future demand for V8 engines.

Some designers of modern even firing V6 engines (even firing especially being desirable where turbocharging is featured) have in fact reverted to a 60° included angle between the cylinder banks, but have retained a three-throw crankshaft supported in four main bearings. The required angular spacing for the crankpins of each throw is then obtained by using what is termed a '*flying arm*' (Figure 1.77). This can be likened to a crankweb that does not connect with a main journal and which may also be utilized for balancing purposes.

Cylinder numbering sequence
The numbering sequence for the cylinders of motor vehicle engines has long been defined by the various standards organizations. In British and European practice, as defined by British Standards BS 5672: 1991 and International Organization for Standardization ISO 1204:1990(E) respectively, the cylinders of an in-line or single bank engine are numbered consecutively as viewed from the flywheel end of the engine (which is in contrast to earlier established practice where the cylinders were numbered consecutively as viewed from the nose of the crankshaft). Both a capital letter and an identical system of numbering is used to designate the cylinders of V and horizontally opposed or double bank engines, the prefix letters A and B denoting the left- and right-hand banks of cylinders respectively as viewed from the flywheel end of the engine. In American practice, the suffix letters R and L are preferred to denote the right- and left-hand banks of cylinders respectively. Owing to the differences that can arise, past and present, the service engineer should always consult the manufacturer's engine specification for confirmation of the cylinder numbering system that applies.

1.6 CYLINDER BLOCK, CRANKCASE AND HEAD

Cylinder block construction

Cylinder block construction takes two forms. *Closed-deck* construction (Figure 1.27a) represents long-established practice and resembles a deep box-like enclosure for the cylinder barrels that also serves as a coolant jacket (Figure 1.28). Transfer ducts are provided in the top face or closed deck of the cylinder block, so as to permit the circulation of coolant to the cylinder head. With the *open-deck* construction (Figure 1.27b) the cylinder barrels are free-standing in that they are attached only to the lower deck of the cylinder block, which in past applications utilized detachable cylinder liners that tended to result in a less rigid construction. By dispensing with a continuous top face, the open-deck construction nevertheless reduces the complexity of the cylinder block casting. Where gravity sand casting is used it facilitates the coring for the mould into which the metal is poured. However, the increasing preference for using aluminium alloy, rather than grey cast iron, for the cylinder block and crankcase of modern lighter weight engines, has led to their manufacture by high-pressure die casting in the interests of economical mass production. Since this method of casting necessarily

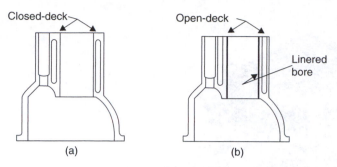

Figure 1.27 Cylinder block construction: (a) closed deck (b) open-deck

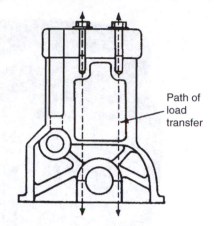

Figure 1.29 Direct load transference between the studs of the cylinder head and main bearing caps

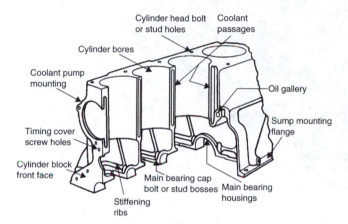

Figure 1.28 Cut-away view of cylinder block

involves the use of steel instead of sand moulds, the need for an open-deck construction to allow withdrawal of the steel cores becomes mandatory. Also, the liners may be cast directly into the cylinder block to restore structural rigidity. An open-deck construction further allows inspection of the coolant jacket for accumulated deposits. To perform this operation in a closed-deck cylinder block requires the addition of detachable cover plates.

In both closed-deck and open-deck cylinder blocks numerous internal supporting bosses must be provided for either the studs, or the setbolts, which clamp the cylinder head to the block. Wherever practicable, these bosses are aligned with the crankcase bulkheads that support the main bearings. This is to secure a direct path of load transference between the cylinder head and main bearing caps and thus minimize bending stresses within the cylinder block and crankcase structure (Figure 1.29). A noteworthy modern example of engine construction where this principle is carried to its logical conclusion can be found in the K Series engine originally developed by the Rover Group (Figure 1.30). In this design ten long bolts pass down through the cylinder block so that they clamp together in sandwich fashion the cylinder head, cylinder block, main bearing ladder and main bearing oil feed rail; all of these layers are located relative to one another by tubular dowels. This type of through-bolt construction therefore makes for a very even distribution of clamping load virtually from top to

bottom of the engine, which is particularly beneficial in view of its structural material being aluminium alloy.

For closed-deck constructions, the internal supporting bosses for the head studs are arranged symmetrically about the cylinder walls, so that the cylinder head clamping load is as evenly distributed as possible around them. The object is, of course, to avoid any tendency towards cylinder bore distortion, which could lead to blow-by of the combustion gases and increased oil consumption. In the case of open-deck constructions, the supporting bosses for the head studs are usually disposed along the walls of the coolant jacket.

Cylinder bores

Clearly, the cylinder bores constitute the most important feature of the cylinder block. Since they act as a guide and a sealing surface for the sliding piston and rings, their accuracy of machining must be such as to minimize any out-of-roundness and taper effects, and to ensure that they are truly at right angles to both the crankshaft and the top deck of the block.

The cylinder bores must also be given a carefully controlled surface finish, because too rough a surface would cause wear, and too smooth a surface would hinder the running-in process. A suitable surface finish is usually obtained by final honing to give a cross-hatched finish (Figure 1.31), which retains the oil in the bores to lubricate the pistons and so reduces friction losses. The question of the most suitable surface finish for new cylinder bores is one of long standing, and it is perhaps significant that an American engineer once observed that somehow the engine knows how to finish the bore better than we do!

Cylinder bore wear in service may be attributed to various factors, which will be reviewed later in connection with the associated wearing of piston rings.

As the degree of wear varies in different parts of the cylinder, an overall assessment of bore wear is most conveniently made by using a dial gauge equipped with an extension spindle, so that wear readings can be observed outside the cylinder. The gauge is self-aligning in the cylinder bore and in use is rocked over the cylinder axis from side to side, the maximum dial reading corresponding to the diameter being measured.

Figure 1.30 A modern example of a through-bolted engine construction

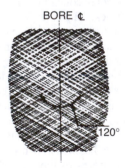

Figure 1.31 Honing pattern for cylinder bore

Crankcase construction

For in-line and V-formation engines, the most commonly used form of crankcase resembles a tunnel structure which extends downwards from the cylinder block (Figure 1.32). The roof is formed by the lower deck of the cylinder block and it is closed off at the base by either a detachable sump or a transmission housing. The crankshaft is underslung in the crankcase and supported by front, intermediate and rear main bearing bulkheads that form a series of crank chambers.

Until recently, it was customary for the side walls or *skirt* of the crankcase to be extended below the axis of the crankshaft, so as to increase the resistance to bending of the structure in the interests of engine smoothness, and also to simplify

Figure 1.32 Underside view of an aluminium alloy crankcase for a four-cylinder engine with five main bearings (*Chevrolet*)

the attachment of the sump. With the aid of modern computer stress analysis techniques, however, the present trend is to return to the much earlier practice of ending the crankcase side walls at the same level as the crankshaft axis, without sacrificing rigidity and with a consequent saving in weight.

For horizontally opposed engines, the cylinder blocks are again cast integral with the crankcase, which is usually divided on its vertical centre line (Figure 1.33). The crankcase halves are then clamped together by through-bolts on either side of the crankshaft main bearings. A series of bolts is also fitted around the peripheral joint faces of the crankcase, it being usual for the sump to be made integral with each half. Less commonly, a one-piece construction is used with an open underside, detachable main bearing caps and oil sump (Figure 1.33).

Main bearing locations

For in-line and V engines, the upper main bearing halves are carried direct in saddles formed in the crankcase bulkheads, whilst detachable inverted caps of great rigidity accommodate the lower main bearing halves. These bearing caps are usually recessed into the underside of their respective bulkheads and secured to them by either studs or setbolts, which thus support the maximum combustion loads imposed upon the crankshaft (Figure 1.34a). In some designs where the crankcase-to-sump joint face is at the same level as that of the crankshaft axis, the bearing caps may be located laterally by a pair of dowels, since it is no longer expedient to recess them into the lower face of the crankcase (Figure 1.34b).

Of more recent application is a form of crankcase construction that embodies a one-piece main bearing ladder or deck (Figures 1.30 and 1.35). This in effect integrates the main bearing caps into a single rigid structural element, rather similar to the bed plate construction found in very large diesel engines.

In some later designs of V eight-cylinder engine with aluminium alloy crankcases, the fastening arrangements for the main bearing caps may be duplicated, so that each cap is retained by four nuts and studs threaded vertically into the cylinder block above (Figure 1.36). As a further contribution to crankcase rigidity and bearing support, transverse bolting for the main bearing caps may also be employed (Figure 1.36), a feature that has previously been found in diesel engine practice (Figure 2.6). To complement these more comprehensive main bearing cap fastening arrangements, aircraft style 'bihexagon' or 'twelve-point' headed bolt and nut forms may

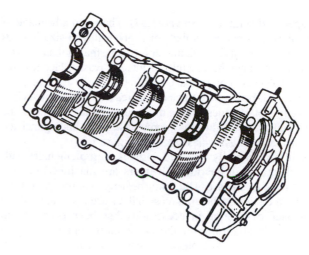

Figure 1.35 A one-piece main bearing deck (*Renault*)

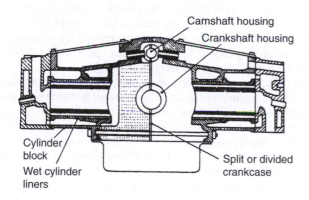

Figure 1.33 Horizontally opposed cylinders with divided crankcase

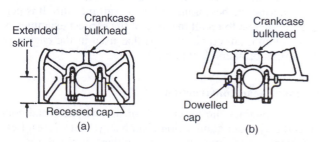

Figure 1.34 Crankcase bulkhead and main bearing cap locations

Figure 1.36 Vertical and transverse fastening for the main bearing caps in a high-grade V eight-cylinder engine (*Lexus/Toyota*)

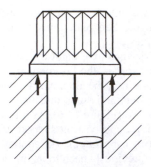

Figure 1.37   Bihexagon or twelve-point headed bolt

be specified (Figures 1.36 and 1.37). The main advantage of the bihexagon pattern is that for a given thread size it saves weight, because its head diameter is much smaller than a corresponding hexagon head, whilst its integral washer face restores an equivalent clamping area. Comparable torque tightening values can still be applied, since the increased number of socket-engaging flats compensate for the reduced turning radius of the bolt head or nut. This form of screw fastener is finding increasing favour among engine designers (Figure 1.30).

With horizontally opposed engines, complementary saddles for the main bearing half-liners are machined in each half of the crankcase. The main bearings cannot, therefore, be changed unless the crankcase halves are taken apart. This drawback of internal inaccessibility has been overcome in a more recent version of the horizontally opposed engine which features a one-piece construction for the cylinders and crankcase, as described in the previous section.

Camshaft bearing locations

A crankcase-mounted camshaft is supported either to one side of, or directly above, the crankshaft, depending on whether an in-line or a V cylinder arrangement is used. With both locations the camshaft bearings are usually carried in webbed extensions of the corresponding main bearing bulkheads. Spanning the underside of the camshaft may be a lubrication trough, which is formed as an integral part of the crankcase. The camshaft followers or tappet barrels slide in bores machined either directly in the material of the crankcase (Figure 1.17), or in bolted-on tappet blocks. In the case of horizontally opposed engines, the bearing bores for a central camshaft are formed similar to those for the main bearings beneath them (Figure 1.33).

Cylinder block and crankcase materials

The combination of cylinder block and crankcase is the single largest and most expensive component of an engine and may be produced from either cast iron or aluminium alloy.

Cast iron is an alloy of iron and carbon; there are two general classifications, known as white cast iron and grey cast iron. Although both varieties have a carbon content between 2.5 and 4.0 per cent, the difference between them is concerned with the condition in which the major portion of the carbon exists in the metal structure. Grey cast iron (so called owing to its grey rather than white appearance when fractured)

is used for cylinder block and crankcase manufacture, because most of the carbon is present as flakes of graphite. This feature not only makes the material more readily machinable, but also provides a satisfactory wear- and corrosion-resistant bearing surface for the cylinder bores. The rigidity of cast iron is such that it exhibits very little tendency towards distortion under the loads and temperatures encountered in the highly stressed engine structure. In addition, it possesses useful sound-damping properties.

Apart from its low cost, an outstanding characteristic of grey cast iron is the ease with which it can be cast into intricate shapes of thin section. Using modern casting techniques, the thickness of the cylinder block walls can therefore be minimized to save weight, which otherwise is the only real disadvantage with the cast iron cylinder block and crankcase.

The term 'aluminium' is generally used to describe not only the very soft and ductile commercially pure variety, but also the numerous aluminium alloys that comprise aluminium with usually more than one element added to it. Only the latter are of interest for cylinder block and crankcase construction, since they can be made harder and more readily machinable than commercial aluminium, which tears badly and poses screw-threading problems. The main attraction of using an aluminium alloy for casting the engine structure is the saving in weight that it affords; its density is about one-third that of cast iron. On the debit side, its strength is about two-thirds that of cast iron. This means that metal sections have to be thickened to compensate for the lower strength, so that in reality the saving in weight is generally nearer to one-half that of the cast iron version.

There are two main classifications for aluminium alloys: those which can be hardened by cold-working processes, and others that can be heat treated to obtain the desired mechanical properties. It is the latter alloys of the aluminium-silicon type that find the widest application for cylinder block and crankcase manufacture, because they retain their strength at moderately high temperatures, possess good casting fluidity and are the most resistant to corrosion.

Although an appreciably lighter construction can be obtained by using aluminium alloy, its wear-resistant properties are less acceptable for the cylinder bores, so that pistons of similar material cannot run directly in them. To overcome this disadvantage, either cast iron cylinder liners, iron-coated pistons or, more recently, silicon-carbide particles dispersed in nickel-plated cylinder bores may be employed. An aluminium alloy engine structure is also less tolerant both of careless handling, especially in respect of screw-thread connections, and of accidental over-heating through loss of coolant.

Historically, aluminium alloys were introduced extensively during the beginning of the aircraft industry. It is perhaps of interest to recall that one of the earliest examples of a cast aluminium crankcase was that used in the engine of the Wright brothers' first aircraft.

Cylinder head construction

In petrol engines the lower deck of the cylinder head contains the cylinder combustion chambers (Figure 1.38). Less commonly, these may be formed either within the piston heads, or by the combination of an inclined top deck for the

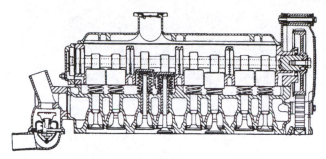

Figure 1.38 Sectional view of a cylinder head (*Fiat*)

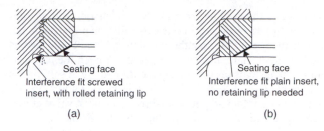

Figure 1.39 Valve seat inserts: (a) early practice (b) later practice

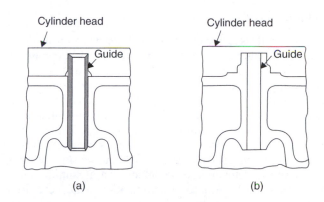

Figure 1.40 Valve guides: (a) removable (b) integral

cylinder block and specially shaped piston crowns. With the latter arrangements, the valves seat directly in the flat under-side of the cylinder head.

The top deck of the cylinder head provides a platform for mounting the overhead valve mechanism and, in the case of some overhead camshaft installations, it may be extended forwards to enclose the upper timing drive system. The walls of the cylinder head form the coolant jacket and provide attachment faces for the intake and exhaust manifolds. An upper continuous flange is formed by the walls for mounting the valve cover and serves to raise the sealing joint face above the level of oil draining from the valve mechanism. The coolant outlet connection at the front part of the cylinder head may also serve as a housing for the thermostat, this being described later. Sandwiched between the upper and lower decks of the cylinder head and surrounded by coolant are the inlet and exhaust valve ports. These are necessarily curved and kept as short as possible, the latter to avoid excessive heat transfer both from the coolant to the induction system and from the exhaust system to the coolant. If one port serves two adjacent cylinders it is said to be *siamesed*.

Valve seats

The valves may seat either directly in the material of the cylinder head or, in harder-wearing rings, inserted therein. Valve seat inserts are usually confined to engines where the cylinder head material is aluminium alloy. Modern valve seat inserts usually take the form of plain rings of greater depth than width, their proportions being such as to confer adequate resistance against distortion (Figure 1.39b). In earlier applications, the more expensive screwed-in type of insert was sometimes used, but it could suffer from inferior heat transfer (Figure 1.39a). For both types it is necessary that they be made an interference fit in the cylinder head. An infrared oven is a feature of the production line for the cylinder head of the Jaguar AJ6 engine, this being used to heat the aluminium alloy cylinder head before the valve seat inserts and valve guides are pressed into place.

Valve seat inserts for aluminium alloy cylinder heads are produced from either a nickel alloy iron casting, or a sintered powder metal the composition of which can be tailor-made to impart the desired hot strength, hardness and corrosion resistance, together with a matching coefficient of expansion. Metal sintering is later explained in Section 1.7 in relation to engine connecting rods.

The use of hard alloy inserts or of local heat treatment to harden the exhaust valve seat areas is also required where cast iron cylinder heads are retained for modern engines designed to run on unleaded petrol. This is because the lubricating effect of lead compounds, formed by the combustion of lead anti-knock additives, previously tended to prevent valve seat recession caused by high temperatures and severe mechanical stress. Although the progressive loss of performance associated with the early stages of exhaust valve recession is hardly likely to be noticed, when the point is reached that valve clearance is lost and one or more valves are prevented from closing properly (Section 1.10), the engine will begin to misfire and potentially serious valve and engine damage can occur.

Valve guides

Coaxial with the valve seatings are the valve guides, which are carried in bosses extending from inside the valve ports to the top deck of the cylinder head. Their length must be such as to present an adequate bearing surface to resist any side loading on the valve stems, and also to provide a ready path of heat transfer from the exhaust valve head. The working surface of the valve guide must possess low friction to minimize resistance to sliding of the valve stem. They are further arranged to project above the level of oil draining from the valve mechanism on to the top deck of the cylinder head.

The valve guides may be either removable from the cylinder head (Figure 1.40a) or, as is sometimes the case in American practice, cast integrally with it (Figure 1.40b). Separate guides are, of course, required with aluminium alloy cylinder heads. Removable guides are always made an interference fit in the cylinder head material, thereby assisting heat transfer from the valve to the cooling medium, which surrounds their supporting bosses. Grey cast iron is generally the preferred material for the valve guides, although sintered metal may also be used, which is hardened and tempered to match the hardness of the valve stems (as previously mentioned metal sintering is later explained in Section 1.7). It may be of interest to add that cast iron inlet and phosphor-bronze exhaust valve guides were specified at the design stage of the Rolls-Royce V8 engine, the latter material being chosen to avoid valve stem scuffing.

Cylinder head materials

The material used for cylinder head construction is either cast iron or aluminium alloy, as in the case of the cylinder block and crankcase. In favour of the much widely used cast iron cylinder head is its greater rigidity and better noise damping properties. However, apart from effecting a saving in weight, the greater heat conductivity of aluminium alloy is beneficial in maintaining a more uniform temperature throughout the cylinder head. This may sometimes permit the compression ratio of an engine to be raised slightly without incurring detonation or knocking. An aluminium alloy head has often been used in combination with a cast iron block and crankcase, and it is usually considered essential for air-cooled engines to ensure efficient heat transfer.

A service difficulty sometimes encountered with aluminium alloy cylinder heads is that they can prove obstinate to remove if corrosion deposits build up around their fixing studs, and it is for this reason that some designers specify setbolts. Owing to the greater thermal expansion and contraction of an aluminium alloy head, a smoother surface finish is generally specified for its mating face to prevent a ratcheting action against the cylinder head gasket.

Tightening down the cylinder head

Before carrying out this operation it is always advisable to consult the particular manufacturer's service instructions (Figure 1.41), especially in respect of the following:

1 Check whether the screw threads and washer faces require lubrication, and if so the type of lubricant to be used. The point here is that the presence or otherwise of a lubricant on screw-thread assemblies will affect the clamping load they exert for any given value of torque tightness.
2 Establish the correct sequence of nut or setbolt tightening, the number of stages in which it is to be achieved and the final torque value to be attained. For improved accuracy in tensioning the cylinder head fastenings a turn-of-nut method (as it is known to structural engineers) may be used in conjunction with initial torque tightening. Following the latter, the setbolts are marked on their heads with paint spots that all face in the same direction; then each nut is given a further specific amount of turn, perhaps in two stages of a quarter-turn each. With the setbolts equally tightened the paint spots will all be facing in the same but of course new direction. Whichever tightening procedure is used, the numerical sequence of tightening specified usually involves starting from the centre and working alternately towards each end.

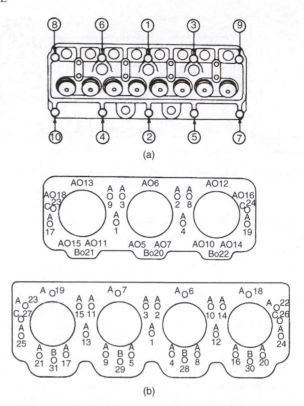

Figure 1.41 Examples of cylinder head nut or bolt tightening sequence: (a) car petrol engine (*Toyota*) (b) commercial vehicle diesel engines (*Gardner*)

Any retightening of the cylinder head after the engine has been run should be done strictly in accordance with the manufacturer's recommendations. Its purpose is to compensate for slight settling of the gasket material after initial exposure to heat and vibration and therefore restore the required sealing pressure. The consequences of incorrect tightening of the cylinder head can prove very expensive to rectify. They may include at least coolant and compression leaks via the head gasket, and at worst the failure of screw-thread fastenings and distortion of the cylinder bores.

Cylinder head gasket

Although the mating faces of the cylinder head and block are machined smooth, flat and parallel, in reality there are always minute surface irregularities and structural deflections to be accommodated. A static seal or head gasket is therefore required, which possesses the necessary degree of both plasticity (pliability) and elasticity (springiness). The effect of tightening down the cylinder head on to the block places the gasket under compression, so that sufficient friction is created between the sealing surfaces to resist extrusion of the gasket by cylinder gas pressure.

However, there is a great deal more to gasket application than this may suggest, because varying amounts of sealing pressure between the cylinder head and block are required at different locations. Sealing against the combustion gases clearly demands far higher unit pressures than are required to seal against the passage of coolant and lubricant through the gasket, while on the other hand very high unit loads are

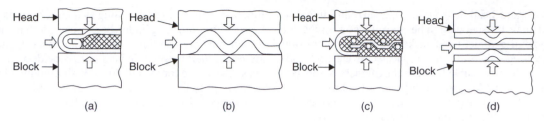

Figure 1.42 Types of cylinder head gasket: (a) metal and asbestos layers (b) embossed steel sheet (c) metal-cored composition materials (d) multi-layer steel

concentrated on those areas immediately surrounding the cylinder head studs or setbolts. The gasket must therefore be designed so that its distribution of sealing pressure conforms to an optimum pattern for any given engine. This requirement has tended to become even more demanding with modern engine design, because not only has there been a trend towards increased peak combustion pressures, but also in the interests of saving weight the structural rigidity of the cylinder block may be compromised, which makes clamping loads more difficult to control.

Apart from maintaining a seal by compensating for the non-uniform loading between the cylinder head and block, the gasket material must also be able to withstand the intense heat of combustion, the penetrating effect of the coolant and the chemical effect of the lubricant. It is for these reasons that the cylinder head gasket has always been regarded as the most critical sealing application on any engine.

Types of cylinder head gasket
In earlier practice, cylinder head gaskets for engines of moderate size and power were traditionally of sandwich construction. They compromised a centre layer of asbestos millboard between two sheets of either copper or tin-plate steel, together with eyelet reinforcements around the combustion chamber openings, which protected the edge of the gasket and increased the sealing pressure in these regions (Figure 1.42a). The embossed steel sheet type of gasket was later introduced for heavier-duty applications, because it better retained cylinder head torque tightening owing to its all-metal construction (Figure 1.42b). A plastics resin coating enabled it to accommodate the minute irregularities of the mating surfaces. By the early 1970s a different type of gasket construction had gained acceptance, which comprised a metal core of tanged or perforated steel sheet faced with fibre composition materials such as asbestos and plastics (Figure 1.42c). In a later version the fibre material facings are bonded to a plain steel sheet to provide a laminated construction, instead of being clinched to the metal core. Eyelet reinforcements around the cylinder bore areas are again used in both forms of composite gasket.

Since exposure to asbestos is now recognized as being a health hazard (and is a topic that will receive further mention in connection with friction materials), the manufacture of non-asbestos cylinder head gaskets has now become established. Gaskets of this type are typically based on a steel core with either aramid fibre, glass fibre or graphite facings with zinc-coated or tin-plated steel beading for protecting their sealing edges.

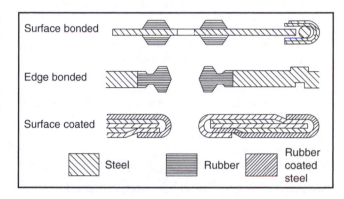

Figure 1.43 Steel and rubber cylinder head gasket constructions (*Cooper Payen T&N*)

A further development has been the introduction of steel and rubber gasket constructions. Three types of steel and rubber gasket that have been developed by Cooper Payen Ltd are shown in Figure 1.43. In the surface-bonded type a rubber bead is bonded on to a stainless steel core around the oil and water ways. A wire and ring eyelet assembly provides the cylinder bore sealing system. Since the rubber requires relatively little load to seal effectively, the surface-bonded gasket confers effective sealing with moderate overall clamping load. It is specified primarily for petrol engine applications. The edge-bonded type comprises a relatively thick steel plate, its edge being made sufficiently wide to allow the direct bonding of a rubber grommet, which provides sealing against oil and coolant. As this type of gasket is associated with medium to heavy diesel engine applications, the cylinder bore sealing system is generally obtained by deformation of the steel plate between the cylinder head and liner protrusion. The surface-coated type is of laminated construction, which incorporates several laminations of rubber-coated and plain steel, the rubber coating providing interlaminar and surface sealing against oil and coolant. Another feature of this particular construction is that it allows a preferential thickness of the gasket in the cylinder bore area. The various types of elastomer used with these steel and rubber gaskets, in ascending order of temperature requirements, are nitrile butadiene, silicone and fluorocarbon.

Since the late 1990s the multi-layer steel or MLS type of cylinder head gasket has gained widespread application. It is constructed from two or more layers of steel laminate, one or more of which will include embossed sealing beads (Figure 1.42d). The gasket may also receive a special surface treatment. In a typical application for an engine with an aluminium cylinder block and head, a 0.5 mm (0.02 in) ultra-thin two layer gasket may be specified. The advantages claimed for using this type of cylinder head gasket are those of improved sealing, enhanced reliability and reduced crevice or 'dead' space. This becomes possible because the thinness of the gasket allows it to be made closer fitting around the tops of the cylinder bores. By reducing crevice space here the accumulation of unburned gases and hence their contribution to hydrocarbon emissions is discouraged.

Cylinder head gasket misbehaviour

A cylinder head gasket that is leaking or 'blowing' may do so in various ways, as follows:

Internal leaks   Between two adjacent cylinders, as evidenced by weak compression pressure in both; through to the coolant jacket, causing bubbling in the radiator and loss of coolant; and through to the coolant jacket, allowing coolant into the cylinder to cause misfiring and steam issuing from exhaust.
External leaks   Outwards from the joint to produce a spitting noise when opening the throttle.

Cylinder liners

During an earlier era of rapid bore wear, cylinder liners were quite widely used either as original equipment or as an overhaul feature. This was because the particular grade of iron from which they were cast centrifugally could be selected for its wear-resistant properties, rather than for the free-flowing characteristics required of an iron for casting the cylinder block and crankcase. However, later developments in the fields of piston ring coatings, lubricating oil formation, oil and air filtration equipment and cooling system control have all combined to minimize cylinder bore wear, so that the need for detachable liners on this score seldom arises. Cylinder liners are now generally specified either to provide a suitable wear-resistant surface for the cylinders of aluminium alloy engines, or to simplify the production of cast iron engines by permitting an open-deck form of cylinder block.

Dry cylinder liners

A detachable dry liner takes the form of a plain or a flanged sleeve (Figure 1.44a), the entire outer wall of which is maintained in intimate metal-to-metal contact with the cylinder block. This is of closed-deck construction and may be of either cast iron or, less commonly, aluminium alloy.

Non-detachable dry liners have been cast integrally with aluminium alloy cylinder blocks of both closed- and open-deck constructions (Figure 1.44b). In a recent V eight-cylinder engine of aluminium alloy construction produced by General Motors, the cast iron liners are retained in their respective positions while the molten alloy is injected into the die cavity surrounding them. Modern cast-in iron liners have a wall thickness that is typically in the region of 3 mm (0.12 in).

Dry liners generally contribute to the rigidity of the cylinder block, but tend to introduce a barrier to heat flow at the adjoining surfaces. This effect is minimized where the cylinder block is made from aluminium alloy, as a consequence of its good heat conductivity.

Wet cylinder liners

The wet type of liner always takes the form of a flanged sleeve, the outer wall of which is largely exposed to the coolant in the cylinder jacket. It may be incorporated in both closed- and open-deck cylinder block constructions. Clearly, the wet cylinder liner is better cooled than the dry type and can more easily be renewed when worn, as will be explained later. It contributes little to the rigidity of the cylinder block, however, and there is always the possibility that coolant leaks into the crankcase may occur.

Two distinct methods of locating wet liners may be used, according to whether they are being installed in closed- or open-deck cylinder blocks, as follows:

Closed-deck   The cylinder liner is provided with a top flange only and is suspended through the coolant jacket from where it is clamped between the cylinder head and the upper deck of the cylinder block (Figure 1.45a).

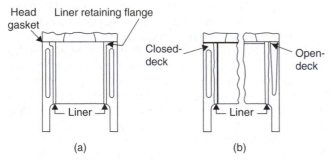

Figure 1.44   Dry cylinder liners

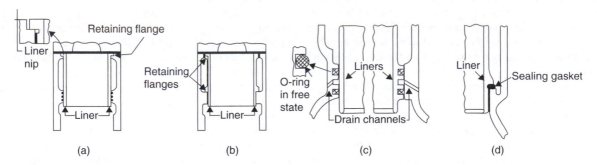

Figure 1.45   Wet cylinder liners in (a) closed-deck and (b) open-deck cylinder blocks. Sealing arrangements for (c) suspended wet liners and (d) wet liner held in compression

Open-deck Here the cylinder liner must be provided with a top and a lower flange and it is held in compression within the coolant jacket between the cylinder head and the lower deck of the cylinder block (Figure 1.45b).

An advantage of the first arrangement is that the cylinder block is relieved of stresses that would otherwise be imposed by the axial expansion of the liner upon heating. With the second arrangement, the less intrusive top flange generally permits better cooling around the upper part of the liner.

For closed-deck cylinder blocks, two oil-resistant (synthetic) rubber O-ring seals encircle the lower part of the liner and are deformed into grooves where it passes freely through the lower deck of the cylinder block. The sealing rings may be grooved into either the cylinder block or the liner itself (Figure 1.45c shows both forms), and a third unfilled groove between them communicates with a drilling in the block that leads to atmosphere. This drilling serves as a drain channel for any coolant and, similarly, oil that may have seeped past the top and bottom sealing rings, respectively.

With open-deck cylinder blocks, a compression sealing gasket is generally used between the flange towards the bottom of the liner and its seating in the lower deck of the block (Figure 1.45d).

In both types of liner installation the cylinder head gasket completes the sealing arrangements for the top end of the liner.

A compromise arrangement of cylinder liner is the so-called 'damp liner', which is part wet and part dry. The upper part of the liner is made thicker in section than the lower part, so that it acts as a wet liner, while the thinner lower part is made a slip fit in the cylinder block and is therefore dry. Since a step is formed between the upper wet and lower dry sections of the liner, it is retained by being nipped between the cylinder head and the abutment in the cylinder block on which the step rests. Although this construction results in a shorter coolant jacket, the direct cooling is nevertheless concentrated where it is most needed in the higher temperature region of the cylinder. There is, in fact, a tendency to shorten the coolant jacket in modern design, because it not only contributes to the rigidity or stiffness of an aluminium alloy cylinder block, but also accelerates the warm-up process to assist emission control.

Cylinder liner installation
It will be evident from the previous descriptions that the dry liner is usually (although not invariably) made an *interference* fit in the cylinder block. Typically, the block is bored out to provide an interference fit of 0.06–0.09 mm (0.0025–0.0035 in) between the cylinder and the liner, which will then need to be lubricated and pressed in under a load of about 2000–3000 kg (2–3 tons) (Figure 1.46). To avoid any possibility of liner bore distortion, the cylinder block studs are usually refitted before the liner bore is honed to final size.

The production method of installing dry liners into the aluminium alloy closed-deck cylinder block of the Jaguar AJ6 engine is of interest, since it does not involve the use of a press. A two-stage infrared oven is used to heat the entire cylinder block for three minutes at a time in each stage, so that when it emerges from the oven the bores have expanded sufficiently for the liners to be slid into position by hand. Immediately all six

liners are in place, the cylinder block enters a special cooling tower where its cooling is rigidly controlled.

In contrast, wet liners are generally made a *slip fit* in the cylinder block (Figure 1.47). A typical cylinder liner-to-block clearance would be 0.05–0.15 mm (0.002–0.006 in). Even so, a manufacturer may recommend that the cylinder block be preheated, so that there is no hindrance to correct insertion of the liners and seals. There is always the danger that if an engine with wet liners is cranked over with the cylinder head removed, the liners could be dragged clear of their locations by the rising pistons. The temporary fitting of retaining clamps on the liners is therefore the safest practice in these circumstances.

Figure 1.46 Installing a dry cylinder liner with a Flexi-Force hydraulic press (*Brown Brothers*)

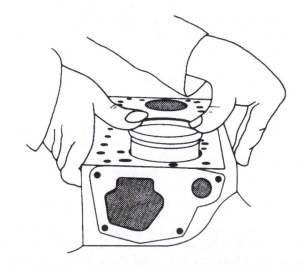

Figure 1.47 Inserting a wet cylinder liner by hand (*Perkins*)

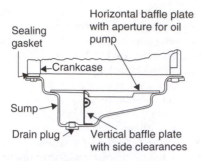

Figure 1.48 Crankcase sump

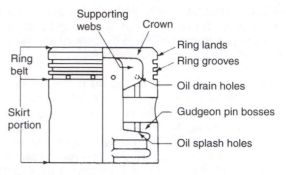

Figure 1.49 Piston construction and nomenclature

A final consideration in fitting flanged liners, either wet or dry, is the provision of a small amount of *nip*. This refers to the amount the liner top flange protrudes above the top deck of the cylinder block, so as to promote an efficient gasket seal when the cylinder head is clamped on to it (Figure 1.45a). In practice, a liner nip of 0.05–0.12 mm (0.002–0.005 in) is fairly typical.

Crankcase sump or oil pan

This unit acts as a reservoir to store the oil that is required by the engine lubrication system. It further serves as a vessel in which any sludge, water and metal particles in the oil can settle out, and also provides an opportunity for any entrained air to escape from the oil.

The sump is either of pressed steel construction, or produced from an aluminium alloy. In its latter form it better contributes to the rigidity of the crankcase and also assists with heat dissipation. Attachment of the sump to the crankcase is usually by means of setscrews through mating flanges, between which is sandwiched a flexible packing or gasket.

Baffle plates are normally fitted in the sump to minimize both oil surging and agitation, the former arising from the changing motion of the car and the latter from the oil flung from the crankshaft bearings (Figure 1.48). A screwed plug is incorporated at the lowest point in the sump for draining the oil in service.

In a recent Japanese application the oil sump is of two-piece construction, comprising an upper cast aluminium section that also houses a harmonic balancer system chain driven from the crankshaft, and a lower stamped steel section.

1.7 PISTONS AND CONNECTING RODS

Piston construction and nomenclature

The piston crown may be either flat topped or specially shaped in order to conform to the particular design of combustion chamber of which it forms one wall. Combustion loads are transmitted directly from the crown to the gudgeon pin bosses through intermediate supporting webs, which also facilitate the flow of heat to the encircling piston rings and thence to the cylinder walls. The ring belt immediately below the crown is thus largely relieved of loads that would otherwise tend to deform its grooves. Any closing in of the grooves would, of course, prevent the free radial movement of the rings and thus impair their sealing ability.

The main part of the piston below the ring belt is termed the skirt, and this is made as close fitting as practicable in the cylinder, thereby ensuring quiet operation and the maintenance of the rings at their most favourable attitude to the cylinder wall. Furthermore, the skirt must present an adequate bedding area to the cylinder wall, not only to minimize contact pressure, but also to assist with heat dissipation. It should be appreciated though that the piston skirt is not normally in direct contact with the cylinder wall, but is separated from it by a film of oil. Recessed flats may be provided at the termination of the gudgeon pin holes.

Modern practice favours the use of a *solid skirt* piston of rigid construction, because of the high combustion loads now encountered (Figure 1.49). Basically, its advantages are that it can be made thinner in section to withstand a given loading, so that it affords a saving in weight. It does, however, need a good deal of modification to provide acceptable expansion control of the skirt, as will be explained later. Another type of skirt which was once widely used is that known as the *split skirt* (Figure 1.50a). This incorporated a near-vertical slot extending from the centre of an upper horizontal slot down to the base of the skirt on the non-thrust side of the piston. Here it should be explained that the thrust side of the piston reacts against the side force arising from the angular motion of the connecting rod on the power stroke, while its non-thrust side reacts against the lesser side forces on the compression and exhaust strokes. (The two sides are also known as the major and minor thrust faces.) The split skirt piston was originally introduced to provide quiet running and, by virtue of its skirt flexibility, to accommodate a certain degree of cylinder bore distortion where this was prone to occur.

Piston materials and expansion control

In low-speed engines of early design, the material from which the pistons were made was cast iron to match that of the cylinders. With increasing engine speeds and output, however, it has long since become established practice for the pistons to be cast from aluminium alloy, materials of this type combining lightness in weight with high thermal conductivity. They have a moderate silicon content so that their mechanical strength is better maintained at high operating temperatures, which may now exceed 300°C, whilst their coefficient of thermal expansion is lower than that usually associated with aluminium alloys (Section 2.4). The addition of silicon also increases their resistance to corrosion and wear.

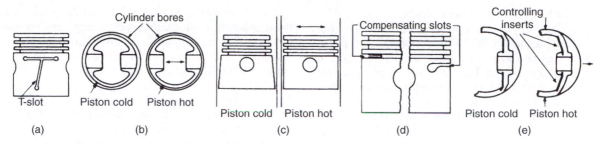

Figure 1.50 Methods of expansion control for pistons. (a) split skirt piston with compensating T-slot. Solid skirt pistons with (b) ovality (c) taper (d) compensating slots (e) controlling inserts

A more recent development has been the *squeeze cast* piston, which was introduced by AE Piston Products in the late 1980s. The squeeze casting technique retains the forming capabilities inherent to the casting process, while demonstrating the strength and integrity of forging. In practice a predetermined amount of molten aluminium is poured into the die, which is then closed and applies pressure to the casting as it solidifies, hence the term squeeze casting.

The associated advantages that accrue from this process include the following:

1 More economic use of aluminium, because the risers and runners needed for gravity diecasting are no longer required.
2 Porosity is minimized by the application of pressure, which compensates for the normal shrinkage that accompanies cooling and solidification of the molten aluminium.
3 Greater refinement of material structure, which arises from the increased heat flow at the interface where the aluminium is squeezed against the die.
4 Improved accuracy and surface finish, which again derives from the more intimate contact between the aluminium squeezed against the die.
5 Alternative materials can be readily introduced into the casting, such as the expansion control inserts or struts mentioned below and piston ring groove carriers (Section 2.4), with freedom from microporosity and hence improvement in the integrity of the bond.
6 A high top ring for better emission control (as explained later) can also be accommodated, by virtue of the enhanced material properties and strength of the squeeze cast piston.

Although the aluminium alloys chosen for the piston material expand less when heated than most other light alloys, they nevertheless expand at nearly twice the rate of the cast iron used for most engine cylinders. In the absence of special expansion control features, this relative difference in expansion rates would result in seizure of the piston when hot, and could be avoided only by tolerating a fitting in the cylinder when cold that would be too slack and would result in piston slap.

A basic method of controlling thermal expansion is to machine the piston to a special form, which is both oval in contour and tapered in profile (Figure 1.50b and 1.50c). The ovality is such that when the piston is cold, the minor axis of the skirt lies in the direction of the gudgeon pin. When the piston is hot the skirt then assumes a circular shape, because of the greater expansion occurring in the mass of metal comprising the gudgeon pin bosses.

The direction of taper allows for additional clearance when cold near the top of the skirt, since this part ultimately attains a greater running temperature and thus expands more than the cooler running lower portion. Immediately below the ring belt, where the skirt temperature is greatest, the degree of taper may be intensified to produce an overall barrelled shape. Similarly, the clearances around the ring lands are progressively increased towards the piston crown, thereby avoiding contact when hot with the cylinder wall.

In order that the piston can be made as close fitting as possible in the cylinder, either compensating slots or controlling inserts or both may be incorporated in its construction. With split skirt pistons, circumferential expansion of the skirt is simply absorbed by temporary closing in of the near-vertical compensating slot. For solid skirt pistons it is usual for part-circumferential slots to be located within or, in some designs beneath the oil control ring groove (Figure 1.50d). The angular disposition of these horizontal slots is such that the flow of heat from the piston head is diverted from the thrust faces of the skirt, thereby reducing expansion across them.

In modern practice, widespread use continues to be made of controlling inserts or struts, which are stamped from alloy steel and cast parallel with the piston axis and integrally around each gudgeon pin hole. Each insert is pre-heated and placed in the die before pouring; no bond is attempted as they are not intended to add strength to the piston. The inserts are covered with aluminium only on the side away from the piston axis, so that they act in the manner of bimetallic strips when the piston is heated. Since the steel inserts have a lower coefficient of expansion than the aluminium piston, they are compelled to bow outwards and thus restrain expansion of the skirt across the thrust axis (Figure 1.50e). It is perhaps of interest to recall that this principle of controlling thermal expansion of the piston skirt was originally incorporated in the American Nelson-Bohnalite piston of the early 1920s, which was also widely used elsewhere. It was regarded by one prominent automobile engineer of the time as being 'the first piston to have been designed with an intelligent appreciation of the difficulties to be overcome'.

To ensure freedom from piston scuffing on cold starting or hot running conditions, it is important that the surface finish of the piston skirt is oil retentive. Various surface treatments may be applied to aluminium alloy pistons, including tin plating to assist the running-in process, graphite coating to avoid scuffing under borderline conditions of lubrication, and PTFE coating (Section 13.2) to reduce power losses through friction. Further developments in the surface treatment of pistons for modern high-performance engines include molybdenum spraying of the skirt, and nickel coating of the piston to combat erosion damage where it occurs in engines that are continuously operated close to their knock limit (Section 3.2).

The fitting clearance for a piston must always be sufficient to maintain an oil film during engine operation and thereby prevent scuffing or seizure of the piston in the cylinder bore. It rarely exceeds 0.05 mm (0.002 in) measured at the top of the skirt across the thrust axis, although it is unlikely to be less than this in the case of turbo-charged engines. The fitting clearance should always be in accordance with the manufacturer's specification. If an engine is fitted with wet cylinder liners, it is usual for matched liners and pistons to be made available in service.

Gudgeon pins and their location

The gudgeon pin is the vital mechanical link that hinges the piston to the connecting rod and, although it is of deceptively simple appearance, it must be recognized as being a precision engineered component. This is because it has to satisfy several conflicting requirements; namely, it must combine strength with lightness, be close fitting but with freedom to move, and resist wear without promoting scuffing. The gudgeon pin is of hollow construction and typically produced from a fine-grained plain carbon steel with controlled hardenability. It is lapped to a mirror finish of 0.05–0.10 (2–4 μin). The diameter of the gudgeon pin may be up to 40 per cent of the piston diameter, so that maximum bearing pressure in the piston bosses does not exceed 55 MN/m^2 (8250 lbf/in^2). Under load its ovality and longitudinal bending are not expected to exceed 0.025 mm (0.001 in) and 0.075 mm (0.003 in) respectively. The methods used for gudgeon pin location depend on whether the arrangement is a semi-floating or fully floating one.

Semi-floating gudgeon pin
This is held rigidly in the connecting rod eye and oscillates only in the piston bosses (Figure 1.51a). A semi-floating gudgeon pin is matched by grading to give a clearance of 0.0075–0.0125 mm (0.0003–0.0005 in) in the piston pin holes. Current practice is for the gudgeon pin to be retained by an interference fit, rather than by clamping it in a split eye with a pinch-bolt as in earlier designs. The latter method introduced a discontinuity into the connecting rod eye that

could be a source of weakness. Two advantages of the semi-floating arrangement are that there is less length of gudgeon pin subject to bending loads, and it eliminates one potential source of noise that can otherwise develop in the small-end bearing of a fully-floating arrangement.

Fully-floating gudgeon pin
This is able to oscillate not only in the piston bosses, but also in the connecting rod eye that is bushed for the purpose, as described later. In order to prevent it from escaping sideways and contacting the cylinder wall, the fully-floating gudgeon pin must be located axially in the piston bosses. This is because it is free to oscillate both in the connecting rod eye and the piston bosses at normal operating temperature. Location is usually provided by spring retaining rings or circlips, which are expanded into grooves near the outer end of each boss and thus act as removable shoulders (Figure 1.51b). A fully floating gudgeon pin is matched by grading to give a clearance of 0.0025–0.0075 mm (0.0001–0.0003 in) in the piston pin holes.

Offset gudgeon pin
When the power stroke begins, the piston is forced by a combination of gas pressure and connecting rod angularity to move laterally across the cylinder; this effect is more pronounced with shorter centre distances for the connecting rod, as found in modern engines. The manner in which this movement occurs from the minor to the major thrust side can be such as to cause piston knocking. To minimize noise from this source the axis of the gudgeon pin can be slightly offset from that of the piston and in the direction of the major thrust side. The effect is to tilt the piston during the compression stroke, so that contact is first established between the lower part of the skirt and the major thrust side of the cylinder. At the beginning of the power stroke it then only remains for the upper part of the piston to move across the cylinder to establish full skirt contact with the major thrust side, this movement being beneficially damped by the friction of the rings in their grooves. The amount of gudgeon pin offset used is generally in the region of 1.5 mm (0.06 in) but it can also be influenced by piston clearance and a 'listening session' on the engine development test bed. A stiffer design of piston skirt has sometimes been necessary where an offset gudgeon pin is used.

Types of piston ring and nomenclature

There are two basic types of piston ring used in petrol and diesel engines. These are designated compression and oil control rings, and in each case the nomenclature is most conveniently presented by illustration (Figures 1.52 and 1.56).

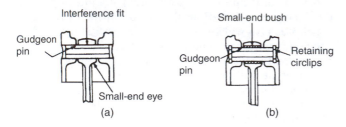

Figure 1.51   Methods of gudgeon pin location: (a) semi-floating (b) fully-floating

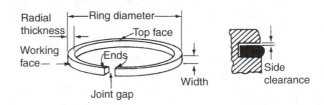

Figure 1.52   Piston ring nomenclature

For modern petrol engines two compression rings and one oil control ring are used on each piston. In earlier practice it was customary for more than three rings to be used. For example, the engines with a fairly low compression ratio used in Rolls-Royce cars of the late 1920s originally had five-ring pistons. The basic reason why fewer rings can now be used is that the higher mean cylinder pressures better augment the sealing action of the rings. Various forms of compression ring may be specified, the differences between them lying in their cross-sectional shapes and the application of wear-resistant surface treatment. Top compression rings are usually of plain rectangular section, their inner and outer edges being slightly chamfered to prevent sticking in the groove (Figure 1.53a). The working surface of the ring may also assume a barrel form, instead of being flat and parallel to the cylinder wall, so that it is better able to accommodate any slight piston rock where the skirt length may be limited (Figure 1.53b).

The second compression ring principally serves to reduce the pressure drop across the top ring, and it can with advantage be made relatively more flexible so as also to assist oil control. Various departures are therefore made from the basic rectangular cross-section, the main object being to compensate for torsional deflection of the ring under combustion pressure, so that top edge contact with the cylinder wall is avoided. Otherwise, the ring tends to pump oil towards the combustion chamber and therefore opposes the action of the oil control ring. To reverse this effect taper-faced and stepped torsional rings are widely used, the two features being combined in some designs (Figures 1.54 and 1.55). Another form of second compression ring with efficient oil control properties is the Napier type, which has a reduced width of taper face with a hook-like undercut below, this being where the bulk of the oil has to be controlled.

The basic slotted type of oil control ring is simply a modification of the plain rectangular-section compression ring. It takes the form of an outward-facing channel section from which collected oil escapes through slots machined radially to the back of the ring (Figure 1.56a). The oil then returns to the sump via communicating holes drilled through the piston wall at the back of the groove. A fairly high radial pressure is exerted against the cylinder wall by virtue of the narrow working surfaces of the ring, so that any tendency for it to ride over the oil film is counteracted.

Oil control requirements have become more exacting for modern engines because the use of high compression ratios increases the depression acting on the rings during the induction stroke, especially on the overrun with a closed throttle. To improve oil control under these conditions various forms of composite rings are commonly used (Figures 1.56b and c). They generally feature two or more independent working faces, comprising flexible rails with rounded wiping edges, which act in conjunction with a spring expander. This presses the rails axially against the sides of the ring groove and radially against the cylinder wall. Since the flexible rails can act independently of one another, they do not lose their effectiveness in the presence of any piston rock.

Piston ring materials and expansion control

Compression rings and slotted oil control rings are produced from descending grades of cast iron, because this material provides a satisfactory wear-resistant surface and retains its elasticity at high temperatures. Composite oil control rings for petrol engines utilize thin steel rails with chromium-plated

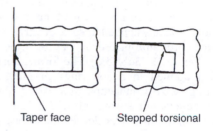

Figure 1.54 Second compression rings

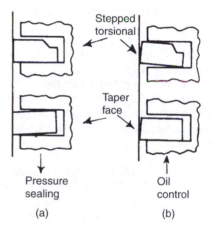

Figure 1.55 Action of second compression rings

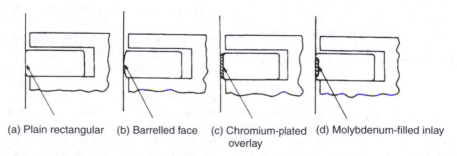

(a) Plain rectangular (b) Barrelled face (c) Chromium-plated overlay (d) Molybdenum-filled inlay

Figure 1.53 Top compression rings

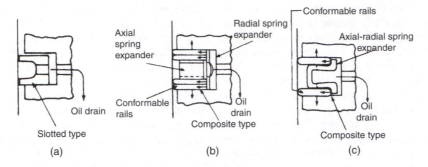

Figure 1.56   Oil control rings: (a) slotted type (b) and (c) composite types

working edges, whilst in the spring-loaded conformable ring used in diesel engines a cast iron slotted ring is expanded by a stainless steel backing spring.

Since the top ring not only is subject to the immediate effects of combustion pressure and temperature, but also must withstand abrasive and corrosive conditions, a wear-resistant coating is now generally applied to its face. This may comprise either a hard chromium-plated overlay or a molybdenum-filled inlay (Figures 1.53c and d). The former treatment was actually first applied to piston rings of military vehicle engines during the African campaign in World War II, and is especially beneficial in resisting abrasive and corrosive wear. Molybdenum inlay was a later development for piston rings, following its introduction on Chevrolet and Cadillac engines in 1963, and is particularly resistant to scuffing wear. Similar benefits are, of course, conferred on the cylinder wall itself.

A piston ring must be provided with a gap at one point on its circumference for three reasons:

1   So that it can be expanded over the piston head and then released into its groove.
2   To allow it to be compressed into the cylinder and thus exert an initial sealing pressure, which is then greatly augmented by gas pressure when the engine is running.
3   To accommodate circumferential expansion of the ring when it is hot.

With regard to point 3, although the coefficients of expansion of the cast iron and alloy steels from which the piston rings are made relate to that of the cylinder material, the circumferential expansion of the rings is greater because they attain a higher mean temperature, otherwise there would be no heat flow from one to the other. It is also for this reason that a piston ring, which maintains a uniformly distributed pressure against the cylinder when cold, is not likely to do so when hot, and this has therefore to be taken into account during manufacture.

The fitting clearances for piston rings are generally as follows:

Piston ring closed gap   To compensate for circumferential expansion this should usually be not less than 3/1000 of the cylinder bore diameter, and is measured with the ring installed in its cylinder. In practice, a gap of between 0.30 and 0.35 mm (0.012 and 0.014 in) is likely to be specified.

Piston ring side clearance   To allow for free radial movement of the ring in its groove this should usually be not less than 0.035 mm (0.0015 in) for cylinder diameters of between 50 and 100 mm (2 and 4 in) (Figure 1.52).

Piston ring misbehaviour

This is indicated by loss of cylinder compression, poor oil control and noisy operation of an engine. These complaints can result from wear, sticking or breakage of a ring, although the fault may not necessarily lie with the ring itself. The faults can be summarized as follows:

Abrasive wear   This has been much reduced by modern air filtration equipment and closed-circuit crankcase ventilation, which prevent airborne dust particles from entering the cylinders and the crankcase, thereby protecting the compression and oil control rings, respectively. Without some abrasive wear the piston rings would never, of course, bed in at all!

Scuffing wear   A definite explanation for this particularly severe type of wear has long been the subject of engineering research (since it can also occur, for example, between gear teeth), but it is generally thought to result from the formation and tearing of tiny welds on the sliding surface of the ring, under the effects of pressure and temperature. If it occurs it does so rather suddenly, typically during the running-in of an engine.

Corrosive wear   This is attributed to chemical attack by the products of combustion and occurs when these are able to condense on the ring surfaces at low temperatures. It is now minimized by better regulation of the cooling system, so that the engine warms up more rapidly and takes longer to cool down, and also by applying various surface treatments to the ring face.

Sticking   This occurs when the ring groove temperature becomes excessive and causes breakdown of the lubricating oil reaching the ring, so that solid products build up in the groove and prevent free radial movement of the ring. The role played by modern lubricating oils in protecting against this condition will be referred to again later.

Breakage   Apart from the breakage that can result from the sticking of a ring, another condition known as ring flutter may cause the breakage of rings in all cylinders if an engine is persistently over-revved. Following experiments made in the late 1940s by Dr P. de K. Dykes at Cambridge University, England, it is now recognized that if the reciprocating inertia

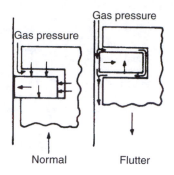

Figure 1.57   Piston ring sealing action

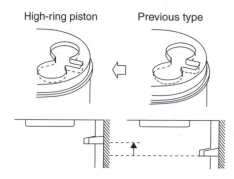

Figure 1.58   Comparison between previous and high-ring pistons (*Toyota*)

force acting upwards on the ring exceeds the gas pressure that is forcing it downwards and outwards, then during the end of the compression stroke and the beginning of the power stroke the ring will lose contact with the lower side of its groove (Figure 1.57). The consequent release of gas pressure from behind the unseated ring results in it collapsing radially inwards, and this can lead to failure.

Further developments in piston and rings

Since the normal operation of the pistons and their rings accounts for about 75 per cent of the total friction losses in an engine, it is perhaps to be expected that more recent developments have sought to reduce these losses in the interests of improving fuel economy.

To reduce the viscous friction that is caused by shearing of the oil film between the piston and its cylinder during operation, there is a trend towards lessening the contact area of the piston skirt, preferably without shortening the skirt as this could result in undesirable tilting of the piston and unsatisfactory ring behaviour. This conflict has been resolved in the case of the AEconoguide type of piston by specially contouring the skirt so as to form three low-area contact pads on each side.

The advent of direct fuel injection for passenger car petrol engines (Section 6.9) can lead to a requirement for relatively complex shaped piston crowns. Since these tend to raise the centre of gravity of the piston, this could be detrimental to its quiet operation, unless countered by an adequate length of skirt to restore accurate guidance. Hence, the piston can become heavier for this type of application. Also, the presence of a bowl-like depression in the piston crown can lead to a concentration of thermal stresses around the rim of the bowl and any projections therefrom (similar to that encountered in the piston crowns on direct-injection diesel engines). This therefore has to be taken into account during the design and manufacturing processes to avoid crack formation.

A reduction in friction of the piston rings, which receive uncertain lubrication especially when sliding against the upper cylinder wall, can be attributed to the recent Japanese development of the two-ring piston, which carries one plain compression ring and one three-piece oil control ring. Associated advantages include a decrease in piston *compression height*, that is the distance from the gudgeon pin axis to the piston crown, so that not only can the piston be made lighter but also the height of the cylinder block can be reduced. An

improvement in both fuel economy and engine manufacturing costs is therefore made possible.

Other developments in connection with piston rings mainly concern their angular and axial dispositions on the piston and their surface treatment. In the early 1980s a trend began in the use of piston ring grooves that were machined with a very slight upward tilt from the piston axis, thereby creating a similar effect to an internally stepped (or bevelled) piston ring. The purpose of this feature is therefore to ensure that when expansion and bending are taken into account, the ring groove and ring are presented squarely to the cylinder wall and not deflected downwards.

Beginning in the late 1980s it was recognized that by moving the piston rings nearer to the piston crown in both petrol and diesel engines, consistent always with maintaining durability, because it increases mechanical and thermal loading of the top ring groove and ring lands, the efficiency of combustion could be improved for increased power and reduced emissions of hydrocarbons. The reason for this is that the 'dead' space above the top ring and between the piston land and cylinder wall is useless for combustion of the gases compressed therein. Thus, if this space is reduced with what is sometimes called a 'high-ring' piston (Figure 1.58), smaller amounts of unburnt gases are expelled from the exhaust. In a recent example the top ring land height has been optimized at 3.0 mm (0.12 in).

Carbonitriding may be used as a less expensive alternative to chromium plating for cast iron top compression rings. This is a surface hardening treatment that involves the simultaneous diffusion of carbon and nitrogen into the ring material. Top compression rings produced from high alloy steel have also been recently introduced, which can allow ring widths as low as 1 mm (0.04 in) to be achieved, as compared with the more usual 1.5–1.75 mm (0.06–0.07 in) found in modern practice. The advantages associated with this type of compression ring include the following:

1 Improved fatigue resistance by virtue of their higher tensile strength.
2 Less tendency towards flutter conferred by their reduced inertia.
3 Increased conformability to improve control of blow-by and oil consumption.
4 Readily incorporated in 'high-ring' pistons for better emission control.

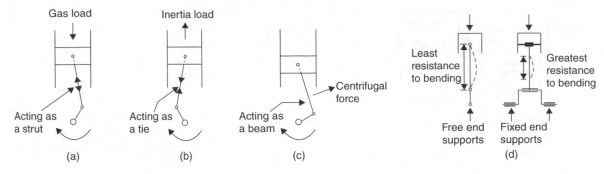

Figure 1.59 Forces acting on the connecting rod: (a) compression (b) tension (c) bending (d) concentration of bending load

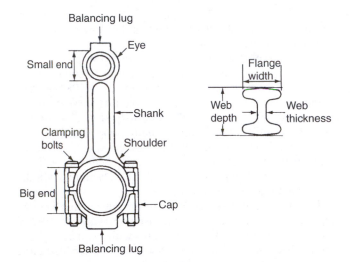

Figure 1.60 Connecting rod assembly

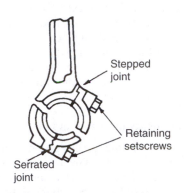

Figure 1.61 Types of angled big end

The Nitrolite steel top compression rings produced by Hepworth & Grandage are hardened on all surfaces for reduced side face wear and have smooth blended edges, instead of the chamfers required on chromium-plated rings, to provide improved control of blow-by and oil consumption.

Connecting rods

The nature of the loading on the connecting rod is such that it is subjected to a combination of axial and bending stresses (Figure 1.59), the former arising from reciprocating inertia forces and cylinder gas pressure, and the latter from centrifugal effects. The shank of the connecting rod is provided with an I cross-section for maximum rigidity with minimum weight (Figure 1.60). Since the end supports for the rod are free in the plane of rotation and fixed in the plane containing the crankpin and gudgeon pin axes, the largest dimension of the I-section is disposed in the plane of rotation to resist the greater bending effect therein. The depth of web in the I-section varies in accordance with any taper of the shank, but the flange width and web thickness remain constant over this length. To reduce stress concentrations the big-end arch and the small-end eye are merged very gradually into the shank portion of the connecting rod.

In some designs the big-end bearing parting line is arranged diagonally (Figure 1.61), because otherwise the width of the housing would be such that the connecting rod could not be passed through the cylinder for assembly purposes. To resist the greater tendency for the cap to be displaced sideways relative to the rod, either a serrated or a stepped joint is generally preferred for their mating faces. Hence, the securing setscrews in their clearance holes are relieved of all shear loads. Where the parting line between the rod and cap is arranged at right angles to the axis of the shank, the cap may be secured by either bolts and nuts, studs and nuts, or setscrews. They are produced from high-tensile alloy steel with special care being taken in their detail design to avoid stress-raising corners, which would lower their fatigue resistance. Their clamping load must always be such as to exceed the inertia forces acting on the rod. Reference has already been made to bihexagon or twelve-point headed screw fasteners in Section 1.6 and for the same reason they may be advantageously used to retain the big-end bearing cap (Figures 1.30 and 1.36).

A more recent development in connecting rod fabrication is for a forged sintered metal rod and cap to be produced in one piece and then forcibly separated. Basically, the sintering process involves highly compressing a metal powder, the composition of which can be tailor-made for the application, in a slim I-section mould. Next, the already formed connecting rod is heated to an elevated temperature in a sintering oven, which fuses together the metal particles to complete the process. The manufacturing advantages of what have become known as 'sinter-forged, fracture-split' connecting rods, are generally those of consistency in fabrication, minimal machining requirements and optimum material properties. In production a fully automatic process allows the cap portion, which is initially defined by a fine groove that corresponds to the parting line, to be split apart from the arch of the rod. This process results in the fractured halves of the rod and cap presenting an ideal mating surface for a perfect fit and improved accuracy of location, as compared to that possible with individually machined parts and, of course, it saves machining time. Although novel to the motor industry in the early 1990s, this type of 'fracture-split' connecting rod construction was originally developed in the late 1960s by the McCulloch Corporation in America, who had investigated various means of improving the structural integrity and fatigue resistance of connecting rods used in their very high speed chain-saw engines.

When a new engine is being developed it is now usual for a component supplier to design the piston and connecting rod as one interacting system, so that the lowest possible reciprocating weight can be attained.

Connecting rod materials

A prime requirement for a connecting rod material is that it should possess a high strength-to-weight ratio. Connecting rods produced from aluminium alloy were used to a limited extent in earlier practice, the primary object being to reduce the loadings on the big-end and main bearings. Also they offered the possibility of being run directly on the crankpins without an intervening separate bearing material. However, their use declined mainly because of an unsatisfactory and unpredictable fatigue life, which made them liable to failure in service. Connecting rods are therefore either forged from a high-strength alloy steel or, as in later American practice, cast from a high-duty iron. With the latter method of manufacture a closer weight tolerance can be maintained than is possible with forgings. However, forged sintered metal connecting rods have more recently found favour for the reason mentioned earlier.

Small-end bearing

Although the gudgeon pin and small-end bearing directly react against the combustion load, they are nevertheless made appreciably smaller in diameter than the crankpin and big-end bearing. This difference in size is not simply to accommodate the small end of the connecting rod within the piston, but can be justified by the small-end bearing benefiting from a much reduced share of the total reciprocating and rotating inertia forces created by the connecting rod and piston.

Where a fully floating type of gudgeon pin is used, a separate bearing bush is pressed into the eye of the connecting rod, its length being such as to limit the maximum bearing pressure to 62 MN/m^2 (9300 lbf/in^2). Since this bearing is difficult to lubricate effectively, because of its oscillating rather than rotating motion, it must possess a high degree of durability. It is now generally of composite construction and comprises a steel backing lined with a hard lead-bronze alloy. The assembly fit of the gudgeon pin in the small-end bush tends to be critical, since too much clearance can produce a small-end tapping noise. Too little clearance when cold can result in the piston being rocked by the angular motion of the connecting rod, thereby causing a temporary piston knocking noise.

Typical pin-to-bush clearances would be 0.005–0.010 mm (0.0002–0.0004 in) for a high-grade petrol engine and 0.012–0.030 mm (0.0005–0.0012 in) in the case of a heavy-duty diesel engine.

Big-end bearing and nomenclature

Since the early 1930s it has become established practice for the big-end (and main) bearings to take the form of thin, flexible half-liners, their nomenclature being most conveniently presented by illustration (Figure 1.62). They are of composite construction and consist of a preformed thin steel backing or shell to which is bonded one or more very thin layers of relatively soft bearing material. Multi-layer bearings of this type therefore became known as thin-wall bearings, and were originally introduced by the Cleveland Graphite Bronze Company in America. Before the advent of thin-wall bearings, the housing was often lined direct with the bearing material, which then required boring and hand scraping to achieve a satisfactory fitting. Thin-wall bearings possess several important technical advantages, including a much improved fatigue resistance, a more compact installation and better suitability to mass production requirements. *Fatigue resistance* concerns the ability of the bearing to withstand fluctuating loads at fairly high temperatures, and is the single most important property required of a bearing. It is achieved in the thin-wall bearing because there is less deformation taking place in the much thinner layer of bearing material.

Big-end bearing materials

A heavier-duty bearing material may be selected for the connecting rod big-end bearing than is used for the crankshaft main bearing, because it is subjected to centrifugal loading by the partly rotating motion of the lower part of the connecting rod. This effect is absent in the case of the main bearings supporting a counterbalanced crankshaft. The big-end (and main) bearings must not only possess adequate fatigue resistance, but also provide satisfactory wear qualities. These requirements tend to conflict in practice, since the relatively soft materials that have the best anti-wear properties are

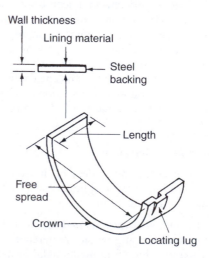

Figure 1.62   Big-end half-liner

usually the least resistant to fatigue. The various bearing lining materials used can be listed as follows:

White metal alloys   These are either tin based or lead based and are sometimes referred to as Babbitt metals, although strictly speaking this description should be confined to the tin-based alloy first developed in 1839 by Isaac Babbitt of Massachusetts. They contribute to a low rate of wear on the crankpins because of their good *embeddability*, which means they can readily bury any unfiltered wear particles entering the bearing oil film clearance and thus prevent scoring. The good *conformability* of these alloys also means they can tolerate any slight misalignment and deflections that may affect a bearing. Their load-carrying capacity is, however, limited by a modest fatigue resistance that decreases rapidly with increasing temperatures.

Copper-lead mixtures   Harder materials of this type were originally developed to meet the combination of increased loads and higher operating temperatures. They possess a higher fatigue resistance than white metal alloys, but show inferior embeddability and conformability. These disadvantages were largely overcome by plating the lining material with a very thin overlay of a softer lead-based alloy, which also improves its corrosion resistance. It is usual for copper-lead bearings to require surface-hardened crankshaft journals.

Aluminium-tin alloys   Bearing alloys containing 20 per cent tin and the balance aluminium, which compared favourably with copper-lead bearings in respect of fatigue resistance, were pioneered by the Glacier Metal Company in the 1950s. Their wear performance is usually such that a plated overlay is seldom required, since they are entirely compatible with both forged steel and the later nodular cast iron crankshafts. Furthermore, their corrosion resistance and thermal conductivity are both high for this type of bearing.

However, in more recent years not only has there been a trend towards reducing the overall length of engines, especially in relation to those transversely mounted, but also power outputs have increased as a result of pressure-charging, multiple valve cylinders and direct fuel injection, all of which have raised combustion pressures. The net effect of these changes has been a trend to shorter and more heavily loaded bearings. To meet this trend aluminium-tin-silicon bearing alloys have been developed to provide improved fatigue strength and anti-seizure characteristics, whilst remaining highly resistant to corrosive attack by the acidic products present in used engine oil. In the case of the aluminium-tin-silicon bearings produced by the Glacier Metal Company, their patented alloy designated Glacier AS124 has a nominal composition of 12 per cent tin and 4 per cent silicon in a matrix of aluminium containing 2 per cent copper.

Location of big-end bearings

Since these bearings are made detachable from the big end of the connecting rod, they must be located in both the rotational and axial senses, as follows:

1 To prevent rotational movement, the two half-liners are retained in their housing by an interference fit which, by virtue of maintaining an intimate metal-to-metal contact between liners and housing, also facilitates heat flow from the bearing. The interference fit is obtained by extending the half-liners a few hundredths of a millimetre beyond their true parting line, so that they are compressed into their housing when the bearing cap is tightened down. This difference in circumferential length between the pair of abutting liners and the closed bore of the big-end housing is known as the bearing *nip* or *crush* (Figure 1.63).

2 To prevent axial movement the two half-liners incorporate lugs that register with offset notches in both the connecting rod and its cap (Figure 1.62). The latter is fitted so that its notch faces the same side as the notch in the rod.

Checking bearing liner nip

The following precautions should generally be observed when carrying out this operation, although it is always advisable to consult the particular manufacturer's service instructions:

1 Ensure that the bearing seatings are absolutely clean.
2 Check that the correct replacement bearing half-liners are being fitted.
3 Note that the half-liners have a certain amount of free spread, so that they can be sprung and retained in position during assembly.
4 Position the half-liners so that their lugs register correctly with the locating notches in the rod and cap seatings.
5 Oil the bearing working surfaces and fit the cap the correct way round.
6 Check that the specified bearing nip or crush is present by first tightening the cap nuts or setscrews to the torque value specified by the manufacturer, then slackening one side to finger tightness and inserting a feeler gauge of appropriate thickness between the joint faces of the rod and cap. This procedure may be repeated on the other side. A bearing liner nip in the region of 0.08–0.10 mm (0.003–0.004 in) is fairly typical. Finally, of course, the cap is retightened.

Where a check on the actual bearing clearance is called for, this should usually be about 1/1000 of the crankpin diameter. An insufficient clearance space for the oil film could lead to an excessive rise in bearing operating temperature, which lowers the fatigue resistance of its material. The bearing clearance may conveniently be established by using the proprietary *Plastigage* method. This involves placing cross-wise in the bearing a suitable length of the thread-like plastics material, so that when the bearing cap is replaced and fully tightened down the material flattens out, because it is initially

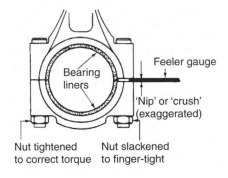

Figure 1.63   Checking bearing liner nip or crush

larger in diameter than the clearance in the bearing. The bearing cap is then removed and the width of the flattened strip of Plastigage is matched against a graduated gauge supplied with the material, so that the corresponding clearance value for the bearing can be checked against that specified by the engine manufacturer. Since the Plastigage material is soluble in oil, care must be taken to wipe clean the bearing surfaces before checking their clearance and, of course, the crankshaft must not be turned during the checking procedure.

Controlling axial movement of the connecting rod

It was once conventional practice to control the axial movement of the connecting rod from the crankpin, so that when the big end came into contact with either shoulder of the crankthrow there still remained a clearance between the small end and the adjacent gudgeon pin boss in the piston. In 1984 the Swedish Volvo company introduced a new method of controlling the axial movement of the connecting rod from the piston instead of the crankshaft, which has now been adopted by other manufacturers in the interests of reducing engine friction losses. Friction is reduced by virtue of the larger-diameter rotating location area being replaced by the smaller-diameter oscillating location area.

1.8 CRANKSHAFT ASSEMBLY AND MAIN BEARINGS

Crankshaft construction and nomenclature

A one-piece, as opposed to a built-up, construction is most commonly used for motor vehicle crankshafts. It consists of a series of crankthrows connected together by the main bearing journals. Each crankthrow is formed by a pair of webs, these being united by the crankpins to which the big ends of the connecting rods are coupled (Figure 1.64). As mentioned in Section 1.5, the angular spacing of the crankthrows is related to engine balance and firing intervals.

The proportions of petrol engine crankshafts are usually such that the crankpin has a diameter of at least 0.60 of the cylinder bore dimension and a length of not less than 0.30 of the pin diameter. Web thickness of the crankthrow is generally in the region of 0.20 of the cylinder bore dimension. The main bearing journal is made larger than that of the crankpin,

with a diameter of up to 0.75 of the cylinder bore dimension and a length of about 0.50 of the journal diameter.

Adequate crankshaft rigidity to resist both bending and twisting is a major requirement for smooth operation. With current short-stroke engines, the proportions of the crankshaft are generally such that in themselves they contribute to greater rigidity. This results from the combination of a smaller crankthrow radius and larger bearing diameters, which permit a beneficial overlap between the main journals and the crankpins (Figure 1.65).

Since the crankshaft is subjected both to bending and to torsional load reversals, it must also be designed to resist failure by fatigue. This condition may be initiated at any point where there is a concentration of stress or, in other words, a heavy loading confined to a very small area. In practice, it may occur at any abrupt change of cross-section, or from the sharp edge of an oil hole or a corner of a keyway. To avoid such stress raisers and therefore extend the fatigue life of the crankshaft, the areas in question are provided with carefully controlled small radii. For example, the corners of each main bearing journal and crankpin may be subject to what is termed 'cold rolling' to a specified fillet radius. This confers a beneficial compressive stress on the crankshaft material. The process of cold rolling basically involves rotating the crankshaft against small hardened steel rollers, which are forced against the corners of the crankshaft journals with a pressure sufficient to cause local plastic deformation and therefore compression of their surface layers. This widely used process actually dates back to 1938, when J.O. Almen of General Motors in America suggested its use to restore the durability of a Chevrolet truck crankshaft following an increase in engine piston stroke.

When the crankshaft is rotating, centrifugal force acting upon each crankthrow and the lower part of its associated connecting rod tends to deflect the crankshaft. Since this deflection is resisted by the main bearings, their loading is correspondingly increased. To reduce these loads, counterbalance weights are either formed integrally with, or separately attached to, the crankthrow webs (Figures 1.66a and b). The former arrangement is now most commonly used, the crankwebs being extended opposite to the crankpin and spread circumferentially.

Crankshaft flywheel

A one-piece construction in cast iron is generally employed for flywheels used in conjunction with friction clutches. For

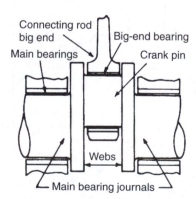

Figure 1.64 Basic crankthrow arrangement

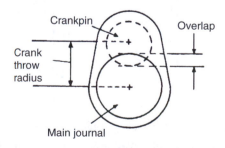

Figure 1.65 Overlap between main journal and crankpin

motor car applications the flywheel is typically of the plain type (Figure 1.67a), although in a few designs a relatively thin disc or *flex-plate* is used to mount a separate cast iron rim (Figure 1.67b). The object of the latter construction is to minimize any disturbance of the flywheel that could result from bending vibration of the crankshaft (Section 1.9) and thus promote smoother running. A similar construction may also be adopted for use with a torque converter automatic transmission, where the flywheel effect derives from the impeller casing, which is mounted from a flex-plate bolted to the end of the crankshaft (Figures 1.67c and 17.6). In the case of some commercial vehicles a pot, instead of a plain, type of flywheel is required to accommodate a heavy-duty, twin-plate, friction clutch (Figures 1.67d and 13.30).

Radial location of the flywheel hub is afforded by a spigot on the rear end of the crankshaft. Owing to its appreciable inertia, the flywheel is located in the rotational sense by dowel pins and clamped firmly to the rear face of the crankshaft by a ring of bolts. The rim of the flywheel provides a mounting

for the starter ring gear and may also bear timing marks for checking the valve and ignition settings, relative to prescribed positions of the crankshaft.

Flywheel with torsional vibration damper

Originally developed in the mid-1980s by Toyota for application to a motor car turbocharged diesel engine, the flywheel with torsional vibration damper or *dual-mass* flywheel as it is now often termed, has in more recent years become increasingly adopted for petrol engines where manufacturers seek additional refinement for the transmission system. The purpose of the dual-mass flywheel is to reduce the extent to which periodic fluctuations in engine torque are passed on to the transmission system, which otherwise create vibration, noise and can lead to wear of components. Typically noticeable with a dual-mass flywheel installation is therefore a reduction in transmission gear noise at low engine speeds. In this context there is a greater opportunity with a modern five-speed and reverse, all-synchromesh, gearbox for light load rattles to occur between the teeth of the more comprehensive train of constant-mesh gears (Section 18.3).

As its name suggests, a dual-mass flywheel basically comprises a two-piece flywheel with an *engine-side mass* and a *transmission-side mass* (Figure 1.68). The latter is supported from the former by an interposed ball-bearing race and its relative oscillatory movements are cushioned by a series of circumferentially spaced compression springs, which are retained in windows shared by the two masses. Frictional resistance to dampen the oscillatory movements between the two masses is supplied in a similar manner to that later

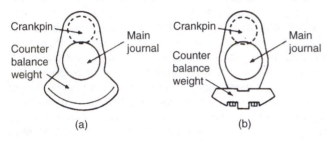

Figure 1.66 Counterbalance weights: (a) integral (b) separately attached

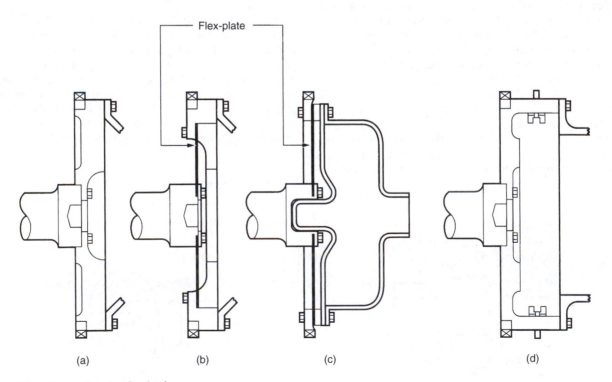

(a)   (b)   (c)   (d)

Figure 1.67 Types of engine flywheel

(a) Plain type for mounting friction clutch
(b) Flex-plate type for mounting friction clutch

(c) Flex-plate type for mounting torque converter
(d) Pot type for mounting heavy-duty friction clutch

described for the centre-plate of a friction clutch (Section 13.3). In an alternative construction hydraulic damping is used between the two masses, fluid-filled chambers with restricted egress and ingress being arranged circumferentially. The changes in volume and restricted fluid movements within the chambers that accompany the relative oscillations between the two masses, therefore, confer the necessary damping properties.

Crankshaft timing wheel and pulley

Ahead of the front main journal of the crankshaft, a cylindrical extension is machined to accept the driving wheel of the timing drive (the various arrangements of which are described at a later stage), an oil flinger (where fitted) and the driving pulley for the belt-driven engine auxiliaries.

These components are made close fitting on the shaft extension or nose and are prevented from turning relative thereto by either a single parallel-faced key, or a series of Woodruff keys, the latter being self-aligning by virtue of their semi-circular form (Figure 1.69).

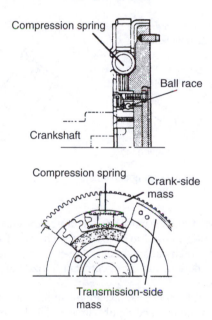

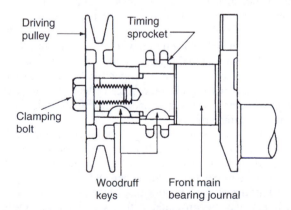

Figure 1.68 Flywheel with torsional damper (*Toyota*)

Figure 1.69 Attachment of timing wheel and pulley to nose of crankshaft

The complete assembly of timing wheel, oil flinger and driving pulley are retained endwise on the shaft extension by means of a large setscrew which enters a hole tapped axially in the nose of the shaft and exerts its clamping load via a thick plain washer. In some engines, a driving gear may additionally be sandwiched either between an inner timing wheel and the pulley for a low-mounted oil pump, or between the front main bearing journal and the outer timing wheel for a concentric oil pump (Figure 1.30). In modern engine design a further requirement can be for a crankshaft mounted sprocket wheel, which provides a chain drive to harmonic balancer shafts housed below the crankshaft.

Crankshaft materials

Until the early 1960s petrol engine crankshafts were traditionally forged from high-strength, low-alloy steels, and indeed these are still used for heavy-duty applications. Since then, however, the majority of car manufacturers have favoured the use of crankshafts produced from iron castings of the spheroidal graphite type, or SG iron as it is commonly termed (Figure 1.70).

High-strength cast irons of this type were first developed in the late 1940s, both in Britain and in America, and their distinguishing feature is that the graphite structure takes the form of spheroidal nodules. This feature confers higher strength, better ductility and greater toughness than the flake graphite structure of normal grey cast iron. It is obtained by injecting a trace of magnesium into the iron melt, which causes the graphite flake to gather into little balls or, more technically, spheroidal nodules, that greatly strengthen the grain structure of the material.

The material composition for a high-strength cast iron crankshaft would generally be as follows:

Carbon	3.50–4.20%	Sulphur	0.03% max.
Manganese	0.30–1.00%	Chromium	0.20% max.
Silicon	1.80–2.75%	Nickel	1.00% max.
Phosphorus	0.08% max.	Brinell hardness no.	217–286

The Brinell hardness test, so named after J. A. Brinell who introduced it in 1900, measures the resistance to penetration of a material by a harder one in the form of a steel ball, and gives a useful indication of the tensile strength of the material being tested.

Crankshaft manufacture

A forged crankshaft, like any other forged metal component, is manufactured by a process in which the metal in a more or less plastic, rather than molten, state is forced to flow into the

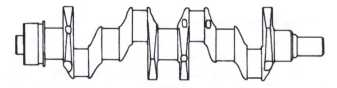

Figure 1.70 General form of a high-strength cast iron crankshaft for a four-cylinder engine

Figure 1.71 Crankshaft turning and grinding (*Laystall*)

desired shape by means of hammering, squeezing and bending. In the case of motor vehicle crankshafts produced in large quantities, the actual shaping is performed by drop-hammer forging in closed dies. The latter are upper and lower blocks of metal in each of which an impression has been formed of the crankshaft.

A cast crankshaft is, in contrast, one that is manufactured by a process in which the metal in a molten state is poured into a mould and allowed to solidify. Motor vehicle crankshafts are not, however, produced in conventional sand moulds, but are cast vertically by the shell moulding process. In this technique, a thin shell-like mould of sand and synthetic resin is made by bringing these materials into contact with a heated metal pattern, the contours of which are exactly reproduced in the shell mould. Two such shells clamped together then form a complete mould.

Among the important advantages offered by the modern shell moulding process is that castings can be produced to much closer tolerances. This in turn reduces the amount of machining required afterwards, and in the case of the crankweb faces can eliminate it altogether.

After heat treatment to remove residual stresses and to give the specified tensile strength of material, usually about $63 \, \text{kg/mm}^2$ (40 tons/in^2), the crankshaft must be machined to its final dimensions (Figure 1.71). This involves principally the rough turning, finish grinding and final lapping of the main journals and crankpins. With SG iron crankshafts it is good practice that the bearing journals and pins be final lapped with the crankshaft rotating in the same direction as it does in the engine.

Crankshaft main bearings and nomenclature

In modern high-compression-ratio engines, the number of main bearings employed to support the crankshaft has tended to increase. This is because the crankshaft is subjected to greater bending loads, resulting from the higher peak gas pressures acting upon the pistons. Hence, the crankthrows must receive adequate support from adjacent bearings to

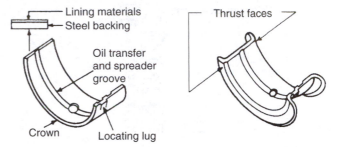

Figure 1.72 Main bearing half-liners

minimize shaft deflections. It is therefore now customary for a main bearing journal to be used at each end of the crankshaft and between each cylinder of in-line engines, and each pair of cylinders in the more compact horizontally opposed and V engines.

The thin-wall main bearings are of similar form to the big-end bearings and must be rigidly supported in the crankcase so as to preserve not only the geometrical truth of their working surfaces but also their correct relative alignment with one another. These requirements must be met to avoid high localized pressures, which may result in destruction of a bearing through breakdown of its oil film with consequent overheating. Thicker than normal main bearing half-liners may be specified for V cylinder engines with aluminium alloy crankcases, so that optimum running clearances are better maintained even though the crankcase and crankshaft differ in their thermal expansion characteristics.

Similarly, the nomenclature of thin-wall main bearings follows that of the big-end bearings with the addition, of course, of an oil groove (Figure 1.72), the purpose of which is explained in Section 4.4. As an alternative to separate crankshaft thrust washers, one pair of the set of main bearing half-liners may be provided with integral flanged ends to serve as thrust faces. Such an arrangement simplifies the production build of an engine but does, however, lack the

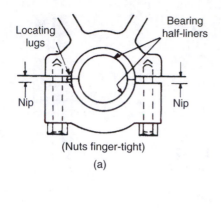

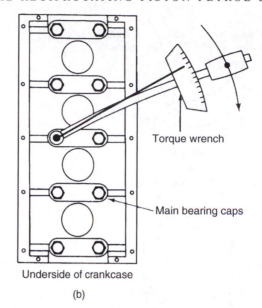

Figure 1.73 Installation of crankshaft: (a) main bearing nip (b) torque tightening the main bearing caps

inherent self-aligning and good heat-conducting properties of separately mounted thrust washers.

Main bearing materials and location

Again, the choice of materials and the methods of location for the main bearing half-liners correspond to those of big-end practice. However, another method of providing lateral location that is sometimes used for heavy-duty main bearings is the dowel-and-hole arrangement. In this method the dowels are fixed securely in the crankcase bearing saddle and cap, and register with circumferentially elongated holes in each half-liner. A disadvantage of this particular method of location is that the presence of the dowel hole disrupts the oil film in the bearing.

Main bearings installation

The fitting precautions to be observed are basically similar to those mentioned earlier for the big-end bearings (Figure 1.73a). It is, of course, necessary to install the crankshaft thrust washers prior to fitting the appropriate main bearing cap, taking care that the oil grooves face the thrust surfaces on the crankshaft. The bearing caps are then torque tightened in sequence to their specified value, starting from the centre and working alternately towards each end (Figure 1.73b). Any undue resistance to rotation of the crankshaft by hand should be checked after finally tightening each bearing cap. The actual main bearing clearances may be checked using the Plastigage method, as previously described for the big-end bearings.

Crankshaft thrust bearings

Before this particular bearing application is discussed, an engineering distinction must be made between two types of bearing. A bearing that is intended to resist a load applied perpendicular to the axis of the shaft is termed a *radial* or *journal* bearing. In contrast, a bearing that is intended to resist a load applied along the axis of the shaft is termed an *axial* or *thrust* bearing.

The crankshaft is located axially in the crankcase by plain (as opposed to rolling) thrust bearings which restrain it against endwise movement from loading imposed mainly by the transmission system. This loading may be in a forwards direction during release of a friction clutch and in a rearwards direction when a fluid coupling is in operation (Sections 13 and 15).

Crankshaft thrust bearings are generally of the same composite construction as that employed for the half-liners of the big-end and main bearings, as described earlier, although bronze bearings may also be used. They often take the form of separate semicircular thrust washers, which are installed either in pairs or singly on each side of one of the main bearing housings. Where thrust washers of smaller size are used in pairs, each lower half is keyed to the bearing cap, thus preventing both lower and upper halves from rotating once the cap is fitted (Figure 1.74a). If lower-half thrust washers only are used, rotation is prevented by their upper ends abutting the joint faces of the crankcase bearing saddle (Figure 1.74b). Since the thrust surfaces of the bearings must be separated by an oil film, the washers are provided with grooves or pockets to distribute the oil reaching them from the main bearings they embrace.

To accommodate both the thickness of the oil film and the thermal expansion of the parts concerned, a small clearance is provided on assembly between the thrust bearing surfaces and the adjacent contact faces of the crankshaft. This clearance must, of course, always be less than that existing between the crankweb faces and the sides of all the other main bearings, in order to relieve them of any thrust loads. The required clearance typically falls within the range 0.05–0.15 mm (0.002–0.006 in) and is termed the *crankshaft end float*.

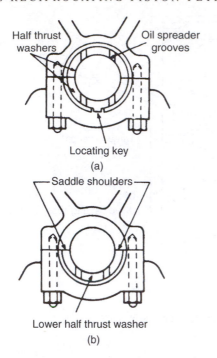

Figure 1.74   Crankshaft thrust bearings: (a) paired (b) single

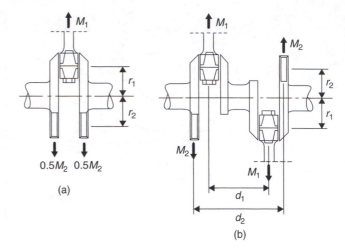

Figure 1.75   Simple examples of force and couple balance in crankthrows:

(a) force balance in single crankthrow when $M_2 = M_1 r_1/r_2$
(b) force balance in opposed crankthrows when
$M_1 r_1 d_1 = M_2 r_2 d_2$
M_1 rotating mass at crankpin
M_2 rotating mass of counterbalance weight
r_1 radius at which M_1 acts
r_2 radius at which M_2 acts
d_1 distance between offset cylinders
d_2 distance between counterbalance weights

Crankshaft balancing

In Section 1.5 our attention was directed in simple terms to those factors which make for acceptable overall balance of a reciprocating piston engine. Similarly, we have now to consider the basic principles involved in balancing the crankshaft itself, which compared with such items as the cooling fan and the flywheel is the most difficult part of an engine to correct for imbalance. At the heart of crankshaft balancing is the requirement for counterbalance weights. These are incorporated by the engine designer for one or more of the following purposes.

Static balancing of the rotating parts
Partial balancing of the reciprocating parts
Dynamic balancing of the rotating parts
Reducing bearing pressures.

Static balancing of the rotating parts

As with all rotating parts used in engineering, the unbalanced rotation of a crankshaft becomes more significant as speed increases. Indeed, the centrifugal forces arising from any such imbalance increase with the square of the engine speed – that is, doubling the speed makes the vibration effect four times worse. It is therefore perhaps stating the obvious that a crankshaft, like a road wheel, must at least be in a condition of *static balance*. This is achieved when the centre of gravity lies on the axis of rotation of the crankshaft, which implies that there is an equal distribution of mass about this axis. To verify this in practice requires the crankshaft to be supported with two of its main journals resting on knife edges; then whatever position the crankshaft is rolled to it should show no tendency to oscillate when coming to rest.

Of all the various arrangements of cylinders used in automotive practice only the single and in-line twin ones require the addition of counterbalance weights to correct for inherent lack of crankshaft static balance. This arises simply because the mass of their crankthrows is not distributed symmetrically about the axis of crankshaft rotation. It therefore becomes necessary to provide counterbalance weights on the opposite side of the axis of rotation to the unbalanced crankthrow(s). Since a single counterbalance weight cannot be placed in the same plane as that occupied by the crankpin and connecting rod, the crankwebs on each side are extended in the opposite direction to the crankpin to form a divided counterbalance weight (Figure 1.75a). In actual practice this weight is calculated so as to counterbalance not only the crankthrow but also the lower portion of the connecting rod, which may be considered as revolving with the crankpin.

For all other multicylinder arrangements, where the crankthrows are symmetrically spaced about the crankshaft axis, there is no inherent static imbalance and any correction needed is related to the machining tolerances on the crankshaft assembly. Counterbalance weights may still be required, however, but from other considerations of balance, as we shall presently find.

Partial balancing of the reciprocating parts

A force must be applied to accelerate a piston over the first half of its stroke, and similarly a force is developed by the piston as it decelerates over the second half of its stroke. As explained in Section 1.5, these reciprocating forces are termed primary inertia forces. Again, it is only with single- and in-line twin-cylinder engines that some attempt must be made to compensate for these forces by adding counterbalance weights to the crankshaft, since in all other multicylinder engines they

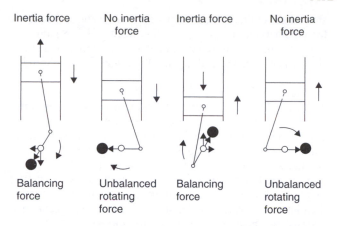

Figure 1.76 Partial balance of reciprocating and rotating parts

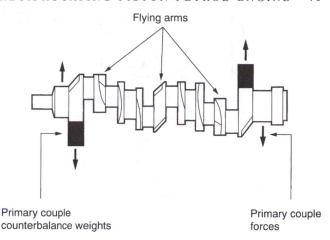

Figure 1.77 Balancing arrangements for a 60° V6 engine crankshaft with four main bearings

are inherently balanced. This is because the arrangement and number of cylinders are so contrived that the primary forces generated in one cylinder are directly in opposition to, and therefore cancelled by, those in another cylinder.

With regard to the rogue single- and in-line twin-cylinder engines, their lack of primary balance can be partially alleviated by increasing the counterbalance weighting referred to in the previous section. The words 'partially alleviated' have been deliberately chosen, because a reciprocating mass can never be completely balanced by a rotating mass on the crankshaft owing to the resultant forces acting along different paths (Figure 1.76). Any attempt to counterbalance the reciprocating forces is always made at the expense of introducing an unbalanced rotating force, the centrifugal effect of which becomes the greatest nuisance at right angles to the cylinder axis. A compromise solution is therefore usually sought by increasing the mass of the divided counterbalance weights such that they produce an opposing force equal to one-half the maximum reciprocating force to be cancelled. In other words, the counterbalance weights added to the crankshafts of single- and in-line twin-cylinder engines generally over-correct for static balance and under correct for reciprocating balance.

Dynamic balancing of the rotating parts

Even though a multicylinder engine crankshaft might be in a state of 'perfect' static balance, it by no means follows that it is also in a similar state of *dynamic balance*. The reason for this is that in static balancing we are dealing with *forces*, whereas in dynamic balancing we are neutralizing *couples* – that is, opposing forces which are not acting in the same plane. An unbalanced couple can impose a rocking effect on the rotating crankshaft and therefore on the engine as a whole.

A ready impression of the manner in which an unbalanced couple acts may be gained by considering the crankshaft layout of a horizontally opposed twin-cylinder engine. Here the crankthrows lie on opposite sides of the crankshaft rotational axis, so that the symmetrical distribution of their masses allows static balance to be achieved without the addition of counterbalance weights. As a result, however, of each piston connecting to an individual crankthrow, the corresponding cylinders cannot be made truly coaxial. It is the presence of this offset distance between the nearly opposite cylinders which causes

the otherwise statically balanced crankthrows to introduce a rocking movement or unbalanced couple on the rotating crankshaft, arising from the centrifugal forces generated.

To attain dynamic balance it is therefore necessary to impose an equal but opposite couple lying in the same plane as the unbalanced one. In practice this can be arranged by extending the end crankwebs in opposite directions to their crankpins to form counterbalance weights (Figure 1.75b). Note that the addition of these counterbalance weights does not destroy the static balance of the crankshaft, so that whereas a statically balanced crankshaft is not necessarily in dynamic balance, a dynamically balanced crankshaft is always in static balance! The requirement for counterbalance weights to achieve dynamic balance of the crankshaft can also arise in other multicylinder engines, notably those with V cylinder arrangements. Taking for example the popular 60° V6 engine, there is an unbalanced primary couple that has to be neutralized by two widely separated and relatively large counterbalance weights, which act in opposition to each other (Figure 1.77). In actual practice a fairly thick counterbalance weight forms part of the foremost web of the crankshaft and a thinner one forms part of the rearmost web, the action of the latter being supplemented by an auxiliary counterbalance weight in the flywheel. A modest unbalanced secondary couple is also present, but this can be comfortably absorbed by the resilience of the engine mountings.

Reducing bearing pressures

This would not be the important balancing consideration that it is but for the fact that a crankshaft cannot be made perfectly rigid. So although a crankshaft might give every appearance of being in 'perfect' dynamic balance, the centrifugal forces developed at what may otherwise be inherently balanced crankthrows are still encouraging the shaft to bend as it rotates. In reality, of course, the crankshaft is being restrained from such bending deflection by the main bearings, but in performing this duty it will be clear that their loading and hence working pressures are being increased.

Another reason for adding counterbalance weights to a crankshaft is therefore to counteract bending deflection and thereby reduce bearing pressures. The crankshaft layout of

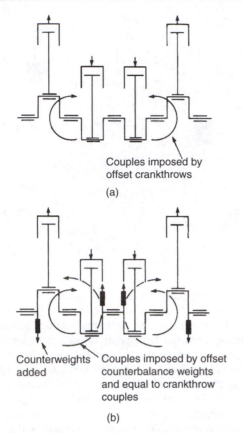

Couples imposed by
offset crankthrows

(a)

Counterweights   Couples imposed by offset
added   counterbalance weights
                            and equal to crankthrow
                            couples

(b)

Figure 1.78 Balancing conditions for an in-line four-cylinder engine: (a) dynamic balance (b) dynamic balance with reduced bearing pressures

modern in-line four-cylinder engines affords a good example of this principle being applied in practice. First, it is useful to consider the state of balance of such a crankshaft without the addition of counterbalance weights. Then we find that it possesses inherent static balance, because the two outer crankthrows lie on one side of the crankshaft rotational axis and the two inner ones on the other side. Also we find that although an unbalanced couple acts about the front half of the crankshaft in the same manner as that described earlier for the two crankthrows of a horizontally opposed twin-cylinder engine, this couple is counteracted by a similar one acting about the rear half of the crankshaft (Figure 1.78a). Therefore on the basis of providing a dynamically and by the same token statically balanced crankshaft, the need for counterbalance weight is non-existent; in earlier practice this was how the in-line four-cylinder crankshaft was designed.

With increasing engine speeds, however, the magnitude of the couples acting about each half of the crankshaft, even though self-cancelling, become such that they are trying very much harder to bend the crankshaft. Since this is being resisted by the main bearings, their working pressures are increased. For this reason it is now necessary to provide counterbalance weights for each half of the crankshaft and thereby cancel the individual couples so that bending deflection is minimized (Figure 1.78b).

A disadvantage of adding counterbalance weights to any crankshaft, especially if it is of the longer and more tortuously

shaped six-cylinder kind, is that they lower the resistance to torsional vibration. This topic will be dealt with in Section 1.9.

During manufacture, final imbalance correction in a crankshaft is performed by drilling into the counterbalance weights.

Harmonic balancers

In Section 1.5 it was mentioned that there has been a revival of interest in the use of twin counterbalancing shafts, to cancel out the unbalanced secondary inertia forces of in-line four-cylinder engines. It was the highly esteemed automobile (and aeronautical) pioneer Dr F. W. Lanchester (1869–1946) who first applied this classic solution to the problem when in 1911 he patented what was originally known as an 'anti-vibrator', then later as a 'harmonic balancer' and in more recent years variously as 'counterbalancing shafts' and 'silent shafts'.

Technically speaking, though, we should perhaps prefer to use the description 'harmonic balancer'. This is because if we were to consult a treatise on engine balancing, it would become evident that the secondary inertia forces of an in-line four-cylinder engine result from the fact that the common centre of gravity of the four pistons, instead of remaining stationary as would be the case if the connecting rods were infinitely long, has a small harmonic motion of double frequency. It is this unwanted small harmonic motion that must be cancelled or balanced out, and explains the designation 'harmonic balancer'. The term harmonic may be thought particularly apt, because as in music when two notes sound right and pleasing together (Dr F. W. Lanchester was also an authority on music theory!), so we can contrive agreement between two moving forces.

However, it also seems to follow that a harmonic balancer would be unnecessary if only we could go out and buy ourselves a set of infinitely long connecting rods! Since the notion of infinitely long connecting rods may seem rather obscure to the motor vehicle service engineer, let us now try to understand its significance in simple terms. Referring to Figure 1.79a it will be seen that instead of using a conventional hinged connecting rod between the piston and crank, we have substituted a crank and slotted-bar or 'Scotch yoke' mechanism, which was sometimes employed in early steam and pumping engines. Comparing this arrangement with the conventional one shown in Figure 1.79b, it will be noticed that there is no inequality of piston travel towards the dead centres for corresponding angular movements of the crank. In fact by conferring a straight-line, instead of a swinging, motion on the connecting rod we have arrived at the practical equivalent of an infinitely long connecting rod. As a result the piston acquires a simple, as opposed to non-simple, harmonic motion. That is, it now oscillates about its equilibrium or mid-stroke position in such a way that its acceleration (positive or negative) towards this position is directly proportional to its displacement therefrom. So in the case of our in-line four-cylinder engine, the common centre of gravity of the four pistons would remain stationary and there would be no unbalanced secondary inertia forces.

Unfortunately a crank and slotted-bar mechanism would be quite impractical for a high-speed internal combustion engine, because of its unwieldy nature and the difficulty in arranging effective lubrication. We are therefore left with little choice but to use a hinged connecting rod, which imposes

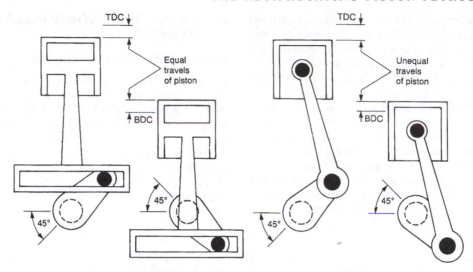

Figure 1.79   Comparison between (a) slotted-bar and (b) hinged connecting-rod and crank mechanisms

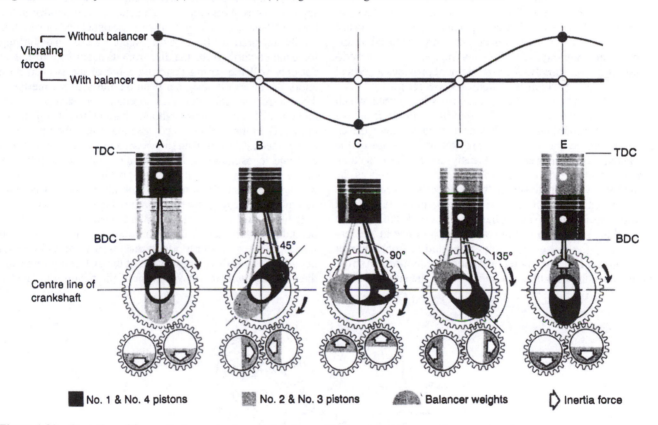

Figure 1.80   Operation of four-cylinder engine harmonic balancer (*International Harvester*)

a non-simple harmonic motion on the piston and is responsible for creating the unbalanced secondary inertia forces with an in-line four-cylinder engine. The purpose of a harmonic balancer is therefore to cancel out these forces by introducing equal and opposite ones.

In Lanchester's original application of the harmonic balancer, two parallel shafts were fitted below and equidistant from the crankshaft axis. These shafts were geared together to run in opposite directions, with one of them being chain driven at twice the crankshaft speed, and they were provided

with bob-weights in or about the plane of symmetry of the engine. The effects of these bob-weights were such that they neutralized each other in respect of their lateral motion, but combined in respect of their vertical motion to give a small harmonic one whose phase was contrary to that created by the error in motion of the pistons. Therefore when the pistons were at their top and bottom dead centres, the bob-weights were always at their lowest position. This sequence of events will be made clear in Figure 1.80, which applies to a heavy-duty diesel engine. In this installation the geared-together bob-weights are in turn driven from a gear shrunk on to the crankshaft and so positioned that they align with the exact centre of the engine. The balancer gears are matched and timed to each other by suitable markings and also to the crankshaft driving gear. Lubrication for the assembly is provided from the engine main oil gallery.

Despite its ingenuity, few designers of a past era in motoring thought it worthwhile to incorporate a harmonic balancer in their four-cylinder engines for passenger cars, their objections being the mainly practical ones of the additional faster working parts and weight and also the theoretical one of creating two otherwise useless lateral forces. It was not until the mid 1970s, when Mitsubishi Motors in Japan introduced their refined version of the harmonic balancer, that interest in its application was revived. Mitsubishi had found that by minimizing engine vibration in this manner, they were able to achieve a reduced level of booming in the body interior and also less intrusive noise at high engine speeds, together with improved durability for the engine auxiliaries and equipment.

The Mitsubishi installation of harmonic balancer, or 'silent shafts' as they call it, is of particular interest because of the way in which the balancer shafts are disposed at different levels on either side of the engine to perform an additional function. Namely, this arrangement provides a vertical separation about which the lateral forces generated by the bob-weight can be utilized to create a resisting couple, which assists in opposing sympathetic rocking vibrations of the engine on its mountings. Both balancer shafts are chain driven at twice engine speed from a sprocket on the nose of the crankshaft. The upper shaft is directly driven from its chain sprocket and therefore rotates in the same direction as the crankshaft, while the lower shaft is indirectly driven from its chain sprocket through a pair of gears, so that its direction of rotation can be opposite to that of the crankshaft (Figure 1.81). Each shaft with its bob-weights positioned in the centre plane of the engine is rigidly supported in front and rear bearings, the latter receiving lubrication under pressure from the former via holes through the centre of the shafts.

1.9   CRANKSHAFT TORSIONAL VIBRATION DAMPERS

The causes of crankshaft vibration

When, say, a heavy and very rugged-looking engine crankshaft is being physically handled in the workshop, the impression given of rigidity is such that the possibility of it being set into a state of vibration might seem unlikely. However, the crankshaft is viewed rather differently by an engine designer, who recognizes that it does in fact behave as an elastic body. This is because in practice the crankshaft cannot be made perfectly rigid, and during engine operation it can be subject to three modes of vibration: torsional, axial and bending.

Torsional vibration
A ready impression of what is meant by torsional vibration may be gained by considering the antics of the simple torsional pendulum used in physics experiments. This piece of apparatus consists essentially of a metal disc suspended in the horizontal plane by a length of wire, the upper end of which is rigidly anchored to a fixed beam (Figure 1.82a). If a disturbing force momentarily turns the disc from its position of rest, then not only will the spring restoring force of the twisted wire return the disc to its original position, but also the inertia of the disc (or natural reluctance to change its state of rotary motion) will cause it to overshoot. This in turn sets up restoring forces in the again twisted wire that cause the process to be repeated in the opposite direction, and so on. The disc will continue to oscillate with diminishing amplitude or rotary movement either side of its original position of rest, until what we can now identify as torsional vibration of the pendulum dies away or decays (of which more will be said later).

If the disturbing force is applied repeatedly, so that the pendulum is excited into a continuous state of torsional vibration, then the amplitude of disc oscillation will depend upon the frequency of application of the disturbing force and the *natural frequency* of vibration of the pendulum. Here it

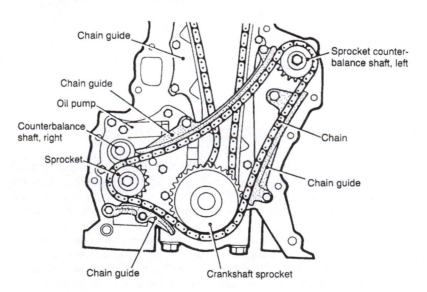

Figure 1.81   Driving arrangements for 'silent shafts' harmonic balancer (*Colt-Mitsubishi*)

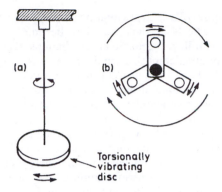

Figure 1.82 Crankshaft torsional vibration: (a) simple torsional pendulum (b) torsional vibration superimposed upon crankshaft rotation

should be appreciated that every body possesses one or more natural frequencies of vibration (the term *frequency* means a certain number of vibrations performed per second) at which it will continue to vibrate when disturbed, until the energy it receives has been dissipated. A few simple experiments conducted with our torsional pendulum would also yield the further useful information that its natural frequency of vibration could be raised by any one or all of several methods. These are to make the wire shorter, to use a thicker wire, and to reduce the weight of the metal disc.

By applying our knowledge of the manner in which the simple torsional pendulum behaves, we are now better able to visualize how torsional vibrations can arise with the much more complex engine crankshaft and how they can be minimized. The disturbing forces applied to the crankshaft are, of course, derived from the pulsating gas and inertia torques acting via each piston, connecting rod and crankthrow combination, as explained in Section 1.4. If the crankshaft were perfectly rigid, then the only effect of these pulsating torques would be to cause some irregularity in its speed of rotation, which could be smoothed out by the action of the flywheel. However, as mentioned earlier, the crankshaft cannot in reality be made perfectly rigid, with the result that the torque pulses are capable of twisting it and therefore of exciting it into a state of torsional vibration. This vibration is superimposed upon the continuous rotation of the shaft (Figure 1.82b).

If the frequency of the disturbing vibrations should coincide with one of the natural frequencies of crankshaft vibration, then a condition known as *resonance* will occur. A danger of resonant vibration is that the energy of the disturbing vibrations may be greater than that lost by the twisting and untwisting of the crankshaft, so that the amplitude of torsional vibration builds up to such a degree that the crankshaft can be over-stressed and eventually suffer a fatigue fracture.

In practice, of course, the design of the crankshaft system is contrived such that its natural frequency of vibration is raised as high as possible. On a similar principle to our simple torsional pendulum, the natural frequency may be raised by making the shaft as short as possible, increasing its diameter and using a lighter flywheel. A resonant or *critical order* vibration would therefore only be expected to occur beyond the normal speed range of the engine, or in other words if the engine is over-revved. However, it may still be necessary in

the interests of both engine smoothness and satisfactory operation of the timing drive to suppress the less critical orders of torsional vibration, which do occur within the normal speed range. For this purpose the crankshaft can be fitted with some form of torsional vibration damper.

Axial and bending vibrations
Axial vibration of a crankshaft is of lesser significance than torsional and bending vibration and is generally regarded as being a by-product of the former. That is, the twisting and untwisting of the crankshaft is accompanied by alternate decreases and restorations in length, which would of course apply to any other body subjected to the same treatment.

Bending vibration of the crankshaft when considered as a beam also results from the pulsating gas and inertia forces imposed upon it via the piston and connecting rod assemblies. As might be expected, this form of vibration is strongly reacted against at the main bearings. In high-compression-ratio engines, the maximum rate of pressure rise during combustion can be such as to cause undesirable vibration of the engine structure and mechanism, especially bending vibration of the crankshaft if it and its bearing supports are not rigid enough.

Types of torsional vibration damper

When describing the vibration of a torsional pendulum, it was implied that in the absence of further disturbing forces the vibration eventually dies out or decays. It does so, of course, because of the natural damping effects of air friction on the surfaces of the oscillating parts and of internal friction in the spring material itself. The internal friction or *damping capacity* of a material is that property which relates to vibration energy being absorbed and subsequently dissipated in the form of heat. Apart from external air and internal material friction there is a further 'apparent' damping effect present in the case of crankshaft operation, this being the energy absorbed and again dissipated as heat by the shearing of the oil films that lubricate the crankshaft, connecting rod and piston bearing surfaces.

Even when the total effect of these natural and apparent damping forces is taken into account, it is by no means sufficient to suppress resonant vibrations of a crankshaft. For this reason is may be necessary to include in the crankshaft assembly a device known as a torsional vibration damper or simply crankshaft damper. The concept of the torsional vibration damper as a means of introducing additional friction to damper vibration of a crankshaft is quite old; C.H. Bradbury, a one-time leading diesel engine specialist, described it humorously as 'a corrective device for engines designed to run at one speed and sold to run at some other speed'.

Two types of torsional vibration damper may be encountered in petrol engine practice. These are generally classified as the slipper damper that was used in earlier engine designs, and the rubber damper that is in current use.

Construction and operation of the slipper damper

This type of damper is basically a device for separating two masses of different inertia by frictional means, one being the crankshaft and the other usually a pair of small flywheels. The latter are not keyed to the crankshaft, but are simply spring loaded apart against friction surfaces on the damper hub, which is rigidly attached to the nose of the crankshaft (Figure 1.83a). So although the damper flywheels normally rotate in unison with the crankshaft, it is also possible for them to slip – hence the description slipper damper – relative to the crankshaft, once the friction torque that restrains their angular movement is overcome. Loosely fitting dowel pins are used to prevent relative angular movement between the flywheels themselves, whilst at the same time permitting them axial freedom under the influence of their spring loading.

The concept of the slipper torsional vibration damper is generally credited to Dr F. W. Lanchester, who developed such a device for the Daimler Company just before World War I, although Henry Royce arrived at very much the same solution to the problem also during this period. Early versions of what was once generally referred to as the Lanchester damper tended to be of elaborate construction, since they virtually amounted to a small multiplate clutch assembly with lubricated metal-to-metal rubbing surfaces. Later examples were of much simpler construction, as originally described and in the first instance pioneered by the American Chrysler Corporation in 1925. Non-metallic friction discs were incorporated in these later dampers, their rubbing surfaces being unlubricated.

At the onset of a critical vibration period, the inertia of the damper flywheels is such that they become increasingly reluctant to follow the torsional oscillations superimposed upon the rotating crankshaft, and ultimately slippage occurs between the faces of their webs and the friction discs. That is, the friction torque of the damper that normally restrains the flywheels from angular movement relative to the crankshaft is exceeded, so that the damper is in effect behaving like a slipping clutch. As a result the energy absorbed in overcoming friction at the rubbing surfaces and dissipated in the form of heat is abstracted from the energy stored in the vibrating crankshaft. A serious amplitude of torsional vibration is therefore never allowed to build up. It should be realized, however, that a slipper damper can only really be effective in dealing with the particular frequency of torsional vibration for which its spring loading and hence slipping torque has been predetermined.

Construction and operation of the rubber damper

One of the shortcomings of the slipper damper was that its performance did not always remain consistent over an extended length of service, as a result of deterioration of its friction surfaces. The ensuing engine vibration periods could also be accompanied by noisy operation of the slipper damper, or what was sometimes referred to by service engineers as slipper roar. Put another way, a misbehaving slipper damper can have a worse effect on crankshaft torsional vibration than no damper at all. Recognition of these difficulties led to the development of a much simpler type of torsional vibration damper in which the two masses of different inertia, represented by the crankshaft and in this case a single small flywheel, were separated by both frictional and elastic means through the medium of rubber.

In long-established practice the rubber damper essentially comprises three concentric parts, these being a carrier cum hub assembly that is rigidly attached to the nose of the crankshaft, a ring-shaped flywheel or inertia ring that may be grooved to accept a V-belt, and a layer of rubber which is either bonded to each of these components (Figure 1.83b) or sandwiched between them under precompression, as in later designs (Figure 1.83c). It will be appreciated that there is no other connection between the carrier and the ring apart from that established by the intervening layer of rubber. The advantages of the later non-bonded version are that it is less costly to manufacture and allows a wider choice of rubber specification. Coincidentally, it was also Dr F. W. Lanchester who first patented a rubber type of crankshaft torsional vibration damper in 1928, and again the American Chrysler Corporation who, in the mid 1930s, developed the idea in the general form that we know it today.

Although the construction of the rubber damper is simpler than that of slipper and viscous dampers (Section 2.5), its operation is rather more complex because it acts as both a damper and a detuner. So far as the former duty is concerned, the onset of a critical vibration period causes the inertia ring to behave in a manner similar to the flywheels of a slipper damper, except that in this case the rubber layer is continually being twisted back and forth (Figure 1.83d). Since the rubber possesses a useful amount of internal friction, the

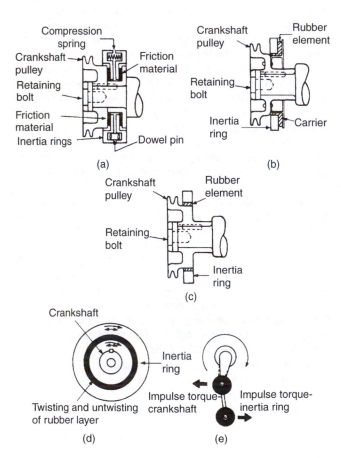

Figure 1.83   Types and action of crankshaft torsional vibration damper: (a) slipper (b) bonded rubber (c) unbonded rubber (d) damping (e) detuning

energy it absorbs and rejects as heat is abstracted from the vibrating crankshaft.

So much for its friction damping function. The *detuning* function of a rubber damper is perhaps not so readily visualized until it is appreciated that, unlike the slipper and viscous dampers, its inertia ring forms part of a torsional spring drive system with its own natural frequency of vibration. It is, in fact, analogous to that of the torsional spring drive used in clutch centre plates described later, except that the wind-up action of the circumferential coil springs is replaced by the rubber layer being deflected in torsional shear. By careful tuning of the damper vibration characteristics, its inertia ring can be induced to move in anti-phase with the torsional oscillations of the crankshaft and thus oppose their build-up. As a further simplification, the action of the damper as a detuner may be compared to that of a double-jointed pendulum, where the application of a disturbing force to the upper half can cause both halves to swing in opposite directions; in other words, their masses will move in anti-phase (Figure 1.83e).

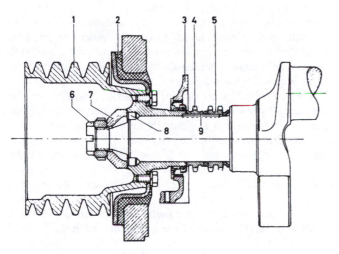

Figure 1.84 Example of attention to detail in rubber type vibration damper installation (*Mercedes-Benz*)

1 V-pulley
2 vibration damper
3 cover with radial sealing ring
4 oil pump sprocket
5 camshaft sprocket and distributor idling gear
6 clamp nut
7 pressure flange
8 dowel pin
9 Woodruff key

Methods of mounting torsional vibration dampers

For maximum effect a torsional vibration damper is always mounted on the front end of the crankshaft (Figure 1.84), since it is this end that suffers the greatest amplitude of vibration as a result of being furthest away from the *nodal point* (position of no vibration) of the crankshaft. The fact that the nodal point lies much closer to the rear end of the crankshaft can be explained by the greater inertia of the nearby flywheel and hence its reluctance to vibrate. If flywheels were mounted at both ends of the crankshaft then the nodal point would lie midway along the length of the shaft. It has sometimes been observed that it is one matter to design a damper of adequate capacity, but quite another to make its hub stay put on the nose of the crankshaft, because any joint transmitting a pulsating torque is inherently trying to fidget itself loose. A mechanically efficient joint is therefore essential for securing the damper and may be achieved by several methods, which most conveniently lend themselves to illustration (Figure 1.85). Split wedge rings and flange fittings are usually associated with heavy-duty diesel engine practice. An interference-fitting hub is sometimes favoured in American passenger car engine design.

1.10   VALVE TRAIN

General background

Chiefly for reasons of accessibility, the inlet and exhaust valves in very early motor vehicles were arranged in two separate rows, one on either side of the cylinders and operated from beneath by similarly positioned camshafts. This long since obsolete arrangement of side valves provided what was known as a *T-head engine* (Figure 1.86a). Its combustion characteristics were later recognized as being poor, and the contrasting hot and cold (exhaust and inlet) sides of the engine could lead to cylinder distortion problems.

The T-head engine was then superseded by the *L-head engine* (Figure 1.86b), in which the inlet and exhaust valves were arranged in a single row on one side of the cylinders and again operated from beneath by a similarly positioned camshaft. Engines with this particular arrangement of side valves underwent considerable development and for many years provided a power unit which was generally cheap to produce. It became obsolete during the mid 1950s because its power output was somewhat limited by space restrictions on the usable size of inlet valves and by difficulties encountered in adequately cooling the exhaust valves.

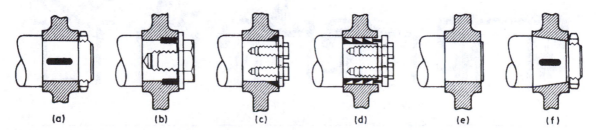

Figure 1.85 Flangeless methods of hub mounting for crankshaft vibration dampers: (a) parallel with key (b) parallel with dowel pins (c) parallel with wedge ring (d) parallel with wedge rings (e) parallel interference (f) taper with key

An early compromise between positioning the valves along-side the cylinders and in the head above them was a combination of these two locations. This resulted in the occasional *F-head engine* being built with overhead inlet and side exhaust valves (Figure 1.86c), both sets of valves being operated from a single camshaft mounted in the crankcase. Each overhead inlet valve was operated by a push-rod-and-rocker system. The main advantage of this type of layout was that larger inlet valves could be used, but being heavier they also placed limitations on maximum allowable engine speed. F-head engines were relatively expensive to produce and have been obsolete since the early 1960s.

While these various developments of the side valve engine had been applied to many touring cars, the designers of racing and sports cars had fairly early recognized that better engine performance would be more readily obtained by placing both the inlet and the exhaust valves over the cylinders, albeit with a certain amount of mechanical complication and less quiet operation. In varying degrees such arrangements allowed for more efficient shapes of combustion chamber and for a less tortuous and therefore faster route to be taken by the ingoing mixture and the outgoing exhaust gases.

Furthermore, the inlet and exhaust valves could either be arranged in a single vertical or near-vertical row, or be separated into two rows and mounted at an included angle to each other. These two arrangements of in-line and inclined valves are thus said to provide *I-head* and *V-head engines*, respectively (Figures 1.86d and e). Their mode of operation can be either directly from a single or a pair of cylinder-head-mounted or *overhead camshafts*, or indirectly through a push-rod-and-rocker system acting upon one, and in some earlier cases two, crankcase-mounted or *side camshafts*.

The present-day requirement for high performance from a medium-power engine has resulted in generally higher maximum crankshaft speeds, typically in the region of 5500–6000 rev/min. This can be explained by recalling the factors governing the power output of an engine since, if neither piston displacement nor mean effective pressure can be further increased, then the only other way to raise maximum power is to permit higher crankshaft speeds. For this reason, it becomes increasingly more important to avoid erratic operation of the valves at high engine speeds. As a result, about 80 per cent of car manufacturers now offer models with

the engine valves operated directly from an overhead camshaft, rather than by the less rigid push-rod-and-rocker system.

Commonly used abbreviations in connection with these valve layouts are as follows:

SV   side valves
IOE   inlet-over-exhaust valves
OHV   overhead valves
SOHC   single overhead camshaft
DOHC   double overhead camshafts

Side camshaft, push-rods and rockers

Camshaft

This serves to open the engine valves positively and to control their closing against the return action of the valve springs. In motor vehicle practice, a one-piece construction is almost invariably used for the shaft and its cams (Figure 1.87). Camshafts are generally produced from hardenable cast iron, which has replaced the case-hardened forged steel material used formerly. The angular spacing of the integral cams is such as to impart the required motion, in correct sequence, to the inlet and exhaust valves in each cylinder. To preserve accuracy of valve motion, the camshaft must be rigid enough to resist deflection under the alternating torsional and bending loads imposed upon it by the valve operating mechanism.

The camshaft journals are supported radially in the crankcase by a series of plain bearings. Since the loading on the camshaft bearings is generally not heavy and the operating speed of the camshaft is only one-half that of the crankshaft, the bearings are sometimes machined direct in the engine structure, especially where this is of aluminium alloy. Otherwise, established practice is to use separate steel-backed bushes that are pressed into machined bores. These bushes are lined with either a white metal alloy or one of the other bearing materials referred to in connection with the big-end

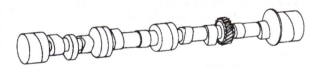

Figure 1.87   Camshaft for a four-cylinder engine

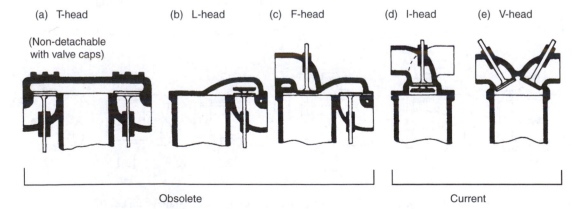

   (a) T-head   (b) L-head   (c) F-head   (d) I-head   (e) V-head

(Non-detachable with valve caps)

Obsolete   Current

Figure 1.86   Identification of basic valve arrangements

and main bearings (Section 1.7). To enable a camshaft to be inserted endwise through its bushes, the radius of its bearing journals is made slightly larger than the operating radius of its cams. There may also be a progressive, but albeit very small, decrease in radii for successive journals and bushes in the direction of camshaft insertion, so that it does not have to be threaded through close-fitting bearings of all the same size when being inserted (or withdrawn).

Cam followers

To convert the radial motion of the cams into the reciprocating motion necessary for opening and closing the valves, cam followers or tappets must be used (Figure 1.88a). This is because the force exerted by a cam acts perpendicular to its contact surface and therefore does not remain in the direction of follower travel. In other words, whatever mechanism bears directly on the cam is subject to a certain amount of side thrust. Sliding followers are almost invariably used in conjunction with crankcase-mounted camshafts. This type of follower is termed a tappet barrel (Figure 1.88a), and is typically produced from hardenable cast iron. However, since certain combinations of camshaft and tappet materials behave better than others, great care has to be exercised by the engine designer to ensure their compatibility and avoid scuffing. Various surface treatments are also applied to the cams and tappet barrels, in order to assist the running-in process of their highly stressed contacting surfaces.

Push-rods

These are required to transmit the reciprocating motion of the cam followers to the valve rockers (Figure 1.88a). Both ends of the push-rod form part of ball-and-socket joints which accommodate the angular movements of the push-rod arising from the straight-line motion of the tappet barrel on the one hand and the arcuate motion of the valve rocker on the other.

Since the push-rod is part of a valve train that constitutes a vibrating system, it must combine maximum rigidity with minimum weight. Push-rods are generally produced from steel and may be of either solid or tubular construction. In comparing solid and tubular push-rods of the same strength, the latter (Figure 1.88b) usually offer some reduction in reciprocating weight and can also serve as an oil conduit in the valve train lubrication system.

Valve rockers

The function of these is to cause both a reversal and a magnification of the motion imparted by the cam and follower to the valve. The valve rocker, or rocker arm, is a short rigid beam that oscillates about an offset pivot of either the journal bearing or the ball-and-socket type. An advantage of the latter type is that it makes the rocker inherently self-aligning, but it is necessary to introduce some means of restraining lateral movement. In cross-section the depth of the rocker arm greatly exceeds its width, since the bending loads imposed upon it are mainly within the plane of oscillation, with very little side loading.

Valve rockers may be of either solid or hollow construction (Figure 1.100a and b), and may be produced from steel forgings, iron castings, steel pressings, or in some later applications from aluminium alloy. Solid rockers oscillate about a stationary hollow shaft known as the rocker shaft. This is supported by a series of pedestals mounted on the top deck of the cylinder head, each pair of rockers generally being separated from the next pair by one of the pedestals (Figure 1.89). Sideways location of individual rockers is against adjacent rocker shaft pedestals, the rockers in each pair being held apart by means of a compression spring, spring clips or spacer tube. For some designs a separate bronze bushing is pressed into the rocker bearing bore, while in others the rocker is hardened all over with a plain bore bearing directly on the rocker shaft. A curved pad is machined on the end of the rocker where it contacts the valve stem tip, so as to allow the partly

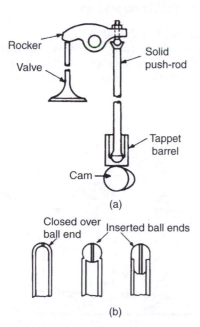

Figure 1.88   (a) Push-rod valve operation (b) types of pushrod

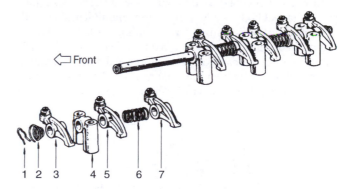

Figure 1.89   Rocker shaft assembly for a four-cylinder engine (*Toyota*)

1 retainer clip
2 conical spring
3 valve rocker arm, no. 1 exhaust
4 rocker shaft pedestal
5 valve rocker arm, no. 1 inlet
6 compression spring
7 valve rocker arm, no. 2 inlet

rolling and partly sliding motion that occurs between them. Although they have been used with conventional rocker shafts, hollow rockers are now usually mounted on individual pivot posts with either hemispherical or part-cylindrical seatings (Figure 1.100b), the latter being currently favoured in American practice. The pivot posts may be pressed and in some cases screwed into bosses on the top deck of the cylinder head.

Overhead camshaft and sliding followers

Camshaft

For a cylinder-head-mounted camshaft, modern practice tends to favour an increase in the number of intermediate bearings to one between each pair of cams. The bearing installations for overhead camshafts are currently similar to those of crankcase-mounted camshafts. An exception is where the bearings are supported in pedestals with detachable caps, which permit the camshaft to be inserted from above the engine. Also, this particular arrangement enables a reduction to be made in the diameter of the journal bearings, since the cams no longer have to be assembled through them (Figure 1.90).

Cam followers

The use of direct-acting sliding followers with a cylinder-head-mounted camshaft demands relatively large-diameter tappets (Figure 1.90). This increase in their base area is dictated by the absence of a multiplying leverage (otherwise provided by rocker arms) in the valve train, so that larger cams have to be used to give the desired amount of valve lift. Since an overhead camshaft is in even closer proximity to the valves than it was the obsolete side valve layout, the tappets must necessarily be both hollow to fit over the valve springs and short in length. For this reason they are aptly termed 'inverted-bucket tappets', although they may sometimes be referred to as Ballot-type tappets, having originally been introduced by the French Ballot company for its straight-eight engine in 1919.

In a recent Toyota development of the inverted bucket tappet, the tappet body is cold forged from aluminium alloy with a sprayed-on iron coating for its sliding surface. A tappet setting steel shim is recessed into its head portion and a stiffening steel disc is caulked beneath its head where contact is made with the valve stem tip (Figure 1.36). This construction promotes better fuel economy by virtue of generally lightening

the valve train, because the valve spring load can be reduced so that less torque is required to drive the camshaft. Furthermore, the valve train is less prone to chatter, since the tappet is lighter in weight and not so hard as steel and is therefore quieter in operation.

Overhead camshaft and pivoting followers

Camshaft

Similar remarks apply as in the previous system regarding the number of installation of the camshaft bearings.

Cam followers

These are known as the pivoting type because they swing between the camshaft and the valve tips and act as either a simple (straight) lever arm (Figure 1.100g and h) or 'bell-crank' (angled) lever (Figure 1.100i).

Modern applications of the former type generally feature a forged steel lever arm, or finger rocker as it is often called, which is self-aligning on a ball-and-socket pivot mounting (Figure 1.91), although journal bearing pivots have been used in the past. To restrain lateral movement of the ball-and-socket-mounted finger rocker, the valve stem tip may be recessed into the curved contact pad on the end of the arm (Figure 1.100g). To reduce valve train friction in some more recent applications, the sliding contact between the cam and finger rocker is replaced by a rolling one, the rocker being slotted in its central portion to embrace a roller mounted on needle bearings, the modern Ford V6 engine being an example of this practice.

The bell-crank type of pivoting rocker is supported on either plain journal or, more recently, needle roller bearings from a conventional rocker shaft. This is mounted above the camshaft, so that the downwards-extending lever arms of the bell-cranks bear on the cams and the outwards-extending arms contact the valve tips (Figure 1.100i)

Comparison of different systems

Side camshaft, push-rods and rockers

The advantages of this method of valve operation are generally that the timing drive to the camshaft is uncomplicated and therefore less expensive to produce, and that tappet clearance

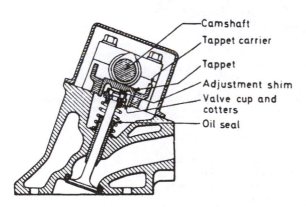

Figure 1.90 Overhead camshaft and sliding followers

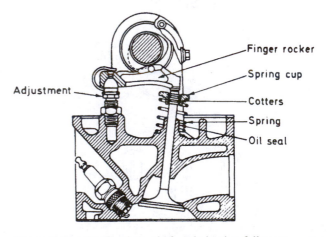

Figure 1.91 Overhead camshaft and pivoting followers

adjustment can be conveniently performed. Disadvantages are concerned with its greater tendency towards vibration, which is due to flexing of the push-rods and rockers and creates false motions of the valves. There may also be the need for frequent tappet clearance adjustment because of the number of points subject to wear.

Overhead camshaft and sliding followers

For this method of valve operation the advantages are chiefly its lower inertia and greater rigidity in the interests of high-speed operation and a reduced requirement for tappet clearance adjustment. Its disadvantages include the need for a more elaborate and therefore expensive timing drive to the camshaft(s); the engine in general tends to be more difficult to service, although tappet clearance adjustment in particular need not be laborious in modern versions, and in any event should seldom be required; and the engine height is increased.

Overhead camshaft and pivoting followers

Here the advantages are that it can provide a further reduction in inertia over that of sliding followers, together with some magnification of the motion imparted by the cam to its valve. It is also simpler to incorporate a ready means of tappet clearance adjustment. Disadvantages are generally that finger rockers do not possess the same degree of rigidity as sliding followers; there must inevitably be some side thrust transmitted to the valves, and effective lubrication can be more difficult to attain for minimum wear between the cams and followers.

Further developments in valve actuating systems

The advent of four-valve combustion chambers for passenger car engines, as discussed in Section 3.3, has necessarily led to some revision of the actuating mechanisms that have traditionally been used with two-valve installations. An interesting comparison between the various mechanisms that may be used, and which were investigated by the Toyota company during the development of their own four-valve engines, is shown in Figure 1.92. It will be noticed that in the second and third examples of SOHC layouts, a single cam and forked

rocker is used to actuate one of the two pairs of valves. The final choice made by Toyota on the basis of valve train rigidity and cost effectiveness was the direct-acting second example of DOHC. This less complex system also minimized friction losses in the valve train, which contributed to improved fuel economy in the engine speed range most used during everyday driving. Furthermore, this particular system had already proven virtually maintenance-free in high-performance engines.

A new approach to camshaft manufacture has been adopted by Ford for their V six-cylinder engine, which was introduced in 1995 and features DOHC for each bank of cylinders. As a weight saving measure and also to reduce inertia loading on the timing drive, each camshaft is of tubular construction with pressed-on sintered metal cam lobes. In the continuing pursuit of engine refinement with multiple valve cylinders an interesting development by BMW has been to counterbalance the eccentricity of the paired cam lobes by offsetting, in the opposite sense, the shaft sections between adjacent cams.

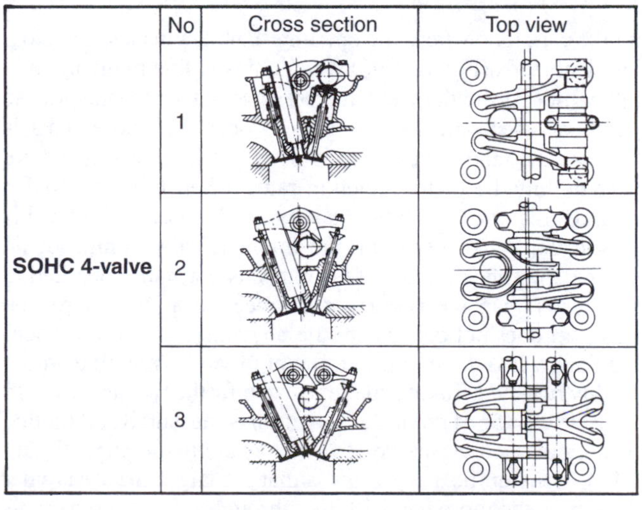

Identifying the parts of a cam profile

At this stage we must consider some of the whys and wherefores of what is probably the most versatile of all machine elements, namely the cam itself. According to their construction, cams used in engineering are usually classified into three types, known as disc, face and cylindrical cams. In the valve train of the automotive engine we are concerned only with the first-mentioned type. To convert rotary motion into a reciprocating movement, the basic parts of a simple disc cam profile comprise the base circle, the flanks and the nose (Figure 1.93). The last two parts constitute the cam lobe or lifting portion of the cam, since they determine the extent to which the follower can be lifted clear of the base circle. The period of valve opening, or valve event, is derived from the angular relationship between those points on the cam lobe where the opening flank begins and the closing flank ends. It therefore includes any interval during which the valve dwells in the fully open position, the governing factor here being the length of the nose arc. Conversely, that portion of cam rotation during which the base circle is operative with respect to the follower determines the valve closed period.

	No	Cross section	Top view
SOHC 4-valve	1		
	2		
	3		

Figure 1.92 Examples of actuating mechanisms for four-valve layouts (*Toyota*) HLA, hydraulic lash adjuster (hydraulic tappet)

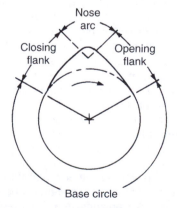

Figure 1.93 Basic parts of a simple cam profile

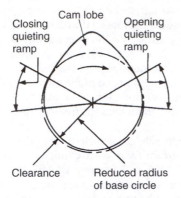

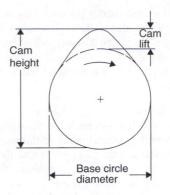

Figure 1.94 Simple cam profile modified by addition of quieting ramps

Figure 1.95 Extent of cam lift

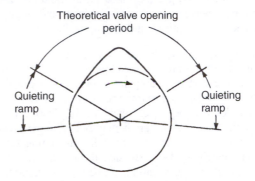

Figure 1.96 Cam valve opening period

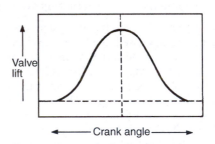

Figure 1.97 Valve opening diagram

The cam profile must also take into account the effects of valve or tappet clearance, so as to minimize noise from this source. For this purpose the simple cam profile is modified to include quieting ramps between the base circle and the flanks (Figure 1.94). These quieting ramps take the form of inclines that occupy 15 to 30° of camshaft rotation with a rise equal to the tappet clearance. Geometrically they are blended into the cam profile by reducing the radius of the base circle to correspond with this clearance and then connecting the undercut base circle to the opening and closing points on the flanks by means of the ramps. Dynamically these ramps serve to reduce the velocity at which the valve initially leaves and finally returns to its seating. Impact loading is thus reduced throughout the valve train, which contributes both to quieter operation and to extended valve life. Ramp heights are reduced for engines that feature hydraulic tappets because these virtually maintain a zero clearance throughout the valve train as explained later.

Valve lift and open period

From the description of the basic parts of a cam profile just given, it should be evident that a complete rotation of the cam momentarily lifts its follower by an amount equal to the overall cam height minus the base circle diameter dimension, the amount of lift being known as the cam lift (Figure 1.95). Here we must be careful to distinguish between cam lift and valve lift, since they are not necessarily one and the same thing. This is because the cam lift may be either less than, or equal to, the actual valve lift, depending upon whether or not a multiplying leverage is provided by valve rockers. The various arrangements of valve rockers were described earlier in this section, and it only remains to add that they usually confer a multiplying leverage in the region of 1.5:1. A further point is that valve lift will always be reduced in the presence of tappet clearance, so in the absence of valve rockers it would be more correct to the state that valve lift is nearly equal to cam lift.

The maximum lift of a valve generally approximates to one-quarter of its head diameter, since any greater opening than this gives diminishing returns in gas flow capacity. That is, the restriction to flow in the port and through the valve opening becomes equal when optimum lift occurs.

The period of valve opening or valve event has already been defined as the angular relationship between those points

on the cam lobe where the opening flank begins and the closing flank ends (Figure 1.96). Whilst this is fine in theory, it is rather more difficult in actual practice to determine the precise points at which the valve begins to open and close. This is because the tappet clearance is usually taken up somewhere near the middle of the quieting ramp when follower lift for each degree of cam rotation is very small indeed. It is for this reason that when verifying the nominal valve timing of an engine, the checking tappet clearance is normally specified to be greater than the running clearance, so as to ensure that valve lift does not occur until the beginning of the cam opening flank when the initial movement of the valve will be more readily detected. Taking matters a step further, if the valve lift for every say 5° of crankshaft rotation is measured and plotted on a graph then we arrive at a *valve opening diagram* (Figure 1.97). The significance of this is that the larger the area under the curve, the more gas will flow through the valve during its open period. A valve opening diagram is therefore of immediate interest to an engine designer and also anyone involved in preparing or supertuning competition engines. It is of much less interest to an automotive service engineer, however, who is more likely to be concerned with measuring cam height during engine overhaul to check whether it lies within the wear limits laid down by the manufacturer.

Types of cam profile

If each valve could be opened to its full lift instantaneously, dwell in the fully opened position for the desired period of opening and then be closed instantaneously, then clearly the area under the curve of its opening diagram would be at a maximum. Equally clearly, too, these ideal conditions of valve operation are unattainable in practice, since the infinite acceleration and deceleration of the valve train components would subject them to mechanically destructive forces. In the real-life automotive engine, it is therefore necessary to employ a cam profile that will open and close the valves in a more considerate manner. Although the theoretical considerations underlying the design of modern cam profiles are far removed from the realms of automotive service engineering, it may nevertheless be of interest to at least identify and briefly describe the various types of cam profile that have been used past and present, as follows.

Concave cams

A roller follower has to be used in conjunction with this early type of cam profile, where each concave flank is an arc of a circle that is tangent to both the base circle radius and the nose radius (Figure 1.98a). Although this type of profile lifts the follower with constant acceleration and deceleration, the transition from one to the other occurs very suddenly. Another disadvantage lies in the need to use a roller follower, which being heavier in construction creates larger inertia forces than a simple flat-faced one and thus makes the concave cam unsuitable for application to high-speed engines.

Tangent cams

Either a roller or a curve-faced follower is required for this again early type of cam profile, which is very simple in construction since each straight-line flank is a tangent joining the base circle radius with the nose radius (Figure 1.98b). It is sometimes described as a fast cam, because so long as the follower remains on the flank it continues to be accelerated. Then follows a very sudden transition from acceleration to deceleration, which is confined to that period of cam rotation when the follower is being lifted by the nose radius and is consequently very high in value. This in turn demands the use of a much increased valve spring force and generally makes the tangent cam again unsuitable for high-speed operation of an engine.

Convex cams

An essentially flat-faced follower is used in conjunction with this type of cam profile, in which each convex flank is an arc of a circle that is tangent to both the base circle radius and the nose radius (Figure 1.98c). A characteristic of this form of cam is that the follower is lifted for only a short period at high acceleration, this being followed by a much longer period of relatively leisurely deceleration, so that the controlling force exerted by the valve spring need not be too

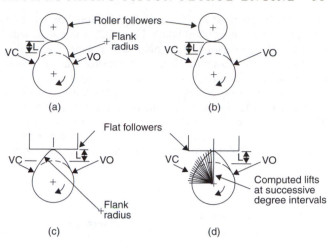

Figure 1.98 Types of cam profile: (a) concave (b) tangent (c) convex (d) non-geometric convex. L lift; VO valve opening; VC valve closing

great. As with the previously described concave and tangent cams, the convex cam still suffers from the disadvantage that the transition from acceleration to deceleration occurs very suddenly. However, the three-arc convex cam was widely used prior to the advent of the modern convex cam of non-geometric type described next.

Non-geometric convex cams

Here again an essentially flat-faced follower is used with cam profiles of this type, where each convex flank is not simply an arc of a circle, but has been tailor-made from a series of computed lift values at successive degree intervals of cam rotation (Figure 1.98d). By this means it becomes possible to make less sudden the transition from acceleration to deceleration during lift of the follower and also, of course, likewise during its return. Furthermore, cam profiles of this type can be mathematically developed to compensate at certain critical operating speeds for the effects on valve lift caused by any lack of rigidity in the valve train components.

Wear considerations related to cam form

The precise alignment required to avoid edge loading or digging in between nominally parallel cam and tappet contact faces is, in practice, very difficult to guarantee. Since edge loading could lead to excessive surface stresses, the contact faces of the tappet barrels may be made very slightly convex and are thus better able to accommodate any such discrepancies in alignment. This particular feature is usually combined with cams that are both tapered across their width and offset axially relative to the axes of the tappet barrels (Figure 1.99). The effect of this geometrical relationship between the cam and tappet contact faces is to induce rotation of the tappet barrel as it reciprocates, thus securing a more satisfactory distribution of wear over its working face. In modern OHC practice only the offsetting of the cams to induce rotation of inverted-bucket tappets is usually necessary, the cam and tappet contact surfaces being ground parallel.

In more recent American engine designs there has been a revival of interest in the roller type of sliding cam follower, the roller being used at the foot of a hydraulic tappet (this type of tappet is described later). The roller is provided with needle rolling bearings at its pivot and operates against a nodular iron camshaft with modern cam forms. Careful design is necessary to ensure that the roller is of optimum size to minimize contact stress. Also, the roller track may require a crowned profile, so as to compensate for any slight misalignment between the roller and its cam. Apart from conferring greater durability on the valve train, it is claimed by one manufacturer that this arrangement reduces friction by about 8 per cent and is therefore a contributory factor in improving the all-important fuel economy. A roller may similarly be applied to the pivoting rocker of an overhead camshaft installation, as described earlier.

Valve clearance: manual adjustments

The design of an engine must take into account the effects of thermal expansion and contraction of the valve train components, relative to that of the engine structure. Since these effects could result in the valves being held off their seats when they should be firmly closed, a small operating clearance has to be introduced into the reciprocating parts of the valve train. This is generally termed the valve or tappet clearance, and provision for either its manual or automatic adjustment is incorporated in all engines. The latter form of adjustment will be dealt with later.

For engines with push-rod overhead valves and solid rockers, valve clearance is usually adjusted at the rockers by means of a locking nut and hardened screw, the ball end of which registers in the push-rod socket (Figure 1.100a). The clearance may be checked by inserting a feeler gauge blade of appropriate thickness between the rocker pad and the valve stem tip. (A useful practical hint is not to slacken completely the locking nut while turning the adjusting screw.)

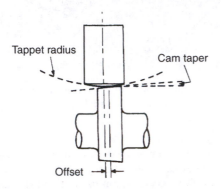

Figure 1.99   Cam and tappet contact geometry

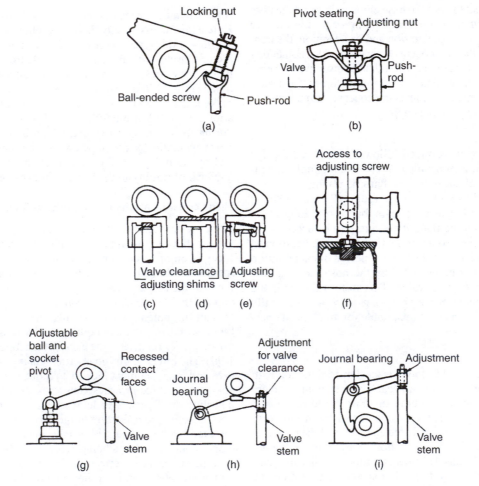

Figure 1.100   Valve clearance adjustments

Where manual adjustment of the valve clearance is employed for push-rod overhead valves and hollow rockers, a screw-threaded portion at the upper end of the pivot post receives a self-locking nut, against the underside of which the pivot seating abuts (Figure 1.100b). Turning the self-locking nut thus allows the valve clearance to be set by virtue of either lowering or raising the rocker pivot. In the case of an overhead camshaft and inverted-bucket tappets, the adjustment can be a time-consuming operation in those installations where valve clearance is set by the selective assembly of graduated-thickness shims, which are inserted between the valve stem tip and the underside of the tappet head (Figure 1.100c). Other less inaccessible means of adjustment have, however, been introduced in recent years for inverted-bucket tappets.

In one now widely used system discs of appropriate thicknesses are located in recesses in the tappet heads and are directly contacted by the cams, so that if a valve is held depressed after being operated by its cam the disc may easily be changed (Figure 1.100d). Another system utilizes the wedging action of a taper-faced adjuster cum locking screw, which is located transversely in the tappet and forms an abutment for the valve tip (Figure 1.100e). Yet another and particularly ingenious system incorporates a self-locking screw in the head of the tappet. Access to this screw, which bears directly against the valve stem tip, is gained through a divided cam and cross-drilling in the camshaft (Figure 1.100f).

For manual adjustment of valve clearance with an overhead camshaft and self-aligning finger rocker, the ball pivot is usually screwed into a boss in the top deck of the cylinder head, so that raising or lowering the ball pin alters the valve clearance accordingly (Figure 1.100g). In the case of journal bearing pivots, a screw-type adjustment for valve clearance is generally provided at the valve tip end of the rocker (Figure 1.100h). The same applies to the bell-crank type of pivoting rocker (Figure 1.100i). An interesting point to note in connection with these various methods of valve clearance adjustment is that those where the screw adjuster remains stationary serve to reduce the inertia of the valve train.

The effects of incorrectly setting tappet clearances
Accurate setting of the valve or tappet clearances in accordance with the manufacturer's specification is of vital importance, otherwise at least the valve timing of the engine can be affected and at worst valve failure may result. In giving incorrect valve timing, should the clearance be set smaller than specified, the valve will tend to open earlier and close later than intended, and vice versa should the clearance be greater.

The effect of setting the tappet clearance much too close, perhaps in a misguided attempt to gain quieter running, can result in the valve not fully closing on to its seating. Apart from loss of cylinder pressure, this will allow the high-temperature combustion gases to blow past the valve face and, since no heat can escape to the valve seat, the valve head temperature becomes so high that rapid burning and destruction of the face occurs. This situation is usually aggravated by charring of the oil film on the valve stem, which causes the valve to stick in its guide. The effect of setting the tappet clearance much too wide is to produce excessively noisy operation and also wear caused by pounding of the valve against its seating. Furthermore, the amount by which the valve is lifted from its seating will be reduced.

Hydraulic tappets

Since their introduction to the American motor industry on the Pierce-Arrow car of 1932 (a make of car that we shall be meeting again later in connection with early power-assisted steering), self-adjusting hydraulically operated tappets are often employed in preference to the simple mechanical type. Although the latter can periodically be manually adjusted to compensate for changes due to wear, they are unable to accommodate the effects of thermal expansion. Therefore a running clearance must be established in the valve train, which not only creates a potential source of noise, but also can be subject to error in its setting. Hence, the reasons for using hydraulic tappets include quietness of valve train operation, constant valve timing and the elimination of valve clearance adjustments in service. These advantages are obtained because tappets of this type automatically maintain zero clearances throughout the valve train, under practically all operating conditions, because their action is such as to compensate for the differences in thermal expansion of the valve train components and the engine structure. They also simplify assembly of the engine and therefore reduce cost of production.

In construction, the hydraulic tappet is essentially a telescopic device that receives a pressurized supply of oil from the engine lubrication system. The engine valve opening load is imposed upon a spring-loaded inner plunger, which has either a ball or a disc type of non-return valve that normally acts to close off the foot of the plunger. An exceedingly fine working clearance is provided between the plunger and the tappet body in which it is free to slide.

Its operating cycle begins each time an engine valve closes, at which point the tappet dwells upon the base circle of the cam and the plunger return spring extends the tappet assembly to absorb any clearances that exist in the valve train. As any extension of the tappet will create a depression beneath the non-return valve of the plunger, the valve opens so that oil under pressure from the engine lubrication system is admitted to the tappet compression chamber (Figure 1.101a). This is formed between the closed end of the barrel and the underside of the plunger. As soon as the cam begins to lift the tappet, the increased pressure on the trapped oil maintains the non-return valve closed, so that the column of oil behaves like a rigid strut to transmit the opening load developed between the cam and the engine valve (Figure 1.101b). There is an intentional slight leakage of oil from the compression chamber, which takes place between the plunger and its operating bore in the tappet barrel. This controlled leakage is known as tappet *leak-down*, and its purpose is to ensure that the engine valve always returns fully to its seating, as once again the tappet returns to the base circle of the cam ready for the next operating cycle.

Although once generally applied to the valve trains of push-rod-operated overhead valve systems, hydraulic tappets are now increasingly used to similar advantage as self-adjusting pivots for the finger rockers in modern overhead camshaft installations. That is, the hydraulic tappet body remains stationary and its plunger provides a pivot mounting for the finger rocker. This arrangement has the advantage of minimizing the inertia of the valve train, since the adjustment mechanism does in effect remain stationary. However, there has been a recent tendency in some multiple valve DOHC engines to incorporate a hydraulic self-adjusting facility in direct-acting

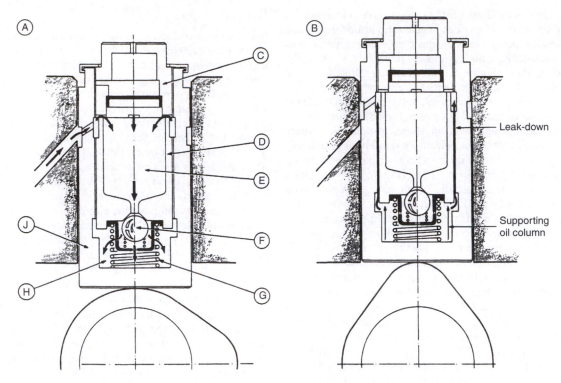

Figure 1.101   Operating principle of hydraulic tappet: (a) tappet absorbing valve train clearances (b) tappet acting as rigid strut to open valve

A Valve closed   D Tappet cylinder   G Tappet spring
B Valve open   E Feed chamber   H Pressure chamber
C Tappet plunger   F Non-return valve   J Tappet body

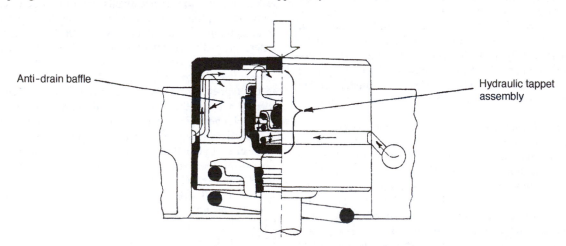

Figure 1.102   Schematic arrangement of a hydraulic bucket tappet showing method of oil feed

sliding type followers, thereby producing a very compact 'hydraulic bucket tappet'. Its operating principle is identical to that already described for other hydraulic tappet applications, but an 'anti-drain' function may also be included. The relevance of the latter is that the nearer to the vertical that the tappet operates, the greater is the chance that air may be induced to enter its compression chamber when significant expansion of the tappet occurs. In other words, following the leak-down of a tappet under the spring load of an engine valve that remained partially open when the engine was switched off and then gradually closed, thereby leaving the tappet in a compressed state prior to the engine being started again. The purpose of an anti-drain baffle in the tappet (Figure 1.102) is therefore to ensure that, when the tappet expands on starting the engine, the level of oil retained in the tappet is high enough to exclude air from its compression chamber, which otherwise results in a noisy tappet until the air is eliminated. From similar considerations it must also be arranged that oil does not drain away from the hydraulic tappets supply gallery in the cylinder head when the engine is not running.

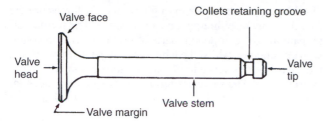

Figure 1.103   Terminology of the poppet valve

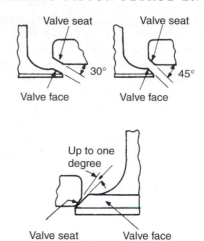

Figure 1.104   Valve seating angles

Engine valves

It has long been established practice to employ what are termed *poppet valves*, which comprise a disc-shaped head with a conical seating and a stem that acts as a guiding surface (Figure 1.103). Their main advantages over other possible alternative valve forms are as follows:

1 They are self-centring as they close on to the cylinder head seating.
2 They possess freedom to rotate to a new position.
3 It is relatively easy to restore their sealing efficiency in service.

For mechanical strength and to assist gas flow, the valve stem is blended into the head portion to form a neck of fairly generous radius, but a certain amount of relative flexibility between them is generally permitted. Under the influence of cylinder pressure, the valve head may therefore better conform to its seating should there be any distortion present. The valve stem is provided with a few hundredths of a millimetre working clearance in its guide; this is usually increased slightly for the hotter exhaust valves to allow for their greater expansion.

It is usual for the head diameter of the exhaust valve to be made less than that of the inlet valve, since cylinder gases may be more easily evacuated at exhaust pressure than admitted at induction depression – truly, as Spinoza observed, 'Nature abhors a vacuum'! Another reason for favouring a smaller-diameter exhaust valve is generally to reduce its thermal loading by virtue of the shorter path of heat flow.

The conical facing of the valve makes an angle of either 45° or 30° with the plane of the head. Although the former angle provides a higher seating pressure for a given valve spring load, the latter angle permits a greater gas flow for a given valve lift. Hence in some engines the face angles of the exhaust and the inlet valves may be 45° and 30° respectively. To improve further their seating conditions, the exhaust valves may be installed with a differential face-to-seat angle of up to 1° (Figure 1.104). The effect of this is initially to concentrate the seat bedding towards the larger-diameter and combustion side of the valve face, so that any subsequent dishing of the valve head then tends to centralize the seat bedding.

In practice, the face angle of the valve may therefore either match, or have a slight positive interference, with the seat angle. Also, the actual bedding area should be neither too narrow, which would hinder heat dissipation, nor too wide, which would reduce seating pressure and be less effective in breaking up deposits.

Valve materials

The valves must endure a particularly arduous existence, since they are subjected to severe mechanical and thermal loading. Exhaust valves in particular have to withstand mean operating temperatures that may approach 900°C, with about 75 per cent of this heat being transferred to the cooling system through the valve seat and the remainder through the valve guide. Seat face burning or 'guttering' must therefore be resisted.

Since the thermal loading on the inlet valves is less severe, they are generally produced from low-alloy silicon-chromium steel, which is also known as Silchrome from its American origins dating back to 1926. Its specification combined 3 to 3.5 per cent silicon (which until then had always been looked upon as an unavoidable impurity) with 8 to 9 per cent chromium and 0.4 to 0.5 per cent carbon. For light-duty applications a 1 per cent chromium steel may instead be chosen. A one-piece valve construction is used with these materials, since they can be locally hardened for tip, groove or seat wear resistance, while the stem may be chromium plated for the same reason.

The most widely used material for the exhaust valves is chromium-manganese-nickel steel, such as that designated 21-4N and referred to again later in connection with diesel engine valves (Section 2.6). This high-alloy material has a combination of hot strength and corrosion resistance that meets the requirements of most engines at temperatures up to over 800°C. The durability of the exhaust valves has sometimes been enhanced by the application of an aluminium coating to their heads, this being a General Motors process to form a tough corrosion resistant layer with valve steel and improve heat conductivity. Bimetal two-piece exhaust valves are also now employed, because they are more economical to manufacture. A necessarily expensive high-alloy steel need to be used only for the head portion, where the maximum value of hot strength is required; a less expensive low-alloy steel can be used for the stem, which confers the best guide wearing properties without the aid of chromium plating.

It may be of interest to mention that the joining of two-piece valves is accomplished by a process known as friction welding. This process is performed by the heat liberated when one of the parts being butted together is rotated against the other stationary part under an axial load, the final weld occurring when rotation is stopped. Friction welding was originally exploited by German engineers to join plastics components during World War II, but was afterwards vigorously developed by Russian engineers to join metal components. In the case of two-piece valves the friction weld in the stem is positioned to remain in the guide during valve travel.

Truly has it been said that the engine valve represents 'a metallurgical triumph over a mechanical monstrosity'!

Internally-cooled exhaust valves

Since the early 1980s the increased thermal loading that has been imposed on high-performance engines, first by turbocharging (Section 9.2) and then more recently by multiple valves (Section 3.3), has led to a modest revival of interest in the 'internally-cooled' exhaust valve. This function is accomplished by providing a hollow interior for the valve which, before its tip is sealed, is partially filled with a material that rapidly transfers heat from its exposed head surface to the stem portion and thence to the valve guide and cylinder head cooling jacket. The purpose of only partially filling (about half) the valve interior is to ensure that the enclosed material is shuttled to and fro by the opening and closing motions of the valve, as though shaking a cocktail drink, which therefore facilitates the transfer of heat from its head to the stem.

This type of exhaust valve was originally developed to meet the arduous duty encountered in aircraft piston engines, the first heat transferring material used being mercury, which rather interestingly was also favoured by Mercedes-Benz for the very powerful supercharged engines of their legendary Grand Prix cars of the 1930s. However, the real breakthrough in the development of the internally-cooled exhaust valve can be attributed to S.D. Heron, a British engineer working in America, who during the same period developed the sodium-cooled valve that posed less of a sealing problem. The advantages of using metallic sodium are several, not least among them being that as a conductor of heat it is over six times better than the valve steel and is also lighter in weight. Since it has a low melting point of 97.5°C it is liquid at valve operating temperatures and therefore when shaken it transfers heat by convection as well as conduction, whilst it has a high boiling point of 880°C that avoids vaporization. The sodium-cooled exhaust valve is usually furnished with a Stellite seat facing alloy (Section 2.6) to complement its high-duty application, so not surprisingly this type of valve construction has only received limited use for automotive engines owing to its greater cost. The Nissan high-performance 4.5 litre V8 engine with four-valve cylinders, which was introduced in 1990, is a modern example where sodium-cooled exhaust valves have been found advantageous in improving knock resistance and increasing torque when the engine is operated under high speed, heavy load, conditions.

Valve springs

As mentioned in Section 1.4, the valve springs are required to ensure that the motion of the valves and their operating mechanism follows faithfully that of the cams. Since this motion is a constantly changing one, inertia forces are created that may be either positive or negative. The former tend to maintain the component parts of the valve train in contact with one another, whilst the latter act to separate them. Hence, the valve springs serve to counteract the unwanted negative inertia forces. The valve springs must also maintain adequate sealing pressure for the valves during their intended closed period, in which respect they are assisted by the cylinder gas pressure acting upon the valve heads during the compression and power strokes.

In conventional practice, the valve springs consist of wire wound in the form of a helix or coil. The valve closing load is conveyed axially along the spring, which stresses the material principally in torsion. For the valve open and the valve closed conditions, the ratio of spring loads is usually in the region of 2:1. Since the valve spring is compressed between parallel abutments, its end coils are ground flat and square with the spring axis. The coil ends are also diametrically opposed, so as to minimize an inherent tendency towards bowing of the spring as it is being compressed. Because they are subjected to severe service, valve springs are produced from high-duty materials, these generally being either hard-drawn carbon steel or chrome-vanadium steel. To reduce stress concentration in the spring wire and thus make it less liable to fail by fatigue, the valve spring may be shot peened. In this process the wire surface is bombarded at high velocity with metal shot, which induces a residual compressive stress that discourages crack propagation in the material.

Valve spring surging

If the frequency of load application in the valve train should happen to coincide with the natural vibration frequency of the valve spring itself, a condition known as spring 'surge' may develop as a result of resonance effects. This phenomenon occurs as a wave motion, which is generated by the inertia of the individual spring coils, such that successive compressions and extensions travel along the spring from coil to coil, causing them to be unequally loaded. The effect of this surging is to oppose normal valve spring loading, so that true valve motion is disturbed.

To lessen any tendency towards surging within the operating speed range of the engine, the valve springs are designed to have a high natural frequency of vibration. Various additional measures may be taken to minimize surge, such as the use of double springs, mechanical spring dampers, and progressive rate springs. Double springs are arranged concentrically about the valve stem, each having a different natural frequency of vibration. Furthermore, the outer diameter of the inner spring may be made equal to the inner diameter of the outer spring, so that (lubricated) rubbing occurs between them (Figure 1.105a). This rubbing contact promotes a friction damping effect, which suppresses surge by dissipating its energy as heat. In some earlier designs the inner coil spring was made from rectangular section wire, so as to obtain the maximum load capacity in the minimum space (Figure 1.105a). The incidental advantages of using double springs include greater spring stability and less risk of engine damage in the event of a spring breaking. Another method of promoting friction damping is to encircle the valve spring

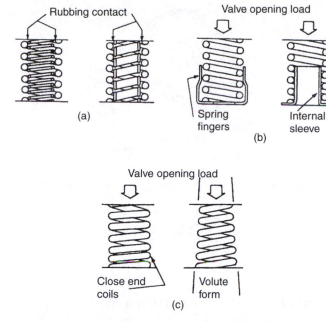

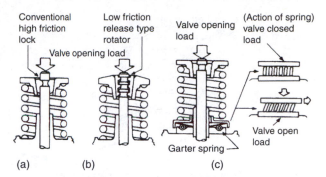

Figure 1.106 Methods of valve spring retention

Figure 1.105 Valve spring arrangements to combat surging: (a) double valve spring assemblies (b) valve spring dampers (c) progressive rate valve springs

with spring fingers. These are located from the stationary end of the valve spring and press inwards to make rubbing contact with its centre coils (Figure 1.105b). Alternatively, an internal sleeve may be installed within the spring to achieve a similar effect (Figure 1.105b).

Progressive rate springs are also commonly used to suppress surge. Springs of this type either are close coiled at their stationary end or, less commonly, utilize a volute form with the smallest diameter and hence stiffest coil at the moving end (Figure 1.105c). In both cases, the effect is to vary the number of active coils in the spring during valve lift, which produces changes in its natural frequency of vibration so that resonance is avoided. Space limitations generally preclude the use of surge-free low-inertia springs of the hairpin and torsion bar types, as alternatives to the helical valve spring in automotive engine practice.

For readers who may be interested in the engine technology of Grand Prix cars, where engine speeds of around 18 000 rev/min are now being achieved, the motion of their valves is controlled not by steel but by pneumatic springs. This concept of using valve stem plungers moving in chambers of compressed air to maintain the valves in contact with their cams, was devised by Bernard Dudot in 1986 for Renault Grand Prix engines.

Methods of valve spring retention

Before leaded petrol was phased out, the different methods of valve spring retention were related to the varying extents that rotation of the exhaust valves was to be encouraged. It was then generally recognized that gradual rotation of the valves during engine operation was beneficial to valve life. This is because it promotes a more uniform temperature distribution around the valve seat, as well as improving heat

transfer from the valve by virtue of more effective removal of seat deposits. The valves are generally credited with an inherent tendency towards rotation, which is variously attributed to engine vibration and the slight winding up of the valve springs as they are compressed when the valve is opened. Several devices of American origin may be employed, however, to impart either a non-positive or a positive rotational movement to the valves.

High-friction-lock retention
This was once the conventional method of retaining a valve spring, the moving end of which encircles a spigot washer with a central conical recess. Seating in the latter is a pair of split-type wedge collets which in turn engage a shallow groove machined near the top of the valve stem (Figure 1.106a). The pressure exerted by the valve spring therefore holds the collets tightly in place, yet they can readily be removed (albeit with a little persuasion from a soft-faced mallet in some cases!) by depressing the spring while holding the valve closed. This method of valve spring retention makes no pretence at encouraging rotation of the valves during engine operation.

Release rotators
This method of valve spring retention confers a non-positive rotary movement to the valves by merely permitting them to rotate relatively freely under the influence of engine vibrations. Such action is made possible by dispensing with the radial clamping exerted by the high-friction-lock method of retention and transmitting the spring load to the valve stem through a multiple-thrust collar arrangement, which is provided by abutting collets (Figure 1.106b). This is in contrast to the gap that must necessarily exist between the sides of wedge collets.

Positive rotators
In this method of valve spring retention, valve rotators of the positive type are incorporated as intermittent motion devices. A modern version consists of an enclosed garter spring, which is sandwiched between a thrust washer and the lower valve spring washer, there being the normal high-friction lock against the upper end of the valve stem (Figure 1.106c). During operation the increase in valve spring load that accompanies valve opening causes the coils of the garter spring to lean over. In so doing they act as friction sprags and transmit a torque to the valve spring and its upper retaining washer, which in turn imparts a small rotational movement to the valve itself.

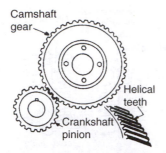

Figure 1.107   Arrangement of gear timing drive

With the requirement to use unleaded petrol, there is less incentive to impart a rotational movement to the valves. In the absence of what was previously a 'lubricating' film of lead oxides and lead sulphides forming on the valve seating, it is now thought that valve rotation could lead to scuffing wear.

1.11   TIMING DRIVE

Methods of driving camshafts

In conventional four-stroke engine practice one or more camshafts, according to valve train requirements and arrangement of the cylinders, are driven from the front end of the crankshaft. Since the axes of the crankshaft and the camshaft(s) are parallel to one another, a positive drive (non-slipping) between them may be afforded by any of the following means of power transmission: gear, chain, and toothed belt.

Gear train
Gear drives were once widely favoured for both crankcase- and cylinder-head-mounted camshafts. Except in the case of competition engines their use is now confined to the former application, helical teeth being generally specified in the interests of smooth and quiet operation (Figure 1.107). These qualities derive from the progressive engagement characteristics of their inclined teeth, one pair remaining in mesh until the following pair are partially engaged (Section 14.1). To reduce further any noise from gear-type drives, a non-ferrous acoustically dead material such as aluminium alloy may be specified for the camshaft gear. A steel crankshaft pinion is retained in this event.

Chain and sprockets
Until the advent of the toothed-belt timing drive for cylinder-head-mounted camshaft(s), the chain and sprockets drive was the most widely used for both crankcase- and cylinder-head-mounted camshafts. The two types of chain drive employed are known as the roller chain and the inverted-tooth chain, each being designed to avoid sliding contact between the teeth of the sprockets and the chain.

Historically, the bush roller chain was invented and patented as a means of power transmission by Hans Renold in 1880. In construction, the side plates of the inner link pairs are rigidly connected by bushings, within which oscillate the bearing pins connecting the side plates of the outer link pairs, and upon which rotate the rollers that engage the sprocket teeth (Figure 1.108). Any wearing of the pivot joints will tend to increase the pitch of the chain, that is the linear distance

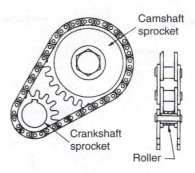

Figure 1.108   Arrangement of roller chain timing drive

Figure 1.109   Inverted-tooth chain timing drive

measured between adjacent roller axes. To accommodate this apparent stretch of the chain, the sprocket teeth are curved in profile so that, as the rollers ride higher on the teeth, contact is still maintained between each roller and tooth included in the arc of chain wrap.

According to the power to be transmitted, roller chains may consist of either a single or two or more strands; the multistrand chains provide a greater drive capacity for a given pitch of chain. In a recent example of a roller chain timing drive for a double-overhead camshaft engine, the sprocket wheels are provided with plastic inserts to support the chain and reduce any tendency to whine.

The inverted tooth chain drive (Figure 1.109) was another Renold invention and, although originally used in Britain until the mid-1920s, it subsequently found widespread application for driving a crankcase-mounted camshaft during the side-valve engine era in America. This type of timing chain comprises a multiplicity of overlapping toothed links, which are connected together by means of either bearing pins and bushes or special rocker pivots. The teeth formed both on the chain links and the sprockets are of tapered profile, so that full-length bedding is established along their contact faces. Any wear occurring in the pivot points of this type of chain is accommodated by the link teeth moving higher up the sprocket teeth. To prevent the inverted-tooth chain from sliding endwise off its sprockets, either the outermost links are extended radially inwards so as to embrace the sides of the sprocket teeth, or central guide links engage a groove in the sprocket. Similar to the roller chain, the drive capacity of the inverted-tooth chain increases in proportion to its width. For both roller and inverted-tooth timing chains the sprockets are usually made from either steel or cast iron. In American practice an inverted-tooth timing chain may be used in conjunction with an aluminium alloy camshaft sprocket that has nylon teeth.

Toothed belt
Originally adopted in 1963 by the Hans Glas Company in Germany as a method of driving an overhead camshaft for a

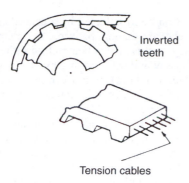

Figure 1.110 Arrangement of toothed-belt timing drive

motor car engine, the inverted-tooth belt was actually invented by an American, Richard Case, of the US Royal Company some 25 years previously and first found application to the synchronous needle and bobbin drive of sewing machines.

To provide a positive timing drive, as opposed to the non-positive friction drive conferred by an ordinary fan V-belt, the belt teeth engage with axial grooves in the peripheries of the crankshaft and the camshaft pulleys (Figure 1.110). Tooth profiles are based on either trapezoidal or semi-circular forms, which permit a smooth rolling motion upon engagement and disengagement of the belt with its pulleys, similar to the action of gearing. The shear strength of the teeth in mesh always exceeds the tensile strength of the belt. In this context the tension loads are transmitted through the neutral axis of the belt by a layer of either fibre-glass cords or steel cables, their strands being spirally wound in alternate directions to stabilize the running of the belt. The main body of the belt is constructed from neoprene synthetic rubber, and to protect the teeth it is faced with a hard-wearing polyamide (nylon) material that also has a low coefficient of friction against the pulley grooves. With some products the intended direction of travel for the timing belt is arrow-marked on its back surface. Axial location of the belt may be effected by a flanged jockey pulley. Lubrication is not, of course, required for a toothed timing belt, which is generally enclosed for protection against the elements and as a safety precaution.

Camshaft drives compared

Apart from quietness of operation, a timing drive must further meet the all-important criteria of low cost, ease of assembly and maintenance, and a long trouble-free life. In purely general terms, the relative merits of the various methods of driving camshafts may be summarized as follows.

Gear drive

This better maintains the accuracy in timing of the camshaft and, especially important in the case of diesel engines, the fuel injection pump. In contrast significant changes in timing can occur owing to apparent stretch in chain drives over an extended period of service, since this particular effect cannot be compensated for by any tensioning device. Gear drives, as always, must be carefully engineered and therefore tend to be expensive, but in high-grade practice they can be made both quieter and longer lived than chain drives. Their application is, however, mainly confined to timing drives for crankcase-mounted camshafts, since roller chain and toothed-belt drives offer a much simpler layout for driving overhead camshafts.

Chain drive

This can readily accommodate either short or long centre distances between the crankshaft and camshaft(s) and is therefore equally suitable for both crankcase- and cylinder-head-mounted camshafts, although the heavier inverted-tooth chain is less well suited to the latter application. Unlike a helical gear timing drive, no end thrust forces are generated at the sprockets of timing chains, which are also less intolerant of any slight misalignment. On the debit side, a timing chain is subject to the cumulative effects of slight wear occurring at the joints of its many links that results in apparent stretch. For most chain runs it therefore becomes necessary to incorporate either a manual or an automatic adjuster to maintain correct chain tension.

Furthermore, at certain critical engine speeds fluctuations in torque can arise in the rotating elements of the timing drive, which can excite the chain into vibration and require the addition of rubber-faced damper shoes to bear against the 'tight' side of the chain (Figure 1.114 and 1.115).

Toothed belt

As earlier inferred, this form of timing drive is complementary to the modern revival of cylinder-head-mounted camshafts. Its advantages include inherent quietness and smoothness of operation, lightness in weight and no lubrication requirements. Hence, this type of timing drive can be accessibly mounted external to the engine main structure. On the debit side, the greater width of the toothed belt tends to demand more installation space at the front of the engine. Also whereas the gear and the chain types of timing drive are usually expected to last the normal service life of an engine, the toothed belt has to be considered a consumable item, a typical early recommendation being that it should be inspected with a view to replacement after 30 000 miles (48 000 km). Higher engine speeds and the adoption of four-valve layouts have both increased the duty imposed on the toothed belt timing drive, even so the present-day replacement interval has tended to become longer and in some cases doubled to 60 000 miles (96 000 km). Whilst this reflects the improvements that have been made in timing belt construction, it is still essential to observe the particular manufacturer's recommendations in this respect.

There has nevertheless been a trend by some manufacturers to revert to a chain timing drive, because they represent a more compact installation and can offer greater durability with lower maintenance. Also, there can be the additional duty to drive a high-pressure fuel delivery pump where diesel and direct injection petrol engines are used.

Timing belt misbehaviour in service

The mileage interval specified by the vehicle manufacturer for replacing a timing belt is intended to provide a margin of safety, before the belt ultimately fails by fatigue of its constantly flexing tensile members. However, there are other factors that can arise in the operation of a timing belt, which at least can prematurely shorten its useful life and at worst lead to breakage, with potentially major damage to the engine where piston to valve contact occurs through loss of timing.

As a general guide to some of the problems that may be encountered with timing belts in service, the following is a list of their likely causes:

Undue wear on belt teeth
Belt tension either insufficient or excessive causing incorrect meshing with toothed pulleys
Undue wear between belt teeth
Belt tension excessive causing too deep meshing with toothed pulleys
Belt abraded by rough tips on toothed pulleys
Undue wear on edge of belt
Misalignment between the toothed pulleys
Damaged flange on toothed pulley that provides axial location for belt
Teeth cracking at roots and peeling from belt
Belt tension insufficient to prevent undue bending of the teeth as they ride too high on the toothed pulleys, and can lead to the belt jumping teeth
Group of teeth sheared from belt
Sudden overload imposed on the belt by seizure of a driven auxiliary
Structural deterioration of belt
Oil or diesel fuel contamination of the belt causing swelling, separation and ultimate destruction of the rubber composite material
Back surface crazing of belt
Excessive frictional heat caused by seizing of either an idler pulley or driven auxiliary acting against the back surface of the belt
Belt torn apart
Acute bending of the belt either before or during installation causing a local weakness and tearing between adjacent teeth
Rogue object caught up between the belt and a toothed pulley causing local over-stretching and tearing across a group of teeth.

Since a toothed belt timing drive is normally quiet in operation, any noise that develops can provide an audible warning of distress. A whining noise can indicate excessive tensioning of the belt, while a slapping sound is associated with insufficient tensioning.

Attachment of the camshaft timing wheel

The once traditional method of attachment was to key the timing wheel on a reduced-diameter extension of the camshaft,

endwise retention being effected by a central clamping nut. To prevent relative rotation between the clamped parts a 'Woodruff' key was used, this being close fitting only on the sides of its keyway and self-aligning by virtue of its semicircular form (Figure 1.111a). In modern practice, the function of the Woodruff key is generally replaced by that of an off-centre dowel pin (Figure 1.111b). The dowel pin is simply a plain cylindrical pin that is made close fitting in corresponding holes in the timing wheel and the end face of the camshaft, these two members being spigotted together for radial location. Their endwise clamping is effected by a central bolt.

For heavier-duty applications, a ring of close-fitting bolts may perform both the clamping and the driving functions, radial location being supplied by a spigot fitting for the timing wheel (Figure 1.111c). This particular method of attachment also lends itself to a fine adjustment feature for setting the valve timing, since if two sets of holes are provided in the timing wheel it is then possible to make incremental adjustments of half a tooth pitch to the timing. Alternatively, if there is no requirement for such fine adjustment, the correct positioning of the timing wheel may be guaranteed by offsetting one of its ring of bolt holes to correspond with a similar offset hole in the camshaft end.

The attachment of the crankshaft timing wheel has been dealt with in Section 1.8.

Controlling end float of the camshaft

In arrangements where the oil pump (together with ignition distributor) is still provided with a right-angle drive from the engine camshaft, via a pair of crossed-helical or skew gears, the obliquity of their teeth is such that an end thrust must be reacted against by the camshaft bearing arrangements. Similar considerations apply, of course, where a helical gear timing drive is used. For these reasons it therefore becomes necessary to control any endwise movement of the camshaft.

Axial location for a crankcase-mounted camshaft is commonly effected by the parallel surfaces of a collar thrust bearing, which acts on the timing wheel end of the camshaft. For some applications, a thrust plate bolted to the front wall of the crankcase is provided with a semicircular opening, the edge of which engages with a clearance groove around the camshaft front journal (Figure 1.112a). In other designs, the thrust plate may completely encircle a similar clearance groove that is formed partly by a shoulder machined on the front end of the camshaft, and partly by the inner end face of the timing wheel boss. Less commonly, a true collar thrust

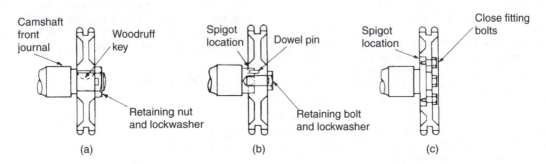

Figure 1.111   Camshaft timing wheel mountings

bearing in the form of a flanged front journal may be employed (Figure 1.112b). The flange registers in an annular clearance groove in the forward part of the bearing housing, which is counter-bored and provided with a closing thrust plate.

Cylinder-head-mounted camshafts are located axially by collar thrust bearings similar to those just described, except where detachable bearing caps are used. Here, the end thrust loads may be reacted against on either side of the front bearing housing by the inner end face of the timing wheel boss on the one hand and by a flange formed immediately behind the camshaft front journal on the other (Figure 1.113a). Where an inverted-tooth belt drive is used for overhead camshafts, axial location may be effected from the rear end of the camshaft, again by methods similar to those already described (Figure 1.113b).

For all these various thrust bearing arrangements the camshaft end float or end play typically falls within the range of 0.15–0.38 mm (0.006–0.015 in).

Timing drive installations

Whereas only a one-stage timing drive is demanded by a crankcase-mounted camshaft for push-rod-operated overhead

valves (Figure 1.114), either one or more stages of chain drive may be needed for cylinder-head-mounted camshafts. For in-line cylinder engines with a single overhead camshaft a one-stage timing drive suffices (Figure 1.115), but where double overhead camshafts are used a two-stage drive is required (Figure 1.116). The latter entails the use of an idler

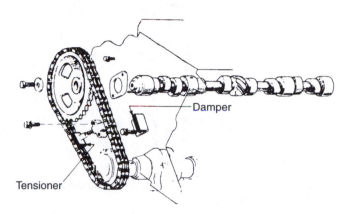

Figure 1.114 Exploded view of a one-stage roller chain timing drive (*Toyota*)

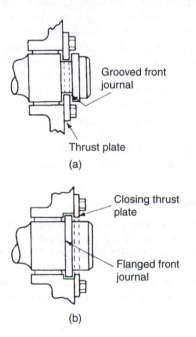

Figure 1.112 Collar thrust bearings for crankcase-mounted camshaft

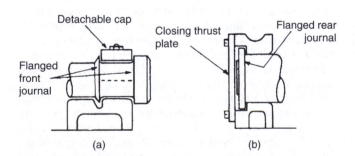

Figure 1.113 Thrust bearings for cylinder-head-mounted camshaft(s)

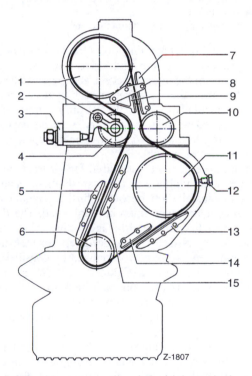

Figure 1.115 A one-stage roller chain driving a single overhead camshaft (*Mercedes-Benz*)

1 camshaft sprocket
2 sprocket bearing and tension sprocket
3 chain tensioner
4 tension sprocket
5 chain guide, outside
6 crankshaft sprocket
7 chain guide, outside
8 support bracket for chain guide, inside
9 chain guide, inside
10 reversing sprocket
11 intermediate gear
12 locking screw
13 chain guide, outside
14 chain guide, inside
15 chain guide, inside

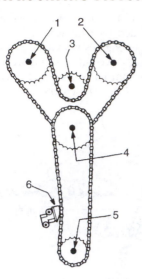

Figure 1.116   A two-stage roller chain driving double overhead camshafts (*Renold*)

1,2 camshaft sprockets
3 secondary stage manually adjustable idler sprocket
4 tandem idler sprockets (one hidden behind the other)
5 camshaft sprocket
6 primary stage slipper shoe chain tensioner

sprocket assembly comprising two sprockets in tandem, so that the primary stage transmits drive from the crankshaft sprocket to one of the idler sprockets, and the secondary stage is from the other idler sprocket to each camshaft sprocket. This arrangement also enables the 2:1 timing drive reduction ratio to be shared between the primary and secondary stages, thus permitting the use of smaller-diameter sprockets than would otherwise be possible.

A similar arrangement is required for V-formation cylinder engines with double overhead camshafts on each bank, although the installation becomes further complicated in that the secondary stage must be duplicated. The detailed manner in which such a multi-stage timing drive has been accomplished by Nissan engine designers for their modern V six-cylinder engine is best visualized by considering the arrangement in its entirety (Figure 1.117). Where a single overhead camshaft is used on each bank of a V engine, a much simpler one-stage drive is possible.

For in-line cylinder engines utilizing a toothed-belt timing drive, a one-stage drive only is required for both single (Figure 1.118) and double overhead camshaft installations (Figure 1.30). Note also in the latter example that the coolant pump is positively driven by the timing belt. Prior to the advent of direct type of ignition systems that do not need a distributor (Section 10.5), it was established practice for the distributor to be driven in tandem from the rear of an overhead camshaft. This source of drive may now be utilized to power a high-pressure pump, which delivers fuel to the common rail of a modern direct injection petrol engine (Section 6.9). An auxiliary or jack shaft that is mounted in the crankcase in a similar manner to that of a camshaft may be included in the timing drive to operate the oil pump and also, in earlier practice, a mechanical fuel pump (Figure 1.118). Otherwise, the oil pump is driven directly from the front of the crankshaft

(Figure 1.30 and 3.10) and an electrical fuel pump is used. A one-stage toothed belt timing drive may also be used for V engines with a single overhead camshaft on each bank, but separate belts with side-by-side drive pulleys on the crankshaft are required for horizontally opposed cylinder engines (Figure 4.13).

Further developments in timing drive installations
In the case of engines with four-valve combustion chambers where a narrow, rather than a wide, included angle is adopted for the two pairs of valves, the closeness of the two camshafts may preclude the use of a one-stage toothed-belt drive to both. That is, a toothed-belt drive can be taken to one camshaft and some other form of inter-camshaft drive must be added to complete the timing drive.

When Toyota were faced with this particular engine development problem, they found that the use of a chain drive over such a short distance between centres and relatively small sprockets was neither quiet nor reliable enough, because of the torque fluctuations imposed by changes in valve spring load as the valves opened and closed. Instead they chose to use a helical gear drive between the two camshafts (Figure 1.119a and 1.36) and mades special provision for ensuring its quiet operation. Since noise can arise from the backlash that results from the tolerance necessary for correct meshing of the teeth in a pair of gears, the driven gear is of divided construction with one part being torsionally spring-loaded against the other to cushion the effect of the backlash (Figure 1.119b and 1.22). Although this method of quietening a pair of gears has been used in earlier timing drive systems, this is the first time it has been applied with such simplicity and to a double overhead camshaft layout. It is referred to by Toyota as their 'inter-camshaft scissors gear drive'.

Timing drive tensioners

Various forms of jockey sprocket, slipper shoe and spring blade tensioners may be arranged to bear against the slack (non-driving) side of roller chain runs. The jockey sprocket type of tensioner is mounted on either a sliding plunger, a swinging arm, or an eccentric bushing. Slipper shoes may similarly be mounted on either a sliding plunger (Figures 1.114 and 1.116) or, less commonly, on a swinging arm (Figure 1.115). Blade-type tensioners generally take the form of rubber-faced spring steel strip, which is so mounted that it bows towards the chain under the influence of an extension spring connecting its ends (Figure 1.120a). A further refinement is for a semirigid blade to be produced from a thermoplastics material, which when hot can deform under the influence of its spring loading and readily adopt the natural contour of the chain run to exert optimum tensioning.

For their automatic operation, all the various types of chain tensioners rely primarily on spring loading (Figure 1.115), which may or may not be adjustable externally. In one version, the spring loading is augmented by hydraulic pressure derived from the engine lubrication system (Figure 1.120b). The oil is admitted behind the spring-loaded plunger of the tensioner mechanism and escapes via a bleed orifice to confer a measure of built-in hydraulic damping action for resisting chain flutter. Chain tensioners of the jockey sprocket and

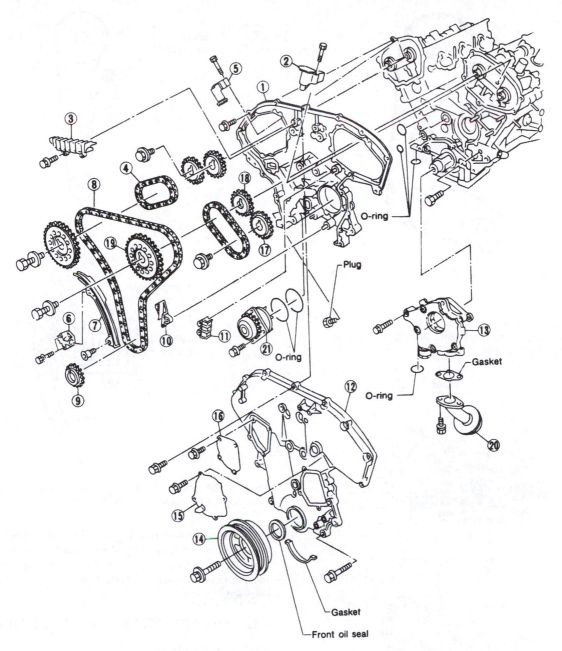

Figure 1.117   A multi-stage roller chain timing drive chain and sprockets for a modern V six-cylinder engine (*Nissan Motor GB*)

1 Rear timing chain case	8 Timing chain	15 Water pump cover
2 Left camshaft tensioner	9 Crankshaft sprocket	16 Chain tensioner cover
3 Intake chain tensioner guide	10 Lower chain tensioner guide	17 Exhaust cam sprocket, 2nd
4 Camshaft chain	11 Upper chain tensioner guide	18 Intake cam sprocket, 2nd
5 Right camshaft tensioner	12 Front timing chain case	19 Camshaft sprocket, 1st
6 Chain tensioner assembly	13 Oil pump assembly	20 Oil strainer
7 Chain slack guide side	14 Crank pulley	21 Water pump

slipper shoe types generally incorporate a non-return mechanism. Its purpose is to prevent any significant loss of adjustment under adverse conditions of chain loading, which may force back the tensioner an undue amount. The non-return or locking feature usually takes some form of pawl and ratchet device (as in a handbrake mechanism), in which the movement of the tensioner in taking up chain slack is followed up by the pawl engaging successive teeth on the ratchet. Over-tensioning of the chain is avoided by the teeth on the pawl and ratchet assuming a trailing angle, such that with each increment of travel they always allow the chain tensioner a limited return movement against its spring loading.

With toothed-belt timing drives for overhead camshafts, a jockey pulley is usually arranged to guide the back of the

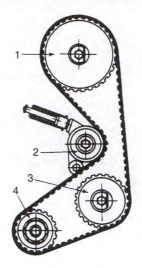

Figure 1.118 A toothed belt driving an overhead camshaft (*Lancia*)

1 camshaft toothed pulley
2 spring-loaded jocket pulley
3 toothed pulley for auxiliary units drive shaft
4 crankshaft toothed pulley

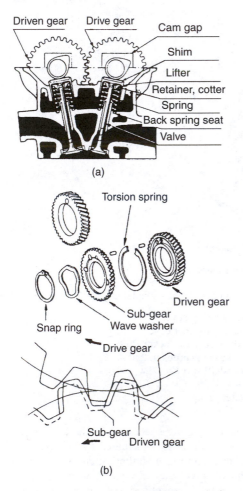

(a)

(b)

Figure 1.119 Upper timing drive for narrow-angle four-valve layout: (a) geared-together camshafts (b) scissors gear mechanism (*Toyota*)

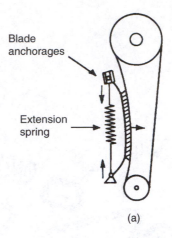

(a)

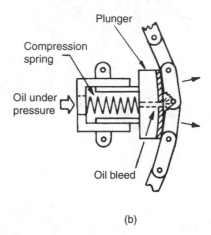

(b)

Figure 1.120 Types of automatic timing chain tensioner: (a) spring loaded blade (b) spring loaded plunger

slack side of the belt run (Figure 1.118). It is either manually tensioned or automatically spring loaded, so as to absorb manufacturing tolerances in the timing drive and to facilitate installation of the belt when the tensioning arrangements are released.

1.12 THE PRINCIPLES OF VALVE TIMING

Basic considerations

So far in our understanding of the four-stroke principle, we have accepted that the opening and closing of the inlet and exhaust valves is timed to coincide exactly with the beginning and ending of the induction and exhaust strokes. For a variable-speed motor vehicle engine such a tidy approach to valve timing would result in very inefficient operation. In actual practice a departure must therefore be made from the basic valve timing implied by the four-stroke principle, this being early acknowledged by the French automotive pioneer Fernand Forest who originally patented the idea of valve timing in 1906. Let us now examine how this departure can be accounted for in relation to the factors involved, which may be listed as follows:

1 Inertia effects of the incoming and outgoing cylinder gases
2 Elastic nature of the incoming and outgoing cylinder gases

3 Mechanical stresses imposed by rapidly opening and closing the valves.

In physics we learn that all bodies possess inertia, no matter whether we are talking about a cannonball or a column of gas, and that the effect of inertia is to resist a change of motion. The incoming and the outgoing flows of cylinder gases are therefore subject to inertia effects. As a consequence there is an unwelcome reluctance of the gases to commence flowing, followed by a welcome reluctance of them to cease flowing.

Also in physics we discover that the volume of a gas, unlike a liquid, is not fixed and that this property of great compressibility is possessed by all gases. The elastic nature of any burnt cylinder gases remaining in the combustion chamber at the end of the exhaust stroke can, therefore, be such that they expand and impede the flow of incoming air and petrol mixture at the beginning of the induction stroke.

In engineering we appreciate that when a material is acted upon by a force, every particle in the material acts upon every other particle with which it is in contact and sets up what is termed a state of stress. This imposes definite limits as to the rapidity with which a valve can be opened and closed, before breakdown of lubrication and overstressing of the contact surfaces of its cam and tappet occurs. Due allowance must therefore be made for the short time it necessarily takes to open and close each valve during its open period.

How then can the basic valve timing of the four-stroke principle be modified in practice to accommodate the various effects mentioned? It is done by providing for the *lead* (advance time) and *lag* (delay time) of the inlet and exhaust valve periods of opening, or *valve events* as the American industry prefers to call them.

Lead, lag and overlap

The inlet valve is given a lead in opening before the piston reaches top dead centre on the exhaust stroke (Figure 1.121a), so that least resistance is offered to the incoming flow of air and petrol mixture as the piston begins its induction stroke. It is also provided with a lag in closing after the piston reaches bottom dead centre and is beginning the compression stroke (Figure 1.121a), so as to take advantage of the reluctance of the incoming mixture to cease flowing as the piston ends its induction stroke. The maximum amount of air and petrol mixture is therefore induced to enter the cylinder, upon which depends the power developed by the engine.

Similarly, the exhaust valve is given a lead in opening before the piston reaches bottom dead centre on the power stroke (Figure 1.121b), so that the burnt gases are already leaving the cylinder under their own pressure as the piston begins its exhaust stroke. Therefore the engine expends less energy on expelling the exhaust gases than would otherwise be the case. The exhaust valve is also provided with a lag in closing after the piston reaches top dead centre and is beginning the induction stroke (Figure 1.121b). This not only better scavenges the combustion chamber of exhaust gases, but also lowers the cylinder pressure to facilitate flow of the incoming air and petrol mixture.

By modifying the basic valve timing of the four-stroke principle in the manner described, it necessarily follows that the opening of the inlet valve before top dead centre on the

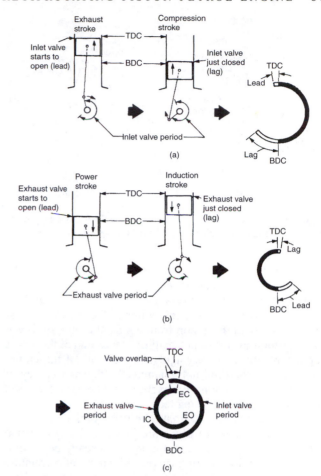

Figure 1.121   The principles of valve timing: (a) exhaust and compression (b) power and induction (c) valve timing diagram

IO inlet valve starts to open
IC inlet valve just closed
EO exhaust valve starts to open
EC exhaust valve just closed

exhaust stroke, and the closing of the exhaust valve after top dead centre on the induction stroke, result in a period when both valves are either partially or fully open. The superimposing of that part of the valve periods where the inlet valve opens before the exhaust valve closes is termed the *valve overlap* (Figure 1.121c).

Valve timing diagrams

For convenience of expression the opening and closing points of the valves have traditionally been shown in the form of a valve timing diagram (Figure 1.121c), although the trend in present-day service manuals is merely to tabulate this data, or simply to quote the number of degrees before top dead centre when the inlet valve begins to open. For tabulating valve timing information the commonly used abbreviations BTDC and ATDC denote before top dead centre and after top dead centre, and refer to the positions of the crankshaft as the piston is respectively advancing towards, and retreating from, the combustion chamber. Similarly, BBDC and ABDC denote before bottom dead centre and

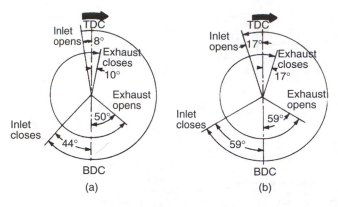

Figure 1.122 Typical valve timing diagrams for four-stroke petrol engines: (a) medium-performance car (b) high-performance car

after bottom dead centre, and correspondingly relate to the opposite sense of piston travel.

An early method of expressing valve timing was simply to relate the opening and closing points of the valves to measured distances of the piston from top dead centre. However, since piston travel is quite small near each end of the stroke for comparatively large angular movements of the crankshaft, this method did not guarantee sufficient accuracy of setting the valve timing in practice, so angular rather than linear values are always quoted.

In everyday engine overhaul the valve timing details are really only a matter of academic interest, since the actual process of setting the valve timing can usually be accomplished simply by ensuring correct alignment of correlation markings on the relevant parts of the timing drive mechanism. However, the need for actually verifying the valve timing values of an engine can arise, for example, in specialist tuning or supertuning of a competition engine where the camshaft(s) has been reground to a specification different from that issued by the original manufacturer.

When comparing the valve timing diagrams for the petrol engine of medium- and high-performance cars (Figures 1.122a and b), it will be noticed that with the latter the valve periods of opening (or events) are extended to greater values of crankshaft angularity. This is especially true in relation to the later closing of the inlet valve, so that at high operating speeds the increased lag allows as much pressure energy as possible to be generated in the cylinder by the incoming air and petrol charge, prior to its further compression by the rising piston. There will also be an increase in the value of valve overlap for the high-performance engine. This means, in effect, that at the top dead centre of the piston both inlet and exhaust valves will be unseated to a greater extent (neither the exhaust valve fully closed nor the inlet valve fully opened) and thereby encourage better breathing of the engine, again at higher operating speeds.

For the medium-performance engine any such increase in the value of valve overlap would be unrealistic, since it could promote a reverse flow of exhaust gases in to the induction system, especially with the increase in valve area of four-valve combustion chambers. This would, of course, tend to occur when manifold depression was high during light and closed-throttle running conditions. The ensuing contamination of the

fresh air and petrol charge by exhaust gases would result at least in poor low-speed performance and unstable idling, and at worst in actual misfiring, if the oxygen content of the contaminated charge became insufficient to support combustion.

Variable valve timing

Although the basic valve timing implied by the four-stroke principle was early modified for better engine performance, as long as the valve timing is fixed with respect to crankshaft rotation it must still remain something of a compromise. More specifically, it would be an advantage if the valve overlap period could be varied, so that its duration could be shortened for engine starting and during low-speed running and lengthened for high-speed running. By varying the valve timing in this manner it should therefore be possible not only to optimize engine performance at all running speeds, but also to improve fuel economy and reduce harmful exhaust emissions. In this context it should be noted that any overlap dilution of the air and fuel mixture with exhaust gas constitutes a simple form of exhaust gas recirculation (EGR) (Section 11.2).

The concept of variable valve timing has long proven an attraction to engine designers, but until recently its practical realization was always considered to be fraught with mechanical complication. Credit is therefore due to the Italian Alfa-Romeo company which, in 1982, was the first to introduce a relatively simple form of variable valve timing on a production car engine. This particular system, known as variable inlet valve timing or simply VIVT, is applicable only to a double overhead camshaft valve train and essentially involves mounting the drive sprocket on the inlet camshaft through an axially movable sleeve or timing piston. This is splined both internally and externally where it engages the camshaft and sprocket respectively, the internal splines being helical and the external ones straight. When the timing piston is moved axially against its return spring, it therefore causes the camshaft to advance rotationally in relation to its sprocket by virtue of the helical spline connection. The timing piston is hydraulically actuated by engine oil pressure, which is admitted via a solenoid valve controlled from the electronic management system for the engine.

A further development of this type of variable valve timing system is the Renold 'Camphase' camshaft angle phase controller. Instead of providing alternative angles of either advance or retard for the inlet camshaft, the purpose of the Renold system is to vary continuously the angular relationship between the inlet camshaft and crankshaft, termed the 'phase angle', over a range of 30° crank angle according to the dictates of the engine management system. Hence the optimum phase angles can be pre-programmed into the electronic control unit and the actual phase angles adjusted to these values. The device is hydraulically actuated by engine oil pressure via solenoid valves, a closed-loop control system being used to monitor the phase angle of the camshaft.

At the heart of the phase angle controller is a linear-to-rotary slidable piston, the bore of which is formed to engage a helical shaft that is mounted as an extension to the camshaft (Figure 1.123). The piston is externally of square form and engages a complementary bore in the phase controller outer body, which carries the camshaft driving sprocket. Therefore

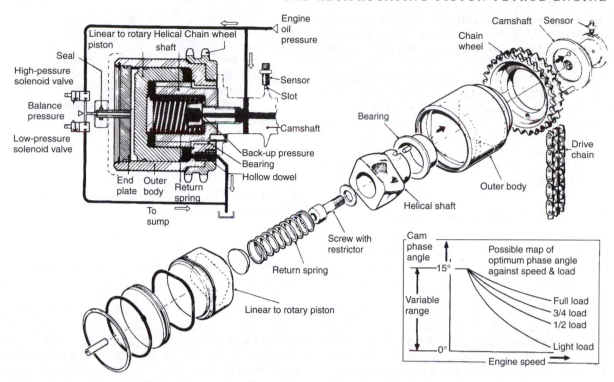

Figure 1.123   Camphase camshaft angle phase controller (*Renold Automotive*)

any enforced endwise movement of the piston over the helical shaft will cause partial rotation not only of the sliding piston, but also of the outer body of the phase controller through which it is sliding. By this means the camshaft can be partially rotated in relation to its driving sprocket.

The piston is actuated hydraulically by engine oil pressure controlled by two solenoid valves. When the high-pressure (HP) valve is opened it admits a balancing oil pressure against the head of the piston, the force on which exceeds that exerted by the return spring and moves the piston endwise to a new position towards the camshaft. Closing the high-pressure (HP) valve and opening the low-pressure (LP) valve releases the oil under pressure to the sump, so that the return spring causes the piston to retreat away from the camshaft. A moderate back-up pressure is maintained within the piston for lubrication and vibration damping purposes. The operation of the phase controller is therefore such that by opening and closing the HP and LP valves, the endwise position of the piston can be adjusted and hence the phase angle between the inlet camshaft and crankshaft varied continuously. Any misbehaviour of the camshaft phase controller in service will be detected and indicated by the engine management system, the device itself possessing a failsafe characteristic that allows continued running of the engine.

Another approach to variable valve timing is that pioneered by the Honda company, which is designated 'VTEC' for variable valve timing and lift electronic control system. Instead of allowing a controlled rotational movement for the camshaft relative to its driving sprocket, it utilizes three cam lobes and three rocker arms for each pair of valves in the four-valve combustion chambers (Figure 1.124). The two outer

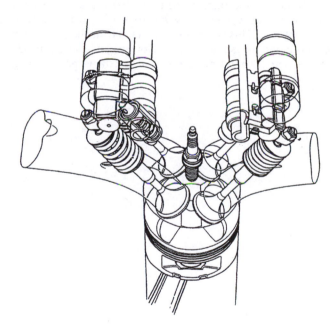

Figure 1.124   Variable Valve Timing and Lift Electronic Control (VTEC) (*Honda*)

cam lobes are active for the low and middle engine speed ranges, while the central cam with an increased lift profile becomes active for higher engine speeds to optimize breathing and power development.

In the lower- and mid-rpm ranges, the two outer rocker arms actuate each pair of valves in the conventional manner, while the central rocker arm swings freely against a spring-returned plunger. At high rpm, the two outer and central rocker arms are locked together by a laterally engaging hydraulic plunger assembly, which is pressurized from the engine lubrication system via an electronically controlled spool valve. When this valve opens the central cam lobe with its increased lift profile becomes dominant, since its rocker arm compels the outer rocker arms to follow the same motion. The increased lift and longer duration of opening for the valves therefore ensure optimum timing for the gases that flow into and out of the cylinders, which contributes to greater power output.

Port timing for the two-stroke petrol engine

The port timing of a two-stroke petrol engine is determined by considerations similar to those affecting the valve timing of a four-stroke unit. That is, the port timing must take into account the time lapse before the ports are either fully uncovered or fully covered, and also the inertia effects of the incoming and outgoing flows of the crankcase and the cylinder gases. Some compromise is inevitable, however, since the ports are necessarily uncovered and covered by the piston at equal angles on either side of the crankshaft dead centres (Figure 1.125). The port timing diagram of a conventional two-stroke engine is therefore said to be symmetrical.

In order to avoid dilution of the crankcase charge, the exhaust port is uncovered at about 15° of crankshaft angularity in advance of the transfer port, so that the latter remains covered until the cylinder pressure drops below that existing in the crankcase. This is generally referred to as the *blow-down* period in the two-stroke engine. However, as a consequence of symmetrical port timing, this lead in exhaust port opening

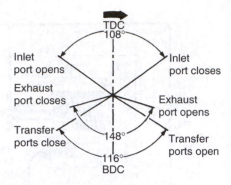

Figure 1.125 Typical port timing diagram for a motor car two-stroke petrol engine

conversely becomes a lag in closing. A part of the scavenge charge transferred from the crankcase therefore tends to short-circuit across the cylinder and be lost through the exhaust port before it closes, which is naturally detrimental to fuel consumption. This condition was especially prone to occur in early two-stroke engine designs that utilized a cross-scavenging effect on the burnt gases leaving the cylinder. A notable improvement in this respect was made possible by the development of a loop-scavenging system and later by the addition of either a reed or a disc valve to control inlet port opening and closing. The operating principles of engines incorporating these features were explained in Section 1.3. Values for the valve, rather than port, timing of an advanced concept supercharged two-stroke petrol engine for a passenger car are included in Figure 1.10, which shows its operating principle.

2 The diesel engine

2.1 SUITABILITY FOR ROAD TRANSPORT

It is a matter of common knowledge that diesel engines have found widespread application in stationary power-generating units and as prime movers for marine propulsion, rail locomotion and especially, by virtue of their more economical operation than the petrol engine, road transport.

The first four-stroke diesel engine to achieve a commercial success in road transport was that introduced by the German firm of MAN in 1924, but it was the intensive research and development carried out on the diesel engine by such British firms as Gardner and AEC during the 1930s and not least by H.R. Ricardo with his 'Comet' swirl combustion chamber (Section 3.5) which probably made the greatest contribution to this major advance in heavy-vehicle technology and its beneficial effect on the economics of road transport operation. Lighter-duty diesel engines later found modest application as power units for passenger cars, following the fuel crisis of the mid-1970s. More recently, the considerably improved driveability of diesel-engined passenger cars in terms of performance and refinement has been such that they now account for about one-third of new cars sold in Europe. This is not difficult to understand because when compared to a petrol-engined car, a diesel-engined version can reduce fuel consumption by about 30 per cent and similarly increase the range between refuelling. By the nature of its construction and operating characteristics, there can also be potential advantages in terms of greater durability, better reliability and the use of a safer non-volatile fuel in the event of an accident. As in the case of modern heavy vehicle diesel engines, the direct-injection combustion chamber used in conjunction with advanced fuel injection systems has pointed the way forward for passenger car diesel engines (Sections 3.5 and 8.11).

Since the mid-1960s the diesel engine for heavy vehicles, and later for passenger cars, has increasingly been developed in turbocharged form. This was originally as a means of improving power output in the medium- to high-speed range, but in more recent years the improved combustion associated with turbocharging has been exploited to produce an environmentally cleaner engine. That is, the turbocharger increases the amount of air delivered to the combustion chamber, so that the injected fuel can be burnt more efficiently. The benefits are a lower fuel consumption for the power developed and, as a consequence, better emission control. Furthermore, the engine can be quieter running because the turbocharger has a silencing effect on the induction and exhaust processes. Similar considerations apply to passenger car diesel engines.

For about the first decade of its life the diesel engine was constructed solely in four-stroke form, following which fairly rapid progress was made in developing two-stroke versions for large, slow-speed marine engines. Notable early examples were those built by the German firm of Krupp and the Swiss firm of Sulzer, the latter being first in the field in 1905. The two main reasons for this development were first that the two-stroke diesel offered a reduction in weight and installation space, and second, that it was possible to reverse easily its direction of rotation – an important consideration for the marine engineer.

Although the two-stroke diesel engine was later developed for other marine, stationary power-generating and rail locomotion purposes, its application to motor vehicles (at least in the United Kingdom) has been on a relatively limited scale. This can almost certainly be accounted for by its fuel consumption being generally less favourable than that of a corresponding four-stroke diesel engine, a matter of prime consideration in the road transport field. Nevertheless, there have been several high-speed two-stroke diesel engines successfully developed for automotive applications. Notable among these have been the uniflow-scavenged engines introduced by General Motors in America and Fodens Limited in England. Another interesting English design was the Commer TS3 two-stroke diesel engine, which had opposed pistons (in common cylinders) with rocker and link connections to the crankshaft. This layout had long been associated with high scavenging efficiency, albeit at the expense of greater mechanical complication.

2.2 OPERATING PRINCIPLES

The four-stroke diesel engine

The essential difference between the petrol and the diesel engine (Figure 2.1) is that the former relies on *spark ignition* (SI) and the latter on *compression ignition* (CI). More specifically, the combustion process in the diesel engine is initiated by spontaneous ignition of the fuel when it is injected into a highly compressed charge of air, which has reached about 800°C. Diesel engine combustion also tends to occur at constant pressure rather than at constant volume as in a petrol engine. This means that in the diesel engine the combustion pressure continues to rise steadily as the piston retreats and the cylinder volume increases, whereas in the petrol engine the combustion process is so rapid that there is very little movement of the piston while it occurs and hence very little increase in cylinder volume. Strictly speaking, though, neither engine fits exactly into either of these categories.

During its early development this type of engine was described variously as compression ignition, Diesel, oil and heavy oil, but it has long since come to be known generically as the diesel engine (with a small 'd'). This acknowledges the major personal contribution made to its development by Dr Rudolph Diesel (1858–1913), who was born in Paris of German parents. He became a student of mechanics and later entered the well-known engineering works of Sulzer Brothers in Winterthur, Switzerland. It was in the early 1890s that he developed his theories on what we now know as the diesel engine principle and subsequently took out various patents, including a British one granted in 1892. A few years later his theoretical work was embodied in a working engine of practical form built by the famous firm of MAN at Augsburg.

In fairness, however, it must be added that Diesel's concept of sparkless ignition was actually predated by the pioneering work of an English engineer, Herbert Ackroyd Stuart (1864–1927). In 1890 he patented an engine operating on a similar principle, but which required a vaporizer surface at the end of the cylinder. For starting the engine, the vaporizer required the application of external heat. Hence, the first true compression ignition engine is generally attributed to Rudolph Diesel.

In the four-stroke diesel cycle the following sequence of events is continuously repeated all the time the engine is running (Figure 2.2):

1 The *induction* stroke, during which air only is taken into the combustion chamber and cylinder, as a result of the partial vacuum or depression created by the retreating piston.
2 The *compression* stroke, in which the advancing piston compresses the air into the very small volume of the combustion chamber and raises its temperature high enough to ensure self-ignition of the fuel charge. This demands compression pressures considerably in excess of those employed in the petrol engine.
3 The *power* stroke, immediately preceding which the fuel charge is injected into the combustion chamber and mixes with the very hot air, and during which the gases of combustion expand and perform useful work on the retreating piston.
4 The *exhaust* stroke, during which the products of combustion are purged from the cylinder and combustion chamber by the advancing piston and discharged into the exhaust system.

As in the case of petrol engine, the timing for the opening and closing of the inlet and exhaust valves, and also that for injecting the fuel, departs from the basic four-stroke operating cycle.

Comparison of diesel and petrol engines

The following generalizations may be made on the use of diesel versus petrol engines in commercial vehicles and, more recently, cars:

1 The diesel engine has better fuel economy than the petrol engine. This is because its thermal efficiency is 30–36 per cent, compared with the 22–25 per cent of a petrol engine. As indicated in Section 1.3, thermal efficiency is the ratio of useful work performed by the engine to the internal energy it receives from its fuel.
2 The diesel engine has generally proved to be more reliable, to need less maintenance and also to have a longer life than an equivalent petrol engine. These advantages derive mainly from its sturdier construction and cooler running characteristics.

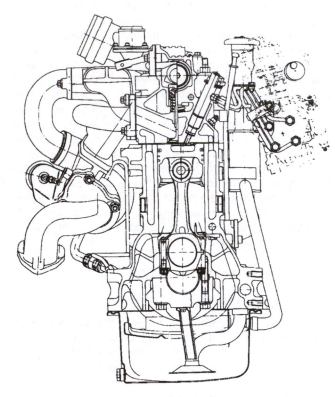

Figure 2.1 Cross-section of a four-stroke turbocharged direct-injection diesel engine (*Perkins*)

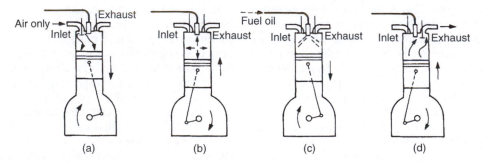

Figure 2.2 The four-stroke diesel engine cycle: (a) induction (b) compression (c) power (d) exhaust

3 Although a petrol engine develops its maximum power at higher rotational speeds than an equivalent diesel engine, the latter can provide better pulling power. This is because the maximum turning effort or torque exerted by the crankshaft of the diesel engine is greater and also better maintained over a wider range of engine speeds.

4 A disadvantage of the diesel engine is that it tends to be heavy and bulky in relation to its power output. This is explained by the greater operating pressures and loads that have to be catered for in the construction of the diesel engine.

5 The noise and vibration level of the diesel engine, especially under idling and low-speed operation, compares unfavourably with the petrol engine. Again, this is chiefly a function of the much higher cylinder pressures in the diesel engine. More recent diesel engine motor cars have nevertheless been praised for their low level of noise at motorway cruising speeds.

6 The diesel engine is sometimes criticized for having smoky exhaust of unpleasant odour, although it is the invisible products of combustion in the exhaust gases of a petrol engine that are more harmful to the environment. Avoidance of a smoky exhaust with a diesel engine is largely a question of good driving technique, regular maintenance and proper adjustments.

7 A safety consideration is that the fuel oil used in motor vehicle diesel engines is far less dangerously flammable than petrol, thus reducing the fire risk in the event of an accident. For taxation purposes, the fuel oil used in automotive diesel engines is referred to as DERV (an abbreviation of diesel engine road vehicle).

8 Finally, the basic cost of the diesel engine, together with its associated fuel injection equipment, is generally higher than that of an equivalent petrol engine.

The two-stroke diesel engine

Since this type of engine may take several practical forms, our attention here will be concentrated on the high-speed version suitable for application to road transport (Figure 2.3).

Apart from the expected difference that air only is introduced into the engine cylinder prior to the injection of fuel oil, another departure from two-stroke petrol engine practice is that, instead of using crankcase compression, a rotary blower is used to charge the cylinder with low-pressure air. This type of blower is sometimes also used for supercharging four-stroke cycle engines (Section 9.2). The distinction that must be made here, however, is that whereas a supercharger is used simply to increase power output of a four-stroke engine, a similar blower is essential for a two-stroke diesel in order that it shall work at all.

Furthermore, a pair of exhaust valves is located in the cylinder head to provide a uniflow system of scavenging. This means that there is no change in direction for the cylinder air stream, which is in contrast with the loop system of scavenging described for the two-stroke petrol engine in Section 1.3. The two-stroke diesel engine is therefore mechanically more complicated.

In the basic two-stroke diesel cycle, the following sequence of events is continuously repeated all the time the engine is running and while the rotary blower is supplying air to the inlet ports of the cylinder (Figure 2.4):

1 The *induction-exhaust* event. Air only is admitted to the cylinder during the period the inlet ports are uncovered by the piston, which occurs towards the last quarter of the power-exhaust stroke and about the first quarter of the induction-compression stroke. During this part of the cycle, the exhaust valves are opened just before the cylinder inlet ports are uncovered and then closed just before the ports are covered again. This sequence of exhaust valve events not only ensures that the exhaust gas pressure falls below that of the scavenging air supply, and thus prevents any return flow of exhaust gases, but also leaves the charge in the cylinder slightly pressurized prior to final compression. Hence, the combination of uncovered inlet ports and open exhaust valves allows air to be blown through the cylinder, which removes the remaining exhaust gases and, by the same token, fills it with a fresh charge of air. Since neither the air nor the exhaust gases change direction in passing through the cylinder, the term uniflow scavenging can justifiably be applied.

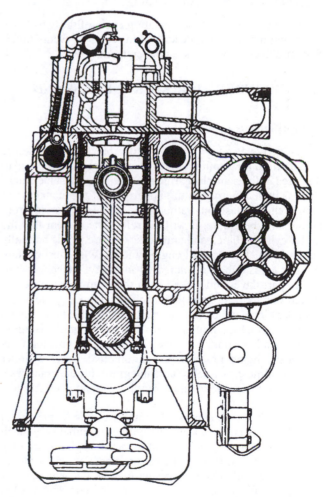

Figure 2.3 Cross-section of a blower-charged two-stroke diesel engine (*General Motors*)

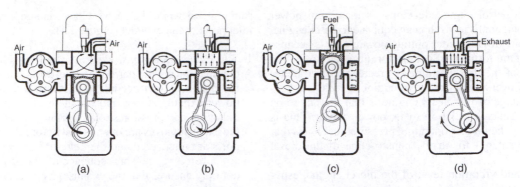

Figure 2.4 The two-stroke diesel engine cycle: (a)/(b) induction-compression (c)/(d) power-exhaust (*General Motors*)

2 The *compression-power* event. The remaining three-quarter portions of the *induction-compression* and *power-exhaust* strokes occur in a very similar manner to that of the four-stroke diesel engine; that is, the advancing piston compresses the air into the lesser volume of the combustion chamber and raises its temperature high enough to ensure self-ignition of the fuel charge. This is injected into the combustion chamber just before the piston begins to retreat on its power-exhaust stroke.

It should be noted that the operating cycle of the two-stroke diesel engine has been described in terms of events rather than strokes in order to assist understanding.

2.3 CYLINDER BLOCK, CRANKCASE AND HEAD

Cylinder block and crankcase

A distinctive feature of the cylinder block and crankcase construction for an automotive diesel engine, as compared to that used in for example marine applications, is that the crankshaft is underslung from the upper part of the crankcase in petrol engine fashion, rather than being supported from below on a bed plate. It will, therefore, be evident that the diesel cylinder block and crankcase (Figure 2.5) is basically similar to that of a petrol engine and likewise must contain the gas loads, albeit of increased intensity, which strive to force apart the cylinder head from the main bearings below. In addition to the appreciable mechanical loading, the diesel cylinder block and crankcase must also be able to withstand severe thermal loading and, in view of current legislation, radiate the least possible noise during engine operation. As in petrol engine practice, the automotive diesel cylinder block is almost invariably cast integral with the crankcase, so as to combine the required rigidity with a reasonably low weight.

Although the thickness of metal sections in the diesel cylinder block and crankcase are unlikely to differ very much from those of a modern high-output petrol engine, since they are largely determined by the requirements of foundry technique, other differences in design may be encountered. For example, a feature that is rarely found in petrol engines but is fairly common in diesel practice, especially with V cylinder arrangements, is the provision of transverse tie bolts as well as vertical studs for securing the main bearing caps (Figure 2.6). This system serves to reinforce the crankcase and confer the necessary rigidity of mounting for the main bearings, because their supporting walls have to react against the dynamic loads imposed by two inclined and offset cylinders. Furthermore, the vertical main bearing studs may extend upwards through the cylinder block of in-line cylinder engines, to act also as cylinder head studs and thus place the entire engine structure in a beneficial state of compressive stress, thereby countering the opposite effect created by cylinder gas pressure. A rather obvious difference in construction is noticeable at the front end of many diesel engines, where provision must be made for mounting a much larger size of timing cover, since the timing chest usually has to accommodate a more comprehensive gear train, as described later.

Although not constituting a basic difference in design, it should be noted that in long-established diesel engine practice for commercial vehicles either wet (Figure 2.5) or, less commonly, dry cylinder liners are employed in order to extend effective engine life for the several hundred thousand miles demanded by vehicle fleet operators. For endwise location it is usual for both types of liner to have a top flange, which is recessed in the closed deck of the cylinder block. In the case of wet liners a narrow 'fire-ring' spigot to protect the adjacent edge of the cylinder head gasket may be formed around the top surface of the flange and engage freely with a circular groove in the mating face of the cylinder head (Figure 2.7). Also, a bead of special silicone sealant may be specified for application to the underside of the liner top flange on engine build. Once the liner has been driven fully home, a clamp is temporarily applied to the liner thereby squeezing out the excess sealant for removal and then allowing that remaining to dry.

Owing to the greater mechanical loading and increase in noise level encountered with the automotive diesel engine, as compared to the petrol engine, aluminium alloy is rarely used as a cylinder block material but a cast iron of similar composition to that mentioned later for the cylinder head is generally specified and typically possesses a slightly higher tensile strength. For greater resistance to impact loading the main bearing caps may be produced from malleable or nodular cast iron. A strongly constructed aluminium alloy sump may nevertheless be used in conjunction with the cast iron cylinder block.

Figure 2.5   The Cummins in-line engine with four valves per cylinder and panelling of the crankcase to reduce noise

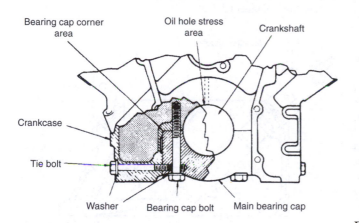

Figure 2.6   Diesel main bearing cap with transverse tie bolts (*International Harvester*)

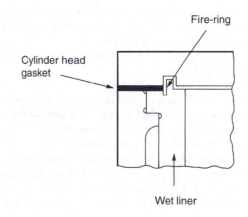

Figure 2.7   Diesel engine wet cylinder liner with fire-ring

For passenger car diesel engines increasing use is now being made of a lighter weight cast iron known as 'compact graphite iron', usually abbreviated to CGI. This material offers about a one-third greater tensile strength and an improved fatigue resistance, as compared to grey cast iron. However, it does call for more expensive high-speed machining techniques to be adopted on production. The relevance of CGI to modern diesel engine design is that it either allows a lighter construction that bears comparison with an aluminium alloy, or its greater strength can be utilized to contain higher cylinder pressures in the interests of efficient combustion and hence improved emissions control.

Cylinder head

The cylinder head for a diesel engine is generally more complex in construction (Figure 2.8) and has a more severe cooling requirement than its petrol engine counter-part. A one-piece cylinder head for an in-line cylinder engine, and similarly one for each bank of a V cylinder layout, can be regarded as established modern practice; although a two-piece cylinder head (two separate heads) for an in-line cylinder engine may still be used in the case of very large commercial vehicle applications, mainly to ease the problem of handling heavy units during service operations. Individual heads are, of course, employed for the cylinders of air-cooled diesel engines (Figure 5.3).

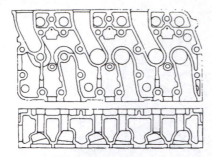

Figure 2.8 Section through a diesel cylinder head casting (*Midcyl*)

Structurally, provision must be made in the diesel engine cylinder head for incorporating a passage for each fuel injector, either through the medium of an inserted metal tube where direct injection is utilized, or by way of an integrally cast boss to suit indirect injection. The latter system also requires the upper part of the combustion swirl chamber to be formed in the cylinder head, whilst the lower half comprises an inserted machined plug that contains the throat portion connecting the chamber to the cylinder. Since this component acts as a hot-spot in the system, as explained later, it is installed with an annular air gap to minimize heat transfer to the cylinder head metal. From the viewpoint of maintaining cylinder head gasket sealing efficiency, it is essential that no settling should occur at the abutment face of the chamber insert, because any sinkage of its lower face into the sealing surface of the head would locally relieve gasket pressure and result in gas leakage (Figure 2.9). To avoid interference with the piston at top dead centre the valves are typically recessed into the surface of the cylinder head (Figure 2.18), and also they are generally provided with inserted valve guides and seats in diesel practice.

Although in common with the petrol engine the very hot metal on the gas side and the less hot metal on the coolant side of the combustion chamber results in a private tug-of-war, owing to the difference in expansion across the thickness of metal wall, the problem becomes more acute in the diesel cylinder head because of a greater lack of uniformity in heat distribution across that part of the head covering the cylinder. This chiefly arises from there being less latitude in positioning and spacing of the inlet and exhaust valve seats, and similarly the swirl chamber where this is applicable. To avoid fatigue cracking in the thermally stressed regions bridging these components, it is necessary to ensure an adequate flow of coolant through the cylinder head; in some designs, coolant entering the head is directed through channels to impinge first above the combustion chamber roof (Figure 2.10).

The materials used for the cylinder heads of motor car engines have already been described in Section 1.6. Except for some motor car diesel engines and also the individual cylinder heads of air-cooled diesel engines, which utilize an aluminium alloy, the cylinder head material for commercial vehicle diesel engines is a cast iron. A high-strength chromium-iron alloy, such as that developed by Midcyl and known as Chromidium, may typically be specified. This type of material is characterized by an extremely high resistance to wear and corrosion, and careful control of metal composition enhances the pressure

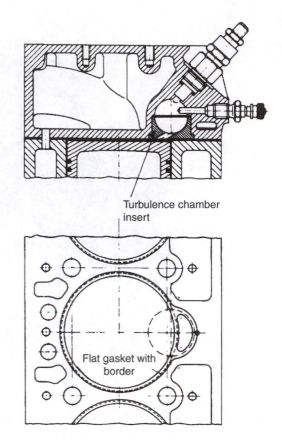

Figure 2.9 Gasket sealing arrangement for a diesel swirl-type combustion chamber (*Reinz*)

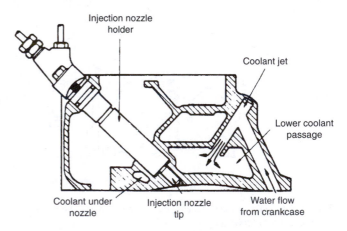

Figure 2.10 Directed coolant flow in a diesel cylinder head (*International Harvester*)

tightness of the finished casting. The following material specification for Chromidium cylinder head iron is given by way of interest:

Carbon	3.10–3.50%	Manganese	0.50–0.80%
Silicon	1.85–2.40%	Chromium	0.20–0.30%
Sulphur	0.15% max.	Brinell hardness no.	190–240
Phosphorus	0.20% max.		

British Standards BS 1452 Grade 14.

2.4 PISTONS AND CONNECTING RODS

Pistons and rings

A diesel engine piston (Figure 2.11) differs from its petrol engine counterpart in the following basic respects.

Increased sectional thickness of metal
This is concentrated in the upper portion of the piston for better heat dissipation and to afford the maximum possible support for the more heavily loaded gudgeon pin bosses, in which the pin is made fully floating. The shape of the piston crown is, of course, largely determined by the design of the combustion chamber, since with a direct-injection system the latter is to all intents and purposes formed within the piston head. In any event, the crown section of the piston has a generous metal thickness to withstand the higher cylinder gas pressures without distorting and to provide a ready path for heat flow to the rings without cracking. To reduce thermal loading on the top compression ring, the piston top land is of increased depth. Temperatures must generally be kept below about 200°C in the top ring groove, in order to avoid ring sticking and scuffing problems due to carbon deposits and lacquer formation. A second land of increased depth is also provided to contain the higher mechanical loading on the top ring.

Longer in relation to diameter
It may be recalled (See Section 1.4) that pistons assume a trunk form to present a sliding bearing surface against the cylinder wall, which thus reacts against the side thrust arising from the angular motion of the connecting rod. Making the piston skirt relatively longer spreads the greater side thrust forces developed in the diesel engine over an increased surface area. Hence, the unit pressure is reduced so that separation between the piston and the cylinder wall is better maintained by the film of lubricating oil, and wear through direct contact is minimized.

May carry more rings
In long-established practice, the diesel engine piston has generally been equipped with three instead of two compression rings and two rather than one oil control ring. More recently, however, there has been a definite trend to reduce the numbers of both compression and oil control rings to the same combination as that used in modern petrol engine practice. Nevertheless, an unfilled groove may be provided below the gudgeon pin to allow the addition of a second oil control ring in later service. Technically, the number of rings used depends upon the severity of application and the projected design targets for oil consumption, blow-by and friction effects, the latter being related to fuel consumption. Where ring-belt temperatures tend to be unusually high, which can promote ring sticking and blow-by, a traditional remedy in diesel engine practice generally has been to use a top ring of 'keystone' or wedge-section form. It acts on the principle that the slightest movement of the tapered ring in its correspondingly shaped groove will vary its side clearance, therefore discouraging any build-up of deposits. An included angle of 10° for the ring faces is typical. Wedge-section top rings may also be adopted for highly rated passenger car

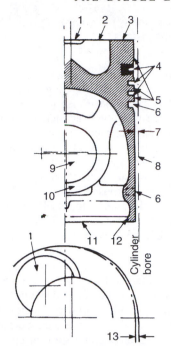

Figure 2.11 Terminology for a diesel engine piston with bonded insert (*GKN Pistons*)

1 valve pockets	8 skirt
2 combustion chamber	9 piston pin hole
3 crown	10 piston pin boss
4 compression ring grooves	11 open end
5 lands	12 register
6 oil control ring grooves	13 ovality
7 minimum clearance	

diesel engines, although in this case a single-side wedge form may be preferred, so as to combine the advantages of this type of ring with those of a conventional rectangular section ring.

Need for top ring carrier
For naturally aspirated (non-pressure-charged) diesel engines where cylinder pressure and temperature conditions are not too demanding, the piston may be simply of monometal (all aluminium alloy) construction. Otherwise, where a more hostile cylinder environment is expected, the piston is generally of bonded insert construction, since the principal limiting factor of the monometal piston is ring groove wear. The insert is bonded into the aluminium alloy and forms a wear-resistant carrier for the top and sometimes also the second compression ring. It is made from a nickel cast iron alloy with hot expansion characteristics that closely match those of the aluminium alloy used in the main body of the piston. Even so it will be noticed that the insert is of tapered form, which ensures that the piston alloy grips it on cooling and contracting in diameter. An aluminium alloy having a 12 per cent silicon content and usually described as a eutectic aluminium alloy (eutectic signifies an alloy composition that solidifies at one temperature rather than over a range of temperatures, and

ensures good castability) is commonly used both for diesel and petrol engine pistons.

Requirement for oil cooling of piston

The importance of avoiding top ring sticking in the diesel engine piston has earlier been mentioned. Clearly this could become a problem with the ever increasing performance of the turbocharged engine, especially in relation to the current trend of reducing the depth of the piston top land, so that its surrounding dead volume is minimized for more efficient combustion. A significant proportion of the combustion heat to which the piston is exposed flows through the rings to the cylinder cooling jacket, the piston itself being cooled only by contact with the cylinder wall and a constantly renewing thin film of oil, which has originally been pumped from the sump. The reservoir of oil in the sump therefore acts as a medium for transferring excess heat away from the piston. Further use can be made of this particular heat transfer medium by arranging for a standing jet of oil to be directed against the underside of the piston crown. For this purpose a series of strategically sited nozzles receive oil under pressure from the main lubrication system of the engine and spray it up into the pistons (Section 4.4). This also permits closer fitting pistons for quieter running.

Two-piece articulated pistons

Although two-piece pistons with steel crowns and aluminium skirts are not unfamiliar to the marine diesel engineer, it has only been in recent years that pistons of this configuration have been adopted in preference to monometal pistons in some high-output turbocharged diesel engines for heavy vehicles such as the Scania DSC 12 engine (Figure 2.20).

Higher power ratings inevitably result in greater thermal and mechanical loadings being imposed on the cylinder components, which have therefore had to be further developed to remove any undue limitation on engine performance and reliability. In particular, higher power ratings generally attract increased piston temperatures and these can have serious practical disadvantages in engine operation. Apart from promoting a tendency towards ring sticking and scuffing due to carbon deposits and lacquer formation, at elevated temperatures the hot strength of aluminium alloys becomes significantly less than their cold strength, which could potentially lead to failures in the highly loaded regions of the piston crown boss supports and the gudgeon pin boss bearings.

To overcome the problems that can otherwise appear with a conventional monometal piston, even though it may incorporate a bonded insert ring carrier and benefit from oil cooling, engineers resorted to a two-piece piston construction. This comprises a forged steel crown that carries the piston rings and an aluminium skirt and therefore combines the best of both worlds, because the steel crown better tolerates the higher temperatures and also expands less to maintain a more constant operating clearance for the rings, while the aluminium skirt still provides a light-weight means of guidance for the piston. However, unlike the two-piece piston used in marine diesel engine practice where the crown is typically bolted to the skirt, those designed for heavy vehicle engines are of more compact articulated construction (Figure 2.12). That is, the connection between the crown and skirt is made solely through the gudgeon pin, so that the crown and semidetached skirt are mutually self-aligning with advantage being taken to minimize the running clearance of the cooler operating skirt. The latter may also incorporate an upper oil collecting gallery to shake cooling oil supplied by sprayer nozzles (Section 4.4) up between the crown and its ring carrier, especially since the steel crown has a lower heat conductivity than if it were made from an aluminium alloy. An oil drain from the gallery is provided through return ducts situated remotely from the oil spray. Another feature of construction, again borrowed from marine diesel engine practice, is that the inner faces of the piston crown gudgeon pin bosses are tapered outwards towards their lower ends and

Figure 2.12 A dismantled two-piece articulated piston and connecting rod assembly (*Scania*)

embrace the matching tapered sides of the connecting rod small-end boss, the purpose of this arrangement being to present an increased bearing area against the gudgeon pin on the firing stroke.

Connecting rods

The connecting rod has sometimes been described as one of the most important components in conventional engine design. Attention to detail in its design has always been of prime importance, and can perhaps be judged by recalling one aspect of the wartime development of the famous Rolls-Royce Merlin aero engine. Here, the alteration from a stepped to a generously radiused shoulder at the junction of the shank to big-end bolt lugs significantly reduced stress concentration in the forked connecting rods and was one of the important modifications that ultimately made possible a safe doubling in power output.

Returning to our comparison of petrol and diesel engine connecting rods, as far as the latter are concerned there is currently no departure from what has long been the established form for this component. Rather the difference lies in its more rugged proportions to withstand the increased compressive loading resulting from the higher combustion pressures in the diesel engine. More specifically, the shank of the connecting rod is often quite noticeably flared out at the junctions respectively with the small-end boss and the big-end bearing housing (Figure 2.13), again to spread the load over these parts with the minimum of stress concentration and to ensure that the thin-wall bearings are rigidly supported so they can retain their roundness.

As in petrol engine practice, the joint between the connecting rod and cap is sometimes made at an angle, so that the big end can accommodate the maximum diameter of crankpin, whilst still allowing the connecting rod and piston assembly to be withdrawn through the top of the cylinder during overhaul. In any event, the lateral location of the cap must be sufficiently positive to avoid mismatch at the joint faces of the big-end bearing, which may entail the use of either serrated or stepped joints (Figure 2.14). With an angled joint, setscrews (instead of bolts and nuts) are normally required to clamp the cap to the rod. For heavy-duty applications two bolts may be used on each side of the big-end bearing assembly (Figure 2.15).

The materials from which connecting rods are forged generally comprise a carbon-manganese steel for passenger car petrol engines and a low-alloy 1 per cent chromium steel for commercial vehicle diesel engines, the latter material being the harder after their heat treatment.

Big-end bearings

Here again, the connecting rod big-end bearings of a diesel engine perform the same function as those in a petrol engine and likewise are more heavily loaded than the main bearings, where the equivalent loading is shared between two adjacent ones. Also, the big-end bearing is subjected to centrifugal loading enforced by the circular path of the lower part of the connecting rod, an effect that is absent in the case of main bearings supporting a fully counterbalanced crankshaft.

However, a basic difference does arise between the loading characteristics of diesel and petrol engine big-end bearings. In transmitting the reciprocating forces from piston to crankshaft and vice versa, it is the upper half-liner in the connecting rod that bears the brunt of the loading in a diesel engine, whereas it is the lower half-liner in the connecting rod cap that is on the receiving end in a petrol engine. The reasons for this contrary behaviour are explained later where the distinction will be made between the predominant sources of loading on diesel and petrol engine bearings. These are cylinder gas pressures and inertia forces of the reciprocating parts respectively, always remembering that the inertia forces vary

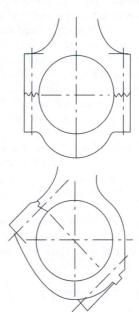

Figure 2.14   Examples of diesel connecting rod cap location (*Glacier*)

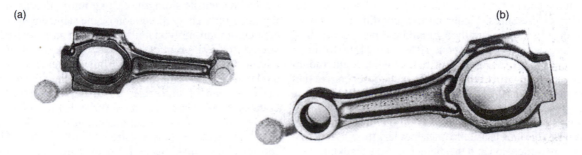

Figure 2.13   Comparison of integral connecting rod cap and forgings for (a) petrol and (b) diesel engines (*GKN-SDF*)

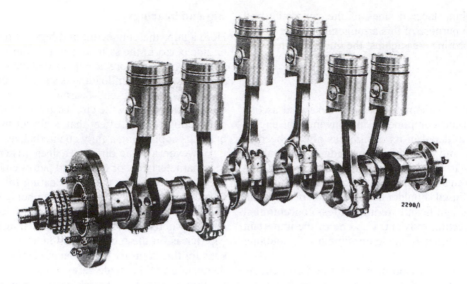

Figure 2.15 Crankshaft assembly of a long-established design of diesel engine (*Gardner*)

as the square of the engine speed and are virtually unopposed by cylinder gas pressure on the exhaust strokes.

Until recent years, the lining material requirements for diesel big-end and main bearings could usually be satisfied by one and the same material. Traditionally this has been either a copper-lead mixture or a tin-aluminium alloy. With the advent of pressure charging, however, the loading on the big-end bearings can rise above the performance capabilities of these materials, since the comparative bearing pressures under maximum-torque running conditions can be in the order of $35 \, MN/m^2$ ($5250 \, lbf/in^2$) for naturally aspirated and $50 \, MN/m^2$ ($7500 \, lbf/in^2$) for turbocharged diesel engines. To provide the increased load-carrying capacity therefore requires the highest-duty bearing materials available, such as lead-bronze and aluminium-tin-silicon compositions with softer overlays of either lead-tin or lead-indium.

In modern engine production, it is not unusual for a specialist manufacturer to develop and supply what is termed a 'piston module', which comprises the piston, piston rings, connecting rod and bearings.

2.5 CRANKSHAFT ASSEMBLY AND MAIN BEARINGS

Crankshaft

The predominant source of loading on the bearings of a diesel engine and hence on the crankshaft itself derives from high cylinder pressures. Under normal conditions of combustion, even with a naturally aspirated engine, peak cylinder pressures can be up to about $10 \, MN/m^2$ ($1500 \, lbf/in^2$), which is therefore in the region of twice that encountered in the petrol engine, whilst for a turbocharged diesel engine the cylinder pressure may be up to $14 \, MN/m^2$ ($2100 \, lbf/in^2$).

Clearly then a diesel engine crankshaft (Figure 2.15) must be of robust proportions, and to this end the diameter of the main bearing journals is usually made not less than 0.75 of the cylinder bore dimension and the diameter of the crankpins not less than about 0.65 of this dimension. These values are, of course, at least on the upper limit for those quoted in Section 1.8 relative to petrol engine crankshafts, and contribute to a substantial and beneficial overlap between main journal and crankpin of typically around 20 to 35 mm (0.8 to 1.4 in). In fact the rigidity of the crankshaft can be such that for shorter versions used in V engines no torsional vibration damper may be needed. Naturally a main bearing has always been used on either side of each crankthrow, as is now the case with most high-compression petrol engines, so that the crankshafts for in-line six-cylinder and V eight-cylinder diesel engines are carried in seven and five main bearings respectively with more recent V six-cylinder versions having four main bearings. The lengths of both main and big-end bearings generally approach 0.30 of the cylinder bore dimension.

When describing the materials and manufacture of petrol engine crankshafts for motor cars, it was mentioned that although the majority of their crankshafts are produced from iron castings of the spheroidal graphite type with a tensile strength of about $63 \, kg/mm^2$ ($40 \, tons/in^2$), crankshafts forged from higher-strength low-alloy steels are still used for heavy-duty applications. Included in this category are, of course, commercial vehicle diesel engines. Although there is no hard and fast rule regarding the material preferences of different engine manufacturers, a few examples can be given. The material used for the crankshaft of one medium-capacity, naturally aspirated diesel engine is a 1 per cent chromium steel with a tensile strength of $71 \, kg/mm^2$ ($45 \, tons/in^2$), while that for a turbocharged version of the same engine is a 1 per cent chromium-molybdenum steel with an increased tensile strength of $79 \, kg/mm^2$ ($50 \, tons/in^2$). The effect of a small addition of molybdenum as an alloying element is to impart a relatively uniform hardness and strength to the material. At the other end of the scale, the material specified for the crankshaft of a large-capacity turbocharged diesel engine can be typically a 3 per cent chromium-molybdenum nitriding steel with a tensile strength of up to $94 \, kg/mm^2$ ($60 \, tons/in^2$). Nitriding or nitrogen hardening is a process performed by heating the material to a temperature of 500°C and passing ammonia gas over it. On contact with the hot steel the ammonia gas breaks down into hydrogen and nitrogen, so that the latter diffuses into the steel to form nitrides at the surface. The benefits conferred by this treatment are an extreme surface hardness coupled with a high fatigue resistance.

Main bearings

Although the crankshaft main bearings of a diesel engine perform the same function as those in a petrol engine and, in established modern practice, are also of the same thin-wall construction, their loading characteristics can be somewhat different. This is because in the diesel engine, the predominant source of loading derives from high cylinder pressures generated at relatively modest operating speeds; whereas in the petrol engine, it is imposed by high inertia forces arising from acceleration and deceleration of the reciprocating parts at much increased operating speeds. Furthermore, the duration of loading for each complete working cycle of the engine is brief for cylinder pressure and protracted for inertia forces. If excessive the former type of loading tends to promote fatigue failures and the latter type wiping failures of a bearing. In the unlikely event of such failures occurring in normal service, they would be characterized by random cracking and superficial melting of the bearing respectively.

It will therefore be recognized that in modern practice both diesel and petrol engines can pose problems of severe bearing loading and heating with their consequent adverse effects on oil film thickness and bearing durability. For this reason an identical choice of high-duty lining materials is often now made to satisfy the main bearing requirements of both types of engine. Materials in current use are copper-lead mixtures and tin-aluminium alloys with typically 70 per cent copper and 20 per cent tin contents by weight respectively, the relative merits of these two classes of bearing material having been compared in Section 1.7. This common requirement for high-duty bearing materials was rarely in evidence some forty years ago, when copper-lead mixtures were confined almost exclusively to diesel engine applications and all but large-capacity petrol engine bearing requirements could still be satisfied by white metal alloys. At that time the tin-aluminium bearing as we know it today had yet to become established and, of course, petrol engine operating speeds were much lower.

Torsional vibration damper

For particularly arduous duty such as may be encountered in the larger sizes of automotive diesel engines, the rubber damper can have certain limitations in respect of its durability, since the high damping forces involved could result in overheating of the rubber element. A viscous damper, so called because the damping forces are generated by the drag of a very heavy or viscous fluid, is of generally more rugged construction and may therefore be the preferred choice for these applications.

This type of damper basically comprises a small fly-wheel that is completely enclosed in a hermetically sealed (airtight) casing, which is rigidly attached to the nose of the crankshaft via a flange fitting against the front pulley (Figure 2.16a). The flywheel ring is not in any way attached to the correspondingly

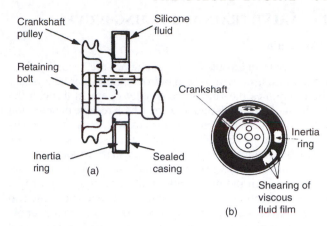

Figure 2.16 Torsional vibration damper: (a) construction (b) action

annular-shaped casing, but is completely separated from its interior walls by an all-enveloping film of silicone fluid, albeit only a few hundredths of a millimetre thick, so that to some extent it can also damp axial vibration of the crankshaft. It is therefore purely the drag of the silicone fluid, which characteristically retains a nearly constant and high viscosity (resistance to flow) over a wide temperature range, that tends to cause the casing and enclosed flywheel to revolve in unison.

In the absence of crankshaft torsional vibration the enclosed flywheel tends to rotate in unison with its casing, because the force required to shear the viscous film of silicone fluid that separates them is quite appreciable. Here it should perhaps be explained that just as one solid body resists the action of sliding when in physical contact with another, so do the multiplicity of molecular layers in a fluid mass resist sliding over one another in the manner of a pack of playing cards pushed endwise. With the onset of a critical vibration period, however, the inertia of the damper flywheel is such that it cannot follow the rapid oscillatory movements of its casing attached to the crankshaft (Figure 2.16b). The relative motion thereby created between the flywheel and its casing results in the molecular layers of the fluid film undergoing a shearing action, which process abstracts energy from the vibrating crankshaft and dissipates it in the form of heat from the damper. An important consideration in the fitting of a torsional vibration damper to the crankshaft of a modern heavy-vehicle diesel engine is that of protecting the comprehensive gear train of its timing and auxiliary drives from vibrational disturbances.

Historically, it was once again Dr F.W. Lanchester who first proposed the idea of a viscous damper to control crankshaft torsional vibrations, and it was the subject of a patent by him as long ago as 1910. However, it was the further research by the Americans G.J. Dashefsky and Capt H.M. Jensen at the New York Naval Shipyard in 1930 that produced the first really practicable viscous damper as we would recognize one today. Rather interestingly these early researchers used furniture glue as the viscous medium, so it was the World War II development of silicone fluids and their application to the viscous damper by the American Houdaille-Hershey Corporation in 1946 that led to finality in design.

2.6   VALVE TRAIN AND TIMING DRIVE

Valve train

The valve train layouts for diesel engines are generally similar to those used with I-head petrol engines. For heavy-duty diesel engines the valves are usually operated through the medium of push-rods and rockers, which receive their motion from a crankcase-mounted camshaft with either tappet barrel or finger rocker followers. In both cases these followers may have a roller contact with the camshaft. In some designs where four smaller valves per cylinder are used and typically those in which the more bulky unit injector system of fuel injection (Section 8.9) would be unable to accommodate two larger valves, each pair of valves is operated from a single overhead rocker that acts upon a vertically sliding cross-head or bridge piece which straddles their stem tips. The cross-head is provided with an adjusting screw beyond the rocker tip, so as to ensure equal operation of both valves (Figure 2.17). In practice, the adjusting screw is slackened sufficiently to allow the other end of the sliding cross-head to be brought into contact with its valve tip under finger pressure. The adjusting screw is then advanced into contact with its own valve tip and held while the locknut is tightened. Valve adjustments proper are carried out in the conventional manner, but with the difference that the clearances are measured between the rocker lever pad and cross-head platform for each pair of valves.

Less commonly, the valves of a heavy-duty diesel engine may be operated from an overhead camshaft, which either attacks them directly via interposed bucket tappets or indirectly through rockers with the object of gaining easier valve clearance adjustment. However, the overhead camshaft has been widely adopted as part of the valve train for light-duty

diesel engines such as used in passenger cars. In late 1996 General Motors Europe introduced a bridge type of four-valve actuation for their direct-injection passenger car diesel engines of 2.0 and 2.2 litre capacities, a point of interest being that the bridge piece sits directly on the hydraulic bucket tappets that surmount the paired valves. By virtue of requiring only one instead of two overhead camshafts, a 30 per cent reduction in valve gear friction is claimed to the benefit of engine fuel consumption.

Comparison of diesel and petrol engine valves

In vehicle diesel engines the poppet valves used are of the same established form as those described for petrol engines, but with more generous metal sections for the heads and with noticeably longer stems (Figure 2.18). The former is provided in the best interests of heat dissipation and mechanical strength and the latter arises from the valves seating in the bottom face of the cylinder head, rather than in the roof of a combustion chamber as for most petrol engines. Although for normally aspirated diesel engines the head diameter of the inlet valve conventionally exceeds that for the exhaust valve, both valves are made equal in diameter in the case of turbocharged versions, where energy from the exhaust gases is utilized to pressure charge the air induction system. To reduce seat wear, the face angle of the inlet valves for turbocharged versions is usually reduced from 45° to 30°, since the inherently lower seating pressure with the latter angle tends to compensate for the higher cylinder pressure. For some direct-injection diesel engines a shroud or masked portion is forged integral with the inlet valve head (Figure 2.19b), the purpose of departing from a plain head (Figure 2.19a) being to promote induction swirl (as explained later). This type of valve does, of course, require

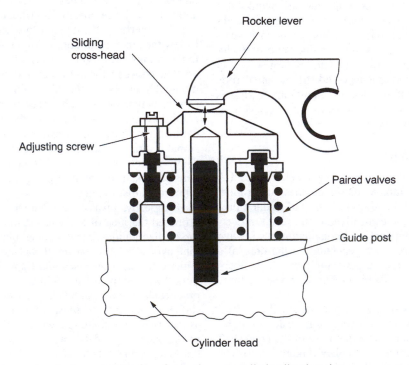

Figure 2.17   Rocker and cross-head valve actuation for a four-valves per cylinder diesel engine

a flat or keyway to be machined on the stem to locate the shroud at the optimum position in the intake port. As in the petrol engine, the limiting factor in the durability of the diesel engine valve train is usually the exhaust valve.

As compared with petrol engines, the materials used for the inlet and the exhaust valves of diesel engines must cater for a much wider range of application and power outputs, as summarized below.

Diesel engine inlet valves

For moderate-duty applications silicon-chromium steel is again commonly used, although for indirect-injection engines where the valves tend to be hotter running a chromium-manganese-nickel steel is often demanded. The latter material has long been known by its original designation 21–4N, and is a high-alloy stainless steel of the austenitic group with a nominally 21 per cent chromium content. Steels in this group are the most resistant to corrosion of any of the stainless steels. For large-capacity turbocharged diesel engines a variety of materials may be used. These range from 1.5 per cent nickel-chromium-molybdenum steel to austenitic chromium-nickel steel, according to running temperature and mechanical loading. Stellite hard facings may be employed to resist

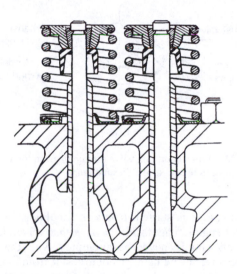

Figure 2.18   Arrangement of diesel engine valves (*Perkins*)

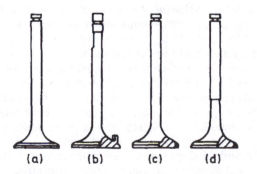

Figure 2.19   Types of diesel engine inlet and exhaust valves: (a) plain inlet (b) shrouded inlet (c) hard-faced exhaust (d) hard-faced bimetal exhaust

wear on the seats of some inlet valves and will be described later in connection with exhaust valves (Figure 2.19c). The significance of their use for inlet valves is that with turbocharging the valve seats can run very dry, because the pressure at the inlet ports inhibits any oil leakage down the valve guides.

Diesel engine exhaust valves

For moderate-duty applications, the two most widely used exhaust valve materials are the high-alloy stainless ones, chromium-manganese-nickel (21–4N) and chromium-manganese-nickel-niobium (21–42) steels. Niobium alloys have a great resistance to heat and oxidation, a point of interest being that niobium metal itself has the very high melting point of 2415°C. The stems of valves made from these materials are usually chromium plated for durability. As in petrol engine applications, the use of a bimetal two-piece exhaust valve construction is becoming more common (Figure 2.19d).

In the case of large-capacity turbocharged diesel engines there are several high-alloy materials in use. These include, apart from chromium-manganese-nickel and chromium-nickel austenitic steels, a material known as Nimonic 80A. This is one of the Nimonic series of nickel-chromium (rather than chromium-nickel) alloys having a basically 80 per cent nickel and 20 per cent chromium composition, which were developed during World War II to withstand the most severe conditions of stress and temperature encountered in aircraft gas turbine engines. Stellite seat facing alloys are generally employed for high-duty applications and contain a high proportion of cobalt. This material has a particular relevance to diesel valves, since it can resist corrosion from the combustion of the sulphur content in fuel oils.

Timing drive

A once typical front mounted timing drive for a diesel engine comprises four gears: one each for the camshaft and fuel injection pump drive, a crankshaft pinion, and an idler gear that engages the other three (Figure 5.3). All four gears have teeth of helical form and are produced from steel for this type of heavy-duty application. Alternatively, a triangulated layout of double-row or triple-row roller chain has been used for the timing drive, the length of the chain being kept to a minimum so that the effects of wear have the least disturbance on the timing of the fuel injection pump. In more recent practice, there has been a need to ensure greater precision of operation for the valve train and unit injectors, which has resulted in the camshaft for in-line cylinder engines being mounted either high in the cylinder block, or on the cylinder head itself, and in both cases this inevitably entails using an extended (Figure 2.5) and sometimes a compound (Figure 2.20) train of gears. Furthermore, the gear train may be mounted at the rear of the engine with the input gear being sandwiched between the crankshaft flange and the flywheel (Figure 2.20). The purpose of this reversal in arrangement of the timing drive will be understood by recalling that when a crankshaft suffers a period of torsional vibration, its maximum amplitude will be greatest at the front end (Section 1.9). Therefore to reduce vibration, wear and now importantly noise in the timing drive, it is clearly advantageous to mount

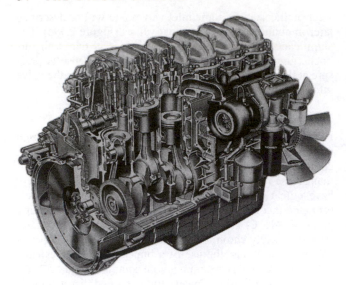

Figure 2.20 The 12-litre Scania DSC 12 in-line engine with four valves per cylinder and rear mounted timing gear (*Scania*)

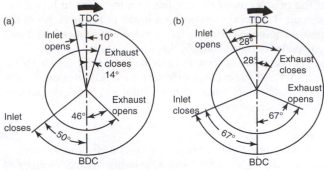

Figure 2.21 Typical valve timing diagrams for four-stroke diesel engines: (a) commercial vehicle, medium speed (b) passenger car, high speed

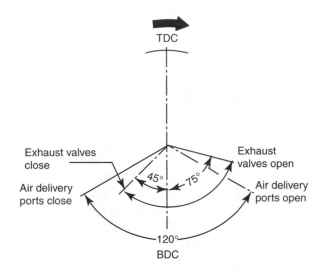

Figure 2.22 Typical port and valve timing diagram for a commercial vehicle two-stroke diesel engine

the gear train at the opposite end of the crankshaft where the torsional vibration is much less. An ideal but hardly practicable course of action would be to site the timing gear train at the nodal point (position of no vibration) of the crankshaft, but this unhelpfully lies somewhere between the rear crankthrows. Another advantage of a rear mounted timing drive is that it can provide a suitably geared power take-off facility (Section 14.5).

As a matter of historical interest, it is worth mentioning that some designers of passenger car engines had early recognized the technical merit of a rear mounted timing drive, notable examples that may be familiar to classic car enthusiasts being those found in the in-line six-cylinder engines of the high-grade AC and Alvis cars of a past era. Other designers objected to this arrangement on the grounds of its inaccessibility in service, but such criticism proved to be largely unfounded because the less highly stressed timing drive seldom required attention.

In the case of light-duty diesel engines used in passenger cars, a toothed belt is now generally used for the timing drive to the overhead camshaft and also includes the fuel injection pump, which is mounted at about mid-height on the engine.

Valve timing for the four-stroke diesel engine

As in the case of the four-stroke petrol engine, the timing for the opening and closing of the inlet and the exhaust valves departs from the basic four-stroke operating cycle. The valve timing values may similarly be presented in the form of a diagram (Figure 2.21). There is a greater likelihood that actual verification of these values may be called for, at least in respect of inlet valve opening and exhaust valve closing points, when reassembling a diesel engine after overhaul. The practical implication of this is that since the piston approaches as closely as possible to the cylinder head, then any deviation from the valve timing specified by the manufacturer can result in the very real danger of the valves fouling the piston

when the engine is started. It is also for this same reason that diesel engine manufacturers warn that valve clearance adjustments should never be attempted with the engine idling, because severe damage could result from inserting a feeler gauge between the valve and its rocker under these conditions.

A further critical factor lies in the choice of value for the inlet valve lag, so that the all-important pressure energy generated during the compression stroke is always at a maximum. More specifically, the inlet valve must neither close too late, which could be detrimental to easy starting of the engine, nor close too early, which could detract from maximum performance. The value of the exhaust valve lead is generally less critical with respect to diesel engine performance, because of the lower operating speeds. For turbocharged diesel engines a valve timing with an increased degree of overlap is generally specified so that, with the earlier opening of the inlet and the later closing of the exhaust valves, the pressurized charge of air can be utilized for more efficient scavenging of the exhaust gases.

As compared with medium-speed diesel engine practice in commercial vehicles (Figure 2.21a), the valve timing diagrams for high-speed engines used in modern diesel-powered passenger cars (Figure 2.21b) usually reveal an increase in duration of both inlet and exhaust valve events. This is, of course, in the interests of providing similar performance characteristics to those that would be expected from a comparable petrol engined car.

Port and valve timing for the two-stroke diesel engine

With reference to the typical port and valve timing diagram for this type of engine (Figure 2.22), the main point to be noticed in comparison with that for a two-stroke petrol engine is that the exhaust event need no longer be symmetrical. This is, of course, made possible by the use of mechanically operated poppet exhaust valves. A sufficiently early opening of the exhaust valves can therefore be provided, so as to initiate thorough scavenging of the exhaust gases just prior to the air inlet ports being uncovered by the piston on the power-exhaust stroke, without incurring a too late closing of the exhaust valves. Since the latter action would reduce the amount of pressure energy generated during the induction-compression stroke, the exhaust valves are usually timed to close just before the air inlet ports are covered again by the piston.

3 Combustion chambers and processes

3.1 BASIC LAYOUTS OF COMBUSTION CHAMBERS

General background

The general arrangement of the combustion chamber in respect of its geometric shape, disposition of the valves, siting of either the sparking plug or the fuel injector, and cooling provision is always regarded by an automotive engineer as a critical factor in the design of spark-ignition (petrol) and compression-ignition (diesel) engines. An enormous amount of both theoretical and practical research has been devoted to this aspect of engine design, especially since Sir Harry Ricardo in England began his pioneering investigations into the combustion process in the early 1920s. The aim then was to prevent knocking in the engine cylinders, or detonation as it was originally known. Some fifty years later, the research priority shifted to the reduction of atmospheric pollutants emitted from the engine and to improving its fuel economy. In modern research computer simulation programs are widely used to predict and refine the combustion process. For our purposes, therefore, we can only attempt to summarize the basic essentials, past and present, as they are generally recognized today.

Geometric shape

To obtain the highest possible thermal efficiency from an engine it is necessary to minimize heat loss to the combustion chamber walls. For this reason, the geometric shape of the combustion chamber must always be such that the ratio of its total surface area to its volume is kept as low as possible. This factor together with the absence of any recessed areas in the chamber is of particular importance in current petrol engines, since it also contributes to a reduction in harmful hydrocarbon (HC) exhaust gas emissions (Section 11.1), by ensuring more complete combustion of the air and fuel charge.

For both petrol and diesel engines the geometric shape of the combustion chamber is contrived, wherever possible, to promote a controlled degree of charge agitation or *compression turbulence* to the cylinder gases. The reasons for promoting this turbulence are not, however, the same for both types of engine. In the case of the petrol engine, the turbulence is associated with propagating the chemical reaction in the air and fuel charge, whereas for the diesel engine a relatively shorter period of turbulence is required for rapid mixing of the air and injected fuel just prior to combustion, the inherent excess of air reducing harmful exhaust gas emissions (Section 3.5). On the other hand too much turbulence is to be avoided, because it can increase heat loss to the combustion chamber walls. Where compression turbulence is difficult to promote in what otherwise may be efficient designs of combustion chamber, it then becomes necessary to obtain a similar effect by promoting a swirling motion to the incoming cylinder charge (either air and fuel or air only), this being known as *induction swirl*.

Disposition of valves

It may be recalled that it has long been established practice to mount both the inlet and exhaust valves over each cylinder in either I- or V-head manner. The latter arrangement in particular is noted for its 'easy breathing' quality in petrol engines. For diesel engines it would, in any event, be physically impossible to obtain a sufficiently high compression ratio without the compact clearance volume associated with the I-head.

In the interests of achieving a high volumetric efficiency, a single inlet valve must be made of adequate size, typically about 45 per cent of the cylinder bore diameter. It should also be provided with ample clearance around the head, so that the incoming cylinder charge does not impinge on the combustion chamber or cylinder walls. The open areas of both the inlet and exhaust valves must, of course, be such as to give adequate flow capacity for the incoming and outgoing cylinder gases. As earlier noted, there is nothing to be gained by lifting a poppet valve from its seating by an amount exceeding one-quarter of its head diameter. A 30° instead of a 45° angle is sometimes used for the inlet valve seating, because it permits a greater gas flow for a given valve lift.

Siting of sparking plug or fuel injector

The siting of the sparking plug electrodes in the combustion chamber of the petrol engine was early recognized by H.R. Ricardo as being one of the important factors controlling pinking or detonation, this being an abnormal condition of combustion described later. To achieve what Ricardo termed the highest useful compression ratio (usually abbreviated to HUCR) and be free from detonation, the distance of flame travel within the combustion chamber must not be unduly long in any one direction. This ensures that there is least time available for a detonating pressure wave suddenly to build up ahead of the flame front. A short flame travel is beneficial for rapid and efficient combustion with least emission of unburned hydrocarbons.

A further compromise usually sought in siting the sparking plug is that preferably its electrodes should initiate combustion in the hottest and therefore exhaust valve region of the combustion chamber, so that the flame front travels from the hottest to the coolest parts. On the other hand the plug electrodes also need to be cooled and scoured of any residual exhaust products by the incoming fresh cylinder charge.

Any reader who takes an interest in vintage cars is likely to know that a 1920s 3 litre Bentley, for example, has two sparking plugs serving each cylinder with one on either side. In fact it was not uncommon practice during that era for designers of high-grade passenger cars to employ twin sparking plugs, especially those associated with firms that also manufactured aircraft piston engines. For the latter it early became conventional practice to use two sparking plugs per cylinder, each served by its separate magneto ignition system. This duplication was partly for reasons of reliability and partly to promote more rapid combustion by initiating two flame fronts from opposite sides of the combustion chamber.

It is therefore of interest that the idea of using twin sparking plugs was revived in the late 1980s by Alfa-Romeo and applied to their long-acclaimed four-cylinder in-line engine. In this case the main advantage sought was the ability to ignite leaner mixtures, especially at low engine speeds, so that both fuel consumption and harmful exhaust emissions could be reduced without sacrificing the smooth running qualities of the engine. These gains may be attributed to less delayed combustion, which results at best from the ignitable charge having greater exposure to the two separate points of ignition and at worst from the increased likelihood that it will be in proximity to at least one of the points of ignition. A further refinement of the twin sparking plugs concept is for them not to be fired simultaneously, the purpose being to reduce combustion noise by avoiding an excessive initial pressure rise as the air and fuel charge is ignited.

With regard to the siting of the fuel injector nozzle in diesel engines, this depends upon the type of combustion system used, as described later. In general terms, a centrally disposed nozzle is required for direct-injection engines with open combustion chambers, whereas for indirect-injection engines an offset nozzle spraying into a combustion antechamber of either the precombustion or swirl type is employed.

Cooling provision

If a high volumetric efficiency is to be obtained, it is essential that the exhaust valve in the combustion chamber be maintained as cool as possible. Otherwise, the heat radiated from it will reduce the density of the incoming charge, either air and fuel or air only, so that a lesser quantity enters the cylinder for compression. A no less important reason for maintaining adequate cooling of the exhaust valve is that of avoiding detonation. Special provision must therefore be made to direct a vigorous flow of coolant around the high-temperature surfaces of the cylinder head combustion chamber in the region of the exhaust valve.

3.2   COMBUSTION IN THE PETROL ENGINE

Basic considerations

The combustion process is concerned with the chemical reaction that takes place between a hydrocarbon fuel – petrol – and oxygen from the air, and which continues until a point of chemical equilibrium is attained. For the moment, we can say that this simply means that the hydrocarbon fuel is oxidized or, more specifically, burned (since the latter implies very rapid oxidation) to carbon dioxide and water, which are oxides of carbon and hydrogen respectively. It is, of course, the intense heat that accompanies this chemical reaction which causes the cylinder gases to expand and pressurize the cylinder, such that the piston is driven away from the combustion chamber on its power stroke.

It will be recalled from Section 1.3 that the sequence of events in the four-stroke petrol engine was defined as induction-compression-power-exhaust, but in some popular types of publication the power stroke may be described as the explosion stroke. The latter may therefore give the impression that the operation of such an engine is one 'Of dire combustion and confused events', to borrow an expression from Shakespeare. With a real engine, however, the energy conversion taking place in the combustion chamber is, or at least should be, somewhat better ordered. In this context, we must first recognize that the combustion process can take two basic forms, which in physical chemistry may be classified as follows:

Spontaneous combustion   This has no external cause and occurs purely as a result of a chemical reaction that is simultaneously set up throughout a mixture of combustible gases, and which is accelerated by the evolution of heat until completion.

Progressive combustion   This does have an external cause which, in our case, is an electric spark from the ignition system, and occurs as a *flame front* that radiates from a combustible nucleus developed around the spark. As combustion progresses, the unburnt and the burnt portions of the mixture are separated by a *reaction zone* in the flame front, the flame cooling and slowing down as it approaches the combustion chamber walls, finally to become extinguished (Figure 3.1).

Normal and abnormal combustion

In further considering the combustion process that takes place in the petrol engine and which is normally of a progressive

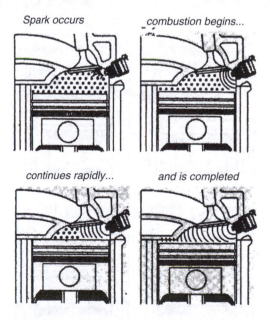

Figure 3.1   Visualizing normal combustion in the petrol engine (*Champion*)

character, we must always remember that we are talking about a phenomenon occupying only a few milliseconds. Paradoxically, it can also be said that probably more time has been devoted by scientific researchers to investigating the combustion process and related combustion chamber design than to any other aspect of automotive engine operation. The reasons for this are basically threefold, the first two being historical in character and the last being of more recent origin: detonation, roughness and pollution.

Detonation

For high-efficiency engines operating at high speeds, it will be evident that a very fast rate of burning of the combustible charge is required. As with any mixture of gases, however, there must be a limit to which the rate of burning may be increased, beyond which the combustion process undergoes a sudden and dramatic change that results in detonation. In normal combustion, the burning progresses in a steady and uniform flame front that travels across the chamber at about 20–40 m/s (65–130 ft/s), this being known as the *combustion speed*. With detonation the flame front similarly advances across the chamber, but about ten times faster, and before it can complete its journey the unburnt gases ahead of it – or end gases as they are usually termed – become heated to such an extent by the overall pressure rise in the chamber that their self-ignition temperature is exceeded (Figure 3.2). When this occurs a condition of spontaneous combustion is created, resulting in a high-pressure detonation wave travelling through the chamber at a speed comparable to that of sound. Quite understandably, this produces a shock loading on the engine structure and components sufficient to generate a clearly audible knocking or pinking noise. If this abnormal combustion is allowed to persist, it can cause serious damage to the engine, such as blowing a hole in the piston crown (Figure 3.3).

Since the early 1920s the control of detonation has been improved in several ways, chiefly by experiment with different combustion chamber shapes, more efficient cooling arrangements for the engine and the production of more knock-resistant fuels. The knock resistance of a petrol is defined in terms of its 'octane rating'. A higher octane rating signifies a greater resistance to engine knocking. There are two different test procedures for determining the octane rating, these being the Research Method giving a Research Octane Number (RON) and the Motor Method giving a Motor Octane Number (MON). Basically, the former rating indicates the resistance to engine knocking during acceleration, whilst the latter rating similarly relates to high-speed running. The octane number of a petrol is established by matching its knock resistance in a test engine against a reference fuel. This comprises a mixture of high knock resistant iso-octane and low knock resistant n-heptane. The actual octane number of the petrol then corresponds to the percentage by volume of iso-octane in the reference fuel, which exhibits the same knock-proneness as the petrol in the same engine running under similar controlled conditions.

In modern practice, combustion knock is often monitored by a knock control system, which comprises a vibration sensor acting in conjunction with the electronically controlled ignition of an engine management system. By relating an adverse signal from the knock sensor to a positional signal from the crankshaft sensor, it becomes possible to retard the ignition timing for any particular cylinder until the knock disappears. A knock sensor is mounted on the cylinder block between pairs of cylinders, where vibration is best amplified by the less stiff walls of the coolant jacket. It therefore converts mechanical vibration into electrical signals and is tuned to detect vibrations typically in the 6–8 kHz (6000–8000 cycles/s) frequency range. A knock control system therefore makes a further contribution to more stable ignition timing and its beneficial effects on engine operation and emission control.

Roughness

Detonation and roughness have sometimes been described as the twin evils of the automotive petrol engine. The reason for this is that whereas better control over detonation allows the use of higher compression ratios to improve thermal efficiency, as described in Section 1.3, it can also result in the maximum rate of pressure rise during combustion being

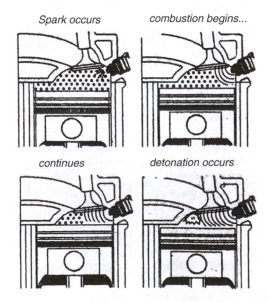

Figure 3.2   Visualizing detonation in the petrol engine (*Champion*)

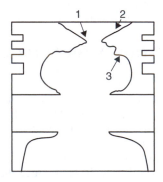

Figure 3.3   Typical piston damage from severe detonation: 1 sharp ragged edges blown out below; 2 occasional metal pitting; 3 possible cracks on underside (*Champion*)

such as to cause undesirable vibration of the engine structure and mechanism, especially bending vibration of the crankshaft. Although this problem of rough operation originates from the combustion process, its practical solution usually lies in providing greater rigidity for the engine parts.

Pollution

As a simplification it was earlier mentioned that during the combustion process the hydrocarbon fuel is burned to carbon dioxide and water. Unfortunately this rather tidy and relatively harmless state of affairs does not actually exist in practice. Evidence of this truth lies not only in the unpleasant odour of the exhaust gases, but also in the common knowledge that they contain poisonous carbon monoxide. It is the presence in fuel of compounds other than hydrocarbons, and also the fact that the combustion process may only be partially completed, which account for the harmful exhaust emissions. More recently a further concern to environmentalists has been the emission of carbon dioxide (CO_2) which, although non-toxic as earlier implied, is an unwanted contributor to global warming. That is, the heat absorbed from the sun cannot leave the atmosphere in the presence of a buildup of carbon dioxide.

The general topic of harmful exhaust gas emissions, their legislated regulation and control is later dealt with in Section 11, and the role played by the geometric shape of the combustion chamber has already been mentioned in Section 3.1.

3.3   PETROL ENGINE COMBUSTION CHAMBERS

The six basic types of combustion chamber employed, past and present, in automotive petrol engines may be classified as follows:

Hemispherical	Bowl-in-piston
Lozenge	Split-level
Wedge	Pentroof

Hemispherical chamber

The hemispherical form of combustion chamber with inclined overhead valves has long been regarded as the classic type for high-output engines (Figure 3.4), especially when used in conjunction with double overhead camshafts after the style introduced as long ago as 1912 by the talented Swiss designer Ernest Henry (1885–1950). Nowadays the hemispherical chamber is also recognised as offering a beneficial reduction in harmful exhaust emissions.

A high volumetric efficiency or 'easy breathing' is afforded by this type of chamber, because not only may large and well-cooled valves be accommodated, but also the radial disposition of the valves ensures that their open areas are not masked at all by the chamber walls. Moreover, the sparking plug is usually sited very close to the central axis of the chamber, so that flame travel distances are minimized and practically equalized for good detonation control (Figure 3.5a).

Since the hemispherical chamber lacks a squish area, charge agitation is normally effected by induction swirl (both these aspects of charge agitation will be described later). A dome-head piston is sometimes required to reduce the combustion chamber clearance volume and obtain a high compression ratio, in which case the dome may be offset slightly to promote compression turbulence. As a consequence of departing from a true hemispherical shape, however, the ratio of total surface area to volume of the combustion space is

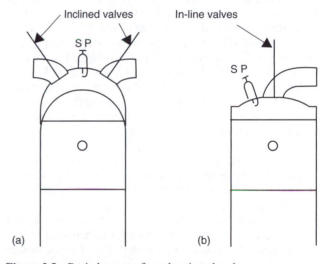

Figure 3.5   Basic layouts of combustion chambers: (a) hemispherical (b) segmental

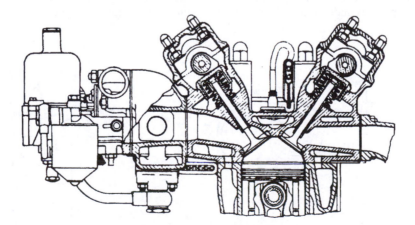

Figure 3.4   The classic Jaguar version of a hemispherical head

increased. This in turn reduces thermal efficiency, owing to the greater heat loss from the combustible charge.

More recently, the basic hemispherical chamber has undergone further changes, such that in some designs it more approaches a segmental form. That is, it assumes a comparatively shallow part-spherical shape with in-line rather than inclined valves (Figure 3.5b). Again, the object is to obtain a favourable ratio of total surface area to volume, thereby improving thermal efficiency and reducing the quantity of incompletely burned products of combustion. Hence, better control is maintained over harmful exhaust emissions.

Lozenge chamber

The lozenge or inverted-bathtub form of combustion chamber is now little used, but it long represented the simplest type for overhead valve engines of moderate output (Figure 3.6a). This type of chamber was evolved in the early 1920s from the plain cylindrical form, as a result of increased compression ratios demanding reduced clearance volumes. The valves are mounted in-line, and in plan view the chamber was originally semicircular at each end with straight sides, its greatest width being such that it overlaps the cylinder bore, but in cross-section it is appreciably narrower than this. To improve volumetric efficiency, which otherwise tends to be a limiting factor with this type of chamber, the valves may be slightly inclined to the cylinder axis and the chamber end walls may be 'undercut' or recessed, where they tend to encroach upon the open areas of the valves (Figure 3.6b).

The chamber side walls may also slope away from the roof, so that the sparking plug entering through one side at a fairly steep angle can be brought closer to the valves. This improves combustion efficiency, in which connection the sparking plug is usually biased slightly towards the exhaust valve. Hence the combustion process is initiated in the hottest, and completed in the coolest, regions of the chamber for good detonation control. A certain amount of compression turbulence is provided by a *squish area,* formed where the flat top of the piston overlaps the combustion chamber opening and approaches very closely the underside of the cylinder head. The squish area, with its high ratio of surface area to volume, also acts as a *quench area,* in as much that it serves to cool the remote parts of the charge or end gases. For this reason the lozenge type of combustion chamber may also be offset laterally from the cylinder axis in the direction of the sparking plug.

A later development of this form of combustion chamber, which was introduced by the British Motor Corporation (BMC) in 1952 for their 'A Series' engine and subsequently adopted for the trend-setting Austin and Morris Mini cars, saw it further modified and made heart-shaped in plan view. The apex portion then directs the incoming cylinder charge towards the sparking plug and also shields it from heat being radiated by the exhaust valve. This version of lozenge combustion chamber is known as the Weslake type (Figure 3.6c), so named after H. Weslake, an English researcher into gas flow through cylinder heads who invented it in the late 1940s.

Wedge chamber

The wedge or ramp form of combustion chamber for overhead valve engines is a comparatively later development. It first appeared in the late 1930s on some American engines, since when it has been widely used for both moderate- and high-output engines. This type of chamber is generally D-shaped in plan view, the circular part of which follows the cylinder opening. In cross-section, this circular portion tapers uniformly away from where the sparking plug is situated in its side wall and terminates at the approach to a very narrow squish area (Figure 3.7). The latter is completed by the top of the piston where it approaches very closely to the underside of the cylinder head. The valves are generally mounted in-line and also inclined to correspond with the angled chamber roof, thus facilitating the flow of gases to and from the cylinder for good volumetric efficiency. Distribution of the clearance volume is therefore such that the greater portion of the combustible charge is concentrated around the sparking plug, so that fairly short and nearly equal flame travels are achieved for smooth combustion. Furthermore, the sloping roof of this part of the chamber reduces its ratio of surface area to volume, thereby improving thermal efficiency as a consequence of less heat loss. The squish area generally covers 25 per cent or more of the piston head surface area and, as in other combustion chamber designs, it also acts as a quench area for good detonation control.

However, in light of current requirements to reduce harmful exhaust emissions, without incurring unacceptable penalties in fuel consumption and power output, the greater ratio of total surface area to volume of the combustion space tends to put the wedge chamber at a disadvantage compared with the hemispherical type.

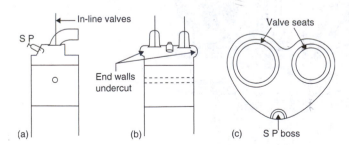

Figure 3.6 Lozenge combustion chamber: (a)/(b) basic layout (c) plan view of Weslake version

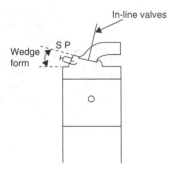

Figure 3.7 Basic layout of wedge combustion chamber

Bowl-in-piston chamber

The bowl-in-piston or piston cavity form of combustion chamber is of more recent origin than the wedge type. Its limited application to petrol engines dates from the early 1960s, following earlier investigations by S.D. Heron in America. In this design the combustion chamber proper takes the form of a cylindrical recess located in the piston head (Figure 3.8a). As compared with other types of chamber utilizing in-line valves, the open areas of the valves are less masked, although they are in themselves limited by the size of valves that can be circumscribed by the cylinder bore (Figure 3.8b). The annular rim of the piston is generally relieved about the gudgeon pin axis, so that a safe working or bumping clearance for the valves is maintained against the piston during their overlap period. A high thermal efficiency is usually claimed for this type of combustion chamber, since its flat roof contributes to a low ratio of total surface area to volume. Furthermore, the geometric symmetry of the bowl-in-piston chamber affords minimum and near-equal flame travels, the sparking plug being positioned as close as is practicable to the central axis of the chamber. Charge agitation is initiated by induction swirl and completed by compression turbulence. The former is derived from the usual angled porting, whilst the latter is promoted by the flat-topped rim of the piston approaching very close to the underside of the cylinder head and thus providing an annular squish area. Again, as in other combustion chamber designs, the squish area also serves as a quench area.

One disadvantage of this type of combustion chamber is that it tends to make for a heavier piston construction. Another, in the context of reducing harmful exhaust emissions, is that the hotter running piston tends to increase the production of unwanted nitrogen oxides.

Split-level chamber

Of further interest is the split-level form of combustion chamber. This is also known as the May fireball, so named after Michael May, the Swiss researcher into combustion chamber design, who first developed the concept in the mid 1970s. This type of combustion chamber was used in the renowned Jaguar V12 engine.

The essential features of the split-level combustion chamber are a lower-level and therefore shallow collecting zone beneath the inlet valve, and an upper level formed by a deeply recessed exhaust valve that constitutes the combustion zone of the chamber (Figure 3.9). A swirl-promoting guide channel is provided between the collecting and combustion zones with a tangenital approach to the latter. The combustion chamber is completed, of course, by the flat top of the piston and the sparking plug is sited in the wall of the upper-level combustion zone.

As the piston closely approaches the combustion chamber towards the end of its compression stroke, that portion of the air and fuel charge residing in the collecting zone is forcibly ejected, via the tangential guide channel, into the combustion zone of the chamber, wherein a rapid swirling motion is generated in the vicinity of the sparking plug. This action makes possible a fast and efficient burning of air and fuel cylinder charges that are leaner than normal, which in turn allows an increase in engine compression ratio without incurring detonation. Hence, the advantages generally attributed to the split-level type of combustion chamber are those of reduced fuel consumption and increased thermal efficiency. Similarly, there is a reduction in harmful exhaust emissions, as would be expected with any new design of combustion chamber if it is to prove acceptable. However, in the context of present-day engine design that generally favours inclined overhead valves, the split-level chamber was necessarily limited to in-line vertical or near vertical overhead valves.

Pentroof chamber

In engine terminology, the designation 'multiple valves' signifies that each cylinder has more than one inlet and one exhaust valve. Since the early 1980s there has been a trend in combustion chamber design towards providing two inlet and two exhaust valves for each cylinder (Figure 3.10), instead of the customary one of each. Four-valve combustion chambers were pioneered by the earlier mentioned Ernest Henry and were actually race proven even before World War I, but thereafter were mainly confined in application to racing car and a few high-performance sports car engines such as for the vintage Bentley cars of the 1920s. The reason for their

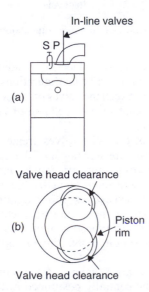

Figure 3.8 Basic layout of bowl-in-piston combustion chamber

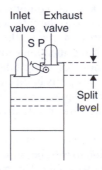

Figure 3.9 Basic layout of split-level combustion chamber

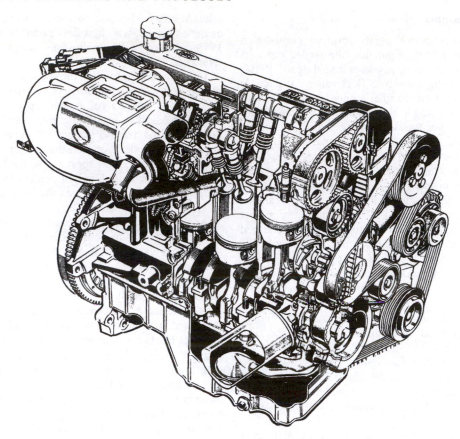

Figure 3.10 A modern four-valve cylinder engine with pentroof combustion chambers and double overhead camshafts (*Ford*)

use was to maximize power output whilst avoiding destructively high valve temperatures.

What then is the reason for now using four-valve combustion chambers in the engines of many volume-produced cars, especially considering the additional complication and cost incurred with such an arrangement? The answer to this question has to be more comprehensive and lies in the requirements of the modern petrol engine given in Section 1.1, namely to provide an optimum performance with high fuel economy and low pollution.

The advantages and disadvantages of the four-valve combustion chamber, which is aptly named the pentroof type, may be summarized as follows:

1 The combined area of the four largest valves that can be accommodated in a given combustion chamber always exceeds that of the two largest which may be used. Doubling the number of valves therefore increases the area through which the incoming mixture and outgoing exhaust gases can flow. As a result the engine can breathe more freely at low valve lifts in the interests of better performance, especially with regard to improving mid-range torque characteristics and hence drive-ability of the car (Section 1.1).

2 The opening and closing motion of the valves is easier to control at high engine speeds, because each valve is smaller and therefore lighter in weight. This can permit an increase in engine operating speeds, which again contributes to

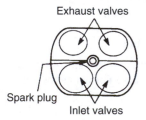

Figure 3.11 Plan view of four-valve combustion chamber

better performance. To reduce the bending moments in the camshafts of four-valve configurations, their bearings may be located between the valves of a pair as in Cosworth designs rather than between each pair of valves as in conventional practice.

3 The provision of two inlet valves creates greater turbulence for the mixture flowing into the cylinder, so that better mixing leads to more efficient burning of lean mixtures and hence contributes to better fuel economy.

4 The provision of two exhaust valves minimizes the intensity of hot-spots in the combustion chamber, thereby avoiding too high combustion temperatures which can promote the formation of polluting nitrogen oxides in the exhaust gases.

5 The four-square arrangement of the valves allows the spark plug to be centrally positioned in the combustion chamber (Figure 3.11), so that flame travels can be both

equalized and minimized for good detonation control. An increase in compression ratio may then be possible to provide better performance and fuel economy.

6 The four valves generally and the two exhaust valves particularly should run cooler, since not only are they individually smaller in size with shorter heat flow paths, but also their seatings are better spaced for more uniform cooling. This reduction in valve operating temperatures should therefore make for greater durability. Even so, the higher combustion temperatures encountered have necessitated the use of sodium-cooled exhaust valves by some manufacturers (Section 1.10).

7 The modern four-valve arrangement inherently demands a V-head with its greater mechanical complexity and increased friction losses. Apart from doubling the number of valve components, there is almost invariably a requirement for twin overhead camshafts to operate them together with their associated additional journal bearings, whereas with a more ordinary two-valve I-head engine only one camshaft is necessary. Similarly, the extra demands placed on the engine lubrication system calls for an increased capacity oil pump.

8 The engine is more costly to manufacture owing to the increased number of valve components used, together with the additional machining required on the cylinder head for the extra valve seats, valve guides and camshaft bearings.

9 The engine may prove more costly to service, but this will to some extent depend on whether greater durability is realized in practice with the four-valve arrangement, such that optimum engine performance can be maintained for extended mileages.

An interesting comparison between two combustion chambers of the pentroof type with four valves, which have been developed by Toyota, is shown in Figure 3.12. The sports type is of classic layout as would be expected for a responsive high-performance engine where maximum power is developed at high revolutions. The high-efficiency type is intended to produce better torque in the low to medium speed range together with more economical operation in everyday town driving. With the latter type it will be noticed that in addition to the short flame travel distance derived from the centrally disposed sparking plug, a narrow included angle has been adopted between the pairs of inlet and exhaust valves to allow the design of a very compact combustion chamber. Furthermore, the larger squish areas at the outer edge of the cylinder are conducive to quicker combustion.

Finally, it should be mentioned that it is by no means unusual to encounter two inlet and two exhaust valves in each cylinder of commercial vehicle diesel engines (Figure 2.5), and even three of each in larger-capacity diesel engines used for rail traction purposes. It may also be of interest to recall that, in the mid-1930s, Rolls-Royce developed a form of pentroof or ramp type of combustion chamber to improve turbulence in their four-valve aero engine cylinder heads. Curiously though, the encouraging results achieved on a single-cylinder test engine did not successfully translate to the war-time Merlin engine, which had a more conventional type of combustion chamber.

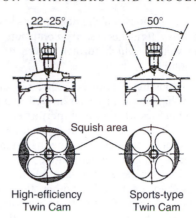

Figure 3.12 Comparison of four-valve combustion chambers with narrow- and wide-angle valves (*Toyota*)

Three- and five-valve combustion chambers

The less common present-day use of three-valve combustion chambers is again a revival of much earlier design practice, which sought to improve engine performance. In modern applications there are the further aims of better controlling exhaust emissions and reducing fuel consumption.

Historically, engine designers originally preferred an arrangement of two exhaust valves and one inlet valve, based on the supposition that more resistance was offered to the exhaust gases leaving the cylinder than there was to the fresh mixture entering it. However, it was Ettore Bugatti the designer of the once famous sports and racing cars that bore his name who, in the early 1920s, introduced the converse arrangement of two inlet valves and one exhaust valve. The two smaller inlet valves and passages serve to increase the velocity of the incoming charge for more efficient filling of the cylinder. Modern versions of the three-valve combustion chamber continue to follow this practice, and it is used, for example, by Mercedes-Benz for their V-cylinder engines (Figure 3.13), including the engine specially developed for their prestigious Maybach car introduced in 2002.

As compared to the more widely used four-valve combustion chambers, the advantages generally claimed for modern three-valve layouts are as follows:

1 Dispensing with one exhaust valve reduces heat loss from the engine cylinder, which not only promotes a rapid warm-up, but also the higher exhaust temperatures enable a catalytic converter to reach its 'light-off' temperature more quickly and reduce exhaust emissions (Section 11.2).

2 The extra space created in a three-valve combustion chamber readily allows the use of twin sparking plugs. These can provide a shorter flame travel and less delayed combustion, so that both fuel consumption and exhaust emissions can be reduced (Section 3.1).

3 For a three-valve combustion chamber it is only necessary to employ a single overhead camshaft to operate both the inlet valves and the exhaust valve, thereby reducing frictional losses in the valve train and hence improving fuel economy.

4   A three-valve combustion chamber and its associated less complex valve train necessarily contribute to a saving in engine weight, which further serves to reduce fuel consumption.

Five-valve combustion chambers with three inlet and two exhaust valves have typically been limited in application to turbocharged engines of ultra high-performance cars, such as certain models of Audi and Ferrari. Although paradoxically the Mitsubishi company was the first to mass produce a small capacity, three-cylinder engine with a five-valve arrangement for a Japanese minicar in 1989.

As compared to a four-valve combustion chamber, the main advantage claimed for the five-valve arrangement is that the increase in total area of the valves allows the maximum benefit from turbocharging to be realized. Induction swirl may be increased by arranging for the outer inlet valves to open slightly earlier than the one offset between them, thereby ensuring more rapid combustion, for better performance and lower emissions.

Altering the compression ratio

The ordinary motorist usually associates the term *engine tuning* simply with the various items of preventive maintenance that seek to restore, as far as possible, the economical running qualities of the engine originally intended by the manufacturer. To the dedicated sporting motorist, however, engine tuning can involve dimensional modifications being made that depart from the original mechanical specification of the engine, in order to increase power output. One such modification is to raise the compression ratio as high as possible without incurring detonation. For obvious reasons, this type of work is normally undertaken by specialist engine tuning establishments.

It may be recalled (Section 1.3) that the compression ratio was defined as the swept volume plus the clearance volume divided by the clearance volume. The only practicable approach to raising the compression ratio of an existing engine is therefore to reduce the clearance volume, which may be accomplished in several ways:

1   Fitting a thinner cylinder head gasket.
2   Skimming metal from the cylinder head face.
3   Installing raised-dome high-compression pistons.
4   Skimming metal from the base of detachable cylinder barrels.

The third method is generally preferred, because it does not alter the vertical distance between the cylinder head and the crankcase, thereby preserving the original operating geometry of the valve train. Otherwise, the removal of metal from say the cylinder head face may require that either steel shims be added beneath the rocker pedestals, or the push-rods be shortened in engines so equipped.

Variable compression engines

Altering the compression ratio of an engine while it is actually running, that is creating a 'variable compression' engine, has in the past been associated solely with single-cylinder test engines used in research laboratories for assessing combustion characteristics and the knock rating of fuels. The earlier mentioned Sir Harry Ricardo pioneered this type of research engine, which took the form of raising and lowering the complete cylinder and head assembly in relation to the crankshaft. In construction the lower part of the cylinder jacket was screw threaded so that the cylinder could be slidably positioned in the crankcase by means of a rotatable ring nut, thereby varying the compression ratio.

It is therefore of interest to record that Saab Automobiles have experimentally applied variable compression principles to a supercharged multi-cylinder engine. In this case it allows the compression ratio to be optimized by the engine management system, based on inputs that relate to engine speed, load and fuel quality. Structurally, one side of the integrated five-cylinder block and head is hinged from within the crankcase, the opposite side being linked to a computer controlled hydraulic actuator system. By this means the compression ratio of the engine can be varied between 8:1 and 14:1 by tilting the block and head in relation to the crankcase and hence the crankshaft. This therefore has the effect of altering the clearance volume remaining above the pistons at TDC. An oil-tight enclosure between the hinged upper and lower fixed parts of the engine is formed by a special rubber bellows seal.

3.4   COMBUSTION IN THE DIESEL ENGINE

Basic considerations

The distinguishing feature of the diesel engine is that its combustion process is initiated by spontaneous ignition of the fuel when it is injected into an agitated, highly compressed and therefore exceedingly hot charge of air. Here we are thinking in terms of pressures and temperatures in the region of $5\,MN/m^2$ ($750\,lbf/in^2$) and 800°C respectively, the temperature rising much faster towards the end of the compression stroke than it does at the beginning. Since the compressed cylinder air attains a temperature that is higher than the self-ignition temperature of the fuel, this being 350 to 450°C and governed by a number of complex factors of which not the least is cylinder pressure, a condition of spontaneous combustion is created.

With the diesel operating cycle it thus becomes possible to ignite a spray of finely atomized, but non-volatile, fuel without the aid of an electric spark. It does, however, mean that the compression ratios for diesel engines must be made appreciably higher than those adopted for petrol engines, and they generally lie between 16:1 and 22:1. If compression ratios as high as these were attempted in petrol engines, then the highly volatile charge of air and fuel admitted during the induction stroke would detonate and burn uncontrollably. Fortunately, no such limitation applies to the diesel engine, because it breathes only 'pure' air. The actual compression ratio chosen for a diesel engine is influenced by combustion chamber design and fuel quality, but in any event it is generally made no higher than is necessary for reliable cold starting.

Although it is a topic that has received much attention by researchers, the precise nature of the combustion process in diesel engines is generally recognized as being complex and as yet not completely understood, since it involves a number of interrelated physical and chemical changes. In simple

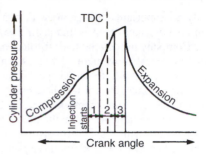

Figure 3.14 The three phases of combustion in a diesel engine

burning almost as soon as it enters the high-temperature combustion space.

In practice, too rapid a combustion period (second phase), with its accompanying high rate of pressure rise, can result in objectionable diesel knock, the sound of which is not unlike detonation in a petrol engine but of much greater intensity. To limit the pressure rise during the second phase of combustion, the delay period preceding it is kept as short as possible, which among other factors depends upon good ignition quality of the fuel. Efficient completion of the third phase of combustion is then achieved by providing for rapid mixing of the injected fuel and the cylinder air.

The three phases of diesel engine combustion can be readily visualized by defining their transition points on a graph as in Figure 3.14, where cylinder pressure has been ploted against crank angle.

The ignition quality of a diesel fuel relates to its ability to minimize the ignition lag during the delay period. It is expressed as a Cetane Number (CN), the higher the number the greater is the tendency for the fuel to ignite and reduce ignition lag. Somewhat similar in principle to determining the octane rating of a petrol, the cetane number of a diesel fuel is established by matching its readiness to ignite against a reference fuel in a test engine. In this case the reference fuel comprises a mixture of fast-burning n-hexadecane and slow-burning methyl naphthalene.

3.5 DIESEL ENGINE COMBUSTION CHAMBERS

Direct-injection combustion chambers

With direct-injection diesel engines the fuel is sprayed through a multihole injector nozzle into an open type of combustion chamber. This is formed partly by a bowl in the piston head and partly by the flat underside of the cylinder head. The basic idea of the bowl-shaped cavity in the piston head is to promote agitation of the cylinder air charge towards the end of the compression stroke, so that combustion is initiated by a process of induction swirl followed by compression turbulence (Section 3.6). Considered in cross-section, the internal shape of the bowl is generally such that it gives a ready impression of conforming to the conical spray pattern of the injector nozzle, which is mounted above in the cylinder head. The raised rim of the piston head prevents the fuel spray from reaching the cooler cylinder walls.

Figure 3.13 Three-valve combustion chamber with twin Sparking plugs (*Mercedes-Benz*)

terms, however, it would appear that each minute droplet of fuel injected immediately absorbs heat from the high-temperature cylinder air and begins to vaporize on reaching its boiling point. Then as long as the cylinder air temperature appreciably exceeds the self-ignition temperature of the injected fuel, the vapour thus formed will spontaneously ignite and each droplet will burn uniformly from the surface. The combustion of the air and fuel does, of course, raise further the temperature and hence pressure in the cylinder, as a result of which the piston is driven away from the combustion chamber on its power stroke.

Phases of combustion in the diesel engine

An overall impression of the combustion process that takes place in the diesel engine may be obtained by considering it as occurring in three phases, a concept that was originally proposed by H.R. Ricardo in this country over 60 years ago. The three phases may be summarized as follows:

First phase This is known as the delay period, and represents the time between the start of fuel injection and the onset of combustion. The delay period is also sometimes referred to as ignition lag.

Second phase A rapid combustion period, as it is termed, constitutes this phase, corresponding to when combustion is initiated and spreads at a rapid rate through the air and fuel that have come into contact during the delay period.

Third phase This final phase is designated the controlled combustion period, and is characterized by the injected fuel

Historically, an important advance in the design of the open type of combustion chamber was made in 1934 when the Swiss Saurer Company introduced their toroidal piston chamber to promote a spiral movement of the air. Various detail refinements have been made to the shape of the bowl since then and these mainly account for the differences in chamber design from one manufacturer to another (Figure 3.15).

Although the combustion process in the diesel engine poses a less severe problem with respect to harmful exhaust emissions than does a petrol engine, since there is a reduction in carbon monoxide by virtue of the excess air involved in the process, increasingly stringent international legislation has demanded that continuing attention be given to reducing exhaust pollutant limits for diesel engines. One area of conflict lies between the requirement for raising combustion temperatures to reduce black smoke or carbon particulates and lowering them to reduce nitrogen oxides. However, the latter can be reduced by retarding the fuel injection timing and, to compensate for the otherwise loss of engine performance, 'square' shaped piston bowls have been developed for direct-injection combustion chambers. These can shorten the ignition delay period by enhancing swirl and squish effects. An example of this type of combustion chamber is that introduced by Perkins and known as the Quadram, where the toroidal recess in the piston crown assumes a four-lobed form. The pockets so formed at each corner of the square bowl then serve to agitate further the swirling vortex of air that becomes trapped in its central cavity and into which the fuel is sprayed. Another advantage associated with the square combustion chamber is that it confers a smoother rise in cylinder pressure for quieter combustion and lower mechanical stresses. A contrasting approach to accommodating retarded injection timing can take the form of a much wider and shallower piston bowl, which is used in conjunction with less air swirl but a greater number of finer fuel sprays injected at higher pressure. Hence there is an increased tendency for the injected fuel to seek out the air rather than vice versa, which again shortens the ignition delay period while avoiding the excessive temperatures associated with too rapid combustion.

These two differing approaches to making the combustion process of direct-injection diesel engines more environmentally friendly, have resulted in what may now be classified as 'high swirl toroidal' and 'low swirl quiescent' forms of combustion chamber (Figures 3.16a and b). In general terms, the large capacity diesel engines used in heavy vehicles require less air swirl than the smaller capacity engines for passenger car application. This difference can be accounted for by the combustion chambers of the latter engines being more restricted in size, together with an increase in engine operating speeds.

Advantages and disadvantages of direct injection

The direct-injection combustion chamber offers the advantages of easy starting and high thermal efficiency with correspondingly low fuel consumption. These advantages derive from a compact combustion space that reduces heat loss and also a method of charge agitation that incurs minimal power loss.

A disadvantage is that the combustion process is not very rapid, owing to the rather long distances the fuel sprays have to penetrate the cylinder air charge. It is also for this reason that what were once thought to be very high injection pressures in the region of 80 MN/m^2 (12 000 lbf/in^2), but which are now often more than double this value, are required to provide a sufficiently 'hard spray', as the fuel injection engineer terms it. These higher injection pressures relate to the increasingly energetic fuel sprays demanded by modern low swirl quiescent combustion chambers, wherein a greater number of finer fuel sprays must be injected with considerable rapidity, typically from an eight-hole injector nozzle with hole sizes of only 0.22 mm (0.009 in). Furthermore, the centrally disposed injection nozzle tends to impose physical restrictions on the sizes of the inlet and exhaust valves that can be accommodated, which is to the detriment of volumetric efficiency.

However, the advantages of the direct-injection diesel engine are such that it continues to be used over an ever widening range of commercial vehicles, both light and heavy, and is acknowledged as pointing the way ahead for future development of the diesel engine in road transport. Although its disadvantages initially made it less suitable for diesel engined passenger cars, it is nevertheless becoming established practice for their modern counterparts to use direct injection systems of the common rail type (Section 8.11)

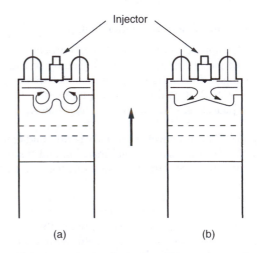

Figure 3.16 Basic layout of direct-injection diesel combustion chambers (a) high swirl toroidal (b) low swirl quiescent

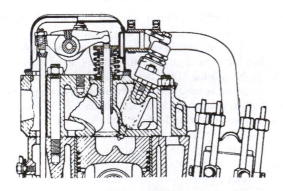

Figure 3.15 A practical application of the direct-injection open chamber

together with turbocharging, thereby taking advantage of the better fuel economy that can be obtained as compared with indirect injection.

Indirect-injection combustion chambers

For indirect-injection diesel engines the fuel is sprayed from an essentially single-hole injector nozzle into a divided type of combustion chamber, which is formed partly in the cylinder head and partly by a saucer-shaped (as opposed to a bowl-shaped) depression in the piston head. Divided combustion chambers actually fall into two quite distinct categories: precombustion, and swirl.

Precombustion chamber

This represents the earliest form of divided combustion chamber, and its originator was Dr Prosper L'Orange of the German Benz Company who patented the idea in 1909. Typically pear-shaped in appearance, the precombustion antechamber is approached from the cylinder through a somewhat restricted passage, or burner, as it is appropriately termed (Figure 3.17a). A narrow cone of fuel is sprayed down towards the burner by the injector nozzle. It should be noted that the whole of the fuel is sprayed into the antechamber. The manner in which the combustion process is initiated in this particular form of divided chamber is based on what is known as *combustion swirl*, which is described later.

Swirl combustion chamber

This form of divided chamber was pioneered by H.R. Ricardo (mentioned earlier), who during the 1930s developed several versions of what have long been known as Comet swirl combustion chambers. The swirl chamber differs from the precombustion type in two clearly identifiable respects. First, the antechamber is of basically spherical form with a relatively unobstructed and tangential passage or throat leading from the cylinder (Figure 3.17b). Second, although the injector

nozzle sprays its narrow cone of fuel wholly into the antechamber, it does so in a direction away from the approach passage. Another interesting feature is that the lower portion of the antechamber containing the throat is separately inserted into the cylinder head and retained with minimal metal-to-metal contact to limit heat transfer to the cooling system (Figure 2.9). It thus forms a hot-spot over which air flows during the compression stroke, thereby raising the temperature of the air to a level higher than that to be expected from compression alone and beneficially shortening the combustion delay period. A process known as *compression swirl* and described later is responsible for initiating combustion in this type of divided chamber.

Advantages and disadvantages of indirect injection

It was early found by P. L'Orange (introduced in the previous section) that the indirect-injection diesel engine ran more smoothly, and less noisily than its direct-injection counterpart. Indeed, it is these self-same operating characteristics that account for the indirect-injection system in either precombustion (Figure 3.18a) or, more commonly, swirl chamber form (Figure 3.18b) also finding early acceptance in diesel engined motor cars. These characteristics may be explained in simple terms by the combustion process being initially confined within an antechamber, so that the rate of pressure rise in the cylinder is reduced. From this it also follows that the engine structure is not so highly stressed, so it can therefore be made less heavy than is possible with direct-injection.

In order to produce sufficient heat for self-ignition of the fuel, an indirect-injection engine requires a higher compression ratio, usually in the region of 20:1, mainly to compensate for the greater opportunities of heat loss from the cylinder charge. Even so the indirect-injection diesel engine can be more difficult to start from cold. For this reason the antechamber of such systems is generally provided with an electrically heated glow-plug, so that when switched on its incandescent electrode instantly ignites any fuel impinging upon it (Figure 8.48).

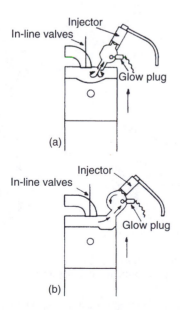

Figure 3.17   Basic layout of indirect-injection diesel combustion chamber: (a) precombustion (b) swirl

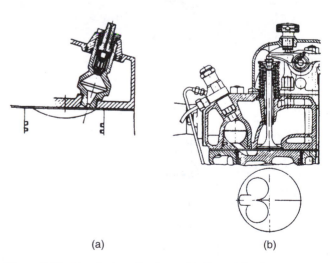

(a)   (b)

Figure 3.18   Practical applications of indirect-injection divided chambers: (a) precombustion chamber of the Caterpillar Tractor Co. (b) Ricardo Comet Mk III swirl-type combustion chamber

The use of such high compression ratios in the indirect-injection diesel engine naturally demands a heavy-duty battery and starter motor installation.

3.6   CYLINDER CHARGE AGITATION

Basic requirements

The need for a controlled degree of agitation for the cylinder charge of either air and fuel in the petrol engine, or air only in the diesel engine, was referred to in connection with the geometric shape of combustion chambers (Section 3.1). It now remains to examine how compression turbulence and induction swirl may be generated in both petrol and diesel engines, and also compression swirl and combustion swirl in diesel engines only (although compression swirl does occur in the more recent split-level type of combustion chamber for petrol engines).

Compression turbulence

In petrol engines this is promoted by arranging for the piston crown to be brought into very close proximity with a limited portion of the combustion chamber roof. By this means a squish area is formed, which again is a term that was first applied by H.R. Ricardo. The effect of this squish area is such that, as the piston approaches the end of its compression stroke, the air and fuel charge trapped in the diminishing space is forcibly ejected across the combustion chamber. In so doing it enters the enlarged and, in some cases, streamlined portion of the chamber, which imparts a turbulent motion to the charge (Figure 3.19a).

Compression turbulence in direct-injection diesel engines is similar in principle, except that the squish area is actually formed by the annular rim of the piston bowl, so that air is ejected radially inwards to the centre of the piston as it approaches the end of the compression stroke (Figure 3.19b). By virtue of the toroidal form of the piston bowl, the air that is forced to enter therein is set into turbulent motion and thereby assists the injected spray of fuel to seek out its oxygen supply.

Induction swirl

For both petrol and direct-injection diesel engines, induction swirl may be generated simply by arranging for the incoming charge to enter the cylinder in a tangential direction. An angled bias may therefore be adopted for the intake port in the cylinder head, so that the entering charge is deflected by the cylinder walls (Figure 3.20a). As a consequence of the angular motion imparted to the charge, a beneficial vortex (swirling) action is set up on the induction stroke and continues during the compression stroke. A characteristic known as the 'swirl ratio' is equal to the swirl speed divided by the engine speed with both being expressed in rev/min. Another method of promoting induction swirl for direct-injection diesel engines is to mask one side of the inlet valve, so that the incoming air is provided with a similar directional control (Figure 3.20b). An additional mechanical complication is introduced with this particular method, since the valve stem must be keyed against rotation. In the now established four-valve pentroof form of combustion chamber for petrol engines (Figure 3.22a and b), research into the burning of leaner mixtures by the Gaydon Technology company in the mid-1980's showed that on the induction stroke a beneficial swirling action can be created across the cylinder, this being termed barrel swirl or 'tumble' to distinguish it from the axial swirl described earlier (Figures 3.21a and b). A tumble motion is an important factor in the induction and compression events of the Mitsubishi Gasoline Direct Injection GDI engine (Section 6.9).

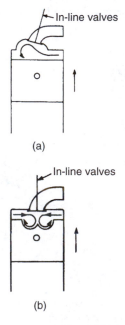

Figure 3.19   Compression turbulence in (a) petrol only (b) mainly diesel engines

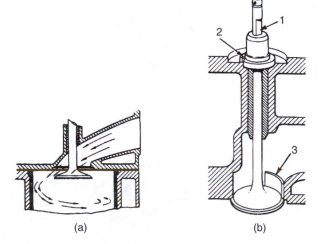

Figure 3.20   Induction swirl in petrol and diesel engines: (a) directed inlet port for petrol and diesel (b) shrouded inlet valve for diesel only.

1 flatted stem to prevent valve rotation against thimble key
2 thimble locating peg
3 shroud

The concept of induction swirl actually dates back to before World War I, when Dr K. Hesselmann of the Swedish Atlas Company first developed it for submarine diesel engines.

Compression swirl

This effect is present with indirect-injection diesel engines where a swirl combustion chamber is used. A characteristic

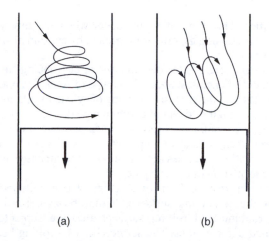

(a)   (b)

Figure 3.21   Identifying induction swirl patterns (a) axial (b) barrel

of this form of chamber is that during the compression stroke about three-quarters of the cylinder air is forced to enter it, via a communicating passage or throat (Figure 3.17b). The latter is relatively unobstructed in its tangential approach to the swirl chamber, which is of spherical form to promote rapid swirling of the air past the fuel sprayed from the injector nozzle. Hence, the oxygen in the air can seek out its fuel supply, which is of course the converse situation to that found with compression turbulence in direct-injection diesel engines.

Combustion swirl

Again applicable to indirect-injection diesel engines, this effect is present with the precombustion chamber type, which receives only about one-third of the cylinder air during the compression stroke (Figure 3.17a). As mentioned earlier, the precombustion chamber is approached through a relatively restricted passage, down towards which the fuel sprayed from the injector nozzle is directed. Ignition of the fuel in the precombustion chamber causes a sudden increase in pressure therein, which results in partially burned fuel and gas being forcibly ejected into the cylinder clearance space and setting up a rapid swirling motion for completion of the combustion process.

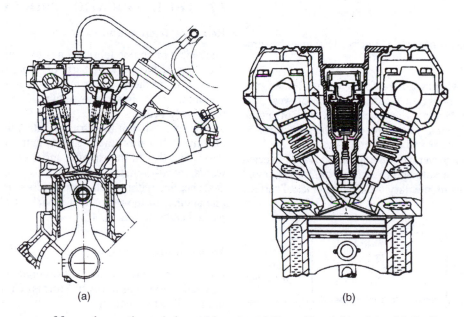

(a)   (b)

Figure 3.22   Arrangements of four-valve porting to induce (a) barrel swirl (Renault) and (b) axial swirl (Saab)

4   Engine lubrication

4.1   FRICTION AND WEAR

Basic considerations

The main factors that can reduce the performance and life expectancy of the engine and other parts of the motor vehicle are the closely related phenomena of friction and wear.

Friction can be described as the resistance that a material develops against sliding contact. When an apparently smooth surface is viewed through a powerful microscope, innumerable tiny projections and depressions are seen; these are termed surface *asperities*. They may be likened to a mountain chain, so that when two surfaces are rubbed together their action is analogous to that of two mountain chains sliding across one another peak-to-peak (Figure 4.1a).

Without lubrication this friction between the rubbing surfaces results in the removal or transfer of solid material from one to the other and constitutes what is known as wear (Figure 4.1b).

Forms of wear

Several different forms of wear can arise. They may be summarized as follows:

Abrasion   occurs whenever hard foreign particles are present between the rubbing surfaces. This type of wear was once a contributory factor in the excessive wearing of engine cylinder walls and piston rings, until it was better controlled by the efficient use of air and oil filters.

Adhesion   arises whenever the rubbing surfaces are subjected to extreme contact pressure and the frictional heat generated between them causes momentary welding together of surface asperities. This type of wear occurs when the tiny welds between opposing asperities are broken by continued sliding of the surfaces.

Corrosion   refers to the deterioration of the rubbing surfaces as a result of chemical attack. This type of wear was once the major factor in the excessive wearing of cylinder walls and piston rings, until the advent of chromium plating for the latter.

Fretting   is confined to rubbing surfaces where the motion between them is almost imperceptible and typically arises from vibration. It causes deterioration of the rubbing surfaces by the formation of corrosive debris.

Fatigue   takes the form of cracking that is initiated just below the contacting surfaces. It may be associated with both rubbing and rolling surfaces that are subjected to repeated deformations. The service life of a rolling bearing can be ultimately limited by this form of wear.

4.2   THE LUBRICATION PROCESS

Reducing friction and wear

The object of engine lubrication is to reduce friction, heating and wear of the working parts or, more precisely, wherever interacting surfaces are in relative motion. It is, of course, accomplished by introducing a film of lubricating oil between the various bearing surfaces to keep them apart, thereby providing a condition of fluid friction rather than dry friction. For effective lubrication the shear strength of this film must be lower than that of the rubbing surfaces it separates. In other words, metal-to-metal contact is avoided by the film of lubricant shearing within itself, as its layers of molecules slide over one another in the manner of a pack of playing cards being pushed endwise (Figure 4.2).

Forms of lubrication

Lubrication is a relatively well-ordered process which is recognized as existing in several forms. These are summarized in the following sections.

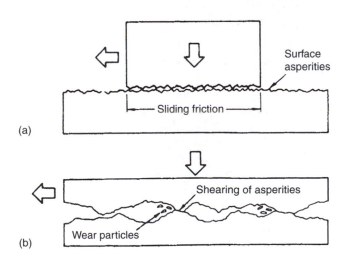

(a)

(b)

Figure 4.1   Sliding friction causing wear

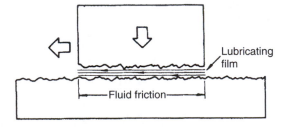

Figure 4.2   Lubrication preventing wear

Hydrodynamic lubrication

This occurs when the clearance condition between the load-carrying surfaces of a bearing favour the formation of a wedge-like film of oil (Figure 4.3). The moving surface of the bearing then acts as its own oil pump to crowd more oil into the converging clearance space occupied by the oil wedge, finally to establish a relatively thick film of oil throughout the entire bearing clearance. This process can occur not only in journal bearings with cylindrical surfaces, but also in slider bearings with plane surfaces. It should be realized that an appreciable oil film pressure can be generated in the engine bearings by hydrodynamic lubrication, which may be more than fifty times the oil supply pressure, or what is commonly termed the engine oil pressure. To maintain a condition of hydrodynamic lubrication it is, of course, essential that a sufficient quantity of oil is always supplied to the bearings by the engine lubrication system.

Elastohydrodynamic lubrication

This represents a modified form of hydrodynamic lubrication. It occurs under conditions of extremely high pressure between the moving surfaces, such as can arise between cams and followers, gear teeth and in rolling bearings where little more than a line or a point contact may exist to support the working load. The viscosity or resistance to flow of the oil trapped between such small interacting surfaces may then increase to such an extent that the oil film momentarily becomes more rigid than the metal surfaces themselves, which then elastically deform to allow the formation of an extended oil film and thus provide lubrication (Figure 4.4).

Boundary lubrication

This exists in a bearing when either the speed of the moving surface is too low for hydrodynamic lubrication, the load acting is too high, or there is an insufficient supply of oil. It therefore relies essentially upon the adhesion of oil molecules to the load-carrying surfaces, so that they are separated by a relatively thin film of lubricant (Figure 4.5). There is, however, always the danger that this thin film may be penetrated by surface asperities, the resulting metal-to-metal contact then leading to wear unless the oil has been fortified by suitable additives.

Hydrostatic lubrication

This can occur in the form of a squeeze film of oil in a bearing if the load momentarily reverses in direction and the speed of the moving surface is very low. Under these conditions the load-carrying surfaces are initially separated by a relatively thick film of oil, which resists being squeezed out from the diminishing bearing clearance during the short interval that the load reverses in direction (Figure 4.6). A cushioning film of oil can therefore be maintained and then restored to full thickness before the next load reversal occurs.

Mixed-film lubrication

A familiar example of a change from one state of lubrication to another occurs in the engine main and big-end plain journal bearings, which are subjected to a complex system of loading. These bearings experience partly hydrodynamic and partly hydrostatic lubrication during normal running, but develop boundary lubrication when starting and stopping (Figure 4.7).

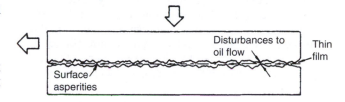

Figure 4.5   Boundary lubrication

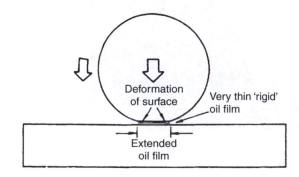

Figure 4.3   Hydrodynamic lubrication

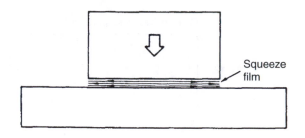

Figure 4.6   Hydrostatic lubrication

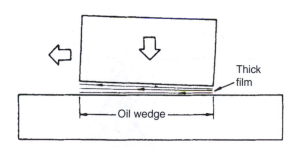

Figure 4.4   Elastohydrodynamic lubrication

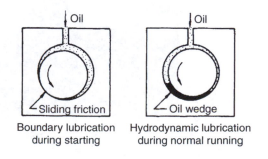

Figure 4.7   Mixed-film lubrication in journal bearing

During the latter stages of operation the shaft journal attempts to climb the side of the bearing wall, either prior to the generation of an oil wedge or after its collapse, so it is under these conditions of starting and stopping that frictional wear of engine bearings is most likely to occur. The transition from one form of lubrication to another is generally referred to as a state of mixed-film lubrication.

Oil film thickness in engine bearings

From our simple explanation of hydrodynamic lubrication, it might appear that so long as a film of oil exists between the bearing surfaces, it does not greatly matter how thick it is. In reality, however, the minimum oil film thickness must be such as to accommodate less than ideal operating conditions, which may exist in a bearing. These can include minute deflections under load, slight distortions arising from thermal effects and insufficient smoothness of the bearing journal. In the latter case, it must be recognized that the highest load capacity of the bearing is developed when the thickness of the oil film is least.

Although the prediction of oil film thickness in a hypothetical bearing, based on the theory of hydrodynamic lubrication proposed by Osborne Reynolds in 1886 (Beauchamp Tower had discovered the practice a few years before), has long been a mathematician's delight, the automotive service engineer is more likely to be curious about the variations in oil film thickness that occur in a real-life engine bearing. By using advanced technology instrumentation, the Castrol oil company can in fact measure the variations in oil film thickness in a big-end bearing with the engine running (Figure 4.8). Castrol refer to this as their BOFT (bearing oil film thickness) test, which forms part of their extensive research programme

to demonstrate the ability of an engine oil to provide effective protection and smooth running power output.

4.3   ENGINE LUBRICATING OILS

Mineral oils

Lubricants can be produced from animal, vegetable and mineral oils; it is the last of these, refined from petroleum and treated with additives, that are suitable for the lubrication of motor vehicle engines. Petroleum is generally thought to have been formed by the decomposition of tiny marine organisms that died, sank to the bottom of the sea and became mixed with sediment. Over many millions of years the accumulating layers of sediment subjected the trapped organic matter to immense pressure and heat, which converted its softer parts into droplets of crude oil mixed with natural gas.

Crude oils are essentially hydrocarbons – compounds whose molecules comprise hydrogen and carbon atoms. The basic types of hydrocarbon found in crude oils are classified as paraffins, naphthenes and aromatics. Oils of predominantly paraffinic type are preferred for the production of automobile engine oils, since they are best suited to the extremes of operating temperature involved. Conventional engine oils therefore consist largely of mixtures of hydrocarbons, but also require treatment with certain carefully chosen chemicals that are known as 'additives'.

Synthetic oils

Although the so-called synthetic oils that do not have a lubricant base of mineral oil form have been developed since the 1920s – indeed, a significant proportion of the lubricants

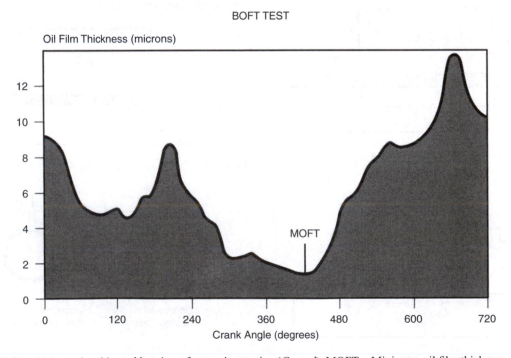

Figure 4.8   Oil film thickness in a big-end bearing of a running engine (*Castrol*). MOFT – Minimum oil film thickness

used by the German armed forces during World War II were of synthetic origin – it has only been in recent years that oils of this type have become available, chiefly to meet the more severe demands of high-performance vehicle engines, albeit at a higher cost than conventional mineral oils. They are generally classified as either 'synthetic-based' or 'full-synthetic' oils, the former containing at least 10 per cent synthetic base stock. A chemical compound that has been found suitable for this purpose after processing is polyalpha-olefin (PAO), which may be classed as a synthetic hydrocarbon base stock. It is produced by polymerization (the combination of smaller molecules to form large ones) of ethane gas, to formulate an engine oil with particularly favourable viscosity-temperature characteristics. Such a product is sometimes described by the oil technologist as synthetically engineered polyalphaolefin.

It is now generally recognized that some synthetic oils will out-perform mineral oils in low-temperature applications and that most also possess better volatility characteristics than mineral oils with equivalent viscosities. In the latter respect this means that thinner oils may be used without necessarily incurring a higher consumption. Other perceived benefits generally depend upon additive composition, as is the case with mineral oils.

Functions and properties of engine oil

The importance of using the correct type and grade of engine oil will be better appreciated if first we consider the essential functions it must perform. These may be summarized as follows:

Lubrication The *viscosity* of the oil is its single most important property, since it represents the internal friction in the film of oil and is caused by the resistance to shearing motion of the oil molecules. More simply, it may be regarded as the resistance to flow of an oil, and it is this property that makes it possible to separate the bearing surfaces in relative motion. To cater for the more difficult conditions of lubrication, such as may arise between the highly stressed cams and tappets, extreme pressure (EP) additives, can be incorporated in the oil. These improve its oiliness by forming a chemical film between the rubbing surfaces.

Entrainment Under less than perfect conditions of lubrication, the flow of oil through the bearings of an engine serves to capture and remove abrasive particles, which otherwise would remain and increase friction and wear.

Sealing To assist in sealing against the leakage or blow-by of the cylinder gases, which is known to cause camshaft wear in some modern overhead camshaft engines, the engine oil must be physically capable of filling the minute leakage paths and surface irregularities of the cylinder bores, pistons and rings. It must therefore possess adequate viscosity stability because of the high-temperature conditions to which it is exposed.

Cooling The conditions under which the oil functions in a cooling capacity generally tend to promote oxidation. This is because the oil is exposed not only to heating – the temperature in the top compression ring groove of the piston may reach 250°C, while even that in the crankcase is typically around 125°C – and agitation, but also to the effects of contact with metals and contaminants. Therefore, to avoid the formation of gummy deposits and piston ring sticking, engine oils are treated with anti-oxidant additives.

Cleaning The lubricating oil must also contribute to the internal cleanliness of the engine, and it is therefore treated with so-called detergent additives. These can be more accurately described as dispersants because they retain insoluble contaminants in fine suspension, which are removed when the engine oil is changed. Similarly, the corrosive action of acidic products that can contaminate the oil may be neutralized by treating it with anti-corrosion additives. A phenomenon of more recent origin is that known as 'black sludge', which manifests itself as a thick tar-like deposit that forms rippling layers in the cooler areas of the engine, especially inside the valve cover, but can flake off to cause obstructions in the lubrication system and at worst can lead to serious engine damage. It first began to appear during the late 1980s in engines designed to minimize harmful exhaust emissions, and has been attributed to a combination of factors that include engine design, operating conditions, fuel quality, engine oil quality and oil change intervals.

Compatibility The exhaust system catalytic converter must be protected against phosphorus induced damage, so to keep the catalyst working cleaner for longer the phosphorus content of the oil must be minimized.

Environmental To reduce the risk of harmful exhaust emissions and also to make for easier and safer disposal and recycling, the amount of chlorine in the oil formulation must also be minimized.

Engine oil additives

Additives are usually chemical compounds of non-petroleum origin, and normally only very small amounts are needed to achieve the desired effect. Their purpose is to confer a service life of acceptable economic length, either by improving a property already possessed by an oil, or by introducing a new characteristic. The different types of engine oil additive and their reason for use, as concisely listed by the Caltex Petroleum Corporation, are as follows:

Viscosity index improvers Reduce viscosity change with temperature; permit reduced fuel consumption; maintain low oil consumption; allow easy cold starting.

Detergents, dispersants Keep sludge, carbon and other deposit-forming material suspended in the oil for removal from the engine with drains.

Alkaline compounds Neutralize acids; prevent corrosion from acid attack.

Antiwear, friction modifiers Form protective films on engine parts; reduce wear, prevent galling and seizing; reduce fuel consumption.

Oxidation inhibitors Prevent or control oxidation of oil, formation of varnish, sludge and corrosive organic compounds; limit viscosity increase that occurs as oil mileage increases.

Rust inhibitors Prevent rust on metal surfaces by forming protective surface films or by neutralizing acids.

Pour point depressants Lower 'freezing' point of oil, assuring free flow at low temperatures.

Antifoam agents Reduced foam in the crankcase.

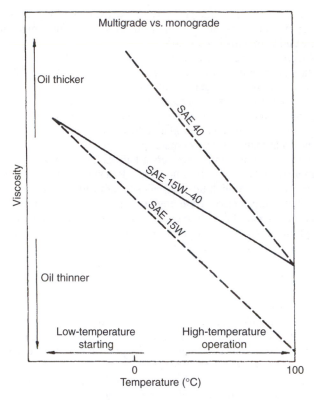

Figure 4.9 Engine oil viscosity change with temperature

Classification of engine oils by viscosity

In the mid 1920s the Society of Automotive Engineers (SAE) in America recommended an arbitrary classification system for engine oils in terms of viscosity only. This system, which has been revised from time to time, has long since been internationally recognized. It established a numerical relationship, defined by SAE numbers, between the viscosities of oils, and specifies minimum and maximum viscosities at stated temperature for an oil to meet a particular grade. In practice this means that the higher the grade number, the higher the viscosity of the oil. On this basis there are now eleven SAE viscosity grades for engine oils, these being SAE 20, 30, 40, 50 and 60 grades, which are intended for normal temperature conditions, while suffix W grades in the range SAE 0W, 5W, 10W, 15W, 20W and 25W are intended primarily for low-temperature conditions.

In common with other liquids, an increase in the temperature of engine oil will cause a decrease in its viscosity (Figure 4.9). As this effect could be quite pronounced with engine oils, which are required to function over a very wide temperature range, they are treated with the earlier mentioned chemical additives known as viscosity index (VI) improvers. The term *viscosity index* expresses the degree of variation of viscosity with temperature, so the higher its value the less will be the change in viscosity of the oil with temperature.

The development of special viscosity index improvers resulted in the introduction of multigrade oils in the early 1950s. These oils are formulated to meet the viscosity requirements of more than one SAE grade and are therefore classified in double SAE numbers, such as 15W–40. The advantages

originally claimed for multigrade oils included easy starting of the engine from cold with minimum oil consumption when hot.

With regard to cold starting, there is a minimum cranking speed below which an engine will not start; this speed is higher for a diesel engine than it is for a petrol engine. A manufacturer therefore determines what is termed the viscosity requirement of an engine; this is the highest value of the sump oil viscosity above which the cranking speed does not exceed the minimum starting speed. Such tests are, of course, related to the cold starting temperatures likely to be encountered in those parts of the world where the car is sold.

Classification of engine oils by performance

In 1947 the American Petroleum Institute (API) introduced a service classification system based on the types and amount of additives used in an engine oil. This was followed in 1952 by a second classification system based on the severity of engine operation. Then in 1970 the API in conjunction with the Society of Automotive Engineers (SAE) and the American Society for Testing and Materials (ASTM) established what are known as 'API engine service categories' to define the overall capability of an engine oil to protect against wear, deposits and self-deterioration. The ability of an engine oil to perform satisfactorily in one or more categories identified by the API is determined by engine tests developed by the ASTM, and the resulting grading system is called 'SAE J183 engine oil performance and engine service classification'. This system can be updated as the need for new categories arises. For petrol engines the existing API service categories and oil descriptions are as follows:

SA	non-additive
SB	anti-oxidant, anti-scuff but non-detergent
SC	protection against high- and low-temperature deposits, wear, rust and corrosion (1964–67)
SD	generally higher levels of protection than SC (1968–70)
SE	high-temperature protection more demanding than SD (1971–72)
SF	improved anti-wear and high-temperature protection again increased (1980–88)
SG	provides improved control of engine deposits such as black sludge, oil oxidation and engine wear relative to oil developed for previous categories; these oils also give enhanced protection against rust and corrosion (adopted 1989)
SH	technically similar to above category, but the validity testing of an oil formulation must now be conducted in accordance with the Chemical Manufacturers Association (CMA) code of practice, which excludes self-certifying a product (adopted 1994)
SJ	describes an engine oil that is subject to more tests than the previous category, and suitable for passenger cars and light commercial vehicles operating under vehicle manufacturers' recommended maintenance procedures (adopted 1996)
SL	similarly describes an engine oil for use in service typical of present and earlier passenger cars, sports utility vehicles and light commercial vehicles. It provides a further advance in tested performance

levels, including improvements relating to oil consumption, engine cleanliness and oil change intervals (adopted 2001)

And for diesel engines:

CA	light duty
CB	moderate duty
CC	moderate to severe duty
CD	severe duty
CE	extra heavy duty for engines manufactured since 1983
CF	introduced in 1994 and intended for service typical of indirect-injection engines.
CF-4	provides improved control of oil consumption and piston cleanliness (adopted 1990)
CG-4	describes an engine oil suitable for severe-duty diesel engine service and designed to meet US 1994 exhaust emission standards
CH-4	superior in performance to service categories API CF-4 and API CG-4, and suitable for engines designed to meet US 1998 exhaust emission standards
CI-4	provides a further advance in severe-duty performance levels, especially in relation to engine durability where exhaust gas recirculation is used to meet US 2004 emission limits (adopted 2002)

(Suffix '4' denotes four-stroke diesel engine categories)

The distinction between rusting and corrosion is that *rusting* occurs from the presence of moisture on those ferrous parts not in continuous contact with the lubricating oil, while *corrosion* arises from the oil becoming contaminated with the acidic products of combustion and attacking various structural and bearing metals.

Since the early 1960s it has been recognized that European requirements for engine oil performance can in certain respects differ from those specified primarily for American operating conditions. Therefore the Committee of Common Market (Automobile) Constructors (CCMC) incorporated additional tests to define the minimum quality for lubricants acceptable to its members. In 1996 the CCMC was superseded by the Association of European Automobile Manufacturers, abbreviated from the French to ACEA. As a result a new classification system for oil performance standards was introduced, which defines the minimum quality level for service fill oils. The system is similar to that of the American Petroleum Institute or API, as it is based on a schedule of physical, chemical and engine tests. A significant difference between the earlier CCMC and current ACEA test sequences is that in the latter case all performance claims must be supported by data generated under the European Engine Lubricant Quality Management System (EELQMS). The ACEA categories or sequences are prefixed by the letters A, B or E to denote petrol, light duty and heavy duty diesel engines respectively, followed by numbers that currently range from 1 to 7 to rate ascending oil performance, and suffixed by the year of issue. Taking a few examples: ACEA A3-02, ACEA B4-02 and ACEA E5-02 category oils relate to recent petrol-engined passenger cars, diesel-engined passenger cars and diesel-engined commercial vehicles. To meet the more demanding 'Euro-4' exhaust emission limits for heavy duty diesel engines (Section 11.3), ACEA categories E6

and E7 have now been introduced; the former to reduce ash accumulation and make the oil compatible with exhaust after-treatment systems, and the latter to minimize abrasive wear of the engine parts including the valve train.

In briefly summarizing the classification and requirements for engine oils, it has to be stressed that a service engineer must always consult the engine manufacturer's recommendations, to ensure that the correct grade and type of oil is used at oil change intervals.

Oil sump capacity and oil level

The amount of oil carried in the sump necessarily represents a compromise, since too great a capacity would hinder warm-up of the engine, while too small a capacity would hasten its contamination. In practice, sump oil capacities generally range from about 3.5 to 7.0 litres (6 to 12 pints) for passenger car engines and in the region of four times these quantities for heavy-vehicle diesel engines. A simple dipstick is all that has traditionally been provided to indicate the minimum and maximum oil levels on most vehicles, although in the past an oil level indicator gauge on the instrument panel was electrically connected to a float unit fitted in the sump on some high-grade passenger cars. In a modern electronically controlled version of an instrument panel low oil level warning system, actuation is by two thermal sensors housed in the sump. One sensor is heated electrically to a temperature higher than that of the oil, while the other sensor continues to register the actual temperature of the oil. When the former sensor is allowed to cool down to the temperature of the latter, a certain time must elapse that will be influenced by the extent of its exposure above a less than required oil level. The actual level of oil in the sump can then be electronically computed based on the signals received from the two sensors.

Too high or too low an oil level will have damaging effects, as indicated in the following sections.

Oil level too high

If the rotating crankshaft approaches too close to the surface of the oil, then the windage it creates can produce foaming of the oil, so that air enters the engine lubrication system and causes inconsistent oil pressure. Should the big ends of the connecting rods actually dip into the oil, then both its temperature and consumption will increase. The latter is because the amount of oil splashed on to the cylinder walls could exceed the controlling ability of the piston rings, so that some of it enters the combustion chambers and is burnt.

As an interesting aside, it was once reported that in oil consumption tests on a fleet of Paris taxis, the average oil consumption was greater for those vehicles where the operator added oil more often and nearer to the dipstick maximum mark, than for those where the operator added oil less often and below the maximum mark.

Oil level too low

This is potentially a more dangerous state of affairs, since if the oil level falls completely below the intake of the oil pump, then obviously no oil will reach the engine working parts, with disastrous results. Even if the oil level has not fallen seriously below the maximum mark, engine oil pressure can still become erratic with any surging of the oil, caused by the changing motion of the vehicle. Furthermore, not only will a reduced amount of oil in circulation be less able to prevent engine deposits, but also the rate at which the oil deteriorates will be increased owing to its reaching higher temperatures.

Oil deterioration and changing

There are two main causes of engine oil deterioration in service: these are oxidation and contamination, as already mentioned. Suffice it here to summarize matters as follows:

Oxidation
Occurs in the hottest parts of the engine
Thickens the oil and makes cold starting difficult
Reduces oil flow until the engine has warmed up
Produces lacquer deposits that can obstruct fine clearances
Corrodes the internal surfaces of the engine.

Contaminants
Are formed of soot and water from combustion
Are composed of lead compounds from fuel additives
Include abrasive particles admitted through air intake
Tend to clog the oil filtering arrangements
Form lacquers independent of those from oxidation.

The reason for periodically changing the engine oil is therefore not because it simply wears out, but as a result of the oxidation and contamination accompanying the depletion of some (not all) of its additives. These are, of course, gradually destroyed in the process of doing their intended job, which in modern oils they do exceedingly well.

Either the depletion of additives, or the actual engine condition, forms the basis upon which the vehicle manufacturer will specify the oil and filter change interval in service, which may also depend upon the annual mileage and whether the vehicle is subjected, for example, to a fair amount of stop-start driving. Also, a longer than average recommended interval between oil changes may apply only if a high-quality synthetic oil is used. For passenger cars the current practice is typically to change the oil in naturally aspirated petrol engines at intervals ranging from 6000 to 12 000 miles (10 000 to 20 000 km) and for turbocharged versions from 3000 to 6000 miles (5000 to 10 000 km). With passenger car diesel engines the intervals are typically in the range from 5000 to 7500 miles (8000 to 12 000 km). In the case of turbocharged diesel engines for heavy vehicles the intervals typically range from 18 000 to 36 000 miles (30 000 to 60 000 km). It is, of course, always necessary to establish the recommended oil change intervals from the vehicle manufacturer's own service schedule.

Since frequent and prolonged contact with used engine oils may cause serious skin disorders, the Health & Safety Executive have issued precautionary guidance in their publication HS(G)67 titled 'Health & Safety in MOTOR VEHICLE REPAIR'.

Excessive oil consumption

Service complaints of excessive oil consumption may be considered under the following general headings.

Oil leakage
External oil leakage may arise at various locations around the engine unit. It is usually caused by a defective sealing gasket, a discontinuity in a vulcanized sealant joint, an ill-fitting blanking plug, a faulty connection in external pipework, or worn oil seals.

Oil burning
This can be attributed to excess oil reaching the combustion chambers. Its possible sources include worn inlet valve guides and stem seals, combined wear in the cylinders, pistons and rings, or increased oil fling-off from worn bearings.

Oil vaporization
This is concentrated in the high temperature regions of the piston rings and valve guides. It is therefore important to use a high quality oil that maintains its thermal stability at elevated temperatures.

The very fine production clearances built into modern engines, together with the closely controlled formulation of their oils, have greatly contributed to a reduction in complaints of this nature, as compared to earlier service experience.

4.4 ENGINE LUBRICATION SYSTEMS

Wet sump system

A splash type of lubrication system was used for the engines of early motor vehicles. Lubrication of the working parts was initiated simply by allowing the connecting rod big ends to dip into a trough of oil in the crankcase each time they passed through their BDC positions. To improve the lubrication of their bearings, the big ends were often provided with oil scoops (Figure 4.10). Surplus oil that was flung upwards into the crankcase was then collected in galleries from whence it gravitated to the main bearings, camshaft bearings and timing drive. The cylinder walls were similarly lubricated by the oil that was splashed on to them.

Following the purely splash system of engine lubrication came the combined pressure-and-splash or semi-forced system. The essential features of this system were that an engine-driven oil pump delivered oil only to the main bearings; the big-end bearings and other working parts were lubricated in the manner of the purely splash system. An attraction

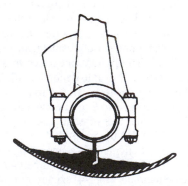

Figure 4.10 Early splash system of engine lubrication

of this system was that no oilway had to be drilled in the crankshaft, as necessary with the fully forced system described below.

The eventual necessity to provide the later high-speed engine with a fully forced lubrication system, arose from the ever-increasing loads and speeds at which the bearings were expected to operate. This meant that the bearings had to be not only lubricated, but also cooled, by the oil circulating through them. We therefore arrive at the long-established fully forced lubrication system, in which oil under pressure is supplied to the crankshaft main bearings, connecting rod big-end bearings and camshaft bearings. Strictly speaking, a fully-forced lubrication system originally included the small-end bearing bushes, which were supplied with oil under pressure from the big-end bearings through drilled connecting rods. The first fully forced lubrication system for motor vehicle engines is generally attributed, like so many features of automobile practice, to the early work of Dr F. Lanchester and was found in the cars that once bore his name. Even so, the basic idea of forced-feed lubrication to all bearings was applied as long ago as 1890 in the Bellis and Morcom steam engine.

With the fully forced oil circulation system, a major contribution to cooling of the bearings is therefore made by passing a large quantity of oil through them, in excess of the amount needed for lubrication alone.

Oil circuit in the wet sump system

This consists of a network of oilways embodied in both the structure and the mechanism of the engine, the extent and complexity of this network being mainly influenced by the number and arrangement of the engine cylinders and their valve train (Figures 4.11, 4.13 and 4.14a). A more convenient representation of an engine lubrication system may be made with the aid of a block diagram (Figures 4.12 and 4.14b). Extensive use was once made of separate pipes and connecting

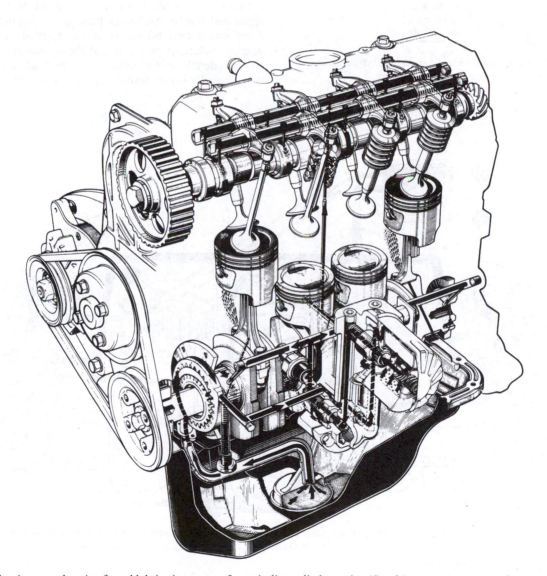

Figure 4.11   A comprehensive forced lubrication system for an in-line cylinder engine (*Suzuki*)

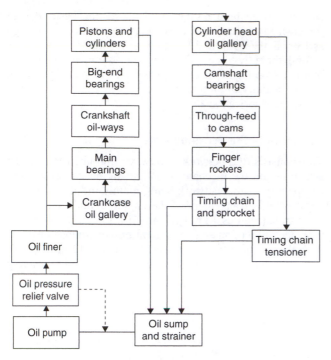

Figure 4.12 Block lubrication diagram for a typical in-line cylinder engine with single overhead camshaft

T-pieces to convey the oil to the various working parts, but since fracturing and loosening of such piping was always a possible source of unreliability, drilled pressure conduits have long since been preferred.

In operation, the engine-driven pump delivers oil from the sump and upwards through a duct in its body, whence it passes into a full-flow filter mounted on the outside of the crankcase. From the filter the oil is directed to either a single or a pair of main oil galleries which extend practically the full length of the crankcase at about the level of the cylinder skirts. A single main gallery is incorporated in the crankcase side wall of inline cylinder engines, while a pair of centrally disposed galleries may be required in the crankcase of some V and horizontally-opposed cylinder engines. The purpose of the main gallery is to distribute the oil to the principal working parts of the engine.

Of major importance is the oil supply to the crankshaft main bearings and their dependent connecting rod bearings. Oil leaving the engine main supply gallery is therefore normally routed directly to the main bearings through communicating ducts in the crankcase webs and main bearing housings. From these ducts the oil passes through an inlet hole in the upper half-liner of each main bearing. At this point, the oil flows into a circumferential groove formed in the actual bearing surface. A proportion of the oil leaving the groove spreads laterally to lubricate and cool the main bearing, whilst the remainder is transferred to perform a similar duty in an

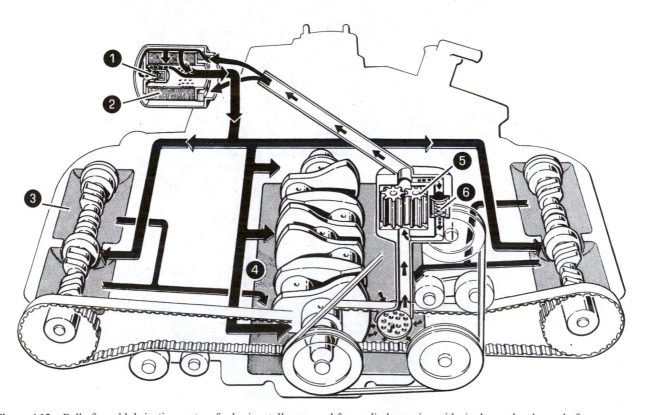

Figure 4.13 Fully forced lubrication system for horizontally opposed four cylinder engine with single overhead camshafts (*Alfa Romeo*)

1 By-pass valve   4 Oil sump
2 Oil filter element   5 Gear pump
3 Camshaft housing   6 Pressure relief valve

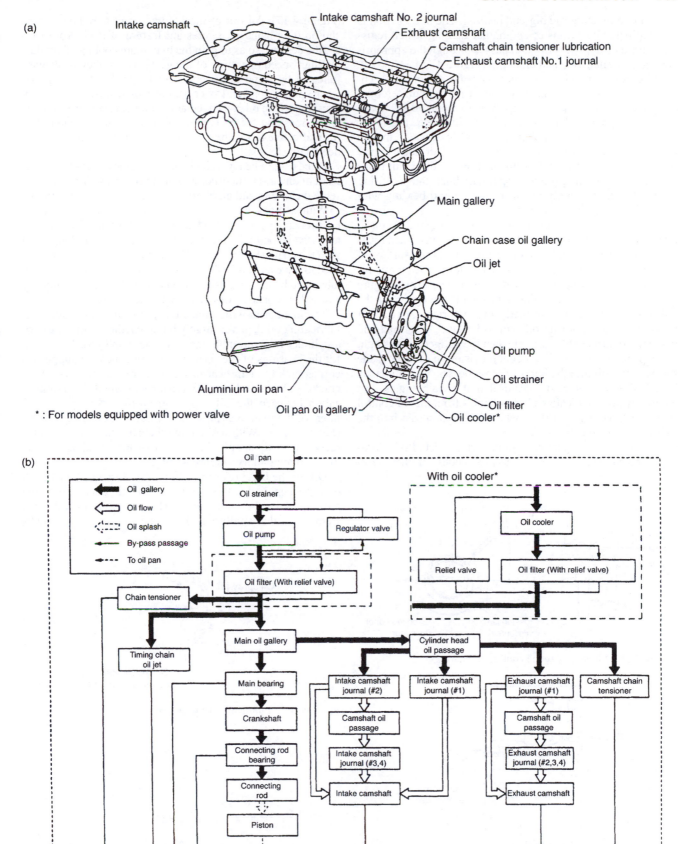

Figure 4.14 Fully forced lubrication system for modern V6 cylinder engine with double overhead camshafts: (a) schematic (b) block diagram (*Nissan*)

adjacent connecting rod big-end bearing. The latter action is accomplished by means of an angular drilling which registers with the supply groove and passes through the appropriate main journal, web and crankpin (Figure 4.15).

Oil grooving in a bearing liner is always a necessary evil. For some high-performance engines a part-circumferential oil supply groove may be employed to improve the load-carrying capacity of the bearings. In this case, the lower half-liner of each bearing is made grooveless, which in turn may require the main journals to be cross-drilled (Figure 4.15b). Hence, either end of this drilling is always in communication with the grooved upper half-liner, thus ensuring an uninterrupted supply of oil to the big-end bearing with each full turn of the crankshaft.

The connecting rod small-end bearing is more difficult to lubricate effectively, because of its oscillating rather than rotating motion. To assist lubrication, the upper bearing surface of a bushed small-end may be grooved to form a reservoir for collecting oil entering through a drilling in the rod eye (Figure 4.16). With the wider use of the semi-floating gudgeon pin the means of its lubrication assumed more importance, because it became necessary for an oil film to be quickly formed and maintained during cold starting when the gudgeon pin is tight in the piston bosses. Various detail provisions have been made for directing oil into the bores of these bosses, which include broached axial slots and holes drilled at certain angles. The latter may communicate with drain holes from the oil control ring groove (Figure 4.16), or with an oil scoop cast into the four corners of the piston skirt.

For lubrication of gear-type timing drives a jet of oil is usually sprayed directly at the in-to-mesh point of their teeth (Figure 4.17). An out-of-mesh spray point may however be chosen (this being an interesting feature of Rolls-Royce V8 engine design), so as to gain the maximum cooling effect during high-speed operation of the gears. This policy is likewise followed by some designers of high-speed industrial gears, where power loss and over-heating could arise from heavy churning of the oil if sprayed at the immediate approach zone to the meshing teeth. The actual amount of oil required to form effective lubricating films between gear teeth is in practice relatively small.

In the case of heavy-vehicle diesel engines an extended network of oilways is incorporated in the crankcase front cover to serve the timing and auxiliary gear drives and their bearings.

With a crankcase-mounted camshaft, driven by either a roller or an inverted-tooth chain, an oil jet is typically arranged to discharge on to the crankshaft sprocket at its meshing point with the chain (Figure 4.18). Where the roller type of chain is used to drive a cylinder-head-mounted camshaft, it is typically lubricated by the controlled leakage of oil from a hydraulically assisted tensioner, as described in Section 1.11.

The manner in which the camshaft bearings receive their lubricating oil depends largely upon whether a crankcase- or a cylinder-head-mounted camshaft is employed. It is usual for the bearings of the former to receive their oil supply via corresponding main bearings. Ducts are incorporated in the crankcase webs and bearing housings to establish communication between the oil supply grooves in the upper half-liners of the main bearings and the camshaft bearing bushes (Figure 4.19a). With a V engine the camshaft bearings may receive their oil supply through cross-ducts leading directly off the main gallery (Figure 4.19b).

Oil for the bearings of a cylinder-head-mounted camshaft may be supplied either directly through ducts extending upwards from the main gallery (Figure 4.20a) or indirectly from a subsidiary gallery in the cylinder head, or similarly by

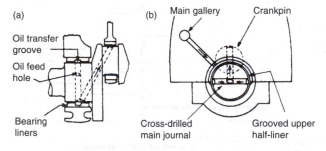

Figure 4.15   Oil supply to the crankshaft bearings

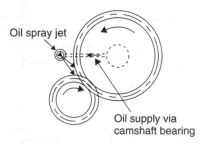

Figure 4.17   Lubrication of gear timing drives

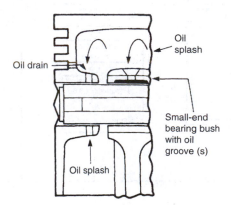

Figure 4.16   Lubrication of small-end bearings

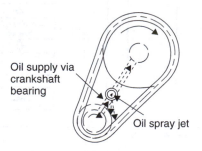

Figure 4.18   Lubrication of chain for crankcase-mounted camshaft

an end-to-end feed from within the camshaft itself (Figure 4.20b). For the now widely used double overhead camshafts, separate oil passages are generally provided to feed the bearings for each camshaft (Figure 4.14). It is usual for the oil supply to overhead camshaft bearings to be restricted, so as to avoid an increase in oil consumption through flooding of the valve mechanism. A further consideration relates to the use of hydraulic tappets in some double overhead camshaft engines, since it must be arranged that oil does not drain away from their supply galleries in the cylinder head when the engine is not running (Section 1.10). To prevent this happening an upper oil retaining valve may be incorporated in the system, which is also provided with a vent hole to facilitate the release of any trapped air when the engine is restarted.

Proper lubrication of the highly stressed cam and tappet contact faces is vitally important. For a crankcase-mounted camshaft, the cams and tappets may be lubricated partly from the oil being flung off from the connecting rod big-end bearings, and partly by the oil draining down the push-rods from the overhead valve rocker mechanism (Figure 4.21a). Alternatively, the cams and tappets may be provided with an oil bath instead of a splash system of lubrication. The cams then dip into a crankcase trough situated beneath the camshaft (Figure 4.21b). Oil escaping from the camshaft bearings constantly replenishes this trough, its level being controlled by a series of overflow holes, the surplus oil draining back to the sump.

For engines with cylinder-head-mounted camshaft(s), the cams and tappets may similarly be oil bath lubricated, the trough being formed either in a detachable camshaft and tappet carrier, or directly in the top deck of the cylinder head. In other designs, a splash system of lubrication is employed for the cams and tappets. Here the subsidiary oil gallery supplying the camshaft bearings may incorporate squirt holes, so that oil is sprayed on to the cams (Figure 4.22a). This function may also be performed by an external gallery pipe, which extends alongside the camshaft and receives a restricted supply of oil from the main gallery. Where an end-to-end oil feed is adopted for overhead camshaft bearings, the cams and tappets are usually lubricated through additional holes drilled radially inwards from the base circle of each cam to the interior passage of the shaft (Figure 4.22b).

Lubrication of the cylinder walls and piston skirts may be effected either wholly, or partly, by oil being flung off from the big-end bearings of the connecting rods. The cylinders in modern engines of high bore-to-stroke ratio tend to receive more oil from this source, since they are both larger in diameter and closer to the crankshaft.

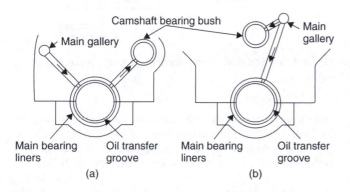

Figure 4.19 Typical oil supply arrangements for crankcase-mounted camshafts

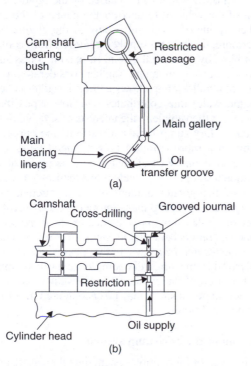

Figure 4.20 Typical oil supply arrangements for cylinder-head-mounted camshafts

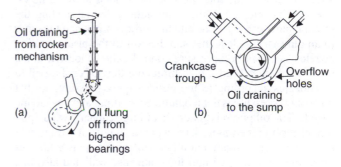

Figure 4.21 Typical lubrication arrangements for crankcase-mounted cams and tappets

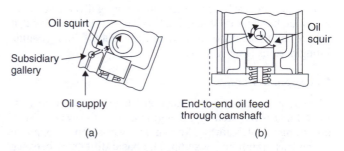

Figure 4.22 Typical lubrication arrangements for cylinder-head-mounted cams and tappets

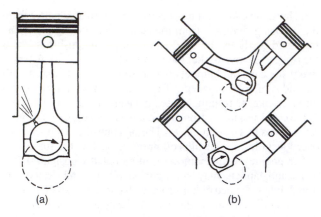

Figure 4.23   Oil spray lubrication of the cylinder walls

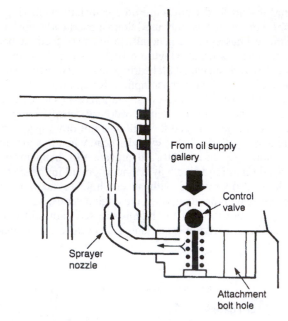

Figure 4.24   Schematic arrangement of oil spray cooling for pistons

In some engines provision is made for the cylinder wall to receive additional lubrication by a fine oil spray, which is directed from a squirt hole drilled in the shoulder of the connecting rod and big-end bearing. The oil spray is usually, but not invariably, directed at the major thrust face of the cylinder (Figure 4.23a). With a V engine it is normal for the oil to be sprayed from the connecting rod of the corresponding cylinder in the opposite bank (Figure 4.23b), so the oil first reaches the upper sides of the inclined cylinders and then gravitates to their lower sides.

Reference has earlier been made to the oil cooling of pistons (Section 2.4), which can lower critical piston temperatures by up to about 30°C. For heavy-vehicle diesel engines it is usually arranged for the separately cooled and then filtered oil leaving the filter head to be divided into two flows; one flow is directed to the main gallery for lubricating the moving parts of the engine and the turbocharger, and the other to an auxiliary gallery that supplies oil to the piston cooling nozzles. A second oil pressure control valve is located in the base of the filter head, which limits the flow of cooling oil to the nozzles according to the pump supply pressure, so that normal oil pressure is not jeopardized at idling and low engine speeds. The oil pump is generally of large capacity where piston oil cooling is employed. On a practical note, it is important to ensure correct rotational alignment of the sprayer nozzles when they are inserted into the crankcase wall and also that their O-ring seals are not twisted in the grooves. For lighter-duty diesel engines such as used in passenger cars, individual nozzle blocks may be bolted to the underside of the cylinder block and directly beneath their oil supply gallery. Each nozzle block has a spring-loaded ball valve at its entrance, so that oil cooling to the piston is automatically cut off during engine idling periods (Figure 4.24).

Dry sump system

The sump was described in Section 1.6 as a reservoir to store the oil that is required by the engine lubrication system. Such an arrangement is further known as a wet sump, as opposed to the less common dry sump. The basic difference between them is the provision of a separate oil tank and the addition of a crankcase scavenge oil pump for the dry sump system. In other respects the engine lubrication arrangements are similar for both systems. The various types of vehicle in which the dry sump system of engine lubrication may be found, together with its related advantages, may be summarized as follows:

Racing cars   The principal advantage here is that even should the oil level in the separate tank become low, it is impossible for surging of the oil content (caused by the rapidly changing motion of the vehicle) to uncover the intake of the oil pump and thereby interrupt the oil supply to the engine bearings. Furthermore, the bulk oil supply is removed from contact with the hot blow-by gases in the crankcase, so it is less subject to oxidation and contamination. Another consideration is that the absence of a bulky sump allows a lower installation height for the engine, which thus contributes to the lowest possible centre of gravity for the vehicle in the interests of safe handling.

Rally cars   Dry sump lubrication affords the opportunity of carrying, in a more compact boot-mounted tank, an appreciably larger quantity of oil than would be practicable in the conventional sump. Hence, there is less tendency towards overheating of the greater amount of oil in circulation. Other advantages are similar to those for racing cars, except that the absence of a bulky sump is welcomed for the increased ground clearance it confers beneath the engine over rough roads.

Other vehicles and tractors   In these cases, the main benefit to be derived from using a dry sump system is to maintain normal lubrication of the engine when the vehicle is negotiating very steep slopes, such as may be encountered in off-the-road operating conditions.

Operation of the dry sump system

With this type of lubrication system the oil pressure and scavenge pumps are engine driven, either separately at different speeds – the scavenge pump at a higher speed for the reason

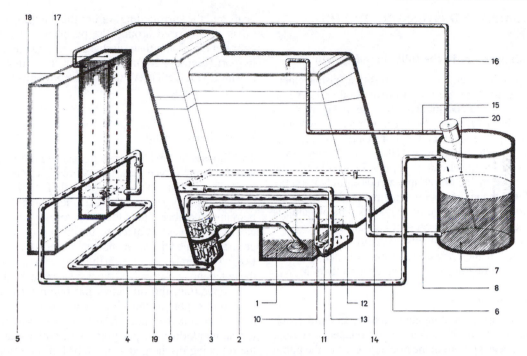

Figure 4.25 Diagrammatic layout of a dry sump lubrication system (*Mercedes-Benz*).

1 oil pan
2 suction pipe
3 return pump
4 oil line from return pump to oil cooler
5 by pass valve in oil cooler
6 oil line from oil cooler to oil tank
7 oil tank

8 oil line from oil tank to pressure pump
9 pressure pump
10 oil line from pressure pump to oil filter
11 oil relief valve in oil filter
12 oil filter element
13 oil line from oil filter to engine
14 pressure retaining valve in oil main duct

15 vent line from oil tank to cylinder head cover
16 vent line from oil cooler to oil tank
17 oil cooler
18 radiator
19 main oil duct
20 oil dip stick

mentioned later – or together in tandem. If they are driven in tandem the pressure pump is sealed from the scavenge pump below it. In operation, the pressure pump receives oil from the separately mounted tank and distributes it to the working parts of the engine in a manner similar to that of the wet sump lubrication system.

The oil returning to the base of the crankcase is then collected by the scavenge or return pump and transferred back to the tank. An important consideration here is that the pumping capacity of the scavenge pump must be appreciably greater than that of the pressure pump. The reason for this is that the inevitable presence of entrained air in the returned oil increases its volume. In other words, the scavenge pump must be capable of handling the combined volume of oil and air.

The actual manner in which a dry sump lubrication system is plumbed-in to both engine and vehicle varies, of course, according to the type of application. Since this is a topic that more readily lends itself to illustration than detailed description, a schematic drawing of a dry sump lubrication system once used in a German high-performance car shows a fairly typical installation (Figure 4.25). It will be noted that in this particular case a tandem oil pressure and scavenge pump was employed (Figure 4.26) and also that an oil cooler was included in the scavenge return to the oil tank. Engine oil coolers receive further mention later.

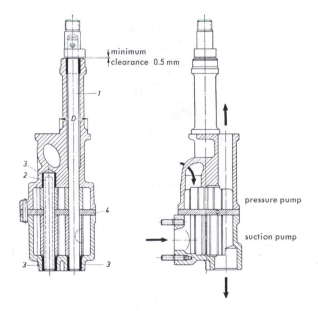

Figure 4.26 Tandem oil pressure and scavenge pump for dry sump system (*Mercedes-Benz*)

1 drive shaft
2 shaft
3 bearing bushing
4 intermediate plate

4.5 OIL PUMPS AND PRESSURE RELIEF VALVES

Developing oil pressure in the fully forced system

In modern practice, engine-driven positive displacement pumps of the rotary type are used to deliver oil and develop pressure in the fully forced lubrication system (Figure 4.27a and c). The term *positive displacement* is used in hydraulic engineering to describe any pump that displaces a definite volume of fluid with each revolution of its drive shaft, regardless of pressure conditions in the circuit it supplies. In other words, the pressure developed by the oil pump depends entirely upon the resistance offered to the flow of oil in the rest of the lubrication circuit. The principal advantage of using this type of pump is that it is compact in design. Because it is not self-priming, the oil pump with its inlet arrangements is usually bolted to a boss on the lower face of the crankcase, so that it is partially submerged in the oil sump but supported independently of it.

External gear pump

This comprises simply an elongated housing and end cover, which enclose a pair of either spur or, for quieter operation, helical gears. A bearing bore in the housing receives the pump drive shaft, rigidly secured to which is the driving gear of the pair. These parts may be either keyed together or machined as an integral pair. A mounting for the driven or follower gear is provided by a spindle supported from the pump housing. The gears generally have a coarse pitch tooth form, so that the least width of gear is required to pump a given amount of oil.

In operation, the oil entering the pump through the inlet port becomes trapped between the teeth of the contra-rotating gears and the surrounding wall of the pumping chamber. The oil is thus carried around the periphery of the gears (not between them) and then discharged through an opposite outlet port (Figure 4.28a). At this point the action of the intermeshing gear teeth prevents the oil from returning to the inlet side of the pump.

Internal gear pump

This houses a smaller driving gear mounted eccentrically within a larger driven gear. The externally toothed driving gear is rigidly secured to the pump drive shaft, while the internally toothed driven gear runs directly in the cylindrical cavity of the pumping chamber. Since the tooth form of these gears more closely resembles that of a lobe, the driving and driven gears are generally termed the inner and outer rotors of the pump respectively. The number of lobes on the inner rotor is one less than on the outer and so determines the displacement of the pump.

In operation, the oil enters the pump through a kidney-shaped inlet port, as the lobes on the rotors are moving out of mesh. Continued turning of the rotors serves to transfer the oil to a similar kidney-shaped port on the outlet side of the pump, the oil being discharged as the lobes of the rotors move into mesh (Figure 4.28b). The advantages usually claimed for this type of pump are that it is still more compact than the external variety and, because there are fewer teeth in mesh for each revolution, it can be quieter running.

More recently, there have been several examples of crescent-type internal gear pumps, encircling and driven directly from the nose of the crankshaft (Figure 4.27c). The advantages offered include simplicity of installation with particular

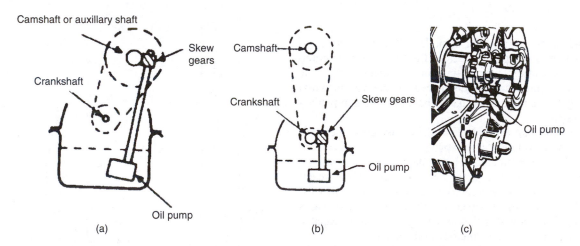

Figure 4.27 Typical pump drives: (a) from side camshaft (b) from crankshaft (c) encircling crankshaft

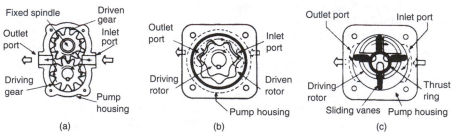

Figure 4.28 Types of engine oil pump: (a) external gear (b) internal gear (c) sliding vane

reference to the overhead camshaft engines and maximum oil pressure can be obtained at half the corresponding speeds of other engines where their oil pumps are camshaft driven, so that substantial oil pressure is maintained during idling.

Sliding vane pump

This takes the form of a driven rotor that is eccentrically mounted within a circular housing. The rotor is slotted and equipped with vanes equally spaced around its periphery, these vanes being free to slide radially within the slots. A pair of centring thrust rings maintains the vanes in close clearance with their track in the housing.

In operation the vanes are, of course, pressed outwards against the track by their centrifugal action. The oil inlet port is sited where the vanes begin to move away from a point of minimum volume between the eccentric rotor and its housing. After the vanes have passed the point of maximum volume and are again approaching a point of minimum volume, the oil is discharged through an outlet port (Figure 4.28c). This type of pump has the characteristic of giving a continuous rather than a pulsating oil flow, the latter usually being associated with gear-type pumps.

Pressure relief valves

The purpose of the pressure relief valve is to act both as a pressure regulator and as a safety device in the engine lubrication system (Figure 4.29).

A means of pressure regulation is required, since at higher engine speeds the increase in the quantity of oil delivered by the pump is not matched by a corresponding increase in the amount of oil that is able to flow through the bearings. This is because of centrifugal action on the oil already in the ducts of the crankshaft main journals, which imposes an increasing resistance to further oil flowing into them *en route* to the connecting rod big-end bearings. (Figure 4.30).

Under low-temperature starting conditions any increase in oil viscosity or reduction in bearing oil film clearances, or both, imposes an appreciable resistance to the oil flowing in the engine lubrication system. The pressure relief valve must, therefore, act as a safety device to prevent the oil pressure building up to a dangerous level. Otherwise, at worst the pump driving mechanism could be overloaded to its detriment, or the oil filter gasket could be damaged, and at least the excessive pressure drop across the oil filter would cause its own internal pressure relief valve to open and bypass unfiltered oil to the engine bearings.

Types of pressure relief valve

The pressure relief valve consists of either a ball, a poppet or a plate valve that is spring loaded on to an orifice seating and opened by oil pressure.

A disadvantage of the simple ball valve (Figure 4.31a), which has complete freedom to turn in any direction, is that random bedding marks on its working surface can form leakage paths across the valve seating. Leakage arising from this source can be avoided with a guided poppet valve, because this can turn only about its lift axis, so that the bedding condition between the valve and its seating is better preserved.

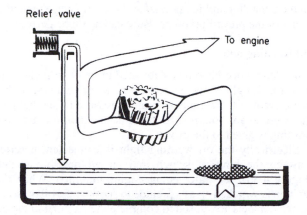

Figure 4.29 Basic action of the engine oil pump and relief valve (*Alfa-Romeo*)

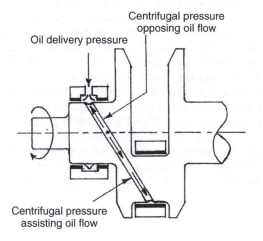

Figure 4.30 Oil flow conditions in a crankshaft

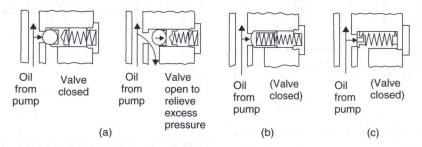

Figure 4.31 Pressure relief valves for engine lubrication systems: (a) ball type (b) poppet type (c) plate type

The poppet valve takes the form of a plunger with either a conical or a hemispherical sealing surface (Figure 4.31b). To minimize friction of the valve sliding in its bore, the plunger guide surface may incorporate annular grooves for oil retention.

A plate-type valve possesses the least inertia and is comprised simply of a thin steel disc (Figure 4.31c). Its edge is usually relieved in several places to prevent cross-binding of the valve in its guide bore.

Pressure relief valve setting

The actual setting of the pressure relief valve is generally related to the rate of oil flow required for adequate cooling of the bearings. In practice, it is typically set to open when the delivery pressure exceeds $280 \, kN/m^2$ ($40 \, lbf/in^2$), although higher maximum opening pressures of up to about twice this value may be specified for heavy-vehicle diesel engines. The maximum delivery pressure does not, as explained earlier, bear any relationship to the pressures generated in the bearing oil films.

Oil pressure indicator

In present-day practice it is usual for an oil pressure indicator, as distinct from gauge, to be featured in the comprehensive instrumentation of a high-grade motor car. An oil pressure indicator is electrically operated and comprises two units known as the transmitter and the indicator (Figure 4.32). The body of the former communicates with the engine lubrication system and encloses a flexible diaphragm. One side of this diaphragm is exposed to engine oil pressure, while the other side attacks a flexible bracket carrying one of a pair of sprung-together electrical contacts. The mating contact is carried by a superimposed bimetallic strip, which is also encircled by a heater winding. Within the dashboard indicator unit, the needle derives its arc of movement from another bimetallic strip similarly encircled by a heater winding, both transmitter and indicator windings being connected in series.

When the ignition is switched on to start the engine, it completes an electrical circuit through the indicator and the earthed transmitter units. The current flowing in the transmitter winding heats the bimetallic strip, thereby causing it to bend and open the contacts. Since this action immediately interrupts the supply of current, the strip will cool and the contacts will close again to restore the circuit. In reality, of course, the process is

being continually repeated such that the bimetallic strip oscillates to provide pulses of electric current. However, the effect of rising oil pressure on the transmitter diaphragm is to press the contacts more firmly together, so that their rate of oscillation is reduced. In other words, the current pulses sensed by the indicator unit will be longer in duration, thus creating a greater heating effect on its bimetallic strip. The bending of the latter then occurs in such a direction as to cause the indicator needle to move over its scale and record the prevailing oil pressure.

Another form of electrically operated oil pressure indicator similarly utilizes a transmitter diaphragm, but this acts instead against a piezoelectric crystal. Since a physical relationship can be established between the mechanical pressure on the crystallographic material and the voltage generated, a signal of the prevailing oil pressure can again be derived.

Oil warning light

It has long since become established practice to include an oil warning light on the instrument panel of vehicles, so that in the event of the engine oil supply pressure falling below a safe working minimum an illuminated and therefore positive warning is given to the driver.

Electrically the oil warning light is connected in series between the ignition switch and an oil pressure switch (Figure 4.33), the latter being engine mounted. When the ignition is switched on as a prelude to starting, the oil pressure switch contacts are closed to complete the circuit and the oil warning light will be illuminated. After the engine has been started and oil pressure is building up, the warning light will be extinguished by the action of the oil pressure switch opening its contacts and breaking the circuit.

The mechanism of an oil pressure switch typically comprises a reinforced rubber diaphragm that is spring loaded in opposition to the engine oil supply pressure acting on its inner face. Interposed between its outer face and the spring is a moving electrical contact of inverted top-hat form, the rim of which can seat upon a similarly formed stationary contact held within the switch body (Figure 4.34). The effective operating pressure of such a switch is generally somewhere in the range of 35 to $70 \, kN/m^2$ (5 to $10 \, lbf/in^2$).

Since under normal running conditions pressures of this order are exceeded several times by the engine oil supply pressure, the deflection of the switch diaphragm against its spring

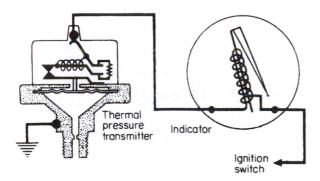

Figure 4.32 Mechanism and circuit of an electric oil pressure indicator (*Smiths Industries*)

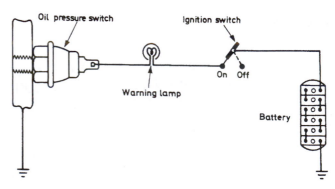

Figure 4.33 Electric circuit for an oil warning light

loading is sufficient to hold apart the switch contacts, so that the oil warning light is no longer illuminated. Conversely, if the engine oil supply pressure falls below the effective operating pressure of the switch, then the contacts close to complete the electrical circuit and the warning light is illuminated.

4.6 OIL FILTRATION AND COOLING

Engine oil filtration

Apart from corrosive wear arising from the acidic products of combustion, the other major form of engine wear is that caused by the presence of abrasive particles in the lubricating oil. This is especially true of those particles of a certain critical size that can span the oil film in the working clearances of the engine parts. Although manufacturers take immense care to build 'clean' engines, traces of casting sand and machining swarf are not unknown sources of abrasive particles. The running-in process can also be responsible for generating wear particles. During normal running any soot particles contained in the cylinder blow-by gases and hardened by high temperatures can enter the crankcase oil as abrasive particles.

Since the engine lubrication system can never be any better than its oil, it is therefore necessary to include some means of filtering the oil circulating in the system, so as to remove the abrasive particles before they reach the sensitive parts of the engine.

Full-flow filtration

An absorbent filter of the type described later may be arranged to arrest abrasive particles from either a small proportion, or practically all, of the oil being delivered by the pump to the principal working parts of the engine. Modern practice favours the full-flow system in which all the oil is filtered continuously, except that discharged from the main pressure relief valve (Figure 4.35a). Exceptionally this valve may be situated on the outlet side of the filter for better protection from contaminants, in which case the filter takes the full delivery from the pump. The full-flow filter is additionally equipped with its own internal relief valve. Its purpose is to bypass an emergency supply of unfiltered oil to the engine bearings, should the filter element become choked through neglect.

Partial-flow filtration

In this system, which can now be regarded as obsolete, an absorbent type of filter is arranged in the engine lubrication circuit such that 5–10 per cent of the oil delivered by the pump is diverted through it and returned to the sump (Figure 4.35b). Hence, the supply of oil to the bearings is not directly filtered or, in other words, only a small proportion of the circulating oil is filtered. To maintain an adequate supply of oil to the engine bearings the flow through the filter is further limited by a restricting orifice. This is usually situated on the outlet side of the filter, so that it is less likely to become obstructed. An internal relief valve is not required for a partial-flow filter, since if the element becomes choked the oil that would otherwise pass through it simply continues with the main oil flow to the bearings.

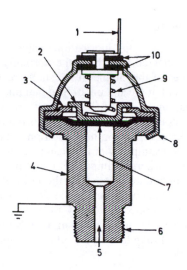

Figure 4.34 Cross-sectional view of an oil pressure switch

1 electrical terminal	6 screw-thread fitting
2 moving contact	7 diaphragm
3 stationary contact	8 metal casing
4 metal switch body	9 compression spring
5 oil pressure	10 insulation washers

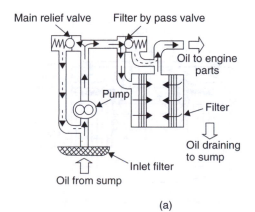

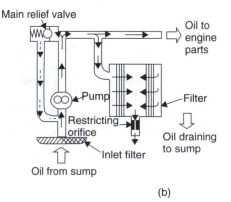

(a) (b)

Figure 4.35 Comparison of (a) full-flow and (b) partial-flow systems of oil filtration

It is possible to employ a very fine degree of filtration with the partial-flow system, on account of the slower rate at which oil is passing through the filter. However, the widespread adoption of the full-flow system in which practically all the oil delivered by the pump is subject to immediate filtration, seems to confirm that it performs the filtration process with more certainty. Indeed, it is now recognized in the field of hydraulic engineering that it is statistically impossible to clean the whole fluid content of a system by partial-flow filtration.

Types of oil filter

Various methods of oil filtration may be encountered in the lubrication system of the motor vehicle engine, and they can be conveniently summarized as follows:

Surface filtration   For this a coarse type of filter is used, which simply comprises a wire gauze screen such as is fitted over the inlet to the oil pump. Its filtering action is in the manner of a sieve and it is, of course, cleanable upon removal of the sump.

Depth filtration   This implies a fine type of filter that once typically comprised a specially processed paper element, but more recently a synthetic fibre material may be used. The replaceable element can be accommodated within either a metal canister made detachable from the engine crankcase, or a housing that is integrated with the crankcase. Because of its lower structural rigidity, the element may be provided with a supporting cylindrical screen. It also generally assumes a pleated form to present the largest surface area of filtering medium for a given size of element. A depth filtering action is obtained as contaminants are caught, and retained, as the oil passes through from the outside to the inside of the porous material.

In modern practice, the pore size of the element may be controlled to arrest particles down to within the region of 5 μm in size. (A micrometre (μm) is equal to 1/1000 of a millimetre. For purpose of comparison, an ordinary grain of table salt measures 100 μm.) As deposits accumulate, the absorbent filter offers increasing resistance to the passage of oil and must therefore be replaced at appropriate servicing intervals, when the oil is changed. Current (full-flow) depth filters are generally sealed and have a spin-on screw fitting, so that service replacement of the element itself is no longer possible, the entire filter being fully disposable.

Another feature that may be found in the modern full-flow filter is the positioning of the element relief valve at the inlet end (Figure 4.36a). The purpose of this is to ensure that any oil bypassed by the relief valve does not pass over contaminated surfaces within the filter casing (Figure 4.36b).

Centrifugal filtration   The centrifugal type of filter acts by subjecting the oil entering it to a high centrifugal force. In so doing it separates the heavier contaminants from the lighter oil, depositing them in sludge traps formed in the perimeter of a rotating filter bowl. The latter may be mounted on the nose of the crankshaft and is of large capacity.

Magnetic filtration   In practical application this generally comprises a plug that replaces the ordinary sump drain plug and incorporates a powerful permanent magnet. It is thus capable of attracting and retaining ferrous metal particles only, although other contaminants may also be retained by agglomeration (collecting in a mass).

Engine oil cooling

The oil circulating in the lubrication system receives heat that has been rejected from both the normally hot and the friction surfaces of the engine. It is usually recognized that 3.5 to 4.5 per cent of the amount of heat that may be obtained from combustion of the fuel received by water-cooled engine cylinders is rejected to the oil in this manner. This can mean that under severe operating conditions, such as may be encountered for example with turbocharged and also in air-cooled engines where cylinder temperatures are always higher, the temperature reached by the bulk of the oil may exceed 100°C. Since oxidation is one of the main causes of deterioration in engine oil, clearly there is justification in such cases to include an oil cooler in the lubrication system with the object of maintaining the oil temperature at a more normal 80 to 90°C.

Types of oil cooler

Oil coolers may utilize either the passage of air (Figure 4.37) or the flow of water (Figure 4.38) through the engine cooling system proper as the heat exchange agent. Where the latter is water we should not place too much emphasis on the word

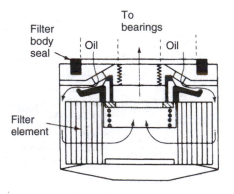
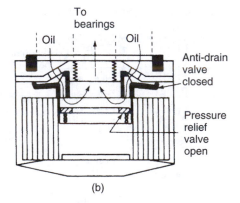

(b)

Figure 4.36   A modern full-flow filter of the spin-on type: (a) normal filtering action (b) filter choked and oil bypassed over non-contaminated surfaces

'cooler', since in this type of application an oil cooler does in fact perform two functions:

1 Brings the oil rapidly to its most efficient working temperature, because as long as the temperature of the oil is below that of the water the oil will be heated.
2 Maintains the oil at its most efficient working temperature under all operating conditions, because as soon as the temperature of the oil exceeds that of the water the oil will be cooled.

Where air is used as the heat exchange agent, necessarily so of course in the case of air-cooled engines, the oil cooler acts predominantly as its description suggests. However, it is then usually required to include an oil cooler bypass valve in the system, which allows the oil to heat up more rapidly from cold by initially restricting its circulation to the engine only. An oil cooler of this type is separately mounted from the engine and comprises a number of finned tubes that are plumbed into the lubrication system. These tubes are exposed to a cool stream of air by virtue of their being mounted across and forward of the engine radiator (Figure 4.39). With an air-cooled engine the oil cooler is mounted directly on the engine and adjacent to the outlet ducting of the cooling fan.

In the case of water being used as the heat exchange agent, the oil cooler is again mounted directly on the engine. It comprises an encased bundle of finned tubes through which the water is circulated from the engine cooling system and over which the oil is pumped in the opposite direction to confer a beneficial counter-flow, suitable piped connections being provided in each case (Figure 4.40).

For a high-performance passenger car oil coolers of both types may be used in series. In a recent Audi installation an oil-to-water cooler is mounted directly between the engine crankcase and oil filter, and is supplemented by an oil-to-air cooler sited as usual beneath the engine radiator. A thermostat control in the oil-to-water cooler limits the amount of oil flowing to the oil-to-air cooler while the engine is idling with the car stationary. When the temperature of the oil exceeds a predetermined limit as power output rises with car performance, the thermostat opens to admit a full flow of oil to the oil-to-air cooler so that both oil coolers are brought into action.

4.7 OIL RETENTION AND CRANKCASE VENTILATION

Static oil seals

These are required because to permit assembly of the working parts into the engine, its main structure must necessarily be fitted with detachable oil-tight covers. Although the mating faces of these parts are produced flat and parallel, in practice there are

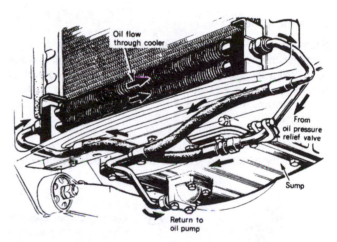

Figure 4.39 Installation of an air-cooled type of engine oil cooler (*Jaguar*)

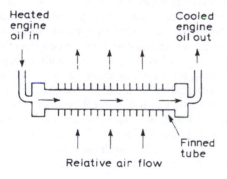

Figure 4.37 Basic principle of oil-to-air cooler

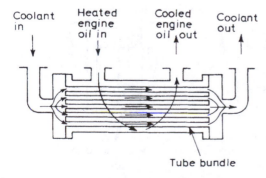

Figure 4.38 Basic principle of oil-to-water cooler

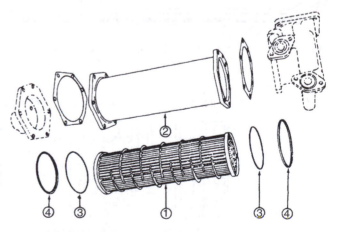

Figure 4.40 Exploded view of a water-cooled type of engine oil cooler (*Cummins*)

1 element   3 O-ring
2 housing   4 retainer

always minute surface irregularities present. If then a metal-to-metal closure was relied upon, these surface irregularities would offer potential areas of oil leakage. Suitably shaped static or gasket seals are therefore inserted between the various joint faces to prevent oil leakage. The gaskets are manufactured from fibre composition, cork with rubber binders and synthetic rubber materials, so that when compressed by bolt loading they tend to fill any leakage paths.

Typical static sealing applications are: the timing cover, for which a fibre composition gasket may be used; the oil sump, which usually requires a cork-based gasket to accommodate the relative expansion and contraction between itself and the crankcase; and the valve cover, where a synthetic rubber seal is generally appropriate in allowing effective sealing for minimum screw tightness (Figure 4.41).

A modern, less costly alternative to these precut gaskets are room temperature vulcanizing (RTV) gaskets or formed-in-place gaskets, which are applied as a silicone sealing bead of toothpaste consistency. Re-tightening of the joint may be required a few minutes after the sealing compound has been allowed to settle.

Dynamic oil seals

Seals are fitted where each end of the crankshaft emerges from the engine structure. Their purpose is to prevent leakage of oil from the adjacent lubricated areas and also to protect the engine mechanism against the ingress of foreign matter.

The sealing devices used for this purpose fall into two categories:

Clearance seals   Although no actual rubbing contact exists between the shaft and seals of this type, they nevertheless impose a hydraulic resistance against oil leakage by virtue of their screw return pumping action.

Contact seals   Their object is to provide a controlled rubbing pressure against the shaft, thereby creating a positive sealing action.

These seals are often used in conjunction with a supplementary oil flinger ring which rotates with the crankshaft. Its action is such that centrifugal force tends to throw off any excess amounts of oil creeping along the shaft towards the main seals. The oil flinger rotates within an annular grooved housing, this in turn being provided with a drain passage for returning oil to the sump.

Contact seals are generally required in modern engines with positive systems of crankcase ventilation, so that advantage can be taken of the latter to maintain a negative pressure in the crankcase and therefore discourage oil leaks.

Sealing the rear end of the crankshaft

Here the sealing requirements are particularly exacting, since the seal has to deal with the relatively large flow of oil from the adjacent rear main bearing. In many constructions the diameter of the crankshaft is substantially increased behind the rear main bearing to form an integral oil flinger, which is followed by a slightly reduced diameter sealing portion.

For clearance seals, either an oil return thread or a band of helical serrations is usually provided on the sealing surface of the shaft (Figure 4.42a and b), which runs within a close-clearance plain bore housing.

With the simple contact seal a flexible packing material is used, the rubbing portion of the shaft being left as a plain journal (Figure 4.42c). The sealing strip is pressed into upper and lower grooves machined in both the rear wall of the crankcase and an extension of the rear main bearing cap. In modern practice, the radial lip type of contact seal made from synthetic rubber has come into widespread use (Figure 4.42d). This type of seal is provided with a spring-loaded flexible sealing lip, which maintains a light rubbing contact with the shaft sealing surface. The sealing surfaces do, in fact, run with a very thin film of oil between them, the positive sealing action being generally attributed to the surface tension effect of the oil film at the exit side of the seal. A press fit must be provided for the seal in its housing and the wiping lip is positioned facing the lubricated bearing.

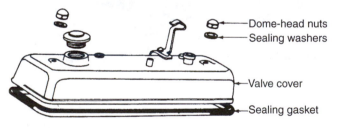

Figure 4.41   A typical static sealing application (*Toyota*)

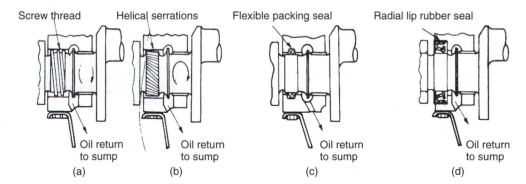

Figure 4.42   Sealing the rear end of the crankshaft

Sealing the front end of the crankshaft

The sealing requirements here are less severe, because the seal is shielded from the oil flowing from the front main bearing by the timing drive sandwiched between them, although this does not apply where an external toothed-belt timing drive is used.

The front end sealing surface is usually furnished by an inner extension of the crankshaft pulley boss. This is left as a plain journal when used in conjunction with either a threaded oil return bush (Figure 4.43a) or a radial lip seal (Figure 4.43b) inserted in the timing cover. Alternatively, an oil return thread may be provided on the pulley boss, which then works within a close-clearance plain bore in the timing cover. Where an oil flinger is also used, it is separately attached to the crankshaft behind the pulley boss.

Valve stems seals

The purpose of these seals is to control the passage of oil between the stems and the guides of overhead valves, especially the oil spill from valve rocker lubrication (Figure 4.44). Any excessive leakage of oil into the cylinders from this source can account for at least half the total oil consumption of a worn engine. If the leakage is bad enough it can produce blue exhaust smoke and hence cause air pollution by emission of hydrocarbons. Detonation or pinking can also arise from carbon build-up in the combustion chambers, and power output can suffer as a result of carbon build-up on the valve stems restricting their free movement.

Types of valve stem seal

It is fairly common practice to use either a simple oil deflector, or a positive seal, or both. The oil deflector or umbrella seal comprises an inverted synthetic rubber cup, which grips the stem immediately below the valve spring retaining collar. Its depth of side is such that the valve guide entrance is shielded from oil mist when the valve is open, but is exposed to it when the valve is closed (Figure 4.45a)

The positive type of seal again takes the form of an inverted rubber cup, but it is smaller in diameter and secured to the upper end of the valve guide where it seals directly against the valve stem moving through it (Figure 4.45b)

The need for crankcase ventilation

In the similar sense that we open the windows of a room in order to ventilate its interior or, in other words, promote a circulation of fresh air to replace that which has become stale, so too does this basic need arise in the case of a vehicle engine, albeit with additional considerations.

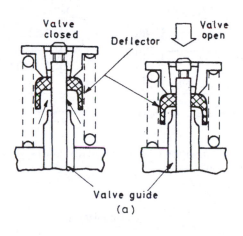

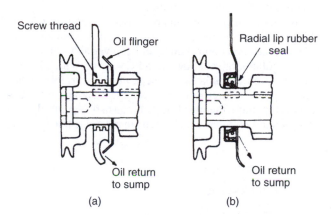

Figure 4.43 Sealing the front end of the crankshaft

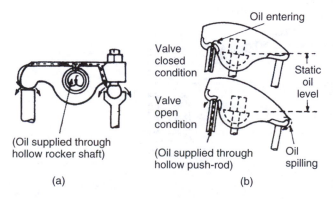

Figure 4.44 Typical methods of lubricating valve rockers

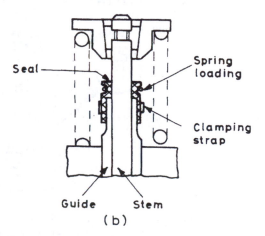

Figure 4.45 Valve stem seals: (a) deflector type (b) positive type

Adequate ventilation of the engine interior may, in fact, be considered as complementary to the satisfactory functioning of the lubrication system for two reasons:

1   The crankcase must be purged of harmful cylinder blow-by gases, since their composition is such as to cause both contamination of the oil and corrosion of the working parts.
2   The crankcase must be relieved of any build-up of internal pressure, because this could cause oil to leak past the crankshaft sealing arrangements.

Methods of crankcase ventilation

Ventilation of the engine crankcase was originally induced by what we would now term an open system, but from consideration of reducing atmospheric pollution the later closed system has long since become established practice. It is nevertheless useful to recall first the operating principle of an open system.

Open system
Here fresh air at underbonnet pressure entered at the highest point in the system, usually through a combined oil filler and

breather cap. It was induced to circulate through the engine interior by a breather pipe, the lower bevelled end of which protruded into the air stream under the car (Figure 4.46a). Purging of the crankcase gases therefore depended upon the motion of the car through the air. The air circulation within the crankcase was also assisted by the rotation or windage of the crankshaft. A basic limitation of this simple system was that ventilation could be inadequate when it was most needed, namely at low car speeds and during idling.

Closed system
Since the gases escaping from an engine with open crankcase ventilation contribute to air pollution, modern practice demands a positive closed system of ventilation, so that pollution from cylinder blow-by gases becomes negligible. Positive crankcase ventilation (usually abbreviated by PCV) was originally conceived by General Motors in America, where it was recognized in the early 1960s that cylinder blow-by gases could account for perhaps 20 per cent of the harmful emissions from a vehicle engine and its fuel system.

In early forms of closed system crankcase ventilation, the depression existing at the caburettor air cleaner was relied upon to maintain an air flow through the engine interior. For this purpose the breather outlet was simply connected by tubing to the air cleaner, through which the crankcase gases were circulated and subsequently burned in the engine cylinders (Figure 4.46b). A baffle plate flame trap was usually included in the system, so as to avoid a crankcase explosion in the event of an engine backfire.

In modern positive systems of closed ventilation (Figure 4.47) the greater depression existing within the intake manifold is utilized and controlled by either a variable- or a fixed-orifice regulator valve, known as a PCV valve, which is sited upstream from a crankcase oil separator. A variable-orifice regulator valve is interposed between the breather outlet tubing and the intake manifold riser. It consists essentially of a spring-loaded plunger, which opens in response to the intake

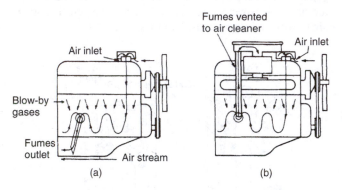

Figure 4.46   Comparison of (a) open and (b) closed systems of crankcase ventilation

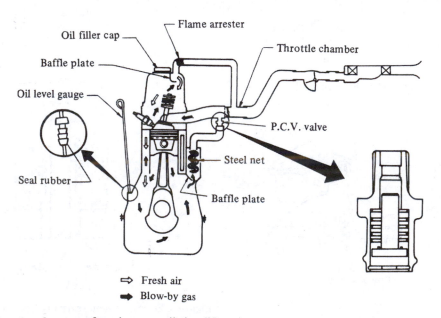

Figure 4.47   A positive closed system of crankcase ventilation (*Nissan*)

manifold depression acting upon it (Figure 4.48a). When the engine is not running the valve is completely closed. Under normal running the valve opens sufficiently to permit crankcase gases to enter the intake manifold with little restriction. At high manifold depressions during idling and overrun, further travel of the valve plunger against its spring loading closes it in the opposite direction, and the gas flow to the intake manifold is then restricted by an internal bleed orifice in the plunger. This particular feature thus ensures that the carburation process is not upset under these conditions. The valve also acts as a safety device in the event of an engine backfire, since any pressure created in the intake manifold automatically closes the valve completely, thereby isolating the engine interior.

With the other type of regulator valve, a fixed-orifice connection to the intake manifold riser is used in conjunction with a check valve connection to the clean side of the air cleaner (Figure 4.48b). At high manifold depressions the pressure difference across the check valve is such as to hold it against its seating, so that the crankcase gases can only pass through the fixed orifice and into the intake manifold. As engine speed rises, however, the reducing manifold depression behind the check valve and the increasing air cleaner depression ahead of it cause the valve to open from its seating. Hence, the crankcase gases may then enter both the intake manifold and the air cleaner to be subsequently burned in the engine cylinders.

Although PCV valves are designed to minimize the chances of blockage, hence the rattling sound made by their loose-fitting valves when shaken, it is nevertheless important that a PCV valve and its connecting hoses do not become clogged with deposits in extended service. If this does occur the crankcase would become congested with blow-by gases, thus hastening contamination of the engine oil. Also, an

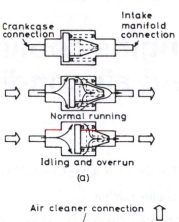

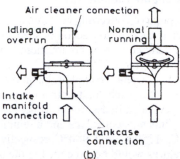

Figure 4.48 Comparison of (a) variable- and (b) fixed-orifice regulator valves

excessive build-up of pressure in the crankcase can jeopardize oil retention to the detriment of oil consumption. It is therefore essential to check for correct functioning of the PCV system at the manufacturer's recommended service intervals, typically every 30 000 miles (48 000 km) or 3 years.

5 Engine cooling, vehicle heating and air conditioning

5.1 HEAT TRANSFER AND COOLING MEDIA

The need for engine cooling

From the viewpoint of converting heat into mechanical energy, it follows that if each piston accomplished its power stroke starting at the temperature of combustion, which for spark-ignition is in the region of 2500°C, such an engine would in theory be highly efficient. To exploit this in practice, however, would impose unacceptably high temperatures on the materials of the working cylinder. In descending order the melting points of cast iron cylinders, steel valves and aluminium pistons are about 1800°C, 1350°C and 900°C respectively, hence all these temperatures would be exceeded in the absence of any cooling.

The behaviour of metals at elevated temperatures also differs from that at normal temperatures. Apart from losing some of their strength they can exhibit a condition known as *creep*, in which the metal deforms slowly and continuously at a constant stress. Excessively high temperatures would in any event cause breakdown of the cylinder lubricating oil films, resulting in the overheated working parts losing their running clearances and seizing up.

For these reasons the engine must be provided with a system of cooling, so that it can be maintained at its most efficient practicable operating temperature. This generally means that the temperature of the cylinder walls should not exceed about 250°C, whereas the actual temperature of the cylinder gases during combustion may, as previously noted, reach many times this figure.

Conversely, there is no merit in operating the engine too cool since this would reduce thermal efficiency and therefore increase fuel consumption. It would also increase oil dilution and hasten corrosion wear of the engine.

It is perhaps worth recalling an observation made many years ago by D.R. Pye, a one-time leading engine specialist, that 'One may go so far as to say that if an engine is not within a narrow margin of overheating at some point or other it is not working at its maximum capacity; or, expressed otherwise, that a change of condition which reduces the cylinder temperatures at critical points under given conditions of running, will raise the limit of power output of which the cylinder is capable.'

Methods of heat transfer

It is a common experience that heat always flows from a hotter to a cooler substance, and a physical principle that the greater the temperature difference between them the more rapidly is the heat transferred, until both substances attain the same temperature. That is, the loss of heat from one substance is equal to the gain in heat by the other. The process of heat transfer can take place in three ways, all of which are encountered in the engine cooling system, and they may be summarized as follows:

Conduction occurs most readily when heat energy is supplied to a solid substance, through which it is transmitted by direct contact or collision of molecules in the substance itself. The resulting molecular vibration increases as the temperature of the substance is raised and decreases when the temperature is lowered. An *insulator* is a substance that is a relatively poor conductor of heat and therefore acts as a thermal barrier.

Convection occurs when heat energy is supplied to a liquid or a gas and produces density changes in it. These promote *natural convection* currents, because the lower-density warmer matter rises to replace the higher-density cooler matter which descends. The process becomes one of *forced convection* with enhanced heat transfer when the circulation of the liquid is assisted by a pump, or that of the air by a fan.

Radiation occurs when heat energy is transferred by wave motion, there being neither contact of the substances nor molecular vibration within them. Radiant heat energy is emitted by all substances and may be either absorbed or reflected by others. Their ability to emit, absorb or reflect heat depends upon the colour and nature of the surfaces concerned, black rough ones being best for absorption and light polished ones best for reflection.

In practice, motor vehicle engines are designed either for indirect cooling by air through the medium of water or, less commonly, for direct cooling by air. Expressed in everyday language, these two systems are known simply as water cooling and air cooling respectively. Each system possesses certain advantages over the other.

Advantages of water cooling

1 Cooling is more uniform because heat is transferred with greater rapidity from the engine metal surfaces to water than it is to air. This can be explained by the specific heat of water being higher than that of air, so that if equal masses of water and air each receive the same quantity of heat, the water will experience the least rise in temperature. Consequently there will be a more rapid transfer of heat as a result of the higher mean temperature differences between the metal surfaces and the water in contact with them.

2 Cooling is more constant because the time taken for the water to rise through a given temperature range is longer than that for the same mass of air. This property of the

water possessing a greater *thermal inertia* than the air can be advantageous in maintaining a more nearly constant operating temperature – that is, when the rate of heat emission equals the rate at which heat is generated in the engine.

3 Interior heating for the vehicle is improved because outside air may be directed through a heat exchanger matrix, which can be conveniently heated by circulating a portion of the engine coolant through it.

Advantages of air cooling

1 Warm-up of the engine is more rapid because of the lower specific heat of the air cooling medium; heat is less readily transferred from the engine metal to the air being circulated around it, as a result of their lower mean temperature difference.

2 The system is inherently more reliable because air cooling is immune to either freezing or boiling of the coolant around the cylinder heads and cylinders, and to the loss of coolant. It is also free from any build-up of corrosive products that can restrict coolant passages.

3 Less maintenance is required in service because there is neither the requirement to check the cooling medium for level and condition, nor the need to inspect rubber connecting hoses for signs of leakage.

Certain criticisms that are customarily levelled at air cooling should be treated with some reserve: these include less reliability for heavy-duty application, greater power consumption and noisier operation. In the long-established and highly regarded range of heavy-duty Deutz air-cooled diesel engines, produced by the German firm of Klöckner-Humboldt-Deutz AG, the fully integrated air-cooling system has been found to have a substantially lower energy consumption than any water-cooled unit of equivalent output. Independently conducted scientific tests have also determined that these engines are among the quietest power units of equivalent output. Air cooling has also long been successfully used on the compact V eight-cylinder engine used in the Czechoslovakian Tatra car. It is perhaps of interest to note that when tested in the 1980s by a British motoring journalist, this car was praised for its 'remarkable silence'.

5.2 ENGINE AIR-COOLING SYSTEM

Circulation of cooling air

With air cooling, the engine structure is directly cooled by inducing air to flow over its high-temperature surfaces. These are finned to present a greater cooling surface area to the air (Figure 5.1), which in non-motor-cycle applications is forced to circulate over them by means of a powerful fan (Figures 5.2a and b). The car engine structure is almost entirely enclosed by sheet metal ducting, which incorporates a system of baffles. A similar arrangement is used with the diesel engine (Figure 5.3).

These baffles ensure that the through flow of air is properly directed over the cooling surfaces of the cylinders and cylinder heads. To maintain uniform temperatures, the air is forced to circulate around the entire circumference of each cylinder and its cylinder head, the direction of flow being along

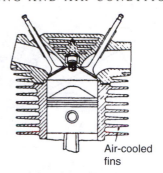

Figure 5.1   An air-cooled cylinder and head

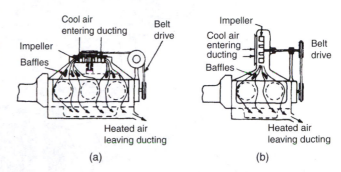

Figure 5.2   Air-cooling systems with (a) vertical (b) horizontal bearing axis for impeller

the cooling fins. These are greatest in number, consistent with providing a sufficient area of flow, on the high-temperature surfaces of the cylinder head in the region of the exhaust valve. The complete system forms what is known as a *plenum chamber* in which the internal air pressure is higher than that of the atmosphere.

Finally, the heated air is discharged from the plenum chamber to the atmosphere, or redirected to heat the car interior.

Air-cooling fan assembly

The forced circulation of air around the engine is generally provided by a centrifugal fan or impeller, which rotates in a spiral-shaped housing. This type of fan is capable of overcoming the appreciable resistance offered to the air as it flows around the ducted and finned cylinders and cylinder heads. The fan is driven by a V-belt and pulley system from the engine.

During operation the air enters at the eye of the impeller via an inlet housing, and flows between each pair of blades. Centrifugal force thus acts upon the enclosed air, which is then discharged under pressure in a radial direction (Figure 5.4). To increase fan efficiency the impeller blades may be curved backwards instead of being straight. The fan housing is also fitted with an inlet ring to minimize any recirculation of air at the point of entry.

Air-cooling throttle valve

This regulates the quantity of air entering the cooling fan, in accordance with engine cooling requirements. It usually takes the form of a movable throttle ring, which acts as a baffle surface at the entrance to the inlet housing (Figure 5.4).

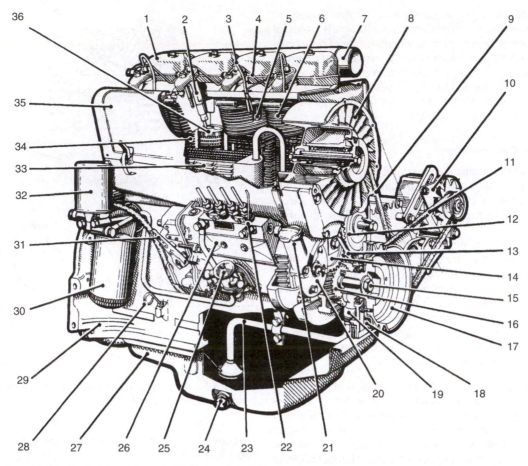

Figure 5.3 Cut-away view of modern air-cooled diesel engine (*Deutz*)

1 rocker chamber cover
2 injector
3 injection line to no. 3 cylinder
4 back-leakage line
5 cylinder head anti-fatigue bolt (four bolts securing each cylinder head with cylinder to crankcase)
6 cylinder head (light alloy)
7 air intake manifold
8 cooling blower (V-belt driven)
9 cooling blower V-belt
10 generator (dynamo or alternator)
11 generator V-belt
12 camshaft gear
13 oil gallery

14 idler gear (driving injection pump and camshaft)
15 anti-fatigue bolt (securing V-belt pulley to crankshaft)
16 crankshaft gear
17 V-belt pulley
18 vibration damper
19 oil pump
20 injection pump drive gear with advance/retard unit
21 oil filler neck
22 overflow line
23 oil suction pipe
24 oil drain plug
25 fuel feed pump

26 Bosch in-line injection pump with mechanical centrifugal governor
27 oil sump (Sheet metal or cast iron)
28 oil dipstick
29 crankcase (cast iron)
30 oil filter
31 speed control lever
32 fuel filter
33 integral oil cooler
34 finned cylinder (grey cast iron), separately removable
35 removable air cowling
36 piston

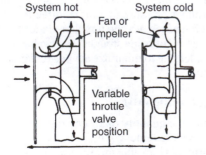

Figure 5.4 Air-cooling fan and throttle value

At normal running temperatures, the throttle ring is retracted from the eye of the impeller to allow free access for the incoming air. When the engine is cold, the throttle ring is advanced towards the eye of the impeller, thus reducing the quantity of air circulated around the engine.

Air-cooling thermostat

Automatic operation of the cooling fan throttle valve is effected by a thermostat, which therefore serves to regulate the rate of engine cooling. It is linked mechanically through a simple leverage system to the throttle valve, so that adequate

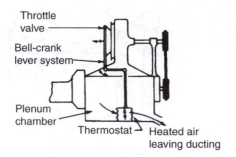

Figure 5.5 Air-cooling thermostat

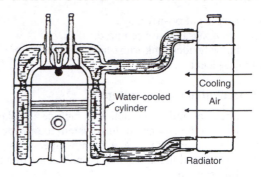

Figure 5.6 The thermosyphonic system of water cooling an engine

movement is transmitted to the ring within the available working stroke of the unit. The thermostat is generally installed in the hot air duct leading from the lower part of the plenum chamber (Figure 5.5).

5.3 ENGINE WATER-COOLING SYSTEM

Circulation of cooling water

A very simple form of engine water-cooling system, known as natural or *thermosyphonic* circulation (Figure 5.6), was used in many early motor vehicles up to the late 1930s and provides a useful introduction to the topic. This type of water-cooling system consisted of the following components:

Water jacket This was formed partly in the cylinder block and partly in the cylinder head and surrounded the cylinder walls, combustion chambers and valve ports. It was provided with a lower inlet and an upper outlet connection, the latter sometimes being called the header pipe.

Radiator The purpose of the so-called radiator is to dissipate the engine heat rejected to the coolant. This it does mainly by conduction and convection, rather than by radiation. It originally consisted of an upper header tank that received the heated coolant from the engine; a matrix or core that served to disperse the down flow of coolant into fine streams and also provided for a through flow of cooling air; and a lower collector tank from which coolant was returned to the engine jacket.

Flexible connections Rubber hoses with clips were used to connect the cooling jacket header pipe to the inlet pipe of the radiator header tank and the outlet pipe of the radiator collector tank to the inlet elbow of the cooling jacket.

Other fittings The radiator header tank was fitted with a filler cap and an overflow pipe and the collector tank with a drain tap. Suitable mounting lugs were also provided at the top and bottom of the radiator.

Radiator fan In many, but not all, early thermosyphonic cooling systems a fan was mounted behind the radiator and driven by belt and pulleys from either the crankshaft or the camshaft. Its purpose was to assist the through flow of air normally resulting from the forward motion of the vehicle and also create an air flow when the vehicle was stationary with the engine idling.

Operating principle of the thermosyphonic system

Upon starting the engine from cold, the following sequence of events occurs:

1 Cold water in the jacket begins to absorb surplus heat from the engine and expands slightly.
2 Since the heated water becomes less dense than the cold water in the jacket, it rises to the header pipe and then flows upwards into the header tank of the radiator.
3 This heated water then begins to be forced down in fine streams through the radiator matrix, as a result of more water being heated in the jacket and rising behind it.
4 The heated water passing through the radiator is then cooled by the throughflow of cooling air, so that it becomes denser again and readily sinks to the collector tank of the radiator.
5 Cooled water is then raised from the collector tank to the inlet of the jacket as a result of the pressure difference created by other cooled water sinking in the radiator and heated water rising in the jacket again.

The rate at which water circulates in the simple thermosyphonic cooling system is in proportion not to engine speed but to the heat output or load placed on the engine. As the engine load increases the circulation becomes more rapid, mainly because the formation of small steam bubbles have the effect of further reducing the density of the coolant being heated.

Purely as a matter of historical interest, perhaps the best remembered application of thermosyphonic cooling was to the legendary Model T Ford car, which remained in production for nearly twenty years up to 1927. Of great simplicity in construction, this was the first car to be mass produced in the true sense of the word. It was also the one about which Henry Ford made his famous remark that any customer could have a car painted any colour that he wanted so long as it was black.

Advantages and disadvantages of thermosyphonic cooling
1 It is a simple arrangement of low cost.
2 Warm-up is fairly quick.
3 Circulation is proportional only to engine load.

4 Circulation is not positive.
5 Large water passages are required.
6 The radiator header tank must have a high location relative to the jacket.

Forced-circulation cooling system

It has long been established practice to include a coolant (or 'water') pump in the engine cooling system to provide what is termed forced circulation. We must now attempt to justify in a little more detail the additional complication of a pump in the cooling system.

With forced circulation it is possible to direct the principal cooling effect to the high-temperature surfaces in the cylinder head, whilst maintaining the cylinder block at acceptable working temperatures by natural circulation. If a relatively high rate of coolant flow is promoted around the exhaust valve seats, exhaust ports and sparking plug bosses, then any tendency towards localized overheating is minimized. From considerations of engine durability it must be appreciated that both the mechanical strength and the corrosion resistance of valve materials decrease with increased operating temperatures. A sufficiently high coolant pressure must also be maintained in these critical cooling regions, so as to avoid the formation of steam pockets that would restrict circulation and impede heat transfer.

The increased rate of coolant flow provided by forced circulation not only improves cooling efficiency, but also permits a reduction in the quantity of coolant required. Hence, the coolant passages in the cylinder block and head may be made correspondingly smaller in size, thus reducing both the bulk and the weight of the coolant jacket. Furthermore, the radiator can be reduced in size, because a higher mean temperature can be maintained between the faster flowing coolant and the air passing through its matrix. This in turn makes for greater convenience of installation, especially where the requirements of modern styling demand a low bonnet line, and also in designs where the radiator is sited alongside a transversely mounted engine.

Advantages and disadvantages of forced-circulation cooling
1 It is more complicated and costly arrangement.
2 Warm-up is quick as long as a thermostat is used in the system.
3 Circulation is proportional to both engine load and speed, the former as long as a thermostat is used in the system.
4 Circulation is positive.
5 Smaller water passages and radiator are permissible.
6 Radiator header tank may be lower with respect to the jacket.

Forced-circulation flow patterns

In practical application the forced-circulation cooling system consists of numerous communicating passages formed within the engine coolant jacket, through which the coolant is circulated around the cylinder walls and combustion chamber regions. To minimize thermally induced stresses in these parts, it is usual for the lower-temperature coolant entering the jacket to be initially circulated over the cooler metal surfaces of the engine. The coolant flow is therefore directed around the

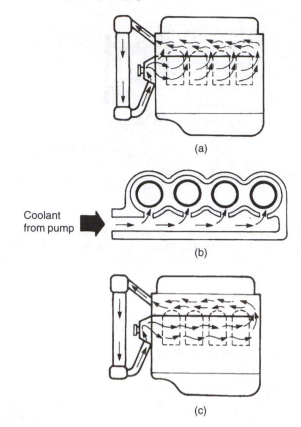

Figure 5.7 Comparison of (a) parallel flow (b) independent parallel flow (c) series flow of coolant between cylinder block and head

cylinder block and then upwards into the cylinder head. In the case of V and horizontally opposed engines, the pump delivers coolant simultaneously to both banks of cylinders. To maintain uniform temperatures throughout the engine, the flow pattern provided between the cylinder block and head can take three forms: parallel, series and reverse.

Parallel flow
In this widely used pattern each cylinder receives a small proportion of the coolant entering at the front end of the cylinder block. After flowing around the 'separately' served cylinders, the coolant then passes upwards through ducts to the corresponding cylinder head regions and thence to the outlet (Figure 5.7a).

More recently there has been a trend to ensure that the separate cooling of the cylinders is done more positively, by directly routing the flow or coolant to each cylinder (Figure 5.7b). This provides for more uniform cooling of the cylinders to reduce any tendency towards detonation and is sometimes known as independent cylinder cooling. The coolant may be arranged to enter each cylinder chamber in a tangential direction, which causes it to swirl around the cylinder barrels as it travels upwards into the cylinder head. This swirling motion of the coolant discourages the accumulation of sludge and rust, since these would greatly reduce cooling efficiency. Mention has earlier been made of directed coolant flow in the cylinder head.

In a variation of this arrangement a dual parallel flow pattern is used, where the coolant from the pump is directed into the front ends of both the cylinder block and head. The parallel flows then continue along the length of their jackets before combining and returning to the radiator, via a rear mounted thermostat. This arrangement is intended to provide improved cooling for the cylinder head, since about 70 per cent of the heat rejected by the engine to the coolant originates from this source.

Series flow

With this less commonly used pattern, all the coolant entering at the front end of the cylinder block circulates around all the cylinders in turn. It then passes upwards through ducts at the rear end of the cylinder block, whence it flows back along the entire length of the cylinder head to the outlet (Figure 5.7c). The advantages of this particular arrangement are that the cylinder head gasket is required to function only as a gas seal and the absence of coolant passages between the main portion of the cylinder block and the head makes for greater rigidity of these components.

Reverse flow

In the early 1990s General Motors in America, contrary to earlier established principles of engine cooling, introduced a reverse flow system in which the coolant is directed first into the cylinder head and then down into the block. The advantages claimed for this system are cooler cylinder head temperatures and warmer cylinder block temperatures, which respectively raise the detonation limit monitored by the knock control system (Section 3.2) and lower the frictional drag of the piston rings (Section 1.7), thereby reducing the fuel consumption. To minimize thermally induced stresses that could otherwise result from the lower temperature coolant being directed first to the much hotter surfaces of the cylinder head, the thermostatic control is arranged to act on the inlet rather than the outlet side of the coolant pump.

It should be appreciated that the cooling systems developed for modern passenger cars must not only dissipate the greater amounts of heat generated by multi-valve cylinders and turbocharger installations, but also be accommodated in the limited space available in the increasingly crowded engine bay.

Location of the coolant pump

It has long been established practice for the coolant pump to be located on the front face of the cylinder block (Figure 3.10). Until the more recent advent of the radiator mounted and electrically driven fan, it was usual for both the pump and the fan to share a common belt-driven spindle. An advantage of this particular location is that the pump output can be conveniently directed through an internal ducted passage, which not only distributes the coolant flow as evenly as possible among the cylinders, but also directs a vigorous flow upwards to the high temperature regions of the cylinder head.

Coolant pump construction

A non-positive pump of the centrifugal type is invariably used to circulate the coolant, its advantages for this application being as follows:

1 It characteristically provides a continuous rather than a pulsating flow.

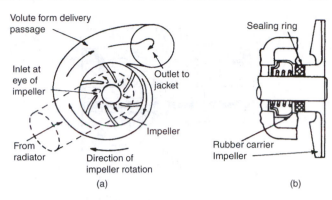

Figure 5.8 Centrifugal coolant pump: (a) action (b) sealing

2 Being non-positive (unlike a gear-type oil pump) it neither builds up an excessive pressure, nor overloads its driving mechanism, if resistance to coolant flow is increased anywhere in the system.

3 It continues to provide a limited passage for thermosyphonic circulation of the coolant in the event of failure of its driving mechanism.

An engine-driven pump of this type circulates the coolant by means of either a metal or a plastics impeller, which revolves in a spiral chamber (Figure 5.8a). To provide a uniform cross-sectional area for the spaces between the impeller vanes, their height is gradually reduced from inner to outer edges. Further the vanes may be curved backwards instead of being straight, since this reduces turbulence as the coolant enters and leaves the impeller. Also, the spiral chamber may be of volute form to obtain a delivery passage of progressively increasing cross-sectional area. This feature serves to reduce the velocity of the discharging coolant, so that its delivery pressure is correspondingly increased.

To prevent leakage of coolant and to exclude air from the system, the now long-established face seal is used (Figure 5.8b). Mechanical seals of this type originally consisted of a stationary counterface of either cast iron or stainless steel, spring loaded against which was a sealing ring of hard carbon material revolving with the impeller. The sealing ring was bonded to a synthetic rubber carrier to prevent coolant leakage along the impeller spindle. In a modern inverted version of this sealing method, a semi-cartridge construction is employed to support and guide the spring loaded sealing ring or primary seal ring as it is sometimes termed, which is then integrated with a synthetic rubber sealing bellows to become the stationary member, the bellows readily allowing the sealing ring freedom of axial movement to compensate for any misalignment, or wearing, against the counterface or mating ring that now rotates with the impeller (Figure 5.9). A further development of the cartridge type of construction is the 'unitized' assembly, which is handled as a one-piece item on build. It differs only in that a pair of lipped telescopic members are used to support and guide both the stationary sealing ring and its rotating mating ring, the inner supporting member for the latter being made an interference fit on the impeller shaft.

The materials used for the pump seal counterface or mating ring are generally harder than that for the primary seal ring.

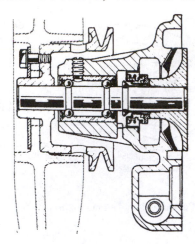

Figure 5.9   Cross-section through a modern coolant pump with a face seal (*Alfa-Romeo*)

minute for every kilowatt of power (3.3 to 3.9 pints per minute for every horsepower) developed by the engine is typical for motor car petrol engine practice. These flow rates are generally lower at 1.75 to 2.25 litres per minute for every kilowatt of power (2.3 to 3.0 pints per minute for every horsepower) developed in the case of diesel engines, because they reject less heat to the coolant. A relatively greater rate of flow is required for indirect- as compared with direct-injection diesel engines, owing to the more severe heating of their cylinder head valve bridge portions.

Coolant pump misbehaviour in service

Modern coolant pumps are usually maintenance free, and in the event of their unsatisfactory operation, causing engine overheating, it is rare for a manufacturer to advise any other course of action than fitting a service replacement unit. The pump may develop such faults as the following:

Leakage
Abrasive particles originating from metal erosion carried around by the circulating coolant and scoring the seal contact faces.
Worn spindle bearing allowing misalignment to occur between the seal contact faces.
Inefficiency
Metal erosion gradually destroying the form of the impeller vanes.
Excessive end float of the impeller spindle in its bearing increasing the working clearance between the impeller vanes and the pumping chamber.
Apparent inefficiency of the pump caused by a worn or excessively slack driving belt.
Inoperative
Mechanical derangement such as a cracked impeller loosening on its spindle or the pulley adaptor doing likewise owing to loss of interference fit.
Failure of the pump driving belt.
Noisy
Wear in the impeller shaft bearing or seal assembly; also air ingress via the latter.

Cooling fan

The purpose of the fan is to maintain an adequate air flow through the radiator matrix, especially at low vehicle speeds and under engine idling conditions with the vehicle stationary. In other words, the fan serves to speed up the natural draught through the radiator or, alternatively, to create a draught where none would otherwise exist. At high vehicle speeds, the ram effect created by the rapid motion of the car through the air usually promotes sufficient air circulation through the radiator.

A multibladed propeller fan is used to promote an axial flow of air through the radiator of a water-cooling system, its blades normally being irregularly spaced to reduce noise level as fan speed rises. Conventionally, the fan is of the puller type; that is, it is mounted between the radiator and engine so that air is pulled through the radiator as a result of the depression created on the inlet side of the fan. Less commonly, a pusher fan with

They must be resistant to corrosion and currently may be produced from either sintered or cast metals, or an aluminium oxide ceramic material, and with sprayed coatings also sometimes being used. For the primary seal ring a carbon graphite material, which may be impregnated with a metallic content, continues to offer an acceptable service life with the required resistance to wear that derives from its inherent self-lubricating property, good thermal conductivity and resistance to physical degradation when exposed to the hot coolant. It nevertheless requires sympathetic handling on build, because it is a material that can be easily damaged.

To compensate for high coolant pressure tending to oppose that exerted by the spring on the seal, the sealing ring assembly is also now designed so that the pressure at the eye of the impeller has a limited access to the nonsealing face of the ring. Hence an increase in coolant pressure then assists the spring loading applied to the sealing ring instead of opposing it, thereby reducing the possibility of leakage across the sealing faces. In this context, it should be appreciated that an extremely thin film of coolant must exist between the rubbing faces of the seal to act as a lubricant. A small drain hole is provided in the coolant pump housing between the seal and spindle bearing, so that the latter is protected against any coolant weeps.

Coolant pump operation

In operation, it must always be arranged for the coolant level to be above that of the pump inlet, since this type of pump is not self-priming. The coolant enters the pump near the drive spindle and is admitted to the eye of the revolving impeller, following which it becomes trapped in the spaces between the vanes and pumping chamber wall. A whirling motion is thus imparted to the coolant, which is accelerated to a high velocity and discharged under centrifugal action in the delivery passage. Since the delivery pressure is higher than that existing at the eye of the impeller, a continuous flow of coolant is maintained through the pump. Coolant delivery is not quite proportional to pump speed, since some of the coolant will recirculate to the inlet side via the internal working clearances of the pump. A maximum flow rate of 2.5 to 3.0 litres per

its blades set at an opposite angle may be required, typically to push the air through the radiator of a rear-engined bus.

Fixed-drive fan

Although now mainly confined to commercial vehicle engines, the once conventional engine-driven fan is generally operated in tandem with the coolant pump, both sharing a common drive spindle that is rotated by a V-belt and pulley system from the crankshaft. The relative speed of the fan ranges from slightly less to rather more than that of the engine, depending on the particular speed ratio chosen for its pulley drive arrangements. Where space limitations sometimes demand a very low positioned radiator, such as in the case of a forward-control heavy vehicle, the fan may be directly mounted on the crankshaft nose (Figure 5.27).

The fan assembly has often comprised a one-piece steel pressing clamped to the face of the coolant pump drive pulley and spigot located on the end of its spindle. In modern practice the fan may be produced from either an aluminium alloy casting or a plastics moulding with a metal boss insert, to provide a lighter, more easily balanced construction with an improved blade form for greater efficiency. The fan is made as large as practicable, not only for efficient utilization of the available radiator matrix area, but also because the volume of air it displaces increases as the cube of its diameter; thus (in an exaggerated sense), a fan of double the diameter would displace eight times as much air.

Variable-drive fan

Although a fixed-drive fan can be designed to provide an adequate air flow through the radiator under arduous operating conditions, such as during prolonged traffic crawling, its continued operation under more normal running conditions can bring several disadvantages. First, the noise level rises. Second, the engine power absorbed in driving the fan will be greater, especially when it is considered that this increases as the cube of fan speed. There may also be a tendency to overcooling; the latter is to be avoided because engine life, lubricating efficiency and vehicle heater performance can suffer as a result. To minimize these disadvantages an automatic control is sometimes employed to confer a variable-speed operation on the fan.

The variable-speed fan may be of either the torque-limiting or the temperature-sensitive type, their common objective being to ensure that the fan is not driven at a speed in excess of that required to maintain an optimum cooling system temperature. Both types of fan embody a viscous coupling, which permits relative motion or slip to occur between the driving and the driven members. (It may be recalled here that we have previously encountered this type of coupling in the viscous torsional vibration damper, as fitted to some diesel engine crankshafts.) A driving disc is mounted directly from the coolant pump pulley and revolves within a sealed coupling chamber to which the fan is attached (Figure 5.10a). The coupling chamber is freely mounted on the coolant pump spindle and partially filled with a highly stable silicone fluid, which characteristically retains a nearly constant viscosity with varying operating temperatures.

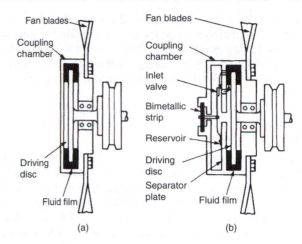

Figure 5.10 Variable-drive fans (a) torque limiting and (b) temperature sensitive

During operation of the simpler *torque-limiting fan* drive the driving disc initially revolves in the stationary mass of fluid and sets up a viscous drag, so that the fluid begins to circulate within the coupling chamber and owing to centrifugal action, moves outwards to fill the narrow gaps between the driving faces of the coupling members. Since these faces are not only arranged to be in close proximity to one another, but also may have mating grooves, torque is transmitted between them by the viscous drag of the silicone fluid, and hence a drive to the fan is established. This drive is sustained with increasing engine speed, until the resisting torque imposed upon the fan by its air flow rises to an extent that overcomes the viscous drag of the silicone fluid and allows the coupling to slip. Maximum fan speed is therefore governed by the ultimate slipping action of the coupling, the latter being predetermined by the viscosity of the silicone fluid and the degree to which the coupling is filled.

The more comprehensive *temperature-sensitive fan* drive differs from the torque-limiting type in that provision is made for automatically varying the degree to which the active part of the coupling is filled. This overcomes the disadvantage that the fan has to operate within a fixed speed range, regardless of cooling system demands. Otherwise, the air flow generated by the fan may be inadequate when vehicle speed is low and engine speed is high, while at low engine speed the uninterrupted drive to the fan may prolong the engine warm-up period.

In construction the coupling chamber of the temperature-sensitive fan is divided into two by a separator plate, thereby forming a working chamber for the driving disc and a reservoir (Figure 5.10b). The fluid in the working chamber is provided with a permanent escape route from an outlet port near the periphery of the separator plate, so that by natural inclination the coupling does not 'want' to drive the fan. However, the working chamber is also provided with an inlet port that is less distant from the centre of the driving disc. This port can be opened and closed by a thermostatically actuated valve, which moves in response to the flexing of a bimetallic strip mounted on the front of the coupling and therefore exposed to the air that has been heated in its passage through the radiator airways. By sensing the coolant temperature in this manner, the bimetallic strip can be arranged to open the inlet port to the working chamber when the engine coolant becomes too hot. Fluid being expelled to the reservoir can then be continuously readmitted to the working chamber to restore viscous drag against the driving disc and therefore bring the fan into operation.

Off-on drive fan

This type of fan drive is sometimes used on the diesel engines of heavy vehicles to assist quick warm-up of the coolant and maintain a constant engine temperature. It basically comprises a small, spring-loaded, single- or multiplate friction clutch that is built into the pulley drive for the fan. While the coolant remains below a predetermined temperature, a thermostatic control valve located in the radiator top hose directs air from the vehicle's compressed air supply to the fan hub, which overcomes the spring loading on the clutch and disengages the fan from its pulley drive. The fan will now idle until the coolant temperature rises above a predetermined level, at which point the air supplied under pressure to the fan clutch is cut off by the control valve and the spring engages the friction elements to drive the fan for normal cooling.

V-belt drives

It has long been established practice to employ either a single, or a matched pair, of endless V-belts for transmitting drive from the engine crankshaft to the coolant pump and where applicable the cooling fan. The drive also includes various other engine-driven auxiliaries such as the alternator, which usually incorporates in its mounting an adjustment facility for initial tensioning of the belt(s). Driving power is transmitted by the friction forces arising from the wedging action of the V-belt in its pulley grooves. The advantages of employing this type of belt drive include minimum space requirements, quietness of operation at relatively high speeds, and the ability to absorb shock loads.

In cross-section the V-belt may be divided into a neutral load-carrying zone, a lower compression zone and an upper tension zone (Figure 5.11a). The load-carrying members comprise a layer of high-tensile cords of a textile material, this typically being polyester, which thus locate the neutral axis of the belt. These cords are embedded in a synthetic rubber core that both protects and supports them. The core consists of a lower supporting cushion of relatively stiff rubber and an upper layer of more resilient rubber. This particular construction enables the tensile loading in the cords to be transmitted as groove sidewall pressure in the compression zone of the belt, whilst also being able to accommodate the repeated flexing of the belt in the tension zone as it bends around the pulleys. A wear-resistant covering of rubberized fabric material completes the build-up of the V-belt.

Since it first appeared on an American commercial vehicle engine in the late 1970s, the multiribbed V-belt has become increasingly widely used (Figure 5.11b). The advantages associated with this type of belt construction are longer service life, better maintenance of tension, higher transmission efficiency and greater flexibility, which allows the use of smaller-diameter pulleys.

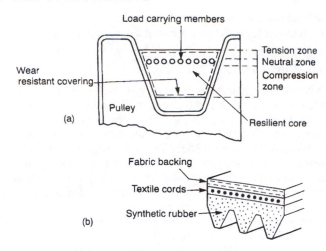

Figure 5.11 V-belt constructions: (a) single (b) multiribbed

Electrically driven fan

The separate electrically driven fan is now widely used in passenger car engine cooling systems. This can be directly mounted from the radiator and switched in and out of operation in accordance with cooling system demands. The switching is thermostatically controlled in response to changes in engine coolant temperature, a sensor being installed for the purpose in the lower half of the radiator. Normally the fan will continue running for a few minutes after the engine has been switched off, this period sometimes being referred to as 'post ventilation'. For greater flexibility of control a two-speed fan and for some high-powered engines twin fans may be used, the latter being individually controlled according to cooling demand. Fuel economy with an electrically driven fan should therefore be improved because the engine better maintains an optimum running temperature, the fan only operates above a preset temperature so that warm-up is quicker, and the fan consumes less energy. The intermittent operation of the electrically-driven fan also makes for generally less noise, although when brought into action at lower engine speeds it may actually be noisier than an engine-driven fan if it has a fixed higher operating speed. An interesting feature of the twin-fan installation for the cooling system of the 2000 Cadillac Northstar V8 engine is that their blades have been acoustically developed with a special curvature, so that operation of the fans is not heard by the car occupants. Not the least advantage of an electrically driven fan is its greater convenience of mounting position; this, of course, accounts for its wide-spread use with modern transversely mounted engines (Figure 5.12).

Fan cowl

A sheet metal or plastic cowl is often made an integral part of the radiator and cooling fan installation. The cowl is funnel shaped to surround the rear face of the radiator matrix and encircle the tips of the fan blades. Its purpose is to contribute to effective operation of the fan by preventing heated air from recirculating through the radiator matrix, which could otherwise lead to engine overheating.

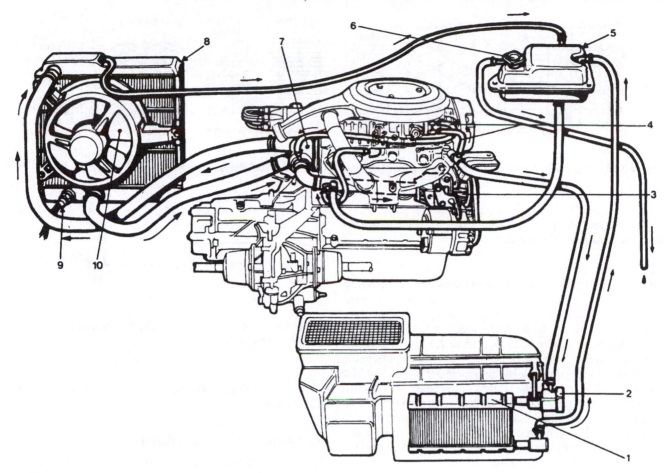

Figure 5.12 Electrically driven fan installation (*Fiat-Lancia*)

1 heater core	6 cap
2 heater core cock	7 blending thermostat
3 water pump	8 cooling radiator
4 semi-automatic choke water hoses	9 electrofan control thermoswitch
5 filling tank	10 electrofan

Radiator construction

The purpose of the so-called radiator is to dissipate the engine heat rejected to the coolant. This it does mainly by conduction and forced convection rather than by radiation. A cooling effect from the radiator is achieved by arranging for the entering heated coolant to be dispersed into fine streams through its matrix or core. Relatively small quantities of coolant are therefore brought into contact with large surface areas of metal. Even greater surface areas of metal must be exposed to the through flow of cooling air, since there is appreciably greater resistance offered to the transfer of heat from the metal surfaces to the air, than from the coolant to the metal surfaces. For this reason the matrix waterways are provided with secondary heat transfer surfaces in the form of metal fins. They are generally shaped to promote turbulence in their airways, which further assists the process of heat transfer between the matrix metal surfaces and the air flowing over them.

Types of radiator matrix

In early practice the construction of the radiator matrix resembled that of a honeycomb. It consisted basically of a stack of air tubes, the ends of which were bulged and sealed together, so that continuous waterways were formed between them (Figure 5.13a). Although this form of construction lent itself to a variety of pleasing radiator shapes now much admired by historic car enthusiasts, in practice its cooling efficiency was limited by the absence of secondary heat transfer surfaces. The honeycomb radiator was therefore eventually displaced by other forms of matrix construction of which there are three basic types, as described in the following sections.

Cellular or film

With this type of matrix the airways and secondary heat transfer surfaces are formed by a series of zigzag ribbons of copper (Figure 5.13b). Each row of airways is bounded along its sides by a corrugated copper strip. The inward facing peaks of the corrugations act as spacers for the air fins, whilst the outward-facing peaks are relieved across part of their width to form waterway channels. Therefore by interlocking successive rows of complete airways any desired number of continuous waterways may be built up, through which the coolant flows in thin films. This type of matrix construction was once widely

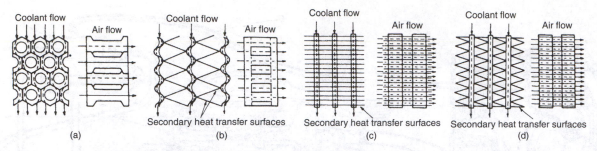

Figure 5.13 Types of radiator matrix: (a) honeycomb (b) cellular or film (c) tube and flat fin (d) tube and corrugated fin

used and generally provided a high rate of heat dissipation for a minimum weight of radiator.

Tube and flat fin

For this type of matrix, also known as the tube and plate type, a series of either in-line or staggered coolant tubes are employed. These in turn are assembled through a stack of continuous air fins, which are spaced fairly closely together and constitute the secondary heat transfer surfaces (Figure 5.13c). The coolant tubes are commonly produced from thin brass strip and assume a flattened oval section, whilst thinner copper strip is used for the air fins. Louvres may be pierced in the latter, so as to promote turbulence in the airways that they form with the coolant tubes. Greater structural strength is usually claimed for this type of matrix, as compared with the cellular or film type, this being an important requirement with the advent of pressurized cooling systems in the late 1940s.

Tube and corrugated fin

Also known as the tube and serpentine fin or the pack type this matrix construction is widely used in modern practice. It comprises alternate rows of waterways and airways, which are provided by brass tubes and zigzag ribbons of copper respectively (Figure 5.13d). Again, the coolant tubes are of flattened oval section and the air fins may be louvred to promote turbulence. There may be two or more rows of tubes, but in more recent practice the requirement for lighter and smaller radiators has led to the development of single-row versions. These have a finer pitch (less separation) for the corrugated fins to maintain an equivalent heat exchange capacity, since the heat radiated from the fins is considerably greater than that from the tubes in this type of radiator. This type of matrix represents a compromise between the cellular or film and the tube and flat fin constructions, in that it generally combines the more efficient heat transfer of the former with the greater structural strength of the latter. An increase in the size of the matrix is generally required for the cooling system of a turbocharged engine, owing to the greater amount of heat that has to be dissipated under maximum torque conditions.

For all types of radiator construction so far described, the sealing joints for the matrix and its header and collector tanks are solder bonded. Since a soldered joint has relatively little strength and both structural integrity and pressure tightness are prerequisites of radiator durability, the necessary strength must be provided by mechanical interlocking

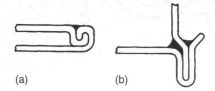

Figure 5.14 Mechanical interlocking of soldered radiator joints: (a) tubes (b) tank

of the parts with the solder connections being used just as a seal (Figure 5.14).

Aluminium and plastics radiators

The increasing use of aluminium alloy in automotive radiator construction, at least as original equipment, is by no means of recent origin. Bolted-on header and collector tanks in cast aluminium were early used in commercial vehicle radiators, and in 1960 the Harrison Radiator Division of General Motors introduced an all-aluminium radiator for a popular American sports car. The mid 1970s saw the introduction of an aluminium and plastics radiator developed by SOFICA in France and now adopted by several German and French car manufacturers. A feature of its construction is that a purely mechanical interlocking process, used in conjunction with synthetic rubber sealing gaskets, is employed for the aluminium alloy fin and tube joints in the matrix and also for joining the matrix and the nylon plastics header and collector tanks.

The advantages claimed for the aluminium and plastics radiator construction are generally those of a saving in weight, lower material costs with no painting requirement, resistance to corrosion and, if required, complex shapes. A possible disadvantage is that in the event of leakage it generally requires service replacement rather than repair.

For all-aluminium radiator construction the methods of joining used, apart from solder bonding, involve either forcible expansion of the tube ends where they enter the header and collector tank plates to provide a simple interference fit, or the application of modern adhesive bonding techniques at these junctures. Both methods lend themselves especially to automated production and the build consistency that it confers. However, the production techniques for copper/brass radiators have also been improved. These have allowed the greater

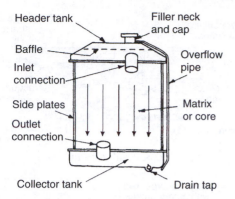

Figure 5.15 Down-flow radiator assembly

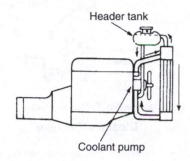

Figure 5.16 Down-flow radiator with separate header tank

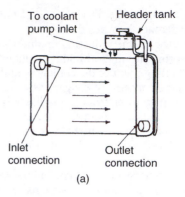

(a)

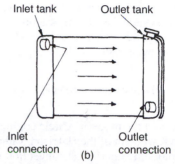

(b)

Figure 5.17 Cross-flow radiators (a) with (b) without separate header tank

strength of the brass tubing to be exploited, especially in reducing tube wall thickness for better heat transfer.

Down-flow and cross-flow radiators

The heated coolant entering the radiator may be dispersed through the matrix in either a down-flow or a cross-flow direction. The differences in radiator installation are described in the following sections.

Down-flow radiator

In this long-established form of radiator construction the matrix is attached to an upper header tank and a lower collector tank (Figure 5.15). The header tank performs several functions because, apart from receiving the heated coolant from the engine, it accommodates expansion of the coolant and provides a reserve against any loss of coolant. As its name would suggest, the purpose of the collector tank is to return to the engine the coolant from which heat has been abstracted.

In modern practice, the header tank proper may be mounted separately from the radiator. An advantage of this is to ensure that the radiator is always maintained 'solid' with coolant (Figures 5.12, 5.16 and 5.27). This reduces any mixing of air with the coolant entering the radiator, since aeration of the coolant has an adverse effect on heat transfer efficiency. For the same reason, radiators with an integral header tank may incorporate a submerged perforated baffle that is horizontally positioned just above the inlet pipe orifice, so that the circulating coolant is effectively separated from the air space in the upper part of the header tank. An incidental advantage of using a separate header tank, where its installation is such that

coolant is returned to the pump inlet, is that any tendency towards cavitation in the pump is reduced by virtue of the increase in pressure head at the pump inlet. *Cavitation* may occur in a pump where a restricted inlet passage promotes the formation of bubbles in the circulating coolant. Upon collapse these bubbles containing vapour and gas implode at extremely high pressures and cause erosion damage. A filler neck and cap, together with an overflow pipe connection, is provided on either the integral or the separate header tank.

Cross-flow radiator

In order best to utilize the available air flow through modern radiator grilles, the matrix itself needs to be low and wide. This requirement led to the development in the mid 1960s, notably by General Motors in America, of the cross-flow radiator, which is flanked on one side by an inlet tank and on the other by an outlet tank. To provide the cross-flow radiator with a reserve of coolant and a means of preventing aeration, a separate header tank is generally connected to the upper end of the outlet tank (Figure 5.17a). In some American applications, however, the separate header tank is dispensed with and the reserve of coolant is accommodated within the capacity of the outlet tank (Figure 5.17b). To prevent aeration, the filler cap and overflow arrangements are located at the upper end of the outlet tank, because it is in this region that any air entering the matrix tends to collect.

The pressurized cooling system

The advantages of pressurizing the cooling system may be summarized as follows:

1 It allows the coolant to circulate at a higher temperature without boiling, so that heat will be transferred more rapidly

from the radiator to the air by virtue of the greater temperature difference between them.

2 It compensates for the reduction in atmospheric pressure when motoring in high-altitude mountainous regions, where the boiling point of the coolant would otherwise be lower, with the consequent danger of overheating.

3 Since the heat transfer from the radiator is directly proportional to the temperature difference between the coolant and the air, a pressurized system allows the radiator to be reduced in size (although it must also be made stronger).

4 The increase in pressure at the coolant pump inlet reduces the possibility of cavitation damage, as earlier mentioned in connection with the separate header tank.

5 Because the system has to be sealed to pressurize it, instead of simply being vented to the atmosphere, coolant losses through evaporation and surging are minimized.

Construction and operation of the radiator pressure cap

The unsung hero of the modern vehicle cooling system must surely be John Karmazin of the Harrison Radiators Division of General Motors in America. His invention was the radiator pressure cap, which was first used in Buick cars in 1939, and has long since been taken for granted.

The radiator pressure cap is a combination of filler cap and pressure control valve. It is installed at the highest point in the cooling system and seals against a seating in the filler neck of the radiator. With the engine running and the cap in position, the cooling system is allowed to become pressurized. This occurs automatically because, as the temperature increases, the coolant is expanding in a closed system. The reason for using a radiator pressure cap is therefore to maintain the cooling system at a pressure above that of the atmosphere.

The cap consists essentially of a spring-loaded plate valve with a rubber facing, which is preloaded on to a fixed seating. According to whether an open or a closed type of construction is used, the fixed seating is provided in either the filler neck or the cap itself (Figure 5.18a and b). If the pressure in the system rises above the limit for which the control valve is rated, the spring loading on it is overcome so that the valve is lifted upwards off its seating and vents the system to atmosphere.

The pressure control valve is in turn fitted with a similar acting and concentric recuperation valve. This is arranged to open in the opposite direction against light spring loading. Air can therefore be admitted to the radiator to relieve any depression as the coolant temperature falls, otherwise at worst there could be a danger of the header tank collapsing.

The pressure control valve in the radiator cap is usually set to open at a specified value in the range of 28 to 105 kN/m² (4 to 15 lbf/in²). For each 7 kN/m² (1/lbf/in²) the system pressure is increased above atmospheric, the boiling point of the coolant is raised by about 1.5°C. The recuperation valve in the radiator cap is set to open whenever the system is subject to a depression higher than about 7 kN/m².

Caution must be observed in checking the coolant level of a hot engine. The radiator pressure cap should first be covered with a protective cloth and then part undone to release steam

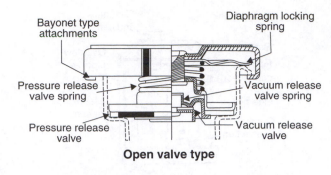

Open valve type

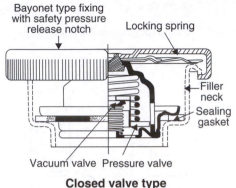

Closed valve type

Figure 5.18   Types of radiator pressure cap and their actions: (a) open (b) closed construction (AC Delco)

and all pressure, before it is completely removed from the filler neck.

Sealed cooling systems

The attraction of the modern sealed cooling system is that it reduces both the need for frequent inspection of the coolant level and the risk of weakening the antifreeze solution by topping up. Although the advent of the radiator pressure cap provided what amounted to a semisealed cooling system, the modern so called fully sealed system constitutes in effect a coolant recovery system, and indeed is known as such by the American motor industry. It basically comprises a reserve tank that is independently mounted from the radiator and vented to the atmosphere (Figure 5.19). A flexible hose connects the overflow stub pipe of the radiator to a dip tube in the reserve tank, which is nominally one-third filled with a coolant or thereabouts.

The action of the system is such that when the coolant reaches a sufficiently high temperature and exerts enough pressure to open the control valve in the radiator cap, the excess coolant will flow into the reserve tank from which it merely displaces air. This is in contrast to the earlier non-sealed system where the coolant would simply drain to the ground. When the engine later cools and the reduced volume of coolant in the system loses its heat, the accompanying drop in pressure will cause the recuperation valve in the radiator cap to open. Since the coolant in the reserve tank is vented to the atmosphere and is now at a greater pressure than that existing in the cooling system, an appropriate amount will flow back to fill the radiator again.

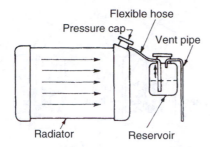

Figure 5.19 Fully sealed cooling system

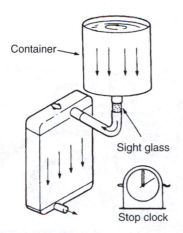

Figure 5.20 Radiator flow testing rig

Radiator inefficiency

The faults that can render a radiator inefficient and lead to engine overheating may be summarized under the following headings.

Partially blocked waterways

The presence of corrosive products on the internal metal surfaces acts as a thermal insulator, so that heat is less readily conducted from the coolant to the metal. A matrix flow testing rig may be used to ascertain *in situ* the extent of any internal restriction to coolant flow through the radiator following removal of its hoses. Briefly such a rig comprises a large cylindrical container that is supported above, and connected via a sight glass to, the inlet pipe of the radiator (Figure 5.20). A bung is temporarily inserted in the outlet pipe. After the radiator and container have been completely filled with water, the bung is removed from the outlet pipe and the time that it takes for the known quantity of water in the container to drain through the radiator is recorded and then compared against the manufacturer's specification. Remedial action usually takes the form of reverse flushing the radiator, in which a waste pipe is fitted to the inlet and water at mains pressure is applied to the outlet for a specified time. Alternatively a special flushing gun is used.

Partially blocked airways

If the external metal surfaces of the matrix are obstructed by foreign matter then the air flowing over them is given less opportunity to absorb their heat. The airways are cleaned out by applying air pressure from the engine side of the radiator; that is, in the reverse direction to normal air flow through it.

Malfunctioning pressure cap

Any malfunctioning of this item that causes a loss of cooling system pressure can result in overheating and give an impression of radiator inefficiency which is more apparent than real. A simple method of testing for cooling system pressurisation is *partially* to release the cap when the engine is hot, following which the release of pressure should be clearly audible. If this does not occur, it then becomes necessary to establish whether malfunctioning of the pressure cap is responsible for the absence of pressure or if this is being lost elsewhere in the system, including of course the radiator itself. To isolate the source of leakage generally requires the application of a proprietary pressure cap and cooling system tester, which combines a pump and gauge together with suitable adaptors for attachment to the pressure cap and the filler neck of the radiator.

Thermostatic control of the water-cooling system

It is worth reflecting that the function of the cooling system is not so much to keep the engine cool, but to prevent it from overheating. In this respect the introduction of a thermostatically regulated cooling system by the Cadillac Motor-Car Company of America for their 1914 V eight-cylinder engine, proved to be an important early step in the development of the engine water-cooling system.

The basic purpose of the thermostat is therefore to regulate the flow of coolant through the radiator in accordance with the cooling requirements of the engine. In modern practice, the reasons for using a thermostat are to:

1 Reduce engine warm-up time.
2 Maintain optimum running temperatures.
3 Meet the requirements of the car interior heating system.

The thermostat consists essentially of a poppet control valve, which is opened by a thermal expansion device or element and is closed by a return spring, or return spring action. Movement of the valve is therefore directly related to the temperature of the coolant surrounding its operating element. The thermostat has traditionally been installed between the coolant outlet from the cylinder head jacket and the inlet to the radiator header tank (Figure 5.25), but the modern tendency is to position it at the inlet to the coolant pump (Figure 5.26). These two arrangements are known respectively as 'outlet-side' and 'inlet-side' thermostats. A disadvantage of an outlet-side thermostat is that when it opens during engine warm-up, it can result in too much low temperature coolant entering the cylinder block, thereby causing unwanted fluctuations in engine running temperature. In contrast an inlet-side thermostat more quickly senses the temperature of the incoming coolant and limits its flow accordingly. Hence, the better controlled coolant flow from the radiator into the cylinder block and head, via the pump, provides a more even warm-up for the engine and improves driveability.

At normal operating temperatures the control valve is open and coolant circulates through both the engine jacket and the radiator, but when cold the valve is closed and circulation through the radiator is prevented, or at least very

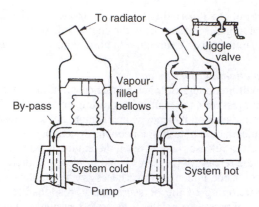

Figure 5.21 Action of the aneroid or bellows thermostat

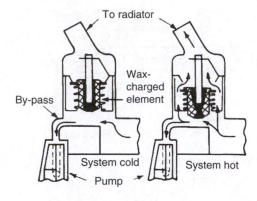

Figure 5.22 Action of the hydrostatic or wax thermostat

restricted. The engine is thus allowed to warm up much more quickly, since a smaller quantity of coolant is then being heated by it.

The two thermostats that may be used in motor vehicle cooling systems are the aneroid or bellows and the modern hydrostatic or wax types.

The aneroid thermostat
This type of thermostat consists of a control valve energized by a vapour-filled metal bellows (Figure 5.21). Control valve movement depends upon the difference between the vapour pressure in the bellows, at any given temperature, and the pressure in the cooling system. The control valve remains closed until the thermostat reaches a predetermined temperature, usually in the range of 75 to 80°C. At this point it begins to open in the direction of coolant flow. It then progressively opens further as the coolant temperature rises, until it is opened fully at around 90 to 95°C. As long as the system is operating, the thermostat continues to control the flow of coolant through the radiator in accordance with engine cooling requirements. The purpose of the jiggle valve is to facilitate refilling of the system by acting as a self-clearing vent hole. Once the engine is running the circulation of the coolant forces the loose-fitting jiggle pin against the vent hole to close the aperture.

The hydrostatic thermostat
With this type of thermostat, the control valve is energized by an element charged with a wax substance having a high coefficient of thermal expansion. The element consists of a cylindrical metal body containing the wax substance which surrounds a rubber insert, this in turn embracing a central operating thrust rod (Figure 5.22). As coolant temperature rises, the wax melts and compresses the rubber insert; but, since this rubber is constrained to act like an incompressible hydraulic fluid, it displaces the thrust rod. The control valve is thus opened against its separate return spring arrangement. For the modern cooling system, the hydrostatic or wax thermostat has the important advantage that it is relatively insensitive to pressure variations and it is also of more robust construction.

A more sophisticated application of the hydrostatic type of thermostat has appeared since 2000, wherein the wax filling

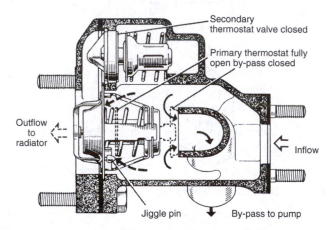

Figure 5.23 Twin-thermostat arrangements (*Gardner*)

is acted upon not only externally by the coolant temperature, but also internally by an electrically heated element controlled from the engine management system. The purpose of this additional control is two-fold. For normal running it allows the thermostat valve to open at a relatively higher temperature in excess of 95°C, so that both thermal and mechanical efficiencies of the engine are increased for improved fuel economy. However, when maximum performance is demanded, the heater element is activated to increase expansion of the wax filling and open more widely the thermostat valve. By virtue of the greater through-flow of coolant, the risk of localized overheating in the region of the combustion chambers is avoided, so that an optimum ignition advance can be signalled from the engine management control system.

Twin-thermostat systems
Two thermostats acting in parallel are sometimes used in the cooling systems of heavy-vehicle diesel engines (Figure 5.23). They are designated primary and secondary elements, with each having a different temperature rating. The primary element is the larger of the two and serves as the principal control in maintaining the operating temperature of the engine at acceptable levels. The secondary, smaller element functions as a supplementary control which opens to allow an increased flow of coolant to the radiator if abnormally high temperatures

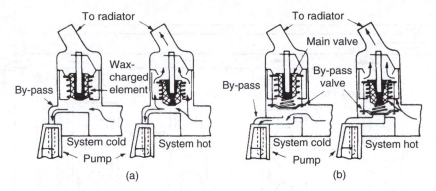

Figure 5.24   Comparing the action of (a) permanent (b) variable radiator bypass arrangements

are threatened. In the event of any misbehaviour of the primary element, the secondary element affords an alternative passage for the coolant flow to the radiator and thus provides a failsafe feature.

More recently, dual-valve thermostats have found application in some high-output passenger car petrol engines. In this case their purpose is to confer a larger open area for the thermostat, so as to allow a smoother increase in coolant flow as the engine warms up.

Radiator bypass arrangements

Rather than attempting to blank completely the coolant flow from the engine during warm-up, either a permanent or a variable radiator bypass arrangement is used with the thermostat. For this purposes a transfer connection must be provided between the underside of the thermostat control valve and the inlet side of the coolant pump. This ensures that when the control valve is closed to the radiator, the pump can recirculate the coolant around the engine to avoid any local overheating during warm-up.

In the case of the simple permanent bypass the transfer connection is no more than a bleed hole, since it must not be so large as to reduce the normal flow of coolant through the radiator when the thermostat control valve is open (Figure 5.24a). With a variable bypass the transfer connection takes the form of a large-bore duct, which is controlled by the lower of two thermostat control valves that operate in unison (Figure 5.24b). Therefore as the upper valve opens to the radiator so the lower one closes the duct to the pump, the net effect of this gradual transition in coolant flow being to maintain more stable the temperature of the coolant entering the engine jacket.

5.4   ENGINE COOLING SYSTEMS FOR PASSENGER CARS AND HEAVY VEHICLES

Typical installations

The Society of Automotive Engineers in America concisely defines the engine cooling system as 'A group of interrelated components to effect the transfer of heat'. Since we have already dealt with the basic principles and individual components of the engine water-cooling system, it now remains to relate this knowledge to examples of actual installations found in modern practice. Once having become familiar with

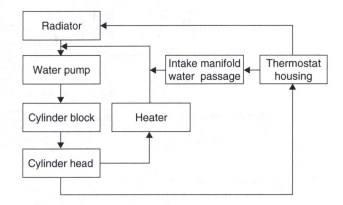

Figure 5.25   Block diagram of an in-line cylinder engine cooling system (*Nissan*)

typical installations by referring to the schematic drawings sometimes issued by vehicle manufacturers, then rather like the engine lubrication system, it often becomes expedient to represent a particular installation in the form of a block diagram (Figures 5.25 and 5.26).

An in-line cylinder engine furnishes a straightforward example of a forced-circulation cooling system, such as that shown for a Nissan six-cylinder engine (Figure 5.25). In contrast an equivalent V-cylinder engine cooling system requires the part-duplication of the circuit, so that each bank of cylinders is separately served (Figure 5.26). Two points worthy of note in the latter system, which shows the cooling circuit for the advanced V6-engined Nissan QX model, are that simultaneous flows of coolant are pumped into both the cylinder block and head in each bank of cylinders, and that the thermostat is positioned on the inlet side of the coolant pump for an even warmup. Another feature of interest in this Nissan installation is that the coolant pump is built into the engine timing case and chain- rather than belt-driven, thereby allowing a more compact design of engine. A similar part-duplication of the cooling circuit is likewise required for the less common horizontally-opposed cylinder engine.

In a typical arrangement of a cooling system for a heavy-vehicle diesel engine (Figure 5.27), the belt-driven pump draws coolant from the bottom of the radiator, via an oil cooler, and circulates it through the engine cylinder block and head. From the latter the coolant enters an externally mounted collecting

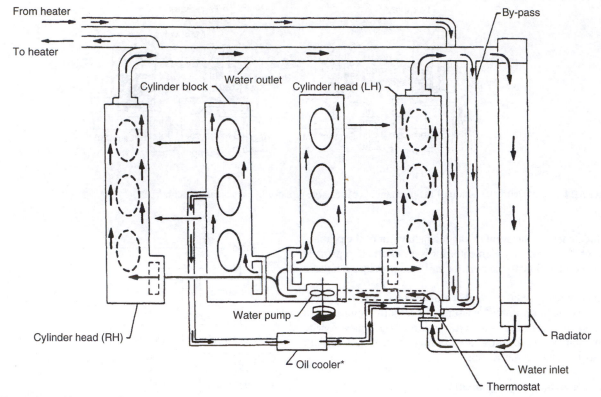

* : For models with power valve

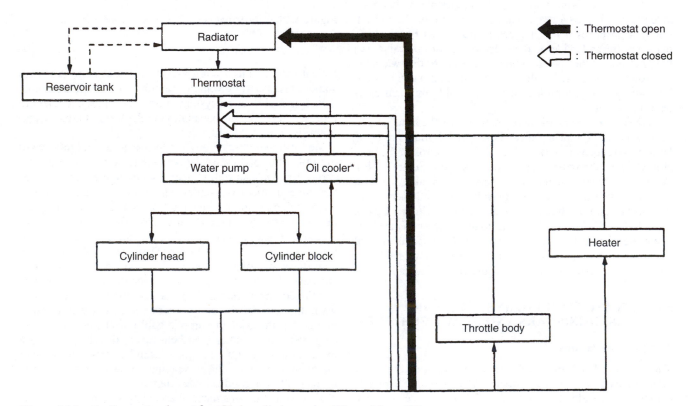

Figure 5.26 Cooling system layout for a V six-cylinder engine (*Nissan Motor GB*)

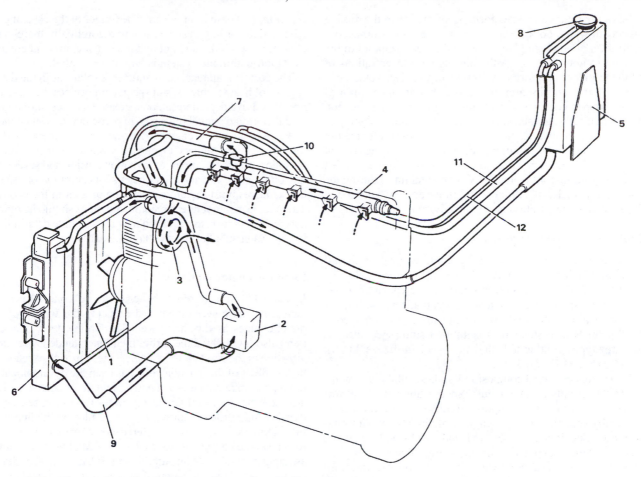

Figure 5.27   Heavy-vehicle diesel engine cooling system (*Seddon Atkinson*)

1 radiator	5 header tank	9 bottom connecting hose
2 oil cooler	6 radiator side tanks	10 thermostat value
3 water pump	7 top engine hose	11 expansion pipe
4 collecting pipe	8 pressure cap	12 expansion pipe

pipe from where it is returned to the top of the radiator, with the usual intervention of a thermostatically controlled bypass. The coolant level in the radiator is maintained by a separate header tank that is mounted on a cross-member behind the cab and has a pipe connection to the radiator top hose. This tank is also interconnected, via expansion pipes, with the side tanks of the radiator and the outlet from the engine to prevent any air pockets forming in the system. The cab heater receives a supply of coolant from the collecting pipe through a controllable flow valve, the coolant being returned to a connector located between the radiator bottom hose and the oil cooler.

5.5   ENGINE COOLANT

Antifreeze solutions

The use of an approved antifreeze solution with corrosion inhibitors, all the year round and in all parts of the world, is generally recognized as offering the following advantages:

1 It protects the cooling system against the damaging effects of frost in cold climates.

2 The coolant passages are protected against corrosion in both cold and hot climates.

3 For hot climates there is the incidental advantage that the boiling point of the coolant is raised.

Chemical composition

Of all the various antifreeze preparations that were once used in motor vehicle cooling systems, a water-soluble liquid known as ethylene glycol has long been accepted as the most satisfactory, no matter how severe the operating conditions. The adoption of ethylene glycol and water solutions really followed from their earlier use as high-boiling-point coolants in piston-type aircraft engines, especially during World War II. Pure ethylene glycol has a boiling point nearly twice that of plain water and this does, of course, explain why only the water evaporates away in an antifreeze solution. For the same reason, ethylene glycol is sometimes described as a permanent antifreeze material.

As far as the motor vehicle cooling system is concerned, when ethylene glycol is added to plain water it lowers the freezing point of the solution to well below 0°C. The actual freezing point, at which either the water content becomes solid or the first ice crystals form, is a function of the mixing ratio with water. For example, a 50 per cent by volume concentration of ethylene glycol lowers the freezing point of the coolant to about $-37°C$. Although higher concentrations of ethylene glycol are normally irrelevant to motor vehicles, it was in fact found in aircraft engines that an upper limit of about 70 per cent was set by the engine running too hot, because of reduced thermal conductivity of the coolant.

In practice many vehicle manufacturers now prefer the use of a 50 per cent concentration, but others generally advise not less than a 30 per cent concentration to take into account fast motorway driving in winter. For Continental travel during the winter, especially if Alpine conditions are encountered, a 40 per cent concentration is usually advised.

Protecting against corrosion

It was also from experience of the liquid cooling of aircraft engines, which posed problems of metallic corrosion, that the need to include a corrosion inhibitor in the coolant became evident. A similar recommendation was made by the Society of Chemical Industry to the motor industry in 1956 and led in 1959 to the British Standards Institution issuing specifications for three types of inhibited ethylene glycol antifreeze, known as BS 3150, 3151 and 3152.

The actual chemical composition of these inhibitor systems is of interest only to the chemist. Specification BS 3150 was based on a Ministry of Supply aircraft specification and is especially suitable for use in engines of aluminium alloy construction. Specifications BS 3151 and 3152 (the latter being based on American practice) are generally suitable for engines of cast iron construction. These formulation specifications were withdrawn in 1983 on the grounds of obsolescence, and in 1985 were replaced by a performance specification designated BS 6580. This prefers the more embracing description of engine coolant concentrate to that of antifreeze, and covers the minimum performance requirements of such a product in terms of boiling point, freezing point, foaming tendency and corrosion inhibition.

Testing antifreeze solutions

Since the degree to which an antifreeze solution can furnish protection depends upon the mixing ratio of ethylene glycol to water, a hydrometer test is necessary to indicate whether ethylene glycol or water, or both, should be added in the event of the radiator requiring topping up.

A hydrometer is an instrument for determining the relative density of a liquid, or the number of times a volume of that liquid is as heavy as an equal volume of water. It will, no doubt, be more familiar as a convenient method for testing the state of charge of a car battery. In a similar manner, it may be necessary to apply a temperature correction to the reading obtained with an antifreeze testing hydrometer, otherwise an appreciable error can arise in determining the freezing point of the solution. It is therefore imperative to follow the instructions issued with the hydrometer.

It should be appreciated that this type of test provides no indication of the condition of the corrosion inhibitors, so the vehicle manufacturer's recommendations as to the intervals when the coolant should be changed must be followed accordingly. These intervals may be longer in the case of cars with sealed cooling systems, where theoretically there is no contact between the coolant and the outside atmosphere and oxidation within the system is largely prevented.

The correct maintenance of antifreeze solutions in the diesel engines of heavy vehicles is of special importance, because the rate of additive depletion can be increased as a result of higher load factors and annual mileages. In particular the wet cylinder liners used in many engines (for the reasons mentioned in Section 2.3) are subject to intense vibration during normal operation, which can promote cavitation erosion and severe pitting on their coolant side. A sufficient concentration of protective additives must therefore always be present in the coolant. The vehicle manufacturer's recommendations should be rigidly adhered to in this respect, which may also include determining if local water supplies meet certain standards.

Long-life engine coolants

In the 1970s much research began to be concentrated on increasing the power output and reducing the weight of the diesel engines used in heavy vehicles. On the debit side, this trend also highlighted the earlier mentioned tendency towards vibration of their wet cylinder liners. One approach to reducing the ill effects of this phenomenon has been the development of more protective coolants. An early example of such a coolant utilized water for heat transfer, glycol for its antifreeze characteristics and nitrites to form a protective film on the liner surface. However, the nitrites suffered depletion and had to be replenished on a regular basis by adding what became known as Supplemental Coolant Additives or SCAs. A further development was the introduction of virtual nitrite-free coolants in the late 1980s, these using instead a carboxylate to confer the necessary protection for the liners.

In the mid-1990s the Texaco company introduced their 'Extended Life Coolant', which was developed jointly by their research facilities in Europe and America, with the object of providing an improved and easier to use heavy-duty coolant that would prove universally acceptable. Technically it is a coolant concentrate that is based on a synergistic combination of carboxylates, these being inhibitors that do not deplete like nitrites, and is generally referred to as 'Carboxylate Technology' or 'Organic Acid Technology' (OAT), although nitrite (NOAT) may be added as a secondary inhibitor. It is used as a 50 per cent concentration with water for all applications to provide maximum corrosion and frost protection. Unlike conventional coolant/antifreezes it does not contain silicates, nitrates, nitrites, borates, phosphates or amines, which has beneficial effects on how the coolant performs. In the absence of silicates, silicate gel or 'greengoo' as it is often known, does not form and silicates do not 'plate-out' in the engine cooling system. By eliminating phosphorus it is more difficult for hard water scale to form. Without silicates, phosphates, nitrates, nitrites and borates, there is a reduction in dissolved abrasive solids that can cause distress to coolant pump seals. As a result it is claimed that Texaco Extended Life Coolant, with no inhibitor testing, may be kept in service for 300 000 miles (480 000 km) or two years. This compares to traditional coolant concentrates that rely solely on nitrites to form a protective film on the cylinder liners, which require

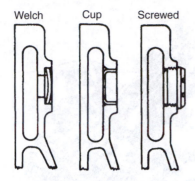

Figure 5.28   Types of engine core plug

the addition of Supplemental Coolant Additives at initial fill and thereafter at 12 000 to 20 000 mile (20 000 to 33 000 km) intervals. Carboxylate coolant technology has more recently been applied to passenger cars, where a typical recommendation is that uncontaminated it may be kept in service for 150 000 miles (240 000 km) or 5 years.

Engine core plugs

Core plugs are present in the engine cooling jacket for the following reasons:

1 They may blank off the holes left by the jacket cores during casting.
2 They may be removed for cleaning out corrosive deposits from the jacket.
3 In the event of the coolant freezing it will expand and force out the core plugs, thereby reducing the risk of cracking the jacket.

They may be of either the Welch plug, drawn steel cup or, less commonly, the screwed plug variety (Figure 5.28). The first and second types are respectively expanded and pressed into core holes, that have been machined to size.

Fail-safe engine cooling

In the event of a serious loss of coolant some manufacturers of V-cylinder engines, notably in America, have provided a 'limp-home' facility. This enables the car to be driven at reduced speed for a limited distance specified by the manufacturer, for example 25 miles (40 km), so as to avoid irreparable damage being done to the engine, especially in view of the modern trend towards aluminium alloy construction. A coolant sensor detects when the engine begins to overheat and signals this information to the driver. The engine management system then alternately delivers fuel to half the number of cylinders, the other half are therefore not firing but continue to pump air thereby helping to cool the engine.

5.6   INTERIOR VENTILATION AND HEATING

Ventilation systems

Among the many features that contribute to the comfortable travel of the occupants of a motor car is the modern ventilation

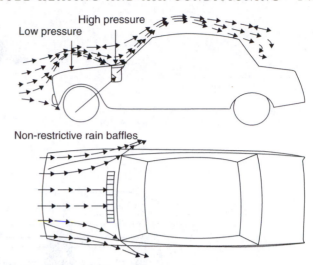

Figure 5.29   Position of ventilation and heating system plenum chamber in high-pressure zone of body air flow

and heating system and, in some cases, a full air conditioning system where refrigeration cooling is also included. First, we shall examine the provision for fresh air ventilation under the headings of direct and indirect ventilation.

Direct ventilation
Fairly obviously the body interior can be directly ventilated simply by opening one or more of the car windows. Whilst this may be perfectly acceptable say to the driver or the passenger alongside, the other passengers in the car may well experience uncomfortable draughts. An improvement in this direction was the once popular feature of the swivelling vent window. With the other windows closed this usually had an optimum degree of opening that provided reasonable ventilation without causing undue wind noise. From a service point of view vent windows could be notoriously difficult to seal against rain when closed, because of the complex shape of their sealing rubber.

Indirect ventilation
Since the early 1960s the approach to interior ventilation has been to provide a through flow of fresh air even with all windows closed. The air admitted to the car interior is usually routed through a plenum chamber, which is located beneath a cowl at the base of the windscreen. It may be recalled from the description of engine air cooling in Section 5.2 that a plenum chamber is one in which the internal air pressure is higher than that of the atmosphere. Hence, the particular location of the plenum chamber and its entrance is chosen to coincide with a high-pressure zone of air flow over the car body, and also where it is reasonably free from engine fumes (Figure 5.29).

The ventilating air flow through the body can be derived simply from the ram effect of air passing over the car and spilling into the plenum chamber, or at low car speeds it may be boosted by means of an electrically driven blower motor connecting with the plenum chamber. Directional control for the ventilating air flow is typically provided by readily adjustable deflectors where it leaves at facia outlets. Of

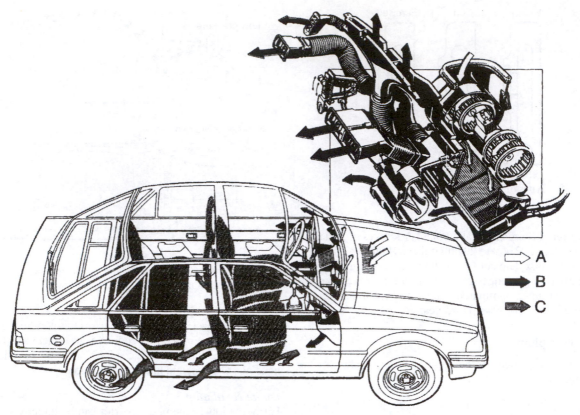

Figure 5.30 Ventilation and heating system air flow inside car

A fresh air into plenum chamber
B warm air into car interior
C stale air out

equal importance to admitting air when it is fresh is that of evacuating it on becoming stale; in other words, a through flow of ventilating air must be possible. Provision for doing this is by means of either extractor grilles incorporated in the rear quarter panels of the body or, as presently favoured, less visible outlet ducts situated in the lower body structure (Figure 5.30). In either case the outlets are sited to coincide with fairly neutral pressure zones of air flow over the car.

Healing systems

As earlier inferred, the ventilating and heating arrangements for a vehicle interior are usually combined into a single system. In conventional practice, the air taken into a plenum chamber reaches the interior through two independent circuits. One of these supplies fresh air and has both flow and direction controls, and the other supplies air at a controllable temperature. It is the latter or heating part of the system that we must now consider and especially in relation to the various sources from which the required heat energy may be obtained: coolant, exhaust, fuel and electrical.

Coolant
The utilization of heat rejected by the engine to its (liquid) coolant is not only the most commonly used source of energy for the heater system, but also is generally regarded as the most effective. As explained in Section 5.1, water

unlike air possesses a useful measure of thermal inertia, so that the heat input to the heater system can be maintained more or less constant during those periods when less heat is being rejected by the engine. For this type of system the basic units therefore comprise an electrically driven booster fan, a heat exchanger matrix plumbed into the engine cooling circuit, and a suitable control system (Figures 5.31 and 5.32). In early practice, the temperature of the air at the heater outlet was controlled by either a manually or a thermostatically set water valve. Although the latter method of control was generally to be preferred, neither version possessed the responsiveness of control provided by the air blending method favoured in modern heater systems. In these the stream of fresh air entering the heater system is divided by a controllable flap valve, so that a proportion flows through the airways of the heater matrix and the remainder bypasses the matrix. The final mixing or blending of the hot and cold air streams then takes place before the heated air enters the car interior. A further arrangement of internal flap valves provides directional control for the heated air in respect of the heating and demisting functions of the system (Figure 5.33).

It has long been recognized that it is more difficult to provide satisfactory interior heating for a car where its engine is directly air cooled rather than water cooled. Mention was made in Section 5.2 that, with air-cooled engines, the heated air discharged from the cooling system plenum chamber could be redirected to heat the car interior. This apparently

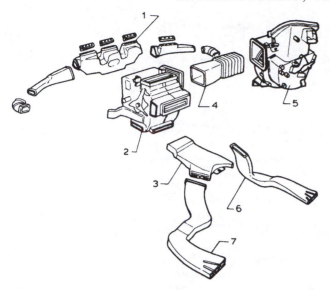

Figure 5.31 Exploded view of a heating and demisting system (*Nissan*)

1 defroster nozzle
2 heater unit
3 rear heater centre duct
4 heater duct

5 intake unit
6 rear heater duct, right hand
7 rear heater duct, left hand

neat solution to providing interior heating does, however, suffer from several drawbacks. First, the amount of air flowing from the heater can vary noticeably according to engine fan speed and hence throttle opening. Second, the temperature of the heated air is generally too low. Third, the system carries the liability that engine fumes could enter the car interior.

Exhaust

A relatively more effective interior heating system for cars with air-cooled engines utilizes exhaust system heat. For this purpose the exhaust pipes near the cylinder head ports are surrounded by muffs that form heat exchangers. Fresh air from the engine cooling fan is passed through the heat exchangers and thence to control boxes. From these the heated air is ducted to various outlet points in the body interior (Figure 5.34). Since the amount of engine heat rejected to the exhaust system varies with throttle opening and is much reduced on the overrun, this type of heater system also tends to give a less consistent performance than that derived from the coolant in water-cooled engines.

Fuel

Under this heading comes the rather specialized combustion heater, which, although utilizing the heat energy of the engine fuel, operates quite independently of the engine itself either as an air or a water heater (Figure 5.35). The mode of heat exchange with the former version is that fresh air destined

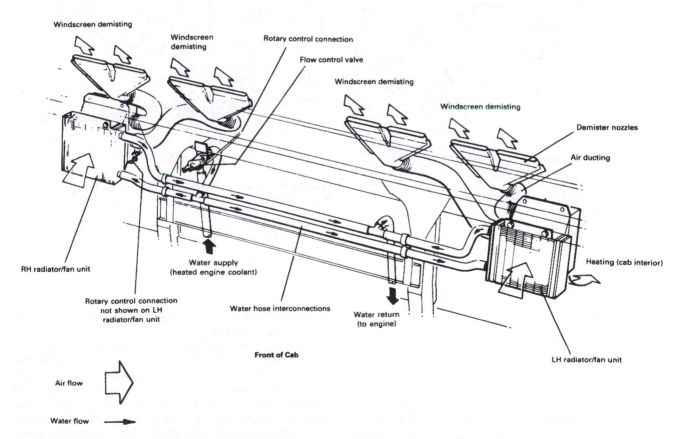

Figure 5.32 Arrangement of commercial vehicle cab heating and demisting system (*Seddon-Atkinson*)

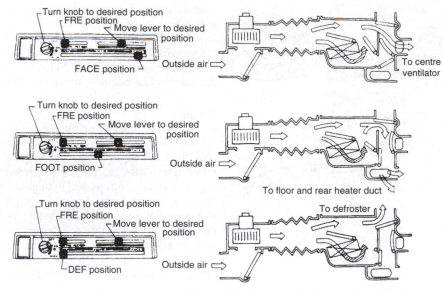

Figure 5.33 Typical control positions with a modern blending method of heating and demisting (*Nissan*). *Note:* FRE = fresh air and DEF = defrost

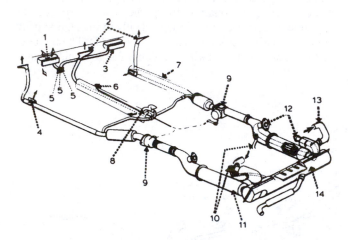

Figure 5.34 Interior heating derived from exhaust system

1 cold demist nozzle
2 hot demist nozzles
3 cold air to feet
4 foot outlet
5 feet/demist outlet controls
6 on/off heater control
7 lower sills
8 rear heater vent
9 heat control boxes
10 carburettor preheat pipes
11 heat exchangers
12 exhaust ports
13 clean cold air from main cooling fan
14 silencer

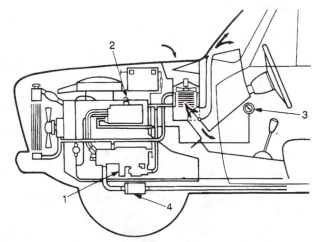

Figure 5.35 Interior heating derived from fuel system (*Smiths Industries*)

1 combustion heater
2 supply unit
3 time switch
4 exhaust silencer

for the vehicle interior is driven by a blower over the outer walls of a hot combustion chamber. Another blower supplies air to the interior of the combustion chamber, which from a pump receives atomized engine fuel and is also provided with an electrical means of igniting it. The products of this continuous combustion process pass into the vehicle exhaust system. Apart from the advantages of the alternative water heater version in being able to preheat both the engine and body interior of a vehicle, once a time switch has been set, combustion heaters of either type are often beneficially applied to such commercial vehicles as ambulances, mobile libraries and luxury coaches. With the latter, for example, the interior heat loss from the very large areas of window glass can be considerable. Of course, a disadvantage of any combustion heater lies in the additional fuel consumption incurred.

Electrical

The role of electrical energy in the interior heating of vehicles, apart that is from the secondary function of driving blower motors and operating the various control units, is confined to the demisting of rear windows. Rapid heating of the window glass is performed either by a flat strip heater element directly attached to the surface of the glass, or by a resistance wire element actually embedded within the glass. Since the current consumption of such heating devices is fairly high, their circuit is usually wired through the ignition switch so that they can only be activated when the engine is running to charge the electrical system. More recent applications of electrical energy in the role of heating have also included the windscreen, and the addition of warming elements in the seats or 'seat heaters' as they are generally known.

5.7   INTRODUCTION TO PASSENGER CAR AIR CONDITIONING SYSTEMS

General background

The term 'air conditioning' does not readily lend itself to restricted definition, because it is widely applied to systems that are needed for example in industrial cooling, where suitable temperatures have to be maintained during the manufacture and storage of certain products, and to systems that provide more comfortable conditions in public and domestic buildings and in our case, of course, the interior of motor vehicles. However, a definition of air conditioning that was proposed many years ago by the American Society of Heating and Ventilating Engineers, will probably best serve our purpose. It stated that 'Air conditioning is the science of controlling the temperature, humidity, motion and cleanliness of the air within an enclosure'.

Quite apart from the expected effects of motoring in hot climatic conditions, if a car is driven with the windows closed and the heating and ventilation system turned off, then irrespective of conditions outside the car the humidity and temperature within will gradually increase. This is because of the evaporation of perspiration from the car occupants and the heat radiated to the passenger compartment from the power unit and its exhaust system. Clearly the car occupants will experience discomfort, since the human body is not very tolerant of variations in atmospheric conditions. It is therefore beneficial if close control can be exercised over the condition of the air circulating inside the car. Although the ventilation system alone can go some way towards reducing the temperature of the car interior nearer to ambient, there is a limit to which the through-flow of air can be increased before it becomes obtrusive, especially with the windows opened, and in any event the ventilation system cannot act as a dehumidifier. The practical answer therefore is to provide a more comprehensive or full air conditioning system, where refrigeration cooling is also incorporated. Such a system allows the air in the car interior to be circulated through a cooling matrix, so that the reduction in temperature also serves to condense the moisture, the water then being drained away. Hence by controlling the temperature of the cooling matrix and the flow of the dry cooled air, the humidity and temperature in the car interior can be maintained at comfortable levels. Since a full

air conditioning system achieves maximum efficiency when all the windows are kept closed, the occupants further benefit from a reduction in general noise level and freedom from draughts and the ingress of dust.

Although the practicability of installing full air conditioning systems in passenger cars received consideration by American manufacturers in the late 1930s, it was not until twenty years later that they first appeared in their larger models and usually took the form of 'boot unit' refrigeration systems, which were so-called because the cooling matrix was installed in the boot and the cooled air admitted to the car interior from ducts in the roof cantrails or the rear parcel shelf. An alternative layout was pioneered by the once well-known American Nash Motor Company, who installed the cooling matrix under the scuttle and admitted the cooled air from ducts in the facia, and it was this type of installation that later became established practice and continues to form the basis of the modern and more sophisticated automatic climate control systems.

5.8   PRINCIPLES OF REFRIGERATED AIR CONDITIONING

The physical principle at the heart of the refrigerating function of a full air conditioning system is concerned with the enforced changes of state of the circulating refrigerant and the latent heat involved, the refrigerant being the term applied to the gas with which the system is charged and the medium by which heat is extracted from the air to be cooled. Latent heat (or hidden heat) is the amount of heat required to cause a substance to change its state without changing its temperature. Taking a familiar example, a good deal of heat or latent heat is taken in by water when it changes to steam and conversely is given out when the steam changes back to water, even though the temperature remains at 100°C during these changes of state. The significance of this to us is that when a liquefied refrigerant evaporates, it must extract its latent heat of vaporization from the air surrounding its container or, in our case, the air from the car interior. If the vaporized refrigerant is next compressed by a pump in order to liquefy it, then it can be arranged to give out its latent heat to the outside air, and the heat exchange cycle between the car interior and exterior can begin again. A suitable refrigerant must therefore possess a high latent heat, although not too high since less refrigerant can then be circulated. By the same token the refrigerant must have a low boiling point, but here again an extremely low boiling point is unnecessary and for automotive applications the temperature is about minus 26°C when it changes from a liquid to a gas.

Until the early 1990s a chlorofluorocarbon (CFC) based refrigerant, which was originally known under its American trade name of Freon-12 and often abbreviated to F-12, had been used to charge automotive air conditioning systems. Recognized by chemists as dichlorodifluoromethane, it was one of a group of commercially used Freon refrigerants that were also manufactured in other countries under different trade names and sometimes with unrelated numbering systems. To avoid this confusion, the American Society of Heating, Refrigeration and Air Conditioning Engineers (ASHRAE) later introduced a system of designating refrigerants in this particular group by 'R' numbers, so that the original Freon-12 and its equivalents have long been generally referred to as 'R-12' refrigerants.

In the mid-1980s there was mounting concern that when CFCs are discharged into the atmosphere and then enter the stratosphere, the planet is subject not only to depletion of the ozone layer that allows an increase in ultraviolet radiation to reach the earth, but also to a larger amount of the heat radiated from the sun to be absorbed by the earth. The damaging effects that these phenomena can have on the relation between living things and their environment or, in other words, the ecology of the planet are now expressed in terms of Ozone Depletion Potential (ODP) and Greenhouse Warming Potential (GWP). It was significantly to reduce these factors that from the early 1990s a replacement hydrofluorocarbon (HFC) based refrigerant, known to chemists as tetrafluoroethane, and designated HFC-134a has been used to charge automotive air conditioning systems. Since this has involved design modifications to these systems, the presently used HFC-134a or R-134a as it is now known cannot be substituted for the R-12 refrigerant used in earlier systems, because the two refrigerants cannot be mixed. The service recommendations of the vehicle manufacturer should therefore be sought in this matter.

Caution In the absence of proper training or qualified supervision, no attempt should ever be made to service refrigeration systems, since there are inherent dangers present in dealing with high-pressure equipment and low temperature refrigerants.

5.9 REFRIGERATED AIR CONDITIONING SYSTEM COMPONENTS

The basic components of a refrigerated air conditioning system (Figure 5.36) may be listed as follows:

Evaporator
Compressor
Condenser
Receiver drier
Expansion valve

Evaporator In refrigeration engineering the earlier mentioned cooling matrix is actually termed an evaporator (Figure 5.37). It is suitably encased and generally produced from aluminium alloy and comprises an arrangement of serpentine tubing through which the refrigerant is pumped at reduced pressure, together with a network of finned airways

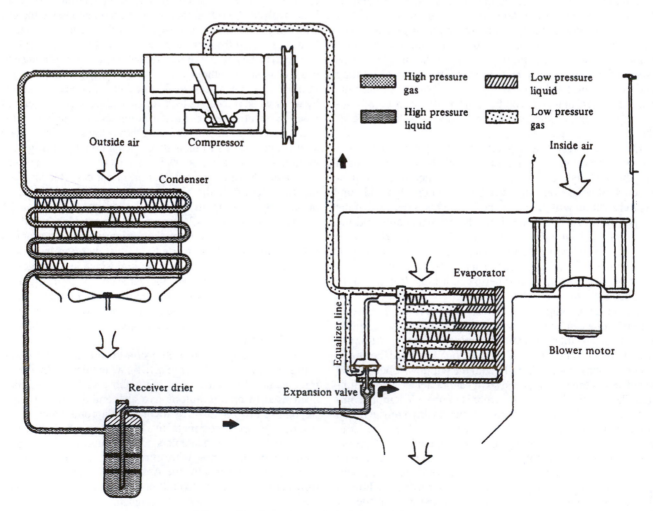

Figure 5.36 Schematic arrangement of air conditioning refrigeration system (*Nissan*)

that considerably increase the effective surface area of the evaporator, over which a flow of ventilating air is directed to the car interior by electrically driven blowers. In operation an atomized flow of refrigerant is delivered to the evaporator and owing to its low boiling point, the evaporator readily absorbs heat from the through-flow or ventilating air. In changing its state from a liquid to a vapour, the boiling refrigerant therefore takes up an appreciable amount of latent heat, which represents the heat loss from the ventilating air. Furthermore this air is not only cooled, but also dehumidified, because its moisture condenses over the large surface of the evaporator and is led away by a collector tray and drain tube to underneath the car. A thermostatically controlled switch may be used to sense the surface temperature of the evaporator, so that if it falls too near to freezing point the system compressor can temporarily be declutched to stop it pumping. If the evaporator airways became frosted, then clearly its cooling efficiency would be reduced and also the system could be damaged.

Compressor   In order that the heat-laden vaporized refrigerant leaving the evaporator can be returned to its liquid state again, the heat that it has absorbed from the ventilating air flow must be removed. The only practicable way of doing this is to introduce a compressor into the system. Its function is to draw off the vapour from the evaporator and then raise its pressure to about $1.5\,\text{MN/m}^2$ ($225\,\text{lbf/in}^2$), so that the corresponding condensing temperature is higher than that of the ambient air, which acts as a readily available cooling medium outside the car. In other words, the heat from the car interior is absorbed by changing the refrigerant from a liquid to a gas and then dissipated to the outside air by changing the refrigerant from a gas back to a liquid again, which explains why the system is sometimes referred to in thermodynamic terms as a 'heat pump'.

The compressor is mounted on the engine and incorporates an electromagnetically controlled clutch, which is belt-driven from the engine crankshaft and allows the compressor to be cut in or out of action according to system (and engine power) requirements. Constructionally the compressor may

be of either the single-acting reciprocating piston(s) and crankshaft type, which may be likened in operation to the compressor of an air brake system (Section 28.2) and was much favoured for early air conditioning systems; the double-acting reciprocating pistons and swash-plate type (Figure 5.38), which is widely used in modern practice and relies on the rotation of a tilted disc to oscillate the pistons, in the manner once found in specialized internal combustion engines of the 'axial' type that were known for their inherently good balance; or the purely rotary vane type, whose action may be compared to that of a similar type of engine supercharger (Section 9.2). In all cases these compressors must be equipped with suction and discharge valves, which are of the reed type and made from high-grade steel. Valves of this type are especially suited to refrigeration compressors by virtue of their simplicity, efficiency, longevity and quietness in operation.

Condenser   Although acting in the opposite sense to the evaporator, the condenser is of similar construction but mounted in front of the radiator for the engine cooling system (Figure 5.39). The condenser represents the second heat exchanger in the refrigerating system, because it dissipates the heat absorbed from the car interior, together with that added by the compressor, to the atmosphere. This occurs when the heat-laden vapour pumped from the compressor into the condenser gives up its heat and the refrigerant condenses to its liquid state. Although the condenser is designed to offer the minimum resistance to the through-flow of air, its presence in front of the radiator can nevertheless jeopardize engine cooling at slow speeds and when the car is stationary. It therefore early became necessary to adopt a pressurized cooling system (Section 5.3) where this had not previously been the case, and to increase the air flow through the radiator by fitting a coarser pitched (and potentially noisier) fan together with a radiator cowl. In modern practice, the advent of the independently mounted and electrically driven fan offered a more flexible solution to the problem. This type of fan can be mounted behind the radiator and additionally in front of the condenser, either singly or in pairs, thereby catering for higher than normal coolant temperatures and refrigerant

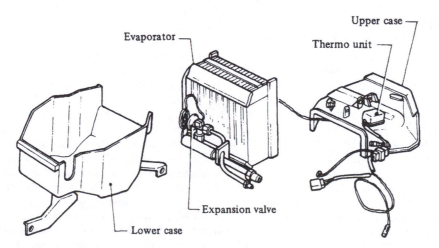

Figure 5.37   Refrigeration system evaporator assembly (*Nissan*)

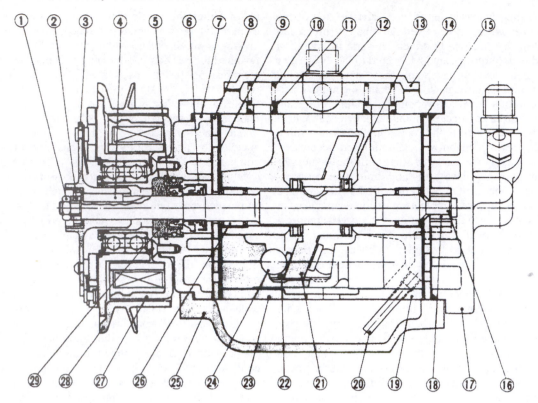

Figure 5.38 Cross-section through a swash-plate refrigeration system compressor (*Nissan*)

1 Shaft nut	9 Suction valve plate	17 Rear end cover	25 Shell
2 Clutch hub nut	10 Silencer spring	18 Oil pump	26 Needle bearing
3 Clutch hub	11 Silencer piece	19 Rear cylinder	27 Clutch coil
4 Key	12 Discharge valve	20 Oil pipe	28 Pulley and bearing
5 Shaft seal seat	13 Thrust bearing	21 Swash plate	assembly
6 Shaft seal	14 Side cover	22 Shoe disc	29 Felt
7 Front end cover	15 Rear cylinder head	23 Front cylinder	
8 Front cylinder head	16 Compressor shaft	24 Drive ball	

pressures respectively. Furthermore, their electrical circuits can be arranged so that lower fan speeds are provided when the thermal loading on the systems is normal and higher speeds when it becomes greater, thereby contributing to reduced noise levels.

Receiver drier Situated below and connected to the outlet of the condenser is the receiver drier (Figure 5.40). Its purpose is to store excess amounts of refrigerant when the cooling demand and hence the thermal load on the air conditioning system is low, and conversely to return stored refrigerant when additional cooling of the car interior is required, thereby maintaining an optimum flow of refrigerant within the system. It also includes a strainer and a desiccant, so that foreign particles and moisture respectively can be removed from the circulating refrigerant.

In operation the condensed refrigerant enters an upper inlet connection on the receiver drier body and then passes down through the strainer and desiccant. There will be a certain amount of vaporized refrigerant above the liquid in the receiver drier, which maintains a high pressure in the system. A central dip tube extending almost to the bottom of the receiver drier is therefore required to convey the liquid refrigerant up again to an outlet connection, which leads to

an expansion valve in the system. The head portion of the receiver drier also incorporates a sight glass and a melting or fusible plug, which communicate with the outgoing refrigerant. In the former case it enables a service engineer to observe the condition of the refrigerant, noting for example if bubbling indicates an insufficiency of refrigerant; whilst the latter fitment acts as a safety device in the unlikely event of an excessive build-up of pressure in the line to the expansion valve. Should this occur then the plug melts at about 105°C and discharges the high pressure refrigerant into the atmosphere.

Expansion valve It should now be evident that the refrigerating cycle of the air conditioning system depends on maintaining a high pressure at the condenser to liquefy the refrigerant, and a low pressure at the evaporator to vaporize it. In order to create this pressure differential it becomes necessary to introduce what is termed an expansion valve or, more specifically, a thermostatic expansion valve, commonly abbreviated to 'TXV'. This valve acts not only as a restrictor to atomize the liquefied refrigerant entering the evaporator, but also as a variable orifice to regulate its flow, so that it is delivered at exactly the same rate as it is being vaporized (Figure 5.41). To meet

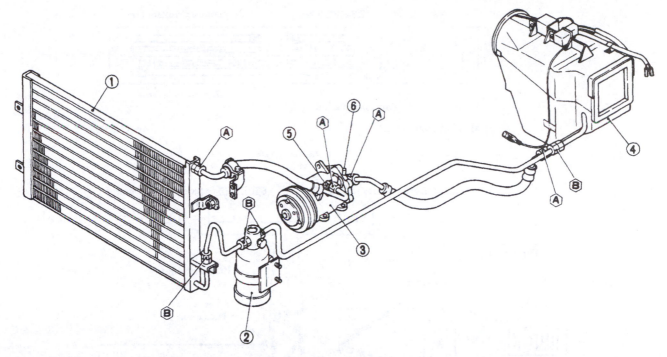

Figure 5.39   Refrigeration system condenser installation (*Nissan*)

1 Condenser
2 Receiver drier
3 Compressor
4 Cooling unit

5 High pressure (discharge)
   service valve
6 Low pressure (suction)
   service valve

Tightening torque N-m (kg-m), ft-lb

Ⓐ: 39–40 (4.0–5.0, 29–36)
Ⓑ: 29–30 (3.0–4.0, 22–29)

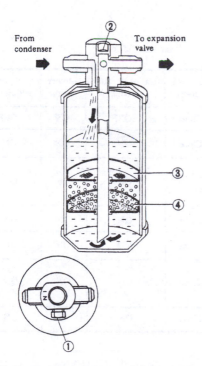

Figure 5.40   Refrigeration system receiver drier (*Nissan*)

1 melting plug
2 sight glass

3 strainer
4 desiccant

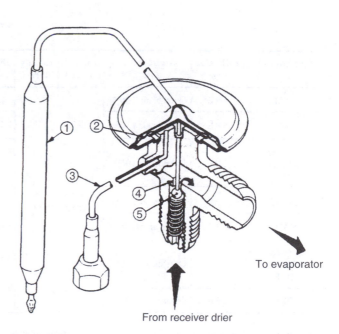

Figure 5.41   Refrigeration system expansion valve (*Nissan*)

1 sensing bulb
2 diaphragm
3 equalizer line

4 orifice
5 valve ball

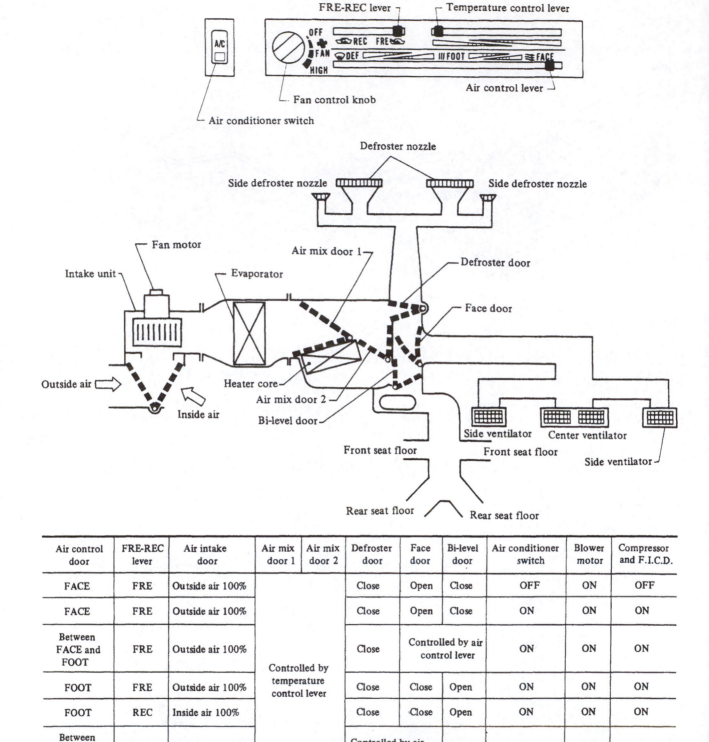

Air control door	FRE-REC lever	Air intake door	Air mix door 1	Air mix door 2	Defroster door	Face door	Bi-level door	Air conditioner switch	Blower motor	Compressor and F.I.C.D.
FACE	FRE	Outside air 100%			Close	Open	Close	OFF	ON	OFF
FACE	FRE	Outside air 100%			Close	Open	Close	ON	ON	ON
Between FACE and FOOT	FRE	Outside air 100%	Controlled by temperature control lever		Close	Controlled by air control lever		ON	ON	ON
FOOT	FRE	Outside air 100%			Close	Close	Open	ON	ON	ON
FOOT	REC	Inside air 100%			Close	Close	Open	ON	ON	ON
Between FOOT and DEF	FRE	Outside air 100%			Controlled by air control lever		Close	ON	ON	ON
DEF	FRE	Outside air 100%			Open	Close	Close	ON	ON	ON

Figure 5.42 Schematic arrangement of full air conditioning system (*Nissan*)

these requirements the TXV essentially comprises a spring-returned ball-valve, the extent of its opening being controlled by a push-rod and superimposed diaphragm flexing in a chamber, which has connections above and below the diaphragm that reflect changes in refrigerant temperature and pressure respectively at the outlet of the evaporator. The pressure acting above the diaphragm is derived from a thermal element termed a 'sensing bulb', which is charged with refrigerant and strapped to the outlet pipe of the evaporator. It communicates with the diaphragm chamber through a capillary tube, so that pressure changes within the bulb resulting from temperature variations at the evaporator outlet are transmitted to the diaphragm.

If for example the temperature of the vaporized refrigerant leaving the evaporator falls too low, then it means that too much refrigerant is being delivered, but since this will also result in a pressure drop within the sensing bulb, there will be less pressure acting above the diaphragm and the ball-valve will be urged closer to its seating by the return-spring. Therefore this reduces the amount of refrigerant delivered to the evaporator. The pressure acting below the diaphragm corresponds to that existing within the outlet pipe of the evaporator, and assists the action of the temperature sensing bulb in controlling the amount of refrigerant delivered to the evaporator. If for example the pressure of the vaporized refrigerant leaving the evaporator rises too high, then again it means that too much refrigerant is being delivered, but since there will also be more pressure acting beneath the diaphragm, it will allow the ball-valve more closely to approach its seating under the influence of the return-spring and thereby reduce delivery of the refrigerant.

5.10   FULL AIR CONDITIONING SYSTEMS

The overall concept of a full air conditioning system for a passenger car therefore features a heater matrix supplied with hot water from the engine cooling system and a cooling matrix circulated with refrigerant from a refrigeration system, either system being capable of independent operation and both possessing sufficient capacity to meet maximum heating or cooling requirements. Full air conditioning systems naturally vary in detail layout, according to a particular manufacturer's preferences, so that the manner in which the heating and cooling matrices together with the necessary flap valves or blending doors to control the ventilating air flow are combined in a system, best lends itself to illustration and tabulation (Figure 5.42).

Later developments of full air conditioning systems, known variously as 'comfort control' and 'automatic climate control', have taken advantage of modern electronic control systems, so that once the driver has preselected a temperature for the car interior, an electronic control unit (ECU) will automatically maintain this temperature regardless of changes in the ambient air temperature. The ECU receives current outputs from a series of strategically positioned temperature sensors, which enables it to build up a complete picture of ambient, car interior, system evaporator, engine coolant and solar radiation

temperatures and then, in accordance with the driver's switch setting, will control the functioning of the system in its entirety from the flap valve actuators to the blower motor and the compressor clutch. Since by definition air conditioning is also concerned with cleanliness of the air entering the car interior, a pleated paper air filter element or 'pollen filter' is now generally housed at the intake of the ventilation system, and typically requires renewal at 20 000 miles (33 000 km) intervals.

5.11   COMMERCIAL VEHICLE REFRIGERATION UNITS

Cold storage is widely used for keeping food fresh during transportation by road. Refrigerated vehicles or trailers are specially designed with insulated bodies and equipped with either overcab or front mounted trailer refrigeration units, which are capable of maintaining the required temperatures for chilled foods at above 0°C and deep frozen foods down to minus 30°C.

The desirable features of a commercial vehicle refrigeration unit may be listed as follows:

1 It must be of sufficiently high capacity to provide very low container temperatures and enable these to be rapidly recovered following door openings.
2 Low weight of the unit is an important requirement in refrigerated distribution to reduce vehicle front axle loading.
3 Quiet operation is essential, even at idle speed, so that early morning deliveries can be made without noisy disturbance.
4 Minimum energy consumption by the use of a low speed compressor and large evaporator and condensing surfaces.
5 Simple and durable construction for reliable operation and least maintenance, thereby reducing vehicle downtime.
6 An external evaporator to avoid loss of load space and the risk of drivers accidentally hitting their heads when sorting the load.
7 Rapid and effective defrosting in 10 to 15 minutes by electric or full reverse cycling of the unit, thus maximizing the refrigeration cycle.
8 An added advantage is a tilting facility for overcab units that permits a lower body height, whilst still enabling the vehicle cab to be tilted.

The manner in which these various requirements are met by a typical overcab unit from the range of German Frigoblock transport refrigeration units is shown and its salient features highlighted in Figure 5.43, whilst the installation arrangements for overcab and front mounted trailer units are shown respectively in Figure 5.44a and b. Refrigerant R-22 is currently permitted to be used in these units, because with longer running times and higher capacity requirements it offers important advantages with regard to energy consumption and refrigeration capacity, as compared to the earlier mentioned HFC-134a refrigerant.

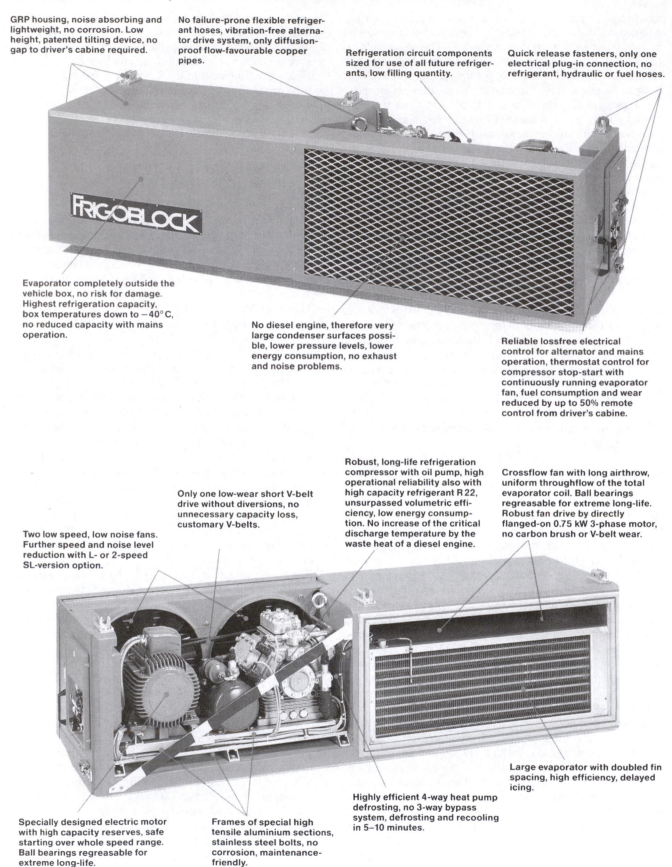

GRP housing, noise absorbing and lightweight, no corrosion. Low height, patented tilting device, no gap to driver's cabine required.

No failure-prone flexible refrigerant hoses, vibration-free alternator drive system, only diffusion-proof flow-favourable copper pipes.

Refrigeration circuit components sized for use of all future refrigerants, low filling quantity.

Quick release fasteners, only one electrical plug-in connection, no refrigerant, hydraulic or fuel hoses.

Evaporator completely outside the vehicle box, no risk for damage. Highest refrigeration capacity, box temperatures down to −40°C, no reduced capacity with mains operation.

No diesel engine, therefore very large condenser surfaces possible, lower pressure levels, lower energy consumption, no exhaust and noise problems.

Reliable lossfree electrical control for alternator and mains operation, thermostat control for compressor stop-start with continuously running evaporator fan, fuel consumption and wear reduced by up to 50% remote control from driver's cabine.

Only one low-wear short V-belt drive without diversions, no unnecessary capacity loss, customary V-belts.

Robust, long-life refrigeration compressor with oil pump, high operational reliability also with high capacity refrigerant R 22, unsurpassed volumetric efficiency, low energy consumption. No increase of the critical discharge temperature by the waste heat of a diesel engine.

Crossflow fan with long airthrow, uniform throughflow of the total evaporator coil. Ball bearings regreasable for extreme long-life. Robust fan drive by directly flanged-on 0.75 kW 3-phase motor, no carbon brush or V-belt wear.

Two low speed, low noise fans. Further speed and noise level reduction with L- or 2-speed SL-version option.

Specially designed electric motor with high capacity reserves, safe starting over whole speed range. Ball bearings regreasable for extreme long-life.

Frames of special high tensile aluminium sections, stainless steel bolts, no corrosion, maintenance-friendly.

Highly efficient 4-way heat pump defrosting, no 3-way bypass system, defrosting and recooling in 5–10 minutes.

Large evaporator with doubled fin spacing, high efficiency, delayed icing.

Figure 5.43 Salient features of a modern overcab commercial vehicle refrigeration unit (*Frigoblock*)

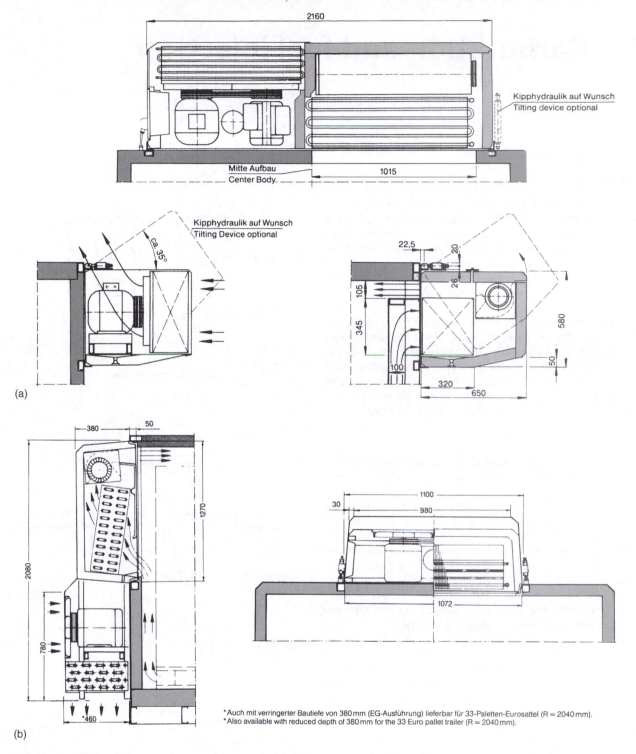

Figure 5.44   Installation differences between (a) overcab and (b) front mounted trailer refrigeration units (*Frigoblock*)

6 Carburation and fuel injection

6.1 FUEL SUPPLY SYSTEM

General layout

The purpose of the fuel supply system for a petrol engine is to store, transfer and filter the petrol required either by the float chamber of a carburettor, or by the pressure regulated circuit of a fuel injection system. It comprises a suitably vented fuel tank with level indicator, either a mechanically or an electrically operated fuel pump, fuel filters and the necessary pipelines to connect these services to their point of delivery.

Caution It perhaps hardly needs stating here that, from a servicing point of view, responsible supervision and extreme care are of paramount importance when working on the automotive fuel system with its attendant risk of petrol fire. Further reference should be made to the precautionary guidelines issued by the Health & Safety Executive in their publication HSG67 titled 'Health & Safety in MOTOR VEHICLE REPAIR'.

Fuel tank

In a conventional front-engined car the fuel tank is located remote from the engine, and its installation is designed to protect against damage in the event of a rear end collision. Fuel tanks are traditionally fabricated from sheet metal, this being either corrosion-protected steel or aluminium alloy. The advantages of using the latter material include a saving in weight and inherently good resistance to corrosion, but any repairs are more difficult to effect. A later German development has been the moulded plastics fuel tank, which is now widely used and not only weighs less than a corresponding sheet metal tank, but possesses freedom from corrosion and also complies with modern safety requirements.

Most fuel tanks are fitted internally with a number of baffles which serve to minimize any violent surging of the fuel content during acceleration, braking and cornering. Similarly, an anti-surge tower with restrictor holes typically surrounds the outlet pipe to the fuel system (Figure 6.1a). Otherwise, the fuel supply to the engine could be interrupted as a result of the fuel surging away from either the outlet pipe to the fuel pump or, where an electrical fuel pump is mounted inside the tank, the pump inlet port. There are cutaway portions at the lower corners of the baffles, which enable the fuel levels in the tank compartments to equalize under non-surging conditions. The pivoted float of the fuel level indicator is generally shielded within a separate compartment formed by the baffle system. In some modern installations where the fuel tank is both wider and shallower, transverse collector funnels with non-return, inwardly opening, flap valves may be used instead of a simple anti-surge tower (Figure 6.2a). Again this is to ensure that the fuel supply cannot be interrupted and allow air to enter the

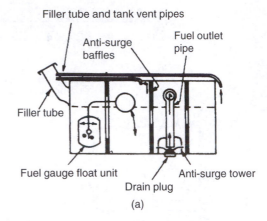

(a)

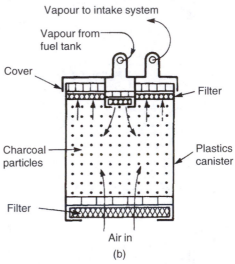

(b)

Figure 6.1 Fuel storage: (a) basic tank construction (b) vapour recovery canister

system, when corner surging or camber tilting occurs with a low level of fuel in the tank. Another feature of modern fuel tank construction concerns the need to cool the return flow of surplus fuel from the fuel injection system (Sections 6.7 and 6.8). For this purpose the returning fuel is directed tangentially into the lower part of a swirl chamber equivalent of an anti-surge tower (Figure 6.2b). This arrangement therefore provides the opportunity for both cooling and deserating the fuel returned to the tank, prior to it being delivered again to the fuel injection system. To prevent any accumulated sediment from entering the fuel supply system proper, the depth to which the outlet pipe enters the tank is always such that a low level of fuel remains at the bottom. For periodic removal of the sediment, a screwed drain plug is generally provided in the base of the tank.

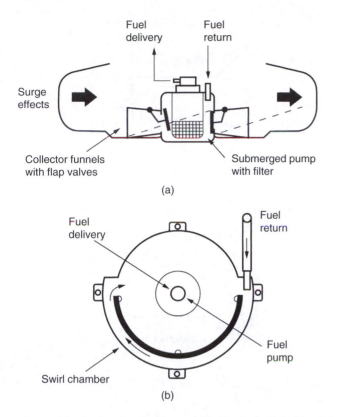

Figure 6.2   Fuel tank refinements: (a) collector funnels (b) swirl chamber

A fuel filler tube with cap is connected towards the top of the tank and must be suitably vented, so that air can enter and take the place of fuel pumped from the tank. In the absence of such venting, a depression would be created in the tank that could not only interfere with normal fuel delivery, but also tend to collapse the tank. Another vent pipe may also be connected from the top of the fuel tank to a point near the top of the filler tube, so that air can leave the tank as fuel is pumped into it and thereby prevent an air lock. As a fire safety precaution an anti-spill device, such as a rubber flap valve, may be incorporated in the fuel filler tube to prevent fuel escaping from the tank in the event of the vehicle overturning in an accident.

In the early 1970s it was recognized in America that an important source of air pollution caused by the motor car was the evaporation of petrol from its fuel tank, especially on hot days. As a result, many cars now have their fuel tanks vented not to the atmosphere, but to an 'on-board vapour recovery (OBVR) system'. When the engine is restarted following a period of shut-down, a mixture of filtered air and the previously stored vapours is then metered into the intake system for burning in the cylinders (Figure 6.1b). For modern fuel injection systems, another reason for using a non-vented filler cap is to maintain a small pressure in the fuel tank, thereby assisting the action of an in-tank fuel delivery pump.

Fuel pump

A positive displacement pump is required to transfer petrol from the fuel tank to the point of delivery, since the installation level of the former is invariably lower than that of the latter and therefore precludes the use of a gravity-feed system, as was once possible with the scuttle-mounted tanks on early motor vehicles. In most carburettor fuel supply systems the pump is of the diaphragm type and operated either mechanically or electromagnetically, the latter usually being described simply as electrical pumps. Electrically driven rotary pumps of the sliding roller-vane type are used to meet the higher pressure demands of fuel injection supply systems. These demands have increased considerably for more recent direct, as distinct from indirect, fuel injection systems, which require high-pressure plunger type pumps driven from the engine camshaft. An important feature of all these fuel delivery pumps and their installation is, of course, that they must be self-priming.

Fuel filters

Filtration in the fuel delivery system to the carburettor or injectors serves two purposes:

1 It must prevent foreign particles from becoming lodged in and interfering with the normal action of the fuel pump valve mechanism.
2 Similarly, it must protect not only the float valve mechanism of the carburettor, but also its fuel metering devices and internal passages. The latter must likewise be protected in the fuel injection systems, especially in the case of direct fuel injection systems that demand greater wear protection and hence finer filtration.

In carburettor installations, fuel filtration is often performed by using a combination of coarse filters on the inlet side of the pump, and a fine filter on its outlet side. Fine filtration is not usually employed on the inlet side, because it may result in an increased tendency towards vapour locking in the fuel line. Here it should be explained that vapour locking is most likely to occur at high operating temperatures, when any overheating of the fuel creates an excess of vapour in the supply system. This causes either partial or complete starvation of the fuel supply to the carburettor with accompanying misfiring and stalling of the engine.

A relatively coarse surface filter or strainer, in the form of either a circular screen or an extended sleeve of wire gauze, is sometimes located in the fuel tank itself at the entrance to the fuel line inlet (Figure 6.3a). All fuel leaving the tank and entering the delivery system must therefore first pass through this filter. The fuel pump typically incorporates a built-in surface filter at the inlet to its pumping chamber. In mechanically operated pumps, the filter often takes the form of a circular wire gauze screen.

In past practice, it was customary to afford additional protection for the carburettor by merely providing a relatively less coarse strainer at its fuel inlet, such as either a sleeve or a thimble gauze screen. More recently, the trend has been towards finer filtration for the fuel entering the carburettor, with the object of minimizing flooding complaints. A fine depth-type filter of specially processed, water-resistant, pleated paper may therefore be incorporated before the outlet of the pump itself (Figure 6.3b), or in the pipeline between the pump and the carburettor (Figure 6.3c). The inline type of filter is usually replaced in its entirety at appropriate servicing intervals.

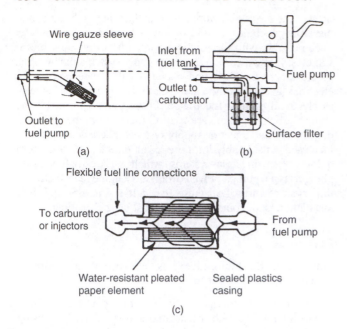

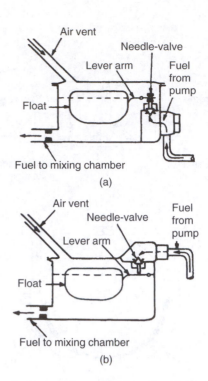

Figure 6.3 Examples of fuel filters: (a) coarse filter at tank outlet (b) fine filter at pump outlet (c) in-line fuel filter

Figure 6.4 Carburettor float chambers: (a) with bottom fuel feed (b) with top fuel feed

In the case of fuel injection system filters, these are mounted in-line between the delivery pump and the pressure regulated circuit (Figure 6.28). The pleated filter medium typically has had a pore width of $10\,\mu m$, but the more stringent filtering requirements imposed by direct injection systems have resulted in the pore width being reduced to $5\,\mu m$ or less. A further development has been the in-tank 'lifetime' filter, the shape of which is tailored to suit the available space in the fuel tank.

Carburettor float chamber

The carburettor float chamber was originally conceived as long ago as 1886 by the German pioneer automobile engineer Karl Benz. The float chamber acts as a constant-level reservoir for the fuel required by the mixing chamber. To regulate the amount of fuel admitted to the float chamber, a float valve mechanism is used. This consists of either a brass, a rubber or a plastics float that actuates a needle valve, the seating for which is contained in the chamber body or its cover. For accurate control of the fuel level, the needle valve has a tapered tip to allow progressive opening and closing of the fuel inlet passage. In some cases a light compression spring is interposed between the needle valve and its end, which contacts the float arm, so as to reduce vibration effects. A rubber-tipped needle valve may sometimes be used instead of the all-metal type, the resilient tip permitting any fine particles to become temporarily embedded. Hence by ensuring that the needle valve does not leak when closed, the carburettor is less likely to suffer from flooding.

The needle valve is sited either coaxially above the float with which it is in direct contact or, more usually, alongside the float from which it is indirectly actuated by a lever arm. An advantage of the latter arrangement is that it provides a leverage in favour of the float, so as to enable a larger needle valve area to be used where a greater through flow of fuel is demanded. The lever arm is pivoted between the float and the needle valve to suit a bottom fuel feed to the chamber (Figure 6.4a), whereas it is pivoted at the opposite end to the float for a top fuel feed (Figure 6.4b).

In operation, the float descends to open the needle valve when the fuel level in the chamber falls. Conversely, as additional fuel flows into it, the float ascends by virtue of its buoyancy. When the desired level is reached, the needle valve closes firmly on to its seating so that no further fuel can enter the chamber. To meet a continuous demand for fuel, the float mechanism remains open by an amount just sufficient to maintain the fuel level practically constant. The float chamber is ventilated either externally to the atmosphere or, as is now generally preferred, internally to the air intake of the carburettor.

Fuel pipelines

An obvious requirement of the fuel pipelines connecting the tank, pump, filter and carburettor is that they must be pressure tight. If this condition is not maintained, any leakage of petrol on the delivery side of the pump presents a fire hazard, while the entry of air on the intake side will reduce pumping capacity. Rigid metal pipelines are used for those parts of the system attached either to the body structure or to a separate chassis frame. Their installation is generally such that they are positioned away from the heat of the exhaust system and also afforded some protection from stones flung up beneath the car from the wheels. Usually the fuel pipelines are supported in small metal clips that may be lined with rubber, which not only protect the pipes against vibration, but also isolate the car interior from transmitted pump noise. A flexible plastic hose is used for that part of the fuel pipeline attached between the car structure and the engine, because of the

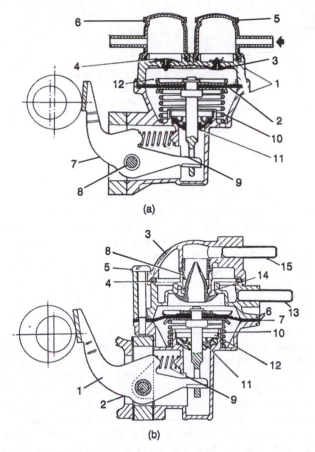

Figure 6.5 SU mechanical fuel pump valve arrangements

(a) Type AUF 800

1 body	7 rocker lever
2 diaphragm	8 pivot pin
3 inlet valve	9 lever return spring
4 outlet valve	10 diaphragm spring
5 inlet nozzle	11 crankcase seal
6 outlet nozzle	12 diaphragm chamber

(b) Type AUF 700

1 rocker lever	9 lever return spring
2 pivot pin	10 diaphragm spring
3 outlet cover	11 crankcase seal
4 sealing washer	12 pressed steel cup
5 cover retaining screws	13 inlet nozzle
6 body	14 annular passage
7 diaphragm	15 outlet nozzle
8 outlet valve	

latter being resiliently mounted. In more recent practice, plastic hose has received wider application in other sections of the fuel system pipeline.

Mechanical fuel pump

This type of fuel pump comprises a pumping chamber, which contains either separate or combination inlet and outlet valves (Figures 6.5a and b) and is sealed by a spring-returned reinforced diaphragm constructed from rubberized fabric layers or laminae. Possessing very low inertia, the separate

valves may be either the early spring-loaded bakelite disc or the later elastomeric moulded types. A moulded construction is also used for the combination inlet and outlet valve, where its lower flap deflects downwards and its upper lips spread outwards respectively to admit and deliver the fuel. An installational advantage possessed by the chamber construction shown in Figure 6.5a is that the two turrets with their moulded nozzles may be splayed in any required direction to provide convenient inlet and outlet connections.

The oscillating or pumping motion of the diaphragm is derived from an eccentric cam, which is usually embodied on the engine camshaft. This eccentric motion of the cam is converted into one of oscillation at the diaphragm, through the medium of a pull-rod, since the latter is working in opposition to the diaphragm compression return spring. At its lower end the pull-rod is attached to a rocker arm follower, which is always in contact with the eccentric cam. This permanent contact is obtained by separately spring loading the rocker arm on to the cam and makes for quiet operation of the pump. In earlier practice, the rocker arm usually took the form of an articulated lever assembly, but in modern examples the rocker arm is often made rigid with a forked end that simply intercepts a clearance collar at the lower end of the pull-rod (Figure 6.5). The reason for the apparent complication of the connection between the rocker arm and pull-rod is that although the pump must be capable of delivering sufficient fuel under all conditions of engine operation – indeed, it must be more than capable so that it can provide quick priming and also handle any vapour present in the fuel pipeline – the quantity of fuel delivered must be only as much as the carburettor needs. This means that the pump must cease to deliver fuel when the needle valve closes in the carburettor float chamber. At this point, the existing pressure in the fuel pipeline between the pump and the carburettor can produce an opposing force sufficiently high to balance the spring load on the diaphragm.

It is for this reason that the rocker arm must embody some form of free-wheeling facility, so that its normal motion can continue without disturbing the retracted diaphragm. This is achieved by ensuring that only on the intake stroke of the diaphragm is there a positive connection between the pull-rod and the rocker arm.

In operation, the intake stroke is performed by the diaphragm being retracted from the pumping chamber by the motion of the rocker arm and pull-rod. The inlet valve thus opens as a result of the depression created in the pumping chamber. Since the fuel tank is vented either to the atmosphere or to a vapour recovery system, the resulting pressure difference in the fuel pipeline causes fuel to flow into the pump from the tank (Figure 6.6a). At the end of the intake stroke the inlet valve closes, following which the outlet valve opens as the diaphragm advances into the pumping chamber on the delivery stroke. This is accomplished solely under the influence of the diaphragm return spring. Fuel is then being supplied under pressure to the carburettor float chamber (Figure 6.6b).

Pump delivery pressure

From the description just given of fuel pump operation, it will be evident that the main factor controlling delivery pressure is the strength of the diaphragm return spring, which is therefore chosen to suit the carburation requirements of a

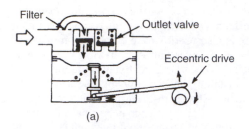
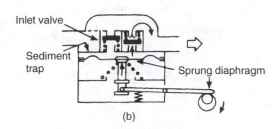

Figure 6.6 Operating principle of mechanical fuel pump

particular engine. In this context it is important that the diaphragm is not subjected to any stretching on its downward stroke, since this action would add to the force exerted by the diaphragm spring.

Too high a fuel pump pressure can result in the needle valve being held off its seat in the carburettor float chamber, which in turn raises the fuel level and increases fuel consumption. In practice, the fuel pump delivery pressure normally lies in the range $25–50 \text{ kN/m}^2$ ($3.5–7/\text{lbf/in}^2$). It is generally highest at engine idling speed, when flow is least, and lowest at maximum speed, when flow is greatest.

Misbehaviour of mechanical fuel pump

Possible faults that can occur with mechanical fuel pump installations may be summarized as follows:

Difficult starting

Air leaking into pipeline connections between petrol tank and pump
Split or over compressed cover gasket
Obstruction in pipeline between petrol tank and pump
Restricted filter in pump or line
Worn actuating mechanism
Defective operation of valves or springs
Incorrectly fitted diaphragm.

Starvation at high speed

Air leaking into pipeline connections between petrol tank and pump
Split or overcompressed cover gasket
Worn actuating mechanism
Fuel vaporization caused by fuel pipe being too exposed to exhaust heat
Fuel vaporization arising from lack of pump heat shield.

Failure to deliver fuel

Damaged diaphragm
Sticking valves.

Noisy operation

Worn actuating mechanism
Broken rocker arm spring.

Electrical fuel pump

As in the case of the mechanical fuel pump, the electric fuel pump used in carburettor installations is also of the positive displacement type. That is, it displaces a definite volume of fuel with each reciprocation of the pumping element, whether this be a diaphragm or (less common) a plunger. We shall consider only the former type (Figure 6.7). An advantage common to all types of electric fuel pump is that they can prime the system from the moment the ignition is switched on, so that petrol can be delivered to the carburettor before the engine is cranked over by the starter motor.

The pumping action of a diaphragm version of what is actually an electromagnetic pump is similar to that of the mechanical type. Whereas the spring-returned diaphragm of the latter is flexed by mechanical interconnection with the engine, the former has its diaphragm flexed by means of an armature being magnetically attracted to a solenoid coil. Either an encircling ring of spherically edged rollers or, as in later versions, a plastic guide plate is fitted between the armature and the coil housing, their purpose being to centralize the armature in its housing whilst also allowing it freedom of axial (pumping) movement. The solenoid coil is energized from the electrical system of the vehicle, via a contact-breaker toggle switch that is actuated by armature travel.

Whenever the toggle switch closes the contact-breaker points and restores the circuit for the intake stroke, the energized solenoid magnetically attracts the armature to which the diaphragm center portion is clamped. Hence, the diaphragm is retracted from the pumping chamber and the inlet valve opens as a result of the depression created (Figure 6.8a). Since the fuel tank is vented to the atmosphere, the pressure difference acting in the pipeline causes fuel to flow into the pump from the tank. Towards the end of the intake stroke, the toggle switch throws over and separates the contact-breaker points, thereby interrupting the circuit and de-energizing the solenoid coil. This then releases the spring returned armature, which thus allows the diaphragm to advance into the pumping chamber on its delivery stroke. The ensuing build-up of pressure causes the inlet valve to close and the outlet valve to open (Figure 6.8b). When the armature approaches the end of its delivery stroke, the toggle switch once again throws over, this time to restore the circuit to the solenoid coil by closing the contact-breaker points, so that the pumping cycle can be repeated. Again, the pump must cease to deliver fuel when the needle valve closes in the carburettor float chamber. At this point, the existing pressure in the fuel pipeline between the pump and the carburettor can produce an opposing force sufficiently high to balance the spring load on the diaphragm, thereby preventing the diaphragm and its armature from travelling fully to the end of their delivery stroke. Since this will leave the contact-breaker

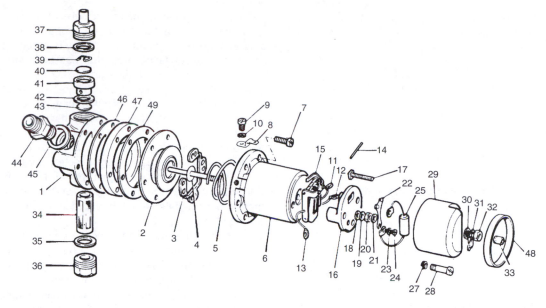

Figure 6.7 Exploded view of a diaphragm-type electric fuel pump (*SU Butec*)

1 pump body	9 earth screw	19 lead washer
2 diaphragm and	(4 BA–H,	20 terminal nut
spindle assembly	2 BA–L)	21 end cover seal
including armature	10 spring washer	washer
3 armature	11 terminal tag, 5 BA	22 contact blade
centralizing plate	12 terminal tag, 2 BA	23 washer, 5 BA
4 impact washer	13 earth tag, 2 BA	24 contact blade
5 armature spring	14 rocker pivot pin	screw, 5 BA
6 coil housing	15 rocker mechanism	25 condenser
7 screw securing	16 pedestal	(or diode
housing, 2 BA	17 terminal stud	resistor)
8 earth connector	18 spring washer	26 –

27 spring washer,	36 filter plug
2 BA	37 outlet
28 screw, pedestal to	connection
housing, 2 BA	38 fibre washer
29 end cover	(medium)
30 shakeproof	39 spring clip
washer, 2 BA	40 outlet valve
31 lucar connector	41 outlet value cage
32 nut, 2 BA	42 fibre washer
33 insulating sleeve	(thin)
34 filter	43 inlet valve
35 washer	44 inlet connection

45 fibre washer
46 joint gasket
47 sandwich plate
48 sealing band
49 diaphragm gasket

points in the open position, there will be no electric current flowing to energize the solenoid coil so that the pumping action temporarily ceases.

Pump delivery pressure

As in the case of the mechanical fuel pump, the main factor controlling the delivery pressure of the electric pump is the strength of the diaphragm return spring. Cut-off delivery pressures for diaphragm-type electric pumps usually lie in the range of 10–30 kN/m² (1.5–4.5/lbf/in²). A pump with a cut-off delivery pressure at the lower limit of this range would be intended for underbonnet mounting, whereas one with a delivery pressure at the upper limit would be suitable for mounting over the rear fuel tank. The advantage of the latter mounting position is that it renders the fuel line less prone towards vapour locking.

Misbehaviour of electrical fuel pump

Possible faults that can occur with electric fuel pump installations may be summarized as follows:

Reduced fuel delivery

Needle valve assembly of carburettor float chamber obstructed or gummed up
Inadequate venting of petrol tank
Restricted inlet filter to pump.

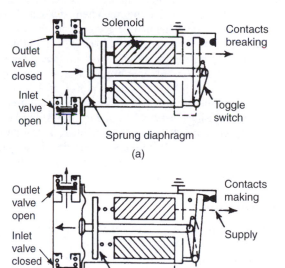

Figure 6.8 Operating principle of electric fuel pump

Failure to deliver fuel

No electric supply to pump
Deterioration or maladjustment of pump contact-breaker points
Obstruction in pipeline between petrol tank and pump

Undue stiffening of the pump diaphragm material

Abnormal friction in the contact-breaker toggle switch mechanism.

Pump operating but not delivering fuel

Air leaking into pipeline connections between petrol tank and pump

Similarly, air entering at coil housing to pump body joint

Pump valves, and especially the inlet one, held open by lodged foreign matter.

Noisy operation

Air leaking into suction side of pump system.

6.2 FIXED-CHOKE CARBURETTORS

Basic requirements

The process of delivering a combustible charge to the cylinders of a petrol engine is termed *carburation*. Strictly speaking, it begins at the air and the fuel intakes and ends at the exhaust outlet, but it is customary to consider the fuel and the exhaust systems as separate entities. The purpose of the carburation system is therefore to meter fuel into the incoming air stream, in accordance with engine speed and load, and to distribute the mixture uniformly to the individual cylinders. Until the 1970s carburation had long been performed by one or more venturi spray carburettors, but since then for reasons later explained in Section 6.6, fuel injection systems in one form or another have become established practice.

For theoretically complete combustion, the engine must be supplied with a homogeneous mixture of air and petrol in the chemically correct proportions of approximately 14 parts to 1 part by weight. This would normally be expressed as an air–fuel ratio of 14:1, and is what the fuel chemist would describe as the stoichiometric (chemically correct) air–fuel ratio. In actual practice, it becomes necessary to deviate from this ideal mixture to compensate for certain shortcomings inherent in petrol engine operation. Suitable provision must therefore be made for an excess of either air or petrol in the mixture, according to engine and, not least, environmental requirements (Section 11.1). Since service engineers are still likely to encounter spray carburettors on older vehicles, they provide an opportunity to introduce the topic of carburation in the general sense, before discussing the fuel injection systems that have superseded them (Section 6.6).

Mixing chamber

Various methods of carburation were tried and eventually rejected in the early development of the motor vehicle, notable among these being the use of surface and wick carburettors. The *surface carburettor* was the simplest and earliest form of mixing chamber, wherein air was drawn by the intake depression of the engine over the surface of the petrol or was caused to bubble through it, thereby forming a petrol and air mixture. Unfortunately, the splashing about of the petrol in the chamber due to vehicle motion caused very erratic vaporization, which demanded frequent correction to the mixture by hand-controlled air valves. In the *wick carburettor* the induced air passed around wicks dipping into a reservoir of petrol.

Capillary action maintained those parts of the wicks exposed to the passage of air in a state of saturation, so that the air evaporated sufficient petrol to form the required mixture. One drawback of this type of carburettor was its considerable bulk.

The shortcomings of the surface carburettor led to the general adoption of the *spray carburettor*, which was first developed in 1892 by another German pioneer automobile engineer, Wilhelm Maybach, (a name that we shall meet again later in connection with vehicle gearboxes). After much further development the spray carburettor came to be the universal vaporizing device for motor vehicle petrol engines.

In the spray carburettor the purpose of the mixing chamber is to measure the incoming air flow and to meter into it the appropriate amounts of fuel. Reduced to its simplest form, the mixing chamber could consist merely of a plain air or choke tube into which is placed a fuel delivery jet, so that the incoming air to the cylinders is induced to flow past it whenever the engine is running (Figure 6.9a). Since a moving stream of air exerts less sideways pressure than when it is stationary, the fuel entering the delivery jet from the float chamber is forced out into the air stream. This is because the pressure acting on the fuel supply in the float chamber remains at atmospheric and is therefore greater than that surrounding the jet. It is misleading to think in terms of the fuel being sucked from the jet; rather it is pushed from the jet owing to a difference in pressures. However, the pressure difference so created would be hardly sufficient to overcome the resistance of the fuel in the jet to flow. In order to increase the depression over the jet, the air tube must be provided with a constricting passage known as a *venturi*. This depends for its action on the Bernoulli effect, so named after the eighteenth-century Swiss scientist Daniel Bernoulli, who wrote extensively on the subject of hydraulics. The principle of the venturi was actually discovered by Bernoulli's contemporary Giovanni Venturi, the Italian physicist.

Stated in simple terms, *Bernoulli's principle* is that the relationship between pressure and velocity in a flowing fluid (in our case air) is such that the total energy possessed by the air flow is the same at every point along its path. The pressure of the air will therefore be reduced where its velocity is increased, since the gain in kinetic energy must be offset by a loss in potential energy. Because the same volume of air must flow through all portions of the carburettor air tube, it will be evident that the effect of introducing a venturi is to increase locally the air speed and cause a corresponding reduction in pressure over the fuel delivery jet.

Verification of this Bernoulli effect could be made by connecting a series of U-tube manometers into the venturi (Figure 6.9b). Pressure gauges of this type work on the principle that the pressure in a column of liquid is directly proportional to its height. Hence, the difference in level of the liquid in the limbs of each U-tube is a measure of how much less the air pressure in the venturi is than atmospheric pressure. It is the pressure drop between the air intake and venturi which, in effect, enables the carburettor to 'measure' the incoming air flow for the purpose of metering the fuel flow. Furthermore, the pressure drop serves to ensure that the jet delivers a finely atomized spray of fuel into the air stream. To promote the best conditions for air flow, the venturi is provided with a smooth, rounded entrance and a gently tapering exit (Figure 6.9c).

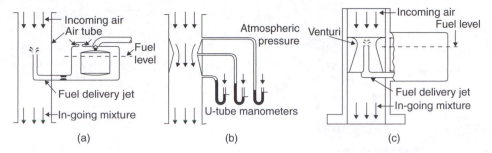

Figure 6.9 (a) Mixing chamber of simple carburettor (b) pressure variations in venturi tube (c) mixing chamber with venturi

The mixing chamber also incorporates a throttle valve, which is positioned on the engine side of the venturi and is linked to the accelerator pedal by either a cable in a conduit or, now less commonly, a rods-and-levers system. This valve serves to regulate the amount of air and fuel mixture admitted to the engine cylinders and hence the power they develop.

A simple mixing chamber of the form just described does, in practice, need considerable refinement if it is to satisfy the mixture requirements of the variable-speed engine. With increasing air flow through the venturi, a plain metering jet would tend to supply disproportionately greater quantities of fuel, resulting in an increasingly richer mixture. This is because the density of the air will vary accordingly to the pressure, whereas the fuel is virtually incompressible and does not share this characteristic. While this is not disadvantageous in supplying the rich mixture required for full-throttle performance, it is opposed to the requirement for a relatively weak mixture suitable for economical part-throttle cruising. To meet the latter requirement demands some method of correcting mixture strength.

Mixture strength correction

There are three basic methods that may be used to correct mixture strength in the fixed-choke carburettor for general performance requirements:

1 Unrestricted air bleed and compensating fuel jet
2 Restricted air bleed and correction jet
3 Variable fuel restriction.

Unrestricted air bleed and compensating fuel jet
This method represents an early and once widely adopted approach to mixture correction. It is sometimes simply referred to as the Bavery system, so named after François Bavery the French scientist who originated it in 1907. The fuel metering orifices comprising the main and compensating jets are submerged in the float chamber and communicate with a common discharge nozzle (Figure 6.10). Whereas the fuel delivered by the main jet is subject to venturi depression, that leaving the compensating jet is exposed to atmospheric pressure via an unrestricted stand-pipe or capacity well. Unlike the main jet, the compensating jet thus delivers a constant amount of fuel regardless of increasing venturi depression, which can result in the fuel level within the capacity well falling to such an extent that air is permitted to enter the metered fuel supply to the discharge nozzle. Hence, as depression upon the discharge nozzle increases, the greater fuel flow from the main jet tends to enrich the mixture, whilst the constant fuel flow

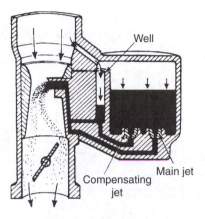

Figure 6.10 Bavery system of mixture correction

from the compensating jet together with any air entering from the capacity well tends to weaken it. By suitable matching of these opposed characteristics, a substantially uniform or corrected mixture strength may be obtained.

Restricted air bleed and correction jet
With this system of air bleed mixture correction, which continues to be used in many modern carburettors, a single fuel metering orifice is employed to serve as the main jet. It is generally sited near to the base of a discharge nozzle and is thus submerged in the fuel supplied to the float chamber. A perforated emulsion tube is either incorporated within, or located remote from, the discharge nozzle and terminates in a restricted air bleed or correction jet (Figure 6.11). This is exposed to the incoming air flow, so that the fuel flow metered by the main jet is subject to a lesser depression than that existing at the venturi.

In operation, an increasing depression in the venturi promotes a greater flow of fuel from the discharge nozzle, the level in which falls accordingly. As a result, the fuel stream becomes emulsified by air admitted through the emulsion tube perforations via the restricted air bleed of the correction jet. Since more of the holes distributed over the length of the emulsion tube will be uncovered as the pressure drop across the venturi increases, a variable air bleed to the main metering circuit is thus effected. In other words, the mixture enriching effect of the main fuel jet is balanced by the weakening effect of the restricted air bleed, so that a substantially uniform mixture strength is obtained.

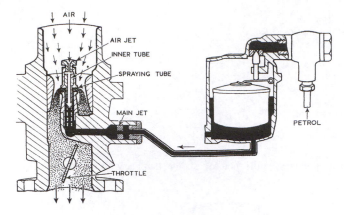

Figure 6.11   An air bleed system of mixture correction

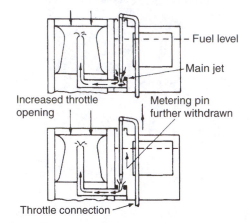

Figure 6.12   Variable fuel restriction system of mixture correction

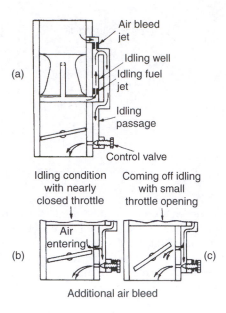

Figure 6.13   Fixed-choke carburettor slow running systems

Variable fuel restriction

This approach to mixture correction is widely favoured in American practice. It involves a variable fuel restriction to the main metering circuit. This is accomplished by means of a stepped and tapered metering pin control, which operates in the fixed orifice of the submerged main jet (Figure 6.12). In simplest form, the metering pin is interconnected solely with the throttle valve linkage, so that for any given throttle opening the metering pin is withdrawn a proportionate amount from the jet orifice. The fuel flowing to the discharge nozzle is therefore metered by the annular opening between the metering pin and the main jet orifice. For economical part-throttle cruising, the relative positioning of the metering pin is such that its largest cross-sectional area remains in the jet orifice to reduce fuel delivery.

Slow running

For slow running at normal operating temperature the engine still demands a fairly rich mixture, the air–fuel ratio usually being in the range of 11:1 to 13:1. In this case, an excess of fuel is demanded owing to the comparatively small amount of incoming fresh cylinder charge becoming diluted by unscavenged and inert exhaust gases. A further weakening of the fresh charge may also occur as a result of slight air leakage

into the induction system, usually via the working clearance between the inlet valve stem and its guide. The degree of mixture enrichment required to compensate for these effects and secure reliable slow running also generally suffices for warm starting conditions.

The main fuel metering circuit of a fixed-choke carburettor cannot meet this slow running requirement, since the depression acting upon it with the throttle valve nearly closed is insufficient to cause the fuel to flow. This difficulty arises not only from surface tension effects to which the liquid fuel is subject, but also because its static level is maintained slightly below the point of delivery in the discharge nozzle to prevent internal spilling when the vehicle is parked off level ground. A separate slow running or idling circuit that is practically independent of the main metering circuit therefore becomes necessary.

Typically, the slow running circuit receives its fuel supply from the base of the main metering circuit, via a metering tube or a well and jet (Figure 6.13a). A controlled air bleed communicates with the upper end of the tube or the well, the air and fuel mixture then passing into the idle passage proper. This in turn conveys the mixture down to the discharge port in the throttle valve body, where it emerges into the air stream just beneath the throttle valve edge. The whole process is facilitated by the high depression acting below the nearly closed throttle valve, which causes the restricted flow of incoming air to attain a high velocity and hence a low pressure.

Manual adjustment of the slow running mixture strength is usually provided by a tapered needle valve, which is screwed into the throttle body at the junction of the idle passage with the delivery port. To avoid too rich a slow running mixture, an additional air bleed or progression port communicating with the idle passage is often incorporated just above the throttle edge (Figure 6.13b). Conversely, to avoid too weak a mixture when coming off idle, this additional port ceases to act as an air bleed and instead provides a further outlet to increase mixture delivery from the slow running circuit (Figure 6.13c).

Rapid acceleration

In order that the engine will accelerate unhesitatingly and smoothly with rapid opening of the throttle, an accelerating pump is often incorporated in the carburettor. Its function is to inject a controlled and metered quantity of additional fuel into the incoming air stream. This additional fuel requirement arises because fuel tends to be deposited on the interior walls of the intake manifold, as the depression within it falls with sudden throttle opening. Since the ingoing mixture would then be robbed of some of its fuel content, it may become momentarily too lean for normal combustion and cause hesitant acceleration. Furthermore, the fuel flow in the mixing chamber is subject to greater inertia than that of the air flow, which adds to the problem.

The accelerating pump receives its fuel supply from the float chamber and is actuated either mechanically or pneumatically. It is equipped with gravity-returned check valves, usually of the ball type, although disc and poppet types have also been used. Their purpose is to prevent fuel returning to the float chamber on the delivery stroke and air from being admitted to the pump on the intake stroke. The outlet valve also normally serves to isolate the pumping chamber from air intake depression, so that fuel cannot be delivered except during pump operation.

With the mechanically actuated pump, either a plunger sliding in a cylinder or a diaphragm flexing in a chamber may be used to generate the necessary hydraulic pressure (Figure 6.14a). Opening movement of the throttle valve depresses the pumping element so as to expel fuel from its chamber and discharge it into the choke tube, via the outlet check valve. A restricting orifice or jet is situated before the point of discharge to meter the additional charge of fuel. Since its delivery is required to be maintained during the period of momentary leanness, rapid opening of the throttle does not positively depress the pumping element. Instead, the connection between them is made flexible by interposing a spring, so that the energy stored in it when deflected prolongs the pump delivery, even though the throttle has ceased to open further. To avoid unwanted mixture enrichment with slow throttle opening, the fuel under pressure is provided with a controlled leakage return path to the float chamber, such as via a deliberately loose fit for the plunger sliding in its cylinder. Closing movement of the throttle valve results in the pump element being retracted either positively in the case of a plunger, or by a return spring where a diaphragm is employed.

In the case of the pneumatically actuated pump, the pumping element is usually a spring-loaded diaphragm, which divides the pumping chamber into two parts (Figure 6.14b). The diaphragm is exposed to air on its spring abutment side and to fuel on its other side. Manifold depression is arranged to act upon the air side, as a result of which the diaphragm is retracted from the fuel side of the chamber. In so doing, fuel is caused to enter the pumping chamber from the float chamber, via the inlet check valve. Under this condition, the energy stored in the compressed spring is available to generate hydraulic pressure on the fuel side of the diaphragm. This occurs with rapid opening of the throttle, since the manifold depression is then no longer sufficient to overcome the spring loading. As a result, the diaphragm is forced to maintain a pressure on the fuel in the pumping chamber. Hence, an additional charge of fuel is expelled from the chamber

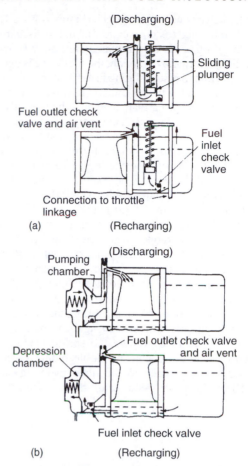

Figure 6.14 Action of fixed-choke carburettor accelerating pumps: (a) mechanically operated (b) pneumatically operated

and discharged into the choke tube, via the usual valve and metering arrangements.

Economical cruising

This is perhaps more conveniently considered in the negative sense of how a relatively richer mixture is supplied for full-throttle performance, either by providing an additional source of fuel supply or by reducing the air bleed facility of the main metering circuit. Devices to accomplish this are therefore referred to by the American industry as power systems, since they enrich the mixture for maximum power. Conversely, they may also be termed economizer systems, which is the preferred description in the United Kingdom, because they weaken the mixture when maximum power is no longer required. It is, however, easier to understand their action on the main metering circuit when considered in the context of a power system rather than an economizer one.

The additional source of fuel supply is normally introduced through a submerged fuel valve, which bypasses the main jet and communicates with the discharge nozzle. This bypass valve is spring loaded on to its seating and is provided with an extended operating stem. It is opened either mechanically or pneumatically. With the former arrangement, the fuel valve spring loading is overcome by mechanical interaction with a lost-motion linkage, which takes up only when the throttle valve is nearly fully open (Figure 6.15a). The fuel valve is then opened to admit extra fuel to the discharge nozzle, but at all other times it remains closed for economical running. For pneumatic operation, the fuel valve is opened by contact with either a piston sliding in a cylinder (Figure 6.15b) or a diaphragm flexing in a chamber. These devices are superimposed upon and spring loaded against the fuel valve stem. In operation, the fuel valve opens only when the manifold depression drops so low that it is no longer able to overcome the spring loading on the piston or the diaphragm. When this occurs, the lesser closing load exerted by the fuel valve spring is overcome and the valve is allowed to open so that additional fuel is admitted to the discharge nozzle.

The alternative means of enriching the mixture for full-throttle performance, whilst retaining a leaner mixture for normal running, is to reduce the air bleed facility of the main metering circuit. For this purpose a pneumatically operated economy air valve may be employed, which is normally retained open against a spring-loaded diaphragm, the chamber for which communicates with the engine side of the throttle valve (Figure 6.16). In operation, the air valve closes only when the manifold depression again drops so low that it is no longer able to overcome the spring loading on the diaphragm. When this occurs, an air bleed duct leading from the carburettor air intake to the main metering circuit is closed off by the valve. As a result, the air bleed facility is reduced and the mixture temporarily enriched for maximum performance.

Cold starting

When starting from cold, the engine demands a very rich mixture. This is because under low-temperature conditions there will be less flammable vapour per unit weight of fuel, so that an excess of fuel must be provided to obtain a readily combustible charge. In terms of mixture enrichment this may require an air–fuel ratio of about 4:1, or even less than this for very low-temperature starting conditions. Cold starting mixture enrichment is then progressively reduced as the engine warms up, this being accomplished by means of either a manual or an automatic choke control system.

Manual choke control

To supply the very rich mixture required when the engine is started from cold, it is necessary to increase to a greater extent than usual the pressure drop (difference in pressure acting on the fuel in the float chamber and in the discharge nozzle) across the main metering circuit. This is generally accomplished by providing a strangler flap, which is similar to the throttle valve but situated between the air intake and venturi of the carburettor. Manual operation of the strangler flap is usually effected by a flexible cable connected to a dashboard control. The strangler flap is mechanically interconnected with the throttle valve, so that when it is fully closed, and also during its initial stages of opening, the throttle valve is opened slightly to admit the enriched mixture to the cylinder (Figure 6.17a). This opening of the throttle also allows faster warming-up of the engine in the interests of reducing engine wear.

In the strangler flap is retained fully closed once the engine has started, the prevailing high depression on the main jet system would cause stalling of the engine from flooding, with consequent difficulty in restarting. To avoid this condition the strangler flap is provided with an offset spindle; the spindle itself is not rigidly connected to its operating lever, but connects via a torsion coil spring. Too high a depression acting on the unbalanced areas of the flap will therefore automatically cause it to swing partly open against the opposing torque of the spring and hence relieve the excess depression.

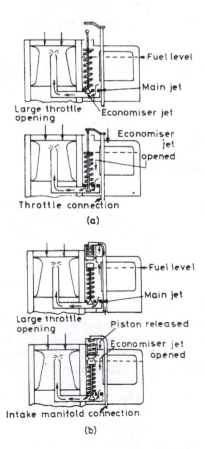

Figure 6.15 Action of fixed-choke carburettor power/economizer fuel jet systems: (a) mechanically operated (b) pneumatically operated

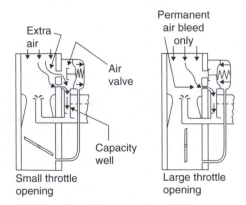

Figure 6.16 Action of fixed-choke carburettor power/economizer air valve system

In operation, the strangler flap is completely closed during initial cranking of the engine. Once the engine fires consistently, the strangler flap is opened slightly to maintain the engine running. As the engine warms up to its normal operating temperature, the strangler flap is gradually opened to its fullest extent.

An alternative method of providing an enriched mixture for cold starting is to use a separate air and fuel metering circuit, amounting in effect to a smaller auxiliary carburettor. In one such design the starter circuit communicates with the engine side of the throttle valve via a manually operated rotary disc valve, which is drilled with a series of different diameter holes to vary the mixture richness (Figure 6.17b). With this arrangement there is, of course, no need for any interconnection between the choke mechanism and the throttle valve, which remains closed during starting and warm-up.

Automatic choke control

An automatic choke was first introduced about 1932 by the Stromberg Carburettor Division of Bendix Products Corporation to meet the demands of the American motorist. It established a pattern of automatic choke control in which some form of thermostatic operation is superimposed on the strangler flap. Modern systems of this type can vary in detail application, but essentially comprise the following units acting in combination (Figure 6.18):

Bimetal spiral spring, known as a thermostat coil, which when cold winds up to exert a closing torque on the strangler flap and when heated gradually unwinds to allow opening of the strangler flap. The latter process is facilitated by engine depression acting on the unbalanced areas of the flap, as earlier described for a manual choke. At its outer end the thermostat coil engages a striker pin carried by a lever arm on one end of the strangler flap spindle, while its inner end is furnished with an adjustable anchorage via a metal heat mass. This is to prevent the coil from cooling too rapidly and overchoking the engine. *Thermostat coil chamber* with an inner heat-insulating disc, which is provided with a kidney-shaped slot to permit arcuate movement of the flap lever striker pin that protrudes through it. Holes drilled through the centre of the metal heat mass allow the passage of heated air over the thermostat coil, the heat being received from either the exhaust

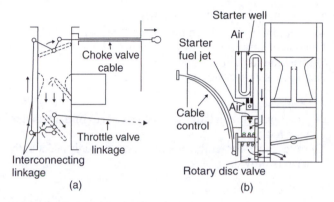

Figure 6.17 Manually operated choke systems

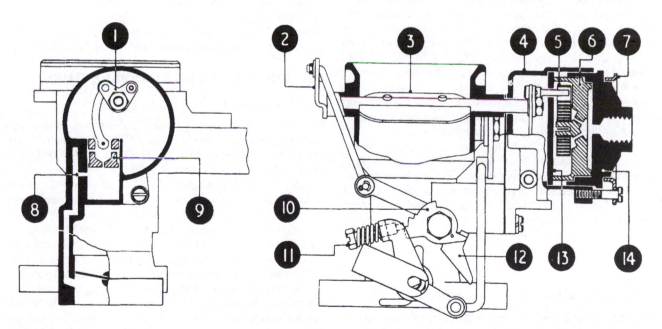

Figure 6.18 Diagrammatic section of carburettor automatic choke system (*Zenith*)

1 bell-crank lever	8 vacuum channel and port
2 strangler spindle lever	9 vacuum-kick piston
3 strangler flap	10 four-step fast-idle cam
4 die cast body	11 fast-idle adjusting screw
5 bimetallic thermostat coil spring	12 interconnection lever
6 die cast heat mass	13 heat-insulating disc
7 flanged clamping ring	14 moulded casing

manifold, the coolant system or an electrical element. The flow of heated air is maintained by engine depression acting via a drilled passage to the coil chamber.

Depression chamber, which communicates with the engine side of the throttle valve and contains either a sliding piston or a flexible diaphragm, the former being known as the vacuum-kick piston. Taking this as an example, the piston is connected by a short curved link to the same lever arm that carries the striker pin for the thermostat coil. The effect of engine depression on the vacuum-kick piston is such that it attempts to open the strangler flap in opposition to the closing torque exerted by a cold thermostat coil. In so doing it is assisted by the engine depression acting on the unbalanced areas of the strangler flap. The latter is therefore forced to open a predetermined amount to maintain the engine running once it has started. Further movement of the vacuum-kick piston then closes off the depression port, so that the movement of the strangler flap will be governed solely by the thermostat coil. If the engine depression is reduced, as a result of an increased load or sudden acceleration, the return motion of the vacuum-kick piston will allow the thermostat coil to close the strangler flap sufficiently to enrich the mixture and prevent stalling.

Fast-idle mechanism, which consists of a multistep cam that is lightly spring loaded into contact with a rocking lever; this in turn carries an interconnection rod to the other end of the strangler spindle. With the engine cold and the accelerator pedal having been depressed once to disengage the cam, the tip of the fast-idle adjusting screw will rest on the highest step of the cam to hold the throttle valve slightly open. This action does not interfere with the gradual opening of the strangler flap as the engine temperature rises and heated air is drawn over the thermostat coil to unwind it. If the accelerator pedal is depressed, the fast-idle adjusting screw will lift clear of the cam and allow it to turn, so that the screw will next engage with a lower step and the idling speed be accordingly reduced. At the end of the warming-up period, the strangler flap will assume its wide open position and control of the engine idling speed will then be resumed by the normal throttle valve slow running screw.

Throughout their long history automatic choke control systems have tended to be somewhat fickle in their service behaviour, and even in recent years it has not been uncommon for a manufacturer to revert to a manual choke control. It is therefore important that the adjustment and setting of an automatic choke be strictly in accordance with the manufacturer's service instructions, and equally important that the adjustments should not be tampered with in any attempt to improve the original build settings.

Misbehaviour of fixed-choke carburettor

The following is intended to be purely a general guide to possible complaints and their related causes. It must always be remembered that a carburettor is both a precision and delicate piece of equipment and must therefore be serviced strictly in accordance with the particular manufacturer's instructions.

Difficult starting from cold
Needle valve in float chamber sticking in closed position
Choke strangler flap not completely closing when dashboard control is fully extended.

Difficult restarting when hot
Worn needle valve assembly in float chamber causing flooding and overrichness
Damaged or sticking float assembly similarly allowing flooding and overrichness.

Poor slow running
Obstruction to slow running petrol feed and air bleed circuits, including progression port
Tapered end of slow running screw damaged by being forced against seating
Slow running adjustment not being retained owing to ineffective spring loading under screw head.

Acceleration flat spot
Too low fuel level in float chamber
Accelerating pump piston not moving freely in cylinder
Pump inlet valve not closing properly
Obstruction to pump discharge jet
Damaged accelerating pump diaphragm.

Lack of power
Obstruction to fuel jet and air bleed passages of main metering circuit
Throttle valve not fully opening even though accelerator pedal is fully depressed.

High fuel consumption
Choke strangler flap not completely opening when dashboard control is released
Bypass valve not closing properly
Damaged economy air valve diaphragm
Worn needle valve assembly of float chamber
Leakage from any joints below fuel level in float chamber.

6.3 VARIABLE-CHOKE CARBURETTORS

Basic construction

Historically, the operating principle of this type of carburettor can be attributed to G.H. and T.C. Skinner, the two brothers whose London firm of Skinners Union (SU) first produced such a carburettor in 1905.

With the variable- (or expanding-) choke carburettor, which is also known as the constant-depression type, mixture correction is obtained by utilizing the depression existing on the carburettor side of the throttle valve to control, simultaneously, the choke tube and jet areas. Whereas most fixed-choke carburettors are now used in down-draught form, conventional variable-choke carburettors are almost always of the side-draught variety. The fuel is metered into the incoming air stream from a single jet, the area of which is made variable by a calibrated tapered needle attached to a moving air valve of piston form. This in turn is provided with an extended lower shank portion, which intersects the axis of the choke tube and thus serves to vary its effective area.

The air valve floats within a superimposed depression chamber, the underside of which is vented to the carburettor air intake. Choke tube depression is conveyed to the topside of the air valve, via a communicating duct in its shank. Sealing of the moving air valve is accomplished either by making its head portion grooved and a close working fit in the depression

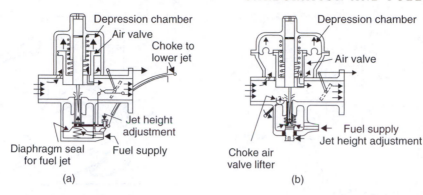

Figure 6.19 Two forms of variable-choke carburettor: (a) sliding piston air valve (b) diaphragm-sealed air valve

chamber (Figure 6.19a), or by employing a flexible diaphragm (Figure 6.19b). To preserve correct alignment between the jet assembly and the needle moving in it, a tubular extension of the air valve slides within a similar bearing member located from the depression chamber roof. This telescopic bearing arrangement is vented to the atmosphere and is partially oil filled, because it embodies a dash pot damping device for the air valve, as explained later. It is also encircled by a return spring, which augments the weight of the air valve as it moves to maintain a constant depression in the choke tube.

Mixture strength correction

As the throttle is opened, the momentary increase in choke tube depression also acts upon the piston head of the air valve, via the communicating duct mentioned earlier. Since the carburettor air intake pressure acts beneath the valve head, the resulting pressure difference forces the air valve to rise against the combined effect of its own weight and return spring load, until equilibrium is restored. The choke area thus becomes enlarged and the depression over the fuel jet is maintained at its original value. In response to changes in engine load and throttle opening, the air valve therefore floats up and down in its depression chamber and in so doing maintains a practically constant depression over the fuel jet. Hence, by selecting a suitable profile for the air valve needle, the mixture strength may be varied in accordance with engine requirements as the needle retreats and advances within the fuel jet.

Slow running

For slow running at normal operating temperatures the air valve remains at rest slightly above the jet bridge, its base being partly undercut for this purpose. Hence, with a nearly closed throttle valve and minimum choke area, adequate depression is available over the jet to promote fuel flow. Since the largest cross-sectional area of the air valve needle remains in the jet, an appropriate amount of fuel is metered into the reduced air supply to provide the desired mixture strength.

Rapid acceleration

For sudden opening of the throttle valve on acceleration, a temporarily richer mixture is obtained by preventing the air valve from rising too rapidly. This is accomplished by the air valve dash pot device, which hydraulically damps and slows its upward motion (Figure 6.20). As a result, an increased

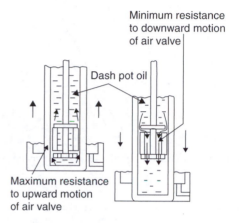

Figure 6.20 Action of air valve damper in variable-choke carburettor

depression acts over the open area of the fuel jet for a given choke area, thereby producing a richer mixture. Since the dash pot damper incorporates a one-way valve, it offers little resistance to the downward motion of the air valve, which is assisted by the return spring.

Cold starting

With manually operated cable control systems, the required mixture enrichment is obtained by either lowering the jet tube beyond its normal setting, raising the air valve and tapered needle from the slow running position, or employing an auxiliary carburettor. All these arrangements are interconnected with the throttle valve, so that it opens a small amount when the choke control is operated. In the first-mentioned system, either an external linkage acting on the jet tube specifically for starting purposes (Figure 6.21), or a similarly acting rocking lever that is otherwise used for normal mixture strength adjustment (Figure 6.19a), may be utilized to lower the jet tube. For the second system a flated rotatable bar can serve to lift the air valve and therefore its needle from the jet tube (Figure 6.19b). Hence when the engine is cranked over, the enforced increase in fuel jet area provides the very rich mixture required for cold starting. Where an auxiliary starter carburettor is employed, still with manual operation, it is of the rotary disc valve type and similar in principle to that incorporated on some fixed-choke carburettors, as mentioned in Section 6.2.

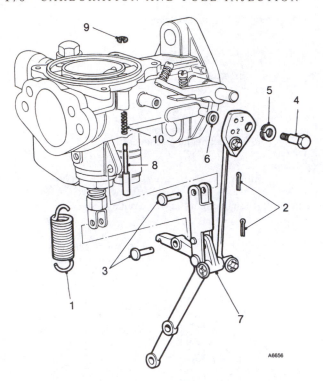

Figure 6.21   Exploded view of a jet lowering mechanism for cold starting (*SU Butec*)

1 lever return spring	6 spacing washer
2 split pins	7 lever assembly
3 clevis pins	8 piston lifting pin
4 cam plate pivot bolt	9 circlip for pin
5 spring washer	10 spring for pin

Mixture strength adjustment

For initial setting of the mixture strength with variable-choke carburettors of the sliding air valve type, provision is made for the jet assembly to be raised or lowered by a manual screw adjustment. This is arranged to act either directly beneath the jet tube or indirectly through a rocking lever pick-up on the jet tube. The adjustable jet assembly is sealed against external fuel leakage by means of a static seal or, alternatively, a small flexible diaphragm may be used. With the swinging air valve type of variable-choke carburettor (Figure 6.23), the mixture strength can be set by altering the lateral position of the tapered needle relative to the jet, manual screw adjustment being incorporated at the connection between needle and air valve.

Although incidental to mixture strength adjustment facilities, it should be mentioned here that one of the detail improvements made to the modern variable-choke carburettor is the spring-biased mounting of the jet needle in the moving air valve (Figure 6.22). This is to provide intentional eccentricity of the needle sliding within its jet tube and is in direct contrast with earlier practice. The explanation for this apparent lack of nicety in alignment between needle and jet is that the crescent-shaped orifice so created offers less viscous drag to the flow of fuel than a perfectly annular one. As a result, there is now less variation in the rate of fuel flow during slow running and at small throttle openings, which contributes to better control over exhaust emissions.

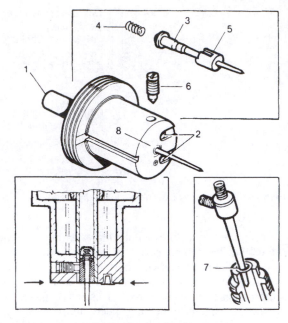

Figure 6.22   Spring-loaded jet needle (*SU Butec*)

1 piston rod	5 needle guide
2 transfer holes	6 needle locking screw
3 jet needle	7 needle biased in jet
4 needle spring	8 etch mark

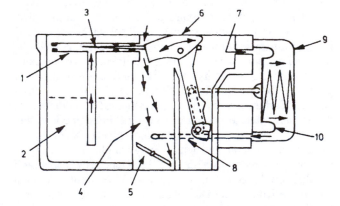

Figure 6.23   Basic layout of a variable-choke carburettor with swinging air valve

1 jet tube	6 swinging air valve
2 float chamber	7 air vent
3 tapered needle	8 vacuum passage
4 choke tube	9 depression chamber
5 throttle valve	10 spring-returned diaphragm

Misbehaviour of variable-choke carburettor

As in the case of the fixed-choke carburettor in Section 6.2, the following is intended to be purely a general guide to possible complaints and their related causes:

Poor slow running
Incorrect fuel level caused by maladjusted floats, wear or
    dirt in needle valve assembly of float chamber

Air valve sticking

Damaged diaphragm seal for air valve

Obstruction to vent holes in air valve and float chambers

Air valve needle incorrectly fitted or not correct to specification.

Acceleration flat spot

As those mentioned above

Air valve damper inoperative due to lack of oil

Air valve return spring not fitted or incorrect to specification.

Lack of power

Throttle valve not fully opening even though accelerator pedal is fully depressed.

High fuel consumption

Any of the causes given for poor slow running or flat spot

Fuel leakage from static seals such as float chamber joint or O-ring seals for jet plug.

Variable-choke carburettor with swinging air valve

A new concept in variable-choke carburettor construction is to be found in some later instruments of Ford manufacture. Although at first glance their interior design strongly resembles that of a fixed-choke down-draught carburettor, they differ in that a horizontally disposed air valve is arranged to swing across the top opening of the throttle bore. Furthermore, the swinging air valve is linked to a calibrated tapered needle that slides within a horizontal main jet. In operation, the movements of the swinging air valve and its associated tapered needle are controlled by choke tube depression, acting via an interlinked and spring-loaded diaphragm housed within a separate depression chamber (Figure 6.23).

Therefore it follows that, as the throttle is opened and the choke tube depression momentarily increases, the swinging air valve will retreat across the throttle bore to enlarge the effective choke tube area and thus maintain constant the depression over the fuel jet. Simultaneously, the tapered needle will be partially withdrawn from its jet so that more fuel is supplied to the engine, thereby complementing the greater air flow. A steady-state condition is reached when the opposing pneumatic and spring forces acting on the control diaphragm become equal in magnitude. In reality, of course, the self-adjustment of this type of carburettor to changes in engine load and speed can be regarded as a continuous process.

6.4 MULTIPLE AND COMPOUND CARBURETTORS

Multiple carburettors

Since the size of the mixing chamber venturi determines the maximum quantity of mixture admitted to the cylinders, large-capacity engines generally demand either multiple single-choke or, as in modern practice, multiple-choke single carburettors to satisfy their requirements.

At first thought it might seem best to provide one carburettor per cylinder and simply attach it direct to the inlet port, as is done in the case of a single-cylinder motor cycle engine and as also was found in the past on multi-cylinder racing car engines, where maximum power was the first need and fuel consumption was of lesser consequence. In classic high-performance cars of the 1930s, it was usual to find two or twin carburettors on four and six-cylinder engines and occasionally three or triple carburettors on the latter, such as used on the high-grade six-cylinders Alvis car engines of that era.

Although this compromise of feeding either two or three cylinders, rather than all the cylinders, from one carburettor certainly resulted in better and quicker mixture distribution for improved and more responsive performance, there were objections to the use of multiple carburettors. Apart from considerations of higher cost, both in respect of initial installation and increased fuel consumption, another problem was the difficulty of keeping each carburettor correctly balanced with the other, even though balance pipes were used to connect their mixing chambers. It also meant that throttle valve connections with varying degrees of complexity had to be introduced into the accelerator control linkage, so that each carburettor could be set at the desired idling speed.

The time-honoured method of checking that idling speeds are correctly synchronized is to use a listening tube which enables the intensity of intake hiss on all carburettors to be compared and their throttle screws adjusted until the hiss is the same. Mixture strength is then adjusted for each carburettor until enrichment produces the fastest engine speed, and then weakened until this speed just begins to fall. If necessary the idling speed of the engine can then be readjusted by equal turning of the throttle screws.

Twin-choke carburettors

It was the mainly American development of V eight- and V twelve-cylinder engines in the 1930s that originally created a demand for a compact twin-choke or, as it was then known, twin-barrel type of carburettor, which could be conveniently mounted between the two banks of cylinders. In this type of carburettor there is a single throttle spindle upon which both throttle valves are mounted and therefore move in unison. Each mixing chamber has its own choke tube and jet system, but is fed by a float chamber that is common to both. A single accelerating pump discharges through a separate jet above each choke tube. The twin-choke carburettor can also be installed on in-line cylinder engines, especially in conjunction with a modern one-piece dual-intake manifold as described later. More than one twin-choke carburettor may be used on some high-performance engines. For example, three twin-choke carburettors are used on the V twelve-cylinder engine of the Ferrari 330 GTC model.

It should be mentioned here that this type of multi-choke carburettor should not be confused with a multi-venturi single-choke carburettor. The latter refinement takes the form of stacking one or two venturis of decreasing size in series with the main venturi, so that each venturi opens above and discharges below the restricted throat of the next. It is then arranged for the main metering circuit to discharge into the smallest venturi where the air depression is greatest, the mixing process of the air and fuel then being intensified in the succeeding venturis. This arrangement of venturis also tends to centralize the mixture stream within the choke tube, because it is shrouded by air that bypasses the additional venturis; therefore liquid fuel is discouraged from depositing on the walls of the throttle body.

Compound carburettors

A compound system of carburation first appeared in America during the 1930s on the in-line eight-cylinder or straight eight engine fitted to Buick cars. In this early installation, two separate twin-choke carburettors were used, the front one being complete in all respects while the rear one contained an idling system and main jet only. The accelerator pedal was connected directly to the throttle valves of the front carburettor, but those of the rear carburettor were opened via a lost-motion linkage only when the throttle valves of the front carburettor were nearly wide open. This sequence of operation had the advantage of maintaining a high air velocity through one carburettor at cruising speeds and during acceleration, yet retaining the effect of two carburettors for increased power at full throttle when high speeds were required.

By 1952 the continued development of V eight-cylinder engines in America saw the introduction of the four-choke or four-barrel type of carburettor, which essentially compounds the action of two smaller or primary choke tubes with those of two larger or secondary choke tubes. The primary choke tubes provide for starting, warming-up and part-throttle economy operation, whilst the secondary choke tubes combine with the primary ones to increase the air capacity of the carburettor for full-throttle running conditions. Again, the secondary throttle valves only begin to open, and then quite rapidly, when the primary throttle valves are approaching their full opening. Typically the secondary throttle valves remain closed until the primary ones are opened within the region of 40 to 60 degrees depending on the installation. Following this all throttle valves reach their fully open position simultaneously, this generally being accomplished by an interconnecting linkage that provides first for lost motion and then a rapid toggle action. A similar linkage is used on smaller twin-choke versions (Figure 6.24a).

A disadvantage of this purely mechanical interconnecting linkage is that full-throttle acceleration from low speed could result in loss of power and possible stalling, because of too large a combined choke area. In some twin-choke compound carburettors used on smaller-capacity engines, this effect is overcome by providing pneumatic operation for the secondary throttle valve. The latter is linked to a spring-loaded diaphragm, the depression chamber for which communicates with the primary choke system (Figure 6.24b). In operation, the depression in the primary choke system must reach a value high enough to overcome the diaphragm spring loading before the secondary throttle valve is opened, which then brings into action the secondary choke and jet system (Figure 6.25).

Compound carburettors that utilize either mechanical or pneumatic operation for delayed movement of the secondary throttle valve are also known as progressive choke carburettors or, in modern terminology, sequential carburettors.

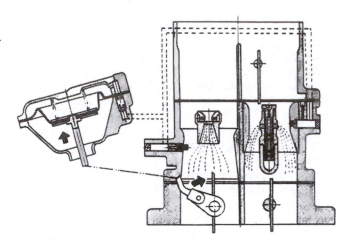

Figure 6.25 Pneumatic operation of compound carburettor secondary throttle: full throttle at high speed (*Nissan*)

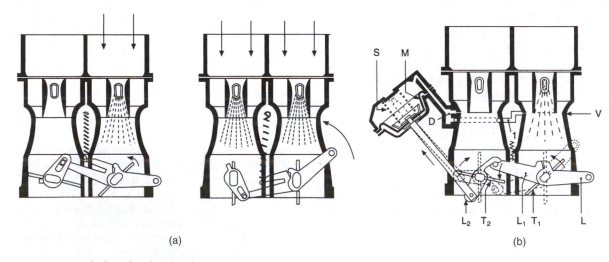

(a) (b)

Figure 6.24 (a) A twin-barrel carburettor in which the secondary throttle is opened by a lost motion mechanisms interconnecting it with the primary throttle (b) Here, the secondary throttle is pneumatically actuated

L Intermediate lever	L$_1$ Primary throttle lever	L$_2$ Secondary throttle lever	M Diaphragm
S Diaphragm return spring	T$_1$ Primary throttle	T$_2$ Secondary throttle	

6.5 ELECTRONICALLY CONTROLLED CARBURETTORS

In the mid 1980s Bosch and Pierburg in Germany further refined the concept of the compound carburettor by superimposing an electronic control over certain functions of its primary barrel, which provided comparable advantages to a more costly fuel injection system. Apart from the normal mechanical connection between the accelerator pedal and the throttle valves, including delayed opening of the secondary valve for full power, all the usual external mechanical controls are replaced by two actuators, these being of the electropneumatic plunger and electrical servo motor types for the throttle and choke valves respectively. The plunger of the former actuator is spring extended and retracted by intake manifold depression acting on its diaphragm. During normal driving the plunger is retracted clear of the spindle lever connecting the throttle valve. Both actuators receive appropriate commands from an electronic control unit, which in turn is supplied with information from various sensors that detect engine speed, coolant temperature, throttle valve position, throttle valve actuator travel and idle switching (Figure 6.26). The signals received from sensing the travel of the throttle valve actuator determine the degree of intake manifold depression to which it is subjected.

The electronic control over the carburettor is exercised in four basic ways, which may be summarized as follows:

Starting and warm-up Cold starts are automatically controlled without intervention from the driver's accelerator pedal because, based on signals received by the electronic control unit relating to the closed position of the throttle valve, engine cranking speed and coolant temperature, appropriate commands are issued to the throttle and choke actuators for rotating their valves to an optimum angular relationship to permit starting. To assist mixture enrichment the full closing of the choke valve also serves to depress a needle valve, which closes off one of the air bleed orifices in the main fuel metering circuit. Once the engine fires and begins to warm up, the commands from the electronic control unit to the actuators are amended, so that the choke valve is progressively opened to weaken the mixture and the throttle valve gradually closed to provide the most economical idling mixture for the prevailing engine temperature.

Idling speed This is held constant by the electronic control unit, via the throttle valve actuator, regardless of any disturbing factors, and therefore makes practicable a lower stable idling speed.

Acceleration The usual accelerator pump for discharging fuel into the choke tube is dispensed with, and instead any rapid opening of the throttle valve is signalled by its sensor to the electronic control unit, which then commands partial closing of the choke valve by its servo motor. This action therefore provides a temporary enrichment of the mixture for maintaining acceleration.

Overrun When the throttle valve closes as the driver decelerates, its spindle lever contacts and triggers the idle switch at the tip of the plunger retracted by the throttle valve actuator. This allows a further retraction of the plunger so that the throttle valve can rotate beyond its normal idle position, which results in the edge of its plate isolating the idle mixture discharge orifice from manifold depression and hence cutting off what would otherwise be an unnecessary delivery of fuel. However, when the engine speed drops to 1400 rev/min the actuator plunger restores the throttle valve to its idle position to prevent stalling.

6.6 PETROL ENGINE FUEL INJECTION

General background

In order to gain a general insight into the different types of fuel injection system used on motor vehicle petrol engines, it is first necessary to review briefly their various stages of development since the early 1950s. Historically though, the first requirement for a petrol fuel injection system arose in the early development of the aeroplane, because the very basic designs of carburettor then available did not take kindly to the climbing, banking and diving manoeuvres inherent to flight; there was a risk of fuel starvation or flooding. It is therefore a matter of interest that the system of fuel injection used, such as the one fitted to Wilbur and Orville Wright's historic aeroplane of 1903, involved delivering the fuel under pressure into the intake ports ahead of the inlet valves, and thus anticipated the modern practice of *port injection* for the motor vehicle engine.

However, a different system of fuel injection was used in earlier motor vehicle applications, beginning in 1951 with the small two-stroke engined Goliath and Gutbrod cars then being made in Germany. These were followed in 1954 by the adoption of a fuel injection system for the now legendary Mercedes-Benz 300SL gull-wing sports car. The system used in these cases was of the *direct-injection* type, where the petrol was sprayed from nozzles directly into the cylinders in diesel engine fashion and delivered from a similar multi-plunger pump as described later. This type of system had been developed from the extensive experience gained by the German aircraft industry working in conjunction with the

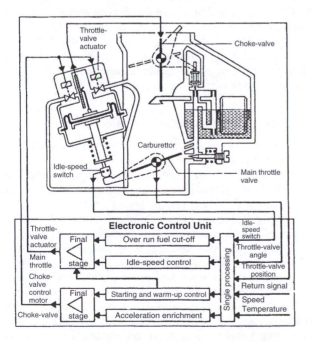

Figure 6.26 Bosch Pierburg electronically controlled carburettor

Robert Bosch company, which resulted in nearly all German aircraft engines during World War II being equipped with petrol fuel injection rather than advanced carburettor systems. In service, though, the direct injection of petrol into the engine cylinders did not prove entirely satisfactory for motor vehicle application, because it tended to cause dilution of the engine oil during the frequent warming-up periods encountered in everyday motoring.

To counter this problem the next significant development was the revival of the earlier mentioned port injection system – again by Mercedes-Benz, this time for their 300d model in 1957. In this application a metered and timed injection of petrol into the separate intake ports was provided, so that the injectors were no longer subjected to combustion pressure and temperature and could therefore be made simpler in design. This system of what is termed *timed injection* into the intake ports still involved the use of an expensive diesel-style fuel injection pump, but at least the delivery pressures could be reduced from the $10.5\,MN/m^2$ ($1500\,lbf/in^2$) previously associated with the cylinder injection system to $0.7\,MN/m^2$ ($100\,lbf/in^2$) for port injection. A simplification of this type of system, again by Mercedes-Benz for their 1958 220SE model, no longer involved a timed and metered quantity of petrol for each of the six cylinders. Instead, the petrol was injected into the intake ports of three cylinders simultaneously, the two groups of three fuel injectors being fed at every revolution of the crankshaft by a twin-plunger pump, thereby providing what is termed *semi-continuous flow* fuel injection.

Meanwhile in America, the General Motors Rochester system of *continuous flow* port injection had been introduced on the Chevrolet Bel Air model of 1957. In this system of continuous flow injection, which was inspired by the simpler Hilborn system developed some years earlier for racing car engine application, it was considered that timed injection of petrol into the intake ports was not really necessary. This was because tests had shown that the speed of engine operation is such that each cylinder cannot distinguish one injection pulse from another. It was, of course, still necessary to meter the quantities of petrol delivered to the open nozzle injectors in accordance with the engine air flow.

A new era in the development of petrol engine fuel injection began in 1968 with the introduction of an *electronic* system, which was the work of R.W. Sutton of the Bendix Corporation in America. The essential difference between this type of system and the others was that it provided for electronic control and timed electro-magnetic operation of the nozzle valves, instead of utilizing pulsations in fuel pressure to open them against their spring loading. In this type of system, as in other port injection systems, the injectors are mounted so as to spray petrol into the intake ports about 100 to 175 mm (4 to 7 in) behind the valve heads. This arrangement had been found to give the best results in terms of engine performance, fuel economy and rapid warm-up. The spring-loaded nozzle valves are opened inwards by solenoid operation, the solenoids being energized through an electrical distributor in conjunction with an electronic control system. This serves to time the duration of nozzle valve opening, so that the correct quantity of fuel is delivered during each operating cycle. For this purpose, the electronic control system receives input signals from the intake manifold of the pressure

and temperature of the air supplied to the engine. Fuel is supplied at a relatively low pressure of 245 to $360\,kN/m^2$ (35 to $50\,lbf/in^2$) to a ring main or, in fuel injection terminology, common-rail circuit leading to all the port injectors. A positive displacement pump is used as the pressure source, together with a control valve for limiting the pressure in the circuit and returning fuel in excess of requirements to the fuel tank. This continual recirculation has a cooling effect on the fuel supply, which therefore avoids vapour locking in the fuel line and improves hot starting behaviour.

The Bendix electronic fuel injection system was later developed under licence by Robert Bosch in Germany, who then produced their D-Jetronic system that was first used on a Volkswagen 1600 model in 1968. By taking advantage of the rapid progress that was being made in electronic circuit technology, where higher precision and greater reliability were being combined with fewer components, the D-Jetronic system was developed into the more sophisticated L-Jetronic system. This has received widespread application following its introduction in 1973, a Nissan installation of this type of system being later described. Since the early 1980s the L-Jetronic system itself has been further refined to incorporate supplementary functions for more precise control over the air–fuel ratio in relation to three-way catalytic converter operation (Section 11.2), overrun fuel cut-off and engine speed limiting. Also in 1973 Bosch introduced their non-electronically controlled K-Jetronic fuel injection system of the hydromechanical type, which again has been widely used and further developed.

A later development in electronic fuel injection (or EFI, to use its modern abbreviation) was the introduction by the American Chrysler company of single-point, as opposed to multipoint, fuel injection on their 1981 Imperial model. This system was shortly followed by similar ones not only from other American manufacturers, but also from Bosch with their Mono-Jetronic system, which differs from the Chrysler system in providing an intermittent- rather than a continuous-flow fuel injection. The aim of developing a single-point fuel injection system was to combine simplicity and reliability with low production cost. A more recent development in electronic fuel injection for petrol engines has been the revival of interest in the earlier mentioned direct injection systems, as described in Section 6.9.

Advantages of petrol engine fuel injection

It was once humorously said that if the motor vehicle petrol engine had begun its life with a fuel injection system and then someone had later invented the much simpler and less expensive carburettor, it would have been hailed as an engineering miracle! So how can the use of a petrol injection system be justified in motor vehicle practice?

The most convincing answers to this question were provided in a 1978 review of the reasons for the market success of petrol injection systems by the Robert Bosch company in Germany. In this review it was pointed out that when the first passenger cars were fitted with fuel injection systems in the 1950s the result was an increase in engine power, but that since the end of the 1960s another feature of the fuel injection system had become more significant. This concerned the introduction of legislation to limit the maximum levels of toxic substances in the exhaust emissions of motor vehicles. Fuel injection systems lend themselves particularly well to meeting these legal requirements, which began in America and have since been introduced in modified form in Australia, Canada, Europe and Japan.

It was also pointed out that another particularly topical reason for the introduction of fuel injection systems was the need to improve the fuel economy of cars. In America, legislation provided for the average fuel consumption of the vehicles produced by any one manufacturer to be reduced to 65 per cent of the 1978 value by the 1985 model year. Again fuel injection systems play an important role when it comes to meeting these requirements.

In addition to the three advantages mentioned, namely those of increased engine power, reduced exhaust emission and improved fuel economy, it was further mentioned that the fuel injection system results in a substantial improvement in driveability, which is particularly reflected in the good cold starting behaviour and warm-up of the engine.

The basic reason why the fuel injection system should possess these fundamental advantages over the conventional carburettor is that the various engine conditions are measured by different sensors, so they can also be considered independently of one another for optimum fuel metering. Other reasons concern the more uniform distribution of mixture to the engine cylinders and the absence of fuel wetting on the walls of the intake manifold, particularly when cold.

6.7 MULTI-POINT FUEL INJECTION

The advantages usually associated with a multipoint fuel injection (MPI) system, where the fuel is sprayed into the inlet port for each cylinder, are a more uniform distribution of mixture to the engine cylinders, an absence of fuel wetting on the walls of the intake manifold, particularly when cold, and the need to design the manifold only for the most efficient flow of air rather than air and fuel (Section 7.2). Another potential advantage is that port injectors can be used with intake manifolds where the separate branches have been tuned in length, so as to exploit inertia ramcharging effects for maximum power (Section 9.2).

Electronic MPI systems

We take as an example a Nissan installation of EFI. The basic features are most conveniently described in terms of the fuel flow, air flow and electrical flow systems, as follows.

Fuel flow system
Fuel is drawn from the fuel tank into a roller vane pump driven by an electrical motor of the wet type which means that the motor runs filled with fuel for greater safety in case of fire. The pump incorporates a relief valve that is designated to open in the event of malfunction in the pressure system, and also a check valve to prevent an abrupt drop of pressure in the fuel pipe when stopping the engine (Figure 6.27). Should the engine involuntarily stop during driving, the electrical circuit to the pump is so arranged that the delivery of fuel ceases in the interests of safety, even though the ignition switch remains in the on position.

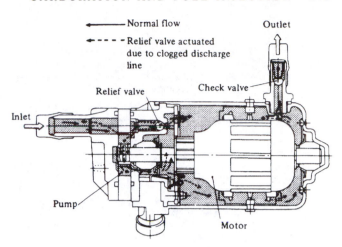

Figure 6.27 EFI fuel pump (*Nissan*)

Fuel discharged under pressure from the pump flows through a mechanical fuel damper, wherein any pulsations in fuel flow are suppressed by a spring-loaded diaphragm that divides the damper into a fuel chamber and an air chamber. After leaving the damper the fuel then passes through a filter unit to remove foreign matter and also any water, the latter being collected at the bottom of the filter casing. The filtered fuel is finally delivered to the common-rail fuel line from which it is injected into the engine intake ports, via the electromagnetically operated fuel injectors (Figure 6.28).

Surplus fuel is directed from the common-rail fuel line through a pressure regulator, which also contains a spring-loaded diaphragm but in this case divides the regulator into a fuel chamber and a depression chamber. The diaphragm actuates an outlet valve through which fuel is returned to the tank and the depression chamber is connected to the intake manifold. In operation, the regulator maintains a constant pressure difference of 255 kN/m^2 (36.3 lbf/in^2) between the fuel delivery pressure and the intake manifold depression, thereby providing optimum fuel injection under all running conditions.

The fuel injectors operate on the solenoid valve principle (Figure 6.29). When an electrical pulse signal is applied to the electromagnetic coil built into the injector, a spring-returned plunger is pulled into the solenoid and withdraws the needle valve from its nozzle seating. The period during which the needle valve remains open governs the quantity of fuel injected and is determined by the duration of pulse signal received from the EFI control unit. When the electromagnetic coil is de-energized, the needle valve is returned to its seating by the return spring and system fuel pressure. During cold starting an additional amount of fuel is injected into the intake manifold independently of the main injectors by a cold start valve. This operates in response to low cooling system temperature via a thermotime switch, which also stops excessive injection of fuel in the event of repeated operation of the ignition switch in the start position.

Air flow system
The engine intake air first passes through an air cleaner and is then measured for quantity by an air flow meter (Figure

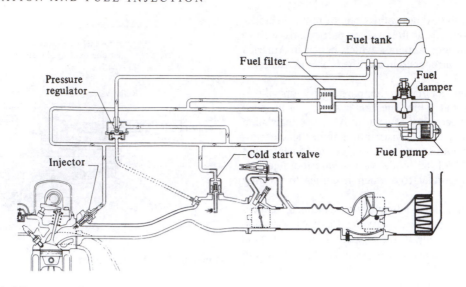

Figure 6.28   EFI fuel flow system (*Nissan*)

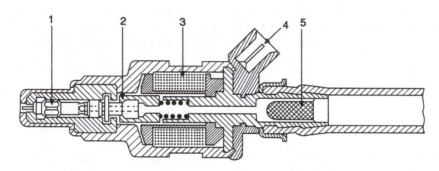

Figure 6.29   Cross-section of a fuel injector

1 nozzle valve   3 solenoid winding   5 filter
2 solenoid armature   4 electrical connection

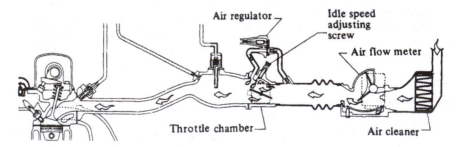

Figure 6.30   EFI air flow system (*Nissan*)

6.30). This incorporates a spring-returned rotatable flap in its air passage. As the air flows through the meter it deflects the flap, until a condition of equilibrium is maintained between the air pressure and the closing torque on the flap. The angle through which the flap is rotated is electronically monitored, so that the EFI control unit can compute the quantity of fuel to be injected. As the flap of the air flow meter is rotated a complementary damper plate, working within a close-clearance air chamber, ensures that the flap is not disturbed by pulsations in the intake manifold.

After passing through the air flow meter the intake air then enters a throttle chamber, which controls the intake air flow while the car is being driven and therefore contains a throttle valve that is rotated in response to accelerator pedal movement. It remains almost closed during idling when the air is diverted through an adjustable bypass port in the throttle chamber. During the warming-up period, the intake air again bypasses the throttle valve and flows through another port via an air regulator. This comprises a thermostatically controlled shutter valve that meters the quantity of air required for fast

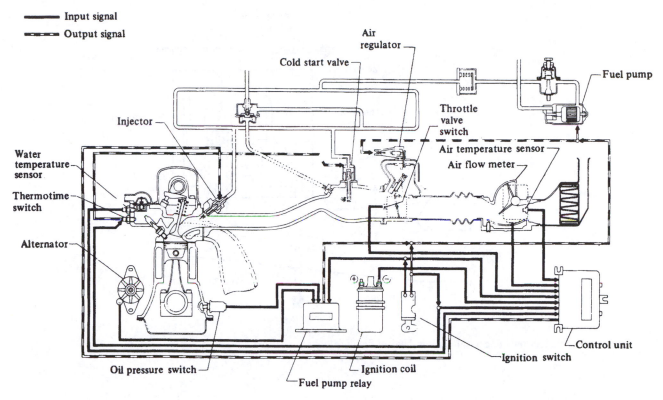

Figure 6.31 EFI electrical flow system (*Nissan*)

idle. Beyond the throttle chamber the intake air flows through each manifold branch and into the cylinders.

Electrical flow system
At the heart of the electrical flow system is the EFI control unit (Figure 6.31). This receives electronic signals from various types of sensor that continually monitor engine operating conditions. These input signals relate to the quantity and temperature of intake air, opening of the throttle valve, engine speed, temperature of coolant and starting operation. Armed with this flow of information, the electronic control unit is then capable not only of computing from information stored in its memory the quantity of fuel that needs to be injected into the engine intake ports, but also of providing the necessary output signals to activate the fuel injectors.

All injectors receive their activating signal from the control unit simultaneously. The required fuel quantity is attained after fuel injection is made twice during one complete four-stroke cycle of the engine, that is once for every revolution of the crankshaft.

Further development of the electronic MPI system
A significant development by Bosch was the introduction in 1983 of an all-electronic air mass meter to supersede the mechanical air flow meter. This new feature first appeared in their LH-Jetronic version of EFI, and permits direct measurement of the volume of air induced by the engine regardless of changes in barometric pressure and hence density.

The meter simply comprises a very fine platinum wire that is suspended in keystone fashion across the measuring venturi. When the engine is running this wire is electrically heated by a circuit that maintains it at a temperature which is always above that of the incoming air by a constant amount. Since the passage of air through the venturi has a cooling effect on the wire, the amount of electrical current required to maintain this temperature differential between wire and air flow can be used as a signal of the mass of air being induced by the engine. Each time the engine is stopped the platinum wire is briefly heated to a higher self-cleaning temperature, because any permanent contamination of the surface could alter its measuring characteristics. The other advantages associated with this type of air mass meter are those of negligible obstruction to the incoming air, a more compact construction and, of course, no moving parts.

Hydromechanical MPI system

Although the earlier mentioned Bosch K-Jetronic fuel injection system bears a certain physical resemblance to the electronically controlled system previously described, it differs in operating on the hydromechanical principle of fuel metering and injects fuel continuously, instead of intermittently, into the inlet port of each cylinder. The injectors therefore neither require any electrical connections, nor serve any metering function. When the injection pressure in the system exceeds a value of typically $360 \, \text{kN/m}^2$ ($50 \, \text{lbf/in}^2$), the poppet valves of the injectors open outwards to create a finely atomized spray of fuel, assisted by their inherent tendency to oscillate or 'chatter', a phenomenon that will be mentioned again later in relation to diesel fuel injectors (Section 8.7).

Confining our attention now to the heart of this particular system, it essentially comprises a combination air flow meter

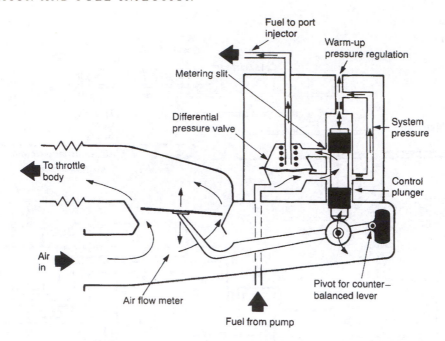

Figure 6.32　Schematic arrangement of hydromechanical fuel injection

and fuel distributor (Figure 6.32), the construction and functioning of these units being as follows.

Air flow meter

The purpose of this unit, like that of its counterpart in the EFI system, is to measure the quantity of air being induced by the engine, since this is an indication of its operating load. However, in this particular system a direct mechanical link is established between the air flow meter and the fuel distributor, so that the latter is signalled the correct fuel quantities required by the injectors. The air flow meter takes the form of a circular plate moving in a venturi of conical shape. A counterbalanced second system of levers supports and conveys the arcuate movements of the plate, as a measure of the air flow rate, to the control plunger of the fuel distributor. A centralizing spring restores the plate to its zero position when the engine is not running. The conical shape of the venturi is so contrived that the movement of the plate, relative to a given air flow, can be varied to obtain different air–fuel ratios according to engine load.

Fuel distributor

As already inferred, this unit serves to meter the fuel quantities required by the injectors, as signalled by the air flow meter. It basically comprises a control plunger of the spool valve type sliding within a barrel, which has lower inlet holes and upper metering slits, there being one pair of each through which the fuel passes *en route* to a port injector. Each slit communicates directly with its own differential pressure valve, which regulates the pressure drop at the slit to a precise and constant value, so that a definite relationship is always preserved between the air induced and the quantity of fuel injected. The differential pressure valves each comprise a chamber that is divided by a diaphragm. This is spring loaded towards the

lower chamber, through which fuel at system pressure passes to the control plunger, and away from the outlet port of the upper chamber, from which fuel at injection pressure is directed to the port injector. The spring loading on the diaphragm is therefore responsible for creating a small pressure differential in favour of the lower chamber.

When the rising plate of the air flow meter transmits a corresponding but albeit reduced movement to the control plunger, which then uncovers an increasing length of metering slit in the barrel, the larger quantity of fuel flowing into the upper chamber tends to increase its pressure relative to the lower one. However, this is countered by the spring-assisted diaphragm deflecting further away from the mouth of the outlet port in the upper chamber, thereby lessening its restriction so that the quantity of fuel directed to the injector is increased but its pressure is maintained constant. A controlling force derived from a restricted source of system pressure acts above the metering plunger and in opposition to that imposed by the plate and lever of the air flow meter. It not only serves to dampen any oscillation of the plate, but also influences the fuel metering during engine warm-up. The control pressure on the plunger is then reduced so that the air induced by the engine increases the movement of the plate. That is, the further rising of the control plunger uncovers a greater length of the metering slits and allows an increased quantity of fuel to be delivered to the port injectors.

6.8　SINGLE-POINT FUEL INJECTION

It was earlier mentioned that the advantages sought with a single-point fuel injection (SPI) system were those of combining simplicity and reliability with low production cost. More specifically, the air–fuel ratio can be controlled to a greater

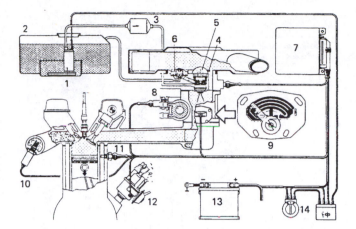

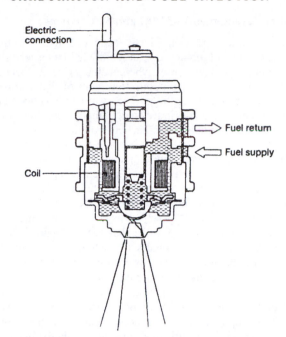

Figure 6.33   Bosch Mono-Jetronic SPI system

1 electric fuel pump	8 throttle valve actuator
2 tank	9 throttle valve potentiometer
3 filter	10 lambda sensor
4 air temperature sensor	11 engine temperature sensor
5 single-point injector	12 ignition distributor
6 pressure regulator	13 battery
7 electronic control unit	14 ignition and starter switch

Figure 6.34   Bosch low-pressure throttle body injector

degree of accuracy than is possible with a conventional carburettor, while realizing a cost advantage over a multipoint fuel injection system. Another technical advantage of the SPI as compared with the MPI system is that the operating temperature of its electromagnetically actuated fuel injector assembly is reduced, thereby allowing the system to work at a lower pressure without the risk of petrol vaporizing in the nozzle. If the latter occurs in a fuel injection system there is less control over fuel metering, with consequent poor engine performance. The lower working pressure of the SPI system, which is typically in the range of 70 to 105 kN/m^2 (10 to 15 lbf/in^2), also permits the use of a non-positive displacement fuel pump of the impeller type, which is mounted inside the tank for quieter running. This type of pump incorporates a check valve at its outlet to keep the fuel supply system primed.

Electronic SPI systems

The two major components of this type of EFI system are the central injection and electronic control units. Since we are already familiar with the necessity for the latter in other EFI systems, we shall confine our attention to the central injection unit. This in effect replaces a carburettor on the intake manifold riser and sprays fuel upstream of the throttle valve (Figure 6.33), instead of downstream as with MPI systems. The fuel is therefore injected into the throat of the throttle body at the point of maximum air flow, thereby enhancing fuel atomization and hence mixture preparation. In the case of V-cylinder engines with two-tier intake manifolds (Section 7.2), the fuel spraying and throttle body arrangements are duplicated in the central injection unit.

Although a pintle valve injector nozzle is favoured by one American manufacturer, a ball valve type is more commonly used as it widens the range of flow rate. This can be an important requirement for modern engines, because at one extreme

their lower friction permits reduced idling speeds with a smaller minimum fuel demand, while at the other the application of turbocharging imposes a greater maximum demand. In the injector of the Bosch Mono-Jetronic system a solenoid attracts a flat armature that carries a part-spherical ball valve, which is spring loaded against its nozzle seating and receives axial guidance from a diaphragm. The angular disposition of the multiple orifices in the nozzle is such that they produce a spray of conical form into the throat of the throttle body (Figure 6.34). An intermittent injection of fuel is triggered to accompany the ignition system pulses. A pressure regulator of the preloaded diaphragm and ball valve type is integrated with the central injection unit, its purpose being to maintain constant the pressure drop across the injector valve regardless of the quantity of fuel injected. It will be noticed that, unlike a port injector with a top fuel feed, the bottom feed of a central injector allows a continuous through-flow of fuel, which ensures efficient cooling of its nozzle region and contributes to good hot starting.

To cater for the varying operating conditions of the engine, the electronic control unit processes the input signals from the usual system sensors. From these it computes the quantity of fuel that needs to be injected for either enriching or weakening the mixture and then commands the appropriate duration of nozzle valve opening for each injection of fuel. An electrical servo motor controls the throttle opening to maintain a consistently low idling speed.

In a Rover system of single-point fuel injection, the intake manifold is provided with a source of electrical heating during the engine warm-up period, so that the air–fuel mixture flowing downstream of the throttle valve is efficiently atomized and does not cause wetting on the manifold walls. The manifold heater is activated once the engine has started and remains in operation until the coolant temperature reaches 50°C.

Further developments of MPI and SPI systems

By virtue of being electronically triggered, fuel injection systems of the multipoint type can readily lend themselves to either continuous flow control, wherein all injectors are opened simultaneously; semi-continuous flow control that involves the injectors being opened in groups; or timed injection control where each injector is individually timed to open just prior to the induction stroke of the particular cylinder concerned. The latter version is tending to become established modern practice in the interests of optimum management of the engine, and is sometimes referred to as 'sequential electronically-controlled fuel injection' (also abbreviated to SEFI) or just simply as 'sequential injection'. For multipoint fuel injection systems, the advent of four-valve combustion chambers (Section 3.3) with two inlet valves has seen a departure from a single tapered spray to a dual-spray pattern. A spray of fuel can therefore be directed towards each inlet valve by a directional orifice disc at the mouth of the port injector.

Other features typical of modern multipoint and single-point fuel injection systems, include the temporary interruption of fuel supply to the injectors or injector not only during engine overrun down-hill, thereby reducing both fuel consumption and exhaust emissions, but also to limit engine speed to the manufacturer's permitted maximum. A safety circuit in the electronic control system, which is responsive to lack of intake air flow, prevents the electric fuel pump from continuing to deliver fuel in an accident situation with the engine stopped and the ignition still switched on.

6.9   DIRECT FUEL INJECTION

As mentioned earlier there has been a revival of interest in directly injecting the fuel into petrol engine cylinders, albeit now taking advantage of modern sophisticated control systems, together with advances in combustion chamber and intake system design. Unlike earlier attempts to adopt a direct-injection system, where the chief objective was to increase engine power, the present aim is to improve fuel economy without sacrificing car performance. This revival of interest in petrol direct-injection by various manufacturers can be attributed to the pioneering work done by Mitsubishi in Japan. After many years of research they introduced their 'Gasoline Direct Injection' or 'GDI' engine in 1996, the designation GDI being a registered trademark of the Mitsubishi Motor Corporation.

Before defining the two combustion modes of the Mitsubishi GDI equipped engine, it is first necessary to identify the relevant features of its construction, which are highlighted by the amusing characters in Figure 6.35. These features may be summarized as follows:

1 Unlike an indirectly injected petrol engine with substantially horizontal intake ports, those of the GDI engine are virtually upright and straight. Their purpose is to direct a greater volume of air smoothly down to the pistons.
2 Since the fuel is being injected directly into the cylinders of the GDI engine, including during the compression stroke when it is operating in the ultra-lean combustion mode, a high-pressure pump is necessary to deliver the fuel at a pressure of 5 MN/m² (725 lbf/in²) to the common rail.

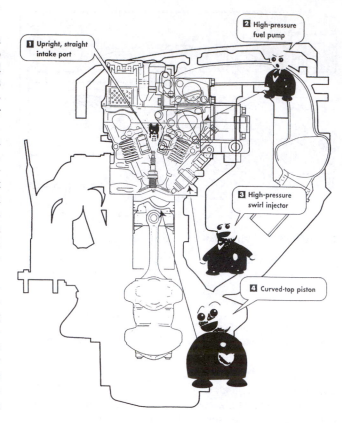

Figure 6.35   Key technical features of the Mitsubishi GDI engine (*Mitsubishi*)

1 Upright, straight intake port
2 High-pressure fuel pump
3 High-pressure swirl injector
4 Curved-top piston

3 High-pressure, swirl action, injectors are similarly employed to control vaporization and dispersion of the fuel spray inside the cylinders. The spray pattern is monitored as necessary by computer control to match each combustion mode of the GDI engine, thereby allowing precise control over the amount of fuel burned.
4 The piston crowns of the GDI engine are curved instead of being flat-topped as in conventional practice. This curvature is such as to form an asymmetrical spherical cavity, which organizes the incoming charge of air into a vigorous tumble or barrel motion and maintains it throughout the compression stroke (Section 3.6). The injected fuel can therefore be concentrated immediately beneath the sparking plugs, which results in clean, complete combustion.

The Mitsubishi GDI engine operates in two distinct combustion modes, these being designated 'Ultra-Lean Combustion Mode' and 'Superior Output Mode'. This particular strategy aims to reduce overall fuel consumption, as compared to an equivalent indirect-injection petrol engine, and therefore provides a more environment-friendly engine but without sacrificing car performance. The Mitsubishi company has best described in words and pictures the two modes of combustion, as shown in Figure 6.36.

Translated into terms of 'driveability', the computerized switching between the two modes of combustion may be related as follows:

Ultra-Lean Combustion Mode – This caters for city driving or high-speed cruising on the motorway, when power

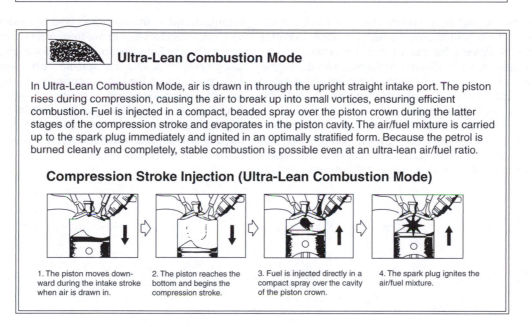

Superior Output Mode

In Superior Output Mode, air is drawn in through the upright straight intake port. Fuel is injected in a wide, conical shower during the piston's intake stroke, when it is descending toward the bottom of the cylinder. The vaporisation of the fuel cools the air inside the cylinder, causing the cylinder itself to cool and thus increasing the compression ratio. The intake port assists by functioning like a supercharger, smoothly drawing in additional air and improving volumetric efficiency. The rising piston carries the homogenous air/fuel mixture up to the spark plug for ignition, turning the GDI engine into a formidable power plant capable of producing 10% higher torque than a conventional engine of similar displacement.

Intake Stroke Injection (Superior Output Mode)

1. The piston moves downward during the intake stroke when air is drawn in.
2. Fuel is injected directly in a wide, conical shower and the air and the cylinder are cooled.
3. The air/fuel mixture is compressed.
4. The spark plug ignites the air/fuel mixture.

Ultra-Lean Combustion Mode

In Ultra-Lean Combustion Mode, air is drawn in through the upright straight intake port. The piston rises during compression, causing the air to break up into small vortices, ensuring efficient combustion. Fuel is injected in a compact, beaded spray over the piston crown during the latter stages of the compression stroke and evaporates in the piston cavity. The air/fuel mixture is carried up to the spark plug immediately and ignited in an optimally stratified form. Because the petrol is burned cleanly and completely, stable combustion is possible even at an ultra-lean air/fuel ratio.

Compression Stroke Injection (Ultra-Lean Combustion Mode)

1. The piston moves downward during the intake stroke when air is drawn in.
2. The piston reaches the bottom and begins the compression stroke.
3. Fuel is injected directly in a compact spray over the cavity of the piston crown.
4. The spark plug ignites the air/fuel mixture.

Figure 6.36 Two combustion modes of the Mitsubishi GDI engine (*Mitsubishi*)

requirements are least demanding. Fuel efficiency can then be comparable to a diesel-engined car.

Superior Output Mode – This allows the responsiveness of a petrol engine to be retained, when more power is required to maintain speed up gradients and for acceleration, such as passing another car on the motorway.

Toyota and other manufacturers have since introduced direct-injection petrol engines or, as they are now sometimes termed, direct-injection spark ignition (DISI) engines. These differ from the Mitsubishi GDI engine chiefly in respect of their intake systems and forms of piston crown recess, which promote modified axial, rather than barrel, swirl motion of the mixture in the cylinders (Section 3.6).

Continuing developments would seem to indicate that a carefully matched combination of direct injection and turbocharger, together with variable valve timing, may allow a beneficial reduction in engine size without loss of power output.

6.10 ELECTRONIC THROTTLE CONTROL

Another innovation is a 'drive-by-wire' control for fuel injection systems, generally termed electronic throttle control. Its main purpose is to eliminate the need for a cable control to connect the accelerator pedal with the system throttle valve. Otherwise this involves routing the cable through the bulkhead

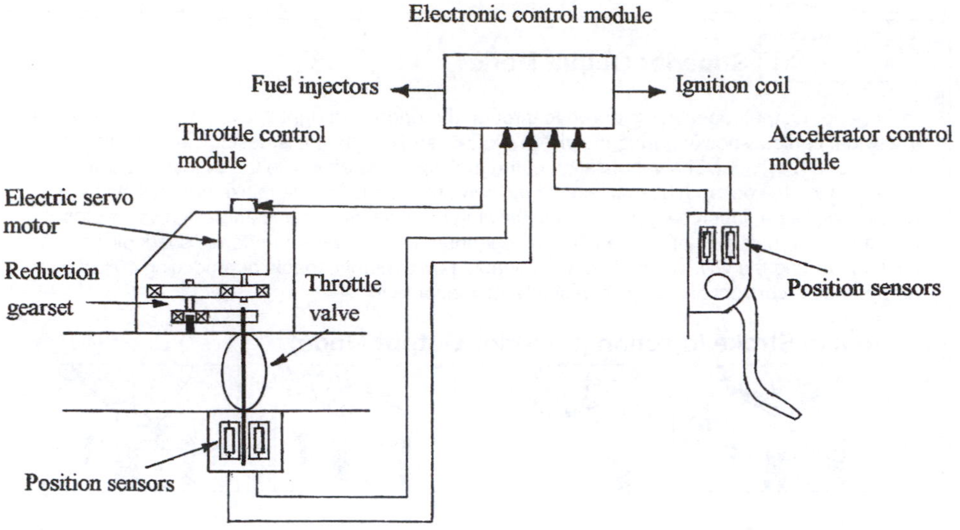

Electronic control module

Fuel injectors ← [ ] → Ignition coil

Throttle control
module

Accelerator control
module

Electric servo
motor

Reduction
gearset

Throttle
valve

Position sensors

Position sensors

Figure 6.37   Schematic layout of an electronic throttle control system

to the body interior, making it a potential source of noise trans-mission. A further benefit of using an electronic throttle control is to enable the driver's commands to be moderated, via an electronic control module. This receives inputs from the engine management and other systems, such as traction and stability controls, and can therefore contribute to improved driveability, stability and also emissions reduction.

In practice, the electronic throttle control system comprises three modules: accelerator pedal, throttle servo motor and electronic controls (Figure 6.37). Accelerator pedal move-ments are monitored by position sensors and the information

communicated to the electronic control module. The throttle valve plate receives its turning movements from a DC electric motor, via a double-reduction gearset that engages with a quad-rant gear on the throttle plate spindle. This combination of elec-tric servo motor and reduction gearing provides responsive and precise angular positioning of the throttle plate to regulate air flow, together with sufficient torque capacity to overcome any sticking tendency. The throttle valve positioning is also moni-tored by sensors, which likewise communicate the information to the electronic control module.

7 Intake and exhaust systems

7.1 AIR CLEANER AND SILENCER

General requirements

Efficient removal of dust particles from the considerable quantities of air flowing into the engine is of vital importance. Otherwise, their presence would contribute to internal abrasive wear, especially between the working surfaces of the cylinders, pistons and rings. An intake air filter is therefore mounted either directly on the engine, or separately in the engine compartment via a flexible connection.

The air cleaner is also required to act as a silencer for the intake system; that is, it must suppress engine induction noise to an acceptable low level. With small throttle openings, induction noise is generally of a high-frequency character. At medium to large throttle openings and especially with four-cylinder engines, the major source of induction noise occurs as low-frequency or boom periods, which arise from the implosion of the air and fuel charge into the cylinders.

Another function performed by all air cleaners is to act as a flame arrester, in the event of the engine backfiring through the induction system.

Methods of air cleaning

The attainment of a satisfactory engine life can be regarded as a function of what is sometimes termed total engine cleanliness. Therefore the air filter may be bracketed with the oil filter as a means of achieving this end, a point of difference in their operation being that the contaminated air passes through the filtering medium of the air cleaner once only, whereas the engine oil is continually recirculated through its particular filtering medium. Modern air cleaners incorporate at least one of the following physical methods of filtration: sieve, impingement and separation.

Sieve

This type of replaceable filter medium has generally comprised a cellulose fibre material of controlled porosity, but more recently a synthetic fibre material may be used. The element assumes a pleated form not only to improve its rigidity, but also to present the largest possible surface area to the through flow of air. In the first instance the element provides surface filtration on the same principle that a gardener will use a wire mesh sieve to remove stones from the earth. However, the thickness of the filter material is such as to provide, in effect, successive layers of sieves. The air flowing through the filter pores is therefore compelled to follow an exceedingly tortuous path, as directed by the fibrous structure of the material. By this process of 'depth' filtration, dust particles of a size much smaller than the mean pore size in any one layer of the element material can be excluded from the air stream.

Impingement

A larger pore size is used with this type of filter medium, which may take the form of either a dry fibrous or an oil-wetted mesh surface. In the former case, a filtering action arises from impact of the dust particles against the surface of the element material, which as a result undergoes minute deformations and physically prevents the dust particles from continuing along with the air stream. With an oil-wetted surface, a filtering action is obtained by the dust particles being held by adhesion after impact.

Separation

In this method of filtration enforced changes in the direction of flow are utilized to separate the dust particles from the air stream. That is, the heavier dust particles are separated from the air by virtue of their greater inertia or reluctance to change direction; centrifugal effects thus become the major factor here. The incoming air is spun around a fixed element in cyclonic fashion to provide a prefiltering action before the air flows through the element proper, thereby prolonging its service life.

Methods of air silencing

Airborne radiated induction noise is mainly reduced in intensity or attenuated by incorporating in the air cleaner the principle of the Helmholtz resonator. This device is so named after H.L.F. von Helmholtz, a noted German physicist of the nineteenth century, who discovered the physical significance of tone quality apart from making many other contributions to science. In its original scientific form, the Helmholtz resonator comprised simply a spherical casing to enclose an air cavity, which was provided with a neck portion on one side and a plain opening on the opposite side (Figure 7.1a). By pointing the

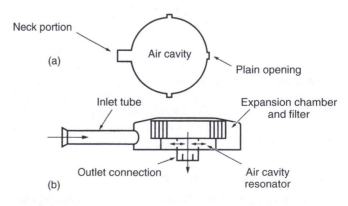

Figure 7.1 Simple analogy between (a) a scientific Helmholtz resonator and (b) an air intake cleaner and silencer

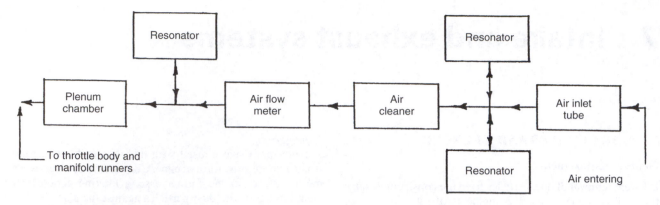

Figure 7.2   Block diagram of air intake system

neck of the resonator towards the source of the sound and then listening at the plain opening, very feeble sounds of the appropriate frequency could be detected from a musical note being played. Helmholtz in his experiments actually constructed a whole series of such resonators, their significance being that each one would respond only to a note of a particular pitch.

The basic structure of an air cleaner is analogous to that of a Helmholtz resonator, because the air flowing through the main expansion chamber and filter also communicates with an annular air cavity in which there is a purely oscillating movement of air (Figure 7.1b), as in a simple Helmholtz resonator. If a suitable length of inlet tube is then matched to the large volume of the main chamber, the air cleaner in acting as a Helmholtz resonator can be tuned to respond to an unwanted peak of induction noise, which may be occurring in the important rural and motorway speed ranges of the car. Since the resonator must owe the energy necessary to excite it into vibration to the source of the unwanted sound, it follows that this source must lose its energy more rapidly.

In modern installations several resonators may be included in the air intake system (Figure 7.2), these being produced from plastics materials. For example, the air inlet tube may be provided with a pair of branched resonators, which are of different size and tuned to suppress troublesome frequencies of induction noise. They may be conveniently mounted under a front wing and the incoming air then routed to an underbonnet air cleaner assembly. From here the filtered air passes through the air flow meter of the fuel injection system. Before being collected by a plenum chamber, the air is routed through intermediate ducting that also has a branched resonator, which further reduces induction noise levels. Both the plenum chamber that connects with the throttle body of the fuel injection system and the branched resonator are mounted directly above the engine. For a V-cylinder engine two plenum (or collecting) chambers may be employed, one for each cylinder bank, in which case a Y-shaped connecting duct forms their entrance.

Air cleaner types and applications

The various types of air cleaner, past and present, that may be encountered in motor vehicle practice are described in the following sections.

Cellulose fibre/synthetic fibre
Air cleaners incorporating these replaceable depth filter elements represent established practice. The element itself is contained in either a cylindrical (Figure 7.3a) or a flatter shape of housing, according to space restrictions imposed by present-day more crowded engine compartments. This type of air filter confers a very high filtering efficiency typically is excess of 95 per cent, and generally requires replacing every 24 000 miles (40 000 km), although some manufacturers may specify much longer intervals. The latter also applies to the larger capacity elements that can be accommodated in the air filters of heavy vehicles, where they may be combined with cyclonic filtration.

Oil-wetted mesh
Prior to the advent of the fibre filtering element, this relatively early type of air cleaner once provided sufficiently good filtration for most UK conditions, but it was not uniformly efficient. The typically wire mesh element was oiled when new and after cleaning. It had the advantage of allowing a fairly generous air flow for its size (Figure 7.3b).

Oil bath and mesh
This type of air cleaner was one widely used on commercial vehicle and large-capacity motor car engines, especially in America from where it originated in the early 1920s, but its installation requirements eventually became incompatible with the low-bonnet styling of modern cars. However, since it is both cleanable and rechargeable with fresh oil, an oil bath air cleaner may still be specified for vehicles operating in dusty countries and where also it may be difficult to guarantee the ready availability of replacement fibre material elements. A filtration efficiency only slightly lower than that for a fibre material element can be obtained, together with a generally high dust retention capacity (Figure 7.3c).

Cyclone and fibre material
Combination air cleaners of this type provide a longer service life for the fibre element even in severe dusty conditions and are particularly suitable for the protection of commercial vehicle diesel engines. In this context it must be remembered that a diesel engine always requires a full charge of air regardless of its operating load, so that the temperature of

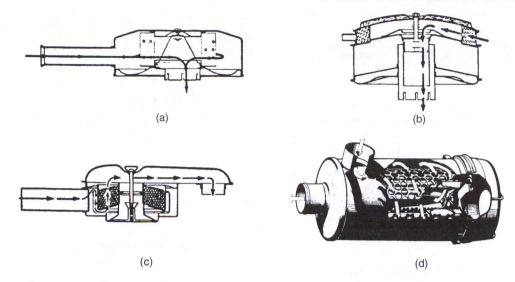

Figure 7.3 Types of air cleaner: (a) fibre element (*AC Delco*) (b) oil-wetted mesh (*AC Delco*) (c) oil bath and mesh (*AC Delco*) (d) cyclone and fibre element (*Mann*)

compression is high enough for efficient combustion. A partially clogged air filter can therefore have a more detrimental effect on the performance of a diesel engine than it would with a petrol engine. Prefiltering of the air is typically accomplished by an integral cyclone unit, the tangential vanes of which impart a spinning motion to the through flow of air and cause the heavier dust particles to be separated therefrom. They are then either retained in a collection chamber or discharged through an outlet port. Up to 85 per cent of the incoming dust particles may be extracted from the air before it finally passes through the fibre element proper (Figure 7.3d).

7.2 INTAKE AND EXHAUST MANIFOLDS

Intake manifold

The function of the intake manifold in its original form was to receive the mixture from one or more carburettors and to distribute it evenly with least variation in air-fuel ratio to the inlet valve ports. It was later adapted to perform the same function for single-point fuel injection systems (Section 6.8). To assist vaporization of liquid fuel in the incoming mixture, it is necessary to maintain a high manifold depression with high mixture velocities. Hence, the cross-sectional area of the manifold passages is kept to a minimum, but not such as to impair volumetric efficiency. Although a circular cross-section for the intake manifold offers the least resistance to mixture flow, this shape may be modified to a rectangular one for the reason of providing a flat internal floor. Therefore instead of any liquid fuel particles precipitated out of the mixture stream becoming channelled, they are spread over the increased surface area of the flat floor to assist evaporation and entrainment (return to the mixture stream).

Further to reduce any tendency for liquid fuel to be deposited in, and retained by, the manifold its bends are generally made fairly sharp in radius rather than being fully streamlined (Figure 7.4). This is because the effect of the

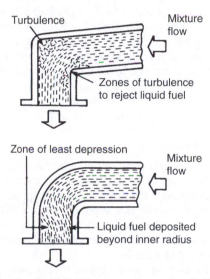

Figure 7.4 Good and ill effects on mixture flow for intake manifolds with sharp and streamlined elbows

latter would be to cause a pressure difference across the outer and inner radii of the bend or elbow, resulting in the mixture stream oversteering round the bend and depositing any liquid fuel particles on the manifold wall beyond the inner radius of the bend.

Various intake manifold arrangements were used while carburettors remained established practice (Figure 7.5). Intake manifold design has to some extent been simplified with modern multipoint fuel injection systems, because the air manifold basically comprises a series of downswept passages termed 'runners', which individually connect the inlet valve ports with the system throttle chamber (Figure 3.10).

Intake manifolds were originally produced from cast iron, but in later practice where there has always been a requirement to save weight, aluminium alloy has long since become

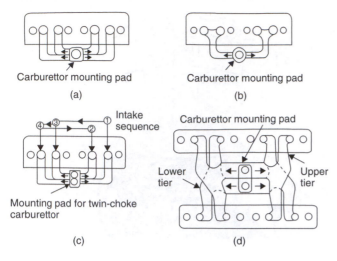

Figure 7.5   Intake manifold arrangements: (a) individual ports (b) siamesed ports (c) dual manifold (d) two-tier manifold

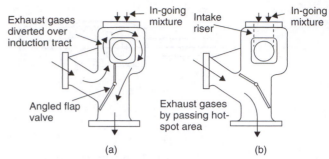

Figure 7.6   Action of exhaust-heated hot-spot with thermostatic control valve during (a) warm-up (b) normal running

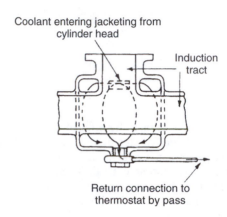

Figure 7.7   Coolant-heated intake manifold and riser

the preferred material. Of more recent origin is the moulded plastics intake manifold, which is produced from DuPont Zytel 33 per cent glass reinforced nylon and can withstand under-bonnet temperatures in the region of 175°C. Apart from achieving an all-important saving in weight of about 60 per cent, as compared to an equivalent aluminium alloy intake manifold, the plastics alternative further benefits from smoother interior walls to improve air flow and hence engine performance, and the material is also recyclable.

Intake manifold hot-spot

To improve vaporization of liquid fuel in the incoming mixture, the intake manifold used with a carburettor system is heated either by the exhaust gases or by heat rejected from the engine coolant. For single-point fuel injection systems where similar considerations dictate the need for an intake manifold hot-spot, either coolant or electrical heating (Section 6.8) methods are favoured.

Exhaust heating has been applied by exposing the underside of the manifold riser portion to the outgoing exhaust gases. This form of localized hot-spot is most conveniently arranged with a counter-flow cylinder head, or in other words where the intake and exhaust manifolds are situated on the same side of the engine. Since the mixture heating requirements are greatest during engine warm-up and any subsequent overheating of the mixture would reduce volumetric efficiency, a thermostatic control valve was sometimes employed (Figure 7.6). In the case of V-cylinder engines, it is usual for a proportion of the exhaust gases to be routed through passages beneath the intake manifold branches.

Coolant heating of the intake manifold may be used in conjunction with a cross-flow cylinder head: that is, where the intake and exhaust manifolds are situated on opposite sides of the engine. Similarly, it can be readily applied to the manifold on V engines. With this method of mixture heating, the heated coolant leaving the cylinder head flows through the partially jacketed intake manifold before entering the thermostat housing (Figure 7.7). Since the returning coolant bypasses the

radiator when the engine thermostat is closed, the flow of coolant and heating of the intake manifold are greatest during engine warm-up. Although this method of heating is less rapid in the first instance, it subsequently tends to minimize variations in temperature of the intake manifold. Furthermore all the exhaust heat remains available for rapidly warming-up a catalytic converter.

Variable-length intake systems

In discussing inertia ramcharging in Section 9.2, mention is made of how it is possible to exploit a pressure wave effect, which accompanies the oscillating motion of the charge in the intake system due to the closing and opening of the inlet valve. That is, the length of the intake manifold pipes can be so contrived that the pressure wave is approaching its crest just as the inlet valve is opening, which therefore improves the volumetric efficiency of the engine, but only over a narrow range of its operating speed.

An ideal, although more complex, solution for improving the behaviour of the intake system would be to use manifold pipes that could be varied in length from being fairly long to relatively short. However, an improvement over a fixed-length intake pipe can at least be obtained by utilizing a dual-chamber intake air passage, as originally developed by Toyota in the mid 1980s. In its simplest form this arrangement comprises a vacuum servo-operated control valve, which alters the effective length of the intake passage by either closing or

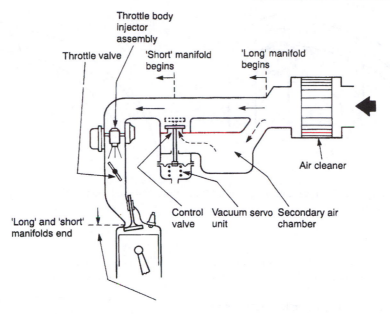

Figure 7.8 Schematic arrangement of a variable-length intake system for a four-cylinder engine with throttle body fuel injection

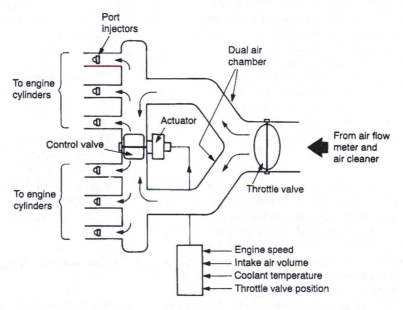

Figure 7.9 Schematic arrangement of a variable-length intake system for a six-cylinder engine with port fuel injection

opening a secondary air chamber (Figure 7.8). At low engine speeds the intake depression communicated to the vacuum servo unit is such as to retain the control valve in its closed position. This serves to maximize the length of the intake passage, thereby creating the most favourable pressure wave effects on the incoming charge of air for improved volumetric efficiency and hence engine torque at low engine speeds. With large throttle openings and reducing intake depression, the vacuum servo unit allows the control valve to open in sympathy with its spring loading, which causes the air now flowing through the secondary air chamber to interfere with that flowing in the primary one, so that the effective length of the intake passage commences at this point and is thus

shortened. Again this creates the most favourable pressure wave effects on the incoming charge of air, but this time to improve engine torque characteristics in the middle speed range.

The principle of the variable-length intake system is especially relevant to in-line and V six-cylinder engines, where it can provide a steadier flow of incoming air by treating the engine as two separate three-cylinder ones. In this case the dual-chamber intake air passage is formed by two relatively large-diameter pipes, each of which feeds three smaller-diameter pipes (Figure 7.9). The control valve is placed between the two main pipes and actuated from the electronic control unit of the engine management system. It is closed at lower engine speeds to allow the air to be drawn from separate

(a)

(b)

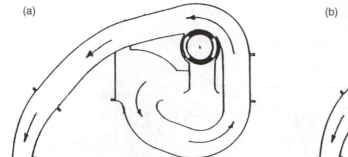

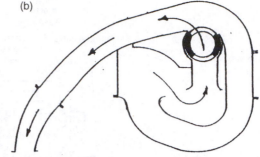

Figure 7.10   Schematic arrangement of a variable-length scroll type intake system: (a) air route longest for maximum torque (b) air route shortest for maximum power

sides of the intake system for improved torque, but opened at higher speeds to provide increased air flow for maximum power.

A more recent development of the variable-length, or perhaps in this case it would more accurately be described as variable-geometry, intake system is the 'scroll' or 'folded' type. It is so called because the manifold is in effect wrapped around a rotary selector cylinder. This can either present a barrier to the incoming air flow and compel it to take the longest route through the manifold (Figure 7.10a), or be turned until cross-channels in the cylinder allow the air flow to be short-circuited within the manifold (Figure 7.10b). That is, the route taken by the incoming air is longest for maximum torque and shortest for maximum power. The angular position of the selector cylinder is servo controlled from the engine management system, based on engine load, torque and temperature inputs.

Exhaust manifold

The function of the exhaust manifold is to connect the exhaust ports in the cylinder head to the downtake pipe, which leads the exhaust gases to the silencing arrangements. In contrast to the form of intake manifold once associated with a carburettor, the branches of the exhaust manifold are generously proportioned and have large-radius bends of fully streamlined form, which merge as smoothly as possible into the downtake pipe (Figure 7.11a). By offering the minimum resistance to gas flow, an efficient design of exhaust manifold contributes to improved power output from the engine.

To promote further the free flow of gases from the cylinders, the exhaust manifold for an in-line cylinder engine sometimes embodies two outlets. These are connected to separate downtake pipes, whence the exhaust gases usually enter a large single pipe via a breeches piece. In the case of in-line four-cylinder engines, the outer pair of exhaust ports communicates with one manifold outlet and the inner pair with the other outlet (Figure 7.11b). For in-line six-cylinder engines, a pair of three-branch manifolds may be employed to serve the exhaust ports of the front and the rear halves of the engine. In all cases the purpose of a divided manifold is to ensure that the gas flows from nearby cylinders, wherein the valve openings overlap, do not merge too soon. This avoids imposing a back-pressure upon a cylinder in which the exhaust

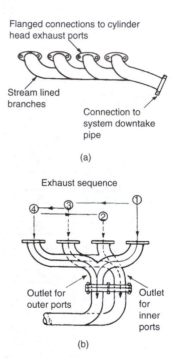

Figure 7.11   Exhaust manifold-arrangements

valves are in the act of closing, since this would be detrimental to its volumetric efficiency. Two separate exhaust manifolds are necessarily demanded in the cases of horizontally opposed and V engines.

A completely separate exhaust manifold produced from cast iron is employed on modern cross-flow cylinder heads, where the intake and the exhaust manifolds are sited on opposite sides. The exhaust manifold may be provided with either a central or an end outlet, as dictated by installation requirements, which then connects directly with the downtake pipe of the exhaust system proper (Figure 7.12) or indirectly via the turbine housing of a turbocharger. Its wall thickness must be sufficient to withstand the burning effect of the high temperature exhaust gases, which may reach 850°C. However, its thickness must not be excessive otherwise this could promote a tendency towards warping and cracking, arising from the effects of uneven heating and cooling. It may

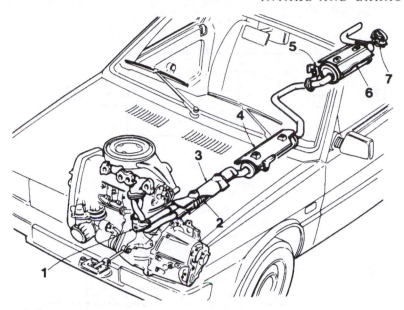

Figure 7.12 A modern exhaust system layout (*Lancia*)

1 front mounting bracket
2 flexible pipe
3 front heat shield
4 intermediate heat shield

5 intermediate pipe mounting bracket
6 rear heat shield
7 rear pipe mounting bracket with flexible pad

also be ribbed to reduce stresses and its fixing holes be slightly elongated with the exception of a centre locating pair. This latter feature better accommodates lengthwise expansion and contraction of the manifold relative to its cylinder head mounting, thereby minimizing any distortion. To avoid assembly distortion it is important that, when tightening the manifold fastenings in service, this is done progressively starting from the centre and working alternately towards each end. A heat resistant gasket provides a seal between the joint faces of the exhaust manifold and the cylinder head. Pressure tightness around the mating ports is especially significant with turbocharger installations and also where the turbine housing is flange mounted to the manifold outlet, because any leakage would cause a reduction in boost pressure at high engine speeds.

A recent trend in exhaust system design is to reduce engine weight by replacing exhaust manifolds of the cast type with ones fabricated from stainless steel tubing. These can then be covered to form a double-layered construction, which improves both thermal insulation and noise isolation, the former promoting quicker 'light-off' for catalytic converters (Section 11.2). Also, their independent branches can better accommodate thermal expansion and contraction between each exhaust port and flange.

7.3 EXHAUST SYSTEM

General requirements

An exhaust system is installed for the purpose of collecting the burnt gases issuing from the engine cylinders and unobtrusively discharging them outside the vehicle (Figure 7.12). To accomplish this satisfactorily the exhaust system must

fulfil certain requirements. It must be quiet in operation, absorb the least possible engine power, minimize heat transference to the car interior and isolate the car structure from vibrations set up in the system. Provision may also have to be made for the exhaust system to supply heat to the induction system, in order to facilitate the process of carburation as already explained. Furthermore, the increasingly stringent international regulations to reduce harmful exhaust emissions have led to the inclusion in the exhaust system of additional sophisticated equipment, such as the catalytic converter as described in Section 11.2.

Exhaust silencer

Since all sound is due to vibrations, exhaust noise arises from the sudden release of compressed gas travelling at a speed in the region of 60–90 m/s (200–300 ft/s), as each exhaust valve is opened. The purpose of a motor vehicle exhaust silencer is therefore to suppress this noise to a legally acceptable level. It does this by breaking up the intermittent stream of high-velocity exhaust gases into numerous small streams, which has the effect of reducing both their velocity and their temperature so that they leave the tail pipe of the exhaust system in a more nearly constant flow at low velocity.

In early practice, the exhaust gases expanding into the silencer were compelled to pass through a series of chambers separated by perforated flat or conical plates, before leaving from the other end of the silencer (Figure 7.13a). Although this simple form of silencer construction could be quite effective in suppressing exhaust noise, it often created too much back-pressure in the system, which could further increase if the small perforations became partially obstructed with carbon deposited by the exhaust gases. The effects of too much

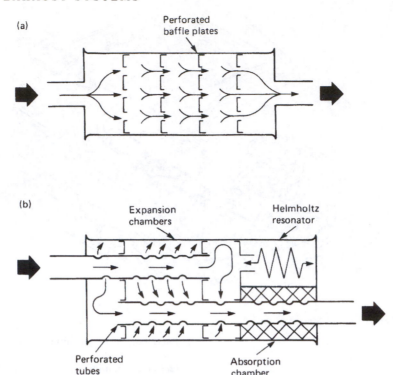

Figure 7.13 Noise reduction features of early and modern silencers

back-pressure in an exhaust system for whatever reason are that the fresh mixture entering the engine cylinders will be diluted by unexpelled burned gases, so that engine performance and fuel economy will both suffer, and the engine will become prone to overheating.

It later became evident that a more scientific approach to silencer design was required, so that the greatest possible degree of noise suppression could be obtained with the least possible back-pressure. This is now achieved by combining several methods of noise reduction, which generally include those of expansion (which is fundamental to all silencer design), resonance and absorption (Figure 7.13b).

Taking first the *expansion* method, the incoming gases from the downtake pipe are initially allowed to expand into the main chamber of the silencer to reduce both their pressure and their temperature. Following this their direction of flow is changed by passing them through side-by-side perforated tubes, in the process of which more of their energy of vibration is converted into heat energy by frictional loss, before they pass into the tail pipe. This method of silencing is best suited to suppressing low and medium-frequency exhaust noise.

In the same way that particularly troublesome peaks of induction noise can be suppressed by making the air cleaner and silencer act as a Helmholtz resonator, as explained in Section 7.1, so can an undesirable boom period of exhaust noise be suppressed by applying the principle of *resonance* tuning. For this purpose the silencer incorporates a separate chamber of calculated volume that has a narrow neck or inserted tube, across which the pulsating flow of exhaust gases

can pass. When the resonant frequency of vibration created in the Helmholtz chamber abstracts sound energy from vibration in the main silencer, the latter will be much reduced and the noise suppressed.

With the *absorption* method, the sound is damped through the medium of an acoustically dead material of a soft and porous nature. As applied to an exhaust silencer, a heat-resistant material such as glass fibre mat or stainless steel wool is packed into the annular space between a perforated or a louvred central tube and the expansion chamber through which it extends. Sound waves leaving the central tube and striking the absorbent material have their energy of vibration partially converted to heat energy. As this method of noise reduction generally offers a low resistance to the through flow of exhaust gases, silencers that work solely on this principle have long been referred to as the straight through type. They are mainly effective in damping out high-frequency exhaust noise.

The general construction of a silencer of the type widely used on heavy vehicles is shown in Figure 7.14.

It has long been established practice to fabricate exhaust silencers from mild steel sheet, although in high-grade practice stainless steel sheet may be used. The piping associated with the exhaust system is commonly produced from seamless steel tube. Since exhaust silencers are very prone to corrosion attack from the outside, as well as from within, various surface treatments may be applied to mild steel construction, including zinc coating, aluminium coating, vitreous enamelling and coating with a refractory material. To improve the life of some replacement silencer systems, a mix of aluminium-coated

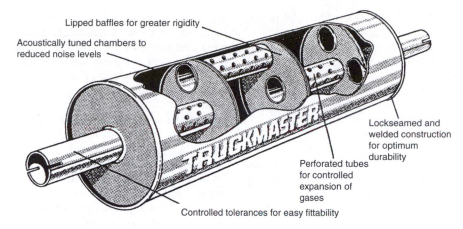

Lipped baffles for greater rigidity

Acoustically tuned chambers to reduced noise levels

Lockseamed and welded construction for optimum durability

Perforated tubes for controlled expansion of gases

Controlled tolerances for easy fittability

Figure 7.14 Construction of an exhaust silencer for heavy vehicles (*Bainbridge*)

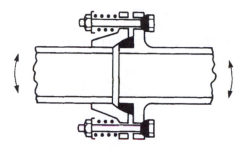

Figure 7.15 Spring-loaded ball-and-socket exhaust joint

mild steel and stainless steel may be employed for the various components, the choice of material depending on which parts of the system have normally been found most vulnerable to corrosion attack. The exhaust system components may be either joined together by welding to form a one-piece assembly, which is eventually renewed in its entirety, or coupled together by means of slip joints, flanged joints, or ball-and-socket joints. An important advantage of the latter type of joint is that it facilitates correct alignment of the system, thereby relieving the support mountings of unnecessary strain. A spring-loaded version of the ball-and-socket joint has been more recently used at the lower end of the downtake pipe on some transverse-engine cars, so as better to accommodate the rocking movements of the power unit (Figure 7.15). Self-locking nuts are provided for the coupling bolts, which are usually shouldered to prevent over-compression of the springs when the nuts are tightened.

The modern exhaust system is very much part of the engine emission control arrangements. It is therefore imperative that gas tightness is maintained at all joints, to prevent not only the leakage of untreated exhaust gas, but also the intake of air and hence oxygen that will adversely affect catalytic conversion (Section 11.2).

Although it may be stating the obvious, before removing exhaust system components in service it is vital from safety considerations that the system should be allowed sufficient time to cool, this is especially true for catalytic converters that can remain much hotter for longer. Reference should also be made to the Health and Safety Executive publication HSG62 titled 'Health & Safety in TYRE & EXHAUST FITTING PREMISES'.

7.4 VACUUM GAUGE AND EXHAUST GAS ANALYSER

Vacuum gauge

An old-established item of engine tune-up equipment is the vacuum gauge, which in the hands of an experienced operator becomes an exceedingly versatile instrument. The vacuum gauge measures engine intake depression, and this can vary with different operating conditions or with different engine defects.

In construction, the vacuum gauge comprises a spring-returned flexible diaphragm that encloses a sealed chamber, which is provided with a detachable hose connection to a suitable offtake point on the intake manifold. The other side of the diaphragm is exposed to atmospheric pressure and linked to a quadrant and pinion mechanism, so that any deflection of the diaphragm causes the indicator needle to swing over the calibrated dial. As the air is drawn into the manifold from the sealed chamber, atmospheric pressure acting on the other side of the diaphragm forces it to move inwards against its spring loading. This movement of the diaphragm is therefore converted into a reading on the vacuum gauge dial, which is expressed in either or both millimetres of mercury (mm Hg) and inches of mercury (in Hg). These measures of intake depression are related to the equivalent height to which a column of mercury would rise if the same extent of depression were to act above it.

Although an engine in a good state of tune would be expected to record a fairly high and steady reading of 430–530 mm Hg (17–21 in Hg) for a specified idling speed and at normal operating temperature, it is more important to observe the behaviour of the indicator needle rather than its actual reading. The following are examples of vacuum gauge indications for different engine conditions:

Leaking engine valve A slow and regular drop of about 25 mm Hg (1 in Hg) from the maximum reading.

Sticking engine valve   An irregular drop of 50–100 mm Hg (2–4 in Hg) from the maximum reading.

Broken valve spring   A rapid fluctuation between 255 and 585 mm Hg (10 and 23 in Hg) with increasing engine speed.

Air leaks into intake system   A steady drop of 75–200 mm Hg (3–8 in Hg) with a tendency to drop lower.

Retarded ignition   An almost stationary reading of 50–100 mm Hg (2–4 in Hg) below normal.

Exhaust back-pressure   Slow return of reading after first fully opening and then rapidly closing the throttle.

Piston ring blow-by   Failure to obtain a momentarily high reading of 585–635 mm Hg (23–25 in Hg) after first fully opening and then rapidly closing the throttle.

Exhaust gas analyser

Another long-established item of engine tune-up equipment is the exhaust gas analyser, its purpose originally being to eliminate the guesswork from carburettor adjustment. This type of instrument provided a meter reading of the approximate air–fuel ratio by weight of the mixture supplied to the engine, based on measuring the relative amounts of carbon dioxide and hydrogen present in the exhaust gas. It operated on the principle that these particular constituents of the exhaust gas possessed markedly different thermal conductivities from that of air. By comparing the thermal conductivity of an exhaust gas sample with that of an air sample, an indication of the air–fuel ratio for the ingoing mixture could therefore be obtained.

In modern practice the exhaust gas analyser, in determining what percentage of the fuel has not been burned in the engine cylinders, has assumed more the role of exhaust emission tester. Such an instrument now provides meter readings for the amounts of either carbon monoxide (CO) or, as in later practice, both carbon monoxide and hydrocarbons (HC) present in the exhaust gas sample. These more sophisticated types of exhaust gas analyser utilize either infrared or ultraviolet, radiation detection units to sample the exhaust gas and, since the test conditions must be properly arranged, it is important to observe the manufacturer's operating instructions. For the MoT emission testing of catalytic converter equipped vehicles in the UK, the actual type and manufacturer of the exhaust gas analyser must be approved by the Vehicle Inspectorate.

8 Diesel fuel injection systems

8.1 FUEL INJECTION SYSTEMS

Basic considerations

It will be recalled from Section 3.4 that the combustion process in the diesel engine cylinder is initiated by spontaneous ignition of the fuel, which is injected into an agitated, highly compressed and therefore exceedingly hot charge of air. To accomplish this demanding task, diesel fuel injection systems may be classified as follows:

1 In-line fuel injection pump (Section 8.3)
2 Distributor fuel injection pump (Section 8.5)
3 Unit fuel injection (Section 8.9)
4 Common rail fuel injection (Section 8.11)

8.2 FUEL SUPPLY SYSTEM

General layout

The motor vehicle diesel engine fuel supply system (Figure 8.1), comprises the fuel tank, preliminary filter, fuel lift pump, main filter(s) and fuel pipelines, as follows:

Fuel tank
This is supported from the side of the chassis frame. It is located in a position as sheltered as possible from the cooling air flow past the vehicle, so as to afford some protection against wax crystals separating from the fuel oil and blocking the fuel system during severe winter operation.

The fuel tank is usually provided with a coarse strainer gauze at its filler neck, a drain plug in its base, and a fuel gauge with either a remote cab reading or a direct tank reading, to indicate tank content. Connections are provided at the top of the tank for the fuel delivery pick-up and return leak-off pipes. Internal baffles are fitted to minimize surging of the fuel. The capacity of the diesel fuel oil tank is usually several times greater than that of the car petrol tank. For medium to heavy commercial vehicles it is typically in the region of 300 litres (65 gallons).

Preliminary filter
This is situated in the fuel pipeline between the tank and the fuel lift pump. Current thinking favours only a simple sedimenter filter (Figure 8.2), since the presence of a fine wire gauze at this point could aggravate any blocking tendency caused by fuel oil waxing. The purpose of a sedimenter filter is to remove the larger droplets of water and abrasive matter that may be present in the fuel oil being pumped from the tank. Water in the fuel oil is to be avoided because it can interfere with the proper lubrication of the fuel injection equipment. The undesirable effect of abrasive matter in the fuel oil is dealt with later.

Fuel lift pump
Since the fuel tank is mounted below the level of the engine fuel injection pump, a fuel lift (or transfer) pump is required to maintain a small constant head of fuel oil in the feed gallery for the injection pumping elements. Although various types of positive displacement lift pumps may be encountered in diesel engine practice, those used for automotive applications are typically of either the plunger or diaphragm variety. Both are provided with hand priming levers so that fuel oil can be forced into the system to vent it of air, without turning the engine.

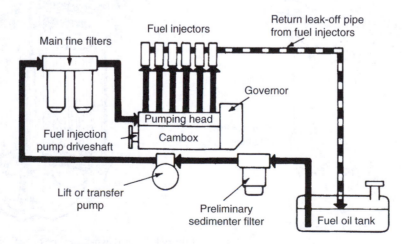

Figure 8.1   Simple representation of diesel engine fuel injection system (in-line pump)

Main filter(s)

Contamination of the fuel oil by abrasive matter is the great enemy of the diesel fuel injection system, and, for that matter, any other piece of hydraulic equipment. The effect of abrasive matter in poorly filtered fuel oil reaching the injection pump and injectors is to damage their highly finished and selectively assembled mating parts. This can result in reduced performance, poor starting and irregular idling of the engine, because of decreased delivery from the injection pump. In the case of the injectors, their faulty spraying and leakage can increase both fuel oil consumption and exhaust smoke. It is therefore necessary to incorporate one or more fine filters between the lift and injection pumps. The main filter(s) is deliberately sited so that it can receive heat from the engine, again to minimize fuel oil waxing problems.

Fuel pipelines

The fairly generous bore size of the low pressure tubing used on the supply side of the injection pump is intended to minimize both pressure drop and the energy required to operate the lift pump. On the delivery side of the injection pump the exceedingly small bore size of the high-pressure tubing is related to the diameter of the pumping elements, and its greater wall thickness to the high injection pressure being handled. It is also important that the delivery tubing should be of minimum, but nevertheless equal, length from the pump to the individual injectors. Seamless steel tubing is used for the fuel pipelines, which should have only large-radius bends and also be clamped to avoid fracture from vibration. The end connections may be effected through either conical sleeves or banjo unions.

Types of fuel filter

The preliminary filter has already been dealt with (Figure 8.2). It now remains to describe the main filters used between the lift pump and the injection pump. These may be classified, according to the direction of fuel oil flow through their filtering media, as down-flow, up-flow and cross-flow, filters.

Down-flow filter

Filters of this type are arranged to provide what is termed an agglomerator flow. This is based on the principle that when fuel oil containing any fine water droplets is passed through a porous filter element, the drops will join together, or agglomerate, into larger drops which can then be collected by sedimentation. In operation the incoming unfiltered fuel passes downwards through the filter element, which comprises a V-form paper roll construction, and into the base of the filter body (Figure 8.3). It is then displaced upwards through the annular space created by the central tube of the filter element and the filter body securing bolt. Finally, the filtered fuel leaves at the outlet connection of the filter head. In this manner, the abrasive particles are retained by the porous filter element and the water content is collected in the base of the filter.

Up-flow filter

A purely filter rather than agglomerator flow is provided with this filtering arrangement. Its action is such that the incoming unfiltered fuel first passes downwards through the annular space created as before by the central tube of the filter element and the filter body securing bolt (Figure 8.4). The fuel is then displaced upwards through the filter element to leave immediately from the outlet connection of the filter head.

Cross-flow filter

This corresponds to the more familiar filtering action described in connection with the engine lubrication system (Section 4.6); that is, the incoming unfiltered fuel passes inwards through the filter element, which in this case assumes a felt tube form (Figure 8.5). The filtered fuel from the interior space of the element then leaves via the outlet connection of the filter head.

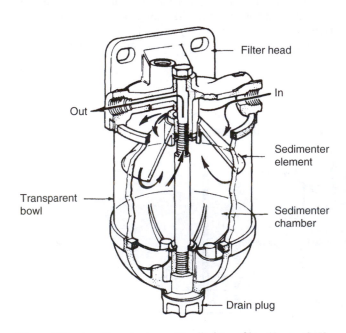

Figure 8.2   A sedimenter type of preliminary filter (*Lucas CAV*)

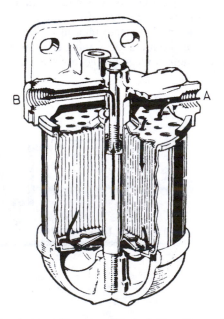

Figure 8.3   A down-flow type of fuel oil filter (*Lucas CAV*): unfiltered fuel oil inlet; B filtered oil outlet

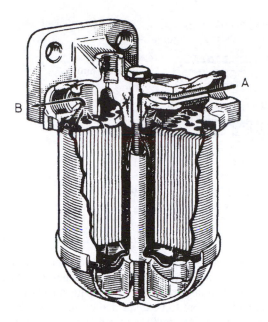

Figure 8.4 An up-flow type of fuel oil filter (*Lucas CAV*): A unfiltered fuel oil inlet; B filtered fuel oil outlet

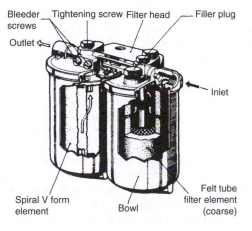

Figure 8.5 Two-stage fuel oil filter incorporating a cross-flow coarse element

Types of lift pump

The single-diaphragm type of lift pump is very similar to that found in petrol engine fuel systems (Figure 8.6). It may be mounted on the engine crankcase or on the fuel injection pump and takes its drive from the camshaft of either. For greater capacity a twin-diaphragm arrangement can be used, which provides a double acting pump and embodies two sets of inlet and outlet disc valves in a central partition. This type of diaphragm pump is driven directly from the camshaft of the fuel injection pump on which it is mounted.

The plunger type of lift pump comprises a spring-energized piston that reciprocates in a pumping chamber containing the inlet and outlet valves (Figure 8.7). A diaphragm is arranged in series with the pumping piston, its purpose being to seal against any fuel oil that leaks past the piston. In a more recent version with improved sealing, a second diaphragm has been introduced for retaining oil in the fuel injection pump cambox.

The pumping piston is operated either through a bell-crank lever or, as in current practice, a direct-acting tappet with roller follower, which bear against an eccentric on the camshaft of the fuel injection pump. This type of lift pump generates a higher pressure. A lost-motion facility is incorporated in the operating mechanism of these different types of lift pump so that they can free-wheel when the force on the diaphragm or the plunger due to back-pressure exceeds that of the energizing spring. In earlier practice this feature was not present and a pressure relief valve had to be included in the system, together with additional piping, so that surplus fuel could be returned to either the tank or the inlet side of the lift pump.

As in the case of the petrol engine fuel pump, the controlling factor for delivery pressure in the diesel lift pump is the strength of the energizing spring acting on either its diaphragm or plunger. The former operates at a lower pressure than the latter with maximum values being in the regions of 30 and 100 kN/m^2 (4 and 14 lbf/in^2) respectively. There is, of course, a reduction in pump output as the back-pressure on the pump increases.

Air venting the fuel system

If air locks are present anywhere in the fuel injection system of a diesel engine, it either may not start or will only run

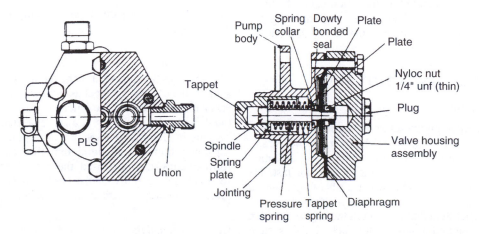

Figure 8.6 Cross-section of a diaphragm-type lift pump (*Lucas CAV*)

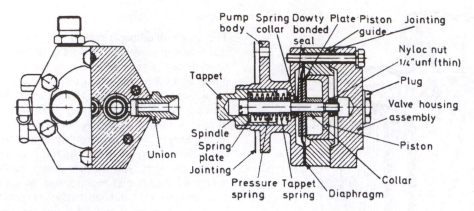

Figure 8.7   Cross-section of a plunger-type lift pump (*Lucas CAV*)

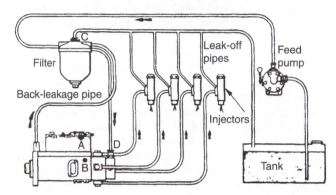

Figure 8.8   Air venting points (A–D) in typical fuel system for distributor-type injection pump (*Lucas CAV*)

spasmodically. Therefore whenever any part of the system has been disconnected during the course of service, such as the renewal of a filter element, or similarly if the fuel tank has been inadvertently run dry, steps must be taken to bleed or vent all air from the system. For this purpose screw fitting vent plugs are provided at various high points in the system and their locations are usually illustrated in the engine manufacturer's operating instructions (Figures 8.8 and 8.9). The supply side of the fuel injection system is vented by partly unscrewing the appropriate plugs and hand priming the lift pump, whilst the delivery side is vented by partly uncoupling the union nut of the pressure pipe for each nozzle holder in turn and cranking the engine. Venting is accomplished when the fuel oil flows freely without any air bubbles from the various points.

8.3   IN-LINE FUEL INJECTION PUMP

General background

Early attempts by Rudolph Diesel to use mechanical methods of injecting fuel into the combustion chamber provided unsatisfactory, no doubt because of the relatively crude injection equipment then available to perform such a precision task. He therefore resorted to what was known as air-blast injection. In this system a compressor and storage tank supplied air at a pressure much higher than that existing in the engine cylinder, as means of injecting and atomizing the fuel oil. This system proved quite satisfactory and remained accepted practice for large marine and stationary diesel engines until the early 1930s.

It was also about this period that the diesel engine was being developed for higher speeds and lighter weight to make it suitable for road transport applications. The air-blast injection system had therefore to be discarded on the grounds that not only was it too heavy, bulky and costly, but also because the air compressor robbed the engine of up to 10 per cent of its power output. A return was therefore made to mechanical methods of injection, albeit of more sophisticated form, in which extremely small amounts of fuel oil (in the order of $0.1 \, \text{cm}^3$) are forced through spray nozzles under high hydraulic pressures from a fuel injection pump. The first commercially successful pump of this type was the port-controlled jerk pump produced by Robert Bosch of Germany in the late 1920s; this was later adopted with detail variations in design by other manufacturers in both Britain and America.

The purpose of the in-line fuel injection pump is therefore to deliver fuel oil at higher than combustion chamber pressures, in minute quantities exactly related to the amount of power required. Furthermore, the delivery of the fuel oil must be timed to occur at precisely the required moment in the engine operating cycle. As in the case of a petrol engine ignition distributor, the fuel injection pump in the four-stroke cycle must also be driven at half engine speed. It is therefore usually gear driven from the engine camshaft through a 1:1 ratio drive.

Construction and operation

The in-line fuel injection pump comprises the following parts (Figure 8.9).

Pump housing
This includes the cambox and the pumping head, the former to support the camshaft, tappet assemblies, control rod and governor mechanism, and the latter to contain the pumping elements and delivery valves. It may be produced as a one-piece housing in aluminium alloy, or a steel pumping head may be separately bolted to an aluminium alloy cambox. Whichever form of construction is used, the overriding consideration is for maximum rigidity of the pump housing, since this can affect both pumping efficiency and operating noise.

Governor housing
This usually takes the form of an enlarged extension of the cambox and is closed by a suitable end cover.

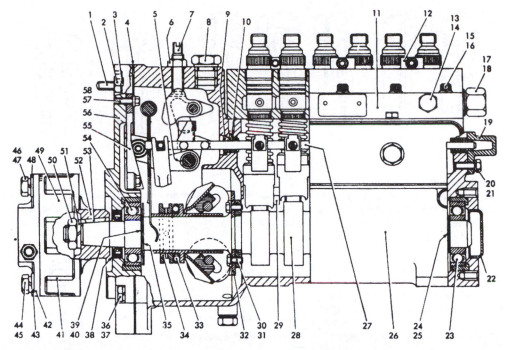

Figure 8.9   An in-line fuel injection pump for a six-cylinder diesel engine (*Lucas CAV*)

1 stud	13 air vent screw	25 shim (0.2 mm)	37 spring washer	49 dog flange
2 nut	14 joint washer	26 pump unit housing	38 oil seal	50 camshaft nut
3 spring washer	15 screw	27 control rod	39 shim (0.1 mm)	51 spring washer
4 gasket	16 spring washer	28 camshaft	40 shim (0.2 mm)	52 pump flange
5 Woodfuff key	17 fuel inlet adaptor	29 tappet locating T-piece	41 insert	53 Woodruff key
6 nut	18 joint washer	30 back plate screw	42 driving flange	54 ball bearing
7 maximum fuel stop screw	19 control rod cover	31 tab washer	43 clamp plate	55 ramp
8 oil filler plug	20 screw	32 back plate	44 dowel screw	56 governor cover
9 control rod bush	21 spring washer	33 thrust bearing	45 spring washer	57 screw
10 Groverlok pin	22 expansion plug	34 thrust pad	46 screw	58 lock washer
11 pump body	23 ball bearing	35 baffle washer	47 spring washer	
12 clamp	24 shim (0.1 mm)	36 screw	48 clamp plate	

Camshaft and tappet assemblies

These operate in conjunction with the plunger return springs to provide the reciprocating motion for the pumping elements. The camshaft is supported by heavy-duty rolling bearings, ball or opposed tapered rollers, and is of rugged proportions to withstand without deflection the heavy loads imposed upon it. Either or both ends may have standard tapers with keyways which provide connections for a driving coupling or gear and the governor. Each spring-returned pump plunger is lifted by its cam through a tappet assembly, incorporating a roller, bush and pin to ensure smooth rolling contact with the cam. A tappet adjustment for phasing the pump is provided; this is an operation that will be described at a later stage.

Pumping elements

Each of these comprises a barrel and plunger assembly, both components being made from steel. In providing a sliding fit, the barrel and plunger are matched to such fine limits as to eliminate the need for any other means of sealing between them. Here it should be appreciated that we are thinking in terms of a working clearance of about 1.5 μm! The barrel is located in the pumping head by either a locking screw or serrations and is retained endwise by the delivery valve holder. Sandwiched between these parts is the delivery valve guide.

Fuel oil inlet and spill ports are provided in the upper portion of each barrel and register with corresponding galleries in the pumping head.

Each plunger is shaped with a sloping edge or helix at its head portion, so that although the plunger operates over a constant stroke, the quantity of fuel delivered can be varied by partial rotation of the plunger. That is, it depends on the radial position of the plunger helix relative to the spill port in the barrel (Figure 8.10a). Hence, the earlier the helix registers with the spill port, as the cam lifts the plunger, the lesser will be the quantity of fuel delivered. No fuel is delivered at all, of course, if the helix uncovers the spill port over the entire stroke of the plunger. Since it is necessary to connect the space above the top of the plunger to the spill port, the plunger head must also be provided with either a vertical slot in its sidewall, or a central hole with a radial duct (Figure 8.10b). An advantage of the latter method is that there is less reduction in contact area, so that wear and leakage are reduced. The actual method of fuel metering by port control was not invented by Robert Bosch, but was conceived earlier by another German, Carl Pieper, in 1892.

Pumping element control

There are two basic methods used for simultaneously turning the reciprocating plungers to control fuel delivery, these

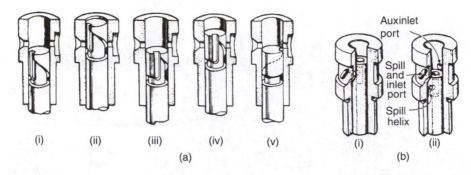

Figure 8.10   Action of two types of pumping element
(a) Helix and slot: (i) commencement of delivery stroke (ii) termination of fuel injection under full load (iii)/(iv) commencement of delivery stroke and termination of injection under partial load (v) position of plunger for stopping engine (b) Helix and drillings: (i) plunger at commencement of spill (ii) plunger at bottom of stroke

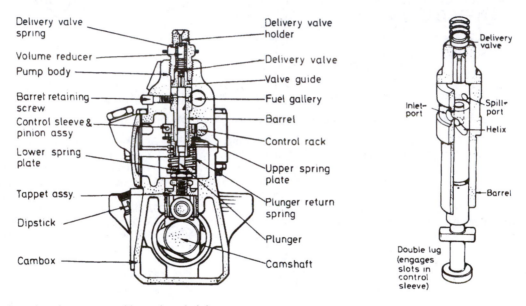

Figure 8.11   Pumping element control by rack and pinion

being known as the rack-and-pinion and fork-and-lever methods, which may be attributed to Robert Bosch in Germany and Simms in Britain respectively.

The *rack-and-pinion* method was the earliest, and for many years the most widely used, pumping element control. In this construction each plunger can be partially rotated through the medium of lugs formed near its lower end. These engage with slots in an extended sleeve that surrounds, and can turn on, the lower part of the barrel. Clamped to the upper part of the sleeve is a quadrant pinion that meshes with a toothed portion of the sliding control rod (Figure 8.11). It thus follows that any endwise movement of the latter will produce a partial rotation of the quadrant pinion, together with slotted sleeve and likewise the reciprocating plunger.

With the *fork-and-lever* method of pumping element control the slotted sleeve, quadrant pinion and toothed control rod are dispensed with and replaced simply by a radial lever attached to either the foot or the waist of the plunger and engaging a vertical fork carried on a square section control rod (Figure 8.12). Therefore any sliding movement of the latter will cause the forks to swing the outer ends of the plunger

levers and thus partially rotate the plungers. The depth of slot in the fork is, of course, sufficient to accommodate the reciprocating movement of the plunger and its lever.

The main advantage generally claimed for this particular method of pumping control is that the sources of backlash in the mechanism are reduced from two with the rack, quadrant and slotted sleeve arrangement to one between the fork and plunger lever, and even this is acting at a large radius to reduce the effect of backlash. There is also less possibility of any misalignment friction occurring and causing sluggish control rod movement. However, it does mean that there is no longer a constant relationship between the linear movement of the control rod and the angular displacement of the plungers, owing to the swing of their radial levers or arms. To compensate for this effect the plungers used with fork-and-lever control are provided with an inclined slot rather than a truly helical one, although the term 'helix' is still commonly applied (Figure 8.12).

In a more recent design of Bosch fuel injection pump, the fuel delivery is controlled by a ball-ended lever, which

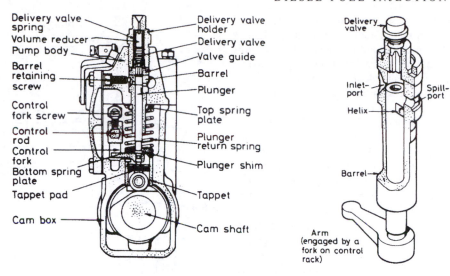

Figure 8.12 Pumping element control by fork and lever

replaces the former quadrant pinion and engages an in-ward-facing fork connector made integral with the control rod.

Pumping element delivery valve

A spring-loaded delivery valve is fitted at the head of each pumping element and serves a two-fold purpose:

1 To act as a non-return valve that isolates the high-pressure from the low-pressure regions of the pumping element, thereby preventing fuel from re-entering the barrel when the plunger retreats on its intake stroke.
2 To serve as an unloading valve that creates additional space or dead volume on the delivery side of the pumping element, thereby ensuring a sudden drop in line pressure and clean termination of spray from the injector.

Although the delivery valve may appear to be a fairly ordinary sort of non-return valve, it does in fact represent a very elegant solution to what was originally a difficult problem in hydraulics, and all credit for this must go to Axel Danielson of the Atlas Diesel Company in Sweden. In 1924 he introduced a retraction-type delivery valve basically resembling a taper-seating poppet valve, but which had its guide stem effectively divided into two parts of the same diameter. The shorter upper part was formed as a separate collar, while the lower part was fluted to allow the passage of fuel (Figure 8.13).

It acts in a manner similar to an ordinary non-return valve, but with the important difference that, when closing, the effect of the collar entering the valve seating bore is to produce an automatic increase in volume in the pipeline to the injector. Hence, the accompanying sudden drop in pressure ensures that the fuel spray from the injector is cleanly terminated when its needle valve closes. Otherwise, there is the likelihood of dribble occurring from the fuel injectors, which was one of the problems encountered with early jerk pump fuel injection systems.

Need for phasing an in-line fuel injection pump

We are now required to examine the provisions for adjustment of timing and of balancing the fuel deliveries from the

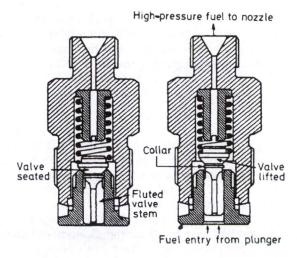

Figure 8.13 Atlas type of pumping element delivery valve

individual pumping elements, these two processes being known as phasing and calibrating respectively. More specifically, *phasing* may be defined as the procedure for checking and adjusting an inline pump to ensure that the interval between successive injections exactly corresponds to the firing interval of the engine cylinders. On production build, the need for phasing a pump arises from the inevitable machining tolerances on the component parts; whilst in service, the need for phasing similarly arises if the pumping elements and tappet assemblies require replacement.

Phasing an in-line fuel injection pump

The phasing operation is performed on what is generally termed an injection pump test bench (or stand), this being a highly developed piece of specialist equipment of which there are several versions. Although these vary in size and individual design features they do, in fact, offer the same basic test facilities. It is, however, essential that before any such testing is performed not only must the test bench manufacturer's operating instructions be consulted, but also those relating to the particular make and model of fuel injection pump being tested, the latter usually including an appropriate fuel setting data sheet. Finally, it must be stressed that this type of work must not be attempted without adequate supervision from someone who is familiar both with the correct use and, most importantly, the safe operation of the equipment concerned.

For the purpose of phasing a pump the tappet body of each pumping element is provided with either a screw-threaded or, as in later practice, a graded thickness of spacer form of adjustment. This tappet adjustment provides the means of altering the overall lengths of the roller tappet and plunger combinations, so as to ensure the correct point of injection for each engine cylinder. If for example the injection is occurring too early in one cylinder, then from the action of the pumping element as described earlier it will be evident that the barrel ports are being closed too soon. In other words the top of the plunger has risen too high in relation to the angle of cam lift and therefore requires lowering, either by turning the adjuster screw further into the tappet body or by fitting a thinner spacer. For the reasons mentioned earlier, it is not proposed to include here the necessarily lengthy detailed instructions that relate to the use of a particular version of test bench and the testing of a given make and model of fuel injection pump. However, as far as the phasing operation is concerned the practical objectives sought are generally twofold:

1   The application of a special plunger stroke measuring tool to check the distance that the plunger of no. 1 pumping element travels from the bottom of its stroke to the closing of the barrel inlet port and hence the commencement of injection (Figure 8.14). When this distance is correct to specification, a small bumping clearance will remain between the head of the plunger at the top of its stroke and the (refitted) delivery valve holder.

2   The use of a graduated degree plate that can be set to zero on the test bench, so that the commencement of injection for no. 1 pumping element will serve as a datum for all the others, in order of cylinder injection sequence. The phase angle or angular difference between each commencement of injection reading should be 60° for a six-cylinder engine or, of course, equal increments of 360° depending upon whatever other number of cylinders is involved. Phase angles normally have to be held within the limits of ±0.5°.

Need for calibrating an in-line fuel injection pump

If full advantage is to be taken of the diesel engine operating cycle, in which fuel oil is delivered by positive means directly into each cylinder or – as it has sometimes been said – 'delivered not on the doorstep but in the entrance hall of the cylinder', then much depends upon the equality of the fuel oil amounts distributed to each cylinder; this explains the need for pump calibration. The *calibration* of an in-line fuel injection pump may therefore be defined as the procedure for adjusting the fuel oil deliveries of the pumping elements so that each will deliver the same specified quantity. It should be appreciated that the calibration of an injection pump is a most important operation, especially because of environmental requirements for a clear exhaust with no unburnt fuel present and, of course, economic considerations as related to the cost of saving in fuel oil.

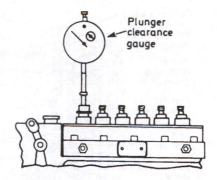

Figure 8.14   Plunger stroke measuring tool used in phasing operation

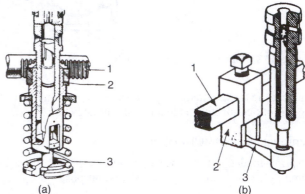

Figure 8.15   Adjustment provisions for calibrating in-line fuel injection pumps:
(a) 1 control rack; 2 quadrant pinion; 3 control sleeve
(b) 1 control rod; 2 engaging fork; 3 plunger arm

Calibrating an in-line fuel injection pump

Injection pump calibration is also performed on a fuel pump test bench. For this purpose it is motorized and equipped with a bank of specially calibrated test injectors, the outputs from these being collected in graduated glass measuring cylinders known as graduates. Facilities are also provided for automatically counting the required number of test oil (not ordinary fuel oil) deliveries or what are termed shots into each graduate, while the pump is being run at a specified speed.

Provision for adjustment of output from each individual pumping element is made either by unclamping and turning the quadrant pinion relative to the plunger control sleeve (Figure 8.15a) or, as in later practice, unclamping and altering the position of the plunger arm engaging fork along the control rod (Figure 8.15b).

8.4   GOVERNING THE IN-LINE FUEL INJECTION PUMP

Reasons for governing a diesel engine

Until the recent introduction of direct-injection for petrol engines (Section 6.9), an indirect fuel injection system or, as in earlier practice, a carburettor has been used to meter fuel into the incoming air stream, which is then throttled in accordance with power requirements. The marriage of fuel and air thus

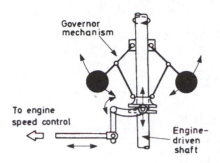

Figure 8.16   Basic principle of a centrifugal governor

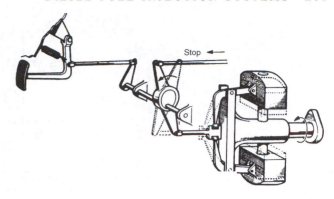

Figure 8.17   Operating principle of a minimum and maximum speed governor for diesel engine (*Lucas CAV*)

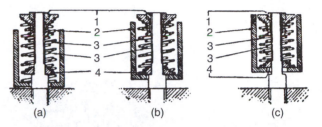

Figure 8.18   Operation of flyweight springs in minimum and maximum speed governor: (a) idling speed (b) normal range (c) speed exceeding predetermined maximum

1 adjusting nuts
2 outer idling springs
3 stiff springs
4 spring plate

takes place outside the engine cylinders and the process may be regarded as essentially self-regulating. This characteristic is not shared by the diesel engine, since its fuel injection system is entirely divorced from the incoming air stream, and power is controlled by varying the quantity of fuel oil injected into the cylinders. Furthermore, it must be appreciated that there is only a relatively small difference between the quantity of fuel oil required to run a diesel engine at idling speed and that used at maximum speed with no load. For these reasons it becomes necessary to superimpose a governor control to provide automatic speed regulation, relative to any set position of the accelerator pedal.

An ungoverned diesel engine could therefore become rather neurotic and tend to race or 'run away' at higher speeds or stall when idling. These two tendencies alone can be countered by using what is termed a minimum and maximum speed or speed-limiting class of governor, but in later commercial vehicle diesel practice an all-speed or variable-speed class of governor was adopted. Mechanical versions of both classes act in the manner of a centrifugal governor (Figure 8.16) and their mechanism is combined with the fuel injection pump.

Minimum and maximum speed governor

Following its incorporation in the original design of German Bosch fuel injection pump of the late 1920s, the Harzmann type of minimum and maximum speed governor was for many years preferred for its simplicity in commercial vehicle diesel engine applications. Its construction and operation were such that radial movement of the flyweights attached to the pump camshaft was transmitted through bell-crank levers to the lower end of an eccentrically pivoted floating lever, via the arms of a cross-head assembly (Figure 8.17). The upper end of the floating lever actuated the control rack for the pumping elements. Almost buried within each sliding flyweight was a trio of concentrically mounted and already compressed (or preloaded) governor coil springs (Figure 8.18). These resist radial movement of the flyweights as they tend to climb outwards under the centrifugal effect created by an increase in speed of the fuel pump camshaft and hence that of the engine itself (Figure 8.19).

Minimum or idle speed governing of the engine during traffic stops was therefore effected by the lesser centrifugal effect on the flyweights being opposed only by the lighter loading of the outer governor coil springs (Figure 8.18a). With any unwanted increase in engine speed, the action of the flyweights moving further outwards and compressing these springs is to pull the crosshead assembly inwards, as a result of which the floating lever will move the control rack to reduce fuel oil delivery and slow the engine. The converse sequence of events occurs, of course, with any tendency of the engine to stall. Maximum speed governing to prevent engine damage through overspeeding was similarly imposed when the greater centrifugal loading on the flyweights eventually exceeded the heavier loading of the inner springs, the outer idling springs already being compressed (Figure 8.18c). In the intermediate speed range between these two extremes of minimum and maximum speed governing, the engine speed was controlled solely by manual repositioning of the floating lever, through the medium of the accelerator pedal linkage and control lever acting about the eccentric pivot. Hence, the control rack could be moved in a direction to either increase or decrease the output from the pumping elements, according to the dictates of the driver and quite independent of the restrained governor flyweights (Figure 8.18b).

All-speed governor

For modern commercial vehicle diesel engines an all-speed governor is generally used to control the output from the pumping elements and is typically of the Hartnell type. In practical application the flyweights may be carried on

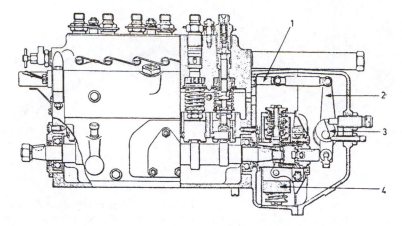

Figure 8.19 Practical application of a minimum and maximum speed governor

1 control rack   3 control lever eccentric pivot
2 floating lever   4 governor flyweights

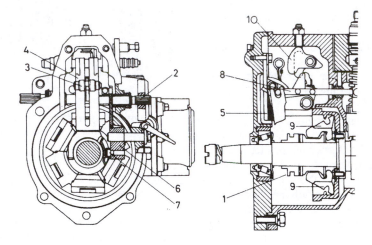

Figure 8.20 An all-speed mechanical governor of the leaf spring type in detail (*Lucas CAV*)

1 thrust pad   6 rocking lever
2 accelerator shaft   7 slipper pin
3 roller   8 rack
4 leaf spring   9 flyweights
5 ramp   10 maximum fuel stop control

bell-crank levers, the inwardly pointing arms of which transmit end thrust to a simply pivoted rocking lever that connects at its upper end to the control rod of the pump. The end thrust derived from the pivoting flyweights is opposed by an externally acting governor spring, in contrast to the internally acting governor springs of the Harzmann minimum and maximum speed device. Either a coil or, as in later practice, a leaf type of governor spring may be used (Figure 8.20), and in both cases the spring constitutes the only mechanical connection between the pump control lever, which is linked to the accelerator pedal, and the control rod for the pumping elements.

In the case of a coil-type governor spring, this may simply be hooked between cranked arm extensions of the rocking and control levers, so that the opposing load on the flyweight assembly is varied according to the relative movements of the two levers (Figure 8.21). With a leaf-type governor spring the variation in opposing load is achieved by altering its effective bending stiffness. For this purpose the spring contacts a roller, which can be moved up or down over a vertically positioned and inclined ramp by means of a forked end on the pump control lever (Figure 8.22). It thus follows, for example, that whenever the rise in engine speed is such that the centrifugal loading on the flyweights becomes sufficient to overcome the load exerted by the accelerator-controlled governor spring, the rocking lever will be swung in a direction to decrease the output from the pumping elements through appropriate movement of their control rod and thereby reduce engine speed.

Misbehaviour of the mechanical governor

As with all mechanisms that incorporate pivoting and sliding components, the mechanical governors employed with diesel fuel injection pumps are subject to gradual wearing of their working parts. This can result in undue sloppiness or sticking

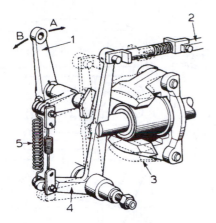

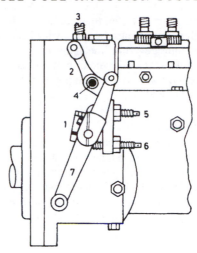

Figure 8.21 Operating principle of an all-speed mechanical governor of the coil spring type

A accelerator pedal depressed
B accelerator pedal released

1 control lever
2 control rack
3 governor flyweights
4 rocking lever
5 coil-type governor spring

Figure 8.23 Typical arrangement of governor controls

1 stop quadrant	5 idling speed stop screw
2 stop control lever	6 maximum speed stop screw
3 maximum fuel stop screw	7 speed control lever
4 excess fuel plunger	

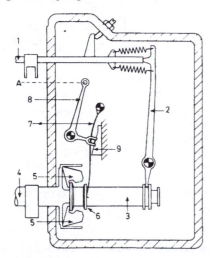

Figure 8.22 An all-speed mechanical governor of the leaf spring type in simplified form

A to accelerator pedal

1 rack control rod	6 shoulder
2 lever	7 leaf spring
3 sleeve	8 bell-crank lever
4 camshaft	9 ramp
5 flyweights	

tendencies of the mechanism, which will upset the sensitivity of the governor action or the difference in engine speed that occurs before a correction is made to the pump output. The following is intended only to be a general list of possible defects that may arise with the two basic types of governor in extended service:

1 Wear on the spring plates and in the bottom of the recesses or pockets of the flyweights used in minimum and maximum speed governors.

2 Wearing of the pivot pins in the bell-crank levers and cross-head assembly of minimum and maximum speed governors.
3 Partial seizing of the eccentric bearing in the floating lever of minimum and maximum speed governors.
4 Wearing of the flyweights bushes and weight carrier pins of all-speed governors.
5 Wear on the leaf-type spring of all-speed governors.
6 Wearing of the roller and ramp assembly of leaf spring all-speed governors.

Adjustment provisions for mechanical governors

Three separate adjustments are usually related to the operation of mechanical governors, as described in the following sections.

Maximum speed stop screw
This is typically mounted external to the governor housing and acts against the stop quadrant of the pump speed control lever and below its pivot axis (Figure 8.23). It is set, locked by a nut and sealed with wire and a lead tag by the engine manufacturer, and must not be tampered with unless appropriate authority is first obtained. When fitting a replacement injection pump, or in the event of an unsatisfactory readjustment having been made to the original, then the maximum no-load speed must of course be checked and set strictly in accordance with the manufacturer's specification with the engine at its normal operating temperature. Under no circumstances should an engine be allowed to operate at a higher speed than that specified or severe damage to it may result.

Maximum fuel stop screw
In effect this is the second stop (once known as a smoke stop) provided for limiting the maximum amount of fuel oil

delivered by the pump, but in this case it acts directly against one end of the control rack or rod, instead of via the speed control lever and governor mechanism (Figures 8.20 and 8.23). The adjustment of the maximum fuel stop screw is set during test bench calibration of the pump, following which it is locked by a nut and sealed.

Idling speed stop screw

This adjustment takes a similar form to the maximum speed stop screw, except of course that it acts above rather than below the pivot axis of the pump speed control lever (Figure 8.23). The adjustment is set, locked by a nut and sealed to provide an engine idling speed in accordance with the manufacturer's specification again with the engine at its normal operating temperature.

The pneumatic all-speed governor

This type of all-speed governor for in-line fuel injection pumps was originally devised by Robert Bosch of Germany in the mid-1930s. It was later adopted elsewhere and, until the advent of the distributor type of fuel injection pump, found ready application on small-capacity, high-speed diesel engines used in light commercial vehicles and also in early diesel cars and taxis.

With a pneumatic governor an accelerator-controlled butterfly valve is used to regulate the air flow through a large venturi unit situated at the entrance to the intake manifold (Figure 8.24). Leaving at right angles from the venturi unit is a pipe connection that is sometimes referred to as a pitot tube (so named after H. Pitot, an eighteenth-century inventor who devised it as a scientific means of measuring air velocity), which communicates with a depression chamber mounted at one end of the fuel injection pump housing. A spring-loaded diaphragm is enclosed within the depression chamber and is directly linked to the fuel control rack or rod for the pumping elements.

In operation, the spring loading on the diaphragm moves the fuel control to its maximum delivery position for starting purposes. Once the engine begins idling, the high depression at the throttled venturi is communicated to the spring side of the diaphragm. The atmospheric pressure acting on the other side of the diaphragm thus overcomes the spring loading and the diaphragm deflects to move the fuel control to its minimum delivery position. So as better to balance the high depression that exists under idling conditions, an auxiliary spring and idling plunger are also mounted from the spring abutment for the diaphragm and thus impose an additional restraint on the fuel control movement.

As the accelerator is depressed and the butterfly valve opens, the depression at the venturi unit gradually reduces, which thus permits the spring-loaded diaphragm to move the fuel control in the direction of increased delivery. Conversely, if when the accelerator is held steady there is any tendency towards an increase or decrease in engine speed with changing load, the resulting variations in governor system depression serve to reposition the fuel control until the desired engine speed is restored. At this point the difference in pressure across, and the spring loading on, the governor diaphragm balance each other out.

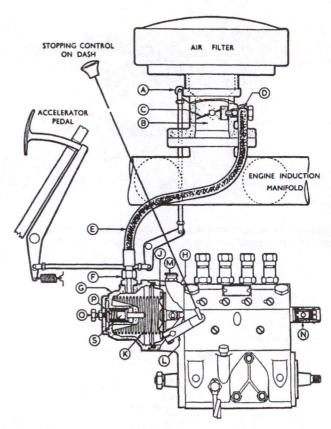

Figure 8.24   Diagrammatic and sectional views of the pneumatic governor with single pitot venturi flow control unit (*CAV*)

A venturi valve control lever
B venturi throat
C venturi butterfly valve
D vacuum pipe union
   vacuum pipe
F diaphragm housing union
G diaphragm housing
H main housing

J diaphragm
K main diaphragm spring
L stop lever
M oil cap
N control rod stop
P auxiliary idling spring
Q auxiliary idling set-screw
S auxiliary idling plunger

The advantages of the pneumatic governor have generally been regarded as: good sensitivity for stable idling control; a useful saving in weight and installation space; and lower-cost construction and maintenance. The main disadvantage is that it is oversensitive at high speeds, because small angular movements of the butterfly valve produce large linear movements of the fuel control. The sudden changes in engine torque that result can make smooth driving difficult.

Misbehaviour of the pneumatic governor

Clearly any defect that allows an air leak into the depression side of a pneumatic governor will affect its normal operation. The following is therefore a list of possible defects that may arise in extended service:

1 Securing unions of vacuum pipe connecting the venturi unit to the governor depression chamber not air-tight.
2 Leaking diaphragm in a depression chamber.
3 Defective sealing surface at the attachment flange of the venturi unit.

4 Excessive clearance between worn butterfly valve shaft and bushes.
5 Butterfly valve loose on its control shaft so that it is neither seating properly on the shaft nor in the bore of the venturi unit.

Caution The engine must not be started before the venturi unit, flexible pipe or pipes, or the air filter are fitted, and the diaphragm unit is checked for air leaks in accordance with the manufacturer's service instructions. Failure to observe these precautions may result in serious damage to the engine caused by overspeeding.

Adjustment provisions for pneumatic governors

Similar to the mechanical governor there are three separate adjustments usually related to the operation of a pneumatic governor, as described in the following sections.

Maximum speed stop screw
This is located from a lug on the body of the venturi unit and acts against the operating lever for the butterfly valve (Figure 8.25). It is provided to ensure that the butterfly valve cannot be turned over centre beyond its fully open position and, if necessary, to restrict the valve opening to less than this amount in order to comply with the manufacturer's specification for maximum no-load engine speed. The absolute maximum speed is ultimately limited by the diameter of the venturi opening, so the maximum speed stop screw is really a speed reducing adjustment.

Maximum fuel stop screw
With a pneumatic governor this is typically carried in a separate small housing mounted at the opposite end of the fuel injection pump to the depression chamber (Figure 8.24). Less commonly, it may act against a rocking lever mounted just ahead of the depression chamber, the other end of the lever having access to the fuel control rack or rod.

Idling speed stop screw
This is also located from a lug on the body of the venturi unit but acts, of course, in the opposite sense to the maximum speed stop screw against the butterfly valve operating lever (Figure 8.25). Readjustment of the auxiliary spring loading for the idling plunger may also be required in conjunction with idling speed positioning of the butterfly valve, if the idling behaviour of the engine becomes unstable. It should be appreciated that idling speed adjustment is impossible in the event of air leaking into the vacuum side of the pneumatic governor, because the pressure difference across the diaphragm would then be insufficient to move the fuel control towards the stop position.

8.5 DISTRIBUTOR FUEL INJECTION PUMP

General background

As the advantages and disadvantages of the distributor fuel injection pump are summarized later, suffice it to say here that this type of pump has materially contributed to the wide acceptance of relatively small high-speed diesel engines for

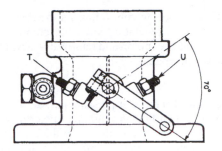

Figure 8.25 Maximum and idling speed controls for pneumatic governor (*Lucas CAV*): T = adjustable screw for maximum speed; U = adjustable screw for idling

vehicles since the mid-1950s. Apart from commercial vehicle applications such pumps are now specified almost exclusively for modern diesel cars.

Distributor fuel injection pump mechanisms basically comprise a single pump and hydraulic metering distributor combined into a single unit and driven at half-engine speed. In one version of distributor pump, introduced by Robert Bosch of Germany in 1963, a ported rotor is reciprocated through the medium of an integral multilobed face-cam riding against fixed rollers, so that the combination of reciprocating and rotating motions provides for pressurization and distribution of the fuel-oil respectively (Figure 8.26a).

Another version of distributor pump, devised in 1939 by Vernon Roosa of America and originally manufactured there by the Hartford Machine Screw Company, utilizes a ported rotor through which radially opposed plungers are reciprocated (Figure 8.26b). The motion of the plungers is derived from their roller ends riding against lobes of a stationary internal cam-ring. Hence, the combination of rotating and reciprocating motions again serves to distribute the fuel-oil under pressure to the cylinder injectors. Since 1956 the Roosa operating principle has been incorporated under licence in distributor pumps manufactured by CAV in the United Kingdom and it is this type of pump that we shall now examine in more detail (Figure 8.27).

Construction and operation

This is more conveniently considered under several headings, as follows:

Transfer pump
In the first instance fuel-oil is delivered from the tank and via a filter to the distributor pump by a conventional engine-driven lift pump. On entering the distributor pump through the main inlet connection, the fuel-oil is further pressurized by a two-bladed, sliding-vane, 'transfer pump' (similar to a corresponding type of engine oil pump as described earlier). This is housed in the 'hydraulic head' of the distributor pump and driven from the end of the rotor remote from the pumping plungers. The purpose of the transfer pump is to create what is termed 'transfer pressure' in the distributor pump proper.

Regulating valve
This is a spring-loaded piston type valve, which regulates the pressure created by the transfer pump not by maintaining it at

a constant value, but by controlling its increase with rising engine speed. The 'regulating valve' assembly is mounted adjacent to the transfer pump and returns by-passed fuel-oil to the inlet side of the pump.

Metering valve

Before the fuel-oil at transfer pressure passes to the pumping plungers, it must first flow through the 'metering valve'. The positioning of this valve and therefore the quantity of fuel-oil it meters, depends upon a combination of pump control lever and governor imposed loads, as later described. Since the extent of outward travel of the pumping plungers on their charging strokes is determined by the quantity of fuel-oil transferred to them, via the metering valve, it thus follows that the fuel-oil output on their injection strokes is related accordingly.

Pumping plungers and cam

These perform the charging and injection strokes of the distributor pump. As each charging port of its driven rotor aligns with the metering port in the hydraulic head, fuel oil is admitted into the central passage of the rotor to push apart the pumping plungers, which are then between successive lobes on the internal cam ring. This constitutes the charging stroke of the pump. The injection stroke occurs when continued turning of the rotor covers the metering port and then aligns its single distributor port with one of the outlet ports in the hydraulic head. At this point both pumping plungers are forced rapidly inwards by the rollers contacting the lobes of the internal cam ring, so that fuel oil under injection pressure is displaced from the central passage of the rotor and thence via the outlet port to the cylinder fuel injector.

An internal adjustable maximum fuel stop acts by limiting the outward travel of the pumping plungers. For this purpose the sliding shoe interposed between the rollers and plungers have cam-shaped lugs on their sides, which engage with corresponding slots in two adjusting plates (Figure 8.28). These

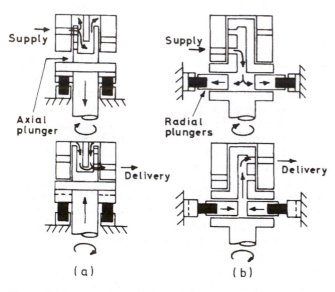

Figure 8.26 Operating principle of distributor fuel injection pumps with (a) axial and (b) radial plungers

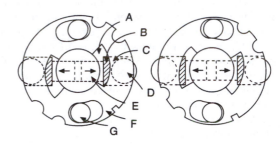

Figure 8.28 Method used to adjust maximum travel of plungers

A cam-shaped slot
B lug on roller shoe
C roller shoe
D roller
E plunger
F adjusting plate
G locking screw hole in shoe carrier

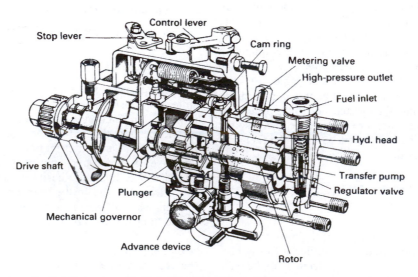

Figure 8.27 Cut-away view of a distributor-type fuel injection pump (*Lucas CAV*)

can be rotated slightly after first slackening the clamping bolts for the rotor drive plate.

Fuel oil distribution

The rotor and hydraulic head are provided respectively with the same number of equally spaced charging and outlet ports as the engine has cylinders. By connecting in turn the outlet ports to the cylinder fuel injectors in the same sequence as the engine firing order, the distribution of fuel oil at injection pressure is thus accomplished.

Pump drive

With commercial vehicle diesel engines the pump drive is usually taken via a gear train from the crankshaft, as described earlier. For a diesel car the pump may be driven in a similar manner to that of an electrical ignition distributor. A designer will always seek to minimize backlash in the pump drive, since otherwise the fluctuations in drive torque imposed by the pumping process can produce troublesome torsional oscillations. The final connection to the pump drive shaft is made through a splined or a keyed shaft, according to the engine manufacturer's requirements.

Advantages and disadvantages

As compared with the in-line fuel injection pump, these may be generally summarized as follows.

Advantages

1 It has an inherent ability to operate reliably at higher speeds, because of lighter reciprocating masses and the absence of any plunger return spring surging problems.
2 It is lighter and more compact in construction owing to the concentric arrangement of all working parts, which remain the same in number regardless of the number of engine cylinders being served.
3 It is less costly to manufacture owing to its simpler construction involving fewer high-precision parts.
4 With only one pumping element the pressure and amount of fuel oil delivered is identical for all engine cylinders served; pump calibration is thus not required.
5 An automatic injection timing control can be readily incorporated (to which further reference will be made).
6 It is more convenient to install since it can be mounted at any desired angle.

Disadvantages

1 There is a greater limitation on the amount of fuel oil that can be handled, which confines its application to engines of up to about 2 litres capacity per cylinder.
2 It is not capable of generating the very high injection pressures demanded by larger diesel engines with direct injection; this topic has been mentioned earlier.
3 The concentric layout places a physical limitation on the number of charging and outlet ports that can be accommodated and hence limits the number of engine cylinders that can be served usually to six.
4 For turbocharged engine applications it is more difficult to incorporate a boost control, the purpose of which will be described later (Section 9.2).

8.6 GOVERNING THE DISTRIBUTOR FUEL INJECTION PUMP

For the type of distributor fuel injection pump described in Section 8.5, governing may be accomplished either mechanically or hydraulically.

Mechanical governor

A convenient point at which to begin a description of this device is the governor arm, which is free to rock about a knife-edge pivot in the manner of a first system of levers (Figure 8.29). One end of the arm receives the end thrust generated by the centrifugal action of the governor weights, which are rotated from the pump drive shaft. The other end of the arm receives the pull from a tension spring connected to the pump control lever. Actually the tension spring is not directly anchored to the end of the governor arm, but is cushioned against it by a light compression spring that provides sensitive control during idling. Mounted in parallel with the tension spring of the pump control lever is a spring-loaded telescopic link, which at its end remote from the governor arm connects with the lever arm of a rotary metering valve. The purpose of the metering valve was explained earlier.

In operation, accelerator enforced movement of the pump control lever influences the tension load exerted by its spring on the governor arm, which in turn transmits an opening or a closing motion to the rotary metering valve, thereby admitting either more or less fuel oil to the pumping element. This action is modified, however, owing to the end thrust exerted by the centrifugal governor opposing the pull from the tension spring of the pump control lever. When a condition of equilibrium is established between the two opposing forces, the engine is then being controlled at the desired running speed. For stopping the engine, a shut-off lever actuates a bar that pushes against the opposite end of the metering valve lever arm, the swing of the arm being accommodated by the telescopic link mentioned earlier.

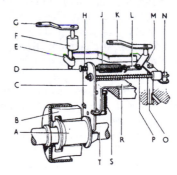

Figure 8.29 Mechanical governor linkage of a distributor-type fuel injection pump (*Lucas CAV*)

A thrust sleeve	L connecting link
B governor weights	M metering valve lever
C governor arm	N spring-loaded hook lever
D idling spring guide	O metering valve
E shut-off bar	P hook lever spring
F shut-off lever	R pivot plate
H idling spring	S retaining spring
J governor spring	T splined driver shaft
K control lever	

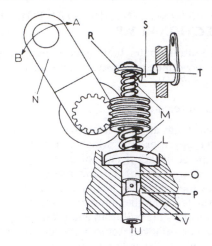

Figure 8.30 Hydraulic governor mechanism of a distributor-type fuel injection pump (*Lucas CAV*)

A accelerator pedal depressed	P metering port
B accelerator pedal released	R shut-off washer
L damping collar	S shut-off cam
M compression springs	T shut-off shaft
N control lever	U transfer pressure
O metering valve	V to pumping element

Hydraulic governor

The most convenient point at which to begin a description is the metering valve itself, which in this case is of the axial sliding type (Figure 8.30). One end of this valve has a stem on which is mounted a slidable toothed rack of cylindrical form, the rack being engaged by a pinion actuated from the pump control lever. Location of the rack on the valve stem is performed by two compression springs, the lighter one of the pair providing sensitive control during idling. The other end of the valve is acted upon by fuel oil at transfer pressure and, since this end of the valve is hollow up to an undercut portion that is also cross-drilled, the fuel oil must flow through the restriction offered by the valve and metering port before continuing on to the pumping element. To prevent unwanted oscillation of the metering valve its stem is provided with a damping collar and cylinder.

In operation, accelerator enforced movement of the pump control lever influences the compression load exerted by the heavier spring on the metering valve, thereby altering the position of the latter relative to its metering port and admitting either more or less fuel oil to the pumping element. The position assumed by the metering valve is modified, however, by the presence of transfer pressure opposing the spring load until, similar to the mechanical governor, a condition of equilibrium is established between the opposing forces and the engine is controlled at the desired running speed. For stopping the engine, a shut-off lever actuates a cam that can be turned into engagement with a thrust washer at the end of the metering valve stem. This results in the valve being lifted to an extent that it completely covers the metering port and prevents any fuel oil from reaching the pumping plungers.

It will be appreciated that both the mechanical and hydraulic governors described are of the all-speed type, the former

generally offering greater precision of control at the expense of increased complication and bulk.

8.7 TIMING IN-LINE AND DISTRIBUTOR FUEL INJECTION PUMPS

Timing an in-line fuel injection pump

First it must be stressed that reference to the engine manufacturer's service instructions for this operation is always advisable, since there is no golden rule as such for carrying it out. However, the following three sections summarize the basic steps to be taken.

Setting commencement of injection for engine
The basic requirement here is for the piston in no. 1 cylinder to be on its compression stroke and set at a position that corresponds exactly with the specified commencement of injection, or injection advance angle. This may be accomplished by observing either angular or linear setting dimensions. In the former case, the crankshaft is first turned in its normal direction of rotation until the piston in no. 1 cylinder is at TDC on its compression stroke, as indicated by alignment of the TDC markings on the flywheel and its housing and the inlet and exhaust valves being closed. The crankshaft is then turned backwards until the commencement of injection or INJ reference mark on the flywheel exactly coincides with the fixed TDC mark or pointer on the flywheel housing. In the second case, the piston in no. 1 cylinder is also first set at TDC on its compression stroke, typically by a timing indication on the crankshaft pulley. The collets, spring cap and spring from the inlet valve of no. 1 cylinder are then temporarily removed and the valve is allowed to rest on the top of the piston. A dial indicator is next suitably mounted and then zeroed with its stylus in contact with the tip of the valve. Finally, the crankshaft is turned backwards for about one-eighth of a turn and then forwards again until the specified piston travel BTDC, corresponding to the required injection advance angle, is recorded on the dial indicator.

Setting commencement of injection for pump
This is usually simply a matter of ensuring that the timing mark on the pump coupling is brought into alignment with the timing indicator on the governor cover (Figure 8.31a).

Synchronizing the engine and pump settings
Connecting the pump to its drive shaft from the engine timing gear is facilitated by slotted bolt holes in the engine side of the exposed pump coupling. Angular correlation between the timing gear shaft and pump side of the coupling is thus automatically conferred and the commencement of injection settings described above are sustained as the coupling bolts are secured.

Timing a distributor fuel injection pump

Again, it must be stressed that reference to the engine manufacturer's service instructions is always advisable before attempting this job. The following is therefore intended only as a basic guide relative to the particular type of distributor fuel injection pump described in Section 8.4. As in the case of an in-line pump, the timing process may be considered under three headings.

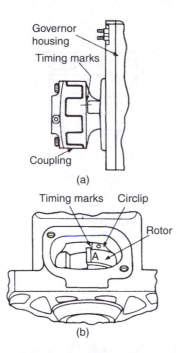

Figure 8.31 Examples of timing marks on fuel injection pumps: (a) in-line pump (b) distributor pump

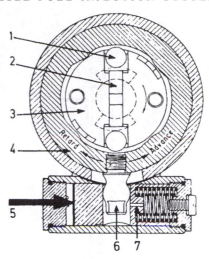

Figure 8.32 Automatic injection timing control for distributor pump

1 plunging rollers	5 transfer pressure
2 pumping plungers	6 ball screw connection
3 rotor	7 spring-loaded piston
4 cam ring	

Setting commencement of injection for engine
The choice of methods that can be adopted here is, of course, the same as that just described for the in-line pump.

Setting commencement of injection for pump
An inspection cover on the pump allows access to internal timing indications. When a specified engraved line on the pump rotor aligns with either another engraved line on the adjacent timing circlip or, as in later practice, that end of an unmarked timing circlip which has a straight edge (Figure 8.31b), it denotes commencement of injection from no. 1 cylinder outlet on the pump.

Synchronizing the engine and pump settings
Again this entails connecting the pump to the engine such that their commencement of injection settings are not disturbed. Since the flange-mounted distributor pump has no visible drive coupling that lends itself to convenient adjustment, other means of accommodating any angular mismatch between the engine timing gear and the pump drive shaft connections must be provided. These may take the form of either slotted bolt holes for the pump drive plate connection to the timing gear, or similarly slotted bolt holes for the pump mounting flange connection to the timing chest. An inspection plate on the timing cover is, of course, required with the former arrangement.

To cater for the wide speed range over which the modern motor vehicle diesel engine operates, an automatic injection timing control facility is built into the distributor pump. Injection timing can therefore be progressively advanced with increasing speed to maintain optimum engine torque. The timing control acts upon the internal cam ring for the pumping plungers (Figure 8.32). This is accomplished by a ball screw connection between the cam ring and a spring-returned piston sliding in a cylinder, which is subject to increasing transfer pressure with rising engine speed. As a result the cam ring is turned in a direction opposite to that of the rotor during pumping, so that the plunger rollers contact their cam lobes earlier and advance the injection timing.

8.8 FUEL INJECTORS FOR JERK PUMP SYSTEMS

General background

Since the pumping elements in combination with the delivery valves not only supply the necessary pressure for injecting fuel oil into the engine combustion chambers, but also control both the timing and period of injection, it might seem at first that a fuel injector need comprise nothing more than a simple spraying hole. This would be feasible were it not important that full pressure is required at the beginning of fuel injection and, as previously mentioned, clean termination of fuel spray at the end of injection. For these reasons the fuel injector must incorporate a spring-loaded valve.

The inward-opening differential needle valve type of fuel injector has long been established practice and was originated by Thornycroft in England as long ago as 1908. In present-day constructions, the larger-diameter guiding portion of the needle valve is lapped into the nozzle body to render the bearing surfaces virtually self-sealing, while its smaller-diameter lower end has a conical seating to seal the nozzle orifice.

Operation of the fuel injector is such that the rising delivery pressure of the fuel acts initially upon the annular area above the valve tip, until the spring preload on the needle valve is overcome. At this stage, the delivery pressure additionally acts upon the valve tip, so that the needle valve rapidly and unhesitatingly attains its full lift and opening (Figure 8.33b). Fuel is then sprayed from the nozzle until the

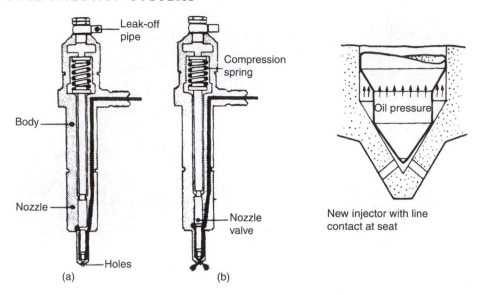

Figure 8.33 Construction and operation of a fuel injector (*Lucas CAV*): (a) fuel spray terminated (b) fuel spray initiated

pressure falls and the spring loading returns the needle valve smartly to its seating (Figure 8.33a). This action is assisted by the delivery valve, as mentioned earlier (Section 8.2).

Since there must inevitably be some leakage past the sliding needle valve, and indeed this is necessary for lubrication purposes, a leak-off pipe is provided to return surplus fuel to the supply tank.

It should be appreciated that the fuel injector leads a particularly arduous life. A leading manufacturer of diesel engine fuel injection equipment, Lucas CAV Limited, state in connection with fuel injectors that to perform their duty consistently and efficiently, nozzles are manufactured to tolerances as small as 1.5 μm in materials which can withstand high pulsing fuel pressure of up to 60 MN/m^2 (8700 lbf/in^2) and temperatures of over 400°C.

Basic requirements

The fuel injectors represent the final link in the diesel fuel injection system. More specifically, they have two functions to perform:

1 To break up the stream of injected fuel oil into minute liquid particles suitable for rapid and complete combustion. This spraying process is often referred to as atomization, although in a scientific sense the description pulverization would be more appropriate.
2 To direct the spray in such a manner and with sufficient penetration that, in conjunction with the movement of air for combustion, the maximum possible quantity of oxygen is consumed without leaving any unburned or partially burned fuel oil.

We must therefore recognize that the design of an actual injection nozzle represents a compromise between providing proper atomization and sufficient penetration – the latter process once being likened by an American diesel research engineer to 'drilling a hole in the air'! This need for compromise arises because the smaller the size of sprayer hole, the better will be the atomization but the poorer will be the penetration.

Types of fuel injector nozzle

The various types of injection nozzle in current use (Figure 8.34) may be classified and their applications identified as follows:

Multihole nozzle, short stem

This type of injection nozzle may have up to six spray holes drilled in the bulbous end under the needle valve seating, their size and angle depending upon combustion chamber requirements. Its application is to direct-injection diesel engines, wherein the fuel oil sprays have to seek out the air for combustion. In this respect an advantage of the multihole nozzle is the versatility with which changes can be made to its spraying characteristics.

Multihole nozzle, long stem

The significance of the extended body is that the nozzle can be used on direct-injection engines where, owing to limited space between the valves in the cylinder head, it is not possible to provide adequate cooling for a short-stem version. It should be appreciated that with a high-speed diesel engine, the rate at which heat is liberated in the vicinity of the nozzle can be exceedingly high.

Pintle nozzle

This type of injection nozzle has long since been preferred to a single-hole short-stem version by virtue of its self-cleaning tendency. It is distinguished by its needle valve having a profiled extension or pintle, which protrudes into the valve body orifice to form an annular spray hole. A hollow cone-shaped spray is produced that makes the nozzle suitable for application to indirect-injection diesel engines of the swirl chamber type, wherein the air for combustion has to seek out the fuel oil, rather than vice versa.

Delay pintle nozzle

Also known as the throttling type of pintle nozzle, this differs from the ordinary version in having an increased length

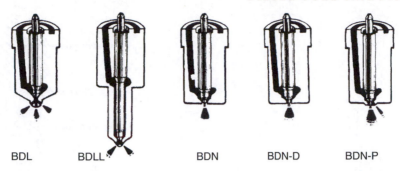

BDL   BDLL   BDN   BDN-D   BDN-P

Figure 8.34   Identification of different types of fuel injector nozzle (*Lucas CAV*)

BDL   multihole nozzle, short stem   BDN-D   delay pintle nozzle
BDLL   multihole nozzle, long stem   BDN-P   pintaux nozzle
BDN   pintle nozzle

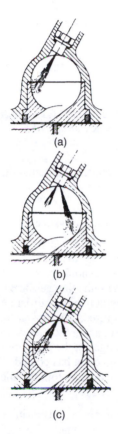

(a)

(b)

(c)

Figure 8.35   Action of the CAV-Ricardo pintaux nozzle: (a) normal pintle nozzle spray in Ricardo Comet chamber (b) pintaux nozzle operating at low speed with bulk of fuel injected towards centre of cell (c) the same at high speed with small spray towards centre and bulk injection following normal pintle nozzle characteristics

of orifice and a greater protrusion of the pintle. It is intended for application to indirect-injection diesel engines of the pre-combustion chamber type. The purpose of this modification is to reduce the initial rate of fuel oil injection during the first phase or delay period of combustion, so as to give smoother combustion and quieter running at idling speed.

Pintaux nozzle
Developed jointly by the CAV Company and H.R. Ricardo in the early 1950s, this type of pintle nozzle (Figure 8.35) was

designed to improve the cold starting behaviour of swirl chamber indirect-injection diesel engines. It does so by means of an auxiliary spray hole (thus the name pintaux, for pintle/auxiliary) such that under starting conditions, when the main spray hole is obstructed by the pintle, practically the whole of the fuel oil injected issues from the auxiliary spray hole. The spray is directed towards the throat of the ante-chamber to attack the air for combustion entering therein, before it has had a chance to lose any of its heat to the cylinder head. At the higher fuel injection pressures that accompany normal operating speeds, the pintaux nozzle behaves in the manner of a normal pintle nozzle, since the pintle valve is then lifted higher to clear the main spray hole.

Poppet nozzle
This small injector nozzle of the single-hole poppet valve type was introduced by Lucas CAV in 1981 and was principally intended for diesel engined passenger cars, where for easy installation it is screwed directly into the cylinder head and utilizes a standard 14 mm thread (the same as a sparking plug). Known as the Microjector, it differs from the other types of injector nozzle listed in having an outward-instead of inward-opening valve (Figure 8.36), this form of construction enabling the nozzle to be much reduced in both size and weight as compared with the smallest size of pintle nozzle. More specifically, the opening of the poppet valve in the same, rather than opposite, direction of fuel oil flow eliminates the need for a leak-off or back-leak pipe for the injector, and also the cylinder compression pressure assists, rather than opposes, the spring load acting to close the valve. The latter feature results in the effective opening pressure being increased over its atmospheric one of 8 MN/m^2 (1150 lbf/in^2) to 11–12 MN/m^2 (1570–1700 lbf/in^2) in the engine cylinder.

In operation, fuel oil from the injection pump passes through the centre of the holder (via an edge filter if specified) and the annular passage created by the valve collar, then between the coils of the spring and through the cross-holes in the nozzle body. The fuel oil under pressure finally acts on the seat diameter of the suitably proportioned valve to open it at the predetermined pressure, the opening continuing until the stop fitted below the collar abuts on the nozzle body. A beneficial degree of swirl is imparted to the fuel oil spray by helical grooves machined in the valve stem, so that

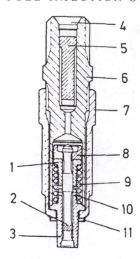

Figure 8.36 General arrangement of the CAV Microjector

1 lift stop
2 nozzle valve
3 nozzle body
4 inlet
5 edge filter
6 nozzle holder
7 capnut
8 collar
9 spring
10 feed ports
11 sealing washer

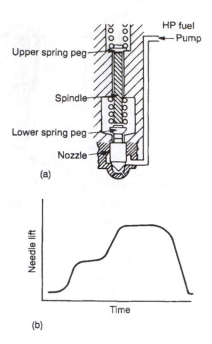

Figure 8.37 Two-stage injector: (a) schematic arrangement
(b) operating characteristic (*Lucas CAV*)

the spray is finely and widely distributed to minimize the delay period of combustion (Section 3.4) for smoother running. The Microjector is essentially a non-serviceable unit with an anticipated life of 50 000 miles (80 000 km).

Two-stage nozzle

The continuing research by Lucas CAV engineers into the combustion process of direct injection diesel engines has indicated that if the fuel oil is injected in two stages it can significantly reduce noise levels at idling and low speeds, provide better control over harmful exhaust emissions and improve driveability in terms of smoother power delivery. All these factors are important in the successful application of the more economical direct-injection diesel engine to passenger cars.

Unlike a conventional single-stage injector nozzle, in which the needle valve unhesitatingly attains a full lift when the rising delivery pressure overcomes its spring preload, the needle valve in a two-stage nozzle is controlled by two springs with each exerting a different preload on the needle valve. The arrangement of these springs, which includes a take-up clearance for the needle valve before the second spring becomes active (Figure 8.37), is such that initially the rising delivery pressure overcomes the preload of the lower spring, so that the needle valve opens by an amount limited to the take-up clearance. This is called the 'prelift' stage and provides the first nozzle opening pressure (designated NOP1) when a small quantity of fuel oil is injected. Full opening of the needle valve is then delayed until the delivery pressure has risen much higher to overcome the preload of the upper spring, which similarly provides the second nozzle opening pressure (NOP2) and allows the remaining larger quantity of fuel oil to be injected. It should always be borne in mind, of course, that the whole process is only occupying a fraction of a second! Since the two-stage injector nozzle reduces the initial rate of

fuel oil injection, it provides a smoother start to the combustion process and hence reduces noise.

Injector nozzle opening pressures

The nozzle opening pressure is usually specified in the engine manufacturer's operating instructions, to which reference must always be made. It should be noted that the nozzle opening pressure is not the same thing as nozzle injection pressure, which is of a considerably higher order and related to the rate of fuel injection required. The nozzle opening and closing pressures are chosen such as to ensure that the needle valve can seal the injector against the high pressure in the combustion chamber after injection has ceased. Otherwise, the combustion gases would enter and foul the injector. Typical nozzle opening pressures are as follows:

1 Multihole injector nozzles used in small diesel engines of the direct-injection type: 17.5 to 21 MN/m^2 (2500 to 3000 lbf/in^2).
2 Multihole injector nozzles used in large diesel engines of the direct-injection type: 21 to 24.5 MN/m^2 (3000 to 3500 lbf/in^2).
3 Pintle injector nozzles used in small diesel engines of the indirect-injection type: 10.5 to 14 MN/m^2 (1500 to 2000 lbf/in^2).

There is currently a trend towards increasing the rate of injection to compensate for the earlier mentioned need to retard injection timing (Section 3.5) and reduce particle emissions, which has led to the use of higher injection pressures. For example, those used in two 1991 versions of large-capacity direct-injection diesel engines for heavy vehicles have been increased from a previous 80 MN/m^2 (11 500 lbf/in^2) to 112 MN/m^2 (16 000 lbf/in^2).

Fuel injector misbehaviour

As noted earlier, the fuel injector leads a particularly arduous life in its handling of fuel oil under conditions of elevated pressure and temperature. It is therefore to be expected that certain faults can arise in extended service, the commoner forms of which may be considered under the following headings.

Dribble
This condition refers to any slight leakage of fuel oil from an injector when its needle valve is supposed to be closed. It is usually caused by the presence of carbon accumulation on the needle valve seating or by actual damage to the seating. Its effect on engine operation is the emission of black exhaust smoke.

Distorted spray pattern
This fault is self-descriptive and arises from fouling of the injector spray hole(s) by ridges of accumulated carbon, a condition that is aggravated by the presence of dribble. Multihole injectors with their very small-diameter spray holes are naturally more sensitive to this type of fouling. A distorted spray pattern with its accompanying poor atomization makes it difficult to obtain complete combustion of the fuel oil and air, which has the effect of increasing fuel consumption.

Reduced spray penetration
If the needle valve opens before the pressure of fuel oil in the injector delivery pipe has risen to the intended value, then the spray will be less penetrative, because its intensity is proportional to the difference in pressure that exists between the fuel oil being injected and the air compressed in the combustion chamber. This can be caused by any falling-off in spring preload on the needle valve. The effect is such that the injection timing will be early and combustion will be incomplete, to the detriment of engine performance. A similar effect can be created by a sticking needle valve, which allows an early injection of an excess quantity of fuel oil and is the classic cause of diesel knock complaints.

Excessive back-leakage
Any increase in normal working clearance due to wear on the sliding surfaces of the needle valve in its nozzle can allow an excessive back-leakage of fuel oil from the injector: that is, a greater back-leakage than is required for internal lubrication purposes. This results in a reduced quantity of fuel oil being injected into the combustion chamber, and is therefore another cause of impaired engine performance.

Testing a fuel injector

In diesel engine servicing the working condition of a fuel injector can be satisfactorily evaluated only by using a piece of test equipment known variously as a nozzle testing outfit or an injector test unit. This is complementary to the fuel pump test bench mentioned in Section 8.2 and must likewise be operated strictly in accordance with the manufacturer's operating instructions. Otherwise, false evaluations can easily be made, especially in the case of delay pintle nozzles.

A hand-lever-operated nozzle tester essentially comprises a single pumping element and delivery valve unit, pressure gauge, check valve, spray chamber and supply tank. The test facilities it provides for an injector may be summarized as follows:

Spray formation   With the check valve closed to isolate the pressure gauge, the hand lever is operated smoothly at a specified rate of strokes, while the fuel oil sprayed into the chamber is observed for any distortion of pattern or coarseness of atomization.

Chatter   Again with the pressure gauge isolated, the hand lever is this time slowly depressed to its full extent, when the nozzle should be heard to make a soft chattering sound. Chatter is a characteristic sound produced by a healthy nozzle during this slow opening and closing. It is explained by the injection process actually taking place in a rapid series of sharp bursts, and is indicative that the fuel oil issuing from the sprayer hole(s) is at a velocity high enough to keep the nozzle tip dry.

Opening pressure   For this test the pressure gauge is, of course, brought into operation and carefully observed while the hand lever is slowly depressed, the highest pressure recorded on the gauge before the pointer flicks indicates the opening pressure of the injection nozzle. This is a function of the spring load on the needle valve and is initially set by either screwed plug or shim washer adjustment.

Leakage   Again with the pressure gauge in operation, the hand lever is slowly depressed to build up a specified pressure that is lower than the opening pressure of the nozzle. A timer is then used to compare, against specification, how long it takes in seconds before wetting becomes visible at the nozzle tip, and also the rate in number of drops per minute at which back-leakage is occurring from the nozzle return line orifice.

A note of caution is necessary concerning injector nozzle testing: *in no circumstances* must a bare hand be allowed to get in the way of a fuel oil jet, which can penetrate deeply into the flesh and destroy tissue. Fuel oil that enters the bloodstream may cause blood poisoning.

8.9   UNIT FUEL INJECTION

General background

Now becoming more widely used by virtue of its lower internal volume and greater rigidity, this type of system can allow much higher injection pressures. These pressures can be up to 203 MN/m^2 ($29\,000 \text{ lbf/in}^2$), their purpose being to promote more rapid injection of the fuel and also to spray it as a finer mist over a larger area through smaller holes in the nozzle. Hence, it becomes possible to meet stricter emission limits, reduce noise levels and improve fuel consumption. Unlike the jerk pump system described in Section 8.3, where a separately mounted multi-element pump provides the injection pressure and regulation of fuel quantity for the fuel injectors, each cylinder-head-mounted unit injector combines the functions of metering, timing and high-pressure fuel injection. These functions are performed by the movement of a plunger actually in the unit injector, via a push-rod and rocker mechanism from the engine camshaft. Each engine cylinder is therefore provided with a self-contained injection pump-and-nozzle unit, which receives fuel at relatively low pressure from a common supply system (Figure 8.38).

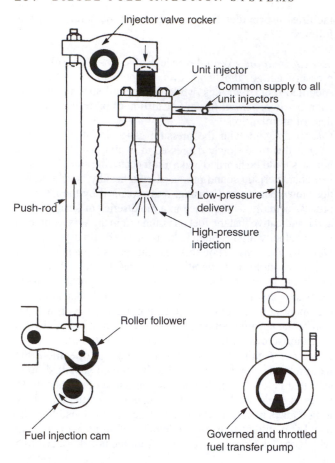

Figure 8.38 Basic layout of unit injector system

A notable pioneer in the application of the diesel engine to heavy vehicles in America was C.L. Cummins, who founded the well-known company that bears his name, and it was he who first successfully developed the unit injector system in the late 1920s for automotive use. This system was originally developed to overcome the unreliability of early high-pressure jerk pumps and to avoid the stresses associated with the high pressure generated in the pipe lines between the jerk pump and the cylinder-head-mounted fuel injectors. Perhaps it should be mentioned that during the early development of the automotive diesel engine, it was not uncommon for engine manufacturers themselves to design and build their own fuel injection equipment.

The pressure-time unit injector system

A basic distinction between the unit injector and the conventional jerk pump systems of fuel metering is that the former is inlet metered while the latter is outlet metered by spill control of output. An important later development of the unit injector system produced by the Cummins Engine Company was the introduction in 1954 of their pressure-time concept of inlet metering for the fuel entering the injector.

With a unit injector operated from the engine camshaft, it will be evident that the time interval available for metering the fuel through a fixed-size orifice into the injector will vary

inversely as the engine speed. Therefore if the fuel supply pressure to the injector remains constant, the quantity of fuel metered for injection would decrease with increase in engine speed. In order to vary and control the quantity of fuel that is metered for injection, so that it is correctly related to engine speed and load, the Cummins pressure-time (PT) system superimposes a governor and throttle control on the engine-driven fuel transfer pump that supplies fuel to the unit injectors. Hence, the quantity of fuel that is metered into them is subject to two variables, namely fuel pressure and metering time.

The Cummins PT system of fuel injection has been continuously refined since its inception, but essentially it comprises the following components:

1 An engine-driven positive displacement fuel transfer pump of the external gear type with a pulsation damper to smooth the fuel flow. The pump draws fuel from the tank via a protective filter.
2 A throttle control, which is combined with the fuel transfer pump for the driver manually to control the engine speed above idle, as required by varying operating conditions of speed and load. Rotation of the throttle shaft via the accelerator pedal varies the amount of flow and therefore the pressure of the fuel delivered to the unit injectors.
3 A centrifugally operated governor, which is similarly combined with the fuel transfer pump. The governor maintains sufficient fuel for idling with the throttle control in the idle position, and it cuts off fuel to the injector units above maximum rated engine speed. During operation between idle and maximum speeds, the fuel flows through the governor to the injector units via the throttle control.
4 A shut-down valve, which is also combined with the fuel transfer pump, so that fuel delivery to the unit injectors can be terminated to stop the engine.
5 Unit injectors, which provide the means of introducing fuel into the engine combustion chambers and combine the functions of metering, timing and high-pressure injection (Figure 8.39). They receive fuel at a relatively low controlled pressure from the common supply system. Except during the metering stage, there is a through-flow of fuel to lubricate and cool the unit injectors, the unused fuel being returned to the tank.

Advantages and disadvantages of unit injector systems

The unit injector system has generally involved the expensive complication of additional push-rods with lower and upper rockers to actuate the injectors from the engine camshaft, although more recent versions may benefit from the simplification of an overhead camshaft and single rocker actuation of the injectors.

However, the system has the advantages (noted earlier) of eliminating the necessity for high-pressure fuel pipe-lines between the pump and injectors and of combining all metering and injection functions. Other advantages generally claimed are those of enabling a wide range of engine speeds to be attained with quick throttle response and low idling speeds, and minimum maintenance requirements. These typically include adjusting the unit injectors at the same time as

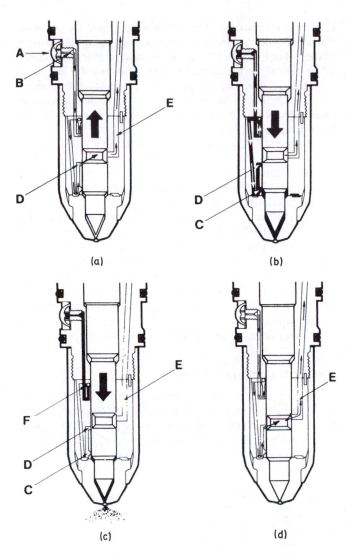

Figure 8.39 The Cummins PTD fuel injector in operation

(a) Start upstroke (fuel circulates). Low-pressure fuel enters injector at A, flows through inlet orifice B, internal drillings, around annular groove in injector cup, and up passage D, to return to tank, Engine speed, governor and throttle determine fuel pressure which, before inlet orifice B, determines amount of fuel flowing through injector.

(b) Upstroke complete (fuel enters injector cup). Injector plunger moving upwards uncovers metering orifice C and fuel enters injector cup. The amount is determined by fuel pressure. Passage D is blocked, momentarily stopping fuel circulation and isolating metering orifice from pressure pulsations.

(c) Downstroke (fuel injection). Plunger moving downwards closes metering orifice and cuts off fuel supply to cup. Continuing down, plunger forces fuel at high pressure from cup through tiny holes as fine spray. This ensures complete combustion of fuel in cylinder. When plunger undercut uncovers passage D fuel again flows through return passage E to tank.

(d) Downstroke complete (fuel circulates). After injection, plunger remains seated until next metering and injection cycle. Although no fuel is reaching injector cup it flows freely through passage E. This cools the injector and warms fuel in the tank.

the engine valves every 60 000 miles (96 000 km) or one year and cleaning and calibrating the unit injectors and fuel pump every 180 000 miles (288 000 km) or two years.

8.10 INTRODUCTION TO ELECTRONIC DIESEL CONTROL

General background

It is perhaps only to be expected that just as sophisticated electronic controls have been applied to manage the fuel injection systems of petrol engines, so they are now being adapted to perform a similar role with diesel engines. Their main purpose is to make it less difficult for manufacturers to comply with stricter exhaust emission regulations and achieve better performance.

The development of electronic controls for diesel engines (often abbreviated to EDC) began in the mid 1970s, when Detroit Diesel of General Motors in America applied them to experimental injection systems used in the course of research into diesel engine combustion. However, it was not until some ten years later that electronic management controls for distributor, in-line and unit injector fuel injection systems first began to appear after further research by Lucas CAV in this country, Bosch in Germany and Detroit Diesel in America.

Basic features of an electronic diesel control system

We take as an example the application of electronic control to the in-line fuel injection pump for a heavy vehicle. The essential differences between a conventional injection pump of this type and one that is subject to electronic control (Figure 8.40) are as follows:

1 The elimination of a mechanical governor, its place on the pump camshaft being taken by an impulse disc that rotates about a stationary speed sensor.
2 The provision of a linear solenoid actuator, which constantly adjusts the position of a spring-returned control rod for the pumping elements.
3 The addition of a sensor that detects the changing position of the pump control rod.

The electronic control unit receives signals from various sensors, this information being processed and compared with preprogrammed values, so that the appropriate commands can be sent to the control rod actuator and the required quantities of fuel oil are delivered to the injectors.

In the case of the Scania electronic diesel control system for a turbocharged heavy-vehicle engine (Figure 8.41), the significance of the various sensors may be summarized as follows:

Amount of acceleration The accelerator pedal is connected to a sensor that converts pedal travel into an electrical signal, there being no need for a mechanical connection between the pedal and injector pump.

Charge air pressure This information is required so that the quantity of fuel oil delivered to the injectors can be accurately matched to the available amount of air.

Charge air temperature It is important to relate this to the quantity of fuel oil delivered to the injectors, because cold air contains more oxygen than warm air and allows more fuel to be burnt.

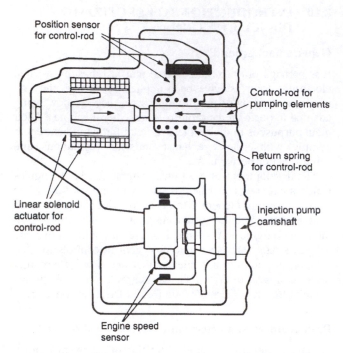

Figure 8.40 Schematic arrangement of EDC for an in-line fuel injection pump

Coolant temperature This needs to be sensed for automatically controlling the cold start function, since the quantity of fuel oil delivered to the injectors must be precisely related to the prevailing temperature whilst allowing the engine to be protected from overspeeding. For example, if the coolant temperature is 0°C the engine cannot be speeded up to more than 1000 rev/min for the first 30 seconds. Information from this sensor also controls idling speed.

Engine speed A sensor is incorporated in the fuel injection pump for constantly monitoring the engine speed, so that it can be maintained at present values, regardless of whether the cruise control is engaged to retain a chosen road speed or the power take-off is being operated.

Injection quantity This information is derived from another sensor in the fuel injection pump that detects the position of the control rod for the pumping elements, thereby indicating if the required quantity of fuel oil is being delivered to the injectors.

As with other modern electronically controlled management systems, several functions for self-diagnosis can be incorporated, which indicate when a fault occurs and can also specify where it has occurred. A light on the driver's instrument panel warns of any misbehaviour in the system.

Electronic unit injector system

A modern example of electronic control applied to unit injectors is the Lucas EUI (Electronic Unit Injector) System (Figure 8.42), which employs solenoid actuated spill control valves in the high pressure injection circuit. It will be seen that various sensors provide the electronic control unit

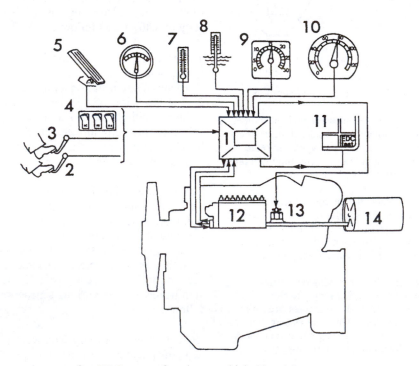

Figure 8.41 Schematic arrangement of an EDC system for a heavy vehicle (*Scania*)

1 microprocessor
2 brake pedal
3 clutch pedal
4 cruise control
5 accelerator pedal
6 atmospheric pressure
7 outside temperature
8 coolant temperature
9 tachometer
10 speedometer
11 fault monitor
12 injection pump
13 electric fuel cut-out
14 fuel tank

(ECU) with the necessary information on engine operation, this relating to accelerator pedal position, engine speed, camshaft position, air intake manifold temperature and pressure, and coolant temperature. The information received is then continuously compared with optimum values stored in the memory of the ECU. As a result the latter can issue command signals to the solenoid actuated spill control valves of the unit injectors, so that their nozzles can deliver fuel to the cylinders according to engine requirements.

The unit injectors are supplied with fuel from a low pressure pump and are mounted above the combustion chambers in the cylinder head, which is furnished with integral passages for the fuel feed and spill functions (Figure 8.43). Each unit injector is of the piston type with a spring-returned pumping plunger (Figure 8.44). It operates through a fixed stroke and receives its motion from the engine camshaft via an interposed rocker and tappet assembly (Figure 8.43). The optimum timing and the fuel quantity required for each individual injection, corresponding respectively to start of injection and governing in a conventional system, is determined by the commencement and duration of a command signal from the ECU, which is computed just prior to injection and then transmitted to the stator of the solenoid actuated spill control valve. This has the effect of activating the solenoid armature to close the spill control valve, so that as the pumping plunger continues its downward stroke the opening pressure of the injector nozzle valve is rapidly overcome and fuel is sprayed into the cylinder. When the ECU command signal deactivates the solenoid armature and allows the spill control valve to regain its open position, there is a collapse in pressure and the injection terminates as the nozzle valve closes in the usual manner.

The basic operating sequence of the unit injector is best visualized with a series of plunger displacement diagrams

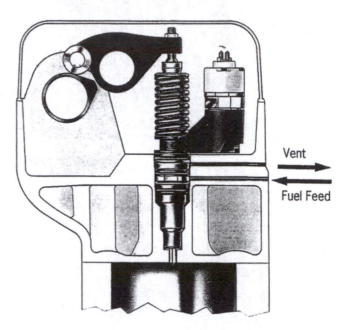

Figure 8.43 Typical installation of electronic unit injector (*Lucas EUI Systems*)

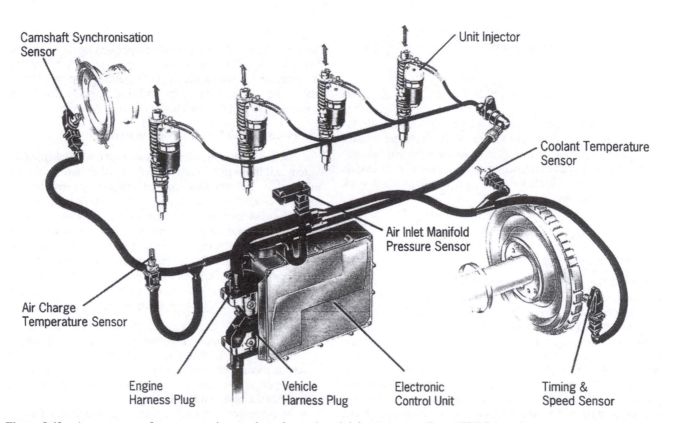

Figure 8.42 Arrangement of components in a modern electronic unit injector system (*Lucas EUI Systems*)

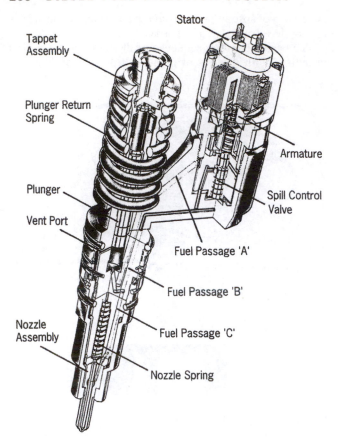

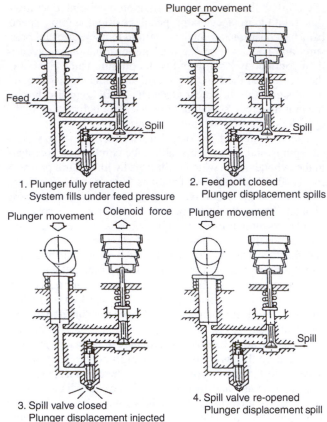

Figure 8.44   Cutaway view of electronic unit injector (*Lucas EUI Systems*).
Fuel passages: 'A' to spill control valve
'B' to pumping chamber
'C' to injector nozzle

Figure 8.45   Basic operating sequence of electronic unit injector (*Lucas CAV*)

(Figure 8.45). It should be noted that the solenoid force labelled in this Figure refers to the action of the high-speed solenoid actuator, which was developed by Lucas CAV for this type of application. By virtue of its electronic control the Lucas EUI System can, of course, be arranged to communicate and act in conjunction with not only other engine systems, but also vehicle systems such as those for anti-lock brakes and traction control.

8.11   COMMON RAIL FUEL INJECTION

General background

Although the so-called common rail fuel injection system has only become established practice for light to medium duty automotive engines since the late 1990s, following the lead set by the Fiat company, its basic operating principle is of much earlier origin. Indeed, the system was one of the first to be successfully applied to industrial and marine diesel engines. The British firm of Vickers Ltd pioneered the concept in 1913, which in practice comprised a high-pressure pump that delivered fuel to a hydraulic accumulator and header, or what we would now term a 'common rail'. From this the fuel at a constantly maintained pressure was sprayed

into the engine cylinders, via cam-operated injector nozzles. Perhaps of equal historical significance in relation to present-day practice was the application of solenoid controlled injectors to the common rail fuel injection system, which was proposed also in 1913 by Thomas Gaff in America.

The common rail fuel injection system

The essential difference between this type of fuel injection system and those previously described is that in the common rail system the two functions of pressure generation and injection control are separately performed. In other words, the high-pressure pump that charges the common rail plays no part in the metering and timing of the injection process, the opening and closing of solenoid or more recently piezo activated injectors being solely controlled from the electronic control unit (ECU) of the engine management system.

A common rail fuel injection system (Figure 8.46) basically comprises the following components:

1   A common rail that takes the form of a manifold-like tube, which is usually mounted at cylinder head level and connected via short piping runs to the fuel injectors. It acts as a constant pressure hydraulic accumulator, and is equipped at one end with a pressure limiting valve and towards the other end with a pressure sensor.
2   A high-pressure plunger pump, which delivers fuel oil through an end connection to the common rail and its

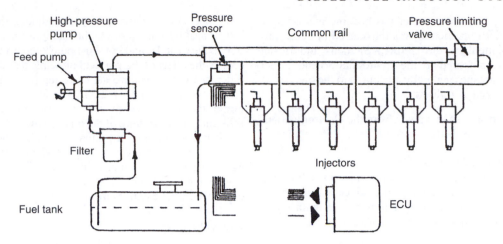

Figure 8.46 Schematic arrangement of a diesel common rail fuel injection system

associated piping (Figure 8.49). It serves only to maintain a constant pressure in the common rail, since all metering and timing functions for the fuel injection are performed separately at the fuel injectors themselves.

3 A feed pump, which transfers fuel oil from the supply tank to the high-pressure pump. These two pumps are typically combined into one unit to provide a compact installation.

4 A protective filter unit, which acts in series with the supply tank and the feed pump.

5 The cylinder head fuel injectors, which are activated by fast-response solenoid or piezo valves and connected hydraulically to the common rail and electrically to the ECU of the engine management system.

6 An electronic control unit (ECU), which serves the two-fold purpose of controlling the pressure of fuel oil in the common rail through the pressure sensor and limiting valve; and signalling the opening and closing of the solenoid or piezo activated fuel injectors to perform the injection process. It can also provide an interface with other vehicle systems.

Advantages and disadvantages of common rail systems

From an installation point of view, common rail fuel injection offers a more compact and less complicated system as compared to other modes of fuel injection. Functionally, since both metering and timing of the fuel injection are performed independent of engine speed, their electronic control can be programmed for optimum engine performance and reduced emissions. Furthermore, each injection event can be multi-phased by virtue of the fuel injectors being activated by fast-response solenoid or piezo valves. This allows minute quantities of fuel oil to be advantageously injected immediately before and after the main injection. The former or 'pilot' injection provides a smoother start to the combustion process, thereby reducing both noise and emission levels, particularly nitrogen oxide (NOx); while the latter or 'post' injection contributes to lower particulate matter (PM) emissions.

Last but not least, a common rail system can deliver peak injection pressures in the region of 160 MN/m^2 (23 000 lbf/in^2). These very high pressures, although not quite so high as may be attained with unit fuel injection, are now required for efficient

atomization of the injected fuel oil, to achieve ultra low emissions and further improve fuel economy. A possible disadvantage of the common rail fuel injection system is that the fuel injectors have continually to seal against these very high pressures, whereas in other systems they are subject only to intermittent pressurization just prior to fuel injection.

Fuel injectors for common rail systems

As mentioned earlier the injectors used with common rail fuel injection systems are of either the solenoid or, more recently, the piezo valve types. Both types of injector operate on a similar principle. Between injections fuel at common rail pressure acts above and below the nozzle needle valve, which remains spring-loaded onto its seating and thereby cutting off fuel delivery. When fuel injection is signalled by the system electronic control unit, a release valve is activated which allows the pressure acting above the needle valve to be reduced via a restricted passageway. Since fuel at high common rail pressure is still acting below the shoulder of the needle valve, a pressure differential is created that overcomes the spring loading on the needle valve. It is therefore compelled to lift from its seating in the injector nozzle, so that fuel at high pressure is forced through the nozzle holes and sprayed into the cylinder combustion chamber.

The difference in construction between the two types of injector is concerned with the method used to activate the release valve. In the case of the solenoid valve type (Figure 8.47a), the spring controlled release valve also acts as an armature. When the solenoid coil situated above it becomes electrically energized, the armature is attracted towards the coil thereby allowing the release valve to open upwards against its spring loading. The injection process can then take place as previously described. For this to occur in the piezo valve type (Figure 8.47b), the spring controlled release valve is opened downwards, via a coupler piston. This acts against the underside of a piezo ceramic element, which extends in length when electrically energized. (It may be recalled that the converse effect occurs when a piezo electric material is subject to compression, as explained in Section 4.5 concerning an engine oil pressure indicator.)

The piezo type of fuel injector has become increasingly adopted in modern practice. It was introduced in 2002, following development by the German Siemens and Bosch companies. As compared to the traditional solenoid fuel injector, the advantages generally attributed to the piezo type may be summarized as follows:

1 Much faster response to signals from the electronic management system for better overall engine performance.
2 Can operate under higher pressure for more rapid injection and better fuel atomization.
3 More precise control over multiple injections for each cycle to improve fuel economy and reduce emissions.
4 Increased number of pilot injections before main injection for quieter and smoother combustion.

A further development by the Bosch company is to incorporate a stepped-piston intensifier in each fuel injector, which can significantly amplify the common rail pressure. Hence, the even better atomization of the fuel spray promotes more complete combustion, to the benefit of fuel economy and reduced particulate emissions.

8.12   COLD STARTING DEVICES

The need for cold starting devices

Although a diesel engine is inherently more difficult to start from cold than its petrol engine counterpart, various measures can be taken in practice to minimize the problem. The basic reason why a diesel engine should pose this problem at all is that even under normal air temperature conditions, there is a greater loss of heat from the more highly compressed air for combustion to the cylinder and combustion chamber walls. This is, of course, the effect of the physical principle that the greater their difference in temperature the more rapidly is heat transferred from a hotter to a colder region. Therefore if the final temperature of the compressed air for combustion remains below the self-ignition temperature of the injected fuel oil, the engine will not start. Furthermore, the situation can be aggravated by too slow a cranking speed for starting, which only serves to encourage heat loss in the manner described. The fact that a diesel engine may eventually start with prolonged slow cranking is explained by the engine warming itself up by its own heat loss!

Excess fuel device

A basic means of improving the cold starting characteristics of a diesel engine is that known as an excess fuel device. This can be conveniently incorporated in an in-line fuel injection pump and simply comprises a spring-loaded plunger (Figures 8.20 and 8.23) that pushes the maximum stop lever clear of the path of the stop fork for the pumping element's control rod. It thus permits overriding of the maximum fuel stop by the control rod, so that during starting the pump can deliver a quantity of fuel oil in excess of the normal maximum delivery. Once the engine starts, the spring-loaded excess fuel device is automatically tripped out of action as the governor exerts its control.

Operation of the excess fuel device in the first instance usually must be performed at the pump itself, because it is undesirable (and in some countries, including Britain, illegal) to

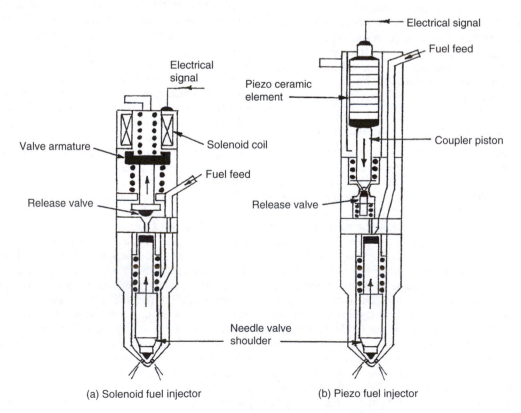

(a) Solenoid fuel injector   (b) Piezo fuel injector

Figure 8.47   Schematic arrangements of solenoid and piezo fuel injectors for common rail systems

allow excess fuel oil to be available while the vehicle is actually being driven. Perhaps it should be explained here that in past years, when an excess fuel device could be operated from the driver's cab, there was a temptation for it to be used as a means of increasing engine power on hills, albeit accompanied by an offensive cloud of dense exhaust smoke. In this context, the following are the provisions of the Motor Vehicles (Construction and Use) Regulations 1978, as amended, relating to emissions:

1 Regulation 33 requires every motor vehicle to be so constructed that no avoidable smoke or visible vapour is emitted therefrom, and Regulation 109 makes it an offence to use a vehicle which emits substances likely to cause damage to property or injury to person.
2 Regulation 34 requires the excess fuel device (for starting) on a diesel engined vehicle to be fitted so that it cannot be used while the vehicle to be fitted so that it cannot be used while the vehicle is in motion if its use increases the smoke emitted. Regulation 111 makes it an offence to use the excess fuel device while a vehicle is in motion and requires the device to be maintained in proper working order.

Other cold start devices

Glow plugs

Also known as heater plugs, these are an accepted starting aid for indirect-injection diesel engines, which tend to be more difficult to start than direct-injection ones. This is because of the increased heat losses from the divided combustion chambers of the former, owing to their greater ratio of surface area to volume. A glow plug acts less as an extraneous source of heat to raise the temperature of the compressed air for combustion than as an incandescent point that will ignite any fuel oil striking against it.

A glow plug is of similar basic construction to a sparking plug, but differs in featuring a resistance wire filament instead of a pair of electrodes at its combustion chamber end (Figure 8.48). Each cylinder glow plug receives energy from the vehicle battery system, via a driver-operated 'ignition' switch. During the preheating period the wire filament reaches about 900 to 1000°C but, once the engine has started and preheating is no longer required, the temperature of the filament reduces to that of the combustion heat.

Decompression control

This type of starting aid really dates back to the pre-electric starter era, but may still be retained so that emergency hand cranking of the engine is practicable. It can also be utilized to facilitate turning of the engine during adjustment of the valve tappets. In a later version of the well-known Gardner system of decompression lift for the inlet valves, the decompression levers at the rear of each cylinder head operate small shafts in which are fitted adjustable screws with locknuts. These are located beneath and towards the push-rod ends of each inlet valve rocker. By turning the decompression levers to their upright position, the adjustable screws of the shafts come into contact with the undersides of the

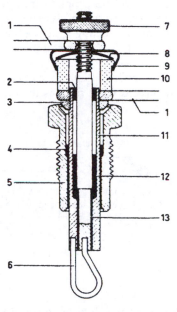

Figure 8.48 Cross-section through a diesel glow plug (*Mercedes-Benz*)

1 contact bar	8 convex washer
2 plastics insulator	9 metal cap on connection insulator
3 collar (on outer electrode)	10 connection insulator
4 insulator	11 outer electrode
5 plug body	12 insulator
6 filament	13 centre electrode
7 knurled nut	

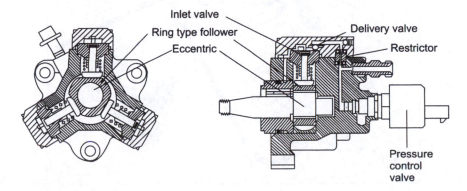

Figure 8.49 High-pressure plunger pump for common rail fuel injection

rockers such as to prevent the inlet valves from reseating, actually holding them open to an extent of 0.50 mm (0.02 in). It thus allows the engine to be cranked smartly at the starting handle, and when maximum speed is attained the decompression levers are released. The kinetic energy stored in the rotating flywheel will then crank the engine against its normal compression pressures for starting purposes. Decompression control is only suitable for application to direct-injection diesel engines, which as earlier mentioned are less difficult to start.

Intake manifold heater
Again intended for application to direct-injection diesel engines, this device heats the intake air by means of what is called a flame glow plug, which is mounted in the intake manifold and burns fuel oil supplied from the engine fuel system. In one version, the fuel oil is metered by a ball valve that is released from its seating as the valve body expands under the influence of the heater coil action (Figure 8.50). The latter receives its energy from the vehicle battery system as in the case of combustion chamber glow plugs. Fuel oil vaporized in the heated valve body is then ignited within a

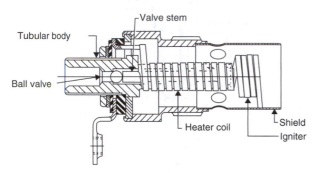

Figure 8.50   Intake manifold heater (*Lucas CAV*)

shielded portion by a glowing filament, thus providing a combustible mixture with the intake air. Cold starting devices of this type not only facilitate starting where the ambient air temperature is below 10°C, but also quickly allow smokeless and smooth engine operation, because of after-flaming at the glow plug during the initial idling period.

Starting fluids
It is not difficult to produce a fluid which, even at very low ambient temperatures, will ignite readily under compression in the cylinder of a diesel engine. A number of common chemicals have been found effective for this purpose, but they can tend to be very violent and uncontrollable in operation, which has been known to cause mechanical damage to an engine. Engine starting fluids are often loosely referred to as ether products because in general they are based on sulphuric ether; indeed, this chemical accounts in some cases for over 90 per cent of the total mixture.

A starting fluid is introduced into the intake system of a diesel engine in the form of an ether spray, either by a hand-held aerosol can which involves lifting the bonnet and spraying into the air cleaner inlet, or by an engine-mounted or cab-operated carburetted system that includes a hand pump, reservoir, tubing and spraying nozzles for the intake manifold. A widely approved example of the latter system is that known as Start Pilot Viso F (Figure 8.51). Although the starting fluid used in it contains sulphuric ether, it does so in relatively small proportion and in a modified form, and solely to act as a pilot to the main combustion process – hence the name Start Pilot. Other constituents of Start Pilot fluid, which also contains an upper cylinder lubricant and an anticorrosion product, cover a wide range of self-ignition temperatures so that each is burned progressively to provide a smooth speed of flame, in place of the total and violent conflagration normally associated with ether. The use of an ether spray carburetted directly into the intake manifold can provide reliable starting down to temperatures as low as −40°C.

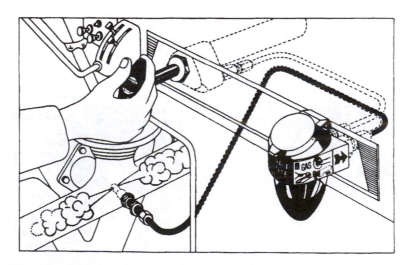

Figure 8.51   Intake manifold starting fluid spray (*Start Pilot*)

9 Forced induction

9.1 NATURAL ASPIRATION AND FORCED INDUCTION

General considerations

Both petrol and diesel engines for motor vehicles may be further classified according to the manner in which their induction process is accomplished, namely natural aspiration or forced induction.

Natural aspiration
So far consideration has been given only to those engines that fall into this particular classification, because their induction process relies solely on the pressure difference that exists between the retreating piston and the air intake arrangements. The density of charge that ultimately fills the cylinder owing to the pumping action and velocity of the piston, therefore, depends on how freely the engine is allowed to breathe. This in turn is related to the design characteristics of the intake system considered in its entirety. The greater the density of charge that fills the cylinder, the higher will be the mean effective pressure developed by the engine, and therefore the greater its torque and power output.

Forced induction
To increase further the mean effective pressure developed by a given engine, it becomes necessary to resort to *pressure charging* of the cylinders. By this means the charge is forced to enter the cylinder at a pressure substantially above that of the atmosphere, although the mean gas velocities through the intake port remain unchanged for the same engine speeds. Since the cylinder ultimately becomes filled with a charge of greater density than is possible with natural aspiration, an increase in torque and power can be obtained from an engine of given displacement. With forced induction or pressure charging it is important to recognize that it is not the accompanying rise in pressure of the charge, but the greater density of the charge with its increased amount of oxygen, which accounts for the improved performance of the engine.

Forced induction for petrol engines

With forced induction it is necessary to avoid any tendency towards excessive cylinder pressures and temperatures, which could lead to destructive detonation. Fortunately, a small reduction in the compression ratio of a petrol engine allows it to tolerate a relatively greater degree of pressure charging without incurring higher maximum cylinder pressures. Any reduction in the engine compression ratio nevertheless implies a lower thermal efficiency and hence an increased fuel consumption. Although forced induction systems were once confined to certain high-performance sports

car and, of course, racing car engines solely to increase their maximum power, there are now additional reasons to account for their modern revival, as mentioned later.

Forced induction for diesel engines

The diesel engine is actually a more deserving case for forced induction than the petrol engine. This is because of the much shorter period available for mixing of the air and fuel, which results in only part of the oxygen in the air received by natural aspiration being utilized for combustion. Forced induction can therefore overcome this shortage of oxygen by allowing the diesel engine to breathe an artificially dense atmosphere, thereby increasing power output. Furthermore, the greater density of the air charge also tends to reduce the delay period of the combustion process, so that the rate of pressure rise is better controlled for smoother and quieter operation.

Modern advantages of forced induction

For both petrol and diesel engined vehicles the main benefits to be derived from forced induction, which in modern practice generally means turbocharging as described later, are that an increased torque and higher power output can be achieved for a given swept volume of engine compared with natural aspiration. These advantages can be exploited by a vehicle manufacturer either to uprate an existing naturally aspirated engine in small or large-volume production, or to equal the performance of an existing naturally aspirated engine with a new engine of smaller swept volume, so that engine installation space, total vehicle weight and fuel consumption can all be reduced. Other advantages of the modern forced induction system are those of compensating for loss of power resulting from stricter exhaust emission controls, and contributing to a quieter exhaust system.

9.2 METHODS OF PRESSURE CHARGING

Inertia ramcharging

This represents a modest form of pressure charging in which no resort is made to the use of a separate compressor, driven either directly or indirectly from the engine. It does in fact exploit a pressure wave effect, which accompanies the oscillating motion of the charge in the intake system due to the closing and the opening of the inlet valve. By suitable adjustment of the length of intake pipe, it can be so contrived that the pressure wave is approaching its crest just as the inlet valve is opening.

Better filling of the cylinder is therefore obtained from this rise in charge pressure at the valve port. However, the pressure wave effect can only be beneficial at certain engine

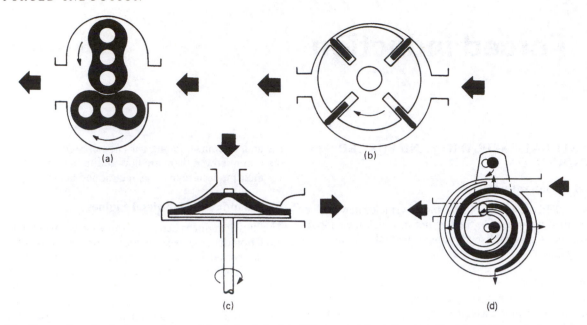

Figure 9.1   Types of mechanical supercharger: (a) rotating lobe (b) sliding vane (c) rotating impeller (d) orbiting spiral

speeds, since it is essentially a matter of resonance between these two factors so the overall gains available are modest. The more recent advent of variable-length intake systems (Section 7.2) has nevertheless increased the useful range of engine speeds, albeit in a stepped rather than a truly variable manner, over which the beneficial effects of inertia ram-charging can be exploited by engine designers.

Mechanical supercharging

This is accomplished by means of an engine-driven compressor. It represents the earliest approach to pressure charging, its potential having been explored before the end of the 19th century by such pioneer designers of the internal combustion engine as Gottlieb Daimler and Wilhelm Maybach in Germany, and at the beginning of the 20th century by Louis Renault in France and Lee Chadwick in America. However, it was not until the early 1920s that firms like Mercedes began to adopt mechanical superchargers for high-performance sports cars. Since World War II they have enjoyed only a limited popularity as conversion rather than original equipment.

For motor vehicle application there have been three long-established basic types of mechanical supercharger, which are positively driven through either gears, chains or belts from the engine crankshaft. These are classified conveniently as rotating lobe, sliding vane and rotating impeller superchargers.

Rotating lobe

This type of supercharger is often referred to as a Roots blower, so named after J.D. Roots, who in 1866 first devised a rotary displacement blower that operated on the same principle for mine ventilation purposes in America. It comprises a meshing pair of either two or three lobed rotors, which are mounted on parallel shafts and rotate in opposite directions on ball bearing races (Figure 9.1a). To maintain the correct relative motion of the rotors, a pair of timing gears is located at one end of the shafts. These gears are an essential feature of this type of supercharger, because the rotating lobes must not be allowed to contact each other, or for that matter the casing in which they are confined. The volumetric efficiency therefore depends on maintaining the smallest practicable working clearances between the pumping elements.

In operation either air or an air and petrol mixture, according to whether the supercharger is mounted upstream or downstream of the carburettor, is simply drawn into the space between the rotating lobes and casing and then carried from the inlet to the outlet. Originally this type of supercharger was mounted upstream of the carburettor, which meant that the carburettor became pressurized and therefore needed to be modified to equalize the air pressure acting on the fuel in the float chamber. It was the once-famous British Sunbeam company who in 1924 first fitted the supercharger downstream of the carburettor, which not only overcame the disadvantage mentioned, but also realized two other advantages. These were concerned with the more thorough mixing of the air and fuel as it passed through the supercharger before reaching the engine cylinders, and the beneficial cooling effect on both the supercharger and the mixture by the heat absorbed as the fuel vaporized.

Since very little compression of the mixture occurs actually within this type of supercharger, the delivery pressure depends upon the back-pressure in the intake system. Except for some falling-off in performance at very low and very high speeds of rotation, its output is roughly proportional to the speed at which it is being driven by the engine. A pressure relief or blow-off valve protects the installation against backfiring and at the same time serves to limit the degree of pressure charging. This is typically in the region of $56 \, \mathrm{kN/m^2}$ ($8 \, \mathrm{lbf/in^2}$) above atmospheric pressure, although much higher boost pressures (to use an aviation term) were once common in motor racing. The rotating lobe supercharger was once used to a limited extent on four-stroke diesel engines, but is more

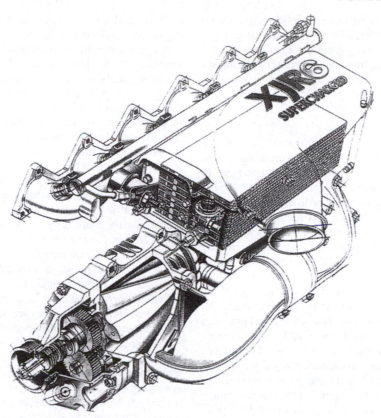

Figure 9.2 A modern supercharger installation (*Jaguar Cars*)

commonly associated with two-stroke diesel engine applications, where it is in any event needed for scavenging purposes (as described in Section 2.2).

The recent modest revival of interest in mechanical supercharging is largely concerned with certain advantages possessed by the rotating lobe or Roots type of supercharger, as compared with the more widely used exhaust turbocharger described later. In particular it can provide a crisper throttle response from low engine speeds, which is especially difficult to attain with high-performance engines unless resort is made to twin turbochargers and then to the likely detriment of part-throttle fuel economy and exhaust emissions. The Roots type of supercharger also creates fewer difficulties in terms of high-speed balancing, high-temperature operation and bulky installation, particularly the latter when twin turbochargers may otherwise be required. On the debit side, it is inherently prone to a pulsating air flow that arises from the periodic discharges of displaced air, which can result in noisy operation. However, it was discovered long ago that by adopting such measures as increasing the number of lobes on each rotor, incorporating helical ports in the casing and providing carefully profiled rotor lobes of helical form, the noise could be reduced to acceptable levels. Hence in modern installations as typified by the Ford Thunderbird and Jaguar XJR high-performance cars, the Eaton supercharger of the Roots type is furnished with three instead of the once traditional two lobes per rotor, each lobe having an involute profile and a helical twist along its length. These features contribute not only to reducing the noise level, but also to increasing the volumetric efficiency of the system. An interesting feature of the Jaguar installation is the compact manner in which the air-to-water heat exchanger of the charge cooling system has been integrated with the intake manifold (Figure 9.2). In these more recent installations the supercharger does, of course, operate in conjunction with a modern fuel injection system.

Sliding vane

Various proprietary superchargers of the sliding vane type were once produced to rival the rotating lobe or Roots-type supercharger. The sliding vane supercharger is essentially another form of positive displacement compressor, but differs from the rotating lobe construction in that it comprises a cylindrical rotor along which a series of radial slots are cut to accommodate sliding vanes (Figure 9.1b). This rotor is also mounted eccentrically in its casing, such that at one point there is the smallest practicable working clearance between them.

In operation the effective space between each pair of adjacent vanes alternately expands and contracts as the rotor turns. The displacement of either air or mixture through the supercharger then occurs, because the inlet port is located where the inter-vane spaces increase in volume and the delivery port where they decrease in volume. Centrifugal force usually suffices to maintain the vanes in their outward position. Ingenious means have sometimes been adopted to eliminate the friction losses and hence the lubrication requirement that results from the vanes being pressed against the casing wall in this manner. These include pivoting the vanes from a centre spindle, so that a very small clearance exists between their tips and the vane tracks. In contrast to its rotating lobe counterpart, the delivery pressure of the sliding vane type of supercharger derives from the compression of air or mixture actually within it, so that its performance characteristics can be superior if perhaps not always its balance and durability.

Rotating impeller

This type of supercharger differs significantly from both the rotating lobe and sliding vane superchargers in being a non-positive displacement compressor of the centrifugal type. In fact it is often referred to as a centrifugal supercharger. It basically comprises a rotating member with radially disposed blades, termed an impeller, which turns within a casing that contains a series of fixed guide vanes acting as a diffuser (Figure 9.1c). For past applications of the centrifugal supercharger the impeller was driven at about five times engine speed.

In operation either air or mixture enters near the eye of the impeller, from whence it is accelerated along the blades by centrifugal action. It then leaves the tips of the impeller blades at a high tangential velocity and is received by the stationary diffuser vanes, which are suitably angled to avoid turbulence effects. The diffusion process so created slows down the gas flow, as a result of which its kinetic energy is converted into pressure energy at the delivery port of the supercharger. Hence, the mixture is delivered to the cylinders under a positive pressure, which with this type of compressor rises substantially as the square of its rotational speed. In other words, doubling the speed of the centrifugal supercharger roughly quadruples its boost pressure.

It is this particular performance characteristic that explains its lack of popularity for automotive applications, where the driver demand for good acceleration favours a torque curve which, if anything, rises with decreasing rather than increasing engine speed. However, the centrifugal supercharger enjoyed a vogue of popularity in America during the 1930s, where it was used on both passenger car and racing car engines, the first popular application being on the once well-known Graham-Paige car in 1934. This type of supercharger was originally pioneered by Lee Chadwick in 1907, who is credited with having humorously observed that if you removed the drive belt the engine seemed to have as good if not better power than without the charger! A further point of interest is that the centrifugal supercharger was widely used during the era of the piston aero engine. In this case the main reason for pressure charging was to maintain sea-level power output through a given altitude range, since atmospheric pressure decreases with increasing altitude.

Orbiting spiral

This unusual type of mechanical supercharger was introduced by Volkswagen in 1986 after a long period of development, and represents a compromise between the ruggedness of the rotating lobe and the efficiency of the sliding vane types. Designated the G-Lader supercharger by Volkswagen, it is so named after the G shape of its compression channels (Figure 9.1d). The basic concept of an orbiting spiral compressor dates back to 1905, when it was devised by a Frenchman,

L. Creux, and was then considered as a possible form of steam engine, although nothing further came of the idea.

In construction the internal moving element of the G-Lader supercharger comprises a disc-shaped displacer, which has identical spirals or scrolls projecting from each side and is made from a light alloy to reduce inertia. Interleaved with these moving scrolls is a pair of fixed scrolls, each scroll being supported from an end wall of the pumping chamber (Figure 9.3). Instead of simply rotating, the displacer orbits in a hula-hoop manner to create a pumping action between the two pairs of scrolls. To confer this orbiting motion the larger central eye of the displacer is driven by an eccentric throw on the main shaft of the supercharger, while a smaller offset eye is guided by an eccentric throw on an auxiliary shaft. The two shafts are synchronously coupled by a toothed-belt drive with the main shaft being driven from the engine.

Although the interleaved scrolls closely approach one another in the radial sense during the orbiting motion of the displacer, they do not actually touch, and so in this respect the G-Lader supercharger may be compared to the rotating lobe type where similarly the lobes are not in contact. An axial sealing requirement arises at the edges of the scrolls and is satisfied by the inclusion of strip-type seals. Both the main and auxiliary shafts are counterbalanced for their eccentric throws and supported in rolling bearings, the bearings used at the displacer eyes being of the needle roller type. The G-Lader supercharger is connected to the engine lubrication system and, since the charge air must not be contaminated with oil, the bearings on the main shaft are provided with efficient radial seals.

When the supercharger is being driven and the outermost parts of the scrolls separate to create a depression, an air charge is induced to enter it through a peripheral inlet port. This air then becomes trapped and squeezed as the scrolls approach one another to form narrowing compression channels. Finally, the air is displaced under pressure through a cluster of central ports at the outlet of the supercharger, the actual output being related to the width employed for the scrolls. The particular

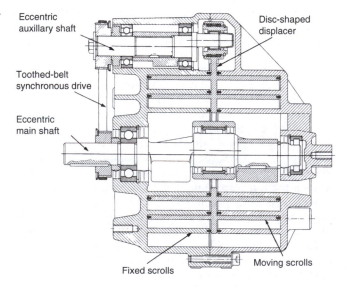

Figure 9.3   Cross-section through G-Lader supercharger showing bearing arrangements (*FAG Kugelfischer*)

advantages associated with this type of mechanical super-charger are those of quick response and low noise level.

Exhaust turbocharging

It was early realized that the amount of heat energy contained in the exhaust gases could be greater than the fuel energy utilized to develop actual engine power. This led to the concept of the exhaust turbocharger, which comprises a centrifugal compressor driven solely by an exhaust gas turbine, there being no mechanical drive from the engine crankshaft. Two notable pioneers of exhaust turbocharging in the period immediately following World War I were A. Rateau in France and A. Buchi in Switzerland, their respective fields of endeavour being in aero petrol engines and marine diesel engines. However, it was not until 1945 that a turbocharger was fitted to a production heavy vehicle diesel engine by the Swiss firm of Saurer, and not until 1962 that one was fitted to a production passenger car engine by General Motors in America. Furthermore, the present-day widespread use of the turbocharger on both diesel and petrol engines only really began in the early 1970s, when comparatively smaller, more efficient and less costly turbochargers first became available.

The construction and operating principle of a centrifugal compressor has already been dealt with under the heading of the rotating impeller type of mechanical supercharger. It therefore remains to describe the other part of the turbocharger, which is the exhaust-gas-driven turbine. For automotive engines the turbine is usually of the radial flow type, which is in contrast to the axial flow turbine found in the larger exhaust turbochargers used on diesel engines for marine, traction and industrial applications. The shape of the passages leading into the turbine are such that the exhaust gases enter tangentially from the periphery of the rotor and then travel radially inwards along its specially profiled blades (Figure 9.4). One advantage of this inward direction of gas flow is that an increasing centrifugal back-pressure is created, which gives the turbine more stable operating characteristics and minimizes any tendency to overspeeding. Another advantage is that of allowing better cooling for the bearing between the compressor and the turbine. This feature of operation, together with a suitable choice of turbine rotor material,

typically a nickel-based alloy that is highly resistant to the effects of heat and corrosion, is an important consideration where temperatures up to 1000°C may be encountered.

The use of an aluminium alloy for the impeller of the centrifugal compressor minimizes the rotary inertia of the turbocharger. Again this is an important consideration, together with precision balancing, when related to the extremely high speeds of rotation attained by the turbine and compressor assembly, which may reach 100 000 rev/min. The greater the rotary inertia of these parts, the longer it will take for the turbocharger to catch up with the engine when the throttle is suddenly opened, to the detriment of acceleration. In more recent Japanese practice, the introduction of a ceramic material for the turbine rotor has significantly reduced the rotary inertia of the turbocharger.

The turbocharger bearings are usually of the plain sleeve type in view of the extremely high-speed operation involved. They are allowed to float within their housing and thus provide an anti-vibration mounting for the shaft that carries the impeller and rotor. Sealing of the bearings is accomplished by mechanical seals of either the labyrinth, piston ring or face type, which are the only sort compatible with the temperatures, pressures and rotational speeds encountered. The lubrication requirements of the turbocharger on the hot side of the engine are quite different from those of the supercharger on the cold side. Whereas a supercharger of, say, the rotating lobe or Roots type can be provided with a self-contained system of lubrication for its rolling bearings and timing gears, the turbocharger requires a continuous flow of clean engine oil under pressure with gravity return to the sump (Figure 9.5). This supply of oil not only lubricates and cools the floating bearings, but also cushions them within their housing. To enhance the reliability of the turbocharger its turbine bearing housing may embody water cooling, again as in more recent Japanese practice.

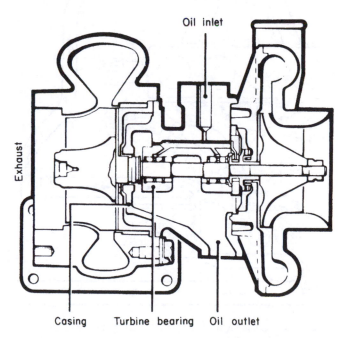

Figure 9.5 Lubrication of turbocharger bearings (*Smallman Lubricants*)

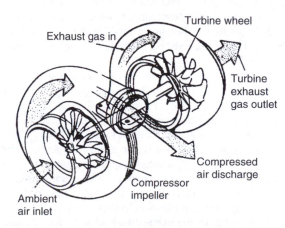

Figure 9.4 Inlet and exhaust gas flow through a turbocharger

Since the temperatures reached in the turbocharger bearings are higher than those encountered in the actual engine, certain precautions are normally advised when starting and stopping a turbocharged engine. For example, a diesel engine manufacturer will typically advise that the engine should be allowed to idle for 3 to 5 minutes before shutting it down, so that the circulating oil will carry away sufficient heat for the oil remaining in the bearings not to reach a charring temperature, which could lead to seizure. Similarly, the turbocharger bearings should be protected during the start-up by not opening the throttle or accelerating above 1000 rev/min until normal engine idle oil pressure registers on the gauge.

Clearly some form of control system is necessary for the automotive turbocharger, so that it can maintain constant the required degree of pressure charging over a suitable range of engine speeds. Otherwise, the output characteristics of its centrifugal compressor would place it at a similar disadvantage to that of a mechanical supercharger of the rotating impeller type. Of the various forms of control that can be applied to a turbocharger, the conventional method is to incorporate a turbine bypass valve. This is now better known as an exhaust wastegate, and it is usually controlled by intake manifold pressure. It basically comprises a spring-loaded poppet valve, which seals against a seating in the entry passage to the exhaust turbine. However, the wastegate is not a simple exhaust blow-off valve that could be liable to flutter, because interposed between the valve spring and valve is a diaphragm that is acted upon by intake manifold pressure (Figure 9.6). The effect of this is such that when the intake manifold boost pressure builds up to about $35 \, kN/m^2$ (5 lbf/in^2), the diaphragm exerts an opposing force that assists exhaust gas pressure to overcome the valve spring load, thereby allowing the opening of the wastegate. When this occurs the excess exhaust gases bypass the turbine rotor and are transferred directly to the silencer system (Figure 9.6).

A refinement of this system is to provide the driver with a dual-mode selector switch for controlling the turbocharger: either 'normal', where the wastegate remains closed until a predetermined boost pressure is reached, thereby directing all the exhaust gas to drive the turbine; or 'low', where the wastegate is partially open to release a portion of the exhaust gas, so that boost pressure is reduced. Partial opening of the wastegate is accomplished via the electronic control unit of the engine management system, which signals a solenoid valve to allow an earlier application of manifold pressure to the wastegate diaphragm.

Turbocharger lag

During normal driving conditions the exhaust gases do not contain sufficient energy to drive the turbine rotor at turbocharging speeds. When the accelerator is depressed to signal an increased demand on the engine, then as with a naturally aspirated engine more air and fuel enters the cylinders and is burned. This in turn increases the amount of exhaust gases passing through the turbine, thereby causing both turbine and compressor to speed up and the latter to provide the required pressure charging, albeit with some delay. An inherent disadvantage of the turbocharger is that of delayed throttle response, generally referred to as throttle lag or turbo lag, which is caused by the inertia of the turbocharger rotating parts (mentioned earlier) and to a lesser extent by the longer and more devious flow path for the gases.

One approach to reducing turbo lag is to shorten the column of gas flowing between the engine exhaust valves and the turbine, which involves using a common casting for the turbocharger housing and the exhaust manifold. A further advantage of this arrangement as pioneered by Opel in Germany is, of course, that it produces a very compact installation. Turbo lag is similarly reduced by shortening the column of pressurized air flowing between the centrifugal compressor of the turbocharger and the engine inlet valves. This may be sought by using an air-to-water rather than an air-to-air heat exchanger for charge cooling, because in the latter arrangement the charge air has to be routed to and from a front mounted heat exchanger sited in the vehicle air flow, whereas an air-to-water heat exchanger can be sited more directly between the turbocharger and the intake manifold.

The effect of turbo lag has also to be taken into account with the turbocharged diesel engine, because if the engine is rapidly accelerated the immediate increase in the amount of fuel injected into the cylinders will initially not be matched by a boosted supply of air. For this reason most turbocharged diesel engines are equipped with a boost control or smoke limiter, which comprises a spring-loaded diaphragm that is responsive to intake manifold pressure. It acts via a bell-crank linkage to prevent full movement of the control rod in an in-line fuel injection pump during initial acceleration.

On the credit side, the exhaust turbocharger has the advantage that it utilizes some of the energy that would otherwise be wasted in the outgoing exhaust gases, and is therefore able to

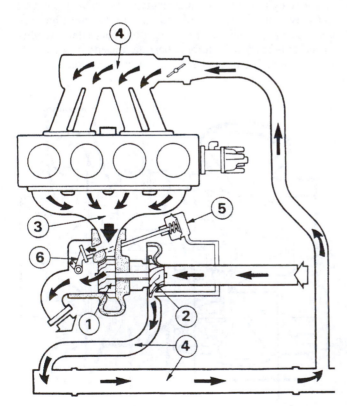

Figure 9.6   Basic layout of a turbocharger system (*SAAB*)

1 turbocharger turbine	4 intake air
2 turbocharger compressor	5 wastegate actuator
3 exhaust gases	6 wastegate valve

avoid the complication of a mechanical drive from the engine crankshaft. This does not mean, of course, that one is getting something for nothing, because the turbocharger must inevitably impose a certain backpressure in the exhaust system, but the loss in engine power is considerably less than that absorbed by a mechanically driven supercharger. Furthermore, the turbocharged engine can be quieter running than a naturally aspirated one, because the turbine itself tends to dampen the pressure pulsations in the flow of exhaust gases. In contrast, one of the main drawbacks of mechanically driven superchargers was often their noisy operation. As compared with a mechanically driven supercharger of the centrifugal type, the turbocharger can be more readily controlled to maintain a substantially constant boost pressure over a fairly wide range of engine speeds.

In 1997 a pioneering development in turbocharging by SAAB has been the 'assymetric' turbocharger installed on their 9–5 3.0 litre V6 engine. With this arrangement the exhaust from one bank of three cylinders is utilized to deliver a charge to all six cylinders, so that the engine can still develop the required power but with a reduced weight penalty.

Variable-geometry turbochargers

To improve throttle response in passenger car petrol engines, another feature of Japanese practice has been the introduction of the variable-geometry turbocharger. This incorporates a set of individually pivoted guide vanes around the periphery of the turbine wheel which move together to vary the gas flow area into the turbocharger turbine. By closing down the guide vanes at low engine speeds the expansion ratio across the turbine is increased giving better response and a higher air flow enabling more low speed torque to be developed (Figure 9.7).

More recently variable-geometry turbochargers have also been introduced for heavy vehicle diesel engines. Their construction differs from that described above by the substitution of the pivoting guide vanes for sliding ones encircling the turbine wheel. An established example of this practice is the Holset HX40V turbocharger.

The principle of this type of turbocharger is shown in Figure 9.8. A nozzle ring with fixed vanes is positioned in

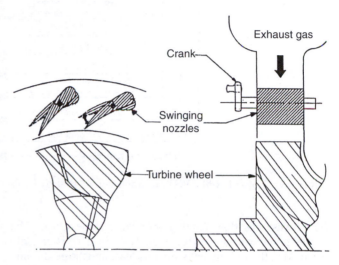

Figure 9.7 Variable-geometry turbocharger for petrol engines (*Holset Engineering*)

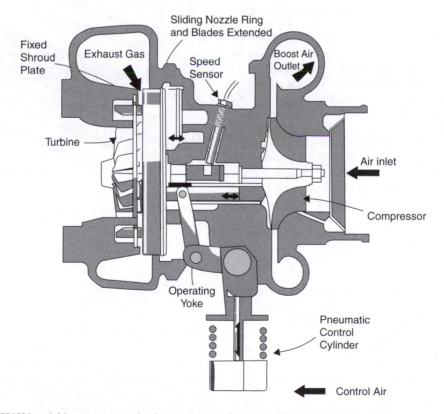

Figure 9.8 Holset HX40V variable-geometry turbocharger for diesel engines (*Holset Engineering*)

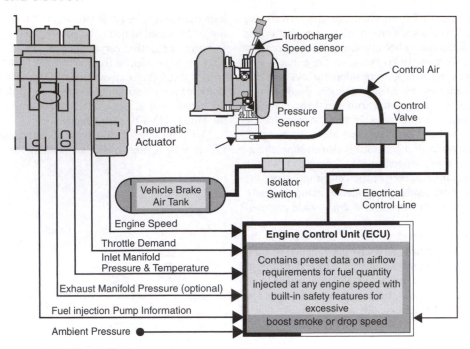

Figure 9.9 Holset HX40V VGT electronic control strategy (*Holset Engineering*)

the exhaust gas flow leading into the turbine wheel. The nozzle guide vanes pass through slots in a register plate mounted in the turbine housing, this plate and the guide vane carrier define the axial width of the gas passage through the nozzle. By moving the nozzle ring axially, the passage width can be varied from a narrow gap to the maximum width, thus controlling the flow area. The nozzle ring movement is provided by a pneumatic actuator, which is supplied with high pressure air from the brake air supply on the vehicle. This air supply is modulated by an electronically controlled valve in order to actuate the nozzle ring movement as required.

In order to behave in the most effective way the control of the VGT has to be integrated into the engine electronic controller. Some of the inputs from the engine and from sensors incorporated in the VGT turbocharger which are necessary for good control under all engine conditions are shown diagrammatically in Figure 9.9. The effect of the variable geometry at small nozzle openings is to allow generation of higher levels of boost pressure at lower engine speeds than would be possible with a fixed geometry turbocharger, and thus the ability to extend the torque curve over a wider speed range. Not only that, but the efficiency of the turbocharger is maintained at large nozzle openings corresponding to higher engine speeds which enables good specific fuel consumption to be achieved over a wide range.

Transient engine performance is also greatly enhanced by controlling the VGT to close during a sudden load demand. This provides additional boost air and enables a higher fuelling rate and torque response without excess smoke.

A further use of VGT is to enhance engine decompression braking systems. Use of these systems is becoming widespread in heavy duty diesel engines to supplement the vehicle service brakes (Section 28). Using the VGT in its closed position to produce a high exhaust manifold pressure and at

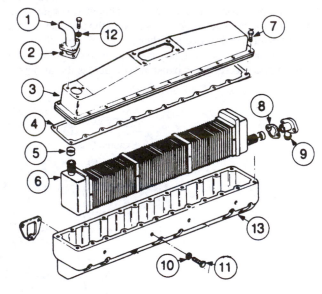

Figure 9.10 Construction of an air-to-water charge cooler (*Cummins*)

1 water outlet connection
2 gasket
3 cover
4 cover gasket
5 O-ring
6 core
7 capscrew
8 inlet connection gasket

9 inlet connection
10 hardened washer
11 cross-bolts
12 copper washer
13 housing

the same time a high air flow through the engine results in braking powers which can exceed the rated power of the engine. Again, control of the VGT under these conditions has to be through the engine ECU.

Charge cooling

It will be recalled that the basic aim of forced induction is to supply the engine with a greater density of charge, which implies that the cooler the charge the better the engine will like it. In practice, though, the charge must inevitably become heated and expand (that is, become less dense) following its passage through whatever type of pressure charger is used. It is therefore to reduce the temperature of the pressurized charge from in the region of 120°C to about one-half this value that a charge cooler may be employed. This takes the form of either an air-to-water or an air-to-air heat exchanger, and may be likened in principle to the engine oil coolers described in Section 4.6. An example of the former type as used with a turbocharged diesel engine for heavy vehicles is shown in Figure 9.10. The air-to-air charge cooling system used by SAAB for their turbocharged petrol engined passenger car is of interest, because it features a heat exchanger constructed from a chromated aluminium core with plastics side headers for the air inlet and outlet connections, together with a bypass valve to avoid damage from any blockages (Figure 9.11). In comparison with a conventional copper construction, this type of charge cooler both reduces weight and increases efficiency. The disadvantages of charge cooling are that it makes for a bulkier and more expensive installation of the turbocharger.

Charge cooling is also known as after-cooling and inter-cooling, but strictly speaking the last mentioned should only be applied where there are two stages of pressure charging. For example, an inter-cooler heat exchanger may be located between the turbocharger and the rotary blower of a two-stroke diesel engine.

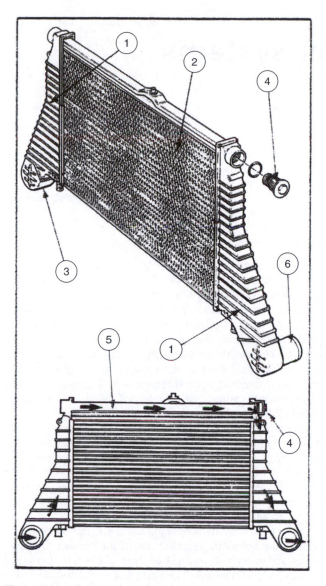

Figure 9.11   Construction of an air-to-air charge cooler (*SAAB*)

1 plastic headers
2 chromated aluminium core
3 inlet from turbocharger compressor
4 bypass valve
5 bypass airflow
6 outlet to engine

10 Ignition and starter systems

10.1 COIL IGNITION EQUIPMENT

General background

To ignite the combustible charge of air and fuel in the cylinder, and thereby initiate the power stroke, some form of ignition system is necessary in the petrol engined motor vehicle. In fact, it is probably no exaggeration to say that the modern high-speed petrol engine would never have been possible were it not for the remarkable efficiency of its ignition equipment.

The earliest source of ignition was, interestingly enough, high-tension electricity, this being used by the Frenchman Etienne Lenoir in 1860 on his famous gas engine, which thus qualifies as the first spark-ignition internal combustion engine. However, the then primitive state of development of high-tension electrical equipment did not make for reliability of this type of ignition system. Various other methods of ignition therefore came into use, notably the self-descriptive hot-tube system, which utilized a red-hot platinum tube to initiate combustion in the cylinder.

Quite naturally, the early automobile engineers attempted to adapt for motor vehicle use the existing stationary engine ignition systems, but they soon ran into difficulties. For example, it proved difficult to keep the lamp alight for heating the hot-tube ignition system once the vehicle was in motion! Furthermore, the timing of the ignition control could not be varied, nor could the same time of firing be obtained in each cylinder. Fortunately, by the time the shortcomings of the early ignition systems became evident, the advances that had been made in the design and construction of high-tension electrical equipment were such that more reliable spark-ignition systems became available.

For many years there were two systems of electrical ignition used for motor vehicle engines. These were the self-contained high-tension magneto and the battery-and-coil systems. The former was pioneered in 1901 by Robert Bosch, who founded the famous firm bearing his name. However, the advent of World War I resulted in much further valuable development work being done on the magneto in Britain, notably by E.A. Watson and also by L. Griffiths, H.W.F. Ireland and J.D. Morgan. The battery-and-coil system was developed in 1912 by Charles F. Kettering, the American engineer who also introduced the electric self-starter and was later president of the world-renowned General Motors' Research Laboratories.

Although it is now a matter of history that the battery-and-coil ignition system was eventually adopted for all petrol engined motor cars and commercial vehicles, it may nevertheless be useful to recall the reasons for this:

1 The intensity of the spark is reasonably consistent regardless of engine speed, and this makes for easy starting.

2 It has few moving parts, thereby reducing possible sources of unreliability.
3 There are no heavy components, if it is accepted that a battery must in any event be carried for other electric services on the vehicle.

Basic requirements of ignition system

An electric spark forms a very convenient means of producing a rapid temperature rise in the engine cylinder, but in practice it requires a very high voltage from the ignition system. For example, with a typical air gap of 0.50 mm (0.020 in) at the sparking plug electrodes, the voltage required across them to produce a spark in the engine cylinder is about 10 000 V (10 kV). The qualifying statement 'in the engine cylinder' is made deliberately, because the voltage requirement would be only a few hundred at normal atmospheric pressure. It thus follows that the voltage requirement of the ignition system is raised as the compression ratio of an engine is increased.

Other factors that raise the voltage requirement include surface deterioration of the sparking plug electrodes, the burning of weak mixtures and cold starting of the engine. The latter is partly explained by the lower temperature of the cylinder gases surrounding the plug electrodes because, unlike a metallic conductor, the electrical resistance of a gas decreases when it is heated.

Apart from actually producing the spark, the next requirement is the not inconsiderable one of the number of sparks necessary in a given interval of time. Take, for example, a four-cylinder (four-stroke) engine where there are two power strokes for every complete revolution of the crankshaft. This means that the ignition system must likewise fire two sparking plugs for every crankshaft revolution. At, say, 4500 engine revolutions per minute (75 per second), the ignition system must therefore produce 150 high-voltage sparks every second. This figure would, of course, be doubled in the case of an eight-cylinder engine running at the same speed.

Furthermore, it is an important requirement of the ignition system that the high-voltage spark must be timed to occur in each cylinder at an optimum point in the engine operating cycle.

The mechanically switched coil ignition system

This once universally used system of ignition provides a useful basis of understanding for later electronically switched systems, since all battery and coil ignition systems share certain features in common. A mechanically switched system comprises the following components:

Battery and charging system This is the source of supply of electrical energy for the low-tension primary circuit of the ignition system.

Ignition switch   The purpose of this is to connect and disconnect the ignition system from the battery and charging system so that the engine can be started and stopped by the driver.

Ignition coil   The purpose of this is to transform voltage from a low-tension source, that is the battery and charging system, into a high-tension voltage sufficient to promote an electrical discharge across a fixed air gap at the sparking plug.

Ballast resistor   This may be added to the ignition coil for the purpose of improving the life of the contact breaker points, since except under starting conditions the resistor reduces the current flowing through the ignition coil primary circuit.

Contact breaker   This is a cam-actuated interrupter switch contained in the ignition distributor, its function being to open and close the ignition coil primary circuit.

Capacitor   This is connected across the contact-breaker points and provides temporary storage for electric energy as they open, thereby minimizing arcing that would otherwise shorten their life.

Rotor arm and distributor cap   In combination these provide a rotary switch that receives the high-tension current from the ignition coil each time the primary circuit is interrupted, and then distributes it to the appropriate cylinder sparking plug.

Ignition distributor   Apart from including the contact breaker, rotor arm and distributor cap, it must also incorporate a mechanism for automatically varying the ignition timing in accordance with engine operating requirements.

Ignition leads   These are heavily insulated cables conveying the high-tension current from the coil to the distributor and thence to the sparking plugs.

Sparking plugs   Their purpose is to conduct the high-tension current from the ignition system into the combustion chambers of the engine cylinders, wherein they promote an electric discharge across a fixed gap between their electrodes to ignite the combustible charge.

The coil ignition circuit

A general understanding of the coil ignition system may be gained by referring to the typical circuit diagram for a four-cylinder engine application (Figure 10.1). The first point to

notice here is that the ignition system comprises two interacting circuits, these being known as the low-tension primary and high-tension secondary circuits. Their meeting place is in the ignition coil itself, where their respective windings of a few hundred and many thousand turns of wire act in the manner of a transformer and provide the necessary step-up in the battery and charging system voltage to fire the sparking plugs.

Unlike an ordinary transformer, however, which utilizes an alternating current, the ignition coil is supplied with a pulsating direct current from the battery and charging system. This is because, as the engine rotates, the contact breaker in the ignition distributor switches on and off the current flowing through the primary circuit winding in the coil. The important effect of this sudden switching action is to produce a corresponding reaction in the secondary circuit winding in the coil, such that if a relatively small voltage is applied to the primary winding a voltage of many thousand times this value will occur across the secondary winding.

A further point to notice in the circuit diagram is that the secondary winding of the ignition coil is connected to the central electrode of the distributor cap and thence to what is termed the rotor arm which, like the contact breaker, is driven at half engine speed. At the same instant as the high-tension voltage occurs across the secondary circuit, the conducting part of the rotor arm also becomes aligned with one of the ring of four internal electrodes in the distributor cap. This immediately results in the high-tension current jumping the small air gap to the electrode and then travelling through an insulated ignition lead to fire the appropriate sparking plug.

10.2   IGNITION COIL AND CAPACITOR

The principles of electromagnetic induction

In Section 10.1 the purpose of the coil was stated to be that of transforming voltage from a low-tension source, namely the battery and charging system, into a high-tension supply sufficient to promote an electric discharge across a fixed gap at the sparking plug electrodes. To understand how this is achieved we must first delve a little into the early history of electrical developments, relative to what is scientifically termed 'electromagnetic induction'.

Electromagnetic induction
Following the discovery in 1819 by Hans Christian Oersted, professor of physics at the University of Copenhagen in Denmark, that an electric current could produce magnetic effects – in his case the opposite deflections of a compass needle placed first above then below a wire carrying a current – scientists of that era tackled the problem of how to bring about the reverse effect: in other words, how to create an electric current in a wire brought near to a magnetic field.

It was not until 1831 that Michael Faraday, who was one of the greatest of English experimental scientists, found that by moving a magnet to and from the end of cylindrical coil of copper wire the presence of an electric current was indicated on a galvanometer connected across the two ends of the wire (Figures 10.2a and c). However, there was no such indication of current flowing when the magnet was simply allowed to remain stationary within the coil (Figure 10.2b).

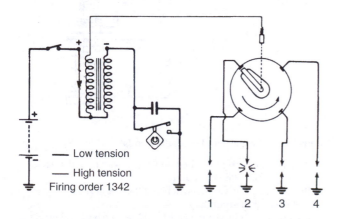

Figure 10.1   Four-cylinder coil ignition system

These observations led Faraday to suggest that an electromotive force (or the force measured in volts that makes an electric current flow) is induced whenever the lines of force of a magnetic field are being cut by a moving wire or conductor. It does not matter, of course, whether the magnetic lines of force are moving across a conductor or vice versa; in both cases an induced voltage causes a current to flow in the conductor. Faraday was further able to demonstrate that the magnitude of the induced voltage could be increased by winding more turns of wire on the coil, using a stronger magnet, moving the magnet more rapidly, and inserting a soft iron core in the coil. The purpose of the last mentioned was to produce a large concentration of magnetic lines of force through the coil and, indeed, the effect of all these modifications was to increase the rate at which the lines of force were being cut by the turns of the coiled wire for a given movement of the magnet. This led Faraday to generalize that the induced electromotive force or voltage is proportional to the rate at which the magnetic lines of force are being cut.

Mutual induction

Before we can relate the foregoing to the operation of an ignition coil, matters must be taken a step further to consider the action of what are termed primary and secondary coils. These comprise entirely separate cylindrical coils of wire that can be brought together end to end or placed one inside the other. The primary coil is connected to a source of electricity supply and the secondary coil to a galvanometer. Their action is such that whenever the current supply to the primary coil is

switched on and off, thereby creating and then destroying a magnetic field that cuts across both the primary and secondary coils, the galvanometer will indicate that a current has been caused to flow in the secondary coil by the voltage induced across it (Figures 10.3a and c). The direction of current flow in the secondary coil will be opposite to that in the primary coil when the latter is being switched on and in the same direction when it is being switched off. This phenomenon is known as mutual induction. If the primary coil is allowed to remain switched either on or off then, of course, there will be no mutual induction (Figure 10.3b).

Self-induction

Consider further the action of these primary and secondary coils. When the current supply to the primary coil is switched on and off, a voltage will be induced not only across the secondary coil but also across the primary coil. Unfortunately, the voltage induced across the primary winding always opposes the action producing it, so that it tends to delay the build-up of current when the primary circuit is switched on and to prolong the flow of current when it is switched off. As will be evident later, the phenomenon of self-induction presents a certain nuisance value to the operation of the coil ignition system.

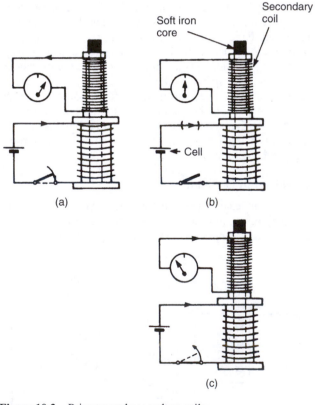

Figure 10.3   Primary and secondary coils

(a) Primary coil being switched on and secondary coil current flowing in opposite direction
(b) Primary coil remaining switched either on or off and no current flowing in secondary
(c) Primary coil being switched off and secondary coil current flowing in the same direction

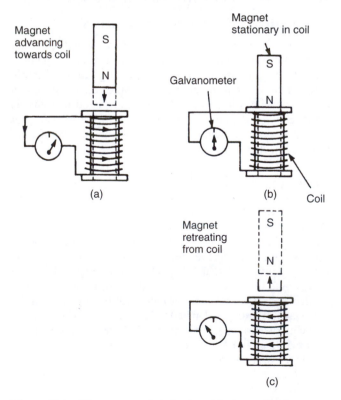

Figure 10.2   Electromagnetic induction: (a) current flowing (b) no current flowing (c) current flowing in opposite direction

Transformer action

There would, of course, be very little point as such in simply causing a current to flow in one coil by switching on and off the current supply to another. The reason for so doing is, in fact, to provide an opportunity for altering the voltages across (the ends of) the primary and secondary coils. This transformer action is made possible because the voltage induced across the secondary coil depends upon the voltage across the primary and also upon the ratio between the number of turns of wire in the two coils. For example, if the secondary coil has 1000 times as many turns of wire as the primary, then the voltage induced across the secondary coil will be nearly 1000 times as great ('nearly' because any such arrangement of coils can never produce 100 per cent efficiency). Conversely, the current flow or amperage in the secondary coil will only be about 1/100 of that in the primary, since electric power is the product of voltage and amperage.

The arrangement of primary and secondary coils just described provides the basic ingredients for the automotive ignition coil (Figure 10.4), since the voltage induced across the secondary coil can be made of sufficiently high tension to generate a spark at the plug gaps and thus ignite the combustible charge in the engine cylinders.

Operation of the ignition coil

In operation the current flow through the primary winding of the ignition coil is being continually interrupted by an engine-driven contact breaker, which therefore acts as a mechanical switch (Figure 10.5). Hence, the ignition coil is supplied with a pulsating direct current as previously explained. Each time the primary current is interrupted, the magnetic field that has previously been built up by the current flow through the primary winding suddenly collapses. It therefore cuts across the conducting coils of both primary and secondary windings and induces a voltage therein. Since the secondary winding of the ignition coil has a considerably greater number of turns than the primary, typically in the order of 20 000 to 300, the voltage induced in the secondary circuit is stepped up from the few hundreds induced in the primary to very high values of 20 000 V (20 kV) or more.

As we know from the principles of electromagnetic induction discussed earlier, this high-tension voltage is theoretically proportional to the rate of collapse of the magnetic field and to the ratio of secondary to primary turns on the coil windings. In practice, however, the voltage will be limited to the lesser amount that is actually required to make the current jump the sparking plug gap, this being termed the breakdown voltage.

Ignition coil selection

This is mainly concerned with the ratio of turns in the secondary and primary windings and also the resistance value of the latter winding. The correct choice of ignition coil for any given vehicle model is therefore decided between the manufacturer and the equipment supplier. It is generally based on what are termed the engine voltage requirements under the extreme load condition, which will vary with compression ratio, engine temperature and sparking plug gap. A reserve of at least 2000 V is then included in the recommended coil output. From this it will be appreciated that the performance of an engine will not necessarily be improved by replacing the standard coil with say a higher-output sports coil, unless a suitable degree of tuning is also applied to the engine itself.

Ballasted coil ignition circuit

Many ignition systems have what is termed a ballast resistor connected in series with the ignition coil primary winding (Figure 10.6). It usually comprises a coil of resistance wire retained within a porcelain holder and provided with suitable end connections, and has a resistance value similar to that of the coil primary winding. Used in conjunction with an ignition coil that is designed to operate with a lower voltage than that of the vehicle electrical system, the ballast resistor is bypassed when starting the engine, but switched into the ignition circuit during normal running. In the former case, the whole of the available battery voltage is applied to the

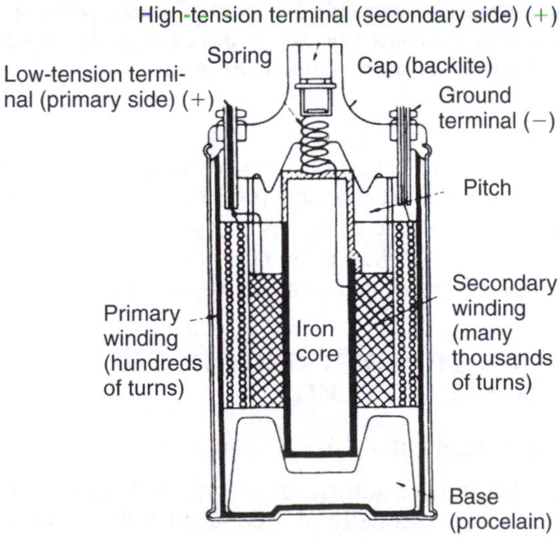

High-tension terminal (secondary side) (+)

Spring
Cap (backlite)
Low-tension terminal (primary side) (+)
Ground terminal (−)
Pitch
Primary winding (hundreds of turns)
Iron core
Secondary winding (many thousands of turns)
Base (procelain)

Figure 10.4 Sectional view of ignition coil (*Yamaha*)

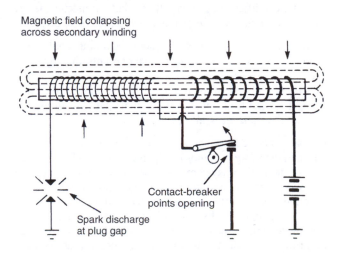

Magnetic field collapsing across secondary winding

Contact-breaker points opening

Spark discharge at plug gap

Figure 10.5 Simplified representation of mechanically switched ignition coil action

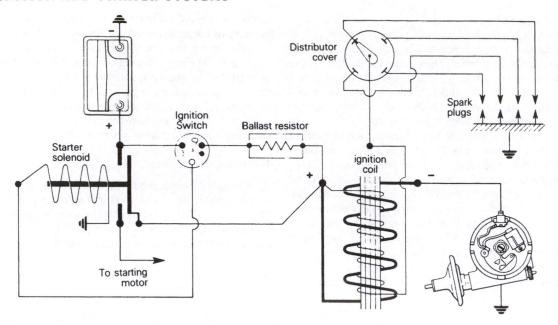

Figure 10.6 Ballasted ignition circuit (*Lucas*)

coil, which then operates temporarily but not harmfully at a voltage slightly above its normal to improve the engine cold starting performance. Under normal running conditions, with the ballast resistor switched into the circuit, the battery voltage is divided between the coil primary winding and the external resistor, so that less heat is generated within the ignition coil itself to provide maximum spark efficiency at high engine speeds.

The need for a capacitor

When the effects of self-induction were described earlier, it was mentioned that the voltage induced across the primary coil always opposes the action producing it. With the coil ignition system it thus follows that this effect will tend to delay the build-up of current in the primary winding when the contact-breaker points close, and prolong the duration of current flow after they open. If the latter tendency is left unchecked, then it would produce two undesirable effects:

1   An electrical discharge with associated arcing would occur across the separating contact-breaker points, thereby shortening their useful life.
2   A rapid collapse of the magnetic field would be prevented and thus the high-tension voltage available from the coil secondary circuit would be reduced.

A capacitor (or condenser, as it was once known in the motor industry) therefore becomes indispensable to satisfactory operation of the ignition system. Since the function of a capacitor is to provide temporary storage for electrical energy, it is connected in parallel with (across) the distributor contact-breaker points (Figure 10.7). Its action is such that when the points begin to separate, it provides an alternative and easier route for the primary current to take, thereby reducing (but not eliminating) arcing across them and also contributing to a rapid collapse of the magnetic field. After the spark has occurred at

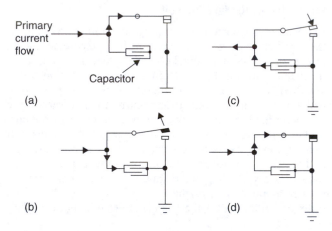

Figure 10.7 Simplified representation of capacitor action: (a) contact-breaker points closed (b) contact-breaker points opening, plug firing (c) contact-breaker points closing, plug ceased firing (d) contact-breaker points closed again

the sparking plug gap, the electrical energy received by the capacitor is dissipated in the process of being discharged against the resistance of the primary circuit, in which the oscillating action of the current just dies out when the contact-breaker points close again and a new ignition cycle begins.

10.3 IGNITION DISTRIBUTOR AND SPARKING PLUGS

Basic arrangement

The mechanism of a non-electronic type of distributor (Figure 10.8) consists essentially of a contact-breaker assembly to make and break the low-tension primary current, a rotating switch to distribute the induced high-tension secondary current

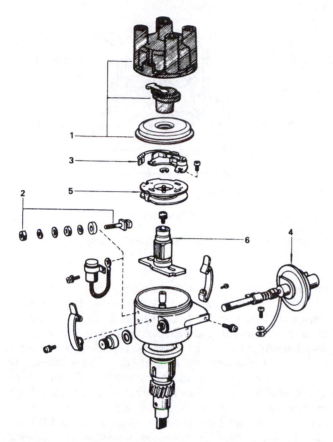

Figure 10.8   Exploded view of ignition distributor (*Toyota*)

1 cap, rotor and cover	4 vacuum advancer
2 terminal	5 breaker plate
3 breaker point	6 cam

whilst the other is attached to the extremity of a lever arm that is pivoted on a base plate. Anchored to both the lever arm and its base plate, but electrically insulated from the latter, is a curved spring blade that maintains the lever arm in contact with the distributor cam. For a single contact breaker the number of lobes on the cam must necessarily correspond to the number of cylinders served. A non-metallic heel pierce or rubbing block is either made integral with, or separately attached to, the lever arm and acts as a cam follower. The lever arm is electrically insulated from its pivot mounting on the base plate, either by virtue of the material from which it is made, or by means of a non-metallic insulating bushing. A light-weight construction with minimum inertia is essential for the lever arm assembly, so that its motion accurately follows that of the cam contour and prevents misfiring at high engine speeds.

Contact-breaker gap size and dwell angle

The correct setting of the contact points gap became very critical as engine speeds increased, since it is directly related to the number of degrees of distributor cam rotation during which the contact points remain closed. Here it must be appreciated that when the primary circuit of the ignition system is restored by the contact-breaker points closing, a finite interval of time, known as the saturation time, is required in order that the current flow and hence the coil magnetic field can build up to their maximum intensities. This time lag again arises from the effects of self-induction on the primary winding, which as mentioned in Section 10.2 opposes any change in the amount of current flowing through it. Since the time during which the contact-breaker points actually remain closed is in the order of milliseconds, clearly one cannot afford to reduce this all too fleeting interval any more than is absolutely necessary. Otherwise the efficiency of the ignition system, and likewise that of the engine, will be impaired. On the other hand, we cannot allow the contact-breaker points to remain closed for too long, because the duration of the primary current flow could then become excessive and lead to their burning.

Although we cannot conveniently record the actual time, or dwell time, during which the contact-breaker points remain closed (and in any event this would become shorter with increasing engine speed), it is practicable to measure with the engine running the number of degrees of distributor cam rotation occupied by their closure. This angular measurement is termed the *cam angle* or *dwell angle* of the distributor (Figure 10.9), and is obtained by connecting a tach-dwell meter into the primary circuit of the ignition system, the instrument being so called because it relates the dwell angle to engine speed.

A little further consideration will show that the dwell angle of the distributor will be increased by reducing the static setting of the contact points gap and vice versa. For example, if the gap is reduced then clearly the heel piece of the contact-breaker lever arm will lose contact with the cam sooner after one lobe has passed and regain contact later before the next lobe arrives. In other words, the contact points will have remained closed over a greater angular movement of the distributor cam and the dwell angle has been increased. The dwell angle (as opposed to dwell time) should remain reasonably

to the correct sparking plug at the right moment, and a means of automatically varying the ignition timing from its static setting, according to changes in engine speed and load.

Since in the four-stroke cycle each cylinder needs a spark only once in every two revolutions of the crankshaft, the distributor is always driven at one-half engine speed. With in-line cylinder engines it is therefore generally expedient to provide either a vertical or a near-vertical side mounting for the distributor and drive it through skew gearing of 1:1 ratio from a crankcase-mounted camshaft or jackshaft, or a horizontal mounting for the distributor and drive it direct from the rear end of an overhead camshaft, as found in some transverse-engine installations. For V and horizontally-opposed cylinder engines, the distributor is vertically mounted on top of either end of the crankcase, because the unavoidable presence of the intake manifolding precludes a central mounting position. In these cases the distributor drive may be taken from the camshaft or the crankshaft, the latter via a similar arrangement of gearing but with a 2:1 reduction ratio.

Contact-breaker mechanism

This comprises a pair of tungsten contact points, one of which remains stationary but with provision for adjustment,

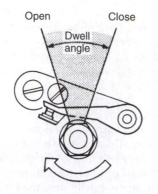

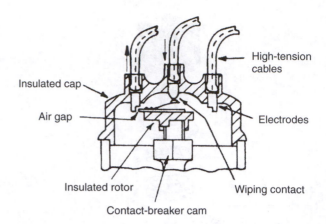

Figure 10.9   Distributor cam dwell angle

Figure 10.11   Distributing the high-tension pulses

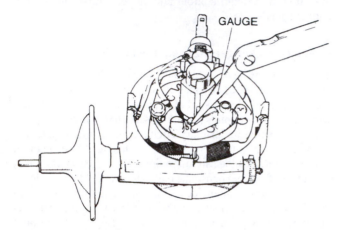

Figure 10.10   Checking the contact-breaker points gap (*Lucas*)

constant over the normal operating speed of the engine, unless there is an unacceptable amount of wear in the distributor shaft bearings that is allowing the cam to float relative to the contact-breaker assembly.

Checking the contact points setting can therefore be interpreted in two ways:

1 Measuring the actual points gap, typically of 0.35 to 0.40 mm (0.014–0.016 in), which should appear between the contact points when the heel piece of the lever arm is resting on the peak of any one of the cam lobes. This is accomplished by the time-honoured method of inserting a feeler gauge blade of appropriate thickness between the contacts points (Figure 10.10).
2 Measuring the distributor dwell angle by connecting a tach-dwell meter and cranking the engine over on the starter motor. The indicated reading on the meter should then be compared to that specified by the car manufacturer. This reference to specification is most important, as evidenced by a random choice of two different makes of four-cylinder engine – one fitted with a German- and the other with an Italian-made distributor – where the dwell angles called for are 40 ± 2 and 55 ± 3 degrees respectively.

Distribution of high-tension pulses

Another essential part of the distributor mechanism is a rotating switch, which receives the high-tension pulses from

the ignition coil each time the primary circuit is interrupted and directs them to each cylinder sparking plug in correct sequence of firing order. It is positioned above the contact-breaker mechanism and consists of a rotor that operates within the distributor cap (Figure 10.11). The former is constrained to rotate with the contact-breaker cam, whilst the latter is mounted on the distributor body; a combination of spigot, tongue and slot locations are used in each case. Both the rotor and the cap with its spring retaining clips are made readily detachable from their locations, so as to facilitate access to the contact-breaker mechanism. They are moulded from an insulating material in which are embedded their respective switch electrodes. The distributor cap contains as many fixed brass electrodes as there are sparking plugs to be served, together with a central carbon electrode that receives the high-tension pulses from the ignition coil. To provide a radial conducting link between the central electrode and each in turn of those surrounding it, the rotor embodies a brass strip electrode.

A wiping contact is invariably employed to provide an electrical path between the central input electrode of the cap and that of the rotor. It takes the form of either a spring-loaded carbon brush in the cap that presses on to the brass electrode of the rotor, or a spring strip attached to and doubling back along the rotor electrode, so that its free ends bears against a carbon insert in the cap. To complete the electrical circuit between the tip of the rotor electrode and those in the cap, as it passes under each in turn, a small air gap is employed across which the high-tension pulse is fairly happy to jump, a wiping contact being unsatisfactory for this particular application. Owing to the internal sparking that occurs across these air gaps, nitric oxide and ozone gases are produced. In order to prevent corrosive condensates from these gases attacking the insulated parts and also the surface of the contact-breaker cam, which in turn increases wear on the heel piece of the lever arm and thus affects the ignition timing and dwell angle, the distributor cap is provided with ventilation holes.

The high-tension impulses are conveyed from the coil to the distributor and finally to the sparking plugs by means of ignition cables, their sequence of connection to the distributor cap electrode terminals depending upon the particular

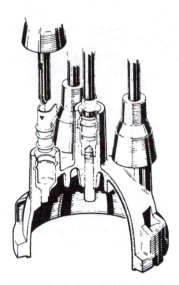

Figure 10.12 A modern distributor cap for push-in type cable terminations (*Lucas*)

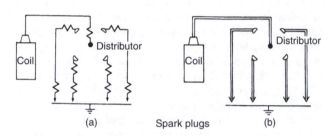

Figure 10.13 Suppression of interference: (a) cable resistors (b) resistor cables

firing order of the engine cylinders (Figure 10.1). Methods of securing the ignition cables in the cap have included screw-in, pierced and push-in terminations, the latter being more frequently used with modern resistive cable (Figure 10.12).

Types of high-tension ignition cable

All vehicles manufactured in the United Kingdom must comply with regulations designed to ensure that there is minimum interference with domestic radio, television and radio communication from ignition equipment. Radio interference suppression is also the subject of ECE (Economic Commission for Europe) and equivalent EEC (European Economic Community) regulations. The interference waves from the high-tension circuit of the engine ignition system may be reduced in range or suppressed by the application of either cable resistors or resistor cables, the latter being the preferred method to meet present regulations (Figure 10.13).

Cable resistors

These are generally less effective than resistor cables. Their application requires that individual wire-wound resistors be fitted at all high-tension cable connections to the distributor and sparking plugs. The type of high-tension ignition cable associated with distributed resistors was the once conventional copper conductor construction (Figure 10.14a).

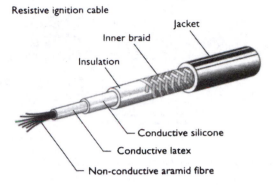

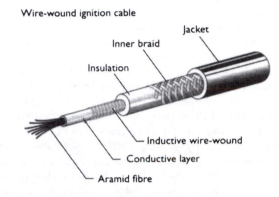

Figure 10.14 Types of ignition cable (Relphi)

Resistor cables

In modern practice the high-tension or HT ignition cables must now conform to stringent requirements, which may be summarized as follows:

1. Suppress radio frequency interference
2. Insulated to withstand 40 000 volts
3. Resistant to temperatures ranging from minus 40°C to 260°C
4. Resistant to effects of ozone, corona and contaminating fluids
5. Possess a service life of 160 000 km (100 000 miles)
6. Remain durable over a 10 year period

The construction of these ignition cables typically features either high-temperature resistive or inductive wire-wound cores, which are sheathed in specialized elastomeric materials that meet the above mentioned requirements (Figure 10.14).

Speed-sensitive advance and retard mechanism

It has long since become established practice for the distributor to incorporate some means of automatically varying the ignition timing, in accordance with engine operating requirements as established on the test bed of the manufacturer. This variation in ignition timing may be accomplished either by speed-sensitive mechanical or load-sensitive pneumatic controls, or more usually by a combination of both. Mechanical control systems operate on the principle of the centrifugal governor (Figure 8.17). In this application it

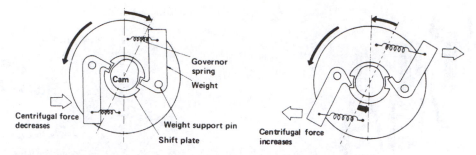

Figure 10.15   Operating principle of a speed-sensitive advance and retard mechanism (*Yamaha*)

serves to advance the angular movement of the contact-breaker cam, relative to the coaxial distributor drive shaft upon which it is freely mounted. Hence, as the engine speed increases, the ignition timing is advanced beyond its static setting, as later described. Compensation is thus obtained for the effect of the combustion process tending to occupy a constant time, regardless of engine speed, by allowing the spark to occur a little earlier on the compression stroke.

In construction this type of mechanism consists of two masses termed the flyweights, which are separately pivoted upon a supporting plate, this in turn being rigidly secured to the distributor drive shaft so that the flyweights circulate about its axis. The controlling force is provided by a tension spring restraint for each flyweight (Figure 10.15). Therefore at any given engine speed, the radius of rotation of the flyweights will be such that the influence of centrifugal force on them is just balanced by the spring controlling force. The arcuate movements of the flyweights are transmitted to the foot of the contact-breaker cam either by toggle, slotted or rolling linkages, or simply by contoured projections on the flyweights. These respectively connect with, or contact, a yoke piece attached to the cam, which is thus capable of a limited angular movement in advance of its drive shaft. Maximum relative angular movement of the cam is limited typically by an extension at its foot, which comes into contact with one of the spring anchor posts on the supporting plate for the flyweights.

The actual ignition advance characteristics are therefore determined by the combined effect of the pivoted flyweights acting against their control springs, together with the variations in leverage exerted through their connection with the yoke piece of the cam. Since the forces involved in the operation of a centrifugal mechanism vary as the square of its rotational speed, one flyweight spring is generally fitted under a certain amount of initial tension, so as to avoid unwanted advance of the ignition timing at engine starting and idling speeds, whilst the other spring is loosely connected through an elongated looped end. This differential spring arrangement acts such that as engine speed increases, the restraining effect of both springs is brought into play to provide the final controlling force. At first, the amount of ignition advance increases fairly rapidly from just above idling speed to a predetermined point in the intermediate speed range of the engine. It then proceeds to advance more gradually until reaching a fixed value as the maximum speed of the engine is approached. Depending upon engine design requirements, the maximum ignition advance provided by the centrifugal

mechanism may be up to about 20° of distributor cam rotation, the actual value being doubled of course in terms of crankshaft rotation or crank angle.

Load-sensitive advance and retard mechanism

To satisfy the ignition advance requirements of the engine under various loads, as distinct from speeds, a pneumatic timing control subject to intake manifold depression is often embodied in the distributor. This requirement arises especially under part-throttle cruising conditions when power output is sacrificed in the interests of fuel economy. The ignition timing must then be advanced sufficiently to compensate for the slower burning of the leaner mixture. Although a vacuum advance unit, as it is generally termed, has sometimes been used alone, it is more widely employed to augment the action of the mechanical control system.

In modern construction the vacuum advance unit is mounted directly on to the distributor body. It comprises a depression chamber that is sealed by a flexible diaphragm, which is spring loaded in opposition to the manifold depression acting upon it. The depression side of the diaphragm does not actually communicate directly with the intake manifold, but is connected to a point just above the carburettor throttle valve in its closed position. This arrangement again avoids unwanted advance of the ignition timing from its static setting at engine idling speed. The other side of the diaphragm is vented to the atmosphere and is provided with a connecting link to the contact-breaker base plate, so that the latter can be partially rotated within the stationary distributor body. Hence, as intake manifold depression increases, the atmospheric pressure acting upon the other side of the diaphragm forces it to move in opposition to its control spring (Figure 10.16). The resulting linear motion of the diaphragm is then converted into one of rotation at the contact-breaker base plate by the connecting link between them. For vacuum advance the contact-breaker mechanism must, of course, be partially rotated in a direction opposite to that of cam rotation.

Under light-load running conditions, the vacuum timing control can therefore advance the ignition over and above that provided by the centrifugal timing mechanism (Figure 10.17). Conversely, under acceleration and full-load running conditions, the vacuum timing control can reduce the amount of ignition advance that would otherwise be imposed by the mechanical control system. This reduction in ignition advance is quite logical, since it complements the faster burning of

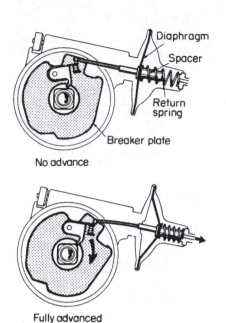

Figure 10.16 Operating principle of a load-sensitive advance and retard mechanism (*Lucas*)

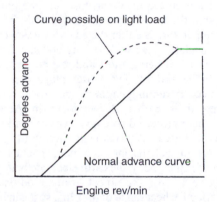

Figure 10.17 Vacuum advance curve (*Lucas*)

the richer mixture under these conditions. The maximum amount of ignition advance provided by a vacuum timing control may be up to about 15° of distributor cam rotation and therefore twice this amount in terms of crank angle. Neither the centrifugal nor the vacuum timing controls can allow the ignition to become retarded beyond its static setting, which is maintained by the distributor body clamp fitting. A form of micrometer screw adjustment is typically incorporated in the depression chamber mounting for the purpose of allowing slight rotational repositioning of the contact-breaker base plate relative to the cam. This feature is sometimes termed an octane selector, since within limits it facilitates fine adjustment of the ignition timing according to the quality of fuel available.

The action of the sparking plugs

It is always worth remembering that the efficiency of the ignition system and hence that of engine operation itself is ultimately dependent upon components that have no working parts and are usually taken for granted – namely the sparking plugs. In fact, the sparking plugs may be said to represent the final link in a chain of three prime requisites for satisfactory operation of the engine, these being fuel of good quality, full cylinder compression and a healthy spark.

The basic action of the sparking plug is simplicity itself, and remains the same as that conceived by the Frenchman Etienne Lenoir, who made the first-ever sparking plug in 1860 for his famous gas engine (referred to in Section 10.1). More specifically, the insulated central electrode of the sparking plug receives a pulse of high-tension current, which in jumping across the air gap to an earthed side electrode creates a spark discharge to ignite the combustible mixture in the engine cylinder.

To maintain this simplicity of action when exposed to the aggressive environment of extreme and rapidly changing cylinder pressures and temperatures is quite a different matter, since little encouragement is needed for the high-tension current to flow in any direction other than across the spark gap between the electrodes. The various service factors that can provoke this wanderlust of the high-tension current will be listed at a later stage. Suffice it to say here that the reliability of the modern product derives from continuing research into the features of construction that improve mechanical strength, electrical insulation, pressure sealing, resistance to spark erosion and chemical corrosion of the sparking plug. Examples of these constructional features are most conveniently presented by illustration (Figure 10.18).

Heat range of a sparking plug

Apart from their size designation, sparking plugs are further classified in terms of their heat range. This classification was first started by the Champion Sparking Plug Company in 1923 when developing plugs for racing, and it refers to the ability of a plug to provide a controlled rate of heat flow from its firing end to the cylinder head cooling arrangements. The satisfactory operation of a sparking plug represents a compromise between avoiding the two extremes of pre-ignition at high speeds and fouling at low speeds.

Pre-ignition intervenes if the insulator nose of the plug becomes hot enough to ignite the combustible charge before the spark is timed to occur. This condition can rapidly lead to one of detonation, with its accompanying loss of power and mechanically destructive effects on the engine. *Fouling* occurs if the insulator nose runs too cool, so that it is unable to burn off the combustion deposits it receives. An accumulation of these deposits lowers the electrical resistance of the plug, since they can form a conducting surface layer on the insulator nose, across which the high-tension current can leak to the plug body and earth. The result of the plug short-circuiting in this manner is either to weaken its spark, or to prevent it from sparking at all.

To avoid pre-ignition on the one hand and fouling on the other, it has long been recognized that ideally the temperature of the plug insulator should neither rise much above 850°C, nor fall below the minimum self-cleaning temperature of about 350°C. In an endeavour to meet these requirements, sparking plugs are constructed with different heat values, ranging from what are termed cold (or hard) through

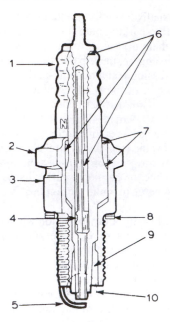

Figure 10.18   Anatomy of a sparking plug (*NGK*)

1 corrugation for flash-over prevention: keeps a distance between terminal and shell, and prevents electrical leakage which would otherwise result from water attached; also provides larger contact area with air to help heat dissipation
2 flat
3 shell: maintains a good seal with the cylinder head hole; good heat conductivity is also needed
4 centre electrode: certain spark plugs have a centre electrode with a copper core to increase heat dissipation and electrical conductivity, thus making the heat range wider
5 ground electrode
6 filler: glass seals are most widely used; excellent heat dissipation and good sealing are required
7 caulking ring
8 gasket
9 plug nose: nose length affects heat dissipation, and thus determines the plug's heat range; cold type has shorter, hot type longer nose
10 centre electrode: a plug with a longer projected part of centre electrode is more exposed to vortex flow of air–fuel mixture; therefore it is little burned under heavy-load operation and less sooty under light-load operation because of greater self-cleaning property

intermediate stages to hot (or soft). Their running temperatures may therefore be controlled as nearly as possible to suit the normal operating requirements of any given engine.

The heat range of a sparking plug is predetermined in practice by varying the length and form of its insulator nose, thereby altering the seating height and gas capacity respectively. If the insulator is provided with a high seating in the plug body, the total length of heat path between the central electrode tip and the cylinder head cooling medium is increased, so that heat dissipation is hindered to provide a hot plug. Conversely, the provision of a low seating for the insulator assists the process of heat dissipation to give a cold plug. Although the insulator nose is always tapered with the object of avoiding excessive heating at its tip, it may also assume either a slender or a stubby form. Since the annular

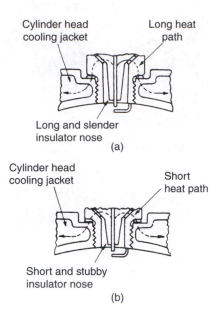

Figure 10.19   Heat range of sparking plugs: (a) hot (b) cold

gas space between the insulator nose and the plug body is increased with the former and decreased with the latter, the insulator temperature is either raised or lowered accordingly. Hence the insulator nose will be long and slender for a hot plug as opposed to being short and stubby for a cold plug (Figures 10.19a and b). The hotter plugs are, of course, installed in cooler running engines, and vice versa.

Following their earlier development in America, the extended nose or projected core nose type of sparking plug is now widely specified. As its name implies, it differs from the ordinary plug in that the insulator nose extends below the underside of the plug body, the earth electrode being reshaped to suit (Figure 10.18). The object of this modification is to widen the effective heat range of the plug, so it can better cope with the conflicting requirements of slow-speed traffic crawling alternating with high-speed motorway running. In operation, the more exposed insulator nose further hinders heat dissipation from the plug to prevent fouling under slow-speed running, but is cooled sufficiently by the incoming rush of mixture to avoid pre-ignition at high speeds.

Importance of correct plug gap setting

The distance, or gap as it is termed, between the tips of the central and side electrodes of a sparking plug determines the value of high tension current or voltage that is required to produce a spark discharge and the available voltage must, of course, exceed this requirement. For any given plug application the actual gap setting is dependent upon the following factors:

Engine compression ratio
Combustion chamber characteristics
Air–fuel ratio
Ignition system design.

In practice the gap setting now usually lies within the range of 0.7 to 1.1 mm (0.035 to 0.045 in) with a gap of 1.0 mm

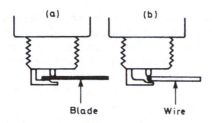

Figure 10.20 Correctly setting gap of worn plug electrodes (shown exaggerated): (a) false setting with blade feeler gauge (b) true setting with wire feeler gauge

(0.040 in) typically being specified. The gap is set by bending the side (never the central) electrode and then checked preferably with a round wire rather than a flat feeler gauge (Figure 10.20). Since the matching of the right sparking plug to a particular engine and the optimization of its gap setting involve close cooperation between the manufacturers concerned, it will be evident that any departure from the gap specified is likely to reduce the efficient operation of the sparking plug and hence that of the engine itself. For example, engine misfiring can result from either extreme of gap setting. This is because too small a gap may reduce the intensity of spark discharge to the extent that the combustible gases are not properly ignited, whilst too large a gap may result in no spark discharge at all if the available voltage is not high enough to cope with hard acceleration when there is a rapid increase in cylinder pressures.

Although the sparking plug electrodes are typically produced from an extremely durable nickel alloy material, their gradual deterioration arises from attack by the corrosive products of combustion and the erosive effects of the spark discharge. It is these factors and also the condition of the insulator nose that ultimately decide the useful life span of the plug. It was once usual for a manufacturer to include in the maintenance schedule a recommendation for the sparking plugs to be removed, cleaned and regapped after 6000 miles (10 000 km) service and finally to be replaced after 12 000 miles (20 000 km), whereas with modern plug design and materials, periodic cleaning may no longer be recommended and their replacement interval extended to every 30 000 miles (50 000 km) or 3 years. Any unnecessary attention to the plugs is to be avoided, because if this should result in engine misfiring then the release of unburnt fuel into a catalytic converter will overheat and damage its monolith (Section 11.2).

Since the early 1990s platinum tip electrodes have become increasingly specified for the sparking plugs of high-performance car engines. Historically, it was during much earlier research in the field of aircraft piston engines, when it was recognized that platinum tip electrodes showed a greater resistance to spark erosion. The advantages of using platinum tip electrodes for the sparking plugs of automotive engines are now perceived as being longer replacement periods, improved cold starting, more tolerant of a low battery charge and, last but not least, greater reliability. This is an important environmental consideration because, as previously mentioned, if an engine misfires it will emit unburned fuel, which will have a detrimental effect on the catalytic converter. In common with the latter component, platinum tip sparking plugs require the use of unleaded petrol for their efficient operation. From the viewpoint of longer life, it is of interest to recall that when the Jaguar V6 engine was introduced in 1999, the replacement period for the platinum tip sparking plugs was given as 100 000 miles (160 000 km).

Sparking plug misbehaviour in service

Often through no real fault of their own sparking plugs can suffer various health problems in service, and to assist in their visual diagnosis sparking plug manufacturers have traditionally supplied troubleshooting charts that depict in colour what to look for. However, a brief summary of sparking plug conditions and their probable causes may prove useful here.

Normal operation
Core nose thinly coated with light grey to tan deposits, which indicates normal high and low speed operation.

Fuel fouling
Dry fluffy black deposit that can short circuit the firing end, which may be caused by an over-rich mixture delivered either by the carburettor or fuel injection system, faulty choke operation or a clogged air cleaner filter.

Oil fouling
Heavy sludgy deposit that can short circuit the firing end, which may be caused by worn valve guides, cylinder bores or piston rings, allowing poor oil control.

Overheating
This can range from a chalky white core nose colouring and excessive electrodes erosion, to burnt electrodes and a blistered or split core nose, according to the severity of overheating. These conditions are associated with over-advanced ignition timing, a fuel with insufficient octane rating, a too weak mixture or a defective engine cooling system, all of which can lead to excessive detonation and pre-ignition.

An anaysis of sparking plug condition can therefore indicate to a service engineer the source of an engine malfunction and the course of action required to rectify it.

10.4 INTRODUCTION TO ELECTRONIC IGNITION SYSTEMS

General background

The concept of introducing electronic components into the electrical ignition system of motor vehicles, originally to improve the life of the contact-breaker points and later to eliminate them altogether, is by no means of recent origin. Joseph Lucas in England developed a breakerless electronic ignition system for racing car engines in the early 1960s, and General Motors in America offered such a system as optional equipment on their 1962 Pontiac models. About the same time a number of accessory fitting systems of various types were also made available. Unfortunately, the reliability of some of these early systems was not all that it might have been, because the electronic components then available had been developed for radio communications work and were therefore not entirely suited to the underbonnet environment of motor vehicles.

However, the demand for better performance from the ignition system, as a contribution to reducing harmful exhaust emissions and improving fuel consumption, led to more vigorous development of the electronic ignition system. By 1977 all American cars were using ignition systems in which the contact-breaker points had been eliminated entirely. This trend was followed elsewhere, and since the early 1980s electronic ignition systems have virtually become established practice on British, European and Japanese cars.

Comparing in simple terms the earlier described mechanically switched ignition system (Figure 10.5) with an electronically switched system (Figure 10.21), the action of the latter is such that when a small trigger current is applied to the base circuit of the transistor, the collector and emitter circuits act as a conductor and allow system current to build up in the primary winding of the ignition coil. Conversely, when the trigger current is interrupted, the collector and emitter circuits act as an insulator, thereby suddenly stopping the current flow in the primary winding of the ignition coil and causing the desired collapse in its magnetic field (Figure 10.21). It is, of course, the extreme rapidity of this collapse made possible with an electronically switched ignition system, which shows to advantage by inducing very much higher secondary voltages to fire the sparking plugs. The timing of the trigger current for the transistor, depends on the sensor system used to monitor the rotational position of the crankshaft and other aspects of engine operation.

Advantages of electronic ignition systems

Modern electronic ignition systems can be expected to confer the following advantages on engine operation:

Increased energy and duration of spark
As compared with a non-electronic ignition system with mechanical contact breaker, the amplifier module of an electronically switched system can utilize a higher value of primary current. This allows a greater amount of energy to be stored in a suitably matched coil, so that when the amplifier module interrupts the primary circuit, a much higher voltage can be induced in the secondary circuit to fire the sparking plugs and therefore provide better combustion. In practice the operating voltage can be in the region of 30 000 V (30 kV) or more, and for this reason it represents a safety hazard to service personnel. It is therefore most important to observe the manufacturer's recommended testing procedures and precautions when working on these systems.

More consistent sparking
This is especially beneficial in relation to better starting in cold weather. It derives from the use of electronic switching so that the rate at which the ignition coil primary current is interrupted does not vary with engine speed. A disadvantage of the mechanical contact-breaker system is that, during starting and idling of the engine, the points are separating relatively slowly. This tends to encourage arcing between them despite the presence of the capacitor, which in practice can only be selected on the basis of an average current flow. Consequently, there is a less rapid collapse of the magnetic field existing in the ignition coil. This in turn has the effect of diminishing the voltage induced in the high-tension secondary circuit, which may then be insufficient to produce a healthy spark at the plugs.

Less restricted rate of sparking
In describing earlier the mechanical contact breaker it was mentioned that the spring-loaded lever arm of the moving contact is designed to have minimum inertia, so that its motion accurately follows that of the cam contour. Even so, a bounce condition is ultimately reached that reduces the contact dwell time and leads to engine misfiring. The onset of this erratic behaviour imposes a practical upper speed limit for such systems in the order of 400 sparks per second. This restriction on rate of sparking therefore places a limitation on the development of ignition systems with mechanical contact breakers for really high-speed engine applications. However, this limitation does not exist with the electronic ignition system, where considerably higher sparking rates are possible as long as it is matched with an ignition coil that can store more energy in a shorter time.

More stable ignition timing
This is an important factor in maintaining optimum engine performance, minimizing fuel consumption and keeping exhaust emissions within specified limits. It is difficult to achieve with a mechanical contact-breaker system, because changes in the points gap occur in service that affect the accuracy of the ignition timing. In the relatively short term these changes are caused by deterioration of the points contact faces, which upon separation are always subject to a certain amount of arcing. Further changes are liable to occur in the long term, since the points gap setting is also affected by wearing of the rubbing block or heel piece of the moving contact and of the distributor drive spindle bearings, the latter condition allowing radial float of the contact-breaker cam. In contrast, there is no contact-breaker gap to alter with an electronically switched system, which thus confers stable ignition timing.

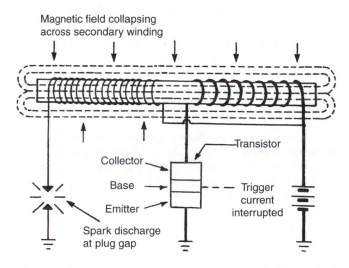

Figure 10.21 Simplified representation of electronically switched ignition coil action

Reduced maintenance requirement

A minimum service requirement for ignition systems incorporating a mechanical contact breaker is that the points have to be either cleaned, or renewed, and then reset at regular intervals. In the modern electronic ignition system, the absence of a contact breaker renders the installation to all intents and purposes maintenance free, except of course for attention to the sparking plugs. This much reduced service requirement for the breakerless electronic ignition system, together with its technical advantages, accounts for its earlier adoption in America, where manufacturers are responsible for the durability of emission controls up to 50 000 miles (83 000 km).

Ready incorporation in engine management systems

An important modern advantage of the electronic ignition system is that it can be integrated with electronic microcomputer control systems, which provide overall management of air–fuel ratio, spark timing and other engine functions to reduce harmful exhaust emissions and improve fuel economy.

10.5   TYPES OF ELECTRONIC IGNITION SYSTEM

Apart from the retention in early transistor-assisted contact (TAC) systems of contact-breaker points, which merely

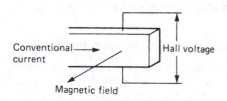

Figure 10.22   The Hall effect

switched transistors in and out of conduction so that the points themselves were no longer subject to high current loading, modern types of electronic ignition system utilize either a breakerless distributor or no distributor at all. Their general features of operation may be summarized under the following headings.

Basic electronic with breakerless distributor

This type of system comprises a breakerless distributor, an amplifier module, and a matched ignition coil. Various types of pick-up unit may be used to replace the mechanical contact breaker. One that gained favour is based on the Hall effect. This is so named after an American physicist E.R. Hall, who in 1879 discovered that a voltage is created across a current-carrying conductor when a perpendicular magnetic field is applied (Figure 10.22).

In the Hall effect distributor this principle is incorporated by interrupting the magnetic field to produce a voltage change in a pulse signal. This signal is then used to trigger the amplifier module (Figure 10.23), wherein the coil primary current is switched on and off by transistor action. A rotor vane assembly in the distributor is provided with a number of vanes, corresponding to the number of engine cylinders, and these vanes are arranged to pass through a gap in the vane switch (Figure 10.24). The latter comprises an inner pick-up coil and an outer permanent magnet with a gap between them. As the rotor vane assembly turns, its vanes successively interrupt the magnetic field in the vane switch and therefore produce the required changes in pulse signal voltage to the amplifier module. An advantage of the Hall effect pick-up unit over the other types is that the pulse signal voltage does not change as the engine speed changes.

In a basic electronic ignition system of the type described, the dwell angle of the distributor depends on the width of

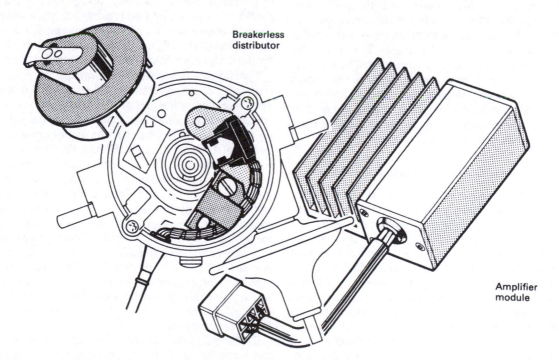

Figure 10.23   Major components of breakerless electronic ignition (*Lucas*)

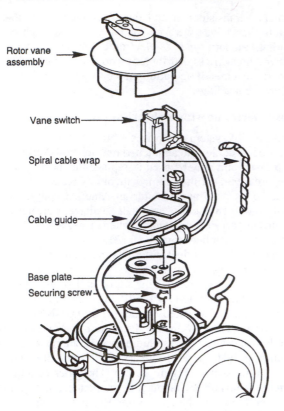

Rotor vane assembly

Vane switch

Spiral cable wrap

Cable guide

Base plate
Securing screw

Figure 10.24 Exploded view of a Hall effect distributor (*Lucas*)

A manifold absolute pressure (MAP) sensor may be used for signalling engine load to the electronic control unit, by responding either to depression with a naturally aspirated engine or to pressure in the case of a turbocharged one, and supplying this information in the form of a voltage signal. Alternative methods of signalling the engine load are to utilize the air flow and air mass meters employed with electronic fuel injection systems (Section 6.7), as part of a combined fuel and ignition engine management system.

Although the engine speed can be additionally signalled by the Hall pick-up unit in the distributor, a more accurate method involves the use of a speed sensor on the crankshaft. As an example, the engine flywheel may carry a toothed reluctor ring, which operates in conjunction with a stationary signal generator. This can then signal the passage of each passing tooth as a voltage pulse to the electronic control unit, so that the number of pulses received in a given time can be interpreted in terms of engine speed. It can also monitor the angular position of the crankshaft, relative to a reference point of typically 90° before top dead centre (BTDC), since by omitting an appropriate tooth from the reluctor ring the consequent absence of voltage pulse reveals the position of the crankshaft and hence that of its pistons to the electronic control unit. This is preprogrammed to take into account a variety of engine operating conditions, so that by comparing these data with the information received from the sensors, it can compute and command an optimum timing for the ignition.

the rotor vanes so that, similar to the once conventional contact-breaker system, the dwell angle remains constant but the dwell time becomes shorter with increasing engine speed. A later refinement to the basic electronic ignition system was to compensate for this effect, by incorporating additional circuitry that allowed the dwell angle to be reduced at low speeds and increased at high speeds. In practice this meant that the primary current was interrupted by the electronic switch of the distributor, but restored at the correct interval by the electronic control unit. The corresponding variations in dwell time therefore eased the current load on the coil at low speeds and created a longer time for maximum buildup of coil magnetic field at high speeds.

Programmed electronic with breakerless distributor

By the mid 1980s a further development of the breakerless distributor system allowed a more precise control over the ignition timing than had so far been possible with mechanical speed-sensitive and vacuum load-sensitive advance and retard mechanisms. This became necessary in the interests of reducing both harmful exhaust emissions and fuel consumption, while preserving overall engine performance. It was achieved by what is termed a programmed electronic ignition system, where optimum ignition timing characteristics are imposed by the electronic control unit. The signals required by the latter for this purpose are the load on the engine, its speed and the angular position of the crankshaft so as to provide a reference point before top dead centre (TDC).

Programmed electronic without distributor

A notable advance in automotive electronic ignition systems is represented by the direct type, which was introduced by SAAB in the late 1980s. Although the concept of an ignition system without a distributor had been pursued some ten years earlier by Citroën for their horizontally opposed twin-cylinder engine, the SAAB system is applied to an in-line four-cylinder turbocharged engine and also features individual small ignition coils mounted directly on the sparking plugs, which accounts for the designation SAAB direct ignition (DI) system.

Functionally this system is of the capacitor discharge type (Figure 10.25). It differs from the conventional or inductive discharge type earlier described in that when a sparking plug needs to be fired, a capacitor that has been charged via a step-up transformer is discharged very rapidly into the primary circuit of the ignition coil. Hence, the step-up in voltage required from the coil itself can be significantly reduced. In other words, the high voltage necessary to fire the sparking plugs is stepped up in two stages, but with greater rapidity and consequently less opportunity for losses to occur. The capacitor discharge type of ignition system therefore applies the primary current, as opposed to interrupting it in the case of a conventional system. A very large sparking plug gap, between 0.9 and 1.5 mm (0.035 and 0.060 in), can be used with this ignition system where 40 000 volts may be reached, which gives the required duration of spark to fire the mixture under all operating conditions and also provides a longer life for the sparking plugs.

The installational features of the direct ignition system are that it requires no distributor, moving parts or ignition leads, because each sparking plug is equipped with its own overhead

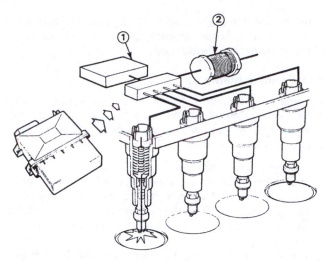

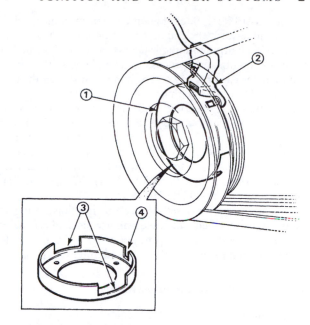

Figure 10.25   Firing the spark with DI system (*SAAB*)

1 capacitor
2 transformer

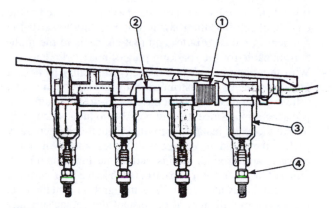

Figure 10.26   Ignition cartridge of DI system (*SAAB*)

1 transformer
2 capacitor
3 ignition coil
4 sparking plug

Figure 10.27   Crankshaft sensor signal (*SAAB*)

1 slotted rotor
2 hall sensor
3 wide slots
4 narrow slot

small ignition coil, the whole system being totally enclosed in a sealed metal cartridge of SAAB patented design (Figure 10.26). This cartridge not only protects the system from flashover that can result from damp, dirty or cracked high-tension surfaces (Section 10.7), but also provides effective shielding against radio interference. Since the system has no distributor, it utilizes the rotation of the crankshaft to signal engine speed and trigger the firing of the sparking plugs at exactly the right instant. For this purpose a slotted rotor is mounted behind the crankshaft pulley and rotates through the gap of a Hall sensor; it has three slots, of which two are wide and one is narrow (Figure 10.27). The wide slots are oriented so that one of them provides the piston position signal for cylinders 1 and 4, and the other similarly for cylinders 2 and 3, while the narrow slot is included to enable the Hall sensor to distinguish between the two wide ones.

In operation the narrow slot will provide an indication before one wide slot passes the Hall sensor, whereas no such indication will be given before the other wide slot passes. This therefore enables the electronic control unit to distinguish between the signals for cylinder pair 1 and 4 and for pair 2 and 3. Although it now becomes possible to identify which pair of pistons is travelling towards top dead centre, a further signal is needed to establish which of the two cylinders concerned is ready for firing. This signal is derived from applying a continuous low voltage to all the sparking plugs, and is based on the principle that the current will only flow in which cylinder combustion has started, because (as earlier stated) the electrical resistance of a gas decreases when it is heated. Hence, by a simple process of elimination the electronic control unit can establish the next cylinder that is ready for firing. A pressure sensor is, of course, used to provide information about the engine load by measuring the pressure in the intake manifold.

As with other programmed electronic ignition systems, the information received from the various sensors is compared with data preprogrammed in the control unit and the necessary commands are issued to the system components. In the SAAB application the electronic control unit is associated with a comprehensive engine management system, which includes a fuel injection and turbocharger operation.

10.6   IGNITION TIMING

The need for correct ignition timing

In the description of the operating cycle of the four-stroke and two-stroke petrol engines (Section 1.3), mention was made that in actual practice the spark to ignite the combustible charge is not timed to occur exactly at the beginning of the power stroke. In fact, it is timed to occur rather earlier than this and towards the end of the compression stroke. This advance of the ignition timing is necessary to allow sufficient time for the combustion process to take place, so that maximum cylinder gas pressure can be attained just after the piston passes TDC and is beginning its power stroke.

As mentioned in Section 10.3, the combustion process tends to occupy a constant period between spark discharge and completion, actually in the order of two milliseconds. Although this is an exceedingly brief interval to imagine in perspective it can be compared say to the five milliseconds that it takes for a piston to make a single stroke when an engine is running at 6000 rev/min.

The static (or initial) setting for the ignition timing is therefore a matter of considerable importance, and is usually such that the spark occurs up to about 10° of crank angle before the piston reaches TDC on the compression stroke.

The effects of incorrect ignition timing

Timing retarded

In this event the combustion process may have only begun just as the piston leaves TDC and is retreating on its power stroke. From this it follows that the maximum attainable cylinder pressure will be less than it would otherwise have been with optimum ignition timing. This is because not only have the gases expended some of their energy against the already retreating piston while they were still burning, but also the volume occupied by the gases when finally burned is appreciably greater than that of the cylinder clearance volume with the piston at TDC. As a result of the accompanying reduction in maximum attainable cylinder pressure, the thermal efficiency of the engine is lowered.

It may be recalled that thermal efficiency represents the ratio of useful work performed by the engine to the internal energy it receives from its fuel. Hence any lowering of the thermal efficiency implies that the power output of the engine will be decreased and its fuel consumption increased. There will also be a greater amount of heat rejected to the cooling system, this increase in operating temperature of the engine typically leading to complaints of overheating with retarded ignition timing.

Timing too advanced

Here it is possible for the combustion process to be completed just as the piston approaches TDC. Although it follows from this that some energy must be expended by the engine for the advancing piston to compress the burning gases, the work performed in this way also increases the pressure and temperature to which they are finally subjected. If this results in the rate of burning of the end gases increasing to such an extent as to cause spontaneous combustion, then as explained in Section 3.2 the outcome will be detonation together with its attendant ills.

As always, of course, in engineering the name of the game is compromise. Therefore the ignition timing recommended by the vehicle manufacturer represents an optimum in the amount of gases burned before and after the TDC position of the piston, so that energy losses are minimized on both the compression and power strokes. The relationship between optimum advance of the ignition timing and that required to avoid detonation finally depends upon the knock resistance of the particular fuel being used.

Static ignition timing

The general procedure for setting the static ignition timing of an engine (after first removing the high-tension cable from the ignition coil as a safety precaution) may be summarized for a non-electronic system as follows:

1 Slowly turn the crankshaft in its normal direction of rotation to bring the piston in no. 1 cylinder on to its compression stroke. This may be ascertained by removing the sparking plug from the cylinder and placing a thumb over the plug hole to feel when the internal pressure is building up.
2 Continue slowly turning the crankshaft until the appropriate timing mark – and here prior reference may be necessary to the vehicle manufacturer's service instruction – on either the front pulley or the flywheel becomes aligned with the timing mark pointer on the engine structure.
3 If the distributor has been removed and then refitted to the engine, check that the rotor arm is pointing towards the segment in the distributor cap that connects to the high-tension cable for the sparking plug of no. 1 cylinder.
4 With the clamping plate loosened sufficiently to permit gentle turning of the distributor body, the latter is turned a little in either direction until the contact-breaker points are just separating. Since it is not possible to judge accurately by eye the exact instant when they separate, the generally preferred method is to connect a 12 V test lamp with a pair of leads and crocodile clips across the low-tension terminal of the distributor and earth (such as the nearby rim of the vacuum advance unit). With the ignition switched on, the test lamp will light at the instant the contact-breaker points are separating, because then the ignition low-tension current is no longer being short-circuited by them but is returning to earth through the test lamp (Figure 10.28). Finally, tighten the distributor clamping plate and restore the entire system to normal.

Dynamic ignition timing

The static ignition timing can also be checked with the engine running at a specified idling speed by the use of a stroboscope (flashing lamp) or, as it is better known for this application, timing light.

If a stroboscope is directed at a rotating disc on which a clear mark has been made and its flashing rate is adjusted so that there is one light flash for every revolution of the disc, then the mark on the disc will appear to become stationary or 'frozen'. This same principle is used when a checking the dynamic ignition timing of an engine with a timing light. In this case the timing light is equipped with a trigger lead that connects to the HT cable of no. 1 cylinder sparking plug (Figure 10.29). The engine is then run at the specified idling speed and the timing light is aimed at the crankshaft pulley on which there is the manufacturer's top dead centre (TDC) timing mark. As the timing light flashes each time no. 1 sparking plug fires, it follows that the timing mark on the

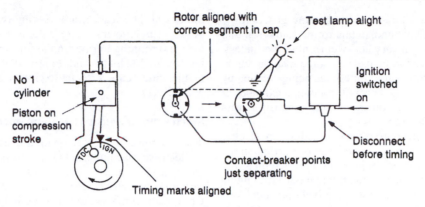

Figure 10.28 Static ignition timing and its related factors for a non-electronic system

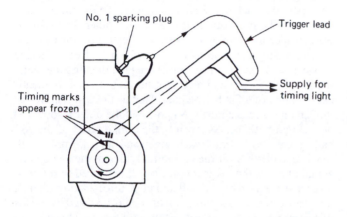

Figure 10.29 Dynamic ignition timing

rotating crankshaft pulley should to all intents and purposes appear to be standing still or 'frozen' in line with the appropriate reference timing mark on the engine timing cover, which then indicates that the ignition timing is correctly set.

10.7 DAMP IGNITION IN SERVICE

Effects of dampness

Dampness on the high-tension surfaces of the ignition system coil, distributor where still used, HT cables and sparking plugs is a common cause of failure to start or difficult starting, because a film of moisture presents an alternative leakage path to earth for electrical energy that would otherwise be available to fire the sparking plugs. This form of dampness arises from condensation of moisture from the atmosphere during periods of high humidity. In wet weather driving conditions the ignition system can also be affected by water entering the engine compartment from road splash or spray from other vehicles, which can at least cause the engine to misfire and at worst to stop it altogether. Even when an ignition system has dried out there can be after-effects from long exposure to damp operating conditions. These effects can arise from localized arcing or flashover across the high-tension surfaces of the system, as the film of moisture begins to break up, such that a conductive carbon track may be gradually burned into the distributor cap or the coil tower. Since this track by natural inclination only wants to form in one direction, that is to earth, the consequent reduction in electrical energy to fire the sparking plug again results in engine misfiring.

Protecting against dampness

The long-recognized problem of dampness has resulted in various detail features of design being incorporated in the ignition system. For example, the distributor where still used can be less affected by moisture if it is provided with the maximum space between each of the electrodes, and also includes a series of sharp-cornered ribs to increase the length of any potential leakage path, since then a higher flashover voltage is demanded before tracking can occur. Additionally, the distributor cap may be sprayed with an anti-track paint. External protection for the distributor cap and its HT cable connections may be afforded by a tight-fitting plastic cover to render it splash-proof. Another example of damp protection is the fitting of a splash-proof cover to the ignition coil tower. Similarly, the plug ends of the distributor HT cables are fitted with protective sealing boots, and a familiar feature of sparking plug design is the prominent anti-flashover ribbing on the ceramic insulator. Anti-damp sprays are also available for treating the ignition system; these are intended to act as either damp prevention or damp dispersion agents, according to the type used. Basic cleanliness of the ignition system equipment is always, of course, an important factor in its reliable operation.

10.8 THE STARTER SYSTEM

General background

The first successful electric self-starter system was that fitted to the 1912 Cadillac. It took the form of a dynamotor that connected directly to the engine crankshaft: in other words, the functions of dynamo and starter motor were combined into a single unit. Such a system would appear to have its attractions in providing a quiet means of starting the engine, but in practice the dynamotor was both cumbersome and expensive to make, because it had to meet the conflicting requirements of being a fast-running dynamo generating a relatively moderate current and, in this form, a slow-running starter motor consuming a very heavy current. Nevertheless, the dynamotor continued to be used by some manufacturers for about another twenty years, sometimes being coupled by a chain drive to the gearbox layshaft, a notable example of this system being that used on the famous English Morris Cowley car of the vintage era.

An entirely different approach to providing an electric self-starter system was that in which a pinion carried on the shaft of an electric motor was brought into mesh with a ring of teeth on the flywheel and then withdrawn once the engine had started. This rather crude but undeniably practical system, albeit much refined, will of course be recognized as that which is still universally used on the modern motor vehicle. However, not all designers of the past were admirers of the system, and the legendary Gabriel Voisin, the French pioneer aircraft and automobile designer, is said to have remarked of its action 'the noise that disgraces the automobile'. Henry Royce was another designer who preferred to modify the idea, and on his Rolls-Royce Phantom I model of the late 1920s, he provided a chain drive from the starter motor to the gearbox layshaft, which was engaged via an electromagnetically controlled jaw clutch.

Two notable advantages were made in 1913 and 1914 that contributed significantly to the early development of the pinion and ring gear starter drive as we know it today. These were the inertia drive, or Bendix pinion as it became popularly known, and the pre-engaged drive, which were introduced by the American Bendix Corporation and the German Bosch Company respectively. Although by the late 1960s the pre-engaged system had become established practice in the American motor industry, it was not until some ten years later that it was generally adopted on passenger cars and commercial vehicles in Britain.

Basic requirements

In order to start a petrol or a diesel engine it must be cranked or turned over fast enough to meet the following requirements:

1 to produce a combustible mixture through agitation of the cylinder charge
2 to develop sufficient energy to fire the mixture through heat of compression
3 to enable the engine to develop enough power to continue firing.

Whether or not an engine can be cranked at sufficient speed therefore depends on the cranking power available and the resistance to cranking offered by the engine, which depends largely upon the viscosity of the oil at the cranking temperature. In practice, the cranking speeds required to start a petrol engine are in the region of 60–100 rev/min and for a diesel engine 80–200 rev/min, although the latter may be reduced where starting aids are used. During initial starting of the engine the starter motor must first overcome the breakaway torque of the engine to get it turning and then the resisting torque to keep it turning at the required cranking speed. In terms of starter motor performance, these requirements are related to what are termed the lock and the running

torques of the starter motor. Taking two examples of starter motor installations on larger petrol and diesel engines, the lock and running torques in the former case would typically be 30 N m (22 lbf ft) and 12 N m (9 lbf ft) respectively and in the latter case 85 N m (62 lbf ft) and 35 N m (26 lbf ft) respectively.

Starter motor construction and operation

The dynamo and the electric motor, are really the same form of electrical machine used in two different ways. If driven by an external source, which in our case is the engine, the machine generates electric current and behaves as a dynamo. Conversely, if the machine is supplied with electric current it can drive another machine, again in our case the engine, and behaves as a motor or a starter motor.

In principle, the operation of an electric motor is dependent upon the fact that unlike magnetic fields attract each other and like magnetic fields repel each other. Therefore if we consider an elementary form of electric motor as comprising a single rectangular coil, which is free to rotate between the poles of a permanent magnet, then the magnetic field created by passing a current through the coil is identical with that of a flat bar magnet. That is, the coil behaves as though one face is a north (N) pole and the other a south (S) pole (Figure 10.30). As a result, the unlike poles of the coil and magnet will attract each other and the like poles will repel each other, so if the coil initially lies in the same horizontal plane as the magnet poles it will be forced to rotate to a vertical position by the pull and push forces acting on it. If now the current passing through the coil is reversed by means of a commutator, the polarity of the coil faces will change to provoke further attraction and repulsion between the poles of the coil and magnet, thereby imposing a continued rotation on the coil, and so on.

Similar to the dynamo, such an elementary form of direct current electric motor requires a good deal of modification before it can perform the arduous duties of say a starter motor. To increase its power it is therefore necessary to provide a large number of coils wound on to a slotted armature with an appropriate commutator, and to replace the permanent magnet between the poles of which the armature is free

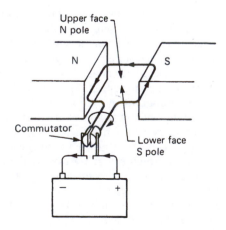

Figure 10.30 Schematic arrangement of a simple direct current motor

to rotate by electromagnetic pole pieces. The manner of winding these differs from that of a dynamo, since the latter is shunt-would whereas the electric motor is series-wound (Figure 10.31). This means that instead of the field windings being connected directly across the brushes, the armature coils and field windings are connected in series. The significance of this change is that the maximum current passes through the armature coils and the field windings, and since the turning effort or torque developed by the motor is proportional to the product of the armature and field currents, a large starting torque can be obtained. However, as the opposing torque on the motor reduces and its speed begins to increase, a smaller current will flow because of the greater voltage induced in the rotating armature coils and opposes that supplied from the battery. In a sense the motor is therefore self-regulating, as it takes automatically just the amount of current it requires, which initially is several hundred amperes.

Although two-pole starter motors were once used for light-duty applications, it has long been established practice to provide four poles and the armature with four brushes to increase power output (Figure 10.32). This may be further increased by connecting the pairs of field windings in parallel and then connecting them in series with the armature, this being known as a series-parallel connection as opposed to series only. In order to cope with the very heavy starting current and its heating effect, flat copper strip of substantial cross-section is typically used for the armature coils and field windings, the latter being wound in opposite directions over adjacent pole pieces to obtain the required difference in magnetic polarity. For light-duty application on some series-only wound starter motors, the field coils are produced from aluminium. Carbon-copper brushes are used with flexible pigtail connectors. An axial instead of a radial brush contact is used with some commutators.

Starter pinion drives

As earlier inferred, the starter motor is equipped with a small drive pinion that is brought into engagement with a large ring gear carried by the engine flywheel, which provides a reduction ratio in the range of 10:1 to 15:1 in favour of the starter motor. This enables the size of both battery and starter motor to be kept within reasonable bounds, while still allowing the starter system to develop the necessary torque for cranking the engine. Less commonly, a further gear reduction may be built into the starter motor drive itself to increase cranking torque. Starter switches are of the ignition key or the press button controlled solenoid type, these having long since replaced the earlier cable hand-pull and plunger foot-push varieties.

The three basic types of starter pinion drive may be conveniently classified as follows: inertia drive pinion, pre-engaged sliding armature and pinion, and pre-engaged sliding pinion.

Inertia drive pinion

In one form of the obsolete inertia or Bendix drive pinion, an extension of the armature shaft carried a spring-retained splined sleeve with an external quick-acting screw thread, which engaged an internally threaded and loose-fitting pinion (Figure 10.32). When the starter motor was operated, the loose-fitting pinion initially (owing to its inertia or reluctance to rotate) revolved more slowly than the armature shaft and as a consequence it was propelled along the threaded sleeve. This process was originally assisted by making the pinion deliberately unbalanced. The resulting endwise movement of the pinion therefore brought it into mesh with the teeth of the flywheel ring gear, engagement being facilitated by generous chamfering of the leading edges of the pinion and ring gear teeth. Once the pinion reached the end of its allowed travel and was forced to rotate bodily with the armature shaft, the action became that of a normal pinion and gear to crank the engine.

Various provisions were made for cushioning the shock of pinion engagement and disengagement. These included the arrangement described in which the endwise reaction of the splined sleeve on pinion engagement and the endwise arrest of the pinion on disengagement were absorbed by a compression spring, and the use of a torsion spring coupling between the armature shaft head and a non-splined threaded pinion sleeve. An overload friction clutch could be incorporated in heavy-duty applications which was capable of slipping and protecting the starter drive under adverse conditions of operation. According to installation requirements, the inertia drive pinion could be arranged to travel either towards or away from the starter motor, these being identified as inboard and outboard types of mounting respectively.

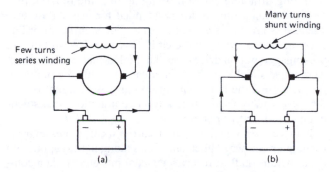

Figure 10.31 Basic difference between starter motor and dynamo field windings: (a) starter motor (b) dynamo

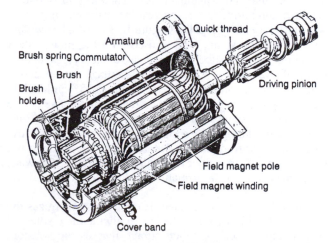

Figure 10.32 Construction of a starter motor with inertia engaged pinion drive

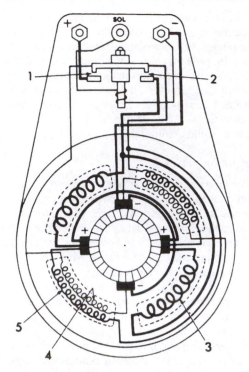

Figure 10.33 Typical internal wiring of an axial starter motor (*Seddon Atkinson*)

1 first-stage contacts
2 second-stage contacts
3 main series winding
4 shunt winding
5 auxiliary series winding

Pre-engaged sliding armature and pinion
Starter motors using this system of pinion drive are based on the original Bosch development earlier mentioned and are also known as axial starter motors. They are now generally confined in application to the larger diesel engine used in heavy vehicles, where owing to the high inertia of the flywheel and crankshaft assembly, the starter pinion is required to engage the ring gear before the starter motor develops full cranking torque, thereby avoiding heavy engagement shock and excessive wear of the teeth. Since engagement between the pinion and the ring gear is effected by a sliding movement of the complete assembly of armature and pinion, a longer than usual commutator is necessary to remain in full contact with the brushes. To protect the pinion teeth against excessive loading, an overload friction clutch is interposed in the drive between the armature and the pinion. It is of the multiplate type, which has a slipping torque above the lock torque of the starter motor but below the shear strength of the pinion teeth.

In operation, the pre-engagement motion of the pinion depends upon a combination of sliding and slow rotation of the armature. For this purpose the field windings of the starter motor consist of a main series winding and auxiliary series and shunt windings (Figure 10.33). The control system for the pinion drive is completed by a solenoid-actuated two-stage switch, which forms an integral part of the starter motor. When the starter is operated, the first of the two-stage contacts on the solenoid switch close and a small current is passed through the auxiliary field windings, causing the armature and pinion assembly to rotate slowly. Simultaneously, the assembly is attracted towards the driving end of the starter motor by the magnetic field created in the windings, which then serves to bring the pinion into mesh with the ring gear. As the sliding armature approaches the end of its travel, a trip plate located just beyond the commutator actuates the solenoid tripping lever mechanism and causes the second of the two-stage contacts to close. This completes the circuit to the main series winding so that the starter motor can now exert its full cranking torque. When the starter is released, the armature and pinion assembly is returned to its disengaged position by a return spring acting inside the partly hollow armature shaft. The auxiliary windings may be arranged either to hold the pinion in mesh until the starter is released, thereby reducing the number of engagements used despite irregular firing of the engine, or (less commonly) to disengage the pinion automatically once the engine fires.

Pre-engaged sliding pinion
This was another early American development of the starter pinion drive. In its original Westinghouse pinion shift form the initially slow rotating pinion was pre-engaged with the flywheel ring gear either electromagnetically or mechanically. It is perhaps of interest to recall that in the electromagnetically operated version, the pinion was carried on a spindle slidably mounted through the armature shaft. This operating principle can still be found in some modern pre-engaged sliding pinion drives intended for heavy-duty use, where it is known as a coaxial drive. The significant difference, therefore, between the early Bosch and Westinghouse pre-engaged pinion drives was that both armature and pinion were made slidable in the former case but only the pinion in the latter; it is the latter type of pre-engagement that is now the most widely used, especially in conjunction with solenoid and shift lever operation.

In this construction, the pinion carrier incorporates an overrunning or one-way clutch and slides along helical splines on the armature shaft. It receives its axial movement via a compression spring and sliding collar, which is engaged by a solenoid-actuated shift lever. The purpose of the helical splines is first to facilitate the meshing process by turning the pinion as it slides and then to augment the engagement thrust once it is transmitting drive to the ring gear. An overrunning or one-way clutch of either the roller (Figure 10.34), or in heavy-duty applications the sprag, type is used with the pinion, their operating principles being identical to the corresponding types of overrunning clutch later described in connection with epicyclic automatic gearboxes (Figures 15.6 and 16.17). The overrunning clutch is incorporated as a protective device for the starter because, although it maintains a positive drive to the pinion during engine starting, it also prevents the flywheel ring gear from driving the armature if the starter is not released immediately after the engine has started. Otherwise, the overspeeding of the armature could dislodge its coils by centrifugal effect.

When the starter is operated its circuit first energizes the pull-in and hold-in coils of the externally mounted solenoid switch (Figure 10.35), which attract inwards the solenoid plunger against its return spring. An opposite movement is therefore transmitted, via the centrally pivoted shift lever, to

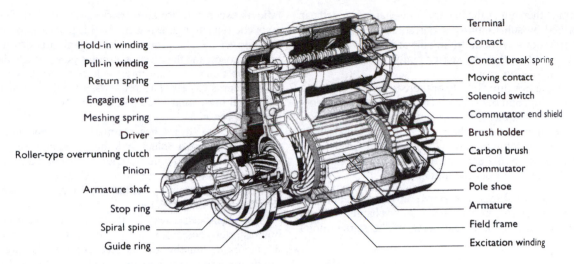

Terminal
Contact
Contact break spring
Moving contact
Solenoid switch
Commutator end shield
Brush holder
Carbon brush
Commutator
Pole shoe
Armature
Field frame
Excitation winding

Hold-in winding
Pull-in winding
Return spring
Engaging lever
Meshing spring
Driver
Roller-type overrunning clutch
Pinion
Armature shaft
Stop ring
Spiral spine
Guide ring

Figure 10.34   Cutaway view of a light vehicle starter motor with pre-engaged drive (*Bosch*)

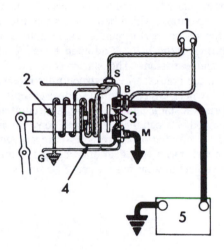

Figure 10.35   Basic solenoid circuit (*Seddon Atkinson*)

1 starter switch	5 battery
2 hold-in windings	S switch
3 contacts	B battery
4 pull-in winding	M motor

the collar of the pinion carrier. This allows the pinion to slide into mesh immediately, always providing it meets a tooth space. Shortly before the end of the pinion engagement travel, the contact moving with the plunger closes the starter switch and completes the circuit to the series windings of the starter motor, so that it can now develop full cranking torque. Since the hold-in coil alone can now exert sufficient attraction in the plunger, the pull-in coil, which is wired in parallel with the solenoid switch contacts, is de-energized as soon as they close. If during its engagement travel the pinion abuts against a tooth, instead of finding a tooth space, the continuing movement of the shift lever causes the collar to slide along the pinion carrier and compress the pinion spring, until the plunger closes the solenoid switch. The starter motor then turns the pinion until it is sprung into the nearest tooth space and engagement is effected. When the starter is released, the hold-in coil is de-energized and the return

spring of the solenoid plunger causes the shift lever to disengage the pinion. To ensure that the starter motor comes to a standstill promptly, so that if necessary a further attempt can be made to start the engine, an armature brake may be incorporated. For this purpose the opposite face of the sliding collar for the pinion carrier may be utilized as a simple disc brake, when it is urged towards its rest position against a suitable thrust surface on the adjacent end cover of the starter motor.

Advantages of the pre-engaged starter pinion drive

As compared with an inertia starter pinion drive, the advantages of the pre-engaged type of drive may be summarized as follows:

1 A smaller pinion can be utilized to provide a higher reduction ratio in favour of the starter motor, so that cranking torque is increased to improve cold starting capability.
2 The pinion drive is not so highly stressed, because less energy has to be absorbed during pinion engagement, especially under difficult starting conditions.
3 Prolonged cranking of a diesel engine is possible without premature disengagement or the pinion repeatedly travelling in and out of mesh as resisting torque fluctuates.
4 Repeated travelling in and out of mesh of the pinion is also avoided when starting modern petrol engines with high compression ratios, and also those where flywheel inertia is reduced with automatic transmission installations.
5 By virtue of advantages 3 and 4 a major source of excessive wearing of the starter pinion and ring gear teeth is eliminated.

In situ checking for starter motor faults

A simple procedure is to switch on the headlamps or connect a 0–20 V voltmeter across the battery terminals, operate the starter motor and check for the following symptoms:

1 If the headlamps dim noticeably or there is an appreciable drop in voltmeter reading, but the starter motor is not heard

to operate, then either the battery is discharged (which can be verified by substitution) or current is flowing through the starter motor without driving the armature. One reason for this could be that the starter pinion has jammed into mesh with the ring gear.

2 If the headlamps dim slightly or there is little drop in voltmeter reading, and the starter is heard to operate but does not crank the engine, the overrunning clutch is slipping.

3 If the headlamps retain their full brilliance or the voltmeter reading remains unaffected, but the starter is not heard to operate, the starter circuit should be tested for continuity and the solenoid unit checked.

The usual cause of sluggish or slow cranking of the engine, apart from the battery being in a low state of charge, is a loose connection causing a high resistance in the starter motor circuit.

Further developments of the starter motor

The continuing development of permanent magnetic materials that posses a greater energy content and are insensitive to demagnetizing influences led to the introduction in the mid 1980s of starter motors from which the electromagnetic pole pieces and field windings had been eliminated. This simplification of the electrical circuit therefore allowed the starter current to flow directly to the motor brushes and armature. The rated output of this type of starter motor is related to its length of armature and associated permanent magnets.

For application to engines of over 2 litres capacity there is a requirement for six permanent magnets, and it also becomes necessary to incorporate a reduction gear-set of about 4.5:1 ratio between the armature and starter pinion shafts. An epicyclic reduction gear is used for this purpose; the armature shaft drives the sun gear, the starter pinion is driven by the planet carrier, and the plastics annulus gear, which also combines the function of an intermediate bearing, is anchored in the motor casing. The principle of epicyclic gearing is explained in Section 16.2, with Figure 16.3a being applicable in this case.

Among the advantages claimed for permanent-magnet starter motors are a reduction in both size and weight for an equivalent or higher rated output as compared with a conventional field-wound machine, and greater reliability in service by eliminating potential short-circuits between field windings and motor frame.

There has been a revival of interest in developing combined starter-generator systems. This has been in response to the increasing number of electrical services that are appearing on modern passenger cars, which may ultimately call for higher voltage systems and hence greater power generation. Such systems can also provide unobtrusive and vibration-free starter operation, which better lends itself to quick and reliable stop-restart performance, where automatic engine cut-out during traffic halts is adopted to reduce fuel consumption and emissions. The starter-generator is directly driven from the crankshaft and structurally integrated with either the flywheel, clutch or torque converter units. It basically comprises a stator and rotor with a circumferential air gap maintained between them. Unlike a conventional starter and generator installation, there is no longer a requirement for drive belts, starter ring gear and engaging pinion, all of which are subject to wear in service.

11 Engine emission control

11.1 PETROL ENGINE POLLUTANTS

The nature and sources of pollutants

In reviewing the modern requirements of the motor vehicle petrol engine (Section 1.1), it was mentioned that since the late 1960s (beginning in America) increasingly stringent legislation has been applied to limit the levels of atmospheric pollutants emitted from it. It may be recalled that the basic pollutants were identified as being carbon monoxide, hydrocarbons and nitrogen oxides, all of which can pose a health hazard.

Carbon monoxide (CO) is a product of incomplete combustion which is difficult to disperse in air and can aggravate respiratory ailments. It is well known as being a dangerously poisonous gas to breathe and for this reason one should never run an engine with the vehicle in a small closed garage, since a high concentration of carbon monoxide in the air can prove fatal in a very short time.

Hydrocarbons (HC) are also products of incomplete combustion and are generally referred to as unburned hydrocarbons. Another source of hydrocarbon pollution is that arising as vapour vented from a fuel system. It is their photochemical reaction in the presence of sunlight that can lead to the formation of dangerous smog, especially in still air, which promotes irritation of the eyes and lungs.

Nitrogen oxides (NO_x) tend to be formed at high combustion chamber temperatures and can take several forms, depending on the amount of oxygen that unites with the nitrogen. Some oxides of nitrogen are major participants in photochemical smog reactions that again aggravate respiratory ailments, and others form acids in the droplets of water in the air that cause irritation to the eyes and throat.

Lead compounds also occur in the exhaust gas in a variety of chemical forms. Ever since the early 1920s when Thomas Midgley of General Motors in America first discovered the value of tetra-ethyl lead as an anti-knock additive for petrol, it was recognized that a substantial part of the lead would be emitted with the exhaust gas and its toxicity could pose a health hazard. Many authorities did in fact study this problem but concluded that this was not likely, for example in 1972 the National Academy of Sciences of America reported that 'lead attributable to emission and dispersal into general ambient air has no known harmful effect'. However, this does not concur with present day thinking on the subject, which accounts for the now established use of unleaded petrol as defined by the European Union emission regulations and similarly by the American Society for Testing and Materials. Apart from health considerations, the presence of lead in petrol would also render inoperative the exhaust system catalytic converter and the lambda oxygen sensor of the engine management system.

Permitted levels of pollutants

Since the early 1990s the European Union has issued Directives that specify maximum permitted levels of pollutants for petrol and diesel engined vehicles. These Directives are periodically amended and defined as 'Euro' standards, which so far have progressed through Euro 1 (1992), Euro 2 (1996), Euro 3 (2000) and Euro 4 (2005). Accompanying these increasingly stringent limits on exhaust pollutants have been further requirements, such as long-term durability of the emission control system together with an onboard diagnostic facility to detect any malfunctioning.

Reference should always be made to the appropriate EU Directives for up-to-date information on exhaust emission standards and their approved test cycles. For background information the Euro 4 maximum permitted levels of exhaust pollutants for passenger car petrol engines, expressed in g/km, are as follows:

Carbon monoxide (CO) 1.0, Hydrocarbons (HC) 0.10 and Nitrogen oxides (NO_x) 0.08.

Conflicting requirements and the need for emission controls

The exhaust emissions from an engine are mainly influenced by the region of air–fuel ratio in which it is operating. To achieve an acceptable compromise between best economy and maximum power, engines are usually tuned to operate in the stoichiometric region which, as explained in Section 6.2, implies an air–fuel ratio that corresponds exactly with the right amount of oxygen from the air to burn the fuel. However, this does not mean that the levels of pollutants emitted from the engine exhaust are environmentally acceptable. Furthermore, if either best economy or maximum power is a priority requirement for an engine, then the air–fuel ratios must be respectively quite lean and just rich of stoichiometry.

If in the interests of maximum power the air–fuel ratio is quite low and there is an excess of fuel, the quality of combustion will be poor and the emissions of both hydrocarbons and carbon monoxide will be high. These can be reduced by moving the air–fuel ratio to a region that is just rich of stoichiometry, which in fact better enables maximum power to be attained, but also incurs a penalty of increased emissions of nitrogen oxides. Should the primary aim be best economy and the air–fuel ratio moved across the dividing line to a region just lean of stoichiometry, then combustion temperatures will be raised to a maximum and produce the highest emissions of nitrogen oxides. These can be much reduced and the emissions of carbon monoxide kept to a minimum by further weakening the mixture until it comes into the 'lean-burn' region, which has been defined as one where the air–fuel ratios lie between 18:1

and 21:1, but the gains mentioned are bought at the expense of increased emissions of hydrocarbons and the possibility of rough running, if there is insufficient fuel for the mixture to burn properly.

Summing up, it will be evident that no matter whether the air–fuel ratio is lean of stoichiometry, stoichiometric or rich of stoichiometry, the engine will still produce environmentally unacceptable levels of pollutants. Since the predominance of any one or more of these is akin to a swings and roundabouts situation, it explains the need for a comprehensive system of emission control, which in modern practice takes the form of a catalytic converter.

11.2 PETROL ENGINE EMISSION CONTROL

General background

The many and varied methods that have been used to control harmful emissions from petrol engines since the early 1960s now virtually constitute a separate branch of automotive engineering, and indeed the subject is sometimes referred to as emission control technology. It will be recalled that references have been made to certain features of emission control, such as fuel vapour recovery (Figure 6.1b), positive crankcase ventilation (Figure 4.46) and those related to combustion chamber design, carburation and fuel injection.

A turning point in emission control came in 1976, when the increasing severity of legislation in America resulted in catalytic converters becoming a vital part of the exhaust emission equipment of all American cars by the late 1970s. This situation was to repeat itself on European cars.

Pre-catalytic-converter emission controls

Of the pre-catalytic-converter emission control systems, the air injection, exhaust gas recirculation, and thermal reactor systems are deserving of brief mention.

Air injection system
Its purpose was to promote the oxidation of any residual hydrocarbons or carbon monoxide. A typical installation comprised an engine-driven compressor that delivered filtered air at low pressure to each exhaust valve port. This was accomplished by means of individual nozzle and tube assemblies that connected to a distribution manifold, which in turn was connected to the air compressor via a supply hose. The oxygen in the air so delivered therefore combined with the unburned combustion gases to promote more complete combustion and less air pollution. A check valve was included at the entrance to the distribution manifold, so that in the event of the exhaust gas pressure exceeding that of the air delivery, it would close and prevent exhaust gases flowing back to the compressor. An anti-backfire valve was also incorporated in the system. This method of control was successful but expensive to apply and also incurred a penalty on fuel consumption, owing to the necessity of using a rather richer than normal air and fuel mixture and the power required to drive the compressor.

Exhaust gas recirculation system
An effective method of reducing the emission of nitrogen oxides is to dilute the ingoing air and fuel mixture with relatively inert exhaust gas, which is bled from the exhaust manifold and routed to the intake manifold. The purpose of recirculating about 15 per cent of the spent exhaust gases in this manner is to reduce the initial formation of nitric oxide, the principal oxide of nitrogen exhaust emission, by lowering both the flame speed and the peak temperature attained in the engine combustion chambers. In practice, it was found necessary to modify the system so that exhaust gas recirculation could be eliminated below normal running temperatures to improve engine response and also during idling to avoid rough running, and reduced under full-throttle running to obtain maximum engine performance. To meet these requirements a variable-flow metering valve, usually referred to as the EGR valve, was included in the system and made responsive to intake manifold depression and coolant temperature (Figure 11.1). Used alone this system was eventually unable to achieve the low level of NO_x emission demanded by American legislation.

To make acceptable the level of NO_x emissions from some recently introduced direct injection petrol engines (Section 6.9), which can operate in an ultra lean combustion mode, it has been found necessary to recycle a large volume of exhaust gas back into the engine to reduce peak temperatures. An EGR system is therefore used in conjunction with a specially developed lean catalytic converter, because conventional three-way catalytic converters can only be effective when an engine is operated within a very narrow margin of its stoichiometric point as later explained.

Thermal reactor system
Another method of limiting the amount of unburned hydrocarbon and carbon monoxide gases emitted from the engine is to replace the conventional exhaust manifold by a heavily insulated and larger-capacity thermal reactor, which acts in effect as a secondary combustion chamber. That is, it allows further combustion of the exhaust gases to take place by increasing the effects of both temperature and time on their

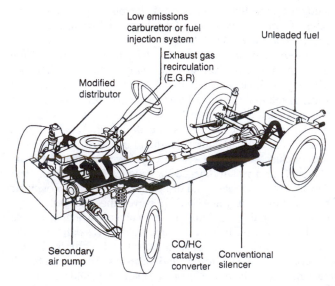

Figure 11.1 Installation of early oxidation catalytic converter (*Johnson Matthey*)

passage from the engine to the exhaust system proper. To assist further the oxidation of the hydrocarbons and carbon monoxide remaining in the exhaust stream, an engine-driven air injector compressor was sometimes used to force fresh air into the thermal reactor; alternatively the engine was run on a very lean mixture. This led to thermal reactors being classified as rich or lean respectively. The internal temperatures of the thermal reactor could be up to 1000°C and therefore demanded expensive materials of construction and also led to various problems arising from very high underbonnet temperatures.

Catalytic converters

Although the possibility of using catalytic converters was investigated early in the history of emission control systems, it was not until the mid 1970s that the increasing severity of American legislation led to a revival of interest in their application.

As we learn in chemistry, *catalysis* refers to the acceleration (or retardation) of a chemical reaction by a substance identified as a catalyst, which in itself undergoes no permanent chemical change or which may be recovered when the reaction is completed. It is a surface action phenomenon, and for this reason the maximum possible surface areas are utilized to attain the highest efficiencies of conversion.

In construction a catalytic converter is similar in external appearance to an ordinary silencer, but inside the catalytic system comprises either a ceramic or a more robust metallic honeycomb element, through which the exhaust gases pass en route to the silencing part of the exhaust system. This element is variously termed the 'substrate' or 'monolith' and its exceedingly large surface area is further increased by the application of a rough textured washcoat, before being coated with a fine 40–50 microns layer of the catalyst. The latter is normally a mixture of noble metals (not easily corroded) with either platinum, or palladium, being used to accelerate the oxidation of hydrocarbons and carbon monoxide and rhodium to reduce the oxides of nitrogen. A ceramic substrate is supported within a sleeve of stainless steel wire mesh, which not only cushions the element against exhaust system vibrations, but also accommodates its rapid thermal expansion and contraction characteristics. The substrate and its supporting medium is then enclosed within a casing fabricated from stainless steel. In contrast a metallic substrate may be welded to the stainless steel casing for greater durability. The casing itself is formed with conical end portions where these connect to the exhaust system piping and their purpose is to assist gas flow through the converter.

We also learn in chemistry that the addition of heat speeds up a chemical reaction and this is certainly no less true of a catalytic converter. In fact it shows little enthusiasm for starting work until a temperature of about 250°C has been attained, this often being referred to as the 'light-off' temperature and is generally reached within a few minutes of starting from cold. To assist this process the catalytic converter is mounted as near to the exhaust manifold as practicable, while the intervening downtake pipe can be thermally insulated by either encasing it with a heat-resistant lagging material, or making it double-skinned to provide an air-gap barrier against heat loss in the manner of a double-glazed

window. During normal running conditions, the catalytic converter maintains an operating temperature of about twice its light-off temperature and for this reason it is generally equipped with protective heat shields. Potential service problems with catalytic converters have earlier been indicated (Figure 7.12).

The earliest type of catalytic converter to be used acted solely as an oxidizing catalyst and was fitted on passenger cars to meet the 1975 emission standards in America. It therefore controlled the emissions of hydrocarbons and carbon monoxide and was usually combined with air injection from an engine-driven compressor (Figure 11.1). The purpose of the added or secondary air was to provide sufficient oxygen to complete the combustion of air–fuel mixtures that were rich of stoichiometry. It left the nitrogen oxides to be controlled separately, either by exhaust gas recirculation or by limiting their formation in the engine itself. The latter may be accomplished by generating a high degree of induction swirl for the incoming mixture (Section 3.6), so that much weaker mixtures with air–fuel ratios in the lean-burn region can be utilized.

A two-way catalytic converter system was the next type to be used and was so called because it provided a reducing catalyst to control the oxides of nitrogen and an oxidizing catalyst to control unburned hydrocarbons and carbon monoxide. Since the former reaction by its nature requires an oxygen-free atmosphere and the latter reaction an excess of oxygen, the two systems were separately housed so that additional or secondary air could be introduced between them. In later practice a dual-bed system was used in the same housing. Hence, the two catalytic systems are arranged in series such that the exhaust gases first pass through the reducing catalyst and then through the oxidizing catalyst, and finally leave the exhaust system via a conventional silencer (Figure 11.2). A disadvantage of the two-way catalytic converter was that it required the engine to operate on a richer than normal mixture, with an adverse effect on fuel consumption.

By 1980 a three-way catalytic converter system had been developed jointly through Volvo in Sweden and Bosch in Germany. This system differs from the two-way catalytic converter by virtue of it promoting reactions actually among those constituents of the exhaust gases that are subject to control. In effect, the unburned hydrocarbons and carbon monoxide remove the oxygen from the oxides of nitrogen, thereby forming harmless water, carbon dioxide and nitrogen. However, a three-way catalytic converter only possesses this capability in full when the engine is being operated within a very narrow margin of its stoichiometric point, so that an extremely precise means of monitoring and adjusting the air–fuel ratio must be used. In practice, this means introducing a regulating oxygen sensor into the exhaust stream that responds electrically to changes in oxygen content of the gases or, more specifically, measures the amount of excess oxygen in the exhaust relative to that in the air. The oxygen sensor continually signals this information on the completeness of combustion to the electronic control unit of the engine management system, which then issues the appropriate commands for adjusting the mixture to either the fuel injection system or, less commonly, an electronically controlled carburettor. This provides what is sometimes referred to as a 'closed-loop' control system (Figure 11.3). The oxygen sensor is also known as a Lambda probe

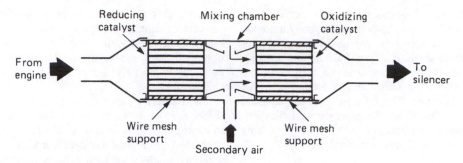

Figure 11.2 Schematic arrangement of a dual-bed catalytic converter

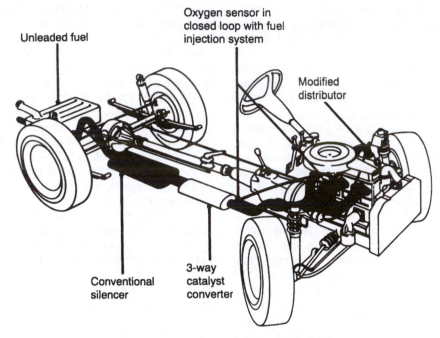

Figure 11.3 Installation of modern three-way catalytic converter (*Johnson Matthey*)

from when it first appeared in the earlier mentioned Volvo installation of a three-way catalytic converter, it being so called after the Greek letter λ (lambda) which is used by combustion specialists as a symbol to denote relative deviation from stoichiometry.

11.3 DIESEL ENGINE POLLUTANTS

Causes of smoke emission

The maximum power developed by a diesel engine is limited in practice to that which can be obtained without the emission of significant amounts of smoke. It may be recalled that in the diesel engine there is a much shorter period available for mixing the air and fuel as compared with the petrol engine. If then some of the injected fuel fails to find sufficient oxygen for complete combustion before the exhaust valve opens, the engine will emit a certain amount of black smoke, because fuel burned with a deficiency of oxygen produces soot. From the design point of view, the efficiency of a diesel engine

combustion system can therefore be judged by its tolerance of a high air–fuel ratio for better mixing, without the onset of black exhaust smoke that becomes especially noticeable when the engine is under load. Apart from readily visible black smoke, which is particularly objectionable when continuously emitted and has been linked to adverse health effects, other forms of diesel exhaust smoke are identifiable as white/grey and blue. The former may be attributable to slow and incomplete combustion of the fuel at low temperatures when the engine is first started, but it can be caused by worn piston rings or valve seating deficiencies that allow a loss of cylinder compression, and also by late injection timing. Blue smoke that is still noticeable when the engine has reached its normal operating temperature can arise from engine mis-fire and incomplete burning of the fuel, which tends to indicate faulty spraying of the injectors, while another cause is over-lubrication if the amount of oil gaining access to the cylinders is excessive. However, once an engine has attained its normal operating temperature, there is generally little tendency for it to emit white/grey or blue smoke.

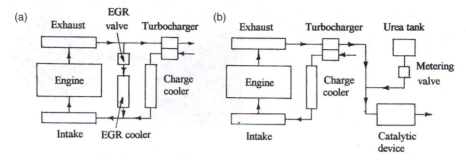

Figure 11.4   Nitrogen oxide emission control systems for heavy-duty diesel truck and bus engines: (a) exhaust gas recirculation (b) selective catalytic reduction

In service, it was once estimated that about 80 per cent of the objectionable black smoke emitted from diesel engined vehicles could be attributed either to the irresponsible use of the injection pump excess fuel device, or to inadequately maintained fuel injection equipment. The latter shortcomings typically include retarded injection timing, a defective fuel injector allowing dribble at the nozzle, and an incorrectly adjusted maximum delivery control stop for the governor. Other possible causes of a smoky exhaust are too restricted an air filter element or high back-pressure in the exhaust system.

It should be noted that the Vehicle Inspectorate Enforcement Group have published a leaflet titled 'How to report smoky diesels', which encourages members of the public to play a part in helping the environment by spotting and reporting large diesel engined vehicles that are extremely smoky.

Assessing the density of smoke emission

Although human observation and judgement are often used to assess the density of smoke emission from a diesel engine, a more accurate and consistent method involves direct measurement of a smoke sample by what is termed a smokemeter. This type of instrument may be of either the opacimeter or the spotmeter variety. In the former case the smokemeter measures the amount of light that fails to penetrate a portion of exhaust gas continuously flowing through a sampling tube, whilst in the latter case it measures the amount of soot particles collected by passing a sample fixed volume of exhaust gas through a filtering medium of controlled density. Since 2000 a smoke opacity test has been included in the European Union Emissions Standards for heavy-duty diesel truck and bus engines.

Permitted levels of pollutants

The Euro standards for maximum permitted levels of pollutants for heavy-duty diesel truck and bus engines have run in parallel with those previously mentioned for passenger car petrol engines. Similarly, the EU standards I through to IV have become increasingly stringent in their requirements to reduce these pollutants. Background information the Euro IV maximum permitted levels of exhaust pollutants for heavy duty diesel engines, expressed in g/km, is as follows:

Carbon monoxide (CO) 1.5, Hydrocarbons (HC) 0.46, Nitrogen oxides (NO_x) 3.5 and Particulate matter (PM) 0.02. As earlier stated a smoke test is also a requirement.

Here it should be mentioned that a particulate is a tiny particle that consists of about one-half carbon or soot which has escaped the combustion process and one quarter sulphate from the fuel, the remainder being mainly lubricating oil and a small amount of fuel.

It is perhaps ironical that the inherently greater efficiency of the diesel engine combustion process, accompanied as it is with intense heat and pressure, is also responsible for producing its major pollutant, namely oxides of nitrogen (NO_x). Although this can be reduced by retarding the fuel injection timing, as earlier mentioned in Section 3.5, this action conflicts with reducing the emission of particulate matter. Also, fuel consumption can likewise suffer through incomplete combustion. To meet Euro IV standards for reducing NO_x, the manufacturers of heavy-duty diesel engines are employing either intercooled exhaust gas recirculation EGR to moderate peak combustion temperatures (Figure 11.4a), or selective catalytic reduction SCR (Figure 11.4b). The latter basically comprises a special catalytic device in the exhaust system, which is activated in the presence of ammonia so that nitrogen oxides in the exhaust gas are converted into nitrogen and water before passing to the atmosphere. In practice, the system requires a solution of urea to be metered into the exhaust gas before it enters the catalytic device, the urea being stored in a separate tank carried on the vehicle.

11.4   DIESEL PARTICULATE FILTERS

The purpose of this type of exhaust after-treatment is to reduce the emission of soot particles, which are released into the atmosphere at full load and during transitory operation of a diesel engine. As previously noted, the emission of particulate matter (PM) is subject to increasingly strict EU legislation on permitted levels of pollutants. To meet this requirement a ceramic filter can be installed in the exhaust system to capture soot particles suspended in the outgoing exhaust gases, but the process is not one of simple filtration. A regenerative process must also be applied periodically to burn off the particles that accumulate in the pores of the filter. Since this process involves generating a temperature of 550°C, or more than twice that normally expected from the exhaust gases of a diesel engine, the shortfall in temperature is sought by briefly altering the characteristics of the fuel injection system. In practice this means adding a post (after

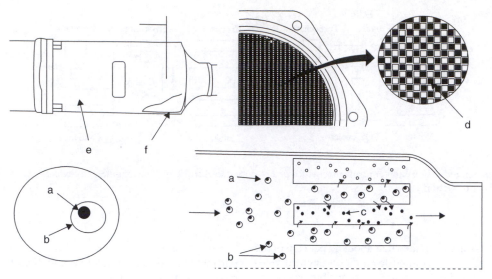

Figure 11.5   Diesel particulate filter (Citroën UK Ltd)
key:

a Carbon particles   d Porous ceramic walls
b Cerine   e Stainless steel casing
c Filtered exhaust gases f Heat insulator

TDC) injection of fuel to raise the combustion temperature and hence the initial temperature of the exhaust gases entering an oxidation catalytic converter, which is located upstream of the particle filter. The chemical reaction in the catalytic converter then further increases the temperature of the exhaust gases passing through the particle filter, thereby attaining the required burn-off temperature to incinerate the trapped soot particles.

An example of a comprehensive diesel particle filter system is that adopted by Citroën for their DW12 TED 4 passenger car engine (Figure 11.5). Although operating in a similar manner to that already described, the regeneration temperature is lowered by an additive injection system in the fuel tank, which is managed by an electronic control unit (ECU). The fuel additive contains Eolys, which chemically is a cerine-based composite that lowers the combustion temperature of the soot particles from 550°C to 450°C. It is stored in an additional tank near to the main fuel tank. The filter should be cleaned and the additive tank replenished every 50 000 miles (80 000 km) by an authorised Citroën dealer.

12 Rotary piston engine

12.1 ROTARY PISTON ENGINE

General background

During the history of the automotive engine many proposals have been made, and as many rejected, to replace the conventional reciprocating motion of the pistons by one of rotation. It was not until the late 1950s that rotary piston engine theory began to be successfully translated into practice, as exemplified by the Wankel engine (Figure 12.1). Even so the only company that has long persevered with a Wankel engined car has been the Japanese manufacturer of Mazda cars, although engines operating on the Wankel principle may be found in other transportation and industrial applications. The Wankel or rotary piston engine owes its original development to the collaboration of two German engineers, Felix Wankel and Walter Froede, following many years of fundamental research by the former after whom, of course, the principle is named. Its development in more recent years, at least as far as its application to the motor car is concerned, owes much to the Japanese engineer Kenichi Yamamoto of Mazda, who has defined this type of engine as 'an internal combustion engine that performs the four strokes of intake, compression, expansion and exhaust while the working chamber changes its volume and the moving parts always rotate in the same direction'.

Operating principle

The Wankel engine essentially comprises a piston of triangular form that orbits within a chamber of approximately figure-of-eight shape, which is embodied in a stationary housing. Sited on opposite sides and adjacent to the minor axis of the chamber are the gas porting arrangements and the twin sparking plugs. The latter are installed in leading and trailing positions with reference to the minor axis, so that the additional trailing one serves to minimize delay in combustion. Since the piston not only rotates about its own axis, but also revolves eccentrically about the output shaft that carries it, a pair of internal phasing gears acting between the piston and chamber is required to preserve correct orbital motion of the piston throughout the operating cycle.

As the piston orbits it creates three small chambers of constantly changing volume, which accommodate the induction, compression, power and exhaust phases of the four-stroke cycle, these being most conveniently presented by illustration (Figure 12.2). On the power phase, a turning moment or torque is imparted to the output shaft by the combustion pressure acting upon the projected area of the piston flank, the resultant force from which is directed through the centre of the piston at the distance of the shaft eccentric throw (Figure 12.3).

It should be noted that the Wankel engine delivers a power impulse for every revolution of its output shaft, in contrast to every other revolution of the crankshaft in a reciprocating engine similarly operating on the four-stroke cycle. On this basis, a single-piston rotary engine may therefore be compared to a twin-piston reciprocating engine. This accounts for the claimed advantages of the Wankel engine that it occupies less space and offers a reduction in weight relative to its power output. On the debit side, the moving combustion space of this type of engine lacks compactness, which is not conducive to rapid combustion and high efficiency and can pose problems relating to fuel consumption and exhaust emissions.

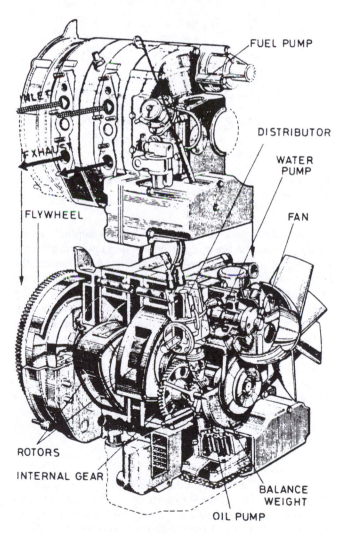

Figure 12.1 Layout of the Wankel engine

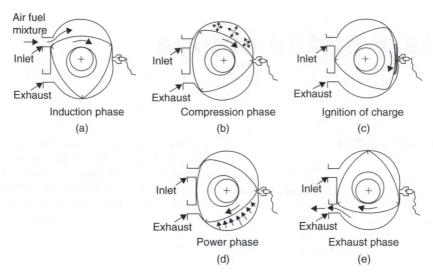

Figure 12.2 Operating cycle of the Wankel engine: chamber size (a) increased to maximum (b) decreasing to minimum (c) at minimum (d) increasing to maximum (e) increased to maximum (note: all chambers are in continuous action)

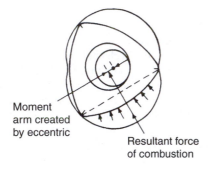

Figure 12.3 Torque generation in the Wankel engine

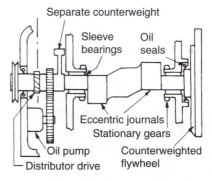

Figure 12.4 Eccentric output shaft assembly in a twin-rotor Wankel engine

Basic mechanism and structure

Engines of this type basically comprise the following components.

Eccentric output shaft

This is the equivalent component to the crankshaft of a conventional reciprocating piston engine. A one-piece steel construction is used, and in the case of a two-piston or twin-rotor engine the eccentric journals are diametrically opposed to provide even firing intervals (Figure 12.4). Counterbalance weights are either formed integrally with the flywheel and the front pulley, or a combination of counterweighted flywheel and separately attached front counterweight may be employed. A spur gear and a skew gear are mounted on the nose portion of the output shaft, as part of a drive system for the oil pump and ignition distributor respectively. Their methods of retention and also that for the flywheel at the rear end of the shaft are similar to those used on a conventional crankshaft assembly.

Output shaft bearings

In twin-rotor engines the rigidity or stiffness of the output shaft is usually such that it can be supported in only two main bearings. Apart from the gas loading applied via the rotary pistons, the centrifugal force acting upon each eccentric journal and piston assembly also tends to deflect the shaft. Since this deflection must be resisted by the main bearings, their loading can be correspondingly reduced by the simple system of counterweighting described in the previous section. Sleeve-type main bearings are assembled on to each end of the output shaft and are of a similar composite construction to the thin-walled half-liners that support a conventional crankshaft. They are retained by interference fit in the central bore of each of the stationary phasing gears, which in turn are mounted through the front and rear covers of the engine. Endwise restraint for the output shaft may be obtained by either plain collar, or rolling contact, thrust bearings.

Rotary pistons

These take the shape of an equilateral triangle with convex flanks, which are pocketed to form part of the moving combustion space and also to assist gas transfer across the minor axis of the chamber. A box-section construction that allows internal oil cooling is used, the material being cast iron for

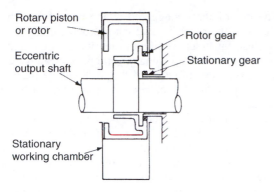

Figure 12.5 Rotary piston and phasing gears

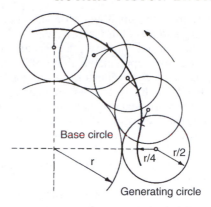

Figure 12.7 Epitrochoid geometry of the Wankel engine

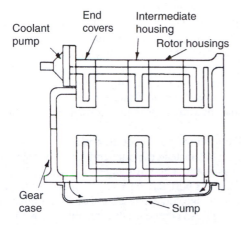

Figure 12.6 Main structure of the Wankel engine

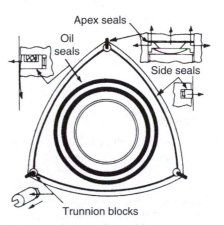

Figure 12.8 Rotary piston sealing grid

durability (Figure 12.5). Each piston embodies a sleeve bearing of the type that supports the output shaft main journals. An internally toothed phasing gear is recessed into the outer side of each rotor and is retained therein, both axially and rotationally, by a ring of spring pins. The sealing arrangements for the rotary pistons are dealt with separately.

Main structure

A twin-rotor engine is built up in the manner of a double-bank sandwich and therefore bears little structural resemblance to its reciprocating piston engine counterpart. It comprises two rotor housings and an intermediate housing, which are flanked by front and rear end covers (Figure 12.6). These parts are located relative to one another by dowels and clamped together by a series of long through-bolts. A double-skinned and internally ribbed construction is used for the main housings and end covers, thereby allowing for the passage of coolant either through or around them. The generally preferred material for the rotor housings is aluminium alloy, which by virtue of its good thermal conductivity minimizes any tendency towards localized overheating. A wear-resistant coating is applied to the working surface of the piston chambers and may be either metallic or non-metallic in character to complement the piston apex seal material.

Fundamental to the Wankel engine is the geometric form of its operating chamber; it is defined as a two-lobed epitrochoid, which perhaps sounds rather off-putting. However, this particular geometric form is simply traced by a given point within a generating circle as it rolls around the outside of a fixed base circle, the radius of the latter being twice that of the former (Figure 12.7). In practice, the given or selected point is located at one-half the radius of the generating circle. The working surface of the rotor chamber is sometimes referred to as the epitrochoid surface.

Sealing the rotary piston

One area that has received intensive research in the short history of the production Wankel engine is the relatively complex gas and oil sealing arrangements of its rotary piston. This is perhaps only to be expected, because the present-day high standard of gas sealing and oil control in the reciprocating piston engine is the result of much sophisticated development work done over a period of many years.

The gas sealing of the rotary piston in its operating chamber is accomplished by what is termed a sealing grid (Figure 12.8). This essentially consists of three apex seals carried radially in trunnion blocks, which are linked by curved sealing strips located in slots on each side of the rotary piston. As with conventional piston rings, the seals are spring loaded against their working surfaces to provide initial sealing pressure, which is then greatly augmented by chamber gas pressure acting behind them when the engine is running. Again similar to the conventional piston ring, the required sealing pressures are derived partly from the small contact area conferred by the narrow working surfaces of the seal elements.

Unlike the piston ring, however, the orbital motion of the rotary piston results in appreciable tilting of its apex seals as they sweep between the major and the minor axes of the operating chamber. The apex seals are therefore the most critical elements of the sealing grid, and current practice is to employ a three-piece construction for each. This comprises a centre section with mitre-fitting end pieces, so that sealing pressure is exerted in both radial and axial directions (Figure 12.8). A radiused nose is applied to the apex seals to accommodate their tilting.

The metallurgy of apex seal and chamber surface compatibility has received considerable attention from Wankel engine researchers, and resulted first in the use of hard carbon and then later special metallic materials for the apex seals. Cast iron piston ring material is employed for the curved sealing strips that bridge the apex seals. This material is also used for the rotary piston oil control rings, which are taper faced and perform the dual function of preventing oil from passing into the combustion space while the engine is running and from draining into it when the engine is stopped.

Later developments

It is perhaps surprising to realize that the rotary piston engine has received continuous development by Mazda engineers since the early 1960s. During this period many improvements have been introduced, chiefly to reduce both fuel consumption and exhaust emissions. As might be expected, some of these improvements have been adopted from contemporary reciprocating piston engine practice. Among the more significant improvements that have been incorporated in the rotary piston engine may be listed the following:

1 The provision of a modern engine management system for electronic coordination of the ignition and fuel injection functions.

2 The use of three staggered intake ports for each rotor chamber, their number and shape providing the best compromise for intake timing of the rotary piston controlled ports. In particular the intake port opening can be retarded to reduce contamination of the fresh charge during low-speed operation, which prevents misfiring and improves torque.

3 The replacement of a compound carburettor by an electronically controlled fuel injection system with twin injectors for each rotor chamber, so as to provide better control over the quantity and timing of the fuel delivered. One injector can be positioned closer to the port than the other and give a finely atomized fuel spray for idling and part-throttle running.

4 The application of turbocharging to improve engine performance, which in this case can take advantage of the smoother exhaust gas flow from a rotary piston engine to reduce turbine lag or delayed throttle response.

5 The adoption first of a thermal reactor and then later a catalytic converter system for more effective control of exhaust emissions, especially by the latter system since it reacts at lower temperatures and makes possible the use of leaner mixtures for improved fuel consumption.

In 2000 a return was made to a natural aspirated version with larger side, instead of peripheral, ports for both intake and exhaust gas flows. The advantages that accrue from this resiting of the gas ports are basically twofold. First, the overlap between the intake and exhaust ports can be eliminated, so that the incoming fresh charge is not contaminated by the outgoing exhaust gases. This allows higher engine speeds, more stable combustion, better fuel economy and a reduction in the emission of unburned hydrocarbons. Second, it promotes more effective lubrication and hence reduced wearing of the apex seals, which no longer have to sweep over an epitrochoidal working surface that is interrupted by gas porting.

13 Friction clutches

13.1 TYPES OF SINGLE-PLATE CLUTCH

General requirements

A clutch is a device used in engineering to engage smoothly two shafts in relative motion, one of which may be stationary, and to release them quickly or slowly at will. The purpose of fitting a clutch between the engine and gearbox of a motor vehicle is to satisfy the following requirements:

1 To connect a running engine smoothly and gradually to the remainder of the transmission system.
2 To permit gear changing when a vehicle is in motion.
3 To allow the engine to continue running when a vehicle is temporarily halted in gear with the clutch pedal depressed to disengage.

A friction type of clutch is used in motor vehicles with manually operated gearboxes. Its functioning depends upon sufficient friction being developed between the contact surfaces of two or more members to transmit the desired torque without their slipping relative to each other. Their ability to slip before full engagement is, nevertheless, a decided advantage because it allows shock-free connection to be made between engine and gearbox. Heat is necessarily generated during any slipping of the clutch, but with fairly infrequent use there is usually sufficient time for it to cool.

An important consideration in the design of clutches is that the driven member connected to the gearbox should behave as little like a flywheel as possible. It should therefore be of light-weight construction and thus possess the minimum of inertia to ensure that it can be rapidly slowed down or speeded up, thereby assisting changing into higher or lower gears respectively. It was chiefly for this reason that the old-fashioned cone-type friction clutch (Figures 13.1a and b), although providing a powerful friction effect by its wedging action, was superseded by the plate-type friction clutch (Figure 13.1c).

The ability of a plate clutch to transmit torque depends upon the following factors:

Effective radius of the friction surfaces
Coefficient of friction acting between the friction surfaces
Number of friction surfaces
Clamping force holding the friction surfaces together.

It is the manner in which the clamping force is applied to the friction surfaces that distinguishes the two basic types of plate clutch used in motor vehicles. They are the multicoil spring clutch and the diaphragm spring clutch.

Multicoil spring clutch

This represents the earlier form of construction (Figure 13.2), and its chief components are as follows.

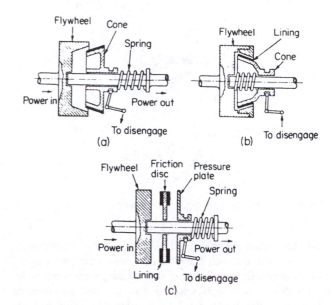

Figure 13.1 Basic forms of friction clutch: (a) cone clutch (b) internal cone clutch (c) plate clutch

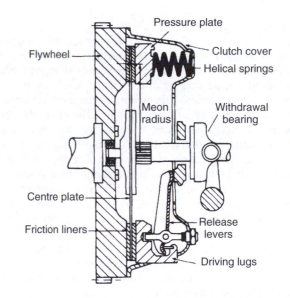

Figure 13.2 General arrangement of a multicoil spring clutch

The driving member

This part of the clutch is driven permanently by the engine crankshaft. It comprises partly the engine flywheel and partly the clutch cover bolted to it. The non-engine side of the flywheel supplies one of the unlined friction contact surfaces,

265

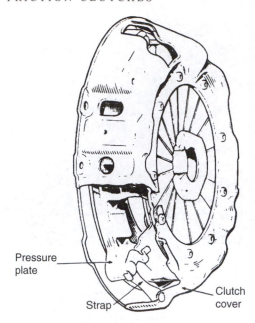

Pressure plate

Strap

Clutch cover

Figure 13.3 Diaphragm spring clutch with strap drive

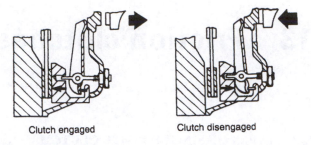

Clutch engaged   Clutch disengaged

Figure 13.4 The operation of a multicoil spring clutch

while the clutch cover forms a housing for the other driving components of the clutch. Produced from a steel pressing, the clutch cover is designed to have appreciable strength and rigidity combined with light construction. It must also provide adequate ventilation for the clutch interior. As previously noted (Section 1.8), the flywheel is produced from cast iron and therefore its graphite content contributes to smooth engagement of the clutch.

The pressure plate

This rotates permanently with the driving member, supplies the other unlined friction contact surface, and conveys the clamping force of the clutch springs to the driven member.

To accommodate wearing of the friction-lined driven member and to allow its release, the pressure plate must be provided with a certain amount of axial freedom. Therefore, the drive between the clutch cover and the pressure plate is obtained either by means of lugs on the latter engaging slots in the former, as in earlier practice, or by tangential spring steel straps secured firmly at their ends to both components and acting in tension. This now long-established form of construction is known as strap drive (Figure 13.3). It not only avoids both friction and wear but also retains the clutch in better rotational balance.

The pressure plate is produced from cast iron and is intentionally of rugged construction, it must neither distort under the loading of the clutch springs nor be unable to absorb and conduct away the heat generated during clutch engagement.

The clutch springs

A series of helical springs is placed around the circumference of the pressure plate, and these springs are compressed between the plate and the inner face of the clutch cover. Their purpose is to maintain a sufficient clamping force between the friction surfaces of the clutch so that maximum engine torque can be transmitted to the gearbox without slip occurring. To disengage the

clutch this force must be counteracted by the driver through a suitable leverage system connected to the clutch pedal.

For heavy-duty applications, insulating washers may be used to reduce heat transmission from the pressure plate to the springs.

The driven member

This is commonly termed the centre plate because it is sandwiched between the friction contact surfaces of the flywheel and the clutch pressure plate.

The centre plate is connected to the input or primary shaft of the gearbox by means of a splined hub so that it can float between the separated contact surfaces of the flywheel face and pressure plate when the clutch is disengaged. Each side of the centre plate is faced with a ring of friction material, known as a clutch liner, which may be secured by either riveting or bonding. Cushioning leaf springs may be provided between the clutch liners and the centre plate to permit smooth, gradual engagement. Cushioning devices acting in the rotational sense may also be incorporated at the hub connection to the centre plate, their purpose being explained at a later stage.

The release levers

These provide leverage to oppose the spring-loading acting on the pressure plate, thereby enabling the driver to disengage the clutch by releasing the clamping force on the centre plate (Figure 13.4).

Around the clutch cover three or more release levers are spaced radially and pivoted from eyebolts on its inside face. The outer, shorter ends of the release levers engage the drive lugs on the pressure plate through the medium of knife-edge struts which minimize friction at these points. An annular release plate is spring retained against the inner, longer ends of the release levers, which locate in channelled recesses to impart rotation to the plate. Facing the release plate is a clutch withdrawal bearing connected through a suitable leverage system to the clutch pedal.

Diaphragm spring clutch

This has long since superseded the multicoil spring type on most private cars and also on many lighter commercial vehicles (Figure 13.5). Originally developed by General Motors of America in the late 1930s, the principle was not adopted in European automotive practice until some twenty years later.

It differs from the multicoil spring type in so far as the spring loading is both applied to and withdrawn from the clutch pressure plate, as described in the following sections.

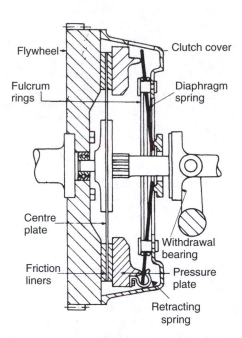

Figure 13.5 General arrangement of a diaphragm spring clutch

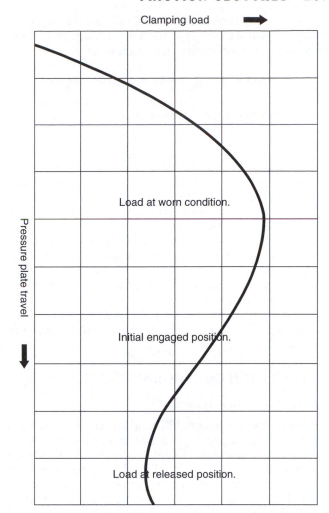

Figure 13.6 Operating characteristics of a diaphragm spring clutch (*Mintex*)

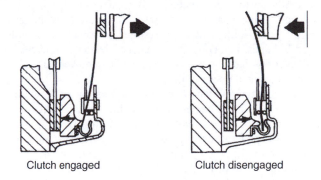

Figure 13.7 The operation of a diaphragm spring clutch

The clutch spring

Because of its shape this is described as a diaphragm spring, but in general engineering it is perhaps better known as a Belleville spring, so named after the French civil engineer Julian Belleville who obtained a British patent for this form of spring as long ago as 1866.

Basically this spring comprises an annular disc initially dished to a conical shape. A compressive load applied to the disc tends to flatten it out, and a spring action is provided as the disc tries to regain its shape. Unlike a coil spring, the compressive load first increases with deflection of the disc and then, as the disc becomes more nearly flat, the load actually decreases with increased deflection, a characteristic unique to this type of spring.

In practice, it means that the load may be held fairly constant over a considerable range of deflections. This is a particularly useful attribute as far as the motor vehicle clutch is concerned, since it can be arranged for the clamping force on the centre plate to be little affected as the liners wear thinner (Figure 13.6).

The fact that this type of spring also possesses a high load-carrying capacity for the space required makes it additionally attractive for application to motor vehicle clutches, which consequently can be of correspondingly shallower form.

The diaphragm spring is incorporated in the clutch in a manner which allows it to flex between a pair of fulcrum rings located on the inside face of the clutch cover. To supply the necessary clamping load on the centre plate, the outer edge of the diaphragm spring bears against the rim of the pressure plate. The outer edge of the diaphragm spring is also embraced by retraction clips secured to the pressure plate so that the latter can be withdrawn from the centre plate when the clutch is disengaged. A strap drive of the type mentioned earlier is used between the clutch cover and the pressure plate (Figure 13.3).

The release levers

To prevent buckling of the diaphragm spring under thermal stress it is provided with radial slits which give rise to the formation of inwardly extending spring fingers. These conveniently act as release levers, because when they are pressed inwards at their free ends the effect is to relieve the diaphragm spring load on the pressure plate (Figure 13.7). The clutch release bearing presses either directly against the diaphragm spring fingers or indirectly through the medium of a release plate that is strap driven from the clutch cover.

Advantages of diaphragm over coil spring clutch

The overall advantages generally claimed in favour of the diaphragm spring clutch, as opposed to the coil type, may be summarized as follows:

1 The higher load-carrying capacity of the diaphragm spring makes for a more compact and lighter construction.
2 The clutch pedal effort can be reduced for the same torque-transmitting ability of the clutch.
3 The torque-transmitting ability can be better maintained as the liners wear thinner in service.
4 The torque-transmitting ability is also less affected by high engine speeds when coil springs can bow and reduce their load.
5 The release mechanism to disengage the clutch can be simplified at source.
6 It readily lends itself to strap drive of the pressure plate for greater mechanical efficiency and better retention of balance.

13.2 CLUTCH CONTROL SYSTEMS

Mechanical control system

On very early cars with rigidly mounted engine, clutch and gearbox units, the clutch pedal was attached directly to the end of a cross-shaft that pivoted through the clutch housing and connected to a withdrawal fork. This in turn engaged a collar on the clutch withdrawal or 'throw-out' bearing, which could slide along the gearbox primary shaft sleeve. Depressing the clutch pedal brought the bearing into contact with the clutch release levers and forced them inwards to disengage the drive.

The advent of flexibly mounted or floating engine and gearbox units meant that the clutch pedal had to be mounted independently on the car structure and provided with a separate linkage to connect an external lever on the clutch withdrawal bearing cross-shaft. This often resulted in somewhat complicated linkages, involving either balance levers or universally jointed cross-shafts, to isolate the pedal from movements of the flexibly mounted power unit.

An ingenious refinement to be found in some of these linkages was the clutch helper or over-centre spring, which was toggle mounted against either the withdrawal bearing cross-shaft or the pedal lever (Figure 13.8). Its action was such that it more strongly resisted initial pedal movement (useful in countering any foot riding of the pedal) but greatly assisted final depression of the pedal to lighten the effort of declutching. Since the modern diaphragm clutch already has a similar built-in characteristic, the helper spring is no longer required in the clutch control system. However, as can be imagined, the less well engineered of these various rod-and-lever mechanisms were apt to wear at the pivots, which not only produced rattles but also could cause clutch judder during engagement.

Cable-operated control system

A later development was to replace the compensating linkage by an enclosed cable and lever system (Figure 13.9). The

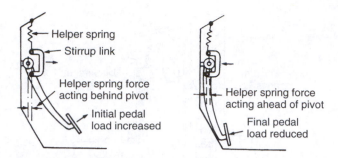

Figure 13.8 Principle of the clutch helper spring (control linkage from pedal omitted for clarity)

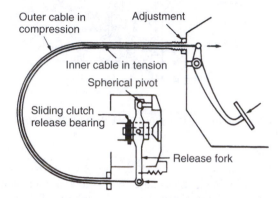

Figure 13.9 Principle of the cable-operated clutch control system

inner cable was connected at its ends to the clutch pedal and withdrawal cross-shaft levers, while the ends of the outer casing were provided with nearby abutments on the bulkhead structure of the body and the clutch housing respectively. Unfortunately, the early cable-operated clutches did not always give consistently smooth operation, especially if their lubrication was neglected in service.

Cable-operated clutch control systems are still widely used in modern practice (Figure 13.10) but their detail application has undergone several changes to improve their efficiency and installation, as follows:

1 The clutch withdrawal bearing still slides on an extension sleeve mounted from the gearbox front cover but is operated by a pressed steel fork which swings from its inner end about a spherical pivot. Its opposite end is connected to the inner cable of the clutch control system, so that the movement of the fork corresponds to the second order of levers.
2 If the force required to initiate sliding of the cable in its casing is much greater than that required to maintain steady sliding, a jerky clutch engagement can result. To reduce these 'stiction' effects the inner working surface of the clutch cable casing may be lined with polytetrafluoroethylene (PTFE). This is a plastics material that is remarkably effective not only in reducing friction but also in reducing it to similar levels under both static and dynamic conditions.
3 The clutch control cable is routed through the engine compartment and terminates at the bulkhead. Its outer casing is anchored usually through a rubber grommet to reduce any noise transmission to the car interior. The alignment of the

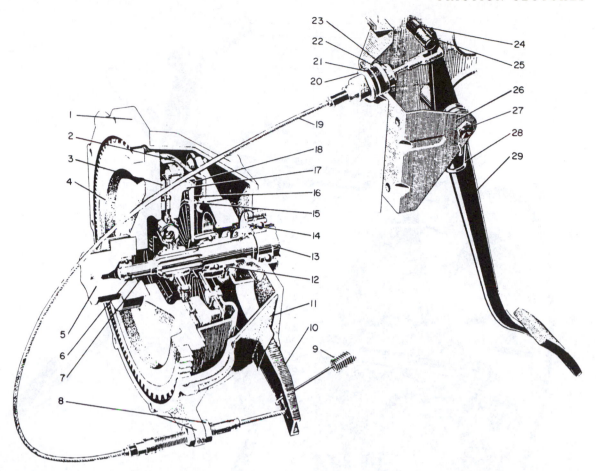

Figure 13.10   The complete installation of a diaphragm spring clutch with cable-operated control system (*Volvo*)

1 flywheel housing	9 return spring	16 thrust spring	24 pedal stop
2 clutch casing cover	10 release fork	17 thrust plate	25 rubber sleeve
3 clutch plate	11 dust cover	18 retainer	26 bracket
4 flywheel	12 release bearing	19 release cable	27 bolt
5 crankshaft	13 plate shaft (input shaft,	20 washer	28 return spring
6 pilot bearing in crankshaft	gearbox)	21 rubber bush	29 clutch pedal
7 circlip	14 cover, gearbox	22 washer	
8 adjusting nuts	15 support rings	23 nut	

outer casing is such that the inner cable runs directly to a hooked connector located a short distance above the pivot of the clutch pedal lever. This arrangement of connecting the clutch control cable to a down-hanging or pendant clutch pedal thus provides an installation comparable to that offered by a hydraulic clutch control system, which is described next.

Hydraulically operated control system

The more general adoption of hydraulic clutch control systems began in the mid 1950s, at which time they offered a simpler and more effective installation than that of the many complicated rod-and-lever, and also indifferent cable, systems then in use (Figure 13.11). A further benefit of hydraulic clutch control is that by suitable relative sizing of the pistons in the master and slave cylinders, an enhanced mechanical

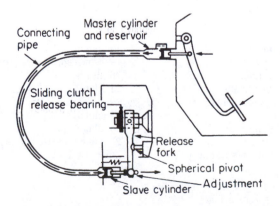

Figure 13.11   Principle of the hydraulically operated clutch control system

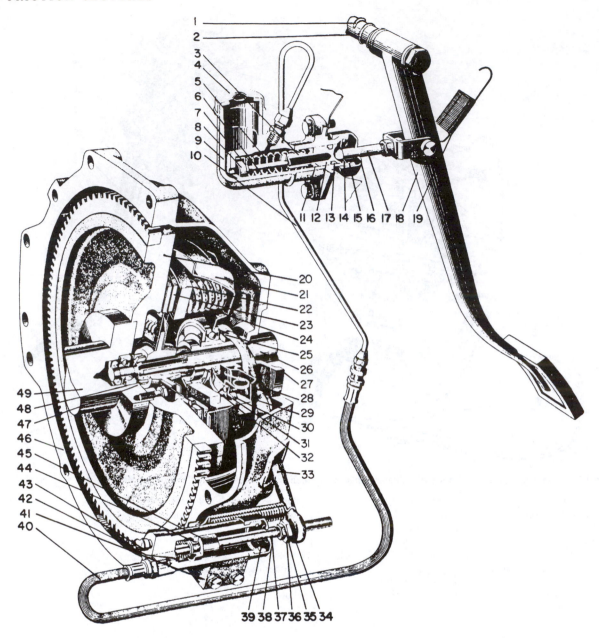

Figure 13.12 Complete installation of a multicoil spring clutch with hydraulically operated control system (*Volvo*)

1 pedal shaft	14 washer	27 cover	39 circlip
2 spring	15 circlip	28 spring	40 hose
3 retainer	16 rubber dust cover	29 pin	41 control cylinder
4 cap	17 thrust rod	30 lip	42 spring
5 spring	18 clutch pedal	31 clutch lever	43 bleeder nipple
6 thrust rod	19 return spring (early prod.)	32 eyebolt	44 plunger seal
7 retainer	20 flywheel	33 clutch release fork	45 plunger
8 check value	21 clutch cover casing	34 return spring	46 flywheel housing
9 master cylinder	22 pressure plate	35 locknut	47 circlip
10 pipe	23 clutch plate	36 adjusting nut	48 pilot bearing in flywheel
11 plunger seal	24 clutch spring	37 thrust rod	49 crankshaft
12 plunger	25 clutch release bearing	38 rubber dust cover	
13 plunger seal (early prod.)	26 input shaft	(early prod.)	

advantage can be obtained in the system, which is not available with a direct-acting cable control.

Apart from completely isolating the clutch pedal from the movements of a flexibly mounted engine and transmission, hydraulic actuation provided another opportunity to exclude engine noise and vibration from the car interior, especially since no clearance slot was required for the pendant mounting of the clutch pedal lever. It was also expected that such systems would function satisfactorily with the minimum of maintenance, because by their nature they are self-lubricating. Basically the hydraulic system comprises the following components (Figure 13.12).

Master cylinder   This unit is mounted on the engine side of the bulkhead, and by means of a push-rod is connected directly to the pendant clutch pedal. By virtue of the sealed piston which slides within it, the master cylinder acts as a hydraulic pressure transmitter whenever the clutch pedal is depressed to disengage the clutch.

Slave cylinder   This unit is mounted on the clutch bell-housing and is connected directly, again by means of a push-rod, to the swinging fork acting upon the clutch withdrawal bearing. Since the outer end of the fork is being pushed from the front, rather than being pulled from the front as in the case of cable operation, its spherical pivot is located between the points of load and effort to correspond with the first order of levers. The slave cylinder acts as a hydraulic pressure receiver because the sliding motion of its sealed piston depends upon a similar action taking place in the clutch-pedal-operated master cylinder.

Connecting pipe   This serves as a hydraulic pressure conduit between the master cylinder and the slave cylinder. It partly comprises a flexible hose consisting of an inner tube, reinforcement, outer cover and screwed end fittings. Its installation must allow it to lie and flex in one plane without suffering any twisting movements.

Fluid reservoir   This contains a reserve of hydraulic fluid to replace any small leakages. It also provides a vented space for expansion and contraction of the fluid due to temperature variations and assists in separating air from the fluid. The reservoir may either form a combined unit with the master cylinder or it may be separately connected to it. In either event, the reservoir communicates with the pressure chamber of the master cylinder through a simple shut-off valve, or orifice, so that as the clutch pedal is depressed the initial movement of the master cylinder piston first shuts off the fluid supply from the reservoir. Further movement of the piston then displaces fluid through the connecting pipe to actuate the slave cylinder piston, which in turn causes the release bearing to disengage the clutch.

It will be evident that to perform their function correctly the hydraulic seals fitted to the sliding pistons of the master and slave cylinders must generate a certain amount of friction. This has sometimes given rise to criticism of the hydraulic type of clutch control system since it can produce an uncertain pedal feel during clutch engagement.

A significant development in hydraulic clutch control systems during the mid 1980s was the introduction of the centrally actuated release bearing unit. With this system the separate hydraulic slave cylinder is removed from outside the clutch bell-housing and mounted concentric with the sliding release bearing itself to provide an integrated unit (Figure 13.13). It therefore eliminates the round-the-house requirement for a push-rod and pivoting release fork mechanism, so that only the clutch pedal, master cylinder and hydraulic pipeline remain external to the clutch bell-housing. To confer automatic adjustment for continuous contact clutch control as described later, a preload compression spring encircles the telescopic members of the slave cylinder.

The advantages to be gained from this more direct-acting system of hydraulic clutch control include a lighter and smoother operation, which is also better maintained in service by virtue of eliminating a mechanism that was partially exposed to the elements, and of course a reduction in the number of parts to save weight and facilitate assembly.

Also during this period, there has been an increasing trend towards producing the master and slave cylinders from a plastics material, such as glass-fibre reinforced nylon, instead of cast iron or aluminium alloy as used in hydraulic brake systems, which originally provided a role model for hydraulic clutch control systems. The advantages generally claimed for these and other detail changes involving the use of plastics materials in hydraulic clutch control systems include a significant reduction in weight, a saving in cost, freedom from corrosion, and consistently smooth operation that is little influenced by temperature effects. From an installation point of view, hydraulic clutch control systems are supplied already assembled, charged with fluid and need no adjustment. In service they are intended to be virtually maintenance-free.

Servo-assisted control system

Although a hydraulically operated clutch control system is generally used in modern heavy-vehicle installations, the withdrawal loads for high-torque-capacity clutches can result in clutch pedal operation becoming unacceptably heavy. To overcome this problem the clutch control system may be provided with servo assistance, by utilizing what is known as an air-hydraulic clutch servo unit. This shares the compressed air supply for the braking system to complement hydraulic pressure transmitted to the clutch withdrawal slave cylinder. However, the degree of servo assistance must be precisely controlled and kept proportional to driver effort on the clutch pedal, because too much assistance could result in an insensitive control with uncertain engagement characteristics.

The air-hydraulic cylinder with servo unit typically comprises a relatively large-diameter pneumatic cylinder with servo piston, which is mounted in tandem with a smaller-diameter hydraulic cylinder with slave piston, this in turn operating a push-rod that is connected to the clutch withdrawal mechanism. Both pistons always move in unison, their connecting spindle passing through a stationary sealing gland assembly. The purpose of the latter is to separate the pneumatic and hydraulic functions of the unit. Bolted to the smaller cylinder is typically a vertical control valve unit, the upper chamber of which contains a hydraulic reaction plunger and the lower one a stepped air control piston. A diaphragm seal is interposed between the plunger and piston, thereby allowing them also to move in unison whilst at the same time separating their hydraulic and pneumatic functions. The air control piston is provided with a central exhaust air passage, and situated a short distance beneath its foot is an air inlet valve. Both piston and valve are spring returned in an upwards direction, the former towards the reaction plunger and the latter against its seating.

When the clutch pedal is depressed, the pressure created in the hydraulic master cylinder is communicated to the slave cylinder via the upper chamber of the control valve unit, so that system pressure is reflected against the reaction plunger. A supply of compressed air is constantly available beneath the inlet valve and can be directed against the servo piston once the inlet valve is unseated. This occurs as a result of the hydraulically actuated reaction plunger pressing down against the air control piston (Figure 13.14). The increasing air pressure acting against the servo piston thus provides a helping hand to the interconnected slave piston, which would otherwise need

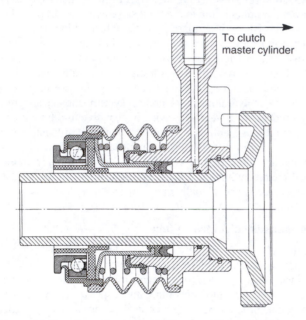

Figure 13.13   Centrally actuated release bearing unit (*FAG Kugelfischer*)

greater hydraulic pressure and hence driver effort at the clutch pedal to effect clutch withdrawal.

If the clutch pedal is held partially depressed, then since the air being directed to the servo piston also has access to the underside of the air control piston, there is an upward force being transmitted to the hydraulic reaction plunger. Once this upward force just exceeds the downward one acting on the reaction plunger, the air control piston lifts just sufficiently to allow the inlet valve to close and shut off further air pressure to the servo piston. In this balanced or 'lapped' condition the clutch is held disengaged with a force that is proportional to the effort being applied at the pedal by the driver. When the clutch is gradually released and hydraulic pressure on the reaction plunger is relieved, the stem of the rising air control piston will lose contact with the already closed inlet valve. This allows the air under pressure in the servo cylinder to be exhausted to atmosphere via the central passage of the air control piston. Simultaneously, the spring load exerted by the clutch mechanism returns both hydraulic slave and pneumatic servo pistons to their off positions. A pre-load spring acting on the servo piston confers automatic adjustment on the clutch control system.

Clutch release bearings

The function of a clutch release bearing was made clear earlier. Its construction may be considered under two headings: metallized carbon ring, and ball bearing race.

Metallized carbon ring

This type of clutch release bearing (Figure 13.15) is of the plain thrust variety and therefore comprises two members. First, there is the rotating member or counterface, this being an annular steel plate carried on the tips of either the release levers or the diaphragm fingers of the pressure plate assembly. Second, there is the non-rotating member, which takes the form of a metallized carbon ring protruding from a cast iron carrier in which it is rigidly held. The pivoting carrier is in turn mounted on a forked withdrawal lever and retained by a pair of spring clips.

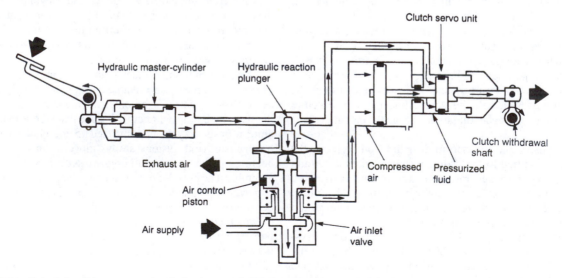

Figure 13.14   Principle of the servo-assisted clutch control system

Although this type of release bearing is (by the nature of its material) inherently self-lubricating, the life that can be expected from it is limited – as in the case of other dry rubbing bearings – by allowable wear. However, the wearing of this type of release bearing did tend to compensate for wearing of the clutch friction linings, since the effect of the latter was to advance the counterface towards the ring. Nevertheless, by the late 1960s it began to be found that the life of the dry rubbing release bearing could become unacceptably short in constant heavy traffic driving conditions, and this led to it being superseded by the ball bearing type.

Ball bearing race

A release bearing of this type must be provided with an axial, rather than an arcuate, withdrawal movement. This is because its rotatable element directly engages either the release levers, or the diaphragm fingers, of the clutch pressure plate assembly. For this reason the bearing must be sleeve mounted and guided along a nose piece extension of the cover for the gearbox primary shaft bearing. The release bearing sleeve must also be provided with either a collar, or a recess, to intercept the withdrawal fork of the clutch control system.

The modern clutch release bearing of the ball race type has passed through several phases of development. Originally a pure thrust bearing was used, which could be likened to a pair of flat washers separated by a ring of ball bearings, their circular path being maintained by matching grooves in the opposing faces of the washers. It was provided with its own shroud and greased for life during manufacture. A drawback with this particular type of release bearing was that its operation at high engine speeds could give rise to destructive skidding between the ball bearings and their raceways owing to centrifugal effects.

A single-row, deep-groove type of ball bearing race was the next choice and is still retained by some manufacturers, especially for heavy-duty clutch release bearings. This type of ball bearing, although having a high radial load rating, also has a substantial thrust resistance, which is of course the relevant factor. It further possesses the well-known features of being simple in design, non-separable and suitable for high-speed operation. Again the bearing is provided with its own shroud or housing and is often greased for life during manufacture, provision for regreasing in service being made in some heavy-duty applications (Figure 13.30).

Finally, we come to the modern single-row, angular-contact ball bearing race. This type of release bearing is distinguished by having a high shoulder on one side of the outer ring and usually another on the diagonally opposite side of the inner ring. The relative positions and steep contact angle of these shoulders give the bearing a high thrust resistance in the required direction of clutch withdrawal. As an alternative to the traditionally machined bearing rings, more recent versions may be produced from pressed steel metal rings and sliding sleeves. The form of the bearing rings imparts the necessary rigidity and also contributes to the permanent sealing arrangements. Since the points of contact between the release bearing and the diaphragm spring fingers must alter as the latter are deflected, it is necessary for either a convex or a flat-faced bearing to be used in combination with flat and curved spring finger ends respectively. To improve the durability of the first arrangement, an innovation by the SKF company is to make the release bearing self-aligning through a rubber sleeve in the bore (Figure 13.16). Hence, any slight eccentricity of the release bearing mounting relative to the rotating clutch assembly can be accommodated, thereby avoiding excessive side loading and wearing of the diaphragm spring fingers.

Release levers setting for coil spring clutches

The satisfactory operation of the multicoil spring type of clutch is very dependent upon the accuracy of the release levers adjustment (Figure 13.17). From what is said later in Section 13.3 in relation to the parallelism of the centre plate friction linings, it should likewise be evident that the pressure plate must always move parallel with the flywheel. In practice, this means that all three or more release levers must be adjusted to the same distance from the flywheel friction face, a uniformity within 0.12 mm (0.005 in) being typically

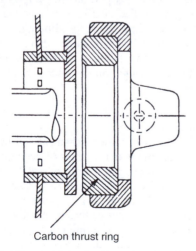

Figure 13.15 Clutch release bearing of carbon ring type (*GKN Laycock*)

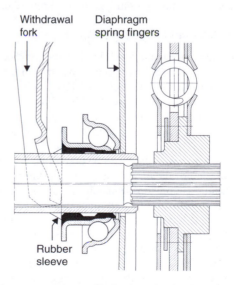

Figure 13.16 A modern self-aligning clutch release bearing of angular-contact ball race type (*SKF*)

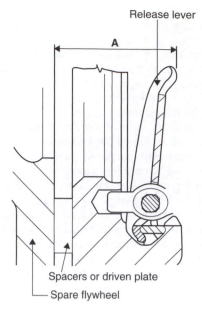

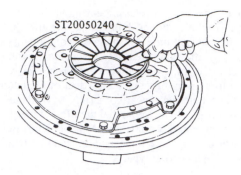

Figure 13.18   Spring fingers setting for diaphragm clutch (*Nissan*)

Figure 13.17   Release levers setting for coil spring clutch: *A* is height of levers (*Seddon Atkinson*)

required. This is achieved by turning the nut on each eyebolt to adjust the height of the levers. When the correct lever setting is obtained, each eyebolt adjusting nut is staked into the slots in the eyebolt. The adjustment is required, of course, only during manufacture or rebuilding of the clutch in service.

Spring fingers setting for diaphragm clutches

A manufacturer may also require that the toe height of the diaphragm spring fingers be checked and if necessary adjusted. The toe height is checked in a similar manner to that already described for the release levers of coil spring clutches, but adjustment is performed by using a special setting tool on the spring fingers (Figure 13.18). Any unevenness of toe height should typically be kept within less than 0.5 mm (0.020 in).

Clutch pedal and linkage adjustment

When the clutch is engaged the clutch pedal must be retained at a certain distance above the toe board, this distance being known as the pedal height (Figure 13.19). It allows the pedal sufficient movement to ensure that the clutch can be completely disengaged, but not so much as to cause overtravel and, by the same token, overstressing of the clutch release mechanism.

There is also another factor to be taken into account. With gradual wearing of the centre plate friction liners, the clutch pressure plate assumes positions closer to the flywheel. This results in an opposite, but albeit magnified, movement of the inner ends of either the release levers or the spring fingers towards the clutch release bearing. Unless their return movement can take place unimpeded by the release bearing and its control system, it will oppose the spring loading on the pressure plate and reduce the all-important clamping effect on the centre plate friction linings (Figure 13.20). The clutch then begins to fight a losing battle with its own control system,

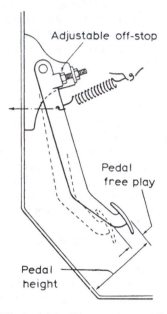

Figure 13.19   Clutch pedal adjustment (control linkage from pedal omitted for clarity)

similar in effect to a driver permanently slipping the clutch. The more it begins to slip, the greater will be the heating and wear and the worse will become the slipping.

In modern transmission practice, there are two basic approaches to accommodating this unwanted movement at the clutch release bearing: non-continuous contact, and continuous contact.

Non-continuous contact

The essential features of both mechanical and hydraulic versions of this once traditional method of accommodating the effects of lining wear are that the clutch pedal height remains constant, but included in the pedal movement is a certain amount of free play (Figure 13.19). The latter is to ensure that a definite clearance exists between the clutch release bearing and either the release levers or the spring fingers, at all times when the clutch is engaged. Hence, the levers or the fingers are permitted a limited amount of unrestricted movement towards the release bearing in its off or no-contact position (Figure 13.21).

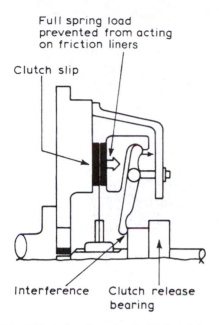

Full spring load
prevented from acting
on friction liners

Clutch slip

Interference   Clutch release
bearing

Figure 13.20   Effect of no free play at clutch pedal

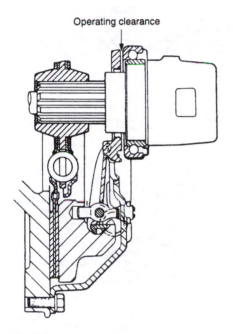

Operating clearance

Figure 13.21   A non-continuous contact clutch release system (*Seddon Atkinson*)

Wearing of the clutch liners is, or at least should be, signalled to the driver by the decreasing free play felt at the clutch pedal. A screw-thread adjustment is provided in the clutch control system to restore the free play, which generally lies in the range of 20 to 30 mm (0.8 to 1.2 in) and is checked in service every 6000 miles (10 000 km).

In the case of a cable-operated system, a screwed sleeve with adjusting nuts is incorporated at one end of the outer cable, usually where it is anchored to the bell-housing (Figure 13.10). For a hydraulically operated system, the screw and adjusting nuts facility is provided on the end of the slave-cylinder push-rod, where it connects to the clutch release fork (Figure 13.12).

Continuous contact
Clutch control systems of this type are of more recent origin and involve continuous contact of the release bearing with the spring fingers for the clutch unit, regardless of whether or not the clutch is disengaged. To all intents and purposes such systems are self-adjusting to compensate for wearing of the clutch lining. The manner in which the release bearing is permitted to retreat, in sympathy with unwanted movement of the spring fingers, differs between mechanical and hydraulic clutch control systems.

With a simple mechanical version the spring loading on the clutch pedal is reversed in direction, so that it tends to move the pedal towards the disengaged position until all free play is taken up. The inner control cable is therefore subjected to a permanent slight tension, such that the release bearing is maintained in continuous contact with the fingers of the clutch unit. As the clutch linings wear, the fingers force the release bearing to retreat. This transmission of force, via the release fork and cable control, opposes the light spring loading on the pedal and raises it. The increase in pedal height thus signals to the driver the state of wear of the clutch linings. An additional refinement comprises a switch on the pedal mounting bracket, so that a warning light operates when the pedal rises to a certain critical height.

In the case of a hydraulic version (Figure 13.22), the installation and operation are similar to that of a non-continuous contact system, except that no return spring is fitted to the release fork. This enables the spring-loaded piston in the slave cylinder to float and retreat in sympathy with the release bearing as the linings wear. It should be noted that the clutch pedal remains spring loaded towards the engaged position, as with a non-continuous contact system, and that a very small amount of pedal free play is retained, typically about 1 to 5 mm. The purpose of this free play is to ensure that the piston in the clutch control master cylinder always fully returns to its off position when the clutch is engaged, so that fluid recuperation can take place unimpeded through an uncovered compensating or bypass port. No indication of lining wear is signalled to the driver with this type of continuous contact clutch control, since there is neither reducing free play nor increasing height of the clutch pedal.

In conclusion it must be said that although the clutch control systems just described are in theory constant contact, in practice they are virtually constant contact. This is because any slight run-out of the spring fingers tends to create a minimal play at the release bearing with the clutch engaged.

13.3   CLUTCH CENTRE PLATE CONSTRUCTION

General requirements

The centre plate was identified in Section 13.1 as the driven member of the plate-type friction clutch, because it is sandwiched between the friction contact surfaces of the flywheel and the clutch pressure plate. At this stage we must consider the construction of the centre plate in a little more detail, especially those features that contribute to the overall smoothness

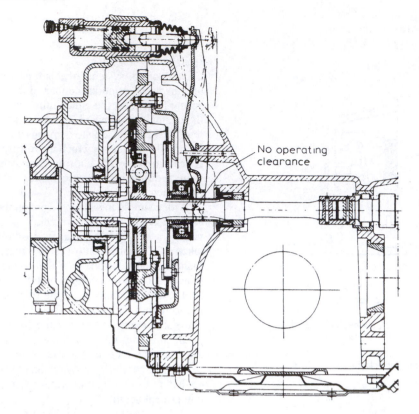

Figure 13.22 A continuous contact clutch release system (*Alfa-Romeo*)

of torque application and transmission. In this context the modern clutch centre plate usually incorporates at least an axial cushioning facility between the friction linings and, except in arduous duty applications, also a torsional cushioning facility between the plate and its hub. It is therefore convenient to classify the two types of centre plate construction as follows:

Solid drive This does, of course, represent the simplest form of centre plate construction. The hub and the plate carrying the pair of friction linings are directly riveted together and thus provide a torsionally rigid drive through the clutch assembly (Figure 13.23).

Cushion drive In this construction the plate carrying the pair of linings is provided with a complementary side plate to form a sandwich with the flanged hub, so although the plate assembly is positively located both axially and radially, in the rotational sense its location is effected through a number of circumferentially spaced compression springs. These are retained in windows formed in the hub and plate sandwich. This feature thus allows a limited degree of angular wind-up of the centre plate about its hub, or in other words torsional cushioning (Figure 13.24).

Functioning of centre plate components

Hub
This is provided with a spline connection to the primary shaft of the gearbox. A sliding fit is permitted so that when the clutch is disengaged, the friction-lined centre plate can float between

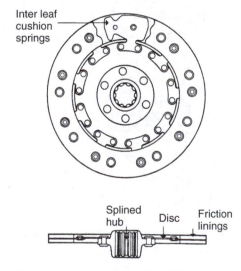

Figure 13.23 Clutch centre plate with solid drive (*GKN Laycock*)

the separated contact surfaces of the flywheel and the pressure plate. Either square-sided or involute-form splines may be specified for this application. An involute spline has teeth similar in form to those of a gear. Since this topic receives attention later, suffice it to say here that the advances of the involute over the square-sided spline are generally those of possessing greater strength and endurance, a better self-centring of the parts, and more economical to produce.

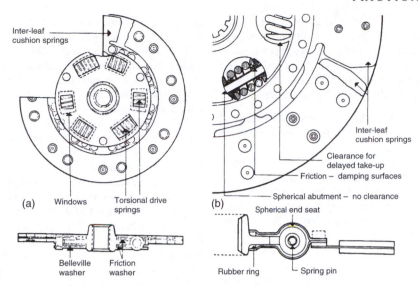

Figure 13.24 Two types of friction damper for cushion drive (*GKN Laycock*)

Torsional spring drive

This feature is incorporated between the centre plate and its hub (Figure 13.24) for the following reasons:

1 To minimize passenger discomfort and reduce shock loading on the transmission components in the event of too sudden an engagement of the clutch. This is achieved by allowing a controlled degree of wind-up between the centre plate and hub, the circumferential springs momentarily acting as energy storage devices.

2 Fluctuations in driving torque are inherent to the operation of the reciprocating piston engine. If they are passed on undiminished by the clutch to the vehicle transmission system, then vibration and noise can result from any source of clearance therein such as splines, dog clutches, synchronizer assemblies and gear teeth. In this case the spring drive is required to act as a torsional vibration damper, which not only involves the previously mentioned wind-up between the centre plate and hub, but also the introduction of a certain amount of frictional resistance to this movement. This is to dissipate the energy of vibration in the form of heat.

The manner in which this feature is built into the centre plate has already been described. As far as the cushioning springs are concerned, it can be added that a greater degree of wind-up tends to be demanded of the modern clutch centre plate. This may entail using springs of different rates (the rate of a spring being the load required to produce unit deflection). Their purpose is to control not only the higher-magnitude vibrations from the engine, but also lower-magnitude vibrations from the transmission drive line. Limiting stops are provided to prevent the springs from being compressed chock-a-block (coils in contact with one another).

Finally, the frictional resistance or damping of centre plate wind-up and recovery motion (referred to earlier) may be accomplished in two ways:

1 By friction washers positioned on both sides of, and spring loaded against, the flanged hub of the centre plate. The spring loading may be supplied by a Belleville spring,

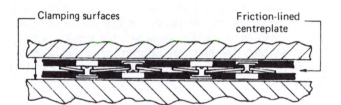

Figure 13.25 Cross-section through axially cushioned friction linings

the principle of which was described in Section 13.1 in connection with the clutch pressure plate (Figure 13.24a).

2 By mounting the torsional drive springs between end caps that slide along specially formed pins, the friction occurring between these parts supplying the necessary damping action (Figure 13.24b).

Dishing and slits

The dishing of a centre plate is concerned with its axial cushioning facility (Figure 13.25), which is provided for the following reasons:

1 In combination with a suitable choice of liner material, the axial cushioning should permit a smooth, gradual clutch engagement free from any judder that may be inherent to the vehicle transmission arrangements.

2 It reduces the initial pedal load required to disengage the clutch, because the spring force due to the axial cushioning opposes that exerted on the liners by the clutch pressure plate.

3 The clutch pedal action or feel is improved by virtue of the engagement process occupying a greater pedal travel, so that a more gradual transition from slip to grip can be made. In some early clutch installations where the pedal travel was very short, the all-or-nothing pedal action could often result in the once well-known 'kangaroo start'!

4 Longer wear life of the friction liners may be obtained owing to their more satisfactory bedding with the friction

contact surfaces of the flywheel and the clutch pressure plate, thereby allowing for a uniform heat generation over the entire friction surfaces. Wear is always more rapid if localized hot-spots occur on friction surfaces with uneven bedding. Also, a flow of air is promoted between the friction linings during intervals of clutch disengagement.

The axial cushioning is provided through some form of dished spring arrangement. For this purpose the friction linings are attached to the centre plate either directly, in which case the outer portion of the plate is provided with a series of crimped tongues, or indirectly through the medium of spring steel, axially curved segments that are separately riveted around a reduced size of plate. The latter method has long since been preferred because it avoids any tendency for the liners themselves to become dished. A parallel approach of the friction contact surfaces is generally essential for judder-free engagement of the clutch.

With regard to centre plate slits, the stresses that are set up in a thin metal disc when heated are such that it tends to distort in a wavy fashion. To relieve a disc of these stresses it is customarily provided with slits or radial expansion slots, which extend to the circumference and terminate at their inner ends in small circular holes, again to avoid stress concentration. It is clearly desirable that the centre plate should remain free from distortion if it is not to drag when the clutch is disengaged.

Centre plate friction linings

It is recorded in a history of the Ferodo Company that when London motor buses first came on the roads fitted with leather clutch linings, these sometimes had to be replaced twice in a day because the drivers had a habit of slipping them and burning them out. It was also not unknown for drivers of fairly powerful cars to have the clutch relined *en route* on journeys of, say, 150 miles; and racing cars almost always had to be relined after a speed test or sporting event. In this field Herbert Frood, who founded the Ferodo Company (Ferodo is an anagram of Frood with one letter added), and one or two colleagues first carried out tests on their own cars by replacing leather with impregnated cotton for the clutch linings. As motoring enthusiasts they were very pleased with the results, which gave long wear, steady friction and smooth engagement. The first known use of impregnated cotton clutch linings on public service vehicles took place about 1905, when they were adopted by the Scottish Motor Buses of Edinburgh and quickly followed by the London General Omnibus Company.

An important point in the development of motor racing in Britain and the British car generally was the opening of the Brooklands racing track in 1909. In those days racing cars gave their constructors the opportunity to test in practice new elements of design, since the conditions experienced in racing accentuated many problems likely to be encountered with standard production cars (although this is less true of racing now). One such problem was bringing into prominence the need for clutch and, of course, brake lining materials with greater resistance to heat and wear than were hitherto available. This led to Herbert Frood experimenting with woven asbestos fabric as a possible basis for high-temperature friction linings. The tests he conducted on this material showed

that a combination of pure woven asbestos spun on brass wire to increase its strength, together with a new bonding agent he had developed, produced a friction material possessing the high-temperature-resisting qualities he was seeking. One of the first users of the new material – Ferodo bonded asbestos, or BA as it was later known – was once again the London General Omnibus Company.

In a typical later development of the clutch lining, the random fibre asbestos base is impregnated with (usually) copper or bronze strands, as these serve not only to assist with heat dissipation but also to 'lubricate' the lining under load and prevent excessive wear. A coefficient of friction (μ) of about 0.4 is fairly representative for a clutch lining.

During the 1980s the growing awareness of the potential health hazards arising from exposure to asbestos fibre led to the increasing use of environmentally safer non-asbestos friction materials for automotive applications, including clutch liners. The general topic of non-asbestos friction materials will receive further attention in connection with vehicle braking systems (Section 27.3).

Sintered metal clutch linings

Sintered clutch linings or segments are generally confined to heavy-duty commercial vehicle applications. The process of sintering may be regarded as the squeezing together and heating of very fine particles of metal, such that very strong adhesion occurs between them. In the case of commercial vehicle clutches a bronze-based sintered metal friction material may be used. Although such materials are initially more expensive than the traditional organic-based ones, the advantages to be derived from their use are generally as follows:

1 Cooler operation and a longer life result from their much higher thermal conductivity.
2 They have more stable friction characteristics under difficult operating conditions.
3 They can withstand high unit pressures without their performance being affected.
4 Their low rate of wear reduces the frequency of manual clutch adjustment.
5 Thinner linings (or segments) can be used to provide a more compact clutch unit, especially if it is of the multi-plate type.

Attachment of clutch linings

Although clutch friction linings are normally attached to the centre plate with rivets, bonding with a thermosetting adhesive may instead be adopted. There are several technical advantages associated with bonded linings, including an increased effective life because of the greater volume of friction material available for wear, and less possibility of scoring owing to the absence of any accumulations of abrasive particles. From practical considerations, however, the greater ease of friction lining replacement made possible with the riveting method continues to guarantee its popularity.

The rubbing surfaces of the clutch linings often have tangentially disposed ventilating grooves moulded into them, their purpose being to act as dust extractors and promote cooling.

13.4 DIRECT-RELEASE CLUTCH

When Alec Issigonis designed the famous transverse-engined Mini car in the late 1950s, the available width in the engine compartment was insufficient for the gearbox mainshaft to be mounted in-line with the crankshaft. This necessitated turning the drive through 180° and burying the gear train in the engine sump or oil pan. It was because of the need to reverse the drive in the least possible space that an unusual form of clutch construction had to be adopted, which would normally be associated with motor-cycle transmission practice and is known as the direct-release type.

With this type of clutch the pressure plate and clutch cover find themselves on opposite sides of the engine flywheel (Figure 13.26), although the centre plate is of course still clamped between the flywheel and pressure plate as in a conventional clutch. Either a series of coil springs or, as in later versions, a diaphragm spring is compressed between the clutch cover and the flywheel flange and supplies, via the pressure plate, the required engagement load for the clutch. The withdrawal bearing is arranged to act on a release plate attached to the centre of the clutch cover. Release of the clutch is thus effected by applying a direct load to the centre of the cover, which in turn moves the pressure plate away from the centre plate so that this is no longer clamped between the flywheel and pressure plate.

Since in this type of clutch the release load is the same as the engagement load it is designated a direct-release type, as distinct from the conventional ratio-release type that embodies a lever withdrawal mechanism as earlier described. Although it is an inherent disadvantage for the release load to be equal to rather than less than the engagement load, the direct-release clutch does possess the advantages of axial compactness and low rotational inertia, which are all-important for its particular application.

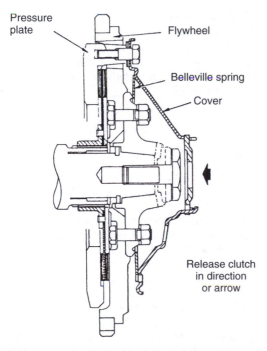

Figure 13.26 Direct release clutch (*Borg and Beck*)

13.5 CENTRIFUGALLY OPERATED CLUTCHES

Source of clamping load

So far we have considered the general construction of conventional plate clutches, which rely entirely on the energy stored in compressed coil and diaphragm springs to supply the necessary clamping load on their friction members. As an alternative construction, the clamping load can be derived either wholly, or partially, from energy generated by a centrifugally operated mechanism, although clutches based on this principle are chiefly associated with past automotive practice. Such clutches may be conveniently classified as centrifugally assisted or fully centrifugal.

Centrifugally assisted clutch

Originally known as the Long clutch (and still so known in America: Long is a registered trademark of the Borg-Warner Corporation), this type of clutch was once widely used on American large-capacity, high-torque engines before the advent of automatic transmission, and also found similar applications elsewhere, such as on the early postwar Rolls-Royce and Bentley cars. It continues to be highly regarded by constructors of competition cars in America.

Clutches of this type represent a fairly simple modification of the conventional coil spring unit. The basis of their operation is in the use of centrifugal assistance acting in parallel with the spring loading on the pressure plate, so as to avoid incipient clutch slip during high-speed gear changes. This effect is achieved by forging offset bob-weights integral with the outer ends of the release levers. The direction of offset is such that, as the clutch revolves, the tendency of the bob-weights to move outwards centrifugally is translated into an additional clamping load at the pressure plate via the pivoted release levers.

In other words, the effect is analogous to that of forcibly pulling the inner ends of the release levers away from the clutch, rather than towards it as when the clutch is being disengaged. It is for this reason that the centrifugally assisted clutch can be critical in respect of pedal free play, because an amount that is just sufficient with the engine idling can be insufficient at higher speeds when the centrifugal assistance is greatest. By virtue of its centrifugal assistance, the spring-loading requirement of this type of clutch can be reduced to a level which only just exceeds that required to transmit maximum engine torque. This criterion has to be comfortably exceeded in a conventional clutch for the reason mentioned later. For ordinary production cars the centrifugally assisted clutch thus required less effort to operate during traffic driving.

Figure 13.27 shows a cross-section through a clutch of this type.

Fully centrifugal clutch

For clutches of this type, the engagement of the pressure plate was effected by the action of bob-weights, each pivoting from the clutch cover and acting in series with a pair of helical control springs. The bob-weights were carried on the free ends of bell-crank levers and swung within clearance holes provided in the flywheel (Figure 13.28). By limiting the outward travel of the bob-weights, and hence the amount

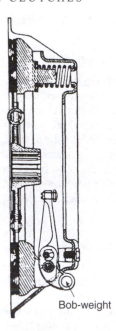

Bob-weight

Figure 13.27 Cross-section through a centrifugally assisted clutch

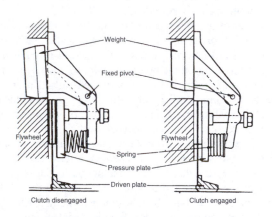

Weight

Fixed pivot

Flywheel   Flywheel

Spring

Pressure plate

Driven plate

Clutch disengaged   Clutch engaged

Figure 13.28 Basic operating principle of a fully centrifugal clutch

of force that could be transmitted through the control springs, it was possible to incorporate a clutch withdrawal mechanism of the form normally associated with a conventional coil spring clutch.

If the bob-weights were allowed to exert, unimpeded, their full centrifugal effect on the engaged clutch, then it could prove wellnigh impossible to disengage the clutch with any sensible amount of effort on the clutch pedal. This is, of course, because the centrifugal loading varies as the square of rotational speed, so that the effort required to disengage the clutch at an engine speed of say 1500 rev/min would be nine times as great at 4500 rev/min. As in the case of a conventional clutch, there was no point in allowing the clamping load to exceed that required for transmitting maximum engine torque by more than about 40 per cent. This extra margin of loading ensures that the clutch springs still exert sufficient clamping load on the pressure plate when the linings are worn, since the compressed length of the springs then becomes greater and their load in position less.

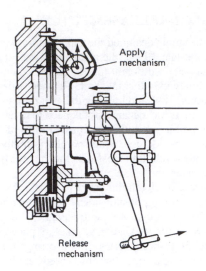

Apply mechanism

Release mechanism

Figure 13.29 Operating principle of roller type of fully centrifugal clutch

A later development was the DAF roller type of fully centrifugal clutch (Figure 13.29), which at engine idle speed is held disengaged by three helical springs. As engine speed is increased above idle, three rolling bob-weights are thrown outwards by centrifugal action. These bob-weights roll in converging tracks between the clutch cover and the pressure plate and force the latter into contact with the friction-lined centre plate, thereby causing the drive to be taken up. An advantage with this particular construction is that the engagement characteristics can be better matched to any given clutch installation, since the roller tracks can be altered in profile to modify the clamping load exerted by the bob-weight rollers.

Among the disadvantages associated with fully centrifugal clutches in the past were the following:

1 Clutch slip is more likely to occur if the engine is allowed to labour at low speeds.
2 Jerky gear changes can result from the less progressive engagement characteristics of the clutch.
3 If clutch judder arises it cannot be minimized by sympathetic driver control.

13.6 MULTIPLATE CLUTCHES

The need for a multiplate clutch

The factors that determine the torque-transmitting capacity of a plate clutch were listed in Section 13.1. So far we have considered the clamping force acting between the friction surfaces and the friction characteristics of the lining materials. Clearly there must be practical limitations to the extent these factors may be increased. Higher clamping forces can demand excessive driver effort to operate the clutch, while materials with higher friction values can tend to make a clutch fierce in engagement.

We are therefore left with the remaining factors of increasing the effective radius of the friction surfaces – that is, fitting a larger-diameter clutch – and increasing the number of friction surfaces in contact. Apart from any space limitation, it is

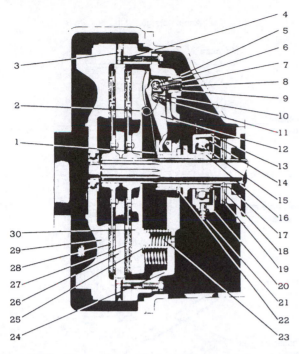

Figure 13.30 Nomenclature of a heavy-duty twin-plate clutch (*Lipe-Rollway*)

1 release lever spider	16 sleeve and bushing assembly
2 retractor spring	17 snap ring
3 intermediate plate drive pin	18 steel disc (clutch brake)
4 socket head (Allen) setscrew	19 friction discs (clutch brake)
5 locknut	20 bearings housing cover
6 eyebolt pin	21 lubrication fitting
7 needle bearing (eyebolt)	22 sleeve locknut
8 eyebolt	23 pressure spring
9 adjusting nut	24 flywheel ring (cover)
10 needle bearing (pressure plate)	25 insulating washer
11 pressure plate pin	26 intermediate plate
12 release lever	27 driven disc assembly (PP side)
13 release bearing housing	28 driven disc assembly (FW side)
14 bearing retaining spring	29 engine flywheel
15 release bearing	30 pressure plate

not always possible simply to increase the diameter of a clutch because of the greater peripheral speed attained by the friction liners. This subjects them to higher centrifugal loading, which could tend to make them fly apart if the clutch were disengaged at high engine speed. One of the more important tests carried out on the physical properties of a clutch lining concerns its ability to withstand bursting stresses.

The final factor to be considered is therefore that of increasing the number of friction surfaces to provide a multiplate clutch. Early clutches of this type utilized a stack of driving and driven thin metal discs and were once widely used as rivals to the early cone clutch, both types being later superseded by the single-plate type. The modern multiplate clutch is, in contrast, usually of twin-plate construction and, apart from being used on competition cars, it is mainly confined in application to heavy-duty commercial vehicles (Figures 13.30 and 13.32).

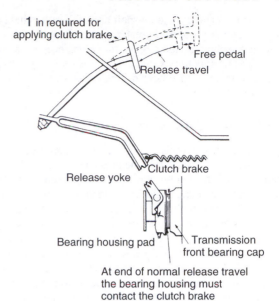

Figure 13.31 Clutch in disengaged position applying clutch brake (*Lipe-Rollway*)

We shall also later encounter the multiplate clutch in automatic transmission applications, but in wet rather than dry form.

A heavy-vehicle twin-plate clutch

In a typical example (Figure 13.30) the clutch cover assembly is produced from cast iron and is attached to the engine flywheel by a ring of setscrews. A pressure plate of the same material is driven by means of lugs extending into mating slots of the clutch cover. Cast iron is also used as the material for the intermediate plate, which is driven by means of six drive pins fitted in the engine flywheel, these being locked in position by Allen screws. The pair of friction-lined driven plates are similar in construction to a conventional clutch centre plate. Three release levers are used, their outer ends pivoting from eyebolts secured to the cover plate. Adjacent to these pivots are those that connect the levers to the pressure plate. To reduce friction, needle roller bearings are incorporated at all six pivot points of the release levers.

It should particularly be noted that the release levers operate on the second rather than the first system of levers, so that the clutch is disengaged by a release bearing that pulls (not pushes) on the release levers. There is a twofold advantage to be gained from this arrangement.

1 The pedal effort is reduced for a given release load at the pressure plate.
2 It allows the ready incorporation of a clutch brake to facilitate upward gear changing.

The clutch brake is brought into action only when the clutch pedal is depressed to the limit of its travel, which results in the release bearing housing pad coming into rubbing contact with the brake disc keyed to the gearbox primary shaft (Figure 13.31). Its purpose is to dissipate in the form of heat the kinetic energy of the spinning clutch driven plates, together with that of the gearbox primary and layshaft assemblies. Inevitably the twin-plate clutch possesses greater inertia than a single-plate one, so that without a clutch brake the reluctance of the driven plates to lose momentum would enforce very slow upward gear changes. It would also render difficult clash-free selection of first and reverse gears.

13.7   ANGLE SPRING CLUTCH

One problem that attends the design of a conventional coil spring clutch is to ensure that the pressure springs have a low enough spring rate (load required to produce unit deflection), because they must maintain an adequate clamping load as they extend in sympathy with gradual wearing of the centre plate liners. Conversely, if the pressure springs have too high a spring rate, either excessive release loads will be imposed when the liners are unworn and in their thickest condition, or the clamping load remaining when the liners are worn and at their thinnest will be inadequate to prevent clutch slip. In other words, it is desirable that the change in clamping load with liner wear should be minimized, and we have already seen how the diaphragm spring clutch can better achieve this aim. However, there are certain limitations to the use of diaphragm spring clutches on heavy vehicles, these being in respect of fully utilizing the increased wear thickness of the liners and the adverse effect of higher temperatures on spring durability.

A different approach to maintaining the clamping load requirements in a coil spring clutch for a heavy vehicle is to be found in the angle spring clutch. This type of clutch was originally developed by the Spicer Clutch Division of the Dana Corporation in America for use on the Ford Transcontinental range of heavy trucks, and since the early 1970s it has been produced under licence in England by the Turner Manufacturing Co. Ltd. It is an indirect-acting clutch in which a series of coil springs are mounted at an angle to the axis of the clutch and transmit a clamping load, via a sliding release sleeve and second order of levers arrangement, to the pressure plate such that the horizontal component of the spring load is multiplied about six times (Figure 13.32). The horizontal component of the spring load therefore changes with movement of the sliding release sleeve, but not in direct proportion to the spring load. As the clutch centre plate liners wear thinner, the sliding release sleeve moves towards the fly-wheel and, although this allows the coil springs to extend and consequently reduce their load, this effect is compensated by the springs assuming a shallower angle to restore the horizontal load component. The engagement load and torque capacity of the clutch therefore remain practically constant. Disengagement of the clutch is effected by a continuously rotating withdrawal bearing that pulls on the sliding release sleeve, thereby removing load from the apply levers and permitting a series of tension springs anchored between the clutch cover and pressure plate to retract the latter. Since the coil springs assume a steeper angle as the sliding release sleeve is being withdrawn, their horizontal load component reduces as does the clutch pedal effort. A servo-assisted clutch control system is therefore not needed with this type of clutch. Another advantage of this particular construction is that the coil springs are less heated, because not only are they removed from direct contact with the pressure plate, but also they are more exposed to the windage in the clutch bell-housing.

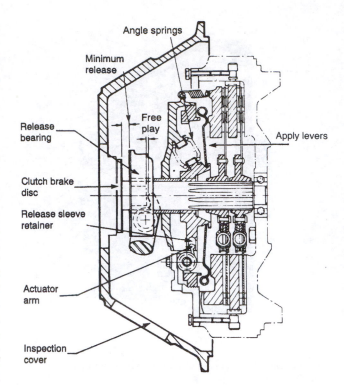

Figure 13.32   Dana Spicer angle spring twin-plate clutch (*Seddon Atkinson*)

For adjustment purposes the apply levers are pivoted at their outer ends from a lockable adjuster ring that is screwed into the clutch cover, so that manual turning of the ring restores the necessary free travel at the clutch pedal. A self-adjusting feature may also be incorporated, which utilizes a worm gear engaging with internal teeth on the adjuster ring; the worm gear is activated by a lost-motion clutch and pick-up lever that respond to increased release sleeve movement. Increments of adjustment take place only when the clutch is disengaged and the load on the apply levers and adjuster ring is relieved. A clutch brake is required where a non-synchronized gearbox is used and may be of the Spicer torque-limiting type. This feature derives from a sandwich construction for the friction elements, which receive their drive via Belleville spring discs from a hub keyed to the gearbox primary shaft. When the friction brake is brought into operation by full depression of the clutch pedal, the braking torque it exerts on the gearbox primary shaft is limited by internal slippage to a predetermined value that minimizes shock loading on the transmission.

13.8   CLUTCH MISBEHAVIOUR IN SERVICE

The service life of a friction clutch must necessarily be influenced by the sympathetic, or otherwise, treatment it receives from the driver and the type of service it is required to perform. It is generally acknowledged that a modern clutch unit with asbestos-free liners should have a life expectancy in the neighbourhood of 50 000 miles (80 000 km) and in some cases this figure may be comfortably exceeded. It is nevertheless important to be able to diagnose clutch problems when they do occur because, despite its basic simplicity, a clutch can sometimes give misleading impressions of where the fault

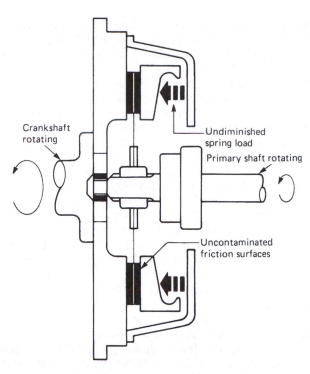

Figure 13.33 Conditions for no slip with engaged clutch

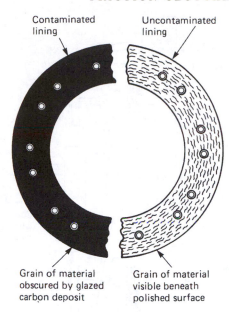

Figure 13.34 Clutch slip arising from contaminated friction linings

actually lies. Another factor that must not be overlooked is the condition of the flywheel face, especially in the case of heavy vehicles. The surface should be examined for unacceptable score marks and heat cracks, since every time the clutch is operated these will cause excessive wearing of the centre plate friction liner.

The symptoms of clutch slip

The term clutch slip should not, of course, be confused with slipping the clutch, since the latter is determined by driver control. As explained earlier, the ability of a friction clutch to slip before full engagement is a decided advantage, because it allows shock-free connection to be made between the engine and gearbox. It was also explained that the functioning of a clutch depends upon sufficient friction being developed between the contact surfaces to transmit the desired torque without their slipping relative to each other (Figure 13.33). If this condition cannot be maintained at all times while the clutch is engaged, then clutch slip is said to be developing. In its early stages clutch slip may not be too obvious under gentle driving conditions, but it will become more noticeable when attempting to accelerate and especially from a standing start.

The probable causes of clutch slip

Lack of free play at the clutch pedal If this condition is present following clutch engagement with a non-continuous contact type of clutch control system, then it implies that either the release levers or the spring fingers of the pressure plate assembly are to some extent pressing against the retracted release bearing. Since this action is tending to oppose the spring loading on the pressure plate, the all-important clamping effect on the centre plate is reduced.

Contaminated centre plate linings One effect of either oil or grease gaining access to the friction linings is that the unwanted lubricant is burnt off, but leaves a very thin deposit of carbon that becomes highly glazed and quite slippery, which naturally reduces the friction acting between the clamped surfaces. Glazing of the friction linings in this manner should not be confused with the normal highly polished appearance of correctly bedded linings, where the grain of the material can still be clearly seen (Figure 13.34).

Excessively worn centre plate linings Even in normal service the effect of slipping the clutch is gradually to wear the friction linings thinner. This reduces the spring load acting on the pressure plate, as its engaged position resides closer to the flywheel. Therefore the effect of excessively worn friction linings is such that the clamping load exerted on them via the pressure plate will ultimately relax to an extent where it will no longer be possible to transmit full engine torque, and the clutch will slip.

Insufficient clamping load Another cause of this is fatigue or gradual weakening of either the coil springs or the diaphragm spring, a process that may be hastened by driver-induced excessive slipping and hence overheating of the clutch. A further possibility is that an incorrect clutch cover assembly has been inadvertently fitted, in which the available spring force is too low for the particular application.

Partially seized clutch control system In this case the incompletely released clutch control system causes either the release levers, or the spring fingers, to remain forcibly in contact with the release bearing, thereby tending to oppose the spring loading on the pressure plate. Examples of this type of complaint include binding in rod, lever and cross-shaft linkages; frayed ends of a release cable jamming in the outer casing; and partial seizing of the slave cylinder piston in a hydraulically operated system.

Partially seized clutch release levers This condition can arise if the release levers are partially seizing or binding about their

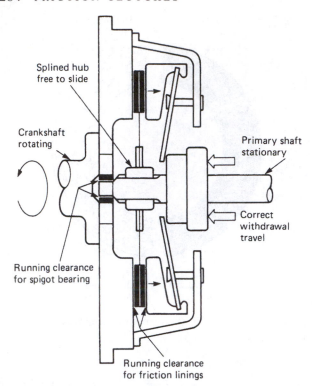

Figure 13.35 Conditions for no drag with disengaged clutch

eyebolt pivot pins. Although this kind of defective clutch action is divorced from the clutch control system proper, in practice its effect on opposing the spring loading on the pressure plate produces the same final result as clutch control system seizure and lack of free play at the clutch pedal.

The symptoms of clutch drag

This type of clutch complaint will become evident when changing gear, especially into low gear. It has the effect of making gear engagement both difficult and noisy, because the centre plate continues to be dragged round by the clutch even though the clutch pedal is fully depressed for disengagement. Needless to say, the presence of clutch drag hardly flatters driving technique and it is mechanically destructive to both the clutch and gearbox, so it must therefore be rectified (Figure 13.35).

The probable causes of clutch drag

Excessive free play at the clutch pedal   This condition of maladjustment is reflected as a reduction in the effective movement of the clutch release bearing against either the release levers, or the spring fingers, of the pressure plate assembly. It results in the pressure plate being incompletely retracted from the cushioned liners of the centre plate, even though the clutch pedal is fully depressed.

Improperly adjusted clutch release levers   This represents the less likely case of internal, rather than external, maladjustment of the clutch control system. If the height setting of

the release levers is not within the limits of equality specified by the manufacturer, then there is a possibility that the pressure plate will tilt as it is retracted, so that the centre plate is not completely released.

Air in hydraulically operated clutch control system   If air is present in the operating fluid then, as with all hydraulic systems, the fluid mass becomes compressible to an extent corresponding to the amount of air entrained. A point can therefore be reached where the effective movement of the piston in the slave cylinder is insufficient to allow complete disengagement of the clutch, via the withdrawal fork and release bearing, even though the pedal is fully depressed.

Splined connection of centre plate partially seized   If the splined hub of the centre plate does not remain free to slide along the splined extension of the gearbox primary shaft, the centre plate cannot nudge itself free of the flywheel friction face when the clamping load is removed from its cushioned linings by depressing the clutch pedal.

Primary shaft partially seizing in flywheel spigot bearing   Strictly speaking this is not a case of clutch drag in the true sense of the description, but as far as the driver is concerned the effect is the same. Relative movement within the bearing only occurs when the clutch is disengaged and the centre plate freed, or if the clutch is slipping. If this movement is prevented by partial seizure of a lubricant-starved bearing, the centre plate will continue to rotate with the clutch assembly even though the pedal is fully depressed.

Contaminated centre plate linings   Although oil or grease contaminants may partially burn off the linings owing to the heat generated by slip, which occurs under normal conditions of clutch engagement, they can leave a resinous and hence sticky deposit on the contact face of the affected lining. When the clutch is disengaged the contaminated lining tends to adhere to the mating friction track of either the flywheel or the pressure plate, causing continued rotation of the centre plate.

Distorted centre plate   Owing to leverage and pedal travel limitations, the clutch designer necessarily has to reduce to a minimum the overall disengagement clearance of the centre plate within the clutch. A distorted centre plate may therefore absorb a normal disengagement clearance of about 2.5 mm (0.1 in) and as a result be dragged round by the clutch assembly. A distorted centre plate is usually caused by careless refitting of the gearbox.

Partial detachment of centre plate linings   Loose or broken linings arising from unusually arduous conditions of service can cause clutch drag, since they create a similar lack of disengagement clearance to that imposed by a distorted centre plate.

Distorted clutch cover   Damage to this component can produce a tilting effect on the pressure plate similar to that created by improperly adjusted release levers. Distortion of the clutch cover can arise if its securing flange has not been evenly bolted to the clamping face of the flywheel. It is for this reason that a manufacturer will recommend the bolts be tightened one turn at a time in criss-cross fashion or, as it is sometimes grandly termed, diagonal selection.

Fatigued diaphragm spring   This condition is not intended to describe one of actual fatigue failure of this component, but rather to indicate a general weakening of the diaphragm material that increases its flexibility. This can arise from normal heating effects repeated over a long period of service and may

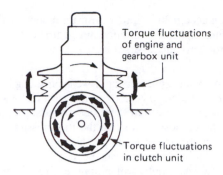

Figure 13.36 Simple representation of clutch judder

Labels in figure:
Torque fluctuations of engine and gearbox unit
Torque fluctuations in clutch unit

be accompanied by wearing thin of the diaphragm spring finger tips. So although the pedal may be fully depressed for clutch disengagement, a complete retraction of the pressure plate is no longer possible and the clutch drags.

Clutch spin

The terms clutch drag and clutch spin are generally used synonymously, but a cautionary note must be sounded here, because it is sometimes necessary to distinguish between real and apparent complaints of clutch drag. In the latter case, for example, it must be recognized that with the engine at idling speed and with the gearbox oil at normal operating temperature, a fully serviceable clutch assembly usually needs 3 to 5 seconds after disengagement or declutching before the spinning centre plate comes to rest. From this it follows that a driver can provoke grating noises in the gearbox if either reverse gear, or an unsynchronized sliding first gear, is selected too hurriedly during vehicle manoeuvring. Before the era of automatic transmission, the professional chauffeur displayed a certain cunning in avoiding such undignified noises by partially selecting or 'stroking' a synchronized gear to kill any spin of the centre plate and then selecting the unsynchronized gear.

The symptoms of clutch judder

This condition describes what a driver senses as a jerky shudder, which can occur during engagement of the clutch as the vehicle moves away from rest, especially on an up-gradient. As a transmission complaint, clutch judder represents a rather special case, since it is influenced by external sources and does not always lend itself to a simple cure (Figure 13.36). In fact a great deal of research into clutch judder problems has been carried out by transmission engineers over the years, and one interesting fact that has emerged is that the frequency of judder is always about 10–13 Hz (hertz or cycles per second).

The probable causes of clutch judder

Distorted clutch cover Damage to this component and the tilting effect it can impose on the pressure plate has previously been mentioned in connection with clutch drag. The pressure plate may be similarly affected if a diaphragm spring clutch is lifted by the release lever plate such that the drive straps become bent.

Distorted centre plate This condition has also previously been mentioned in connection with clutch drag, together with its likely cause. Misalignment implies, of course, a non-parallel approach of the friction contact surfaces as the clutch is engaged, with a consequent tendency towards judder.

Flywheel misalignment Yet another source of misalignment that can provoke clutch judder, flywheel misalignment or run-out can be readily checked with a dial indicator (see Section 13.9).

Defective engine or gearbox mountings Loose mountings should be checked for wear and retightened. Rubber mountings can become softened by oil or grease contamination and these should be replaced. Similar remarks apply to gearbox tie bars, which should also be checked for bending and maladjustment.

Contaminated centre plate linings Clutch judder can be initiated if unusually large changes in friction value occur between the contact faces of the clutch. It is therefore necessary to check for any oil or grease contamination of the friction linings, correct the fault at source and renew the centre plate.

The symptoms of clutch fierceness

As might be expected from its description, this type of clutch complaint is one in which the drive is taken up very abruptly in an all-or-nothing manner, no matter how gently the driver releases the clutch pedal. Apart from causing discomfort to the vehicle occupants, the shock loading on the clutch assembly, engine and transmission system can be mechanically destructive.

The probable causes of clutch fierceness

Distorted centre plate As in the case of clutch judder, any misalignment within the clutch assembly which results in a non-parallel approach of the friction contact surfaces is the enemy of smooth engagement.

Contaminated centre plate linings Again, the effect of this condition on clutch operation has previously been mentioned in connection with judder. If a clutch gradually becomes fierce with use, it can be caused by oil contamination.

Partially seized clutch control system The various ways in which this can occur have also been dealt with in connection with clutch slip. Any excessive friction that spoils the sensitivity of control can provoke fierce engagement.

Excessively worn centre plate Fierce engagement can arise if the friction linings have worn down to the rivets and scored the flywheel or pressure plate, and be aggravated if these surfaces are heat spotted by accompanying slip.

Clutch noises

A rattling noise from the clutch when the vehicle is being driven could indicate distress, such as a broken spring, in the torsional spring drive of the clutch centre plate (Section 13.3). If the rattling noise occurs when the engine is idling and the clutch pedal depressed, it could be caused by wearing of the retracting springs in a diaphragm spring clutch. A clicking noise can arise from a worn ball mounting for the clutch release fork. The clutch release bearing itself can be a potential source of noise if it leaks its lubricant. This becomes evident as a whirring noise whenever the clutch pedal is depressed, which brings the bearing into operation. On the other hand, if a squealing noise occurs when the clutch pedal is depressed and a gear is selected with the vehicle held stationary, this would indicate distress in the flywheel spigot or pilot bearing that supports the overhanging end of the gearbox primary shaft. It should be noted that this test would be invalid unless a gear is selected, since only then can relative movement be enforced between the bearing in the rotating flywheel and the spigot end of the non-rotating primary shaft. Heavy-duty clutch installations can sometimes prove more demanding in terms of durability of their clutch release and spigot or pilot bearings.

13.9   FLYWHEEL AND CLUTCH HOUSING ALIGNMENT

The effects of misalignment

Unless the engine crankshaft and the gearbox primary shaft are maintained in correct alignment or, more specifically,

held within acceptable limits of misalignment by the clutch bell-housing that unites engine and gearbox, the following service problems can arise:

1   It can be responsible for rapid wear on the splines of the clutch centre plate to gearbox primary shaft connection, the consequent looseness of these parts resulting in noticeable 'backlash' or 'lost motion' in the transmission, accompanied by an impact noise on drive take-up.
2   Undue bending stresses can be imposed on the clutch centre plate assembly, which may result in the hub breaking loose from the plate and the consequent total failure of the clutch to transmit drive. In other words, we have what amounts to a do-it-yourself fatigue test!
3   It can promote conditions of clutch drag and judder, since the skewing of the centre plate within the flywheel and pressure plate assembly prevents parallel movement of the friction contract surfaces during disengagement and engagement of the clutch.

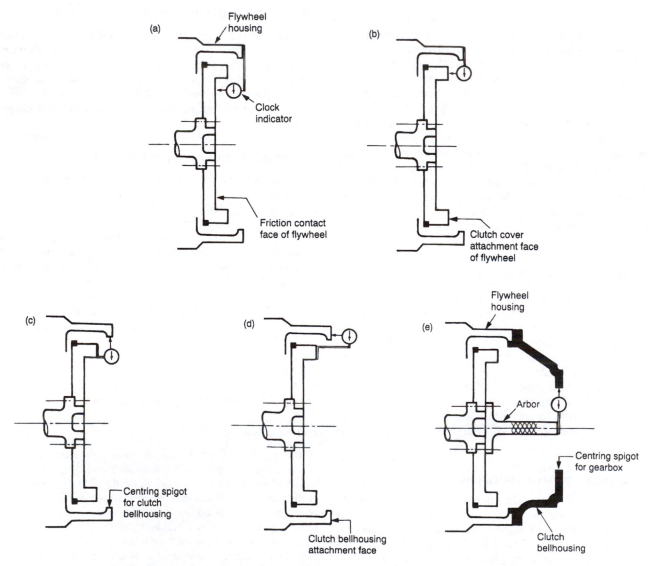

Figure 13.37   Checking flywheel and clutch housing alignment

4 There can be difficulty in gear changing arising from point 3, especially in the case of unsynchronized heavy-vehicle gearboxes, which not only affects the operation and durability of the clutch, but also is detrimental to the gearbox bearings, dog clutches and gears.

Checking alignment

Although normally there need be no concern about fly-wheel and clutch housing alignment in passenger car practice, a manufacturer of heavy vehicles may well deem it advisable to provide service instructions for checking such alignment, in the event of a complaint of clutch drag arising from an undetermined cause or if appreciable backlash becomes evident in the transmission. Checking flywheel and clutch housing alignment involves first the removal of both gearbox and clutch and then the mounting of a clock indicator or dial gauge in various positions to record any run-out as the engine is turned over slowly by hand for one complete revolution.

The stages of checking flywheel and clutch housing alignment to ascertain whether any components need renewing may be summarized as follows:

1 Clock indicator mounted on bracket from the flywheel housing to clutch bell-housing joint face and engaging the friction contact face of the flywheel (Figure 13.37a), which should be pressed forwards to cancel the effect of crankshaft end float. The allowable run-out on the friction contact face should generally be within 0.025 mm (0.001 in) per 25 mm (1 in) of flywheel radius measured from the crankshaft axis to the clock gauge stylus. If not, the fly-wheel should be removed from the crankshaft to check the mating surfaces for burrs and cleanliness.

2 Clock indicator mounted as in stage 1 but engaging the clutch cover attachment face of the flywheel (Figure 13.37b). The allowable run-out should not exceed that for the friction contact face of the flywheel.

3 Clock indicator mounted on bracket from the clutch cover attachment face of the flywheel and engaging the centring spigot for the clutch bell-housing (Figure 13.37c). The allowable run-out generally ranges from 0.15 mm (0.006 in) to 0.30 mm (0.012 in) according to the diameter of the housing, and the manufacturer's specifications should be consulted. Readings taken at the 12, 3, 6 and 9 o'clock positions will indicate in which direction the housing may need small corrective movement at its clamping face.

4 Clock indicator mounted on bracket from the clutch cover attachment face of the flywheel and engaging the fly-wheel housing to clutch bell-housing joint face (Figure 13.37d). The allowable run-out should be within the same limits in relation to diameter as for stage 3. If not, the fly-wheel housing should be detached to check the mating surfaces for burrs and cleanliness.

5 Clock indicator mounted on arbor (special support spindle) from the flywheel attachment bolts and engaging the centring spigot for the gearbox casing after refitting a separate bell-housing (Figure 13.37e). The allowable run-out should generally be within the same limits as for stage 3 except, of course, that for 'housing diameter' read 'housing bore diameter' in the manufacturer's specifications.

All test surfaces should be cleaned thoroughly before carrying out the above inspection. Each test is generally repeated two or three times and the readings compared to eliminate test error.

14 Layshaft gearboxes

14.1 PURPOSE AND ELEMENTS OF THE GEARBOX

The need for a gearbox

A gearbox is incorporated in the vehicle transmission system for the following reasons:

1 To assist the engine to do its work.
2 To enable the vehicle to be reversed.
3 To provide a ready means of disconnecting the engine from the transmission system.

The first reason merits closer examination. In the same sense that man is often required to perform physical tasks beyond his unaided strength, so too can the performance demanded of a motor vehicle exceed the turning effort available at the engine crankshaft to drive the road wheels.

In practice the turning effort, or torque as the engineer prefers to call it, developed by an engine generally rises to a maximum somewhere near the middle of its speed range, following which it begins to decrease. This characteristic of the engine would not unduly concern us if the vehicle were always driven at an appropriate constant speed, but it does pose a problem when starting from rest, accelerating from low speed and climbing a steep hill. In these circumstances, a heavier workload is imposed upon the engine and this clearly demands a multiplication of its torque, a function which is performed by the gearbox.

The gearbox therefore provides the means of altering in stages (hence the different speeds) the relationship between engine speed and vehicle speed, since what is gained in torque is always lost in speed, as will be made evident later. Its chief purpose then is to ensure that within reasonable limits the engine can be allowed to run efficiently and hence economically at a speed that produces sufficient torque to drive the road wheels as the workload on the vehicle varies.

Reduction gearing

Before examining the various types of layshaft or, in American terminology, countershaft gearbox used in light and heavy vehicles, it is necessary first to recall the basic elements of reduction gearing and gear engagement.

Starting from first principles, it is convenient to consider gears merely as an altered form of lever. The lever is, of course, one of the simplest machines known to man and consists of a bar or crowbar free to rock over a pivot. From practical experience it will be known that by suitable adjustment of the pivot position a relatively moderate effort applied downwards at one end of a crowbar can raise a heavy load at the other end. The relationship is such that the nearer the pivot is placed to the load, in comparison with its distance

from the point of effort, the greater will be the multiplication of the effort, although the lesser will be the distance the load is raised. It is thought provoking to recall that the early Greek philosopher Archimedes was so impressed by this fact that he was led to remark, 'Give me a firm place to stand on, and I will move the world.'

Let us next consider a pair of levers arranged end to end with their inner ends overlapping slightly and provided with cam-like projections (Figure 14.1). If one lever is made half the length of the other and they are both centrally pivoted, then any torque applied to the shorter lever will be doubled at the longer lever, as long as the cam-like ends remain in contact. Here then we have the essential ingredients of a pair of reduction gears, since if we take matters another step forward and arrange in each case for a whole series of levers extending to a full circle, we can maintain a multiplication of torque during continuous rotation.

In this analogy we have therefore arrived at the lever equivalent of a simple gear train in which the pinion would be half the size of the gear and would have only half as many teeth. The relationship between the numbers of teeth on the gears of a simple gear train is known as their ratio to one another. In the example just given, the ratio would be expressed as 2:1; similarly, if the pinion had only one-third as many teeth as the gear then the ratio would be 3:1.

At this stage, two important points of terminology should be noted. First, the smaller gear of a meshing pair is always referred to as the *pinion* and the larger gear as the *gear* (Figure 14.2). Second, a *low-ratio gear* refers to a pair of gears of *high numerical ratio* and vice versa. For example, a pair of gears giving a 3:1 reduction ratio is said to be of lower ratio than a 2:1 ratio pair.

It is, of course, desirable that the arrangement of gearing in a gearbox is as compact as possible. If a simple gear train is required to give a fairly large reduction ratio the gear could become unavoidably large and cumbersome. It is usual, therefore, to employ what is known as a *compound gear train*, so that a large reduction ratio may be obtained from gears of small size (Figure 14.3). A single compound gear train comprises two simple gear trains arranged so that the pinion of the second pair rotates at the same speed as the gear of the first pair, their rigid connection being effected through a countershaft or *layshaft*.

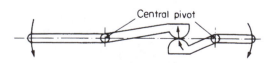

Figure 14.1   The levers analogy of gearing

The final reduction ratio of a compound gear train is equal to the product of the ratios of the simple gear trains from which it derives. For example, the first simple gear train may have a pinion with 25 teeth and a gear with 50 teeth, while the second may have a pinion with 15 teeth and a gear with 45 teeth. Their respective ratios are therefore 2:1 and 3:1 which, multiplied together, provides a final ratio of 6:1. This means that the torque available at the gear of the second train is six times greater than that supplied to the pinion of the first train. By the same token, the rotational speed of the second train gear will be one-sixth that of the first train pinion.

The engagement and disengagement of a gear in a simple or a compound gear train may be arranged in two different ways: sliding mesh or constant mesh.

Sliding-mesh gears

With this method gear engagement is accomplished by sliding the appropriate gear endwise along a splined shaft (in effect a series of keys formed integral with the shaft and spaced evenly around it) until it meshes with its opposite number (Figure 14.4).

The sliding-mesh principle of gear changing was the first to be used in motor vehicle gearboxes. Its introduction is generally attributed to the French automotive pioneer Emile Levassor, who as long ago as 1895 adopted the idea from the lathe type of clash gear.

As can be imagined, the action of changing gears called for a considerable nicety in timing on the part of the driver, since it was no easy matter to bring into mesh a gear and a pinion revolving at different speeds. The inexperienced driver would often produce most unpleasant grating noises from the gearbox as the ends of the gear teeth momentarily ground together, which naturally damaged them. Those readers who are interested in vintage cars will probably have heard sliding-mesh gearboxes referred to as crash gearboxes. Indeed, Levassor is said to have once commented, 'It's brutal but it works.'

Constant-mesh gears

With this arrangement the gear is retained permanently in mesh but is free-running on the shaft. Engagement is effected by locking the gear to its shaft through the medium of a dog clutch sliding on splines (Figure 14.5).

Except usually for reverse gear, gear trains arranged on the constant-mesh principle have long since formed the basis of most layshaft gearboxes, which in the case of all passenger cars and many heavy vehicles incorporate the additional refinement of a synchromesh mechanism to simplify the action of gear changing.

All forward speed gears are therefore retained permanently in mesh with their pinions, although they are free to revolve independently of their shaft when disengaged. As previously explained, their engagement is effected through sliding dog clutches rotationally fixed to the gearshaft (Figure 14.6). This system naturally relieves the gears and pinions of any abnormal wear which might otherwise result from endwise clashing of their teeth during gear changing.

Requirement for helical gears

When two gears meshing with each other are used to connect parallel shafts, and their teeth are straight and parallel to the

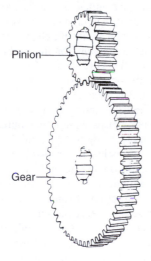

Figure 14.2 A pair of meshing gears

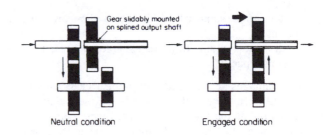

Figure 14.4 Principle of sliding-mesh gear engagement

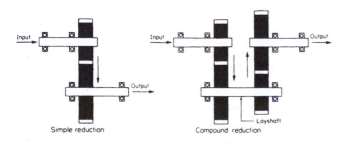

Figure 14.3 A comparison of simple and compound gear trains

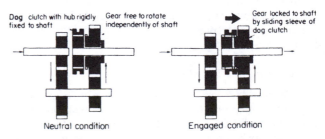

Figure 14.5 Principle of constant-mesh gear engagement

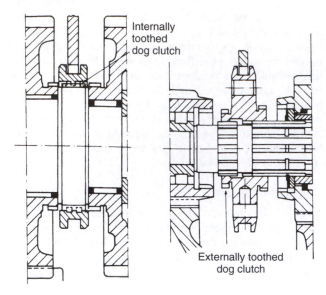

Internally
toothed
dog clutch

Externally toothed
dog clutch

Figure 14.6 Constructional details of two forms of sliding dog clutches (*Eaton*)

shaft axes, we call them spur gears. Such a pair of gears was illustrated in Figure 14.2, wherein their application to sliding-mesh gear engagement was inferred.

If, on the other hand, the pair of gears have teeth that are inclined to the axes of the shafts, forming part of a helix, then we describe them as helical gears. These are normally retained in constant-mesh engagement; it is possible to have a sliding-mesh helical gear, but only by introducing the expensive complication of helical splines to mount it. Since constant-mesh gear engagement is used exclusively for the forward speeds of all car and many commercial vehicle gearboxes, clearly there is a choice of using either spur of helical gears. The reasons for generally preferring helical gears are basically as follows:

1 For the same standard of accuracy they are quieter running.
2 For the same width they are capable of greater load-carrying capacity.

These advantages are realized with helical gears as long as there is sufficient face width available to give a continuous helical action, as the load is transferred from one group of teeth to another. In other words, there is no sudden transference of load from one tooth to the next, as occurs in ordinary spur gearing. The load is gradually put on a tooth and as gradually taken off, so that the stress in any one tooth is maintained practically constant. In contrast, the stress in a straight tooth of a spur gear fluctuates according to its relative position of engagement.

Practical application of helical gears

A problem that arises when single helical gears are employed is that provision must be made to absorb an end thrust force which is imposed by the obliquity of their teeth. This end thrust force is in addition to the separating force that in any event acts between two gears transmitting torque.

As we shall next be dealing with the various types of rolling bearings that are used to contain these forces, it may be useful to consider in simple terms how they are generated.

Let us start with a meshing pair of straight-tooth spur gears. The tooth form of these gears is based on what is known in geometry as an involute curve, which was first applied to the teeth of gears by the French scientist De la Hire as long ago as 1694. It is traditional to imagine an involute as a curve that is described by the knotted end of a cord being unwound from the circumference of a circle. This circle is known as the base circle (Figure 14.7a).

In the case of a meshing pair of gears, their action may be visualized as being the same as if their base circles were simply connected by a crossed-over belt (Figure 14.7b). The point at which the belt crosses over is designated the pitch point of the teeth (Figure 14.7c), and the straight portion of the belt not in contact with the base circles represents the line of action. It thus follows that the pressure on the teeth of involute gears is constantly in the direction of the line of action, and this guarantees their uniformity of motion. In other words, when two meshing gears are transmitting torque, the resultant force acting normal to the profile of each tooth occurs along the line of action (Figure 14.7d). Since the latter lies at an acute angle, termed the pressure angle, to the common tangent to the pitch circles of the gears, it can be resolved into two mutually perpendicular components: a tangential force that is responsible for transmitting torque, and a radial force that is trying to separate the gears (Figure 14.7d). We have therefore accounted for the separating force that occurs between a pair of straight-tooth spur gears and similarly, of course, between a pair of helical gears.

It now remains to account for the additional end thrust force generated by the latter type of gear. This end thrust force arises from the obliquity of their teeth, which is termed the helix angle and is measured from the axis of the gear (Figure 14.7e). The force acting along the helical teeth of the gears can therefore be resolved into two mutually perpendicular components, these being the tangential force (mentioned earlier) that is responsible for transmitting torque, and an axial force that is producing an end thrust. This can act in either direction and depends upon the hand of the helix (right or left), the direction of rotation and whether the particular gear is driving or being driven.

It has sometimes been said that modern gear design in engineering owes much to the development of the motor vehicle. Since the motor vehicle was invented, gears have been one of its most important components as they provide a compact and reliable means of transmitting power and of varying speed. Improved materials and the progress made in heat treatment and machining techniques have increased the resistance of the gear teeth to surface pitting and bending fatigue. Since these factors ultimately determine the gear tooth load capacity, it has allowed the gears to become generally smaller and of finer pitch for quieter operation.

Gearbox rolling bearings

The selection of suitable types of bearing for the motor vehicle gearbox is based on considerations similar to those for any other engineering application. Each type of bearing has

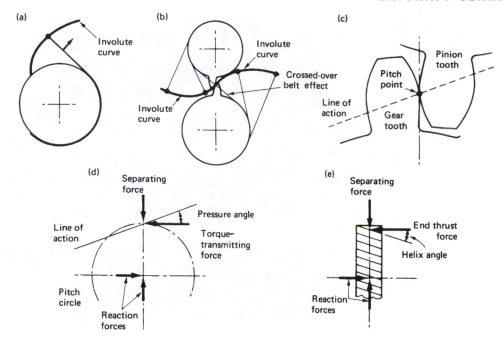

Figure 14.7 Summary of basic gear tooth action: (a) involute curve generated by unwinding a knotted cord (b) action of involute gear teeth (c) close-up of gear teeth action (d) radial separating force acting on a spur or a helical gear (e) radial separating and axial thrust forces acting on a helical gear

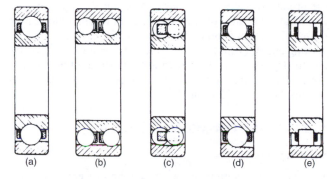

Figure 14.8 Simple ball and roller bearings: (a) standard ball journal (b) double-row standard ball bearing (c) double-row self-aligning ball bearings (d) combined radial and thrust (one direction) (e) standard roller journal

characteristic features which make it particularly suitable for certain duties (Figure 14.8a–e), so its selection must take into account at least the following basic factors:

Available space In many instances at least one of the main dimensions of the bearing (usually the bore) is predetermined by the design of the gearbox shafts. If radial space is particularly limited, such as in the case of the spigot bearing in the constant-mesh pinion of the primary shaft, then a needle roller bearing with a small sectional height may be required.

Magnitude of load This is normally the most important factor in determining the size of bearing required, together with the related factors of life and reliability. Generally,

roller bearings can carry greater loads than ball bearings of the same external dimensions.

Direction of load Cylindrical roller bearings having one ring without flanges and needle roller bearings can carry only radial loads. All other types of radial bearing can carry a radial and an axial load acting simultaneously, or what is termed a combined load. The most important feature affecting the ability of a bearing to carry an axial load is its angle of contact, which was referred to in connection with clutch release ball bearings (Section 13.2).

Axial displacement The normal bearing arrangement consists of a locating (fixed) bearing and a non-locating (free) bearing, so that the latter can be displaced axially. This prevents cross-location of the bearings as a result of shaft expansion or contraction. The bearings can also be arranged so that axial location is provided by each bearing in one direction only, as for example with the tapered roller type (Figure 20.7), which is further discussed later in connection with final drives.

In a conventional double-stage, constant-mesh gearbox, as shown in Figure 14.11, there is virtually no load on the gearbox bearings when direct drive top gear is engaged. It is possible that some load may be imposed on the mainshaft rear bearing, but this arises from propeller shaft operation.

When the indirect gears are selected, it is then that the separating and end thrust forces acting on the helical constant-mesh gears, and whichever other pair of gears happens to be engaged, have to be taken into account. This means that there will be combined radial and axial loading on both the primary shaft bearing and the main-shaft rear bearing, together with a radial load on the mainshaft front or spigot bearing.

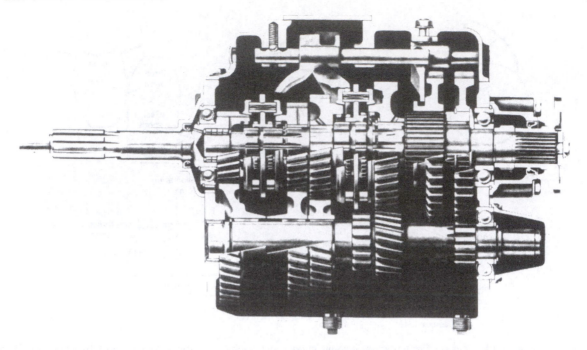

Figure 14.9   Ball and needle roller bearings in a modern five-speed commercial vehicle gearbox (*Turner-Spicer*)

Although the layshaft bearings will be subjected to radial loading, they can be relieved of axial loading (or nearly so) by providing the driven constant mesh gear and the driving pinions with helices of the same hand, so the two end thrusts on the layshaft are in opposite directions. A careful choice of helix angles for the layshaft gear and pinions also plays a part in achieving this balance of loading.

Until recently, it was long-established practice to use single-row, deep-groove ball bearing races on the primary shaft and mainshaft, in conjunction with a needle roller spigot bearing (Figure 14.9). Needle roller bearings were also typically used to support a hollow layshaft with plain thrust washers to take any end thrust loading (Figure 14.18). However, the need for a hollow layshaft could restrict the diameter of its supporting spindle for the needle rollers, which tended to limit the load capacity of the gears owing to deflection. Assembly of the gearbox was also made a fiddly job, because the layshaft could only be located on its spindle after being lifted into mesh with the primary shaft and mainshaft gears, which had first to be inserted endwise into the gearbox casing. A reduced length dummy spindle was therefore required to hold the needle rollers and thrust washers temporarily in place, until the proper length spindle could be pushed through to replace it, once the layshaft gears had been meshed with those on the primary shaft and main shaft.

In current practice, tapered roller bearings are often used instead of ball, roller and needle bearings, if not throughout the gearbox (Figure 14.10), then at least to support the layshaft or its equivalent in a transaxle (Figure 20.12). The reasons for their use include quieter operation, more compact design and simpler assembly. Also, the closer control over endwise movements of the primary shaft and mainshaft is claimed to improve the quality of gearchange. It is nevertheless critical for the 'preloading' of these bearings to be

carefully controlled, especially where an aluminium alloy is the preferred material of construction for the gearbox casing. This is because of the difference in thermal elongation of the alloy casing and its steel gear shafts, between the cold starting and hot running conditions. The operating geometry and the significance of preloading for tapered roller bearings are explained later, in connection with axle final drive applications (Sections 20.1 and 20.2).

14.2   CONSTANT-MESH GEARBOXES

Basic layouts

We take as the example a four-speed (and reverse) gearbox. The two basic layouts of constant-mesh gearing, omitting for the moment any reference to synchromesh mechanisms, are double stage and single stage.

Double-stage constant mesh

This is the conventional layout used where a front engine drives the rear wheels through a gearbox separated from the final drive (Figure 14.11). The following are its chief components.

The primary shaft

This takes from its splined end the drive from the clutch-driven member or centre plate. Supported by a rolling bearing in the front end of the gearbox casing and a spigot bearing in the engine flywheel, it carries the constant-mesh pinion for driving the layshaft. The pinion is also equipped with dog clutch teeth providing direct connection with the mainshaft dog clutch when fourth speed is engaged. This principle of utilizing a direct-drive top gear was first established by another famous French automotive pioneer, Louis

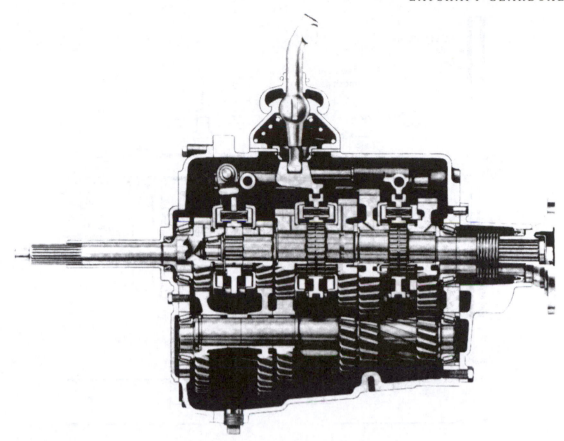

Figure 14.10   Tapered roller bearings in a modern five-speed commercial vehicle gearbox (*Turner-Spicer*)

Renault, in 1899. A spigot bearing is contained within the constant-mesh pinion to support the nose of the mainshaft.

The layshaft
This carries the constant-mesh gear that takes its drive from the constant-mesh pinion of the primary shaft. The layshaft therefore revolves whenever the clutch centre plate is revolving. It is provided with three pinions for transmitting drive to the forward gears and another for reverse gear. These pinions, and the constant-mesh gear previously mentioned, are all rotationally fixed to the layshaft, which is usually supported in rolling bearings carried either on a separate through spindle or directly in the front and rear end covers of the gearbox casing, the latter being found in heavy-duty applications (Section 18.3).

The mainshaft
This takes its drive from either the layshaft or the primary shaft according to whether an indirect gear (first, second or third speeds or reverse) or direct top (fourth speed) is engaged (Figure 14.11). When driven, it conveys the drive to the propeller shaft and thence to the rest of the transmission system. The mainshaft is provided with a series of journal bearing surfaces upon which the indirect gears are free to revolve. Adjacent splined portions are provided to secure rotationally the hubs of the sliding dog clutches and also the propeller shaft flange coupling. Support for the mainshaft is given by a rolling bearing in the rear end cover of the gearbox casing and the spigot bearing in the constant-mesh pinion of the primary shaft. A tail shaft bearing may also be found where a rear extension housing is used on the gearbox, so as to reduce the length of the propeller shaft.

The indirect gears
These gears, with the exception usually of reverse gear, are free to revolve on the mainshaft either with or, less commonly, without separate bearings. In the former case caged needle rollers are now generally preferred to plain bushes, so as to minimize friction and prevent tipping of the gears under load on a plain bearing clearance, and also to make the lubrication arrangements less critical. The indirect gears run permanently in mesh with their corresponding pinions on the layshaft (Figure 14.11) and they are restrained against endwise movement by a suitable deployment of thrust collars on the mainshaft. One gear each serves for the first, second and third speeds, all the gears being provided with dog clutch teeth to enable them to be locked to the mainshaft by an adjacent sliding dog clutch.

The sliding dog clutches
A pair of these are permanently fixed, in the rotational sense, to the splined portions of the mainshaft. Their hubs are provided with external teeth which engage with the internal teeth of the sliding sleeves, these in turn being able to intercept the complementary dog clutch teeth of the indirect gears and the primary shaft pinion. To facilitate engagement of the dog clutches, their teeth are relieved at the ends by

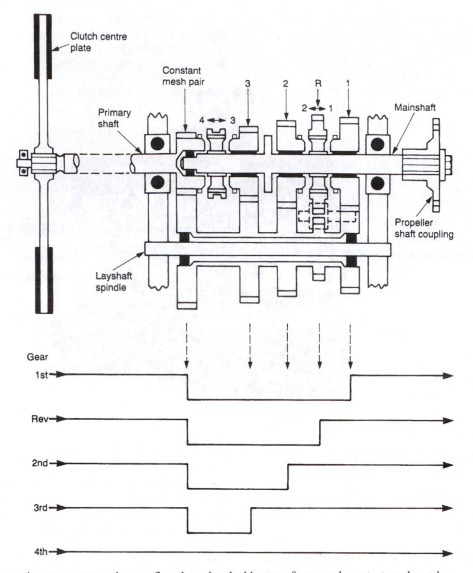

Figure 14.11   Schematic arrangement and power flow through a double-stage four-speed constant-mesh gearbox

chamfering. The mainshaft reverse gear is combined with the sleeve of the sliding dog clutch that engages first and second speeds.

The reverse gear
To enable the vehicle to be driven backwards, an idler gear must be introduced between the non-meshing reverse pinion and gear carried respectively by the layshaft and the first and second-speed sliding dog clutch on the mainshaft. When the reverse idler gear is engaged, it thus causes the mainshaft to revolve in the direction opposite to the primary shaft. The sliding-mesh principle is retained to bring into mesh the reverse idler gear, except in more recent designs where they are in constant mesh and synchronized (Figure 18.5).

The gearbox casing
This provides a rigid support for the bearings of the various shafts and the gear selector mechanism as well as an oil-tight enclosure for the gears. A typical construction allows for endwise insertion of the gearshaft assemblies during build. The casing may be produced from either cast iron or, more commonly now, aluminium alloy to reduce weight and improve heat dissipation. Its surfaces may be ribbed to increase both strength and rigidity and to minimize noise transmission.

Single-stage constant mesh

This type of layout is required where the gearbox is combined with the final drive to provide what is termed a transaxle (Figures 14.12 and 14.13), which is a necessary arrangement for front-wheel drive and also rear-engined passenger cars. It differs from the double-stage version in the following respects:

The mainshaft
This is no longer divided into two parts since the extended primary shaft lies above the final drive gears and merely

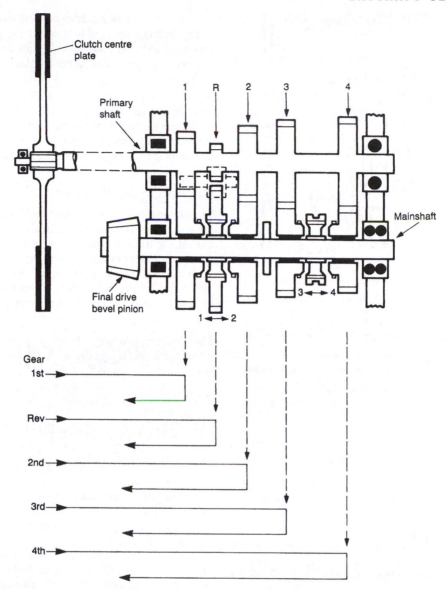

Figure 14.12 Schematic arrangement and power flow through a single-stage four-speed constant-mesh gearbox for longitudinal mounting

conveys the drive from the clutch centre plate to an output gear of the mainshaft below. Support for these shafts is by rolling bearings in the front and rear ends of the gearbox casing.

The indirect pinions

All the pinions serving the four forward speeds and reverse are carried by, and rotationally fixed to, the primary shaft. A direct-drive top speed between the primary shaft and the mainshaft is therefore no longer possible, each forward speed being indirect and obtained through a simple gear train (Figure 14.14). In contrast, the first, second and third speeds in a double-stage gearbox are obtained through compound gear trains. The single-stage gearbox thus reverts in basic principle to the all-indirect gearbox originally designed by the earlier mentioned Emile Levassor, albeit now much less brutal!

The constant-mesh gears

These and their sliding dog clutches, one of which normally carries the reverse gear, were all originally mounted on the output mainshaft (Figures 14.12 and 14.14) in a manner similar to that found on the mainshaft of a double-stage gearbox. In some more recent designs of (synchronized) four-speed transaxle, the sliding dog clutches for the third and fourth gears are carried not by the mainshaft but by the primary shaft, on which the third and fourth gear pinions are then made rotationally free (Figure 14.13). The advantages to be gained from this redistribution of the sliding dog clutches between both mainshaft and primary shaft, which provides a 1–2 on output and 3–4 on input arrangement, are as follows:

1 The rotational inertia of the primary shaft and its associated clutch centre plate is minimized to reduce effort in gear changing, since only first and second gear pairs and

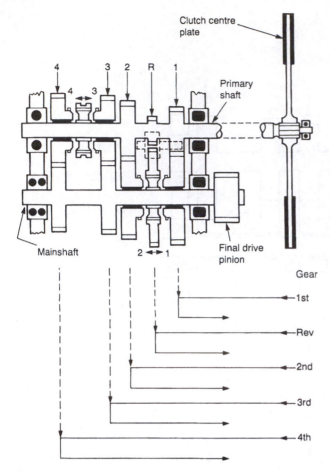

Figure 14.13 Schematic arrangement and power flow through a single-stage four-speed constant-mesh gearbox for transverse mounting

the third and fourth sliding dog clutches permanently rotate with it.

2 Quieter running in neutral is attained because half the number of gears, namely third and fourth, remain stationary.

3 A more compact gearbox becomes possible by virtue of eliminating the need for a mainshaft spacer between the second and third gearsets, and also allowing a tapering gear casing.

The redistribution of the sliding dog clutches in this manner is not an entirely new idea, since what perhaps may be designated a 'dual-line' synchronizer arrangement was originally to be found in the transaxle of the rear-engined Porsche car in the mid 1960s. This same principle may also be extended in application to single-stage five-speed overdrive top gearboxes, where an overhung fifth speed step-up gear is freely mounted towards the end of the input primary shaft and engaged by a synchromesh dog clutch splined to the shaft extremity (Figure 18.7).

The mainshaft also serves as either the bevel pinion or helical pinion shaft for the final drive gears (Figure 14.12 and 14.13), according to whether the engine is mounted longitudinally or transversely. For this reason, the type of rolling bearing selected to support the mainshaft must also accommodate the additional radial and end thrust loads imposed by the final drive pinion (Figures 14.12 and 14.13).

The gearbox casing
In the single-stage gearbox the casing is extended to include the final drive gears and their bearing mountings, the complete unit then being known as a 'transaxle' (Section 19.5).

Gearbox speedometer drive

The older style magnetic speedometer is driven by a flexible drive shaft, or 'speedo cable', from reduction gearing within the vehicle transmission system, such that its driving speed is always proportional to vehicle speed (Figure 14.15a). With the modern car there are generally three sources from which the speedo cable may receive its right-angle drive, depending upon the layout of the transmission system:

Double-stage gearbox From a spiral driving gear that is either machined integrally with, or separately mounted on, a rearward extension of the mainshaft.
Single-stage (transaxle) gearbox In this case the spiral driving gear is clamped over a short rearward extension of the mainshaft.
Transverse gearbox and final drive Here the spiral driving gear is mounted on the opposite end of the differential cage to the final drive helical gear.

It has long been customary to use a nylon material for the driven gear to which the speedo cable is directly connected and, more recently, also for the driving gear where this is separately mounted.

Where a modern electronic speedometer is used, similar arrangements to those just described provide the necessary drive to a speed sensor unit. This converts mechanical rotary motion of the transmission output gear into an electrical frequency proportional to vehicle speed, which is then utilized by the main speedometer system.

Gearbox tachograph drive

The tachograph registers road speed and distance traveled automatically and correct to time. As a member of the EC all vehicles of over 3.5 tonnes gross weight operated in the United Kingdom have, since the end of 1981, been required to have a tachograph fitted and calibrated.

Most mechanically driven tachographs are so designed that for every 1000 turns from the drive cable a distance of 1 kilometre is recorded. To ensure that the tachograph is correctly calibrated in relation to the rear axle ratio, circumference and rotational speed of the vehicle wheels, an adaptor gearbox is required at the speedometer drive take-off connection (Figure 14.15b). Calibration of the adaptor gearbox is obtained by means of interchangeable and very closely graduated sets of gear wheels. Vehicles with two-speed rear axles must also have an electrically operated corrector unit

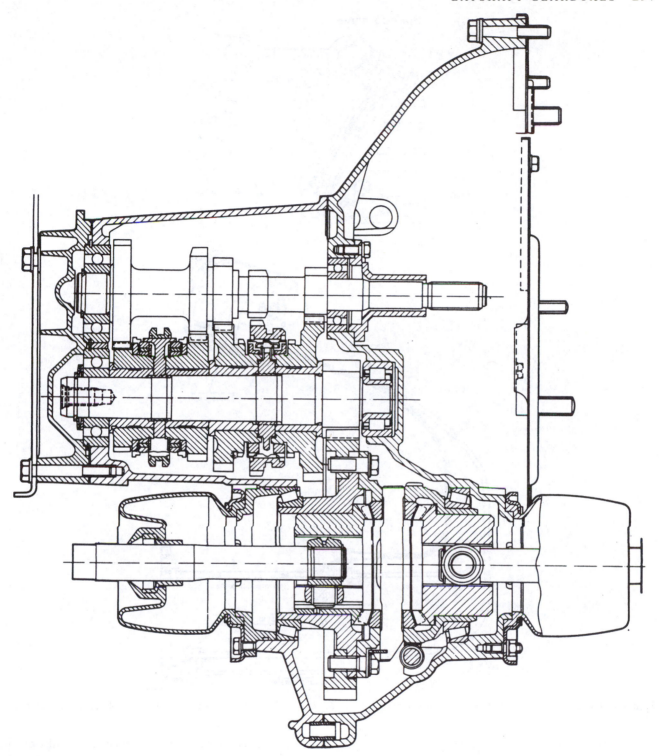

Figure 14.14   Single-stage gearbox and final drive for a transverse-engined car using a combination of baulking and servo ring types of synchromesh (*Lancia*)

built into the tachograph drive, so as to maintain correct speed and distance indications.

With modern electronic tachographs the flexible drive cable is replaced by an electrical cable. An impulse sender unit is then fitted on to the vehicle gearbox and transmits the required signals to the tachograph head.

14.3   SYNCHROMESH GEARBOXES

General background

When previously describing the constant-mesh gearbox, a brief reference was made to the additional refinement of a

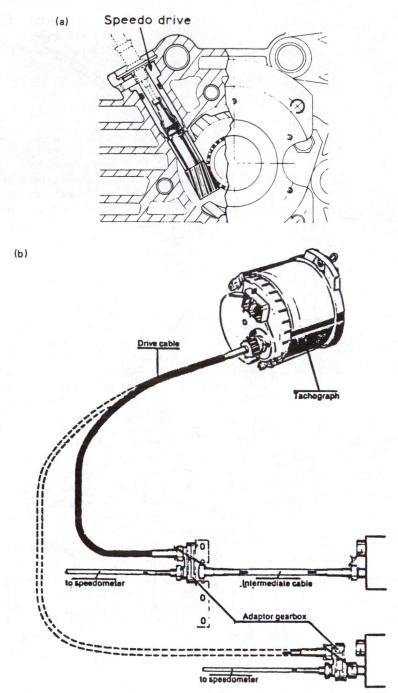

Figure 14.15   (a) Cross-section through speedometer drive (*Alfa-Romeo*) (b) Drive arrangement for a tachograph (*Lucas Kienzle*)

synchromesh mechanism for the purpose of simplifying gear changing.

Historically, this type of mechanism may be included among the many important contributions to more relaxed driving that have been pioneered by General Motors in America. The first synchromesh gearbox appeared on their 1928 model Cadillac, since when there have been many variations on the theme. In all cases the operating principle, which was devised by Earl Thompson of General Motors, is such that gear engagement by sliding dog clutch is preceded by frictional contact between a pair of conical surfaces that

possess a high torque capacity for a relatively small size. The latter thus serve to equalize or synchronize the rotational speeds of the gearwheel and its sliding dog clutch, so that final engagement of their dog teeth can be effected quietly. In the act of changing gear it therefore overcomes the rotational inertia or, in other words, a reluctance to be either speeded up or slowed down, of the clutch centre plate together with those particular elements of the gear train that are permanently connected to it. Since the adoption of synchromesh also encouraged the wider use of helical gearing, it may be said that at the time the synchromesh gear killed

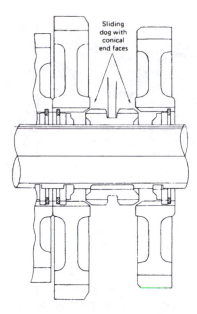

Figure 14.16 Conical dog type of synchronmesh (*Eaton*)

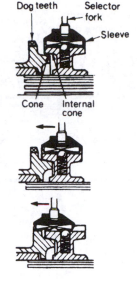

Figure 14.17 Operation of constant-load type of synchromesh

two birds with one stone; it provided easier gear changing and quieter running in the lower gears.

It is our purpose now to describe examples of these particularly ingenious synchromesh devices, which may be classified as follows:

Conical dog
Constant load
Proportional load
Servo ring.

Conical dog synchromesh

From a purely historical point of view, this form of synchromesh device should be last on the list, since it was introduced in the early 1960s by the Fuller Transmission Division of the Eaton Manufacturing Company for application to heavy commercial vehicle gearboxes. The reason for describing it first is because it represents the simplest possible means of obtaining a synchronizing action in the constant-mesh gearbox.

With this form of synchromesh (Figure 14.16) each gear to be engaged is locked to the mainshaft by a dog clutch sliding on splines in the conventional manner, but the ends of the dog teeth both in the gear and on the clutch are provided with matching chamfers at an angle of 35°. Before the dog teeth finally slide into engagement, their conical end faces can therefore act in the manner of a small cone clutch, so that any difference in relative speed between the gear and its dog clutch may be resolved by frictional drag.

In other words, the conical dog teeth confer a synchronizing action to prevent their own premature engagement, which otherwise would result in noisy gear changing together with undesirable shock loading on the parts concerned.

Constant-load synchromesh

It is sometimes wrongly inferred that this now obsolete type of synchromesh was the first to find application in the motor car gearbox. This is possibly because its action was much simpler, but albeit less foolproof, than that of the original proportional load synchromesh introduced by General Motors and described later. The so-called constant-load form of synchromesh was, in fact, a later development by the Warner Gear Division of the American Borg-Warner Corporation, who introduced it in 1931 for their proprietary T 81 model gearbox.

This type of synchromesh (Figure 14.17) differed from the previously described conical dog system by physically separating the synchronizing and engaging functions of the sliding dog clutch. For this purpose the latter was made telescopic with a splined inner hub, each end of which was formed as an internal cone, and a splined outer sleeve that served also as the internal dog teeth. A corresponding male cone and external ring of dog teeth were carried by the gears to be engaged. The synchronizer hub and sleeve assembly were spring loaded together by a series of radially disposed detent springs and balls that located in a groove within the sleeve.

In operation, the initial pressure exerted on the gear lever caused the selector fork to move simultaneously the synchronizer sleeve and the hub along the mainshaft, until the internal cone of the hub was brought into contact with its mating cone on the gear. Hence, the frictional drag so created synchronized the rotational speed of the gear with that of the mainshaft. Further pressure on the gear lever then overcame the restraining action of the spring-loaded balls on the outer sleeve, the resulting movement of which allowed its internal splines cum teeth to engage the dog teeth on the gear wheel.

From the foregoing description it should be appreciated that the designation 'constant load' did not actually relate to the effort exerted by the driver on the gear lever, but referred to the definite load that had to be placed on the friction cones in order that sufficient synchronizing action could take place. The load in question was therefore derived from the

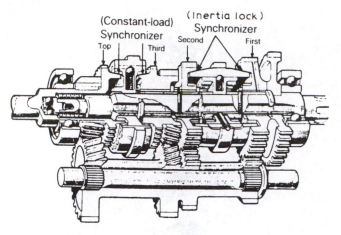

Figure 14.18 An interesting rare combination of constant-load and proportional-load types of synchromesh once used on Rootes Group vechicles

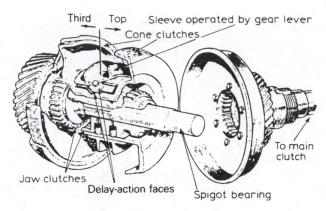

Figure 14.19 A classic example of inertia lock synchromesh with radial blocking pins once used on Rolls-Royce cars

spring-loaded balls acting between the hub and sleeve. A disadvantage of the constant-load synchronizer was that this load had to be fairly high. Even for a light car it could result in a breakaway force of about 200 N (45 lbf) being required to slide the sleeve over its hub before gear engagement could be effected. Furthermore, there was nothing to stop the inexperienced driver from clashing the engaging dog teeth by exerting too much effort on the gear lever, especially during a down-change. This difficulty was recognized in one early design of all-synchromesh gearbox where only the third and top gears had constant-load synchronizers, the more difficult to engage second and bottom gears being provided with proportional load synchronizers.

It is perhaps of interest to recall that for sports cars the constant-load synchromesh gearbox was once thought preferable to the virtually foolproof proportional load type, quite possibly because their drivers were better practised in correctly timing a gear change and had no need for synchromesh assistance anyway!

Figure 14.18 illustrates a combination of constant-load and proportional load types of synchromesh.

Proportional load synchromesh

With the constant-load type of synchromesh it is possible for engagement of the dog teeth to take place before synchronization is complete, as already explained. This is virtually impossible with the proportional load type of synchromesh, because while friction torque exists between the mating cones of the sliding dog clutch and the gear wheel to be engaged, the sliding dog teeth are positively prevented or baulked from engaging those on the gear wheel. For this purpose, delay action inclined surfaces must be incorporated in series with the friction cones and the sliding dog clutch to provide an inertia lock facility between them. The inertia lock remains active until all friction torque between the mating cones disappears, following which the sliding dog teeth are no longer restrained from engaging those on the gear wheel and the gear change is completed.

As indicated earlier, there are many variations on the original inertia lock theme, which may be sub-classified as blocking pin and baulking ring types.

Blocking pin synchronizers

These represent the earliest forms of inertia lock devices used by General Motors of America and also licensed to certain other manufacturers at that time. The first example to be illustrated (Figure 14.19) shows the blocking pins disposed radially between the sliding dog clutch and a slotted outer sleeve that carries the friction cones. Spring-loaded ball detents are fitted between the hub and sleeve. During gear selection the whole assembly of hub and sleeve slides axially along the mainshaft splines until the friction cones engage. If the synchromesh hub and gear are not rotating at the same speed, the friction torque at the cones causes the slotted sleeve to turn slightly relative to the hub, until the delay-action faces of the blocker pins register in the V-shaped recesses of the slots. This effectively locks the sliding dog hub against further endwise travel.

With continued pressure on the gear lever the hub blocker pins produce a centralizing torque on the slotted sleeve, so that the dog clutch teeth may proceed into engagement with those on the gear wheel. This can occur, however, only when the centralizing torque exceeds the synchronizing torque at the friction cones – or in other words, when all slipping between them has ceased and there is no longer any difference in rotational speeds of the synchromesh hub and gear wheel to be engaged. At this point the blocking pins are free to ride out of their recesses in the slotted sleeve, so the spring-loaded ball detents can be overridden and the sliding dog clutch moved into engagement to complete the gear change.

This particular version of blocking pin synchronizer lends itself to a robust form of construction suitable for heavy-duty applications. It may also be regarded by many service engineers as representing something of a Chinese puzzle to assemble! In the second example to be illustrated (Figure 14.20) the synchronizer friction cones receive their support from axially disposed blocking pins spaced around the inner hub comprising the sliding dog clutch. The latter is provided with a relatively large-diameter central flanged portion, this

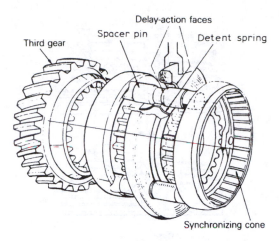

Third gear

Spacer pin

Delay-action faces

Detent spring

Synchronizing cone

Figure 14.20 An example of inertia lock synchromesh with axial blocking pins (*Smiths Industries*)

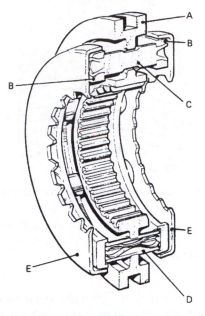

Figure 14.21 A modern inertia lock synchronizer with axial blocking pins (*Turner-Spicer*)

A synchro hub   D split energizer pins
B synchro rings   E synchro cups
C blocker pins

being drilled with a series of holes such that it can slide either way over the blocking pins, as directed by the gear selector fork. A system of detent spring loading is incorporated between the assembly comprising the blocking pins and cones and the sliding dog clutch.

As so far described, such a device would simply provide a constant-load synchromesh action. Therefore in order to introduce an inertia lock effect to avoid premature engagement and clashing of the dog teeth, the blocking pins have a central waisted portion with delay-action chamfers at each end. The corresponding blocking pin clearance holes in the hub flange are similarly chamfered. During gear selection the whole assembly of flanged hub, blocking pins and cones slides axially along the mainshaft splines until the friction cones engage. If the synchromesh hub and gear wheel are not rotating at the same speed, the friction torque at the cones causes the assembly of blocking pins and cones to turn slightly relative to the hub, until the waisted portions of the blocking pins come into contact with one side of their chamfered holes in the hub flange. The sliding dog clutch is therefore effectively locked against further endwise movement. With continued pressure on the gear lever, the chamfers on the blocking pins act against those of the hub flange holes to produce a centralizing torque on the assembly of blocking pins and cones, so that the hub dog teeth may proceed into engagement with those on the gear wheel. As in the previous case, it is not until the friction torque at the cones falls to zero that the centralizing torque can exceed the synchronizing torque. Then the hub flange chamfers riding on those of the blocking pins move the latter aside and allow the sliding dog clutch to override the detent spring loading and engage the dog teeth of the gear.

This type of blocking pin, inertia lock synchronizer continues to find favour mainly for heavy-duty applications, because it permits a construction with compact overall dimensions and has the ability to operate for high mileages without attention. A notable design in this field is that produced by Turner-Spicer and known as the split pin synchronizer (Figure 14.21). The adjacent synchromesh cones are linked to each other by three blocking pins in the usual

manner, but these are supplemented by three split pins that provide the detent spring loading. Each split pin is made in two halves with a square centre section into which two specially shaped springs are assembled back to back. The particular advantage of this form of construction is that it eliminates sliding friction on the detent springs themselves.

Baulking ring synchronizers
These may be said to represent the later forms of proportional load synchronizers, of which the classic baulk ring design, introduced by the American Warner Gear Division of the Borg-Warner Corporation in 1939, remains in production worldwide. An interesting variation on this theme is the inverted-cone type of baulk ring synchromesh introduced by the Italian Fiat Company in 1967.

The first example of a conventional baulk ring synchronizer to be illustrated (Figure 14.22) may appear at first sight to be very similar to the earlier Warner constant-load type. Closer examination will show, however, that the splined outer sleeve actuates three spring-loaded detent shifting plates. These can be moved along the inner hub, which is splined and held axially on the mainshaft. The ends of the shifter plates engage with corresponding slots in the adjacent back faces of externally toothed baulk rings, but the width of the slots is such that the rings may turn to the extent of being a half-tooth out of alignment with the sleeve splines. As in the case of the constant-load synchronizer, the sleeve splines also serve as the internal teeth of the sliding dog clutch, except that now their chamfered ends form delay-action faces with those of the teeth on the baulk rings. The latter are tapered internally to act as the synchronizer

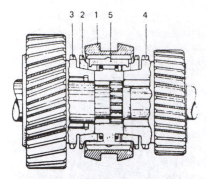

Figure 14.22   A conventional baulking ring type of synchromesh (*SAAB*)

1 synchronizer sleeve
2 bronze baulking ring
3 dog clutch, third-speed gear
4 dog clutch, fourth-speed gear
5 spring-loaded shifting plate

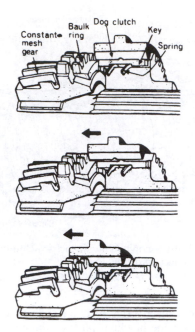

Figure 14.23   Operation of baulking ring synchronizer

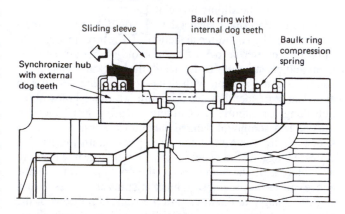

Figure 14.24   Operation of inverted-cone type of baulking ring synchromesh (*Lada*)

friction cones and they are generally manufactured from a copper-based alloy material such as manganese bronze.

During gear selection, the splined outer sleeve together with shifter plates is moved along the inner hub until the plates butt against the bottom of the slots in the baulk ring. This is then brought into contact with its mating cone on the gear wheel to be engaged. If the synchronizer hub assembly and gear wheel are not rotating at the same speed, the friction torque at the cones causes the baulk ring to turn slightly relative both to the hub and its outer sleeve. Since the sleeve dog teeth and those on the baulk ring are now axially misaligned, the sliding dog clutch function of the synchronizer is temporarily halted. With continued pressure on the gear lever, the delay-action chamfers on the teeth of the hub sleeve and baulk ring produce a centralizing torque that is trying to align the two sets of dog teeth. Again, the centralizing torque can only exceed the synchronizing torque when all frictional drag ceases between the mating cones. When this occurs the detent spring loading on the shifter plates is overridden and the baulk ring teeth move aside and allow those on the hub sleeve to slide past and engage the dog teeth on the gear wheel, thus completing the gear change (Figure 14.23).

In another fairly widely used version of conventional baulk ring synchromesh, the back faces of the rings are provided with lug projections rather than slots, the lugs riding in extra width slots in the inner hub. End thrust on the baulk rings is transmitted from the sliding sleeve by detent balls and springs acting via a follower piece that can be moved along the hub. This particular form of construction permits a stronger (non-slotted) baulk ring at the expense of a longer synchronizer.

The third illustration of a baulk ring synchromesh is that of the inverted-cone type mentioned earlier (Figure 14.24). This differs by virtue of the baulk rings having internal, rather than external, teeth and being carried on splined extensions of the gear wheels. Each baulk ring is spring loaded away from its gear wheel and retained against the ends of the splines by a locking ring. The spline ends and the teeth of the baulk ring are chamfered to provide delay-action faces and a limited rotational movement for the ring. No axial freedom is required for the inner hub of the synchronizer, which is a rigidly held on the mainshaft and simply

carries a splined outer sleeve with each end formed as an internal cone.

During gear selection, the splined outer sleeve is moved along the inner hub until its friction cone contacts that of the baulk ring. Since the latter is spring loaded away from the gear wheel to be engaged, any frictional drag between the cones will cause it to turn slightly and bring the delay-action chamfers on the spline ends and the baulk ring teeth into contact, thereby temporarily halting the sliding movement of the hub sleeve. With continued pressure on the gear lever a situation once again presents itself where a centralizing torque is acting in opposition to the synchronizing torque, except that in this case it occurs between the gear spline ends and the baulk ring teeth. When this conflict is resolved the baulk ring teeth move aside and the ring overcomes its spring loading to slide along the gear splines. Synchronization having been achieved, the internal splines cum dog teeth of the hub sleeve follow into engagement with those on the gear extension to

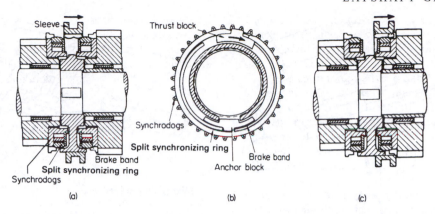

Figure 14.25 Operation of servo ring type of synchromesh: (a)/(b) synchronizing (c) coupled

complete the gear change. The advantages of the inverted-cone type of baulk ring, as compared with the conventional arrangement, are that its larger effective diameter of friction cone not only augments the synchronizing torque, but also minimizes wear and tear.

Servo ring synchromesh

This type of synchronizing device was developed by the German Porsche Company and in its original form first appeared on their 1953 type 356 cars, which were based on Volkswagen mechanical units. Since in the first instance it was intended to occupy no more space than that required for gears without synchromesh engagement, the servo ring synchronizer has always enjoyed the advantage of being a very compact design (Figure 14.14).

The basic operational difference between the servo ring and other types of synchronizer is that the inner cone of each mating pair assumes a split ring from, which is brought directly into contact with the chamfered splines or internal dog teeth of the hub sleeve. If the latter is not rotating at the same speed as the split ring (driven by the gear to be engaged), then the initial frictional drag between them tends to unwrap the ring – rather like trying to push an expanded circlip round in its groove by its leading end – causing yet more friction and an increasingly powerful servo or synchronizing action.

Its operation was not, however, made clash proof until the system was further developed in 1959 to include a baulking feature. This involved the addition of an anchor block located from the gear hub, part-circular brake bands and a thrust block intercepted by the ends of the captive split ring, so that the latter embraced the entire assembly. Its action is such that as the internally toothed hub sleeve moves over for gear engagement it tries to compress the synchronizer split ring (Figure 14.25a). The rotational drag on the split ring then forces the thrust block against the appropriate end of the anchored brake band, which bows outwards to resist further compression of the split ring and thus promotes a powerful synchronizing torque between the split ring and the advancing toothed sleeve (Figure 14.25b). It continues to do this until the hub sleeve is rotating at the same speed as the gear to be engaged. At this instant, the end force of the split ring acting against the thrust block collapses to relieve bowing of

the brake band. This finally allows the toothed sleeve to compress the split ring, pass over it, and effect engagement with the dog teeth on the gear wheel (Figure 14.25c).

A possible disadvantage of the servo ring type of synchromesh for some types of vehicle is that it can demand greater driver effort to operate, because its shift loads tend to be higher than those of a baulking ring system. In this respect a mixed system of synchromesh may uncommonly be used, so that baulking ring synchronizers are used for the lower gears and servo ring synchronizers for the upper ones.

Other considerations of synchromesh

Before leaving the topic of gear synchronizing devices, it should be realized that the quick, positive gear changes taken for granted with the typically modern baulk ring gearbox are the result of much attention to detail by designers, including the following items:

1 Correct choice of friction cone angles. These are usually of 12 or 14° included angle; a smaller angle further increases the synchronizing torque, but could lead to difficulty in separation of the cone surfaces.
2 Optimum choice of delay-action chamfer angles. These usually range from 105 to 125° included angle. Too small a chamfer angle could lead to clashing engagement of the dog teeth, while too large an angle could result in difficult gear changing.
3 Provision of a suitable screw-thread finish on the conical surface of the baulk ring, this being to ensure quick dispersal of the oil film for effective synchronization.
4 Careful determination of detent spring loading in the synchronizer which, if too great, would result in an excessively heavy gear change.
5 Suitable choice of lubricants. In particular, their viscosity should be low enough for them to be readily wiped off the friction cone surfaces, so as to avoid premature gear engagement and clashing of the dog teeth (Section 14.6).

14.4 GEAR SELECTOR MECHANISMS

Multirail selection

It is clearly desirable that any speed in the gearbox should be available for engagement at will without having to pass

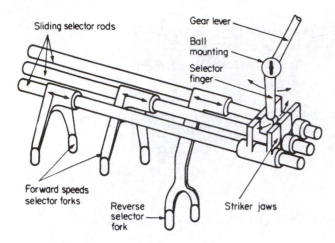

Figure 14.26　Multirail gear selector arrangement

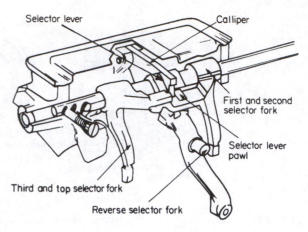

Figure 14.27　Single-rail gear selector arrangement

through intermediate speeds. This requirement was early met by the selective change speed mechanism devised by Wilhelm Maybach, the German automotive pioneer.

In the case of the four-speed constant-mesh gearbox, the once conventional multirail gear selection arrangement (Figure 14.26) would be as follows:

Selector rods　A separate selector rod is provided to control the axial movements of each sliding dog clutch and the reverse idler gear.

Striker mechanism　Each selector rod carries a fork permanently engaged with the grooved collar of the corresponding sliding dog clutch, or the reverse idler gear. The semi-circular ends of each fork terminate in slightly raised contact pads. Striker jaws provided at one and the same ends of the selector rods intercept the selector finger of the gear lever. In neutral, when none of the gears is engaged, the slots in all the striker jaws are transversely aligned.

Gear lever　Fore-and-aft movements of the ball-mounted gear lever cause the selector rods and forks to slide in a direction parallel to the mainshaft, while side-to-side movements of the gear lever merely select which one of them is to be moved. To prevent unintentional engagement of the reverse gear, the gear lever typically has to be lifted against the spring loading of its ball mounting before the selector finger can enter the slot in the reverse striker jaw.

Single-rail selection

A more recent and simpler type of gear selector arrangement (Figure 14.27) is that known as the single-rail system, which operates as follows:

Selector rod　The two selector forks actuating the forward speeds are slidably mounted on a common selector rod, which in its turn is free both to slide and partially rotate in the gearbox casing according to the movements of the gear lever. A rocking selector lever with a crankpin engages the reverse idler gear.

Striker mechanism　The forward speeds selector forks are formed with overlapping cranked arms which extend alongside

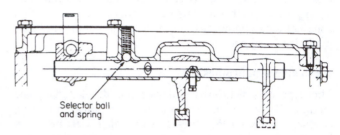

Figure 14.28　Spring-loaded detent for selector rods (*Eaton*)

the selector rod and carry inward-facing striker jaws. In the neutral position these stroker jaws, and that on the other end of the rocking reverse lever, are all in alignment and form an arc about the axis of the selector rod.

Gear lever　A radial selector finger is suitably positioned on the selector rod so that it can intercept the striker jaws. Fore-and-after movements imparted by the gear lever to the selector rod thus cause either one of the forks or the rocking lever to move with it. The choice is decided by partial rotation of the selector rod which derives from side-to-side movements of the ball-mounted gear lever.

Retention of the gears

To hold the sliding dog clutches or the reverse idler gear in the engaged or the neutral position, a spring-loaded detent acts radially upon each selector rod of a multirail selector mechanism (Figure 14.28). This device comprises simply a spring-loaded ball, or conical-ended plunger, which engages with transverse V-grooves into each selector rod. It is paradoxical that the detent force actually becomes less as the ball or plunger enters the selector rod groove, since the spring then extends and loses some of its loading just when it most needed!

In a four-speed gearbox the two selector rods controlling the forward speeds each have three grooves, their spacing corresponding to the neutral and the two engaged positions of the sliding dog clutches. The remaining selector rod has two grooves to provide for the neutral and engaged positions of the reverse idler gear, which may also require a greater detent force to prevent it from jumping out of mesh.

A single-rail selector system does, of course, require only one spring-loaded detent to act upon the common selector rod for the forward speeds.

Another feature of gear retention is the provision of some form of anti-jump-out characteristic, which prevents the dog clutch teeth from disengaging themselves from the teeth of the mainshaft gear, as their relative alignment alters slightly during the change from drive to overrun. Otherwise, the forces generated by this misalignment can sometimes be sufficient to overcome the detent force acting on the selector rod.

To counteract this tendency a simple modification to the sliding dog splines, which was widely adopted for the gearboxes of British military vehicles during World War II, is often used. This consists of slightly undercutting the central portion of the hub splines and chamfering out an equivalent splined portion of the sliding sleeve. When the sliding dog teeth engage those of the gear wheel, the drive is transmitted via the undercut portion of the hub splines. Any tendency of the sliding sleeve to be rejected from the gear dog teeth is resisted by the overhanging sleeve splines, which become slightly out of step with the ends of the hub splines and hence butt against them (Figure 14.29a). However, there is no difficulty in disengaging the dog clutch during normal gear changing, since the amount by which the splines need to be undercut does not exceed 0.25 mm (0.010 in).

In another arrangement that has long been favoured in General Motors gearboxes, the outer ends of the sliding sleeve internal splines are tapered to provide a 'keystone' contour, which matches a similar contour formed on the gear wheel dog teeth (Figure 14.29b). This provides a wedging action against unwanted disengagement of the sliding dog. A somewhat different approach is to use uneven spline widths on the sliding sleeve (Figure 14.29c). This results in there being fewer splines in contact with the gear wheel dog teeth on the overrun, so that contact pressure is increased on the remainder to prevent disengagement of the sliding dog.

Gear interlock devices

An essential feature of all gear selector arrangements is a gear interlock device without which the driver could inadvertently select, or at least attempt to select, two speeds simultaneously. Quite apart from the consequences of any loss of vehicle control owing to locking of the transmission system, the gearbox mechanism would of course suffer catastrophic damage.

Locking devices used for this purpose act either between the selector rods or across the selector rod striker jaws, and they are known respectively as sliding plunger and swinging plate interlocks.

Sliding plunger interlock
In a typical sliding plunger interlock, two balls on each side of the middle selector rod are positioned with their centres aligned with the axis of a short pin or plunger free to slide in a hole drilled diametrically through the middle rod (Figure 14.30). The balls each register in a groove machined in the

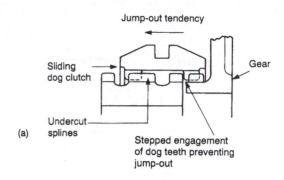
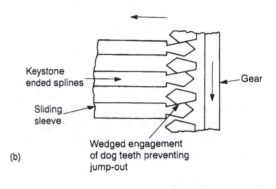
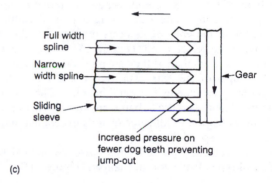
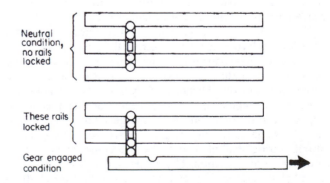

Figure 14.29 Anti-jump-out features of gear engagement: (a) undercut splines (b) keystone splines (c) uneven width splines

Figure 14.30 Operation of sliding plunger type of gear interlock

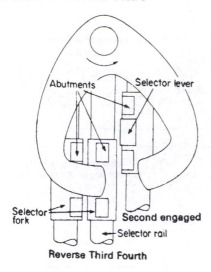

Figure 14.31 Operation of swinging plate type of gear interlock

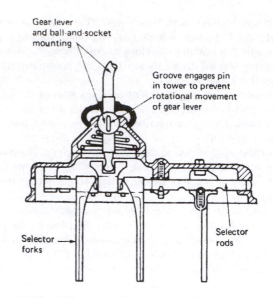

Figure 14.32 Direct type of gear lever control for commercial vehicle

adjacent rod. When one of the selector rods is moved to select a speed it unseats the ball registered in its groove. This ball then forces the other balls and the plunger sideways until they register in grooves in the other two rods, thereby locking them in their neutral positions.

Swinging plate interlock

This type of interlock acts in a rather different manner. A C-shaped pivoting plate resembling a pair of callipers is mounted with the ends of its arms registered in the striker jaw slots of the selector rods (Figure 14.31). These jaws are of a thickness such that, as they are moved fore and aft by the selector finger of the gear lever, only one can pass through the opening presented by the swinging plate. The arms of the swinging plate thus serve to baulk any movement away from the neutral position of all selector rods except that required for selecting a speed.

A swinging plate interlock may also be applied to the single-rail type of selector mechanism (Figure 14.27).

Direct type of gear lever control

This relatively early type of gear lever control was provided with a spring-loaded ball-and-socket mounting, directly in the tower of the gearbox top cover (Figure 14.32). A cross-pin in the gear lever ball engaged a groove in the tower socket to prevent rotational movement of the gear lever. To guarantee against inadvertent engagement of reverse gear, some form of blocking device was placed across the path that the lever had to follow to engage reverse, this obstacle being overcome only by some additional action on the part of the driver. In some cases it involved lifting slightly upwards the whole gear lever before it could be moved in the appropriate direction to engage reverse, while in others the gear lever motion was just the same as for any of the forward gears, but a fairly strong spring loading had to be overcome when moving towards reverse position. The blocking devices used were a guard plate and a spring-loaded plunger respectively, both of which were arranged to act against the selector finger of the gear lever.

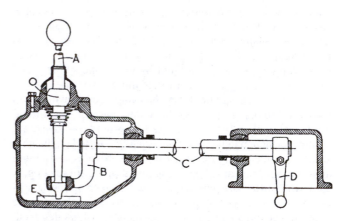

Figure 14.33 Operating principle of remote type of gear lever

A gear lever   D selector finger
B cranked arm   E lift lever reverse guard
C remote control rod C ball-and-socket joint

Remote type of gear lever control

A remote gear lever control is usually associated with the following transmission layouts:

1 Forward-mounted engine and gearbox, rear-wheel drive.
2 Front engine with rear transaxle.
3 Front engine, front-wheel drive.
4 Rear engine, rear-wheel drive.
5 Front engine with forward-mounted cab (commercial vehicle).

A remote type of gear lever control (Figure 14.33) is required where the gearbox is mounted so far forward, or rearward, of the seating arrangements for the driver that it is no longer possible to use a direct type of gear lever. It comprises basically a remote push-pull rod that is usually cranked downwards where it engages the lower end of the gear lever. Alternatively, a toggle connection may be used

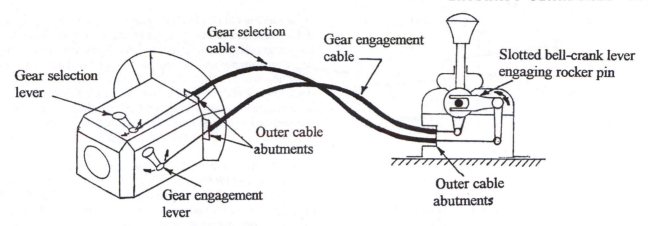

Figure 14.34 Schematic arrangement of enclosed cable and level type gear change for transversely mounted gearbox

between the remote control rod and the gear lever. In some installations it is necessary for the control rod to have universally jointed ends. These various features enable the rod to transmit both axial and rotational movements to the gear selector mechanism, in sympathy with fore-and-aft and sideways rocking movements of the gear lever. At its end remote from the gear lever, the control rod actuates either a multi- or a single-rail gear selection system, as earlier described. A reverse guard similar to those previously described for the direct type of gear lever control is, of course, still required for this and other versions of remote gear lever control.

In the case of a transversely mounted engine and gearbox, the inherent problem of isolating their motions from the remote gear lever control is a difficult one and can only be overcome by using either enclosed push-pull cables (Figure 14.34), or a system of rods and relay levers, to operate the selectors. In the latter case it typically becomes necessary to introduce two bell-crank relay levers, so that the control rod movements can be turned through a right angle. One of the levers serves to engage the selector finger with the appropriate selector rod and the other moves the selector finger and the selector rod to engage the gear.

A remote gear lever control does not necessarily have to be floor mounted, since less commonly it may be operated from the facia or the steering column. For example, the former type of control makes for a less complicated linkage in those versions of front-wheel drive where the transaxle, combining the gearbox and final drive, is mounted ahead of the engine. It comprises simply a 'walking stick' twist and push-pull control rod, which is cranked where it intercepts the top of a direct-mounted gear lever.

The steering column type of gear change control (Figure 14.35) was an American development of the late 1930s, following its introduction by General Motors on their 1938 Pontiac, although the basic idea was pursued by designers of some very early cars. Its purpose was to give the front-seat passengers more space and leg room and also make possible a three-abreast bench seating arrangement.

For these reasons the steering column gear change enjoyed a similar vogue of popularity in Britain during the late 1940s through to the early 1960s. However, by present-

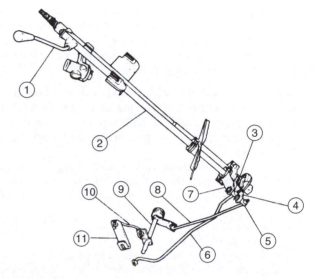

Figure 14.35 A modern example of steering column gear control for three-speed transmission (*Nissan*)

1	control lever	7	first-reverse shift lever
2	control rod	8	second–third upper shift rod
3	trunnion rod	9	cross-shaft
4	second–third shift lever	10	second–third lower shift rod
5	trunnion rod	11	second–third change lever
6	first-reverse shift rod		

day standards of gear control many such installations lacked precision of operation (to put it kindly!) because of the complexity of their linkage and the multiplicity of pivot points that were subject to wear. Also, they are not entirely compatible with the modern safety requirement for a collapsible steering column.

Further developments in gear selector mechanisms

During the 1980s car manufacturers became increasingly aware that their products should be pleasurable or even fun to drive. Not least this meant that the gear change should be

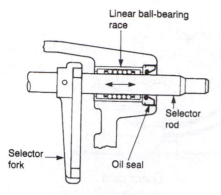

Figure 14.36 Reducing friction in the selector mechanism

effortless, quick and positive in action, or what a past generation of sporting motorists used to describe as a 'slicing butter' gear change. In that past era such refinement in gear changing was usually only to be found in high-grade cars, of which Rolls-Royce and Bentley were notable examples. In order to improve the quality of the gear change in mass-produced cars, various refinements may now be applied to the gear selector mechanism as follows:

1 To reduce friction the selector rods can be allowed to slide either in linear ball bearing races (Figure 14.36), which comprise a multirow ball cage encapsulated in a mounting sleeve, or in sintered bronze and teflon lined plain bushes.
2 Again to reduce friction, roller-ended plungers can be used as the spring-loaded detents for the selector rod(s).
3 To provide positive action the proportions of the selector forks are designed to confer better resistance against tilting, thereby allowing a close-clearance fit for their pads in the grooves of the synchronizer sleeves.
4 Complementary to point 3, positive stops can be used in the dog teeth of the gears rather than at the selector rods.
5 To reduce noise transmission through the selector mechanism and also reduce pad wear, the selector forks material may revert to aluminium bronze, rather than using the less costly steel versions that have tended to replace it.

14.5 HEAVY-VEHICLE GEARBOXES

Synchromesh gearbox developments

It was not until the late 1960s that synchromesh engagement, at least on the upper ratios, became increasingly adopted for the gearboxes of heavy vehicles. Until then haulage operators had tended to put initial cost of the vehicle before ease of driving, although in fairness it must be said that the application of synchromesh to heavy-vehicle gearboxes, as compared with passenger cars, did pose additional technical problems. However, synchromesh engagement had previously found application on many public service vehicles, where the layshaft gearbox was in competition with the pre-selective epicyclic gearbox (Chapter 16), which had already proved so successful on London Transport buses.

For heavy-vehicle gearboxes a greater torque capacity of the synchronizer cone surfaces is required, because the inertia

of the rotating clutch plate(s) together with that of the primary and layshaft assemblies is higher owing to their increased size and weight. Hence, the synchronizer cones are of necessity larger in diameter and comparatively narrow. Their accurate mounting has therefore always demanded careful attention, since any tendency to wobble would result in misalignment of their friction surfaces and reduced synchronizing torque. Proportional load synchronizers are used and correspond to heavy-duty versions of the blocking pin and baulking ring types (Section 14.3).

During their investigations into semi-automatic transmissions in the mid 1960s, S. Smith and Sons (England) Ltd conceived the idea of a multicone baulking ring synchronizer. It provided two additional friction surfaces by the inclusion of two extra cone members between the usual male and female cones. The two members were arranged concentrically, so that the inner one received its drive through tongues and slots from the normal male cone and the outer one similarly from the normal female cone. This arrangement therefore provided three friction surfaces that became loaded in series during synchronizer operation, the total synchronizing torque thus being the sum of the individual torques generated at each friction surface. In practical terms the Smith multicone synchronizer offered the potential advantages of either significantly reducing gear change effort or achieving quicker changes. Furthermore, the higher synchronizing torque allowed the use of more favourable angles for the delay action faces of the sleeve and baulking ring teeth, which less obstructed gear engagement when the oil was cold.

Unfortunately, there was no market interest shown in the Smith multicone synchronizer at the time and the idea quietly died. It has therefore taken some twenty-five years for this basic concept to be revived by other manufacturers and notably by the German firm of Zahnradfabrik Friedrichshafen (ZF) for heavy-vehicle gearbox applications. Their design is known as the 'double-cone' type with two rather than three friction surfaces, and achieves a reduction in both shift force and time. Other commendable features of ZF double-cone and conventional synchronizers (Figure 14.37) are their short travel and reduced width for a more compact installation and their molybdenum-coated steel friction surfaces for long service life. Double-cone synchronizers are now also used in some passenger car gearboxes, where their more powerful action allows a reduction in overall leverage for gear selection. A shorter gear lever with less travel between positions is a particularly attractive feature for the sports car driver.

Twin-layshaft gearboxes

For heavy-vehicle gearboxes where five or more speeds are required, a twin-layshaft construction is sometimes employed, as originally devised by the Fuller Transmission Division of the Eaton Corporation in America and introduced into the United Kingdom in 1964.

This type of gearbox construction utilizes the principle that if torque is transmitted simultaneously from one gear to two or more, instead of to a single gear, the loads imposed on the gear teeth are reduced. It therefore allows the gears to be made narrower and their supporting shafts correspondingly shorter. The advantages that arise from this construction are

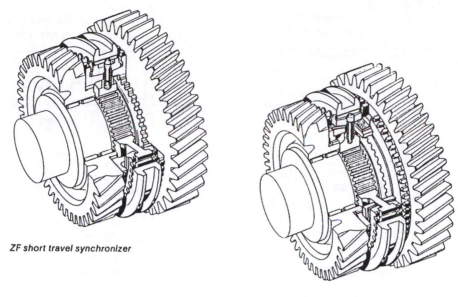

ZF short travel synchronizer

ZF double-cone synchronizer

Figure 14.37 Comparison of single- and double-cone synchronizers for heavy vehicles (*ZF*)

those of a high ratio of torque capacity to weight, and a compact layout.

In practical application an essential feature of this type of gearbox is that the primary and mainshaft gears are permitted a limited amount of radial freedom relative to their nominal axes of rotation, so that they can float between the opposed pairs of layshaft gears. In this manner the primary and mainshaft gears can assume their true meshing condition with the layshaft gears, which thus share the torque loading equally between each cluster. To achieve this floating action for the primary and mainshaft gears, the former is driven through clearance splines on the hub of the primary shaft, while the latter are mounted on a flexibly supported mainshaft. Although rigid within itself, the mainshaft receives its support each end from relatively flexible cantilever spindles, one of which enters the primary shaft spigot bearing and the other a splined sleeve carried by the mainshaft rear bearing.

The engagement of the mainshaft gears is effected through the usual sliding dog clutches, which in Fuller versions have conical end faces to facilitate gear changing (Figure 14.16). Power flow through a twin-layshaft gearbox is identical to that for an equivalent two-stage gearbox, except of course that there are two parallel paths of power flow instead of a single one (Figure 14.38).

Splitter drive gearboxes

The fuel-efficient diesel engines used in modern heavy vehicles are designed to be operated within a specific range of engine speeds for maximum fuel economy. To realize this in practice has led to the increasing use of so-called multispeed gearboxes, in which a two-speed auxiliary gearset is integrated with a typically five-speed main gearbox to provide ten speeds in all. The most suitable choice of gear can then be made relative to either the load carried, or the terrain over which the vehicle is to be operated. This particular form of construction has been found

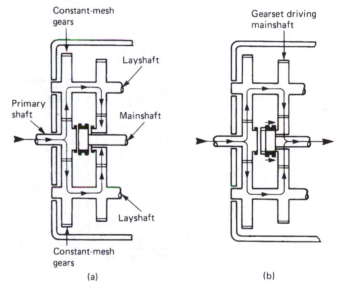

Figure 14.38 Schematic arrangement and operating principle of a twin-layshaft gearbox: (a) neutral (b) engaged

necessary, because unless the main gearbox is to become excessively long and heavy, the number of forward speeds it can provide must be limited to six.

In the case of a splitter gearbox, as it is popularly known, the auxiliary gearset is usually overhung from the front wall of the main gearbox. This gearset may be arranged in different ways to provide the equivalent of an extra pair of constant-mesh gears, so that an alternative lower (numerically higher) ratio underdrive to the layshaft becomes possible. Two methods of achieving this, which clearly show the principle involved, are for either the gears (Figure 14.39) or the pinions (Figure 14.40) of the twin constant-mesh pairs to

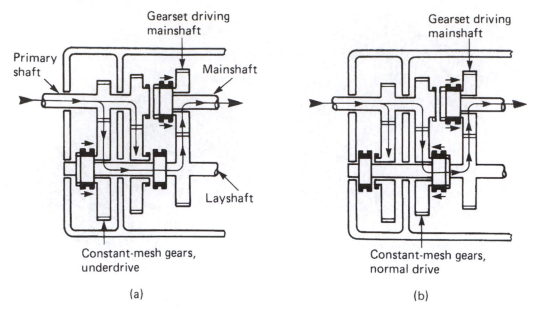

Figure 14.39 Schematic arrangement and operating principle of a splitter drive gearbox with freely mounted gears of the twin constant-mesh pairs: (a) underdrive (b) normal drive

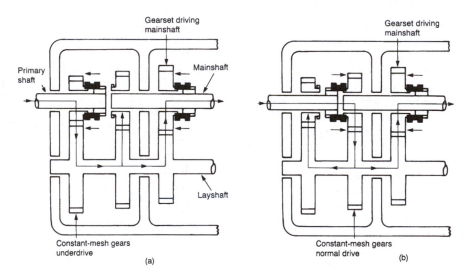

Figure 14.40 Schematic arrangement and operating principle of a splitter drive gearbox with freely mounted pinions of the twin constant-mesh pairs: (a) underdrive (b) normal drive

be freely mounted on their shafts with appropriate engagement through adjacent sliding dog clutches for underdrive and normal drive. The power flow within the main gearbox still remains the same as for a conventional two-stage gearbox.

With a splitter gearbox a close-ratio auxiliary gearset is used in conjunction with wide ratios in the main gearbox, so that upward gear changes, for example, occur successively in underdrive and normal drive for each speed in the gearbox. A separate control for the auxiliary gearset is therefore operated with each main gearbox ratio.

Range change gearboxes

In this type of multispeed gearbox for heavy vehicles, an auxiliary gearset is overhung from the rear wall of the main gearbox. The gearset is arranged to provide either a direct drive or an underdrive between the gearbox mainshaft and tailshaft, so that a high or a low range can respectively be selected. A simple method of achieving this range change, which again clearly shows the principle involved, is to drive a compound reduction gear from the mainshaft with its output gear freely mounted on the tailshaft. A sliding dog clutch

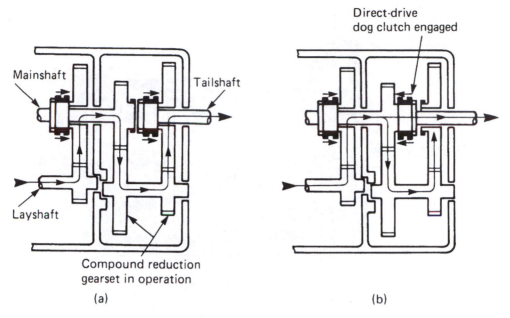

Figure 14.41 Schematic arrangement and operating principle of a range change gearbox with a compound auxiliary gearset: (a) low range (b) high range

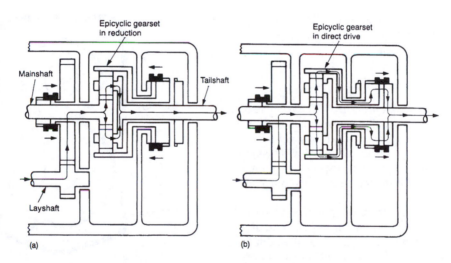

Figure 14.42 Schematic arrangement and operating principle of a range change gearbox with epicyclic auxiliary gearset: (a) low range (b) high range

is then carried on the tailshaft, so that it may engage either the tailshaft gear to allow a double-reduction underdrive or low range between the mainshaft and tailshaft (Figure 14.41a), or the mainshaft pinion to provide a direct-drive or high range from the main gearbox (Figure 14.41b).

An epicyclic auxiliary gearset may be used in other designs. In this case the sun gear is driven from the mainshaft, the planet carrier drives the tailshaft, and the annulus gear is either held stationary for low range (Figure 14.42a) or locked to the planet carrier for high range (Figure 14.42b). The general principles of epicyclic gearing are later explained in detail. Whichever form of auxiliary gearset is used, the power flow within the main gearbox still remains the same as for a conventional two-stage gearbox.

In contrast to a splitter drive gearbox, the range change alternative utilizes a wide-ratio auxiliary gearset in conjunction with close ratios in the main gearbox. The gear change or shift pattern is likewise different from that of the splitter drive gearbox, because upward gear changes (for example) occur progressively through the main gearbox first in low range and then again in high range. With a five-speed main gearbox the two-speed auxiliary gearset is therefore operated only once in each sequence of ten gear changes.

A splitter drive and range change can both be integrated with a four-speed and reverse gearbox with all except reverse gear engagement being synchronized (Figure 14.43). The eight forward speeds provided by the combined four-speed section and the range change group can be doubled by

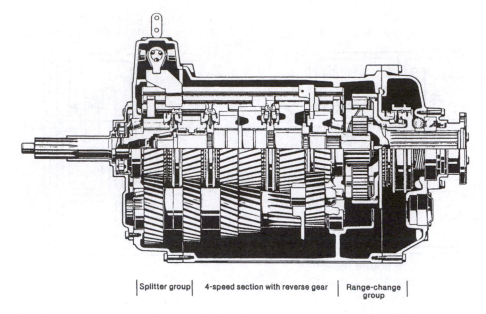

|Splitter group| 4-speed section with reverse gear | Range-change |
group

Figure 14.43   A splitter drive and range change integrated with a four-speed and reverse gearbox (*ZF*)

means of the splitter group so that a total of sixteen ratios are available, which can be selected consecutively. This particular gearbox made by ZF will be referred to again later for its semi-automatic operation (Chapter 17).

Gearbox power take-off

A power take-off unit, usually abbreviated PTO, is an auxiliary drive assembly that conventially is face mounted to, and gear driven from, the main gearbox of a heavy vehicle. Its purpose is to utilize engine power for driving a wide variety of vehicle-mounted ancillary equipment, especially hydraulic pumps that operate tail-gate lifts, tipping gear and other specialized services.

Heavy-vehicle gearboxes are therefore provided with suitable openings in their casings for mounting a PTO unit, these openings having long been standardized in both their dimensions and bolting arrangements by the Society of Automotive Engineers in America.

A sliding-mesh gear is carried on the splined output shaft of the PTO unit and receives its drive either from one of the constant-mesh gears or from an adjacent gear on the layshaft (Figure 14.44). An interposed idler gear may sometimes be used to change the direction of rotation for the PTO unit. Engagement or disengagement of the PTO is effected through either a manual or a power-operated selector fork, which intercepts a collar on the sliding-mesh gear. The splined output shaft is supported at each end on rolling bearings that are protected by oil seals. Lubrication of the PTO unit is shared with the main gearbox.

Forward cab remote gear control

The remote gear control linkage must be so contrived as to create sufficient leverage to engage the gear synchronizers, whilst at the same time be as light as possible commensurate with the required stiffness to transmit the loads involved. If the

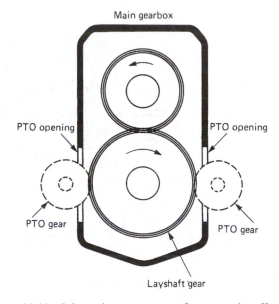

Figure 14.44   Schematic arrangement of a power take-off installation

connecting links together with the associated parts to be shifted represent too heavy a mass, then their inertia will be such as to impose an insensitive action when changing gear. A typical remote control mechanism (Figure 17.3) comprises a combination push-pull and swinging tube. This is clevis jointed at the front to a universally pivoted gear lever, and ball jointed at the rear to an external lever attached to the end of a slidably mounted gearbox selector shaft. To-and-fro engagement movements of the gear lever can therefore be transmitted to the selector rods and forks through partial rotation of the selector shaft and finger. The push-pull tube also carries a relay lever towards its rear end, which is ball jointed to a reaction rod that pivots from the gearbox casing. This arrangement

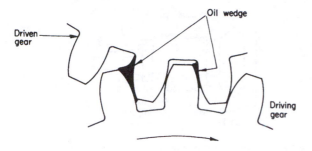

Figure 14.45 Basic lubricating action between meshing gear teeth

allows side-to-side selection movements of the gear lever to swing the push-pull rod laterally about the pivoted end of its reaction lever and thereby impose an endwise movement on the slidable selector shaft, so that the selector finger can intercept the appropriate selector jaw.

14.6 GEARBOX LUBRICATION AND SEALING

General lubrication requirements

The main purposes of a gearbox oil are that it must prevent metal-to-metal contact between the mating gear teeth under high-load conditions by maintaining a sufficiently strong lubricating film; and it must be of an adhesive nature, not only to prevent being wiped off the mating gear teeth by their combined rolling and sliding action (Figure 14.45) but also to resist being flung off the rotating gears by centrifugal effect. Additionally, gearbox oils must act as a coolant and must protect the gears against corrosion. Viscosity is an important factor because it determines the ease of gear changing in a manually operated layshaft gearbox. More specifically, the viscosity of the oil and also the additives it contains can influence the friction torque developed at the synchronizer cones, because if the oil cannot be wiped rapidly enough from their engaging surfaces, premature engagement and clashing of the synchronizer dog clutch teeth is likely to occur. The oil viscosity itself is, of course, affected by gearbox operating temperatures, which may approach or even exceed 100°C under particularly arduous conditions of service. Channelling of the oil at low temperatures is also to be avoided. If a rotating gear sustains a clearly visible channel where it dips into its oil bath, then its teeth may not pick up sufficient oil for their proper lubrication. Furthermore, the drag imposed on the gear train demands greater synchronizer and hence driver effort, to increase gear speeds when making downchanges. The viscosity of a gearbox oil must therefore not be so high as to prevent the channel from collapsing. On the other hand, the viscosity must not be so low as to affect adversely load-carrying capacity, resistance to leakage and noise suppression.

When a manufacturer decides upon the most suitable oil to be used in a gearbox, which may now be filled-for-life at the factory, the lubrication of the gears themselves necessarily places an emphasis on friction reduction and heat removal but, as already mentioned, the performance of the synchronizers must also be taken into account. In fact the requirements for lubricating the gears, selector forks and synchronizers may to some extent be conflicting ones. For double-stage gearboxes used in front engine, rear-wheel drive cars, an engine oil is often specified with a viscosity grade of SAE 5W-30 or 10W-30 (Section 4.3), although an automatic transmission fluid (Section 17.6) may instead be preferred in a modern five-speed transmission, to reduce gear change effort at low temperatures. With front engine, front-wheel drive cars, where a single-stage gearbox is combined with the final drive to form a transaxle, the choice of oil is determined by the lubrication requirements of the final drive gears, which in transverse engine installations are of the helical type as in the gearbox itself. For these applications a gear oil product is typically specified, which may be a synthetic oil conforming to an SAE 75W-80 or 80W-90 viscosity grade and an API GL-5 service category (Section 20.5).

Some manufacturers may specify their own Gear Lubricant Special (GLS) oil for single- and double-stage gearboxes, such as the 'Synchromesh fluids' of General Motors and Chrysler. It is therefore most important to observe the manufacturers' transmission oil recommendations. In the unlikely event of a modern filled-for-life transaxle requiring significant topping up, it should be carefully checked for leakage. Oil specifications for the gearboxes of heavy vehicles vary according to whether their gears are synchronized or not, and also recognize the severity of duty for which the vehicle is designed, especially in terms of gearbox operating temperatures and their effect on oil degradation.

Oil bath lubrication

Oil bath lubrication is commonly used in motor vehicle gearboxes although it is sometimes combined with a forced-feed system in some heavy-duty applications. In either event, the gearbox casing or housing forms an oil-tight reservoir for the lubricating oil. It is generally made as small as practicable, so as to restrict the amount of oil present. The casing is also provided with drain and level plugs. The latter prevent overfilling of the gearbox, which could result not only in excessive churning of the oil and needless loss of power but also overheating and accelerated deterioration of the oil. There would also be a risk of flooding the oil sealing arrangements of the primary and mainshaft bearings. In more recent practice, where a lifetime fill of oil is provided for the gearbox, a large ceramic magnet is provided in the casing to collect any metal particles and prevent their circulation in the maintenance-free lubricant.

The operation of the oil bath system of lubrication is as follows, the term 'gear' being applied both to the pinions and the gears to simplify the description:

1 The layshaft gears run partially immersed in the oil contained in the gearbox casing.
2 The oil is transported by the revolving layshaft gear teeth to their points of mesh with the primary and mainshaft gears.
3 Radial drillings in the primary and mainshaft gears or, in the case of larger gears, angled drillings in their hubs conduct oil inwards to the primary shaft spigot bearing and the bearings, or bearing surfaces, of the mainshaft gears (Figure 14.46). This occurs once per revolution of each drilling as it coincides with the meshing point of the gear teeth.
4 An oil spray is naturally set up by the revolving gears, and this serves to lubricate the primary shaft and the mainshaft

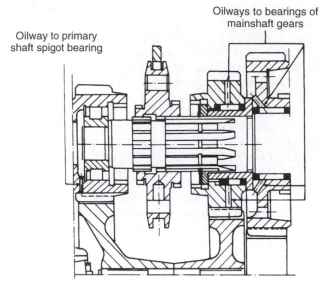

Figure 14.46 Lubrication arrangements for bearings of mainshaft gears and spigot bearing (*Eaton*)

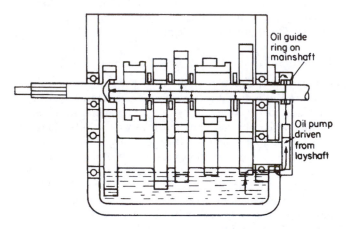

Figure 14.47 Oil flow in a gearbox with forced-feed lubrication system

support bearings, the sliding dog clutches and also the synchromesh units when fitted, and the gear selector mechanism.

Forced-feed lubrication

The addition of a forced-feed system is usually required in gearboxes where the lubrication conditions for the bearings, or the bearing surfaces of the mainshaft gears, are especially critical (Figure 14.47). It is usually arranged in the following manner.

1 A positive displacement pump of the rotary gear type (similar to those used in engine lubrication systems) mounted in the rear of the gearbox takes its drive from the tail end of the layshaft.

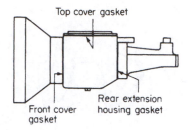

Figure 14.48 Typical arrangement of sealing gaskets for gearbox casing

2 The oil is drawn into the pump from the gearbox reservoir through a filter and inlet pipe.
3 The oil delivered by the pump first enters a passage leading to an oil guide ring encircling the tail end of the mainshaft.
4 From the oil guide ring the oil flows through a crossdrilling and into the centrally bored mainshaft.
5 Finally, the oil is conveyed to the bearings or bearing surfaces of the gears via radial outlet holes suitably positioned along the mainshaft.
6 A restricted supply of oil is also directed from the end of the centrally bored mainshaft into the primary shaft spigot bearing.

Static oil seals

If oil leakage from the gearbox is to be prevented, oil retaining devices of both the static and dynamic varieties must be provided. The former are required because, to permit assembly of the working mechanism into the gearbox, its casing must necessarily be fitted with detachable oil-tight covers (Figure 14.48). These typically comprise the following:

1 A front end cover to retain the primary shaft bearing and which may form part of the clutch bell-housing.
2 A rear extension housing to support the gearbox tail shaft bearing.
3 An upper or a side housing to contain the gear selector control mechanism.
4 A base cover plate in some cases.

Although the mating faces of these parts are machined flat and parallel, in practice there are always minute surface irregularities present which would offer potential areas of oil leakage if a metal-to-metal closure was relied upon. Suitably shaped static or gasket-type seals are therefore inserted between the various joint faces to prevent such leakage. These gaskets are produced from a non-metallic flat sheet material which, when compressed by bolt loading, will tend to fill any leakage paths.

Dynamic oil seals

The need for these arises because in a conventional gearbox two rotating shafts, namely the primary and main shafts, project from bearings mounted in the front end cover and the rear extension housing. Seals of the dynamic type are thus required in each case to act between the rotating shafts and

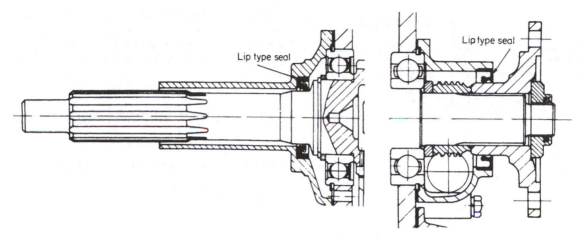

Figure 14.49 Typical oil sealing arrangements for the gearbox primary and mainshaft bearings (*Eaton*)

their stationary supporting members (Figure 14.49). They fall into two categories, generally termed contact and clearance seals:

Dynamic contact seals The object of a contact seal is to provide a controlled rubbing pressure against the shaft, thereby creating a positive sealing action. It takes the form of a radial lip seal made from a synthetic rubber that is not degraded by contact with hot gearbox oil. The flexible sealing lip is spring loaded to maintain a light rubbing contact with the oil-wetted sealing surface of the shaft.

Dynamic clearance seals Although no actual rubbing contact exists between the shaft and a seal of the clearance type, it nevertheless imposes a resistance against oil leakage by virtue of its return pumping action. A seal of this type simply comprises an oil return scroll (a form of screw thread) machined on the sealing surface of the shaft, which runs within a close-clearance plain bore housing.

The application of these seals to a double-stage gearbox is as follows:

Primary shaft This may be equipped with either an oil return scroll, or a radial lip seal, acting within the front end cover. An oil slinger washer may be fitted on the other side of the bearing, where it also serves as a clip shield to prevent any metal particles from entering the rolling elements.

Mainshaft A radial lip seal is installed in the rear extension housing to act against a sealing surface machined on the sleeve of the propeller shaft flange coupling.

In the case of a single-stage gearbox, the dynamic sealing points are reduced to only one, this being where the primary shaft enters the combined final drive and gearbox casing. A radial lip seal is used for this particular application.

An oil breather cap is fitted to the top of motor vehicle gearboxes to relieve any build-up of internal pressure due to heating of the gearbox oil, which would encourage oil leakage past the seals due to accelerated wearing of their lips. Internal deflectors may be used to protect the breather cap

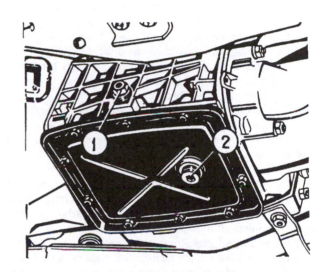

Figure 14.50 Gearbox oil level and drain holes (*Lada*)

1 level/filler plug
2 drain plug

from direct oil spray. The breather cap itself takes the form of a similar application for rear axles (Figure 20.34). It may also incorporate a filter material to prevent the ingress of contaminants, as internal pressure drops upon cooling.

Checking the level and changing the oil of a manual gearbox

The oil level can be checked by first raising the vehicle on a hoist, or placing it over a pit, so as to provide safe access to the gearbox from below with the vehicle maintained in a level position and the handbrake applied. Always ensure that the engine is switched off before checking the oil level.

First identify and then remove the combined oil level and filler plug located in the side of the gearbox casing (Figure 14.50). The oil contained in the gearbox should be either at or

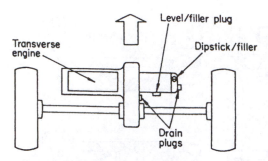

Figure 14.51 Typical alternative locations for transaxle oil level and drain holes

just below the level of the plug hole as detected by extending a forefinger to the inner edge of the hole. If the oil level is correct, or after it has been topped up, it simply remains to replace the plug, for which the vehicle manufacturer may quote a torque tightening value. In the case of older cars where convenient access to the gearbox could be gained from above, a combined filler plug and level dipstick was sometimes provided in the top cover of the gearbox.

The procedure for checking the oil level in a transaxle, whether it be fore-and-aft or transversely mounted, is very similar to that just described for a conventional gearbox. However, if a dipstick is provided instead of a combined oil level and filler plug (Figure 14.51), the check can be conveniently performed from above the engine compartment. If in doubt as to the location of the more usual combined oil level and filler plug, then refer to the manufacturer's service instructions, but the plug is typically to be found in either the front face or the end cover of the gearbox part of a transversely mounted transaxle and in the conventional gearbox position where a fore-and-aft mounted transaxle is used.

Except in certain more recent designs that are provided with a lifetime fill of oil, the gearbox oil must be periodically changed in accordance with the manufacturer's lubrication schedule. Once the approximate refill capacity of the gearbox and the type of oil required are ascertained, the procedure for changing the oil is merely an extension to that of checking its level. The oil must be drained completely into a suitable receptacle following removal of either a drain plug (Figure 14.50) or, less commonly, the lowest retaining setscrew for the gearbox extension housing. To expedite draining of the oil the vehicle should previously have been run so that the oil is at or near working temperature. After the drain plug is replaced, the gearbox is refilled through the combined oil level and filler plug until the oil just begins to overflow, the plug being replaced when the overflow ceases. The vehicle manufacturer may similarly quote a torque tightening value for the drain plug. To feed the oil into the gearbox manually generally requires the use of a syringe, unless the oil is dispensed from a polythene container complete with detachable filler spout.

The procedure for changing the oil of a transaxle corresponds to that already described for a conventional gearbox, once the appropriate refill capacity and type of oil have been ascertained. Access is, of course, necessary from below the

vehicle and the drain plug is typically to be found in either the underside, or the lower end face, of the final drive part of a transversely mounted transaxle (Figure 14.51), and in the underside of the final drive part of a fore-and-aft mounted transaxle.

14.7   GEARBOX MISBEHAVIOUR IN SERVICE

Types of gearbox complaint

Gearbox complaints can usually be categorized as follows:

Difficulty in engaging gears
Faulty synchromesh action
Jumping out of gear
Noisy gearbox
Oil leaks

Owing to the variation in detail design of gearboxes, the following summary of the probable causes of these complaints is intended only as a general guide. It is always advisable to consult the service diagnosis guide that is normally included in the manufacturer's service literature.

Difficulty in engaging gears
Clutch drag, for which the various causes have been given in Section 13.8.
Worn, damaged or maladjusted remote gear lever control mechanism, preventing full engagement.
Faulty synchromesh action
Worn or damaged synchronizer ring and cone, causing insufficient frictional grip
Weak, broken or displaced spring elements in synchronizer hub assembly, allowing crashing through
Worn or damaged teeth on synchronizer baulk rings, again allowing crashing through
Worn or loose blocker pins in other types of synchromesh, similarly allowing crashing through.
Jumping out of gear
Worn, damaged or maladjusted remote gear lever control mechanism, preventing full engagement
Weak or broken detent springs in selector mechanism, or worn detent grooves, not providing sufficient restraint
Distorted, loose or worn selector fork, preventing full engagement
Excessive slackness in synchromesh hub and sleeve assembly
Undue end float of mainshaft, arising from bearing slackness.
Noisy gearbox
Insufficient oil in gearbox
Worn or damaged bearings
Worn or damaged gear teeth
Excessive end float in layshaft gear cluster.
Oil leaks
Worn or damaged oil seals
Damaged joint washer
Front, rear or side covers loose or damaged.

15 Fluid couplings and torque converters

15.1 FLUID COUPLINGS

General background

The basis of most forms of semi-automatic and fully automatic transmission systems used in motor vehicles is a hydrokinetic drive, so called because it operates by means of changes in kinetic energy of a circulating fluid. This type of drive involves the use of either a fluid coupling or a torque converter. Although quite similar in construction, the fundamental difference between the operation of these two hydraulic devices is that the torque converter is capable of multiplying engine torque whereas the simpler fluid coupling is not.

It may be of interest to recall that the torque converter and the fluid coupling both had their early origins in marine engineering and were patented in 1905 by Dr H. Föttinger, who was chief designer at the Vulcan Works of Stettin (at that time in Germany). The torque converter was originally conceived as a device to reduce speed and increase torque between a high-speed steam turbine and the propeller drive for ship propulsion. A few years later the advent of the marine high-speed diesel engine introduced severe torsional vibrations into the propulsion system of ships, and it was as a means of damping out these vibrations that the fluid coupling first found successful application.

Fluid couplings in motor vehicles

It remained for an enterprising British engineer, Harold Sinclair, to perceive that there were wider uses for fluid couplings than for the propulsion of ships. Sinclair had noticed the fierce clutch engagement on some of the London buses he rode in as a passenger in the 1920s. The thought therefore occurred to him that smoother progress should be possible if a fluid coupling could be used rather than a friction clutch, so he took his idea to the Daimler Company who further developed it. In 1930 they introduced a novel transmission system that employed a fluid coupling in conjunction with permanently meshed epicyclic gears, which were preselected by an arrangement of brake bands.

In later alternatives to the Daimler system, both a fluid coupling and a friction clutch have been used in conjunction with a conventional layshaft gearbox, so that the fluid coupling provides a completely smooth getaway from rest while the friction clutch allows the drive to be interrupted for changing gear.

The first fully automatic transmission system to incorporate a fluid coupling was the American General Motors Hydra-Matic gearbox, which was introduced in their 1939 Oldsmobile car. This system comprised a fluid coupling operating in conjunction with an epicyclic gear train, wherein changes of ratio were automatically effected through a hydraulic control system acting upon plate clutches and band brakes.

Advantages of fluid couplings

The purpose of using a fluid coupling in a vehicle transmission system is generally to secure the following advantages:

1 Absence of direct mechanical contact between the driving and driven members minimizes the transmission of shock and torsional vibration between the engine and the drive line.
2 No positive disengagement or engagement of drive allows a smoother starting characteristic, this being particularly advantageous when restarting up a steep hill.
3 Protects against harmful labouring of the engine at low speeds, since the fluid coupling will merely slip and allow the engine to increase speed when overloaded.

Since a fluid coupling cannot provide a positive disengagement of drive, a drag torque must always be present between the driving and driven members of the coupling. It is for this reason that a fluid coupling is not compatible with a layshaft gearbox where there is sliding dog engagement of the gears, unless a friction clutch is interposed between the coupling and the gear-box. This problem is avoided in those semi-automatic and fully automatic transmissions where permanently meshed epicyclic gears are used.

Basic construction of a sealed fluid coupling

The internal appearance of a fluid coupling has traditionally been likened to the two halves of a grapefruit, each facing the other with the pulp scooped out and the cell dividers left intact. Translated into technical language, the fluid coupling may be described as consisting of two toroidally grooved discs facing one another with a small clearance between them. Radial blades are formed across the grooves to divide them into curved cells. These blades also support hollow semicircular cores for guide rings, which reduce turbulence in the coupling. The guide rings are offset within their toroidal cavities so as to equalize flow areas in the cells. One disc is mounted from the engine flywheel via a torus cover and is termed the *impeller*. The other one is enclosed by the flywheel and torus cover but connects to the input shaft of the gearbox and is termed the *turbine* (Figure 15.1).

For the sealed type of fluid coupling a sealing gland is provided where the gearbox input shaft passes through the centre of the impeller. The sealed coupling is only partially filled with fluid and incorporates a reservoir space at the back of the turbine. Fluid couplings are either produced from aluminium die castings or, as in later practice, fabricated from steel pressings. One of the early problems encountered with the fluid coupling was that of balancing it, such that the effect of the fluid rotating with the impeller and turbine could be taken into account. To solve this problem each component was filled with wax of the same density as the fluid used and balanced with the wax in the curved cells, the wax being melted out after balancing.

Operation of a fluid coupling

In considering the operation of a fluid coupling it must first be stressed that the transmission of motion between the impeller and turbine does not occur as a result of viscous drag, such as may be found for example in the action of some automatically controlled cooling fans, but depends upon the kinetic energy of the fluid circulating within the coupling. For this reason the desirable fluid is one of low viscosity and high density; a high-viscosity fluid would in fact increase slip.

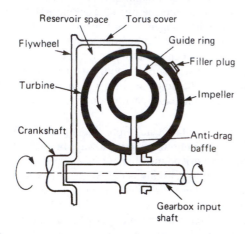

Figure 15.1   Schematic arrangement of a fluid coupling (section)

The circulation of fluid and hence the transmission of energy within the coupling is derived from two distinct although of course not separately acting motions of the fluid, these being known as rotary flow and vortex flow. As would be expected, *rotary flow* occurs in the direction of impeller rotation. However *vortex flow* is set up across the impeller and turbine and deserves further explanation. For simplicity, let us try to imagine swinging a series of buckets partially filled with water in a circle and noting that the water stays in the buckets even when each bucket is momentarily overhead and upside down. The water is, of course being acted upon by centrifugal force and the faster the buckets are swung the greater will be the force holding the water in them (Figure 15.2a). If we can next imagine very finely slicing the buckets down their middle, and allowing one circle of bucket halves to slow down slightly, then the centrifugal force holding the water in the slower-moving bucket halves will be less than for the faster-moving halves. Since the pressure of water at the bottom of the faster-moving bucket halves will now be greater than in the slower-moving halves, a cross-flow of water will be established between them. This will result in the slower-moving bucket halves becoming fuller than the faster-moving ones, so that water will then flow across the tops of the slower- to the tops of the faster-moving bucket halves. A continuous cross-circulation or vortex flow of water is therefore created between the bucket halves rotating side by side (Figure 15.2b).

If we can further imagine the bucket halves as corresponding to the cell halves of the impeller and turbine, then we can understand how a vortex flow is established in a fluid coupling. We must not forget, however, that it is the combined effect of the rotary (Figure 15.2c) and vortex flows that finally determines the angle at which the fluid leaves the

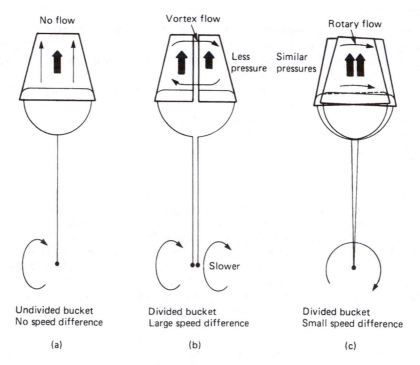

Figure 15.2   Swinging bucket analogy of vortex and rotary flow

vanes of the impeller and strikes those of the turbine to transmit driving torque.

Taking matters a step further, the extent of vortex flow within the coupling is constantly changing while the vehicle is being driven. During acceleration, when there is a considerable difference in rotational speed between the impeller and turbine, the vortices move farther apart to resemble a stretched corkscrew of fluid flow. This condition of high vortex increases the angle at which the fluid leaves the blades of the impeller and strikes those of the turbine, thereby raising the torque capacity of the coupling. Conversely under cruising conditions when the impeller and turbine are rotating at almost the same speed, the vortices move closer together to give a low-vortex flow. The effect of this is therefore to reduce the angle at which the fluid strikes the turbine blades, thereby lowering the torque capacity of the coupling. As the impeller and turbine approach the same speed of rotation, a condition of zero vortex will obtain and the torque capacity of the coupling is sustained only by rotary flow of the fluid. This is known as the *coupling phase* of a fluid coupling.

It thus follows that a limited amount of slip between the impeller and turbine is always necessary for normal operation of a fluid coupling, and this state of affairs is practically guaranteed as long as the engine is driving the vehicle. On the overrun when the vehicle is driving the engine and the turbine is rotating faster than the impeller, the circulation of fluid within the coupling is reversed after passing through zero vortex, so the braking effect of the engine is still retained. Since the slowing down of fluid circulation in one direction and starting it up in the other is a relatively slow process, it provides a cushioning effect against the reversal of forces in the transmission system.

Finally, it must be recognized that a fluid coupling gives out precisely the torque which is put into it, neither more nor less. It is incapable of multiplying its torque input, because the fluid returning from the turbine is flowing in the wrong direction to assist rotation of the impeller. Fluid couplings have maximum efficiencies between 93 and 95 per cent.

15.2 IMPROVEMENTS TO FLUID COUPLINGS

Reservoir space

Fluid couplings used in early motor vehicle applications were of the partially filled sealed type and incorporated a reservoir space at the back of the turbine, as indicated in Figure 15.1. This space was to permit expansion of the fluid as it became heated in circulation and thereby to avoid excessive build-up of pressure in the coupling. It also allowed the circulating fluid partially to empty into the reservoir when the turbine is stalled, thus reducing idling drag. A low drag torque is essential to minimize any tendency for the vehicle to creep forward when temporarily halted with the engine idling and a gear engaged.

Anti-drag baffle

This was another feature introduced for the sealed-type fluid coupling, and comprised a central baffle plate that partially obstructed the exit areas of the turbine cells (Figure 15.1). Its purpose was further to reduce the idling drag torque by impeding the circulation of the remaining fluid, which had not emptied into the reservoir.

Pressure feed

In some early fully automatic transmission systems a constant flow of fluid was maintained through the coupling, which commenced as soon as the engine was started. An engine-driven gearbox pump was used to transfer fluid from the gearbox sump into the coupling and maintain a pressure in the coupling of about $490 \, \text{kN/m}^2$ ($70 \, \text{lbf/in}^2$). Surplus fluid exhausted from the coupling then served to lubricate gearbox parts on its return to the sump. This pressurized through flow of fluid in the coupling served not only to lower operating temperatures, but also to prevent cavitation (the formation of vapour bubbles in the fluid) which would reduce coupling efficiency.

Variable blade spacing

To reduce noise in fluid coupling operation it is necessary to avoid exact coincidence of the blades in the impeller and turbine. For this reason either the blades are unequally spaced, or a different number of equally spaced blades are used in these members.

Lock-up coupling

A centrifugal friction clutch is incorporated in some heavy-vehicle applications of the fluid coupling. This eliminates slip during continuous high-speed running in the interests of better fuel economy. To minimize wear the lock-up clutch operates wet in the coupling fluid.

15.3 TORQUE CONVERTERS

Torque converters in motor vehicles

About the same time that Daimler introduced the fluid coupling for their passenger car and bus transmission systems, A. Lysholm of the Ljungstrom Steam Turbine Co. of Sweden developed a relatively small torque converter drive suitable for use in motor vehicles. This unit could multiply engine torque up to about five times with a peak efficiency of just over 80 per cent. It was further developed under licence by Leyland Motors in England, and in 1933 they began manufacturing such a system for use in buses and railcars, following which it became known in this country as the Lysholm-Smith transmission. A disadvantage of this system was that it increased engine fuel consumption.

During World War II various torque converter installations were developed for American military vehicles, notably in a tank destroyer built by the Buick Company. Here a Lysholm-Smith torque converter was used in conjunction with a three-speed epicyclic gearbox. No doubt it was experience gained in this field which enabled Buick to introduce in 1948 the first production fully automatic transmission embodying the torque converter principle for passenger car use. It is now a matter of history that the torque converter in conjunction with epicyclic gears became established practice for passenger car automatic transmission systems (Figure 17.6).

The use of a torque converter between the engine and a conventional friction clutch and gearbox can also bring

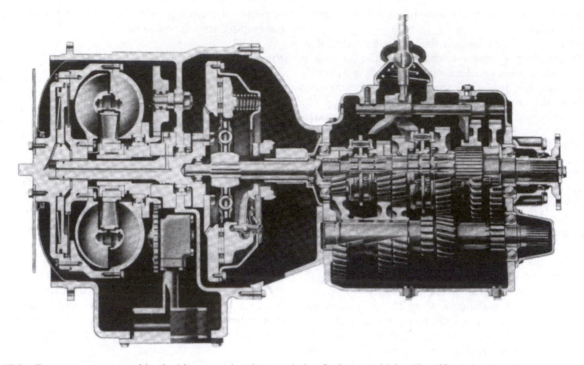

Figure 15.3   Torque converter combined with conventional transmission for heavy vehicles (*Brockhouse*)

operating benefits for heavy vehicles (Figure 15.3), especially in certain applications. Typically these involve refuse collection, skip loading, ready mixed concrete and other localized delivery or collection services, particularly in city centre traffic conditions. The main advantage to be gained from this system is the available surge of torque and hence acceleration of the vehicle at low speed and indeed from rest, which is conferred automatically through the torque converter. This compares with a conventional transmission where such progress demands more frequent gear changing and is often accompanied by prolonged and damaging slipping of the clutch. The torque increase through the converter is such that the lowest gear can be dispensed with except for starting on hills with a heavy load.

Advantages of torque converters

Torque converters share the same general advantages as those previously listed for plain fluid couplings, but in addition they possess the important capability of automatically multiplying engine torque, which enables a vehicle to accelerate smoothly under conditions when conventional gear changing can detract from the driver's concentration.

Basic construction of a torque converter

Since the torque converter was first introduced for passenger car automatic transmissions in the late 1940s many varriations in design have been tried, but established modern practice favours the use of what is termed a single-stage three-element converter. It is so called because in addition to the impeller and turbine, as found in the plain fluid coupling, there is a third member termed the reactor or *stator* (Figure 15.4). This is also provided with blades but is of smaller diameter than the impeller and turbine. It is interposed between the entrance

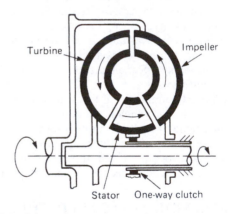

Figure 15.4   Schematic arrangement of a torque converter (section)

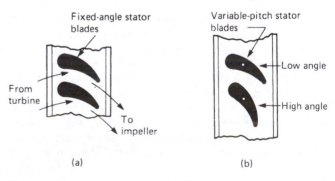

Figure 15.5   Fixed-angle and variable-pitch stator blades

to the former and the exit from the latter (Figure 15.5a). All three sets of blades have a working clearance and are mitred to complete the toroidal form. For quiet operation the turbine has fewer blades than the impeller.

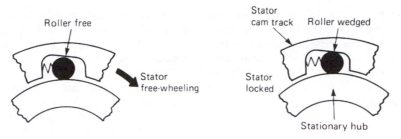

Figure 15.6 Operating principle of a roller one-way clutch

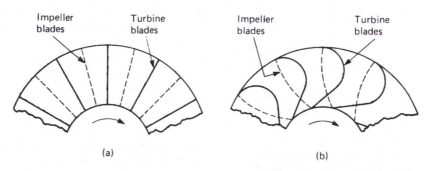

Figure 15.7 Difference between fluid coupling and torque converter blade forms: (a) fluid coupling (b) torque converter

An overrunning or one-way clutch supports the stator from a gearbox reaction shaft to provide it with a free-wheeling facility, but only in the same direction of rotation as the impeller and turbine. The one-way clutch is of the roller type, which comprises a fixed inner race and an outer cam track that is splined to the stator. A series of spring-loaded rollers occupies the converging spaces formed between the inner race and the cam track. In the free-wheeling condition the direction of rotation of the cam track causes the rollers to walk towards their springs and unwedge themselves, while any attempted movement of the cam track in the opposite direction results in the spring-loaded rollers becoming wedged (Figure 15.6).

Apart from the addition of a free-wheeling stator, another important difference between the torque converter and the fluid coupling concerns the curvature of the blades in all three converter elements, as compared with the straight blades of the plain fluid coupling (Figure 15.7a). If we 'look through' the torque converter from the front of the engine, the impeller vanes are curved slightly away from the direction of rotation. The turbine blades have a more pronounced curvature, first in the direction of rotation and then reversing in direction towards the centre (Figure 15.7b). The curvature of the stator blades is such that they act as scoops, to that fluid leaving the turbine and entering the impeller is reversed in direction to help rather than hinder the rotation of the impeller (Figure 15.5a).

Owing to the considerable amount of heat that can be generated during torque conversion, it has long been established practice for the torque converter to be supplied under pressure with a cooling through flow of fluid. For passenger car installations the fluid is usually pumped through a heat exchanger that is combined with the engine cooling system (Figure 15.8). The cooled fluid then returns to the transmission case where it serves to lubricate the gearbox mechanism

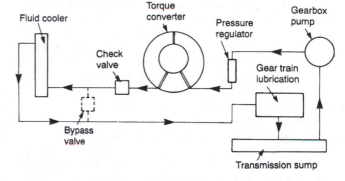

Figure 15.8 Cooling circuit for torque converter automatic transmission

before draining to the sump (Figure 16.17). Converter supply pressure must be sufficient to prevent cavitation, as earlier mentioned in connection with the fluid coupling, but not excessive as to cause distortion and ballooning of the torus cover. In practice the supply pressure is regulated to about $210 \, \text{kN/m}^2$ ($30 \, \text{lbf/in}^2$). Heated fluid leaving the converter does so via a check valve in the stator support, the two-fold purpose of this valve being to maintain a minimum pressure in the converter during operation and to prevent the converter fluid from draining back through the cooler and lubrication circuits to the transmission sump when the engine is not running. If a bypass valve is included in the cooling circuit, it serves to maintain a fluid flow through the heat exchanger when the gearbox pump speeds are low during traffic crawling conditions, since these produce the greatest temperature rise in an automatic transmission. In heavy-vehicle torque converter installations a separate sump may be provided to cool the fluid being pumped (Figure 15.13). Passenger car torque converters are now of welded construction and cannot be dismantled. When used with transversely mounted engines the converter members may be made slightly elliptical in cross-section; this departure from a truly circular form is made to save installation space. Heavy-vehicle torque converters are typically of bolted-up construction (Figure 15.13).

Operation of a torque converter

In essence the torque converter operates in a similar manner to that previously described for a plain fluid coupling, because the transmission of energy is again derived from the rotating and vortex flow motions of the circulating fluid. Where its operation differs from that of a plain fluid coupling is in the torque multiplying properties conferred by the specially curved forms of its blades and the presence of a bladed stator.

The purpose of curving the blades is to take advantage of the physical fact that when a jet of fluid strikes an object, the amount of force transmitted to the object will be in proportion to the extent that the velocity of the fluid has been reduced. Taking a few examples, if a jet of fluid is aimed against a flat plate at a low angle very little force will be transmitted to the plate, because of the small reduction in fluid velocity (Figure 15.9a). Clearly a more definite reduction in velocity of the fluid will occur if the jet is aimed directly at the plate, in which case an increased force will be transmitted to the plate as the fluid is randomly dispersed (Figure 15.9b). If the plate is now modified to assume a curved form and receives a favourably aimed jet of fluid, then the force transmitted to the plate can be further increased, because the plate is not only slowing the jet of fluid but also reversing it in a controlled manner (Figure 15.9c). The more slowly the fluid travels in the opposite direction, the greater will be the force exerted on the plate.

Applying this principle to the elements of a torque converter, it should now be evident that the slight backward curving of the impeller blades is intended to transfer fluid into the turbine with both added force and in the most favourable direction. Once the fluid has entered the turbine, the more pronounced curvature of the turbine blades will cause the path of vortex flow to change direction sharply and hence greatly increase the driving torque exerted on the turbine. Unfortunately, this so far promising state of affairs would result in the vortex flow of fluid leaving the turbine in a direction opposite to that of impeller rotation. Since this would oppose engine torque, the benefits gained from curving the blades would be cancelled out. It is therefore to overcome this problem

that a stator is used: the curved blades of the stator redirect, again with increased force, the vortex flow of fluid so that it re-enters the impeller in a direction that assists rather than impedes rotation (Figure 15.10).

There is a considerable difference in rotational speed between the impeller and the turbine as the vehicle accelerates away from rest and the greatest torque conversion is taking place. As the difference in rotational speed of these elements reduces under cruising conditions, the vortex flow of the circulating fluid changes from high to low and the rotary flow vice versa, similar to that occurring in a plain fluid coupling. When the turbine achieves a speed ratio of about 90 per cent, or in other words about 90 per cent of the running speed of the impeller, the vortex flow is entirely replaced by rotary flow and torque multiplication ceases. The torque converter is then said to have reached its coupling phase.

At this point, and in the absence of vortex flow, the stator is no longer serving a useful purpose and, if it were permanently fixed, would hinder rotary flow and reduce the efficiency of the torque converter. It is therefore for this reason that the stator is mounted on an overrunning or one-way clutch, so that once the coupling phase of the torque converter is reached the stator can free-wheel and merely be swept around by the rotary flow of fluid acting on the back of its blades. The torque converter is thus allowed to act as a plain fluid coupling once torque multiplication ceases, and it is this feature of operation that technically classifies it as a converter coupling.

Modern three-element torque converters attain maximum efficiencies between 87 and 90 per cent. Torque multiplication is generally limited to within the range of 2:1 to 2.5:1, otherwise difficulties can be encountered with overheating under severe conditions of loading.

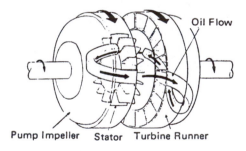

Figure 15.10 Fluid flow in a torque converter (*Toyota*)

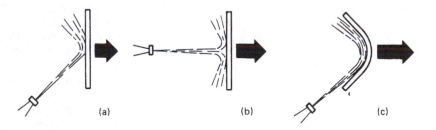

Figure 15.9 Relative force transmission by a jet of fluid directed against a plate

The descriptions 'soft' and 'hard' are sometimes used to describe the performance characteristics of a torque converter in relation to engine power. In simple terms these characteristics are related to the size or, more specifically, the maximum diameter of the flow path of the torque converter. An increase in diameter provides a hard characteristic to the benefit of fuel consumption, whilst a reduction in diameter confers a soft characteristic that improves acceleration times. For the automatic transmissions of modern passenger cars there is a tendency to use a soft characteristic, but to incorporate a lock-up clutch to help restore fuel economy (Section 15.4).

15.4 IMPROVEMENTS TO TORQUE CONVERTERS

Multistage converters

The designers of some early automatic transmissions sought to extend the useful operating range of the torque converter so that it could be used in conjunction with a simple system of epicyclic gearing, which provided only for emergency low- and reverse-gear operation. This made it necessary to employ a multistage torque converter in which two or more turbine members were used. In a typical arrangement of a two-stage unit the stator was positioned between the first- and second-stage turbine members (Figure 15.11). The object of abstracting the maximum amount of energy from the circulating fluid in this manner was to sustain the multiplication of torque over a wider operating range. However, the multistage torque converter fell into disfavour for passenger car transmissions because maximum efficiency was reduced by increased fluid friction losses and this resulted in poor fuel economy.

Variable-pitch stator

In some designs of torque converter the stator blades are pivotally mounted to their hub and may be swivelled to either a low-angle or a high-angle position (Figure 15.5b). This is accomplished by an in-built hydraulic control system, which is responsive to movement of the accelerator pedal. With the accelerator released and the vehicle standing in gear, the stator blades assume the low-angle position. This reduces the multiplication of torque and prevents the vehicle from creeping forward. When the accelerator is depressed and there is a large difference in rotational speed between the impeller and

the turbine, the stator blades are moved to the high-angle position. As a result the vortex of fluid leaving the turbine changes direction more sharply, so that it re-enters the impeller in the most favourable direction for increasing torque. Easing the accelerator for cruising speed causes the stator blades to return to the low-angle position; therefore when the coupling phase of the torque converter is reached it can operate more efficiently.

Lock-up clutch

The use of a lock-up wet clutch has already been mentioned in connection with the plain fluid coupling (Section 15.2). This feature was first used in a multistage torque converter of a passenger car automatic transmission by the American Packard Company in 1949. Since the original oil crisis of 1973 its use has also spread to the modern and more efficient single-stage three-element torque converters for several reasons. First and foremost, it can eliminate the 10 per cent slip that occurs between the turbine and the impeller during the coupling phase of the converter and therefore can reduce engine speed under cruising conditions, which has the all-important effect of reducing fuel consumption. Second, the reduction in engine speed during cruising contributes to quieter running. And third, the absence of rotary flow in the converter when the lock-up clutch is in operation serves to lower the working temperature of the fluid.

The lock-up clutch is built into the torque converter, and in passenger car applications is of single-plate construction. It operates on the same axis as the turbine member of the torque converter and comprises a hydraulically operated clutch piston of annular from which is made from a steel stamping. One side of the clutch piston engages the turbine member through a series of tabs that engage cushioning springs, while the other side has a machined friction track that can be brought into contact with a friction liner bonded to the inside face of the torque converter front cover. Engagement and disengagement of the lock-up clutch is determined, via the hydraulic control system of the transmission (Section 17.5), by changes in the flow direction of the fluid supplied to the torque converter. These cause a difference in pressure one way or the other across the clutch piston, which pushes it either towards or away from the friction liner (Figures 15.12a and b). The purpose of the cushioning springs is to minimize torsional vibrations in the transmission when the lock-up clutch is engaged, since the impeller and turbine members then revolve as one and the torque converter is locked up in the manner of an engaged friction clutch. Multiplate lock-up clutches may be used in heavy-vehicle torque converters (Figure 15.13).

15.5 FLUID COUPLINGS AND TORQUE CONVERTERS IN SERVICE

Fluid couplings

When sealed-type fluid couplings were still being used in the transmission systems of some cars, their only servicing requirement was normally that of topping up the coupling with engine oil every 3000 miles (5000 km). Either of two combined filler and level plugs on opposite sides of the

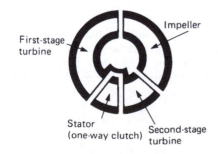

Figure 15.11 Schematic arrangement of toroidal members in a two-stage torque converter (section)

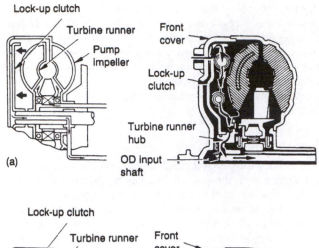

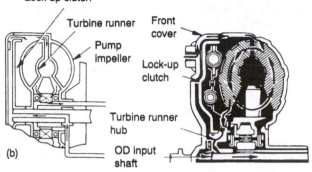

Figure 15.12 Operation of a torque converter lock-up clutch: (a) engaged (b) disengaged (*Toyota*)

coupling could be turned to top dead centre and the oil added until it overflowed. This ensured that the coupling received no more oil than its optimum amount, which was typically about 80 per cent of coupling total volume. In later applications of the fluid coupling in heavy-vehicle transmission systems, the pressure-fed coupling generally shares the same oil supply as the gearbox, so that no separate maintenance is required.

Torque converters

The pressure-fed torque converters used in passenger car automatic transmissions also share the same oil supply as the gearbox, so again no separate provision is made for them to be topped up, although in some cases they may be drained by turning a plug to bottom dead centre. However, it is unusual for modern automatic transmissions to require periodic changes of fluid; all that is necessary is to check and top up the fluid level, typically at every 6000 miles (10 000 km) or 6 months, whichever comes first. As mentioned later, it is most important that certain procedures be adopted that recognize the design differences between the various types of automatic transmission. The particular vehicle manufacturer's service instructions must therefore always be consulted (Section 17.7).

The torque converter of an automatic transmission can also act as an in-built dynamometer and therefore offers a unique facility for testing the general overall condition of both engine and transmission. This is accomplished by carrying out what is termed a stall test, which records the maximum speed (rev/min) that can be attained by the engine under full throttle with the transmission engaged and the vehicle stationary: that is, the engine is driving the impeller of the torque converter while the turbine is being held against rotation. Given a correctly tuned engine, a carefully performed stall test can yield useful information about the condition of an automatic transmission. However, it cannot be over-emphasized that a stall test is particularly severe on the transmission and may not always be recommended by a manufacturer. In any event it must only be performed strictly in accordance with the manufacturer's service instructions. Of no less importance is that proper supervision must always be available to ensure the safety of personnel, since clearly no person must be allowed to stand in front of or behind the vehicle while a stall test is being conducted.

A stall test essentially consists of the following sequence of events:

1 The engine and transmission are brought to their normal operating temperature.
2 A check is made on the transmission fluid level.
3 To record engine speed a tachometer is connected into the ignition system, and must be readily visible from the driver's seat.
4 The road wheels are chocked and the handbrake is firmly applied.
5 With the footbrake also firmly applied and a drive range selected, the accelerator is fully depressed just long enough for the engine speed to stabilize, otherwise serious overheating would occur.
6 The highest speed (rev/min) attained by the engine is noted and compared with the manufacturer's test schedule.

If the stall speed is well above that specified by the manufacturer, it indicates that either the torque converter is not receiving an adequate supply of fluid, or excessive slip is occurring in the friction elements of the gearbox. In the event of the stall speed being well below that specified by the manufacturer, it indicates either a lack of engine performance, or that the one-way clutch of the torque converter stator is slipping when it should be holding. The latter condition reduces torque multiplication in the converter, which therefore results in poor acceleration from rest and difficulty in moving away up a steep gradient. If the stall speed is normal but there is a noticeable reduction in top speed together with poor fuel economy, the one-way clutch of the stator may have seized in the engaged position. Since the stator can then no longer freewheel, the torque converter is prevented from reaching its coupling phase. Torque converter problems in service generally involve replacement of the complete unit.

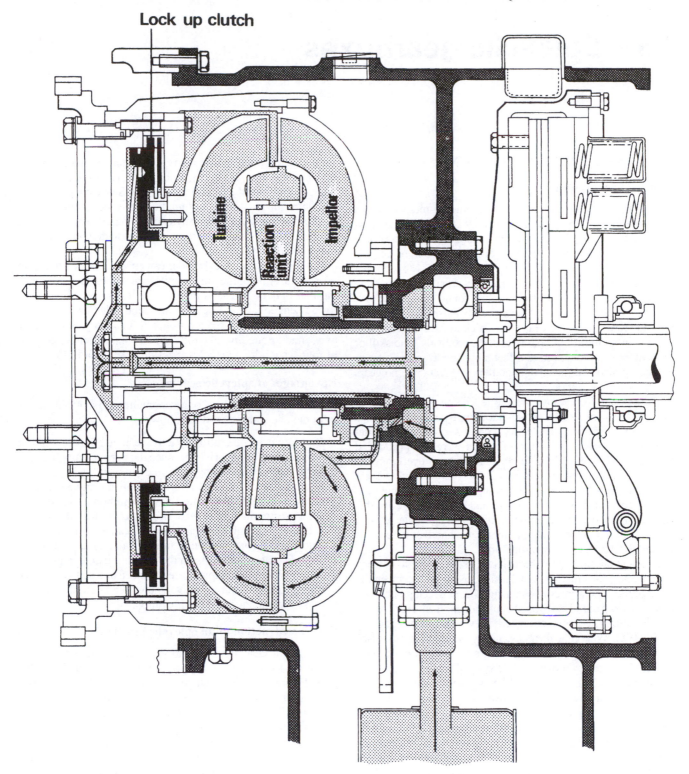

Lock up clutch

Turbine

Reaction unit

Impeller

Sectional view of converter showing oil flow

 **Oil flow**

Figure 15.13 Detail construction (section) of a modern torque converter for heavy-vehicle application, showing oil flow (*Brockhouse*)

16 Epicyclic gearboxes

16.1 BASIC EPICYCLIC GEARING

General background

During the early development of the sliding-mesh or crash gearbox some pioneer designers, notably Dr F.W. Lanchester in England and Henry Ford in America, sought to avoid its crude mode of operation by adopting a different system of gearing known as epicyclic or planetary. The system is so named because the behaviour of its geared elements, which rotate about a common axis, is similar to that found in our solar system. Since all the gears were permanently meshed and engaged by some form of friction brake, the value of the epicyclic gearbox lay in its foolproof characteristics, relatively little skill or judgement being required to effect a change from one ratio to another. It also had the advantage that in direct drive no gears whatever were in action.

Perhaps the best known example of epicyclic gearing in motoring history was the somewhat abruptly acting two-speed and reverse gearbox that Henry Ford fitted to his immortal Model T car, over 15 million of which were mass produced between 1908 and 1928. In this rather primitive epicyclic gearbox the contracting bands of the friction brakes were operated solely by pedals, there being no gear lever, and neutral was obtained by applying the handbrake! However, this early design did at least demonstrate that an epicyclic gearbox could be made to work reliably in service and that it need not be unduly expensive to manufacture.

At about the same time that interest in the epicyclic gearbox was waning in America, not least because of the invention there of the synchromesh gear, a significant development took place in England. This was the introduction in 1927 of the Wilson four-speed epicyclic gearbox, which was designed by Major W.G. Wilson, a former military engineer. It possessed the unique feature of allowing the driver to select in advance the required gear by moving a steering-column-mounted lever, following which the gear could be engaged at will simply by depressing and releasing the equivalent of a clutch pedal. For this reason it was popularly referred to as the Wilson preselector gearbox, and was mainly used in conjunction with a fluid coupling drive. Although this type of epicyclic transmission system has not been used in passenger cars for many years, it has continued in modernized form to find convenient application in buses, where the frequency of gear changing can be very high.

The successful combination of the Wilson epicyclic gearbox and fluid coupling is known to have influenced the early development of the fully automatic gearbox by General Motors in America. Where the modern automatic gearbox differs from the earlier Wilson gearbox is chiefly in the more economic arrangement of its epicyclic gearing and the more

sophisticated manner in which the motion of selected members is controlled. However, a point of further interest is that in the early 1980s the British company of Self-Changing Gears Ltd introduced a fully automatic five-speed transmission system for European and American city buses, which is based on the original Wilson gearbox but is provided with electro-hydraulic control.

Epicyclic gearing has by no means been confined to use in gearboxes. Among the other applications of this type of gearing that may be found in motor vehicles are overdrive units for passenger cars and light commercial vehicles; two-speed final drives and hub reduction gears for heavy-vehicle rear axles; and torque-dividing differentials in the transfer gearing of four-wheel-drive vehicles (Sections 18, 20 and 21).

Advantages of epicyclic gearing

The principal advantages of an epicyclic over a layshaft gear train may be summarized as follows:

1 Changes from one ratio to another can be effected without interruption of the drive, because the permanently meshed gear train is engaged by friction elements.
2 It can be made more compact for a given torque capacity, because its members rotate about a common axis and there are a greater number of teeth in contact to share the load.
3 It is more versatile in application because it makes available a wider combination of ratios from a given gear train, although there are limits on the ratio attainable in practice.
4 The gear separating forces are balanced and self-contained, which together with a relatively low speed of tooth engagement is conducive to quiet operation.

Basic construction of an epicyclic gear train

An epicyclic gear train comprises three essential members, which are known as the sun gear, planet pinions and annulus gear (Figure 16.1). The sun gear is situated on the central axis of the gear train. Surrounding the sun gear are the planet pinions, which can freely rotate on pins mounted in a planet carrier and are equally spaced around the sun gear with which they are permanently meshed. Hence the similarity to our solar system, where the planets rotate around the sun. The annulus gear embraces the whole gearset, and comprises an internally toothed flanged disc that likewise is permanently meshed with the planet pinions. In practical application the output torque is not normally taken direct from the sun gear, because the high tooth loading that this could involve would demand an increase in size of the gearset.

For quiet and efficient operation the epicyclic members may be provided with helical gear teeth and the planet pinions mounted on needle roller bearings. Lubrication of the epicyclic gear teeth is effected from a pumped supply, the oil being directed through radial holes in the shafts. Methods of lubricating the planet pinion bearings include radial drillings in the planet pinions and splash lubrication slots in either the endwise locating washers for the planet pinions or the end faces of the pinions themselves.

Epicyclic gear trains may be classified as being of either the simple or the compound type (Figure 16.2a and b). To widen the range of available ratios it is necessary either to couple together simple trains or to provide a compound train in which the planet pinions may be stepped in diameter or, alternatively, of non-stepped diameter but combined in different lengths.

16.2   OPERATION OF EPICYCLIC GEAR TRAINS

Simple epicyclic gear train

A simple epicyclic gear train can perform the following functions in a motor vehicle gearbox:

Neutral condition   If any one member is rotated and the remaining two members are allowed to run free, no drive will be transmitted through the gear train and the whole unit will merely idle in neutral.

Direct drive   If any two of the three members are locked together, the third member will be carried around by the teeth of the two locked members at the same speed and in the same direction. The gear train has in effect become solidified, and therefore acts in the manner of a direct mechanical coupling with no increase in torque.

Forwards reduction   If the annulus gear is held stationary and the sun gear is rotated, the planet pinions will be compelled to 'walk' in the annulus (Figure 16.3a). The planet carrier will therefore rotate in the same direction as the sun gear, but at a much reduced speed and with a corresponding increase in torque. Conversely, if the sun gear is held stationary and the annulus gear is rotated, the planet pinions will be compelled to walk around the sun gear (Figure 16.3b). The planet carrier will therefore rotate in the same direction as the annulus gear, but at a less reduced speed and with not so great an increase in torque.

Reverse reduction   If the planet carrier is held stationary and the sun gear is rotated, the 'idling' planet pinions will rotate the annulus gear in the opposite direction (Figure 16.3c). The speed of the annulus gear will be reduced and its torque increased.

Coupled simple epicyclic gear trains

When considering the practical implications of the simple epicyclic gear train, it should be evident that since one member must necessarily be permanently connected to the output shaft of the gearbox, there is no possibility of obtaining two forwards and one reverse reduction in speed from a single gear train. An early approach to overcoming this problem was to couple together several simple epicyclic gear trains, so that beginning with a large reduction in speed and increase in torque, each successive gear train engaged was driven more slowly to provide lesser reductions in speed and increases in torque, until finally all the gear trains were locked together to give a direct-drive top gear.

For example, two simple epicyclic gear trains may be coupled together such that both sun gears rotate as one on the input shaft, the annulus gear of the first train is rigidly coupled to the planet carrier of the second, and the planet carrier of the first train is connected to the output shaft

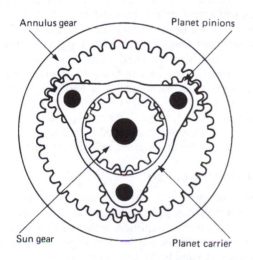

Figure 16.1   Identifying the members of an epicyclic gear train

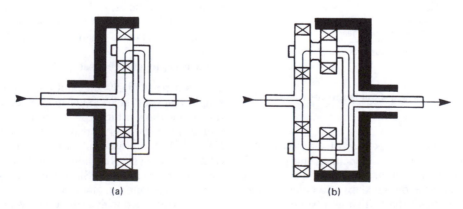

Figure 16.2   Identifying (a) simple and (b) compound epicyclic gear trains

Forward reduction: annulus gear held

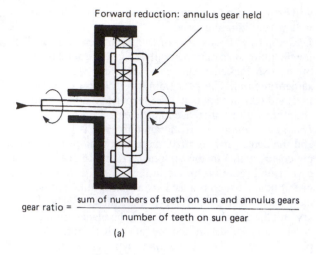

$$\text{gear ratio} = \frac{\text{sum of numbers of teeth on sun and annulus gears}}{\text{number of teeth on sun gear}}$$

(a)

Forward reduction: sun gear held

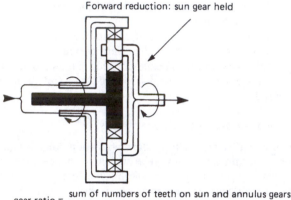

$$\text{gear ratio} = \frac{\text{sum of numbers of teeth on sun and annulus gears}}{\text{number of teeth on annulus gear}}$$

(b)

Reverse reduction: planet carrier held

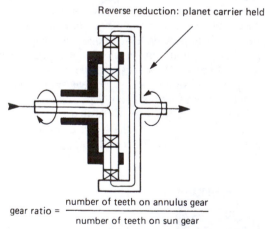

$$\text{gear ratio} = \frac{\text{number of teeth on annulus gear}}{\text{number of teeth on sun gear}}$$

(c)

Figure 16.3 Simple epicyclic gear train in forwards and reverse reduction

(Figure 16.4). Now if the annulus gear of the first train is held stationary, the rotation of its associated sun gear will cause the planet pinions to walk in the annulus gear and the planet carrier will drive the output shaft at a much reduced speed than that of the input shaft. It should be noticed that while the annulus gear of the first train is held stationary, the

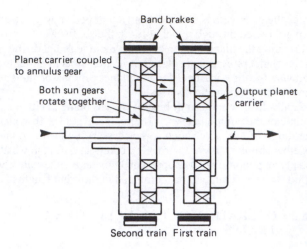

Figure 16.4 Schematic arrangement of simple epicyclic gear trains coupled on Wilson principle

planet pinions in the second train will be idling under the influence of their rotating sun gear. The fact that the annulus gear in the second train is freely rotating in the opposite direction is of no consequence.

If next the annulus gear of the second train is held stationary and that of the first train is released, then likewise the planet pinions will walk in the annulus gear under the influence of their rotating sun gear. However, since the planet carrier of the second train is driving the annulus gear of the first train and the sun gear of the first train is also being driven in the same direction, the net result is that the planet pinions of the first train are speeded up. Hence their carrier will drive the output shaft of the gearbox with a lesser reduction in speed. In other words, first and second gears are obtained by holding stationary the annulus gears of the first and second trains respectively.

Wilson epicyclic gearset

The operating principle of the original and present-day versions of the Wilson epicyclic gearbox is identical to that described above, the coupling together of simple gear trains being extended in some city bus applications to give five forward speeds. A band brake is provided for each gear train, so that each time a brake is applied a gear change is made with the gears always in mesh. When a change into top gear is made, none of the brake bands is applied, but instead a multiplate clutch is engaged to lock together the gear trains and provide a direct drive (Figures 16.5 and 16.6).

Simpson epicyclic gearset

This is a particularly ingenious coupling together of two simple epicyclic gear trains to provide a three-speed and reverse gearbox, which can be used in conjunction with a torque converter. It is so named after its American inventor, Howard W. Simpson, and originally featured in the long-established Chrysler Torque-Flite automatic transmission that was introduced in 1957. Since then this system of epicyclic gearing has featured in many designs of automatic gearbox. It was later combined with a simple epicyclic gearset if an overdrive top gear was required (Section 18.4).

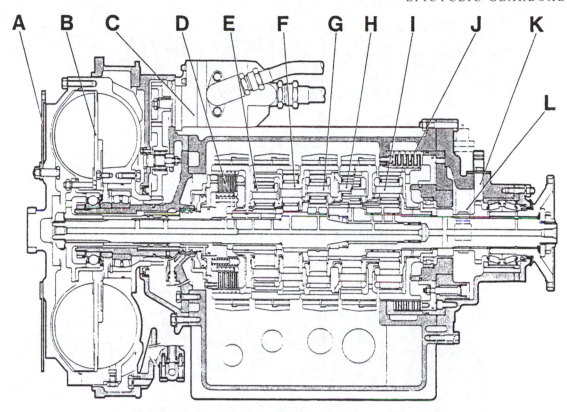

Figure 16.5 A modern Wilson five-speed epicyclic gearbox for bus operation (*Self-Changing Gears*)

A diaphragm plate	E fourth-speed gear assembly	I reverse gear assembly
B fully charged fluid coupling	F third-speed gear assembly	J friction retarder
C hydraulic pump	G second-speed gear assembly	K transducer
D top-speed clutch	H first-speed gear assembly	L speedometer drive

Unlike the compound epicyclic systems used in some automatic gearboxes, the Simpson gear train is designed so that both epicyclic trains have the same numbers of teeth and identical gear specifications. Since both gearsets are the same, the component parts of the planet carriers are all interchangeable with consequent economy in manufacture. Furthermore, a single sun gear of extended length serves for both gear trains. Associated with the operation of the Simpson epicyclic gearset are two multiplate clutches, two brake bands and a one-way or overrunning clutch (Figure 16.7). Hydraulic engagement is employed for the clutches and bands, fluid under pressure being supplied from the control system of the automatic transmission as explained later.

The functions performed by the clutches and brake bands acting on a Simpson gearset may be summarized as follows:

Front clutch This drives the transmission in reverse, but also combines with the rear clutch to provide a direct-drive gear; it is alternatively known as the reverse-high clutch. It transmits torque from the input shaft to the common sun gear.

Rear clutch This drives the transmission in all forward drive situations, and for this reason it is alternatively known as the forward clutch. It transmits torque from the input shaft to the annulus gear of the front train.

Front band This combines with the rear clutch to provide an intermediate gear; it is also known as either the intermediate

or the kick-down band. It serves to hold the common sun gear stationary for both gear trains.

Rear band This combines with the front clutch to provide a reverse gear, and is also used in manual low gear to retain engine braking effect; it is alternatively known as the low-reverse band. It serves to hold the rear planet carrier stationary. In some designs another multiplate clutch replaces the rear band.

One-way clutch This is situated towards the rear of the gearbox. It is not engaged hydraulically, but will automatically lock to hold the rear planet carrier stationary when required.

Power flow through a Simpson gearset

The operating modes of the Simpson gearset as used in modern automatic gearboxes are best visualized with the aid of a series of power flow diagrams (Figure 16.8). The following notes are intended to supplement these diagrams.

Neutral
Whenever the engine is running, the input shaft of the gearset is being driven via the torque converter. However, since there is neither engagement of the clutches nor application of the bands, no drive is transmitted through the epicyclic trains to the output shaft.

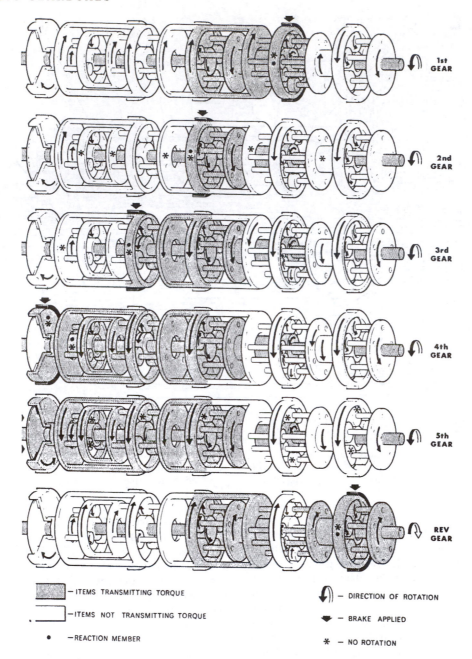

Figure 16.6   Power flow through a Wilson five-speed and reverse epicyclic gearbox (*Self-Changing Gears*)

Low gear

Drive from the input shaft is transmitted through the applied rear clutch to the annulus gear of the front train. Rotation of the annulus gear causes the planet pinions initially to idle, because their carrier is attached to the stationary output shaft. Now as the common sun gear can freely rotate on the output shaft, it is not only being driven by the planet pinions of the front train, but is also driving the planet pinions of the rear train. The effect of this is to cause the planet pinions of the rear train to idle, because their carrier is locked against rotation by the one-way clutch, so that drive is finally conveyed to the output shaft via the annulus gear of the rear train. A

double reduction in speed, together with a corresponding increase in torque, has therefore been obtained between the input and output shafts to provide a low-gear drive.

Manual low gear

When this is selected the power flow through the gearset is exactly the same as in automatic low gear, except that the rear band is applied to provide engine braking on vehicle overrun (Figure 16.8a). This feature is otherwise absent in automatic low gear, because the one-way clutch free-wheels on the overrun in the interests of smooth gear changes.

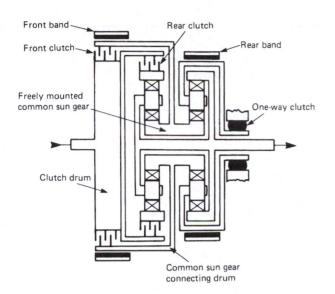

Figure 16.7 Schematic arrangement of simple epicyclic gear trains coupled on Simpson principle

Intermediate gear

The rear clutch remains engaged so that drive from the input shaft is again transmitted to the annulus gear of the front train. In this case, however, rotation of the annulus gear causes the planet pinions to walk around the common sun gear, because it is being held stationary by application of the front band to its connecting drum. This results in the output shaft, which is attached to the rotating front planet carrier, being driven at a reduced speed (Figure 16.8b). The speed reduction and torque increase between the input and output shafts is less than that obtained in low gear, because only the front train is in reduction. It should be noted that the rear annulus gear is also rotating with the output shaft, the one-way clutch free-wheeling to allow the rear planet carrier pinions to walk around the stationary sun gear.

High gear

The rear clutch again remains engaged and the front clutch is also engaged, but the front band is released. By engaging both clutches, the input shaft is now rotating the common sun wheel via its connecting drum and front clutch and also the annulus gear of the front train via the rear clutch. Since both the common sun gear and the annulus gear are being driven by the input shaft, there can be no relative rotation between them while the clutches are engaged. The planet pinions in the front train are therefore prevented from rotating and as a result their carrier is compelled to rotate with the locked train. Since the front train planet carrier is attached to the output shaft, a direct-drive high gear is obtained (Figure 16.8c). In this case it should be noted that the rear train is unavoidably locked against relative rotation of its members, the one-way clutch again free-wheeling to allow the rear planet carrier to rotate with the output shaft.

Reverse gear

This requires engagement of the front clutch and application of the rear band. Drive from the input shaft is therefore transmitted first to the common sun gear via its connecting drum and the front clutch. Now since the planetary carrier of the rear train is being held stationary through application of the rear band to its drum, the rotation of the common sun gear causes the planet pinions to idle and drive the annulus gear in the opposite direction (Figure 16.8d). A reversal in direction of rotation is thus established between the input and output shafts of the gearset, together with a reduced speed and increased torque. Two further points to be noted are, first, that the planet carrier of the rear train is being held stationary by the rear band rather than the one-way clutch, which would otherwise free-wheel and result in no drive; and, second, that although the planet carrier of the front train is being rotated by the output shaft the whole unit will merely idle, because none of its members is being held stationary.

Ravigneaux epicyclic gearset

A compound epicyclic gear train known as the Ravigneaux type, so named after its French inventor, Pol Ravigneaux, who developed the principle in the mid-1930s. Since the late 1940s it has been widely used in conjunction with a torque converter, first in two-speed and then later in three-speed versions. More recently it has been combined with one or even two simple epicyclic gearsets, so as to increase the number of gear ratios up to five including an overdrive top gear. To obtain a basic low, high and reverse two-speed geartrain, the Ravigneaux epicyclic gearset comprises an input shaft and sun gear, co-axial shaft and low sun gear, three equally spaced pairs of long and short planet pinions, output shaft and planet carrier, and a single annulus gear (Figure 16.9). There are two important relationships to note about this gearset, which are that each long planet pinion meshes with the input sun gear at one end and with a short planet pinion at the other end (but not with the low sun gear in close proximity), and that each short planet pinion also meshes with the low sun gear and the annulus gear. The friction elements associated with the forward gears are a combination of band brake and multiplate clutch, where the former serves to arrest the low sun gear and the latter to lock together the shafts of the input and low sun gears (Figure 16.10). For the reverse gear either a band brake, or a multiplate clutch, is used to arrest the annulus gear. In practical application this type of epicyclic gearset is generally regarded as offering a compact construction with simple control.

Power flow through a Ravigneaux gearset

Neutral

Whenever the engine is running, the input shaft and sun gear is being driven via the torque converter, but since none of the friction brakes is applied to the gearset, it is merely exercised without being required to do any work and no drive is transmitted to the output shaft from the planet carrier.

Low gear

This is obtained when the band brake is applied to arrest the low sun gear, so that it can become the reaction member of the gearset. Drive from the input shaft and sun gear then causes the long planet pinions to rotate in the opposite direction, but since they also mesh with the short planet pinions, the latter are compelled not only to turn in the same direction

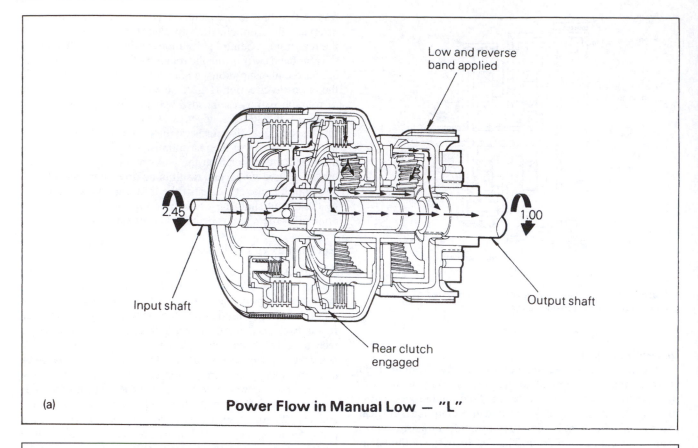

Low and reverse
band applied

2.45

1.00

Input shaft

Output shaft

Rear clutch
engaged

(a)   **Power Flow in Manual Low — "L"**

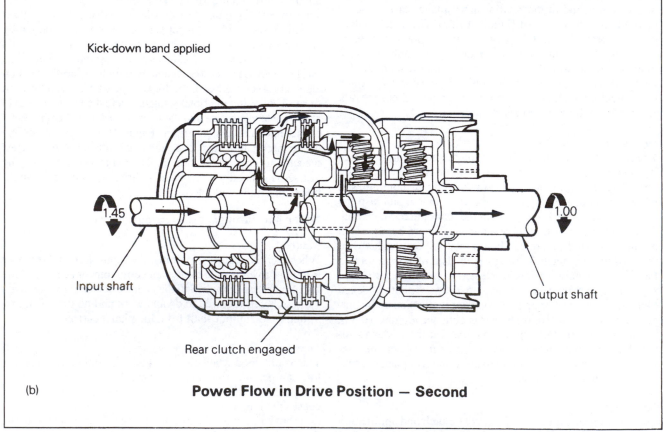

Kick-down band applied

1.45

1.00

Input shaft

Output shaft

Rear clutch engaged

(b)   **Power Flow in Drive Position — Second**

Figure 16.8   Power flow in Simpson gear train (*Colt-Mitsubishi*): (a) manual low (b) drive, second (c) drive, direct (d) reverse

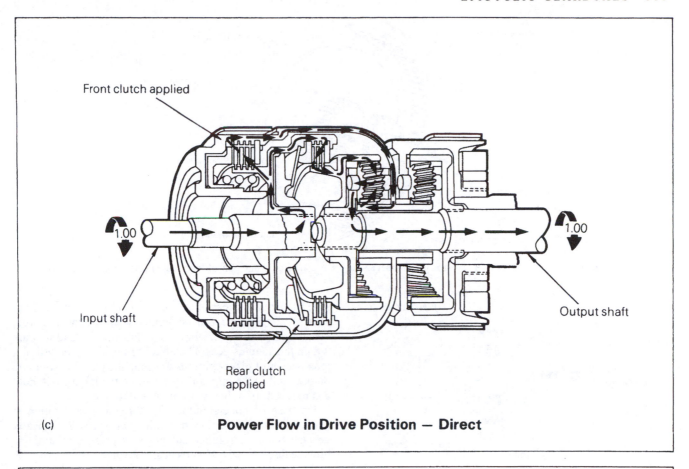

(c) **Power Flow in Drive Position — Direct**

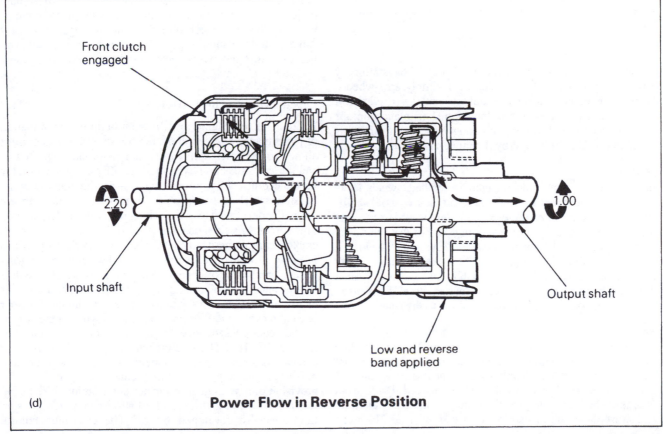

(d) **Power Flow in Reverse Position**

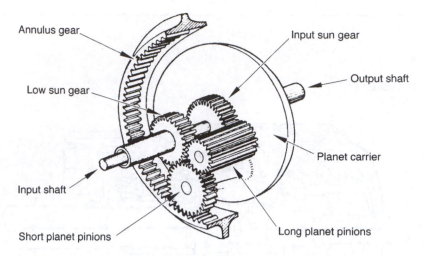

Figure 16.9   General arrangement of a Ravigneaux gearset (Note – For clarity only one pair of long and short pinions is shown)

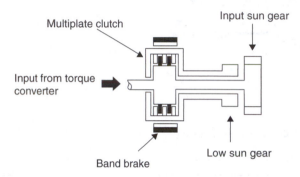

Figure 16.10   Schematic arrangement of forward speeds friction brakes for a Ravigneaux gearset

as the input sun gear, but also to walk around the stationary low sun gear while the annulus gear simply free-wheels. Hence, the planet carrier and output shaft are driven at a reduced speed by the orbiting long and short planet pinions and, of course, with a corresponding increase in torque as compared to that delivered by the input shaft.

High gear
In this gear the multiplate clutch is applied, which locks together the input and low sun gears, via their co-axial shafts. As a result there can be no relative motion occurring between the orbiting long and short planet pinions, so that the planet carrier and output shaft is now being driven through the locked gearset (including the annulus gear being carried around with it). A direct-drive is therefore established and in the same direction between the input and output shafts of the gearset, with no increase in torque being delivered to the latter.

Reverse gear
For reversing purposes the other band brake or multiplate clutch is applied to arrest the annulus gear, which therefore becomes the reaction member of the gearset, the low sun gear now being allowed to free-wheel instead. Drive from the input shaft and sun gear then causes rotation of the long and short planet pinions, as in the case of low gear. However,

since the annulus gear is now being held stationary and the short planet pinions can no longer walk around the free-wheeling low sun gear, they are compelled instead to walk within the annulus gear. The effect of this is for the orbiting long and short planet pinions to rotate the planet carrier and output shaft in the opposite direction to the input shaft, both at a reduced speed and with increased torque.

For ease of understanding the foregoing description is confined to the powerflow through a two-speed Ravigneaux gearset. Some rearrangement of the gearset is required to provide three speeds. This basically involves using a smaller input or primary sun gear that meshes with three short plant pinions. These in turn mesh with three long planet pinions, which in this case mesh directly with the larger secondary sun gear and the annulus gear. The planet carrier is held stationary in first and reverse gears, but is free to revolve in second and third gears.

Lepelletier epicyclic gearset

In order for the engine to deliver an optimum combination of performance and economy, the transmission system must allow it to work within its most efficient torque range. To this end the Transmission Division of the ZF Group in Germany introduced in 2002 a pioneering six-speed automatic transmission, which is based on a form of epicyclic geartrain originally developed by a French engineer Pierre Lepelletier. This ZF transmission, designated 6 HP 26, possesses considerable technical merit, since it is less complex than a previous five-speed automatic transmission and, by virtue of requiring fewer components, offers a more compact construction with a significant reduction in weight (Figure 16.11a). The six speeds or mechanical gear ratios are obtained by linking together one simple and one Ravigneaux gearset, which are engaged in the required sequence by five multiplate wet clutches (Figure 16.11b). These clutches comprise three rotating and two stationary types, there being no requirement for band brakes or one-way overrunning clutches. Apart from the provision of a sixth gear, a further contribution to reducing fuel consumption is the use of a low-viscosity transmission fluid specially developed by ZF. The electronic control

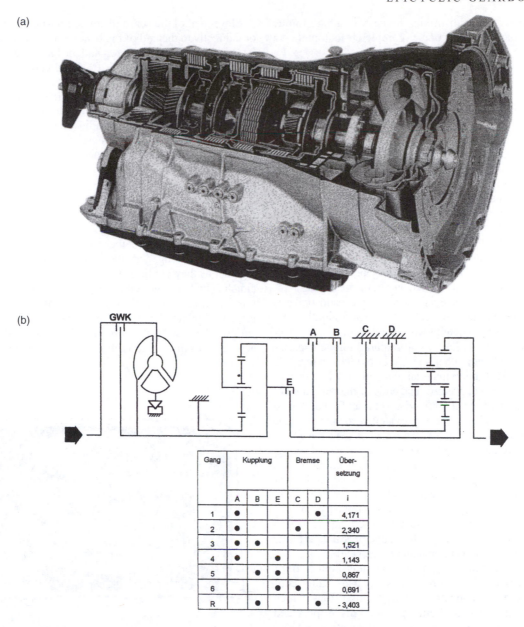

Figure 16.11 Six-speed ZF 6 HP 26 automatic transmission (*ZF Group UK*): (a) cutaway view showing one simple and one Ravigneaux gearset with five multiplate clutches (b) powerflow through the transmission GWK – controlled-slip converter lock-up clutch; Gang – gear; Kupplung – rotating multiplate clutches; Bremse – stationary multiplate clutches (brakes); Übersetzung – gear ratios

system for this advanced transmission is later described in Section 17.5.

16.3 FRICTION BRAKES FOR EPICYCLIC GEARBOXES

Types of friction brake

Three types of brake mechanism may be used to hold stationary the various members of an epicyclic gear train:

Band brakes
Multiplate clutches
One-way clutches.

All of these brakes rely on frictional contact between oil-wetted surfaces for their effectiveness. A band brake possesses the advantages of being able to withstand a high torque loading and requires only a simple anchorage point in the gearbox housing. The multiplate clutch has found increasing use in modern automatic gearboxes, because it can act as both a rotating and a stationary friction brake, whereas a band brake can only perform the latter function. Another convenient feature is that the torque capacity of the multiplate clutch can fairly readily be tailored to suit different installations, either by increasing or by reducing the number of discs and plates. An important service consideration is that the multiplate clutch never requires adjustment for wear. One-way clutches provide a compact form of friction brake with a high torque capacity and they are, of course, completely automatic in operation.

Band brakes

In a simple band brake the band wraps partly around a rotating drum and a braking action is obtained by pulling the band tightly on to it, the actual braking force being equal to the difference in tensions at the two ends of the band. The band brake was a natural choice for the designers of early epicyclic gearboxes, because the annulus gear lent itself to a convenient form of drum construction against which the band could be contracted. However, a band brake can be a fickle device and many early troubles were caused by chattering and grabbing of the bands. It was Major W.G. Wilson (Section 16.1) who later recognized the need for more attention to detail in brake band design, including the fully floating principle of band application that is found in modern automatic gearboxes.

For application to an epicyclic gearset, the band brake is made from either cast iron or steel and is sprung slightly away from the drum to give a natural working clearance in the non-applied condition. According to the required torque capacity the band is formed with either a single or a double wrap. It is lined with a special friction material of a porous nature to control the fluid film that is present. The ends of the band are not actually pulled but rather are pushed to contract the band on to the drum; one end of the band is provided with an anchorage in the gearbox casing. An apply force acts upon the other end of the band and may be derived from either a compressed spring, compressed air or hydraulic pressure. A compressed spring was used in early Wilson gearboxes, and compressed air in later ones intended for heavy-vehicle installations (Figure 16.12). In both cases the apply force was greatly multiplied by a self-adjusting toggle mechanism that finally closed the band on to its drum. A servo unit actuated by hydraulic pressure performs the same function in conventional automatic gearboxes.

A hydraulic servo unit is either separately bolted to, or made integral with, the gearbox housing. It basically consists of a cylinder with one or more fluid connections, which houses a spring-returned sliding piston. This is provided with synthetic rubber lip seals for pressure retention and also an actuating stem. The purpose of the spring loading is not only for returning the piston to its off position when the hydraulic apply pressure is cut off to release the band, but also for cushioning the application of the band in the opposite situation.

In detail design the servo piston can vary according to operating requirements. For example, the piston may have two lands of different diameter (Figure 16.13). A controlled load may then be applied to the band, because the higher apply pressure required to prevent the band from slipping under full-throttle conditions could result in jerky gear changing under light throttle. Therefore with a two-land piston the normal apply pressure is admitted between them and acts on the annular area of the larger one. Then under heavy-throttle conditions an apply pressure is also admitted below the smaller land. On the other hand, to ensure a very rapid release of the band for correctly timing a gear change, a release pressure is admitted above the larger land to assist the action of the piston return spring.

Movement of the servo piston is transmitted either directly or indirectly to the end of the band; in the latter case a lever is interposed between the piston and the band end to increase the application force (Figure 16.13). Final contact with the band is usually made via freely pivoting struts, so that the band is fully floating and therefore self-centring on its drum. This is an important factor in preventing chatter as the band is applied to arrest the drum. A screwed adjustment for setting the band-to-drum clearance is provided either at the fixed anchorage of the band, or in the lever mechanism of an indirect-acting servo (Figure 16.14).

Multiplate clutches

The main use of a multiplate clutch in an epicyclic gearbox is to connect a gearset member to the input shaft, so that when the clutch is engaged the two members remain locked together as they continue to rotate. Unlike the multiplate dry clutch, which is used with some heavy-vehicle manual transmission systems (see Section 13.6), the hydraulically energized multiplate clutch used in epicyclic gearboxes is the wet variety. The term 'wet' is not intended to mean that the clutch actually runs immersed in fluid, but rather that it works in the

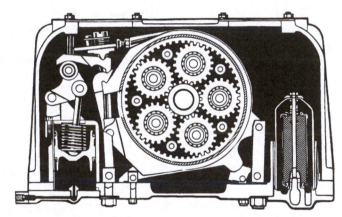

Figure 16.12 Compressed air actuation of band brake (*Self-Changing Gears*)

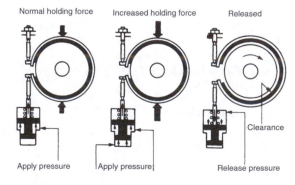

Figure 16.13 Operating principle of direct-acting hydraulic servo for band brake

presence of fluid that could not otherwise be readily excluded and which may beneficially perform a cooling function. Despite the lubricating effect of this fluid on its plates, the multiplate clutch can nevertheless be designed to sustain a high torque loading from compact dimensions and with acceptable pressures.

The friction elements of the clutch, known as the clutch pack, consist of several friction-lined driving discs alternating with (one more in number) unlined driven plates. There can be variations in the total number of discs and plates used depending on the required torque capacity. The flat steel driving discs are lined or faced on both sides with a friction material similar to that used in band brakes and are internally splined to provide a slidable connection with the clutch hub. In contrast, the steel driven plates are not faced with

friction material and have a series of external lugs to effect a slidable connection within the clutch drum. To withstand the thrust reaction of the engaged clutch pack the end plate is made much thicker than the others and is retained in the drum by a large snap ring. Rather than being made completely flat the driven plates may be either coned or waved very slightly, so as not only to cushion the engagement of the clutch but also better to separate the plates and discs on release.

Similar in principle to the hydraulic servo unit employed for band brake application, the multiplate clutch is engaged either directly or indirectly by a spring-returned sliding piston (Figure 16.15). As the piston must necessarily be of annular form to slide within the drum and over the hub, it requires synthetic rubber lip seals for both its outer and inner diameters, although in some arrangements the inner seal may be located from the hub. The rotating clutch drum is provided with inlet ports that communicate with fluid channels in its support, the desired path of fluid flow being established by strategically placed piston-ring-type seals.

In the direct-acting front clutch of a Simpson gear train the hydraulic apply pressure acting behind the piston is transferred directly to the clutch pack, the clutch becoming engaged once the plates and discs are clamped together (Figure 16.16). With the indirect-acting rear clutch a diaphragm or Belleville spring is interposed between the piston and the clutch pack, which acts as a lever to multiply the hydraulic apply force of the piston and clamp together more tightly the plates and discs to increase torque capacity (Figure 16.16). The diaphragm spring also acts as a return spring for the piston. A vent and ball check valve are also incorporated in either the drum or the piston of multiplate clutches. This feature allows the escape of trapped air as fluid is admitted during clutch engagement and also prevents fluid from remaining trapped after clutch release. The effect of such trapped fluid would be to cause drag in the clutch-pack, because the centrifugal action on the fluid builds up pressure behind the piston.

In later designs of automatic transmission, the multiplate clutch has also become used increasingly to hold a member of a planetary gearset stationary (Figure 16.11b). Unlike a

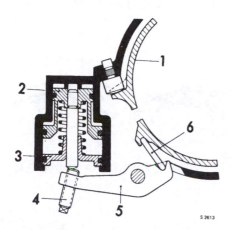

Figure 16.14 Identifying the parts of an indirect-acting hydraulic servo for band brake (*SAAB*)

1 brake band
2 piston, inner
3 piston, external
4 adjusting screw
5 lever
6 push-rod

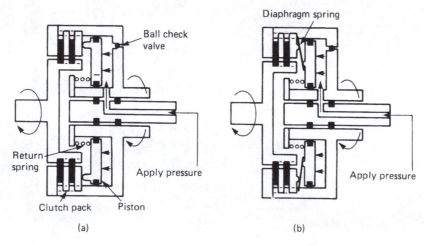

(a)   (b)

Figure 16.15 Schematic arrangement of (a) direct- (b) indirect-acting multiplate wet clutches

band brake performing the same function, the stationary type of multiplate clutch has the advantage that it never requires adjustment.

One-way clutches

In the design of an epicyclic automatic gearbox it is often expedient to substitute a one-way or overrunning clutch for the type of friction clutch previously described. Its ability to permit rotation in one direction and prevent it in the reverse direction can simplify the transition from one gear ratio to another, since there is then no necessity for timing one clutch against another. A one-way clutch can also perform a useful free-wheeling function in low gear by preventing objectionable lurching when the vehicle overruns the engine. Engine braking can then be provided in manual low gear by applying the low-reverse rear band (or clutch) to cancel out the free-wheeling action, as earlier explained for the Simpson gearset.

One-way clutches may be classified as being of either the roller or the sprag type. The roller type has already been described in connection with its use in torque converters where it supports the stator (Figure 15.6). Of later origin than the roller type, the sprag type differs in that the cam surfaces are formed not in one of the tracks but on each end of a series of spring-biased struts or sprags, which act between truly cylindrical races (Figure 16.17). The sprags are evenly spaced and constrained to tilt in unison by inner and outer retaining cages. They are also tilted towards a position of

incipient engagement by an intermediate ribbon spring. Since the diagonal distance across the sprag contacts is least in the direction of their initial tilt, as when the clutch is free-wheeling, any tilting in the opposite direction will cause the cam surfaces of the sprags to wedge tightly against the inner and outer races and lock them against relative rotation, thereby solidifying the one-way clutch.

Lubrication of an automatic gearbox epicyclic gear train

The comprehensive epicyclic gear train of an automatic transmission for a passenger car requires a well-planned system of lubrication. Typically the heated fluid from the torque converter flows through a cooler before it is directed by passages in the transmission case to both the centre and rear supporting members of the epicyclic gear train (Figure 16.18). From these sources the fluid is fed into annular clearance

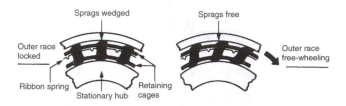

Figure 16.17   Operating principle of a sprag one-way clutch

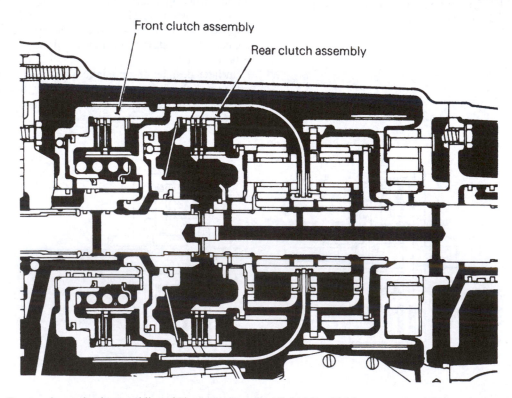

Figure 16.16   Front and rear clutch assemblies of Simpson gear train (*Colti-Mitsubishi*)

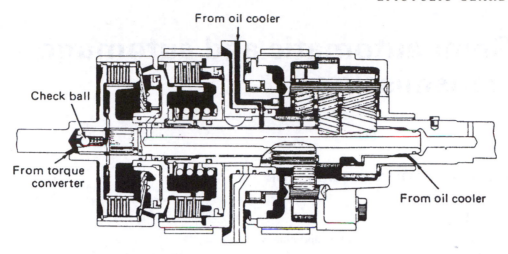

Figure 16.18 Lubrication system for an automatic transmission (*Toyota*)

spaces formed between the intermediate shaft and both the surrounding rear clutch sleeve at its centre and the counterbored output shaft at its rear end. The fluid is then directed either radially through sleeve holes or laterally from bush spiral grooves to lubricate the gear train teeth. Methods of conveying lubricant to the planet pinion bearings were mentioned earlier. The epicyclic gear train receives lubrication in all operating ranges of the transmission.

17 Semi-automatic and automatic transmissions

17.1 SEMI-AUTOMATIC TRANSMISSIONS FOR PASSENGER CARS

General background

The manually operated combination of gearbox and clutch has posed two problems during its long history of development. First, as already mentioned, the act of changing gears called for a considerable nicety of timing on the part of the early motorist. Eventually this problem was to all intents and purposes overcome by the invention of the synchromesh gear. Second, the friction clutch by its very nature has always demanded a certain delicacy of operation, especially when starting the vehicle from rest. Although the experienced driver can undoubtedly derive much satisfaction from skilful use of the clutch and gearbox, there are many drivers who would gladly dispense with the clutch pedal given a choice. In other words, there has always been an incentive for transmission designers to produce a two-pedal accelerator and brake control system for the motor vehicle, as found in modern automatic transmissions.

However, the first step towards providing two-pedal control for passenger cars was the introduction of the semi-automatic transmission, which is defined by the Society of Automotive Engineers in America as 'a transmission in which some of the functions of normal ratio changes are effected automatically'. There have been many versions of semi-automatic transmission, which have basically superimposed a degree of automatic control over layshaft and epicyclic gearboxes used in conjunction with friction clutches, magnetic clutches, fluid couplings and torque converters.

Although many early developments in semi-automatic transmission originated from the American motor industry, interest in this type of transmission waned there after World War II because of the increasing popularity of fully automatic transmission, but it was revived in Europe for smaller-engined cars. In these applications, emphasis was usually placed on retaining a layshaft synchromesh gearbox with conventional means of changing gear, but eliminating the need for a clutch pedal by operating the friction clutch through a vacuum servo unit (Figure 17.1). The British Lockheed Manumatic system was a typical example of this approach. When a change of gear was required the act of gripping the gear lever closed a microswitch contained within the knob. The electrical circuit thus completed served to energize a solenoid and open a control valve, which admitted intake manifold depression to the vacuum servo for the purpose of withdrawing the clutch through a rod and lever connection. Other more sophisticated but albeit elaborate designs of semi-automatic transmission still incorporated a

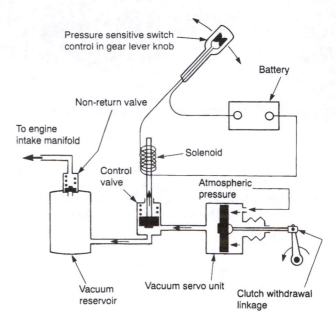

Figure 17.1 Schematic arrangement of an earlier vacuum servo-operated clutch for semi-automatic transmission

layshaft gearbox, but revived the American use of a fluid coupling in series with a vacuum servo-controlled friction clutch, thereby eliminating the need for a clutch pedal and also ensuring smooth moving away. Finally, we must arrive at the early 1990s to see how a relatively simple electronically controlled actuator can eliminate the need for a clutch pedal and provide a semi-automatic transmission, as described below.

Electronic clutch control

With the aid of modern electronics it has become possible to confer a sophisticated automatic control over a conventional friction clutch, which used in conjunction with an existing layshaft gearbox provides a cost-effective semi-automatic transmission that eliminates the need for a clutch pedal. Such a system has been developed by the French Valeo company and is marketed in Britain as the Valeo Electronic Clutch (Figure 17.2).

At the heart of this system is an electromechanical actuator, which converts signals received from an electronic control unit into angular displacement of the clutch withdrawal fork. The actuator basically comprises an electric motor with worm reduction gearing for operating the withdrawal fork, but with the important addition of an integrated overload or

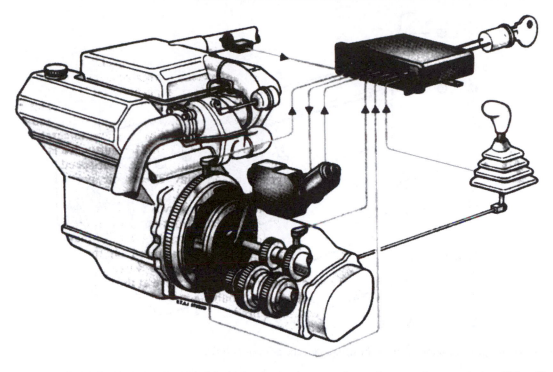

Figure 17.2 General arrangement of the modern Valeo electronic clutch system for semi-automatic transmission (*Valeo Clutches*)

compensator spring. Rather ingeniously this spring stores energy from the clutch diaphragm spring during engagement of the clutch and returns it during disengagement, so that the power requirement and hence current consumption of the electric motor are reduced to a minimum. An automatic wear compensation device is also incorporated in the actuator to maintain constant clutch performance throughout the normal life of the centre plate friction liners.

The electronic control unit comprises a microprocessor to interpret the data received from the system sensors, together with a power stage that transmits control signals to the electromechanical actuator. Data received by the electronic control unit are derived from rotational speed sensors and a potentiometer, which in combination sense engine speed, gearbox input speed, gear shift movements and throttle valve position. It will therefore be evident that the system can be programmed to enhance the performance characteristics of any particular type of vehicle, and indeed it has successfully met the stringent demands placed on the clutch operation of rally cars, as well as being approved by Ferrari for their Mondiale model.

In operation the system is such that with the engine idling in neutral, the actuator is in the disengaged position. When the driver shifts into first gear, the actuator immediately allows incipient engagement of the clutch, the engagement process being completed as the driver accelerates and the speed of the gearbox input shaft becomes synchronized with that of the engine. Each time a gear shift is made, the actuator quickly disengages the clutch and then re-engages it smoothly by taking into account engine speed, gearbox input speed, gear selected and accelerator movement. Since the precision of control is such that there need be no excess travel of the clutch withdrawal fork, a very short clutch engagement time

of about 100 ms is possible with this system, which therefore contributes to quicker gear changing.

17.2 SEMI-AUTOMATIC TRANSMISSIONS FOR HEAVY VEHICLES

General background

Although semi-automatic transmission systems can bring undoubted benefits in simplifying the driving process, there has always been a certain reluctance on the part of the car buyer to pay the extra cost of such an installation. This is no less true for the transport engineer in relation to heavy vehicles, who perhaps understandably may be reluctant to change something that is known for something that is new and not so well understood, and who recognizes that almost anything new will increase the first cost of the vehicle. However, several well-developed semi-automatic transmission systems for heavy vehicles have been introduced since the late 1980s and are being favourably received. These systems are based on well-proven existing manual gearboxes, so that development and production costs are minimized. Also, the gearbox spare parts can be commonized between manual and semi-automatic versions. In general the modern semi-automatic transmission for heavy vehicles utilizes either pneumatic or electropneumatic servo controls, according to the degree of automaticity that is conferred upon the operation of the clutch and gearbox.

Pneumatic servo-controlled transmissions

An example of pneumatic servo control for part operation of the gearbox is that applied to the well-known ZF-Ecosplit

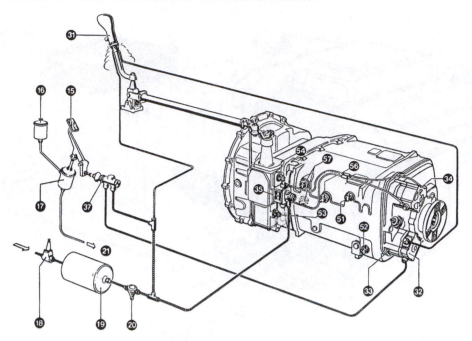

Figure 17.3   Gear change control for ZF-Ecosplit semi-automatic gearbox (*ZF GB*)

15   clutch pedal
16   refill rank
17   master cylinder for hydraulic clutch operation
18   overflow valve without return flow
19   compressed air tank, independent of air brake reservoir
20   compressed air filter with water separator and drain valve
21   to clutch slave cylinder
31   preselector valve for splitter group
32   relay valve for splitter group
33   integrated shift cylinder for splitter group

34   integrated shift cylinder for range change group
35   shift valve for range change group
37   actuating valve for operating the splitter group
51   indicator switch for range change group
52   indicator switch for splitter group
53   indicator switch for neutral
54   indicator switch for reverse
56   indicator switch for first gear
57   indicator switch for fourth gear

gearbox, which has a splitter drive and range change integrated with a four-speed and reverse section of the gearbox (Figure 14.43). Here the normal four-speed and reverse section of the gearbox is controlled entirely mechanically, but the range change group only partly so and the splitter drive not at all. The final control for the rear-mounted range change group is performed automatically by a pneumatic servo system and occurs during gear lever movement, which has a double H or superimposed H shift pattern, from 3–4 to 5–6 or from 5–6 back to 3–4. This particular servo system comprises a shift valve and a double-acting shift cylinder which are integrated with the gearbox (Figure 17.3). The shift valve is controlled by the selector shaft of the four-speed section of the gearbox and only directs air pressure to the shift cylinder in neutral.

A pneumatic servo system is also used to operate completely the front-mounted splitter group, which is controlled by a preselector valve that takes the form of a toggle switch mounted on the gear lever (Figure 17.4). This preselector valve operates in conjunction with an actuating valve at the clutch pedal, a relay valve and a double-acting shift cylinder (Figure 17.3). The required splitter group H (high) or L (low) can be preselected via the relay valve, which is spring loaded and not pressurized in the H range. When the clutch pedal is fully depressed, the actuating valve directs the compressed air to the shift cylinder and the splitter group shifts into the preselected setting.

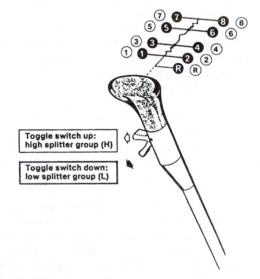

Figure 17.4   Gear lever operation for ZF-Ecosplit semi-automatic gearbox (*ZF GB*)

Electropneumatic servo-controlled transmissions

In 1982 the Swedish Scania company was the first heavy-vehicle manufacturer to superimpose an electropneumatic servo control system on the gear change of a layshaft gearbox.

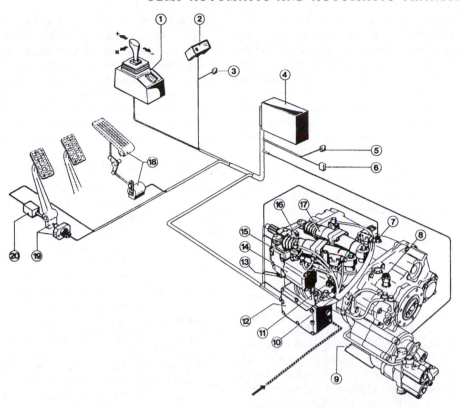

Figure 17.5   ZF-AVS automated preselection gear change system (*ZF GB*)

1 preselector switch	13 indicator switch for 'gear engaged'
2 gear display	14 shift cylinder for reverse
3 main power supply 2	15 indicator switch for neutral position
4 electronic control unit	16 shift cylinder for 3–4 and 7–8
5 diagnostics connection	17 shift cylinder for 1–2 and 5–6
6 main power supply 1	18 accelerator pedal with load sensor
7 speed sensor for input speed	19 clutch pedal with electrical release switch
8 speed sensor for tachograph	20 sensor for return signal 'gear engaged' and valve
9 speed sensor for output speed	'hold down clutch'
10 indicator switch for 'range change group engaged'	
11 indicator switch for reverse gear	Compressed air pipe with water separator and
12 central valve block	dehumidifier to be installed separately

Their computer-aided gear shifting (CAG) system utilizes a computerized electronic control for the pneumatic valves and can be operated in either A (automatic) or M (manual) modes from the driver's selector console. In automatic mode the electronic control unit determines the gear to be selected, according to the control signals it receives from the vehicle and engine speed sensors, but the actual gear change does not occur until the driver fully depresses the clutch pedal. However, the driver can override the automatic choice of gear if it is prudent to do so. Alternatively, in manual mode the driver makes the gear selection by means of a toggle switch on the selector console, but the electronic control unit will still overrule any choice of gear that would result in dangerous overspeeding of the engine.

A somewhat similar approach can be found in the ZF-AVS gear change system, which comprises an electronic control unit that automatically preselects the most suitable gear and indicates it on a display panel. The driver is then given the option of engaging the recommended gear by depressing the clutch pedal, or changing gear manually as conditions demand. The general arrangement of the AVS system (Figure 17.5) can best be considered in terms of gearbox- and vehicle-related components.

The gearbox-related components feature a central valve block that contains the main shift valve and the solenoid valves required for pneumatic control of the shift cylinders. These engage 1–2 and 5–6, 3–4 and 7–8, and reverse gears. In addition there is a range of indicator switches for 'range change group engaged', 'reverse gear', 'gear engaged' and 'neutral position'. Also included are sensors for input and output speeds and another speed sensor for the tachograph. The vehicle-related components comprise a preselector switch, which is a combined tip lever and selector switch with drive modes designated D, N and R; a display panel which not only shows the driver which gear is engaged but also signals the gear preferred by the electronic control unit; a release switch actuated by the clutch pedal; a load sensor actuated by the accelerator pedal; a sensor return signal 'gear engaged'; and a valve for the function 'hold down clutch'. The electronic control unit is programmed with data relating to engine, vehicle and gear ratio steps, so that by taking into account the prevailing running conditions it can determine the most suitable gear for engagement. It therefore processes signals and impulses from the various switches and sensors, as well as monitoring all gear changes to prevent engine overspeeding or gearbox damage. The shift command is passed on to the main shift valve of the central valve block when the clutch pedal is depressed. This actuates the appropriate shift cylinder to engage the most suitable gear. A diagnostic facility is also included for the gear change system.

With the Eaton semi-automated mechanical transmission (SAMT) the electropneumatic servo operation is superimposed on both the gear change and the clutch actions. In response to signals received from the driver's gear selector switch, the computerized electronic unit controls the declutching, accelerator and gear change actions. That is, the driver simply moves the gear selector switch to the upwards position to change up and similarly downwards to change down. The only requirement to use the clutch pedal is during initial gear selection and when the vehicle is brought to rest.

There are clearly installational advantages to be gained with electropneumatic gear change systems, because they eliminate expensive remote-control gear selector linkages, lend themselves to easy fitment in left- and right-hand-drive vehicles, provide a straightforward installation for tilting cabs and present no soundproofing problems.

17.3 AUTOMATIC TRANSMISSIONS FOR PASSENGER CARS

General background

Unlike a semi-automatic transmission system, a fully automatic one completely relieves the driver of the duty of changing gear, while still allowing the driver to override its normal operation if thought desirable. An automatic transmission has therefore been defined by the Society of Automotive Engineers in America as 'a transmission in which ratio changes are effected automatically without manual assist'. The individual roles played by the fluid coupling, torque converter and epicyclic gear train in the development of automatic transmissions have earlier been related in Sections 15.1, 15.3, 16.1 and 16.2. However, it was not until the late 1930s, when a brilliant team of General Motors transmission engineers led by Earl Thompson (who, it may be recalled from Section 14.3, pioneered the synchromesh gear change) added a sophisticated hydraulic control system to the operation of an epicyclic gear train, that the fully automatic transmission became a practical reality.

Advantages of the fully automatic transmission

The advantages of the fully automatic transmission, usually referred to simply as an automatic transmission, may be listed as follows:

1 It minimizes driver fatigue, especially in heavy traffic, by eliminating the need to operate a clutch pedal and gear lever for starting from rest and changing gear.

2 It contributes to safer driving because the concentration of the driver is not disturbed by the need to change gear; also, both hands can remain on the steering wheel.
3 Progress can be smoother under normal driving conditions, because gear changes will occur at the theoretically correct moment in terms of road speed and throttle opening.
4 It allows the driver to override the automatic control and enforce a gear change when desired, since no control system can anticipate road or traffic conditions ahead.

Basic construction of an automatic transmission

In a typical modern construction the components of an automatic transmission are based in three aluminium alloy castings. These comprise the bell-housing for the torque converter, the main casing for the epicyclic gearset together with its associated clutches, bands and servo units, and the rear extension housing that encloses the output shaft with parking sprag and governor. All three castings are mutually located by spigots and bolted together (Figure 17.6).

A stator support for the hub of the torque converter is spigoted into the open front face of the main casing and bolted to the rear face of the adjoining bell-housing. The stator support further incorporates bearing bushes for the input shaft from the turbine and also provides an integral operating chamber for the front pump. Of annular form, the integral rear wall of the main casing serves as both a bearing support and a distributor sleeve for the governor assembly, which is carried on the output shaft. A further bearing support for the output shaft is provided, via its sliding joint connection with the propellor shaft, by the rear extension housing. There are two main applications where dynamic lip-type seals are required to seal rotating parts in a longitudinally mounted automatic transmission, these being the front pump and the rear extension housing. In an automatic transaxle only the front pump seal is required, since the other sealing function is transferred to the output from the final drive gears. The front pump seal is in any event by far the more critical application, because it is subject to fairly high-temperature operation and is relatively inaccessible for service replacement. For this reason synthetic rubber seals are used with carefully controlled physical properties and resistance to ageing. Finally, the control valve body assembly is bolted to the underside of the main casing and is enclosed by a detachable sump that contains the transmission fluid.

Automatic transaxles

The widespread adoption of transversely mounted engines for front-wheel drive cars created a need to combine not only a layshaft gearbox with the final-drive assembly to form a 'transaxle' (Section 19.5), but also to adapt an automatic gearbox in a similar manner and thereby produce an 'automatic transaxle' (Figure 17.7). Several different drive arrangements can be used, which may conveniently be classified as follows:

Input chain drive
Output three-gear drive
Output four-gear drive

Input chain drive This less common arrangement has found particular favour on various General Motors cars, such as

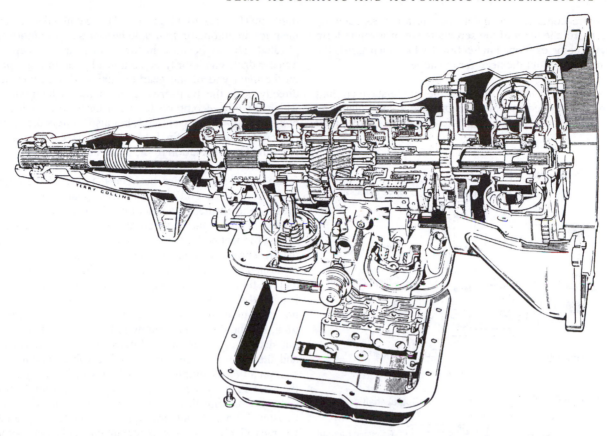

Figure 17.6 Cut-away view of a modern automatic transmission with torque converter and three-speed gearset (*Ford*)

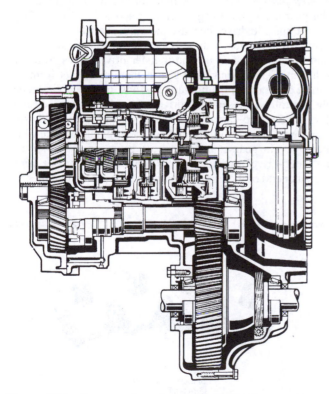

Figure 17.7 Cut-away view of a modern automatic transaxle with torque converter and three-speed gearset for transverse-engined cars (*ZF*)

Vauxhall, since it was first adopted for the Oldsmobile. Toronado in the mid-1960s. Its distinguishing feature is a step-down chain and sprockets drive, which transfers the output from the torque converter to the input side of the automatic gearbox (Figure 17.8a). The drive from the latter is then transmitted to the differential assembly, via a permanently acting epicyclic gearset. This gearset therefore constitutes the final-drive gearing, because it provides an additional and constant gear reduction in the transmission system, regardless of the output from the automatic gearbox itself. The final-drive epicyclic gearset corresponds in function to that shown in Figure 16.3a. That is, the sun gear is directly coupled to the output shaft of the automatic gearbox and the differential case serves as the planet carrier for the pinions, while the annulus gear is anchored to the gearbox casing. To complete the drive-line, one output shaft from the differential backtracks through the centre of the final-drive and automatic gearbox, where it emerges to connect with the far side drive-shaft coupling.

An interesting form of construction is used for the drive chain, which is known as the Morse 'Hy-Vo' type (the name being derived from high-velocity and involute). The chain is of multi-strand pattern with inverted teeth, similar to that earlier described for an engine timing chain, but instead of the links articulating about pins and bushes they are joined through convex-faced divided pins (Figure 17.9). As the chain links articulate about the sprocket teeth, a rocking motion is substituted for a rubbing one at their joints, this being more conducive to quiet and vibrationless running. Furthermore, the teeth of the sprockets are of involute form (this term being explained in Section 14.1), which assists the chain to run without the need for tensioners.

Output three-gear drive Whereas the single-stage layshaft gearbox is intended by design to provide a convenient and what is sometimes termed a 'two-gear' transfer drive for the

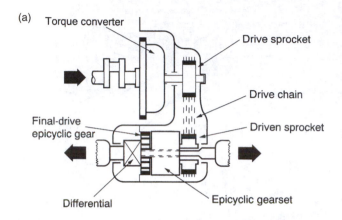

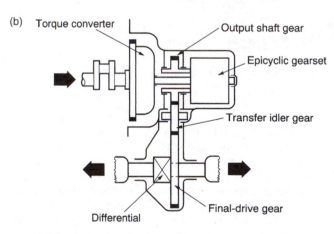

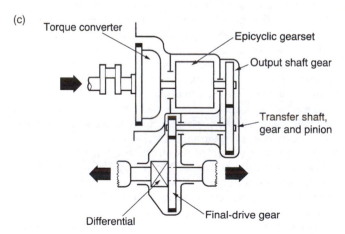

Figure 17.8 Schematic arrangements of automatic transaxles for transverse-engined cars: (a) input chain drive (b) output three-gear drive (c) output four-gear drive

transaxle (Figures 14.13 and 14.14), the equivalent arrangement for an automatic transaxle becomes a little more complicated. This is because in the former case, a step-down transfer drive can simply be obtained by mounting a pinion on the inner end of the gearbox mainshaft and meshing it directly with the final-drive gear of the differential case, which not only accommodates the distance between the gearbox and final-drive assemblies, but also confers the required reduction ratio for the final-drive as well as restoring the correct direction of rotation to the drive shafts.

However, the situation with an automatic transaxle is different, since the distance between the axes of the gearbox epicyclic gearset and the final-drive gear is significantly greater, and further provision must be made in the transfer gearing to restore the correct direction of rotation to the drive-shafts. To meet both of these requirements therefore demands the presence of an intermediate or transfer idler gear, which is meshed between the output shaft of the automatic gearbox and the final-drive gear (Figure 17.8b), thereby giving what is termed a 'three-gear' transaxle drive. The idler gear does not, of course, affect the reduction ratio of this simple gear train. It should be noted that with this arrangement the output shaft gear is sandwiched between the torque converter and the automatic gearbox, which involves mounting the gear on a hollow output shaft that backtracks partly through the gearbox epicyclic gearset, the input shaft from the torque converter passing through it. Helical gears are used in the transfer drive and they are typically mounted in preloaded tapered roller bearings to provide stability and quiet operation.

Output four-gear drive A more orthodox arrangement of the gearbox epicyclic gearset is practicable in this 'four-gear' transaxle drive, because the input shaft from the adjacent torque converter directly enters the gearset at one end and the output shaft for the transfer drive leaves from the other end. The output shaft gear is therefore offset from the torque converter and automatic gearbox. In order to intercept the centrally disposed final-drive gear and differential assembly, it therefore becomes necessary to provide a transfer shaft, gear and pinion, which backtracks parallel to the gearbox epicyclic gearset and produces what is sometimes termed a 'folded-drive' arrangement. To complete the transfer drive, the inner end of the transfer shaft carries a pinion that meshes

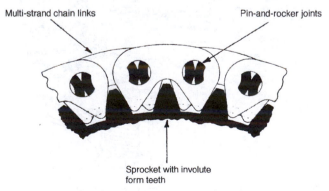

Figure 17.9 Construction of Morse Hy-Vo chain for automatic transaxle

with the final-drive gear (Figures 17.7 and 17.8c). Since we have now introduced a compound gear train, as distinct from the simple gear train for the previously described arrangement, between the gearbox epicyclic gearset and the final-drive gear, the overall ratio for the final-drive will be equal to the product of the ratios for the two pairs of gears.

This particular arrangement of automatic transaxle drive possesses the production advantage of allowing for convenient changes of final-drive ratio to suit different models. As may be expected, helical gears are again used throughout the transfer drive train and all are typically mounted in preloaded tapered roller bearings (the reasons for preloading bearings being later dealt with in Section 20.2).

Since the construction and operation of the torque converter and epicyclic gearset have earlier received individual attention, it now remains to explain the basic operating principles of the hydraulic and then later the electrohydraulic control systems for automatic transmissions.

17.4 HYDRAULIC CONTROL SYSTEMS

Functions of the hydraulic control system

The hydraulic control system of an automatic transmission has to perform the following functions:

1 Fills and then maintains a constant flow of fluid under pressure through the torque converter for efficient transmission of torque.
2 Supplies fluid to a valve control body and governor for regulating the application and release of the gearset clutches and bands, according to road speed and throttle opening.
3 Circulates heated fluid from the torque converter through a separate cooler to lubricate the gears, bearing bushes, thrust washers, multiplate clutches and one-way clutch.

Operation of the hydraulic control system

It has often been said that, like many complex devices, the hydraulic control system of an automatic transmission may be considered as a complicated sum of a number of fundamentally simple units (Figure 17.10). The first step is therefore to gain a basic understanding of the individual units and then to relate their operation to the system as a whole by referring

to the hydraulic circuit or 'bloodstream' diagram issued by the particular manufacturer. The basic hydraulic units with which we are concerned here may be listed as follows:

Hydraulic pump
Pressure regulator and boost valves
Manual valve
Throttle and kick-down valves
Governor valve
Shift valves
Accumulators.

Hydraulic pump

An engine-driven pump provides the energy source for the hydraulic control system. It is mounted at the front of the gearbox main casing and is driven through either two flats or two tags from the hub of the torque converter. A positive displacement pump is used and, as earlier explained for engine oil pumps (Section 4.5), this class of pump can only generate flow: it is the resistance to this flow offered by the circuit that creates the system pressure. An internal gear and crescent type of positive displacement pump is the most widely used in current automatic transmissions, the relatively large size of the pump requiring a separating crescent between the gear teeth not in mesh to ensure the efficient transfer of fluid (Figure 17.11).

Since the pump is driven by the engine rather than the transmission, it will not circulate fluid for gearbox lubrication if the vehicle is being towed. It is therefore necessary to observe the particular manufacturer's recommendations with regard to speed and distance restrictions on towing, it being generally preferred to tow the vehicle with the driven wheels off the ground.

Pressure regulator and boost valves

The pressure regulator valve is included in the hydraulic control system to regulate its primary pressure, which is usually referred to as the line or control pressure. In high gear with the transmission in direct drive, the pressure regulator valve provides for a variable line pressure that will increase

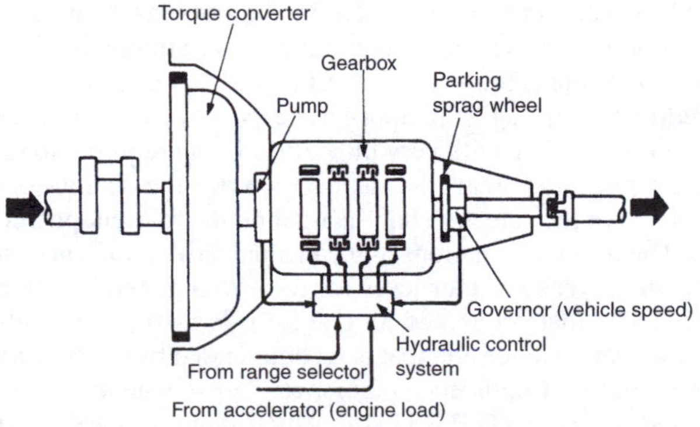

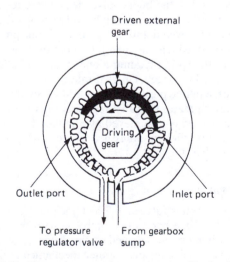

Figure 17.10 Basic hydraulic control system for an automatic transmission

Figure 17.11 Internal gear and crescent type of hydraulic pump

with engine torque and hence throttle opening. Similarly the line pressure is further increased as greater torque is transmitted through the gearbox in intermediate, low and reverse gear ranges. The purpose of varying the line pressure in this manner is to provide the best compromise between always ensuring that it is high enough to prevent slipping and wearing of the clutches and bands, while avoiding unnecessarily high pressures that would produce harsh gear shifts. An additional function of the pressure regulator valve is to deliver fluid at a reduced pressure to the torque converter, oil cooler and gearbox lubrication circuit.

In practical application the pressure regulator valve is installed in a branch of the main supply line from the hydraulic pump and operates in several stages. Initially the valve remains closed under the influence of its spring loading and simply allows the pump to fill the fluid lines of the hydraulic control system. As the pressure in the system in neutral builds up to about $420 \, kN/m^2$ ($60 \, lbf/in^2$), the valve no longer remains bottomed but is forced to move against its spring loading by the line pressure acting on its end. This balancing pressure gains access to the end of the valve via a damping orifice, so that the valve cannot be set into rapid oscillation and cause a buzzing noise. The movement of the valve is then sufficient to uncover a port so that fluid delivered by the pump to the main supply line can be metered into the torque converter circuit and thus pressurize it (Figure 17.12a). Once this has been accomplished, any further rise in line pressure will similarly increase the balancing pressure on the end of the valve and force it to open further against its spring loading. This causes an exhaust port to be uncovered, such that all fluid delivered by the pump in excess of that required by the hydraulic control system and torque converter can be either returned to the sump or bypassed to the pump inlet passage (Figure 17.12b).

When the vehicle is being driven, the balancing pressure acting on the end of the regulator valve has to oppose not only the spring loading but also an additional force exerted by what is called a boost valve. The amount of force exerted by this valve varies according to throttle opening and the particular gear range selected. In the former case a throttle pressure that is proportional to throttle opening is directed against the end of the boost valve. This added resistance to movement of the pressure regulator valve allows the exhaust port to remain covered until a high enough line pressure has been achieved, which may be up to three times greater than that due to the spring loading alone (Figure 17.12c). When intermediate, low or reverse gear ranges are selected, a further increase in line pressure is required which is about three to four times greater than that regulated by the spring loading alone. This can be achieved by allowing line pressure to act on the difference in areas between the stepped lands of the boost valve (Figure 17.12c).

Manual valve

The function of this valve is to direct line pressure to the various circuits of the hydraulic control system in accordance with the driving range selected by the driver. It is connected to the driver's selector lever, either mechanically through a direct-acting system of rods and levers or an enclosed cable,

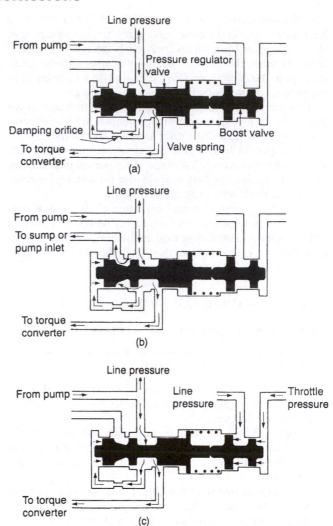

Figure 17.12 Schematic arrangement and operation of pressure regulator and boost valves: (a) line pressure normal (b) line pressure stabilizing (c) line pressure increased

or unusually (as in the Rolls-Royce application) through an electrically operated servo mounted on the gearbox. The manual valve is of the spool type with several equal diameter lands, and its design is such that line pressure is always admitted between the same two lands, thereby ensuring that the hydraulic forces acting along the valve are maintained equal and opposite (Figure 17.13a). Line pressure can therefore be redirected through the appropriate ports of the valve body (Figure 17.13b), with very little effort being required to move the valve other than that of overcoming the spring-loaded detents of the selector shaft that define its different positions.

The different positions of the manual valve and hence those of the driver's selector lever conform to a general pattern or sequence that originated long ago for two-speed gearsets in American practice and that is recommended by the Society of Automotive Engineers in America. This sequence is designated by the letters P R N D L, which stand for park, reverse, neutral, drive and low. Variations on this basic sequence are required for modern automatic transmissions, which usually

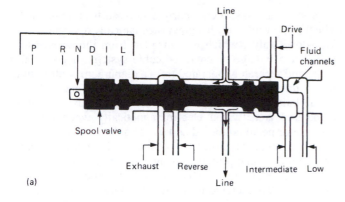

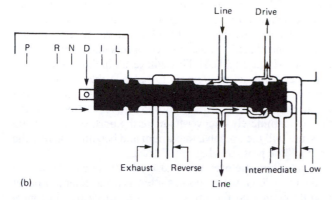

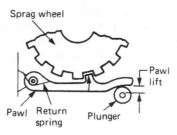

Figure 17.14   Operating principle of automatic transmission parking lock

Figure 17.13   Schematic arrangement and operation of manual valve in two of its ranges: (a) neutral (b) drive

have three- or four-speed (and sometimes now five-speed) gearsets. Typical sequences are: P R N D I L, where I stands for intermediate; P R N D 2 1, where 2 and 1 are equivalent to intermediate and low; and similarly P R N D3 D2 D1.

In practice the control facilities provided by the corresponding positions of the manual valve and its associated mechanism may be summarized as follows:

Park   The early development of the automatic transmission provided an opportunity to incorporate a means of positively locking the transmission when parked, which was in addition to the handbrake control. With the gear control mechanism set in the park position, a sprag wheel on the output shaft of the gearbox is engaged by a plunger-operated pawl that pivots from the casing and prevents rotation of the shaft (Figure 17.14). This is also one of the only two positions of the gear selector in which the engine can be started, the other of course being neutral.

Reverse   This position may be selected when the vehicle is stationary, its purpose being self-evident.

Neutral   When the selector lever is moved to this position the output shaft of the gearbox is unlocked and the engine may be started and run without driving the vehicle, since no gear changing can take place.

Drive   This range is selected for all normal driving conditions and maximum economy. The transmission starts in low gear and shifts automatically to intermediate and high gears as the speed of the vehicle increases. Changing down to the

appropriate lower gears similarly occurs automatically as the vehicle speed decreases. A kick-down facility is also available in the drive range; this is so called because when accelerating to overtake, for example, the driver can enforce downward gear changes by fully depressing or flooring the accelerator pedal.

Intermediate   In this range the vehicle will either start off in low gear and shift automatically to intermediate gear, or start off in intermediate gear, according to design; in either event the gearbox is prevented from shifting above intermediate gear. This range is appropriate for gentle downhill slopes or very winding roads, the continuously engaged intermediate gear contributing a useful engine braking effect for better vehicle control. It is normally advised that intermediate range must not be used above 60 mile/h (100 km/h).

Low   When this range is selected the vehicle will start off in low gear and remain in this gear regardless of vehicle speed or throttle opening. It is particularly appropriate for steep downhill slopes when the quite noticeable engine braking effect reduces the need for severe brake usage. To prevent overspeeding of the engine it is normally advised that low range must not be selected above 35 mile/h (60 km/h).

For safe operation of the automatic transmission, the gear selector mechanism connecting with the manual valve also incorporates both mechanical and electrical inhibitor devices. Apart from visual indication of the gear range selected, the selector lever is also provided with a detent arrangement that enables the driver to detect by feel which range is being selected. For example, a two-stage stop under the selector gate can prevent certain movements of the lever, unless its spring-loaded knob is depressed. Partial depression of the lever knob is then necessary before intermediate or reverse can be selected, while full depression with greater effort is required before selecting low or park.

A neutral safety or inhibitor switch is always incorporated in the electrical circuit for the starter motor, so that as earlier noted the engine can only be started with the gear selector lever in the neutral or park positions (Figure 17.15). The inhibitor switch may be of either the plunger type, which is activated by a striker cam on the driver's selector lever or alternatively from the gearbox selector shaft, or the rotary type, which is confined to the gearbox selector shaft.

Throttle and kick-down valves

The purpose of the throttle valve is to receive fluid at line pressure via the manual valve and meter it as throttle pressure, which is also known as throttle valve or TV pressure, into those circuits of the hydraulic control system that require a signal of engine load or, in other words, an indication of the torque requirements of the engine. As earlier mentioned, throttle pressure is arranged to act on the boost valve of the regulator valve assembly. It is also required to oppose governor pressure which, as later explained, provides a signal of vehicle speed. Otherwise, the gear shifts would occur at exactly the same road speed for identical throttle openings and would therefore take no account of the engine labouring on a steep incline when clearly a downshift is demanded. The throttle valve may be either mechanically operated via a kick-down valve, which is controlled from the accelerator pedal by a linkage or enclosed cable system, or pneumatically and separately operated from a diaphragm control unit that utilizes intake manifold depression to sense engine load. In the latter case, the associated kick-down valve operates alone and is activated electromagnetically whenever the accelerator pedal is floored and closes a detent switch.

In operation, the throttle valve itself is hydraulically balanced on the same principle as the pressure regulator valve. Therefore when fluid is delivered to the throttle valve at line pressure and is metered to throttle pressure, this pressure is also reflected on the end of the valve to oppose its spring loading. Take for example the simpler case of the mechanically operated throttle and kick-down valves. As the accelerator pedal is partly depressed the kick-down valve will move towards the throttle valve and compress the spring acting against it (Figure 17.16a). This will result in the throttle pressure increasing as it attempts to balance the throttle valve against its increased spring force. Now if the accelerator pedal is fully depressed the throttle valve will bottom and thereby completely uncover its metering port, so that throttle pressure will then reach a maximum and become equal to the original line pressure (Figure 17.16b).

Fluid at throttle pressure is also admitted behind the land of the kick-down valve spool, so that as the accelerator pedal is fully depressed the inward travel of this valve becomes sufficient to uncover a further port, which releases the fluid at what is now termed kick-down pressure. Fluid at this pressure

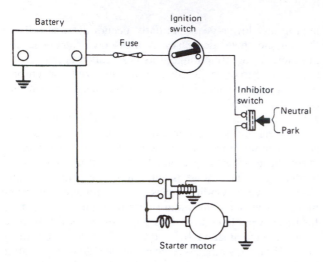

Figure 17.15 Schematic arrangement of neutral and park inhibitor switch circuit

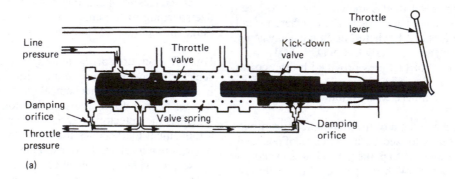

(a)

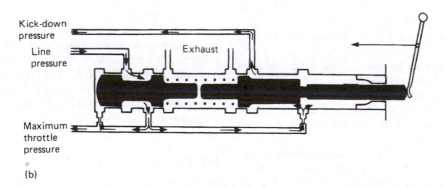

(b)

Figure 17.16 Schematic arrangement and operation of throttle and kick-down valves: (a) part-throttle kick-down (b) full-throttle kick-down

is then delivered to the hydraulic control system for enforcing downshifts.

Governor valve

In contrast to the throttle and kick-down valves, the purpose of the governor valve is to meter fluid under pressure to those circuits of the hydraulic control system that require a signal of vehicle speed rather than engine load. Governor pressure is therefore a fluid pressure that varies in relation to vehicle speed and hence the rotational speed of the gearbox output shaft. For this reason a typical attachment for the governor housing is through a bolted face mounting to the parking sprag wheel, the hub of which can then be provided with internal channels to convey fluid from the distribution sleeve to the governor valve housing via strategically placed piston-ring-type seals. The mechanism of the governor can vary in detail design, chiefly in the manner that movement of its flyweights is transmitted to the regulating valve, either directly through a pull-rod or indirectly via bell-crank levers. Although the governor pressure regulating valve acts in much the same way as the throttle pressure regulating valve, an important difference to note is that as the governor valve moves from rest it gradually uncovers the port admitting fluid at line pressure, whereas similar movement of the throttle valve gradually covers its corresponding port.

A simple governor comprising a regulating spool valve that is being urged to open by the centrifugal action of its flyweight, but at the same time is striving to close under the influence of fluid pressure, would not provide suitable gear shifting characteristics for an automatic transmission system.

This is because, being a centrifugal operated device, the governor would produce a pressure signal that was in proportion to the square of vehicle speed. That is, a doubling in vehicle speed would quadruple the governor pressure to the hydraulic control system, which would have difficulty in utilizing such a wide range of pressures. A refinement to the simple governor is therefore necessary, and takes the form of a two-stage governor valve and flyweight assembly. In a typical construction two coaxial flyweights slide in the governor housing and are flanged to accept a compression spring between their walls, the inner flyweight being connected to the governor valve by a pull-rod that passes freely through the output shaft of the gearbox.

The operation of this type of two-stage governor is such that as the output shaft begins to rotate with movement of the vehicle, both flyweights tend to move outwards because of centrifugal force acting upon them. This action pulls the control valve inwards so that the port admitting fluid at line pressure starts to uncover and meters fluid at governor pressure. As the vehicle continues to gain speed, the flyweights will move further out until the outer one approaches its retaining ring, but the increasing governor pressure that is acting on the larger spool of the valve compresses the spring to restrain the inner one from reaching its full travel (Figure 17.17a). This condition of valve balance represents the first stage of governor operation and prevails at low vehicle speeds up to about 20 mile/h (35 km/h). It provides a fairly rapid rise in governor pressure for early shifting from low to intermediate gear at light throttle. The second stage of governor operation occurs when rising vehicle speed compels the inner flyweight to resume moving outwards and overcome the existing governor

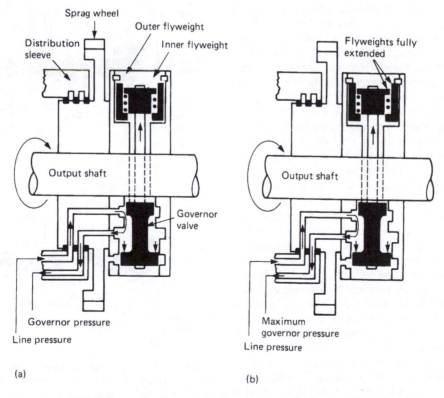

Figure 17.17 Schematic arrangement and operation of governor valve: (a) low vehicle speed (b) high vehicle speed

pressure acting on the valve (Figure 17.17b). This action pulls the governor valve further open and thereby produces an increasing governor pressure, but the build-up of pressure is more gradual than in the first stage and therefore more appropriate for shifting from intermediate to high gear at moderate to high speeds according to throttle opening.

Shift valves

In describing the arrangements of the throttle and kick-down valves and the governor valve, it was indicated that their functions were to meter fluid at appropriate pressures to those circuits of the hydraulic system that require signals of engine load and vehicle speed. These circuits do in fact contain the shift valves, which automatically bring about a change in gear ratio when acted upon by the control forces signalled to them. The shift valves may therefore be regarded as the central nervous system of the hydraulic control valve unit in which they are housed, and which some transmission engineers aptly refer to as the brainbox. These valves, like others in the control valve unit, are precision machined to slide in the close-tolerance bores of their housing. They are also provided with sharp edges to prevent sticking; as an American engineer once humorously remarked, 'The valve can thus act as its own street cleaner.'

Unlike the previously described throttle and kick-down valves and governor valve, the shift valves act primarily as relay rather than balancing valves. This means that their action is analogous to an electrical switch that is either off or on with no intermediate function; similarly, a shift valve is either closed or open in response to the control signals it receives. Each valve therefore has only two operative positions; thus two valves are required to change the gear ratios of a three-speed gearset, these being known as the 1–2 and 2–3 shift valves. An important requirement of shift valve operation is that there must be a sufficient lag, or *hysteresis* as it is technically known, between upshifts and downshifts. In practice, this means that there must be a difference between the speed of the vehicle at which the upshift takes place at a given throttle opening and that at which the downshift occurs at the same throttle opening. If this were not so, the hydraulic control system would develop an anxiety complex and impose an annoying recurrent gear shift, first up and then down, which is known as shift hunting to the transmission engineer.

A shift valve assembly comprises a spool valve with several equal diameter lands, a separate governor plug with larger and unequal diameter lands, and a compression valve spring. Take for example the operation of a 1–2 shift valve. The valve is always being urged to its closed or downshifted position by a combination of spring load and variable throttle pressure, which act on one of its ends. Its natural preference to remain closed is, however, opposed by variable governor pressure directed towards the other end of the valve and acting on it via the movable governor plug (Figure 17.18a). At a given vehicle speed the governor pressure acting against the governor plug builds up to the point where the force transmitted to the shift valve is greater than the opposing one from the spring load and throttle pressure. As this happens the governor pressure will begin to push the shift valve to this open or upshifted position, which not only directs line pressure to

actuate the appropriate friction brake element of the gearset and thus effect a change in gear ratio, but also abruptly cuts off throttle pressure so that completion of valve opening is then opposed only by spring load (Figure 17.18b). This latter feature provides the desired snap action of the shift valve to prevent hunting.

The governor plug for the 1–2 shift valve can also serve another purpose, which is to prevent a 1–2 upshift when the manual valve is positioned in the low range. In this case line pressure is directed between the lands of the governor plug, where it reacts against the larger land to overcome the effect of governor pressure on the other side of the plug. By holding the plug in its bottomed position the fluid at line pressure can then be transferred to the appropriate friction brake element of the gearset (Figure 17.18c). Since line pressure is also directed at both ends of the shift valve, the net effect is that

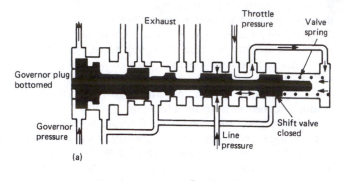
(b)

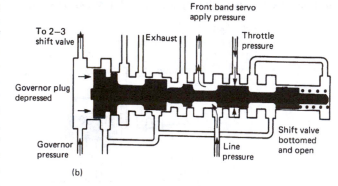
(a)

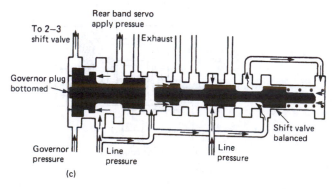
(c)

Figure 17.18 Schematic arrangement and operation of a shift valve: (a) 1–2 shift valve closed in low gear (b) 1–2 shift valve open in intermediate gear (c) 1–2 shift valve balanced in manual low range

the valve is positioned by spring load such that throttle pressure is prevented from acting on it.

As might be expected, the 2–3 shift valve operates in the same manner as the 1–2 shift valve. However, the governor plug is of smaller diameter and the valve spring load is greater, so that a higher governor pressure and hence vehicle speed is necessary to open the 2–3 shift valve according to performance requirements (Figure 17.19).

Accumulators

In the early development of automatic transmissions a distinction was drawn by American engineers between those hydraulic controls that caused a shift to occur and those which controlled the smoothness of its occurrence. For this reason the hydraulic controls are sometimes classified as being either causative or smoothing in their functions.

Clearly a shift valve is a causative hydraulic control, because it brings about a ratio change when acted upon by control forces. As previously inferred, it must exhibit a no-nonsense

snap action in order to avoid the transmission hunting between gear shifts. Paradoxically it is this same action of the shift valve that can provoke too rapid a pressure rise in the fluid directed to the band servos or clutch units. To prevent their harsh application and therefore improve the quality of gear shift, it becomes necessary to include smoothing controls in the apply circuits.

A very simple form of smoothing control is the restrictor orifice, which we have already encountered elsewhere in the hydraulic control system as a damping orifice. The purpose of the restrictor orifice is, of course, to delay the pressure build-up in the band servos or clutch units. Once fluid flow through the restrictor orifice ceases, the pressures either side of it become equalized, so that full apply pressure is then available at the band servo or clutch unit.

In practice it has usually been found necessary to use a restrictor orifice in conjunction with a hydraulic accumulator (Figure 17.20), especially in the case of clutch apply circuits, because the band servo itself provides a cushioning effect. The purpose of the accumulator is not to restrict the flow of fluid, but temporarily to absorb a portion of its pressure energy. This it does by exposing the fluid delivery in the apply circuit to a spring-returned piston acting in a blind-ended cylinder. It therefore delays the build-up of full apply pressure in the band servo or clutch unit until just before the piston has completed its stroke, and ensures their smooth application. Since all gear shifts do not occur at the same engine torque, the piston may be provided with two lands of different diameter and work in a stepped cylinder, so that the spring load can be modified by throttle pressure acting on the difference in piston areas. This means that there can be a greater accumulator or smoothing effect during light throttle shifts and less effect for a more rapid response to heavy throttle shifts.

Other refinements

In order to provide a simplified approach to the study of the hydraulic control system for an automatic transmission, our attention has been confined only to those circuits and their associated valves that are necessary for its basic operation.

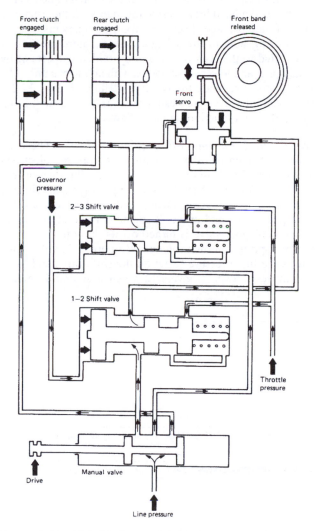

Figure 17.19 Simplified control circuit diagram showing both 1–2 and 2–3 valves upshifted to provide direct drive for a three-speed Simpson gearset

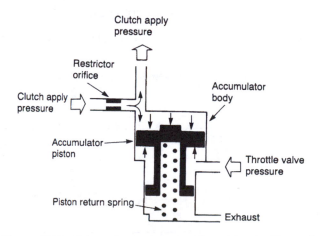

Figure 17.20 Schematic arrangement of a restricted orifice and hydraulic accumulator smoothing control

During the long development history of the automatic transmission many refinements have been introduced into the hydraulic control system for the gearset. Further information on these additional circuits, which vary according to design requirements, is most conveniently obtained from the service manual for the particular type of automatic transmission that is being dealt with in service.

17.5 ELECTROHYDRAULIC CONTROL SYSTEMS

General background

The benefits of using electrohydraulic, rather than purely hydraulic, control systems were first recognized in the late 1960s by American engineers and applied to heavy-duty automatic transmissions for on- and off-highway vehicles. However, it was not until 1978 that such a control system was adopted for a passenger car automatic transmission when Renault introduced it in a three-speed unit, followed in 1982 by Toyota with a four-speed application. The electrohydraulic control system is now generally referred to simply as electronic transmission control.

Advantages of electronic transmission control

The advantages to be expected from this apparent complication, but in reality simplification, of the automatic transmission control system may be summarized as follows:

1 A greater precision of control over gear shift points and torque converter lock-up is possible than with a conventional hydraulic control system, which is beneficial in terms of improving fuel economy, more closely relating engine performance to vehicle speed, better exhaust emission control and reduced noise levels.
2 Preprogramming of the electronic control system can provide the transmission with two shifting and lock-up modes, so that the driver can switch to either a fuel-efficient or a high-performance operating characteristic, usually designated as 'economy' and 'power' or their equivalents.

3 Close communication becomes possible between the electronically controlled engine and transmission management systems, thereby reducing any shock formerly experienced during gear shifting and torque converter lock-up.
4 There is no longer a requirement for a hydraulic governor.
5 The complexity of the hydraulic valve body is reduced.
6 A self-diagnostic facility can be built into the electronic control system, so that it can monitor its own state of health.

Arrangement and functioning of an electronic transmission control system

Taking the electronic control of a Toyota four-speed automatic transmission as an acclaimed example of modern practice then, apart from control by solenoid valves, the hydraulic portion of the control system is arranged similar to that of a non-electronic version and may be simply represented as a block diagram (Figure 17.21). The hydraulic system therefore comprises a hydraulic pump, the control valve body, the solenoid valves, the accumulators, the stationary and rotating clutches and a band brake servo, together with the usual network of fluid passages to connect these components.

Referring now to the solenoid valves, there are four of these fitted to the control valve body, which consists of the usual upper and lower valve bodies with an intervening valve body plate to regulate the passage of fluid (Figure 17.22). The solenoid valves are designated nos 1, 2, 3 and 4, and their individual functions are as follows:

Nos 1 and 2 solenoid valves (Figure 17.23) These are turned either on or off by signals sent from the electronic control unit (ECU) to their solenoid coils, and they operate the 1–2, 2–3 and 3–4 shift valves to perform gear shifting. The line pressure is acting on no. 1 solenoid valve in any forward gear range and on no. 2 solenoid valve in all ranges. When a solenoid valve is turned off its drain port is closed by the spring-returned plunger (Figure 17.24a), which causes the line pressure to act on the shift valve and therefore moves it over. Conversely, when a solenoid valve is

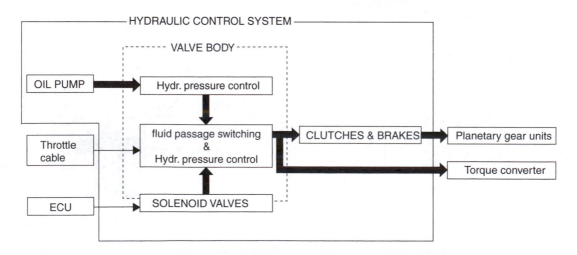

Figure 17.21 Electrohydraulic control system for an automatic transmission (*Toyota*)

turned on the plunger is pulled inwards and opens the drain port to relieve the line pressure (Figure 17.24b), so that the shift valve moves back again under the influence of its own return spring.

No. 3 solenoid valve (Figure 17.25) This is also operated by signals received from the ECU, and its purpose is to engage or disengage the lock-up clutch of the torque converter. It is supplied with a stabilized line or solenoid pressure and modulates this into lock-up control pressure. This is accomplished by means of the magnetic force generated in the solenoid control, which acts upon a spool that moves the control valve

in opposition to a compression spring. The latter provides a force that balances the magnetic force, which is varied in accordance with the amount of current supplied to the solenoid coil from the ECU.

No. 4 solenoid valve (Figure 17.26) Likewise operated from the ECU, this valve serves to control the back pressure in the accumulators for smooth gear shifting. It is constructed and

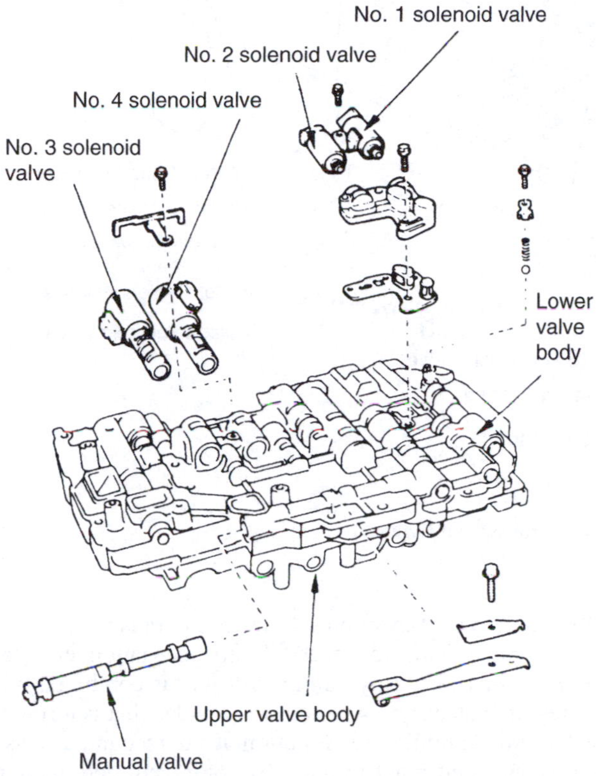

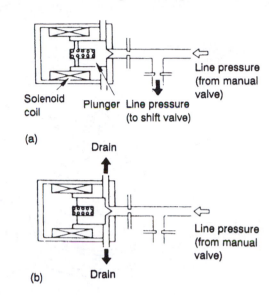

Figure 17.24 Nos 1 and 2 solenoid valves in off and on positions (*Toyota*)

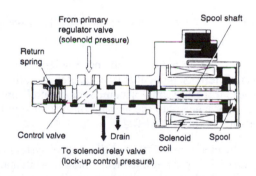

Figure 17.25 No. 3 solenoid valve for torque converter lock-up control pressure modulation (*Toyota*)

Figure 17.22 Valve body for electronic transmission control (*Toyota*)

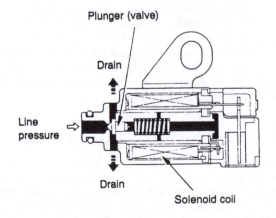

Figure 17.23 Nos 1 and 2 solenoid valves for shift control (*Toyota*)

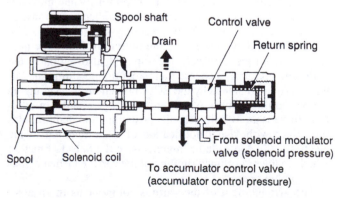

Figure 17.26 No. 4 solenoid valve for accumulator back pressure modulation (*Toyota*)

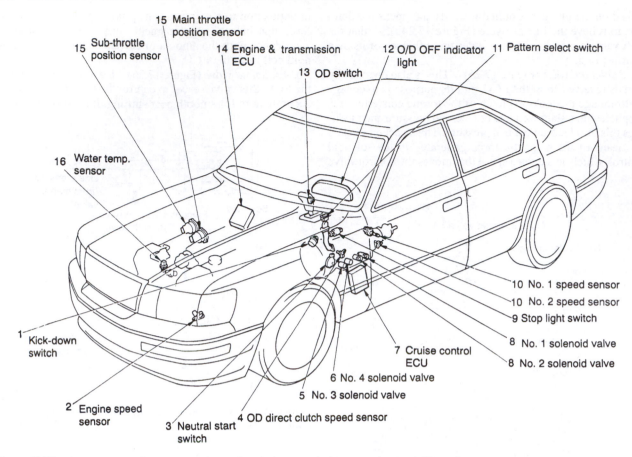

Figure 17.27   Arrangement of components in an electronic transmission control system (*Toyota*)

operates in a similar manner to the no. 3 solenoid valve, the functional difference being that it modulates a stabilized line or solenoid pressure into accumulator control pressure, again in accordance with the amount of current supplied to the solenoid coil from the ECU.

So far as the electronic portion of the control system is concerned, the ECU is programmed with optimum shift and lock-up patterns and selects the appropriate one in accordance with signals received from the 'pattern select' and 'neutral start' switches. The former allows the driver to choose an operating mode in terms of either 'normal' or 'power', while the latter detects the position of the gear shift lever. Then in response to further signals received from various sensors (Figure 17.27), the ECU not only determines the optimum timing for gear shifts and torque converter lock-up, but also regulates the hydraulic pressure to ensure their smooth operation. The latter is further enhanced by the ECU momentarily delaying the engine ignition timing when initiating a gear shift. In addition, there is an in-built self-diagnosis function that provides the driver with visual notification of a malfunction in the electronically controlled components, and a failsafe function which enables continued running of the car even when a malfunction has occurred.

The arrangement of the various components in an electronic transmission control system is shown in Figure 17.27, and their functions are listed in Table 17.1.

In a more recent example of electronic transmission control, as applied to the advanced ZF six-speed automatic gearbox (Section 16.2), a hydraulic shift unit is combined with an integrated electronics system to provide what is termed a Mechatronic module. Its function is to recognize sensor input signals and react to them by conferring optimal shift timing according to the prevailing driving conditions. It must therefore monitor drag resistance as imposed by car weight and road gradient, and interpret driving styles that may place emphasis on either performance or economy. Further sophistication of the electronic control system is provided by a Stand-by-Control (SBC) feature. This controls the input clutch of the transmission and allows the torque converter to be disconnected from the geartrain when the car is stationary, whether or not it is in gear. By virtue of minimizing internal drag in the transmission, a useful contribution is made to lower fuel consumption.

17.6   AUTOMATIC TRANSMISSION FLUID

The main requirements of a suitable fluid for a torque converter automatic transmission are generally that it must act as a power transmission medium; a hydraulic control fluid; a lubricant for the epicyclic gears, wet friction elements and bearings; and a heat transfer medium. To perform these diverse functions an automatic transmission fluid contains

Table 17.1 Functions of the components of an electronic transmission control system (*Toyota*)

Component	Function
Kick-down switch	Detects the accelerator pedal position depressed beyond full throttle valve opening
Engine speed sensor	Detects the engine speed
Neutral start switch	Detects the shift lever position
OD direct clutch speed sensor	Detects OD input shaft speeds from first to third gears
No. 3 solenoid valve	Controls the hydraulic pressure applied to the lock-up clutch and controls lock-up timing
No. 4 solenoid valve	Controls hydraulic pressure acting on accumulator back chamber during gear shifting to smoothly engage clutches and brakes
Cruise control ECU	Prohibits vehicle running in OD gear and lock-up control when vehicle speed drops lower than a predetermined level of the auto drive set speed
No. 1 and no. 2 solenoid valves	Control the hydraulic pressure applied to each shift valve, and control gear shift position and timing
Stop light switch	Detects the brake pedal depression
No. 1 and no. 2 speed sensors	Detect the vehicle speed; ordinarily, ECT control uses signals from the no. 2 speed sensor, and the no. 1 speed sensor is used as a back-up
Pattern select switch	Selects the shift and lock-up timings by the power mode or the normal mode
OD OFF indicator light	Blinks and warns the driver, while the OD main switch is pushed in, when the electronic control circuit is malfunctioning
OD switch	Prevents up-shift to the OD gear if the OD switch is off
Engine and transmission ECU	Controls the engine and transmission actuators based on signals from each sensor
Throttle position sensor	Detects the throttle valve opening angle
Water temperature sensor	Detects the engine coolant temperature

a comprehensive range of additives and is subject to a very closely controlled specification.

When General Motors in America introduced the first fully automatic transmission in 1939 no special fluid was then available, so there was little choice but to use a straight mineral oil. Unfortunately, this often deteriorated because of the high operating temperatures and led to oxidation and the formation of lacquer deposits. Since oxidation caused the oil to thicken, the automatic action of the gearbox could become sluggish, while the presence of lacquer on the control valves caused them to stick and impose unsatisfactory gear changes. So clearly an automatic transmission fluid requires some of the same types of additive that are used in modern engine oils in order to avoid these problems. It has also long been recognized that if an automatic transmission fluid becomes too viscous under low-temperature conditions, its reduced ability to flow can be responsible for less effective action of the torque converter, reduced pumping efficiency and pressure loss in the hydraulic control system. The corresponding pressure losses in the band servo and clutch units can induce slippage of their friction elements, and subsequent transmission failure through burned clutch plates was once not unknown. Conversely, the fluid must possess a sufficiently high viscosity during high-temperature conditions in order to avoid excessive internal leakage, which arises from increased clearances in the hydraulic units and could cause delayed gear changes or none at all. It is for these reasons that, in terms of viscosity, modern automatic transmission fluids have been likened to SAE 20 grade oils with exceptionally good low-temperature properties.

It was not until 1949 that it became feasible for General Motors to issue a specification for a fluid that could satisfy the diverse requirements of an automatic transmission, and this was the original automatic transmission fluid (ATF) type A. Since then, of course, further specifications have been issued by General Motors and also by Ford, which are known

by their designations 'DEXRON' and 'MERCON' respectively. These specifications recognized the trend towards increased severity of operation and the differing properties of the friction elements used in these manufacturers' transmissions. However, there is now a greater accord in respect of the latter requirement and this has led to the widespread acceptance of the General Motors Dexron-11 specification for automatic transmissions of modern design. More recently DEXRON II has been superseded by DEXRON III (GM 6297-M), which especially takes into account the increasingly stringent requirements of compactly designed and more highly stressed automatic transaxles, associated with transversely mounted engines and front-wheel drive. Similarly the Ford requirements are met by MERCON (Ford M2C185A). It is still nevertheless important to refer to the car manufacturer's approved list of lubricants.

17.7 CHECKING THE LEVEL AND CHANGING THE FLUID

Checking the fluid level

Although physically checking the fluid level of an automatic transmission by means of its dipstick is a simple enough task in itself, it is most important that certain procedures be adopted that recognize the design differences between the various types of unit. The particular manufacturer's service instructions must therefore always be consulted.

In general terms, car attitude and fluid temperature are two important factors. Fairly obviously the fluid level can be checked only when the vehicle is on an even and level surface. The fluid must also have reached its normal operating temperature of about 80°C, since the correct fluid level at lower operating temperatures can be below the 'add' (or equivalent) mark on the dipstick and the correct fluid level at higher operating temperatures can rise above the 'full' (or equivalent) mark. At normal operating temperature, the fluid level should register between these marks on the dipstick.

Before the fluid level is checked, a typical procedure calls for the handbrake to be applied and then with the engine idling the gear selector lever is moved through all ranges. This is to ensure that all fluid passages in the hydraulic control units are filled to operating capacity, otherwise the level shown on the dipstick could be artificially high. Finally, with the gear selector lever returned to the park position and the engine still running to ensure that the torque converter is full, the dipstick is removed from its tube, wiped clean and reinserted to its full depth, and then withdrawn again to observe the fluid level indication (Figure 17.28).

Changing the fluid

When the fluid of an automatic transmission is changed, reference must again be made to the manufacturer's lubrication instructions to establish the approximate refill capacity, the type of automatic transmission fluid required, and whether any special points of procedure need to be followed. As with draining the oil of a manual gearbox, provision must be made for safe access to the automatic transmission from below, especially as the fluid may be very hot. The procedure for refilling involves pouring an initial fill of fresh fluid through the filler tube, the actual amount being in accordance with the lubrication instructions. A check on the fluid level is then carried out as described earlier. Additional fluid is next poured in to bring the level on the dipstick either up to, or slightly above, the 'add' mark. After bringing the unit up to its normal operating temperature, a final check on the fluid level is made; further fluid is added if necessary to correct the dipstick indication. The unit should also be examined for any signs of fluid leakage.

It should be realized that an underfilled automatic transmission can at least suffer from delayed engagement of 'drive' or 'reverse' and slipping changes, and at worst from complete loss of drive. These conditions basically arise from the transmission fluid pump taking in air along with the fluid, so that the otherwise incompressible fluid becomes spongy and affects the pressure build-up in the hydraulic control system. Since this supplies the servo units that actuate the friction elements for engaging the epicyclic gear trains, a slipping action occurs. This in turn leads to overheating and rapid wearing of the friction elements, the overheating being worsened by the torque converter also not receiving an adequate supply of fluid.

With an overfilled unit, the end result can be similar although the basic reason is not quite the same. Here the pump still takes in air along with the fluid, but does so because the excess amount of fluid has been churned and aerated by the rotating gear trains. The resulting foaming can therefore produce sluggish and slipping changes and therefore overheating. Also, the combination of foam and overheating will accelerate oxidation of the fluid and lead to varnish formation and sticking of the control valves. Furthermore, the foam can expand such that fluid loss occurs through the air vent of the automatic transmission unit.

17.8   AUTOMATIC LAYSHAFT GEARBOXES

Since the early history of the layshaft gearbox, there have been occasional attempts by designers to eliminate the sliding dog clutch and its potentially noisy engagement with a free-running constant-mesh gear (Section 14.1), and to use instead some form of friction brake or clutch that can lock the gear to its shaft in a more considerate manner. An eventual solution to this problem was found in adopting hydraulic actuation of small multiplate clutches to engage the constant-mesh gears (Figure 17.29). The detail design of these clutches is based on long-established practice for conventional automatic transmissions, where multiplate clutches are used extensively in conjunction with epicyclic gearsets as described earlier. Hence, the advantage of employing hydraulic actuation for the multiplate clutch engagement of constant-mesh gears is not

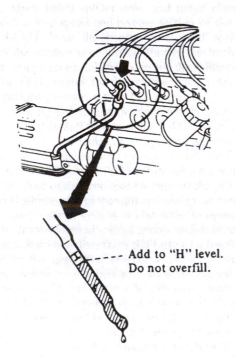

Figure 17.28   Checking automatic transmission fluid level (*Nissan*)

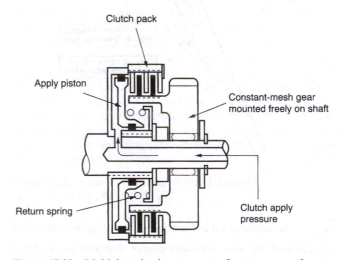

Figure 17.29   Multiplate clutch engagement for constant-mesh gear

only to confer a foolproof means of operation, but also to provide the opportunity for super-imposing a hydraulic control system in the manner of a conventional automatic transmission. A fully-automatic transmission can therefore be produced by combining a torque converter with a hydraulically controlled layshaft gearbox as pioneered by the Honda company in the early 1980s.

17.9 DUAL MODE TRANSMISSIONS WITH SEQUENTIAL GEARCHANGE

General background

When the driver of a Grand Prix car is braking at a seemingly impossible rate of deceleration to take a corner, he is simultaneously making lightening downchanges through the gears to select one that will give maximum acceleration away from the corner. In the modern Grand Prix car these gearchanges are no longer made by the driver using a conventional gear lever and clutch pedal, but by operating what is known as a 'sequential' gearchange system. Although such a system retains a layshaft gearbox with its associated friction clutch, it differs in that the movements of the gear selector forks and rods are controlled by double-acting hydraulic actuators and the clutch withdrawal mechanism by a spring-returned single acting actuator. When changing gear fluid under pressure is directed towards the appropriate ends of the selector rod plungers in these actuators,

via an electronically controlled electrohydraulic valve unit that receives its signals from a driver-operated rocker or 'paddle' lever, which is mounted beneath and straddles the steering wheel (Figure 17.30).

In effect the driver is provided with a 'shift-by-wire' gearchange system, which merely requires fingertip movement of the paddle lever towards the steering wheel, the right-hand side of the paddle being used for successive upchanges and the left-hand side similarly for downchanges. Perhaps understandably, sequential gearchange systems like this are now finding application on high-performance passenger cars, albeit with additional refinements. An acclaimed development in this field has been the Selespeed system adopted by the Alfa Romeo company, which confers dual mode semi- and fully automatic operation.

Selespeed sequential gearchange system

For this system a conventional manual transmission, which combines a synchromesh gearbox with a single-plate friction clutch, is integrated with an electronically-controlled electrohydraulic unit that manages the gear shifts and clutch action (Figure 17.31). It should be noted that the use of this system does not entail any internal alterations to the already proven Alfa Romeo gearbox and clutch assemblies, only the clutch bell-housing requires structural modification to accommodate the hydraulic actuators controlling gear selection

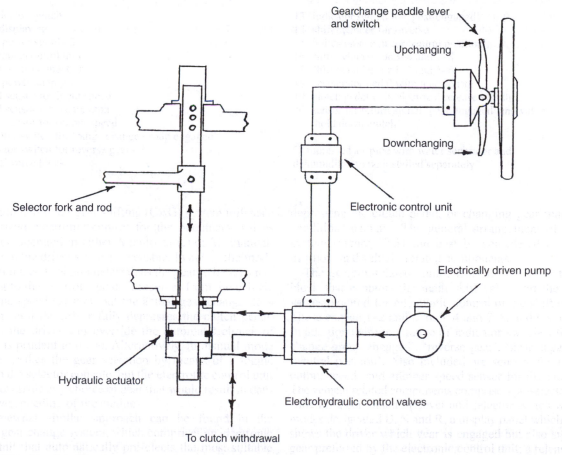

Figure 17.30 Simplified layout of a sequential gearchange system

and clutch action. No clutch pedal is necessary with this system, the accelerator alone being used to move away.

The transmission offers two distinct operating modes:

1 Semi-automatic – where the driver manually requests gearchanges by operating a lever on the central console, buttons on the steering wheel spokes (Figure 17.32), or in a more recent version a Grand Prix style paddle lever beneath the steering wheel.
2 Automatic that is designated 'CITY' – where the system itself decides when best to make a gearchange.

The semi-automatic mode is set automatically whenever the car is started; when the driver operates the gear lever or steering wheel buttons while in CITY mode; and when the driver presses the CITY button again while in CITY mode. During operation the system will accept an upward or a downward shift request with the car in motion and clutch fully engaged, only if it is compatible with the prevailing engine rpm range. Normally a command via the gear lever or a steering wheel button with a car speed more than 6 mile/h (10 km/h) will trigger only one upward or downward gearchange.

In automatic CITY mode the speed is selected on the basis of a single characteristics map pre-programmed into the memory of the electronic control unit. This receives input signals from sensors that recognize accelerator pedal position, car speed, optimal gear (based on accelerator pedal position), and brake pedal position. The system does not enforce an upshift if the accelerator pedal is released, thereby retaining an engine braking effect. Gearchanges occur in exactly the same way as in semi-automatic operating mode, because the transmission and engine control parameters remain unchanged.

Electrohydraulic gear selection

At the heart of the Selespeed system is the electrohydraulic gear selection unit (Figure 17.33), which receives its operating signals from the electronic control unit fitted beneath the

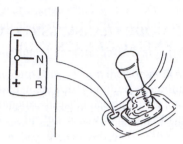

Console speed selection lever

Steering wheel speed selection buttons

Figure 17.32 Dual mode driver controls of Selespeed system (*Alfa Romeo*)

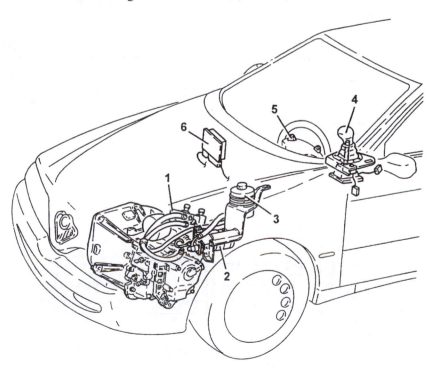

Figure 17.31 General layout of Selespeed dual mode gearchange system (*Alfa Romeo*)

1 hydraulic unit   3 fluid reservoir   5 speed selection buttons
2 pump   4 gear selector lever   6 electronic control unit

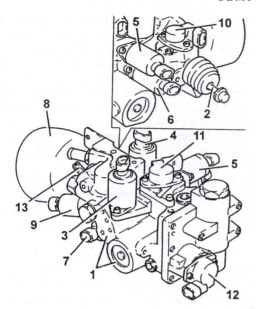

Figure 17.33 Components of the Selespeed electrohydraulic unit (*Alfa Romeo*)

1 cast body fastened directly to the gearbox
2 single acting actuator for operating the clutch lever
3 proportional pressure valve EV1 controlling the engagement actuator
4 proportional pressure valve EV2 controlling the engagement actuator
5 on-off solenoid EV3
6 on-off solenoid EV4
7 proportional flow solenoid EVO controlling the clutch actuator
8 hydraulic gas accumulator
9 pressure sensor
10 sensor for detecting clutch actuator position
11 sensor for detecting engagement actuator position
12 sensor for detecting selection actuator position
13 bleed valve

facia. The hydraulic power for the electrohydraulic unit is supplied by a gear pump driven from a commutator-type electric motor. Three basic functions are performed by the electrohydraulic unit, these being to operate the clutch control lever and the gear control shaft, and to maintain a reserve of energy/hydraulic power for efficient actuator operation. The manner in which the components of the electrohydraulic units are arranged is identified in Figure 17.33 and their functions shown in Figure 17.34.

Advantages of the Selespeed system

The advantages claimed for this type of system may be summarized as follows:

1 The driver no longer needs to operate a conventional clutch pedal and gear lever, but can still enjoy all the driving satisfaction that derives from direct control of the transmission system.
2 Driving safety is enhanced because the electronic control system protects against driver error and improper use of the transmission system.

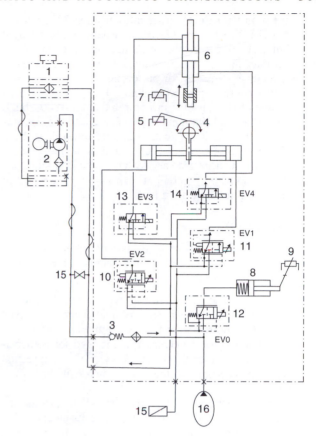

Figure 17.34 Function diagram of Selespeed electrohydraulic gear selection (*Alfa Romeo*)

1 tank
2 electric pump
3 check valve
4 engagement/release actuator (with three mechanically-defined positions)
5 engagement actuator position potentiometer
6 selection actuator (with two fixed positions, the third selection position is defined by the inner gearbox springs)
7 selection actuator position potentiometer
8 clutch actuator (single acting)
9 clutch actuator position potentiometer
10 proportional pressure solenoid EV2 controlling the engagement actuator
11 proportional pressure solenoid EV1 controlling the engagement actuator
12 proportional flow solenoid EVO controlling clutch actuator
13 on-off solenoid EV3 controlling selection actuator
14 on-off solenoid EV4 controlling selection actuator
15 pressure sensor
16 hydraulic gas accumulator
17 bleed valve

3 Improved performance with conventional manual transmission components, including reduced clutch wear.

17.10 DIRECT SHIFT GEARBOX

General background

For the enthusiastic driver who demands a responsive car with competitive acceleration times, the Volkswagen Group in conjunction with BorgWarner introduced in 2003 what is termed their Direct Shift Gearbox, usually abbreviated to 'DSG'. It is claimed to be able to change gear faster and more smoothly than conventional manual and automatic gearboxes, without incurring any loss of performance or fuel economy since it does not require a torque converter. This type of gearbox was first used in an Audi rally car in the mid-1980s, but it had to await the development of reliable electronic controls before its more recent application to production models. In principle the gearbox may be classified as a 'dual-clutch' transmission, which in this case features a divided input primary shaft flanked by two output mainshafts that connect to the final-drive gear and differential.

Basic layout of Direct Shift Gearbox

Before considering its method of operation it is first necessary to identify the interrelated components of the gearbox, these being as follows:

1   Unlike a conventional single-stage manual gearbox (Section 14.2), the primary shaft is divided into two coaxial inputs, each controlled by its own multiplate wet clutch. These clutches are concentrically arranged for compactness, so that the outer clutch connects with the inner section of the primary shaft and the inner clutch with the hollow outer section.

2   The inner section of the primary shaft carries the first/reverse, third and fifth gears, while its outer section carries the second, fourth and sixth gears. All these gears are integral with their respective sections of the primary shaft. Both sections are also provided with pulse wheels, so that their rotational speeds can be signalled by sensors to the mechatronic module mentioned later.

3   Again unlike a conventional single-stage gearbox, two relatively shorter output mainshafts are employed lying either side of the divided primary shaft. The rotationally-free mainshaft gears that mesh with the divided primary shaft gears become active when engaged by their synchronizers in the usual manner. Instead of a mechanical linkage being used to move the selector rods and forks, these are controlled by hydraulic actuation. Separate reverse gears and shaft take their drive from the first/reverse gear on the inner section of the primary shaft. Each mainshaft is additionally provided with an end final-drive pinion, these pinions remaining in permanent mesh with the larger final-drive gear and differential assembly.

4   The hydraulically energized multiplate wet clutches are similar in construction to those found in automatic transmissions (Section 16.3). Both concentric clutches receive their drive from the rotating drum that encases them, which in turn is connected to the engine flywheel through a splined input hub. Either clutch is applied when hydraulic pressure is directed, via a channelled sleeve at the entrance to the drum, against an annular plunger that forces their friction plates together. When hydraulic pressure is released the plunger for the outer clutch is retracted by a diaphragm spring whilst the plunger for the inner clutch is retracted by a series of coil springs.

5   As a consequence of using hydraulic controls for gear selection and wet clutch application, which therefore require a pressure source, an internal gear and crescent type of oil pump (Section 17.4) is driven from a spindle that passes freely through the divided primary shaft and connects to the input hub for the clutches' drum. This pump also has to perform the additional duties of supplying a flow of cooling oil to dissipate frictional heat generated by the clutches, and to lubricate the geartrain. It is not only provided with a pressure control valve but also a cooling control valve, the output from the latter being varied in accordance with the temperature of the oil leaving the clutches.

6   Finally a comprehensive control system is required for the DSG and is styled a 'mechatronic module'. Basically, this combines an internal electrohydraulic control unit with shift valves and modulators to actuate the gear selector mechanism and to apply and release the clutches, and an electronic control unit that processes signals from a range of sensors and also liaises with the vehicle and engine electronic systems. Since it confers an overall control strategy for the DSG, the mechatronic module can be regarded as its nerve centre. In this respect it bears comparison with the electrohydraulic control systems employed for conventional automatic transmissions (Section 17.5)

Operating principle of the Direct Shift Gearbox

In essence the DSG may be considered as two single-stage gearboxes operating in parallel. This arrangement then provides the opportunity for two pairs of gears to be selected, but only one pair actively transmitting drive according to which clutch is applied. The other pair has been preselected by the electronic control system in anticipation of the next gear shift. Changing to this pair of gears can then be rapidly accomplished simply by releasing the applied clutch and applying the other clutch, thereby activating the preselected pair of gears. A simplified representation of this sequence is shown in Figure 17.35. Unlike a conventional manual gearbox and single clutch, there is virtually no interruption of power flow as the gearchanges take place.

The electronic control system for the DSG provides for three modes of operation, which are selected by a floor-mounted gear shift lever and under steering wheel paddle levers. Mode 'D' for drive allows fully automatic gear shifting for relaxed driving. Mode 'S' for sport modifies the gear shift points by delaying them for upshifts and advancing them for downshifts to give maximum acceleration. Sequential gear shifts are made by moving the gear lever across from the automatic to the manual side of the gate and then forwards for upshifts and backwards for downshifts. Alternatively the paddle levers behind the steering wheel may be used, the right-hand one for upshifts and the left-hand one for downshifts.

17.11 CONTINUOUSLY AND INFINITELY VARIABLE TRANSMISSIONS

General background

The notion of providing a continuously variable transmission system for motor vehicles is far from being a new one. Robert Philips, who was an early authority on vehicle transmissions,

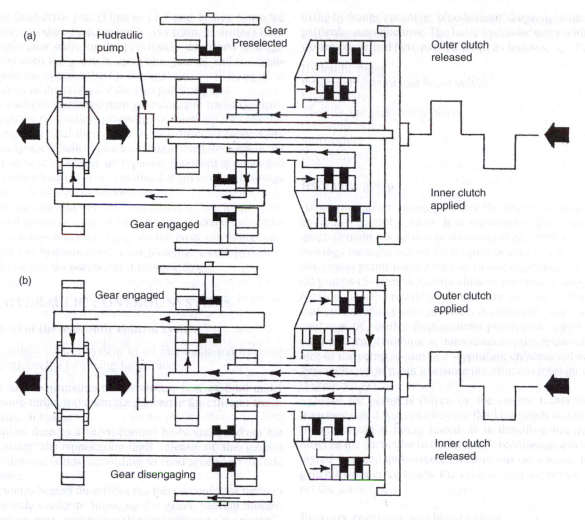

Figure 17.35 Basic operating sequence of a dual-clutch transmission. For simplicity only two gearsets are shown: (a) right-hand gearset engaged and transmitting drive, inner clutch applied; left-hand gearset preselected for next gearchange, outer clutch released (b) left-hand gearset now engaged and transmitting drive, outer clutch applied; right-hand gearset now disengaging, inner clutch released

wrote in 1916 'nothing is more fascinating than the possibility of effecting a smooth and gradual change of speed over the whole desired range, whereby the prime mover can always be kept running at its most effective speed'. It is also interesting that, of the different variable speed mechanisms he listed at the time, the expanding pulleys and belt friction drive system was mentioned first, because it is the modern application of this principle that has so far shown most promise. Another simple form of variable speed friction drive system mentioned by Robert Philips was that employing a flat disc and slidable roller. Although in this early period of automobile development it proved to have more immediate application than the expanding pulleys and belt system, it also had severe limitations in practice. The operating principle of the flat disc and roller friction drive nevertheless offers a convenient introduction to the revived development of the toroidal discs and rollers drive, as explained later.

The expanding pulleys and belt type of variable speed transmission is believed to have first been used around the turn of the last century in the now long-forgotten Swiss Weber car, albeit with manual control of the axially expanding pulleys. Owing to belt wear problems little further progress was made with this type of transmission until 1955, when the Van Doorne brothers introduced their twin-belt Variomatic system for the Dutch DAF car. While the basic principles were the same as the Weber transmission of over fifty years earlier, the significant difference was that the expanding pulleys were automatically operated. It is this feature, together with the replacement of the rubber drive belt by a unique segmented metal one that pushes rather than pulls and other refinements to the control system, which constitute the modern continuously variable transmission or CVT as it is usually abbreviated.

Since early 1987 continuously variable transmissions incorporating these features, which have been developed in conjunction with Van Doorne, have been introduced by the Ford, Fiat, Subaru and ZF companies. To transmit the greater torque of more powerful engines the Audi company in 2000 introduced their Multitronic CVT, wherein the segmental construction for the push-belt is replaced by a more robust multi-strand link plate chain.

The flat disc and roller type of variable speed transmission was used on a number of early light cars, notably the GWK in England. As perhaps can be imagined, the practical difficulties encountered with this early attempt at a variable speed transmission were those of minimizing frictional wear between the contact surfaces of the roller and disc, and avoiding the jerkiness that could occur when easing the pressure of the roller against the disc to change its position. Although simple systems of this type have long been obsolete for motor vehicles, the basic principle of loading a driving roller into slidable contact with an engine driven disc, so that its path of contact could be altered to different radii, remained an attractive idea to transmission designers. In future developments they sought to overcome its disadvantages by using multiple rollers to reduce their individual loading. Then instead of retaining a flat disc they achieved a similar effect by sandwiching tiltable rollers between saucer-shaped or 'toroidal' discs preloaded together to resist slipping. Furthermore, the dry friction between the rolling elements is replaced by fluid friction, so that in fact the drive is transmitted by the shearing of a very thin film of oil that separates them.

This type of drive was originally conceived by W.D. Hoffman in the late 1890s, but it was not until the early 1920s that first Citroën and then a few years later General Motors developed it for automotive application. Although fully developed for production by the latter company, it was not proceeded with on the grounds of extra weight and cost. However, an American engineer, F.A. Hayes, who had been involved in the development of this concept, then persuaded the once long-established Austin company in England to produce such a transmission for their cars in the mid-1930s. Known as the Hayes 'Self-Selector' it was abandoned after being fitted to just a few hundred cars, owing to its greater cost and poor durability. In the early post World War II years interest in this form of transmission was revived in England by Forbes Perry. His company Perbury Engineering, introduced significant detail improvements to the operating geometry of the tiltable rollers and coupled the drive to an epicyclic gear that provided a reverse and a geared neutral. Since 1988 this form of toroidal drive has undergone extensive further development by Torotrak (Development) Ltd in England who has licensing agreements with a number of car manufacturers worldwide. Another approach to a toroidal drive system has been made by the Japanese transmission manufacturer Jatco. In this system half-toroidal cavities embrace the tiltable rollers, the operating loads being higher with this more compact geometry but with less likelihood of slippage. A hydraulic torque converter for starting purposes and an epicyclic geartrain to give forwards and reverse drives, are also features of this particular design.

Operating principles of variable-speed belt drives

Instead of using gears, rotary motion can be transmitted from one shaft pulley to another by an endless flexible connector or driving belt. It is the force of friction that compels the driving pulley to drive the belt and likewise the belt to rotate the driven pulley. This force of friction is derived from the contact pressure of the belt against the rims of the pulleys, which in turn is imposed by initial belt tension. The velocity ratio of a pulley drive is equal to the diameter of the driving pulley divided by the diameter of the driven pulley; in other words, the speeds of the shafts expressed in rev/min are inversely proportional to the diameters of their pulleys.

If we can now imagine a simple transmission system for a motor vehicle, in which a small-diameter driving pulley is attached to the engine crankshaft and is connected by a flat belt to a large-diameter driven pulley on the wheel axle shaft, then we shall have the equivalent of a low-ratio gear. (It will be recalled from Section 14.1 that in automotive terminology a low-ratio gear refers to one of high numerical ratio and vice versa.) Similarly, if we make the driving and driven pulleys equal in diameter then we get the equivalent of a direct drive, while if we go to the opposite extreme and make the driving pulley larger in diameter than the driven one, it will give a high-ratio gear equivalent to overdrive.

Taking matters a step further, let us replace our simple flat belt with a V-belt of trapezoidal cross-section, which connects with pulley grooves of the same shape and transmits drive by virtue of the friction created by the sides of the belt and the pulley flanges. We have now not only considerably increased the torque transmitting capability of the pulley drive system, but also introduced the possibility of using a stepless speed change or variable-speed drive (Figure 17.36). This is accomplished by using a divided pulley construction, so that the two halves of each pulley may be moved axially to change their effective pitch diameters. If one pulley is then provided with a positive axial movement across its width and the other is spring loaded towards its minimum width condition, the latter will automatically adjust both its diameter and axial position so that the belt remains under correct tension at all times. An additional advantage of the spring-loaded pulley is that it can act as a slipping clutch in the event of the drive being overloaded. In this context it must be recognized that, unlike in a gear train, there will be a reduction in the speed of the driven pulley and therefore a proportional loss of power transmitted if the belt slips or creeps. The former is caused by too low a friction force if either the tension of the belt is insufficient or

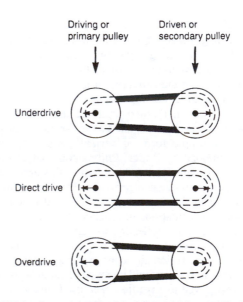

Figure 17.36 Operating principle of a variable-speed belt drive

its arc of contact is inadequate for the application, while the latter arises from the inherent elastic properties of the belt.

Taking the Ford CTX (continuously variable transaxle) as an example of advanced design (Figure 17.37), the transmission comprises the following basic units.

Epicyclic gearset and clutches

A single-stage epicyclic gearset controlled by two multiplate wet clutches is interposed between the engine and the driving or primary pulley of the transmission. Its purpose is to provide a forward and reverse facility (Figure 17.38). For forward drive the stationary clutch is released and the rotating one applied. This locks together the sun wheel connecting the primary pulley and the pinions of the input shaft planet carrier, thereby providing a direct drive from the engine to the belt and pulley system with the annulus gear spinning freely. For reverse drive the rotating clutch is released and the stationary one applied. This compels the pinions of the input shaft carrier to walk in the annulus gear locked to the housing and, since dual planet pinions are used, the direction of drive imparted to the sun gear and hence the primary pulley is reversed. In neutral both clutches are released, so that the planet pinions rotate around the stationary sun gear and the annulus gear spins freely. The multiplate wet clutches engage progressively to start transmitting drive to the wheels. They can also slip without wear or overheating to provide a slight degree of forward creep at idling speed, as an aid to low-speed manoeuvring or driving in traffic.

Belt and variable-pitch pulleys

At the heart of the modern CVT is, of course, the variable-speed belt and pulleys system, and not least the unique construction of its belt (Figure 17.39). Instead of transferring the drive between the pulleys by tension, as in a conventional rubber belt system, the elements of the steel belt work in compression and transmit a peripheral force by thrust. This 'pusher' operation of the belt allows a greater torque capacity for a given size of belt; it also contributes to quieter running and increased durability. The belt is made by threading 320 V-shaped steel elements on to two sets of composite steel bands. Each element is precision made from a special high-technology high-friction steel, and each composite steel band set is itself made up of 10 steel loops. In operation each element of the belt transmits thrust by pushing against its neighbour, while the steel bands hold the elements in place, guide the slack side of the belt and prevent kinking under load.

The variation in axial spacing of the primary or driving pulley halves is controlled by a hydraulic servo cylinder, which is integrated with the hub of the pulley. Changes in hydraulic pressure supplied to the servo cylinder therefore

Figure 17.37 Cut-away view showing the basic elements of the Ford CTX (*Ford*)

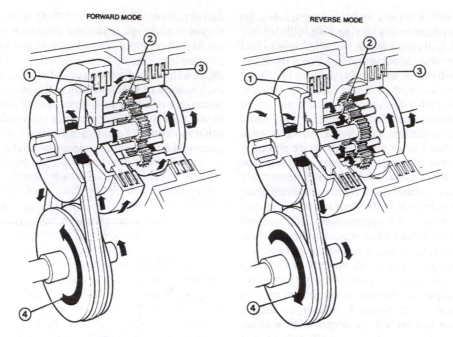

Figure 17.38   Power flow diagram for Ford CTX epicyclic gearset and clutches (*Ford*)

Forward mode

1 Forward clutch engaged; planetary gearset locked to sun wheel, providing drive straight through to primary belt pulley.
2 Planet gears do not rotate.
3 Reverse clutch disengaged.
4 Output pulley rotates in forward direction.

Reverse mode

1 Forward clutch disengaged.
2 Planet gears rotate.
3 Reverse clutch engaged.
4 Output pulley rotates in reverse direction.

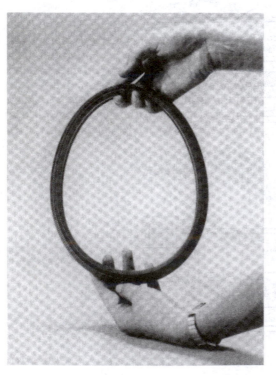

Figure 17.39   High-friction steel thrust belt used in Ford CTX (*Ford*)

determine the radial position of the belt on the primary pulley and, because of the fixed length of the belt, its position on the secondary or driven pulley. In addition, an axial clamping force is applied to the secondary pulley halves by means of a compression spring and a second, but independently acting, hydraulic servo cylinder that is similarly integrated with the hub of the pulley. The hydraulic pressure supplied to this servo cylinder is regulated in accordance with the prevailing drive ratio and load, so that it is not only sufficient to prevent slip in the belt and pulleys system, but also not so excessive as to impose unnecessarily high clamping loads that could waste power. To provide rapid response to changes in their controlling hydraulic pressure, the movable half of each pulley slides on ball bearing splines. The axial expansion and contraction of the primary and secondary pulleys is therefore such that, when the fixed-length belt is acting at a minimum radius on one pulley, it is acting at a maximum radius on the other, and vice versa (Figure 17.40). By this means a continuous variation of drive ratio can be obtained between 'low' and 'high' limits.

Hydraulic control system

This is arranged on generally similar lines to that of a conventional automatic transmission. Hydraulic pressure is supplied by an engine-driven pump to a control unit, which determines the pressure supplied to each of the pulley servo cylinders and also for the stationary and rotating clutches. Two flexible

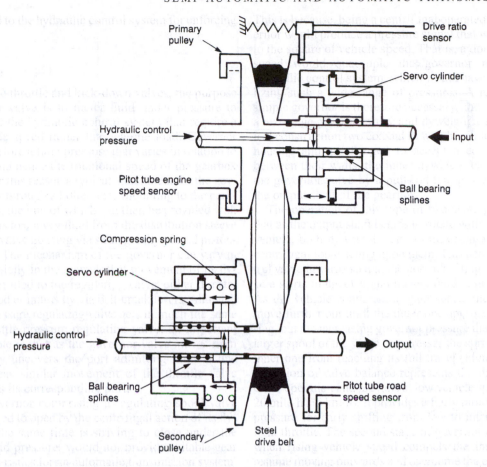

Figure 17.40 Schematic arrangement of belt and variable-pitch pulley transmission

cables connect the hydraulic control unit to the floor-mounted gear selector lever, which has the normal P R N D L positions, and to the accelerator for sensing engine load. Rather interestingly, the engine and road speeds are sensed by means of pitot tubes, which communicate with fluid replenished annular chambers that are made integral with the fixed halves of the primary and secondary pulleys (Figure 17.40). A pitot tube is more usually associated with the measurement of air speed (Section 8.3). The prevailing drive ratio is sensed mechanically by a rod that detects the axial separation of the primary pulley halves and hence the radial position of the belt (Figure 17.40). All these signals are processed by the hydraulic control unit to determine, against a programmed operating strategy, what fluid pressure should be supplied to which parts of the transmission running gear.

Transmission housing
Housed in an aluminium die-casting together with the final drive gears (Figure 17.37), the Ford continuously variable transaxle contains the three basic elements of the transmission system: the epicyclic gearset and clutches, the belt and variable-pitch pulleys, and the hydraulic control system. The reduction ratio of the final drive compound (not simple as in conventional transaxles) gear train is chosen to provide the best overall reduction ratios for any given vehicle application. In the case of the Ford Fiesta application, for example, the

overall ratio on the lowest setting of the system is 14.75:1 and 13.37:1 respectively for the 1.1 litre and 1.4 litre engined models, and similarly 2.52:1 and 2.41:1 on the highest setting.

Other versions
The Subaru variable-speed belt and expanding pulleys CVT differed from the Ford CTX in utilizing an electronically controlled magnetic powder clutch, which was used in conjunction with a forward and reverse synchromesh gearset. In a more recent version Subaru have replaced the magnetic powder clutch with a hydraulic torque converter, which is combined with an electronically controlled lock-up clutch. Similarly the ZF-Ecotronic CVT features a hydraulic torque converter with lock-up clutch, so that it too combines the advantages of smoother starting-off with reduced engine speed and lower fuel consumption during normal driving. It is controlled by an electrohydraulic system in the manner of a modern conventional automatic transmission. For this purpose a microprocessor-controlled ECU continuously determines the most favourable operating mode for the system and selects whatever CVT ratio is appropriate by adjusting the relative diameters of the primary and secondary belt pulleys. The transmission unit is designed for fill-for-life lubrication with ATF (automatic transmission fluid), which minimizes maintenance costs, makes servicing simpler and reduces environmental pollution. For their previously mentioned

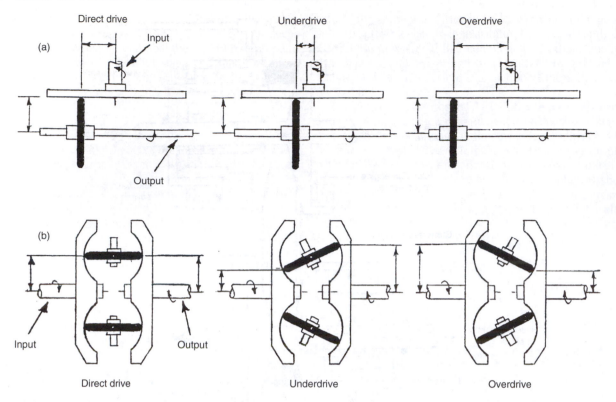

Figure 17.41   Operating principles of (a) flat disc and roller and (b) toroidal disc and rollers variable-speed drives

Multitronic CVT, Audi employ one oil-cooled multiplate clutch for transmitting forward drive and another for reversing, instead of using a hydraulic torque converter that absorbs more power.

Operating principles of variable-speed toroidal drives

For simplicity of understanding the toroidal drive, let us first consider the operating principle of the earlier mentioned flat disc and roller drive. This system of drive typically comprised a roller with a friction material lined rim, which was slidably keyed to a cross-shaft, so that it could be moved at right-angles across the face of an engine driven disc. The cross-shaft for the slidable roller was then connected by a chain and sprockets drive to the rear axle. To obtain a variable-speed drive the lateral position of the roller, which was normally pressed against a driven disc was controlled by a driver operated hand-lever. Hence, the nearer the slidable roller was moved towards the centre of the disc, the greater was the reduction in output relative to input speed of the transmission or, in other words, the lower the effective drive ratio and the greater the torque transmitted. If the slidable roller was moved further so that its position actually coincided with the centre of the disc, then a neutral condition existed, while if it was moved beyond this position to the other side of the disc a reverse drive was obtained. It would therefore be seen that by altering the operating radius at which the roller engaged the disc (Figure 17.41a), the relationship between the input and output speed of the transmission could be varied according to the torque requirements at the wheels.

Let us next imagine two fixed, but tiltable, rollers sandwiched between a pair of discs and engaging with saucer-like or toroidal depressions formed in their faces. With this arrangement we can therefore provide an engine-driven input disc and a driving output disc. A variable-speed transmission may then be obtained by tilting the rollers so that they can operate at different radii against the toroidal discs (Figure 17.41b). It will be noticed that the driving disc is being driven in the opposite direction to the driven one, which would if required need suitable gearing to restore the correct direction of rotation in the drive line. Also, unlike our simple flat disc and roller drive, this arrangement of toroidal discs and rollers cannot in itself confer a neutral position or a reverse drive without additional provision being made, as explained later. In practical application, two sets of three equally spaced rollers are mounted in tandem between two pairs of toroidal discs that are preloaded together. The use of multiple rollers in this manner does of course reduce their individual loading as noted earlier. A further refinement involves the rollers not only being tiltable to alter the drive ratio, but also being steerable by modifying their operating geometry. The purpose of this feature is to prevent the rollers from being scrubbed across the surfaces of the toroidal discs as they are tilted, instead of rolling across. This consideration also raises the matter of effective lubrication between the hardened steel contact surfaces of the rollers and toroidal discs, because again unlike the early flat disc and roller drive where dry friction existed, these operate under fluid friction conditions. In fact the drive is transmitted by the shearing of a very thin film of oil that separates the rolling contacts, which corresponds to a state of elastohydrodynamic lubrication (Section 4.2).

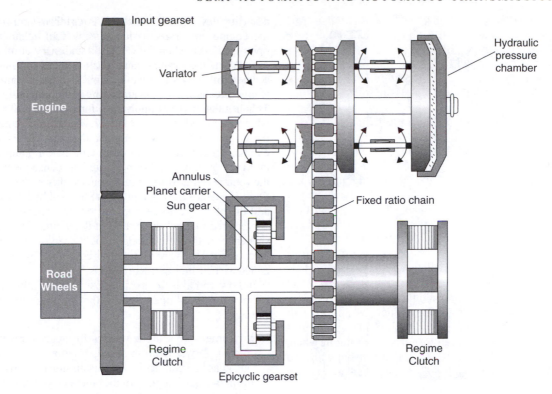

Figure 17.42 Schematic arrangement of the Torotrak IVT (*Torotrak (Development) Ltd*)

This requires the use of a specially developed type of oil, whose molecular structure creates a low friction characteristic under low pressure, but as the pressure increases its molecules begin to interlock until under extremely high pressure a high friction or traction effect is conferred. For this reason toroidal drives are sometimes technically referred to as 'oil film drives' or 'toroidal traction drives'.

Basic layout of an infinitely variable transmission

The unique features of the earlier mentioned toroidal transmission developed by Torotrak (Development) Ltd and known as Torotrak, clearly justify its description as being infinitely variable, as distinct from continuously variable, because it can deliver a continuous range of drive ratios from full reverse through to high overdrive, without the need for any form of slipping clutch or other starting devices.

Before considering its overall operation, it is necessary to identify the interrelated mechanical components of the Torotrak transmission (Figure 17.42), its basic layout being as follows:

1 As in other transmission systems, the input shaft receives its drive from the engine via a torsional vibration damper. The drive is then directed in series to the input gearset and the input toroidal discs. These are positioned outermost in the 'variator', which is at the heart of the Torotrak transmission system. A hydraulic pressure chamber is mounted at the rear end of the input shaft, so an appreciable end-to-end loading can be imposed upon the variator assembly of discs and rollers.

2 The output toroidal discs engage either side with a central toothed drum, which is allowed rotational freedom on the input shaft. An inverted tooth chain drive of fixed ratio is then taken, via a second toothed drum, to a hollow shaft mounted parallel to the input shaft. At its front end the hollow shaft carries a sun gear that engages with an epicyclic gearset, while at its rear end it forms part of a hydraulically-actuated, multiplate, wet clutch known as the 'high regime' clutch.

3 In the cavity between each pair of input and output toroidal discs there are three rollers, each roller being mounted on individual, hydraulically controlled, roller carriage pistons (Figure 17.43). Drive ratio changes are achieved, under computer control, by raising or lowering the hydraulic pressure to enforce an offset condition between the rotational axes of the rollers and the discs. This promotes a self-steering effect on the rollers, thereby inducing them to tilt and change the drive ratio.

4 The planet carrier for the pinions, which mesh with the sun gear of the epicyclic gearset, is mounted on a separate hollow shaft just ahead of the chain driven one. It connects at its front end to the second gear of the input gearset, via a hydraulically-actuated, multiplate, wet clutch known in this case as the 'low regime' clutch.

5 Finally, the output shaft of the transmission not only carries the annulus gear to complete the epicyclic gearset, but also passes with rotational freedom through successively the second gear of the input gearset; the low regime clutch; and the chain-driven sun gear shaft, before emerging from the latter to form part of the high regime clutch.

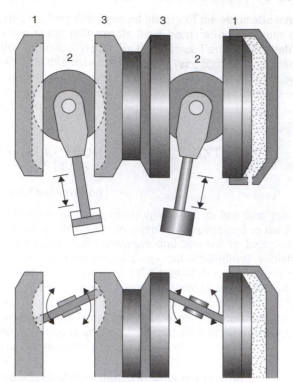

Figure 17.43 Operation of the Torotrak variator (*Torotrak* (*Development*) *Ltd*)

1 The input disc(s) powered by the engine
2 The variator roller(s) which picks up the power and transfers it to...
3 The output disc(s) which transmits the power from the variator to the drive shaft

From the above description it will be evident that one aspect of particular interest concerns the ingenious manner in which the epicyclic gearset is combined with the variator. Unlike in a conventional automatic transmission where it will be recalled (Section 16.1) that, except for the neutral condition, one or other of the three elements of an epicyclic gearset need to be held stationary to obtain forward drive ratios and a reverse, the Torotrak epicyclic gearset functions in effect as a mechanical adding machine. This is because the speed of rotation of the output shaft annulus gear is equal to the sum of the motions of the sun gear driven from the variator and the planet carrier driven from the input gearset. Hence, it follows that by controlling the relative speeds of the sun gear and planet carrier, the annulus gear and its output shaft can be made to rotate forwards, backwards, or stand still to provide what is termed a 'geared neutral'.

Operating principle of the Torotrak transmission

In operation, the combining of the toroidal discs and rollers variator with the epicyclic gearset thus eliminates the need for a starting device such as a clutch. Apart from simplification of the engine and transmission system, fuel economy can be optimized even when the vehicle is stationary or moving from rest. When the driver wants to alter speed, pressure on the accelerator signals a 'drive-by-wire' electronic control system to create a change in pressure of the variator hydraulic system. This has the effect of applying a force on each of the rollers, which is reacted against by the toroidal discs. The net result of these forces is to apply a torque to the road wheels and on the engine. An electronic control system then monitors the optimum performance of the engine and the consequent motion of the vehicle. With further input from the driver, the system continues to determine an appropriate level of hydraulic pressure and therefore the force imposed on the rollers. It should be appreciated that the drive ratio of the variator is a direct consequence of the engine and vehicle speeds, and the tractive force applied to the road wheels is controlled by the force generated at the roller carriage pistons.

It now remains to explain the 'low regime' and 'high regime' modes of operation for the Torotrak transmission as follows:

Low regime – This mode obtains for neutral, reversing, and moving forwards up to speeds roughly equivalent to second gear in a conventional transmission system. It involves the active participation of the epicyclic gearset and engagement of the low regime clutch. As the vehicle is accelerated, the tilting of the rollers between the toroidal discs of the variator imposes a progressive shift to a higher (numerically lower) drive ratio. The planet carrier of the epicyclic gearset, which it will be recalled is connected directly to the engine via the input gearset and low regime clutch, therefore dominates causing the annulus gear to increase the speed of the output shaft and that of the vehicle itself. When the vehicle speed approximates to that of second gear in a conventional transmission, the drive ratio of the variator reaches its highest limit, ready for a smooth transition to be made into high regime mode.

High regime – This mode is intended to cater for all higher forward speeds, including overdrive. Entering this mode of operation coincides with the speeds of the annulus gear, sun gear and planet carrier becoming equal, therefore providing an ideal opportunity for shock-free engagement of the high regime clutch. Both low and high regime clutches then briefly remain engaged, allowing the drive to be maintained while the variator control is set for high regime operation. Finally, the low regime clutch is released, so that the epicyclic gearset no longer participates in the power flow. Since the high regime clutch remains engaged, the variator is now directly connected to the output shaft of the transmission and hence the final-drive arrangements of the vehicle, which can then achieve its optimum performance.

18 Overdrive gears

18.1 PURPOSE AND POSITION OF AN OVERDRIVE GEAR

General background

An overdrive gear may be defined as a means of providing a higher (that is, numerically lower) ratio than that given by the final drive gears, so that engine speed can be reduced in relation to vehicle speed.

Even in the very early years of motor car development, there were some designers who recognized that it could be advantageous to provide a step-up gear within the gearbox, thereby enabling the output shaft to rotate faster than the input shaft. In other words, the engine could be allowed to rotate at a slower speed than the propeller shaft when in top gear. At the time this feature was confined to the more expensive makes of car, mainly to give their owners the enjoyment of effortless top gear cruising on favourable level stretches of road, when less power was needed to propel the car. Therefore we find that Henry Royce incorporated what we would now call an overdrive fourth gear in the sliding-mesh gearboxes of Rolls-Royce cars as early as 1905. Similar prestigious but now long-forgotten American makes of car, such as Locomobile and Lozier, also incorporated an overdrive top gear in their sliding-mesh gearboxes.

The main problem encountered with these early attempts to provide an overdrive gear in an otherwise conventional gearbox was their noisy operation. Indeed, it was for this reason that Henry Royce and others later abandoned the overdrive gear, because so many drivers were tempted to remain in it at lower speeds than were intended. The overdrive gear was then largely forgotten until the early 1930s, when it reappeared in America on Chrysler cars but in a different form and with the aim of improving fuel consumption very much in mind.

In the Chrysler application a Warner overdrive unit was bolted on to the back of their main gearbox and contained an epicyclic step-up gearset. This was semi-automatically controlled so that the tailshaft could be rotated faster than the mainshaft of the gearbox and provide an overdrive ratio of 0.722:1 at speeds above 45 mile/h (75 km/h). Yet another form of overdrive gear, which found favour in America on the high-performance Auburn and other cars of that period, was the dual-ratio Columbia rear axle. This incorporated a driver-controlled step-up epicyclic gearset in the final drive between the crown wheel and differential.

As interest in the overdrive gear declined in America with the arrival of automatic transmission, in Great Britain an epicyclic overdrive unit, originally known as the Laycock de Normanville and based on the inventions of Captain Edgar J. de Normanville, became increasingly popular, not least from considerations of improving fuel consumption. From the early 1950s onwards this bolt-on overdrive unit was adapted to fit the existing gearboxes of many British cars, until the advent of transversely mounted engines. It then became difficult to incorporate this form of overdrive gear in their transaxles. However, the GKN Laycock overdrive still found application in light vans, especially where long-distance motorway haulage was involved.

The position was different with most Continental car manufacturers; they had found it more acceptable to provide a rather higher final drive ratio, so the incentive to incorporate an overdrive gear was much less than in the UK. Those few that did generally preferred to include it in the gear train of a layshaft gearbox, such as the mid-1950s French Peugeot with an overdrive fourth speed, although an epicyclic step-up gearset was earlier used to give an overdrive fifth speed in the high-grade German Horch cars.

Since the mid 1970s history has tended to repeat itself, because it has now become established practice to incorporate an overdrive top gear within the gearbox itself, or at least where this has five speeds. This is also true in the case of those transversely engined cars that use five-speed transaxles, a trend that began with the 1969 Austin Maxi.

Following its introduction by Toyota in the late 1970s, an overdrive top gear for automatic transmissions has become increasingly adopted by other manufacturers. Included by now in the list of objectives are improved fuel economy, reduced exhaust emission, less engine noise and better driveability, so as to complement the advances that have been made in engine technology.

An overdrive step-up gear may also be included in the transmission arrangements of heavy vehicles, so that they may be operated more efficiently at cruising speeds, especially if partially loaded or empty on return journeys. Again the overdrive gearset is usually incorporated in a conventional layshaft gearbox, which has five or more speeds.

Advantages of an overdrive gear

The reduction ratio of the final drive gears, whether these are contained in a rear axle or a transaxle, necessarily represents a compromise. It should be low (numerically high) enough to confer good flexibility in top gear and thus avoid the need for frequent gear changing. On the other hand it should be high (numerically low) enough to prevent over-revving of the engine at the maximum speed of which the car is capable in top gear on a level road – that is, the speed at which aerodynamic and other drag forces acting on the car balance the tractive effort available at the driven wheels.

It is this compromise in final drive ratio that can result in the engine speed being unnecessarily high when cruising at motorway speeds, because to preserve good top gear flexibility it is generally arranged for the maximum speed of the car on a level road to occur beyond the peak of the engine power curve. The choice of step-up ratio for an overdrive gear is therefore such that although the maximum speed of the vehicle is attained at a similar engine power, it occurs before and not after the peak of the power curve so as to reduce engine speed (Figure 18.1).

Utilized to best advantage, an overdrive gear can thus permit any given journey to be completed in the least total number of engine revolutions. This confers the following benefits on vehicle operation:

Improved fuel economy
Reduced exhaust emissions
Quieter cruising speed
Better driveability
Longer engine life.

Position of an overdrive gear

As already indicated, an overdrive gear may occupy one of several positions in the transmission system. These may be identified as follows.

Between clutch and gearbox
The provision of a slimline version of the Laycock type of overdrive, which could be sandwiched between the clutch and transaxle, was once considered for transverse-engined front-wheel-drive cars. In this position the step-up ratio occurred between the primary shaft and the mainshaft of the transaxle gearbox. The overdrive unit was therefore never required to transmit a torque higher than that being delivered by the engine, so there was no need to restrict its engagement to the higher gears. The disadvantages of this position are that the greater inertia of the rotating parts increases the workload on the gearbox synchromesh mechanism, and that any removal and refitting of the unit in service is less convenient, involving as it would the removal of the transaxle.

In the gearbox
Clearly the simplest form of overdrive gear is that incorporated in the main gearbox, since its operation can be integrated with the other gear trains and their associated selector mechanism. In contrast to early practice, the more rigid support provided for the overdrive gearset in modern five-speed

gearboxes is such as to ensure their quiet operation and durability. The advantages now generally associated with positioning the overdrive gear in the gearbox itself are those of least cost, less increase in weight and potentially greater reliability. It can also be used with equal facility in front-wheel-drive and rear-wheel-drive transmission systems.

Between gearbox and propeller shaft
For the once conventional front-engined rear-wheel-drive car, the positioning of an overdrive gear between the gearbox and the propeller shaft offered the advantages of convenient installation and servicing. Furthermore, the space available in this position was usually sufficient to accommodate an overdrive gear with a comprehensive control system, such as the Laycock unit, which to a limited extent could provide two-pedal motoring. In this position the step-up ratio occurs between the gearbox main shaft and the overdrive unit tailshaft, so the unit must necessarily transmit engine torque multiplied by any reduction ratio being used in the main gearbox. For this reason the selection of overdrive had to be confined to the higher gears, in order not to exceed the torque capacity of the unit. This was never considered to be a serious disadvantage. However, more significant was the increased weight, cost and complexity associated with this type of installation, as compared with the now conventional five-speed overdrive top gearbox.

Between propeller shaft and final drive
Another position once considered for mounting an overdrive gear of the Laycock type was on the nose of the final drive unit, where this was attached to the body structure and used in conjunction with independently sprung rear wheels. This position had the advantage of improving the front-to-rear weight distribution of the car and also or contributing to a lower floor height in the passenger compartment. Since the step-up ratio occurs between the propeller shaft and the pinion shaft of the final drive gears, the engine, gearbox and propeller shaft were all reduced in speed when overdrive was selected.

Between final drive and differential
The early use of an overdrive gear positioned between the crown wheel and the differential of a live rear axle likewise permitted the engine, gearbox and propeller shaft all to run at reduced speed when overdrive was selected. This position for the overdrive gear also contributed to a lower floor height in the passenger compartment. Apart from being expensive to manufacture, the two-speed overdrive axle was inevitably heavier than a conventional axle. This proved to be a serious disadvantage, because the increase in unsprung weight of the rear suspension was detrimental to road holding. It should be noted that an opposite function is performed by the two-speed rear axles later described in connection with heavy vehicles, since these provide an underdrive or step-down ratio (Section 20.7).

18.2 EPICYCLIC OVERDRIVE GEARS

Basic considerations

When the properties of a simple epicyclic gear train were examined (Section 16.2), it was shown that a direct drive can be obtained if any two of the three members comprising the sun

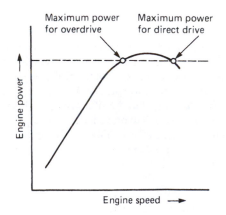

Figure 18.1 Maximum power for direct drive and overdrive

gear, planet pinions and annulus gear are locked together. This is because the third member will then be carried around by the teeth of the two locked members at the same speed of rotation and in the same direction. We can therefore satisfy one requirement of an epicyclic overdrive gear, which is to allow a direct drive to be transmitted when overdrive is not selected.

It was also shown that if either the sun gear was held stationary and the annulus gear rotated, or the annulus gear was held stationary and the sun gear rotated, then two reduction or step-down ratios could be obtained from the gearset. If for our present purposes we now regard these as underdrive gears, then conversely we can obtain overdrive gears with step-up ratios should either the sun gear or the annulus gear be held stationary and the planet carrier rotated. For practical application to an overdrive unit the former combination is always used (Figure 18.2), since it provides the most suitable step-up ratio. For example, the overdrive ratios that have commonly been used in Laycock units for application to British cars are 0.778:1 and 0.820:1.

Laycock overdrive

In the Laycock overdrive a hydraulically engaged epicyclic gearset with electrical control is employed. The epicyclic gearset and its wet double-cone clutch, together with hydraulic system, are housed in a two-piece aluminium alloy casing. This is attached to the back of a conventional gearbox for rear-wheel drive and conveniently takes the place of the rear extension housing. The main-shaft of the gearbox remains extended, but in this case to engage the planet carrier of the overdrive unit via a splined connection. Next comes the sun gear, this being freely mounted on the extended mainshaft adjacent to the planet carrier. Slidably mounted on the externally splined hub of the sun gear is the double-cone clutch with inner and outer friction linings. These linings can respectively engage either the stationary brake ring that is sandwiched between the two halves of the overdrive casing, or the externally coned surface of the annulus gear. The latter completes the epicyclic gearset and is carried on the output or tailshaft of the overdrive unit, this shaft being internally bushed at its forward end to furnish a spigot bearing support for the end of the gearbox mainshaft.

An addition to the basic gearset is a one-way roller clutch that acts between the mainshaft of the gearbox and the tailshaft of the overdrive unit. The inner member of the roller clutch is splined to the end of the mainshaft and provides the cam track, while the outer member has a plain track and is retained within the hub of the annulus gear.

The double-cone clutch is actuated via a thrust ring with ball race, and offers the advantages of high torque capacity for a moderate clamping load and minimum drag when disengaged. It is spring loaded to engage the externally coned surface of the annulus gear and thus lock the gearset to provide direct drive (Figure 18.3a), but can be hydraulically retracted to engage the brake ring and prevent rotation of the sun gear to achieve overdrive (Figure 18.3b). The purpose of the one-way clutch is to maintain drive during the change-over period of the cone clutch, because the overdrive unit is designed to operate when transmitting full power.

The hydraulic retraction of the cone clutch against its spring loading is obtained through a simple built-in hydraulic system. This essentially comprise a plunger pump driven from the gearbox mainshaft, which delivers oil to a circuit that includes a

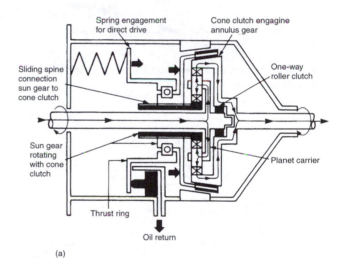

(a)

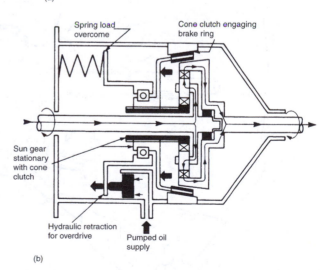

(b)

Figure 18.3 Schematic arrangement and operation of Laycock-type overdrive: (a) direct drive (b) overdrive

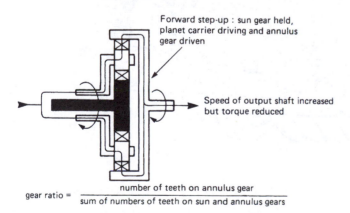

$$\text{gear ratio} = \frac{\text{number of teeth on annulus gear}}{\text{sum of numbers of teeth on sun and annulus gears}}$$

Figure 18.2 Simple epicyclic gear train utilized as an overdrive gear

pressure relief valve, a solenoid-operated control valve and a pair of pistons that act against the thrust ring of the cone clutch. To select overdrive the solenoid is energized from the car electrical system through a driver-controlled switch, which is mounted on either the steering column or the facia panel. This actuates the control valve and allows oil pressure to build up against the pistons, so they can move the thrust ring in opposition to its spring loading. Cushioning springs may be incorporated in the piston assemblies to permit a progressive engagement between the cone clutch and brake ring. When overdrive is selected the hydraulic system is maintained at a pressure of about 2.8 MN/m^2 (400 lbf/in^2), any excess delivery of oil from the pump being directed to lubricate the gearset and

one-way clutch. The overdrive unit typically shares a common oil supply with the main gearbox.

An inhibitor switch is fitted in the electrical circuit between the driver-controlled switch and solenoid for the control valve, the switch being activated by the gearbox selector mechanism (Figure 18.4). Its purpose is to prevent selection of the overdrive in reverse and in some or all of the indirect gears according to the requirements of a particular manufacturer.

18.3 LAYSHAFT OVERDRIVE GEARS

In current motor car applications an overdrive top gear may be incorporated in both double-stage gearboxes and single-stage transaxles, where these have five forward speeds.

Double-stage five-speed gearboxes with an overdrive top gear and a direct-drive fourth gear were originally confined in application to a few specialist high-performance cars, such as the Italian Lamborghini. In this application the overdrive gearset was positioned at the rear of the gearbox and adjacent to a constant-mesh reverse gear train, so affording the opportunity for synchronizing the engagement of not only the five forward speeds but also reverse. This principle has been revived in the latest Ford MT 75 gearbox, which therefore becomes the first volume-produced five-speed overdrive top gearbox to have a synchromesh reverse gear for clash-free engagement and reduced wear (Figure 18.5). There has also been an understandable tendency for volume car

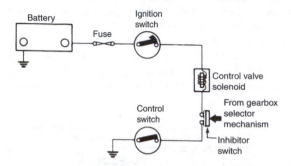

Figure 18.4 Schematic arrangement for overdrive inhibitor switch circuit

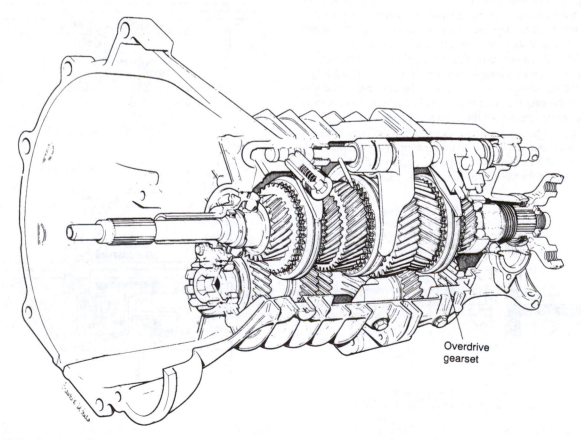

Figure 18.5 Cut-away view of a modern five-speed overdrive top all-synchromesh gearbox (*Ford*)

manufacturers to make certain changes in detail design so that either a four-speed direct top gearbox or a five-speed overdrive top gearbox can be made from the same tooling. This manufacturing economy may be achieved by simply either adding an overdrive gearset, or omitting it and substituting spacer sleeves, near the middle of a lengthened gearbox casing. In other words, the five-speed overdrive top gearbox came to be regarded as a 'four-plus-one' unit from a production point of view.

An alternative approach has been to mount an overdrive gearset in an extension housing behind the rear wall of a four-speed gearbox casing. When the overdrive gearset is mounted in this overhung fashion (Figure 18.6a), it has the advantage of allowing a shorter span between the gearbox main bearings and thereby reducing deflection of the shafts. In this context it should be noted that a live layshaft is required in the construction of five-speed overdrive top gearboxes. The difference between a live and a dead layshaft is that instead of the layshaft gear cluster being made hollow and rotating on bearings at each end of a fixed layshaft spindle that passes through it (Figures 14.11 and 14.18), the layshaft cluster remains solid throughout with a spigot on each end to mount its support bearings (Figure 18.5). Live layshafts have long been used in the construction of heavy-vehicle gearboxes (Figures 14.9, 14.10 and 14.43).

The increasing performance and driveability requirements of modern high-performance cars have led in some cases to the adoption of six-speed, overdrive top, all-synchromesh gearboxes. In a double-stage layout for a front-engine, rear-wheel drive car, the third and fourth gear pairs are typically situated midway between the constant-mesh/direct-drive fifth and adjacent overdrive sixth speed gear pairs and the first and second gear pairs. Since the torque conversion in the gearbox is higher in the lower gears, this arrangement allows the first and second gear pairs to be mounted close to the gearbox rear bearings. Shaft deflections under load and hence gear noise are therefore minimized when either gear is selected. The basic arrangement of the gear pairs may be compared to that of a six-speed, overdrive top, gearbox for a heavy vehicle (Figure 18.6), but usually without the addition of intermediate

support bearings. Furthermore, the pinions of the third and fourth gear pairs may be allowed rotational freedom on the layshaft with synchronizers being mounted between them. The purpose of this departure from the conventional layout of a double stage gearbox (Section 14.2), where all the synchronizers are carried by the mainshaft, is to avoid any tendency towards gear rattle from torsional vibration. To reduce gearchange effort, six-speed gearboxes generally incorporate more powerful synchronizers. These are therefore graduated according to severity of duty, so that single-cone synchronizers are used for the direct-drive fifth and overdrive sixth gears, double-cone synchronizers (Figure 14.37) for the third and fourth gears, and even triple-cone synchronizers for the first and second gears. By this means the gearchange effort can be equalized, regardless of the gear being selected, to confer nicety of control.

For five-speed overdrive top transaxles of single-stage layout (Figure 18.7), which are widely used in modern transverse-engined cars, the overdrive gearset is overhung from the main casing and need only be enclosed by a simple end cover. As in all single-stage gearboxes an exact 1:1 ratio direct drive cannot be provided in fourth gear, because it is necessary to avoid having equal numbers of teeth in the gear pair. This ensures that each tooth of one gear engages every tooth on the other in regular succession for smooth and quiet operation. In practice there is a one-tooth difference between the gears, which means that the fourth gear is made very slightly overdrive.

In the case of heavy-vehicle gearboxes with an overdrive top gear, the pair of overdrive gears may be incorporated in several different ways. For a five-speed gearbox the overdrive gears are either positioned adjacent to the constant-mesh gears and contained within the gearbox main casing (Figure 14.10), or they may be overhung from the rear wall of the main casing (Figure 18.6a). With a six-speed gearbox the overdrive gears can similarly be accommodated, but when an adjacent mounting is used for the constant mesh and overdrive gears they may alternatively be contained in a forward compartment of the gearbox main casing for better support (Figure 18.6b).

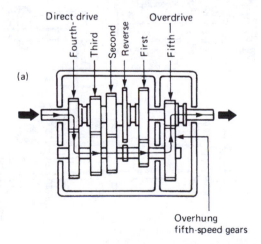

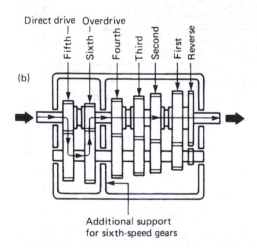

Figure 18.6 Schematic arrangement and power flow through five- and six-speed gearboxes in overdrive top gear

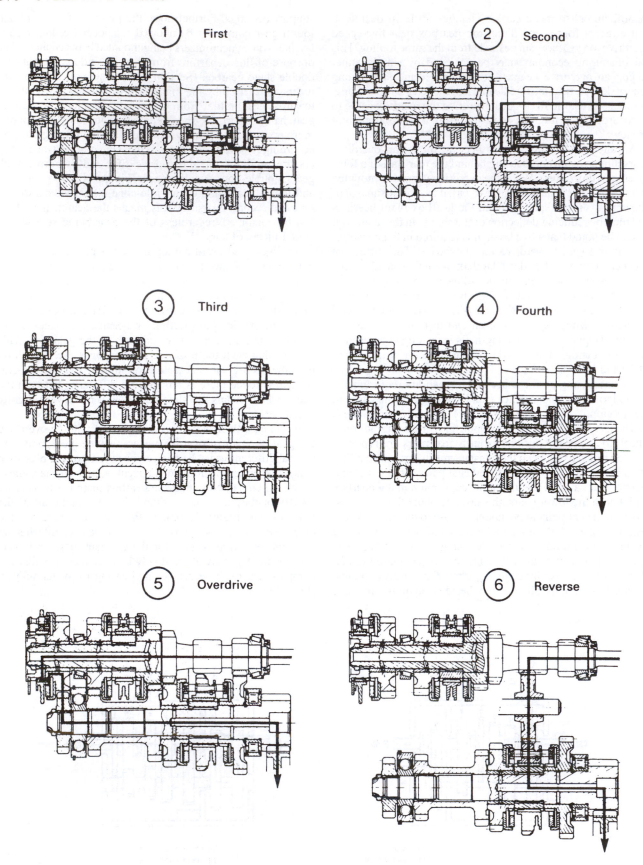

Figure 18.7   Power flow through a modern five-speed overdrive top transaxle (*Citroën*)

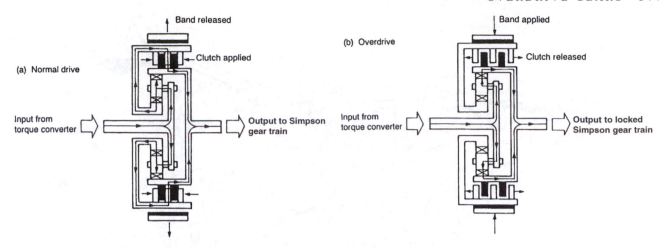

Figure 18.8 Schematic arrangement and operating principle of an automatic transmission overdrive gearset: (a) normal drive (b) overdrive

18.4 AUTOMATIC TRANSMISSION OVERDRIVE GEARS

A four-speed automatic transmission with an overdrive top gear can be obtained by adding a step-up epicyclic gearset between the torque converter and a conventional Simpson-type three-speed gear train. Transmission engineers sometimes refer to this arrangement as a 'Simpson plus simple' or 'three plus one' gear train. The main reason for incorporating the overdrive gearset before, rather than after, the Simpson gear train is that the required torque transmitting capacity of the overdrive gearset is minimized. This arrangement also makes for a more convenient connection between the overdrive hydraulic circuit and the control valve unit.

In practical application, the following basic changes are required to the general arrangement of the transmission:

1 The input shaft from the torque converter is connected directly to the planet carrier of the overdrive gearset.
2 What was formerly the input shaft connecting the torque converter to the Simpson gear train is replaced by an intermediate shaft, which connects the annulus gear of the overdrive gearset to the Simpson gear train. The output shaft from the latter does, of course, remain unchanged in its duty.
3 A direct-acting multiplate clutch is required to act as a rotating brake between the sun and annulus gear drums of the overdrive gearset.
4 Either a band brake or a direct-acting multiplate clutch is required to act as a stationary brake between the sun gear drum and the gearbox casing.
5 A 3–OD (overdrive) shift valve must be added to the hydraulic control system of the transmission.
6 The lubrication arrangements for the transmission gear train must be extended to include the overdrive gearset, especially since its planet pinions can be subject to high rotational speeds during overdrive operation.

It will be recalled from Section 18.2 how a simple epicyclic gearset can be utilized as an overdrive gear, and this same principle is adopted here. During normal drive the overdrive gearset is locked internally by application of its rotating multiplate clutch, so that it transmits drive directly via the intermediate shaft to the three-speed Simpson gear train (Figure 18.8a), which then operates in the normal manner to provide low, intermediate and direct-drive high gears as described in Section 16.2. When there is an up-shift into overdrive top gear the Simpson gear train remains locked as for the previous high gear, the rotating multiplate clutch of the overdrive gearset is released, and either the band brake or the stationary multiplate clutch is applied to arrest the sun gear. This results in the pinions of the planet carrier, which is permanently connected to the input shaft from the torque converter, walking around the stationary sun gear and driving the annulus gear at a faster speed, albeit with reduced torque, than the planet carrier. Since the drive from the annulus gear is directly transmitted to the locked Simpson gear train (Figure 18.8b), the speed of the output shaft will exceed that of the input one. More importantly, the input shaft speed can be reduced whilst maintaining the original speed of the output shaft, thereby fulfilling the main aim of the overdrive gear to reduce engine speed in relation to road speed.

A well-known example of a Simpson plus simple four-speed overdrive top automatic transmission, which also features a lock-up facility for the torque converter on the two higher gears, is the Ford A4LD unit (Figure 18.9), which was developed from the earlier C3 three-speed direct top transmission (Figure 17.6). An overdrive ratio of 0.75:1 has been adopted for the A4LD transmission. In a more recent four-speed overdrive top automatic transmission introduced by Toyota, alternative overdrive gear ratios of either 0.705 or 0.753 may be specified according to the country of destination.

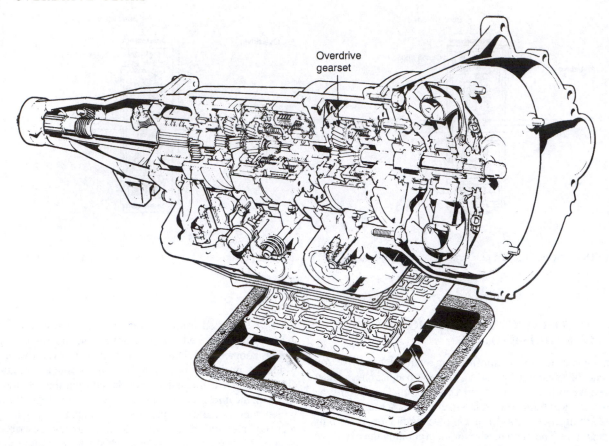

Figure 18.9 Cut-away view of a modern four-speed automatic transmission with overdrive top gear (*Ford*)

19 Drive lines

19.1 UNIVERSAL JOINTS

Flexible element universal joints

Since the early history of the motor car automobile engineers have always been attracted to the idea of a universal joint that does not require lubrication or any other attention during the useful life of the vehicle. In practical terms, this led to the development of various forms of universal joint in which the driving and driven shafts were provided with spider-arm connections to a flexible disc of either metal or leather. It was the British engineer E.J. Hardy who, in 1911, introduced the first really successful flexible disc universal joint. The disc was built up from many layers of rubberized fabric material, the weave of successive layers being staggered to provide a uniform strength of disc. Later developments in this field became concentrated on using rubber as the flexible medium in the form of either bushings or a complete ring.

Before describing the various modern types of rubber element universal joint, it may be helpful to list the looked-for advantages in using such joints:

1 To eliminate lubrication and wearing problems in the propeller shaft assembly.

2 To reduce transmission of vibration and noise.
3 To cushion shock loads on the transmission system.

On the debit side, rubber element universal joints generally do not provide a constant-velocity drive when operated at an angle.

Layrub joint

This represents a once well-known version of the rubber bush type of universal joint, which was widely used on the propeller shafts of commercial vehicles and, rather interestingly, on electrically driven trolley-buses to cushion their suddenly applied and high starting torque.

It had no working parts in the accepted sense of the term, because all relative movement in the joint was accommodated by means of rubber bushings housed between the complementary halves of a circular steel housing (Figure 19.1a). The rubber bushings were under compression in their housing and had specially shaped cavities at their ends. Steel trunnion sleeves having spigoted shoulders were pressed into the rubber bushes. Each trunnion sleeve located in a registered

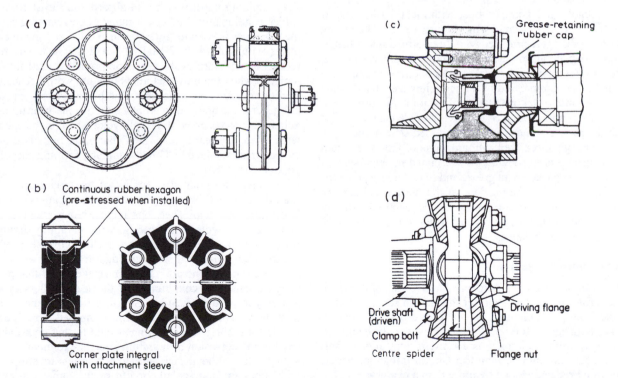

Figure 19.1 Types of rubber element universal joint: (a) Layrub joint (b) Rotoflex joint (c) rubber element joint with centring device (*Alfa-Romeo*) (d) Moulton joint

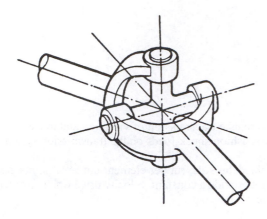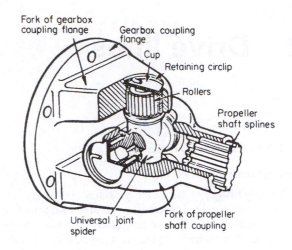

Figure 19.2   Hooke's coupling in principle and in practice

portion of bolt hole in the two-armed spider flanges of the driving and driven shafts.

For the propeller shaft speeds at which the Layrub universal joint was normally operated no centring device was required, because the effect of any imbalance of the propeller shaft assembly would not be sufficient to cause whirling. Its angular capacity was generally limited to about 5° for continuous deflection. The axial flexibility of the rubber bushings was sufficient to make redundant the need for a sliding spline joint in the propeller shaft assembly.

Rotoflex joint

For this version of rubber ring or doughnut type of universal joint, a regular polygon of rubber with metal bushes vulcanized at the corners is employed (Figure 19.1b). The driving and driven shafts are provided with three-armed spider flanges for bolt connections to the metal bushes. An important design feature of this type of universal joint is that the rubber element is supplied precompressed by an encircling steel band, which is then discarded after installation of the joint. Its purpose is to ensure maximum endurance of the rubber element in service, since the rubber remains in a compressed state during operation. For high-speed applications, joints of this type require a mechanical centring device, the associated spherical seat and bush being prepacked with grease and sealed with a protective cap (Figure 19.1c). The purpose of the mechanical centring device is, of course, to restrain any tendency towards whirling of the propeller shaft assembly, for which it is never possible to achieve perfect balance.

Hooke's universal joints

This is the most common form of universal joint used in engineering. It is so named after the British scientist Robert Hooke, who was granted a patent for a cross-pin universal joint as long ago as 1664. In Europe, however, the equivalent form of coupling is known as a Cardan joint after its supposed actual inventor Jerome Cardan, an Italian mathematician who predated Hooke by more than a century.

The simple cross-pin universal joint used in most motor vehicle propeller shafts is therefore a development of the basic Hooke's or Cardan joint (Figure 19.2). It consists of a pair of driving and driven yokes disposed at right angles to each other and pivoting from a four-armed cross-piece or spider. A universal action is thus obtained by virtue of the joint hinging in two planes at right angles.

In early designs of universal joint the cross-pins oscillated in plain hardened steel bushes carried by the eyes of the yokes. At a later stage, the plain bearings were superseded by needle roller bearings to increase the torque capacity and thereby satisfy the requirements of more powerful engines. Needle bearings are roller bearings in which the length-to-diameter ratio of the rollers is greater than two.

The needle roller bearings are assembled into hardened steel retaining cups, which also absorb end thrust from the spider trunnion pins. Endwise location of the retaining cups in the yoke eyes is obtained by either snap rings, staking or cover plates with screws. It should be noted that with the more recent staked bearings the complete propeller shaft must be replaced when the universal joints become worn because the bearing cups are non-removable. One-piece steel forged yokes are generally employed to embrace the needle bearing cups, although constructions of the split yoke type may be encountered where the bearing cups are either of the bolted-on block variety or are retained by U-bolts similarly forming part of the yoke (Figure 19.3).

The taken-for-granted reliability of the modern propeller shaft depends entirely upon the efficient lubrication and sealing of its universal joint bearings. If these requirements are not met, then the needle roller bearings will fail rapidly through a condition known to the engineer as fretting corrosion, which occurs when two metal surfaces fidget together. Two factors are present in the operation of the propeller shaft that potentially can cause this condition. First, the needle rollers are subject only to partial instead of complete rotation, and second, they have vibrational movements imposed upon them. Inefficient sealing of the bearings would further contribute to any fretting corrosion by admitting moisture to them.

In practice, either the universal joint bearings are pre-packed with lubricant and sealed for life or, in some heavy-duty applications, they may be fitted with lubricators for periodic servicing (Figure 19.4). With the former construction, the

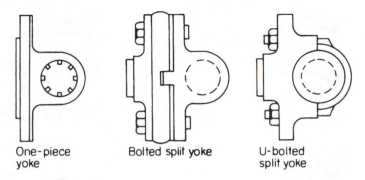

One-piece yoke

Bolted split yoke

U-bolted split yoke

Figure 19.3 Different types of bearing cup mountings for universal joints

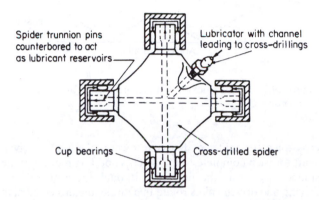

Spider trunnion pins counterbored to act as lubricant reservoirs

Lubricator with channel leading to cross-drillings

Cup bearings

Cross-drilled spider

Figure 19.4 Lubrication arrangements for a heavy-duty universal joint

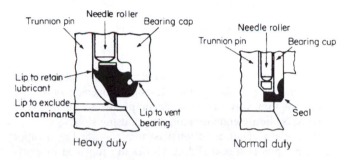

Trunnion pin   Needle roller   Bearing cap

Lip to retain lubricant

Lip to exclude contaminants

Lip to vent bearing

Heavy duty

Needle roller

Trunnion pin   Bearing cup

Seal

Normal duty

Figure 19.5 Sealing arrangements for universal joint bearings

spider trunnion pins are usually counterbored to provide oil reservoirs for each needle roller bearing. If a lubricator is fitted, it is centrally located on the spider and communicates with cross-drillings connected with the trunnion pin reservoirs. A transmission grade of oil is generally specified for relubrication of the universal joint in service. During initial assembly the bearing cups are partially filled with grease to facilitate assembly of the loose needle rollers.

Cork gaskets with retainers were once commonly used on the spider trunnion pins to seal the needle roller bearings, but seals of this type have long been superseded by rubber seals located on the rims of the bearing cups (Figure 19.5). In some heavy-duty applications, multi-lipped seals may be used that not only retain the lubricant and exclude contaminants but also allow the bearings to be vented during their relubrication in service. In connection with the sealing of universal joint bearings, one point that should be appreciated is that there is no tendency for the lubricant to be flung from them when the propeller shaft is revolving. On the contrary, the lubricant is better retained in the bearing cups under the influence of centrifugal force.

Moulton joint

This type of rubber element joint (Figure 19.1d) is so named because it was evolved and patented by Moulton Developments Limited in the UK. It is basically of the Hooke's type, but its four-armed steel spider carries bonded-on part-conical rubber bushes. These are embraced by split sleeves and retained

against the saddle-shaped forked ends of the driving and driven shafts by U-bolts and nuts. The U-bolts are shouldered to ensure that the correct amount of preload is applied to the rubber bushes on assembly of the joint. In this context, it should be appreciated that the fatigue life of rubber is always improved if it is never allowed to relax completely when undergoing changes in loading – exactly the same principle being applied by the civil engineer to prestressed concrete!

Bipot joints

This type of pot joint first appeared as long ago as 1894 on early steam cars and was revived later on the once well-known French De Dion cars. On these the rear wheels were carried by a dead axle beam curving around behind the final drive, and were driven by two drive shafts each equipped with two universal joints. Since the arcuate movements of the short drive shafts mismatched those occurring at the axle beam ends, each shaft together with its two joints had to be capable not only of universal but also of telescopic action. This type of De Dion drive is, in fact, still favoured on certain high-performance cars, because it eliminates the wheel tramping tendencies of a live axle (Figure 20.25).

Early constructions of bipot joint, usually referred to by motoring historians as De Dion and sliding block joints, comprised simply a cross-pin at the end of the drive shaft, each end of the pin being free to pivot within its own square-shaped sliding block. These in turn contacted the flat faces of two slots in the joint housing. The sliding blocks were also provided with radiused ends to allow the shaft to rock in the plane

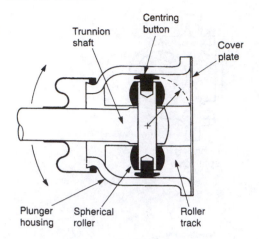

Figure 19.6   A modern bipot universal joint

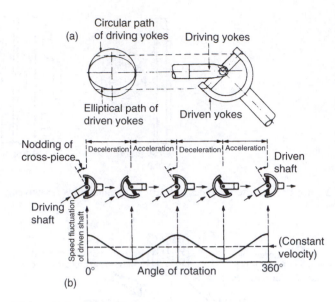

Figure 19.7   Single Hooke's joint: (a) geometry (b) speed fluctuation

of the cross-pin and therefore complete its universal action, in the manner of a Hooke's joint. Telescopic movement or plunge within the joint was, of course, accomplished by the action of the blocks sliding endwise along the housing slots.

In modern versions of what is now classified as a bipot joint, the pivoting and sliding blocks are replaced by part-spherical rollers operating within partly cylindrical bores in the joint housing (Figure 19.6). The part-spherical rollers are themselves supported on the cross-pin by needle roller bearings. Such joints (notably the Detroit type) were once fairly widely used at each end of the normal propeller shaft on American cars and also on certain high-grade cars elsewhere, because their free-plunging capability eliminated the need for a sliding splined joint. It is now more usual to find the bipot joint at the inner ends of drive shafts for independently sprung and driven rear or front wheels. This type of joint is not suitable for the outer ends of front-wheel drive shafts, because of its non-constant-velocity operation.

Hooke's joint in series

The motions of the rear axle occurring in unison with, and relative to, the rear springs in a Hotchkiss drive system are later described in connection with the propeller shaft requiring a sliding joint. It is these same antics of the rear axle that further demand the use of two universal joints, one at each end of the propeller shaft. These are necessary if the propeller shaft is to accept, without bending, the relative angular and parallel movements of the gearbox and rear axle shafts that it connects.

Apart from the physical necessity for both ends of the propeller shaft to be so hinged, the use of two universal joints in series also provides for a more uniform rotation in the drive line. The reason for this is that the simple Hooke's coupling is not a constant-velocity joint. This means that its presence between two shafts, running with angular misalignment, causes a non-uniform rotational speed to be imparted from the driving to the driven shafts.

A simple explanation for this peculiarity of operation may be found by observing closely the universal joint during hand rotation of the shafts. It will then be noticed that the

cross-piece or spider nods twice towards the angled driven shaft for each complete rotation. This nodding motion of the spider arises from the fact that although the arms of both driving and driven yokes rotate in a circle, the projected path of the driven yoke arms describes an ellipse owing to the inclination of the driven shaft (Figure 19.7a). These differing paths of effective movement of the connecting yokes account for there being two positions in which the driven shaft has advanced in rotation relative to the driving shaft and two intermediate positions in which the driven shaft has lagged behind it (Figure 19.7b). Each shaft does, of course, make the same total number of revolutions in a given time despite the periodic fluctuations in angular velocity.

Returning now to the motor vehicle propeller shaft, the effect of this non-uniform rotational speed can be to create an objectional vibration in the drive line, which leads to a general roughness being felt in the car. Fortunately, this peculiarity in operation of the Hooke's joint can to all intents and purposes be countered by the use of two such joints in series, as is found in the conventional one-piece propeller shaft. With this arrangement, the fluctuations in angular velocity of the swinging propeller shaft are still present but they are cancelled by the second universal joint before reaching the rear axle final drive shaft (Figure 19.8).

19.2   CONSTANT-VELOCITY JOINTS

General requirements

As explained in Section 19.1, a simple Hooke's coupling or universal joint of cross-pin construction causes a non-uniform rotational speed to be transmitted from the driving to driven shafts when they are running with angular misalignment. It was further explained that these fluctuations in rotational speed could be cancelled by adding a second Hooke's coupling in series with the first. Although our thoughts were

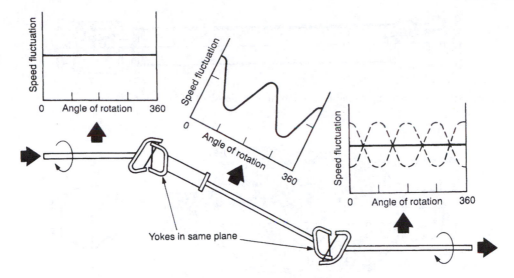

Figure 19.8   Hooke's joints correctly phased to provide cancellation of speed fluctuations

then concerned with a conventional propeller shaft, incorporating a simple Hooke's coupling at each end, it does of course follow that if we place two Hooke's couplings back-to-back between any pair of driving and driven shafts, running with angular misalignment, then these shafts would immediately be free of relative fluctuations in rotational speed.

In other words, the use of two back-to-back Hooke's couplings can be said to create a constant-velocity universal joint, of which there are also several other more modern forms. Another description of a constant-velocity, universal joint is homokinetic joint (literally of constant velocity), a term that was coined in the late 1920s by Automobiles Tracta who were pioneer manufacturers of front-wheel-drive cars in France.

In modern automotive practice the transmission applications of constant-velocity (CV) joints can be summarized as follows.

Front-wheel drive shafts

Certainly the earliest and perhaps the most obvious requirement for a constant-velocity joint is at the outer end of each drive shaft on front-wheel-drive cars (Figure 19.9). The presence of a single Hooke's joint at this point, where large-angle steering movements have to be accommodated, makes for totally unacceptable kick-back reactions at the steering when cornering.

Rear-wheel drive shafts

A later and perhaps not so obvious requirement for a constant-velocity joint is at both ends of each drive shaft to independently sprung rear wheels (Figure 19.10). Although the use of single Hooke's couplings at both ends of each drive shaft does to all intents and purposes give a constant-velocity drive to the road wheels, as they rise and fall with suspension movements there still remain the fluctuations in rotational speed of the drive shaft itself. In a modern car, the inertia effects that arise from this unevenness of drive shaft

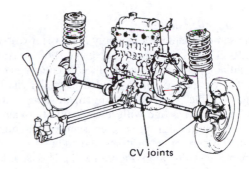

Figure 19.9   Application of constant-velocity universal joints to front-wheel drive (*GKN Hardy Spicer*)

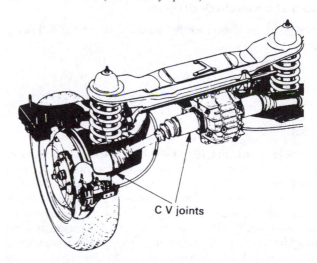

Figure 19.10   Application of constant-velocity universal joints to rear-wheel drive (*GKN Hardy Spicer*)

rotation can be sufficient to produce an unacceptable level of vibration. Hence constant-velocity joints are required to provide an overall uniformity of rotational movement within each drive shaft assembly.

Figure 19.11   Application of constant-velocity universal joints to propeller shaft (*GKN Hardy Spicer*)

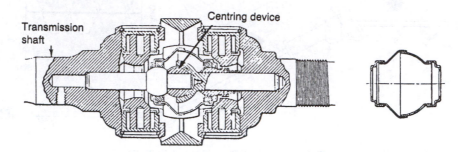

Figure 19.12   Cross-section through a double Hooke's joint (*Citroën*)

Rear axle propeller shafts

Much the same considerations apply here as for rear-wheel drive shafts. This trend began in the early 1960s in America when Cadillac fitted back-to-back Hooke's couplings to both ends of the rear part of their divided propeller shaft. The advantages claimed for this arrangement were smoother operation of the drive line with less sensitivity to any misalignment. More modern types of constant-velocity joint are used in similar current applications and additionally perform the function of a sliding joint (Figure 19.11). This latter feature is also relevant to the first two applications mentioned above.

Types of constant-velocity joint

These may be conveniently classified under the following headings:

Double Hooke's
Rzeppa
Weiss
Tracta
Tripot.

All these joints aim to satisfy one essential requirement of constant-velocity operation, which is that all contact points between their driving and driven members must always lie in a plane that bisects the angle between the shafts (Figure 19.15), this plane being known synonymously as the constant-velocity plane, the bisecting angle plane and the homokinetic plane. In theory the tripot joint does not quite meet the stated requirement, but nevertheless represents an excellent compromise.

Double Hooke's joint

Perhaps the best known version of this particular form of constant-velocity joint was that introduced in 1934 by the French firm of Glaenzer-Spicer for the then new front-wheel-drive model Citroën and used for many years thereafter. An

important constructional feature of this type of joint is the ball-and-socket centring device required between the driving and driven shafts (Figure 19.12). The basic functions of this device are those of maintaining the double cross-pin joint both self-supporting and self-aligning, and also dividing equally (or nearly so) the total joint angle between the driving and driven shafts. In performing these functions the centring device is subject to appreciable relative motion between its socket, ball and stud. Suitable arrangements for lubricating these critical parts of the double Hooke's joint must therefore be made, typically by a grease nipple and communicating channel in the coupling yoke.

Given satisfactory behaviour of the centring device, the advantage of the double Hooke's joint are generally those of rugged construction, durability and the ability to operate at high joint angles (Figure 19.13). Its main disadvantages are those of complicated and cumbersome construction, so that it can now be considered obsolescent for passenger car front-wheel-drive applications, although it may still be used in heavy-duty axles for off-road vehicles.

Rzeppa joint

The operating principle of this type of ball constant-velocity joint was devised in 1926 by the Czechoslovakian inventor A.H. Rzeppa (pronounced 'Sheppa'), a one-time consultant to an American firm of gear-grinding machine manufacturers. With further development the Rzeppa concept has probably become the most widely used design of constant-velocity joint in motor vehicle drive lines (Figure 19.14).

In basic construction it comprises a ball-and-socket joint with both parts grooved to embrace six rolling steel balls, which are held in a common plane by a controlling ball cage. The latter assembly thus constitutes the intermediate member of the joint and must be steered so that the six balls always lie in a plane which bisects the angle between the shafts, this being an essential condition for a constant-velocity drive (Figure 19.15). It is the manner of steering the ball

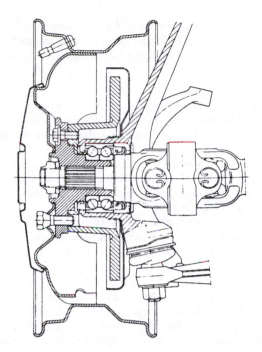

Figure 19.13 Double Hooke's joint and hub bearing arrangement for front-wheel drive (*SKF*)

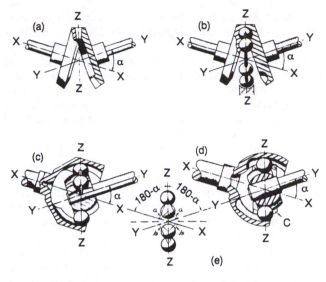

Figure 19.15 Constant-velocity principle of the Rzeppa-type joint (*Con-Vel Dana*)

(a) Constant-velocity bevel gear, fixed angle
(b) Constant-velocity gear, fixed angle: driving balls replace gear teeth and mesh with pockets in gears
(c) Constant-velocity joint, variable angle; transverse grooves in driving/driven members replace pockets
(d) Rzeppa constant-velocity joint: driving balls are engaged in transverse grooves and maintained in an angle bisecting plane *Z-Z* by ball cage C
(e) Constant-velocity principle: driving balls move in a plane bisecting the angle included between shaft axes *Z-Z*: every ball in every position maintains equal distance from both axes *X* and *Y*

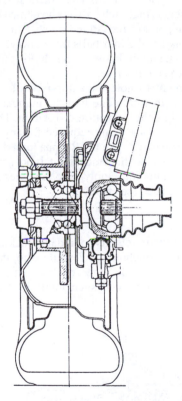

Figure 19.14 Rzeppa joint and hub bearing arrangement for front-wheel drive (*SKF*)

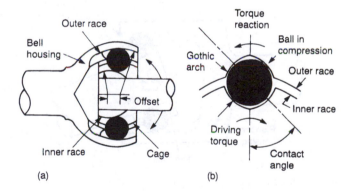

Figure 19.16 Track-steered Rzeppa-type joint and groove form

cage that is one of two features which distinguish the early from the later designs of Rzeppa joint. As originally designed, the ball cage was steered by the toggling action of a pilot pin, the ball ends of which registered in the ends of the driving and driven shafts, with endwise movements of the pin being allowed only in the latter. In later designs, by Hardy Spicer in England and Lobro in Germany, the ball cage was track steered by the ingenious expedient of offsetting the geometric centres of the ball-and-socket grooves by equal amounts from the true centre of the joint (Figure 19.16a). Whenever there is articulation of the driving and driven shafts, the deliberate mismatch in curvature of the ball-and-socket grooves creates a wedging effect on the caged balls, until they retreat into a plane which again bisects the angle between the shafts. The same effect has been produced in a further development of the Rzeppa joint by offsetting the geometric centres of the cage inner and outer surfaces, so that the balls are cage rather than track steered (Figure 19.17). This in turn allows the use of straight instead of curved ball grooves in the socket housing, so that a certain amount of plunging as well as complete universal action of the drive shaft can be accommodated within the joint itself, thereby eliminating any need for a splined coupling. Track- and cage-steered Rzeppa joints are therefore sometimes also referred to as fixed and plunging joints respectively (Figure 19.18).

The second feature which distinguishes the early from the later designs of Rzeppa joint is related to the cross-sectional shape of the ball grooves. Originally these were of purely circular form, so that torque was transmitted with the balls loaded in shear. This could lead to severe edge loading and wearing of the grooves, which then resulted in noisy operation of the joint. To overcome this problem, Hardy-Spicer has long since modified the form of the grooves to an elliptical or Gothic arch cross-section (Figure 19.16b), so that the balls not only assume an angular contact away from the edges of the grooves, but also become advantageously loaded in compression rather than shear. A Gothic arch form of groove is so called after the pointed style of architecture that was prevalent in Western Europe from the twelfth to the sixteenth centuries. The Hardy-Spicer version of Rzeppa joint incorporating this feature and also a track-steered ball cage were introduced in the late 1950s on the Austin-Morris Mini cars. In modern form, the Rzeppa type of joint offers the several

advantages of being structurally self-supporting and compact in relation to its torque capacity, and having the ability to operate quietly at high joint angles up to about 45°.

Weiss joint

So named after its inventor C.W. Weiss, who was a pioneer figure in the American diesel engine industry, it was patented by him in 1923 and represented the first of the ball-type constant-velocity joints. Rather interestingly the Weiss joint (Figure 19.19) was originally intended to replace the enclosed single Hooke's joint used behind the gearbox of the then popular torque-tube-drive rear axles. These will be referred to again later (Section 19.4).

The innovative feature of the Weiss joint was the use of four rolling steel balls interposed between the interlocking forked ends of the driving and driven shafts. Location of the balls is by means of curved tracks machined in the opposed faces of the forked ends, each of which resembles a pair of jaws. These are separated endwise by a centring ball about which the two shafts articulate. At this juncture it is important to appreciate that the geometric centre of the curved tracks in each pair of jaws is again deliberately offset from the true centre of the joint. Hence, the opposing ball tracks cross each other even when the driving and driven shafts are in alignment (Figure 19.19b). The practical significance of this arrangement is that each ball has no choice but to remain at the crossing point of the curved tracks that embrace it, so that in combination the four balls must always lie in a plane which bisects the angle between the shafts, thereby providing a constant-velocity drive.

A further point to note with this type of ball constant-velocity joint is that only one-half the number of balls will transmit drive in either direction of rotation. This can result in a larger size of joint being required for any given torque capacity, in comparison with the Rzeppa joint described earlier. Also, the Weiss, like the Tracta joint, is not self-supporting, so that similar limitations apply as to the types of vehicle on which it may be used (see below).

Tracta joint

This type of constant-velocity joint dates back to 1926 and is attributed to the Frenchman P. Fenaille who, in conjunction with another pioneer designer of front-wheel-drive cars, J.A. Gregoire, incorporated such joints in the (previously

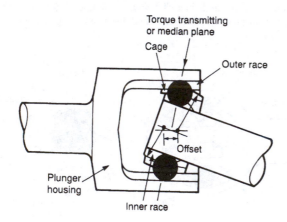

Figure 19.17 Cage-steered Rzeppa-type joint

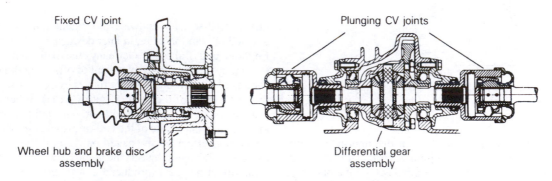

Figure 19.18 Typical application of Rzeppa-type fixed and plunging joints (*GKN Hardy Spicer*)

mentioned) Automobiles Tracta cars: hence the name of the joint (Figure 19.20).

In construction the driving and driven shafts are provided with forked ends of semicircular form, which in turn are free to articulate about a correspondingly grooved intermediate member. The latter is actually a two-piece assembly with a spigot and slotted joint that permits relative circular motion in a plane at right angles to the forked ends of the shafts. It is thus possible for the two-piece intermediate member to adjust itself continuously to their angular displacement, thereby providing a universal action with constant-velocity drive. It will be evident that a distinguishing feature of this particular type of joint is that all relative motions occur as sliding, rather than rolling, contact between the working parts. However, the bearing surfaces are of fairly generous area and should be expected to provide a low rate of wear. On the other hand, the sliding friction at the bearing surfaces can result in overheating of the joint when articulating under high torque loading conditions, and this can impose a limit on its usefulness.

Another disadvantage of this type of joint, at least for the independently sprung driving wheels of the modern car, is that it is not self-supporting. That is, the driven shaft requires the external bearing supports that can only really be furnished by a live axle with steerable wheels. For this reason its later development in the automotive field was mainly confined to the driven front wheels of specialized heavy-duty and military vehicles, where it provided a simple, rugged and compact joint capable of operating at high joint angles.

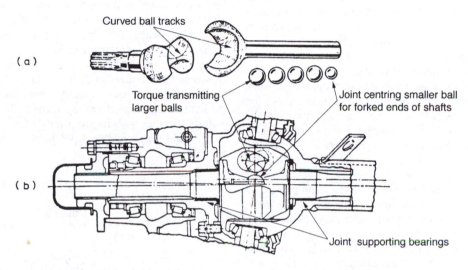

Figure 19.19 Weiss-type joint in front-wheel-drive axle: (a) components (b) practical application

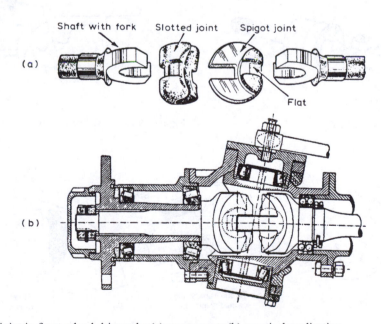

Figure 19.20 Tracta-type joint in front-wheel-drive axle: (a) components (b) practical application

Tripot joint

As its designation would suggest, this type of pot joint features a tripod member comprising three equispaced radial pins and part-spherical rollers (Figure 19.21) and may be regarded as a further development of the bipot joint. It was introduced in its present-day form by the French firm of Glaenzer-Spicer, who in the early 1960s called it the Tripode universal joint. The tripot joint shares the same advantage as the bipot one in that it will accommodate endwise movement within the joint itself with a minimum of frictional resistance, since the partspherical rollers are themselves supported on the tripins by needle roller bearings. Its principal advantage over the bipot joint is that it will transmit a very nearly constant-velocity drive, which perhaps calls for a little more explanation.

From the descriptions given of the double Hooke's, Tracta, Weiss and Rzeppa joints, it may be recalled that in all cases the intermediate member of the joint (such as the ball cage in the Rzeppa joint) is constrained always to lie in a plane which bisects the angle between the driving and driven shafts. It is, therefore, somewhat difficult to visualize how the tripot joint without such an intermediate member can provide this, since the median plane always lies perpendicular to the axis of the drive shaft (Figure 19.22a). In actual practice, however, the variations in angular velocity are largely compensated for by the tripod member weaving very slightly off its own centre three times per revolution, so that to all intents and purposes the tripot joint is a constant-velocity one. This slight weaving motion of the tripod member is, of course, accompanied by small radial oscillations of the spider pins within the part-spherical rollers, the frequency of these movements being twice the rotational speed of the shaft. To minimize not only scuffing but also edge contact between the rollers and their embracing tracks during high torque transmission, each part-spherical roller has an edge radius that is slightly less than that of the track curvature (Figure 19.22b). This same principle was incorporated in later versions of the bipot universal joint (Figure 19.6).

So far we have considered only the original plunging type of tripot joint. In the late 1960s there was a further development of the tripot joint in the form of a fixed type, which was devised by M. Orain of the earlier mentioned firm of Glaenzer-Spicer. This version of tripot joint made it eminently suitable for the outboard station or wheel end of front-wheel-drive shafts, since it can operate at relatively large angles of articulation. The tripod spider is rigidly held in the joint housing and supports the part-spherical rollers. These are engaged by the drive shaft, the forked end of which has tulip-shaped prongs that form the roller tracks. Axial location of the shaft relative to the housing tripod is derived from a retaining clip, which embraces the ends of the shaft prongs and makes part-spherical contact with the central boss of the tripod spider. This location is maintained by an opposing spring-loaded centring button, which emerges from the other side of the central boss and similarly makes part-spherical contact with the root of the shaft prongs (Figure 19.23).

The plunging and fixed types of tripot joint are also known as 'end motion' and 'fixed centre' types respectively. Each has a good torque capacity for its diameter by virtue of the load being divided among three rolling elements. This feature, together with its simplicity of construction, has led to the wide usage of the tripot joint of the drive shafts of front-wheel-drive cars.

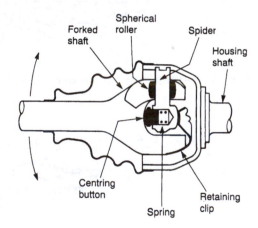

Figure 19.23   Fixed centre tripot joint for steered wheel

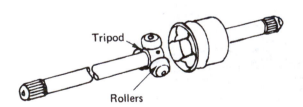

Figure 19.21   Tripot-type joint (*SAAB*)

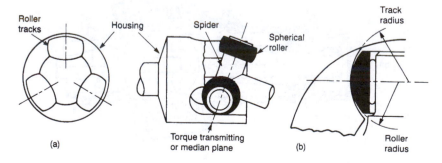

Figure 19.22   End motion tripot joint and track form

Lubrication and sealing of constant-velocity joints

The double Hooke's, Rzeppa and tripot types of constant-velocity joint are generally grease lubricated for life and sealed by a synthetic rubber sealing boot when used on the drive shafts for independently sprung wheels. This form of lubrication and sealing is not required when double Hooke's, Tracta, Weiss and Rzeppa joints are installed in live axles with steerable wheels, because part-spherical enclosures are provided at the steering pivots to seal them and they can receive their lubrication from the same source as that used for the final drive gears (Figures 19.19 and 19.20).

19.3   PROPELLER SHAFT CONSTRUCTION

One-piece propeller shafts

The general construction of the propeller shaft is as follows (Figure 19.24).

Shaft   The main body of the propeller shaft comprises a steel tube into the ends of which are welded steel stub portions forming part of the end joints. Chiefly to reduce the transfer of any noise from the rear axle to the car interior, some designs of propeller shaft simply have inserted cardboard sleeves to dampen vibration of the steel tube, while others may incorporate a rubber sleeve sandwiched between a tubular extension on either, or both, of the stub portions and the inside of the main tube (Figure 19.25). This also introduces a certain amount of torsional resilience into the propeller shaft assembly to cushion shock loads through the drive line. A further development has been the first production fitment by Renault of a composite material propeller shaft. This is of hollow construction and comprises 75 per cent carbon and 25 per cent glass-fibre with bonded steel end fittings, and provides an appreciable saving in weight over an all-metal equivalent shaft.

Hinged joints   Each end of the propeller shaft is provided with a hinged joint of the universal type, so that within practical limitations the shaft can articulate in any direction. The hinged joints thus allow the shaft to adjust its position to the changing angle of the drive line at all times as it rotates.

Sliding joint   Some from of sliding or telescopic joint is required in the propeller shaft assembly, so that it can accommodate itself to small variations in effective length of the drive line, as explained later. A splined coupling is therefore typically incorporated between the front universal joint and either the main body of the propeller shaft (Figure 19.24) or the tailshaft of the gearbox. The latter arrangement is widely used in cars. It requires the addition of a supporting bush acting both as a journal and a sliding bearing, but since this and the splines are continuously lubricated from the gearbox, the need for hand lubrication of the sliding joint is dispensed with in service (Figure 17.6).

In 2001 the Visteon company in America introduced a new concept in propeller shaft construction, known as 'Slip-in-Tube'. This eliminates the need for splined stub portions, because the propeller shaft simply comprises part length inner and outer tubes of either steel, or aluminium. These are free to slide co-axially, their integral male and female splines being cold formed so that no material is removed from the tubes by machining. A convoluted plastics sealing boot protects the emerging portion of the splines. The advantages claimed for this construction include lighter weight, quieter operation, enhanced durability and controlled telescoping to accommodate endwise movement of the power unit safely in a crash situation.

To ensure that a uniform speed is transmitted from the gearbox shaft to the axle shaft, two conditions must be met

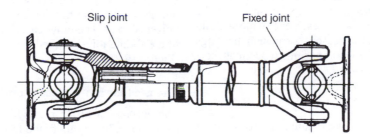

Figure 19.24   General arrangement of a commercial vehicle propeller shaft (*GKN Hardy Spicer*)

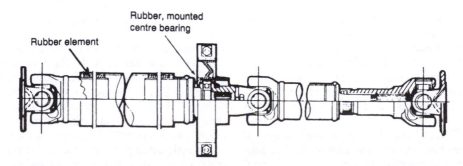

Figure 19.25   Divided propeller shaft assembly (*GKN Hardy Spicer*)

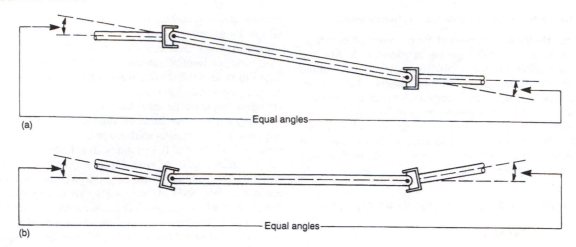

Figure 19.26 Conditions for constant-velocity drive with two Hooke's joints

by a conventional installation of propeller shaft with Hooke's universal joints:

1 The gearbox and final drive shafts must form equal angles with the propeller shaft (Figures 19.26a and b).
2 The axes of the joint yokes at each end of the propeller shaft must lie in the same plane (Figure 19.8).

The first condition is essentially a problem for the designer and must of necessity represent something of a compromise because of the complex axle motions occurring in a Hotchkiss drive system, as described later. The second condition can be guaranteed only by correct initial build of the propeller shaft and, perhaps more to the point, correct reassembly of the splined joint propeller shaft during any subsequent overhaul in service. A source of error here lies in the possibility of randomly inserting the splines of the propeller shaft into the universal joint, such that the yokes at each end of the shaft do not lie in the same plane. For this reason, correlation markings are usually to be found on these components to facilitate correct alignment during assembly.

Two-piece propeller shafts

For reasons that will be discussed later, a two-piece or divided propeller shaft is often employed in the drive lines of front-engine RWD vehicles (Figure 19.25). This type of assembly comprises the following basic components:

Primary shaft This forms the front section of the propeller shaft, its forward end connecting to the universal joint behind the gearbox. At its other end, the primary shaft is shouldered to accept a support bearing. It is also splined either externally or internally, the former to provide a rigid connection for the front universal joint of the secondary shaft, the latter to form a sliding connection for this joint according to design requirements.

Centre bearing Through being located on the rear and of the primary shaft, this bearing serves as an intermediate support for the propeller shaft assembly. In some early commercial vehicle applications, the centre bearing housing was originally attached to the chassis frame and therefore incorporated a self-aligning bearing of the ball or roller types. Later practice for both commercial vehicles and cars saw the introduction of various forms of rubber mountings for the centre bearing housing to dampen vibrations and permit slight misalignments of the propeller shaft assembly (Figure 19.25). A deep-groove ball bearing race prepacked with grease is now generally used, in conjunction with either integral seals or separate lip-type seals and augmented by finger rings.

Secondary shaft This forms the rear section of the propeller shaft and connects to the rear axle final drive shaft. Unlike the primary shaft, the secondary shaft has to accommodate movements of the live rear axle on its springs. It is therefore provided with two universal joints, one at each end. In the absence of a sliding joint at either end of the primary shaft, a sliding joint is provided between the main body of the secondary shaft and either the front or, less commonly, the rear universal joint. The only reason for using a rear end location of the sliding joint is to reduce loading on the centre bearing (Figure 19.25).

The need for a two-piece propeller shaft

The purpose of fitting a divided propeller shaft to a motor car is generally to satisfy the following requirements:

1 To provide the lowest possible transmission tunnel height, thereby increasing rear seat foot room.
2 To raise the critical speed of the propeller shaft assembly, without resorting to tubing of unduly large diameter.

The first requirement was originally enforced by styling considerations and the demand for a reduction in car height, which in turn called for some lowering of the body floor line (Figure 19.27). Unfortunately, this also meant that, in order to maintain an adequate working clearance for the propeller shaft beneath, the height of the transmission tunnel could seriously intrude upon the foot room for the rear seat passengers.

An early attempt to overcome this problem consisted of inclining the axes of the engine and gearbox and the rear axle final drive shaft, so that the lowered propeller shaft

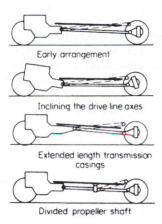

Early arrangement

Inclining the drive line axes

Extended length transmission casings

Divided propeller shaft

Figure 19.27 Comparison of methods used to lower the body floor relative to the transmission drive line

made equal but opposite angles with its connecting shafts (Figure 19.26b). Another development sought to reduce the length of the swinging propeller shaft by providing an extension housing for the gearbox and, in some cases, an extended nose casing for the final drive. The later introduction of the divided propeller shaft proved to offer the best solution to this problem because of the reduced working space required by the shorter moving part of the assembly.

The second requirement calls for some explanation. In common with many other power transmission shafts used in engineering, the propeller shaft is subject to high rotational speeds and, by its nature, is of appreciable length in relation to its diameter. These factors would be of little account were it possible in practice to produce and maintain a shaft in perfect balance, so that its centre of gravity always coincided with its axis of rotation. Any departure from this ideal running condition generates a centrifugal force which tends to bend the shaft. Although the shaft offers increasing resistance to this bending it eventually reaches a rotational speed at which violent vibration occurs and which may lead to the shaft being whirled around in the manner of a skipping rope. This dangerous critical or whirling speed of the shaft is reached when the number of natural vibrations and the number of revolutions per second are equal. If, however, a long propeller shaft is divided into shorter sections, each can then have a much higher critical speed lying safely beyond the upper speed range of the car.

The explanation for this behaviour is that the critical speed of a shaft is inversely proportional to the square of its length. For example, the critical speed will be increased four times if the length of the shaft is halved. On the other hand, an unacceptably large increase in the diameter of the shaft, far exceeding that required to transmit maximum driving torque, would be required to achieve the same gain in critical speed.

So far as commercial vehicle applications are concerned, their wheelbase lengths are often such that a two-piece, or in some cases a similarly supported three-piece propeller shaft, is an essential requirement. Where tandem rear axles are used, a further shaft known as an inter-axle or coupling shaft is required to transmit the drive from one axle to the other (Figure 19.47).

Drive line vibration in service

Complaints of vibration from the transmission drive line may conveniently be categorized as follows:

Propeller shaft vibration at medium to high speeds
Propeller shaft vibration during acceleration and overrun
Propeller shaft vibration on moving away from rest.

Propeller shaft vibration at medium to high speeds
A typical test procedure for confirming this type of complaint is momentarily to slip the gear lever into neutral and release the accelerator, while the vehicle is in the speed range under criticism, then note if the vibration still persists. If it does then the engine and its ancillaries can be ruled out as a possible cause.

The probable causes of this type of propeller shaft vibration may be listed as follows:

1 Worn spider trunnion pins and yoke journal bearings in the universal joints.
2 Worn splined connection of the propeller shaft sliding joint.
3 Worn journal bushing in rear extension housing of the gearbox.
4 Loosening of the bolts and nuts of the propeller shaft coupling flanges.
5 Random substitution of coupling flange bolts, washers and nuts with others that do not have the same weight.
6 Mud or carelessly sprayed undercoating clinging to the propeller shaft tube.
7 Balance correction weights adrift from propeller shaft tube.
8 Damage to propeller shaft caused by accidental denting or bending of tube.
9 Run-out of coupling flange in propeller shaft to rear axle connection.
10 Unserviceable centre bearing in two-piece propeller shaft assembly.

Propeller shaft vibration during acceleration and overrun
A complaint of this type may usually be identified as a rumbling disturbance in the drive line and may also be accompanied by a certain amount of gear rattle. The disturbance is essentially one of torsional vibration arising from non-constant-velocity effects in the drive line. Any tendency towards this type of propeller shaft vibration will be aggravated in the presence of the following factors:

1 Improperly phased universal joint yokes, the significance of which was earlier explained.
2 Incorrect installation angles of the power unit and rear axle. A manufacturer may require these angles to be checked by using a special form of spirit-level protractor gauge or equivalent method (Figure 19.28).

Propeller shaft vibration on moving away from rest
This type of complaint usually manifests itself as a shudder in the drive line that occurs at a speed of 8–12 mile/h (13–19 km/h). It is most likely to be confined to two-piece propeller shaft installations, and arises from shaking forces that become obtrusive when both the angularity and the torque

load present at the rearmost universal joint, connecting the rear section of the propeller shaft to the rear axle, approach their maximum values (Figure 19.33). By its very nature this type of drive line vibration is more amenable to alleviation than to cure, and usually requires attention to the following items:

1 Checking for correct installation angles of the power unit, propeller shafts and rear axle in accordance with the particular manufacturer's instructions.
2 Inspecting for any deterioration in the flexible mounting and security of attachment of the centre bearing for the propeller shafts.
3 Examining for any deficiencies in the rear axle to suspension mountings that would encourage greater rocking of the axle in opposition to drive torque.

Rebalancing a propeller shaft in service

There are two basic approaches to rebalancing a propeller shaft in service, these being 'on the car' by trial and error methods and 'off the car' by using a balancing machine; the latter is usually confined to specialist drive line servicing.

The simplest and in some cases the only corrective action that a vehicle manufacturer may advise is to disconnect the propeller shaft at the companion flange of the rear axle pinion shaft, rotate the latter through 180° and reconnect the propeller shaft. If this simple reindexing and therefore alteration to the relative eccentricities of the drive line components does not reduce any imbalance effects present, then typically the vehicle manufacturer will advise that a replacement propeller shaft assembly be installed.

A basic method of attempting to correct propeller shaft imbalance in service is to introduce a counterbalancing effect, which may conveniently be provided by the lack of symmetry offered by a worm-drive hose clip. The hose clip may encircle either end of the propeller shaft, and its effect on reducing imbalance may be ascertained by moving its heavier head portion to various positions around the tubing. Clearly, the object of the exercise is to locate the head portion of the clip 180° from the 'heavy side' of the propeller shaft. Note that the hose clip should be positioned as close to either end of the propeller shaft as practicable, thereby minimizing any imbalance effects that can result from changes in deflection at the mid-span of the propeller shaft during high rotational speeds.

A final point concerned with the prevention, rather than cure, of drive line vibration is that correlation or match marks should always be made on the various components of the propeller shaft before any disconnection or actual dismantling is begun, thereby preserving as far as possible the original state of balance. From the safety point of view, it is essential that only the correct part number bolts and nuts, properly tightened, should be used when replacing a propeller shaft in service. Any loss of clamping load and separation at the mating faces of the companion flanges can at least introduce vibration into the drive line and at worst result in partial detachment of the propeller shaft, the latter with a potentially catastrophic effect on vehicle controllability.

In established modern design it is the friction forces acting between the mating faces of the companion flanges that are fully exploited to transmit the torque from one shaft to the other. The underlying engineering principle is that embodied in what the structural engineer has long known as a slip-resistant or friction grip joint. Here the externally applied load acts in a plane perpendicular to the bolt axis (as in our propeller shaft flange joint) and is completely transmitted by the frictional forces acting between the contact faces of the joint, which are very tightly clamped together by high-strength bolts (Figure 19.29). In other words, each bolt is subjected only to the torsional and tensile stresses of tightening and 'doesn't know' that it is indirectly resisting a shear load. The absence of a fidget-provoking shear load is an important factor in lengthening the fatigue life of any bolted joint.

It thus follows that the frictional resistance developed in the modern slip-resistant joint depends partly on the bolt preload, which explains why a torque tightening value is now usually specified for propeller shaft flange nuts, and partly on the coefficient of friction existing between the joint interface, which also explains why the mating surfaces of propeller shaft companion flanges must be clean and dry on assembly.

In the case of heavy vehicle propeller shaft installations, the companion flanges are now usually provided with mating sectors of radial serrations, which are confined to areas immediately behind the yokes of the universal joint.

19.4   DRIVE LINE ARRANGEMENTS

The Hotchkiss drive system

With front-engine RWD vehicles a propeller shaft is required to transmit the driving effort from the gearbox to

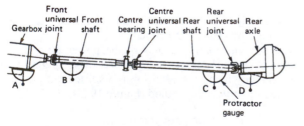

Figure 19.28   Checking drive line installation angles

A power unit angle
B front shaft angle
C rear shaft angle
D rear axle angle

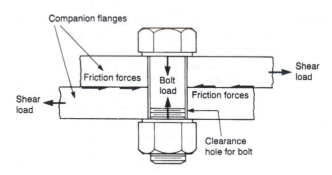

Figure 19.29   Basic principle of slip-resistant joint

the final drive, which in turn is mounted either in a sprung rear axle (Figure 19.30) or, in the case of independently sprung and driven rear wheels, in a separate housing attached to the vehicle structure (Figure 19.10).

At this stage, we shall confine our attention to the former arrangement of a propeller shaft driving a rear axle mounted solely on semi-elliptic leaf springs which thus combine the functions of spring medium and locating linkage. This method of transmitting drive to the wheels, through the cushioning effect of what is termed a live axle on its springs, dates from the early 1900s and is well known as the Hotchkiss drive principle (Figure 19.31). It became so named because it was introduced by the once famous Hotchkiss company in France.

Although simple in concept, the Hotchkiss drive imposes exacting conditions on the operation of the propeller shaft. The shaft must be of the open type and therefore, unlike many other applications of transmission shafting, it is not supported in fixed bearings. The complete assembly of propeller shaft and universal joints must therefore be accurately balanced, otherwise it would excite severe vibrations in the drive line.

It was earlier mentioned that a sliding joint is incorporated in the propeller shaft assembly, so that it can adjust itself to the fluctuations in effective length of the drive line between the gearbox and the rear axle. These lengthwise variations are enforced mainly by the Hotchkiss drive system and require some explanation:

1 With suspension movements of the rear wheels and body, the arc of travel of the propeller shaft is much shallower than that of the rear axle (Figure 19.32). This is because the propeller shaft swings about the centre of its front universal joint at the back of the gearbox, whereas the rear axle swings about an imaginary axis that lies a short distance behind the front pivots of its semi-elliptic leaf springs.

2 During acceleration and braking, the arcs of travel of the propeller shaft and rear axle not only differ in curvature, they also conflicts in direction (Figure 19.33). This arises because the torque reaction on the axle casing tends to make it rock about an imaginary axis lying a short distance behind its attachment to the leaf springs.

3 To reduce the transmission of vibration and noise extensive use is made of rubber mountings for supporting the engine and gearbox units and also for attaching the rear axle and springs to the vehicle structure. These groups of components are therefore permitted a limited amount of fore-and-aft movement relative to each other.

From the above considerations, it will be evident that unless the propeller shaft assembly is allowed a certain freedom of telescopic movement, destructively high end thrust loads would be imposed on the bearings of the gearbox and rear axle shafts to which it is attached. The telescopic joint will, in any event, always transmit some end thrust due to friction between the sliding joint splines.

Fortunately, this friction tends to be lessened by the torque fluctuations present in the drive line and, of course, by proper

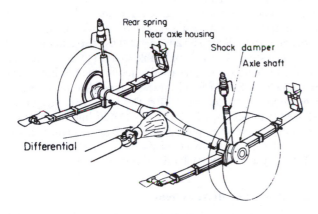

Figure 19.30 A later application of the Hotchkiss drive system with telescopic shock dampers (*Toyota*)

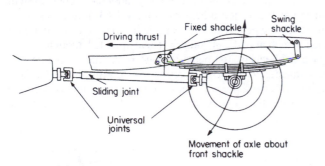

Figure 19.31 Layout of the Hotchkiss drive system

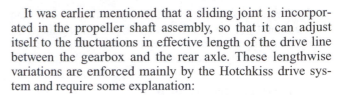
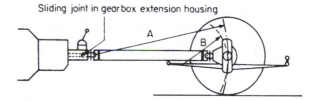

Figure 19.32 Propeller shaft movements due to suspension travel of the rear wheels and axle. The sliding joint in the gearbox extension housing allows the axle to follow its natural arc of movement B instead of conflicting with the natural arc of propeller shaft movement A

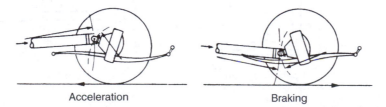

Figure 19.33 Axle casing torque reaction and the need for a sliding joint and rear universal joint in the propeller shaft

lubrication of the splines. To reduce further the friction in the sliding joint, there have been some designs of propeller shaft where the splined connection is replaced by tracks and rollers which afford rolling instead of sliding friction.

Modern live axle drive systems

Where still used in modern practice, the live axle usually forms part of a coil spring rear suspension system, a trend that began on American cars as long ago as the late 1930s. To relieve the coil springs of all axle locating duties, the axle is mounted to the car through a system of pivoted links. These do, of course, exercise a more accurate control over suspension movements of the rear axle than is possible with purely leaf spring installations. Comprehensive locating systems for live axles may be conveniently classified by their number of longitudinally disposed (or nearly so) links, of which there may be either two, three or four. These systems will now be identified by their basic geometric layouts, rather than by relating their geometric effects to vehicle handling qualities.

Two-link system

This was an early development and comprised two trailing arms pivoting at their front ends to the car chassis frame and at their rear ends to the axle casing extremities (Figure 19.34a). Rubber bushings were used at both ends of each arm, but those at the rear were so arranged that the axle was restrained from rocking about the arms under torque reaction. Since the two trailing arms could not provide for lateral location of the axle, this duty was performed by a Panhard rod, so named after the French make of car on which it first appeared in the late 1930s. This consists simply of a rubber-bushed transverse control rod pivoting at one end from the car chassis frame or structure and at its other end to the far side of the axle assembly. It thus confers positive lateral location on the rear axle, albeit at the expense of slight relative sideways movement at the rear end of the car incurred by the arcuate swing of the rod.

Three-link system

This was a later system and again involved the use of two trailing arms, but in this case they are freely pivoted to the axle casing and their action is complemented by a third and shorter trailing arm, such that in combination they form a parallelogram linkage as viewed from the side of the car (Figure 19.34b). The triangulated third arm is usually placed uppermost with a pair of widely spaced rubber bushes at its forward end. It is ball jointed at its apex to the centre part of the axle casing, thereby furnishing the necessary lateral location for the axle. Less commonly, the third arm may take the form of a simple connecting link, in which case a Panhard rod must be added for lateral location.

Four-link system

With the current four-link system, one pair of trailing arms is pivoted below the axle and another pair of either semi-trailing or trailing arms is pivoted above it (Figures 19.34c and d). If in plan view both upper and lower pairs of arms lie substantially parallel to the centerline of the car, then a Panhard rod must be added to locate the axle laterally. Alternatively, the upper pair of arms may be disposed in a semi-trailing fashion, so that by virtue of their diagonal mounting they stabilize the axle in the lateral direction. A quieter ride with reduced harshness has often been attributed to the modern four-link system of axle location, this being derived from the isolating effect of the multiplicity of rubber-bushed pivot connection involved.

In conclusion, it may be added that these two and three-link locating systems are also applicable to dead axle rear suspensions, which are sometimes used to mount the trailing wheels of FWD cars. Similarly all these systems have been used in conjunction with the now rare De Dion rear suspension referred to in Sections 19.1 and 20.4. Multi-link locating systems continue to be employed where coil sprung rear axles are retained for the currently popular 4 × 4 vehicles, typically the four-link system with Panhard rod (Figure 19.34d).

The torque-tube drive system

Although no longer employed in its original form, an early rival to the Hotchkiss drive system (and paradoxically later used by Hotchkiss themselves) was another French system known as the torque-tube drive (Figure 19.35). In this system the revolving propeller shaft was totally enclosed within a rigid tube, which at its front end was usually pivoted about a hollow spherical bearing just behind the gearbox, and at its rear end was attached to the nose piece of the axle. Adjustable diagonal radius rods were generally utilized to maintain the torque tube in proper alignment with the rear axle. Since the relative

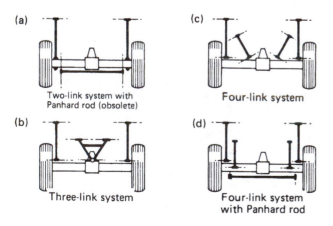

Figure 19.34   (a)–(d) identification of live axle location systems by defining their pivot axes in plan view (*Talbot*)

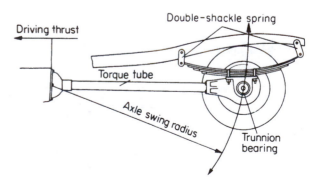

Figure 19.35   Layout of full torque-tube system

movement between the rear axle and the gearbox was confined to a simple angular one, the inner propeller shaft required only a single universal joint at its front end. However, it was later recognized that the fluctuations in rotational speed inherent with this arrangement could be a disadvantage, as inferred earlier in connection with the Weiss constant-velocity joint.

Where semi-elliptic leaf springs were used for the rear suspension they had to be shackled at both ends, so as to accommodate the swinging motion of the torque tube and axle with rise and fall of the road wheels. Trunnion bearings were ideally, but not always in practice, provided for anchoring the axle to each spring. It should be appreciated that the fore-and-aft location of the axle was finally dependent upon the engine and gearbox mountings, since end thrust loads due to wheel traction and braking were transmitted to them via the spherical bearing of the torque tube.

Semi-torque-tube drive

General Motors Opel in Germany introduced this compromise form of torque-tube drive in the early 1960s, following which there were several other examples of its use.

It comprises an abbreviated-length torque tube, or pinion housing extension, which is flange mounted to the nose of the axle (Figure 19.36). The front end of the torque tube is pivoted from various forms of rubber bearing supported from the understructure of the car body. Within the front end of the torque tube is the equivalent of a propeller shaft centre bearing. This carries the input end of either a solid or tubular secondary drive shaft, the output end of which is connected to the final drive bevel pinion shaft through a splined muff coupling. This comprises simply a cylindrical collar with internal splines that is pressed over the splined ends of the two coaxial shafts being connected, approximately one-half of the collar contacting the end of each shaft. In its various forms the muff coupling has its origins in very early engineering practice, since a common problem in mechanical design is that of rigidly connecting together two shafts.

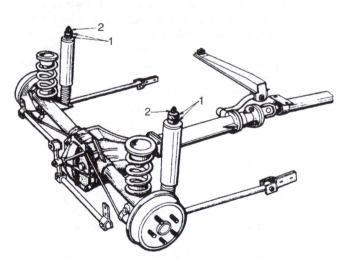

Figure 19.36 A semi-torque-tube system with longitudinal links, Watt lateral linkage, coil springs and telescopic shock dampers (*Fichtel & Sachs*) (1) shock damper mounting rubbers (2) securing nuts

A splined connection is also provided for a coupling flange at the front end of the secondary drive shaft, which in turn connects to the rear universal joint of an open primary propeller shaft. The latter is equipped with either two Hooke's universals and a sliding joint, or simply two constant-velocity joints of the plunging type, since a limited amount of fore-and-aft and vertical movements of the rubber-mounted torque tube and axle must be permitted relative to the power unit. To assist in maintaining alignment between the semi-torque tube and axle, rubber-bushed longitudinal links pivot between the axle ends and brackets attached to the body structure. The radius of articulation of these links generally coincides with that of the axle and torque tube about its forward rubber bearing. To prevent sideways movement of the axle a separate means of control must be used, such as a Panhard rod or Watt linkage (Figure 19.36).

The utilization of torque-tube drive

To appreciate why the full torque-tube drive has long been obsolete and the semi version has received limited acceptance, it is necessary to make a few general comparisons not only between these two forms of torque-tube drive, but also to relate them to the rival Hotchkiss drive system.

Full torque-tube v. Hotchkiss drive

1 The alternative full torque-tube drive immediately relieved the leaf springs of axle casing torque reaction, together with the thrust loads associated with traction and braking of the wheels.
2 Since with torque-tube drive the springs are left to perform only their vehicle supporting function they can be made more flexible to improve ride comfort.
3 Similarly, the adoption of torque-tube drive allowed the early use of friction-free coil springs for non-independent rear suspension, again to improve ride.
4 Full torque-tube drive required the use of only one universal joint in the drive line, albeit at the time a non-constant-velocity one with its usual disadvantage.
5 A major disadvantage of full torque-tube drive was that it increased the unsprung weight of the rear suspension system, which could adversely affect road holding.
6 The axle casing was subjected to greater bending loads in the horizontal plane, because of the traction and braking thrust loads being transmitted to the vehicle mainly by the central torque tube.
7 Undesirable transmission noise could be telegraphed to the car interior by the hollow torque tube, which thus acted as a sound amplifier.
8 The full torque-tube drive generally demanded a more elaborate overhaul procedure with consequent higher repair costs.

Semi v. full torque-tube drive

1 In comprising little more than a pinion housing extension, the semi-torque-tube drive is at much less disadvantage in terms of rear suspension unsprung weight.
2 By virtue of its rubber-mounting system the semi-torque tube with separate locating links provides better isolation from road noise and transmission vibrations.

3 The semi-torque tube permits a reduced height of transmission tunnel in the body floor structure, thereby improving rear seat passenger accommodation.

4 For additional refinement the short primary propeller shaft can be of one-piece construction and embody modern constant-velocity joints of the plunging type.

5 In reacting crown wheel torque (not pinion torque, as is sometimes inferred) the semi-torque tube is of a more convenient length to act as a short lever arm and reduce rear end squat of the car during acceleration. Apart from this anti-squat effect it likewise confers an anti-dive effect when the axle casing is resisting brake torque reaction.

6 The semi-torque tube still remains more complicated than a Hotchkiss drive, but is less complicated than the driving arrangements for independently sprung rear wheels, with which system of suspension it has often been favourably compared.

19.5 REAR-WHEEL DRIVE AND FRONT-WHEEL DRIVE LAYOUTS

General layout of passenger cars

As far as the general layout (as opposed to the styling) of a car is concerned, it is the disposition of the power unit and transmission units relative to either the rear or the front wheels that constitutes its main distinguishing feature. The different arrangements of these units may be conveniently categorized as follows:

Front engine, rear-wheel drive (RWD)
Front engine, front-wheel drive (FWD)
Rear engine, rear-wheel drive (RWD)

Front engine, rear-wheel drive

In long-established practice this takes the form of a fore-and-aft mounting for the engine combined into a single unit with either a manually or an automatically controlled gearbox. A universally jointed propeller shaft is employed to transmit the drive from the gearbox to the final drive, this being embodied in either a sprung rear axle (Figure 19.37a) or a unit separately attached to the car structure. In the latter case, a pair of short universally jointed drive shafts are required to transmit the drive to the rear wheels, which are usually independently sprung (Figure 19.37b).

An alternative and less common arrangement is where a rear-mounted gearbox is combined with the final drive unit and separately attached to the car structure for improved weight distribution. The combination of the gearbox and final drive into a single unit is sometimes referred to as a transaxle. Again, the rear wheels receive their drive from a pair of universally jointed drive shafts and may be mounted on a De Dion dead axle (Figure 19.37c).

The engine, either in unit with or separate from the gearbox, must be supported on a resilient, anti-vibration mounting system. This is necessary because all motor vehicle engines are subject to complex vibratory forces, which can conveniently be reduced in magnitude only by the spring effects of resilient mountings. These are almost invariably of the

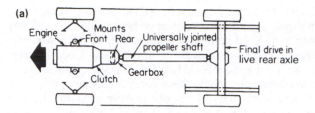

(a)

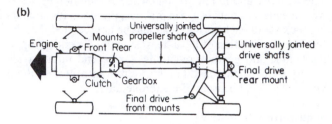

(b)

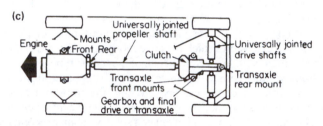

(c)

Figure 19.37 Layouts of front-engine RWD cars

rubber-to-metal bonded type. Various three-point mounting systems are employed, their basic aim being to allow the engine the greatest freedom of movement about its natural axis of oscillation or roll axis (Chapter 30). In a typical installation, the engine is supported at about mid-length by mounts on either side of its crankcase, with a third mount located on a central cross-member to support the rear end of the gear box.

A resilient mounting system is also required for the final drive unit if this is separately attached to the car structure. Again, a three-point mounting system is generally used, in which a pair of front mounts also serve to locate a subframe from which the independently sprung rear wheels are pivoted on links. The single rear mount then completes the location for the combination of final drive unit and rear suspension subframe, relative to the car structure (Figures 24.19 and 30.11b).

Front engine, front-wheel drive

With this particular layout the engine, gearbox and final drive are combined into a single unit, the gearbox and final drive forming a 'transaxle' drive to the independently sprung and steerable front wheels, via a pair of universally-jointed drive shafts. Where a fore-and-aft layout is used, the engine unit may be positioned either in front of or behind the transaxle (Figures 19.38a and b) according to design requirements. A much less common front-engine FWD arrangement is the positioning of the transaxle alongside a fore-and-aft mounted engine, the final drive unit extending ahead of the latter.

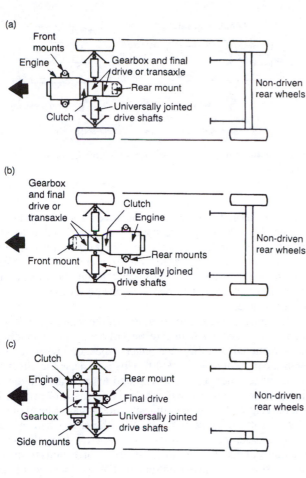

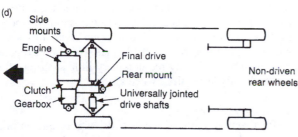

Figure 19.38 Layouts of front-engine FWD cars

already described for front-engine and RWD cars, except that the pair of mounts supporting the engine and the single mount supporting the gearbox are spread much further apart owing to the intervention of the final drive.

A transversely mounted engine and transaxle was originally supported by a lower mount at each end of the engine and gearbox, augmented by a single rubber-bushed upper link pivoting from the bulkhead to control fore-and-aft rocking movements of the entire unit, but current practice favours more sophisticated arrangements of the mounts as described in Section 30.1. It should be appreciated that such a mounting system must be capable of reacting against the full final drive torque, or turning effort, that is transmitted to the front wheels. This is not the case with a conventional front-engine RWD arrangement, where the reaction on the engine and gearbox mounting system is limited to the torque delivered by the gearbox to the propeller shaft before it is further multipled by the final drive gearing.

Rear engine, rear-wheel drive

In the now less popular rear-engined car, the engine is mounted either fore-and-aft behind a transaxle or transversely behind the gearbox and final drive units (Figures 19.39a and b). As might be anticipated, both these arrangements correspond closely to the similarly combined engine and transmission units of front-engine FWD cars. Three-point mounting systems are usually favoured for the engine and transmission systems of rear-engined cars. A single high mounting is used to support the rear end of a fore-and-aft mounted engine, the transaxle being supported at its front end by a widely spread lower pair of mounts. With a transversely mounted rear engine a widely spread pair of mounts supports the engine from behind, the gearbox and final drive units being located by a single front mounting.

It should perhaps be mentioned that in some very high-performance cars, the position of both fore-and-aft and transversely mounted engines may be reversed with respect to their transmission systems, but in these cases the layout is generally regarded as being mid engine rather than rear engine (Figure 19.39c).

Reasons for differences in layout

If it is accepted that most private cars are considered merely as a means of transportation, why then should there be such diversity in their general layout? The answer is that there must be a compromise, and this is not the same for every car because special emphasis must be placed on certain aspects of design according to the type and tradition of the particular manufacturer.

The advantages and disadvantages of the different arrangements of power unit and transmission systems may be considered under the following general headings:

Interior space
Ease of handling
Maintaining traction
Balanced braking
Other considerations.

Two basic arrangements of the widely adopted transversely mounted front engine and transaxle can be identified. In the early Issigonis version, the final drive was positioned alongside and behind the gearbox with the latter residing beneath the engine (Figure 19.38c). For the later Autobianchi and now most favoured version, the engine and gearbox are mounted in-line with the final drive positioned alongside and behind them (Figure 19.38d). In both arrangements the front wheels are driven through a pair of universally jointed drive shafts, which may not necessarily be equal in length.

The resilient mounting systems used for the combined engine and transmission systems of FWD cars similarly feature rubber-to-metal bonded mounts. In the case of a fore-and-aft mounted power unit and transaxle, a three-point mounting system is arranged in a manner very similar to that

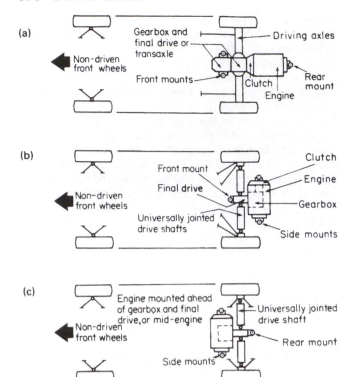

Figure 19.39   Layouts of rear-engine RWD cars

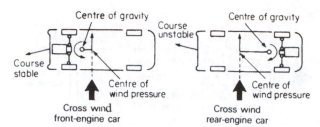

Figure 19.40   Simple representation of course stability and instability

Interior space

For any given dimension of wheelbase (distance between the front and rear wheel centres) the regrouping of the engine, gearbox and final drive into a combined unit mounted at either end of the car should result in more space being made available for passengers and luggage. This is because the underbody no longer requires a tunnel and a raised rear platform to accommodate the propeller shaft and either the sprung rear axle or a body-mounted final drive unit, as is the case with a front-engine RWD layout.

Since it follows that there must be more luggage storage space available between a pair of non-steered and non-driven wheels, the front-engine FWD car is at an advantage as compared to both rear-engine RWD and front-engine RWD layouts. It would therefore appear that a transversely mounted front-engine FWD unit should offer the best compromise in terms of both passenger and luggage accommodation, especially for small to medium cars.

Ease of handling

By virtue of its more nearly balanced laden weight distribution, the modern front-engine RWD car usually provides a very acceptable compromise in respect of general handling, directional stability and parking effort, although power-assisted steering is often required for medium to large cars.

The inherent front end heaviness of the front-engine FWD layout can nevertheless provide the small to medium car with excellent directional stability, and very safe handling derives from the driven front wheels. This is because car instability can usually be corrected by the natural reflex actions of the driver, which are to decelerate in the event of entering a corner too fast

and to steer away from the skid if the driven front wheels begin to slide. Both these actions could prove detrimental to car stability given similar circumstances with driven rear wheels. Parking effort is usually greater with front-engine FWD owing to the front end heaviness, unless either the number of steering wheel turns lock-to-lock is increased, or a power-assisted steering gear is fitted.

In contrast, the rear end heaviness of the rear-engine RWD can put it at a disadvantage in respect of both general handling and directional stability, the latter especially when crosswinds are present (Figure 19.40). This is because the effect of a cross-wind tends to be concentrated towards the middle to front portion of any car owing to its basic shape or profile. If then the centre of wind pressure acts substantially ahead of the vehicle centre of gravity, as must be the case where a rear engine is used, the car will tend to veer away from the wind direction and may need frequent steering corrections to maintain a desired course. (A front-engine FWD estate type of car, in which the centre of wind pressure acts behind the centre of gravity, should possess excellent directional stability, since in trying to steer into the crosswind it better maintains a true course or is said to be course stable.) As would be expected, the rear-engine car usually requires very little steering effort to turn the lightly laden front wheels.

Maintaining traction

The effectiveness with which the driving torque applied to the road wheels is translated into a tractive force to propel the vehicle, so that the road exerts a forwards force on the wheels, depends on the vertical load and the coefficient of friction existing between the tyres and road. Clearly, the coefficient of friction is a variable quantity since it depends upon the state of both the road surface and the tyres. So, too, is the vertical load on the tyres a variable quantity, as will be seen later.

Taking first the simple case of maintaining traction on a level but slippery road, it will be evident that the front-engine FWD and the rear-engine RWD layouts will be at an advantage over the front-engine RWD car, because the major portion of their weight is concentrated over the driven wheels. However, the front-engine RWD car has an advantage over the front-engine FWD car when towing, since the load on the driven wheels is then increased rather than decreased.

Suppose next the friction conditions between the road and tyres are normal but the car is being fiercely accelerated. In these circumstances an inertia force due to acceleration acts rearwards through the vehicle centre of gravity and is equal

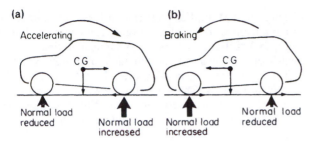

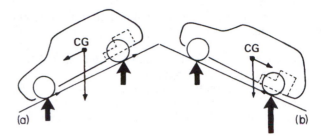

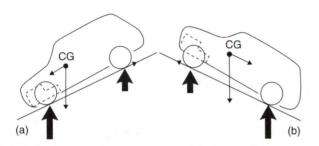

Figure 19.41 Pitching of car during (a) acceleration and (b) braking owing to load transfer between the front and rear wheels, shown for RWD

Figure 19.43 Effects on braking and traction with a rear-engine RWD car (rear end heavy): (a) normal load on front wheels increased, on driven rear wheels decreased, better balanced braking (b) normal load on driven rear wheels increased, better traction

Figure 19.42 Effects on braking and traction with a front-engine FWD car (front end heavy): (a) normal load on rear wheels reduced, brakes lock more easily (b) normal load on driven front wheels reduced, less traction

in magnitude to the tractive forces existing between the tyres of the driven wheels and the road. The effect of this inertia force acting at the height of the centre of gravity is to impose a pitching couple on the car, thus increasing the load on the rear wheels and reducing it on the front ones (Figure 19.41a). As compared with the front-engine FWD layout, both rear-engine RWD and front-engine RWD cars will therefore better maintain traction under fierce acceleration, because of the temporarily increased load on their driven wheels.

The latter is also true, of course, when the car is being driven uphill. Additional weight is then transferred on to the driven rear wheels as a consequence of the gravitational force on the vehicle acting vertically downwards through its centre of gravity and thus moving closer to the rear wheels at ground level. Purely for traction reasons, the rear-engine RWD layout would therefore appear to offer the best compromise (Figure 19.43b), with the front engine FWD layout coming off the worst (Figure 19.42b).

Balanced braking
Unbalanced braking between the front and rear wheels leads to instability of the vehicle. For example, if the rear wheels are locked while the front wheels are still rolling, the car will tend to slew round. On the other hand, if the front wheels are locked while the rear wheels are still rolling, the car tends to pursue its original course but cannot be steered.

Contrary to what occurs when a vehicle is accelerated, the application of its brakes creates a deceleration inertia force acting in a forwards direction through the centre of gravity of the car. This inertia force is, of course, equal in magnitude

to the retarding forces acting rearwards between all four tyres and the road. Again, the effect of the inertia force is to impose a pitching couple on the car, but in this case the load on the front wheels is increased and that on the rear wheels is reduced (Figure 19.41b). From this it follows that an already front-heavy front-engine FWD car, and to a lesser extent a front-engine RWD car, are likely to suffer from premature locking of the rear wheels when even more load is removed from them during heavy braking and especially downhill (Figure 19.42a).

The problem becomes still more acute under slippery road conditions, and especially when braking downhill. It is for this reason that a special control valve is often fitted in the hydraulic braking systems of such cars to prevent the rear wheels from locking.

So, as in the case of maintaining traction, the rear-engine RWD layout would seem to be preferable for balanced braking, since the transfer of load from the heavily laden rear wheels to the lightly laden front ones serves better to equalize their braking effect and especially downhill (Figure 19.43a).

Other considerations
Various other advantages and disadvantages are some-times claimed relative to the different layouts of engine and transmission systems. These range from their effect on engine cooling systems and car interior heating systems to interior noise levels and manufacturing costs, but in most cases it would be very difficult to assess such comparisons in realistic terms.

Other comparisons may no longer be valid in terms of modern design technology. An example of this is that the available steering lock (the angle through which the front wheels may be steered either side of their straight-ahead position) can be unduly restricted by the outer universal joints of the drive shafts on FWD cars, this being true only of earlier products.

However, there would seem to be little doubt that overhaul procedures can become more complicated and time consuming on cars where the engine and transmission system are combined into a single unit, and also that tyre wear tends to be greater on wheels that are both steered and driven.

19.6 FRONT-WHEEL DRIVE SHAFTS

General considerations

Before commenting upon the construction of front-wheel drive shafts, it may be useful first to compare their duty with that of an equivalent single propeller shaft in the Hotchkiss drive system:

1 The location of the drive shafts is such that they are necessarily shorter in length.
2 Since the drive shafts operate directly between the final drive and the road wheels they transmit greater torque.
3 From point 2 it also follows that they rotate at a slower speed.
4 When used in conjunction with inboard-mounted front brakes, the drive shafts must be able to withstand brake torque loading which can be applied more suddenly than that generated by the engine and transmission system.

Naturally the above considerations apply also to the drive shafts of independently sprung and driven rear wheels and De Dion axles.

Equal length drive shafts

In all front-wheel-drive cars with fore-and-aft mounted engines and transaxles, the drive shafts can obviously be made equal in length. For the modern small to medium FWD car they are produced from steel bar and may therefore be described as being of solid as opposed to tubular construction. Their diameter is typically in the region to 20 to 255 mm (0.8 to 1.0 in), and according to the type of universal joint used either one or both ends may be splined to provide an interference fitting for the inner member of the joint. A locking ring generally completes the assembly of shaft end and joint. This is usually supplied as a replacement unit, because the relevant parts are individually matched and are therefore not interchangeable.

Tubular construction drive shafts may be encountered where double Hooke's joints are used, because this type of joint lends itself to easier attachment to a tubular shaft and may also require a sliding spline coupling. A shaft diameter of about 40 mm (1.6 in) would be typical in such cases.

Unequal length drive shafts

For front-wheel-drive cars with transversely mounted engines and transmissions, the final drive housing can be offset appreciably from the mid-track of the wheels, such that the difference in lengths between the long and short drive shafts can be typically in the ratio of 3:2. In practice this may cause a steering pull to one side on hard acceleration for the following reasons:

1 If the long and short shafts are made the same diameter then the difference in their torsional rigidity results in the former taking longer to wind up than the latter, so that power is supplied to one wheel momentarily before the other. To minimize this effect the longer shaft is stiffened by making it of larger-diameter tubular construction, while the shorter shaft remains of smaller-diameter solid construction (Figure 19.44). In some designs the longer shaft may also be fitted with a rubber element torsional vibration damper, which resembles a miniature version of that earlier described for the engine crankshaft.
2 Any difference in the relative droop angles of the drive shafts as the front of the car rises on hard acceleration

(Figure 19.45a) implies that friction losses are increased in the universal joints of the more steeply angled shorter shaft, so that slightly less power is supplied to that particular wheel. This can only be remedied in designs where a jackshaft and steady bearing are added to equalize the lengths of the drive shafts (Figure 19.45b). Either a rigid, or a resilient rubber insert, type of bearing mounting may be used for the jackshaft, the advantages of the latter type being that it compensates for any misalignment between the final drive and the engine-mounted steady bearing, such as may arise from relative thermal expansion effects, and it serves to dampen vibrations for quieter running.

19.7 TANDEM AXLE DRIVES FOR HEAVY VEHICLES

General arrangement

Following the introduction of the rigid-six three-axled heavy vehicle with all four rear wheels being driven, as originally developed by the American firm of Goodyear Tire and Rubber in 1918, it was early recognized that the driving and indeed the suspension arrangements for the two axles in tandem posed certain problems. It will be evident that a conventional bevel gear final drive as described later does not readily lend itself to the tandem axle unit, since it is not practicable to provide a coupling shaft directly between the driving pinions of the leading (foremost) and trailing (rearmost) axles, simply because the differential assembly of the former gets in the way. It was for this reason that worm-and-wheel

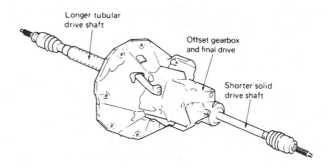

Figure 19.44   Tubular and solid FWD shafts (*Fiat*)

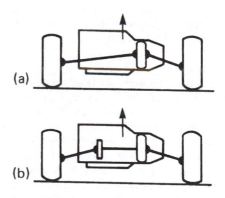

Figure 19.45   Droop angles of FWD shafts during acceleration: (a) unequal angles (b) equal angles

final drives long offered a convenient solution to the problem. That is, the drive was transmitted from the gearbox to a through worm drive on the leading axle by a conventional propeller shaft, and thence carried on to a regular worm drive on the trailing axle via a universally and slidably jointed coupling shaft (Figure 19.46a).

A less common alternative to the arrangement just described was a more complicated one in which side-by-side propeller shafts were used to drive conventional spiral gearsets in the tandem axles. The propeller shafts were driven from transfer gears mounted immediately behind the main gearbox, a single propeller shaft being used to drive the leading axle and a two-piece propeller shaft being routed alongside and over the leading axle to drive down to the trailing one.

Although worm drive tandem axles remained popular in British practice until the late 1950s, they had proved less acceptable in America. By the mid 1930s the American industry had developed more sophisticated versions of the bevel drive axle that incorporated a two-gear transfer train, so that the propeller shaft connection to the leading axle could be made at an upper driving head. This facility provided a convenient through drive connection to an angled coupling shaft, which then transmitted drive to a conventional spiral-bevel gearset in the trailing axle (Figure 19.46b). Note the disposition of the crown wheels relative to their pinions, as enforced by the two direction-reversing transfer gears. This consideration does not arise in later hypoid bevel drives when used in conjunction with a three-gear transfer train, because the idler pinion restores the correct direction of rotation for the crown wheel in the leading axle. Bevel drive axles with transfer gear trains have long since become established practice for tandem axle drives on heavy vehicles (Figure 19.47).

19.8 DRIVE LINES FOR PUBLIC SERVICE VEHICLES

The different arrangements of drive line that may be used on public service vehicles can be summarized as follows.

Offset front engine and gearbox
This arrangement early represented a means of obtaining increased head clearance in the centre gangway of double-deck buses. For this purpose the engine, gearbox, propeller shaft and final drive unit of the rear axle were offset in their entirety towards the nearside of the vehicle (Figure 19.48a).

Amidships engine and gearbox
To provide as flat as possible a floor area, suitable for forward control and one-man operation of single-deck buses, either a vertically or more especially a horizontally disposed

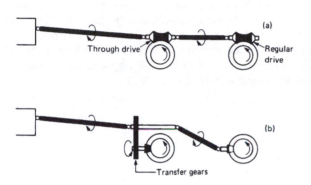

Figure 19.46 Basic comparison between (a) earlier worm drive and (b) later bevel drive tandem axles

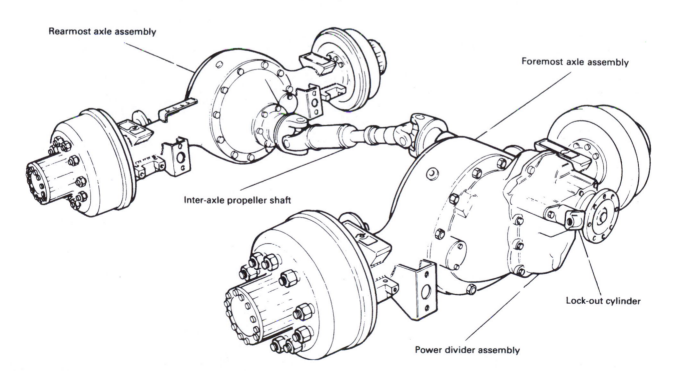

Figure 19.47 General view of a modern tandem axle installation (*Seddon Atkinson*)

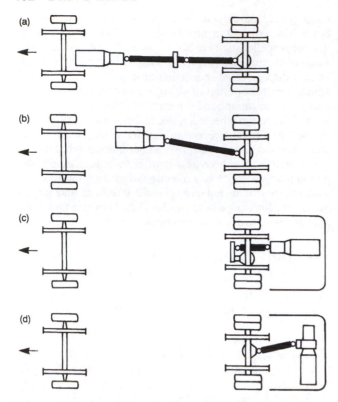

Figure 19.48　Drive lines for public service vehicles: (a) offset front engine and gearbox (b) amidships engine and gearbox (c) rear engine and gearbox, longitudinal (d) rear engine and gearbox, transverse

engine and gearbox may be carried amidships beneath the chassis frame. A single-piece propeller shaft, slightly angled across the vehicle, is then all that is necessary to transmit the drive to the rear axle (Figure 19.48b).

Rear engine and gearbox

Clearly this particular layout offers a low, flat floor area for both single- and double-deck buses, together with greater versatility in choice of location(s) for passenger entrance. This can be ahead of the front wheels, within the wheelbase, or behind the rear wheels; the first two locations are sometimes used in combination. If the rear engine is mounted longitudinally in the vehicle, then either a fairly short universally and slidably jointed propeller shaft simply connects the forward-facing gearbox to a rearward-facing final drive unit of the rear axle, or a rather longer version of propeller shaft passes beneath an axle tube to connect with a transfer gearset on a forward-facing final drive unit (Figure 19.48c). In the case of a transversely mounted engine and gearbox, the transmission arrangements become complicated by the need for a right-angled bevel gear take-off drive from the gearbox, so that a forward-routed propeller shaft can be angled across the vehicle to connect with a rearward-facing and offset final drive unit in the rear axle (Figure 19.48d).

20 Final drives and rear axles

20.1 FINAL DRIVE GEARS AND BEARINGS

The purpose of final drive gears

Final drive gears are incorporated in vehicle driving axles and transaxles for the following reasons:

1 To provide a right-angled drive from either the propeller shaft, or the gearbox layshaft, to the driven wheels.
2 Alternatively, to provide a parallel drive from the gearbox layshaft to the driven wheels.
3 To permit an additional and constant gear reduction in the transmission system.

The first requirement will be fairly evident and it derives from the particular layout of the vehicle. It therefore applies to any vehicle in which a longitudinal mounting is adopted for the engine and gearbox, regardless of whether they are driving the rear wheels or the front wheels (Figures 20.1a–c). Equally clearly, the second requirement applies only in the case of passenger cars with transversely mounted engines lying immediately parallel to the axis of the driven front or rear wheels (Figure 20.1d). An exception to this rule can arise, however, in the case of some buses with a transversely mounted rear engine and gearbox that drives forward to the rear axle via an angled propeller shaft (Figure 19.48d).

The third requirement arises from the practical limitations on the physical size of the driven wheels and hence the distance they travel in the course of a single revolution. If relatively large-diameter wheels and the crankshaft were so coupled that they revolved at the same speed, at least when direct-drive top gear was engaged, then the engine speed could never rise high enough for the maximum performance of the vehicle to be attained.

Types of final drive reduction gearing

Two basic types of final drive reduction gearing have generally been used in motor vehicles with longitudinally mounted engines. These are known as bevel pinion and crown wheel, and worm and worm wheel, usually described more simply as bevel drive and worm drive respectively (Figure 20.2). Although once used quite extensively in both car and commercial vehicle final drives, especially the latter where large reduction ratios can be required, the worm drive based on the screw-thread principle is now obsolete. In the modern vehicle, a bevel drive has proved to be lighter, more efficient, less expensive and equally quiet running. For cars with transversely mounted engines, the final drive reduction gears are of the helical pinion and gear type as described in connection with the gearbox in Section 14.1.

Bevel drive gearsets

Whereas the gears used in conventional gearboxes are of cylindrical form, bevel gears are conical with teeth that taper in both thickness and height. In principle their action may be compared to two friction-driven cone-shaped rollers sharing a common apex and mounted one on each of two shafts at right angle to one another (Figure 20.3).

By altering the relative sizes of the rolling cones various reduction ratios may be obtained, the smaller cone corresponding to the bevel pinion and the larger one to the crown wheel. With actual gears the number of teeth on the crown wheel divided by the number of teeth on the bevel pinion gives the *axle ratio*.

Here again, it should be noted that a high ratio axle refers in fact to one with a low numerical ratio. For example, a final drive with an axle ratio of 4.10:1 is a higher-ratio axle than one providing a 4.56:1 ratio. In this connection, it will be observed when consulting motor vehicle specifications that the axle ratio appears to be a somewhat 'fussy' figure, such as

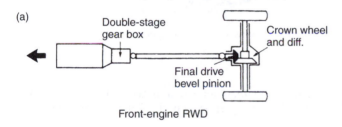

(a) Double-stage gear box — Crown wheel and diff. — Final drive bevel pinion
Front-engine RWD

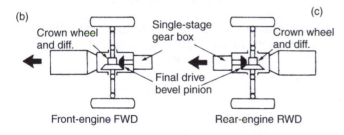

(b) Crown wheel and diff. — Single-stage gear box — Final drive bevel pinion
Front-engine FWD

(c) Crown wheel and diff.
Rear-engine RWD

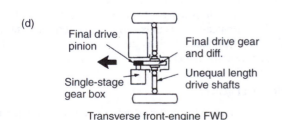

(d) Final drive pinion — Final drive gear and diff. — Single-stage gear box — Unequal length drive shafts
Transverse front-engine FWD

Figure 20.1 Schematic layouts of final drive gearing: in all cases the crankshaft rotates clockwise looking towards the gearbox

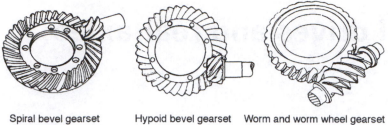

Spiral bevel gearset   Hypoid bevel gearset   Worm and worm wheel gearset

Figure 20.2   Identifying the basic types of final drive gears (*David Brown*)

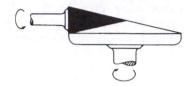

Figure 20.3   Friction cones analogy of bevel gearing

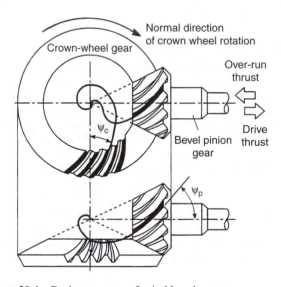

Figure 20.4   Basic geometry of spiral bevel gearset

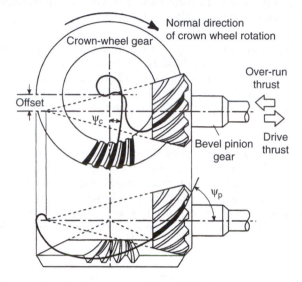

Figure 20.5   Basic geometry of hypoid bevel gearset

those just quoted. This is because the axle designer prefers to use an odd number of teeth on the pinion so that each tooth on the crown wheel engages every tooth on the pinion in regular succession. Known as a 'hunting tooth' gearset, it contributes to the quiet running of the bevel gears.

In the final drive gear arrangements of early motor vehicles, straight bevel gears were often used since these were the simplest form of bevel gear. Unfortunately, it was very difficult to make gears of this type run really quietly, so they were eventually superseded by two further types of bevel gear: the spiral bevel and the hypoid bevel.

Spiral bevel gears

These differ from straight bevel gears by having a lengthwise curvature of their teeth, which also makes an appreciable angle with the axis of the gear (Figure 20.4). They generally

provide a smooth, quiet operation with a high load capacity. These advantages derive from the following features:

1 One pair of teeth begins to engage before another pair disengages.
2 The pressure on the gear teeth surfaces is reduced because the transmitted load is shared by more than one pair of teeth in simultaneous contact.
3 The pressure on the gear teeth surfaces is also reduced by the curved teeth having larger contact areas.

Although spiral bevel gears have long since been superseded by hypoid gears in all car and many commercial vehicle rear axles, they are still employed in the combined gearbox and final drive units, or transaxles, of some FWD and rear-engined cars. In these applications there is generally no requirement for an offset bevel pinion, which is an important feature of the hypoid gear described next.

Hypoid bevel gears

These differ in turn from spiral bevel gears by having the axis of the bevel pinion appreciably offset from the axis of the crown wheel, this offset being in the direction of crown wheel rotation (Figure 20.5). In front-engine RWD cars hypoid gears can thus allow a lowering of the propeller shaft, and this contributes both to a flat floor and a reduced overall body height. Gears of this form can be even smoother and quieter running than spiral bevel gears, and also stronger. These particular advantages arise from the following considerations:

1 The hypoid gear combines the essentially rolling tooth action of the bevel drive with some of the sliding tooth action of the worm drive in which the driving member takes the form of a screw thread.
2 The pressure on the gear teeth can be further reduced by virtue of the greater spiral angle of the pinion teeth.
3 The size of the pinion increases in accordance with its off-set, so that for any given reduction ratio the durability of the final drive gears may be improved.

Although no longer regarded as a real disadvantage, hypoid gears are extremely sensitive to correct lubrication on account of the appreciable sliding action that occurs under pressure between their teeth. For this reason they require extreme pressure (EP) lubricants to provide an increased strength of oil film between their teeth and prevent them from scuffing.

In connection with the special lubrication requirements of hypoid gears it may perhaps be of interest to recall a little of their background history. The first successful hypoid gears were introduced by the Gleason Works of America in the mid 1920s and their first automotive application was made a few years later by the then famous Packard Motor Car Company which manufactured luxury cars. The use of hypoid final drive gears was confined to the products of this company for some ten years, chiefly because the discerning purchasers of these expensive cars could be relied upon to have them maintained by Packard service stations, since only these establishments stocked the special brand of oil then considered suitable for hypoid axle lubrication Needless to say, hypoid gear oils have long since become a highly developed product and are readily available to all.

The basic requirements for the durable and quiet operation of hypoid bevel gears may be summarized as follows:

1 Accuracy and consistency in the manufacture of the gears themselves.
2 Rigidity in mounting of the gears as provided by their housing and bearings.
3 Correct choice and application of axle lubricant.

During their manufacture the crown wheel and pinion are lapped together to a high degree of precision. This means that they have been subject to a final polishing operation, which refines the hardened gear teeth surfaces to improve their contact conditions and thereby promote smoothness and quietness of operation. Alternatively, precision grinding of the hardened gear teeth may be employed that more accurately reproduces optimum tooth geometry. Since the crown wheel and pinion have been matched in this way by the manufacturer, it naturally follows that they must remain matched in service. If either the crown wheel or the pinion is divorced from the other, then any new combination will result in unacceptable noise and wear. It is, therefore, for this reason that service replacement crown wheels and pinions are always supplied in matched sets – 'togetherness' being the key word here!

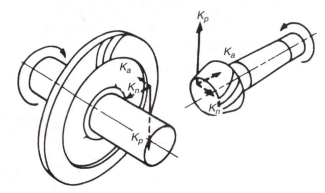

Figure 20.6 Forces in spiral bevel gears resulting from meshing action of crown wheel and pinion: K_a thrust force; K_n separating force; K_p tangential force (*SKF*)

Crown wheel and pinion bearing loads

The loads acting on the crown wheel and pinion bearings are primarily those which result from the meshing action of the gear teeth and are most conveniently presented by illustration (Figure 20.6). Here, in the case of spiral bevel gears, it will be seen that several pairs of opposing forces are created during torque transmission:

1 An end thrust force on the pinion with a corresponding radial separating force on the crown wheel.
2 An end thrust on the crown wheel with a corresponding radial separating force on the pinion.
3 The corresponding tangential forces on the crown wheel and pinion.

Whereas each pair of opposing forces is equal in magnitude with spiral bevel gears, this is not so in the case of the relatively more complicated meshing action of hypoid gear teeth, where the tangential forces responsible for torque transmission are neither opposite in direction nor equal in magnitude. It should be noted that when the engine is being used as an aid to braking on vehicle overrun, the loads acting on the gears are reversed in direction, and similarly, of course, when reverse gear is engaged.

A point of interest in connection with spiral bevel and hypoid gears is that, unlike straight-toothed bevel gears, the curvature of their teeth produces a winding-in or winding-out action. The resulting end thrust on the pinion must therefore be taken into account, along with that arising purely from its conical form. In fact, the high loading imposed on the pinion bearings makes their application the most critical in the axle or its equivalent.

The additional loads that may be imposed on the crown wheel and pinion bearings, at least as far as a live axle is concerned, are respectively radial ones from the inner ends of semi-floating and to a lesser extent three-quarter floating axle shafts, and axial ones from any sliding joint friction in the propeller shaft.

Crown wheel and pinion bearings

It is generally regarded as being less difficult to maintain spiral bevel and hypoid gears in satisfactory alignment with

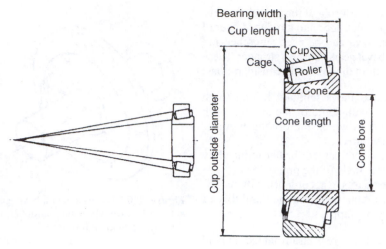

Figure 20.7 Nomenclature and basic principle of the tapered roller bearing: the apices of the tapered surfaces meet at a common point on the axis of the bearing

rolling element bearings than it would be with plain bearings. The use of rolling element bearings, especially those of the tapered roller type (Figure 20.7), has therefore long been established practice for supporting the final drive gears of motor vehicles. Before describing the various arrangements of crown wheel and pinion bearings, it may be of interest to consider the principal advantages of using tapered roller bearings:

1 They possess true rolling characteristics because the projected conical surfaces of the rollers and their raceways all meet on the axis of the bearing. This is in contrast to the combined rolling and sliding that occurs between ball bearings and their raceway surfaces owing to the variations in circumferential speed of a ball rolling freely in a conforming groove. (The sliding effect is sometimes referred to as Heathcote slip, so named after H.L. Heathcote who identified it in the early 1920s.)
2 By virtue of their unique geometry, the alignment of the rollers is not wholly dependent upon either the thrust rib of the cone (the inner race of a tapered roller bearing is termed the cone and the outer race the cup), or the bearing cage.
3 The line contact of the tapered rolling elements, as opposed to the (theoretical) point contact in ball bearings, provides a greater load-carrying capacity for a given size of bearing. This is of particular advantage in the case of the heavily loaded pinion bearings, where there are space limitations on their size.
4 Tapered roller bearings are inherently rigid and have the ability to withstand any combination of radial, thrust and tilting loads within the designed capacity of the bearing. Shallow-angle bearings are specified for carrying heavy radial loads and steeper-angle ones for carrying heavier thrust loads.
5 Owing to their tapered construction, the operating geometry remains undisturbed when bearings of this type are adjusted to remove initial play and also preloaded as described at a later stage.

6 Although the ultimate fate of all types of rolling element bearing is fatigue, in practice the useful life of the final drive bearings is usually determined by wearing of the pinion bearings causing noisy operation of the gears. In this context, the rate of wear for tapered roller bearings is generally lower than that for ball bearings.
7 Tapered roller bearings possess the useful property of acting as their own oil pump, since by virtue of their form oil flows through the bearing from the small end to the large end of the rollers.

Crown wheel and pinion bearing arrangements

In modern automotive practice, the bearings for the crown wheel and pinion may be mounted in either a live rear axle, a separate final drive housing attached to the vehicle structure, or a combined gearbox and final drive assembly known as a transaxle.

For live rear axles and separately mounted final drives (Figures 20.8, 20.9 and 20.10) the arrangement of the crown wheel and pinion bearings is typically as follows.

Pinion

This is made integral with its shaft and is overhung from two tapered roller bearings mounted back to back (small ends of rollers pointing inwards). These are spaced apart by their outer races or cups, which locate against shoulders in the pinion housing. Endwise retention of the pinion shaft is effected through the inner races or cones of the bearings, since these are rigidly held between the pinion and the boss of the companion flange for the rear universal joint of the propeller shaft by a clamping nut on the front end of the shaft.

The bearing cups are pressed into the pinion housing; similarly, the cone of the bearing adjacent to the pinion is made an interference fit on the shaft. A lighter fit suffices for the cone of the other bearing at the less heavily loaded end of the pinion shaft, since this facilitates the build and adjustment of the

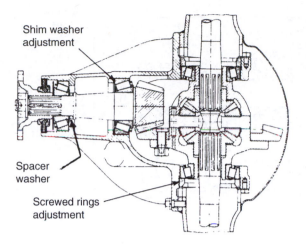

Figure 20.8 Tapered roller bearing mounting for final drive gears in banjo-type axle (*SKF*)

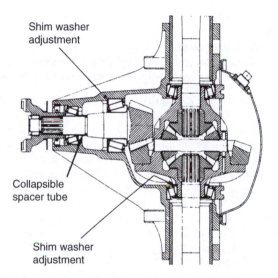

Figure 20.9 Tapered roller bearing mounting for final drive gears in carrier-type axle (*SKF*)

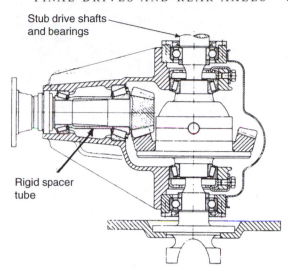

Figure 20.10 Arrangement of separately mounted final drive gears (*SKF*)

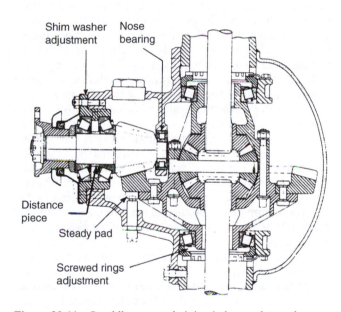

Figure 20.11 Straddle-mounted pinion in heavy-duty axle (*SKF*)

pinion and bearings. The spacing of the pinion bearings is generally 2.5 times the overhang distance of the pinion.

Crown wheel
This is bolted behind its mesh to provide a slip-resistant joint (Figure 19.29) with the flanged differential cage, which in turn is carried by two tapered roller bearings mounted face to face (small ends of the roller pointing outwards) in the final drive housing. On each side of the differential cage bosses are machined to locate the bearing cones, these being made an interference fit on their seatings. For endwise retention of the crown wheel, the cups of the bearings locate against shouldered bores in the final drive housing, in which they are made a lighter fit, again to facilitate build and adjustment. Detachable bearing caps are generally required to permit assembly of the crown wheel and bearings into the final drive housing. The spacing of the crown wheel bearings usually

corresponds to about three-quarters of the mean pitch diameter of the crown wheel.

In heavy-duty rear axles, the pinion may be straddle mounted by having a pair of back-to-back tapered roller bearings in front of it and a single parallel roller bearing immediately behind it (Figure 20.11). The purpose of this arrangement is, of course, to increase the rigidity of mounting for the pinion. Since the crown wheel must likewise be immune from deflection under heavy side loading, it is often provided with a steady pad that is mounted from the final drive housing. This steady pad is either screw or shim adjusted to be just clear of the non-meshing face of the crown wheel under light-load running conditions.

Where the final drive is combined with the gearbox to provide a transaxle, the bearing arrangements for the pinion

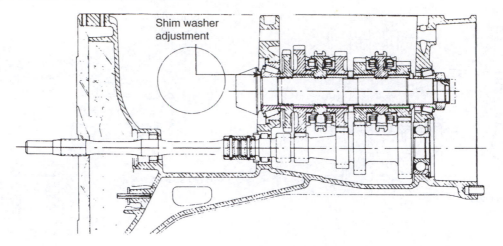

Shim washer
adjustment

Figure 20.12   Typical mounting of bevel pinion in a transaxle (*Alfa-Romeo*)

may differ slightly from those found in the live axle, because the gearbox mainshaft also serves as the bevel pinion shaft, as explained earlier. The pinion cum mainshaft is therefore supported from the gearbox part of the transaxle housing, usually either by two very widely spaced tapered roller bearings mounted back to back (Figure 20.12) or by a single parallel roller bearing at the pinion end of the shaft and a pair of back-to-back tapered roller bearings at the other end.

The final drive reduction gearing for a transversely mounted transaxle usually comprises a simple helical gear train. A driving pinion is overhung from the gearbox mainshaft and meshes with the driven gear of the final drive and differential assembly (Figure 14.14). The driven gear is spigoted and bolted to a flange near the outer end of the differential carrier, which is supported in the final drive housing by two opposed taper roller bearings. To minimize deflection of the driven gear under load and thereby improve not only its durability but also the life of the bearings, the bearing adjacent to this gear is mounted as closely to it as practicable. Great care is also taken in design and production to minimize initial backlash of the gears in this type of final drive, because it has been found that the distance between their centres can be appreciably affected by heat expansion of the housing.

20.2   ADJUSTING THE FINAL DRIVE GEARS

Adjustment sequence

To an engineer one of the pleasant things about bevel gears is that ready provision can be made in their bearing mounting arrangements for adjusting one gear relative to the other, both on initial build and later in the life of the gears. However, it is most important that the necessary adjustments should be performed in a logical sequence.

Taking for example a carrier-type axle, as described later, the adjustment sequence is generally as follows:

1 Preload the crown wheel and differential assembly bearings.
2 Set the pinion mounting distance with the above assembly temporarily removed.
3 Preload the pinion assembly bearings.

4 Set the backlash between the refitted crown wheel and pinion.
5 Check tooth contact.
6 If tooth contact is unsatisfactory, reset pinion mounting distance and restore backlash with crown wheel adjustment.

Before dismantling a final drive unit, it should be evident that the original build adjustments made by the manufacturer must not be disturbed unless absolutely necessary. Due note should therefore be taken of the backlash value. Correlation markings should be made to ensure that the crown wheel bearing caps are not interchanged side-to-side and to index any threaded sleeves relative to the bearing caps; note also the number of their exposed threads. Finally the location and thickness of all shims should be noted. The purpose of taking these various precautions is to ensure that upon reassembly at least the original settings can be quickly restored, even though further adjustments may still be required.

Unless familiarity with a particular final drive assembly has already been acquired, the build and setting instructions issued by the manufacturer's service department should always be consulted. In some cases a manufacturer may advise that, beyond replacement of the pinion oil seal, no adjustment or dismantling of the final drive unit should be attempted in service. This is because of the specialized equipment needed to carry out an overhaul, the policy then being to fit a factory replacement unit.

Preloading the crown wheel and differential assembly bearings

Here we are considering the adjustment of the crown wheel bearings in their purely load-supporting role, rather than in the sense of their allowing sideways repositioning of the crown wheel to secure a correct meshing condition with the pinion. In other words, we must take into account the heavy loads acting on the crown wheel bearings that result from the meshing action of the gear teeth and in particular the end thrust force on the crown wheel. This force is restrained by the bearing adjacent to the crown wheel and, remembering always that we can never achieve absolute rigidity in engineering constructions, it tends to cause elastic deflection of the housing in which the bearing is supported. As a result of this action not only is the gear mesh disturbed, owing to the crown wheel being elbowed away from the pinion, but also the other bearing remote from the crown wheel can become unseated. We have then lost mounting stability for the crown wheel when it is most needed.

Since transmission engineers have long since recognized that, to obtain maximum endurance and quiet operation of the final drive gears, the utmost rigidity is essential in their mountings, we have no choice but to resort to preloading of their bearings, as opposed to adjusting them with zero end play or (even worse) actual end play. The term *preload* signifies a bearing setting that is tighter than metal to metal. A non-technical person may perhaps be forgiven for thinking that preloading simply means that the bearings have been over tightened, whereas an engineer is more likely to conjure up the abstract notion of a 'less than zero clearance'!

To understand what we are trying to achieve by preloading the crown wheel bearings, let us first consider a condition of zero end play, where the bearing remote from the crown wheel becomes unseated owing to the end thrust force on its opposite number. If under this condition of operation we could readjust the bearing adjacent to the crown wheel, so that it is advanced towards the unseated bearing just sufficient to restore therein a metal-to-metal rotating contact, we shall have accomplished two things and to all intents and purposes solved our problem. First, the crown wheel will have been persuaded back to its correct meshing position with the pinion, and second, the mounting stability of the crown wheel and differential assembly will have been re-established.

Under non-operating conditions it follows, of course, that both bearing supports will assume a slightly sprung-apart condition. Instead of merely zero end play we therefore have an assembly condition in which the crown wheel bearings are placed under an initial compressive end load by their sprung-apart supports. The bearings are therefore said to be preloaded.

Methods of preloading the crown wheel bearings

In actual practice the preloading of the crown wheel bearings is not of course set with the unit in operation, but is taken into account during build of the unit. One of the following techniques is generally used.

Shim washers acting behind inner ends of bearing cones
The crown wheel and differential assembly complete with bearings is installed in the housing without any shims (Figure 20.9). A clock indicator is mounted on the housing and its stylus is brought into engagement with the back face of the crown wheel. Two levers are inserted between bearing cup and housing to force the crown wheel assembly over to the pinion side; the clock indicator is then set to zero reading. Next the crown wheel assembly is forced over to the other side of the housing, so that the total clearance between the bearings can be read on the indicator. A certain additional amount specified by the manufacturer is then added to the total clearance just established, and in sum this value represents the final thickness of shims to be divided between sides

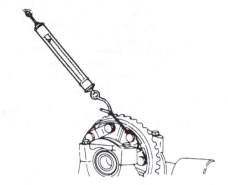

Figure 20.13 Checking preload crown wheel and differential bearings (*Nissan*)

and hence fitted behind the bearing cones. The additional amount of shim thickness beyond that required to give zero end play is, of course, responsible for springing apart the bearing supports to exert a preload on the bearings. A stretching fixture may be applied to the final drive housing to facilitate refitting the crown wheel bearings in carrier-type axles.

Threaded sleeves bearing against outer ends of bearing cups
With this particular arrangement the crown wheel assembly and bearings, together with the threaded sleeves, are installed in the housing and the sleeves are turned hand-tight against the bearing cups (Figure 20.8). A clock indicator is mounted on the housing and its stylus is brought into engagement with the back face of the crown wheel, following which the indicator is set to zero reading. The opposite side threaded sleeve is then slackened back sufficient to notice an end play reading on the clock indicator. Next the same threaded sleeve is advanced towards the crown wheel just sufficient to confer zero end play. Both threaded sleeves are then advanced towards the crown wheel assembly by equal amounts; these are specified by the manufacturer, but typically are one-notch or one-window turns depending upon the form of the sleeves. This further tightening of the threaded sleeves beyond that required to give zero end play is, as before, responsible for springing – or, perhaps more appropriately in this case, jacking – apart of the bearing supports to preload the bearings.

Whichever method of adjustment is employed to preload the crown wheel bearings, a manufacturer may additionally call for a check to verify the correctness of setting by measuring the resistance to steady rotation offered by the crown wheel. This is generally accomplished by hooking a spring balance to one of the crown wheel bolts and then noting whether the tangential force required to maintain steady rotation is correct to specification (Figure 20.13). Nothing is to be gained by increasing the preload beyond that specified, and indeed much could be lost, since any overheating would reduce the life of the bearings.

Setting the pinion mounting distance

We must have a starting point whereby the quiet meshing condition of a crown wheel and pinion, as originally established

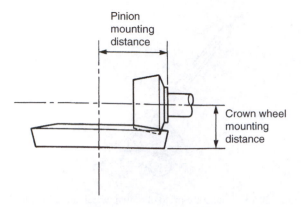

Figure 20.14 Defining crown wheel and pinion mounting distance

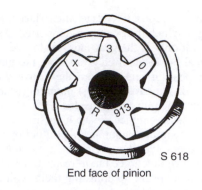

End face of pinion

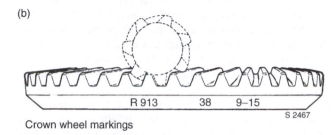

Crown wheel markings

Figure 20.15 Examples of markings on (a) end face of pinion (b) crown wheel (*SAAB*)

(a)	+3	measurement for pinion setting
	R913	mating number, also stamped on crown wheel pinion not offset: shaft centreline intersects crown wheel centreline (all pinions are marked 0, and this datum has no relevance to the adjustment)
(b)	38	no. of teeth on crown wheel
	9	no. of teeth on pinion
	−15	backlash 0.15 mm
	R913	mating number, also stamped on pinion

by the gear manufacturer during noise testing, can be accurately reproduced when these components are assembled into the final drive housing. This starting point is variously referred to as the pinion mounting distance, pinion depth or pinion height, and may be defined as the proper distance from the centre-line of the crown wheel to the bearing shoulder on the pinion (Figure 20.14). From the point of view of a service engineer, however, this dimension does not have direct application because of the practical difficulty in gauging it, once the pinion and bearings assembly is installed in the final drive housing.

What we do in fact measure is the distance from the centre-line of the crown wheel via its bearing bores to the nose of pinion, and we express this distance as the gauge dimension. More specifically, we can say that the nominal gauge dimension is equal to the nominal pinion mounting distance minus the measured pinion head thickness. The term 'nominal gauge dimension' has been used deliberately, because in practice the crown wheel and pinion gearsets for any given type of application cannot all be manufactured exactly alike, so that very slight variations in pinion mounting distance occur. Such variations are etched on the pinion nose during noise testing as plus (+) or (−) limits (Figure 20.15a), which must be either added to or subtracted from the nominal gauge dimension quoted by the manufacturer.

A pinion setting gauge is generally of either the micrometer or clock indicator type and is arbor mounted from a pair of adaptor discs, which fit into the crown wheel bearing bores of the housing. The measuring rod of the gauge is brought into either direct contact with the nose of a spiral bevel pinion, or indirect contact via an offset adaptor plate in the case of a hypoid pinion. With an overhung pinion any adjustment required to achieve the proper pinion mounting distance, as indicated by the nominal gauge dimension plus or minus the amount etched on the pinion, can be effected by either of the following methods:

1. The insertion of an adjusting shim washer of appropriate thickness between the pinion head bearing cone and the adjacent shoulder of the pinion (Figure 20.12).
2. The insertion of an adjusting shim washer of appropriate thickness between the pinion head bearing cup and its locating shoulder in the housing (Figures 20.8 and 20.9).

A straddle-mounted pinion lends itself to a more convenient method of adjusting mounting distance, since a shim washer pack can be interposed at the flange mounting of the pinion bearing cage assembly to the final drive housing (Figure 20.11).

Preloading the pinion assembly bearings

The need to preload the pinion assembly bearings arises from considerations similar to those mentioned earlier in connection with the crown wheel bearings. Indeed, the requirement for preloading the pinion bearings is greater because they are more heavily loaded as torque multiplication occurs in the gearset. In the case of the pinion assembly we are therefore seeking to maintain both its proper mounting distance by reducing endwise deflection at the head bearing adjacent to the pinion, and its mounting stability by preventing the tail bearing remote from the pinion from becoming unseated in sympathy.

Since an overhung pinion is carried by two widely spaced tapered roller bearings mounted back to back (small ends of the rollers pointing inwards), it becomes expedient to preload them by tensioning together their cones using some form of spacer, which acts as a controlling device. The spacer may take the form of either a selective spacer washer, a rigid spacer tube or a collapsible spacer tube. With the first method of adjustment a ground spacer washer, fitted between the tail bearing cone and a shoulder on the pinion shaft (Figure 20.8), is selected for thickness such that the clamped distance between the head and tail cones produces the required preload on the bearings. The rigid spacer tube method of adjustment utilizes either a selective length of spacer tube (that is, a range of spacer tubes of graded length) or a fixed length of spacer in conjunction with shim washers (Figure 20.10) to give a clamped distance between the head and the tail cones of the pinion bearings sufficient to impose the required preload. As the spacer tube spans the entire distance between the bearing cones, there is no need to incorporate a shoulder on the pinion shaft. The collapsible spacer tube takes either a short or a long form; the former is sandwiched between the tail bearing cone and a shoulder on the pinion shaft (Figure 20.9), whilst the latter spans the entire distance between the head and the tail bearing cones. An essential feature of either type of collapsible spacer tube is that its wall thickness and shape are very carefully contrived, so that it can yield a small amount under a specific axial load, which is derived from tightening the retainer nut for the final drive coupling flange. The pinion bearings preload is therefore set by gradually tightening the retainer nut to a torque value specified by the manufacturer, while checking at intervals the actual resistance to steady rotation of the pinion assembly. Since bearing preload is being controlled by using the permanent set of the collapsible spacer tube, it follows that if an excessive preload is measured the spacer tube must be discarded and another one fitted before the preload is reset to its correct value. An important practical advantage of using a collapsible spacer tube is that the bearings can be adjusted for preload without the necessity for a trial assembly. It thus lends itself admirably to large-scale production application, where automated bearing setting equipment can be utilized.

Finally, it should be mentioned that in the case of a straddle-mounted pinion the close proximity of the back-to-back tapered roller bearings, mounted in a separate cage, dictates the use of a very short rigid spacer tube or distance piece to set the bearing preload (Figure 20.11). This can be conveniently set and measured in terms of rotating friction torque with the pinion and cage assembly removed from the final drive housing. A practical method of measuring this torque is to hold the companion flange between the protected jaws of a vice, wrap a length of soft wire around the bearing cage and then pull it tangentially with a spring balance to rotate the cage and bearings (Figure 20.16). The initial reading should be ignored since the starting torque will be higher than when the bearings are moving steadily, as with all preloaded bearings.

Setting the backlash between crown wheel and pinion

To complete the mesh adjustment of the crown wheel and pinion, as distinct from preloading their bearings, we need to restore the crown wheel mounting distance as originally established by the manufacturer. This term may be defined as the proper distance from the centre-line of the pinion to the back or mounting face of the crown wheel gear, and is complementary to setting the pinion mounting distance as previously described

Figure 20.16   Checking preload of straddle-mounted pinion bearing (*Seddon Atkinson*)

(Figure 20.14). In the actual rebuilding of a final drive unit, however, it is neither necessary nor practical to measure the crown wheel mounting distance. This is because it is automatically conferred once the correct backlash has been set between the crown wheel and the pinion teeth.

What do we mean by backlash, and for that matter why should it be required? To answer the first question, *backlash* may be defined as the amount by which a tooth space exceeds the thickness of an engaging tooth. In answering the second question, a gear designer is likely to cite at least a dozen good reasons why backlash is required, many of them being related to manufacturing tolerances, but for our purposes we should at least recognize the following reasons:

1 To provide clearance space for an oil film to be formed between the engaging teeth.
2 To maintain sufficient clearance space in the presence of heating and relative thermal expansion of the gearset.
3 To allow for any slight misalignment within the gearset arising from wear in the supporting bearings.

The actual setting of the backlash between the crown wheel and pinion is obtained solely by lateral adjustment of the crown wheel relative to the pinion. This may be effected either by the transference of an adjusting shim washer of appropriate thickness from behind the bearing cone on one side of the differential cage to behind the bearing cone on the other side (Figure 20.9), or by the equal screwing and unscrewing of threaded sleeves that bear against the outer ends of the bearing cups (Figure 20.8). It will be noted that neither method disturbs the already established preload on the crown wheel assembly bearings, since the distance between their cups remains unchanged and all that we are doing is to reposition the assembly within that set distance.

To verify that the backlash specified by the manufacturer and etched on the crown wheel (Figure 20.15b) has been obtained, it is necessary to secure a clock indicator gauge from the final drive mounting flange and bring its stylus into contact with a tooth surface on the drive side (the convex side of the tooth which contacts the opposite member when the vehicle is

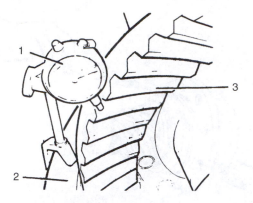

Figure 20.17   Checking the backlash between the crown wheel and pinion (*Seddon Atkinson*)

1 dial gauge
2 driving head assembly
3 crown wheel

driven in the forward direction) of the crown wheel teeth. The crown wheel must then be rocked in both directions against the teeth of the pinion, which is held stationary while the clock indicator reading is observed (Figure 20.17). Any correction will involve either moving the gear away from the pinion to increase backlash, or conversely moving the gear towards the pinion to decrease backlash. Typical values for backlash are 0.15–0.20 mm (0.006–0.008 in) in the case of motor car final drive gears, and 0.20–0.40 mm 0.008–0.016 in) for heavy-vehicle units.

Checking tooth contact

Finally, we must carry out a tooth contact pattern check. The purpose of this is to verify that a correct meshing relationship exists between the crown wheel and the pinion teeth, so as to ensure their quiet operation and long life. A typical method of checking is as follows:

1 The teeth of the crown wheel gear and pinion should be thoroughly cleaned.
2 A section of the crown wheel teeth, say about five and ten teeth respectively for motorcar and heavy vehicle gearsets, should be coated by brush with either the once traditional mixture of red lead or yellow ochre and oil, or engineer's marking blue.
3 When the pinion is rotated the marking material on the crown wheel teeth will be partially squeezed away by the successive line contacts of tooth engagement, thereby leaving bare areas that correspond to the exact size, shape and location of tooth contact. A sharper impression may be gained by manually applying a resistance to rotation of the crown wheel while at the same time turning the pinion.
4 A satisfactory contact pattern on the drive side of the tooth is generally one that is centred between the top and bottom (or in gear terminology the face and flank) of the tooth and also biased towards its inner end (or toe). The tooth contact on the concave or coast side of the teeth should automatically be correct, once a satisfactory pattern has been established on the drive side.

5 In the event of an unsatisfactory meshing condition the pinion mounting distance is likely to require slight alteration in accordance with contact pattern indications (Figure 20.18). It will, of course, then become necessary to restore the correct backlash setting between the crown wheel and the pinion teeth. A point worth remembering is that repositioning the pinion affects the depth of contact, whilst repositioning the crown wheel affects the length of contact.

A visual impression of the satisfactory tooth contact pattern shown in Figure 20.18 might suggest that its coverage and in particular its length are somewhat abbreviated, until it is realized that the successive line contacts occurring between the meshing teeth of the crown wheel and pinion undergo change from the negligible to maximum load conditions. The resulting variation in tooth contact pattern arises from relative deflections of the gear and pinion and their supports, since despite the best intentions of the designer neither the materials of construction nor the bearing mountings can be made perfectly rigid. As a matter of background interest, these deflections are determined experimentally for any newly designed final drive unit by operating it on either a gear or a bearing manufacturer's special test rig, which is instrumented with a multitude of clock indicator gauges mounted at strategic positions around the unit (Figure 20.19). By checking tooth contact patterns at low speed under various torque inputs from negligible up to maximum load at which the unit is expected to operate, it is possible to correlate very accurately the change in tooth contact pattern with the relative deflections of the gear and pinion and their supports.

It should therefore be evident that during original build and any later overhaul service, the aim is to achieve a suitably abbreviated and positioned contact pattern that will progressively spread over the profile of the teeth, as the load on them increases under actual operating conditions. In theory, the crown wheel and pinion are operating under ideal conditions when they exhibit a contact pattern that utilizes the full working profile of the teeth, without load concentration in any particular area and with gradual fading of the contact pattern at the tips and ends of the teeth. Clearly then, if we fail to obtain a satisfactory meshing of the gears during bench rebuild of a final drive unit, their meshing condition can only worsen when the gears are subjected to working loads during vehicle operation.

The most obvious effect of incorrectly meshed final drive gears is that of noisy operation, whether it be the immediate result of defective rebuild or the long-term outcome of bearing wear allowing unwanted deflection of usually the pinion. In either event we have a situation where unsatisfactory tooth contact patterns are not allowing sufficient tooth overlap, so that a uniform transfer of torque load from one tooth to another is no longer possible. The input loading on the teeth of the crown wheel and pinion can then generate a noise corresponding to tooth contact frequency, which is equal to the number of tooth contacts per second. When considering the topic of gear noise it is perhaps worth recalling the words of a one-time American authority on gear design, Professor Earle Buckingham, who observed that in many respects it is not a question of why gears are noisy, but rather why they are ever quiet! This serves to highlight the care that must be taken at all stages of rebuilding a final drive unit in service.

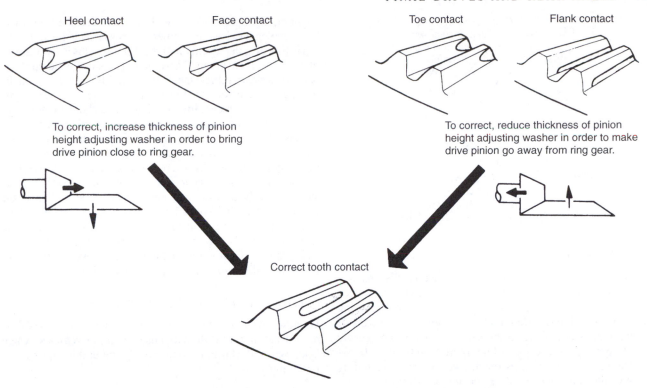

Heel contact

Face contact

Toe contact

Flank contact

To correct, increase thickness of pinion height adjusting washer in order to bring drive pinion close to ring gear.

To correct, reduce thickness of pinion height adjusting washer in order to make drive pinion go away from ring gear.

Correct tooth contact

Figure 20.18 Identifying and correcting tooth contact patterns (*Nissan*)

Figure 20.19 Deflection and contact check by gear manufacturer (*Gleason Works*)

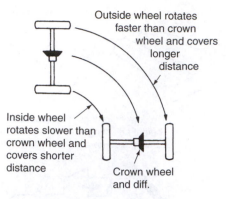

Outside wheel rotates
faster than crown
wheel and covers
longer
distance

Inside wheel
rotates slower than
crown wheel and
covers shorter
distance

Crown wheel
and diff.

Figure 20.20   The need for a differential gear in the final drive

Worm drive adjustments

A few brief notes on the method of adjusting this obsolete type of final drive may be of background interest. In a once typical heavy-vehicle application, the roller bearings carrying the worm wheel and differential were adjusted by means of two threaded sleeves that embraced the ends of the bearing cups, to provide an end play of not more than 0.1 mm (0.004 in). The ball thrust bearings for the worm were adjusted by selective distance piece to allow an end play of 0.15–0.2 mm (0.006–0.008 in). A past generation of service engineers will no doubt recall that it was customary to adjust the worm wheel slightly offset from the central axis of the worm, so that a less pronounced bedding was obtained on the entering side of the wheel teeth. This not only assisted the lubrication process but also allowed for slight movement of the worm wheel under load when the bedding moved towards the entering side of its teeth.

20.3   DIFFERENTIAL GEARS

The need for a differential

When a motor vehicle is steered round a corner, the circular path followed by the inner wheel of an axle is of lesser radius than that of the outer wheel (Figure 20.20). At any given distance through the corner, the inner wheel has therefore turned through fewer revolutions than the outer one. From this it follows that if the driven wheels were coupled directly to the crown wheel or, less commonly, the worm wheel of the final drive, enforced scrubbing of the tyres would occur during cornering as the wheels sought to rotate at different speeds. This would result in not only rapid wearing of the tyres, but also a virtually unmanageable vehicle.

In practice, the driven wheels are coupled to the final drive through separate driving shafts and a differential gear, a device which not only accommodates the differing speeds of the driven wheels, but also equalizes at all times the axle torque distributed between them.

A ready impression of the manner in which the differential gear acts may be gained by jacking the driven rear axle of a vehicle on to stands. After releasing the handbrake, and with the gearbox in neutral, first count the number of turns made by the wheels for a given number of hand turns of the propeller shaft. Then with one wheel held against rotation, repeat the count on the other wheel. It will be found that the free wheel has completed twice the number of turns it made previously.

Historically, the invention of the differential is usually attributed to the Frenchman Onésiphore Pecqueur, who in 1828 patented a steam vehicle that included a differential gear on the driving axle. However, a student of mechanics may well argue that the motor vehicle differential gear is technically not a differential mechanism at all, but a cyclic equalizing gear. This is because its basic function is that of dividing the crown wheel torque equally between the two driven wheels, irrespective of their relative speeds of rotation. A true differential mechanism, on the other hand, is characterized by similar parts moving simultaneously at different rates. An example of this is the familiar differential chain hoist in which the endless chain operates over a compound sprocket wheel having two different radii. Be that as it may, perhaps we should stick to convention and continue to call it a differential gear (or even 'diff-unit' as it is universally called in motor vehicle repair work).

Construction of the bevel gear differential

We now have to consider the mechanical construction of the most usual form of differential used in motor vehicles, which is known as the straight-tooth bevel type and comprises the following parts:

Differential cage   It may be recalled from Section 20.1 that the crown wheel is bolted to the flanged differential cage, which in turn is carried by the crown wheel bearings. To gain maximum rigidity the differential cage is provided with substantial stiffening ribs and is produced from a malleable iron casting. It may be of either one-or-two-piece construction, according to whether it contains two or four planet bevel pinions. The latter number is used in heavy-duty axles so that contact pressure between the differential gear teeth is reduced by distribution (Figure 20.21).

Planet pinions   These bevel pinions are produced from steel and provided with coarse-pitched teeth of stub form. They are free to rotate either on a shaft located and retained by a tapered pin in the cross-bore of a one-piece cage or, in the case of a four-planet pinion design, on a spider-shaped member that is embraced by a two-piece cage. The thrust faces of the planet pinions are of part-spherical form, as are their thrust washers and seatings in the differential cage.

Sun pinions   Also known as side pinions, these are similarly bevel toothed and have internally splined hubs registering in recesses in side bores of the differential cage. Their thrust faces and those of the thrust washers and cage recesses are flat, since they can be made larger in diameter than the planet pinions and hence have sufficient bearing area to accept end thrust loading. The inner splined ends of the axle half-shafts engage with the hubs of the sun pinions.

Operation of the bevel gear differential

This is perhaps best explained by reference to the balance beam and slotted discs analogy, in which the former corresponds to one of the planet pinions (input from the crown wheel) and the latter to the sun pinions (output to the axle shafts).

From this it should be evident that with the balance beam centrally pivoted, the loads transmitted by its ends to the

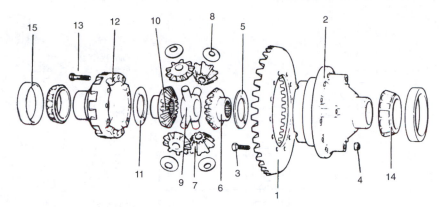

Figure 20.21 Exploded view of a heavy-duty four-planet pinion differential (*Seddon Atkinson*)

1 crown wheel	6 sun pinion	11 flat thrust washer
2 flanged differential cage	7 planet pinion	12 plain differential cage
3 bolt	8 part-spherical thrust washer	13 capscrew
4 nut	9 spider	14 differential bearing race
5 flat thrust washer	10 sun pinion	15 differential bearing track

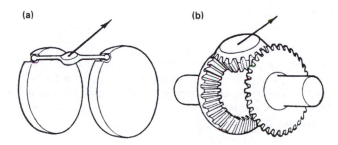

Figure 20.22 Operating principle of the bevel gear differential

rotatable slotted discs will be equal within the available swing of the beam (Figure 20.22a). Such a differential would, of course, permit only a limited relative movement of the two discs, but if we could arrange for a whole series of balance beams extending to a full circle and then provide the discs with a complete set of slots, a continuous relative movement of the two discs can be accommodated. Here we have the essential ingredients of the real-life bevel gear differential, which thus divides equally the turning force imparted by the crown wheel to the axle half-shafts (Figure 20.22b), even though circumstances may dictate that one must rotate faster than the crown wheel and the other an equal amount slower. It should be noted that any alteration to the relative sizes of the planet and sun pinions will have no effect on the 2:1 operating ratio of a bevel gear differential, as long as the axes of the planet pinions remain at 90° to the axis of the sun pinions.

It follows, of course, that if both half-shafts are rotating at the same speed as the crown wheel, there will be no relative movement between the planet and the sun pinions and the differential gearing becomes temporarily redundant.

The purpose of differential locks

An inherent disadvantage of the conventional differential, especially for heavy vehicles operating under difficult off-road conditions, is that if one of the driven wheels encounters a slippery patch of ground a relatively low driving torque will cause it to lose traction and spin. Since the differential always distributes the axle torque equally between the driven wheels, the driving torque delivered to the gripping wheel will be reduced to that of the slipping wheel and as such is unlikely to maintain the vehicle in motion. This only serves to aggravate matters, of course, as the spinning wheel churns itself deeper into the ground and immobilizes the vehicle.

Under these circumstances there is clearly a good case for temporarily locking the differential out of action, so that sufficient driving torque can still be delivered to the gripping wheel and vehicle progress maintained. It is for this purpose that a positive differential lock, more often referred to simply as a differential lock, may be specified. The possible disadvantages of such a device are that it may be engaged too late when the vehicle is already in trouble, and if inadvertently left engaged can impose destructive strains on the axle shafts during cornering on normal dry road surfaces.

Single-axle differential lock

For this application the differential may be locked out of action by a sliding dog clutch, which through a combination of internal spline connection and external tooth engagement couples the crown wheel supporting half of the differential case to the half-shaft that passes through it (Figure 20.23). Since this also prevents the sun pinion of the half-shaft from rotating relative to the differential case, the remaining sun and planet pinions of the differential gearing automatically become locked together. The differential bevel gearing and case will therefore simply act in the manner of one solid lump of metal. No difference in speed can arise between the axle wheels, because both half-shafts are constrained to rotate at the same speed as the crown wheel. By the same token neither wheel will receive a diminished driving torque as a result of the other one slipping, since a locked differential has no means of 'knowing' whether a wheel is slipping or gripping.

Engagement and disengagement of the sliding dog clutch is effected through either a sliding or a rocking selector fork,

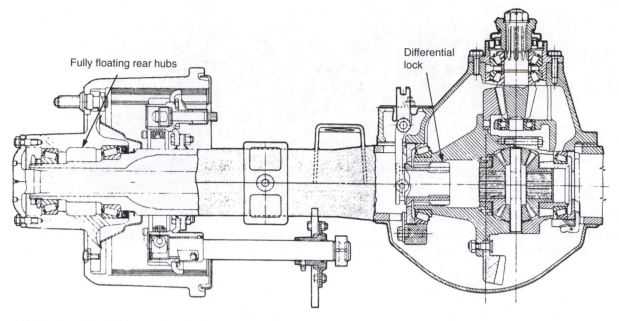

Figure 20.23 Heavy-vehicle rear axle with differential lock (*Eaton*)

which intercepts a collar on the outer end of the dog clutch. Actuation of the selector fork in past practice was manual by either rod or cable, and in present-day installations typically by an air shift unit, the unit being similar to that described later in connection with heavy-vehicle two-speed axles. It should be noted that the differential lock must not be engaged if either driving wheel is already spinning, otherwise unless the clutch pedal is depressed to stop the wheelspin the axle could suffer damage.

Limited-slip differentials

A limited-slip differential is essentially a device that will allow normal differential action of the driving wheels for negotiating corners, but prevents loss of traction in the event of either driving wheel losing adhesion. Since the conventional differential always divides torque equally, it follows that if means can be found to increase the torque needed to turn the wheel having less grip, then the torque delivered to the wheel having more grip can be increased to maintain traction. Although it was in the field of motor racing that the need for some limitation on differential action first became imperative, because otherwise such cars would be virtually unmanageable when accelerating from a standing start or out of a slow corner, it was the ever-increasing power of American cars in the late 1950s that led to the development of the limited-slip or controlled-slip types of differential as we know them today.

What is really required then is an in-built frictional means of automatically conferring resistance to rotation of the differential sun (or side) pinions relative to the differential case, according to the prevailing conditions of traction. This braking effect can be supplied by a pair of preloaded friction clutches of either the multiplate or the cone type. Taking for example the multiplate construction, the friction plates are held alternately on the outside diameter by lugs engaging with slots in the differential cage and on the inside diameter

by splines connecting with the hubs of the sun pinions. Preloading of the friction clutches is achieved by one or more lugged plates being dished to act also as a Belleville spring. The frictional resistance to differential rotation of the sun pinions and hence the road wheels is derived from the following two sources:

1 The spring loading of the friction clutches, which by imposing a fixed braking torque on the sun pinions relative to the differential cage guarantees that a certain amount of traction will always be available at one wheel, no matter how poor the grip may be at the other.
2 The additional thrust loading on the friction clutches due to the separating forces created between the sun and planet pinions during torque transmission, these forces being of a similar nature to those described earlier for the crown wheel and pinion. Resistance to differential action of the sun pinions will therefore increase in proportion to the torque applied and inhibit wheel spin during acceleration.

A long-established and more sophisticated version of the basic multiplate clutch limited-slip differential is that originally developed in the 1950s by the American Dana Corporation, under patents of the Thornton Axle Company, and known as the Powr-Lok. In Britain it is known as the Salisbury Powr-Lok (Figure 20.24) and is manufactured by GKN Salisbury Transmissions Limited. This type of limited-slip differential possesses a unique third source of loading on the friction clutches, which is derived from the opposite pairs of planet pinions being mounted on two separate pins that are stepped to cross one another. On the ends of each cross-pin are machined flats of V form that ride freely within correspondingly shaped recesses in the sun pinion clutch rings, which engage the differential case through lugs and slots. On the outside of each planet

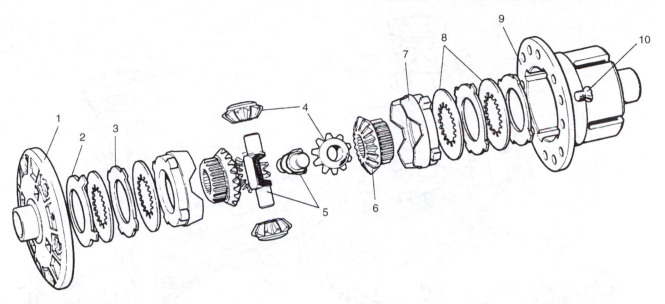

Figure 20.24 The Salisbury Powr-Lok limited-slip differential (*GKN Salisbury Transmissions*)

1 differential cage flange half	6 sun pinion
2 Belleville clutch plate	7 sun pinion ring
3 eared clutch plate	8 splined clutch plates
4 planet pinions	9 differential cage (pot half)
5 cross-pins	10 differential cage securing screw

pinion there is a smooth shoulder that can engage in rolling contact with the mating face of its sun pinion clutch ring.

When torque is applied to the crown wheel and differential case during straight-ahead driving, the reaction on the planet pinion cross-pins is such that they attempt to climb up the V-shaped recesses of the clutch rings in which they reside. In so doing they move fractionally sideways and, since this produces a corresponding movement of the planet pinions, their rolling contact with the clutch rings increases the endwise clamping load on the friction plates of the clutches. Differential action between the sun pinions is thus effectively inhibited.

During cornering, however, when the outer rear wheel has a greater distance to travel than the inner one, the speeding up of its sun pinion is such as to cause relative rotation of the planet pinions about the other sun pinion, which of course is slowing down. This reversal of torque in the plane of the cross-pins allows them to retreat down their ramps and partially de-energize the friction clutches, so that less inhibited differential action can take place.

A technical advantage of utilizing this third source of loading on the friction clutches is that it allows the use of low-friction plates, so that any difference between their static and dynamic (sliding) coefficients of friction is minimized. Otherwise, there is more opportunity for unwanted stick-slip action to occur, which can generate a cornering noise known as chatter. Of related significance is the fact that limited-slip differentials have always been sensitive to axle oil specification, which must include special friction modifying agents. A further point is that under certain conditions the total crown wheel torque may be transmitted by only one half-shaft, so it follows that the differential gears must be made stronger to sustain the heavier loading.

Since the advent of modern traction control systems (Section 29.6), where a driving wheel about to spin can be automatically braked, the limited-slip differential has tended to feature less in rear axle design.

Limited-slip differentials in service

Safety aspects Under no circumstances must the engine be run with the transmission in gear and only one driving wheel jacked clear of the ground, since the inhibited differential action will cause the grounded wheel to move the car forward. For the same reason an on-the-car wheel balancer may only be used as long as the complete axle is raised and the wheel opposite to the one being balanced is removed.

Functioning check A limited-slip differential may be checked for functioning by jacking one driving wheel clear of the ground and then recording the manual torque required to maintain it in steady rotation. This will represent the torque at which the preloaded friction clutches in the differential are slipping, and it should be compared with that specified by the vehicle manufacturer. The check is performed by means of a suitable capacity torque wrench, which engages a special adaptor plate typically held by two of the wheel nuts.

Lubrication requirements This topic has already received mention. It only remains to add that in the event of a limited-slip differential misbehaving in service, a necessary first step in diagnosis is to drain the old oil from the axle and refill it with an oil specially approved by the vehicle manufacturer. This oil will have been formulated with anti-chatter agents, otherwise known as 'LS' (limited-slip) additives, which have been evaluated on rig and vehicle testing. These ensure that the level of friction between the lubricated sliding elements

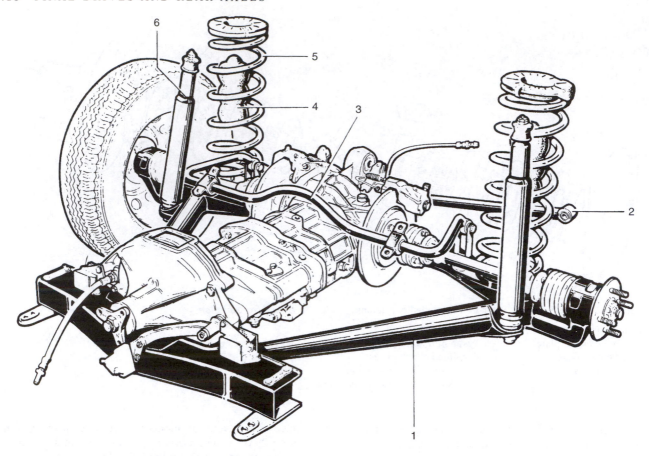

Figure 20.25   A modern De Dion axle (*Alfa-Romeo*)

1 De Dion axle   4 bump rubber
2 transverse link   5 spring
3 anti-roll bar   6 shock absorber

of the differential always remains sufficient to generate the desired partial locking effect.

20.4   REAR AXLE CONSTRUCTION

Live and dead axles

The method of transmitting drive to the rear wheels through a live axle has earlier been described in connection with the Hotchkiss drive principle (Figure 19.31). From this it should be evident that the function of a live axle is not only to provide bearing mountings for the rear wheels, but also to support the rotating elements of the final drive mechanism. A dead axle is employed either in the form of a De Dion drive to the rear wheels, which represents the original use of the term 'dead axle', or simply as a beam of lightweight construction to mount the trailing wheels of modern front-wheel-drive cars, or as a non-driven 'pusher' or 'tag' axle in the tandem axle system of a heavy vehicle (Section 23.5). The De Dion axle has also received earlier brief mention (Section 19.1), and is a form of rear suspension and drive that offers a compromise between the live axle and independent rear suspension (Figure 20.25). Essential features of this system are that the driven rear wheels are mounted on a tubular axle beam that is

cranked to ensure working clearance with the final drive assembly, which is separately mounted and insulated from the car structure. Driving torque is transmitted from the final drive assembly to the rear wheels by universally jointed drive shafts. Location of the De Dion axle may be achieved by a variety of linkage systems, the simplest comprising a pair of radius rods disposed in a V-formation together with a Panhard rod.

Functions of the rear axle casing

The functions performed by the rear axle casing may be summarized as follows:

1 To provide a single rigid means of mounting the road wheels, final drive arrangements and brake assemblies, relative to the vehicle rear suspension system and driveline.
2 To perform a load-carrying function by supporting the rear end of the vehicle on the road wheels.
3 To perform a suspension function by containing those forces arising from the transference of road shocks and cornering forces to the springing system.
4 To perform a tractive function by containing those forces associated with the transmission of driving torque to the road wheels.

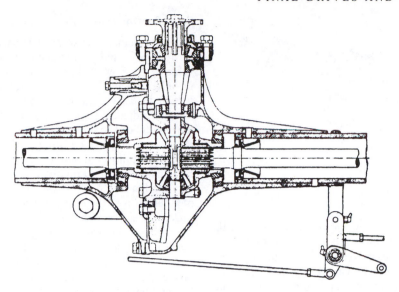

Figure 20.26 An interesting early postwar trumpet-type axle fitted to a commercial vehicle

5 To perform a retarding function by containing those forces associated with the application of brake torque to the road wheels.

Types of rear axle casing

Before examining the different types of casing construction for live axles, it is necessary to identify the two main requirements that an axle casing must satisfy as a load-bearing structure:

1 It must be sufficiently rigid to avoid any distortion of the final drive housing under the most arduous operating conditions and thereby ensure quiet running of the gears.
2 Its rigidity must also be such as to maintain the halfshafts in true alignment with the final drive unit and thus reduce any possibility of their fatigue failure due to bending.

There are three basic types of live axle construction, two of which by virtue of their form (and not because of any lack in quietness of running!) have rather musical designation: trumpet, banjo and carrier.

Trumpet axle

This is constructed from two trumpet-shaped axle halves, which are spigoted and bolted either directly together across their vertical joint faces, or indirectly together across their vertical joint faces, or indirectly together with an interposed aluminium alloy centre section. In the former case one of the axle halves is extended beyond the true central parting line of the axle in order to accommodate the bevel pinion and its bearing mountings (Figure 20.26), while in the latter arrangement both the crown wheel and pinion may be supported independently of the axle halves by the centre section housing, as long found in Rolls-Royce live axle designs.

The trumpet axle can enable the final drive gears to be very rigidly supported and generally provides an unusually robust form of construction. It has, however, long been regarded as being somewhat difficult to manufacture and

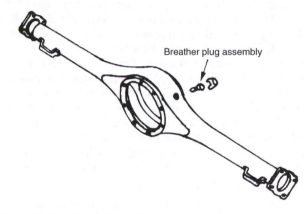

Breather plug assembly

Figure 20.27 Banjo-type axle casing

service, so that its application has generally been limited to a very few high-grade cars, certain specialized 4 × 4 and battery-electric vehicles.

Banjo axle

For normal duty axles this type of casing is of one-piece construction. It is formed from a single tube and is provided with an enlarged open centre section, which accommodates the flange-mounted final drive housing (Figure 20.27). The rear part of the centre section is closed by a domed rear cover that need not be made detachable and can thus be welded on to the casing in the interests of both rigidity and oil tightness.

Although this type of axle construction, is by virtue of its form, well adapted to resist vertical bending loads, it is less able to resist tractive and braking forces that act in the horizontal plane. Also, the crown wheel tends to be less well supported, because its bearings are cantilever mounted (overhung) from the final drive housing. However, this type of axle construction does allow the final drive assembly to become a self-contained unit, which was an important servicing consideration when axles were not so reliable as they are nowadays.

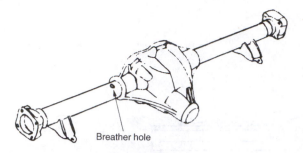

Breather hole

Figure 20.28   Carrier-type axle casing

Carrier axle

The essential feature of this type of axle casing is that the crown wheel and pinion are mounted in a rigid central housing or carrier of malleable cast iron, into which the steel axle tubes are pressed and welded in position (Figure 20.28). This form of construction necessarily requires the crown wheel and differential unit to be installed through a large aperture in the rear of the carrier. Semicircular shouldered recesses are machined in the carrier to locate the bearing cups of the crown wheel and differential unit, which are retained by bolted on bearing caps. A detachable cover is, of course, provided to close the rear aperture of the carrier.

Until the widespread adoption of independent rear suspension, the carrier type axle had become established practice for passenger cars. However, it is still used in the construction of some 4 × 4 and many light commercial vehicle live axles and so for that matter is the banjo type axle. In Britain it began to find increasing acceptance in the 1950s, but its American origins date back to the early 1930s. The important advantages conferred by this form of axle construction are therefore worth enumerating.

1 The carrier can be provided with a substantial stiffening rib that completely spans the bearing mountings for the crown wheel and pinion, so that the gears are given maximum support.
2 In the carrier axle the crown wheel and differential bearings can be widely spaced, which further enhances the stability of the crown wheel.
3 Maximum lateral support is also provided for the crown wheel and differential bearing cups, which butt against continuous surfaces in the carrier.
4 Oil circulation channels can be readily incorporated in the carrier, a feature that is of particular importance for effective lubrication of the pinion bearings, as mentioned later.
5 From a manufacturing point of view, the carrier-type axle can be readily adapted to a wide range of wheel track widths by alteration in length of the axle tubes and half-shafts, while retaining a common carrier assembly.

Rear hub mountings

In addition to their other features of general construction, driven rear axles are classified as to the type of bearing mounting used to support the hubs. This is because the nature of the loading on the half-shafts and the hub bearings varies in accordance with the particular arrangement adopted (Figure 20.29). There are three distinct arrangements of rear hub: semi-floating, three-quarter floating and fully floating.

Semi-floating rear hub

This arrangement became established practice for passenger car rear axles and continues to be used for light commercial vehicles. Each wheel hub is either formed integral with or, as in earlier practice, separately attached to the outer end of the half-shaft. This in turn is supported by a rolling bearing mounted adjacent to the hub and registered inside the end of the axle tube. The inner end of the half-shaft is supported by the corresponding side bearing of the final drive crown wheel.

It thus follows that the half-shaft must withstand not only the torsion load involved in driving the wheel, but also bending loads in the vertical and horizontal planes. The former derive from the effect of vehicle weight acting about the offset wheel bearing, and also from the side thrust forces imposed on the wheel during cornering. In the horizontal plane, the bending loads arise from traction and braking forces. However, it should be realized that these bending movements on the half-shaft are being restrained by the very widely spaced hub and final drive bearings. From a practical point of view, it is not possible to withdraw a semi-floating half-shaft from the axle without removing the wheel and disturbing the hub bearing assembly.

Noteworthy constructional features of modern semi-floating axle arrangements (Figure 20.30) are as follows.

Half-shafts

Each of these may be forged integrally with its wheel hub which is then formed by an upset flange, there being a generous radius at this juncture to minimize stress concentration. Alternatively a taper and key fitting may be used to secure the wheel hub. The inner end of the shaft is splined to engage the differential gearing of the final drive.

Hub bearings

A single-row, deep-groove, ball bearing race is usually chosen to support the outer end of each half-shaft and its associated hub. This type of rolling bearing comprises an inner ring, caged ball bearings and an outer ring. The outer stationary ring of the bearing registers in the counterbored end of the axle tube and is held therein by a bolted-on retainer plate. Endwise location of the bearing inner ring is provided by a shoulder near the outer end of the half-shaft, against which it is held by an interference-fitting retaining sleeve. The latter is thus subjected to thrust loads acting in the outward direction on the wheel.

A similarly mounted single-row, tapered roller bearing has sometimes been used in preference to a ball bearing. The tapered roller bearing comprises an inner ribbed cone, caged tapered rollers and an outer cup. In this case, the thrust loads acting on a wheel in the inward direction are freely transmitted through both half-shafts, via a central spacer, to the bearing on the other side of the axle. The reason for this side-to-side transfer of loading is that, acting alone, the conventional tapered roller bearing is capable of accepting axial thrust loading in one direction only.

More recently, however, a modified form of single-row tapered roller bearing taking axial thrust loading in either direction has been applied to motor vehicle axles by the Timken Company. This type of bearing is mounted inverted in the axle tube so that the heavy cornering loads acting on the outside wheel are not taken by the cone and cup ribs.

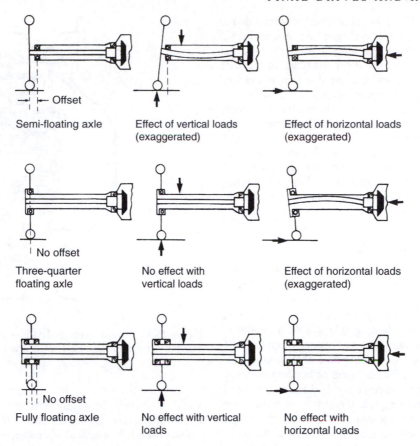

Figure 20.29 Schematic arrangements of semi-, three-quarter and fully floating rear axles, also showing bending effects on axle shaft arising from vertical and horizontal loadings

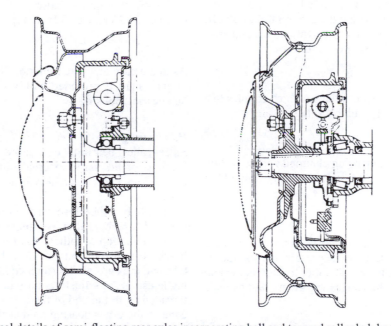

Figure 20.30 Constructional details of semi-floating rear axles incorporating ball and tapered roller hub bearings (*SKF*)

Hub oil seals

Where ball bearing races are used they generally have integral seals and are grease lubricated for life. As their name implies, integral seals are built-in seals forming part of the bearing itself, which may be extended in width to accommodate them. A radial lip oil seal is normally used on the inner side of the bearing to prevent axle oil from surging into the bearing and gaining access to the rear brake assembly. The outer side of the bearing is usually sealed by a dust shield which is intended to retain the lubricating grease and exclude contaminants. Provision may also be made for sealing between the outer diameter of the bearing and the inner diameter of the axle tube. In this case an O-ring seal is located within an external groove on the outer ring of the bearing race.

Tapered roller bearings do not lend themselves so readily to integral sealing arrangements. It is therefore usual for separately mounted seals of the radial lip type to be used on each side of the bearing and acting against suitably machined sealing surfaces on the half-shaft itself. The sealing lips of both seals point inwards towards the final drive, but the outer seal may also have an extended wiper lip pointing in the opposite direction to exclude contaminants.

Three-quarter floating rear hub

This type of bearing support for the hubs was once widely used in the rear axles of both cars and light commercial vehicles, but its popularity has declined in favour of the less costly and simpler semi-floating arrangement. An essential feature of the three-quarter floating axle was the mounting of each hub bearing on the axle tube end instead of on the half-shaft, and its location on the central plane of the road wheel. This arrangement relieves the half-shaft of bending moments due to the vehicle weight, traction and braking forces, but it must still withstand cornering side thrust forces on the wheel and of course transmit driving torque.

A safety feature of this arrangement was that the wheel could not become detached through fracture of the splined end of the half-shaft. Whereas this is also true in the case of the modern semi-floating axle, it was not always the case with some earlier versions, where endwise location of the half-shaft was effected from within the differential assembly.

The constructional features of a typical three-quarter floating rear axle (Figure 20.31) are as follows.

Half-shafts
On the outer end of each shaft is a flange which couples to a similar flanged portion of the hub bearing housing.

Hub bearings
A ball bearing race has its outer ring clamped within the hub bearing housing and its inner ring retained on the end of the axle tube by a locking nut.

Hub oil seals
Later designs incorporated two radial lip seals, one located within the end of the axle tube to prevent the escape of axle oil along the half-shaft, and the other located in the hub bearing housing to retain the lubricating grease and exclude contaminants.

Fully floating rear hub

This arrangement represents standard practice in the mounting of rear hubs on commercial vehicles and was also used in the past on some heavier cars. With this arrangement, the axle tubes entirely relieve the half-shafts of bending moments

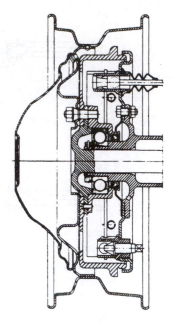

Figure 20.31 Constructional details of a three-quarter floating rear axle incorporating a ball hub bearing (*SKF*)

arising from vehicle weight, traction, braking and cornering thrust forces. Hence, the half-shafts are subjected only to torsional loading. A practical advantage of the fully floating axle is that a broken half-shaft can be withdrawn without removing the road wheel and, of course, the vehicle remains towable in these circumstances.

The constructional features of a fully floating rear axle (Figures 20.23 and 20.32) are typically as follows.

Half-shafts
As in the case of a three-quarter floating axle, the outer end of each shaft is provided with a flange for coupling to the hub bearing housing.

Hub bearings
Since roller bearings can carry greater loads than ball bearings of the same external dimensions, the former are the obvious choice for heavily loaded commercial vehicle hubs.

Each hub is therefore carried on a widely spaced pair of opposed tapered roller bearings which can accommodate any combination of radial and thrust loads. The cones of the bearings are made a push fit over the stepped end of the axle tube and the cups an interference fit in the rotating hub bearing housing. An adjusting nut with locking arrangements is threaded on to the end of the axle tube and bears against the cone of the outer bearing so that the correct running clearance of the bearings can be accurately set.

It will be noticed that the arrangement and adjustment of fully floating hub bearings are very similar to those often found in the mounting of front hubs on their spindles.

The late 1980s saw the development of more compact and integrated designs of bearing unit for this application to heavy

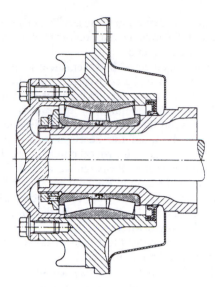

Figure 20.32 Heavy-vehicle compact wheel bearing unit (*FAG Kugelfischer*)

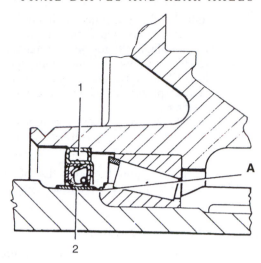

Figure 20.33 Oil seal and wear sleeve (*Seddon Atkinson*)

1 oil seal
2 wear sleeve
A location

vehicles. In the German FAG design the integrated bearing unit comprises a double-row tapered roller bearings with one common outer ring forming the cups, which is pressed into the hub. The two inner rings or cones contact each other with their small faces and slide over the axle tube to be clamped tightly against its shoulder by a retaining nut. It does not matter which side first the bearing unit is fitted because it is symmetrical in construction (Figure 20.32). The components of the bearing unit are adapted to each other in such a way that after mounting the appropriate operating clearance is automatically obtained. Conventional setting of the bearing clearance is therefore no longer necessary. The tapered roller bearings themselves are of modern low-cross-section form and have cages of either glass-fibre-reinforced polyamide or sheet steel. For lubrication purposes the wheel bearing space and the oil space of the axle can be combined. Several advantages are claimed for this type of integrated rear-wheel bearing unit, which include simpler overall construction, smaller mounting space, lower weight, increased capacity, easier installation and reduced cost.

Hub oil seals
These are generally of the radial lip type, and their arrangement depends upon whether the hub bearings are grease lubricated or, as in more recent practice, oil lubricated from the axle (Figure 20.32). In both cases, a seal with its lip pointing outwards and bearing against the axle tube is installed in the hub adjacent to the inner bearing. For oil lubrication the hub may be provided with a drain hole that leads from a catcher behind the oil seal, so that in the event of a seal failure the leaking oil is prevented from reaching the brake assembly.

An associated feature of the oil sealing arrangements for the fully floating hubs of some heavy vehicles is an interference-fit 'wear sleeve'. This fits over the end of the axle tube and thus permits easy replacement of the axle sealing surface whenever the oil seal requires renewing (Figure 20.33).

20.5 FINAL DRIVE LUBRICATION AND SEALING

Classification and grades of oil

Similar to the earlier described way in which engine oils have been classified by the Society of Automotive Engineers in America, transmission gear oils are classified into their various viscosities by assigning them a different series of SAE numbers. The system was first adopted in 1931.

The current SAE viscosity classifications for gear oils are 70W, 75W, 80W, 85W, 90, 140 and 250, these viscosities being measured at low and high temperatures as appropriate. It should be noted that these numbers follow on from those of engine oils, so that it is immediately apparent the product is a gear oil, even though a gear oil and an engine oil may in fact have similar viscosity characteristics. For example, an SAE 90 gear oil can have viscosity characteristics similar to those of an SAE 40 or SAE 50 engine oil.

Multigrade gear oils of SAE 80W–90 and 85W–140 again follow the same system as for multigrade engine oils; that is, an SAE 80W–90 gear oil conforms both to the SAE 80W requirement at low temperature and the SAE 90 requirement at high temperature.

As in the case of engine oils, the selection of a gear oil demands careful consideration of the performance characteristics required of it, and again the American Petroleum Institute (API) has been responsible for developing since 1957 an API classification system for these oils based on the severity of transmission gear operation. Briefly, this widely accepted system of gear oil designation comprises the following API service categories and oil descriptions:

GL1, GL2   non-additive for mild conditions of tooth contact pressure and sliding speed

GL3   mild EP for moderate conditions of tooth contact pressure and sliding speed

GL4   medium-high EP for severe conditions of tooth contact pressure and sliding speed

GL5   high EP for very severe conditions of tooth contact pressure, sliding speed and shock loading

GL6   high EP for very severe conditions, originally intended for high-offset hypoid gears. This service category is now obsolete.

Most final drive gears require the protection of extreme pressure (EP) oil additives, to prevent surface deterioration of their teeth.

Lubrication of the gears and bearings

It is perhaps a tribute to the well-planned lubrication and sealing of the modern rear axle that ultimately it wears generally rather than locally. The special lubrication requirements of hypoid gears were mentioned in Section 20.1, and it is now necessary to consider first the mode of application and then the sealing arrangements.

Oil bath lubrication is employed for motor vehicle final drives, the axle housing forming an oil-tight reservoir for the lubricating oil (Figure 20.34). A drain plug is usually provided on the underside of the housing and a combined filler and level plug on the rear cover. The oil level is generally made just high enough to cover the full-face width of the crown wheel teeth at bottom dead centre.

At comparatively low road speeds lubrication of the gears relies on the transport of oil by the crown wheel gear, as teeth that have dipped into the oil travel round to the point of mesh with the bevel pinion. With increasing road speed oil splash or spray begins to be thrown up, which also serves to lubricate the differential gear assembly.

Whereas the crown wheel bearings may be splash lubricated, the chief concern is to ensure that the heavily loaded pinion bearings receive adequate lubrication under all running conditions. This is because under low-speed operation there is a tendency for the oil to cling to, and build up on, the rim of the crown wheel so that oil in the reservoir can be drawn away from the rear of the housing to the detriment of pinion bearing lubrication at the front end.

As the road speed rises, however, there is an increasing tendency for the oil to be centrifuged from the rim of the crown wheel and thrown against the housing wall. This oil can then be collected to lubricate the pinion bearings by directing it between them via a suitably positioned passage. Once it arrives between the bearings their tapered form enables them to act as their own oil pump to circulate the oil over the rolling elements. Research work carried out by the SKF Laboratory at Luton in England has shown that for the most effective lubrication of the pinion bearings their oil supply passage should be aligned with the rim, rather than the teeth, of the crown wheel.

In the case of heavy-vehicle tandem drive axles (Figure 20.43), a separate oil pump may be driven from the final drive pinion shaft of the foremost axle. Its purpose is to circulate oil to the bearings that support the upper transfer gear and to the working surfaces of the inter-axle differential.

Final drive oil sealing arrangements

The purpose of these is not only to seal in the lubricant but also to prevent any foreign matter from entering. In modern practice, the sealing arrangements must take into account final drives incorporated in live rear axles, those separately mounted on the car structure, and others that form part of transaxles.

Live rear axles
Here we may confine our attention to the sealing of the bevel pinion shaft, since the hub sealing at the axle ends was dealt with in Section 20.4. An oil sealing arrangement for the bevel pinion shaft typically comprises a combination of the following:

Lipped metal dust shield   pressed against a shoulder on the companion flange and working in conjunction with the lipped casing of the oil seal (Figure 20.8).

Radial lip oil seal   pressed in a recess in the pinion housing and acting against a sealing surface machined on the boss of the companion flange (Figures 20.9 and 20.34). In this context, it should be appreciated that for proper functioning and satisfactory life of the seal, a very thin film of oil must always be maintained between its lip and the boss, the film thickness being controlled by the force exerted by the lip on the sealing surface. A certain tolerable leakage of oil therefore occurs, even though it should be hardly measurable in service.

Oil slinger washer   sandwiched between the rear end of the companion flange boss and the pinion front bearing cone (Figures 20.8 and 20.9). Its purpose is to prevent flooding of the radial lip oil seal at high rates of oil circulation to the pinion bearings. However, an oil slinger washer is not always included.

Separately mounted final drives
The sealing arrangements for these are twofold:

Bevel pinion shaft   The arrangements for sealing this are identical to those previously described for the live axle.

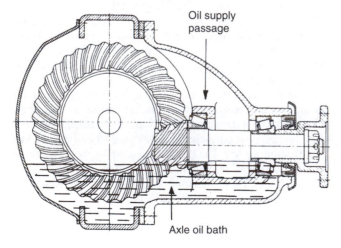

Figure 20.34 Lubrication of final drive gears and pinion bearings (*SKF*)

Oil supply passage

Axle oil bath

Stub drive shafts These connect with the differential sun pinions and are usually carried in rolling bearings mounted independently of those supporting the crown wheel and differential unit. A need is therefore created to seal the stub shafts where they emerge from the sides of the final drive housing (Figure 20.10). For this purpose radial lip seals are pressed into recesses machined in the separate bearing housings of the stub shafts and act against sealing surfaces provided near the flanged ends of the shafts. An additional sealing consideration arises where inboard brakes are located on each side of a separately mounted final drive, since the stub shaft seals then have to be capable of withstanding limited periods of very high-temperature operation.

Transaxles
In the now widely used transversely mounted transaxle (Figure 14.14), the radial lip sealing arrangements for the stub drive shafts are similar to those previously described for the separately mounted rear final drive unit (Figure 20.10). This also holds true for sealing the stub drive shafts in the less common longitudinally mounted transaxle, but another seal is required where the primary shaft enters the transaxle casing (Figure 20.12).

Axle casing breather

An appreciable temperature rise accompanies the normal operation of the final drive, this being especially true since the advent of motorways where vehicles can run at sustained high speeds. Axle oil temperatures can in fact exceed 100°C, and it is significant that among the various tests applied by research laboratories to axle lubricants there is one that evaluates their thermal stability at a controlled temperature of typically 135°C.

Clearly, then, some means of venting the axle interior to the atmosphere must be provided, otherwise there would be a danger of pressure building up in the axle and forcing oil past its seals. For a motor car rear axle this venting may be arranged either simply by a small-diameter hole in the axle tube (Figure 20.28), or by the once traditional drilled and shielded breather plug screwed into the final drive housing (Figures 20.27 and 20.35).

For heavy vehicles a manufacturer may recommend that the breather valve be removed and thoroughly cleaned at regular intervals, typically every 5000 miles (8000 km) or monthly, although more frequently if the vehicle is operating in extremely dirty conditions. On some heavy vehicles a breather valve may be fitted that is so constructed as to allow

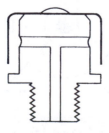

Figure 20.35 Shielded axle breather plug

heated air to pass out of the axle, but at the same time to prevent the ingress of dust and moisture.

20.6 REAR AXLE MISBEHAVIOUR IN SERVICE

Final drive noise

Noisy operation is the most likely type of complaint to arise in a final drive unit, as a result of either defective build or long-term wear. It is fairly readily distinguishable to the vehicle occupants, because its frequency is usually much higher than the other potential sources of noise in a motor vehicle. Before attempting to diagnose any such complaint of noise, the final drive must be not only checked for correct oil level, but also allowed to attain a normal running temperature before testing proper begins. Furthermore, the possibility of the noise being caused by the tyres must be eliminated by noting during road-test if the character of the noise changes on different road surfaces.

The following is intended to be a purely general guide to possible complaints of final drive noise and their related causes:

Drive noise This may be reproduced when the vehicle is being accelerated and therefore corresponds to a higher than normal torque load on the final drive gears. Any condition of maladjustment or of wear that allows the contact pattern on the drive side of the teeth to become concentrated towards their heel or wider portion (Figure 20.18) can generate a resonant whining noise. Typically this begins at a certain road speed, reaches a maximum intensity and then fades out at a higher speed.

Coast noise Conversely, this may be reproduced when the vehicle is being decelerated, or in other words when the engine is being driven by the vehicle. In this case any condition of maladjustment or wear that allows the contact pattern on the coast side of the teeth to become concentrated towards their toe or narrower portion (Figure 20.18) can similarly generate a resonant whining noise. As might be expected, this begins at a certain road speed, reaches a maximum intensity and then fades out at a lower speed.

Float noise This may be reproduced while maintaining a throttle opening just sufficient to prevent the vehicle from driving the engine. As a continuous whirring noise at low speeds it is usually symptomatic of wear in either the pinion or, less commonly, the crown wheel bearings. The former tend to generate a relatively higher-pitch noise than the latter, since they rotate at a faster speed in accordance with the final drive ratio.

Shunt noise This may be reproduced by 'blipping' the throttle at a relatively low road speed, which creates a sudden torque reversal on the final drive gears. It usually occurs as a slapping noise and can arise from any condition of maladjustment or of wear that results in excessive backlash between the teeth of the crown wheel and pinion.

Take-up noise This may be reproduced on moving away from rest, as torque load is initially applied to the final drive gears, and also repeated during gear changing. Heard as a single clunking noise, it is usually associated with general wear in the differential mechanism and may include excessive clearance between the teeth of the sun and planet pinions, slack fitting of the sun pinions in the bearing bores of the differential case, and looseness of the planet pinions shaft in the differential case.

Cornering noise This is self-explanatory and, as would be expected, it emanates from the differential mechanism and not from the final drive gears themselves. A moaning noise may arise from several conditions including partial seizing of the sun pinions in the bearing bores of the differential case, partial seizing of the planet pinions on either their cross-shaft or spider, and worn thrust washers allowing excessive backlash between the teeth of the sun and planet pinions. A chattering noise may be encountered with a limited-slip differential, the source of which can be attributed to either an incorrect type of lubricant or wear of the friction clutches.

Hub bearing noise

During road test it is usually possible to distinguish between hub bearing and final drive noise, because when running straight ahead a noisy hub bearing is likely to be heard only on the overrun from modest vehicle speeds, and a gentle application of the brakes will often reduce the level of noise under these conditions. During cornering a noisy hub bearing will become noisier when the direction of turn is such as to transfer more weight on to that particular wheel and hub, so a righthand turn will intensify the noise from a left-hand hub.

After road test the wheel with the suspect hub bearing is jacked clear of the ground and slowly spun by hand to verify the source of noise. Unless it is simply a matter of readjusting the hub bearings, as is sometimes the case with the fully floating rear hubs of heavy vehicles (Figure 20.23), there is normally little choice but to replace a hub bearing that becomes noisy in service.

Oil leaks

Leakage of oil past the hub bearings or the final drive pinion shaft of the conventional rear axle occurs when the casing oil seals reach the end of their useful life. This is determined by age hardening and embrittlement of their sealing lips due to heating, or by wearing of the shaft and sealing lip to an extent that effective seal contact is prevented. Other factors that may cause oil leaks are overfilling of the final drive casing, such that oil can surge along either axle tube during cornering and flood past the hub bearing seals, or a blocked breather, which allows pressure to build up in the hot axle and disturb seal contact. Damaged sealing gaskets offer another source of oil leakage.

The installation of a new oil seal of the non-integral type demands care, since it can be rendered unserviceable if distorted out of round or tilted in its counterbore, or if the sealing lip is torn. The use of a special guidance tool may be advised by the vehicle manufacturer to protect the seal against physical damage during installation.

20.7 HEAVY-VEHICLE REAR AXLES

Single-speed double-reduction axles

The development of higher-speed engines for heavy vehicles, which began in the early 1950s in America and then spread elsewhere, generally dictates proportionally lower (numerically higher) axle ratios. These were necessary to provide a greater final reduction in speed between the engine and road wheels in direct drive, since there were already restrictions on maximum vehicle speed and minimum size of tyre equipment.

In earlier British practice, worm drive axles had been widely used to achieve reduction ratios greater than about 7:1. This type of final drive gearing (as related in Section 19.7) was never popular in America, where it became established practice to use double-reduction gearing. This enabled greater reduction ratios to be obtained in two stages, or what are termed primary and secondary stages, and it was this system that was later adopted in Britain and elsewhere.

Double-reduction axles may be arranged in several different ways, which are chiefly as follows:

Spiral or hypoid bevel gears for primary stage, spur or helical gears for secondary stage
The crown wheel of the primary stage and the pinion of the secondary stage are carried on an intermediate shaft (Figure 20.36). This may be positioned towards the top of the gear-driven differential unit to reduce the overhang of the final drive housing. Three pairs of bearings are required to support the primary- and secondary-stage reduction gears, and as in conventional single-reduction axles these bearings are usually of the preloaded taper roller type. Since the primary-stage reduction gears are less heavily loaded than the secondary-stage ones, a simple overhung mounting for the bevel pinion can be used. It should be noted that in comparison with a conventional single-reduction axle, the crown wheel is carried on the other side of the bevel pinion, so that the correct direction of rotation is restored for the road wheels.

Spur or helical gears for primary stage, spiral or hypoid bevel gears for secondary stage
In this case the gear of the primary stage is carried on the bevel pinion shaft of the secondary stage (Figure 20.37). The input pinion shaft of the primary stage is mounted above the bevel pinion shaft. Three pairs of bearings are again required to support the primary and secondary-stage reduction gears,

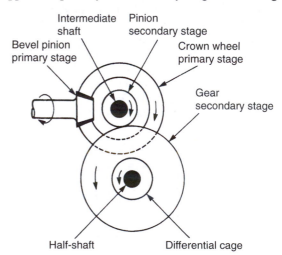

Figure 20.36 Schematic arrangement of a single-speed double-reduction axle with spur or helical gears for the secondary stage

and similarly the crown wheel is carried on the other side to normal of the bevel pinion. This reversed arrangement of double-reduction gearing before the differential unit is generally not thought to be at any particular disadvantage to the previous form.

Spiral or hypoid bevel gears for primary stage, epicyclic gears for secondary stage

At first glance this arrangement of double-reduction gearing gives every appearance of being a conventional single-reduction axle. This is because a normal mounting is adopted for the crown wheel and pinion, while the secondary-stage epicyclic gearing is hidden adjacent to the differential gearset (Figure 20.38). The epicyclic gearing is interposed between the ring gear of the crown wheel and the differential cage. This is accomplished by forming the annulus gear teeth on the inside of the ring gear, attaching the carrier for the planet pinions to the differential cage and providing a tubular stub mounting for the stationary sun gear. A secondary-stage reduction is therefore obtained as the combined crown wheel and annulus gear rotates and compels the planet pinions to walk around the stationary sun gear. Since the planet carrier is directly coupled to the differential cage, the half-shafts and hence the road wheels will rotate in the same direction as the crown wheel but at a reduced speed. This arrangement of double-reduction gearing reduces the loading on the primary-stage crown wheel and pinion and produces a compact design of axle.

Spiral or hypoid bevel gears for primary stage, hub epicyclic gears for secondary stage

An alternative approach to providing double-reduction gearing is to separate the secondary stage from the axle final drive and incorporate it in the wheel hubs (Figure 20.39). In a typical installation the sun gear is attached to the end of the half-shaft, the hub is extended to form a carrier for the planet pinions and the annulus gear is held stationary via a connecting plate to the axle tube. Therefore as the sun gear rotates with the

half-shaft it compels the planet pinions to walk in the stationary annulus gear. Since the hub acts as the planet carrier it will rotate in the same direction as the half-shaft but at a reduced speed. Hub reduction gears must necessarily be of compact design to allow for fitting of the road wheels. They typically receive their lubrication from the axle oil supply, the level of which is then controlled by inspection plugs at the hub centres. An important advantage of using a hub reduction gear is that it makes for a less bulky axle, because of the reduced torque being transmitted by the final drive gears, differential and half-shafts, which can be reduced in size accordingly. A disadvantage is naturally that the secondary-stage reduction has to be duplicated for both hubs.

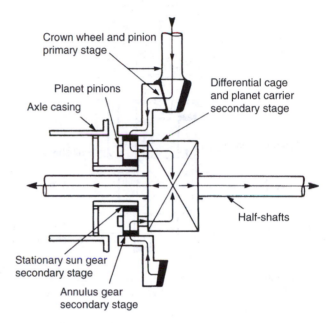

Figure 20.38 Schematic arrangement of a single-speed double-reduction axle with epicyclic gears for the secondary stage

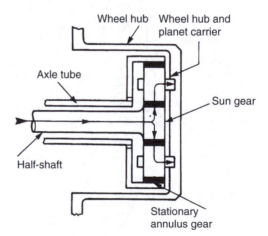

Figure 20.39 Schematic arrangement of a single-speed double-reduction axle with hub epicyclic gears for the secondary stage

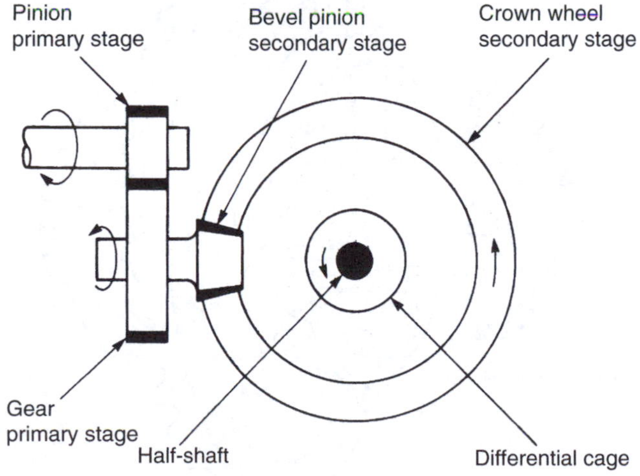

Figure 20.37 Schematic arrangement of a single-speed double-reduction axle with spiral or hypoid bevel gears for the secondary stage

Two-speed axles

It was again in America that the two-speed axle first became popular for heavy vehicles, especially when used in conjunction with a five-speed gearbox. The advantage of providing an axle with alternative reduction ratios is that they can be chosen to straddle the optimum ratio for a single-speed axle. As either ratio can be selected readily by the driver, variations in load or road condition can more economically be accommodated.

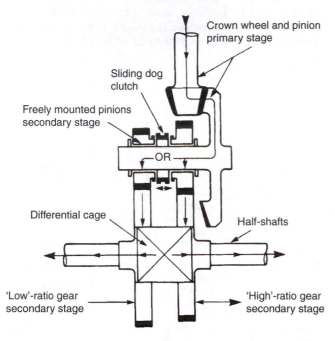

Figure 20.40 Schematic arrangement of a two-speed axle with two alternative pairs of spur or helical gears for the secondary stage

Two-speed axles are essentially derived from certain forms of single-speed double reduction axle, as follows.

Spiral or hypoid bevel gears for primary stage, two alternative pairs of spur or helical gears for secondary stage
With this arrangement the crown wheel of the primary stage and two different size pinions of the secondary stage are carried on the usual intermediate shaft (Figure 20.40). Both pinions are free to rotate on this shaft until either is engaged by a centrally disposed sliding dog clutch, the hub of which is made integral with the shaft. These pinions are in constant mesh with two gears that differ correspondingly in size and are rigidly attached to the carrier for the differential gears. A choice of two double-reduction ratios is therefore made available by dog clutch engagement.

Spiral or hypoid bevel gears for primary stage, lockable epicyclic gears for secondary stage
This type of two-speed axle is virtually identical in construction to its single-speed double-reduction counter-part, the essential difference being in the mounting of the sun gear for the secondary stage. For the purpose of locking the epicyclic gearing to give a direct drive from the crown wheel to the differential, the tubular stub mounting for the sun gear is made slidable and provided with dog teeth, so that it can be disengaged from its axle tube support and then engage internal teeth formed at the outer end of the planet carrier. Since this action results in two members of the epicyclic gearing becoming locked together, these being the sun gear and planet carrier, no secondary-stage reduction is possible. This type of two-speed axle therefore provides either a single-reduction high ratio with no secondary-stage reduction through the epicyclic gearing, or a double-reduction low ratio with the crown wheel and pinion acting in combination with the epicyclic gears (Figures 20.41a and b).

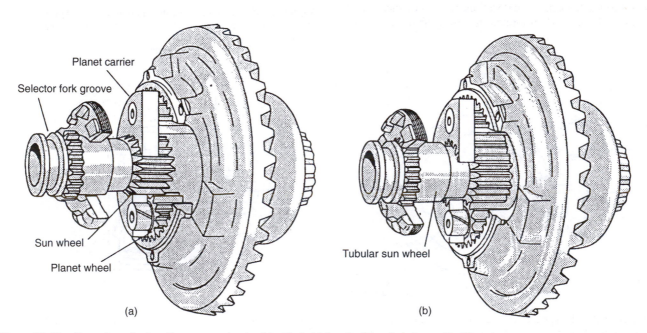

Figure 20.41 Operation of epicyclic two-speed axle: (a) axle in high ratio (b) axle in low ratio (*Eaton*)

In modern practice the two-speed axle has a power-operated driver-controlled selector mechanism that utilizes either a pneumatic or an electric servo unit. For example, the sliding sun gear of an epicyclic secondary-stage reduction can be operated by a forked rocking lever that intercepts a collar on its hollow stub support. The other end of the rocking lever receives its motion from a striker pin, which is activated from the rocker shaft of the servo unit via a torsion spring cushion drive (Figure 20.42). Finally, the rocker shaft is acted upon by either a spring-returned piston in a compressed air cylinder, or a screw and nut push-pull mechanism driven from an electric motor. The servo unit also incorporates an electrical control switch that signals when a change of drive ratio is required to a speedometer adaptor, so that speedometer readings are corrected to suit whichever axle ratio is being used.

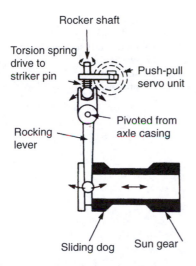

Figure 20.42 Schematic arrangement of a servo-controlled selector mechanism for lockable epicyclic gears at the secondary stage

Tandem axles

The problem involved in transferring the drive between two axles suspended in tandem has been mentioned in Section 19.7. This type of installation also requires the use of a third or inter-axle differential, the function of which is to prevent any wind-up effects between the two axles arising from slight differences in rolling radii of the tyres. It is further necessary to make provision for locking the inter-axle differential, because otherwise its action would be self-defeating when the vehicle encountered slippery conditions. That is, the drive would be limited to the traction ability of the wheel having the least grip. With a tandem axle drive an inter-axle lockable differential is therefore embodied within the driving head of the foremost axle, so that the torque input from the propeller shaft can be routed directly to its four planet pinions. Two equal torque outputs from the inter-axle differential are then taken from its sun pinions; one output is indirectly routed via transfer gears to the final drive pinion of the foremost axle, whilst the other output is received more directly via output and inter-axle propeller or coupling shafts by the final drive pinion of the rearmost axle. A convenient method of locking the differential out of action when required is to employ a dog clutch, which slides on the splines of the input shaft to the differential and has face-type dogs to engage the upper transfer gear (Figure 20.43). As in similar applications for heavy-vehicle axles, the actuation of the locking clutch is usually performed through an air shift unit (Figure 20.44). On a slippery road surface the driver will primarily engage the inter-axle differential lock, but if this does not prove adequate then the differential locks on each axle are also engaged. In the latter case this will tend to make the vehicle reluctant to being steered away from the straight ahead direction of travel. Tandem axles may also incorporate double-reduction gearing in the form of an epicyclic gearset for the secondary stage of each final drive unit, the principle of which has earlier been explained (Figure 20.38).

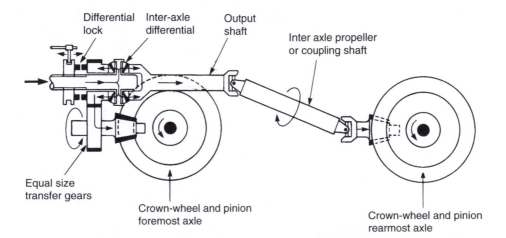

Figure 20.43 Schematic arrangement of tandem drive axles and power flow through inter-axle differential

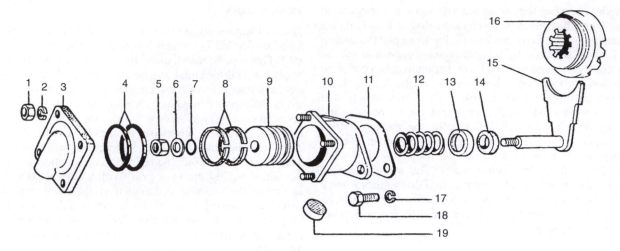

Figure 20.44 Exploded view of differential lock-out mechanism (*Seddon Atkinson*)

1 nut	8 piston felt oiler	14 felt seal
2 washer	9 lock-out cylinder position	15 shift fork
3 lock-out cylinder cover	10 lock-out cylinder body	16 input shaft sliding clutch
4 grommet	11 gasket	17 washer
5 nut	12 compression spring	18 setscrew
6 washer	13 seal retainer	19 breather
7 push-rod grommet		

21 Four-wheel-drive systems

21.1 TYPES OF FOUR-WHEEL DRIVE

General background

Like certain other modern developments in automotive engineering, the origins of four-wheel drive may be traced back to the early history of the motor vehicle and in this case particularly to its military applications. The basic concept of providing the once almost universal front-engine rear-wheel-drive vehicle with a central transfer gearbox from which the drive could also be transmitted to the front wheels is generally attributed to the Dutch Automobile Works who incorporated such a system in their Spyker car of about 1905. Several years later the American Four Wheel Drive (FWD) Auto Company introduced a car with a transmission system of similar layout, this car being somewhat formidably known as their Battleship model and for which they coined the amusing slogan 'It's a whole new haul game.' More significantly the FWD Auto Company later and successfully applied the same concept to their 3 ton commercial vehicles, some 20 000 of these four-wheel-drive trucks being produced to satisfy American military requirements during World War I.

In relation to modern practice, the FWD Auto Company pioneered two important technical features. First and foremost the transfer gearbox was equipped with an inter-axle or centre differential, which not only compensated for the momentary difference in the speeds of turning of the front and rear axles when rounding corners or travelling over uneven ground, but also could be locked out of action by a sliding dog clutch when traction could only be sustained by one pair of wheels. Second, a chain drive rather than a gear drive was used for the connection between the gearbox mainshaft and the differential carrier for quieter operation.

After World War I the demand for four-wheel-drive vehicles became limited to specialist requirements, and it was not until the advent some twenty years later of World War II that interest in vehicles of this type was revived. In particular it saw the arrival of a versatile military vehicle designated ¼ ton 4 × 4, which had originally been developed by the American Bantam Car Company and then built by the Ford Motor Company and Willys Overland Motors. This vehicle later became universally known as the Jeep (A name derived from the initials of 'general purpose') and over 600 000 of them were built. A simplifying technical feature of its four-wheel-drive system was that either the rear wheels only, or for better traction all four wheels, could be driven, so that in the absence of permanent four-wheel drive there was no need to incorporate a centre differential in the transfer gearbox. This constitutes what is now generally referred to as a 'part-time' system of four-wheel drive, in contrast to the permanent or 'full-time' system which applied in the case of the earlier mentioned FWD Auto Company truck.

The Jeep was therefore suitable both for open-road rear-wheel-drive and cross-country four-wheel-drive operation, and it is now a matter of automotive history that it set the pattern for many other similar four-wheel-drive (normally abbreviated to 4WD) vehicles beginning, of course, with Britain's now equally famous Land-Rover.

The British Jensen C-V8 FF high-performance car was the first modern, albeit limited production, passenger car as opposed to cross-country vehicle to be made available with four-wheel drive. It had a permanently engaged or full-time system that incorporated many ingenious design features developed by the Harry Ferguson Research Company. However, it was not until the early 1980s that a continuing interest in the development of four-wheel-drive passenger cars really began, and it coincided with the introduction of two significant cars, the Audi Quattro in Germany and the American Motors Eagle in America. Both had full-time four-wheel-drive systems, the former being advanced in having essentially a front-wheel-drive transaxle from which the rear-wheel drive had been taken, and the latter more conventionally at the time having a central transfer gearbox from which the front-wheel drive had been routed forwards. These two distinct drive arrangements are sometimes designated 'All-wheel drive' (AWD) and 'Four-wheel drive' (4WD) respectively. However, it was the inclusion of a viscous coupling to control the centre differential of the Eagle system which was later to prove an important step forward in the successful development and wider acceptance of the four-wheel-drive passenger car. Another significant development was made in the mid 1980s by the Austrian firm of Steyr-Puch, who developed for Fiat the first four-wheel-drive system adapted to a transverse-engined car, namely the Fiat Panda 4 × 4.

A further requirement was for the operation of four-wheel-drive vehicles to be compatible with modern anti-lock brake systems. The transmission arrangements must therefore avoid simultaneous interlocking of all four wheels, otherwise an anti-lock brake system is unable to comprehend whether the vehicle is merely stationary, or sliding uncontrollably with all wheels locked. In other words, an anti-lock brake system must be given the opportunity to detect any significant variation of wheel speed relative to vehicle speed before it can act (Section 29.2).

Advantages and disadvantages

To some extent the advantages of four-wheel drive have to be related to the type of vehicle that is so equipped, and in general terms may be summarized as follows:

Off-road utility vehicles The optional engagement of four-wheel drive, together with a two-speed transfer gearbox, confers superior traction for climbing or descending steep slopes, negotiating really rough ground and travelling in mud or snow. Vehicles of this type are therefore admirably suited for general farming, construction site and light industrial duties.

Off-road motor cars and estate vehicles   These vehicles usually have similar part-time four-wheel-drive arrangements to those of off-road utility vehicles, but otherwise offer a better compromise in terms of all-weather capability both on and off the road, combined with a more acceptable level of comfort for private use.

On-road motor cars   In modern practice these have usually been developed from basically front-wheel-drive layouts, their transaxles having been adapted to provide an offtake for rear-wheel drive, which is permanently engaged and gives a full-time system. The advantage sought here is not so much concerned with the ability to cross rough ground, since their ground clearance is in any event much less than on off-road vehicles, but chiefly that of best exploiting available engine power by improving traction and hill climbing on slippery roads. Theoretically the fact that the tractive forces are being shared between four, rather than two, driven wheels should also mean that a four-wheel-drive car can be accelerated harder out of a corner before any tyre loses its adhesion. That is, the disadvantage of two-wheel drive in this respect is based on the premise that the greater the tractive force imposed on a tyre, the lesser will be its sideways grip. However, the potential advantage in cornering power with four-wheel drive is not necessarily realized in practice, according to some comparative tests conducted by the motoring press.

The disadvantages associated with four-wheel drive, at least for everyday motoring, are those of increased cost of vehicle; greater complication, weight and bulk; and higher fuel consumption. With the continued development of on-road four-wheel-drive passenger cars these disadvantages are, of course, being minimized.

21.2   BASIC CONSIDERATIONS OF FOUR-WHEEL DRIVE

Transmission wind-up

A condition known as transmission wind-up may be defined as one in which a positive torque occurs in one axle and a negative torque in the other. The possibility of this condition occurring between tandem drive axles for heavy vehicles has already received brief mention in Section 20.7, and it is now necessary to consider it in relation to four-wheel-drive transmission systems. Unless suitable means are taken to prevent it, transmission wind-up can occur in a four-wheel-drive vehicle for the following reasons:

1   During cornering the front wheels will rotate faster than the rear ones, because the mean radius of the front wheel paths will always be greater than that of the rear wheel paths in deference to the Ackermann steering geometry, the principle of which is described in Section 24.1.
2   Any difference in front to rear inflation pressures for equal tyre loading will produce different rolling radii of the front and rear wheels and thus make them want to rotate at different speeds, this effect being less significant with radial-ply than cross-ply tyres.
3   Any difference in front to rear tyre loading at equal inflation pressures will produce the same effect as described in point 2, which again will be less significant in the case of radial-ply tyres because they better maintain their rolling radius under these conditions.
4   Any difference in front to rear types or sizes of tyre can similarly produce a tendency for them to want to rotate the wheels at different speeds, and different types of tyre can also vary in their ability to scrub over the ground and thus relieve any transmission wind-up.

If transmission wind-up is allowed to occur in a four-wheel-drive vehicle then it follows that the transmission system propeller and drive shafts will be subjected to unusual amounts of torsional deflection or twisting, and indeed it is not unknown in service for propeller shafts to suffer permanent distortion for this reasons. Since such wind-up will only be relieved by the tyres scrubbing over the ground, or in the event of one wheel momentarily losing its traction, it also follows that the most severe wind-up will occur when the vehicle is heavily laden and being operated on dry roads. Under these circumstances it is likely that the driver will become aware of unpleasant vibration and snatching from the transmission system, accompanied by a noticeable increase in rolling resistance on sharp corners and even stalling of the engine. On stopping it may also be difficult to disengage the drive to either the front or the rear optionally driven wheels, as a direct result of the torsional wind-up imposed on the transmission system. Practical proof that this condition has really occurred may be obtained simply by jacking up any one wheel very gradually until its tyre skids over the ground, as the wheel smartly rotates to release the wind-up energy stored in the transmission system.

Torque distribution

It may be recalled from Section 19.5 that the effectiveness with which the driving torque applied to the road wheels is translated into tractive force to propel the vehicle depends on the vertical load and the coefficient of friction existing between the tyres and road. So in the simplest case of a four-wheel-drive vehicle having an equal distribution of weight between the front and rear wheels, a likewise equal division of driving torque between its axles would seem to be acceptable. In fact whether or not there is an equal distribution of weight between the axles, this equality of torque is automatically imposed by a part-time system when four-wheel drive is engaged, because both output shafts from the transfer gearbox are then positively locked together. Similarly an equality of torque is also imposed by any full-time system, which incorporates an ordinary bevel gear differential between both output shafts.

In actual practice most front-engined four-wheel-drive passenger cars have a decided weight bias to the front wheels, which would therefore suggest that these should receive a greater share of the driving torque. However, there are two other factors that need to be taken into account and which dictate a certain amount of compromise. The first of these is the load transfer that occurs under acceleration when, it will be recalled from Section 19.5, the vertical load is increased on the rear wheels and reduced on the front ones. Second, the distribution of driving torque between the front and rear wheels can influence the handling characteristics of a vehicle. As later explained in Section 22.2, the cornering force developed by a tyre decreases as the tractive force between the tyre and road is increased and vice versa; therefore if either the front or the rear wheels receive too much torque, it can provoke excessive understeering and oversteering tendencies respectively.

The practical means of obtaining an unequal distribution of driving torque between the front and rear wheels involves the use of an epicyclic gear centre differential as later described. In one example of a four-wheel-drive passenger car with a longitudinally mounted engine, the torque distribution is 33 per cent to the front wheels and 67 per cent to the rear ones; while in another example where a transverse engine is used, the torque distribution is 56 per cent to the front wheels and 44 per cent to the rear ones.

21.3 PART-TIME FOUR-WHEEL DRIVE

Purpose of a transfer gearbox

At the heart of the part-time four-wheel-drive system, as usually now confined to off-road vehicles, is the transfer gearbox. Since the basic layout of this type of vehicle had been derived from the once conventional front-engine rear-wheel-drive arrangement, it meant that either the propeller shaft drive to a live front axle, or the final drive for independently sprung front wheels, had to lie alongside the engine and main gearbox. This feature therefore introduced the requirement for a transfer gearbox. In fact the transfer gearbox also served another purpose, because it housed an additional pair of gears that could provide a 'wall-climbing' emergency low gear when four-wheel drive was engaged. So we therefore arrive at the designation two-speed transfer gearbox, which generally implies the following duties:

1 Provides a means of offsetting the drive to the front wheels, so that it can run alongside the main gearbox, clutch and engine.

2 Allows the drive to the front wheels to be engaged or disengaged.

3 Offers an emergency low gear with four-wheel-drive operation.

4 Facilitates the fitting of a power take-off if required.

Basic layout of a two-speed transfer gearbox

A two-speed transfer gearbox is either directly attached to the rear of the main gearbox, or separately mounted from it and driven via a short coupling shaft. The object of the latter arrangement is to provide a centrally disposed transfer gearbox, and so to avoid an unduly short front propeller shaft operating with large angular movements of its universal joints where a live front axle is employed.

In the modern two-speed transfer gearbox all the gears have helical teeth and are retained in constant mesh, their engagement being effected through a dog clutch. For earlier designs of transfer gearbox the low gear was usually of the sliding-mesh type with spur teeth (Figure 21.1). Another dog clutch is provided to engage or disengage the drive to the front wheels.

The chief components are as follows:

Transfer input shaft and gear The shaft is a splined extension of the gearbox mainshaft and carries the transfer input gear. It may receive additional support from an outrigger bearing in the rear wall of the transfer gearbox.

Intermediate shaft and gears Similar to the layshaft cluster of some gearboxes, it comprises a fixed spindle with needle roller bearings upon which rotate an integral idler gear and pinion.

Transfer output shaft and gears This assembly comprises a main and an extension shaft in series with a spigot bearing between them. The main shaft is typically supported in tapered roller bearings, which are located either side of a pair of high- and low-speed gears that are normally free to revolve upon the mainshaft. For engagement of these gears the central portion of the mainshaft is splined to carry a sliding dog clutch. Each end of the mainshaft is also splined, the rear end to accept the flange coupling for the rear propeller shaft and the front end to carry a sliding dog clutch for engaging the

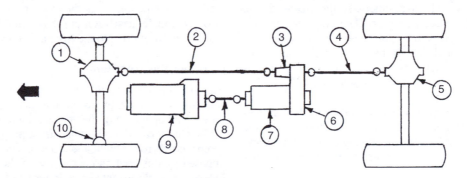

Figure 21.1 General arrangement of a part-time four-wheel drive system for off-road vehicles

1 front axle final drive	6 transfer gears
2 front propeller shaft	7 main gear box
3 FWD dog clutch	8 coupling shaft
4 rear propeller shaft	9 engine and clutch
5 rear axle final drive	10 constant-velocity joints

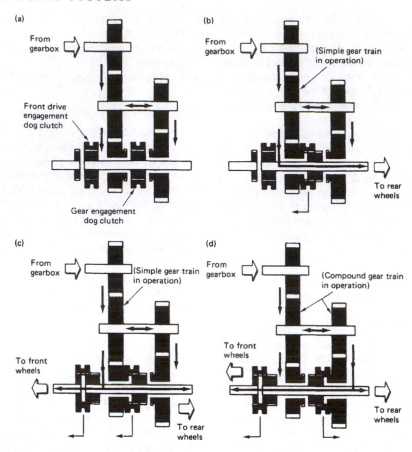

Figure 21.2 Operating principle of the two-speed transfer box: (a) no drive to wheels (b) drive to rear wheels (c) high drive to all wheels (d) low drive to all wheels

extension shaft. This in turn is splined at its front end to accept the flange coupling for the front propeller shaft.

Transfer selector mechanism   A pair of conventional rods and forks is used to perform the separate functions of engaging or disengaging the drive to the front wheels and selecting either high or low gear range. An interlock between the selector rods ensures that the emergency low gear can only be used in conjunction with four-wheel drive, otherwise the greater torque could overstress the rear axle's parts. The driver is provided with two extra shift levers to control the transfer gearbox, and it should be appreciated that shifting from high gear to low gear should not be attempted except when the vehicle either is being operated at very low speeds or is at a standstill.

The operation of a two-speed transfer gearbox lends itself better to illustration (Figure 21.2) than to description, so the modes of power flow have been presented accordingly and should be further related to the arrangement of propeller shafts described below.

Arrangement of propeller shafts

In the conventional type of off-road vehicle, the drive from the transfer gearbox is transmitted to the front and rear final drive units through two propeller shafts, each with Hooke's-type universal joints at each end and incorporating a sliding coupling.

The rear propeller shaft is typically coupled to a live rear axle in the conventional manner. In the case of the front propeller shaft, which runs alongside the engine and main gearbox, this again may be coupled to a live axle or, less commonly, to a final drive unit mounted on the vehicle structure and driving independently sprung front wheels via the usual universally jointed drive shafts. Two arrangements of propeller shafts used in conjunction with directly attached and separately mounted transfer gearboxes are shown in Figure 21.3, in these cases with offset drive to the front axle only.

Free-wheeling front hubs

Although the drive to the front wheels of a part-time four-wheel-drive system may be disengaged at the transfer gearbox for on-road operation of the vehicle, it will be evident that the half-shafts, final drive gears and front propeller shaft are still compelled to rotate with the wheels. The rolling resistance imposed by this continuous rotation of the front-wheel drive line is clearly disadvantageous in terms of fuel economy and also in unnecessary wearing of the components concerned.

A convenient method of overcoming this problem exists in providing free-wheeling hubs for the front wheels. In their simplest form they comprise a splined sliding dog clutch, which is normally spring loaded into engagement between the male splines of a sleeve on the end of the half-shaft and

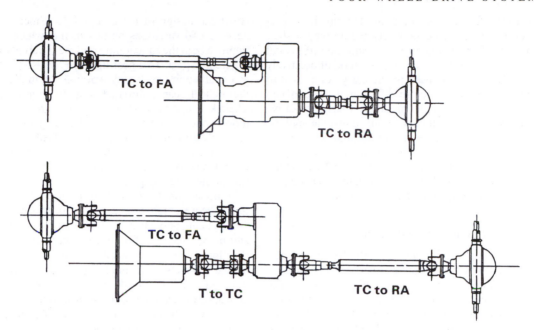

Figure 21.3 Conventional arrangements of propeller shafts used with 4WD (*GKN Hardy Spicer*)

TC to FA: Transfer Case to Front Axle – Two Joint and Shaft Assembly.
TC to RA: Transfer Case to Rear Axle – Short Coupled Joint Assembly.
T to TC: Transmission to Transfer Case – Short Coupled Joint Assembly.
TC to RA: Transfer Case to Front Axle – Two Joint and Shaft Assembly.
TC to RA: Transfer Case to Rear Axle – Two Joint and Shaft Assembly.

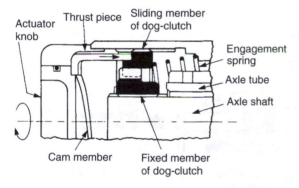

Figure 21.4 Arrangement of free-wheeling hub for part-time 4WD

the female splines in the hub body, thereby transmitting a positive drive when the transmission is in four-wheel-drive mode. For on-road operation when the drive to the front wheels is disengaged and it is also required to unlock the drive between the front-wheel hubs and their half-shafts, an actuator knob in each hub end cover is partially rotated, so that an internal cam member forces the sliding dog clutch out of engagement with the half-shaft end sleeve (Figure 21.4). This then allows the hub to free-wheel and its associated drive line components to remain stationary. It should be noted that the spring loading on the sliding dog clutch also facilitates relocking of the hubs for four-wheel drive, because if the splines do not align the spring remains compressed to urge them into engagement as the half-shaft begins to turn.

21.4 FULL-TIME FOUR-WHEEL DRIVE

Role of the centre differential

As earlier inferred, the transmission arrangements of a full-time four-wheel-drive system are such that the drive is permanently directed to all wheels. From previous considerations it thus follows that if the front and rear drive lines were rigidly connected together and the vehicle operated on dry roads with very little opportunity for wheel slip to occur, a condition of transmission wind-up would soon intrude. In order to accommodate any difference in the mean, rotational speeds of the front and rear pairs of wheels, it becomes necessary to include an inter-axle or centre differential between the front and rear drive lines. Here it should be recognized that in more recent passenger car practice where a transverse engine with adjacent four-wheel-drive transaxle is used, the centre differential is no longer geographically near to the middle of the car. However a centre differential, working in conjunction with the other two differentials in the front and rear final drives, would be self-defeating from the point of view of improving mobility under slippery road conditions. This is because it would only need one of the four driven wheels to slip and start spinning to cause the other three to lose their drive as well; although it should not be thought that this may happen too readily, owing to the frictional drag imposed by the end thrust loading of the bevel gears in a conventional differential (Section 20.3).

To overcome this disadvantage with a centre differential it is necessary to have either a driver-controlled positive lock or an automatically imposed limitation, on the free action of the differential. In the former case a sliding dog clutch is operated to effect a splined connection between an output shaft from one of the differential sun pinions and the differential case itself (Figure 21.5). This particular method of positively locking the differential was used in the early Audi Quattro full-time system, where the centre differential was situated at the rear of a longitudinally mounted four-wheel-drive transaxle (Figure 21.6). It will be recognized that the application of a positive lock to the centre differential is similar in principle to the differential lock used between the rear axles of some heavy vehicles (Figure 20.42). The positive locking centre differential of the Audi system was later replaced by a Gleason Torsen differential, which is described next.

Automatic control of the centre differential

An automatically imposed limitation on the free action of a centre differential may be obtained by either substituting a suitable arrangement of partly reversible crossed-helical gears for the completely reversible bevel gears, or superimposing a viscous coupling across the differential output shafts to the front and rear drive lines. The limited-slip types of differential that depend on preloaded friction clutches, as described in Section 20.3, generally do not allow sufficient freedom of centre differential action for steering the vehicle.

The concept of using crossed-helical gears in the differential of an ordinary rear axle, as a means of increasing internal friction and limiting differential action, dates back to an original American design of the early 1920s. Since then there have been several variations on this theme, which have sought to improve both the functioning and the durability of this type of differential gear. To a gear designer, crossed-helical gears are essentially non-enveloping worm gears, in the sense that neither helical gear is throated so as to wrap around the other. Their key feature lies in the point contact that continually moves in a sloping line across their engaging teeth. It is this action that results in a combination of rolling and sliding friction, which can provide a holding effect on the gearset. For a given load on the gear teeth the resistance to rotation is a function of their helix angles.

A modern application of this principle to a centre differential is represented by the Gleason Torsen unit, the name Torsen being derived from torque sensing. In this unit the sun and planet pinions of a conventional bevel gear differential (Section 20.3) are replaced by two interconnected crossed-helical gearsets. The inter-connection is effected through three pairs of non-meshing but separately geared-together helical gears, which in turn mesh at right angles with the helical gears on the output shafts of the differential. If any wheel suddenly begins to spin, the geometrical characteristics of the crossed-helical gear teeth and the change in loading on them are such that progressively more torque will be directed to the slower turning drive line and hence to the pair of wheels where traction is still being maintained. In the absence of any significant change in torque distribution, the differential acts freely in the manner of a conventional one to allow for steering

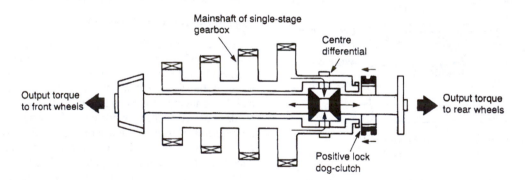

Figure 21.5   Positive lock centre differential for full-time 4WD

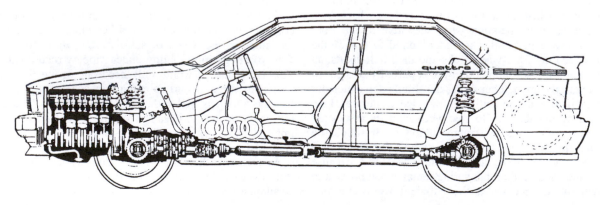

Figure 21.6   A modern 4WD arrangement of front transaxle and rear propeller shafts (*Audi*)

of the vehicle. The internal arrangement of the gearing in a Gleason Torson differential, in this particular case applied to an axle final drive, is shown in Figure 21.7.

There can be little doubt that the use of a viscous coupling to limit the free action of a centre differential has, since the late 1980s, made a substantial contribution to the progress of four-wheel-drive technology. We have previously encountered examples of simple viscous drive couplings in connection with crankshaft vibration dampers and variable-drive cooling fans (Sections 2.5 and 5.3), and it is this same basic concept, albeit more comprehensively applied, that may be incorporated in the centre differential of full-time four-wheel-drive systems. The transmission viscous coupling unit (often abbreviated to VCU) was invented by Derek Gardner and Anthony Rolt of the Harry Ferguson Research Company in the late 1960s, but it was not until some ten years later that it was first applied to the four-wheel-drive system of a production passenger car, namely the American Motors Eagle mentioned earlier.

A transmission viscous coupling unit essentially comprises a stack of interleaved steel discs, one set being splined to an outer drum and the other set similarly attached to an inner shaft (Figure 21.8). The discs of the former set are perforated and those of the latter set slotted to enhance viscous drag in the coupling. Also, the discs of one set are allowed freedom of axial movement on their splines, while the discs of the other set have spacer rings between them. This assembly of closely spaced discs is encased and sealed within the outer drum, which is almost completely filled with silicone fluid having a viscosity about that of honey and remaining stable under wide variations in operating temperature.

When a viscous coupling unit is connected across the output shafts of a centre differential (Figure 21.9), its residual viscous drag will still permit a moderate differential action between the front and rear pairs of wheels, such as required for steering purposes. If there is a sudden increase in differential action caused by any wheel beginning to spin, then the greater relative motion of the coupling discs will result in the fluid film between them being sheared at a higher rate, which will impose an increasing viscous drag and hence limitation on differential action. At this stage the viscous coupling is said

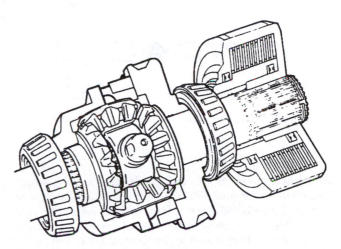

Figure 21.8 Centre differential and viscous control unit (*Colt-Mitsubishi*)

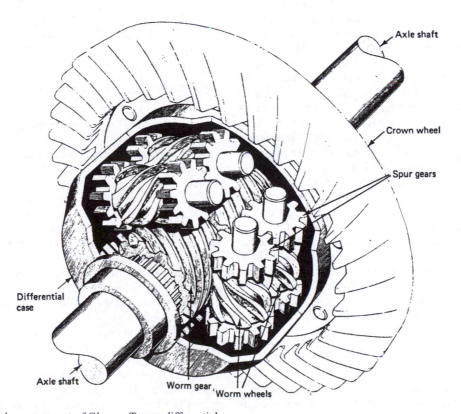

Figure 21.7 Internal arrangement of Gleason Torsen differential

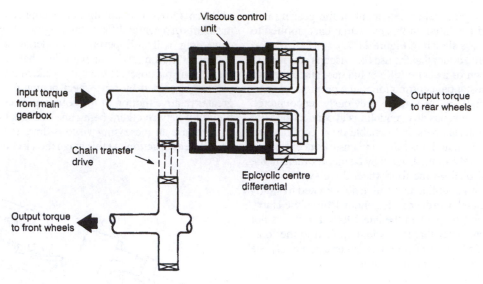

Figure 21.9 Schematic arrangement of epicyclic centre differential with viscous coupling unit

to be operating in its 'viscous' mode and the drive torque through the differential will be progressively directed more to the pair of wheels with the best traction. If under extreme conditions this control proves insufficient to restore the normal distribution of drive torque to the front and rear wheels, the continued operation of the viscous coupling will cause a rapid rise in the temperature of its fluid, which will therefore expand and completely fill the coupling. Although this initially serves to maximize the wetted area of the discs for greater viscous drag, any further rise in temperature will create a build-up of pressure in the coupling. Since the presence of the perforations and slots in the discs modifies the effect of the pressure acting on them, the floating discs are squeezed into contact with the fixed ones. This causes the coupling to act temporarily in the manner of a friction clutch, which further increases the torque directed to the gripping pair of wheels. The viscous coupling is then operating in what is termed its 'hump' mode. As relative motion between the discs ceases, the liquid begins to cool and the coupling will revert to its viscous mode of operation if differential control is still needed.

A more recent alternative to connecting a viscous coupling across the output shafts of a centre differential is to connect a wet multi-plate clutch in a similar manner, which can be engaged electromagnetically by a solenoid control. This arrangement is then capable of modulating the basic torque distribution conferred by an epicyclic differential gear up to the rigidly coupled or locked condition, depending on the electronic sensing of the torque requirements, front to rear, of the drive axles. During normal driving the centre differential does, of course, operate quite freely with its clutch disengaged.

General arrangement of full-time four-wheel-drive systems

For passenger cars there are four basic arrangements of full-time four-wheel-drive systems that may be summarized as follows.

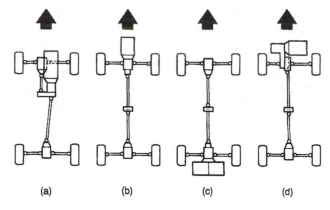

Figure 21.10 Basic arrangements of full-time 4WD

Longitudinal front engine and gearbox with 4WD transfer case (Figure 21.10a)
This represents the earliest but still used system where front-wheel drive has been added to an existing rear-wheel-drive vehicle. The rear propeller shaft is driven directly from the back of the transfer case and the front propeller shaft is offset to run alongside the engine and gearbox and is driven from the front of the transfer case through either a gear or a chain drive. A centre differential is contained within the transfer case (Figure 21.9), but unlike an off-road vehicle there is no requirement for an additional low-range gearset. This type of installation tends towards mechanical complexity and can be difficult to accommodate in the space available.

Longitudinal front engine with 4WD transaxle (Figure 21.10b)
In contrast to the previous arrangement, this system involves the addition of rear-wheel drive to an existing front-wheel-drive vehicle. The centre differential is enclosed at the rear end of the transaxle and receives its drive from the hollow mainshaft of a single-stage gearbox. From the differential

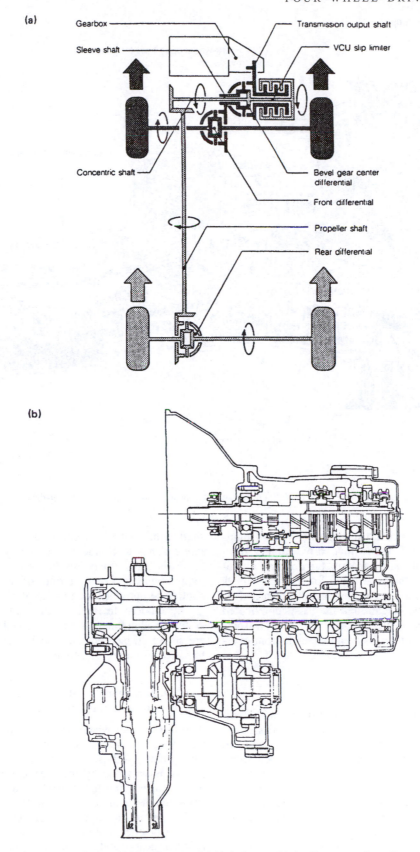

(a)

Gearbox

Sleeve shaft

Transmission output shaft

VCU slip limiter

Concentric shaft

Bevel gear center differential

Front differential

Propeller shaft

Rear differential

(b)

Figure 21.11 Full-time 4WD system of Mitsubishi Galant car: (a) basic layout (b) detail construction of transaxle (*Colt-Mitsubishi*)

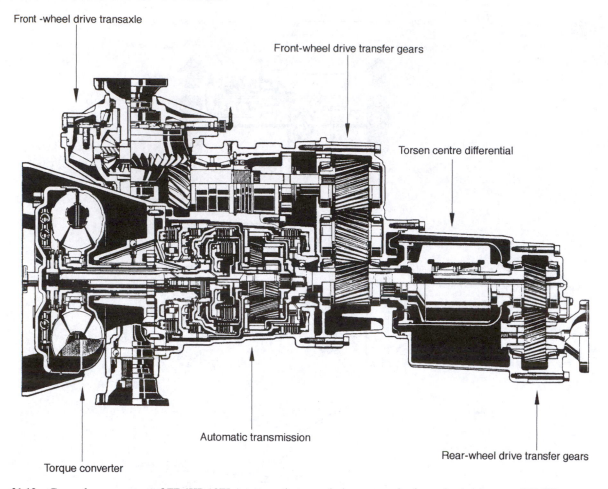

Figure 21.12   General arrangement of ZF 4HP 18FLA automatic transmission system for four-wheel-drive cars (*ZF GB*)

one output shaft passes forward through the hollow main-shaft to drive the pinion of the final drive for the front wheels; the other output shaft transmits the drive via a two-piece propeller shaft to the final drive for the rear wheels (Figure 21.5). This arrangement is generally regarded as being both mechanically elegant and the least complicated conversion to four-wheel drive, but as in the case of a front-wheel-drive only version there is an appreciable front over-hang of the engine.

Longitudinal horizontally-opposed rear engine with 4WD transaxle (Figure 21.10c)
For this uncommon system of four-wheel drive where max-imum traction is being sought for a specialist high-performance car, the transmission arrangements are essentially similar to those described for the second system, but turned back to front.

Transverse engine with 4WD transaxle (Figure 21.10d)
This type of system represents the most modern arrange-ment of four-wheel drive for passenger cars and involves the addition of rear-wheel drive to an existing front-wheel-drive vehicle. It tends towards mechanical complexity because one of the outputs from the centre differential has to be turned through a right angle by a bevel gearset, so that the drive can be transmitted via a two- or three-piece propeller

shaft to the final drive for the rear wheels. Also, a second dif-ferential has to be incorporated near to the centre one to serve the final drive for the front wheels. Despite these requirements a certain amount of design ingenuity usually contrives to produce a transaxle of fairly compact size.

The modern features of the full-time four-wheel-drive system used in the transverse-engined Mitsubishi Galant car are shown in basic layout and in detail construction of its transaxle in Figure 21.11a and b.

General arrangement of full-time four-wheel-drive system with automatic transmission

A full-time four-wheel-drive system that incorporates an automatic transmission can prove an attractive combination for a luxury high-performance car. An example of such a system is provided by the German ZF 4HP 18FLA auto-matic transmission for four-wheel-drive cars, the entire unit being longitudinally mounted in the vehicle (Figure 21.12).

The automatic transmission itself follows established modern practice, comprising a torque converter, four-speed epicyclic gearset and an electronic control system. Structurally, the housing for this transmission is integrated with that for the front-wheel drive transaxle, which embodies a crown-wheel and pinion final-drive differential assembly. The pinion shaft

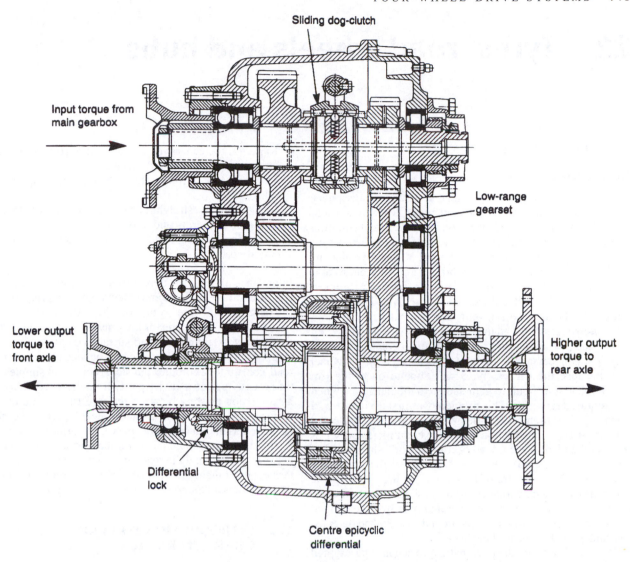

Figure 21.13 General arrangement of running gear and bearings in full-time 4WD transfer gearbox for a heavy vehicle (*FAG Kugelfischer*)

extends alongside the automatic transmission and receives its transfer drive from a train of two helical gears with idler, these gears being supported within a narrow housing sandwiched between those for the automatic transmission-cum-transaxle and the inter-axle or centre differential. However, the transfer gears for the front-wheel drive transaxle do not receive their input directly from the output shaft of the automatic transmission, because this shaft passes freely through the first transfer gear and connects to the cage of the adjacent centre differential, which is of the torque sensing Torsen type (Figure 21.7).

Both outputs are arranged to emerge concentrically from the rear end of the differential, the outer one being conveyed forwards by a driving shell that encircles the differential and connects to the first transfer gear for the front-wheel drive transaxle, and the inner one that directly connects to another transfer gearset carried in a rear extension housing. The driven gear of this pair is coupled to the front end of a propeller shaft(s), which then conveys the drive to the rear wheels via their crown-wheel and pinion final-drive and differential assembly, and it also serves to lower the height of the drive-line.

With this system of full-time four-wheel-drive, there is an equal division of driving torque between the front and rear axles of the vehicle unless a wheel spins.

General arrangement of full-time four-wheel-drive systems for heavy vehicles

For heavy vehicles a full-time drive system to the front and rear axles, together with a positive lock for the centre differential, may be either integrated with the main gearbox or separately housed within an auxiliary gearbox. The centre differential is usually of the epicyclic type, where the input drive torque is applied to the planet pinions carrier, and the output torques to the front and rear axles are taken from the sun and annulus gears respectively, so that a greater proportion of the input torque is distributed to the more heavily laden rear wheels. An additional low-range gearset with dog clutch engagement may also be incorporated in a heavy-vehicle full-time drive system, the arrangement of the gearset and supporting bearings for one example being shown in Figure 21.13.

22 Tyres, road wheels and hubs

22.1 TYRE REQUIREMENTS

General background

The application of a hollow inflatable tube to the periphery of a road wheel was first patented by a Scot, Robert Thomson, in 1845. The object of his invention was to render the motion of a carriage more comfortable, easier and quieter, these attributes being confirmed by practical testing of what he called his Aerial wheels on a horse-drawn brougham in Regent's Park, London. Unfortunately, this admirable invention aroused very little real interest at the time and it remained for another Scot, John Dunlop, to reinvent the pneumatic tyre in 1888 when he successfully adapted it to cycle wheels. However, before the pneumatic tyre could become a commercial success it was necessary to make it detachable from the wheel rim, because the earliest tyres were bound on to the rim which made it difficult to gain access to the inner tube when this required repairing. This problem was first overcome by a young English engineer, Charles Welch, who in 1890 patented a form of tyre construction in which circumferential wires were embedded in the edges of a horse-shoe-shaped rather than circular section tyre cover. The wheel rim was made with a well in the centre of its section, so that in fitting the tyre part of its wired edge could be pushed temporarily into the well followed by working the remainder over the flanged rim of the wheel. Similarly, this procedure could be reversed to remove the tyre from the wheel.

Within a few years the pneumatic tyre began to be adopted for the wheels of the newly introduced motor car, the initiative for this application being credited to André and Edouard Michelin, who first fitted pneumatic tyres to a Peugeot racing car in 1895. Since those early beginnings the motor vehicle pneumatic tyre has evolved through many stages, and the modern product is the result of intensive research, rigorous testing and joint development by the tyre and vehicle manufacturers. The role played by the pneumatic tyre in the development of the motor vehicle cannot be overestimated; indeed, it has sometimes been suggested that without the pneumatic tyre the internal combustion engine itself would not have guaranteed any real future for the motor vehicle.

Tyre requirements

Pneumatic tyres for motor vehicles must basically satisfy the following requirements:

1 To support the weight of the vehicle and distribute it over the road surface.
2 To offer the minimum rolling resistance to the motion of the vehicle and thus reduce power absorption.
3 To contribute to the suspension cushioning of impact forces created by road surface irregularities.

4 To permit the generation of traction, braking and steering forces on dry or wet road surfaces.
5 To confer safe operation up to the maximum speed of the vehicle.
6 To provide quiet straight-ahead running and freedom from squeal on cornering and braking.
7 To realize an acceptable tread life under varied running conditions.

In practice, all tyre design represents a compromise in meeting these various and to some extent conflicting requirements.

The grip of the tyre on the road is generally referred to as 'tyre adhesion', and depends not only on the tyre but also and to a greater extent on the road surface condition. Tyre adhesion, or the frictional effect existing between the tyre and the road, constitutes a complex physical process. In simple terms, the process appears to derive partly from the forces of molecular attraction existing between the tyre rubber and the road surface, and partly from a mechanical interlocking action occurring as the rubber conforms to the surface roughness of the road. For maximum adhesion the normal area of contact between the tyre and the road, or 'footprint area' as it is often termed, must therefore be maintained under all running conditions and is a function of the tyre construction.

22.2 INTRODUCTION TO TYRE CHARACTERISTICS

When a vehicle is being driven along the road, the tyres are acted upon by a complex and variable system of forces that arise in the vertical, longitudinal and lateral directions. The manner in which the tyres react to these forces defines their main characteristics, which may be summarized as follows.

Cushioning ability

The pneumatic tyre contributes to the ride comfort of the vehicle by acting as a spring in series with the suspension system (Figure 22.1). However, since the tyre is much less flexible than the vehicle springing, its cushioning ability is chiefly related to the high-frequency disturbances emanating from the road surface. Radial flexibility of the tyre is mainly derived from the elastic properties of its carcass and tread, which in turn are pretensioned by the inflation pressure of the air in the tyre. The flexibility of the tyre may thus be increased by reducing the inflation pressure, but in order to sustain the same normal load the tyre section must be made larger to accommodate an adequate amount of air. This basic approach to providing greater cushioning ability was first successfully exploited by the American Goodyear Company, who in the mid 1920s introduced their lower-pressure 'balloon' tyres.

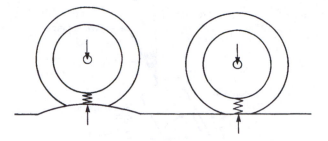

Figure 22.1 Spring cushioning effect of tyre

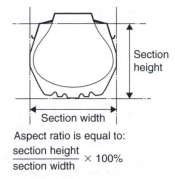

Aspect ratio is equal to:

$$\frac{\text{section height}}{\text{section width}} \times 100\%$$

Figure 22.2 The aspect ratio of a tyre

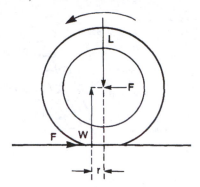

Figure 22.3 Rolling resistance of tyre

L vertical load
W vertical reaction force
F friction force
r offset distance

Resistance torque to rolling $= Wr$

In the later development of the passenger car, the trend has been towards a lowering of its overall height to realize both greater stability and more acceptable styling. This in turn meant adopting tyres of smaller outside diameter, because in order to confer a similar amount of usable interior space the wheel housings must be reduced in size. At the same time, the higher performance of the modern car demands adequate rubbing surface areas for the brakes, which in conventional practice are mounted within the wheels and thus control the inside diameter of the tyre. These conflicting space requirements are met in practice by making the smaller-diameter tyre wider than its height of section, so that its air capacity and therefore load rating remain undiminished. The ratio of the sectional height to the sectional width of a tyre is known as its 'aspect ratio' (Figure 22.2) and is expressed as a percentage, so the lower its aspect ratio the wider a tyre is in relation to its height. An aspect ratio in the range of 60–70 per cent is typical in modern practice and may be compared with a value of 95 per cent for the balloon tyres of an earlier era. A significantly lower aspect ratio of 30–40 per cent may be specified for ultra high-performance cars, but these very low profile tyres are related to larger wheel sizes and can be recognized by their almost rubber band-like appearance.

Rolling resistance

A basic consideration affecting the flexibly tyred wheel, freely rolling on a rigid road surface, is that it must inherently possess a resistance to motion. This effect arises because the deformation of the tyre tread tends to be especially concentrated at the leading edge of its contact area with the road, which produces therein an asymmetrical pressure distribution. As a result the normal reaction force due to vertical load acts

through a point ahead of the rotational axis of the wheel, thereby exerting a turning moment in the opposite sense to that of wheel rotation and thus imposing a resistance to rolling (Figure 22.3). In the presence of a tractive force, as opposed to free rolling, the normal reaction force moves further ahead of the rotational axis of the wheel and therefore rolling resistance increases. This all means, of course, that some of the energy being supplied to propel the vehicle along the road is wasted, actually in the form of heat due to flexing and internal friction of the tyre materials.

Since rolling resistance is closely associated with tyre deflection, it follows that the energy expended in the tyre derives mainly from the structural distortions imposed upon it by the vehicle load and the forces generated by traction and cornering. Rolling resistance is also increased by an increase of speed and a decrease in inflation pressure. To a lesser extent the resistance may be attributed to the creeping or shuffling movements of the tread elements, which accompany their passage through the tyre to road contact area. Apart from minimizing heat build-up in the tyre, a tyre construction that offers a low rolling resistance is always to be desired in the interests of either reducing fuel consumption or increasing performance.

Directional ability

Since a pneumatic tyre is essentially a flexible structure, it travels in a straight path only when it is allowed to roll freely without sideways restraint. In reality, of course, the tyre is always called upon to exert a sideways force either as the car is being steered round a bend, or when it is being maintained in a straight path in the presence of a road camber or against a cross-wind. This sideways force is transmitted through the rotational axis of the wheel to the rim and is resisted by the adhesion forces existing between the tyre tread and the road surface. Considering the effect of these opposing forces in a little more detail, the tyre carcass and tread begin to experience a lateral deformation as the tread elements approach their contact with the road surface. Then as the tread elements

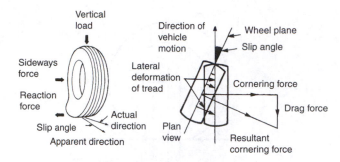

Figure 22.4   Slip angle of tyre

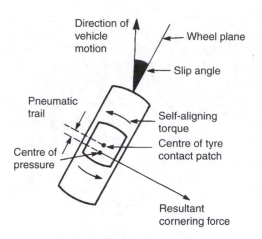

Figure 22.5   Self-aligning torque of tyre

actually establish contact with the road surface, they undergo an increasing lateral deformation until finally they may slide across the remaining portion of their footprint area. Any tread slipping or creeping is accompanied by a reduction in lateral deformation of the carcass, and after the tread elements have passed through their contact area with the road surface the tyre returns to its underformed condition. As a result of the tyre carcass and tread deforming in this manner, an angular difference is created between the direction in which the tyre is being aimed and the direction in which it actually travels. This angular difference is termed the tyre 'slip angle' (Figure 22.4). It may reach a value of between 8° and 12° as the limit of adhesion is approached. The slip angle concept was originally advanced by an early French authority on suspension design, G. Brouhliet, in the mid 1920s, and it is this feature of tyre behaviour that is responsible for generating a cornering force at the road surface. In other words, the reaction of the road against the sideways push from the tyre compels the vehicle to follow a curved path.

For any given vertical wheel loading, the cornering force generated by a tyre is in fairly close proportion to the slip angle up to the limit of adhesion. The cornering force for each slip angle attains a peak value that coincides approximately with the rated load of the tyre, thereby ensuring that the cornering ability of the tyre is least affected by moderate changes in vertical loading. Apart from the influence of tyre construction and road surface condition, the cornering force generated by a tyre is increased by raising inflation pressure and by adopting wider wheel rims and tyre sections. Tyre cornering force is also increased if the wheel leans into a curve in motor-cycle fashion, and conversely reduced if it leans away from the curve; these effects are referred to as camber thrust. Another factor that affects the cornering ability of a tyre is traction; the cornering force decreases as the tractive force between the tyre tread and road surface is increased. The same consideration also applies to braking force. Other terms that describe the cornering ability of a tyre are cornering power and cornering coefficient, the former referring to the cornering force per degree of slip angle and the latter to the cornering power per unit of vertical load.

Self-aligning facility

A characteristic of the pneumatic tyre is that it prefers to roll freely in a straight path, and if compelled to do otherwise it

generates its own self-aligning torque. This effect arises because in the presence of a slip angle the trailing portion of the tyre contact patch experiences greater lateral deformation than the leading portion, so that the shift in centre of pressure causes the cornering force to act at a small distance behind the rotational axis of the wheel. The moment arm distance at which the cornering force acts is known as 'pneumatic trail' (Figure 22.5), and this tends to decrease in length with increasing magnitude of cornering force. Hence, the self-aligning torque tends to remain fairly constant until the limit of adhesion is approached, at which point the pneumatic trail distance diminishes until the self-aligning torque falls to zero. Apart from the influence of road surface condition and tyre construction, the tyre self-aligning torque is increased markedly by greater than normal vertical loading and to a lesser extent by the application of a tractive force. It is also increased if the wheel leans away from, rather than into, a curve and by adopting a wider section tyre. An increase in the inflation pressure and the application of a braking force in both cases reduce the tyre self-aligning torque.

The practical significance of tyre self-aligning torque is that it not only plays a major role in causing the vehicle to straighten up after leaving a curve, but also provides the driver with a certain measure of 'feel' in handling the vehicle. For example, the driver is able to sense an impending loss of tyre adhesion by the steering 'going light' as a result of the diminishing self-aligning torque under these conditions. A reasonable but not excessive amount of tyre self-aligning torque is usually sought, so that the steering retains an acceptable degree of feel without becoming heavy in operation.

22.3   TYRE CONSTRUCTION

Tubed and tubeless tyres

The separate rubber inner tube acting as a flexible bag to retain air introduced under pressure early became an established feature of pneumatic tyre construction. It comprises an endless tube slightly smaller than the inside of the tyre itself, its only opening being an inflation valve projecting through the wheel rim. Inner tubes were originally made from high-quality natural rubber, but this has long since been superseded by a certain type of synthetic rubber able to retain air much better.

Since the mid 1950s it has become established practice for private cars, and to an increasing extent for commercial vehicles, to dispense with the separate inner tube in favour of an inherently air-tight or tubeless tyre.

History often repeats itself in automotive engineering. The concept of a tubeless tyre was certainly not new; it had been tried in some very early versions of pneumatic tyres but had been abandoned. However, its later successful development by Frank Herzegh of the American Goodrich tyre company demanded comparatively little modification to the modern tubed tyre construction. First, it involved vulcanizing a layer of self-sealing synthetic rubber to the inside of the tyre casing. Second, it meant making the tyre valve assembly a sealed interference, instead of an unsealed clearance, fit in the wheel rim. Third, it was necessary to avoid porosity in the bead region of the tyre, which otherwise was well adapted for air sealing because of the 5° tapered bead seat of the wheel rims. For tubeless tyres on commercial vehicle wheels, the bead seat angle is increased to 15° to ensure air tightness.

The advantages of tubeless tyre construction are several:

Increased safety   With tubed tyres there is always the possibility that rapid deflation can occur as a result of air escaping freely from around the loose-fitting valve stem in the wheel rim, either because the inner tube has been punctured by a sharp object penetrating or embedded in the tyre casing, or owing to the inner tube chafing in the bead region of the tyre.
Cooler running   This is made possible because internal friction occurring between an inner tube and its tyre cover is absent.
Improved balance   A contribution to better tyre balance follows from eliminating the additional mass of a separate inner tube.

Basic tyre construction

Before attempting to distinguish between the two principal forms of tyre construction, namely cross and radial ply, it is necessary to describe the basic components of a pneumatic tyre:

Carcass   This unites the tread, side walls and beads of the tyre, and it is made from rubberized cord plies, the material for which is specially chosen to impart the required strength to the tyre.
Beads   These locate the tyre on the wheel rim and provide anchorages for the looped ends of the carcass plies. As there must be no relative movement between the tyre and wheel rim, the bead assemblies incorporate an inextensible wire core.
Side walls   These regions span the distance between the tread shoulders and the beads. They are responsible for imparting the necessary vertical flexibility to the tyre, and together with the tread they share a protective role for the carcass beneath.
Tread   The chief function of the tread is, of course, to establish and maintain contact between the tyre and the road surface. The various patterns of tread used make this the most distinguishable, but by no means the most important, part of the tyre.

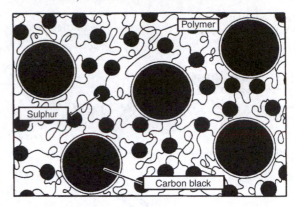

Figure 22.6   Microstructure of tyre compound (*Bridgestone*)

Tyre materials may be classified as rubber compounds, cord materials and bead wire, and it is worth reflecting that although their various properties differ considerably they must work in harmonious physical contact throughout the life of the tyre.

Rubber compounds
These refer to the composition of the rubber mix used in the construction of a tyre. At this point it is first necessary to understand what we mean by the term 'rubber'. A useful engineering definition attributable to the American Society for Testing and Materials tells us that rubber is 'a natural or synthetic material that can be or is already vulcanized to a state in which it has high extensibility and forcible, quick retraction'. For many years it was the generally held opinion of tyre manufacturers that the use of natural rubber was essential for reasons of reliability, but the increase in speed and cornering capabilities of high-performance cars designed towards the end of the 1950s brought about the serious development of 'less bouncy' synthetic rubber materials for tyres. Their eventual acceptance made possible safer braking and cornering on wet roads, a reduction in squeal when braking or cornering on dry roads, and an increase in tyre life, although natural rubber still plays an important role in the tyre construction.

The principal synthetic rubbers or polymers used in the construction of modern tyres are typically those known to the tyre engineer as styrene-butadiene rubber (SBR) and polybutadiene (PBD), their relative proportions determining the compromise between tyre grip and wear resistance. These polymers, combined with other chemical agents such as sulphur to impart elasticity, carbon black for abrasion resistance and oils that confer plasticity, produce the desired properties of the compound. The microstructure of an advanced modern tyre compound developed by the Bridgestone tyre company is shown in Figure 22.6.

Cord materials
These serve as reinforcing components in the carcass plies and tread bracing layers of a tyre. The carcass unites, the tread and the beads and is produced from rubberized cord plies that impart the necessary strength to the tyre while also being highly flexible. Each reinforcing cord consists of twisted

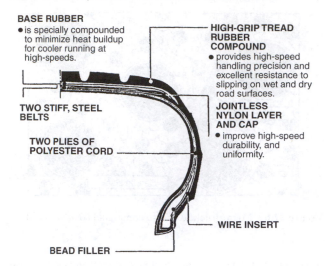

Figure 22.7   Cross-section of a radial-ply high-performance car tyre (*Bridgestone*)

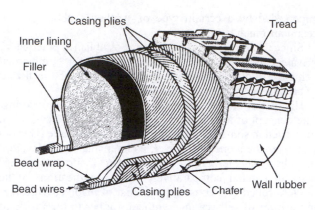

Figure 22.8   Identifying the parts of a cross-ply tyre (*Pirelli*)

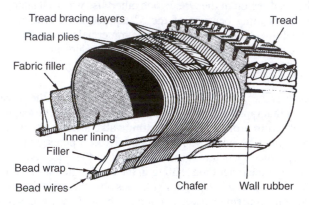

Figure 22.9   The structure of a modern radial-ply tyre (*Pirelli*)

together yarns of either rayon, nylon or polyester fibre, these materials possessing in varying degree superior physical properties to the cotton originally used for this purpose. For current versions of radial-play tyre, the virtually inextensible stabilizer belt typically consists of two layers each of rubberized high-tensile steel and polyester cords (Figure 22.7).

Bead wire
A rubberized high-tensile steel wire, larger in diameter than that used for tread bracing purposes, is used to wind either round, square or hexagonal bead bundles. These are necessarily the least flexible component of the tyre structure, since there must be no relative movement between the inflated tyre and wheel rim.

The least readily distinguishable, but very important, part of tyre construction is the path taken by the cords in the separate plies of the carcass, because this feature governs the behaviour of the tyre on the road in respect of cushioning ability and directional stability. These requirements tend to conflict with each other because cushioning ability depends upon vertical flexibility while directional stability demands horizontal rigidity. The manner in which this conflict is resolved in practice leads us next to consider and distinguish between cross-ply and radial-ply tyres.

Cross-ply tyre construction

Until the late 1960s the cross-ply construction represented long-established conventional practice. Fundamental to this type of construction is the criss-cross arrangement adopted for the several layers of cords (Figure 22.8). The angle at which the cords of each ply intersect the circumferential centre line of the tyre is usually about 40°, this being termed the crown angle. It is chosen to provide the best compromise in the cushioning and directional properties of the tyre.

The cross-ply tyre possesses certain shortcomings. First, the continuous flexing of its carcass imposes slight fidgeting movements in the rubber layers between the criss-crossed ply cords. This activity causes frictional heating which, if

generated to excess, can shorten tyre life by weakening the ply cords. Second, the vertical deflection of the tyre results in contraction of the flattened part of the tread in contact with the road, and this is a prominent source of tyre wear.

Radial-ply tyre construction

The cushioning and directional functions are to all intents and purposes catered for separately in tyres constructed on this principle, which was first successfully applied by the French Michelin Company in the late 1940s. Since the late 1960s the radial-ply tyre has gradually been adopted as original equipment on most cars and many commercial vehicles.

To achieve good cushioning properties, the walls of this type of tyre are made as flexible as possible by arranging the carcass ply cords at right angles, or very nearly so, to the circumferential centreline of the tread. In other words, the ply cords follow the contour of the tyre from bead to bead. To confer the necessary directional properties, the radially arranged ply cords are crowned by a virtually inextensible stabilizer belt beneath the tread (Figure 22.9).

The radial-ply tyre is therefore relatively free from internal friction because of the non-criss-crossing of its carcass plies. This in turn means cooler running and permits reduced inflation pressures for greater cushioning ability. The rigidity of the stabilizer belt effectively resists contraction of

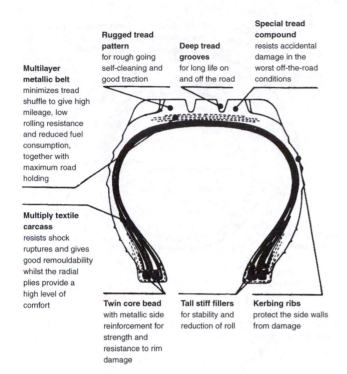

Multilayer metallic belt minimizes tread shuffle to give high mileage, low rolling resistance and reduced fuel consumption, together with maximum road holding

Rugged tread pattern for rough going self-cleaning and good traction

Deep tread grooves for long life on and off the road

Special tread compound resists accidental damage in the worst off-the-road conditions

Multiply textile carcass resists shock ruptures and gives good remouldability whilst the radial plies provide a high level of comfort

Twin core bead with metallic side reinforcement for strength and resistance to rim damage

Tall stiff fillers for stability and reduction of roll

Kerbing ribs protect the side walls from damage

Figure 22.10   Cross-section of a radial-ply lorry tyre (*Pirelli*)

the flattened part of the tread in contact with the road, so there is less tread shuffling and wearing of the tyre (Figure 22.10).

A disadvantage of the radial-play tyre is its tendency towards increased ride harshness at low speeds owing to its vibration characteristics, which differ from those of the cross-ply tyre.

Run-flat tyre construction

The run-flat tyre is concerned with both safety and mobility, since it serves the twofold purpose of maintaining vehicle stability in the event of a sudden deflation and permitting limited vehicle operation after the deflation. Its concept is not new, the principle having been early applied by the Dunlop Company to tyres used for land-speed record breaking when they developed a rubber bead spacer for the tyres of Sir Henry Segrave's Golden Arrow car in 1929, and later to tyres for combat vehicles during World War II. Despite this promising background of development the relatively greater sophistication and higher cost of the run-flat type of tyre, which generally demands a special form of wheel rim, have been such that it has failed to gain wide acceptance by vehicle manufacturers. A further complication is that most tyre manufacturers now consider it essential to combine a deflation warning system with the run-flat tyre.

Tyre manufacturers have adopted various approaches to the design of run-flat tyres, of which the following may be listed as typical examples in recent years:

Internal release of lubricants to minimize heat generation
Specially reinforced tyre side walls

Tyre side walls embracing rim flanges
Internal reinforced-plastics support system
Multi-chambered tyre construction.

Space-saver tyres

The object of the sometimes controversial 'space-saver' tyre is to provide a more compact spare wheel and tyre combination for emergency use, which can bring benefits in boot space and to a lesser extent in weight and cost. This type of tyre is of smaller diameter and reduced section compared to the conventional tyre it temporarily replaces. To compensate for its much reduced size in relation to the same axle loading, the space-saver tyre is inflated to a higher pressure and it is also restricted in speed to 50 mile/h (80 km/h) from safe handling and wear considerations. A particular manufacturer's recommendations concerning the use of a space-saver tyre must therefore be strictly observed.

Load-carrying capacity of a tyre

This may be considered in terms of its size and ply rating, as follows:

Size   The choice of tyre size must primarily take into account the weight per axle of the vehicle. In practice, it is the economic load that is used as the basis for tyre selection. This term is applied to the load capacity beyond which a tyre of larger size or with a stronger carcass will be more economical under medium service conditions.

Ply rating   This indicates the strength of a tyre and is a term used to identify a given tyre with its recommended load and inflation pressure when used in a specific service. At one time the ply rating directly related to the number of cord plies used, but the advent of cord ply materials stronger than the original cotton meant that a tyre with fewer plies could have the same load-carrying capacity. Car tyre equipment is usually of four-ply rating with two actual carcass plies of rayon cord being used in the typical modern radial-ply tyre. In contrast, a radial-ply tyre for a heavy commercial vehicle with a single carcass ply of steel cord, may have a ply rating as high as eighteen.

Tyre sizes and designations

In the early days of motoring it was customary to express the size of a pneumatic tyre by its overall diameter and width, this already being established practice for the solid rubber tyre. With the advent of low-pressure tyres in the mid 1920s, this convention was abandoned in favour of denoting the size of a tyre by its nominal section width and the diameter of the wheel rim at the bead seat. These dimensions were originally expressed in either inches or millimetres, and they may still be found on cross-ply tyres supplied for use on older cars. Nowadays, a combination of both metric and imperial dimensions is used, the sectional width being quoted in millimetres and the rim diameter in inches. This system of marking has become the convention for radial-ply tyres, together with the identifying letter R originally preceded by a speed coding symbol. The symbols used indicate the following maximum speeds that may be safely sustained for any given design of radial-ply tyre:

Speed symbol	mile/h	km/h
P	93	150
Q	99	160
R	106	170
S	112	180
T	118	190
U	124	200
H	130	210
V	over 130	over 210
Z	over 150	over 240

NB: If the V symbol comes after, rather than within, the tyre size markings it indicates a speed capability up to 150 mile/h or 240 km/h. Two other pieces of information that may appear on a radial-ply tyre are the aspect ratio and a load index (LI) number that relates the maximum load a tyre can carry to its speed symbol (Figure 22.11).

Taking an example, a tyre with a designation 195/65R14 89H means that it has a nominal section width of 195 mm, has an aspect ratio of 65 per cent, is of radial construction, fits a wheel rim of 14 in diameter at the bead seats, can carry a maximum load of 580 kg (this being equivalent to its load index number of 89), and is suitable for use on cars having a maximum speed capability of 130 mile/h.

A similar system of designation is now used for the tyres of heavy vehicles, but as would be expected the speed symbols are extended below, and the load index numbers extended above, those for passenger cars. Also the load index number may be expressed as a double one, the first for single tyres and the second for twinned tyres.

Tyre tread patterns

On a wet road surface the water film acts as an excellent lubricant for the tread rubber, so that in the absence of some form of tread pattern to disperse this water the grip or adhesion of tyre to road would be drastically reduced and the safe handling of the vehicle jeopardized. The main purpose of the tread pattern is therefore to provide both a wiping action and a drainage facility between the tyre and road.

A flooded, as opposed to a wetted, road surface poses a much more serious threat to tyre adhesion. It is then possible for the combined effects of speed and depth of water to be such that a worn tyre is no longer able to remove quickly enough the water from its contact area with the road, especially if the remaining depth of tread pattern is approaching the legal limit stated in the next section. In this event, the tyre can become entirely supported by the film of water with consequent loss of control. This condition is known as aquaplaning or dynamic hydroplaning, since a water wedge builds up between the tyre and road to generate a hydrodynamic upthrust on the tyre (Figure 22.12). Indeed, the action can be likened to that of hydrodynamic lubrication of a plain bearing, as described in Section 4.2.

Additional requirements of the tyre tread pattern are that it should provide quiet running and acceptable wear characteristics. A non-linear (not repeating itself at equal intervals)

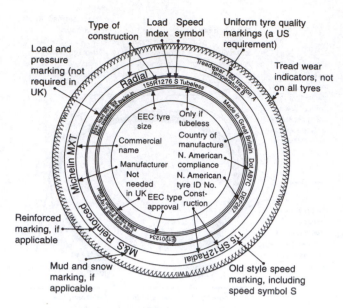

Figure 22.11 Diagram showing the markings to be found on the sidewalls of tyres

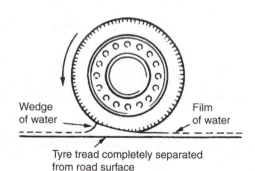

Figure 22.12 Tyre aquaplaning effect

pattern is the most widely used method of avoiding a constant frequency note that could provoke tyre noise. Adequate support for the tread pattern elements is also essential to prevent their excessive distortion, which could otherwise lead to high rates of tread wear (Figure 22.13a and b).

Another phenomenon that may be encountered in tyre operation is that known variously as 'tram-lining' or perhaps more exotically as 'white-line consciousness'. Here the tyre offers an initial and disconcerting resistance to steering as it first attempts to climb over and then suddenly darts across a longitudinal ridge in the road surface. To reduce the sensitivity of the tyre to this type of road surface irregularity, the shoulder regions of the tread incorporate shallow antiwander ribs that allow the tyre to be steered smoothly over such ridges.

To achieve adequate traction in adverse off-road conditions, such as encountered during rallying a relatively deep open tread pattern is required (Figure 22.13c). The concentration of loading on the individual tread blocks then causes them to penetrate the material covering the ground to the

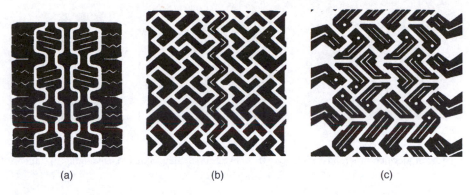

(a)   (b)   (c)

Figure 22.13   Examples of modern passenger car tyre tread patterns: (a) everyday motoring (b) high performance (c) all-seasons mud and snow (*Pirelli*)

firmer surface beneath. At the same time, the wide spacing, shape and flexibility of the tread blocks assist in evacuating the penetrated material, which may be deep mud and snow.

Actual tread patterns fall into three basic categories; circumferential ribs, continuous zigzag and individual blocks. In early narrow tread patterns these features were often used in isolation, but it was later found that whereas plain circumferential ribs gave good steering control and were quiet running, their behaviour was less exemplary when braking in the wet. This shortcoming could be improved by altering the ribs to a continuous zigzag pattern, but usually at the expense of noisy running. Individual blocks gave a similar improvement, but again were noisy and they also tended to wear irregularly. Progress therefore involved compromise by combining more than one of these features into a wider tread pattern. For example, the provision of circumferential ribs flanked by individual blocks or bars. By this means multi-directional drainage channels are created between the ribs and blocks to disperse surface water on the road. It is still necessary, of course, to maximize the area of tread rubber that meets the road surface, not only to give an acceptable tread life, but also to increase tyre adhesion because a rolling tyre does not obey the normal laws of friction, owing to the complex manner in which the rubber conforms to the surface roughness of the road. The ratio between the area of rubber in contact with the road and the corresponding area occupied by the drainage channels is termed the 'void ratio' of the tread, and is also known amusingly as the 'land-to-sea ratio' by tyre technologists. Other detail tread improvements include the provision of rectangular and vee-notches or teeth along the edges of the circumferential ribs, which increase their effective length, and also of miniature slits in the tread surfaces. These slits communicate with the main drainage channels of the tread pattern to produce a squeegee effect on wet roads and therefore make a secondary contribution to water dispersal. Commonly known as 'sipes' they are so called after their American inventor Harry E. Sipes. A more recent version is the Bridgestone dual-width sipe which has a stepped interior to give better stone ejection.

Special tread patterns have been designed to meet the requirements of ultra high-performance cars, these being generally classified as 'asymmetric' and 'unidirectional'. As

Figure 22.14   Bridgestone ultra high-performance sports tyre with super slant tread pattern (*Bridgestone/Firestone*)

its description implies an asymmetric tread pattern differs side-to-side, the purpose being to concentrate more rubber on the outside tread area, because this is the most highly stressed part of the tread during fast cornering and is therefore liable to increased wear. Conversely, the inside tread area has a greater concentration of voids, so as to restore the overall water dispersal properties of the tyre. This type of tyre must, of course, be positioned the correct way round. A unidirectional tread pattern is designed to exploit the forwards rolling motion of the tyre to best advantage, in terms of traction, braking and water dispersal for optimum stability on wet roads. The tread pattern typically features swept-back lateral ribs in place of circumferential ones (Figure 22.14), which provide for more efficient water dispersal to the sides and rear of the tread, thereby improving resistance to aquaplaning. There is also a better distribution of contact pressure between the tread and road surface for enhanced wet and dry handling. An arrow marking on the sidewall of a unidirectional tyre indicates the correct direction in which it must roll during forward travel.

Although it is probably true to say that every conceivable form of tread pattern has, at some time, been evaluated by the various tyre manufacturers, development work still continues

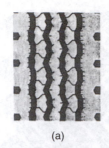  

(a)	(b)	(c)

Figure 22.15 Examples of modern commercial-vehicle tyre tread patterns (a) steer-axle (b) drive-axle (c) trailer-axle (*Bridgestone/Firestone*)

on this theme except that nowadays it is aided by sophisticated computer programming, as indeed is tyre design in general.

Heavy vehicle tyre tread patterns

Maximizing tyre life on heavy vehicles is clearly an important economic consideration for the fleet operator. To help achieve this objective it was once the traditional practice to interchange the wheels according to their constantly observed wear patterns, rather than at any definite intervals. It was also recognized that indiscriminate interchanging of the wheels could actually have an adverse effect on tyre life. More recently, tyre manufacturers have introduced tread patterns that have been tailored to suit specific duties. This is to overcome an inherent tendency towards either shoulder or crown wear, which can occur on steer-axle and drive-axle tyres respectively, thereby avoiding the time-consuming task of periodically changing their positions to even-out wear.

In practice the tread patterns adopted for modern commercial vehicle tyres may be identified as follows:

Steer-axle tyres These essentially feature longitudinally disposed grooves, a modest tread pattern and solid shoulders (Figure 22.15a), the aim being to provide responsive steering and combat shoulder wear arising from camber thrust.

Drive-axle tyres In contrast these exhibit transverse grooving, a more pronounced tread pattern and ribbed shoulders (Figure 22.15b), which in combination with full-width steel stabilizer belts give optimum wet and dry traction and counteract crown wear.

Trailer-axle tyres These generally feature virtually plain longitudinal grooves with minimal tread pattern (Figure 22.15c), the design objective here being to allow sideways yielding and hence reduce scuffing wear when manoeuvring tri-axle trailers in confined spaces.

The law and vehicle tyres

In Britain The Road Vehicles (Construction and Use) Regulations make illegal both the fitting of radial-ply tyres at the front and cross-ply tyres at the rear, and the mixing of radial-ply and cross-ply tyres at the same end of the vehicle.

The reasoning behind this particular regulation will not have been made apparent in the previous description of cross-ply and radial-ply tyres because it concerns their different

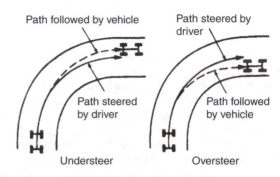

Figure 22.16 A simple representation of understeer and oversteer

behaviour during cornering. A simple explanation of this may be broken down into the following stages:

1 If during cornering the front tyres run at a greater slip angle than the rears then an understeering effect is produced whereby the front of the vehicle tends to run wide (Figure 22.16). This characteristic is not regarded as dangerous because any breakaway is progressive and fairly easily controlled by further turning the steering wheel. On the other hand, if the rear tyres run at a larger slip angle than the fronts then the rear of the vehicle tends to swing wide into the corner and this can result in breakaway that is vicious and difficult to control.

2 For a given sideways or cornering force the greater lateral rigidity in the tread of a radial-ply tyre reduces its slip angle as compared with that of a cross-ply tyre. It therefore follows that with the unlawful combination of radial-ply front and cross-ply rear tyres an unstable oversteering condition could arise on corners (Figure 22.16).

3 With the equally unlawful mixing of radial-ply and cross-ply tyres at the same end of a vehicle it will be evident that there could be disconcerting differences in vehicle handling when cornering in either direction.

The following also constitute illegal practices with respect to the use of vehicle tyres:

Types If there is unsuitable for the use to which the vehicle is being put or is incompatible with the types of tyre fitted to the other wheels.

Pressure If the tyre is not sufficiently inflated to make it fit for the use to which the vehicle is being put.

Tread The grooves of the tread pattern shall be of a depth of at least 1.6 mm throughout a continuous band situated in the central three-quarters of the breadth of tread and round the entire outer circumference of the tyre. On vehicles with a maximum design gross weight exceeding 3.5 tonnes, the tread depth must be at least 1 mm across any three-quarters of the breadth of tread or, if the grooves of the original tread pattern did not extend across three-quarters of the tread breadth, the original grooves must have a depth of a least 1 mm.

The effect of tread wear when braking on a wet road surface should not be underestimated, since stopping distance increases disproportionately to the reduction in tread depth.

Structure If the tyre has a break in its fabric or has a cut that exceeds 25 mm or 10 per cent of the section width, whichever is the greater, measured on the outside of the tyre and deep enough to reach the body cords. Also if the tyre has any portion of its ply or cord structure exposed, and further, if it has any lump or bulge caused by separation or partial failure of its structure.

Recut pneumatic tyres must not be fitted to any wheel of a motor car, unless it comes into the category of a goods vehicle the weight of which unladen is 2540 kg or more and the rim diameter of the wheel is 405 mm or more, or is an electrically propelled goods vehicle. It is not permitted to use on the road any motor vehicle or trailer a wheel of which is fitted with a recut pneumatic tyre where the fabric has been cut or exposed by the recutting process. In America the regulating agency for tyres is the Federal Motor Vehicle Safety Standards (FMVSS).

The tyre valve

The tyre valve that we take so much for granted today, in terms of its standardization and reliability, is unchanged in principle from the original version introduced as long ago as 1898 by the firm. A Schrader's Son of New York. It serves the following purposes:

1 Allows the tyre to be inflated with air under pressure.
2 Prevents air from escaping after inflation.
3 Permits release of air for adjustment of tyre pressure.
4 Readily accepts a pressure testing gauge.

For both tubed and tubeless tyres the tyre valve comprises three basic components:

Body For cars the metal stem of the valve is bonded into a rubber body which is either attached to the inner tube during its vulcanization or, in the case of tubeless tyres, made with a snap-in grommet fitting that provides an air tight seal against the wheel rim hole (Figure 22.17a). Commercial vehicle tyre valve bodies comprise an all-metal stem bonded to a rubber base and vulcanized to the inner tube (Figure 22.17b). To position the valve stem away from the brake drum the metal stem is bent at nearly a right angle where it emerges from the wheel rim hole (Figure 22.17c) and may then be bent again to a lesser angle to permit pressure checking and inflation. A clamp-in rather than a snap-in valve is required for tubeless tyres on commercial vehicles, owing to the higher inflation pressures involved.

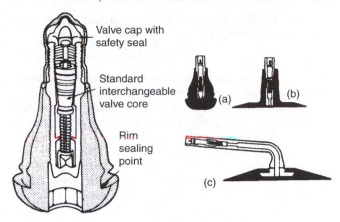

Figure 22.17 Tyre valve equipment (*Schrader*)

Core This is a complete assembly comprising a screw-in hollow plug with a plastics seal making an air-tight contact with a seating in the valve body, and a spring-loaded plunger with a synthetic rubber seal making an air-tight contact against the underside of the hollow plug (Figure 22.17). The spring-loaded plunger acts as a check valve in that it is closed and held on to its seating not only by its spring load but also by the pressure of air it is retaining. It can thus admit air at a higher pressure during inflation, and release air only when one is either pressure checking or purposely deflating the tyre, since both these operations involve forcing down the plunger pin to open the valve.

Cap This acts as an extra air seal and also excludes dust and moisture from the valve (Figure 22.17). It incorporates a floating rubber sealing gasket providing a positive seal against the mouth of the valve body.

Tyre inflation pressures

The flexibility of a pneumatic tyre is derived mainly from the elastic properties of its carcass and tread, which in turn are pretensioned by the inflation pressure of the air in the tyre.

It thus follows that a tyre must be maintained at its correct inflation pressure since this can be related to the following aspects of tyre operation: load capacity, vehicle handling and tyre life.

Load capacity
For any given size and type of tyre the manufacturer recommends a range of inflation pressures commensurate with the load to be carried to ensure that the deflection of the tyre under load does not continually exceed a predetermined safe amount. In the event of the inflation pressure being too low, then the greater rolling resistance will increase fuel consumption, and the Michelin company have found that if a commercial vehicle tyre is under-inflated by 20 per cent it can lose 30 per cent of its potential life. Similarly, under-inflated commercial vehicle twin-tyres can be damaged by excessive flexing, which allows their adjacent side-walls to scrub against each other. Too high an inflation pressure reduces the cushioning properties and also increases the possibility of a

concussion failure of the tyre because its carcass plies are overstressed in tension.

Vehicle handling
The manner in which a pneumatic tyre flexes and develops a slip angle during cornering has already been described. At a constant cornering force the effect of increasing inflation pressure is to reduce tyre slip angle and vice versa. The final choice of pressures can thus be related by the tyre and car manufacturers to the handling requirements of any particular vehicle. Any departure from recommended pressures is therefore likely to affect vehicle handling adversely. For example, under-inflation of the rear tyres would reduce the stable understeering characteristics of the vehicle and, at worst, introduce an unstable oversteering effect.

It is also important that the tyres at the same end of the vehicle are inflated to equal pressures, otherwise the effect on vehicle handling would be similar to that resulting from the mixing of cross-ply and radial-ply tyres.

Tyre life
To extend the life of tyres it is important that their relative positions on the vehicle be changed or, as it is commonly termed, 'rotated' on a regular basis, typically every 6000 miles (10 000 km). This is necessary to even out the tread wear between the front and rear tyres, because of the different duties they perform according to vehicle design and the further influence of driving style. The particular vehicle manufacturer's recommended rotation intervals and sequence must therefore be followed.

Underinflation and over-inflation both adversely affect tread wear as in both cases the area of contact between the tyre and the road changes in shape. With the former the load, and therefore most wear, tends to be concentrated towards each side of the tread, while with over-inflation the load is concentrated around the central part or crown of the tread, which thus experiences the greatest wear. The Bridgestone/Firestone company have found that 80 per cent of premature tyre failures are caused by incorrect inflation, underinflation with its excessive heat build-up being more damaging than overinflation.

An important point in connection with maintaining correct pressures is that these will rise while the vehicle is running, especially during high-speed motoring in hot weather, and under these conditions no attempt should ever be made to readjust them to their cold setting.

A pressure equalizing system may be used to improve the life of twinned tyres on heavy vehicles and also provide a ready visual warning of any unacceptable loss of pressure in the two tyres. Systems of this type incorporate a normally open link between the valve stems of adjacent tyres for pressure equalization, while remaining failsafe by closing the link in the event of either tyre suffering a puncture so that air is retained by its neighbour. Tyre pressure monitoring systems suitable for application to passenger cars are also being developed. One approach has been to utilize the wheel speed sensors of an anti-lock brake system (Section 29.2), which allows the rotational speed of each wheel to be compared to the other three. Hence, a reduction in rolling radius of an underinflated tyre and the associated speeding up of its wheel can be signalled to the driver. A tyre pressure monitoring system is an important requirement for the safe use of run-flat tyres, as earlier mentioned.

22.4 ROAD WHEELS AND HUBS

Basic requirements

Three main types of road wheel are produced for cars and commercial vehicles, and these must be designed and constructed to meet certain essential requirements according to their specific applications. The design features necessary for all wheels may be summarized as in the following sections.

Lightweight construction
Each wheel adds significantly to the unsprung mass of its suspension mechanism. Since it is easier for a shock damper to control the bouncing movements of a small rather than a large unsprung mass, it follows that for good road holding the wheel should be made as light as possible. For cars this requirement has been helped by modern body styling trends demanding smaller diameter wheels than were used in the past.

Strength and rigidity
These are not one and the same requirement. A wheel must possess sufficient strength to withstand all normal service loads imposed upon it, as well as being resistant to accidental damage. It must further be rigid or stiff enough as a structure to minimize bending deflection under steering and cornering loads. Again, the smaller diameter wheels used in modern cars offer an advantage in this respect.

Tyre retention
The design of the wheel rim must permit the mounting and removal of the tyre and retain it securely in position when inflated. As regards the former requirements, heavy-duty tyres used on commercial vehicles generally require a detachable flange rim as compared with the simpler one-piece rim adopted for cars. For tyre retention taper seatings are provided adjacent to the rim flanges so that the beads of the inflated tyre are forced to climb these tapers, thus wedging themselves against the insides of the rim flanges. The rim must also accommodate the tyre valve assembly.

Following the development of the tubeless tyre in the late 1950s, American safety regulations created a requirement for a retaining hump on the outer bead seat and adjacent to the well of the wheel (Figure 22.18), which later became accepted practice elsewhere. Its purpose is to ensure greater safety by preventing the tyre bead from sliding off its seat in the event of sudden deflation or hard cornering with under-inflated tyres.

Wheel retention
It is, of course, an essential requirement for both cars and commercial vehicles that their wheels should be simple to remove and refit (even though the jacking procedure may sometimes leave a lot to be desired!). Established practice is for each wheel to be attached to a flanged hub by an inner ring of either studs and nuts, or screws (Figure 22.19). In a non-spigot type mounting, these serve the dual purpose of

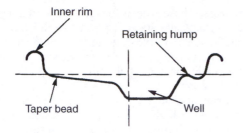

Figure 22.18 Wheel rim with safety retaining hump

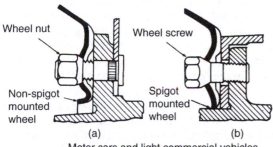

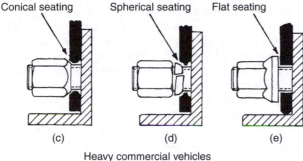

Figure 22.19 Methods of road wheel retention (a), (c) and (d) non-spigot mountings (b) and (e) spigot mountings

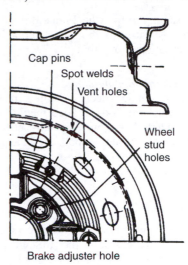

Figure 22.20 Constructional details of a pressed steel wheel (*Dunlop*)

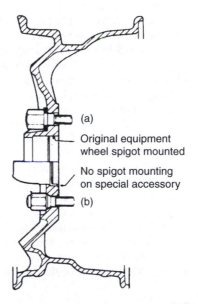

Figure 22.21 Cross-section of a typical cast alloy wheel and methods of fastening (*GKN Kent Alloys*): (a) typical original equipment steel bolt and flat washer assembly clamping into spot face in casting (b) special accessory cast aluminium practice with conical nut seating integral with bolt clamping directly into machined cone seating in casting

retaining the wheel endwise and locating it radially. They are therefore provided with either conical or spherical seatings (Figures 22.19a,c,d and 22.21b), which exert a wedging action to prevent their loosening against the wheel. Under no circumstances must these two different forms of nut or screw seating be interchanged. Less commonly, the nuts or screws may be left-hand threaded on the left (nearside in the UK) of the vehicle and right-hand threaded on the right (offside in the UK). For modern heavy-duty commercial vehicles, and increasingly also for greater accuracy of wheel mounting on some cars, the function of radial location may be performed by the wheel being made close fitting over the end of its supporting hub, it then being known as spigot mounted. In the case of commercial vehicles, flat-faced nuts with captive washers are used to retain the wheel (Figure 22.19e), while for car wheels either conical or flat-faced screws serve to retain pressed-steel and cast alloy wheels respectively (Figures 22.19b and 22.21a).

It is imperative that the wheel nuts or setbolts are tightened progressively in a diagonally opposite sequence and are

neither under- nor over-tightened. Under-tightening is especially hazardous in the case of heavy vehicles, since there is then a distinct possibility of a wheel becoming detached on the road. The manufacturers of heavy vehicles specify the correct torque and re-torque tightening values to be used, which are typically based on lightly oiled threads, so it is most important to consult their detailed instructions and also the British Standards Code of Practice BS AU 50: Part 2: Section 7a: 1995. These are intended to ensure that sufficient tension is being generated in the wheel studs and by the same token clamping force imposed on the wheel, which will prevent any fidgeting movements at its mounting interface and subsequent loosening of the wheel nuts. It therefore follows that the mating surfaces of the wheel mounting must be clean and preferably free of paint.

Economic and accurate manufacture

Obviously, motor vehicle wheels must be suitably designed for mass production methods to keep manufacturing costs down to a reasonable level, but they must also be made to exacting tolerances to minimize both radial and lateral run-out, as measured at their rims. These features contribute to smooth running and satisfactory tyre life.

Types of road wheel

The types of road wheel used on modern cars and commercial vehicles fall mainly into the following categories: pressed steel, aluminium alloy and detachable rim.

Pressed steel

This type is fabricated from a dished and flanged disc pressed into a rolled-section rim and retained by spot welding, a process that has superseded riveting to provide more reliable air sealing for tubeless tyres (Figure 22.20). The joint between the disc and its rim is not continuous, but interrupted by a series of equally spaced ventilation slots derived from the original polygon-shaped disc blank. These may be supplemented by swaged ventilation holes in the dishes portion of the disc, so that in combination they not only induce a flow of cooling air over the brake assembly, but also minimize heat transfer from the latter to the tyre. The wheel fixing holes are formed in the central portion of the disc.

A one-piece construction is used for the well-base rim, so called because a well is formed around the middle of its base to facilitate removal and refitting of the tyre. The tyres are provided with taper seatings, and in the case of pressed steel wheels for cars a 5° angle of taper is used. This type of wheel generally meets all the requirements listed earlier and, compared with its one-time rival the wire-spoked wheel, it possesses the additional advantage of being easy to clean.

Aluminium alloy

The demand for attractively styled wheels manufactured from a high-strength aluminium alloy has continued to grow

in recent years. These wheels are generally produced from one-piece castings and since they must be finish machined, their manufacture is to closer tolerances than steel wheels (Figure 22.21). For high-performance cars a two- or three-piece construction in forged aluminium alloy may be used. This typically comprises a centre section with either a single or divided rim held together by a ring of small diameter, bihexagon headed, high-tensile screwed fastenings. An air sealing gasket is required between a divided rim.

Aluminium alloy wheels possess certain technical advantages. The lighter material permits the use of thicker metal sections for greater rigidity. Similarly, the rims can be increased in width so that correspondingly wider tyres may be used for improved car handling. In this context the lighter wheels also contribute to better road holding, by virtue of reducing unsprung weight of the suspension system. Apart from allowing a saving in the overall weight of the vehicle and therefore the fuel it consumes, the lower rotational mass or 'flywheel effect' of the wheels absorbs less energy in speeding them up or slowing them down during acceleration and braking. Furthermore, the better heat conductivity of this material permits cooler running for the tyres and brakes, also the pattern of the wheels can be designed to induce more air to flow over the brakes. The possible disadvantages of aluminium alloy wheels are their reduced resistance to both accidental damage and corrosion, although a wide variety of protective finishes are now applied to them.

Aluminium alloy wheels of the ventilated disc type are also used to a limited extent on heavy vehicles, where they offer similar technical advantages, especially in terms of heat dissipation from the brakes. However, an important commercial consideration is that they confer valuable weight savings, which can be translated into greater payload, which is the actual useful weight carried by the vehicle.

Detachable rims

This type of construction is confined to commercial vehicle applications and it arises from the need for ease of mounting and removal of their heavy-duty tyres. Apart from rim design, it is basically similar to the welded pressed steel wheel used on cars, although the disc portion may be of either constant or variable-section thickness. This is so that the necessary strength requirements can be met without producing an excessively heavy wheel. It is usually deeply dished not only to accommodate the bulk of the brake drum but also to provide sufficient offset for twin rear wheels to be mounted back-to-back. Ventilation apertures are generally incorporated in the disc.

With detachable rim wheels an integral flange is formed around only one edge of the rim, the other flange being detachable and taking the form of either a split spring ring, which locates directly in a groove around the edge of the rim, or the combination of an endless rigid ring and a split locking ring likewise sprung into the grooved edge of the rim. These two forms of detachable rim construction, known as two-piece and three-piece respectively are used according to whether the wheel loading is medium or heavy (Figure 22.22). In both cases, taper bead seatings are provided for the tyres.

Another type of wheel that may sometimes be encountered on heavy vehicles and particularly those used by the

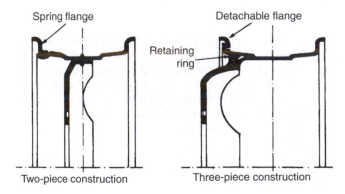

Figure 22.22   Detachable rim wheels for commercial vehicles

military is the divided wheel. It is so called because it is constructed in two main parts, which are clamped together by an outer ring of bolts and nuts (Figure 22.23), The complete wheel then being secured to its hub by another inner ring of nuts. Wheels of this type are typically used in conjunction with massive tyres that are designed to operate at very low pressures. The tyres are so hefty, in fact, that they could not be pulled off an ordinary rim owing to insufficient give. The divided wheel allows the wheel to be taken out of the tyre when the inner and outer halves have been separated, instead of attempting to take the tyre off the wheel. A safety note is given later with regard to handling precautions for divided wheels.

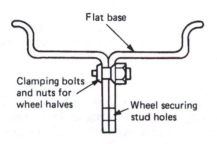

Figure 22.23 Divided wheel construction

Wheel hubs and bearings

The arrangement of hubs and bearings used in live rear axles has been described in Section 20.4. It now remains to consider other arrangements which are associated with independently sprung wheels, as follows:

Independent front suspension (IFS) with non-driven wheels
Independent front suspension (IFS) with driven wheels
Independent rear suspension (IRS) with driven wheels
Independent rear suspension (IRS) with non-driven wheels

IFS with non-driven wheels

The hub bearing arrangements for this system (Figure 22.24a) generally continue to follow those of beam-axle-mounted wheels. That is, each hub rotates about a dead spindle through the medium of an opposed pair of either angular-contact ball, or tapered roller, bearings. However, the requirements in respect of their load-carrying capacity have tended to increase, because of the more forward mounting of the engine permitted by IFS. It is usual for the line of action of the wheel load to pass closer to the inner bearing than to the outer, so that the latter mainly performs a stabilizing function and can be of smaller size and therefore less expensive. To allow for expansion caused by heat dissipated from the brake assemblies, a very small amount of end float

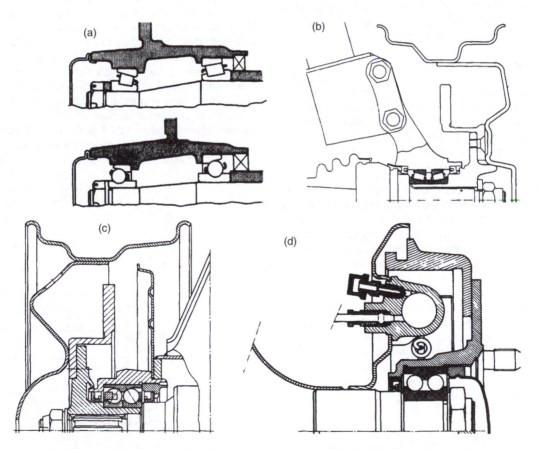

Figure 22.24 Typical hub bearing arrangements: (a) IFS, non-driven wheels (*RHP*) (b) IFS, driven wheels (*Timken, Ford Fiesta*) (c) IRS, driven wheels (*SKF*) (d) IRS, non-driven wheels (*Citroën*)

is usually permitted between the bearings when they are adjusted.

IFS with driven wheels

The development of the front-wheel-drive principle where each wheel is secured to a live spindle, which rotates within a pair of bearings mounted in a hub carrier, early created a demand for bearing arrangements that occupied less width than those previously described. This led to the application of very compact double-row angular-contact ball bearing units in which a split inner ring is used to obviate the need for filling slots for the balls. Hence the optimum number of fairly large-diameter balls may be used to increase radial and axial load-carrying capacities, so that a much reduced axial space is required and the wheel spindle can be made short and rigid. Tapered roller bearings may be similarly adapted to give high load-carrying capacity in the minimum space (Figure 22.24b). These modern types of hub bearing unit are set for correct axial clearance during manufacture and require no adjustment in service.

IRS with driven wheels

For independently sprung and driven rear wheels, the modern approach is to use a double-row angular-contact ball bearing hub unit similar to that described for IFS with driven wheels (Figure 22.24c). Since the bearing arrangements in both cases are basically the same, one manufacturer for example uses a common hub bearing unit for the driven front wheels of one model and for the driven rear wheels of another. For heavy-duty applications, the earlier practice of using an opposed pair of either deep-groove ball, or tapered roller, bearings may still be followed.

IRS with non-driven wheels

The lightly laden rear wheels of FWD cars readily lend themselves to the use of a hub bearing arrangement either similar to that described for non-driven front wheels, or again to a modern double-row angular-contact ball bearing unit (Figure 22.24d). In the latter case it follows, of course, that the bearing unit is housed within the wheel hub and the split inner rings are mounted directly on the dead spindle.

In long-established practice the front wheel hub bearings used on heavy vehicles with beam axle front suspension have featured an opposed pair of tapered roller bearings, their mounting being identical with that shown in Figure 22.24a. The smaller outer bearing is retained on the stub axle by a washer and nut, which is also used for adjusting the bearings to allow for heat expansion under normal operating conditions. Each hub is, of course, fairly massive in comparison with its passenger car counterpart and is provided with ten wheel studs pressed into holes bored in its outer flange. The studs may be knurled on their shanks to enhance the press fit in the flange. In a more recent development the integrated type of wheel bearing unit has been applied to the front hubs of heavy vehicles (Figure 22.25), the advantages being the same as those cited in Section 20.4 for fully floating rear hubs.

Lubrication and sealing of hub bearings

Long-term lubrication of the hub bearings constitutes the most important grease application for the vehicle running-gear.

Since the late 1950s established practice has been to specify a multipurpose lithium complex grease, the characteristics of which combine high temperature and water resistance, together with extreme pressure and corrosion inhibiting properties. The use of disc brakes, particularly in the case of heavier vehicles, introduced more severe temperature demands and required the grease to resist peak temperatures of up to 200°C. In this connection it should be noted that although reference is sometimes made to 'high melting point' grease, this expression has little real meaning. This is because grease is a fluid lubricant blended with a thickening agent, so that it is basically a plastic material that has a melting range, rather than point, during which it becomes gradually softer. For this reason a typical analysis for a hub grease will instead refer to its 'dropping point', which may be defined as 'The temperature at which a quantity of grease first begins to exude oil, when heated in a special apparatus under controlled test conditions'. A dropping point of 300°C generally recognizes modern requirements for hub bearings grease.

In practice, a vehicle manufacturer is usually content simply to quote the required 'NLGI' (National Lubricating Grease Institute of America) grade number, typically NLGI 2 for hub bearings grease. This relates to the degree of stiffness or consistency of a grease, the grade number indicating the depth of timed penetration of a standard weight metal cone dropped into the grease, again using a special apparatus under controlled test conditions. Before testing the grease is mechanically agitated or 'worked', so that consistent results can be obtained. The grease is, of course, subjected to working in the hub bearings, so that any tendency of a grease to leak from them must be avoided in determining its suitability for this application. NLGI grade numbers actually range from 000 through to 6 (almost fluid to extra hard), the commonly specified NLGI 2 grade for hub bearings corresponding to a medium soft grease. This method of classification is therefore analogous to that of SAE grade numbers for engine and transmission oils. It is important that the particular manufacturer's recommendations should always be adopted regarding the assembly greasing of hub parts and the extent

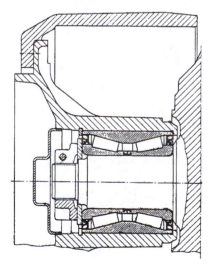

Figure 22.25   Integrated front wheel hub bearing unit for heavy vehicles (*FAG Kugelfischer*)

to which the hub cavity should be charged with grease (Figure 22.26). The modern hub bearing units described earlier are lubricated for life. Established sealing practice for hub bearing arrangements of all types is generally to use some form of radial lip seal, in which the flexible sealing lip is spring loaded to maintain a light rubbing contact with the surface to be sealed. In the case of the integrated front wheel hub bearing for heavy vehicles shown in Figure 22.25, it should be noted that it is sealed on both sides for maintenance-free service. The lip seal between the outer washer and the bearing has sealing lips that act in the radial and axial directions, so as to retain the grease reliably in the bearing and exclude fine dirt and water. In dead spindle applications, a hub cap is used to prevent the direct access of dirt and water to the bearing assembly.

22.5 WHEEL BALANCING

The reasons for wheel balancing

Until the early 1950s very little was heard about wheel balancing, although it was not unknown to those concerned

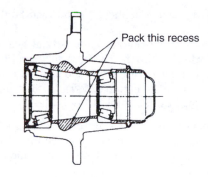

Figure 22.26 Lubrication and sealing of front hub bearings (*Nissan*)

with the servicing of a relatively select few high-performance cars. Since then, of course, the need for balancing the wheel and tyre assemblies has become a fact of life for all concerned. Historically, the reasons for this are basically twofold:

1 It was found that the more softly sprung car with IFS displayed a greater sensitivity to any imbalance of the wheel and tyre than was previously the case with a relatively stiffly sprung beam axle suspension system.
2 The ill effects of any such imbalance became more pronounced with the trend towards smaller-diameter and hence faster-rotating wheels, which from tyre loading considerations were also increased in width. Technically there are two types of wheel imbalance, known as static and dynamic (Figure 22.27).

Static imbalance in principle

It is common experience for a schoolboy to invert his cycle and then proceed to hand pedal it until the back wheel rotates so fast that the rear end of the machine begins to jump up and down of its own accord. This is caused, of course, by the out-of-balance forces generated by the revolving weight of the tyre valve, and provides a simple example of static imbalance of a wheel and tyre assembly (Figures 22.27b and c). The reason why this particular condition of imbalance should be called static can be explained as follows.

If the schoolboy transferred his attention to the more freely running front wheel, he would find that from whatever stationary position the wheel is released it would slowly turn until the unbalanced valve portion finally settled at bottom dead centre. Fairly obviously this state of affairs could be corrected by attaching a counter balance weight on the wheel rim, equal to that of the tyre valve, directly opposite to the latter (Figure 22.27d).

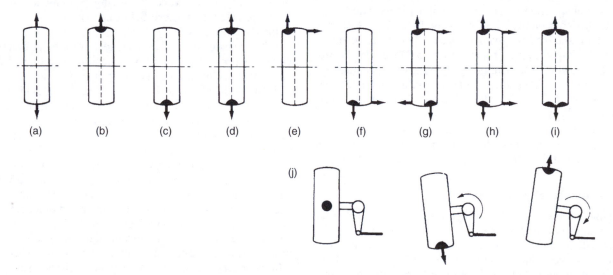

Figure 22.27 Conditions of wheel and tyre balance and imbalance: (a) ideal dynamic and static balance (b)/(c) static imbalance (d) static balance restored (e)/(f) dynamic and static imbalance (g) dynamic imbalance made worse but static balance restored (h) dynamic and static balance restored (i) ideal restoration of dynamic and static balance. Diagram (j) shows in plan form how static imbalance can cause wheel shimmy and steering joggles

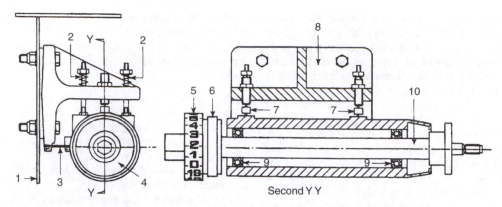

Figure 22.28   Wheel mounting shaft and bearing assembly of an off-the-car balancer (*Churchill*)

1 cubicle	6 shaft drive pulley
2 compression springs (mounting)	7 transducers
3 leaf spring (mounting)	8 bed casting
4 plastic cylinder	9 ball bearings
5 position of shaft indicator	10 mounting shaft for wheel

Summing up, a condition of static balance is achieved when there is an equal distribution of weight about the axis of rotation of the wheel and tyre.

Dynamic imbalance in principle

Returning once again to the schoolboy hand pedalling his inverted cycle, if for some peculiar reason the tyre valve emerged not from the centre of the rim, but was offset to either side of it, then the out-of-balance forces generated would cause the rear end of the machine not only to jump up and down, but also to shake from side to side owing to dynamic imbalance (Figures 22.27e and f).

To cure both these conditions of imbalance it would be necessary to attach a counterbalance weight, equal to that of the tyre valve, directly opposite and offset in the same direction to the latter on the wheel rim (Figure 22.27h). A state of both static and dynamic balance would then be achieved.

On the other hand, if the counterbalance weight was placed centrally on the rim or offset in the opposite direction to the tyre valve, then static balance would still be obtained but not dynamic, and in fact this would be made worse (Figure 22.27g).

Summing up, a condition of dynamic balance is achieved when there is an equal distribution of weight about the centre plane of the wheel and tyre (Figures 22.27a and i).

Wheel imbalance on the vehicle

A glance at any heap of scrapped tyres will reveal that the varied nature of the tyre wear is usually such that the distribution of the rubber mass becomes irregular, so as a tyre wears it is likely that the initial balancing of the wheel and tyre by the manufacturer will be disturbed. Static imbalance alone can make itself evident in two ways:

1 The vertical out-of-balance forces generated can provoke the wheel into bouncing up and down or 'pattering',

which is conducive neither to good road holding nor to long tyre life.
2 The corresponding horizontal out-of-balance forces can promote wheel 'shimmy', because they tend to steer the road wheel first in one direction and then in the other about its offset swivel axis (Figure 22.27j). This is signalled to the driver by 'joggles' felt at the steering wheel.

Dynamic imbalance again has the effect of promoting wheel shimmy, except that in this case it is the transverse out-of-balance forces generated that tend to steer the road wheel about its swivel axis.

Wheel balancing equipment

Modern proprietary wheel balancing equipment falls into two basic categories, these being known as off-the-car and on-the-car balancers.

As their designation would suggest, off-the-car balancers require that the wheel and tyre assembly be removed from the vehicle for balancing. Their operating principle is typically such that the wheel to be balanced is mounted on a motor-driven horizontal shaft. The bearing housing for the shaft is spring mounted in the horizontal direction only against a pair of electrical transducers, which produce voltages proportional to the forces imposed upon them by any imbalance of the spinning wheel (Figure 22.28). This information is interpreted by a small computer so that, after the wheel has been braked to a stop, a meter read-out can be taken of any out-of-balance at the inner and outer rims and the appropriate correction weights attached thereto. It is, of course, necessary in the first instance for the operator to program the rim position, width and diameter of the wheel.

In contrast, the on-the-car balancer checks static and dynamic out-of-balance with the wheel on the vehicle. Hence, the hub and brake drum or disc, as well as the wheel and tyre assembly, are compensated for imbalance. Naturally, this condition of overall balance can be lost in the event, say, of

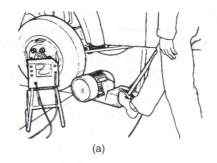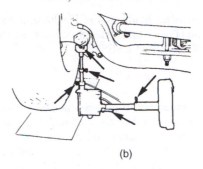

(a)   (b)

Figure 22.29   An on-the-vehicle wheel balancer in operation: (a) spinning the wheel (b) adjusting the pick-up (*Churchill*)

the wheel being removed and then not replaced in exactly the same rotational position relative to the hub. With this type of equipment the wheel to be balanced is jacked clear of the ground and, unless it is normally driven through the vehicle transmission system, it is rotated by a mobile motor-driven, wheel spinner held in contact with the tyre (Figure 22.29a). For sensing any out-of-balance forces acting on the wheel assembly, a transducer pick-up device is attached magnetically beneath the suspension linkage and as near as practicable to the spinning wheel for static balancing (Figure 22.29b), and against the forward part of the brake assembly backplae or splash shield for dynamic balancing. The equipment is completed by an amplifier, stroboscopic flash unit (similar to an engine timing light), weight indicating meter and controls. Any vibration set up by inbalance of the spinning wheel is therefore signalled by the pick-up to the strobe flash and the weight indicating meter. The maximum unbalanced force shown on the meter and the position on the wheel at which a previously affixed piece of white tape appears frozen by the flashing strobe light are noted. Reference is then made to an adjustment of weights chart (provided with the equipment) before attaching correction weights to the wheel rims.

For all types of wheel balancing equipment it is vitally important that the particular manufacturer's operating instructions are consulted.

22.6   SAFETY PRECAUTIONS IN TYRE SERVICING

A comprehensive list of safety points for servicing, removing and fitting tyres of commercial vehicles, which may also be applied in part to motor car tyres, has been drawn up and published by the British Rubber Manufacturers' Association and is reproduced below:

1 It cannot be stressed too strongly that when servicing tyres every precaution must be taken to avoid accident or injury.
2 If a wheel or tyre is damaged, remove that valve core completely to deflate the tyre before removing the wheel from the vehicle.
3 Before removing any tyre from a wheel rim remove the valve core and make sure the tyre is fully deflated.
4 Take care not to distort loose flanges and locking rings when dismantling tyres.

5 Thoroughly clean all flanges, locking rings and rim gutters.
6 Carefully examine all flanges, locking rings and rim gutters; if they are damaged, distorted, cracked or broken they must not be reused.
7 Thoroughly examine the tyre inside and outside to ensure that it is in sound condition.
8 Make sure that the correct size and type of tube, flap, flanges and locking rings are used.
9 Use an approved tyre bead lubricant when fitting tyres.
10 Once a wheel and tyre have been reassembled and before any inflation, check the positioning of loose flanges and locking ring.
11 Before any inflation, check and confirm the correct inflation pressure for the size and ply rating strength of the tyre.
12 The tyre should be gently inflated to not more than $15\,\text{lbf/in}^2$ ($1000\,\text{kPa}$) and checks made to ensure that the tyre is properly centred on the rim and that the flanges and locking rings are correctly seated.
13 Do not inflate above $15\,\text{lbf/in}^2$ ($100\,\text{kPa}$) until you are satisfied by a thorough external examination that the assembly appears to be in a safe condition.
14 Before final inflation above $15\,\text{lbf/in}^2$ ($100\,\text{kPa}$) put the tyre/wheel assembly into a safety cage or use a portable safety device.
15 In exceptional circumstances where no safety device is available (e.g. on site or at a roadside breakdown), position the wheel and tyre assembly with the loose flange and locking ring side facing a protective barrier such as a wall or bank or the side of a vehicle.
16 At all times when inflating assemblies, whether or not mounted on a vehicle hub, stand well clear of the area in front of the assembly.
17 Never lean over, sit or stand on or in front of a tyre during inflation.
18 Check pressure during inflation and never over-inflate above the maximum pressure allowable for the size and ply rating strength of the tyre.
19 An unattended airline should never be left attached to a valve.
20 Ensure that all pressure gauges and air metering devices are regularly checked against a master pressure gauge at least monthly.
21 For workshop use the BRMA poster 'Danger – don't risk a fatal accident' should be prominently displayed.
22 Do not take changes.

Reference should also be made to the Health and Safety Executive publication HSG62 titled 'Health & Safety in TYRE & EXHAUST FITTING PREMISES'.

Divided wheels

Care must always be taken not to confuse the nuts by means of which the wheel is mounted on the hub with those of the clamping bolts that hold the inner and outer halves of the wheel together. It is for this reason that British Standard BS AU 50:Part 2a:1973 specifies that the nuts of the clamping bolts for divided wheels shall be painted for identification – red for wheels painted in colours other than red, and yellow for wheels painted red. Also, this standard calls for divided wheels to be legibly marked with wording that states: Deflate before dismantling. Tighten nuts before inflating.

It may be worth recalling the homely advice offered by a vehicle manufacturer to military personnel during World War II, regarding the significance of the coloured nuts on divided wheels: 'You can take our word for it that it isn't done to make the wheels look pretty when they go round, or because we happened to have a spare pot of red paint.'

23 Suspension systems

23.1 BASIC RIDE CONSIDERATIONS

In the design of motor vehicles in general and passenger cars in particular, the comfort of the occupants is clearly of prime importance. The basic function of the suspension system is therefore to provide a flexible support for the vehicle, so that its occupants ride comfortably isolated from the imperfections of the road surface. An additional and no less important requirement of the suspension system is that of stabilizing the vehicle under all conditions of driver handling, involving as it does cornering, braking and accelerating manoeuvres. These two basic requirements in respect of vehicle ride and handling tend to conflict in actual practice, since 'soft' springing is indicated on the one hand and relatively 'hard' springing on the other. A successful design of suspension system is therefore one that achieves an acceptable compromise between these two areas of conflict.

The basic factors that govern the ride comfort of a vehicle in practice may be defined as vertical flexibility, horizontal flexibility, shock damping and linkage friction. It will later become evident how these factors are related to the design and construction of modern suspension systems, but first it is necessary to understand a little more about vertical flexibility since this is at the heart of any vehicle suspension system.

When the wheels rise and fall over road surface irregularities, the springs momentarily act as energy storage devices and thereby reduce greatly the magnitude of impact loading transmitted by the suspension system to the car structure. The energy of impact loading is related to the product of the disturbing force and the distance over which it is constrained to act by the spring medium. It thus follows that a soft or low-rate spring, which permits a generous deflection for a moderate loading, will reduce to a minimum the impact or disturbing forces transmitted to the vehicle occupants. The rate of the spring is equal to the load per unit of deflection (Figure 23.1a).

To lend real meaning to the commonly used expressions 'soft' and 'hard' springing, the suspension engineer prefers to define vertical flexibility in terms of 'static deflection' (Figure 23.1b). A value for static deflection is obtained from the normal spring load acting at the wheel divided by the effective spring rate measured at the wheel. Here the spring load is less than the actual load at the wheel by an amount equal to the unsprung weight of the suspension mechanism, and the effective spring rate is measured at the road wheel to take into account any leverage exerted on the spring by the suspension linkage.

Static deflection is in turn directly related to the 'natural frequency' of the suspension system: that is, the number of oscillations or cycles per second that the sprung mass would perform if allowed to vibrate freely on its suspension. Since this motion of the sprung mass is comparable with that of a

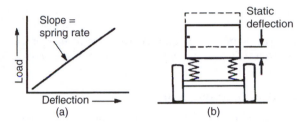

Figure 23.1   Spring rate and static deflection

simple pendulum, the physical relationship between the suspension frequency and deflection may be expressed as follows:

$$N = \frac{1}{2\pi}\sqrt{\frac{g}{D}}$$

where N is the number of cycles per second (hertz, Hz), g is the acceleration due to gravity (9810 mm/s^2) and D is the static deflection (mm). In modern passenger car practice, suspension frequencies generally fall within the range of 1 to 1.2 Hz (60 to 72 cycles/min) with corresponding static deflections of 225 to 175 mm (8.8 to 6.9 in).

The suspension frequency therefore varies in inverse proportion to the spring load and the amount of static deflection produced by that load. This explains why the ride comfort in a heavily laden car is often significantly better than when it is lightly laden, albeit within the limitations imposed by the curtailed bump travel of the suspension due to the greater static deflection. Whereas this limitation may be removed by including a self-levelling facility in the suspension system, the same comfortable ride in both the lightly and the heavily laden conditions can only be achieved by a suspension system that automatically varies its resistance to deflection or stiffness in accordance with load, so that the suspension frequency remains constant. Since there is a fractional time lapse between the front and the rear wheels passing over an obstacle, the suspension frequency at the rear end of the car is usually raised above that chosen for the front end, so that the quicker restoring effect of the rear suspension tends to synchronize the front and rear ride motions and thus minimize pitching.

23.2 TYPES OF SUSPENSION

Motor vehicle suspension systems may be classified as follows:

Dependent
Both wheels of a pair are mounted on a common axle that acts as a rigid beam, which is then connected by springs to the vehicle structure.

461

Semi-dependent

Both wheels of a pair are again mounted on a sprung common axle, but in this case it acts as a beam with limited flexibility.

Independent

Each wheel of a pair is separately linked and sprung from the vehicle structure, so that its movement relative to the latter is the same in roll as it is for ride.

Interdependent

Both wheels on each side of the vehicle have their springs coupled together, either mechanically or hydraulically, so that single wheel bumps are shared between front and rear springs for a softer ride, together with reduced pitching except during acceleration and braking.

Dependent front suspension

In dependent front wheel suspension systems a beam-type axle is used to support the two stub axle pivots at a fixed distance apart. Although beam axles have long since been obsolete for cars, for reasons that will be explained later, this form of front wheel suspension continues to be used on almost all commercial vehicles on the basis of the following considerations:

1 Front wheel alignment is better maintained with a beam axle because, unlike independent linkages, it does not demand great rigidity at the front end structure of the chassis frame. This is an important consideration in the design of a commercial vehicle chassis frame, which is intended to accommodate a certain amount of flexing in response to road shocks transmitted through the suspension springs.
2 Tyre wear is minimized because, first, the front wheels remain perpendicular to the road in the presence of body roll, and second, the track width remains constant with suspension movements of the wheels (Figure 23.2). It should be appreciated that the attainment of satisfactory tyre life is an important factor in the economic operation of a large fleet of commercial vehicles.
3 By virtue of its rugged construction and its mechanical simplicity with a minimum number of wearing parts, the beam axle is also favoured on the grounds of low initial and servicing costs.
4 Finally, the use of a beam axle ensures that the ground clearance of a commercial vehicle remains constant in both its unladen and laden conditions.

Dependent and semi-dependent rear suspension

Both live and dead rear axles are, of course, forms of dependent beam axle suspension and have previously been described in connection with final drive arrangements (Section 20.4). A further development of the type of dead axle beam that is used to mount the trailing rear wheels on some front-wheel drive cars is the semi-dependent H-beam axle system, which was originally introduced by Volkswagen for their 1980 Passat. In this ingenious system each wheel is mounted on a trailing arm that pivots from rubber bushes carried by the body structure, and both arms are connected together by a flexible axle beam that lies between the axes of the pivot bushes and wheel bearings. The axle beam is made flexible in torsion and stiff in bending, such that on the one hand it allows the trailing arms and wheels to move up and down almost independently of each other, while

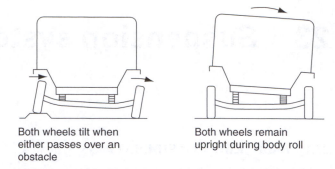

Both wheels tilt when either passes over an obstacle

Both wheels remain upright during body roll

Figure 23.2   The action of beam axle front suspension

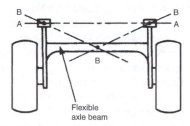

Flexible axle beam

Figure 23.3   Plan view of H-beam axle system

on the other hand it prevents the wheels from tilting to the same extent as if they were attached directly to the ends of the axle beam. That is, both wheels can move in unison about a purely trailing axis A–A (Figure 23.3) while a single wheel can move about a semi-trailing axis B–B.

Lateral location of rear axles

In the study of mechanics a *parallel motion* mechanism is an arrangement for constraining a point to move in a straight line. Two such mechanisms are known as the Watt and the Scott-Russell linkages, so-named after James Watt and F. Scott-Russell, who invented them in the 18th and 19th centuries respectively for use on steam engines. Our interest in these historic mechanisms is concerned with their application to live and dead axle rear suspension systems, and their ability to provide accurate sideways control of vehicle body and axle motions.

The Watt linkage basically comprises two equal or nearly equal length transverse links, staggered in height, so that their outer ends are pivoted from each side of the car body structure, while their inner ends are pivoted to a third vertical link that is centrally pivoted on the axle (Figures 23.4a and 19.36). It will therefore be evident that any upward or downward movements of the axle, relative to the body, must cause the transverse links to describe opposing arcs of motion. Since this has the effect of increasing the lateral separation of their inner ends, the vertical link to which they are connected on the axle turns slightly in sympathy with their motions. Hence, its central pivot has no choice but to continue in a strictly vertical path, which therefore imposes the required sideways control for the axle relative to the body. Notable examples of Watt linkage applications to the rear suspension of high-performance cars are those used to control the De Dion and the live rear axles on Aston Martin and Bristol cars respectively.

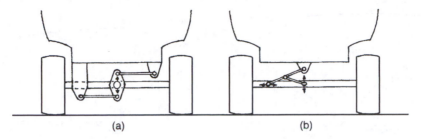

Figure 23.4 Rear axle lateral location linkages: (a) Watt (b) Scott-Russell

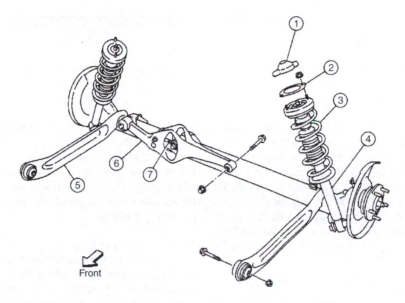

Figure 23.5 Arrangement of Scott-Russell axle linkage on Nissan QX car (*Nissan Motor GB*)

1 Shock absorber cap   5 Torsion beam
2 Shock absorber mounting seal   6 Lateral link
3 Coil spring   7 Control rod
4 Shock absorber

Perhaps an even subtler, but albeit rarer, approach to obtaining a parallel motion for the axle is that conferred by the Scott-Russell linkage. This essentially comprises a transverse link, one end of which pivots from near the center of the body structure, while the other end is allowed a limited lateral freedom where it connects close to the appropriate end of the axle beam. To complete the linkage a shorter link is pivoted both near to the center of the axle and towards the mid-point of the longer link (Figure 23.4b). The parallel action of this linkage is most easily visualized by imagining a pair of scissors resting edgewise on a table and the point of the lower blade held from moving. If then the scissors are opened and closed while still in contact with the table, the point of the upper blade will travel in a vertical path above the point of the lower one. In other words, the upper and lower points of the scissor blades correspond to the upper pivot of the longer link and the lower pivot of the shorter link respectively. A potential difficulty in applying this type of linkage in practice concerns the need to provide lateral freedom of movement where the longer link connects to the axle end, because a metal-to-metal sliding joint would clearly be unacceptable in modern design. This problem has been neatly solved in the sole application of the Scott-Russell linkage by Nissan

for their high-grade QX model (Figure 23.5), and involves the use of a special rubber bush at this point in the linkage. The bush is partially voided so as to offer the lowest resistance to lateral and the highest resistance to vertical deflections, hence the former characteristic accommodates the required lateral freedom where the longer link connects to the torsion beam type of dead rear axle.

For lateral location of rear axles, both the Watt and the Scott-Russell linkages offer a superior geometry as compared to the more commonly used and less complex Panhard rod linkage (Sections 19.4 and 20.4). Two shortcomings of the latter are scuffing and jacking effects, which result from the swinging movement of the transverse rod and its installed angularity. The former hinders the straight-line stability of the car (Figure 23.6a), while the latter imposes either a jacking-up or -down effect according to the direction of lateral or cornering force (Figure 23.6b) and therefore produces inconsistent handling characteristcs.

Independent front suspension (IFS)

A significant step towards reducing the earlier mentioned conflict between vehicle ride and handling was the proper

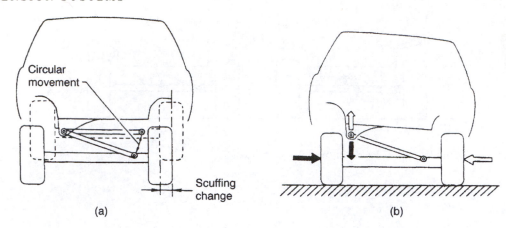

Figure 23.6 Shortcomings of Panhard rod linkage (*Nissan Motor GB*) (a) scuffing effects (b) jacking effects

application of independent front wheel suspension which followed chiefly from the research work done in the early 1930s by Maurice Olley, an ex-Rolls-Royce engineer then working for the Cadillac Motor Car Company in America. With an independent front wheel suspension system the steered wheels are located by entirely separate linkages rather than being united by a common axle beam. The use of some form of independent front suspension (IFS) has long been established practice for all conventional motor cars for the following reasons.

Improved ride comfort
The more precisely controlled location of the front wheels afforded by using an independent linkage system in conjunction with a rigid vehicle structure permits them to have a greater range of suspension movement. This in turn allows the use of much softer springs, which reduce the magnitude of impact loads transmitted by the front suspension to the car structure. Furthermore, the springs themselves are generally no longer required to play any part in locating the wheels, so that leaf springs can be discarded in favour of other types of springs possessing very little internal friction and thereby prevent harshness of ride.

Better roadholding
To some extent the springs can be made softer with an IFS system without reducing the roll resistance at the front end of the car, which otherwise could lead to over-steer on corners as a result of the rear suspension then offering too much resistance to roll. With beam axle suspension the lateral separation of its pair of semi-elliptic leaf springs is restricted to about one-half the wheel track dimension so as to leave sufficient clearance for the wheels to be steered. This narrow spring base compares unfavourably with that of an independent system where it is always equal to the wheel track irrespective of the lateral separation of the springs.

More accurate steering
An independent linkage is better able to ensure that each front wheel follows its prescribed geometrical path relative to the car structure and hence those parts of the steering linkage carried thereon. This can be difficult to achieve with a beam axle which is located solely by semi-elliptic leaf springs. For example, early attempts to increase their flexibility usually required the addition of an axle control linkage to prevent the axle from winding up on its springs and causing instability during braking.

Reducing steering joggles
As compared with a beam axle system, an independent linkage can be arranged to reduce by about one-half the amount either front wheel tilts inwards when passing over an obstacle. This serves to lessen the gyroscopic forces acting on the road wheels, because in tilting inwards they also attempt to steer themselves inwards and this produces an unwanted reaction or joggle at the steering wheel. Furthermore, both wheels of a beam axle system are tilted in unison when either of them passes over an obstacle, a state of affairs that at worst can lead to a wobble or shimmy of the steered wheels.

Increased passenger space
Last, but by no means least, the introduction of IFS made a direct contribution to improved passenger accommodation by having the power unit mounted further forward in the car, an arrangement which removed the need to provide front end clearance for the moving center portion of the axle beam. It thus became practicable to reposition the rear seats from above the rear axle to a lower level within the wheel base. Similarly, the rear-mounted fuel tank could then be moved forward, thereby increasing the capacity of the luggage boot.

The linkages used in modern IFS systems generally fall into two basic categories: the unequal transverse links, or wishbone system; and the transverse link and strut, or MacPherson system.

Unequal transverse links IFS

This system, pioneered by General Motors of America in the mid 1930s, is sometimes referred to as a wishbone system, because in plan view the front suspension links of their Buick models were originally of this form. With this type of IFS, each wheel is guided over obstacles by a short upper and

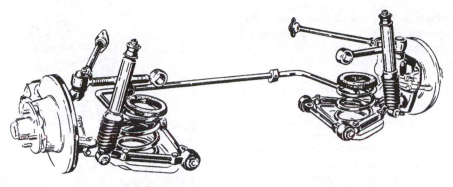

Figure 23.7 A traditional arrangement of unequal length transverse links IFS (*Alfa-Romeo*)

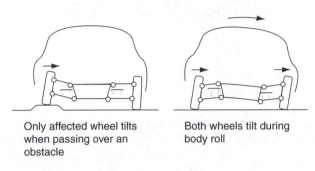

Only affected wheel tilts when passing over an obstacle

Both wheels tilt during body roll

Figure 23.8 The action of unequal transverse links IFS

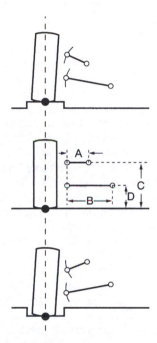

Figure 23.9 Constant wheel track IFS with unequal transverse links where: $\dfrac{A}{B} = \dfrac{D}{C}$

a long lower link, the inner ends of these links being pivoted from the car structure and their outer ends now ball jointed to a stub axle carrier or yoke (Figure 23.7).

As viewed from the front, the relative lengths and angles of these links are chosen so as to offer the following basic compromises:

1 To reduce by about one-half the tilting or camber change as either wheel rises or falls (Figure 23.8), which otherwise would be imposed on both wheels of a beam axle system.
2 To minimize changes in wheel track and thereby reduce any tyre scrub accompanying the rise and fall of the wheels, changes which would be greater if similar links were made equal in length and arranged to lie parallel. If the lengths of the links are made to vary inversely as their height above the ground, then the tyre contact will move up and down vertically without lateral scrub (Figure 23.9). This explains, of course, why the shorter link is always placed above the longer one. However, a small amount of scrub is usually accepted so as to raise the roll-centre above ground level (Section 23.3).
3 To allow the wheels to remain more nearly upright as the car rolls (Figure 23.8) as compared with the use of equal length and parallel links. In recent years, the advent of low-profile tyres of greater tread width has made it even more desirable to maintain the wheels upright so that the treads are kept flat on the road during cornering.
4 To realize certain installation advantages, namely that the short upper links protrude less into the engine compartment and the longer lower links reduce bowing of the suspension coil springs acting against them.

Various forms of springing medium may be used in conjunction with unequal transverse links IFS systems, including coil, torsion bar and rubber cone springs, all of which will be considered in more detail at a later stage, together with a description of shock damper action.

The spring and shock dampers may act against either the lower (Figure 23.7) or, less commonly, the upper link of each wheel linkage (Figure 23.32). In modern practice the transverse links almost invariably pivot from rubber bushings, with the two fold objective of isolating the car from road noise and eliminating the need for lubrication. The widespread use of rubber bushings for the suspension linkage pivots has largely obviated the static friction that could exist with the previously used metal-to-metal ones. These were particularly prone to excessive friction if any misalignment errors were present during initial build and, of course, in the absence of proper lubrication in service. Linkage friction was earlier mentioned in connection with ride considerations and its presence is to be avoided. It could result in the unsprung and sprung masses of the car momentarily locking together, so that the suspension spring ceased to function as such and the car rode solely on the much stiffer spring effect of the tyres. This caused ride harshness that was particularly noticeable when the wheels were running over minor road surface irregularities, since the impact forces were then insufficient to overcome the static friction or 'stiction' in the suspension linkage. The use of rubber bushings also allows for a limited amount of fore-and-aft movement of the wheels which, in absorbing any structural vibration of the tyres, reduces low-speed harshness in the ride. This feature of suspension design therefore provides for horizontal flexibility in the system, which again was referred to earlier in connection with ride considerations and is known to the suspension engineer as 'horizontal compliance'.

Increasing use is now also being made of aluminium alloy to fabricate the transverse links or control arms and, in some designs, its application is also extended to include the steering yokes. The object is further to minimize unsprung weight of the suspension system and hence reduce shock damping requirements, especially now that car bodies are being made lighter to save fuel, otherwise ride comfort could deteriorate.

For reasons mentioned later, many manufacturers gradually abandoned the unequal transverse links IFS in favour of the transverse link and strut type, but since the late 1980s interest in the former system has been revived to exploit its inherently lower friction and more amenable geometry, especially by Japanese designers. One problem that arises in using this type of IFS with a modern transverse engine layout is to find sufficient room between the wall of the engine bay and the wheel to locate an upper link of the required length and with appropriate angle of inclination, which not only maintains the desired camber change characteristics but also preserves an optimum steering offset.

The basic solution to this problem that has been adopted by both Honda and Nissan is to raise the upper link to a much higher level, so that it lies above and partly over the wheel. This involves using either a longer and suitably dished upper extension for a one-piece hub carrier to intercept the outer pivot of the upper link (Figure 23.10), or an additional link between the hub carrier and its upper pivot that enables the steering swivel axis to be separately determined (Figure 23.11). In both arrangements the high-mounted upper link is pivoted at a leading angle, so that as the wheel rises towards its bump position the top end of the hub carrier moves not only inwards but also slightly rearwards. The latter has the effect of compensating for rearward movement at the bottom end of the hub carrier, which accompanies longitudinal compliance of the suspension and braking force, and therefore preserves optimum castor angle for the steered wheels.

It is instructive to consider the manner in which the Nissan multilink version of an unequal transverse links IFS evolved. In Figure 23.12a the difficulty of accommodating the upper link in the conventional position between the wall of the engine bay and the wheel will be evident, and the necessity for raising it above and partly over the wheel made clear. However, it will be seen in Figure 23.12b that repositioning the upper link in this way has considerably reduced the king-pin angle and

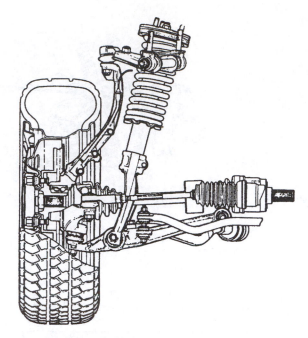

Figure 23.10 Unequal length transverse links with high-mounted upper one, rear view (*Honda*)

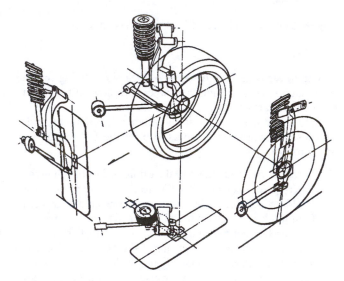

Figure 23.11 Transverse multilink IFS (*Nissan Europe NV*)

produced too great a steering offset. This effect could be avoided by shortening the upper link, but only at the expense of introducing excessive camber changes as the wheels rise and fall. A better compromise has therefore been sought as shown in Figure 23.12c, where the required length of upper link with appropriate angle of inclination has been retained, but instead of a one-piece hub carrier connecting directly to the outer ball-joints of the upper and lower links to establish the king-pin inclination, this is determined independently by introducing an additional vertical link. It is mounted between the outer pivot bushing of the upper arm and the steerable hub carrier, which is still ball-jointed to the outer pivot of the lower link. The lower end of the additional link incorporates a suitably spaced pair of ball bearings, into which is inserted the upper pivot of

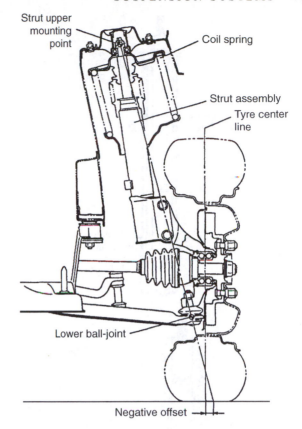

Figure 23.13   Arrangement of a modern transverse link and strut IFS (*Mitsubishi*)

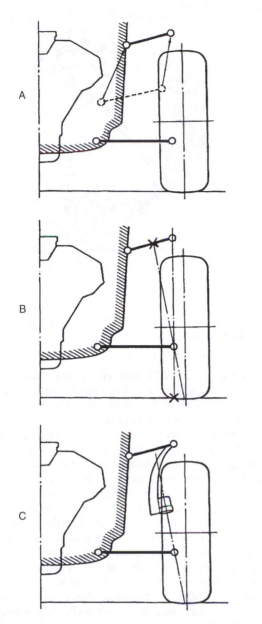

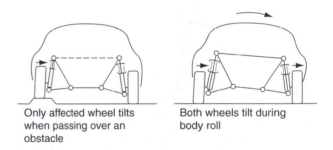

| Only affected wheel tilts when passing over an obstacle | Both wheels tilt during body roll |

Figure 23.14   The action of transverse link and strut IFS

Figure 23.12   Evolution of transverse multilink IFS (*Nissan Europe NV*)

the steerable hub carrier, so that the arrangement may be regarded as a partial king-pin (Figure 23.11).

Transverse link and strut IFS

Since its introduction into Britain in the early 1950s, the transverse link and strut system of IFS (Figure 23.13) has come into increasingly wider use. The system is also known after its originator E.S. MacPherson, sometime chief engineer of the Ford Company in America, although for many years the MacPherson system was not used in that country. With this type of IFS each wheel is guided over obstacles by a strut comprising a telescopic shock damper connected between the ball-jointed outer pivot of the transverse link and an upper flexible mounting on the body structure (Figure 23.14).

In its original and still most widely used form, the suspension coil springs encircle the telescopic shock damper struts. At the upper end of the strut the shock damper piston rod swings from the body structure via a conical rubber mounting, which has always required careful design to confer adequate noise suppression. Conventionally, this mounting incorporates a separate thrust bearing for the suspension spring to seat against (Figure 23.13), so that steering movements of the strut and its hub carrier can easily be performed. Less commonly these steering movements may be accommodated simply by twisting of the rubber mounting itself, which then also assists in returning the wheels to the straight-ahead position after cornering.

Since the guiding surfaces of the strut where it slides over the piston rod assembly must be capable of withstanding high

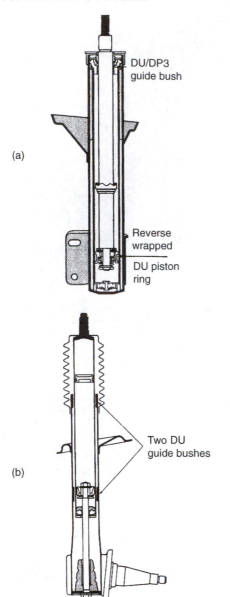

Figure 23.15 Reducing sliding friction in MacPherson struts by using Glacier DU materials: (a) conventional strut (b) inverted strut (*T&N PLC*)

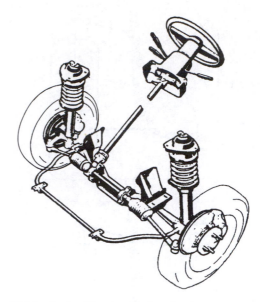

Figure 23.16 Anti-roll bar installed so that it acts as part of the front suspension linkage (*Ford Escort*)

sideways loading, arising from the normal action of the suspension, there is clearly a necessity to reduce friction effects to a minimum. In early designs the use of a sintered iron upper bush was found to produce excessive static friction, which resulted in the strut transmitting vibration to the car structure at slow speeds over rough road surfaces. To overcome this problem Glacier DU guide bushes have been widely used for this application and also in reverse wrapped form in the lower piston ring assembly (Figure 23.15a). Glacier DU is a composite material consisting of a steel backing to which is bonded a porous bronze sinter, this being impregnated and overlaid with a mixture of PTFE and lead. This type of bush may also be used as the upper and lower guide bearings in the inverted form of MacPherson strut IFS mentioned later (Figure 23.15b).

As in the case of some modern unequal transverse links IFS systems, the lower links often take the form of slim track control arms, which receive their fore-and-aft location from either the splayed ends of an anti-roll bar (as in the original Ford application: Figure 23.16) or separate diagonal links. Rubber bushes are similarly used at all pivot points in the linkage. In a few recent designs the lower links assume the form of crescent or sickle arms, so that the positioning and relative flexibilities of their front and rear bushes can be such as to ensure both precise location and optimum compliance for the wheels without adversely affecting the steering. As in the case of unequal transverse links IFS, aluminium alloy is likewise being used in the construction of transverse link and strut IFS and for the same reasons mentioned earlier.

Later developments of the basic transverse link and strut IFS include the notable inverted-strut design by Alfa-Romeo, where the stub axle carrier tube slides on the outer casing of the inverted shock damper to reduce bending and friction effects in the suspension (Figure 23.17a). These effects may also be reduced by angling the coil spring relative to the axis of the strut, so as to counteract the normal tendency for the wheel to tilt inwards during straight-ahead running (Figure 23.17b). In 1977 the Ford Company in America introduced a modified MacPherson system in which the suspension coil springs no longer encircle the struts, but are arranged to act against wishbone-type transverse lower links (Figure 23.17c). The advantages claimed for this modified layout, which was later used by Mercedes-Benz are that it gives more installation space for the engine, it is easier to service because no special spring compressor is required if the shock damper needs to be removed, and it allows for more effective suspension mountings.

Horizontal compliance and steering effects

Since the provision of horizontal compliance in both unequal transverse links and link and strut IFS systems can influence the directional stability of a car, the manner in which this can

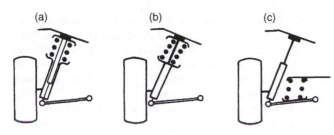

Figure 23.17   Later developments of transverse link and strut IFS: (a) inverted strut (b) angled coil spring (c) modified MacPherson

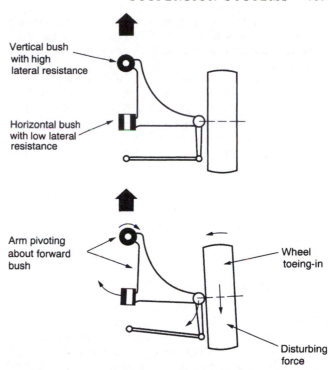

Figure 23.18   Horizontal compliance with stabilizing toe-in effect

occur will now be considered. In the former system, the limited amount of fore-and-aft cushioned movement at hub level for the front wheels is conventionally derived from the rubber bushed inner pivots for the lower link, so that in plan view it can skew backwards (and forwards) about a point contained in the transverse plane of the steering yokes. In other words, when the lower link skews backwards in response to a disturbing force and therefore its outer ball-joint moves back in relation to the corresponding ball-joint of the non-complying upper link, the steering yoke will swing backwards and since the wheel hub is mounted from the yoke the wheel will retreat rearwards. Although in plan view the steering track rod typically shadows the lower transverse link of the suspension (Section 24.1), so that any skewing of the link is complemented by a parallelogram movement of the track rod and should not affect the steering, other factors still have to be taken into account. In particular, when the steering yoke swings backwards as the lower link allows compliance, not only will it temporarily reduce the steering castor angle (Figure 24.9b), but also a rear pointing side steering arm attached to the yoke will tilt upwards and lift the outer ball-joint of the steering track rod, the resulting angular movement of the latter then causing the wheel to toe-out. Both effects, the reduction in castor angle and the toe-out of the wheel, are undesirable in terms of maintaining directional stability and especially during braking on surfaces that offer more grip to the wheels on one side of the car than on the other.

A later development to overcome this destabilizing tendency towards compliance toe-out has been to move forward the effective pivot point about which the lower link skews. This results in a more pronounced inwards movement of the lower link outer ball-joint, which provides a greater incentive for the wheel to self-steer about its track rod connection on the side steering arm and develop a beneficial toe-in to prevent veering off course (Figure 23.18). To obtain this effect in practice, the lower link front and rear bushings may be mounted vertically and horizontally, thereby conferring high and low lateral resistance respectively to compliance forces acting at its pivot points. With this arrangement, the torsional flexibility of the front bushing for normal suspension movements occurs as a tilting deflection of its rubber element. As inferred earlier, the unwanted effects of horizontal compliance are minimized in some more recent designs of unequal transverse links IFS systems with a high-mounted upper link (Figures 23.10 and 23.11), because there is inherently less angular movement of the much longer steering yoke for the same degree of skewing at the lower link.

Unwanted steering effects can also be minimized where a certain degree of horizontal compliance is conferred by mounting an unequal transverse links IFS system on a sub-frame, which is then flexibly mounted through rubber bushings to the main body structure (Figure 30.11), rather than concentrating the flexibility in the suspension linkage pivots themselves.

Horizontal compliance does, of course, play an equally important role in the design of the transverse link and strut or MacPherson IFS system as earlier indicated. However, similar to the unequal transverse links system with a high-mounted upper arm and longer steering yoke, the long length of strut inherently tends to minimize unwanted castor angle and toe changes in the steering, but additional measures may still need to be incorporated in the pivoting arrangements for the transverse link (Figure 23.18). A total compliance of typically 24 mm (0.95 in) is divided equally between the fore and aft movements of the wheels.

Comparing the two types of IFS

It is of interest to compare the installation and operating characteristics of the transverse link and strut and the unequal transverse links system:

1   Whereas the unequal transverse links system was originally developed for installation on a separate chassis frame, the transverse link and strut system was intended from the outset to be used with integral body construction. Advantage was therefore taken to provide widely spaced attachment points for the strut and link, thereby reducing load concentration on the front end body structure which can then be made lighter in weight.

2 It has been found expedient to incorporate the transverse link and strut system in many FWD cars featuring transversely mounted engine and transaxle units, since the absence of upper transverse links contributes to the additional installation space required for these units.

3 An inherent disadvantage of the transverse link and strut system is that frictional forces arising from the guiding function of the strut are much greater than those occurring in purely link systems. This is because the axis of the strut often has an appreciable offset from the centre plane of the road wheel, which also accounts for the steering being more sensitive to any wheel imbalance.

4 With rise and fall of the wheels there is generally less tilting or camber change with a transverse link and strut system, but changes in wheels track tend to be greater, especially as the wheel drop towards the rebound position and the angularity of the initially drooping lower links increases further.

Independent rear suspension (IRS)

Whereas a few car manufacturers continue to mount the rear wheels on a sprung live axle, many others have long since adopted various forms of independent rear suspension (usually abbreviated to IRS). It first became widely used by German and Middle European manufacturers during the 1930s, notably in the designs of Drs Ferdinand Porsche and Hans Ledwinka, but this development did not really gain popularity elsewhere until some thirty years later.

The chief benefits to be expected from using a modern IRS system are generally concerned with the inter-related qualities of ride, handling and, in the case of rear-wheel-drive cars, traction. Ride comfort in particular should benefit from the reduction by about one-half in the un-sprung mass of the suspension mechanism, resulting from the final drive assembly being mounted on the vehicle structure. Also an increase in useful space within the body rear portion is implicit with IRS, since the propeller shaft and final drive assembly do not rise and fall in sympathy with the suspension movements of the rear wheels.

The improvement in traction to be expected with independently sprung and driven rear wheels deserves a few words of explanation, Mention was made in Section 19.4 of the antics performed by the live axle of a Hotchkiss drive system during acceleration and braking. Taking matters a little further, we find that during acceleration the axle casing rocks on its

springs not only in opposition to crown wheel torque, but also to a lesser extent (as related to the final drive gear ratio) in sympathy with pinion torque. In other words, the tendency during acceleration is to press one rear wheel harder against the ground and to lift the other one off it. This effect, combined with the one mentioned previously, can cause the axle and wheels to writhe about a conical path and generate an unstable handling condition known as axle tramp. Although this state of affairs may to some extent be alleviated by additional means of axle control, as earlier described, such misbehaviour is absent from IRS systems. The reason for this is, of course, that the final drive is divorced from the road wheel mountings and is attached to the vehicle structure, the drive to the wheels being transmitted through universally jointed drive shafts.

Comparing different types of IRS

Apart from the need to identify the various systems by their basic geometric layout, a few brief notes on their general characteristics may prove useful to explain their current popularity or otherwise. The systems may conveniently be classified into four types, as in the following sections.

Swing axle: pure and diagonal

A pure swing axle system was once widely favoured by Continental manufacturers (Figure 23:19a), especially for rear-engined cars where its use proved mechanically expedient. Although body roll tends to be less with this type of IRS, hard cornering can produce outward lean of the outer wheel and a smaller inward lean of the inner wheel, the result being that the rear end of the car is lifted. The sudden onset of this jacking effect can lead to an unstable oversteering condition. For normal ride motions of the car there is pronounced tilting of the wheels, with accompanying changes in wheel track as they rise and fall. These undesirable effects can, to some extent, be reduced by using a diagonal swing axle (Figure 23.19b). This system involves less tilting of the wheels and also causes them to steer inwards or toe-in slightly as they rise and fall, which counteracts the oversteering tendency.

Trailing arm: pure and semi

The trailing arm system has long been favoured for the relatively lightly laden rear wheels of front-wheel-drive cars. In its pure form the arms pivot about an axis that lies parallel with

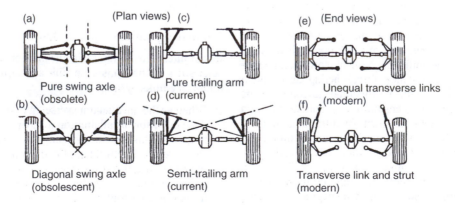

Figure 23.19 Identification of IRS systems by defining their pivot axes

the ground and normal to the centreline of the car (Figure 23.19c). Although the wheels can rise and fall vertically during normal ride motions of the car, they are necessarily tilted to the same angle as the body with cornering roll, which tends to be greater with this type of IRS. This leaning away from the curve of the wheels has the disadvantage of reducing their cornering power. In the case of rear-wheel-driven cars, a departure is usually made from the pure system to one where the pivot axis of each arm is moderately angled in plan view and known as the semi-trailing arm (Figures 23.19d and 24.19). The purpose of this modified geometry is to maintain the wheels more nearly upright during cornering and also to cause them to steer inwards or toe-in slightly as they rise and fall, thereby contributing to a stable understeering condition.

Unequal transverse links

Until recently this system of IRS was comparatively little used, because of its potentially greater intrusion into valuable rear body space (Figure 23.19e). On the credit side, a better compromise with respect to suspension geometry can fairly readily be obtained, a particular advantage being that the heavily loaded outer wheel can be maintained more nearly upright in the presence of body roll during cornering. By allowing each drive shaft to perform a dual role and serve also as the upper link, as so ably demonstrated by the Jaguar Company, the unequal transverse links system of IRS may be simplified, its un-sprung mass lessened and its vertical space requirements reduced.

However, the more recent revival of interest in this type of IRS has produced somewhat more complicated versions, which are intended to exercise better toe control over the rear wheels when they are subjected to driving and braking forces, while still allowing them the required amount of horizontal compliance. Reduced to simple terms, these systems typically seek to replace the single upper and lower transverse links with two pairs of separately pivoting links that are angled together in plan view, which are therefore well disposed to withstand both lateral and horizontal loads. Furthermore, the apices of the upper and lower pairs of angled links meet at points contained within the centre plane of the wheel, so that it is no longer offset from the outer pivots of the links and therefore has no tendency to twist about them and impose toe changes during driving and braking forces (Figure 23.20a). This is difficult to achieve with conventional two-link versions, because the brake disc gets in the way of where the pivots ought to be (Figure 23.20b). To complete the system a fifth pivoting link is added between the upper and lower pairs and connects to either a forward-or a rearward-facing arm on the hub carrier. Its purpose is to act in the manner of a steering track rod (and indeed it does this in the true sense of the description in the case of certain four-wheel steering systems, as we shall see later), so that optimum alignment can be enforced upon the rear wheels under practically all conditions of loading and deflection, such as attained in the Nissan 'multilink' IRS system (Figure 23.21).

Transverse link and strut

Also sometimes referred to as a Chapman strut, so named after the Lotus car designer who first applied the MacPherson

transverse link and strut principle to rear wheel suspension (Figure 23.19f), this type of IRS has since become widely used for the non-driven rear wheels of front-wheel-drive cars. For the lateral and longitudinal location of the non-steered rear wheels, the transverse link may pivot about an axis parallel to the centreline of the car. The link may comprise either a substantial wishbone arm (Figure 24.20), or a track control arm that is located fore-and-aft by a trailing link (Figure 23.34), or a similarly located transverse parallelogram linkage or, as it is sometimes termed, 'tri-link' system that better maintains wheel alignment with optimum compliance as developed by Mazda. Alternatively, the transverse link may be skewed in the manner of a diagonal swing axle, again with the purpose of correcting for any oversteering tendency that may be present, because the geometry of the transverse link and strut is not quite as good as the unequal transverse links system of IRS.

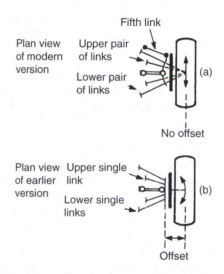

Figure 23.20 Unequal length transverse links IRS: (a) five-link (b) two-link

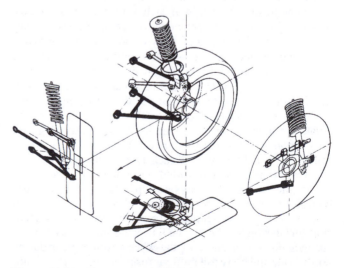

Figure 23.21 Transverse multilink IRS (*Nissan Europe NV*)

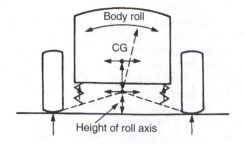

Figure 23.22   Simplified representation of suspension geometry

23.3   BASIC HANDLING CONSIDERATIONS

Every vehicle has two roll centres, which are imaginary points determined by the particular geometrical features of the front and rear suspension linkages. If a line is drawn through these points it establishes the roll axis of the vehicle, because it represents an imaginary hinge between the sprung mass of the vehicle and the unsprung mass of its suspension mechanism and wheels.

In simple terms the significance of the roll axis is twofold (Figure 23.22). Firstly, the amount by which the vehicle will roll when cornering is influenced by the vertical distance or moment arm between the roll axis and the vehicle centre of gravity. The lower the roll axis and the higher the centre of gravity, the greater will be the angle of roll assumed by the sprung mass of the vehicle under a given centrifugal or side force. Secondly, the increase in vertical loading on the outer pair of tyres and the corresponding decrease on the inner pair when the sprung mass of the vehicle rolls on its springs during cornering, partly depends upon the height of the roll axis above ground level. The higher the roll axis the greater will be the load transfer for a given centrifugal or side force. It is therefore possible for a designer to establish an optimum balance between the cornering slip angles of the front and rear tyres by making relative changes to the roll centre heights, although this effect is more conveniently achieved by altering the thickness and hence torsional stiffness of anti-roll bars (Figure 23.53).

Geometrically, the height of each roll centre may be determined by joining the instantaneous centres of the wheel locating linkages to the centres of the tyre contact patches and then producing, as necessary, these lines until they intersect. The point of intersection thus locates the instantaneous roll centre of the linkage (Figure 23.23). It is instantaneous because it can move with respect to the ground and the sprung mass of the vehicle with suspension deflection. It will be noticed that to obtain a roll centre above ground level imposes a swing axle effect upon the wheels as they are deflected upwards or downwards, so that the centre of each tyre contact patch moves sideways to produce a so-called scrub angle. In practice, of course, the rolling tyre develops a slip angle rather than scrubbing sideways over the road surface. Even so, the geometrical characteristics of very high and very low roll centre suspension linkages are to be avoided, because unacceptably large changes in camber angle of the wheels would occur during normal ride and body roll motions respectively. Hence, a compromise as always is sought and the roll centre heights of the

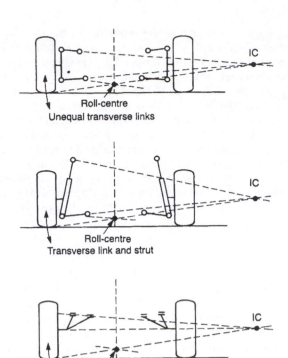

Figure 23.23   Determination of roll centres for independent suspension linkages

front and rear suspension linkages generally lie somewhere between the ground level and the height of the wheel hubs.

23.4   TYPES OF SUSPENSION SPRING

Basic requirements

Springs that utilize the elastic properties of metal, rubber and air are variously employed in motor vehicle suspension systems, the actual choice made being determined largely by versatility in application and best economy of material in terms of energy storage per unit volume.

Semi-elliptic leaf springs

Following the invention of the leaf spring in full elliptic form as long ago as 1804 by Obediah Elliot, the use of leaf springs first on horse-drawn and later on railway rolling stock considerably predates the era of the motor vehicle. Perhaps not surprisingly, the designers of early motor vehicles soon realized that the great practical advantage of this type of spring is that its capacity can easily be changed by varying the number of leaves, and any degree of resilience can be obtained by varying the thickness and width of the leaves. It is therefore interesting to find that leaf springs continue to be used extensively in both the front and rear suspension systems of commercial vehicles and some 4×4 vehicles, but they are now rarely used in the suspension systems of passenger cars.

Although originally used in full elliptic form, it has long been established motor vehicle practice to use the leaf spring in semi-elliptic form, in the interests of preventing excessive sideways movement of the vehicle relative to its axles (Figure 23.29).

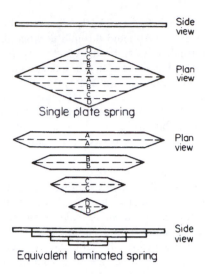

Figure 23.24   Basic development of laminated or leaf spring from double-cantilever single-plate spring

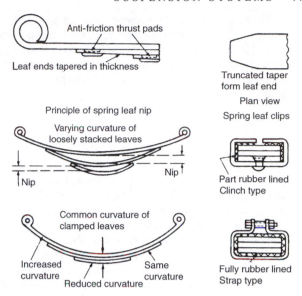

Figure 23.25   Constructional features of the multileaf spring

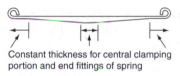

Figure 23.26   Profile of single-leaf spring

The traditional leaf spring may be regarded as being the laminated equivalent of a simple plate spring, which if made constant in thickness would have to be diamond shaped in plan view, so that the maximum bending moment on the spring coincided with its greatest dimension of width for uniform stress (Figure 23.24). Clearly, the overall width of a simple plate spring of this type would pose considerable installation problems in a motor vehicle. It is therefore expedient, in effect, to cut the diamond-shaped plate into strips or leaves of diminishing lengths which are then stacked one upon another to form a much narrower laminated spring. However, modern leaf springs sometimes assume the form of a modified simple plate spring and generally fall into three categories: multileaf, single leaf and tapered leaf.

Multileaf spring
At one time, the conventional multileaf spring was built up from a large number of a narrow, thin leaves, which in rubbing against each other with flexing of the spring exerted an appreciable friction damping effect on suspension movements of the wheels. For other than commercial vehicles, it is no longer considered desirable that the suspension springs should also act as friction shock dampers, so that leaf springs are now designed with fewer leaves of relatively greater width and thickness.

Furthermore, the leaves are separated at their ends by anti-friction thrust pads such as recessed plastics buttons (Figure 23.25). A series of retaining clips positioned along the length of the spring has the twofold purpose of preventing the leaves from separating during rebound travel of the spring and ensuring that the sideways loads imposed on the spring are not borne solely by the uppermost or master leaf.

Another established feature of multileaf spring construction is that of providing nip between the leaves by forming the leaves below the master leaf with successively reducing radii of curvature (Figure 23.25). This gives a beneficial stress reduction for the master leaf, because when the leaves are clamped together it is subject to a bending preload opposite in direction to that caused by the vehicle load.

Helper leaf spring
One of the problems encountered with the suspension system of a heavy vehicle is the wide variation between the unladen and laden running conditions, which often demands the addition of a helper leaf spring. This form of leaf spring is shorter than the main spring and clamped directly above it, but makes no permanent contact with the vehicle chassis frame. The main spring carries the load in the normal manner when the vehicle is moderately laden, but as the load increases to about 75 per cent of maximum the otherwise free ends of the helper spring make contact with their own frame-mounted slipper brackets. Once this contact occurs the load is shared by both main and helper springs, so that their combined spring rate offers considerably greater resistance to further deflection.

Single-leaf spring
In more recent years advances made in spring manufacturing technology have resulted in the limited use of single-leaf springs (Figure 23.26). This type of construction is similar in principle to the previously mentioned simple plate spring, but it avoids the excessive width disadvantage by having its single leaf varying in both width and thickness. Advantages generally claimed for this simplified form of construction are the useful reduction in unsprung weight and the elimination of interleaf friction, both of which contribute to improved ride comfort of the vehicle.

In the late 1980s the GKN company introduced a further development of the single-leaf spring by producing it in a composite material of the glass-fibre-reinforced polyurethane resin type, the spring eyes being fabricated from pressed steel and cradled by the extended ends of the leaf. Apart from a significant saving in weight, the other advantages claimed for this form of construction are good resistance to fatigue, wear, corrosion and noise transmission.

Tapered-leaf spring

This type of spring represents a compromise between the conventional multileaf and single-leaf springs, and it can be described in effect as a stacked single-leaf spring (Figure 23.27). It comprises several full-length leaves of constant width but of tapering thickness towards their ends. At their thick middle portion the leaves are separated by interleaf liners and contact one another only towards their ends. Since the tapered-leaf spring offers advantages similar to those of the single-leaf spring, but with a much greater load-carrying capacity, it has been used to advantage in some commercial vehicles where it gives better cushioning for both cargo and driver.

Leaf spring mounting

Whichever type of leaf spring is used, its mounting must locate it positively with respect to both the axle and the vehicle structure, as follows.

Attachment to axle

The centre portion of the spring is attached to a seating or saddle formed towards either end of the axle. For installation purposes, the leaves are generally held together by a centre bolt the head of which further serves as a dowel for locating the spring relative to its seating (Figure 23.28). To avoid the stress-raising effect of a hole through the leaves, the usual centre bolt may be dispensed with in some heavy-duty applications and replaced by dimples or cups pressed into successive leaves, with a corresponding depression being provided

in the axle seating (Figure 23.28). Final attachment of the spring to the axle is effected through the medium of either U-bolt or normal bolts and nuts, together with a clamping plate if required. For the driven rear axles of cars it has long been established practice for the spring-to-axle clamping arrangements to be rubber lined (Figure 23.28). This not only reduces the transmission of road noise through the spring mounting, but also minimizes stress concentration on the leaves where they emerge from the clamp.

Attachment to vehicle structure

The flexing of a semi-elliptic leaf spring is such that its curvature, and hence its effective length, change constantly with suspension movements of the axle. To locate the axle positively, and at the same time accommodate these lengthwise movements of the spring, the latter is provided with fixed and free end pivots (Figure 23.29). The fixed end of the master leaf, usually the front end, is provided with a rolled eye embracing a suitable pivot connection on the vehicle structure.

For cars the pivot, or spring pin, clamps the inner sleeve of a rubber bushing pressed in to the spring eye, while for commercial vehicles lubricated metal bushings are generally required. The free end of the master leaf is provided with either a similar rolled eye or a plain end. In the former case, the rolled eye and its bushing connect by means of a shackle pin to a shackle link, which in turn is hinged from the vehicle structure (Figure 24.30a).

For commercial vehicles this usually requires a separate shackle bracket fixed to the chassis frame (Figure 24.30b). The shackle link bushings are complementary to those used at the fixed pivot of the spring. A plain end mounting for the master leaf is used in conjunction with a slipper-type bracket,

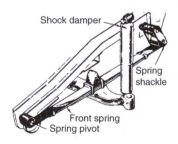

Figure 23.29   Attachment of leaf spring to vehicle structure

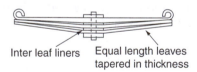

Figure 23.27   Profile of tapered-leaf spring

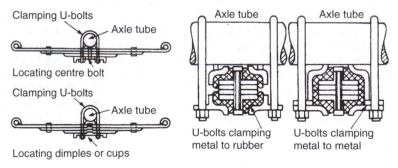

Figure 23.28   Attachment of leaf spring to rear axle

an arrangement now confined to commercial vehicles. Flexing of the spring thus produces metal-to-metal sliding contact of its plain end against the curved underside of the slipper bracket. It also confers a variable-rate effect on the spring installation, because when the vehicle is lightly laden the end of the spring plate contacts the outer end of the slipper bracket, but as the load increases and the curvature of the spring alters the contact moves towards the inner end of the bracket. As a result there is an effective reduction in the length of the spring, which therefore increases its rate for greater resistance against further deflection.

Coil springs

Helical compression or coil springs are probably now the most widely used type of suspension spring for motor cars. In comparison with a leaf spring, the coil spring can store more than twice the amount of energy per unit volume of material, and it possesses minimal internal friction. However, the vertical space requirements of a coil spring are greater and its inherently low resistance to buckling is such that it can function as a spring medium only when used in conjunction with a separate wheel locating linkage.

A coil spring is produced from a length of round wire wound into the form of a helix, and is installed in a manner that loads it as nearly as practicable in the axial direction. When the spring is compressed, a twisting moment is imposed on the wire as a result of the axial load acting at a distance equal to the mean radius of the coil turns. It may be provided with either two flat, one flat and one tangent tail, or two pigtail ends (Figure 23.31). For the second arrangement, the flat end remains stationary relative to the car structure while the tangent tail end seats against a helical-shaped abutment on the moving suspension linkage. This prevents rotational creep of the spring in service, which otherwise can affect the installed rate of the spring. The third arrangement of two pigtail ends is generally used where the ends of the spring are positively clamped to their stationary and moving abutments. Since this form of spring end also permits a reduction in the fully compressed or solid length of the spring, it finds application where the vertical space to accommodate the suspension spring is strictly limited. As in the case of leaf spring installations, rubber isolators are usually incorporated in the coil spring mountings to reduce the transmission of high-frequency noise through the suspension system (Figure 30.11a and b).

Several examples of conventional application of the coil spring to IFS systems have earlier been described (Figures 23.7 and 23.13). A less common application is where the spring is arranged to act against the upper link of an unequal transverse links system (Figure 23.32). An advantage conferred by this overhead mounting of the coil spring is that the suspension loading is distributed over an appreciably larger area of the body front structure, so that the need for a heavy front cross-member is avoided. Another advantage is that it allows the lower link more freedom for longitudinal compliance.

For rear-wheel drive cars the semi-trailing arm system of IRS is generally associated with coil springs (Figures 23.33b, 24.19 and 30.11b) and was first used by Lancia on their Aurelia model of 1951. Coil springs are also usually employed with the increasingly popular transverse link and strut system of IRS,

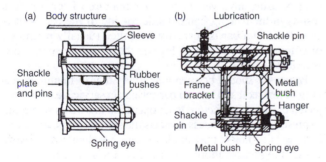

Figure 23.30 Typical constructions of spring shackles

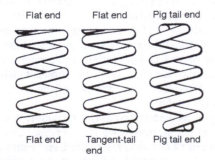

Figure 23.31 Types of suspension coil spring

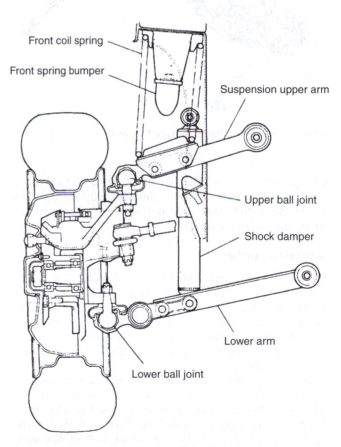

Figure 23.32 Unequal transverse links IFS with overhead coil spring (*Suzuki*)

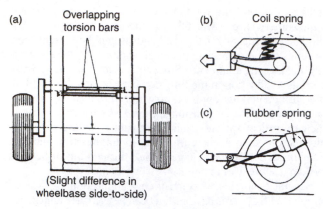

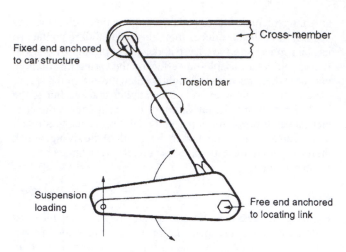

Figure 23.33 IRS with different types of spring: (a) pure trailing arm with torsion bar springs (plan view) (b) semi-trailing arm with coil spring (side view) (c) pure trailing arm with rubber spring (side view)

Figure 23.35 Torsion bar springing in principle

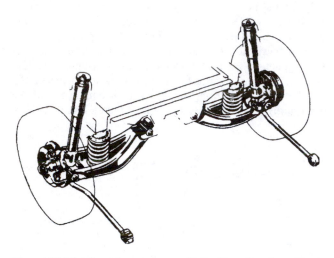

Figure 23.34 A modern strut-type IRS with coil springs (*Ford*)

although instead of encircling the strut they may act against the lower links to reduce intrusion into rear body space (Figure 23.34). This system also lends itself neatly to a transverse leaf spring installation (Figure 24.20).

Torsion bar springs

This type of spring has not been so widely adopted as the helical or coil spring for motor vehicle suspension systems, although it is used extensively in a secondary role as an anti-roll bar (Figure 23.53). In common with the coil spring it possesses minimal internal friction and can only function as a spring medium when used in conjunction with a separate wheel locating linkage.

Its method of application is such that one end of the bar is clamped to the car structure to prevent rotation (except for adjustment purposes) and the other end is connected to a lever arm that receives suspension loading (Figure 23.35). With rising of the road wheels a twisting moment is thus imposed on the bar as it performs its function of energy storage. In this context, the torsion bar offers in the region of a fourfold increase in

energy storage per unit volume of material over that of a leaf spring. This further gain in energy storage is mainly attributable to there being less inactive or dead material in the bar end connections, as compared with that in the end turns of a coil spring.

For anchorage purposes the torsion bar is provided with end connections of either serrated or hexagonal form. An advantage of the former is that it requires the least diameter bar ends, while the latter has the production merit of requiring no further machining after grinding of the bar stock and upsetting the ends.

To reduce stress concentration in the critical regions of the end connections, the main body of the bar merges very gradually into the increased diameter ends. Another stress reducing measure involved in the manufacture of torsion bars results in their being handed side-to-side of the car, so it is essential that a bar is correctly identified (usually by means of colour-coded markings) for whichever side suspension it is intended.

Perhaps it should be mentioned here that to minimize settling of leaf, coil and torsion bar springs in service, each is subjected during manufacture to an operation known as scragging or presetting. This process subjects the spring to repeated overloading of sufficient magnitude to cause yielding in any highly stressed regions. A more uniform distribution of stress is thus obtained upon removal of the load, so that settling is less likely to occur and also the fatigue life of the spring is improved.

For convenience of setting the ride height or standing height of the car, a manual adjustment for winding the bar is provided at either the fixed or, less commonly, the free end (Figure 23.36). It may take the form of a reaction lever bearing against a trunnion-mounted jacking screw, or alternatively against a lockable eccentric cam. Yet another method is to have a different number of serrations at each end of the bar to provide a vernier setting facility.

In IFS systems the torsion bar is used mainly in conjunction with unequal length transverse links (Figure 23.37), where it has the merit of contributing to the basic simplicity of the layout. It is also compatible with front-wheel drive because this type of spring does not get in the way of the drive shaft.

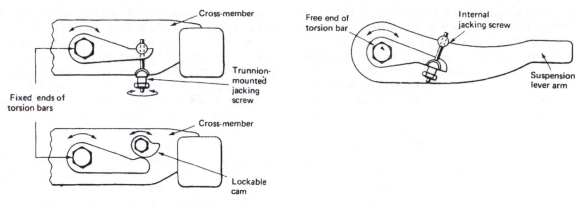

Figure 23.36 Types of height adjuster for torsion bar suspension

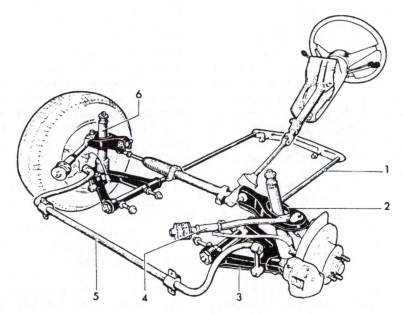

Figure 23.37 Torsion bar springing in practice (*Alfa-Romeo*)

1 torsion bar	4 castor rod
2 upper wishbone	5 anti-roll bar
3 lower wishbone	6 shock damper

The pure trailing arm system of IRS associated with front-wheel drive lends itself fairly readily to transversely mounted torsion bars (Figure 23.33a).

A torsion bar is generally claimed to be less bulky than its near relation the coil spring (the material in each is stressed principally in torsion, and so a torsion bar may conveniently be regarded as an unwound coil spring), although in most installations space must be found under the body floor to accommodate its length. With longitudinal mounting of the torsion bars there is the advantage that the suspension loads are transferred to the main part of the body structure, which is better able to contain them, but torsion bars tend to be noise transmitters so they are usually anchored to a cross-member that is rubber mounted to the body structure. Another advantage of the torsion bar spring is that it contributes to a reduction in unsprung mass of the suspension system (Section 22.4). Where a torsion bar spring is conventionally connected

to the lower link of a transverse links IFS system, the provision for horizontal compliance necessarily has to be made in a splayed upper link (Figure 23.37). The torsion bar type of spring is generally regarded as being more expensive both to manufacture and to build into the car.

Rubber springs

The idea of utilizing the elastic properties of rubber as a spring medium for wheel suspension has long been an attractive, but not always a successful, venture for designers in various fields. In Britain it has ranged from the use of rubber block springs in 1853 by the Bristol and Exeter Railway on their legendary 4–2–4 locomotive tank engines with 9 ft diameter driving wheels, to the use of stacked rubber discs on the undercarriage legs of the famous World War II Mosquito high-speed bomber aircraft. Similarly, the use of rubber has found relatively limited application in the field of passenger car suspension systems, although there have been a few notable exceptions, the best known of these being the Moulton cone spring applied originally to the then BMC Mini in the late 1950s. In a secondary role rubber has, of course, been used extensively as a spring material for suspension bump and rebound buffers (Figure 23.38). These act as progressive limiting stops to prevent metal-to-meal contact between the wheel locating linkages and the car structure at the extremities of suspension travel. However, in modern practice the use of a microcellular urethane elastomer is usually preferred for the suspension buffers, because unlike rubber materials it can provide a more progressive stiffening characteristic over a wider range of deflection and also exhibits reduced bulging so that less mounting clearance is required.

There are two factors that have tended to mitigate against the widespread application of rubber springs in passenger car suspension systems. One concerns their settling caused by what are termed primary creep and secondary creep of the rubber. The former arises from static loading on the spring and the latter from dynamic loading. Primary creep in a rubber spring is mainly responsible for any reduction in ride height or standing height of the vehicle in service, but since this effect tends to occur early in the life of the spring it can usually be catered for by setting or trimming the vehicle initially high on its suspension. The other factor concerned is perhaps less significant and is the stiffening in action of the spring at low ambient temperatures.

On the credit side, a rubber spring has a very high energy storage capability, possesses a useful amount of inherent damping due to the internal friction of its material, and it does not rust. It also contributes to the reduction of noise transmission through the suspension system.

The familiar Moulton cone spring for IFS (Figure 23.39) consists of a natural rubber element bonded between inner and outer pressed steel cones. To provide the required load and deflection characteristics and also to lend inherent stability to the spring the rubber element assumes a form reminiscent of a ring doughnut; although perhaps we should be more formal and describe it as frusto-conical. By virtue of its shape the rubber element is therefore loaded partly in shear and partly in

compression, so as to provide the best compromise between greatest energy storage and long service life. The purpose of the co-operating plate is to stiffen the action of the spring under high loading and thereby eliminate the need for a separate bump buffer. This type of rubber spring is also used in conjunction with pure trailing arm IRS (Figure 23.33c).

The use of rubber as a spring medium in heavy-vehicle suspension systems and in particular for tandem axle bogies is rather a different matter, because like the air spring it is now regarded as a 'road-friendly' type of suspension, which in current international legislation can allow certain payload benefits to be realized, apart from providing better cushioning properties than steel springs for the transportation of fragile goods. One of the qualifying factors that defines a rubber suspension as being equivalent to an air system is that it must provide a laden ride frequency of less than 2 Hz or 120 cycles/min.

Rubber springs lend themselves admirably to the walking-beam type of tandem axle bogie, the principle of which is explained in Section 23.5. Suffice it to say here that each walking beam can be pivoted beneath either a semi-elliptic leaf spring or a pair of inclined rubber springs termed bolsters. Their inclination allows them to work in combined compression and shear for the same reason as mentioned earlier in connection with engine mountings (Section 1.4). They also feature bonded-in metal plate interleaves, which not only restrain excessive bulging of the rubber but also conduct away the heat generated by internal friction in the rubber. To limit the rebound travel of the rubber bolsters and enhance stability, a rubber buffer-ended control rod may be pivoted vertically from the walking beam and act downwards against a stop bracket on the chassis frame (Figure 23.40).

The inclined rubber bolsters are cradle mounted from the chassis frame side member and may be disposed either longitudinally (Figure 23.40) or transversely (Figure 23.41) with respect to the pivot axis of the walking beam. An advantage claimed for the former arrangement is that all normal suspension loads are better distributed into the side members of the chassis frame, without there being any need

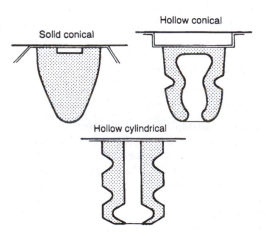

Figure 23.38 Suspension linkage buffers

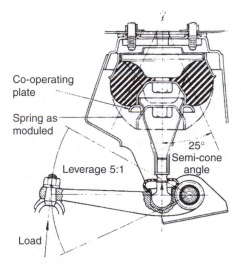

Figure 23.39 Moulton rubber cone spring installation

for reinforced outrigger brackets that increase weight. With the latter arrangement, additional torque radius rods are required to restrain fore-and-aft movement of the walking beams, but it is claimed that there is easier articulation of the walking beam and that the rubber bolsters are better placed to resist vehicle roll. In general the advantages claimed for the rubber suspension of tandem axles are: a better cushioned ride; no metal-to-metal contact of the working parts for quieter operation and the elimination of wear; low maintenance, because there is no need for lubricating or checking the bolt tightness of leaf spring fittings; tyre life can be increased by virtue of the suspension being more compliant to cornering and manoeuvring scrub loads; and the rubber bolsters are failsafe in operation.

Air springs

All gases are readily compressible without ever being damaged, and this property has led to the limited use of either air, or an inert gas, as a spring medium in motor vehicle suspension systems. The attractions of using this type of spring are essentially twofold:

1 The load deflection characteristics of an air spring are such that the ride comfort of the suspension system is little affected by changes in vehicle load.
2 An automatic height control facility may readily be incorporated in the system to ensure that the full range of suspension movement always remains available.

As compared with other spring media the air spring possesses the highest energy storage capability, but in practical applications this advantage can be reduced by the requirement for relatively elaborate and heavy automatic height control equipment. Since an air spring takes the form of either a bellows or a strut unit, it cannot assume any wheel

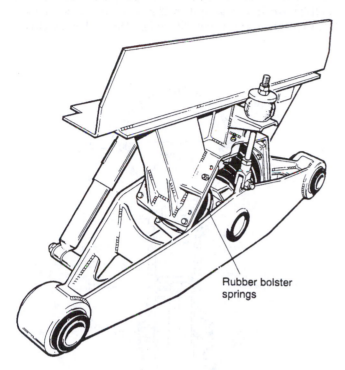

Figure 23.40 Walking-beam tandem axle bogie with longitudinal rubber bolster springs (*Hendrickson Norde*)

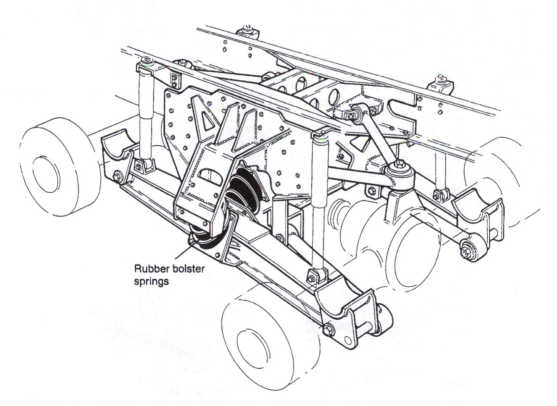

Figure 23.41 Walking-beam tandem axle bogie with transverse rubber bolster springs (*GKN Tadchurch*)

locating function, which must therefore be performed by a separate linkage. Air spring applications fall into two categories, these being generally termed fixed-mass and variable-mass air springs.

Fixed-mass air springs

This form of springing was pioneered by the French Citroën Company in the early 1950s and is universally known as hydropneumatic suspension (Figure 23.42). It has long been acclaimed for the high level of comfort it provides.

In considering the operation and construction of this type of suspension system, it is important first to establish the division of labour between the fluid and the inert gas:

1 The compressible gas constitutes a self-contained springing medium.
2 The incompressible fluid provides the means of transmitting suspension loads to the springing medium.

The Citroën type of hydropneumatic strut (Figure 23.43) is connected between the car structure and the wheel locating linkage. At the head of each strut is a spring unit formed in the upper half of a spherical chamber, which is pressure charged with non-flammable nitrogen and sealed from the lower half by a flexible diaphragm. Communicating with the other side of the diaphragm is a fluid-filled cylinder, in which a piston reciprocates in sympathy with suspension movements. Since the column of trapped fluid is virtually incompressible, its displacement by the piston further compresses the fixed mass of gas in the spring unit to absorb the road shock taken by the rising wheel. A damper valve is incorporated at the base of the lower half of the spherical chamber, thereby creating a resistance to fluid flowing to and from the spring unit. To provide an automatic height control facility, the column of fluid trapped in the strut is adjusted in length by either releasing fluid or admitting fluid under pressure from a separate energy source.

Front and rear height control (or levelling) valves are flexibly connected to the suspension anti-roll bars to signal the flow of fluid to and from the strut units. Each height corrector comprises a three-way hydraulic distributor block which, depending on the position of its suspension activated slide valve, connects the suspension cylinders either to the high-pressure supply inlet or to the reservoir outlet, or isolates them from both the inlet and outlet when the slide valve is in its central position (Figure 23.44).

Interdependent suspension with fixed-mass air springs

Alternatively the fluid sections of the front and rear suspension units on each side of the car may be interconnected in the interests of providing a relatively pitch-free ride, as in the well-known Hydragas system that was introduced in 1973 by Moulton Developments for certain British Leyland cars.

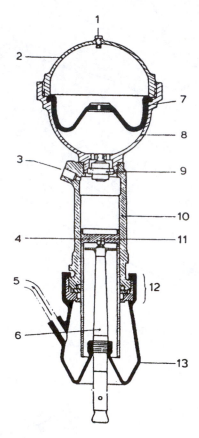

Figure 23.43   Cross-section through a hydropneumatic strut (*Citroën*)

1 filler screw	8 lower half-sphere
2 upper half-sphere	9 damper
3 high-pressure fluid inlet	10 cylinder
4 piston	11 thrust pad
5 overflow return	12 sealing system
6 piston rod	13 dust shield
7 diaphragm	

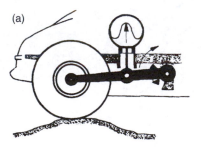

Figure 23.42   Basic action of a fluid gas suspension system: (a) bump (b) rebound (*Citroën*)

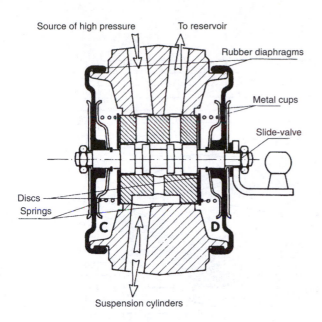

Figure 23.44   Height corrector (*Citroën*)

The Moulton type of hydragas unit similarly has at its head a fixed-mass gas spring, but the fluid acting against it is contained within a displacer body in which a tapered plunger reciprocates according to suspension movements (Figure 23.45). Again, a damper valve is incorporated at the base of the lower half of the gas spring chamber, or egg as it is sometimes called. The sealing function between the displacer body and the tapered plunger is performed by a second diaphragm of the rolling lobe type, such that the extent to which it envelopes the plunger head determines the effective area acting upon the fluid. As mentioned earlier, the Hydragas units may be interconnected on each side of the car, so as to suppress pitching. If this begins to occur, a restoring force is created by virtue of the difference in effective diaphragm areas and hence fluid displacements created by the rising and falling of the tapered plungers, since the excess fluid displaced serves further to compress the gas in both spring units.

Variable-mass air springs

Following earlier successful applications of this principle to motor coaches and heavy vehicles, variable-mass air

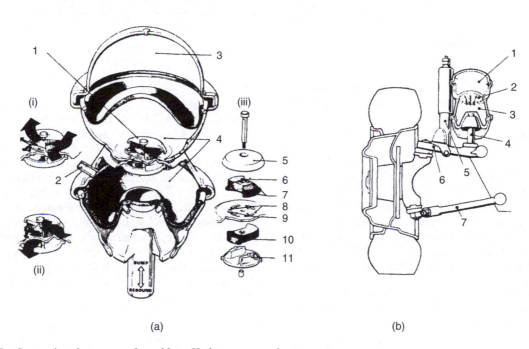

    (a)   (b)

Figure 23.45   Comparison between early and later Hydragas suspension types

(a) Early type: (i) bump flow through damper (ii) rebound flow through damper (iii) exploded view of damper valve

1 damper valve
2 interconnection pipe
3 gas
4 fluid
5 bump housing
6 bump compression block
7 flaps
8 bump bleed
9 main bleed
10 rebound compression block
11 rebound housing

(b) Later type
1 nitrogen spring chamber of Hydragas unit
2 drilled stainless steel diaphragm in lieu of damper valve
3 fluid chamber
4 push-rod
5 external telescopic damper
6 upper suspension forging
7 lower suspension forging

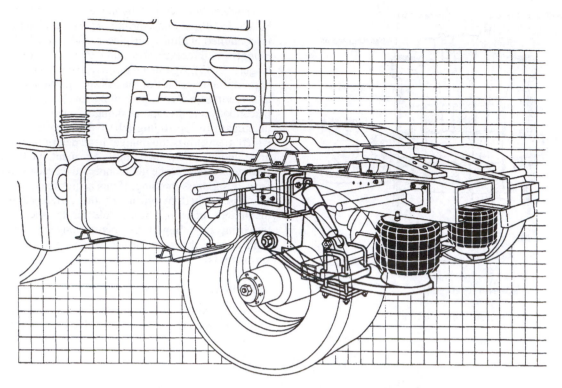

Figure 23.46   Flexolink drive axle air suspension for heavy vehicle (*Dunlop*)

suspension systems originally enjoyed a vogue of popularity on passenger cars built in America during the late 1950s and early 1960s, but with few exceptions were little used elsewhere. Certain problems could be encountered in service with these early systems, especially in climates where winter operation involved high humidity near freezing point. Also, the air suspension could sink onto its bump rubbers when the car was parked overnight. However, since the mid-1980s there has been a revival of interest in the use of variable-mass air springs to improve passenger car suspension characteristics. This time electronic control modules are employed in conjunction with either Hall-effect rotary, or magnetic linear, height control sensors and air flow solenoid valves to increase or reduce pressure in the air springs.

For modern heavy vehicles the use of variable-mass air suspension on their tractive units (Figure 23.46) and semi-trailers (Figure 23.50) brings advantages in respect of smoother riding that reduces driver fatigue and better protects fragile cargo, provides height control for ease of trailer coupling and docking, eases maintenance requirements, and causes less damage to the road. In the case of buses the use of air suspension, apart from improving ride comfort, allows a constant height to be maintained for the entrance platform or step, which has always been an important consideration as far as passenger safety is concerned.

The variable-mass air spring has usually been designed to occupy the same space as that vacated by a steel spring. As a consequence, its practical application typically takes the form of a frame-mounted air chamber, in which the spring plunger and a rolling diaphragm are free to move both axially and radially according to the dictates of the suspension linkage that

retains them (Figure 23.47a). The sealing function between the spring plunger and air chamber is performed by the rolling lobe diaphragm, which is of fabric-reinforced synthetic rubber construction. An important advantage that accrues from using this type of air spring is that the plunger may be specially contoured over its length, which enables the effective area of the spring to be varied over the whole range of suspension travel (Figure 23.48). Therefore the change in rate can be least for moderate deflections of the suspension either side of the normal ride position, followed by a gradual increase in rate to contain greater deflections and culminating in a rapid increase in rate to resist more severe deflections. A fairly generous radial clearance is required between the spring plunger and the air chamber, in order to accommodate both tilting of the plunger and looping of the diaphragm.

In contrast to earlier practice, where variable-mass air springs were applied exclusively to wishbone independent front and live axle rear car suspension systems, springs of this type may now be incorporated in both independent front and rear suspension systems utilizing MacPherson struts. The air spring is again designed to occupy the same space as that vacated by a coil spring (Figure 23.47b), which is traditionally used with a MacPherson strut. In either application the air chamber is pressurized from an external source to support the vehicle load, a self-levelling facility being provided by either releasing or admitting air as signalled by the levelling mechanism.

At the nerve centre of the levelling system are the height control valves, which respond either mechanically, or electrically, to relative movement between the vehicle and its suspension mechanism. The vehicle is therefore maintained at

(a)

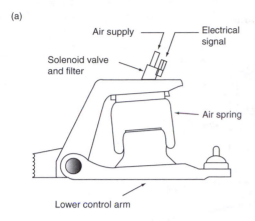

Air supply — Electrical signal

Solenoid valve and filter

Air spring

Lower control arm

(b)

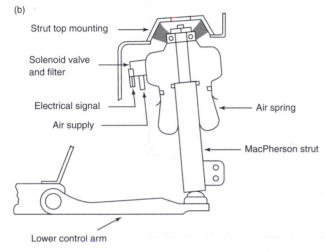

Strut top mounting

Solenoid valve and filter

Electrical signal

Air supply

Air spring

MacPherson strut

Lower control arm

Figure 23.47 Basic arrangements of passenger car variable-mass air springs: (a) wishbone suspension system (b) MacPherson strut suspension system

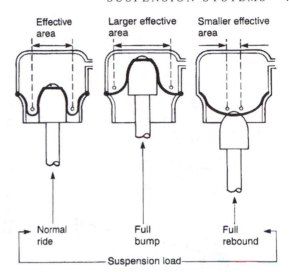

Effective area — Larger effective area — Smaller effective area

Normal ride — Full bump — Full rebound

Suspension load

Figure 23.48 Action of a variable-mass air spring with rolling lobe diaphragm

a constant standing height, regardless of load, by adjusting the quantity of air in the springs (Figure 23.49). A spring-cushioned lost-motion connection is provided between each height control valve and its actuating linkage, so that the system ignores normal ride motions. Based on the principle that a uniformly laden three-legged, as opposed to four-legged, table will provide an equal distribution of load among its legs even though they may not be exactly equal in length, so three and not four height control valves are used for levelling purposes where air suspension is used at both ends of the vehicle. Four height control valves could of course be used, but only if the perfect setting of each valve could always be guaranteed in practice. For heavy vehicle trailers with air suspension systems, one levelling valve is used on each side of the unit, so that the air springs at each wheel station operate at the same pressure. This equalizes axle loads and prevents an axle from being overloaded provided the total load is not excessive. A modern example of trailer air suspension is the Rubery Owen-Rockwell Indair system, which is applied to independently suspended wheels (Figure 23.50). In this installation rubber brake line quality high-pressure hoses are used in the piping system. The filter/pressure protection valve assembly is connected between the air reservoirs for the trailer brake and suspension systems. Its purpose is to ensure

a clean supply of air to the suspension levelling valves and to protect the air brake system from loss of pressure in the event of a failure in the air suspension system.

As mentioned earlier, variable-mass air suspension systems for modern cars utilize electronic control systems as a basis for their operation (Figure 23.51). Such systems may then be extended to recognize an increased number of operating variables and, in some applications, include control over adaptive shock dampers (Section 23.6). Sensors may therefore be installed that not only detect and, if necessary, restore correct standing height of the car by signalling an increase or a decrease in air pressure for the springs, but also can take into account dynamic considerations. For example, an input from a road speed sensor can signal the trim height of the car to be varied automatically according to its speed. This feature is sometimes known as 'speed levelling' and confers the following advantages:

1 On rough road surfaces the car can be raised to provide increased ground clearance.
2 With increasing speed on normal road surfaces the car can be lowered. This enhances stability by reducing the height of its centre of gravity.
3 From consideration of (2) the aerodynamic drag or Cd value of the car is reduced to improve fuel consumption (Section 30.7).

It will be evident that the height control systems used in the fixed-mass hydropneumatic and variable-mass air suspensions are similar in general principle, but with the important difference that in the former system the levelling function is performed by displacing small amounts of incompressible fluid at high pressure, but in the latter by moving large amounts of compressible air at low pressure.

For heavy vehicles the air suspension linkage is conventionally of the simple type where trailing arms that pivot from the chassis frame are rigidly attached to the axle and locate it both longitudinally and laterally. These arms are

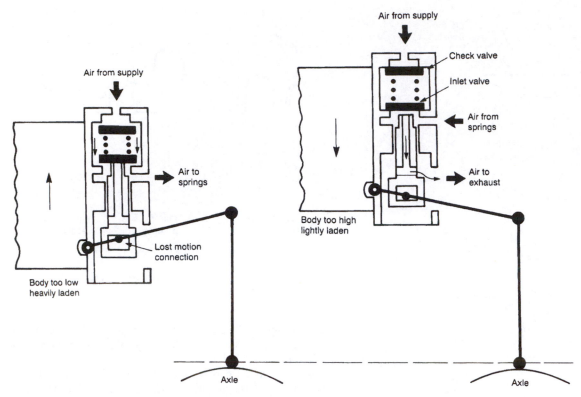

Figure 23.49 Schematic arrangement and operation of a height control valve

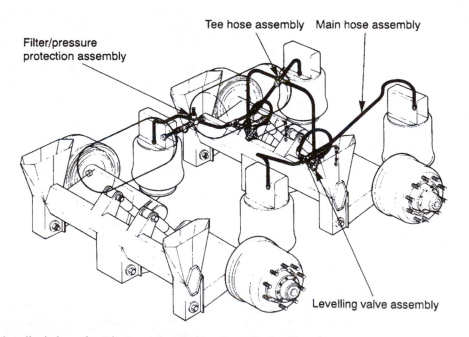

Figure 23.50 Indair trailer independent air suspension (*Rubery Owen-Rockwell*)

extended beyond the axle to mount the air spring and may be of either rigid box-section construction, or comprise a quarter-elliptic leaf spring typically with twin tapered leaves. To allow for rocking movements of the axle during normal ride and when the chassis rolls relative to the axle, the rigid type of arms need to pivot from thick rubber bushings that are flexible enough to prevent overstressing of the axle when the arm on one side swings upwards and the arm on the other side swings downwards. Less flexible rubber bushings can of course be used at the pivots of the leaf spring type of arms (Figure 23.46), since these are able to twist in accommodating axle movements.

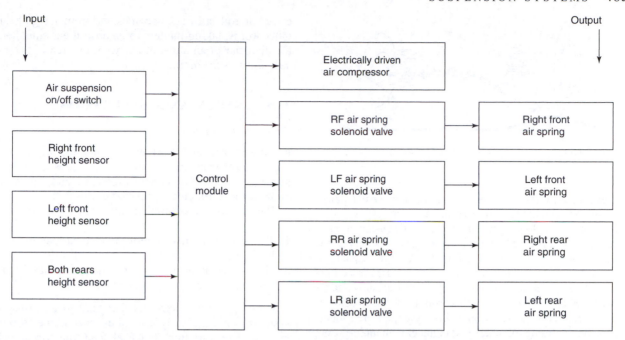

Figure 23.51 Block diagram of a passenger car electronically controlled height control system

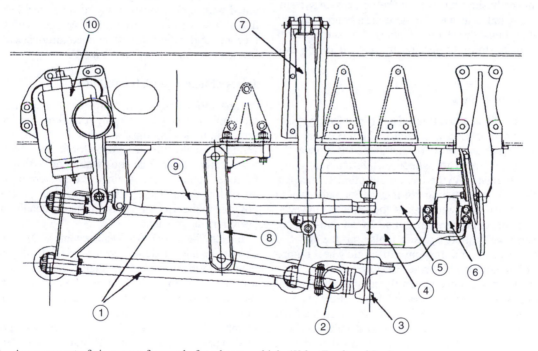

Figure 23.52 Arrangement of air sprung front axle for a heavy vehicle (*Volvo Truck and Bus*)

1 axle parallelogram locating linkage
2 axle mounted anti-roll bar
3 axle beam
4 air spring plunger
5 air spring bellows
6 Panhard rod connection
7 telescopic shock damper
8 anti-roll bar shackle
9 steering drag link
10 steering gear

In Continental heavy vehicle practice air springs may also be applied to the beam axle front suspension of rigid units, especially when they are used for towing drawbar trailers (Section 30.5). Taking for example the general layout of the Volvo system (Figure 23.52), this incorporates two pairs of side-mounted trailing links, each pair arranged in parallelogram fashion, which not only locate the axle in the longitudinal direction, but also restrain it from winding-up under brake torque reaction. For lateral location of the axle a Panhard rod is connected across from one end of the axle beam to the opposite side

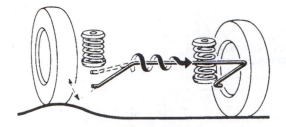

Figure 23.53   Action of anti-roll bar (*Alfa-Romeo*)

member of the chasses frame. In combination, the two pairs of trailing links and the Panhard rod therefore confer physical stability for the bellows type air springs, which are mounted directly above the axle beam and operate at a tank supply pressure of about 1.2 MN/m² (175 lbf/in²). A larger capacity air compressor (Section 28.2) is required when used in conjunction with an air suspension system. An anti-roll bar is attached to the axle beam and the ends of its arms are shackled below the frame side members. Telescopic shock dampers are mounted alongside the frame members and connect with multi-purpose axle brackets, which also serve as anchorage points for the locating links and mounting platforms for the air springs. Pivot points are rubber bushed throughout the suspension system. The steering drag link is arranged to describe a sympathetic arc with the trailing links, thereby minimizing any disturbance of the steering arising from suspension movements of the axle (Section 24.1).

Suspension anti-roll bars

An anti-roll bar is basically a torsion bar type of spring with angled ends, and its action is shown in Figure 23.53. It is mounted transversely and in such a manner as to connect the locating linkages of opposite wheels, while remaining torsionally free with respect to the vehicle structure. It thus performs the function of resisting rolling, as its name would suggest, but not bouncing motions of the vehicle. However, the amount of roll restraint contributed by an anti-roll bar must necessarily be limited, because it also has the effect of stiffening the suspension in response to single wheel bumps and can therefore compromise ride comfort.

The anti-roll bar was once generally confined in application to the front suspension, because it had the effect of redistributing the overall lateral weight transfer between the front and rear pairs of tyres in favour of promoting understeer. The influence of tyre slip angles on understeer was described in Section 22.3. The slip angle of a tyre can also be increased if the vertical load becomes greater, which is the effect that the addition of an anti-roll bar has on the already heavily loaded outer wheel of a pair during cornering. In more recent years there has been a trend to include an anti-roll bar in the rear suspension system as well, partly to correct any condition of excessive understeer and partly to reduce roll further.

It will sometimes be seen mentioned that to improve car handling, a manufacturer has increased the diameter of an anti-roll bar by a seemingly insignificant few millimetres. However, when this change is related to the torsional stiffness increasing as the fourth power of the bar diameter, the

effect of roll stiffness becomes far from insignificant. In other words, taking matters to an absurd extreme, doubling the diameter of an anti-roll bar would increase its torsional stiffness by sixteen times!

23.5   TANDEM AXLE SUSPENSION

Basic requirements

It was earlier related (Section 19.7) that the driving and suspension arrangements for the tandem axles used on some heavy vehicles posed certain problems, and we must now examine these from the suspension point of view.

The suspension requirement are basically as follows:

1   To equalize the distribution of vertical load between the leading and trailing axles.
2   To ensure that the equal distribution of vertical load is not disturbed by axle torque reactions.

Modern legislation demands that tandem axle suspension systems comply closely with these requirements because, in respect of variation of wheel load, the Motor Vehicles (Construction and Use) Regulations in Britain for example state that vehicles with more than four wheels 'shall be provided with such compensating arrangements as will ensure that all the wheels will remain in contact with the road surface and under the most adverse conditions will not be subject to abnormal variations in load'.

Types of load equalizing linkage

At first thought it might seem that tandem axles could be mounted in close proximity by providing each with a pair of semi-elliptic leaf springs, and thereby simply duplicate conventional single-axle suspension for heavy vehicles. However, it was early realized that equalization of vertical load between the axles could not be obtained if they were individually suspended in this manner, although in later practice it proved possible to incorporate various types of equalizing linkage to overcome the problem.

In 1926 the British War Department (WD) evolved a classic design of tandem axle suspension, which readily demonstrates the basic principles of an equalizing linkage. With this design, inverted semi-elliptic leaf springs are used in pairs, one above the other, and are clamped on each side of the vehicle to a trunnion block that pivots about a frame-mounted cross-tube member (Figure 23.54a). The eyes of the leaf springs are pivoted in pairs to ball-jointed axle brackets, which allow for tilting of the axles without distortion of the springs.

In operation, it should be evident that each pivoting pair of leaf springs acts in the manner of a balance beam, so as to equalize the vertical loads on the wheels of the leading and trailing axles. Furthermore, as compared with suspending the axles individually, the amount of vertical movement imposed on the vehicle as the wheels of the interconnected axles pass over road irregularities is reduced by one-half. This balance beam effect is perhaps even more visually evident in the equally classic American walking-beam tandem axle suspension (Figure 23.54b), which also appeared in 1926 and was invented by Magnus Hendrickson.

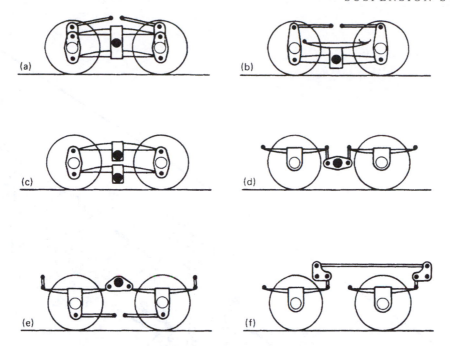

Figure 23.54 Schematic arrangements of leaf spring tandem axle suspension

Referring again to the WD tandem axle suspension, as so far described the system does not take into account the driving and braking torque reactions imposed on the axle casings and their tendency to rotate bodily the pivoted pairs of leaf springs. These effects result in the unwanted transfer of vertical load from the leading to the trailing axles on acceleration and vice versa on braking, and can lead to wheel slip. To eliminate this transference of load within the tandem axle itself, it becomes necessary to add a pair of torque rods or torque reaction links, which are pivoted between an upper frame-mounted cross-member and an upstanding bracket on each axle casing. By arranging for the torque rods to act in parallel with the pivoted pairs of leaf springs (Figure 23.54a), the latter are relieved of the turning effect of axle casing torques and the equality of vertical loading on the leading and trailing axles is no longer disturbed. A similar arrangement of torque rods is used with the earlier mentioned walking-beam tandem axle suspension (Figure 23.54b). Such systems are said to be non-reactive, as opposed to being reactive, where there is no compensation for axle torque effects.

There can be little doubt that 1926 was a very good year for tandem axle suspension systems! Because it was also in that year that yet another notable design appeared, this time from the once prominent British firm of Thornycroft. Although still employing two pairs of inverted semi-elliptic leaf springs, as in the WD design, it differed in the important respect that the springs were individually pivoted at their centres (Figure 23.54c) instead of being pivoted as pairs. Since the eyes of the leaf springs still remained pivoted in pairs to the axle brackets, a parallel motion rather than a balance beam linkage was created. It thus provided an equivalent non-reactive but albeit simpler arrangement than the WD system with added torque rods. In modern versions of the Thornycroft system, the lower of each pair of leaf springs is often replaced by leading and trailing pivoted links.

As previously mentioned, it was later found possible to duplicate a conventional single-axle suspension system, yet still equalize the vertical loads on the leading and trailing axles by using various interconnecting linkages, which have usually been inspired by railway practice. A relatively simple linkage was that originally used by the French Berliet company, where the adjacent eyes of the front and rear semi-elliptic leaf springs were shackled to a frame-mounted swinging equalizer, the springs being pivoted at their other ends directly to the frame (Figure 23.54d). In modern applications, the swinging equalizer is typically positioned above the spring eyes and may be mounted on tapered roller bearings. A variation on this theme involves directly pivoting the adjacent eyes of the springs to the frame-mounted swinging equalizer and shackling the other ends of the springs to the frame (Figure 23.54e). Although both types of linkage serve to equalize the loads on the leading and trailing axles, the suspension is limited to being a reactive one. This is because drive torque reaction on the leading axle will tend to depress the rear eye of its spring, which via the swinging equalizer will then raise the front eye of the trailing axle spring, so that there is a transfer of vertical load from the trailing to the leading axles (opposite to that occurring in the WD design without additional torque rods). Here again, this disadvantage can only be overcome by the addition of torque rods, usually pivoting beneath the springs (Figures 23.54e and 23.55), which then render the system non-reactive.

A more elaborate type of equalizing linkage is that originally associated with the once well-known British Tilling-Stevens company. In this still used arrangement the rear ends of the front and rear semi-elliptic leaf springs are shackled to frame-mounted bell-crank levers, which in turn are interconnected by an overhead push-pull rod, the front ends of both springs being pivoted directly to the frame (Figure 23.54f). This type of equalizing linkage cleverly produces a

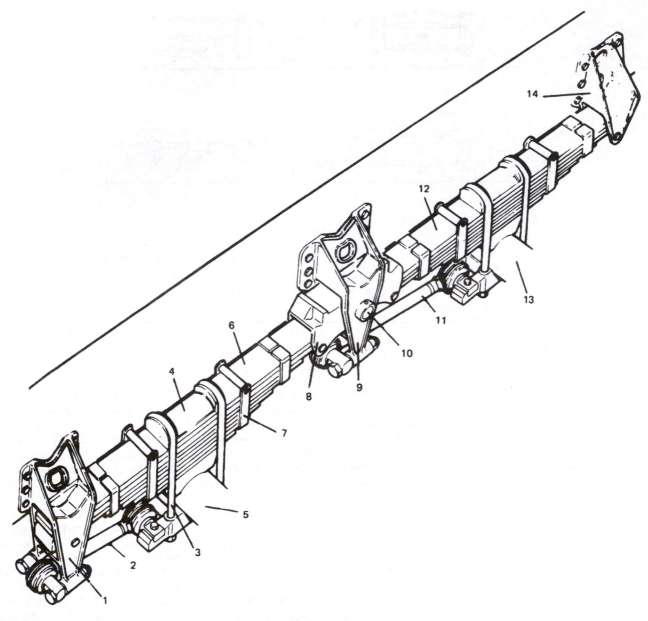

Figure 23.55   Details of a modern tandem axle suspension (*Seddon Atkinson*)

1 foremost rear spring bracket	6 road spring (foremost)	11 torque rod (long)
2 torque rod (short)	7 leaf retaining clamp	12 road spring (rearmost)
3 U-bolt	8 equalizer	13 axle tube
4 U-bolt	9 equalizer bracket	14 rearmost rear spring bracket
5 axle tube	10 shaft (equalizer)	

non-reactive suspension system, because when drive torque reaction on the leading axle tends to depress the rear eye of its spring, this same motion is transmitted via the interconnected bell-cranks to the rear eye of the trailing axle spring, and therefore equalization of loads on the two axles remains undisturbed.

Where twin front axles are used on heavy vehicles, these can still be individually suspended. This is because the centre of gravity of the vehicle is so far removed from the front axles, that variations in their vertical loading remain modest enough to disregard the need for any equalizing linkage.

Later developments

In modern applications of these basic types of load equalizing linkages for tandem axle suspension, the installation details can vary considerably. For example, the replacement of leaf springs by other spring media will entail the use of additional swinging links to restore proper fore-and-aft location for the axles (Figure 23.41). Similarly, these links will be required in leaf spring systems where slipper-ended springs are preferred to pivot-ended ones to reduce maintenance requirements (Figure 23.55). It may further be necessary to employ either

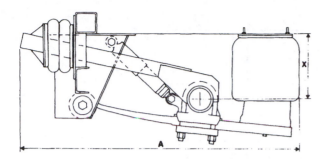

Figure 23.56 Flexolink trailer air suspension with lift/lower facility (*Dunlop*): axle lift system which can be incorporated on 7.5 and 11 tonne systems, manual operation or automatic override

A 1412 mm
X 335/425 mm, 250/315 mm
Weight of axle lift system: 40 kg

triangulated torque rods above the axle centres (Figure 23.41), or Panhard rods, to control the sideways movement of the axles.

Axle lift facility

In a multi-axle installation, the facility to lift and retain a non-driven and solely load-carrying axle clear of the ground when the vehicle is running light, can bring economies in both tyre life and fuel consumption. It can also transfer weight temporarily to the drive axle for increased traction under difficult conditions.

Modern air suspension systems for heavy vehicles offer several convenient means of retracting an axle, including the following:

1 An air spring may be sandwiched between the chassis frame side-member and a semi-elliptic leaf spring, the installation of the latter being such that the axle is held clear of the ground unless the air spring is pressurized.
2 An air lift bellows is mounted vertically from the chassis frame and is provided with flexible stirrup connections to the axle locating arms of an air suspension system, the axle being retracted when the air lift bellows is pressurized.
3 A pair of air lift bellows are mounted horizontally from a chassis frame cross-member and connect to the axle via a T-shaped tension link, the air sprung axle being retracted through a bell-crank motion about its locating arms when the air lift bellows are pressurized (Figure 23.56).
4 Similar in principle to (3), but with a pair of air lift bellows mounted vertically from a chassis frame cross-member, which act upon the ends of cantilever arms rigidly attached to the axle and extending beyond the pivot axis of the axle locating arms. Hence, when the air lift bellows are pressurized the see-saw action of the cantilever arms retracts the axle.

Coil spring actuated mechanical devices have also been used to retract axles in earlier practice.

23.6 SHOCK DAMPERS

The need for shock dampers

Shock dampers play an important role in the motor vehicle suspension system and work in conjunction with the springs.

The need for damping the action of the springs can be readily visualized by considering the gymnastics of a very simple vibratory system consisting of a mass supported by a helical extension spring.

If a disturbing force momentarily displaces the mass from its position of rest, then not only will the spring restoring force return the mass to its original position, but also the inertia of the mass (or reluctance to change its state of motion) will cause it to overshoot. The mass will then continue to oscillate with diminishing amplitude either side of its original position until the vibration eventually dies out or decays. This must occur, of course, because of the natural damping effects of air friction and of internal friction in the spring material. In fact these frictional effects have damped the energy of vibration by dissipating it in the form of heat energy.

With practical anti-vibration systems found in engineering, and certainly in the case of the motor vehicle suspension system, it is necessary to reduce the amplitude of vibration far more rapidly than is possible with the natural damping effects just described, otherwise the vehicle ride and road holding would prove totally unacceptable. Hence, it is usual for some form of deliberate damping to be introduced into the system, and this is the duty performed by the familiar shock dampers. At this point it may be appreciated that the universal description of the shock damper as a shock absorber is really a misnomer, since it is the suspension springs and tyres that actually absorb the road shocks and the shock dampers that damp the oscillations of the springs and their associated masses. However, we must accept that the designation shock absorber has long since come into general usage.

Basic application

Shock damping is performed in the suspension system of the modern motor vehicle through the medium of hydraulic friction, which is created by the resistance offered to a fluid that is forced by movement of a piston to flow through submerged orifices.

Permanent bleed orifices are used to provide an increasing rate of damping for slow movements of the suspension system (Figure 23.57a). Since the motion of the fluid being forced through these small orifices is of a turbulent nature, the resistance to flow increases as the square of velocity or, in other words, the resistance is four times as much if the velocity is doubled. The aim here is to obtain freedom from vehicle body float on the one hand and to avoid normal ride harshness on the other. However, this rapidly increasing rate of damping would become excessive with rapid movements of the suspension, so what is termed a linear blow-off valve is arranged to act in parallel with the permanent bleed orifices (Figure 23.57b). Once the predetermined setting of this valve is exceeded by the build-up of fluid pressure imposed by the bleed orifices, it opens to allow fluid flow through its much larger orifices. Since the fluid motion then becomes of a streamlined nature, its resistance to flow remains in direct proportion to the velocity, thereby moderating the overall damping effect. The amount of damping can therefore be made sufficient to control wheel hop for good road holding without jeopardizing ride comfort.

The telescopic type of shock damper has been used in the suspension systems of most motor vehicles designed from

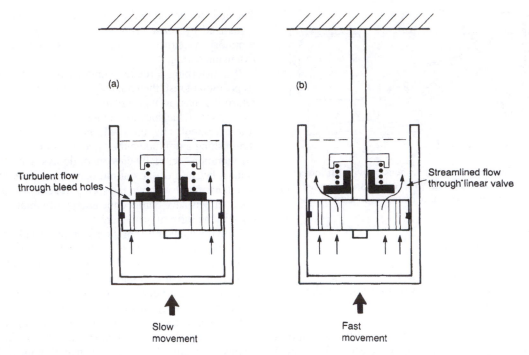

Figure 23.57   Schematic arrangement and fluid flow through hydraulic shock damper bleed orifices and linear valve

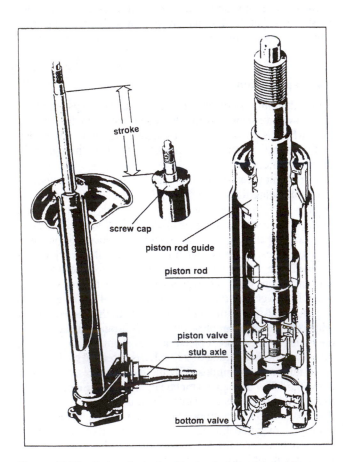

Figure 23.58   Twin-tube hydraulic shock damper insert for MacPherson strut (*Fichtel & Sachs*)

the 1950s onwards. It is usually described as direct acting by virtue of its operating piston and cylinder being directly attached between the vehicle structure and the suspension linkage. In contrast, an earlier form of hydraulic damper was the lever arm type where the piston and cylinder assembly was almost invariably mounted on the vehicle structure and intercepted movements of the suspension linkage via a lever arm and connecting link. An inherent disadvantage of this type of damper was that any reduction in damper resistance was considerably amplified at the road wheel, because of the intervening lever arm arrangement.

Returning to the telescopic damper, this is installed either as an individual unit, or as an integral part or a renewable cartridge of a MacPherson strut (Figure 23.58). It may also be of either the original twin-tube or the later monotube construction, the latter being pressurized internally to prevent aeration of the fluid.

In actual practice there is, of course, a good deal more sophistication in the design of a telescopic shock damper than the foregoing simple description might suggest, especially in respect of valve characteristics and sealing techniques. Indeed, it should be recognized that the hydraulic damper is a precision piece of equipment that may have to withstand up to 20 000 oscillations per mile, and on bad roads the energy of vibration dissipated as heat to the atmosphere may approach 100°C and 200°C for passenger car and heavy duty commercial vehicle applications respectively. The piston rod sealing elements therefore range from simple lip seals in synthetic rubber to multi-lip seals in plastics compound, according to the severity of application, so that a long service life can be achieved.

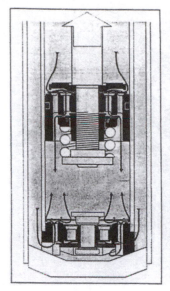
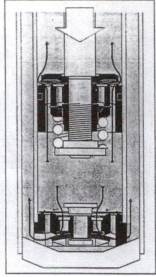

Figure 23.59   Operation of a twin-tube hydraulic shock damper (*Fichtel & Sachs*)

Figure 23.60   Operation of a monotube hydraulic shock damper (*Fichtel & Sachs*)

Twin-tube shock dampers

The basic operation of this type of telescopic shock damper (Figure 23.59) is as follows:

Bump movements of the suspension   As the damper retracts, a two-way flow of oil is established through the hydraulic resistances offered by valves in the piston and in the base of the cylinder. That is, the damper fluid passes upwards through the stationary piston and downwards through the base of the rising cylinder, the volume of fluid passing through the latter to the reservoir being equal to that displaced by the piston rod entering the head of the cylinder. To ensure that the increasing space above the piston is maintained completely full of fluid, the hydraulic resistance in the base of the cylinder must be greater than that in the piston and therefore governs the damper resistance to bump movements of the suspension.

Rebound movements of the suspension   As the damper extends the fluid passes downwards through the piston and upwards through the base of the cylinder, the volume of fluid returning from the reservoir to the cylinder similarly being equal to that of the retracting piston rod. Conversely, in this case, it is the hydraulic resistance in the piston that must be the greater and therefore governs the damper resistance to rebound movements of the suspension.

Monotube shock dampers

Their construction is such that a pressurized gas (usually the inert gas nitrogen) accepts the variations in effective volume of the piston rod as it enters and leaves the cylinder, and also compensates for changes in volume of the damper fluid due to thermal effects (Figure 23.60). This type of telescopic shock damper has received increasingly wider use since the mid 1960s, the concept having been pioneered by De Carbon in France some ten years earlier.

To eliminate aeration the gas reservoir that supplies the necessary pressure, typically in the region of 2.1–3.1 MN/m^2

(300–450 lbf/in^2), is separated from the damper fluid by a floating piston. The hydraulic disc valves and their orifices for both bump and rebound damping are embodied in the main piston, and no recuperation valves are required by virtue of the internal pressurization, which thus simplifies damper construction. In the absence of an annular reservoir, the piston displacement area of the monotube shock damper may be increased over that of twin-tube construction, without increasing the overall diameter of the unit. This in turn leads to greater fluid displacement for the same damper travel and improved heat dissipation. Pressurizing the fluid not only promotes a more responsive action of the valves, but also discourages cavitation thereby eliminating one source of noisy operation. Cavitation may occur in any damper where incomplete recuperation with its accompanying drop in cylinder pressure promotes the formation of bubbles containing vapour and gas, which upon collapse implode noisily at very high pressures. A further point in connection with the pressurized monotube shock damper is that it is necessarily self-extending with a force typically in the region of 250–300 N (57–68 lbf) and therefore contributes to the spring stiffness of the suspension system.

In another version of pressurized gas monotube shock damper, the fluid and gas are not actually separated, but instead a deflection disc is incorporated that is always below the fluid level. Its purpose is to deflect and arrest the upward surge of fluid resulting from piston movement, thereby preventing fluid from entering the gas chamber that would otherwise result in foaming of the entire fluid content. This particular type of monotube shock damper may be installed only with the piston rod pointing downwards, but has the advantages of contributing least to the unsprung weight of the suspension mechanism and permitting a shorter retracted length.

Shock damper ride control

Before World War II it was not uncommon for high performance cars to be provided with either a mechanical or a hydraulic

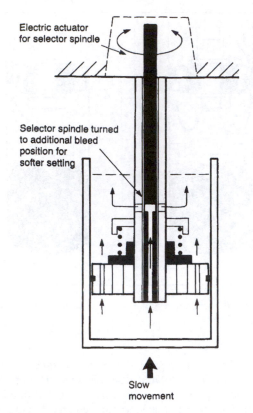

Figure 23.61 Schematic arrangement of ride control to vary shock damping

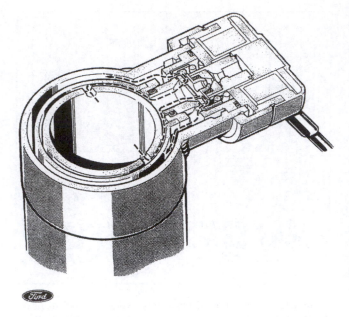

Figure 23.62 Solenoid operated ride control valve (*Ford*)

centralized control system for the shock dampers, so that the resistance they offered to suspension movements could be adjusted by the driver, according to the state of the road surface and the extent to which the car was laden.

Since the mid 1980s there has been a revival of interest in the idea of ride control, or automatic damping control as it is sometimes now known, but in a modern and more comprehensive form that utilizes an electronic control system. Typically this involves using various sensors that transmit signals to a central processor on the angle of the steering wheel and the rapidity with which it is being turned, the angular position of the front anti-roll bar (which reflects pitching), the accelerator position, the braking force and the vehicle speed. This information can then be translated into terms of optimum shock damper adjustment in about 150 ms, thereby better controlling body motions during cornering, acceleration and braking.

The actual adjustment at the shock damper can be performed by an electrical actuator, which acts upon a rotatable selector spindle that passes down through the piston rod. Both the selector spindle and the piston rod are provided with additional bleed orifices above the level of the piston, which can be made to coincide by partial rotation of the selector spindle and hence bypass fluid to provide a softer setting for the shock damper (Figure 23.61). Another method of re-routing fluid is to incorporate a solenoid operated valve and bypass channel at the lower end of each suspension strut-cum-shock damper, as in certain Ford Mondeo models. When the valve opens it allows an increased flow of fluid to soften the action of the damper (Figure 23.62), and vice versa when the valve closes. An

interesting feature of this particular system is that the firmer damper setting can be manually selected, otherwise the process of switching between soft and firm settings is automatically controlled with extreme rapidity and occupies less than 20 ms.

A different form of ride control is that provided by the ingenious Sachs Vario shock damper. In this type of twin-tube but albeit moderately pressurized shock damper, the inner working cylinder is entirely filled with fluid and the reservoir between the inner and outer cylinders is filled about two-thirds with fluid and one-third pressurized gas. The ride control characteristic is derived from one or more shallow grooves with tapering ends formed part way along the working surface of the inner cylinder, this grooving being confined to that portion of the cylinder wall swept by the damper piston during normal ride movements of a lightly laden vehicle. In this condition the resistance to flow of the fluid is therefore reduced and confers softer damping, because some of the fluid bypasses the piston valve and flows instead along the grooving in the cylinder wall. Conversely, when the vehicle is heavily laden and the shock damper working surface is less extended, that portion of the cylinder wall swept by the damper piston during normal ride movements lies beyond the grooving. Since part of the fluid flow can no longer bypass the piston, the full flow is compelled to pass through its valve, so that the increased resistance provides firmer damping.

Magnetorheological shock dampers

Rheology is a term applied by physicists to the study of the flow characteristics of a fluid, which in this particular case can be modified in the presence of a magnetic field. Such a fluid has been specially developed by the Lord Corporation in America as a damping medium for controllable shock dampers. The system is known as Delphi MagnaRide and was introduced by General Motors in 2002 on their Cadillac Seville model and since used more widely by them. In practice the magnetic field is created by an electromagnetic coil, which is embodied in the piston of a monotube telescopic shock damper. Hence, the consistency and resistance to flow of the fluid that governs the amount of energy dissipated during shock damper action can be controlled by varying the current supplied to the piston electromagnetic coil. That is, strengthening or weakening the magnetic field to which the fluid is subjected. Furthermore, the computer control system for MagnaRide can respond to inputs of vehicle speed, suspension, steering and body movements, to provide active dynamic control of the car. Structurally, the shock damper is simplified by virtue of there being no need for electromagnetically operated control valves.

Shock damper misbehaviour in service

With modern designs of relatively frictionless suspension systems, the duty imposed upon the shock dampers is much more arduous than was previously the case when multileaf springs were widely used. Although motor vehicles are originally equipped with shock dampers that give the required control of the suspension, and with the optimum settings having been established by the development engineer, it is the length of time that this control continues to be effective that is the concern of the service engineer, especially since the deterioration in their performance can be so gradual as to pass unnoticed by the vehicle owner. Worn shock dampers can cause the following complaints:

Wheel hop on acceleration
Lurching on corners
Pitching and wallowing
Vibration through the steering
Less effective braking
Increased tyre wear.

Deterioration of the rubber end mountings, significant fluid leakage, badly scored or bent piston rod and damaged dirtshield are among the defects that may be encountered with telescopic shock dampers.

Caution must be exercised in disposing of unserviceable gas pressurized shock dampers. It is therefore prudent to seek advice from the shock damper manufacturer who may, for example, recommend drilling a small vent hole in the pressure chamber.

23.7 ADAPTIVE SUSPENSION SYSTEMS

In an ideal world the suspension of a motor vehicle would probably be an 'active' rather than a 'passive' one, so that instead of using springs and shock dampers which are idly waiting to react against the suspension loads applied to them, their function would be performed by a computer-controlled system of very busy hydraulic actuators that actually anticipated the suspension loads. The hydraulic actuators would therefore positively raise and lower the wheels to conform to road surface irregularities and also restrain any movement of the vehicle body, in accordance with signals received by the computer from strategically positioned sensors. Such a system must necessarily tend to be more elaborate and expensive than a conventional passive one and so far has mainly been the subject of advanced research, notably by the Lotus company.

In the real world an acceptable compromise between these two extremes of active and passive suspension can be obtained by adopting an electronically controlled 'adaptive' suspension system. With this type of suspension system both the spring and shock damper characteristics are subject to constant adjustment to provide optimum ride and road holding. A widely acclaimed example of adaptive suspension is the Citroën Hydractive system introduced in 1989, which represents a further development of their hydropneumatic suspension mentioned earlier. With the Hydractive system an additional gas-spring/hydraulic-damper unit can be switched into operation between the two main front units and another between those at the rear. The purpose of these two additional spring/damper units is to enable the spring flexibility and the amount of damping to be varied in terms of soft and firm modes. This variation in suspension characteristics is computer controlled to match road conditions and driving style, the computer receiving appropriate signals from strategically placed sensors in a manner similar to that mentioned earlier in connection with shock damper ride control systems. The computer exercises its control over a single three-way and spring-loaded solenoid valve, which provides a connection between the central hydraulic power system and two slide valves termed 'firmness regulators', there being one for each of the additional spring/damper units.

The basic operation of the system is such that when the computer switches on the solenoid valve to obtain the soft mode (Figure 23.63a), it admits fluid under pressure to open the slide valves of the firmness regulators, which then allow the movements of the suspension linkages to be hydraulically transmitted to the additional as well as the main gas spring units. This not only has the effect of increasing the quantity of gas under compression to permit greater suspension flexibility, but also allows the fluid that is transmitting the linkage movements to flow through twice the number of restrictor orifices, which acting in parallel reduce the amount of shock damping. Conversely, when the computer switches off the solenoid valve to obtain the firm mode (Figure 23.63b), the fluid previously trapped under pressure returns to the reservoir of the central hydraulic power system and allows the slide valves of the firmness regulators to return to their closed positions. This action therefore isolates the additional spring/damper units from their associated main ones, so that the front and rear suspension linkages are each acting against only two instead of three spring/damper units, thereby reducing spring flexibility and increasing shock damping. Among various other refinements in this system is a manual mode selector, so that the driver can choose between either automatic mode as described above with the suspension normally in soft mode, or sports driving mode when the suspension remains in firm mode except at low speed to balance out the pressure in the four spring/damper units.

The Citroën Hydractive principle has been subject to continuous development and in the current third generation version, introduced in 2001, an electronically controlled variable ride height function has been incorporated in the system. This is accomplished by varying the amount of fluid in the hydraulic levelling circuitry, depending on car speed and road surface conditions, and is signalled by fast-acting electronic sensors. The benefits of a variable ride height system have previously

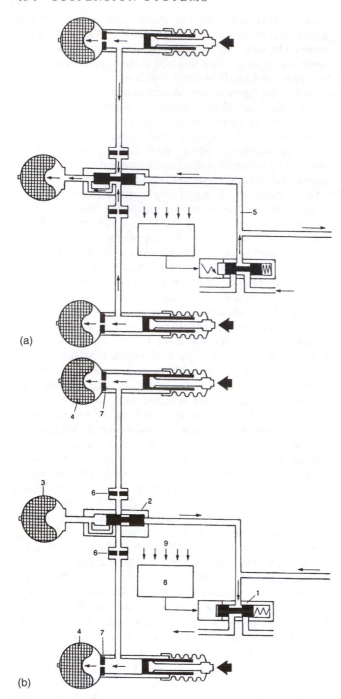

(a)

(b)

Figure 23.63 Schematic arrangement and basic operation of Citroën Hydractive suspension system: (a) soft mode (b) firm mode

1 solenoid valve	6 additional shock dampers
2 firmness regulator	7 main shock dampers
3 additional gas spring	8 computer
4 front gas spring	9 sensors
5 high-pressure fluid	

been explained in connection with air suspension (Section 23.4). Provision is also made for an overriding push-button control of the system by the driver, which caters for the extremes of a high position to facilitate wheel changing and a low position for coupling a caravan.

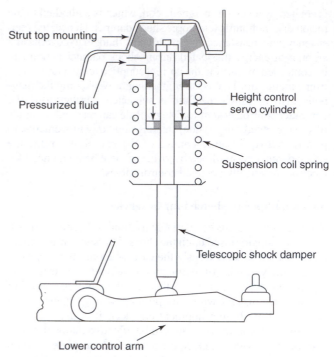

Figure 23.64 Basic arrangement of passenger car active roll control spring mounting

23.8 ACTIVE ROLL CONTROL SYSTEMS

From time to time various ideas have been proposed that sought to eliminate rolling of the vehicle body during cornering and, in some cases, even to make it lean motor-cycle fashion into a bend as in some advanced railway systems. Such systems involved the use of either elaborate mechanical linkages between the body and the suspension arrangements, or hydraulically or pneumatically controlled jacking effects that utilized an existing height-control facility of an air suspension system.

In view of these complications, it is therefore not surprising that the much simpler option of adding anti-roll bars to the suspension system (Figure 23.53) remains a generally preferred compromise, otherwise there are practical limitations on the extent to which the springs themselves can be stiffened or the roll axis be raised (Sections 23.1 and 23.3) to achieve a similar effect. However, the addition of anti-roll bars is still something of a mixed blessing, because the more effective they are in controlling body roll during cornering, the more they tend to spoil the ride when travelling straight ahead, since when one wheel is deflected towards the bump position it also tries to lift the opposite wheel. This can impose rocking movements on the vehicle body over secondary road surfaces, a phenomenon that is sometimes referred to by testers as 'anti-roll bar waddle'.

Active roll control

Before describing a modern active roll control system, it is relevant first to recall that in the mid-1960s Rolls-Royce introduced a hydromechanical, as distinct from hydropneumatic, system of active height control for their cars. Fluid stored under pressure was used to actuate servo pistons, which could raise the car against the upper ends of the suspension coil springs, thereby compensating for their shortening according to the load in the car. Although this system did not confer any measure of active roll control, it will be evident that the same general principle (Figure 23.64) can be applied for this purpose, when the hydraulic servos are controlled by a modern electronic system with fast-acting sensors.

Such an arrangement forms the basis of what Mercedes Benz call their Active Body Control (ABC) system introduced in 1999. At the heart of this comprehensive system are 13 sensors placed at strategic positions on the car, which detect any movement of its body and relay this information to two on-board processors. These compute the optimum suspension setting for each wheel under given conditions within milliseconds. Fluid stored under pressure then acts, via two control and shut-off valves for each axle, against servo cylinders that form movable abutments for the upper ends of the suspension coil springs, so as to counter body movements and keep the car balanced and in even trim. By virtue of the active roll control conferred by this system, that is the hydraulic servo cylinders extend to increase pressure on the outer springs and retract to reduce pressure on the inner springs when the car is cornering, there is no requirement for anti-roll bars. Their absence is of course to the benefit of ride comfort. Mercedes Benz engineers have coined the technical expression 'Mechatronics' to describe the interrelationship between the mechanical, hydraulic and electronic functions of their Active Body Control system.

23.9 SUSPENSION MISBEHAVIOUR IN SERVICE

Suspension and steering troubles are often interrelated. For this reason it is always advisable to consult the particular vehicle manufacturer's trouble-shooting guide or its equivalent, owing to the many variations in modern suspension design that may be encountered. The following brief summary is therefore intended to be no more than a general guide to purely suspension, rather than steering related, problems and their causes:

Hard ride
Tyres over-inflated
Excessive friction in leaf springs, screwed pivot bearings, or ball joints
Stiff operation of suspension struts
Too frequent contact with bump rubbers due to settling of springs
Excessive friction in equalizing linkage of tandem axle suspension.

Excessive pitching
Inefficient shock dampers
Broken rebound clips on leaf springs.

Increased rolling
Slackness in anti-roll bar attachments.

Uneven trim
Vehicle is down at either end or on one side owing to settling of springs
Vehicle remains low on air suspension, visually check air lines and air springs for leakage and then carry out the required diagnostic procedure on the electronic height control system, as specified by the particular manufacturer.
Vehicle is down on either side owing to damaged or worn equalizing linkage of tandem axle suspension.

Noisy operation
Insufficient lubrication of metal pivot bearings or of ball joints
Deterioration of suspension linkage flexible bushings
Looseness of suspension strut mounting
Defective operation of shock dampers including their flexible end mountings.

24 Manual steering

24.1 STEERING PRINCIPLES AND LAYOUT

The Ackermann steering principle

Considered in plan view, a basic feature of the steering geometry of any vehicle concerns the relative angular movements of the front wheels while travelling along a curved path.

Very early motor vehicles had their smaller front wheels mounted on a turntable to provide a single-pivot steering layout after the fashion of horse-drawn vehicles (Figure 24.1a). This simple arrangement suffered from the great disadvantage that shock loads acting in the horizontal direction on either wheel produced a leverage on the steering mechanism equivalent to half the wheel track dimension of the vehicle. This could make the vehicle difficult to control, and chiefly for this reason the double-pivot steering layout later became established practice. In this system the front wheels are mounted on separately pivoting stub axles and linked together by a track rod.

The double-pivot steering linkage was originally proposed by George Lankensperger of Munich for improving the stability of horse-drawn carriages. The system was patented in Britain in 1817 by Rudolph Ackermann, the London agent for Lankensperger, but difficulties were encountered with its practical application and the idea was soon forgotten.

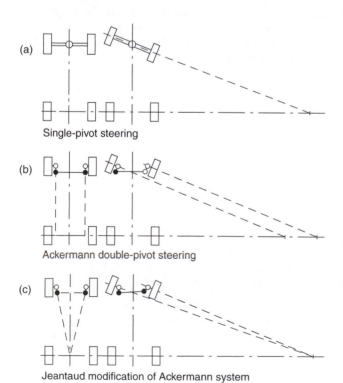

(a)

Single-pivot steering

(b)

Ackermann double-pivot steering

(c)

Jeantaud modification of Ackermann system

Figure 24.1   The evolution of vehicle steering layouts

496

A disadvantage of the Ackermann steering linkage was that the wheels swivelled at equal angles when the vehicle followed a curved path (Figure 24.1b), the effect of this parallel action being to cause scrubbing of the wheels over the ground and heavy operation of the steering. This problem was actually foreseen in the original Ackermann patent, but the geometrical means of overcoming it were only vaguely defined.

It remained for Charles Jeantaud, a French carriage maker and pioneer builder of electric vehicles, to demonstrate in 1878 that this scrubbing of the wheels on turns could be overcome by arranging for the rearward-facing side steering arms to converge towards each other, so that the inner wheel swivelled to a greater angle than the outer wheel. The system required that during cornering the axes of the steered wheels should, when produced, intersect at a common point on the extended centreline of the rear wheels (Figure 24.1c). Confirmation of its success came in the Mechanical Cab Trials in Paris of 1898, when it was said of the Jeantaud electric car – 'the steering is done with a great deal of accuracy'.

In theory this meant that the angular difference between the steered wheels had to be related to the length of wheelbase and width of track. On very early cars of short wheelbase, an approximation to this requirement was usually achieved by arranging for the side steering arms to lie along lines that met at the centre of the rear axle (Figure 24.1c). As the proportions of cars changed and their wheelbase lengthened, it became necessary for the side steering arms to lie along lines that met forwards of the rear axle, the point of intersection with the vehicle centreline being typically at a distance of about two-thirds the wheelbase from the front wheels. In actual practice the Jeantaud system provides for correct angular movements of the steered wheels in three positions only, i.e. straight ahead and one position either side. The intervening errors are due to the track rod moving out of parallel with the centreline of the front wheels. These errors are such that the divergence of the steered wheels is greater than required for small steering angles and smaller than required for large steering angles.

To conclude the historical background to this particular aspect of steering geometry, it must be explained that in English-speaking countries the description 'Ackermann steering' is always taken as including the Jeantaud modification to the Ackermann principle. This modification was originally devised, of course, for rigid tyred carriages operating at relatively slow speeds. It will therefore be evident that Ackermann steering has gradually had to be further modified to accommodate the effects of tyre slip angles and their influence in determining the vehicle centre of turn (Figure 24.2), since any increase or reduction in its effective radius produces understeer and oversteer respectively (Figure 22.15). In particular the over-Ackermann effect that obtains with the traditional layout for small steering angles of the wheels is in direct conflict with the large slip angles that can be developed by the tyres during fast cornering, which is normally accomplished at small steering angles.

It is for this reason that a reduced Ackermann effect is now usually introduced into the steering geometry. This simply involves less divergence or toe-out of the wheels on turns and requires that the steering arms lie along lines that meet behind, rather than ahead of, the rear wheels' centreline. For the average modern passenger car this compromise arrangement of reduced Ackermann steering therefore still provides some relief from tyre scrub during slow-speed manoeuvres, executed with large steering angles of the wheels. The lesser role now played by Ackermann steering accounts for it sometimes being humorously referred to as 'showroom geometry'.

The Ackermann steering principle is also applied to heavy vehicles, and in the case of rigid six- and eight-wheeled types calls for a little further explanation. Taking the latter type, for example, it is usual to arrange for the projections of the axes of all four steered wheels to intersect a line parallel to the rear axles and equidistant from them (Figure 24.3). It will therefore be evident that all four steered wheels will assume different steering angles when turned away from the straight-ahead position. Since the projected axes of the rear wheels cannot pass through the common centre of turning, it necessarily follows that a certain amount of lateral tyre scrubbing will occur as the axles fight each other for position. In actual practice this effect is not quite so bad as it may appear because, apart from being minimized by close positioning of the two axles, in some designs of tandem axle with initially drooping linkage, slight relative steering movements of the axles are allowed, so that when cornering the outer wheels move slightly apart and the inner ones vice versa as the vehicle rolls on its suspension. In other designs of tandem axle a certain degree of flexibility is built into the locating linkage for the axles, so that slight relative lateral movement can occur between the two axles to assist their tracking on corners.

Steering gear layout

The particular form of steering gear layout adopted for a vehicle depends upon the suspension arrangements of the front wheels. As previously mentioned, these are of the non-independent beam axle type for most commercial vehicles and have long since been of the independent linkage type for all private cars. It is therefore necessary to examine the steering gear layout of vehicles under two headings: beam axle steering, and independent front suspension steering.

Beam axle steering

In a commercial vehicle, the driver is usually seated either above or slightly forward of the front wheels so as to provide a minimum overall length of vehicle for a given cargo space and also to ensure that the front axle carries its legally required fair share of the load. These requirements have therefore resulted in the steering box itself being mounted ahead of the front axle (Figure 24.4). Bearing in mind this basic consideration, the overall layout of the steering gear comprises the following components:

Steering box assembly This incorporates some form of worm-and-follower reduction gear which serves to reduce to an acceptable level the driver effort involved in steering. In some installations this may require the addition of a power assistance system. The steering column shaft transmits turning effort from the steering wheel to the reduction gearing in the box, thereby creating a partial rotary movement at the output rocker shaft.

Drop arm As its name suggests, this is simply a down-hanging lever with its top end rigidly attached to the steering

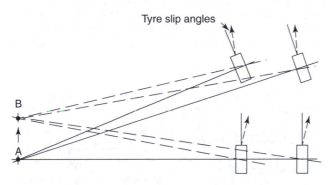

Figure 24.2 Effect of tyre slip angles on Ackermann steering
A apparent center of turn due to Ackermann steering
B real center of turn due to effect of tyre slip angle

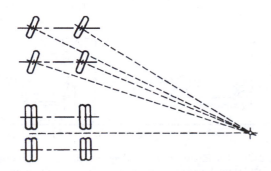

Figure 24.3 Ackermann steering for rigid eight-wheeled heavy vehicle

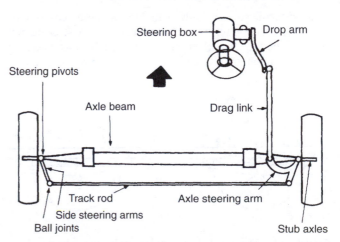

Figure 24.4 Steering layout for commercial vehicle beam axle suspension

box rocker shaft. Angular movement of the latter therefore causes the drop arm to swing in either the fore or aft directions.

Drag link   This is known also as the steering side tube and is fitted with a ball joint connecting socket at each end. The front and rear sockets embrace ball pins retained in the lower end of the drop arm and the outer end of the axle steering arm respectively. The drag link thus acts as a pull-and-push rod to convey steering movements from the chassis-mounted steering box to the spring-mounted axle.

Axle steering arm   In a right-hand-drive vehicle the steering arm is rigidly attached to, and projects inwards from, the offside stub axle. Both stub axles pivot about the ends of the axle beam.

Track rod   This is known also as the steering tie rod and is transversely mounted. Similar to the much shorter drag link, it is fitted at both ends with ball joint connecting sockets which embrace ball pins retained in the axle side steering arms.

Side steering arms   Each pivoting stub axle carries a side steering arm which may project either rearwards or forwards of the axle. Any steering movements conveyed to the offside stub axle are thus transmitted to the nearside stub axle via the side steering arms and their connecting track rod.

An important detail in relation to the steering gear layout used with a beam axle suspension system concerns the arcs of movement of the drag link and the axle steering arm, in the sense of the latter being governed by the action of the road springs (Figure 24.5). If in side view these arcs do not coincide, then unwanted steering movements and vehicle wander will accompany the rise and fall of the axle and road wheels.

In order to preserve the Ackermann effect on heavy vehicles with twin-steer axles, the drag link to the second axle connects to the steering box drop arm at a lesser radius of swing than the drag link connections to the first axle, because the wheels on the second axle are required to steer at reduced angles on turns. Owing to installation length a two-piece drag link is used in practice to connect the second axle to the steering box drop arm, an intermediate drop arm being mounted from the frame on tapered roller bearings to lend the necessary support.

Independent front suspension steering

In a car the driver is usually seated towards the mid-point of the wheelbase; therefore in modern practice the steering box itself can be conveniently mounted between the front wheels to operate a purely transverse linkage. The steering gear layouts

employed with current types of independent front suspension comprise the following components:

Steering box assembly   This incorporates a reduction gearing of either the worm-and-follower or the now more common rack-and-pinion type. In the former case the mounting of the steering box permits its rocker shaft to lie in the vertical plane so that the drop arm can swing to either side. With a rack-and-pinion steering box the steering connections are carried directly by the transversely mounted rack that can slide to either side.

Track rod   In conventional practice a two-piece track rod is used in conjunction with a rack-and-pinion steering box and a three-piece track rod where a worm-and-follower mechanism is employed (Figure 24.6). With the latter arrangement, a pivoting idler arm is required to transfer the motion of the steering linkage to the road wheel on the side of the car opposite to where the steering box is fitted. In the case of a rack-and-pinion steering box, each track rod is connected to either one end or the centre of the sliding rack, as dictated by the layout of the suspension linkage.

Whichever steering layout is used the linkage is typically arranged so that it shadows either the lower or the upper transverse links of the suspension system. The term 'shadow' means that the track rods lie in the same horizontal plane as the appropriate suspension links and swing about the same radius of articulation. If the steering track rod is too long its arcs of movement do not agree with those of the suspension links (Figure 24.7). This results in unwanted steering effects independent of driver control as the wheels rise and fall, which would be worse if in addition the rod and links were not parallel. A track rod of the correct length fitted parallel with the links ensures agreement between their respective arcs of movement

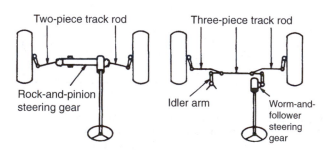

Figure 24.6   Modern two-piece and three-piece track rod layouts used with IFS systems

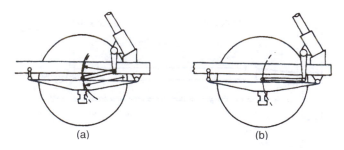

Figure 24.5   The effect of beam axle suspension on steering properties: (a) arcs of movement not coincident, unwanted steering effects (b) arcs coincide, good straight-running properties

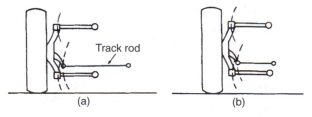

Figure 24.7   The effect of independent suspension on steering properties: (a) track rod too long, arcs of movement disagree (b) correct length rod, agreement of arcs avoids unwanted steering effects

and thus avoids the unwanted steering. Strictly speaking this is true only when the wheels are pointing straight ahead. This is because when they are turned towards either lock the presence of king-pin inclination (Section 24.2) results in the outer ball joints of the track rods rising and falling, which introduces angular errors into the steering geometry relative to that of the suspension linkage.

Where rack-and-pinion steering is used in conjunction with MacPherson transverse link and strut front suspension, a modern trend is to mount the steering gear high on the engine compartment bulkhead for reasons of saving space and easing access to the transverse engine and transaxle. This has meant that the side steering arms have had to be raised from the level of the stub axles, where the track rods could conveniently shadow the transverse links (Figure 24.8a), to a correspondingly higher position on their carrier tubes (Figure 24.8b). Since with rise and fall of the road wheels the motion of the ball-pin connections on the side steering arms now more nearly approaches a straight line rather than an arc, each track rod is arranged to pivot from the centre of the rack to maximize its radius of swing and thereby reduce any unwanted steering effects. Although this high mounting of the rack and pinion may appear to give a less accurate steering control, it does in practice have the advantage of minimizing any self-steering effects arising from lateral deflections of the inner rubber bushing for the transverse link. This is because the high mounted side steering arms experience less sideways movement against their track rods as the strut swings about its upper mounting.

24.2 FRONT END GEOMETRY AND WHEEL ALIGNMENT

General background

In order to take into account the various features that constitute the steering or front end geometry of a vehicle, it is necessary not only to consider the system in plan view as related to Ackermann effect, but also in end and side views. Considered in end or front view, the orientation of the steered wheels both with respect to the road surface and to their swivel axis is defined in terms of camber angle, king-pin inclination and steering offset. In side view, it is the angular disposition of the swivel axis relative to the vertical that we are concerned with, and which is termed castor angle.

Camber angle

This is the inclination of the road wheel centre plane relative to the vertical, as viewed from the front of the vehicle (Figures 24.9a and 24.10a). The camber angle is considered to be positive if the top of the wheel leans outwards and negative if it leans inwards. Historically, a noticeably positive camber was given to the front wheels of motor cars of an earlier era (as any reader who has ever seen one of the legendary prewar Bugatti

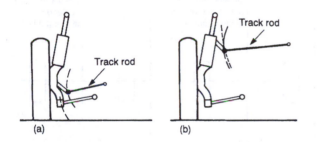

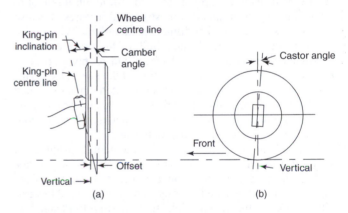

Figure 24.8 Low- and high-mounted rack-and-pinion steering linkages with MacPherson transverse link and strut suspension

Figure 24.9 Front end geometry of beam axle suspension: (a) camber and king-pin angles (b) castor angle

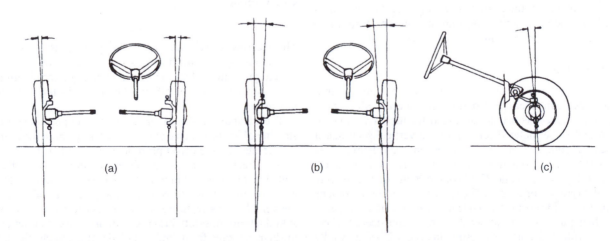

Figure 24.10 Front end geometry of IFS: (a) camber angle (b) king-pin angle (c) castor angle (*SAAB*)

cars head on will agree!), its purpose being to maintain the wheels normal to the surface of the then steeply cambered or crowned single-track roads.

In modern automotive practice, however, the front wheels are usually provided with either a zero or a small camber angle of up to about 1° positive, at the normal standing height of the vehicle. The tendency of a cambered wheel to roll in a direction towards which it is leaning, thereby producing a side force or camber thrust, can have two beneficial effects on the steerable wheels. First, it tends to counteract any small sideways forces imposed on the tyres by ridges in the road, and second, it induces a small lateral preload in the steering linkage. The overall effect is therefore to promote stable, straight-ahead running.

King-pin inclination

This is the inclination of the rod wheel swivel axis relative to the vertical, as viewed from the front of the vehicle. It should be observed that the steering swivel axis coincides with the centreline of the king-pin on beam axle front suspensions (thus the origin of the term king-pin angle: Figure 24.9a), whereas it passes through the centres of the spherical joints on modern IFS systems (Figures 23.13 and 24.10b).

In conventional practice, the angle of the swivel axis is such that it leans inwards at the top. Although this inclination is in any event dictated by structural features, it is also usually desirable from handling considerations. With regard to the former, it is the physical bulk of the brake assembly sandwiched between the road wheel and its pivot arrangements that demands some inclination of the swivel axis. This is because when the swivel axis is projected down to ground level it must either coincide with, or more usually be slightly offset from, the centre of the tyre contact area. Further consideration will be given to this matter later.

From a handling point of view, an incidental advantage of king-pin inclination is that it confers a self-aligning torque on the wheels, since their sloping swivel axes cause the front of the vehicle to be lifted slightly as they are steered towards either lock. As the vehicle sinks to its normal front standing height, the wheels thus have a natural tendency to return to their straight-ahead position, this being effective in both the forward and reverse directions. A king-pin inclination in the range of 6 to 12°, at the normal standing height of the car, is typical of modern practice, as is an inclination of not more than 6° for commercial vehicle beam axles.

Steering offset

The wheels offer their greatest resistance to being steered during parking manoeuvres. To minimize steering effort under these conditions, it is usually arranged for the swivel axis of each wheel to be inclined relative to the vertical such that it intersects ground level at a point either inboard (Figures 24.9a and 24.10b) or, as sometimes found in more recent practice, outboard (Figure 24.11 and 23.13) of the centre of what is aptly called the tyre footprint area. These swivel axis configurations are said to provide positive and negative steering offset respectively. Their purpose is to allow the tyre of the steered wheel partly to roll and partly to scrub, instead of being wholly scrubbed, about the geometrical centre of its footprint area.

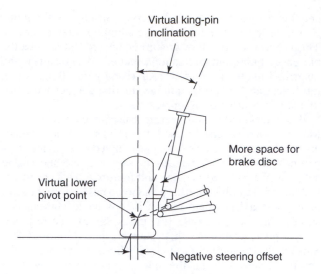

Figure 24.11   Negative steering offset applied to strut suspension with double-joint lower links

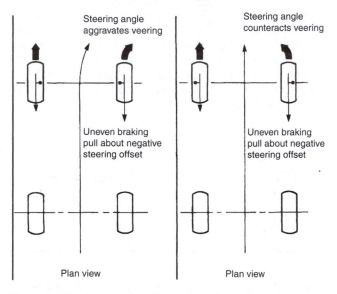

Figure 24.12   Comparative effects of positive and negative steering offset

The steering offset dimension rarely exceeds one-half the width of the tyre footprint area, since greater amounts than this tend to give diminishing returns in reducing steering effort, as well as exaggerating the effects of road shocks felt at the steering wheel.

The advantage claimed for negative, as opposed to positive, steering offset is that tyre drag forces acting about the moment arm created by the offset promote toe-in, rather than toe-out, of the wheels (Figure 24.12). Therefore in the event of a tyre deflation or uneven braking, the wheel turning-in effect of negative offset tends to counteract any veering off course of the vehicle. This feature is also relevant to modern diagonal split dual-circuit safety braking systems, since if either circuit misbehaves one front brake will develop much more drag force than the other.

To achieve a negative steering offset with the widely used MacPherson strut and link suspension can pose a clearance problem between the ball joint mounted from the lower arm of the steering knuckle and the brake disc. In some designs this problem has been overcome by using separately acting transverse links, one of which is diagonally disposed to confer fore and aft location for the strut, and connecting their outer ends to the lower arm of the steering knuckle by individual ball joints. The effect of this departure from using a single ball joint is to confer a virtual pivot point where the extended centre-lines of the two links intersect. It therefore establishes what is termed a 'virtual king-pin inclination' and thus restores the desired negative steering offset (Figure 24.11). However, it should be noted that this offset will vary as the steering is turned towards either lock, but this is not necessarily a disadvantage.

Centre point steering is obtained when the swivel axis of the wheel intersects ground level at the centre of the tyre footprint area, instead of being laterally offset from it as in conventional practice. True centre point steering implies a vertical swivel axis that is contained within the centre plane of the wheel, but this has seldom been found practicable except in the case of some earlier front-wheel-drive cars with brakes mounted inboard on the final drive unit. Otherwise, an apparent centre point steering geometry is obtained by resorting to a fairly generous swivel axis (king-pin) inclination, usually in the region of 10° or more.

An important advantage of centre point steering is that in the presence of braking forces and also traction forces in the case of driven front wheels, acting through the centre of the tyre footprint area, the wheels experience no turning moment or disturbance about their swivel axes. A disadvantage of centre point steering is, of course, the earlier mentioned increased scrubbing effect on the tyres, which demands greater steering effort during parking manoeuvres unless power-assisted steering is used.

Castor angle

Not to be confused with king-pin inclination, this is the angle of the road wheel swivel axis relative to the vertical, as viewed from the side of the vehicle (Figures 24.9b and 24.10c). It is considered to be positive if its intersection point with the ground lies ahead of the vertical centreline through the wheel and to be negative if it lies behind. The effect of a positive castor angle is to create at ground level a moment arm distance or mechanical trail, such that the centre of the tyre contact area always tends to trail behind the swivel axis where it meets the ground (Figure 24.13). Exactly the same principle is involved in the design of the inclined steering head and forks of pedal and motor cycles. A familiar household application of this principle is, of course, found in the construction of furniture castors.

In the case of motor vehicles, however, the mechanical trail due to castor angle is merely additive to the much greater pneumatic trail generated by the tyre itself (Figure 22.5), so that in combination the two trails present a moment arm distance or castor trail at which cornering force acts to provide a self-aligning torque on the steered wheel (Figure 24.13). Castor effect from this source becomes most effective at the relatively small steering angles used during normal and high-speed driving. This is in contrast to the castor effect derived from the previously mentioned king-pin angle, which is most

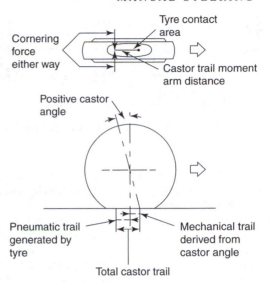

Figure 24.13 The principle of castor action

effective at the relatively large steering angles confined to low-speed driving.

On modern vehicles fitted with radial-ply tyres a positive castor angle in the range of 1 to 3° is typically specified. There are generally two reasons why a castor angle appreciably greater than this may be employed. The first is now the less common one of providing sufficient castor effect for the lightly laden front wheels of rear-engined cars, and the second is the rather subtle one of keeping the steered wheels more nearly upright in the presence of body roll, since there is usually some tilting of independently sprung front wheels under these conditions.

The cambering effect that accompanies high castor angles is, of course, most readily observed during parking manoeuvres, and explains the somewhat drunken posture of the wheels as they lean increasingly towards the direction in which they are being steered. A glance at any 2CV Citroën parked with its front wheels turned to either full lock should make this effect abundantly clear!

Toe-in and toe-out

When front wheel alignment for straight-ahead running is being adjusted during vehicle construction and in service, the wheels are rarely set parallel with each other as viewed in plan, but are provided with a small amount of either toe-in or toe-out (Figure 24.14). Toe-in is an angular misalignment causing the planes of the wheels to converge towards a point ahead of the vehicle, whereas toe-out causes a divergent angle. In practice, toe-in and toe-out are not measured in terms of angular misalignment, but are expressed as the difference in the distances between the rearmost and the foremost points of the wheel rims.

At this juncture it should be appreciated that vehicle suspension and steering mechanisms, in common with other engineering constructions, cannot be made completely rigid, especially where rubber bushings are present at linkage pivot points.

The purpose of deliberate static toe-in misalignment of non-driven wheels is to compensate for small deflection effects which the rolling resistance of the tyres imposes upon the

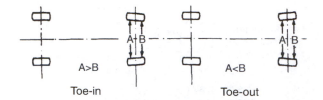

Figure 24.14   Static toe-in and toe-out of the front wheels (grossly exaggerated)

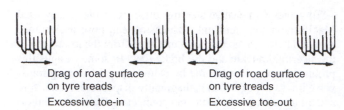

Figure 24.15   Feathered edge tread wear of misaligned tyres

suspension and steering linkages. In other words, the wheels become parallel when the vehicle is in motion by misaligning them when it is stationary.

Although a toe-in setting is the usual requirement to counteract the tendency of non-driven front wheels to toe-out in motion, a toe-out adjustment may be specified for front-wheel drive. This is because the tractive forces generated by driven wheels about a positive steering offset act in the opposite sense to tyre rolling resistance and tend to cause the wheels to toe-in. Since there are always exceptions to the rule, it is vitally important to consult the manufacturer's wheel alignment specification. A toe-in setting of 0–3 mm (0–0.125 in) is typical for commercial vehicle beam axles.

The effects of incorrect steering geometry

With independent front suspension the service engineer is mainly concerned with the effects of the directly adjustable angles of camber and castor, since the king-pin inclination is fixed relative to the camber angle.

Camber angle

This is a fairly critical setting, especially in relation to modern wide-tread tyres that are intended to provide maximum area of contact with the road surface and do not like being tilted. The effects of camber angle settings that are equal but too positive, or too negative, will be to reduce the grip of the tyres on cornering as further changes in wheel camber occur, and to increase their wear on one side of the tread in straight-ahead running. If the camber angles are set outside their tolerance on equality, then directional stability will suffer as the steering tends to pull towards the side with the greatest positive camber. It should be noted, however, that a manufacturer may occasionally specify a fractionally less positive camber angle for the wheel that normally runs along the inner edge of cambered roads to preserve straight-ahead running.

Castor angle

Again this is a critical setting, mainly because of its considerable influence on the directional stability of a vehicle. The effects of castor angle settings that are too positive will be to increase the amount of steering effort required for cornering and also create a tendency for the vehicle to run down a road camber, as well as being more easily deflected off course by cross-winds. Caster angle settings that are too negative will reduce steering effort on cornering, but inhibit the self-centring action of the steering on straightening up. They will also tend to make the vehicle wander, so that constant steering correction may be demanded from the driver to maintain straight-ahead running. If the castor angles are set outside

their tolerance on equality, then the steering will continually pull towards the side with the least positive castor.

The effects of incorrect wheel alignment

To say that the front wheels of a vehicle should be correctly aligned implies that their tyres must roll over the road surface without scuffing, both in straight-ahead running and, with reservations, during cornering. The need for maintaining, at all times, correct alignment of the front wheels is justified on the grounds of economy of operation and safety of control.

Economy of operation

This refers not only to tyre life but also to fuel consumption. A reduction in normal tyre life can result from incorrect alignment during straight-ahead running because the tread is being scrubbed sideways over the road surface. This type of abrasive wear can be identified by the feathered edges of the worn tread pattern (Figure 24.15). Any misalignment in the straight-ahead position will result, of course, in some departure from the Ackermann effect when the wheels are steered. This can be detrimental to tyre life if excessive tread scrubbing occurs towards full steering lock either side during slow-speed manoeuvres.

At normal to high cornering speeds, when only very small amounts of steering lock are used, it is more important that any deviation from the Ackermann effect is consistent in either steering lock, since the slip angles resulting from angular distortions of the tyres assume more importance than the Ackermann angles of the wheels. In any event, it should be remembered that cornering contributes more to the overall wear of the tyres than do harsh braking and accelerating. Fuel economy can also suffer from incorrect wheel alignment, simply because the tyres offer greater resistance to rolling and thus absorb more of the engine power.

Safety of control

This relates to the handling characteristics of the vehicle. In this respect, incorrectly aligned front wheels can be one cause of a tendency to wander, thus obliging the driver to make frequent corrections with the steering wheel to maintain a true course.

With the now widespread use of independent rear suspension systems on cars, the correct alignment of all four wheels in relation to each other assumes even greater importance.

Adjusting and checking the front end geometry

Perhaps surprisingly there are many different forms of adjustment provided to set the camber and castor angles on the two established types of independent front suspension.

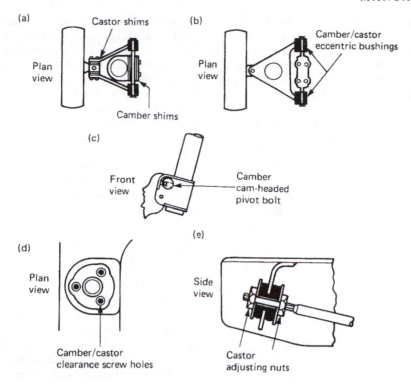

Figure 24.16 Methods of adjusting camber and castor angles

For some applications a separate means of adjustment is incorporated for each of the two angles, which is generally to be preferred, while in others a common adjustment sets both angles. In principle, however, these adjustments normally involve the use of either slotted shims, eccentric bushings, cam-headed pivot bolts, clearance holes and screws, or adjustable length diagonal links, or some combination of these features. Taking one example of each of these should illustrate their basic applications.

Slotted shims In an unequal transverse links IFS, slotted shims can provide for camber angle adjustment by interposing them between the upper wishbone fulcrum shaft and the cross-member spring turret to which it is attached (Figure 24.16a). Castor angle adjustment can similarly be made by incorporating slotted shims on each side of the steering ball-joint housing, where it is embraced by the outer ends of the upper wishbone.

Eccentric bushings These can be incorporated at the front and rear pivots of the lower wishbone of an unequal transverse links IFS (Figure 24.16b). In this application, if both bushings are turned together in either direction of rotation they will enable the camber angle to be set, while if the bushings are turned equally but in opposite directions of rotation they will cause skewing of the lower wishbone so that the castor angle can be set.

Cam-headed pivot bolts In a transverse link and strut IFS, a cam-headed pivot bolt can be used as one of the two bolts that conventionally clamp the strut assembly to the upper arm of the steering knuckle, the corresponding holes in the strut bracket being slotted to allow relative angular movement of the strut during adjustment (Figure 24.16c). Turning of the cam-headed pivot bolt in either direction will then enable the camber angle to be set.

Clearance holes and screws Again with a transverse link and strut IFS, clearance screw holes can be provided at the attachment for the upper strut mounting to its tower, thereby allowing a limited degree of both camber and castor adjustment (Figure 24.16d). That is, moving the position of the strut mounting either inwards or outwards will enable the camber angle to be set, while moving it backwards or forwards similarly sets the castor angle before the nuts are tightened.

Adjustable length diagonal links These can be used in conjunction with the lower track control arms of both unequal transverse links and transverse link and strut systems of IFS. Since either reducing or increasing the effective length of the link will cause skewing of the track control arm, this form of adjustment is used to set the castor angle (Figure 24.16e). Advancing or retreating both nuts that clamp the link front bushing effects the required adjustment.

Various specialized combination gauges have long been available to enable the front end geometry angles of camber, castor and king-pin inclination to be measured accurately and rapidly. Such a gauge may simply comprise a metal post with an adjustable upper foot and a double lower foot that incorporates a spirit level, so that the gauge may be hand held vertically in tripod fashion against the wheel rim. Attached to the post is a moving arm with pointer that reads over fixed scales, the arm also being provided with a spirit level so that when the arm is moved to centralize the bubble a reading can be taken.

With the vehicle on a level surface, the tyres correctly inflated and the wheels in the straight-ahead position, the camber angle is measured first directly from the appropriate scale on the gauge. To measure next the castor angle the wheels require to rest on steering turntables with degree graduations, so that they can be steered through equal specified angles to the left and right. The castor angle is then shown on the appropriate scale as being the difference between the camber readings in each position. King-pin inclination is obtained in a similar manner, but the wheels must be braked while being steered to the left and right and the rim must be marked so that the gauge is reapplied in exactly the same position for the second reading. The purpose of also checking the king-pin inclination is to detect any bending of the hub spindle or its support if this is suspected.

Some front end geometry gauges are held instead against the trailing portion of the tyre wall, so as to eliminate any errors arising from wheel rim damage, while other types of gauge are mounted from the hub spindle or a wheel nut. In all these cases it is important that the particular manufacturer's operating instructions are consulted. Automatically displayed-read-outs of the front end geometry angles are a feature of the more sophisticated types of permanently installed wheel alignment equipment.

Adjusting and checking the wheel alignment

With a commercial vehicle beam axle suspension system, the toe adjustment of the front wheels is made by either lengthening or shortening the effective distance between the track rod ball joints (Figure 24.17a).

For this purpose, the screwed attachment of the ball joint sockets to the track rod ends is also utilized as a convenient method of fine adjustment (Figure 24.18). The ball joint socket housings are threaded either right hand or left hand and used in pairs, so that making the toe adjustment is a matter of trial and error rotation of the track rod tube in the appropriate direction. When the correct toe setting is obtained, the adjustment is locked to prevent the track rod unscrewing from the socket housings. Finally, a check is made to ensure that the assembly of track rod and socket housing remains free to rock about the ball pins in the side steering arms to prevent any binding.

The toe adjustment of independently sprung front wheels is carried out on a similar principle, but the points in the steering linkage at which the adjustments are made are closely related to the type of steering box employed.

In the case of a two-piece track rod used in conjunction with a rack-and-pinion steering box the points of adjustment are usually confined to the outer socket housings. If the inner ends of the track rods are connected by axial ball joints to the ends of the steering rack (Figure 24.17b) then toe adjustment can simply be a matter of unlocking the threaded outer socket housing, slackening the clips securing the rack sealing bellows to the track rods and equally rotating the track rod tubes as necessary. With the alternative arrangement of the inner ends of the track rods connecting to the centre of the rack, toe adjustment can involve detaching the outer ball joints from the side steering levers, so that the threaded socket housings can be rotated relative to the track rods.

In a much less common arrangement, the track rods are fabricated from steel pressings, so that it is no longer practicable to provide screwed attachments for the outer ball joints. In this case the problem of setting toe alignment is solved by moving the steering box rack housing either fore or aft within its mounting clamps.

Where a three-piece track rod is used in conjunction with a worm-and-follower steering box, the points of toe adjustment are confined either to the centre track rod of a forward-mounted linkage (Figure 24.17c) or the outer track rods of a rearward-mounted on one (Figure 23.17d). In the former case, the points of adjustment are the same as those for a beam axle system described earlier. This same arrangement can also be duplicated for the outer track rods of a rearward-mounted three-piece steering linkage, each track rod being separately and equally adjusted.

Perhaps a more convenient but otherwise similar form of adjustment is that where the inner socket housings are made non-adjustable and the outer ones are attached to the ends of the track rods by means of short adjuster tubes. These tubes are threaded right and left hand, so that their rotation makes the required length alteration to the track rods, which again must receive equal adjustment.

In conventional practice, where the steering ball joint socket housings are screw threaded both for retention and adjustment, some form of locking device must be provided to ensure that the pieces cannot unscrew. For this purpose the threaded adjustment is locked either axially with a jam nut or radially by a clamp and pinch bolt (Figure 24.18), or less commonly by a combination of both axial and radial clamping through the medium of a conical compression ring and back-up nut.

The correct alignment of independently sprung rear wheels assumes just as much importance as it does for the front wheels. Here again it must be appreciated that the wheel locating linkages for IRS systems cannot be made completely rigid,

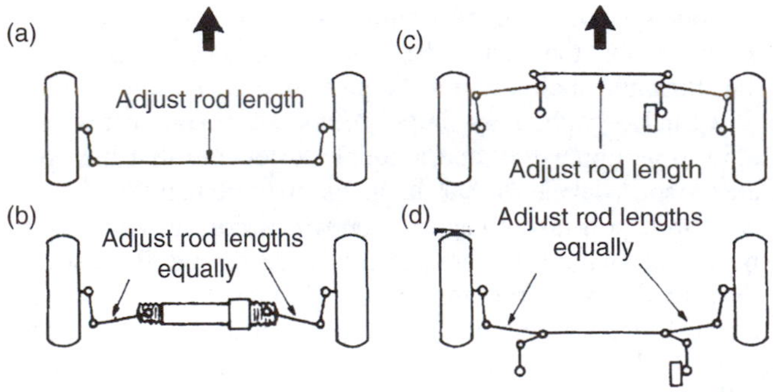

Figure 24.17 Points of toe adjustment in beam axle and IFS

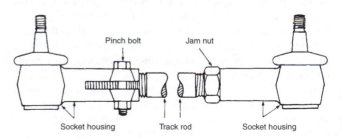

Figure 24.18 Two commonly used methods of locking toe adjustment

especially where rubber bushings are present at the pivot points. As a result there is a tendency for driven rear wheels to toe-in during acceleration and to toe-out during deceleration and braking. (The non-driven rear wheels of FWD cars tend to toe-out from the effects of rolling resistance and braking.) If, as in the case of the front wheels, we are going to set the rear wheels deliberately misaligned when stationary, so that they run parallel when the vehicle is in motion, then which condition are we going to compensate for: toe-in during acceleration, or toe-out during deceleration? In fact it is the latter condition that must be avoided, since it can provoke an unstable oversteering condition when entering a corner. For this reason, it is therefore usual to find that independently sprung rear wheels are provided with a small amount of static toe-in.

The once popular semi-trailing arm type of IRS especially requires accurate toe setting and is generally provided with some means of adjustment at either, or both, the bearing pivots for each arm. Fore-and-aft repositioning of one of the pivot pins may be accomplished by the use of a slot and a fixed

location for its cam-headed pivot pin (Figure 24.19) and also by the use of a sliding bearing bracket with jacking screw. Alternatively, selective shimming may be used between both bearing brackets and the vertical face of the cross-member to which they are attached. This form of adjustment can also be applied to the bearing brackets of a transverse link and strut system, where it serves the dual purpose of setting toe and camber of the rear wheels (Figure 24.20).

Checking for correct alignment of the front (or rear) wheels is accomplished by either measuring their static toe setting with some form of trammel gauge, or verifying that they have no side slip as the vehicle is moved forward over a wheel alignment indicator pad. Trammel-type gauges can vary in construction from a simple base bar and two columns with adjustable pointers to an optically registering instrument that affords a high degree of precision. The required front and rear measurements of toe are made at the height of the wheel centers and taken from either the wheel rims, or band lines chalked on the tyre treads, according to the vehicle manufacturer's preference.

Another alignment check for the steered wheels concerns what is generally termed their toe-out on turns. This is a measure of the extent to which the wheels diverge on part lock owing to Ackermann effect. For this check the wheels are required to rest on steering turntables with degree graduations; one wheel is then turned outwards at 20° (although other angles may be advised) and the lesser turning angle of the other wheel is noted. The significance of any discrepancy between the toe-out on turns measured on the vehicle and that specified by the manufacturer is that the steering arms may be distorted. An alignment 'set-back' check may also be necessary, which determines whether either front wheel is displaced backwards or forwards from its true position.

Incorrect alignment of the wheels on heavy vehicles can prove very costly in terms of both accelerated tyre wear and increased fuel consumption due to greater rolling resistance.

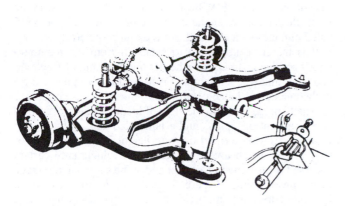

Figure 24.19 An example of toe adjustment for semi-trailing arm type IRS (*Ford*)

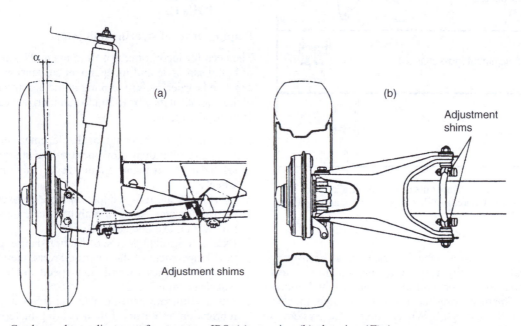

Figure 24.20 Camber and toe adjustment for strut-type IRS: (a) rear view (b) plan view (*Fiat*)

Front wheel toe-in/out	
Front wheel toe-out on turns	
Front wheel camber angle	
Front wheel castor angle	
King pin inclination	
Twin steer alignment	
Steering box centring	
Rear wheel toe-in/out	
Rear wheel camber	
Front/rear alignment (rigid chassis)	
Trailer alignment	

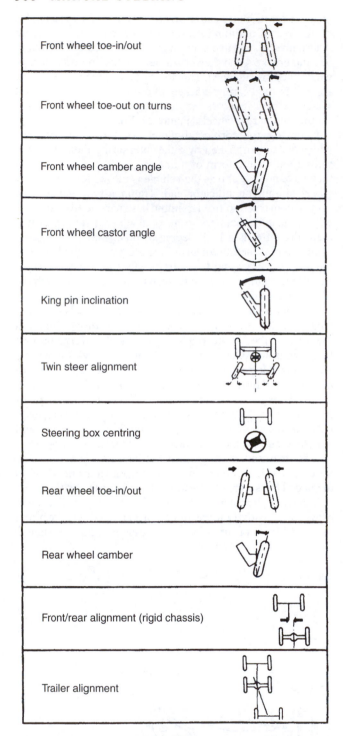

Figure 24.21 Wheel alignment measurements available for heavy vehicles using Churchill Wheel Force 1950 equipment

Steering alignment checks require the use of measuring equipment that is specially adapted to single- and twin-steer three-axled and twin-steer four-axled vehicles and multi-axled trailers. The measuring capability of such equipment as exemplified by the Churchill Wheel Force 1950 System is summarized in Figure 24.21.

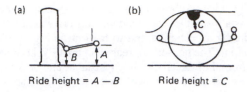

Ride height = *A* − *B*   Ride height = *C*

Figure 24.22 Measuring vehicle ride height

Vehicle loading, front end geometry and wheel alignment

Finally, it must always be borne in mind that service personnel necessarily have to make adjustments to front end geometry and wheel alignment on a stationary vehicle which are intended to ensure correct operating conditions when it is both laden and running. The static settings of independently sprung wheels must therefore take into account geometry and alignment changes that accompany their suspension movements, and which by the same token relate to vehicle loading. In the case of one modern FWD car, the changes for example in the front wheel toe between the laden and unladen condition is 0.5 to 2.5 mm toe-out and 0 to 2.0 mm toe-in respectively.

It is therefore important to establish from the manufacturer's service instructions whether the checking of front end geometry and wheel alignment is to be done with the vehicle in the unladen or laden condition. In the latter case reference is usually made either to the state of loading required, such as two front seat occupants and a half-filled tank of fuel; or to what is variously termed the ride height, standing height or trim height, which involves loading the vehicle until certain specified measurements are attained at the suspension linkage lower pivot points (Figure 24.22a) or, in the case of a Hotchkiss drive system, between the bump rubbers and frame member (Figure 24.22b).

24.3 STEERING AND SUSPENSION BALL JOINTS

Requirements of steering ball joints

The need for incorporating ball joints in the steering linkage of both beam axle and independent suspension systems was explained earlier. Before looking at the constructional details of these joints, it may be useful to summarize what is expected of them. They must:

1 Accommodate both angular displacement of the steering links and rotational movement of the steering levers, as related to the suspension and steering motions of the road wheels.
2 Provide freedom from lost motion or backlash in the steering linkage, so as to maintain accurate steering control over an extended service life.
3 Exert a controlled degree of friction damping on the rotational movement of the steering levers, thereby suppressing any tendency towards low-speed wobble of the road wheels (Figure 24.42).
4 Either eliminate or minimize the frequency of periodic lubrication by virtue of their bearing material properties and sealing capability.

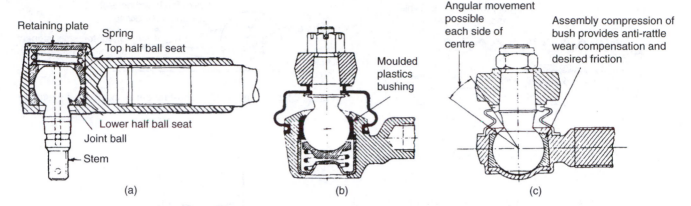

Figure 24.23 The development of steering ball joints: (a) early spring-loaded type with metal seatings (b) later spring-loaded type with plastics seating (c) modern type with compressed plastics bushing (*Cam Gears*)

There are basically two forms of steering ball joint, which may be classified by the direction in which the steering forces are transmitted through the ball pin, these being known as transverse and axial joints.

Transverse ball joints

This type of ball joint is overhung from the steering levers, so that the ball pin is mainly subjected to sideways loading. The connection between the ball pin and its steering lever is effected through a tapered shank and hole, acting in conjunction with a threaded portion and clamping nut.

In its once traditional form for car applications, the spherical surface of the ball pin was embraced by an opposed pair of fixed and floating seatings located within the ball joint socket. These seatings were usually spring loaded together by means of a coil spring compressed between the floating seating and the socket end cap (Figure 24.23a). The latter often took the form of a screwed adjusting plug, so that the build sequence was first to tighten fully the plug to solidify the joint and then unscrew a fraction of a turn to provide the joint with both freedom of movement and a certain degree of frictional resistance.

A metal-to-metal combination was originally employed for the ball pin and its seatings, with provision being made for their periodic relubrication in service. From the late 1950s moulded plastics materials became more widely used for the seatings and generally eliminated the need for any further lubrication after initial build (Figure 24.23b).

Some ten years later a different form of ball joint construction, incorporating new concepts in materials and design, became established to avoid the complication of spring-loaded seats. With joints of this type the ball pin is made of constant spherical radius and is encapsulated in a partially enveloping plastic bushing. This is typically produced from high-density polyethylene material, which is resistant to abrasion, provides a smooth low-friction bearing surface, and is self-lubricating. The bushing is installed under pressure into the tapered bore of the socket and the natural compressibility of the plastic material is relied upon to maintain the required degree of frictional resistance to joint movement (Figure 24.23c). Another form of transverse steering ball joint sometimes used in heavy commercial vehicle practice is that known as the pendant joint, the distinguishing feature of which is that the fixed and floating

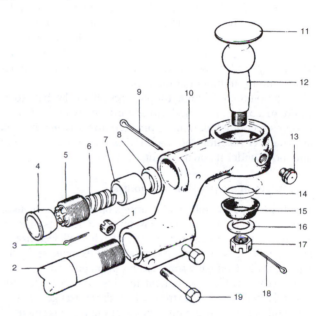

Figure 24.24 Exploded view of pendant-type track rod end (*Seddon Atkinson*)

1	nut	11	plug
2	track rod	12	ball pin
3	split pin	13	grease nipple
4	dirt excluder	14	metal cover
5	adjusting screw	15	rubber cover
6	spring	16	washer
7	ball cup, inner	17	nut
8	ball cup, outer	18	split pin
9	split pin	19	bolt
10	body		

metal seatings are disposed normal to, rather than coaxial with, the axis of the ball pin (Figure 24.24).

Axial ball joints

This type of ball joint (Figure 24.25) is usually confined to applications where the track rods articulate directly from each end of a rack-and-pinion steering gear, as described earlier. Its construction is such that it is integrated with the steering rack, each end of which is counterbored to receive

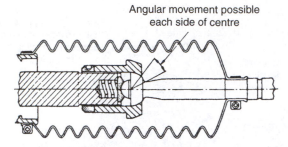

Figure 24.25 Axial ball joint used with rack-and-pinion steering gear (*Cam Gears*)

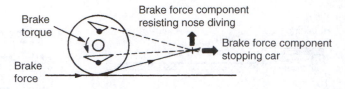

Figure 24.26 Ball joints in anti-brake-dive suspension geometry

the spring-loaded inner seating. The outer seating is formed integrally within a steel thimble housing screwed on to the threaded end of the rack. A convenient means of adjusting ball joint friction is thus afforded by the threaded thimble, which can then be retained by a locking collar. Alternatively, the threaded thimble may be screwed fully home against a hexagonal flange (for a holding spanner) and the ball joint friction adjusted by selective shimming between the inner seating and its abutment in the rack.

The amount of friction torque present in the ball joint should, of course, be strictly in accordance with the manufacturer's recommendation. In practice this is typically such that whatever position the track rod is swung to, it will not quite move under its own weight.

Advantages of suspension ball joints

Since the late 1950s spherical bearings have come to be used almost exclusively for the outboard pivot connections of independent front wheel suspension linkages. This followed from a few earlier designs that were inspired by the example set by the French Citroën Company some twenty years previously, where an almost inevitable requirement for this type of suspension joint arose because of their front-wheel-drive arrangements.

As compared with the once conventional arrangement of trunnion-mounted yokes with separate stub axles hinged from king-pins, the use of spherical bearings or ball joints that accommodate both suspension and steering movements of the yokes offers the following advantages:

1 The simpler construction reduces the unsprung mass of the suspension system and thus contributes to improved road holding.
2 Improved support is provided for the wheels, because the vertical distance separating their pivot points is always at a maximum.
3 Front-wheel-drive universal joints can be accommodated between the span of the suspension ball joints.
4 Fore-and-aft compliance movements of the wheel locating links are readily allowed, which reduces low-speed harshness in the ride.
5 The inboard pivot axes of the suspension links can be angled relative to each other, in side view, so as to reduce nose diving of the car during braking (Figure 24.26).

Although the above points have been made relative to unequal transverse link IFS systems, similar considerations apply to transverse link and strut installations.

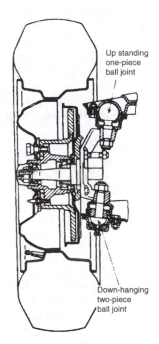

Figure 24.27 Two types of ball joint used with transverse links (*Lada*)

Types of suspension ball joint

Two basic constructions are used for suspension ball joints as follows:

One-piece, in which both suspension and steering movements of the yoke are accommodated by a combination of rocking and partial rotation of the ball pin within its seating.

Two-piece, where the suspension and steering movements of the yoke are accommodated separately by means of a part-spherical bushing drilled to receive a rotatable thrust pin.

An important consideration in the application of ball joints to unequal transverse link IFS systems is that the joint through which the suspension spring load is transmitted should not impose undue frictional resistance to steering movements of the yoke. For this reason either an up-standing one-piece or a down-hanging two-piece ball joint is used at the outer end of the lower link, against which the suspension spring is normally arranged to act. The former choice of transmitting the spring load via the crown of the ball was a Jaguar innovation of the late 1940s which allowed a reduction in effective friction radius at the bearing seating to minimize steering effort. A similar advantage is gained by using a down-hanging two-piece joint at this station (Figure 24.27)

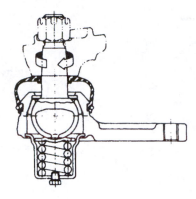

Figure 24.28 An example of ball joint used with transverse link and strut IFS (*Nissan*)

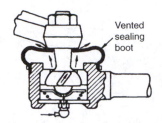

Figure 24.29 Lubrication of steering ball joint

and originally represented the American approach to minimizing steering friction, while at the same time transmitting spring load.

An up-standing one-piece ball joint is conventionally used at the outer end of the upper link (Figure 24.27). This may also be retained for the less common type of layout where the suspension spring acts against the upper link, since again the spring load will be transmitted via the crown of the ball in the interests of reducing steering friction (Figure 23.32).

Early designs of suspension ball joint were usually made adjustable by means of either a screwed cap, or selective shimming against a fixed cap, which set the endwise positioning of the floating seating. Like their steering ball joint counterparts, modern suspension ball joints are rarely provided with means for external adjustment in service (Figure 24.28).

Ball joint lubrication and sealing

From the descriptions already given of car steering and suspension ball joints, it should be evident that in modern practice they generally require neither adjustment nor lubrication during their service life. If the latter is required it is usually at intervals of not less than 12 000 miles or 20 000 km, unless operating conditions are particularly severe, although with some older-established models this interval may be reduced to 6000 miles or 10 000 km.

A major contribution to this desirable state of affairs lies in the more dependable sealing arrangements of the modern ball joint, which is positively sealed by a flexible boot made from rubber or rubber-like plastics materials. An encircling garter spring is typically used to retain the sealing boot in a groove around the socket housing. The neck of the boot must, of course, allow for freedom of rotational movement of the ball pin relative to the socket housing, so it therefore floats against a sealing face machined on the ball pin supporting member. It is also good practice periodically to examine all ball joint sealing boots and gaiters for splits or cuts and, of course, to renew them if necessary.

The steering ball joints of commercial vehicles either are regularly greased under a planned lubrication programme, or receive oil from an automatic chassis lubrication system (Section 30.4). Grease lubricated joints usually feature a large-capacity grease reservoir in their end cap, a grease distributing

slot across the ball that also prevents internal pressurization of the joint, and a vented sealing action that permits excess grease to escape thereby preventing rupture of the boot (Figure 24.29).

As an example of routine maintenance requirements, the heavy commercial vehicles manufactured by Seddon-Atkinson require their steering ball joints to be lubricated after the first 1000 miles or 1600 km of vehicle life and thereafter every 1500 miles or 2400 km or weekly. The grease used is a multipurpose lithium-based compound whose characteristics combine water resistance with extreme pressure- and corrosion-inhibiting properties. If the vehicle is operated under conditions in which corrosive elements such as salt are encountered, then a suitable calcium-lead-based grease should be used.

24.4 MANUAL STEERING GEARS

General requirements

At the heart of all motor vehicle steering systems is, of course, the steering gear itself. This comprises some form of reduction gearing to provide the driver with a mechanical advantage in respect of steering the road wheels. To the gear designer such a mechanism would be classified as a control gear, since it must have the minimum possible free play or backlash. This is because its primary job is that of accurately transmitting motion, rather than conveying power. The general requirements for a motor vehicle steering gear may be summarized as follows:

1 It should provide for accurate and light steering control without involving excessive movements of the steering wheel.
2 The transmission of road wheel disturbances to the steering wheel should be minimized, while still allowing the driver to retain an acceptable feel of the varying road surface conditions.
3 It should not hinder the self-centring action of the steered wheels as the car returns to a straight path.
4 During parking manoeuvres the driver effort required to steer the wheels should be reduced to a minimum, because it is then that their resistance to being steered is greatest.

Translated into practical terms, these requirements mean basically that the steering gear should possess a high mechanical efficiency in the 'forward' direction (steering wheel turning the road wheels), but preferably a lower mechanical efficiency in the 'reverse' direction (road wheels attempting to turn the steering wheel). At this stage we shall confine our attention to manual, as distinct from power-assisted, steering gears.

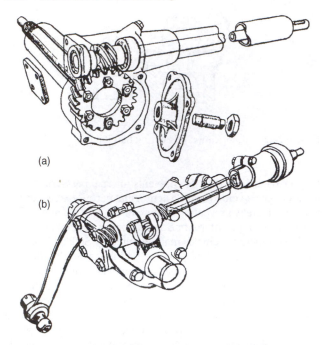

(a)

(b)

Figure 24.30 Comparison between (a) early worm-and-wheel and (b) worm-and-nut steering gears

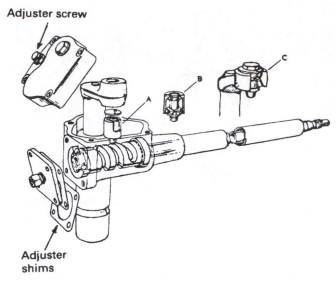

Adjuster screw

Adjuster shims

Figure 24.31 Worm-and-peg steering gear: three basic types of peg assemblies are shown at A, B and C

Worm-and-follower steering gear

The designers of early motor vehicles soon recognized that the steering gear should possess a certain degree of irreversibility or, as previously stated, a lower mechanical efficiency in the reverse direction. In other words they sought to reduce the transmission of road wheel disturbances to the steering wheel. This particular problem could assume quite large proportions, because of the poor state of road surfaces and the limited knowledge then available to relate steering and suspension characteristics to best advantage. It was countered by the general adoption of steering gears that incorporated a simple worm-and-follower reduction gearing of either the worm-and-wheel (or part-wheel known as a sector), or the worm-and-nut variety (Figure 24.30). In both designs the worm may be regarded as acting in the manner of a screw thread against its follower. Their reverse efficiency was governed by the angle of the worm threads, the steering becoming less reversible with smaller thread angles. It should be appreciated that total irreversibility of the steering was not desired since all self-centering ability would then be lost, and in any event the steering gear would have to absorb greater shock loading.

The mid 1920s saw the advent of increased-section lower-pressure tyres for greater ride comfort. This was further improved by the mid 1930s with the gradual adoption of independent front wheel suspension systems, which also contributed to better passenger accommodation by virtue of permitting a more forwards mounting of the engine. In combination, these changes meant that a greater proportion of the car total weight was being carried by the steered wheels on tyres with a substantially increased contact area with the road.

It therefore became increasingly more difficult to keep the driver steering effort within reasonably bounds, especially during parking, without drastically increasing the number of

steering wheel turns required from lock to lock. That is, the steering had to be made lower geared, which in automotive terminology implies a higher numerical ratio for the reduction gearing. This in turn led to the development, especially in America, of various proprietary high-efficiency steering gears in which rolling, as opposed to sliding, motion was provided between the worm and its follower.

These improved mechanisms reduced steering effort without resorting to excessively low gearing, since forwards efficiency was raised by about 50 per cent. The accompanying increase in reverse efficiency now assumed less importance by virtue of there being better road surfaces and better matching of steering and suspension geometries. High-efficiency worm-and-follower steering gears fall into three basic categories: worm and peg, worm and roller, and worm and recirculating-ball nut.

Worm-and-peg steering gear

This particular type of mechanism was originally devised by a prolific American inventor, David Ross of the Ross Gear & Tool Company, who introduced it in 1923. It possessed the twin virtues of being an unusually compact and robust form of steering gear.

The worm-and-peg steering gear basically comprises a worm with a coarse pitch thread, which intercepts a conical-headed peg carried by a side-mounted lever arm on the drop arm rocker shaft (Figure 24.31). (Descriptions were given earlier of the manner in which the drop arm of worm-and-follower steering gears can be connected into the steering layouts of beam axle and independent suspension systems.) Strictly speaking, this type of steering gear did not become a true high-efficiency mechanism until Ross also mounted the peg in rolling bearings.

Although its mode of operation is one of extreme simplicity, in that rotation of the steering wheel compels the peg to follow the helical-thread groove of the worm and thus convert the motion of its line of contact into partial rotation of the rocker shaft, its internal geometry is a little more complex. This is because the arcuate movement of the peg causes it to rise and fall relative to the rotational axis of the worm, so that there is an accelerated movement between the worm and the rocker shaft as the steering is turned towards each lock. In driver terms this reducing angular ratio means that the steering is less direct acting about the straight-ahead position, which is the range of movement most used during normal driving, than it is when manoeuvring on either lock. Also the steering will require more effort towards the locks.

To compensate for this effect, the worm thread may be generated with a variable helix angle, which is greatest towards the middle portion of the worm. In this event, the worm can no longer be regarded as a true gear but rather as a form of cam, which accounts for this type of steering gear also being known as a cam-and-lever mechanism. Further modifications to the worm-and-peg steering gear that not only maintained more nearly constant the effective leverage of the side lever arm about the rocker shaft, but also permitted a substantial increase in angular capacity (drop arm travel), were the introduction in the mid 1930s of twin and, at a later date, triple pegs.

However, the worm-and-peg steering gear is obsolete for motor car practice, having disappeared from American vehicles by the mid 1950s and from vehicles elsewhere by the late 1960s. The reason for this was the increasing popularity of worm and recirculating-ball nut and rack-and-pinion steering gears respectively.

Worm-and-roller steering gear

This type of mechanism was invented by a British engineer, Henry Marles, who in 1919 was the first to substitute purely rolling for sliding contact between the meshing elements of a worm-and-follower steering gear. It evolved through several versions before assuming its most popular form, which comprises an hourglass-shaped worm that engages with a double-toothed roller follower carried between the jaws of the rocker shaft for the drop arm. The periphery of the roller is formed into the contour of a worm gear tooth.

Its mode of operation is such that around the straight-ahead position of the steering the worm thread bears on the inner flanks of the double-toothed roller, but as either lock is approached it begins to bear on the appropriate outer flank of the roller. Since with an hourglass worm the helix angle of the threads reduces as the pitch diameter increases, the internal leverage of this type of steering gear remains substantially constant from lock to lock. A triple-toothed roller is sometimes used to increase the angular capacity of the worm-and-roller steering gear (Figure 24.32a and 24.33). Similar to the case of the worm-and-peg mechanism, the specialized geometry of the hourglass worm really qualified it as a form of cam rather than a gear, which accounts for it also being known as a cam-and-roller steering gear.

The worm-and-roller steering gear continues to find limited application in European and Russian-built cars, but in America it was entirely superseded towards the late 1960s by the worm and recirculating ball nut mechanism.

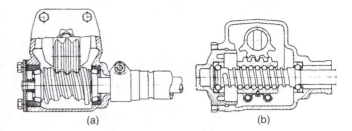

Figure 24.32 Cross-sectional comparison between typical (a) worm-and-roller and (b) worm and recirculating-ball nut steering gears (*SKF*)

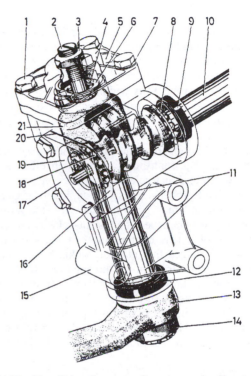

Figure 24.33 Identifying the parts of a worm-and-roller steering gear (*Volvo*)

1 bolt	13 drop arm
2 adjusting screw	14 nut
3 locknut	15 steering box body
4 drop arm shaft bush	16 steering column bearing
5 cover	shell, lower
6 circlip	17 steering column cover
7 tab washer	18 steering column bearing,
8 steering column bearing, upper	lower
9 steering column oil seal	19 washer
10 steering column	20 shims
11 drop arm shaft bush	21 drop arm shaft with roller
12 drop arm shaft seal	

Worm and recirculating-ball nut steering gear

Introduced in 1939 by the Saginaw Steering Gear Division of the General Motors Corporation, this further American development in high-efficiency steering gears represented a return to the basic worm-and-nut principle, but with the important difference that a chain of ball bearings was interposed between the thread grooves of the rotating worm and its translating nut.

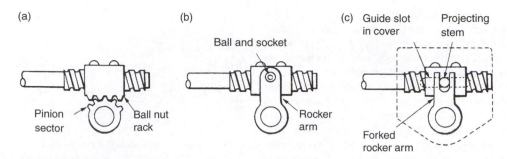

Figure 24.34   Worm and recirculating-ball nut mounting systems: (a) rack mounting (b) rocker mounting (c) fork mounting

Since in this type of construction the ball bearings cannot roll indefinitely around the helical path of the worm, a diagonal return passage must be provided that connects externally the beginning and end of the thread groove in the nut. Hence, the chain of ball bearings rolls within a closed circuit in the nut and replaces sliding by rolling friction against the worm. Either a single, or two independent, ball bearing circuits may be used according to the severity of duty imposed.

A not readily apparent, but nevertheless very significant, design detail that contributes to the high efficiency of the modern worm and recirculating-ball nut concerns the profile of their thread grooves. These are typically of Gothic arch cross-section and may be compared with the form of ball groove used in some types of constant-velocity universal joint (Figure 19.16b). This configuration of non-circular groove therefore minimizes free play or backlash by reducing the radial and angular freedom of the assembly. It also provides clearance space for any small particles of foreign matter that might enter the thread grooves. Differences in the design of present-day steering gears of this type are mainly concerned with the choice of ball nut mounting, which may take the form of either a rack, a rocker or a fork connection (Figure 24.34).

Rack connection

This is widely used, and featured in the original Saginaw arrangement. One side of the ball nut is provided with a toothed rack and meshes with a pinion sector, integral with the rocker shaft for the drop arm (Figures 24.32b and 24.34a). By virtue of this rack-and-pinion type of connection, the angular ratio between the movements of the worm and the rocker shaft remains constant throughout the wide operating range of this type of steering gear (although this characteristic can, in fact, be modified by interposing one long tooth engagement between two shorter ones, thereby varying the internal leverage to match the addition of power assistance). A disadvantage of the rack mounting for the ball nut is that it tends to produce a less compact steering box than the alternative rocker-and-fork mountings. However, this particular form of steering gear still finds application on some American and Japanese cars, and also 4 × 4 vehicles.

Rocker connection

This is effected through a spherical joint and side arm on the rocker shaft (Figures 24.34b and 24.35). The resulting arcuate travel of the joint is reflected in small angular movements of the nut about its own axis. Hence, there is a reducing angular ratio between the movements of the worm and the rocker shaft as the steering is turned away from the straight-ahead position. In this respect it is similar to the worm-and-single-peg steering gear, except that the reduction in angular ratio differs slightly according to which lock is being approached, because in one direction the angular movement of the nut occurs in the same sense as worm shaft rotation, whereas in the other direction it occurs in the opposite sense.

Fork connection

In this case the ball nut is provided with a projecting stem upon which is mounted a roller. This is constrained to run in a straight channel machined in the steering box cover plate, and thus prevents the ball nut from turning as it translates. At the base of the roller stem is a chamfered collar, which engages the open forked end of the rocker shaft side arm (Figure 24.34c and 24.36). As the steering is turned away from the straight-ahead position, the chamfered collar slides along the fork, thereby increasing the effective length of the side arm. Hence, there is an increasing angular ratio between the movements of the worm and the rocker shaft, which is the opposite characteristic to that found in the rocker mounting for the ball nut and makes it suitable for heavy vehicles by reducing steering effort towards full lock.

Rack-and-pinion steering gear

The established popularity of the rack-and-pinion steering gear can be accounted for by its precision of operation, simplicity in construction and low cost. This type of steering gear first became popular on certain French and German cars of small to medium size during the early 1930s. For cars of similar size in the United Kingdom, rack-and-pinion steering was essentially a development of the late 1940s when IFS became more generally adopted there. It is also finding increasing acceptance on much larger cars because power assistance can now be added to overcome the leverage limitations of a simple rack-and-pinion gear.

The operation of a rack-and-pinion steering gear is one of extreme simplicity. Rotary movements of the steering wheel are converted by the pinion and rack into linear push-pull movements of the track rods which in turn connect to the side steering levers and turn the road wheels (Figure 24.37). It should be noted that the pinion may be located either above or below the rack according to whether the side steering levers project rearwards or forwards respectively.

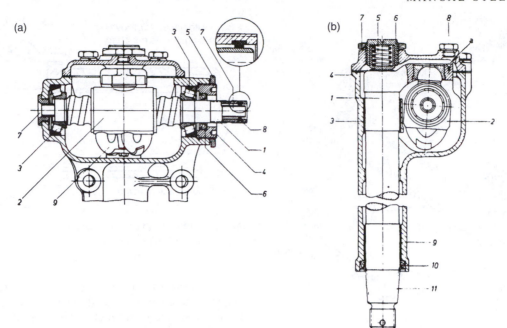

Figure 24.35 Identifying the parts of a worm and recirculating-ball nut steering gear (*Mercedes-Benz*)
(a)   (b) *a* centring bore for the centre position of the steering shaft arm

1 steering worm
2 steering nut
3 tapered roller bearing
4 adjusting ring
5 hexagonal nut for adjusting ring
6 seal ring in adjusting ring
7 cable tube
8 seal ring for cable tube
9 ball guide tube

1 steering shaft
2 steering nut
3 bearing bushing, top
4 pressure sleeve
5 pressure spring
6 pressure screw
7 hexagonal nut
8 screw plug
9 bearing bushing, bottom
10 seal ring
11 steering shaft

Construction of rack-and-pinion steering gear

A typical rack-and-pinion steering gear comprises the following components, but there can be many differences in detail construction (Figure 24.38):

Gear housing   The housing for the rack consists of a steel tube, one end of which is cast in an aluminium alloy housing for the pinion. Alternatively, a one-piece housing construction in aluminium alloy may serve to enclose both the rack and the pinion.

Rack gear   The toothed rack is produced from steel bar, its guiding surface being of either semi-circular or V profile. A V profile may be used to resist any rotary movements of the rack which can be imposed by steering connections that are overhung from its centre section rather than from its ends.

Pinion gear   This is carried on a stub shaft connected either directly, or through the medium of a flexible coupling, to the lower end of the steering column assembly. The pinion gear is usually provided with helical instead of straight-cut teeth, a feature which not only ensures smooth action but also reduces kick back by reacting it as side thrust and increased friction on the rack bearings.

Rack-and-pinion bearings   A two-point support is provided for the sliding rack gear. At the end of its housing remote from

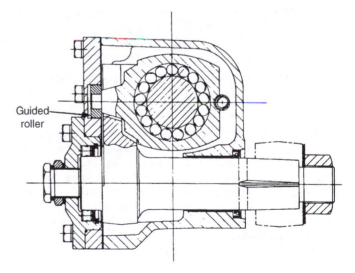

Figure 24.36 Worm and recirculating-ball nut steering gear with forked rocker (*Burman*)

the pinion the rack is guided by an inserted bearing bush which may be of the oil-impregnated, sintered bronze type. The other end of the rack is supported between the pinion gear and an opposed slipper bearing. This is spring loaded against the rack

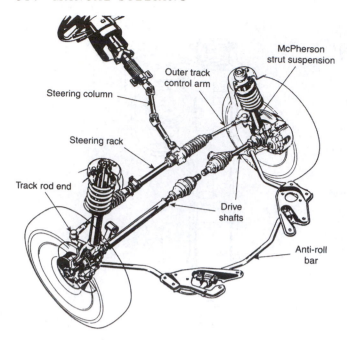

Figure 24.37 Typical installation of a rack-and-pinion steering gear (*Triumph*)

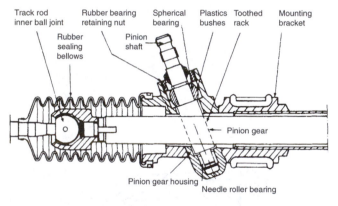

Figure 24.38 Constructional features of modern example of rack-and-pinion steering (*Fiat*)

to exert a friction damping effect on unwanted rack movements caused by road shocks at the wheels. Various combinations of bearings may be used to absorb the radial and axial loads imposed upon the pinion gear. According to design the bearings may be of either the rolling or the plain varieties, or both in combination such as an upper plain bush and a lower needle roller bearing (Figure 24.38). It is now recognized that the pinion must be rigidly supported by its bearings to provide the driver with a good feel of the road.

The lubricant is retained in the steering box by a pair of concertina-like rubber bellows, which form an oil-tight enclosure between either end of the rack housing and the adjacent track rod to which they are clipped. Some form of radial lip seal and dust excluder are fitted at the point where the pinion stub shaft enters the pinion housing. The cover plates for the latter are sealed by simple gasket-type seals.

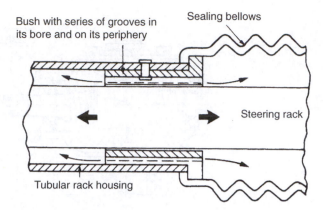

Figure 24.39 Detail of rack bushing to allow air and oil a free passage during sliding of the rack

An important point of design detail concerns the provision made to allow air and oil to pass either side of the rack bearing bush remote from the pinion (Figure 24.39). This is generally achieved by incorporating a series of lateral grooves on the periphery of the bushing where it is inserted into the end of the rack housing.

The complete assembly of rack and pinion is mounted across either the front suspension cross-member, or the engine compartment bulkhead, depending upon the design requirements of any particular car (Figures 23.16 and 24.37). Methods of attachment include outrigger brackets from the rack housing and encircling U-shaped clamps (Figure 24.38). The latter are sometimes rubber lined so that the entire rack-and-pinion assembly can float sideways to a very limited extent in sympathy with any push or pull movements of the track rods arising from road wheel shocks (Figure 24.37). Since the rack is then not sliding relative to the pinion, kick-back is prevented from reaching the steering wheel. For the same reason a flexible coupling is often used between the pinion stub shaft and the lower end of the steering column.

It is important to realize that both these features, together with the friction damping imposed by the rack slipper bearing (Figure 24.46), are intended to suppress an inherent shortcoming of the rack-and-pinion steering mechanism. This concerns its high reverse efficiency or, in other words, the ease with which the road wheels can turn the steering wheel, when they are subject to disturbing forces. It is mainly for this reason that some manufacturers still prefer to use worm-and-follower types of steering box, which possess a certain degree of irreversibility.

A later development for manual rack-and-pinion steering has been to increase its reduction ratio towards the steering locks. This variable-ratio effect reduces steering effort for low-speed manoeuvres, whilst retaining a more direct feel (smaller movements of the steering wheel for given angles of the road wheels) about the straight-ahead position for normal driving. It has been achieved by the ingenious expedient of making all the rack teeth either side of centre with different profiles. The mid-point teeth have a virtually saw-tooth form with a wide pitch, so that contact with the pinion is then near the helix tips to increase the effective pitch circle, but all teeth on either side of mid-point have a gradually narrowing pitch

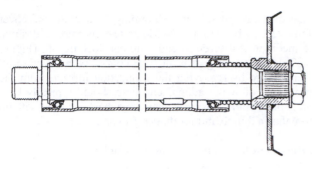

Figure 24.40 Arrangement of upper steering column bearings (*SKF*)

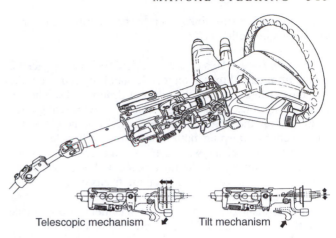

Telescopic mechanism   Tilt mechanism

Figure 24.41 Steering column with tilt and telescopic features (*Honda*)

with the flat flanks moving to increasingly shallow angles and reducing the effective pitch circle. In other words, it may be imagined that the pinion gradually shrinks in diameter as its contact with the rack moves away from mid-position. The overall effect is that the steering becomes lower geared (higher numerical ratio) and therefore less heavy towards the steering locks, albeit accompanied by an increase in the total number of steering wheel turns lock-to-lock.

Steering column bearings and adjustable columns

Before the advent of international safety regulations that required the use of collapsible steering columns to prevent the driver from being impaled in a crash, the upper bearing for a rigid steering column could be either a plain bush or a rolling bearing, the latter usually where it also served as a thrust bearing for the worm shaft.

In modern practice the upper part of the collapsible steering column is generally carried in a pair of rolling bearings, their design being such that the installation can survive crash tests which require steering ability to be retained. A steering column arrangement that utilizes SKF-developed single-row angular-contact ball bearings is shown in Figure 24.40. Here the opposed caged ball bearings are preloaded together by a compression spring beneath the boss of the steering wheel, and for quiet operation the bearing raceways are polished after hardening.

Another feature often found in modern practice is the readily adjustable steering column, either manually or power operated, which can be both telescoped and tilted to suit driver requirements. In principle this is not a new idea, because in earlier practice many British cars were fitted with the Bluemel telescopic steering column. On certain high-grade cars the driver was offered a choice of alternative length steering columns and also bolting arrangements for the steering box mounting bracket to alter column tilt or 'rake', but these settings were of course fixed on chassis build. However, the separate upper column associated with present-day safety steering systems provided an opportunity to incorporate a readily adjustable splined telescopic and pivoted tilting mechanism (Figure 24.41). The tilting function is accommodated by a universal joint enclosed in the articulated steering column, the geometric centre of the universal joint coinciding with the pivot axis of the column. A lever operated toothed locking mechanism establishes alternative angles of steering wheel tilt, above and below the standard position. In the case

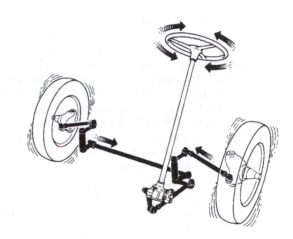

Figure 24.42 The nature of steering shimmy (*Alfa-Romeo*)

of a power operated system, an offset mounted electric motor actuates a jacking rod that lies alongside the articulated column and connects to a projecting lug at its upper end, thereby acting in the manner of a bell-crank adjuster for steering wheel tilt.

24.5 INSPECTING AND ADJUSTING THE STEERING MECHANISM

Wear in the steering mechanism

Wear in the steering mechanism proper, that is the steering gear and its ball-jointed linkage to the road wheels, can be either wholly or partly responsible for a variety of steering complaints that can make driving both difficult and hazardous. For example, the presence of wear in the steering linkage can materially contribute to low-speed steering wobble or shimmy of the front wheels. This may be described as a rapid oscillation of the front wheels, which try to turn in and out alternatively, causing the steering linkage to shake side-to-side (Figure 24.42). Clearly a process of elimination must be adopted to establish whether any free play is confined either to the steering linkage or to the steering gear, or is cumulative to both:

Steering linkage This may be checked for wear by jacking up the front end of the car on to axle stands placed under the suspension links so that they do not droop to the rebound position. In other words, the angularity of the steering track rods remains the same as for the normal ride height of the car. Then by grasping both front tyres at hub level and attempting to force the wheels apart and together again, the presence of any free play in the linkage will be shown up by relative lateral movement between the fronts of the wheels.
Steering gear A visual check on whether the steering gear requires adjustment, or possibly replacement, can be made with the car on a ramp. The drop arm may then be observed for follow-up movement, while an assistant 'waggles' the steering wheel either side of the straight-ahead position against the turning resistance of the tyres.

As far as the Department of Transport's inspection of goods vehicles is concerned, if a point on the rim of the steering wheel moves without the road wheels turning, for a distance more than one-fifth of the diameter of the steering wheel, then this can be a reason for rejection in the case of non-power-assisted steering mechanisms. For example, there should not be more than about 105 mm (4.2 in) free play on a 525 mm (21 in) diameter steering wheel. In the case of cars and light goods vehicles not exceeding 1525 kg (30 cwt), free play at the steering wheel in excess of 75 mm (3 in) is considered unacceptable where worm-and-follower steering gears are employed, and only one-sixth of this amount is permitted for the more direct-acting rack-and-pinion steering gears.

Once having pinpointed the source or sources of free play in the steering mechanism it is then, of course, necessary to carry out a more detailed examination of the affected parts. From this it may be decided by a combination of experience and knowledge of the particular manufacturer's service techniques whether replacement, simple adjustment or overhaul is required.

Adjustment of worm-and-follower steering gears

Two main adjustments are normally incorporated in the modern steering gear, their purpose being to remove all end play from the worm bearings and to establish correct mesh between the worm and its follower. Another related adjustment concerns the steering lock stops, which are intended not only to limit the maximum angularity of the steered wheels, but also to prevent the steering gear mechanism from over-travelling on extreme locks (Figures 24.43 and 24.44). The vehicle manufacturer's recommendations must always be strictly followed, of course, when making these important adjustments to the steering.

Adjustment of worm bearings

This is always the first adjustment to be checked and if necessary performed. The worm is supported between either angular-contact ball, or tapered roller, bearings mounted face-to-face. These bearings are adjusted to give zero end float or, in some cases, a slight preload. Their adjustment is usually effected at the abutment for the lower bearing outer race,

either by means of selective shimming against the end cover of the steering box (Figure 24.31) or, less commonly, through the medium of a screwed end plug and locking nut (Figure 24.32b). The steering gear must, of course, be relieved of all external load by removing the connecting linkage from the drop arm. Following adjustment there should typically be a barely perceptible drag as the steering wheel is turned with no more than a thumb and forefinger grip on its rim.

Adjustment of worm-and-follower mesh

Often this takes the form of a single adjuster that not only establishes the correct meshing condition between the worm and its follower, but also controls the end float of the rocker shaft, which is usually carried in either plain journal bushings or needle roller bearings.

An important point concerning mesh adjustment is that the follower is generally intended to be in closer contact with the worm around the straight-ahead position of the steering, than it is towards either lock. The purpose of this is to compensate for any greater wear occurring over that range of travel where steering movements are most concentrated during normal driving.

Worm and peg
Any lost motion that develops in the straight-ahead position of the steering can be eliminated by altering the depth of engagement of the peg, since a clearance always exists between its truncated end and the bottom of the thread groove in the

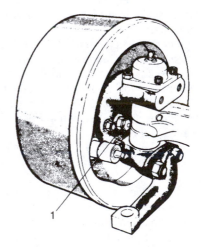

Figure 24.43 Steering lock stop adjusters used with beam front axle (*Seddon Atkinson*): 1 axle stop adjusting screw

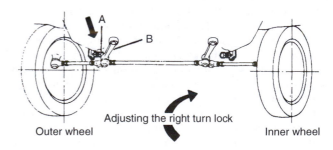

Figure 24.44 Steering lock stop adjusters used with IFS (*Alfa-Romeo*): A = lock stop adjuster; B = steering arm

worm. For mesh adjustment it was therefore found expedient to allow the side lever arm end of the rocker shaft to bear either directly against a selectively shimmed side cover of the steering box, or against the end of an adjuster screw carried by the side cover and secured with a locking nut (Figure 24.31). Hence, the peg can be brought into closer engagement with the worm by the removal of shims or by screwing in the threaded adjuster.

Worm and roller

It has been found expedient in some designs to provide a single adjustment that not only controls rocker shaft end float, but also establishes the correct meshing condition between the worm and roller in the straight-ahead position of the steering. The meshing facility is made possible by the intentional offsetting and raising of the roller axis relative to that of the worm, as viewed endwise. To effect the necessary adjustment, a coaxially mounted threaded stud with thrust collar engages a T-slot in the roller end of the rocker shaft. Screwing in the threaded stud thus brings the roller into closer engagement with the worm thread groove, the adjustment being retained by a locking nut (Figure 24.33). In some earlier designs of worm-and-roller steering gear, notably those of American origin, the roller was located centrally below the worm. Here

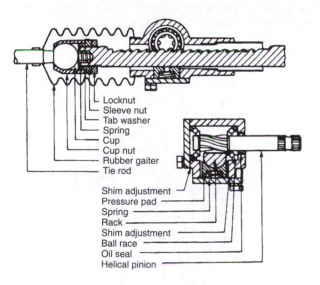

- Locknut
- Sleeve nut
- Tab washer
- Spring
- Cup
- Cup nut
- Rubber gaiter
- Tie rod

- Shim adjustment
- Pressure pad
- Spring
- Rack
- Shim adjustment
- Ball race
- Oil seal
- Helical pinion

Figure 24.45   Typical adjustment facilities for rack-and-pinion steering gear

the adjustment took the form of an eccentric sleeve or lantern for mounting the worm shaft bearings, so that partial rotation of the sleeve in the appropriate direction brought the worm into closer contact with the roller. This form of adjustment was also once used by Rolls-Royce in their version of the Marles worm-and-roller steering gear.

Worm and recirculating-ball nut

All three types of ball nut mounting may be brought into closer radial contact with the worm through endwise adjustment of the rocker shaft. For both rocker and fork ball nut mountings this adjustment can be simply effected by the usual screw and locknut arrangement on the side cover of the steering box (Figure 24.36). In the case of the rack ball nut mounting, the endwise adjustment of the rocker shaft is accomplished by a coaxially mounted threaded stud, as described previously for the worm-and-roller mechanism. Screwing in the threaded stud has the effect of bringing the teeth of the sector into closer mesh with those of the ball nut rack (and therefore the ball nut into closer radial contact with the worm) because the sector teeth are provided with a small wedge angle specially for this purpose. Whichever type of mesh adjustment is provided it is, of course, carried out with the steering in the straight-ahead position.

A final point in connection with the rocker-mounted ball nut is that this particular arrangement readily lends itself to a self-adjustment facility. This may be obtained by incorporating a compression spring and plunger within a hollow threaded adjuster so that once the endwise adjustment of the rocker shaft has been set, the spring will exert a preload on the ball nut and its worm and thus maintain the steering gear free of backlash throughout its full working range (Figure 24.35b).

Adjustment of rack-and-pinion steering gears

The adjustments that can be carried out on a rack-and-pinion steering box are generally those confined to initial build and overhaul, since this type of mechanism is essentially self-adjusting by virtue of the toothed portion of the rack being spring loaded into mesh with the pinion teeth. Typical adjustment facilities may therefore be summarized as follows.

Meshing of rack-and-pinion gears

A cover plate with either packing shims or an adjuster screw is provided to control the correct meshing of the rack-and-pinion gear (Figures 24.45 and 24.46). This is accomplished,

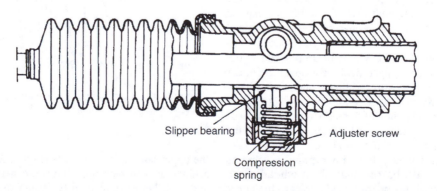

Slipper bearing   Adjuster screw

Compression spring

Figure 24.46   Alternative screw adjuster for meshing of rack-and-pinion gears (*Fiat*)

in effect, by limiting the amount to which the spring-loaded slipper bearing allows the rack to climb away from the pinion teeth, as the turning effort is applied at the steering wheel. The correct setting is usually such that, with the rack centralized, the steering is just free from binding.

Centralization of the rack is necessary because the teeth in its middle portion are normally machined fractionally higher than those towards the ends to compensate for the greater amount of wear suffered around the straight-ahead position of the steering in any steering box.

Pinion bearings preload

This adjustment applies only in cases where the pinion gear is supported between a pair of opposed angular-contact ball bearing races. By reducing their end float to less than zero the bearings become preloaded, the object of which is to increase the rigidity of the pinion mounting. Their adjustment is effected at the abutment for either upper or lower bearing, through the medium of selective shimming against the end cover of the pinion housing (Figure 24.45).

24.6   CONVENTIONALLY STEERED AND SELF-STEERING AXLES FOR HEAVY VEHICLES

Conventionally steered axles

The steering layout for these has been described in Section 24.1. An important constructional feature of this layout concerns the pivot arrangements for the stub axles, which are carried on each end of the axle beam, the latter being of I-section and forged from nickel steel. The basic arrangement is such that the stub axles are straddle mounted on the outer bosses of the axle beam, a construction of steering head that has long been known as 'reverse Elliott'. It is so named after an American engineer, Sterling Elliott, who pioneered the converse arrangement where forked ends of the axle beam embraced the stub axles. It was originally applied to a quadracycle prior to the advent of the motor vehicle. Although this particular arrangement was adopted for early motor vehicles, it was later found to be structurally incompatible with front wheel brake installations, hence the adoption of the previously mentioned reverse Elliott construction.

For medium-duty applications each stub axle pivots about a king-pin or swivel pin, which is made a tight taper fit in the axle boss and passes through plain bushes pressed into the lower and upper jaws of the stub axle. These bushes may be produced from either solid lead bronze or steel-backed lead bronze lined materials. Alternatively a polymer-based composite material such as Glacier DX may be employed (Figure 24.47), which was developed specifically for operation under conditions of marginal lubrication. Either a plain thrust bearing is interposed between the lower end of the swivel pin and a closing plate for the lower jaw of the stub axle, or a tapered roller thrust bearing is sandwiched between the boss of the axle beam and the lower jaw of the stub axle.

In heavy-duty applications the upper plain bearing is replaced by a tapered roller bearing that takes both radial and thrust loads. The cone of this bearing is clamped to the upper end of the swivel pin, while the cup seats against a shoulder in

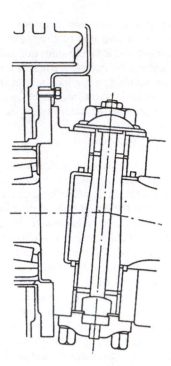

Figure 24.47   Commercial vehicle king-pin design employing two Glacier DX bushes and thrust washers (*T&N PLC*)

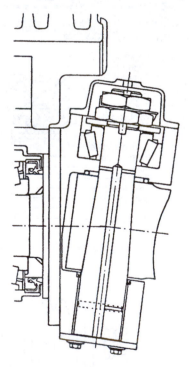

Figure 24.48   Commercial vehicle king-pin design employing a tapered roller bearing and a single Glacier DX bush (*T&N PLC*)

the upper jaw of the stub axle (Figure 24.48). This particular arrangement of bearings allows the steering to be less heavy and with better returnability during low-speed manoeuvres, because it reduces the friction torque that would otherwise be imposed on the steering by the jacking effect of swivel pin inclination. It is recognized that the city bus is a prime example where friction torque has to be reduced, because much of its duty is at low speed where steering loads are high and lock angles frequently greatest. Both oil and grease methods of lubrication are utilized for swivel pin bearings.

Self-steering axles

The concept of the self-steering axle was first developed during the 1970s in Italy, where it was designed to be incorporated as the trailing axle of a tandem group on rigid heavy vehicles. Its purpose was to reduce tyre scuffing on tight turns, since this adversely affected both the vehicle and the road. In this application the load equalization on the tandem axles was biased at least 60 per cent in favour of the leading driven axle, so that the vehicle could always generate the required cornering force. More recently the use of a self-steering axle in the rearmost position of tri-axle trailers for articulated vehicles has proved advantageous in conforming to international legislation on turning circle limits.

A self-steering axle is similar in layout to a conventionally steered one, but the beam is of either round or square tubular form and the stub axles pivot about plain bearings. The wheel spindles are offset to the rear of the stub axle swivel pin axes to provide a castor trail for the wheels. An essential feature of the self-steering axles is the addition of a pneumatically controlled centring system to supplement the castor trail effect. This system acts between the axle beam and the track road, and not only overcomes the internal friction in the axle steering arrangements to ensure that the wheels return smoothly but not suddenly to the straight-ahead position after turns, but also largely counteracts the effects of any uneven braking between the wheels. A separate hydraulic damper may also be connected in a similar manner to prevent any wheel shimmy. Another essential feature of the self-steering axle is the provision of a pneumatically controlled steering lock to facilitate reversing the vehicle.

A modern example of a self-steering axle is the Rubery Owen-Rockwell Stefan design (Figure 24.49), which can be fitted at either the front, the rear or both positions on a tri-axle trailer and may be used in conjunction with steel leaf springs or an air suspension system. The pneumatically controlled centring device that acts between the axle beam and the track rod is termed a Torpress bellows. This is mounted from the axle beam and is connected to an air supply, its operating pressure being regulated through a valve that senses vehicle load. In operation the pivoting arms of the pressurized bellows intercept a striker plate attached to the track rod, so that an additional centring effect can be exerted on the wheels. A hydraulic vibration damper is directly connected between the axle beam and the track rod. For reversing purposes, a pneumatically controlled locking cylinder is mounted from the axle beam and when operated it clamps an arrester plate that is also attached to the track rod. The locking cylinder is actuated with either a manual switch located on the side of the trailer or an electropneumatic switch from inside the vehicle cab. The Stefan self-steering axle has a toe-in requirement of 4–6 mm (0.16–0.23 in), which has to

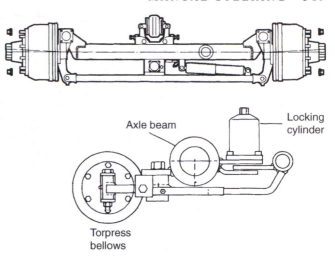

Figure 24.49 General arrangement of self-steering axle (*Rubery Owen-Rockwell*)

be set strictly in accordance with the procedure laid down by the manufacturer.

24.7 STEERING WHEEL AIRBAGS

As in the case of various other passive safety features that have appeared in modern passenger cars (Section 30.1), the now increasing adoption of airbags to protect at least the driver owes much to the pioneering efforts of the Daimler-Benz company in Germany during the late 1960s. Whereas safety belts have long continued to provide the occupants of a vehicle with a primary restraint system, airbags are intended to offer a supplemental restraint system. The distinction between these two restraint systems in practice is that safety belts help to reduce the risk of injury from being thrown about within, or thrown from, a vehicle in a variety of accident situations, while airbags are designed to distribute impact forces more evenly over a restrained occupant's face and chest in moderate to severe frontal or near-frontal collisions. Since the chances of a correctly belted front seat passenger coming into contact with the dashboard are likely to be less than those of a similarly restrained driver hitting the steering wheel, a driver-side airbag naturally received priority installation, to be complemented later with a passenger-side airbag.

Airbags are typically produced from a sturdy nylon material, the driver-side airbag being stored within the steering wheel hub and concealed by a stylized trim cover (Figure 24.50). It deploys when a sensor system detects a very sudden deceleration, the severity of which is associated with a frontal or near-frontal impact at speeds above about 20 mile/h (32 km/h). In this event a propellant chemical in the inflator is triggered by the sensor system and produces a large volume of mainly nitrogen gas, which then inflates the airbag in less than 0.05 seconds (Figure 24.51). It should be appreciated that the airbag does not act merely as an inflated cushion, but has small side vents so that when the driver makes forcible contact with it, the energy of impact is dissipated by the gas escaping from the deflating bag. A more recent refinement in airbag design

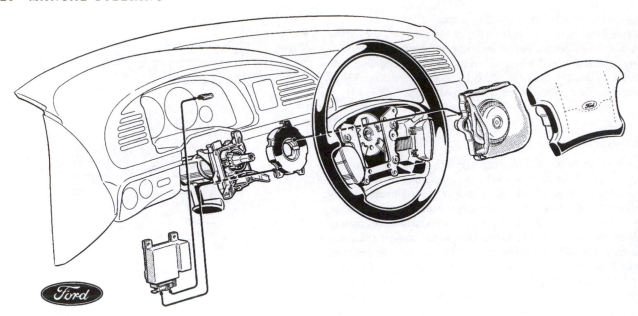

Figure 24.50   Installation of driver's side airbag system (*Ford*)

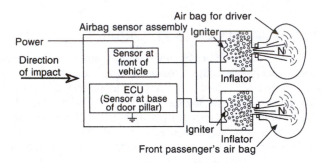

Figure 24.51   Electrically actuated airbag system

includes an additional propellant charge, which is triggered to give a two-stage deployment of the airbag. The time interval between detonation of the first and second propellant charges, albeit measured in milliseconds, depends upon the severity of the impact as detected by the system crash sensors.

Caution   In the workshop the manufacturer's advice should always be followed regarding the precautions necessary to prevent accidental triggering of the airbag, since this is potentially hazardous to personnel working on the vehicle and also involves expensive replacement for the owner.

25 Power-assisted steering

25.1 THE NEED FOR POWER-ASSISTED STEERING

General background

The reasons for the early development in America of high-efficiency manual steering gears to reduce driver effort have previously been explained. It was also in America that Francis W. Davis, who has sometimes been called the father of power steering, first demonstrated in 1926 the benefits of providing power assistance to reduce the steering effort on a once prestigious make of car known as the Pierce-Arrow. However, the problems attending this pioneer effort were considerable, and it was not until twenty-five years later, at least as far as passenger cars were concerned, that Chrysler in America introduced an acceptable power-assisted steering system. Such was the enthusiasm for this new feature, soon popularly abbreviated to PAS, that by 1952 it was introduced in one form or another on the more expensive models of five American car manufacturers. At the time many American cars had grown so large and heavy that power assistance for the steering had virtually become essential, at least to restore earlier levels of steering effort if not significantly to reduce them.

Since the oil crisis of 1973 there has been a worldwide trend towards building smaller, lighter-weight passenger cars, but paradoxically this has not meant any lessening in public demand for PAS. The reason for this lies in the now established preference for front-wheel drive with a transversely mounted power unit, which can result in about 65 per cent of the vehicle weight being concentrated over the steered wheels. When this is further related to the use of modern low-profile wide-tread tyres and a strong castor action, it can similarly lead to unacceptable levels of steering effort unless PAS is made available.

As might be expected, the most acute need for PAS first arose in connection with heavy goods vehicles, buses and coaches, military and various specialized off-road vehicles. In the latter cases true power steering rather than power-assisted steering may be used, since no direct mechanical connection between the steering wheel and the road wheels is necessarily retained. We confine our attention to more conventional heavy vehicles. The power-assisted type of system took various forms from about 1930 onwards and utilized either vacuum, compressed air or pressurized fluid as the energy source. However, it was during World War II that hydraulic PAS received widespread application on American military vehicles, notably for Chevrolet armoured cars, and it was systems of this type that formed the basis of those later adopted for passenger cars. The first British application of hydraulic PAS to a heavy vehicle was the Lockheed system, which was installed on a Daimler double-deck bus chassis in 1948. Since the late 1950s hydraulic PAS has become more widely used, either as standard or optional equipment, for heavy vehicles in the UK and elsewhere.

In design and construction modern hydraulic PAS systems for heavy vehicles are generally similar to the worm-and-follower installations that are still used in some passenger cars, except that they are made necessarily more robust to withstand the larger forces involved. Hydraulic limiters may be required to reduce the high pressure in the system as the steering lock stops are approached, thereby affording protection to the axle, steering linkage and pump. In very heavy-duty applications an emergency steering pump and standby valve are sometimes employed.

Advantages of power-assisted steering

The advantages generally associated with the use of PAS, including perhaps the less obvious ones, may be listed as follows:

1 Minimizes driver fatigue by lightening the action of the steering, especially during parking manoeuvres when the resistance to turning the wheels is greatest.
2 Reduces the required number of steering wheel turns from lock to lock, about three generally being preferred, by allowing an optimum choice of reduction ratio.
3 Lessens kick-back at the steering wheel by counteracting road shocks that would otherwise be transmitted back through the steering gear.
4 Improves safety by better resisting any sudden swerving of the vehicle in the event of a tyre deflating and causing uneven drag forces.
5 Permits heavier loading of the steered wheels to allow a greater freedom in overall vehicle design for obtaining maximum passenger or cargo space.

25.2 PRINCIPLES OF HYDRAULIC POWER-ASSISTED STEERING

Basic requirements

At the heart of a conventional PAS system is a hydraulic servo mechanism, which is brought into operation whenever the resistance to turning the steering wheel exceeds a predetermined amount. It then furnishes additional effort to assist the manual operation of the steering and thus greatly reduces the steering effort that would otherwise be imposed upon the driver.

An engine-driven pump circulates fluid around the closed circuit of the servo mechanism and provides a build-up of hydraulic pressure when needed. As the steering wheel is turned, a hydraulic control valve is displaced by reaction that is derived from the resistance to turning of the steered wheels by the steering reduction gearing. This valve then admits the hydrulic pressure to a double-acting power cylinder or ram, where thrust is developed on one side or other of a servo piston, which is connected at some point in the steering mechanism to augment the driver's effort.

Reasons for using hydraulic power assistance

Hydraulic power offered an attractive choice for PAS operation on motor vehicles and was based on the following considerations:

1. It has inherent self-lubricating characteristic.
2. It can readily develop high pressures.
3. Large forces can be transmitted in small spaces.
4. Large forces can be rapidly applied and removed.
5. The incompressibility of the fluid allows precise control over movements.
6. It provides a closed-circuit system to exclude contaminants.
7. It is readily adaptable to the available installation space.

Desirable features of PAS operation

These generally come under the three headings of safety, sensitivity and servicing, and may be summarized as follows.

Safety

An important feature of any PAS system is that it should be fail-safe. For this reason the power assistance is always superimposed upon the normal steering mechanism, so that in the unlikely event of the power assistance failing or the vehicle having to be towed it can still be steered in the ordinary way, albeit with a somewhat greater effort.

Sensitivity

Apart from providing adequate power assistance under all driving conditions, it is also essential that the system maintains a high degree of feel at the steering wheel. If during hard cornering the front tyres begin to approach the limit of their road grip, their pneumatic trail (see Sections 22.2 and 24.2) greatly diminishes and reduces castor effect. With manual steering this potentially dangerous condition is signalled to the driver by the steering appearing to go light.

It is therefore important that with PAS the control should not be so insensitive as to prevent this early warning signal from reaching the driver. Similarly, the feel from the road wheels should be retained when driving over slippery or icy roads. To improve generally the steering feel with power assistance it is usual for the castor angle of the front wheels to be specified as 1° more positive, since unlike the tyre pneumatic trail the castor angle does not reduce with an increasing cornering force so that less castor effect is lost.

Servicing

Perhaps as a tribute to the precision engineering that is necessarily demanded by PAS systems, their routine maintenance requirements have generally been relatively modest. In a typical modern installation the service schedule calls for the system to be inspected for leaks and the hydraulic pump reservoir to be checked for correct fluid level every 6000 miles

(10 000 km), although this interval may be halved in the case of heavy vehicles. An automatic transmission fluid is normally specified as being most suitable for use in PAS systems. The pump drive belt should be checked for condition and correct tension every 12 000 miles (20 000 km). In some systems there is a renewable filter element in the pump reservoir that usually requires changing every 20 000 miles (32 000 km).

Types of power-assisted steering system

Three main types of hydraulic PAS have been used on light and heavy vehicles: integral gear, semi-integral gear and linkage unit.

Integral gear

The distinguishing features of this type are that the power cylinder, servo piston and control valve are all incorporated in the steering gear housing itself. It can therefore equally well be applied to worm-and-follower (Figure 25.1) and the now more widely used (at least for passenger cars) rack-and-pinion steering gears (Figure 25.2). Other advantages of the integral gear are that being a self-contained unit it provides a more compact and less heavy installation than rival systems and also lends itself to a very sensitive control valve action.

This was the first type of production PAS system to be used on a passenger car, and since 1965 it has become established modern practice in this field. The integral gear has similarly found many applications on heavy vehicles since it first appeared at the beginning of the 1940s in America.

Semi-integral gear

This type is characterized by having the power cylinder and servo piston connected directly to the steering linkage, while the control valve alone is incorporated in the steering gear housing (Figure 25.3a). It offers a good compromise between reducing the forces transmitted through the steering reduction gear and providing a sensitive control valve action. The

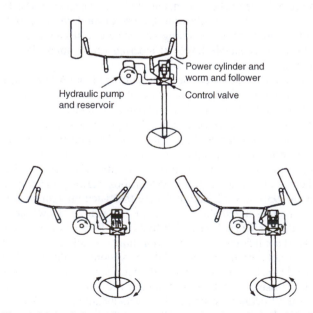

Figure 25.1 Schematic arrangement and action of integral gear PAS with worm-and-follower mechanism

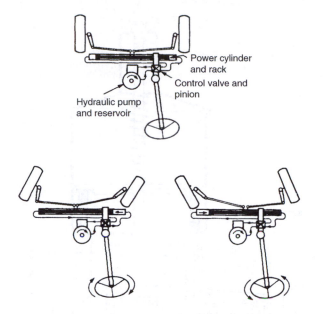

Figure 25.2 Schematic arrangement and action of integral gear PAS with rack-and-pinion mechanism

system was favoured by Rolls-Royce for their first PAS system in 1957, but its application has been limited mainly to heavy vehicles where the required hydraulic power cannot reasonably be exerted within the steering gear housing. It is possible, although uncommon, to incorporate the semi-integral gear principle in rack-and-pinion PAS (Figure 25.3b). In the case of a Citroën application, the control valve is fitted at the lower end of the steering column and the power cylinder is mounted alongside the rack housing with its servo piston rod connecting to one end of the rack.

Linkage unit
With this type of system the power cylinder, servo piston and control valve are all embodied at some point in the steering linkage (Figure 25.4a). It therefore offers a certain versatility of application for both light and heavy vehicles. The system was originally introduced in America during the early 1930s for low-volume production heavy vehicles, where it allowed the simple provision of PAS on otherwise standard chassis. A Bendix linkage unit PAS system was first used on passenger cars by Ford of America in 1954 and a Girling system later featured on the first British car to have PAS, the now classic Armstrong-Siddeley Sapphire of 1956. As compared with a

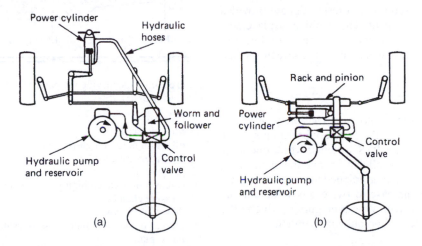

Figure 25.3 Schematic arrangements of semi-integral gear PAS: (a) worm-and-follower (b) rack-and-pinion systems

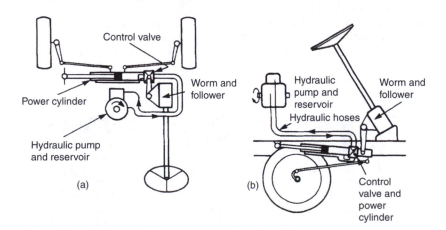

Figure 25.4 Schematic arrangements of linkage unit PAS: (a) light vehicle (b) heavy vehicle

modern integral gear, the linkage unit tends to be less sensitive and the system also requires rather lengthy hydraulic hose runs connecting the pump to the power cylinder. It has rarely been used on passenger cars since 1965, but continues to find convenient application on heavy vehicles (Figure 25.4b). A well-known example of the latter is the Hydrosteer linkage unit, which is made in England under licence from the Ross Gear & Tool company of America.

25.3   HYDRAULIC POWER-ASSISTED STEERING COMPONENTS

Hydraulic pump

This supplies the hydraulic energy for the PAS system and is mounted from and driven by the engine. According to installation requirements the pump may be provided with a V-belt and pulley adjustable drive, taken either directly from the crankshaft pulley or via an intermediate pulley. Alternatively, the pump can be driven in tandem with and flange mounted to one of the engine-driven auxiliaries, or it may be driven from the timing train. A fluid reservoir either is made integral with the pump (Figure 25.5), or for reasons of accessibility may be separately mounted in the engine bay and connected by flexible hose to the pump. Some pump reservoirs are provided with a renewable filter element. The positive displacement pumping elements used are mainly those of internal gear or rotor type and the vane type, which may take several specialized forms.

Rotor type

The operating principle of the rotor pump has previously been explained in connection with the engine lubrication system. It was widely used in earlier PAS systems, where operating pressures were usually controlled to a maximum of about $4.2 \, \text{MN/m}^2$ ($600 \, \text{lbf/in}^2$). In later practice the maximum pressure requirements can increase to twice this value and are generally better obtained with vane-type pumps, which offer higher volumetric efficiency and quieter operation.

Vane type

All forms of vane pump incorporate a driven rotor with a number of radial slots in which slide either blades, rollers, or slippers (Figure 25.6). As the rotor spins these steel vane elements are maintained in contact with the inner wall of the pumping chamber by centrifugal force, which in the case of

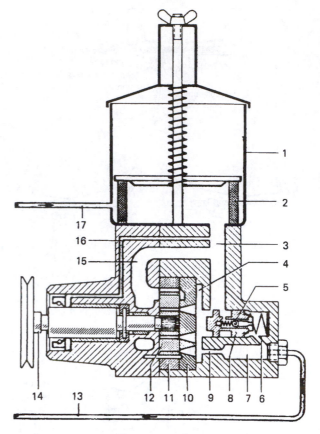

Figure 25.5   Identifying the parts of a high-pressure PAS pump (*Mercedes-Benz*)

 1 reservoir
 2 filter element
 3 oil return bore in high-pressure oil pump
 4 high-pressure chamber
 5 pressure relief valve
 6 connection between high-pressure outlet and spring chamber
 7 pressure chamber
 8 flow control valve
 9 calibrated bore from high-pressure chamber to pressure chamber
10 thrust plate
11 cam ring
12 impeller with pump vanes
13 high-pressure line to power steering
14 pulley
15 intake duct in high-pressure oil pump
16 bore for lubricating the drive shaft
17 oil return line from power steering

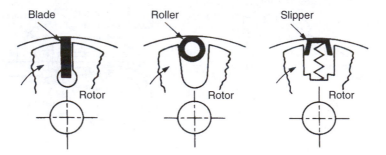

Figure 25.6   Types of sliding element used in vane pumps

the slipper type is augmented by spring loading. The pumping chamber is not circular but assumes an oval form, so that there is only a small clearance for the rotor across the minor axis or narrowest part of the chamber. There are thus two zones where the vanes slide further out of their slots as they approach the major axis or widest part of the chamber and form expanding cavities for the fluid to enter, and two other zones where the blades are pushed back again as they approach the minor axis of the chamber and form contracting cavities to discharge the fluid. Inlet and outlet ports are arranged to coincide with each of the two zones of operation. By virtue of there being two pressure cycles per revolution such pumps are said to be hydraulically balanced (Figure 25.7), so that unequal side loading on the rotor shaft bearing as found with internal gear pumps can be avoided.

The application of a hydraulic pump to supply the energy needs of PAS poses a rather unusual problem, in that the greatest energy demand for steering assistance occurs during parking manoeuvres when engine and hence pump speeds are low, while there is least energy demand at all speeds above this. Since the output of the pump naturally increases with its speed of operation, this would lead in normal driving to an excess of fluid being pumped through the restricted passages of the PAS control valve, which would result in overheating of the circulating fluid and failure of the pump.

It therefore becomes necessary to limit the discharge of fluid from the pump by recirculating the excess flow back to the inlet port. This is accomplished by what is termed a flow valve, which in essence is the opposite of a pressure relief valve because it produces large variations in pressure drop with small variations of flow. However, a pressure relief valve is also needed in the pump to limit to a safe value the maximum pressure that may be built up in the system, should for example an attempt be made to steer the wheels while obstructed by the kerb.

In practice, the two functions of flow control and pressure relief may be combined into a single valve assembly (Figure 25.8). Its operation is such that fluid leaving the pumping chamber passes through metering holes in the flow valve body, which allow fluid to pass into the system at the required rate. As the pump output increases, the volume of fluid will become too great to pass through these holes. This causes fluid from the pumping chamber to build up pressure on the end of the flow valve and push it forward against its spring

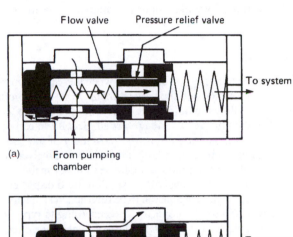

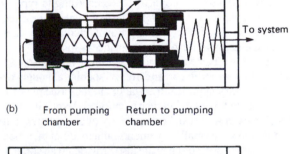

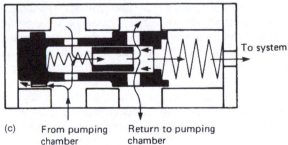

Figure 25.8 Operating principle of flow and pressure relief valves

(a) Pump output low; fluid flows through metering holes of flow valve, which otherwise remains at rest
(b) Pump output increasing: fluid flows through metering holes and bypass passage of moving flow valve
(c) Pump output high: fluid flows through metering holes but is also returned to bypass passage by opening of pressure relief valve

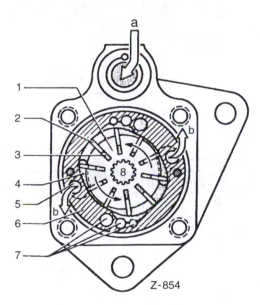

Figure 25.7 Arrangement of hydraulically balanced PAS pump (*Mercedes-Benz*)

a oil supply from reservoir
b outlet for pressure oil from housing ring
1 impeller
2 pump vanes
3 pressure oil under pump vane
4 recess in housing for admission of pressure oil to bore 5
5 pressure oil bore
6 cam ring
7 intake bores
8 drive shaft

loading. As this occurs an annular passage in the valve housing is uncovered so that fluid flow in excess of the required rate can be redirected to the inlet side of the pump. Whenever the fluid pressure beyond the pump reaches a maximum safe value, the spring loading on the relief valve is overcome. The fluid pressure acting on the relief valve then pushes it deeper into the body of the flow valve until release ports are uncovered, thereby allowing fluid to be redirected to the inlet side of the pump and thus reducing system pressure.

A pair of flexible hydraulic hoses are used to connect the pump with the PAS control valve, one being suitable for high-pressure supply and the other for low-pressure return. Not the least of the problems that had to be solved in the early development of PAS systems was that of noise being generated by hydraulic pulsations in the pump supply line. For this reason the high-pressure line from the pump to control valve may comprise a section of large-bore highly expansible hose followed by a section of small-bore non-expansible hose, so that in combination they will dampen pulsation and thus minimize noise.

Control valve

This has three duties to perform. First, when there is no demand for power assistance and the valve remains in its neutral position, it must allow the fluid being delivered by the pump to be recirculated via the valve to the reservoir. Second, in response to an increasing steering effort, the movement of the control valve must be such as to redirect the fluid being delivered by the pump to the appropriate side of the servo piston in the power cylinder, and simultaneously it must allow fluid displaced on the other side of the servo piston to be returned via the valve to the reservoir. Third, the control valve according to type must contrive to give either a natural or an artificial steering feel to the driver, by providing a proportional increase in steering effort as the demand for power assistance is increased.

There are two basic types of PAS control valve, these being generally known as linear and rotary spools, although detail applications vary significantly. The operating principle of both types originated from the earlier mentioned development work of F.W. Davis in America. For sensitivity of control these valves are limited to very small movements either side of their neutral or open-centre position. They are also provided with some form of spring centring, so that not only must the valve be forced to move a predetermined amount before power assistance becomes available, but also it must be readily neutralized to allow the castor return of the steering when straightening up. The actual pull required at the steering wheel rim to overcome the preload of the control valve spring centring can vary from as little as 9–18 N (2–4 lbf) in American practice to the 27–36 N (6–8 lbf) generally favoured for European applications.

Linear type

This form of spool control valve was introduced by the Saginaw Division of General Motors in 1942 for integral gear PAS on heavy vehicles, and ten years later was adapted to passenger car PAS requirements. It is fitted at the lower end of the steering column and is provided with three collars or lands so that there are two grooves or ports for the passage of fluid. These grooves are deliberately mismatched with three

annular ports in the housing within which the spool valve is permitted a very limited sliding movement. The spool valve is held in its neutral position normally by two sets of opposed reaction plungers, which are spring loaded apart against the thrust bearings for the steering gear reduction worm to provide the spring centring effect (Figure 25.9).

As the resistance to turning the steering wheel increases, the steering worm attempts to move endwise against the reaction of its follower and eventually overcomes the preload of the centring springs within the plungers. When this occurs the spool valve is compelled to move axially within its housing, so that the fluid ports move towards alignment in a direction that corresponds to steering lock. The fluid delivered by the pump to the control valve can then no longer be recirculated to the reservoir, but is instead directly to the appropriate side of the servo piston in the power cylinder to render steering assistance. Less commonly, a linear type of

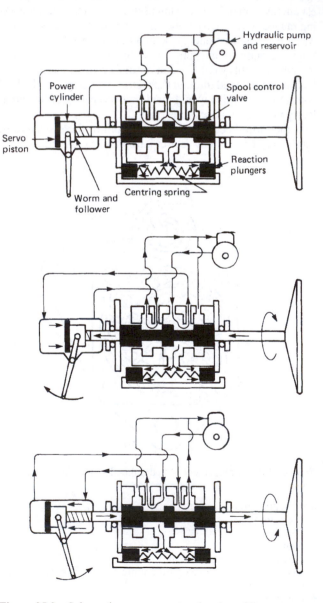

Figure 25.9 Schematic arrangement and action of linear-type spool control valve PAS

control valve may be incorporated in an integral gear rack-and-pinion PAS system. In this case the axial movement of the spool valve is derived from the attempted endwise movement of the helical-toothed pinion against the teeth of the rack, as the latter offers resistance to sliding.

An additional feature of this type of control valve operation is that the fluid pressure in the system also acts against the inner ends of the reaction plungers and thus adds to the separating force exerted by the centring springs. Since the fluid pressure will increase with the demand for power assistance, it follows that the greater centring effect on the valve is attempting to unwind the steering worm against its follower, or helical pinion relative to the rack, so that additional resistance to turning of the steering wheel is sensed by the driver. A useful degree of natural feel is therefore conferred on the steering action, and for this reason a control valve of this type is termed a reactive one.

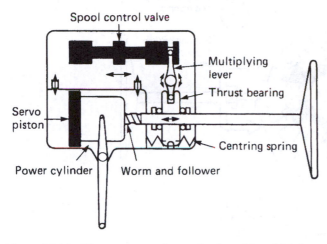

Figure 25.10   Linear-type spool control valve with multiplying lever

In a further development the response of the linear type of control valve was improved by actuating a separately mounted spool valve via a multiplying lever, which engaged with the steering worm thrust bearings (Figure 25.10). Its operating principle was otherwise similar to that just described.

Another quite different application of a linear control valve is that used since 1961 by Mercedes-Benz and which has long been acclaimed for conferring excellent steering feel. In this design the spool valve is mounted tangential to the axis of the steering worm and receives its axial movement from a control edge that extends lengthwise along the recirculating-ball nut, the latter being allowed a limited rocking motion within the servo piston (Figure 25.11). As the resistance to turning the steering wheel increases and the nut is travelling in either direction along the worm, it tries to take the least line of resistance by attempting to rotate with the worm. In so doing the rotary movement of the nut control edge, which is simultaneously sliding through its engagement with the control valve, imparts an axial movement to the spool. Pumped fluid is then directed to the appropriate side of the servo piston in the usual manner. The movement of the spool valve is controlled by a centring spring and also by system fluid pressure acting on each end of it, which thereby makes the valve reactive for natural steering feel.

Rotary type

This form of spool control valve was originally a later development from Saginaw and introduced by them in 1959. It was by then considered that to improve the sensitivity of PAS the control valve should be the first element of the system beyond the steering wheel. In this new arrangement the steering wheel therefore actuated the rotary motion of the spool valve directly. The rotary valve comprises an outer sleeve that is internally splined and rotates with the steering worm, and an inner sleeve that is externally splined and rotates with the steering column. A connection between these coaxially

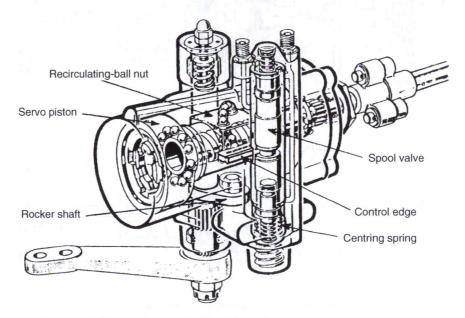

Figure 25.11   Another application of a linear-type control valve (*Mercedes-Benz*)

mounted sleeves is effected not through the splines, which act solely as valve ports, but through a valve centring torsion bar. This is anchored at one end to the steering column and at the other end to the steering worm (Figure 25.12). Internal mechanical stops limit the relative angular movement that can occur between the steering wheel and worm and hence the two sleeves of the spool valve. In some designs the annular space between the inner sleeve and torsion bar serves as an outlet for the return of fluid to the reservoir.

As the resistance to turning the steering wheel increases, it is accompanied by twisting of the torsion bar which allows the splined fluid ports of the valve sleeves to change their relative angular alignment, even though the spool valve assembly as a whole is being rotated. Since the outer sleeve of the valve is also provided with holes that communicate with annular grooves on its external surface, the fluid that is normally being recirculated between the pump and the valve is instead directed to the appropriate side of the servo piston in the power cylinder. As this rotary type of control valve is not subject to hydraulic reaction from system pressure, it is termed non-reactive. The driver feel of the steering is therefore artificial rather than natural and is derived from the resistance to twisting of the torsion bar and the profile of the valve ports. A typical relationship between the pressure of fluid delivered from the spool valve to the servo piston and the angular displacement of the valve is shown in Figure 25.13.

Another version of the rotary control valve principle is that used since 1964 in the German ZF (Zahnradfabrik Friedrichshafen) PAS system. This similarly embodies a valve centring torsion bar but, instead of influencing the

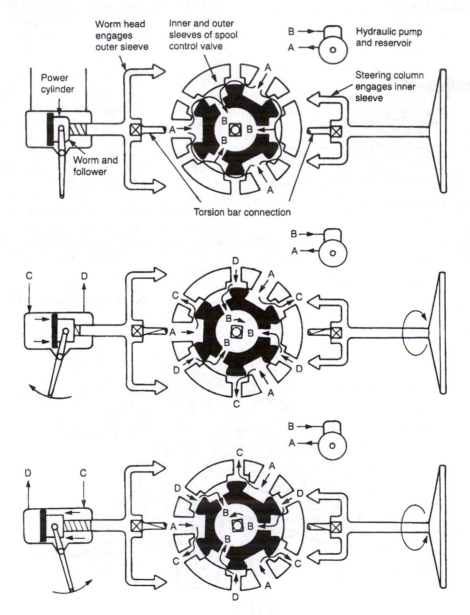

Figure 25.12   Schematic arrangement and action of rotary-type spool control valve PAS

A from pump   C to cylinder
B to reservoir   D from cylinder

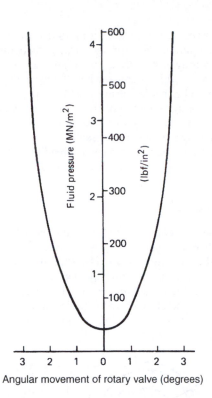

Figure 25.13 Typical operating characteristics of a rotary-type control valve

relative angular position of two ported sleeves, it actuates a cross-valve spool assembly carried in the enlarged head of the worm shaft (Figure 25.14). The head is ported on its external surface and accommodates two transversely disposed piston valves, which receive opposing motions from a cross-beam at the lower end of the steering column. As the resistance to turning of the steering wheel increases, the twisting of the torsion bar causes relative displacement of the two spools, so that instead of the pumped fluid being recirculated through the control valve assembly it is directed to the appropriate side of the servo piston in the power cylinder.

The rotary type of control valve is now widely used in PAS systems, especially as it readily lends itself to modern rack-and-pinion steering installations, where one element of the valve is connected to the pinion shaft and the other to the steering column (Figure 25.15).

However, a criticism that is sometimes levelled at the conventional torsion-bar-controlled rotary valve concerns a feeling of vagueness in the steering for small movements of the steering wheel either side of the straight-ahead position, typically during motorway driving. This condition basically arises from the absence of a preload centring spring, which otherwise would transmit these small movements of the steering wheel directly to the pinion of the steering gear. As it is, these small movements of the steering wheel are subject to the initial wind-up of the torsion bar and associated friction in the mechanism before they reach the pinion. To

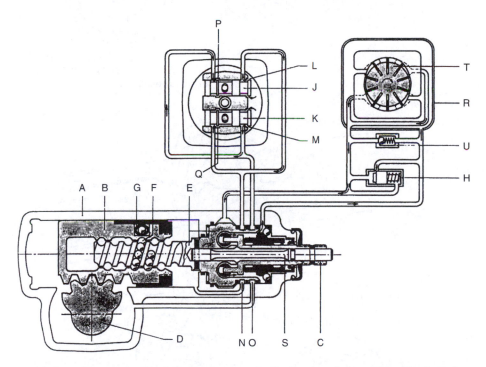

Figure 25.14 Power-assisted steering system with piston-type rotary control valve: the valve is additionally shown in cross-section in order to give a schematic representation of the connection of the valve pistons to the power cylinder and of the functioning of the valve (*ZF GB*)

A steering gear and valve housing	G recirculation tube	O radial groove
B piston	H flow limiting valve	P return groove
C input shaft	J valve piston	Q return groove
D sector shaft	K valve piston	R oil tank
E worm	L inlet groove	S torsion bar
F balls	M inlet groove	T pressure oil pump
	N radial groove	U pressure limiting valve

overcome this problem a TRW Cam Gears design of rotary valve, known originally as Positive Centre Feel and now called Centre Plus, was introduced in the late 1980s which restores a preload into the system. This is achieved by eliminating the torsion bar and transmitting steering torque from a rigid input shaft to the pinion via a spring C-ring, this being preloaded into position against a drive pin (Figure 25.16). Hence, the hydraulic valve only operates when the torque applied to the steering wheel is sufficient to exceed the preload exerted by the spring C-ring. A more positive feel to the system is therefore provided for small movements of the steering wheel, whilst retaining a smooth progression of power assistance once the preload centring spring begins spreading and allows relative movement within the rotary control valve (Figure 25.17).

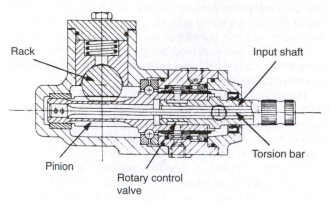

Figure 25.15 General arrangement of a modern rack-and-pinion PAS with rotary valve (*Burman*)

Power cylinder

This is double acting and provides a means of utilizing the available hydraulic energy to produce force and travel at some point in the steering system. The size of power cylinder is typically such that maximum power assistance will enable the driver to steer, relatively easily, the wheels from lock to lock on dry concrete with the vehicle stationary. In the case of semi-integral gear and linkage unit systems, the control

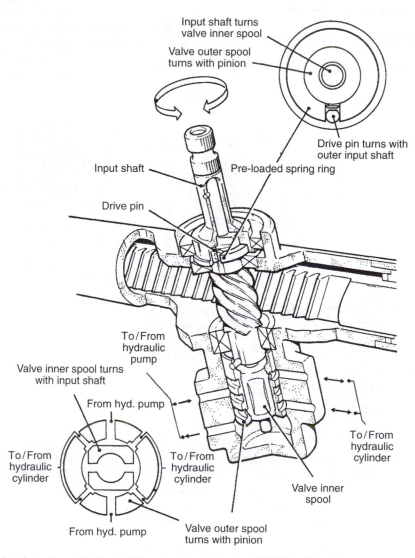

Figure 25.16 Power-assisted steering with Centre Plus rotary control valve (*TRW Cam Gears*)

valve operation must allow for a higher pressure to act on the reduced area of the rod side of the servo piston, so that the output force in each direction of piston travel is balanced. A piston ring type of seal is generally used on the servo piston of all types of PAS. With integral gear PAS of the worm-and-follower type, it has long been established practice for the recirculating-ball nut to act also as the servo piston. The reduction gear housing is then adapted to become the power cylinder and is usually provided with an internal channel to convey fluid to and from the blind end (Figure 25.18a).

As mentioned earlier, hydraulic limiters may be required in the power cylinder of heavy-vehicle applications of integral gear PAS. They take the form of either a two-way relief valve fitted in the blind end of the servo piston, or twin relief valves operating on either side of the reduction gear rocker shaft. The two way valve is opened in either direction by stops provided at the blind end of the power cylinder and

on the end of the worm shaft (Figure 25.18a). A striker cam on the inner end of the rocker shaft servers to operate the twin-valve arrangements, one or other of the valves being opened after the drop arm swings a certain amount in either direction (Figure 25.18b). Both types of limiter valve serve to reduce operating pressure in the power cylinder as the steering lock stops are approached, thereby protecting the axle, steering linkage and pump from overloading.

For integral gear versions of the modern rack-and-pinion type, the rack housing becomes the power cylinder and the rack is adapted to become the servo piston. In designs where the track rods are connected to the extremities of the rack, the servo piston is located on the rack remote from the pinion. Pressure seals are arranged to act on the rack at each end of the servo piston travel and thus provide an enclosed power cylinder (Figure 25.19). If the track rods are connected to the middle of the rack, then the blind ends of the rack housing

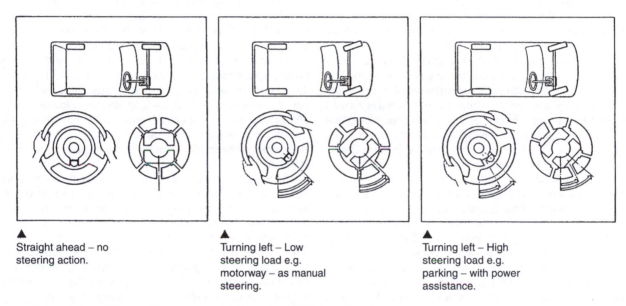

Straight ahead – no steering action.

Turning left – Low steering load e.g. motorway – as manual steering.

Turning left – High steering load e.g. parking – with power assistance.

Figure 25.17 Action of Centre Plus power-assisted steering (*TRW Cam Gears*)

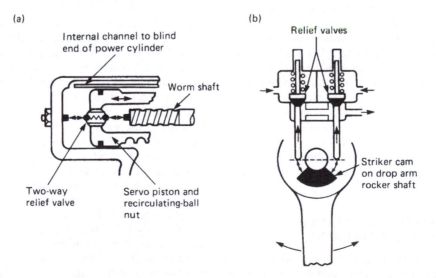

(a)

Internal channel to blind end of power cylinder

Worm shaft

Two-way relief valve

Servo piston and recirculating-ball nut

(b)

Relief valves

Striker cam on drop arm rocker shaft

Figure 25.18 Two arrangements of limiter valves for PAS

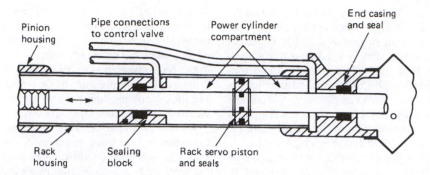

Figure 25.19   Power cylinder and servo piston arrangement for rack-and-pinion PAS with end take-offs for track rods

each form part of the power cylinder and the entire length of rack becomes the servo piston (Figure 25.2). In this case pressure seals are used at each end of the rack. The external fluid connections between the control valve housing and the power cylinder are effected by rigid steel tubing.

With semi-integral gear PAS a simple telescopic power cylinder is pivoted from a flexible bushing at its blind end and at its other end is furnished with a pressure seal through which slides the servo piston rod. The outer end of this rod is pivoted to the steering linkage (Figure 25.3a). Fluid connections are incorporated at each end of the cylinder, flexible hoses being required to convey fluid between the control valve housing on the steering gear and the power cylinder.

In the case of heavy-vehicle linkage unit PAS systems where the power cylinder is combined with the control valve, the internal arrangement of the cylinder becomes more complex. Here the control valve end of the power cylinder or booster is connected to the steering box drop arm and the other end is flexibly anchored to the chassis frame, via the servo piston rod (Figure 25.4b). Fluid connections are furnished at the control valve end of the cylinder for a pair of flexible hydraulic hoses that connect with the pump. The power cylinder is thus free to move in response to increased fluid pressure on either side of the stationary servo piston. A linear type of control valve is clearly relevant to this type of application (Figure 25.20). The valve spool is centred in its neutral position by means of a reaction spring and also by system fluid pressure to provide a natural steering feel. As long as the steering effort is less than that required to overcome the spring loading on the spool valve, the fluid circulates freely between the control valve and pump. As the resistance to turning the steering wheel increases and overcomes the spring loading, the valve spool will be displaced towards either end of its ported housing according to the motion of the drop arm ball pin. Fluid from the pump is then directed to the appropriate side of the servo piston, thereby moving the power cylinder and hence the steering linkage to render power assistance (Figure 25.21).

25.4   SPEED-SENSITIVE HYDRAULIC POWER-ASSISTED STEERING

Ideally a power-assisted steering system should provide a variable degree of assistance depending upon the effort required to turn the steering wheel. That is, the maximum assistance

is required by the driver during slow-speed manoeuvres, but this should be significantly reduced under normal straight-ahead running. It is perhaps of historical interest to recall that in their late 1950s adaptation of the Davis-type linear spool valve, Rolls-Royce recognized this requirement by constraining the axial movement of the valve not only by the conventional preload springs and reaction plungers, which governed the amount of assistance for normal straight-ahead driving, but also a second set of preload springs without reaction plungers, which compressed only under conditions when maximum assistance was required and then allowed the full controlled pressure of the pump to be applied to the power cylinder.

The modern approach to achieving the desired variation in power assistance is to make the system 'speed-sensitive'. Two methods of accomplishing this are termed the flow control and reactive systems, both of which incorporate an electronically controlled variable-orifice valve. The electronic control unit (ECU) is responsive to either vehicle speed alone, or both vehicle and engine speeds. A small electrical stepper motor and linear actuator may be used to position the valve, or alternatively a direct-acting solenoid performs this duty.

Flow control systems

Taking for example a power-assisted rack-and-pinion steering system, the variable-orifice valve can simply be interposed between the two feed ports of the power cylinder (Figure 25.22). Therefore as the road speed increases, the linear actuator progressively opens the valve to reduce the pressure difference across the rack piston, which has the effect of reducing the amount of power assistance. During slow-speed manoeuvres the valve remains completely closed, so that no pressurized fluid bypasses the power cylinder and maximum assistance is obtained. Conversely at high speed the valve is fully open and bypasses the greatest proportion of pressurized fluid to the reservoir for least assistance, so that to all intents and purposes the steering becomes manual rather than power assisted in feel.

In the early 1980s the ZF company further developed their rotary cross-valve type of power-assisted steering to provide a speed-sensitive characteristic, this being known as the Servotronic system (Figure 25.23). Here again the power assistance is varied by means of an electronically controlled

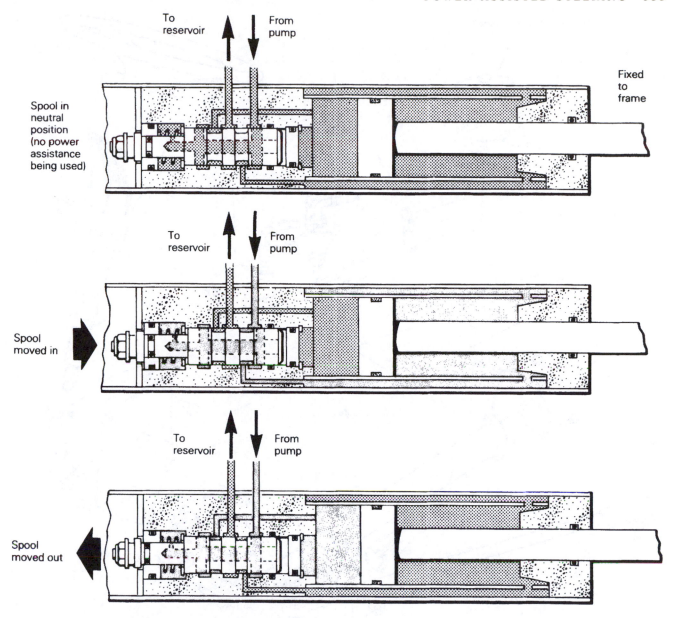

Figure 25.20 Operating principle of linkage unit PAS (*Seddon Atkinson*)

variable-orifice valve that is progressively opened with rising vehicle speed, but the reducing pressure difference across the power cylinder piston is sensed by unbalanced reaction pressures that increasingly oppose turning of the steering wheel (Figure 25.24). During slow-speed manoeuvres the simultaneous closing of one and opening of the other piston valve directs fluid under pressure not only to the appropriate side of the power cylinder piston, but also via a non-return valve to the opposite end of the closed piston valve. With the variable-orifice valve closed the pressure difference across the power cylinder piston is greatest, while the pressure that is created in the reaction chamber of the closed piston valve is also reflected undiminished in the reaction chamber of the open piston valve. By virtue of the see-saw connection between the two piston valves, the equalization of pressures in their reaction

chambers means that there is least opposition to turning the steering wheel for maximum assistance (Figure 25.25a).

As vehicle speed increases the variable-orifice valve progressively opens. This results in a reduced pressure difference across the power cylinder piston and less assistance, because a proportion of the fluid delivered to the cylinder is bypassed to the reservoir via the reaction chamber of the closed piston valve. Since this fluid has to flow through a restrictor of which there is one beyond each reaction chamber, a greater back-pressure is created on the closed piston valve than on the open one. The effect of these unbalanced reaction pressures is to impose a turning moment on the see-saw connection between the valves, which acts in opposition to the driver turning the steering wheel and therefore improves steering feel for high-speed driving (Figure 25.25b). A reaction

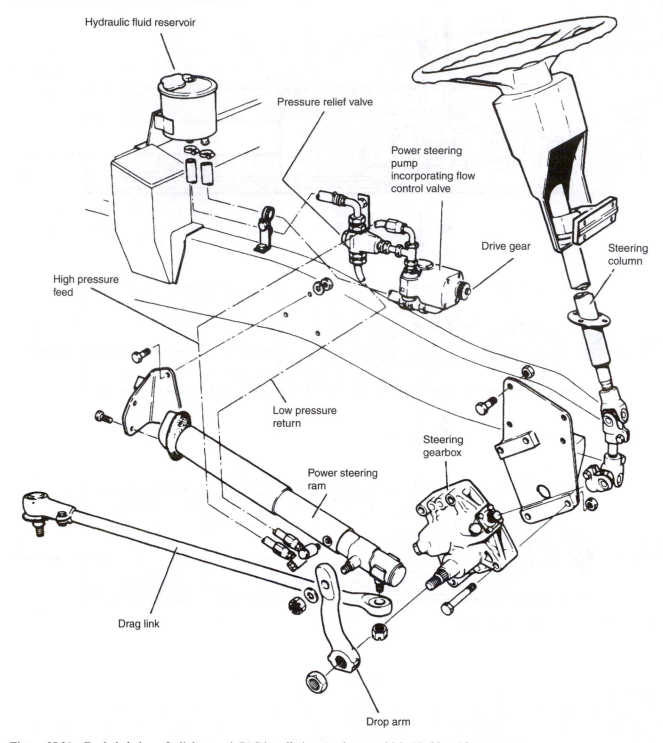

Hydraulic fluid reservoir

Pressure relief valve

Power steering
pump
incorporating flow
control valve

Drive gear

Steering
column

High pressure
feed

Low pressure
return

Steering
gearbox

Power steering
ram

Drag link

Drop arm

Figure 25.21   Exploded view of a linkage unit PAS installation on a heavy vehicle (*Seddon Atkinson*)

pressure limiting valve ultimately determines the extent to which the pressure can drop in the reaction chamber of the open piston valve and therefore the amount of increase in steering effort.

Reactive systems

A reactive control is not only speed-sensitive, but also confers an added torsional preload in the steering system, which

has to be exceeded before power-assistance becomes available. Furthermore, the preload is varied in relation to the speed of the vehicle. In other words, the added preload is reduced almost to zero during parking manoeuvres, so that power-assistance is at a maximum and steering effort is very light. Conversely, as vehicle speed rises the preload is gradually increased and the power-assistance sensed by the driver therefore decreases, so that the overall effect is to

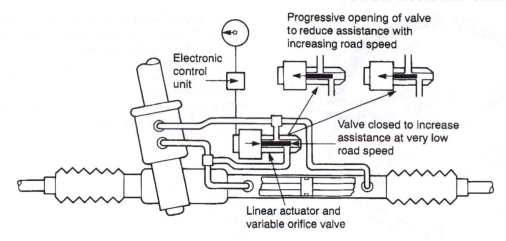

Figure 25.22 Schematic arrangement of a speed-sensitive power-assisted rack-and-pinion steering

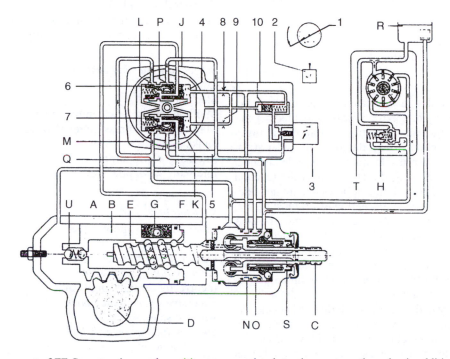

Figure 25.23 Arrangement of ZF-Servotronic speed-sensitive power-assisted steering system: the valve is additionally shown in cross-section in order to give a schematic representation of the connection of the valve pistons to the power cylinder and of the functioning of the valve (*ZF GB*)

1 electronic speed indicator	A steering gear and valve housing	L inlet groove
2 microprocessor	B piston	M inlet groove
3 linear actuator for variable-orifice valve	C input shaft	N radial groove
4 reaction pressure	D sector shaft	O radial groove
5 reaction pressure	E worm	P return groove
6 non-return valve	F recirculating balls	Q return groove
7 non-return valve	G recirculating tube	R reservoir
8 restrictor	H pressure relief and flow control valve	S torsion bar
9 restrictor	J valve piston	T fluid pressure pump
10 reaction limiting valve	K valve piston	U full-lock pressure limiter

provide a more positive feel to the steering when it is most required.

Again taking for example a power-assisted rack-and-pinion steering system, a spring-loaded reaction piston and face cam assembly, which can additionally be subject to system hydraulic pressure, is superimposed on the torsion bar drive between the steering input and pinion output shafts. In the Subaru version a face cam in the form of radially grooved spider arms is rigidly attached to the steering input shaft, the V-shaped grooves being engaged by ball bearings that receive

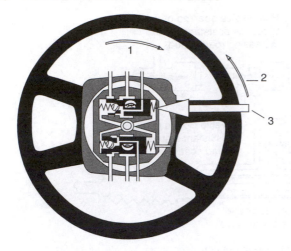

Figure 25.24 Steering wheel feedback with ZF-Servotronic system (*ZF GB*)

1 direction of turn
2 reaction feel at steering wheel
3 hydraulic reaction on spool valve

their location from holes in the flanged connection of the pinion output shaft (Figure 25.26). Slidably mounted beyond this flange is a spring-loaded reaction piston, which urges the ball bearings into contact with the grooved spider arms. Additionally, the modest spring-loading can be supplemented by fluid under pressure being directed against the reaction piston, as vehicle speed rises. It thus follows that for any relative turning to occur between the steering effort must be sufficient to cause the ball bearings to climb the sides of their V-grooves and by the same token force back the reaction piston. A torsional preload is therefore imposed on the steering system and, until it is overcome, the valve centring torsion bar cannot undergo any twisting movement to summon power-assistance in the usual manner.

The supplementary hydraulic pressure directed against the reaction piston is controlled by a valve sleeve, which is positioned as required by a direct acting solenoid whose current supply corresponds to vehicle road speed, as monitored by an electronic control unit (ECU). Hence, when driving at high speeds and the most positive feel is required for the steering, the solenoid moves the valve sleeve into a position where the full pump pressure at the rotary spool valve is also directed against the reaction piston, thereby providing maximum torsional preload in the steering system (Figure 25.26). During slow-speed manoeuvring, of course, the hydraulic pressure is no longer directed against the reaction piston and steering preload is at a minimum and assistance at a maximum for effortless steering.

25.5 HYDRAULIC POWER-ASSISTED STEERING MISBEHAVIOUR IN SERVICE

Before attempting any repairs it is always advisable, unless already familiar with a particular PAS installation, to consult the fault diagnosis and correction chart or its equivalent that is normally included in the vehicle manufacturer's service manual. Similarly, the manufacturer's recommended procedures for checking steering wheel rim loads with the vehicle stationary and engine idling, carrying out a pressure check on the system and bleeding the system of air, should all be carefully observed.

The following list of faults and their possible causes is therefore intended to be purely a general guide to some of the problems that may be encountered with PAS systems in service:

Noisy operation of steering
Low fluid level
Slack drive belt
Malfunctioning flow control valve
Internally worn pump
Over-stroking in linkage unit.
(The familiar slight hissing sound heard during parking manoeuvres does not indicate faulty components.)

Inadequate steering assistance
Low fluid level
Slack drive belt
Malfunctioning flow control valve
Malfunctioning pressure relief valve
Kinks in hoses
Internally worn pump
Internally worn power cylinder.

Steering lacks feel
Internally worn spool valve.
Poor castor return
Malfunctioning spool valve.
Steering pulls to one side
Incorrectly centred spool valve.
Excessive steering kick-back
Air in system.
Steering judder towards lock
Low fluid level
Slack drive belt
Malfunctioning flow control valve.
Fluid leakage
From pump or reservoir
From hoses or connections
From integral gear housing
From linkage unit power cylinder.

25.6 ELECTRO-HYDRAULIC POWER-ASSISTED STEERING

A more recent development by TRW Steering Systems has been the electrically powered hydraulic steering (EPHS) system, where the pump is driven by an electric motor rather than directly from the engine (Figure 25.27). The advantages claimed for this scheme include a more compact and integrated arrangement for simpler installation; a reduction in energy consumption by virtue of less power being taken from the engine; a cost and production effective design; and

(a)

(b)

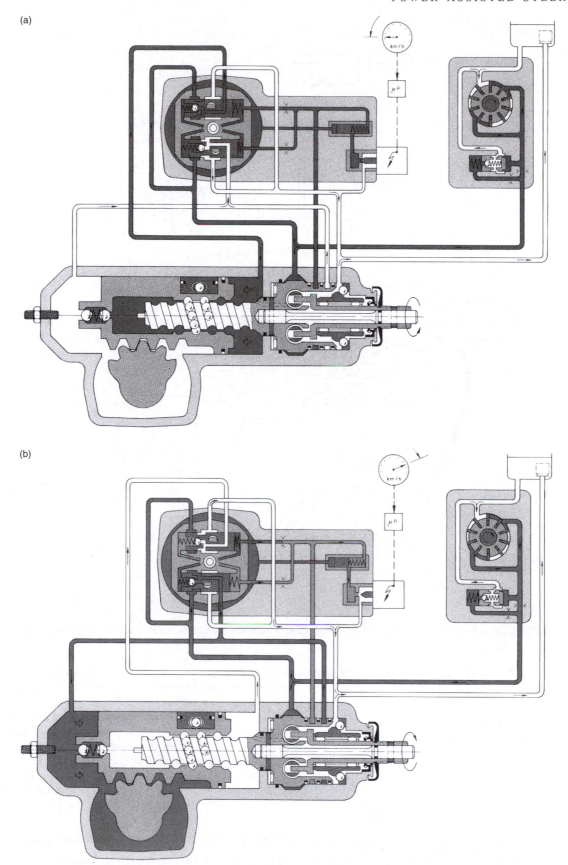

Figure 25.25   Action of ZF-Servotronic power-assisted steering at low and high speeds (*ZF GB*)

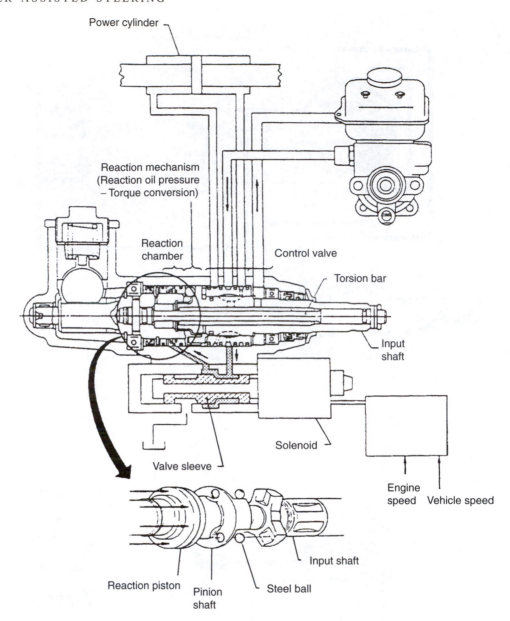

Figure 25.26 Action of a speed-sensitive, power-assisted, steering system with reactive control during high-speed driving (*Subaru*)

power-assistance remains available during towing providing the battery is sufficiently charged.

25.7 INTRODUCTION TO ELECTRICAL POWER-ASSISTED STEERING

General background

Electrical power-assisted steering (EPAS) or electric power steering (EPS) as it is sometimes more simply but perhaps less accurately termed, basically comprises a system that uses electronic control and an electrical power supply, as distinct from the mechanical control and hydraulic power supply long associated with conventional power-assisted steering systems. Electrical power-assisted steering was first used by the Suzuki Motor Company in Japan for their small 3-cylinder engined,

front-wheel drive, city cars in 1988 and they were followed soon after by other Japanese manufacturers of this class of car. Also during this period, the TRW Company in America announced their 'Powertronic' system for more general application. In 1990 the Honda Company introduced their own design of electrical power-assistance for the steering of automatic transmission versions of their advanced mid-engined NSX sports car. More recently, the Rover Company offered the Japanese NSK system, as an option to manual steering on their similarly mid-engined MGF sports car.

Advantages of electrical power-assisted steering

1 Fuel economy can be improved, because to all intents and purposes the system only consumes energy when there is a demand for power-assistance, there being no

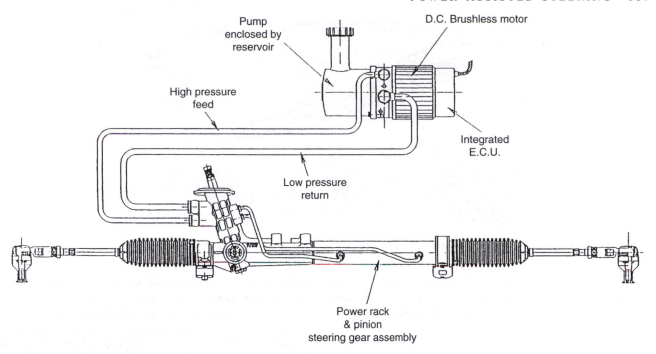

Figure 25.27 Basic arrangement of an electrically powered, hydraulic assisted, steering system (*TRW Steering Systems*)

requirement for a pump to be continuously driven by the engine.

2 A more compact installation can be achieved for the system by virtue of its integrated construction, hence there is less intrusion on engine compartment space.

3 The system is more economic to install, since fewer separate components are involved and it doesn't need filling with fluid and purging of air.

4 A significant saving in weight is generally attributed to electrical power-assistance systems, which again derives from their integrated construction and fewer components.

5 Similar to hydraulic power-assistance, the various systems can be provided with speed-sensitive characteristics and naturally have a fail-safe facility.

6 In some systems power-assistance is still available even though the engine is not running, while in others this is intentionally prevented to avoid discharging the battery.

7 The service nuisance of dealing with fluid leaks is obviously non-existent with electrical power assistance, as is the hazard of fluid spill in the event of an accident.

25.8 ELECTRICAL POWER-ASSISTED STEERING COMPONENTS

Electric motor

This furnishes the motive power for the system and comprises a direct-current machine with the armature rotating between permanent magnets. The current input to the motor is varied in relation to the torque output required for steering power-assistance, and its direction of rotation is controlled through changing polarity of the signal received by the motor. An important design consideration for the motor is to minimize the rotary inertia (or flywheel effect) of its armature, otherwise there could be a possibility of steering overshoot.

Actuating mechanism

The actuating mechanism embodies a reduction gearset and is interposed between the electric motor and whichever part of the steering-gear that receives the assisting force. Therefore its purpose is not only to transmit this force smoothly and efficiently, in either a rotary or a linear sense, but also to amplify the torque output from the electric motor. An electromagnetic clutch may also be included in some actuator mechanisms, so that the motor is disengaged when assistance is not required at higher vehicle speeds, thereby eliminating the effects of armature inertia on the steering.

Steering input sensors

A steering torque sensor is a vital element in the control system for electrical power-assisted steering, because it detects not only the driver effort at the steering wheel in terms of rotary torque being transmitted to the steering input shaft, but also the direction in which the steering wheel is being turned. One method of accomplishing these functions is to introduce a centring torsion bar within the steering input and output shafts to the rack pinion, so that relative angular movements occurring between the shafts can be converted into axial ones by a radial pin engaging a helical grooved slider (Figure 25.30). The slider movements are then intercepted by a potentiometer, whose electrical output signals are proportional to steering input torque. A system may additionally be provided with a steering speed sensor, this detects the rapidity with which the steering wheel is being turned and, in this case, also its

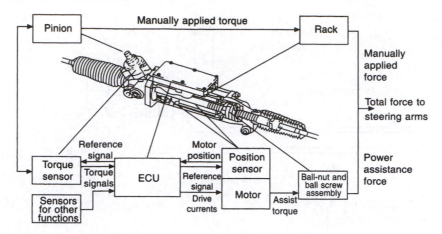

Figure 25.28 General layout of an integrated electrical power-assisted steering system (*TRW*)

direction of turning. The electrical output signals from this sensor are derived from a small direct-current generator, which is driven through step-up gearing from the steering column input shaft.

Electronic control unit

At the nerve centre of an electrical power-assisted steering system is the electronic control unit (ECU), which is typically mounted integral with the motor and actuating mechanism (Figure 25.28). It receives input signals from the earlier mentioned steering sensor(s) and also from vehicle and engine speed sensors. This data is then processed and related to pre-programmed power-assistance characteristics, which have been tailored to suit the specific requirements of the vehicle. From this the ECU is able to control the current supplied to the motor, so that the amount of power-assistance delivered is in accordance with prevailing demands. The electronic control system further possesses both fail-safe and self-diagnosis capabilities.

25.9 TYPES OF ELECTRICAL POWER-ASSISTED STEERING

Electrical power-assisted steering systems may be conveniently classified as either 'column-drive', or 'rack-drive', where the motor and actuating mechanism is applied to the steering input shaft and the steering output rack respectively.

Column-drive type

For light duty applications of this type of system a combination of motor, clutch and actuating mechanism is mounted normal (at right-angles) to the upper steering column, which contains the torsion-bar centred input and output shafts, the former actuating the system torque sensor and the latter being driven by a worm and worm-wheel reduction gearset from the actuating mechanism (Figures 25.29a and 25.30). An established system of this type is that pioneered by the NSK Group in Japan and known as their EPAS (Electrical Power-Assisted Steering) system. Two important factors that account for the smooth and quiet operation of this system are the careful choice of materi-

als for the worm and worm-wheel reduction gearset and an optimum preloading for the worm gear bearings.

As earlier mentioned the NSK EPAS system is an optional feature on the MGF sports car, which has a mid-engined layout and would have therefore required lengthy hose runs from an engine mounted pump to the steering gear, if a conventional hydraulic power-assisted steering system had been installed. Similar considerations also apply to the mid-engined Honda NSX sports car, which uses Honda's own rack-drive type of electrical power-assisted steering system as described later.

For medium duty applications of the column-drive type of system, the power assistance operating mechanism is arranged to act directly on the steering pinion shaft (Figure 25.29b), thereby reducing the torsional loading on the steering column itself.

Rack-drive type

For systems of this type a combination of motor and actuating mechanism is mounted from the steering gear housing, and drives the rack either through a recirculating ballnut and lead-screw arrangement or, less commonly, by introducing an auxiliary engagement pinion.

The application of a recirculating ballnut and lead-screw mechanism, to impose an assisting force on the sliding steering rack, may take two distinct forms. That is, the electric motor can be arranged to drive either the ballnut or the lead-screw to convert rotary into linear motion at the rack (Figures 25.29c and d). In the earlier mentioned TRW Powertronic system (Figure 25.28), the electric motor is mounted concentric with the rack and is provided with a hollow armature, thereby allowing the rack to move freely through it. A splined connection joins the armature to a rotatable ballnut, which engages with the lead-screw portion of the rack remote from the pinion. When the motor is energized for either direction of steering, the rotary motion of its armature with ballnut is translated into linear motion at the lead-screw portion of the rack, and thereby transmits the required assisting force.

In contrast, the recirculating ballnut and lead-screw mechanism introduced by the Honda Company is applied parallel to the steering rack. The electric motor is again mounted

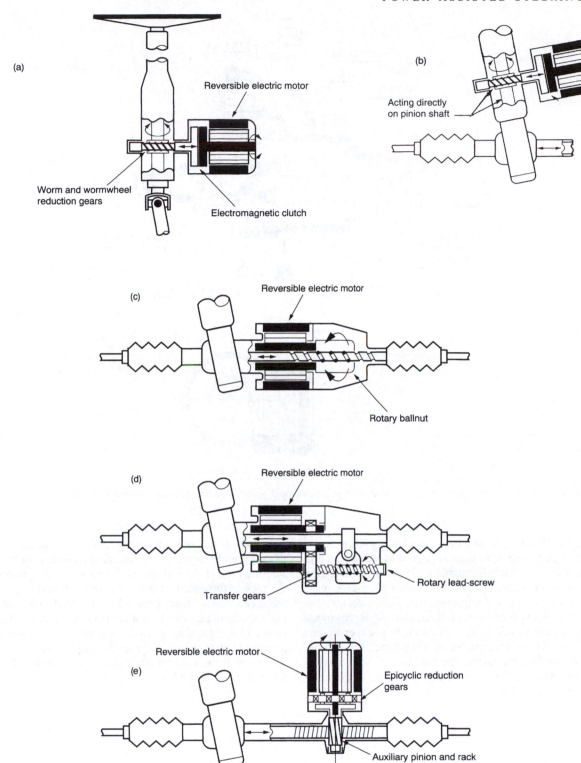

Figure 25.29 Types of actuating mechanism for electrical power-assisted steering systems (a–b) column-drives (c–e) rack-drives

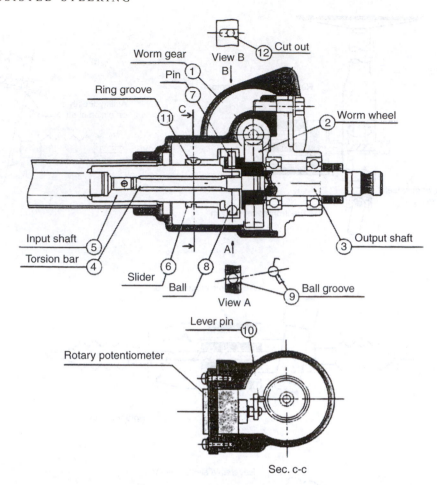

Figure 25.30 Actuating mechanism of a column-drive type of electrical power-assisted steering system (*NSK-RHP*)

concentric with the rack and therefore has a hollow armature, but in this case it is geared to a lead-screw running alongside the rack. A non-rotatable ballnut engages with the lead-screw and is connected to that end of the rack remote from the pinion, via a self-aligning joint. This is designed to absorb any deflection of the rack from steering forces transmitted by the track rods. Hence, when the motor is energized the rotary motion of its armature is transferred through the gears to the lead-screw, so that the resulting linear motion of the ballnut transmits the required assisting force to the steering rack (Figure 25.29d).

In the less common system of rack-drive, the combination of motor and actuating mechanism is mounted normal to the steering rack at the end remote from its column pinion. Here the rack is furnished with another toothed portion, which is engaged by an auxiliary pinion driven from the electric motor, via a reduction epicyclic gearset (Figure 25.29e). When the motor is energized its geared down pinion therefore rotates and exerts an assisting force on the sliding rack, in the same manner that the column pinion transmits a manual steering force at the other end of the rack.

26 Four-wheel-steering systems

26.1 THE NEED FOR FOUR-WHEEL STEERING

General background

The concept of steering all four wheels of a passenger car is another feature of modern automotive technology that has its origins in the early history of the motor vehicle and, similar to four-wheel drive, is not without its military connections. Early attempts at providing four-wheel steering were concentrated on making the rear wheels steer in the opposite direction to the front ones, or what we would now term 'counter' steer. Just prior to World War I a French automotive pioneer, Eugene Brillé of the Schneider company, successfully modified the chassis of a petrol engined Paris omnibus to provide a motor lorry with four-wheel counter steering. His aim was to improve manoeuvrability of the vehicle and thereby simplify and save time in the operation of loading and unloading. This idea has also been applied to small electrically driven motor lorries, and led one authority at the time to make the following observation: 'The man who was first put to work on one of these vehicles had had absolutely no experience in driving a car of any sort, but in less than a week he was seen dodging in between rows of machines in an extraordinary way, largely because the driver had such wonderful facilities under his control owing to the steering on all four wheels.' One can only speculate as to how long this driver's luck held, or that of any other of the firm's employees who failed to get out of his way quickly enough! More realistically a well-engineered 2-ton truck with four-wheel counter steering was manufactured in 1918 by the Walter Motor Company of New York.

In 1923 there appeared what has sometimes been described as a truly remarkable vehicle known as the Holverter car, which was designed by A.A. Holle and built by French's Motor Engineering Works in London. This unique vehicle not only had four-wheel independent suspension and four-wheel drive, but also four-wheel steering. Furthermore, the steering could be set so that the rear wheels could be steered in either the opposite or the same direction as the front ones, the latter facility now being what we would term 'parallel' steer. This vehicle was therefore capable of being turned around in a very small space and also of being steered crabwise, the whole object of the design being to produce a vehicle of great manoeuvrability and mobility. Despite being successfully demonstrated to the military authorities, this ambitious design was no doubt too far ahead of its time and it never went into production. Although the military authorities had made very early attempts at four-wheel steering, it was not until the late 1930s that this feature began to appear on both British and German armoured vehicles. In the former case the Daimler Scout car originally had an unusual system of progressive

four-wheel steering, in which the rear wheels did not start to counter steer until the front ones approached full lock. When the vehicle was put into reverse the steering automatically changed over to back axle only for safety reasons. However, the system proved troublesome during the early part of World War II and it was abandoned on later versions of this vehicle.

Apart from the limited use of four-wheel steering on specialist industrial and military vehicles, there was little further interest shown in applying this concept to conventional road vehicles until the mid 1960s, when Japanese automotive engineers began investigating its potential for improving the manoeuvrability and handling of passenger cars. Over twenty-five years were then to elapse before the Honda Motor Company introduced the first production version of four-wheel steering (often abbreviated to 4WS) on their Prelude model in 1987, which is perhaps a measure of the engineering effort that was involved in translating an old concept into successful modern practice. It is now a matter of history that other Japanese manufacturers quickly followed this led with their own particular versions of four-wheel steering. The most significant difference in principle between these modern and earlier applications of four-wheel steering is that only a relatively small degree of rear-wheel steering is now employed.

Advantages and disadvantages of four-wheel steering

The advantages generally associated with the modern use of four-wheel steering may be summarized as follows.

Improved low-speed manoeuvrability
Considering first the extreme and early application of four-wheel counter steering, where the rear wheels were steered at equal and opposite angles to the front ones, it will be evident that when turning the effective wheelbase of such a vehicle becomes half its actual wheelbase. In other words, a very much smaller turning circle can be performed for the same angle of steering lock, as compared with a vehicle steered by the front wheels only. Alternatively, for a given turning circle the angle of steering lock will similarly be reduced with the rear wheels also being steered. However, an unacceptable problem is created by using such a generous amount of counter steer, because when moving away from a kerb the rear wheel tends to mount it while the front one runs away from it.

The modern approach that, to all intents and purposes, overcomes this problem is to compromise with a much smaller degree of rear-wheel counter steering. In practice this typically means that with the front wheels turned on full lock, the opposite steering angles of the rear wheels are limited to

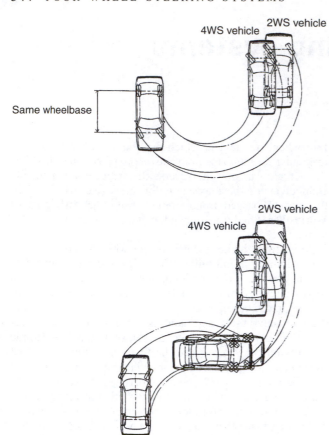

Same wheelbase

4WS vehicle 2WS vehicle

4WS vehicle 2WS vehicle

Figure 26.1 Improved low-speed manoeuvrability with 4WS (*Mazda*)

about 5°, which confers about a 10 per cent useful reduction in turning circle (Figure 26.1). Any sideways push exerted by the kerb against the rear wheels steered to this modest degree can then be absorbed by the flexibility of the tyre wall and suspension bushings. It should be noted that provision for counter steer is not included in all modern four-wheel steering systems.

Enhanced high-speed stability

Here we are chiefly concerned with the effect of steering the rear wheels in the same direction as the front ones to provide parallel steer, but only to a very modest degree. The purpose of this feature expressed in simple terms is to reduce the slip or attitude angle of the vehicle when cornering and during lane changing, thereby minimizing any sensation of tail swing. It was explained in Section 22.2 that a tyre is always called upon to exert a sideways force when a vehicle is being steered round a bend, and that as a result of the tyre and its tread deforming an angular difference is created between the direction in which the tyre is being aimed and the direction in which it actually travels, this effect being known as tyre slip angle. The cumulative effect of the tyre slip angles is that the vehicle itself develops a slip angle in relation to the centre about which it is turning. Hence the slip angle or, as it is also known, 'attitude angle' of a vehicle corresponds to the angular difference between the tangent of its turning circle and the direction of its centreline. The manner in which vehicle posture is affected by changes in slip or attitude angle when turning at low and high speeds is compared in Figure 26.2.

It will therefore be evident that if under high-speed cornering conditions the rear wheels are provided with a modest degree of same direction or parallel steer, so as partly to

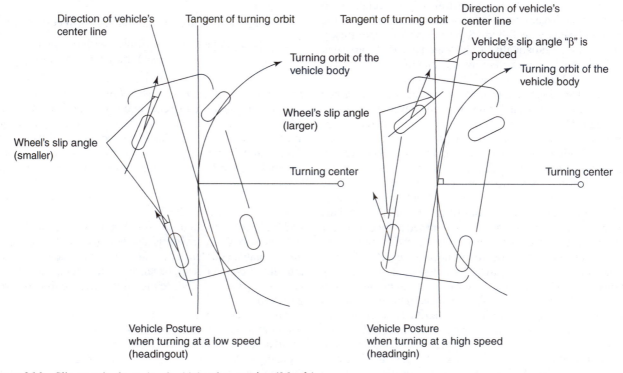

Direction of vehicle's center line

Tangent of turning orbit

Turning orbit of the vehicle body

Wheel's slip angle (larger)

Wheel's slip angle (smaller)

Turning center

Tangent of turning orbit

Direction of vehicle's center line

Vehicle's slip angle "β" is produced

Turning orbit of the vehicle body

Turning center

Vehicle Posture when turning at a low speed (headingout)

Vehicle Posture when turning at a high speed (headingin)

Figure 26.2 Slip or attitude angle of vehicle when turning (*Mazda*)

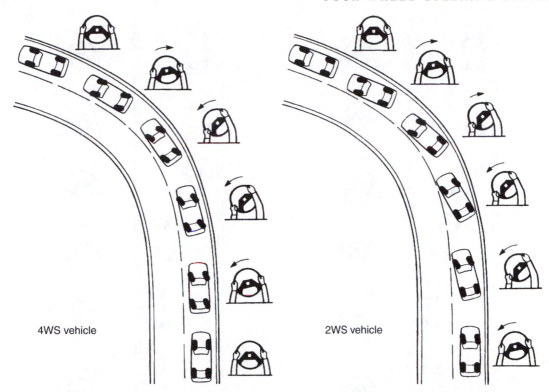

4WS vehicle

2WS vehicle

Figure 26.3 Comparison of cornering behaviour with 4WS and 2WS (*Mazda*)

compensate for the effect of the rear-tyre slip angles, then the attitude angle of the vehicle can be reduced to minimize the sensation of tail swing. That is, the direction in which the rear tyres are actually travelling now more closely coincides with the direction of the vehicle centreline. In practice the maximum amount of parallel steer conferred typically does not exceed 2°. A comparison in vehicle cornering and lane change behaviour with four- and two-wheel-steering systems is shown in Figures 26.3 and 26.4. It should be noted that the provision of parallel steer is common to all modern four-wheel-steering systems.

More responsive handling
When the driver of a conventional front-wheel-steered vehicle turns the steering wheel to enter a corner, two things happen but not simultaneously. First, the tyres of the steered front wheels generate slip angles to create a cornering force, which acts laterally about the centre of gravity of the vehicle and thus initiates the desired change of direction. This is experienced as a yawing movement of the vehicle as it swings slightly in plan view. Second, it is only after this initial change in direction of the vehicle has begun that the tyres of the otherwise directionally fixed rear wheels can develop their slip angles and similarly create a cornering force. The latter then enables the vehicle to complete the steering process and establish a steady state of turning. Hence, there is an inherent delay in the build-up of cornering force between the tyres of the front and rear wheels.

With four-wheel steering it becomes possible to reduce this delay and provide more responsive handling. It is achieved by momentarily applying counter steer to the rear wheels as the vehicle enters a corner, so as quickly to enforce a larger yawing movement. When sufficient this is followed instantly by a change to parallel steer, which then establishes steady state turning with a minimum attitude angle for the vehicle, the process being shown in Figure 26.5.

So far as disadvantages of four-wheel steering are concerned, these inevitably relate to increased cost and weight of the vehicle, greater intrusion into body space that may dictate a smaller fuel tank and wider rear wheel arches, and increased complexity of the steering system from a service point of view.

Passive and active rear-wheel steering

Reference has been made in Section 23.2 to the need to enforce optimum alignment on independently sprung rear wheels, under all conditions of loading and deflection. In practice, optimum alignment can mean that the rear suspension linkages and their flexible bushings must be so contrived as to impose toe-in on the heavily loaded outer wheel in a corner, while allowing its opposite number to toe-out. That is, the rear wheels will automatically provide a beneficial parallel steer effect once their tyres develop a cornering force. Rear-wheel steering conferred in this manner is termed 'passive', as opposed to 'active', since the latter requires the addition of a separately controlled rear-wheel-steering mechanism.

Passive steer has, either by accident or design, almost always been present in motor vehicle rear suspension systems.

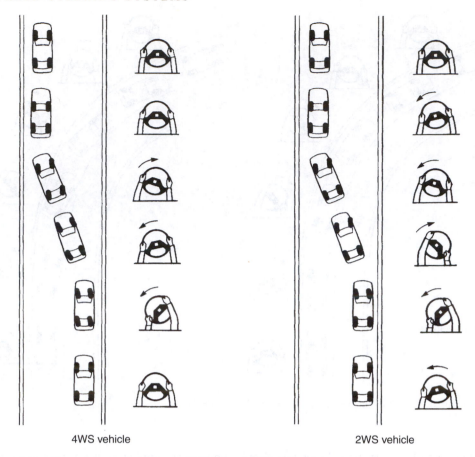

4WS vehicle   2WS vehicle

Figure 26.4 Comparison of lane-changing behaviour with 4WS and 2WS (*Mazda*)

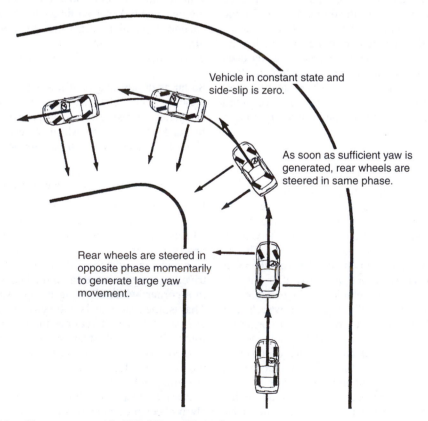

Vehicle in constant state and side-slip is zero.

As soon as sufficient yaw is generated, rear wheels are steered in same phase.

Rear wheels are steered in opposite phase momentarily to generate large yaw movement.

Figure 26.5 Improved handling response with 4WS (*Nissan Europe*)

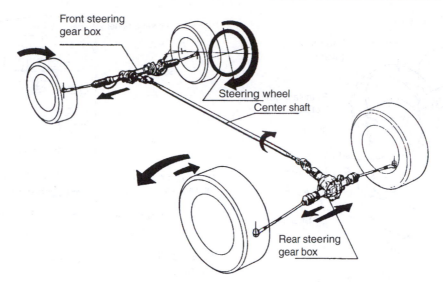

Figure 26.6 *General arrangement of Honda 4WS system* (*Honda*)

Consider for example earlier practice when many passenger cars had a Hotchkiss drive system (Section 19.4). If the semi-elliptic leaf springs with upturned eyes were set horizontally, the equivalent geometry was that of locating the axle by two forward pivoted links that sloped downwards to the rear. When the car was, say, steered to the right and therefore rolled to the left, the contra-movements of these imaginery links were such that the left-hand end of the axle was pushed backwards and its right-hand end pulled forwards. As a result the axle would be steered slightly in plan view to provide an unwelcome counter steer effect. It was later found that this effect could be beneficially cancelled, or reversed, by lowering the front ends of the springs so that the equivalent imaginary links became horizontal. This resulted in both ends of the axle being pulled forwards when the car rolled, thereby maintaining parallelism of axle movement.

Since it is possible to achieve a beneficial passive rear-wheel-steering effect when cornering with a conventional front-wheel-steered car, it may be thought why then bother with the complication of active rear-wheel steering? The answer to this question is that the latter introduces the rear-wheel-steering effect much sooner into the cornering process for more responsive handling, and passive steer cannot, of course, improve low-speed manoeuvrability if this is also sought.

26.2 TYPES OF FOUR-WHEEL STEERING

The physical interconnection between the front and rear steering systems may be effected through either mechanical or hydraulic means and can have a superimposed electronic control. A conventional rack-and-pinion steering system with hydraulic power assistance is employed for steering the front wheels, while the rear wheels receive their steering movements from the stroke rod of a sliding actuator or hydraulic power cylinder. A failsafe feature is incorporated in all systems where there is no direct mechanical interconnection between the front and rear steered wheels. It takes

the form of centring springs in the hydraulic power cylinder, which return the rear wheels to their straight-ahead position in the event of system misbehaviour through fluid loss.

The parallel or counter steering of the rear wheels in all four-wheel-steering systems is necessarily dependent on the angular direction of the front wheels, but this characteristic may be modified by superimposing the effects of either vehicle speed, vehicle speed and steering effort, or vehicle speed and steering rate. In view of this the various types of four-wheel steering may be classified as follows:

Steering angle dependent
Vehicle speed dependent
Vehicle speed and steering effort dependent
Vehicle speed and steering rate dependent.

Steering angle dependent

In the pioneer Honda system of modern four-wheel steering, provision is made for both parallel and counter steer. A hydraulic power-assisted rack-and-pinion steering mechanism is used to steer the front wheels in a conventional manner, but the steering box is modified to incorporate a suitably geared off-take drive for the rear-wheel-steering arrangements. This drive is transmitted to the rear-mounted steering box by a long, universally jointed, centre shaft. The rear steering box takes the form of a sliding actuator, wherein the transverse movement of what is termed a stroke rod steers the rear wheels in a rack-and-pinion fashion (Figure 26.6).

A unique feature of the purely mechanical rear steering box is that it changes the direction of the rear wheels from parallel to counter steer, both gradually and smoothly, when the steering wheel is turned past 140° from the straight-ahead position. The reversal of stroke rod movement, with continued same direction turning of the steering wheel, is accomplished by an epicyclic gearset and eccentric slider mechanism. Steering input from the centre shaft is transmitted to a single planetary gear that walks within a fixed annulus gear. The planetary gear is in turn provided with an offset

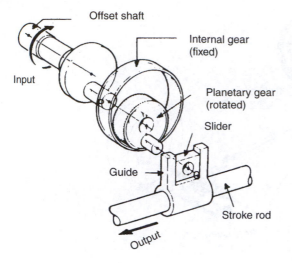

Figure 26.7 Mechanism of Honda rear steering box (*Honda*)

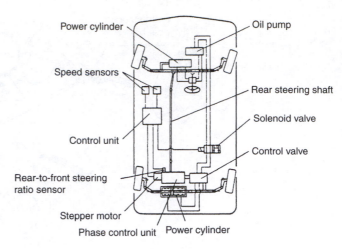

Figure 26.8 General arrangement of Mazda 4WS system (*Mazda*)

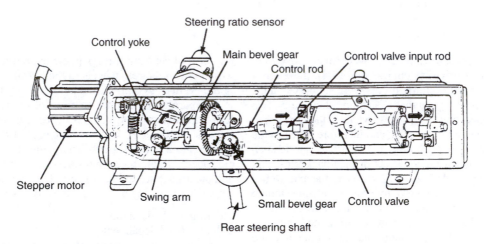

Figure 26.9 Mechanism of Mazda rear steering phase control unit (*Mazda*)

peg that engages, via a sliding block, a slotted guide member attached to the stroke rod (Figure 26.7).

When the front wheels are pointing straight ahead the offset peg of the planetary gear assumes a bottom dead centre position. As the steering wheel is turned towards either lock, the offset peg of the now rolling planetary gear undergoes a sideways displacement first in one direction and then to a greater extent in the other, thereby imparting the required contra-steering movements to the stroke rod and hence the rear wheels. Vertical movements of the peg are, or course, accommodated by its guided sliding block.

Vehicle speed dependent

Similarities between the Mazda system of this type of four-wheel steering and the Honda type described above extend to the provision for both parallel and counter steer, and the use of a hydraulic power-assisted rack-and-pinion steering mechanism for the front wheels, together with a suitably geared off-take drive for the rear-wheel-steering arrangements taken via a long, universally-jointed, centre shaft.

A significant difference is that instead of the centre shaft directly actuating a rear steering box, it transmits the steering angle to a rear-mounted steering phase control unit, which operates in conjunction with a hydraulic power steering cylinder. Steering movements are then conveyed to the rear-wheel-steering arms through track rod connections from the stroke rod of the power cylinder. A hydraulic pump supplies fluid under pressure to both front and rear steering systems. The latter system also incorporates centring spring in its power steering cylinder, which return the rear wheels to their straight-ahead position in the event of a solenoid valve in the hydraulic circuit being activated by misbehaviour of the system (Figure 26.8).

At the heart of the Mazda system is the steering phase control unit, the purpose of which is to control the rear-wheel-steering mechanism according to front steering angle and vehicle speed (Figure 26.9). The basic inputs to the phase control unit are therefore derived mechanically from the rear steering centre shaft and electrically via an electronic control unit that receives signals from two vehicle speed sensors. One sensor is within the speedometer and the other at the transmission output shaft to provide a cross-check for accuracy and as a failsafe measure. The phase control unit contains a hydraulic control valve, which receives and directs fluid under pressure from the pump to the power steering cylinder according to the phase, either parallel or counter steer, and the stroke rod movement required for appropriate steering of the rear wheels.

Actuation of the hydraulic control valve sliding spool for either parallel or counter steer is performed by a ball-jointed connecting rod, which at its other end connects with a short lateral arm hinged from an articulating yoke member. The angular movement of the latter is determined by an electrical servo motor and reduction gearing, which in turn is activated by vehicle speed signals received from the electronic control unit. To relate the parallel and counter steer requirements of the rear wheels to the particular direction in which the front ones are being steered has involved an additional and ingenious controlling feature for the movement of the connecting rod. This takes the form of passing the connecting rod through an offset spherical bearing in a crown wheel, which can be partially rotated by a bevel pinion on the input end of the centre steering shaft.

When the front wheels are in their straight-ahead position, no movement is transmitted by the connecting rod to the hydraulic control valve spool, because the crown wheel has guided the connecting rod to a position where the axes of the lateral arm to which it is ball jointed and the pivot of the articulating yoke member coincide. Rotary movements of the latter are therefore neutralized with respect to the connecting rod. As soon as the front wheels are steered away from the straight-ahead position, the centre shaft pinion partially rotates the crown wheel to impose a conical movement on the connecting rod, which according to the direction of steering either raises or lowers its end attached to the hinged lateral arm of the yoke member. Once this occurs and the axes of the lateral arm and the pivot of the yoke member no longer coincide, the lateral arm can then act as a lever whose effective length may be varied according to crown wheel rotation. This enables the connecting rod to control the position of the hydraulic control valve sliding spool, so that fluid is supplied under pressure to the appropriate end of the power cylinder for steering the rear wheels in the required direction. They are provided with counter steer below 22 mile/h (35 km/h) and parallel steer above this speed.

Vehicle speed and steering effort dependent

The Mitsubishi system of four-wheel steering of this type is significantly different from both the Honda and Mazda type described above. Functionally the steering of the rear wheels is performed almost entirely by hydraulic means, without any intervening mechanical or electrical controls (Figure 26.10). Hence there is no requirement for a rear steering centre shaft or an electronic control unit. Furthermore the system is concerned only with providing parallel steer, which is transmitted to the rear wheels by deflection of their suspension locating linkage. The rear wheels are not steered at all until the vehicle speed exceeds 31 mile/h (50 km/h).

In basic layout the front- and rear-wheel-steering mechanisms have their own separate hydraulic systems, but share

a common fluid reserve tank with a level sensor for dashboard indication of an insufficient fluid level. The power-assisted rack-and-pinion steering for the front wheels receives its supply of fluid under pressure from the front steering pump. This pump additionally operates a hydraulic control valve that regulates the fluid flow from a separate rear steering pump to the power steering cylinder for the rear wheels. Since the vehicle using this system also has four-wheel drive, the rear pump is driven from the final drive for the rear wheels. The stroke rod of the power steering cylinder is furnished with the usual track rod connections to the rear wheels, but in this case they act against the wheel locating trailing arms of the unequal transverse links suspension arrangements, these arms being capable of limited lateral articulation adjacent to their forward pivot points. That is, a small outward or inward movement enforced on the trailing arms causes the wheel to toe-out and toe-in respectively.

When the steering wheel is turned, say, to the left, the power assistance for the rack-and-pinion steered front wheels acts in the usual manner. However, the pressure build-up that occurs in the left-hand chamber of the rack power cylinder serves also to increase the pressure in the left-hand chamber of the rear steering hydraulic control valve. This forces its spool to move over to the right, which then allows fluid under pressure from the rear steering pump to flow into the right-hand chamber. From here the fluid is directed into the right-hand chamber of the rear power steering cylinder. As a result a parallel steer effect is transmitted to the rear wheels, via the track rod connections to their trailing arms (Figure 26.10). Any misbehaviour in either the front or rear hydraulic systems causes the centring springs in the rear power steering cylinder to return the rear wheels to their straight-ahead position.

Since the rear hydraulic steering pump is driven from the final drive for the rear wheels, their steering angle increases or decreases in proportion to vehicle speed. The rear-wheel steering angle is also determined by the amount of power assistance required, so that less steering effort decreases the steering angle of the rear wheels and thereby minimizes understeer.

Vehicle speed and steering rate dependent

The Nissan system of four-wheel steering of this type is known as Super HICAS. Although this is abbreviated from 'high-capacity actively controlled suspension', it is to all intents and purposes a hydraulically operated and electronically controlled four-wheel-steering system with provision being made for both parallel and counter steer. Similar to the Mitsubishi type of system, the front and rear steering mechanisms have their own hydraulic systems, but in this case both share a common pump and fluid reserve tank. The hydraulic power assistance for the rack-and-pinion steered front wheels is arranged in the usual manner. Another supply line from the pump and a return line to the reserve tank are routed, via a hydraulic pressure control valve, to the rear-mounted power steering cylinder (Figure 26.11). Track rod connections are provided to convey the movements of the power cylinder stroke rod to the steering arms for the rear wheels. (In a previous version of the HICAS system the rear

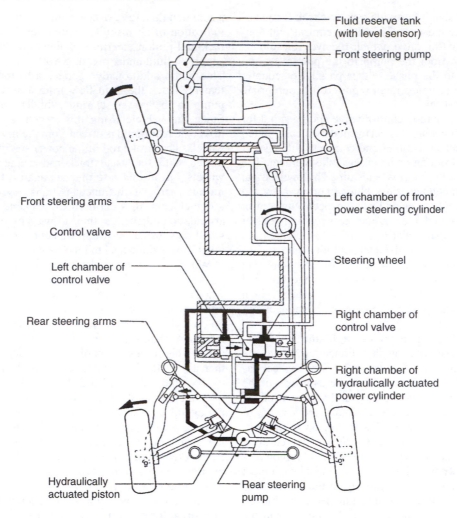

Figure 26.10   General arrangement of Mitsubishi 4WS system (*Colt Mitsubishi*)

wheels were steered by a pair of hyraulic power cylinders, which acted at either end of a flexibly mounted rear suspension subframe, so that the rear wheels could be steered slightly with it.)

An electronic control is superimposed on the hydraulic system and comprises a control unit that receives signals from vehicle speed and steering angle sensors. Based on these input signals that include rate of steering angle application, together with preprogrammed information, the electronic control unit computes the required steering direction and angle for the rear wheels. A control current is transmitted to actuate the hydraulic pressure valve, which then directs fluid at a suitable pressure to the appropriate end of the power steering cylinder. Hence the rear wheels are either parallel or counter steered to an angle determined by the electronic control unit, which in practice confers the mode of handling shown in Figure 26.5. To achieve a smooth control, the timing of the steering of the rear wheels is fractionally delayed with respect to the turning of the steering wheel and the steering of the front wheels. The electronic control unit is self-diagnosing in relation to any misbehaviour of the four-wheel-steering system, which is rendered failsafe by the actuation of a solenoid cut-off valve. In the event of losing hydraulic

pressure the cut-off valve momentarily holds the rear wheels in position, before gradually allowing the centring springs in the power cylinder to return them to straight-ahead running.

Later developments

In more recent years there has been less development of active four-wheel steering systems for passenger cars. This is mainly because of the high standard of handling that can now be achieved by passive means, designed into modern five-link independent rear suspension systems. However, there has been an increasing interest in providing four-wheel steering for sports utility vehicles (SUV) and pickup trucks in America. Such a system developed by Delphi Automotive Systems and known as 'Quadrasteer', was introduced in 2002 on selected General Motors vehicles of the types mentioned.

The Quadrasteer system essentially combines hydraulic power assistance for steering the independently sprung front wheels, and electromechanical power steering for the rear wheels mounted on a live axle. A rack-and-pinion type of steering mechanism is attached directly to and behind the rear axle and linked to the steerable wheels. It is actuated by an electric motor and drives the pinion through an epicyclic

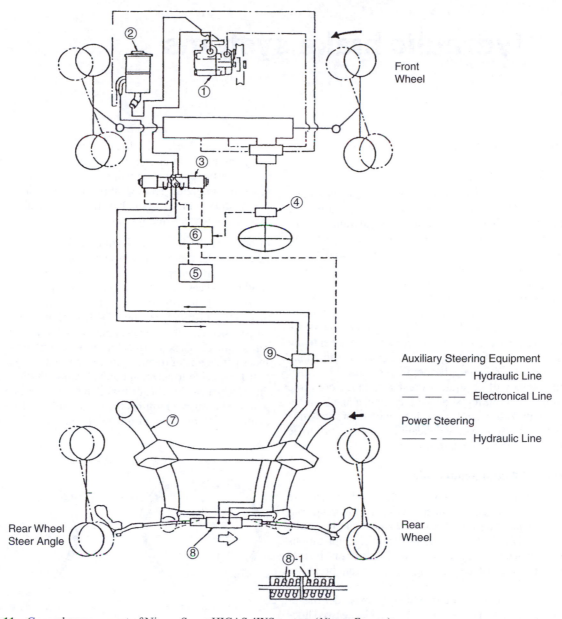

Figure 26.11 General arrangement of Nissan Super HICAS 4WS system (*Nissan Europe*)

1 hydraulic pump	5 speed sensor	8 hydraulic actuator (power cylinder)
2 reservoir tank	6 control unit	8-1 centering spring
3 pressure control valve	7 rear suspension member	9 cut-off valve
4 steering angle sensor		

gearset. An electronic control unit (ECU) for the electric motor receives information from front wheel position and vehicle speed sensors, so that the rear wheels can be steered to an appropriate angle and in the required direction. As for the previously described passenger car four-wheel steering systems, the rear wheels can be either counter or parallel steered and in this case up to an angle of 12° in relation to the front wheels, or retained in the straight ahead position. These three phases of operation are defined as 'negative', 'positive' and 'neutral' in Quadrasteer terminology. Therefore in the negative phase, the rear wheels are counter-steered to increase low speed manoeuvrability, while in the positive phase they are parallel steered to improve high speed handling. At moderate

speeds, the neutral phase retains the wheels in the straight ahead position.

To utilize the Quadrasteer four-wheel steering system to best advantage, the driver has the choice of three 'driving modes' which can be selected using an instrument panel control. The three modes are '2WS' for conventional steering of the front wheels only; '4WS' for four-wheel steering; and '4WS Tow' to modify the amount of rear wheel steering when towing a trailer. In the latter mode, the amount of rear wheel steering is reduced in negative phase at low speeds to assist parking in confined spaces with adjacent walls, and is increased in positive phase at higher speeds to provide enhanced stability by combating trailer sway.

27 Hydraulic brake systems

27.1 DRUM BRAKE ARRANGEMENTS

General background

The earliest and simplest form of hydraulically operated, internally expanding drum brakes is that in which a pair of friction-lined shoes is mounted in leading and trailing fashion on either a common or individual pivots. Before the layout of this type of drum brake assembly is examined in detail, it must first be briefly explained what is meant by the expressions leading and trailing shoes (Figure 27.1).

A leading shoe may be defined as one where the direction of drum rotation is from the applied load end or toe to the fulcrum end or heel. Conversely, a trailing shoe is one where the direction of drum rotation over its liner is from the heel to the toe. The significance of leading and trailing shoes is that they are self-energizing and non-self-energizing respectively. That is, a leading shoe will generate a greater braking drag force on the drum than a trailing shoe for the same applied load at their toes. On the other hand, a leading shoe is more sensitive to variations in friction level with heating and thus tends to be less stable in operation than a trailing shoe.

Fixed-cam and floating-cam brakes

In a leading and trailing shoe drum brake, the pivoting shoes are expanded by what are termed either fixed-cam or floating-cam mechanisms, which can take both mechanical and hydraulic forms (Figures 27.2 and 27.3). Before practical applications are discussed it is necessary to define in simple terms the theoretical significance of fixed-cam and floating-cam brakes:

Fixed-cam brake This is one in which the expander mechanism (not necessarily a cam as such) compels the two shoes to have equal movements and do the same amount of work. However, it does not exert equal loading on the leading and trailing shoes because of their self-energizing and non-self-energizing tendencies respectively.

Floating-cam brake In this case the expander mechanism (very rarely a cam as such) does exert equal loading on the two shoes but permits them unequal movements. This allows the self-energizing leading shoe to follow its natural tendency and do more work than the trailing shoe, but at the expense of greater lining wear.

In practice, the fixed-cam leading and trailing shoe brake is inherently the more stable in operation, but requires a much greater effort to apply. For this reason, its use is now usually confined to heavy commercial vehicles where full power operation of the brakes by compressed air is used, as described later. Since there is usually an appreciable thickness of liner to wear on commercial vehicle brakes, instead of having flat sides the cam is often S-shaped so that the amount of lift it imparts to the shoes per degree of rotation remains constant (Figure 27.4).

It is possible to have an equivalent hydraulic expander to the mechanical fixed-cam brake, but this is rarely found in practice because it requires the use of either a stepped wheel cylinder with the large piston acting on the trailing shoe, or a

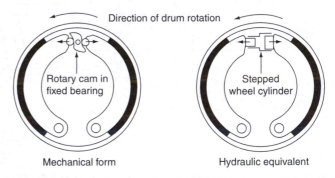

Figure 27.2 The fixed-cam drum brake in mechanical and hydraulic forms

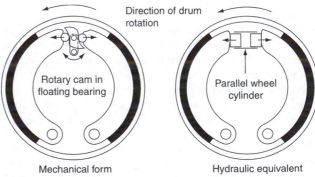

Figure 27.3 The floating-cam drum brake in mechanical and hydraulic forms

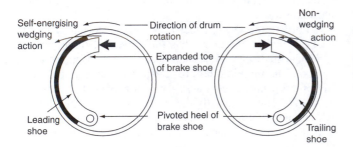

Figure 27.1 The wedging and non-wedging actions of leading and trailing brake shoes

conventional non-stepped wheel cylinder with mechanically interlinked shoes. The object in both cases, of course, is to ensure that both leading and trailing shoes share the workload equally despite their natural inclinations to do otherwise.

For all other classes of vehicle using leading and trailing shoe brakes a floating-cam expander of either hydraulic or mechanical form is almost invariably used and its limitations are found acceptable.

Before the advent of hydraulic brakes, a floating-cam expander usually took the form of either a toggle or a wedge-operated mechanism, but rarely a rotary cam. At first this may seem curious, but the explanation is that it was more difficult to arrange for a cam to float between the shoe tips because this involved mounting the camshaft in a separate swinging link from the back plate (Figure 27.5). Wedge-operated shoe expanders with a floating-cam characteristic were once popular for car brakes, especially since they were operated mainly by a more rigid system of rods in tension, rather than shafts in torsion as in the case of rotary cam expanders. Wedge-operated shoe expanders are still widely used on some commercial vehicles, where they lend themselves to hydraulic and compressed air operation, as described later.

The introduction of the basic hydraulically operated, leading and trailing shoe brake immediately produced a simple hydraulic equivalent of a floating-cam expander. It did so because the opposed pistons sliding in the double-acting wheel cylinder exerted equal loading on the shoes, while at the same time allowing them freedom of relative movement. A leading and trailing shoe brake may also be expanded by a single-acting wheel cylinder free to slide on its back plate. This arrangement likewise produces a hydraulic equivalent of the floating-cam expander.

Two leading shoe brakes

It was earlier explained that a leading shoe will generate a greater braking drag force on the drum than a trailing one for the same applied load at their toes (or tips). We should perhaps now express this difference in braking drag force in terms that are more familiar to an automotive engineer, by saying that a leading shoe possesses a higher shoe factor than a trailing one. That is:

$$\text{shoe factor} = \text{drum drag/shoe tip load}$$

Other things being equal, the shoe factor is about three times as great for a leading shoe as for a trailing one. Considering next the combined effect of both shoes, we can further say (except in the more complicated case of a fixed-cam shoe brake) that the brake factor equals the sum of the separate shoe factors, or simply:

$$\text{brake factor} = \text{total drum drag/total shoe tip load}$$

It was almost inevitable that a demand for a brake factor higher than could be obtained from a simple leading and trailing shoe (L/TS) brake should arise first in the field of motor racing. As a result we find that a further development of the drum brake, known as the two leading shoe (2LS) type (Figure 27.6), was first adopted by the German firms of Daimler-Benz and Auto-Union for their legendary racing cars of the 1930s.

Following the introduction of 2LS front brakes on the 1940 production model Chrysler cars in America, this particular arrangement of front brakes became increasingly popular in the United Kingdom and on the Continent, although it was not further pursued in America. It has the merits of concentrating the greater braking torque on the more heavily loaded front wheels, while allowing the retention of L/TS rear brakes which were equally effective in the forwards and reverse directions and therefore better suited for handbrake operation. For cars the 2LS brake was largely superseded by the disc brake in the early 1960s, but continues to be used on commercial

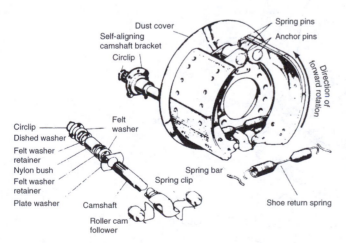

Figure 27.4 A modern fixed-cam drum brake for commercial vehicles (*Girling*)

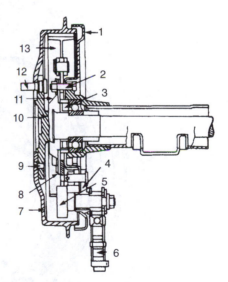

Figure 27.5 An interesting rare example of a mechanical floating-cam brake used on some prewar Vauxhall cars

1 brake drum dust shield	7 brake drum
2 anchor plate bolts	8 axle shaft retaining plate
3 oil retaining washer	9 drum locating screw
4 swinging bearing bracket for floating-cam	10 axle shaft flange
5 brake shoe floating-cam	11 oil deflector
6 brake operating lever	12 wheel attachment bolts
	13 brake shoe

vehicles not only for the front but also for the heavily laden rear wheels.

Types of 2LS brake

The identifying features of any 2LS brake are that the shoes pivot from diametrically opposite abutments; each shoe is provided with a separate means of actuation; and that during forwards braking the direction of drum rotation is from the toe to the heel of each shoe.

We next have to consider the different means by which the two shoes can be separately activated, these being hydraulic and mechanical.

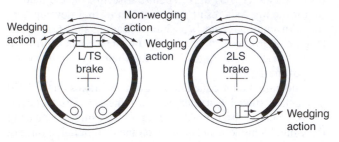

Figure 27.6 The basic difference between leading and trailing (L/TS) and two leading shoe (2LS) brakes

Hydraulic 2LS

This is the original and simplest form of 2LS brake. Each shoe is actuated by a separate single-acting hydraulic wheel cylinder of which the blind end forms an abutment for the adjacent heel of the opposite shoe (Figure 27.7). The two wheel cylinders are linked on either side of the brake back plate by a separate pipe and the brake hose connects to the upper cylinder.

Mechanical 2LS

For this arrangement a wedge and tappets expander is used in conjunction with a bell cranks and strut linkage on what would normally be the trailing shoe of a L/TS brake (Figure 27.8). The trailing shoe is therefore converted into a second leading shoe (Figure 27.9), because the appropriate tappet of the wedge expander does not bear directly against the shoe tip, but acts through the bell-cranks and strut linkage to expand the other end of the shoe.

A perhaps simpler mechanical arrangement for actuating two leading shoes is that found in the front brakes of some motor cycles, where the shoes are expanded by two diametrically opposite cams (Figure 27.10). The camshafts are linked externally by a levers and pull-rod mechanism, one lever being extended for connection to the cable and actuating system.

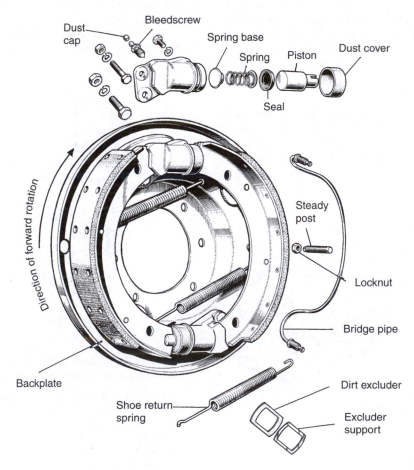

Figure 27.7 Hydraulically operated 2LS commercial vehicle brake (*Girling*)

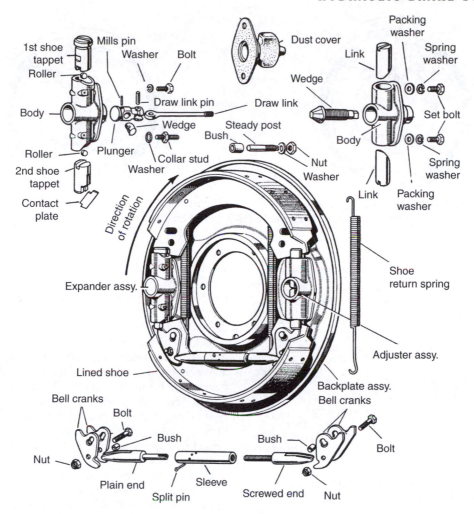

Figure 27.8 Mechanically operated 2LS commercial vehicle brake (*Girling*)

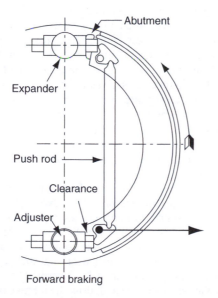

Figure 27.9 Trailing shoe converted to leading shoe by bell-cranks and strut

Advantages and disadvantages of 2LS brakes

As compared with an L/TS brake the 2LS brake offers the following advantages:

1 Greater braking torque is generated by the same operating force or, perhaps more realistically, a smaller operating force is required to achieve the same braking torque.
2 Since each shoe is equally loaded there are no unbalanced loads imposed on the wheel bearings.

As far as disadvantages are concerned these are:

1 When braking in the reverse direction the two leading shoes become two trailing shoes, so that a considerably greater operating force is required to generate the same braking torque.
2 The 2LS brake is very sensitive to changes in friction level between the linings and drum, so that as the brake heats up there is a greater falling off (fade) in brake torque for a given operating force.
3 The purely hydraulic version does not lend itself to additional operation by the handbrake system, but this can

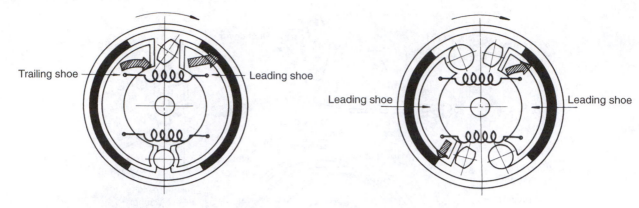

Figure 27.10   Comparison between cam-operated L/TS and 2LS motor cycle brakes (*Yamaha*)

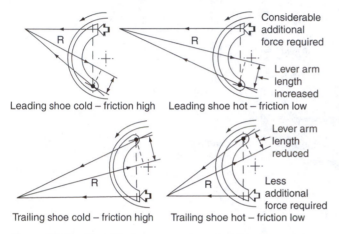

Figure 27.11   Simplified representation of additional operating force for L/T shoes with falling friction

readily be arranged in the mechanical version through the wedge expander mechanism.

Regarding the second disadvantage, this deserves a few words of simple explanation. As mentioned later, brake fade arises from a reduction in the frictional properties of the lining material as its temperature increases. This in turn has the effect of altering the direction of the resultant force on the brake shoe and hence the distance it acts from the shoe pivot. Since this leverage acts in opposition to that of the brake shoe apply force, which also acts about the shoe pivot, then clearly with the onset of brake fade the brake apply force must be increased to maintain the same braking torque. The reason why fade does not greatly increase the effort required to operate a trailing shoe, in comparison with a leading shoe, is that in the former case fade decreases the effective lever arm of the resultant force about the shoe pivot, whereas with a leading shoe it increases the length of the effective lever arm (Figure 27.11). In other words, the operating force on the leading shoe must be increased to a greater extent to maintain a given brake torque. It is, of course, only fair to add that the force needed to operate a trailing shoe is greater in the first place.

Other brake shoe arrangements

The earliest form of internal-expanding shoe brake was that devised by the French automotive pioneer Louis Renault in 1902, in which the two shoes for each rear brake shared a common fixed pivot at their heels and were expanded against the drum at their toes by turning a simple flat-sided cam. In principle this early shoe brake will be recognized as being of the fixed-cam type that is still used in modern form on heavy vehicles. The first significant development of this basic type of brake was made in the early 1920s when another French automotive pioneer Henri Perrot, who had previously been responsible for introducing the first successful mechanically operated four-wheel braking system, invented the self-servo type of shoe brake. Instead of the leading and trailing shoes being provided with fixed pivots at their heels, they were connected together by a floating link and termed primary and secondary shoes.

In the original design only the primary shoe was expanded against the drum, with the secondary shoe being anchored to an adjacent abutment. When the brake was applied in the forwards direction, the rotation of the drum tended to drag the primary shoe around with it and, since this drag was applied directly to the secondary shoe, this in turn shared the self-apply or self-servo effect. The total drag and hence the braking torque generated could be quite considerable for a given size of brake, thereby reducing brake pedal effort, but the resulting high shoe factor could make this arrangement of brake shoes more susceptible to fade. Clearly in its original form with only the primary shoe being expanded, the brake was just as lazy in reverse as it was vigorous in forwards application. This defect was largely overcome by the Bendix company in America who later produced their duo-servo version, wherein both primary and secondary shoes were hydraulically expanded against the drum and provided with an adjacent common abutment for either shoe according to the direction of drum rotation.

The duo-servo arrangement of brake shoes continued to be widely used in America until the advent of disc brakes. Elsewhere it continued to find application on light to medium commercial vehicles, where especially as a rear brake it conferred very effective handbrake operation in either direction of holding. It is also for the latter reason that this type of shoe arrangement is incorporated in what are termed 'drum-in-disc' brake installations, where a small drum handbrake mechanism is built into the hub of a brake disc and is cable operated.

At the opposite extreme to the duo-servo arrangement of brake shoes was the two trailing shoe brake, which represented the final development of the drum brake for high-performance cars by the Girling company, who introduced it in the early 1950s. From its designation it will be clear that during forwards braking, and in contrast to the two leading shoe brake, the direction of drum rotation is from the heel to the toe of each shoe. As a result of having a brake factor that was 50 per cent lower than even a leading and trailing shoe brake, the two trailing shoe brake will be remembered for being not only the least susceptible to fade of all drum brakes, but also the smoothest in operation. By the same token they did need servo assistance to apply them, owing to the much higher operating forces required at the shoe tips to generate a given brake torque.

Finally, it should be mentioned that various attempts have been made in the past to exploit multishoe brakes, which typically employed three or four shoes per brake arranged in either leading or trailing fashion. Although it might seem a good idea to have a 'drumful of shoes', as such brakes were once described, in practice they invariably proved disappointing. This was because the drums were less able to dissipate the greater amount of heat generated during heavy braking, which resulted in excessive wearing of the liners.

Manual adjusters for brake shoes

Since the adjustment facilities required by brake shoes with pivoting and sliding abutments can differ, it is convenient to consider them under separate headings, as follows.

Pivoting shoes
These are usually provided with a centralizing adjustment at their heels and a wear adjustment at or near their toes, although both adjustments can be effected at the heels of the shoes where a wedge-type adjuster is used between them. The purpose of centralizing adjusters is to ensure that during initial build, and later during reline of the brake, there is no localized bedding of the liner at either the heel or the toe of the shoes. This is because a heel bedding shoe will produce an inefficient brake and a toe bedding one a grabbing brake. As its name suggests, a wear adjuster is provided for each shoe so that it can be set closer to the drum to compensate for lining wear.

Sliding shoes
These were first featured in proprietary drum braking systems during the late 1940s and by the nature of their mounting are self-centring within the drum, thereby eliminating any need for a centralizing adjuster. A wear adjuster for each shoe is, of course, still required. The advantages of the sliding shoe over the pivoting type are generally that it allows more uniform wear along the length of the lining and it can be more easily serviced.

Centralizing adjuster
A typical example of this type of adjuster comprises a hexagonally headed eccentric bushing at the heel pivot of the brake shoe (Figure 27.12). Once the shoe has been centralized, the bushing is clamped to prevent further rotation, but the shoe remains free to pivot about the bushing.

Wear adjusters
Several forms of wear adjuster may be encountered, especially those known as snail cam, screwed plug and wedge.

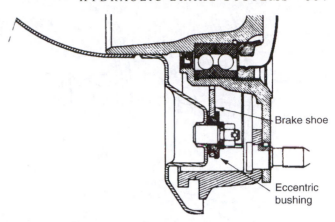

Figure 27.12 Brake shoe centralizing adjuster (*Citroën*)

Snail cams are usually provided with serrated edges and may act either between the brake back plate and a striker pin on the shoe web (Figure 27.13a) or between the wheel cylinder piston and shoe tip. A screwed plug adjuster may be incorporated with the wheel cylinder piston and therefore also acts against the shoe tip (Figure 27.13b). In contrast, the wedge-type adjuster acts between the brake back plate and the heels of the shoes and thus furnishes them with adjustable abutments (Figure 27.13c). Slight repositioning of the wedge adjuster on the back plate, before its retaining bolts are finally tightened, allows for the centralization of pivoting-type shoes when they are expanded within the drum.

Automatic adjusters for brake shoes

Although automatic adjusters for the shoes of drum brakes had earlier been used to a limited extent on cars, it was not until self-adjusting front disc brakes began to be widely used in conjunction with rear drum brakes that a self-adjustment facility for the latter seemed a desirable improvement and led to a revival of interest in automatic shoe adjusters. Since then automatic adjusters have also been increasingly adopted for the front and rear drum brakes of commercial vehicles. It may be useful to summarize the advantages to be gained from the otherwise added complexity of automatic adjusters:

1 A welcome reduction in the amount of routine servicing.
2 Constant travel for the brake pedal to ensure quick response of service brakes.
3 Constant travel for the handbrake lever to safeguard emergency braking.
4 Minimum provision required for wear of linings in reserve pedal travel, so that some of it may be utilized to allow increased hydraulic leverage in the system and thereby reduce pedal effort.

The design of automatic shoe adjusters is carefully contrived to take into account the effects of high braking temperatures on drum expansion, because an adjuster that reacted with an excess of zeal could impose less than normal running clearance between the shoes and drum on cooling of the brake. Equal care is demanded in the setting up of modern automatic adjusters when replacing brake shoes or servicing wheel cylinders. Automatic adjusters for drum brakes may be of either the one-shot or the incremental type, the former

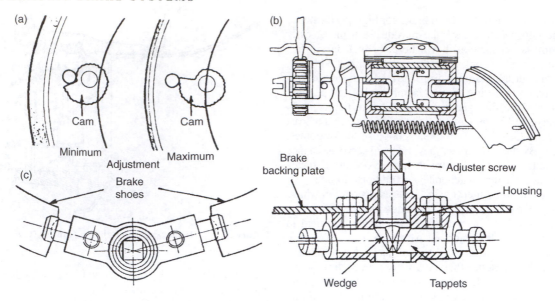

Figure 27.13 Examples of brake shoe wear adjusters: (a) snail cam (*Girling*) (b) screwed plug (*ATE*) (c) wedge and tappets (*Girling*)

having the advantage that only a single brake application is usually required to adjust the brakes after initial setting up. In contrast the latter type restores the normal running clearance between the shoes and drum by a small amount at each successive application of the brakes, until adjustment is completed. Various forms of these basic types of adjuster have been used past and present, but their action is essentially that of either a friction-grip, ratchet-and-pawl or jacking-screw mechanism to reposition the brake shoes as their linings wear, each version incorporating a lost-motion facility to allow the shoes their normal running clearance once repositioned.

An early version of the friction-grip adjuster represents an easily visualized method of automatic shoe adjustment. The toe of the brake shoe is extended downwards with a striker pin at its extremity. This pin engages a clearance slot in a forked lever with a friction pivot, the effect of the latter being such as to resist turning under the influence of the brake shoe pull-off spring. A normal running clearance of the brake shoe with the drum is permitted by the smaller free movement of the striker pin in the forked lever (Figure 27.14a). As the lining wears, the increased outward movement of the shoe forces the lever to rotate slightly via the striker pin and reposition the shoe, which then resumes its normal running clearance. This same principle is incorporated in the shake-back type of automatic adjuster, wherein a flanged sleeve is located in an elongated hole in the shoe web by a pair of spring-loaded friction washers. A striker pin is secured to the brake back plate and engages with the clearance hole in the sleeve to effect the necessary repositioning of the shoe with lining wear and also allow a normal running clearance (Figure 27.14b).

Ratchet-and-pawl and jacking-screw automatic adjusters tend to be more elaborate in their application. With the former arrangement the toothed end of a ratchet strut may engage a spring-loaded pawl secured to the brake back plate, while the other end of the strut is pin jointed with a clearance hole to the brake shoe (Figure 27.14c). The clearance hole allows the normal running clearance of the brake shoe with the drum, but as the lining wears and the shoe moves further outwards the pawl engages the next tooth on the ratchet strut to reposition the shoe. Jacking-screw adjusters may be incorporated either as a strut between the shoes, or as a tappet within the actuating wheel cylinder piston. In a typical example (Figure 27.15) a spring-biased pawl pivots from near the toe of the leading shoe and engages in bell-crank fashion with a ratchet wheel on the jacking-screw strut, which although part of the hand brake operating mechanism also acts as an off-stop for the striker pawl. As the shoe linings wear, the increased arcuate movement of the striker pawl that accompanies the greater separation of the shoes turns the ratchet wheel on brake release and lengthens the jacking-screw strut to reposition the shoes further apart. No movement of the striker pawl is intended to occur when the handbrake is applied and the adjuster strut is under load, so that automatic adjustment can only happen during footbrake release.

Brake shoe return and retention

When the brakes are released, a pull-off tension spring or springs returns the shoes to their off position against the adjusting mechanism (Figure 27.41). In other designs the shoes may be maintained in light rubbing contact with the drum, in which case a bias rather than a pull-off spring is used for the purpose of preventing any tendency towards grabbing-on of the leading shoe. Hold down or 'steady' springs may be used to retain each shoe sideways against the supporting back plate to prevent vibration and potential squeal (Figure 27.16).

Brake drums

These rotate with the road wheels by virtue of being rigidly attached to the wheel hubs. Less commonly, they may be mounted either side of the final drive unit and coupled to the road wheels by the transmission drive shafts. The two arrangements are known as outboard- and inboard-mounted brakes respectively. To resist wear at the rubbing surfaces of the shoe

(a)

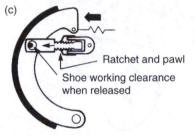

Shoe working clearance
when released

Fork friction
pivot

(b)

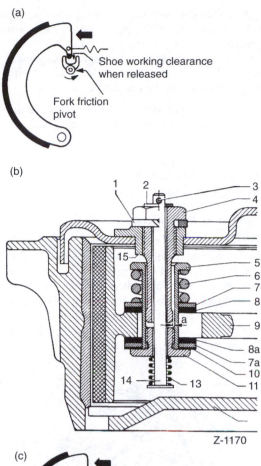

Z-1170

(c)

Ratchet and pawl

Shoe working clearance
when released

Figure 27.14 Principles of automatic brake shoe wear adjusters:
(a) forked pivot (b) shake-back (*Mercedes-Benz*) (c) ratchet and
pawl

1 lock washer	9 brake shoe
2 washer	10 washer
3 cotter pin	11 tensioning screw
4 bolt	12 brake drum
5 adjusting sleeve	13 pressure spring
6 pressure spring	14 guide pin
7 and 7a thrust washers	*a* clearance
8 and 8a friction washers	

liners and drums, the latter are generally cast from an alloy
iron. Circumferential ribs may be provided on their external
rim surface both to assist cooling and to reduce distortion.

The physical dimensions of a drum brake are concerned
with providing an adequate swept area for the rubbing surfaces which will not only allow sufficient brake torque to be
developed for exercising the necessary braking control over the
vehicle, but also be capable of absorbing the heat generated

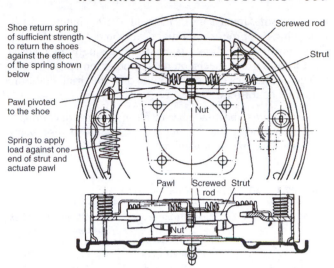

Figure 27.15 An incremental type of drum brake adjuster
(*Girling*)

Figure 27.16 Methods of retaining the brake shoes against the
back plate (*Girling*)

by this braking and then to dissipate it at a rate appropriate to
the operating conditions; thereby contributing to a consistent
brake performance and a reasonable life for the shoe linings.
In practice, of course, the diameter of the brake drum is necessarily limited by the size of road wheel used, less the radial
clearance required to allow the circulation of air through the
wheel for brake cooling. Whilst the width of the brake shoes
that can be accommodated within a drum does not directly
increase the brake torque, wider shoes will serve to lower temperatures for more stable friction values and also to increase
the life of the shoe linings. Here again there are space restrictions on the width of drum that can be used and, in any event,
a too wide drum can suffer from 'bell-mouthing' (distorted in
the manner of a cone) with consequent disturbed bedding of
the shoe linings.

Since all-drum brake systems have long been confined to
commercial vehicles, it may here be appropriate to quote a
couple of examples of heavy vehicle, albeit air braked (Section
28), wheelbrake sizes from the Volvo range of trucks as shown
in Table 27.1.

27.2 DISC BRAKE ARRANGEMENTS

General background

It is now a matter of motoring history that the spot type
of disc brake (so named to distinguish it from a much earlier
and less successful type of disc brake that acted in the manner of a plate clutch) contributed greatly to the motor racing

successes of Jaguar cars in the mid 1950s (Figure 27.17). Following this promising initial development by the Dunlop Company the disc brake became increasingly adopted, at least for the front wheels, on most production cars.

The advantages of the disc brake over the drum brake are generally as follows:

1 Less prone to heat fade because a higher rate of cooling is achieved with the rubbing surfaces of the disc being exposed to the air stream.

2 Any heat fade that does occur is less pronounced because being virtually non-self-energizing reduces its sensitivity to changes in friction level.

3 Inefficiency and apparent fade caused by expansion of the drum, and consequent poor bedding of its shoes, are absent in the case of the disc brake.

It is now proposed to classify the different forms a disc brake may take (Figure 27.18) and to indicate their relative merits, as follows:

Fixed caliper and cylinders
Floating caliper and cylinder
Floating caliper and fixed cylinder.

Fixed caliper and cylinders

This represents the earliest development of the modern disc brake and remained the most widely used type until the early 1970s (Figure 27.19). It comprises simply a mounting half and a rim half, which are bolted together and straddle the brake disc rigidly attached to the wheel hub. One or more cylinders in each half of the caliper house sealed and self-adjusting pistons. These press the friction pads against the respective sides of the disc, each pad being located in a recess

Table 27.1 Wheelbrake sizes of Volvo trucks

Model	GVW (kg)	Brake sizes	Diameter (mm)	Width (mm)
FL7	34 200	First axle	410	200
8 × 4 Rigid		Second axle	410	200
		Third axle	410	200
		Fourth axle	410	200
FH12	26 100	Front axle	410	200
6 × 2 Tractor		Intermediate axle	410	125
		Drive axle	410	225

GVW – Gross Vehicle Weight

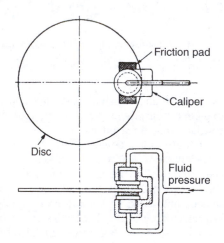

Figure 27.17 The hydraulically operated disc brake in principle

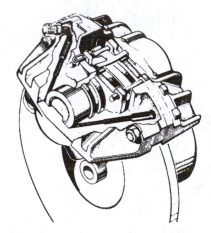

Figure 27.19 Cut-away view of fixed caliper and cylinders disc brake (*ATE*)

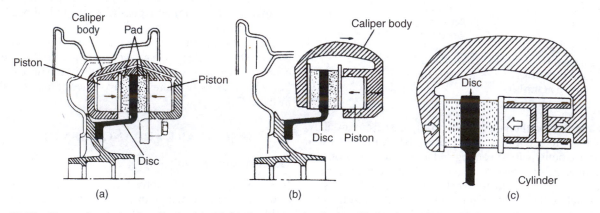

Figure 27.18 Types of non-pivoting disc brake: (a) fixed caliper and cylinders (b) floating caliper and cylinder (c) floating caliper and fixed cylinder (*Suzuki*)

within the caliper. When the brake is applied both pads are moved simultaneously into contact with the disc, because a hydraulic pressure balance between the opposed cylinders is obtained either by an external bridging pipe or internal drillings through the caliper bridge itself.

The drawbacks generally associated with this form of disc brake may be summarized as follows:

1 Since the rim half of the caliper houses one or more pad operating pistons, its bulk can be difficult to accommodate within the confines of the road wheel.
2 Under particularly severe usage the temperature reached by the brake fluid, especially in the transfer passage between the caliper halves and in the outboard cylinder(s) shrouded by the wheel, may exceed its boiling point so that vaporization and possible loss of braking could occur.
3 It is difficult to provide an integral handbrake mechanism with this form of disc brake, which demands an auxiliary pair of pads with their own actuating linkage.

To their credit we can say that:

1 Virtually all the movable parts are lubricated by hydraulic brake fluid.
2 They offer the simplest form of pad location.
3 Four-cylinder versions readily lend themselves to an L-split divided-line brake circuit.
4 Their inherent rigidity makes them especially suitable for high-performance car brakes, where even six-cylinder front calipers may be used as in a Porsche installation.

Floating caliper and cylinder

The distinguishing feature of this type of disc brake is that only the inboard pad is directly operated by hydraulic means, the reaction force from this being transmitted to the outboard pad by an abutment in the floating caliper. A floating action for the caliper is obtained by virtue of either a pivoting or a sliding mounting system.

In one example of the pivoting mounting system, the caliper takes the form of a frame produced from a rigid steel pressing. A single hydraulic cylinder and piston are secured on the inboard side of the caliper frame, which is pivoted at its upper end to the mounting bracket. Therefore when the inboard pad is pressed into contact with the disc by the hydraulic piston, the caliper frame and cylinder swing in opposition and apply the other pad with equal pressure. As the pads wear the angular relationship between the caliper frame and disc gradually changes, so to compensate for this the friction material is initially of tapered section, but approaches parallelism with the disc as wear proceeds (Figure 27.20).

Current developments of the disc brake would seem to be concentrated on the sliding mounting system for the caliper. In a typical modern example a fist-type sliding caliper is used, this being so called because of its similarity in form to a partly clenched fist (Figure 27.21). A hydraulic cylinder and piston (in some cases two are used) are housed in the wrist portion and press the inboard pad into contact with the disc, while the fingers form the moving abutment or claw for the outboard pad. A carrier bracket reacts against brake torque and provides the

necessary support and location for the pads. Also, it is furnished with precision-bored guides to receive the mounting pins of the sliding caliper. (One of the pins is rubber sleeved to prevent rattle in certain types of Girling Colette brake.)

It should be appreciated that the sole purpose of the sliding caliper is to clamp the pads on to the disc, since it plays no part whatsoever in reacting the drag forces on the pads when the brake is applied, this being the function of the carrier bracket mounted from the steering yoke. The sliding surfaces of the caliper mounting pins are greased and sealed to minimize friction (Figure 27.22); otherwise problems can arise such as unequal pad apply forces, false self-adjustment and dragging of the brake.

The advantages of the modern floating caliper disc brake may be listed as follows:

1 Brake fluid operating temperatures can be reduced by virtue of the more remote and unshrouded inboard mounting for the hydraulic cylinder(s).
2 The absence of hydraulic cylinders on the outboard side of the disc makes for easier accommodation within the confines of the road wheel, especially where negative steering offset is used.
3 A sliding version of floating caliper readily lends itself to the addition of an integral handbrake mechanism (Figure 27.55).

On the debit side we may include the following points:

1 Pad replacement can be less easy with a sliding version of floating caliper, because the fist must be either

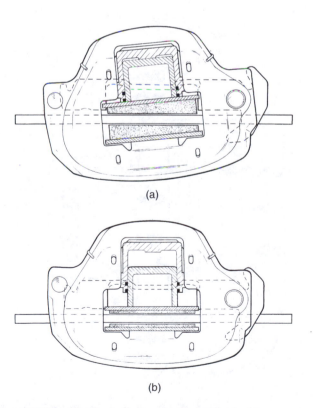

(a)

(b)

Figure 27.20 Pivoting caliper and cylinder disc brake: (a) assembly with new pads (b) pads fully worn (*Lockheed*)

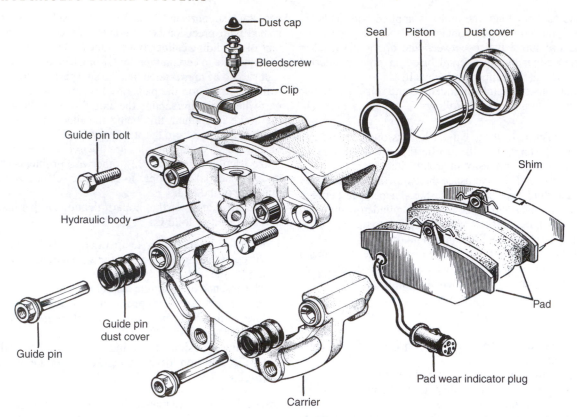

Figure 27.21 Exploded view of a modern floating caliper and cylinder disc brake, showing carrier bracket for reacting against brake torque (*Girling*)

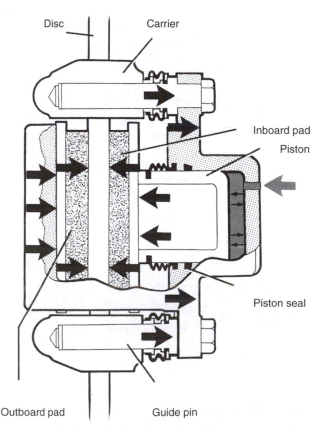

Figure 27.22 Action of Colette caliper (*Girling*)

completely or partially removed from its mounting bracket.

2 There is a risk of corrosion and possible seizure in designs where the caliper floats on open V-slides, which gives rise to the operational problems mentioned earlier.

Floating caliper and fixed cylinder

This design of disc brake represents something of a compromise between the fixed and floating caliper constructions, because although it is basically a sliding caliper type it retains the easy pad replacement of the fixed caliper. Its construction is such that a hydraulic cylinder with opposed pistons is rigidly attached to a mounting bracket on the inboard side of the brake disc. A steel yoke or frame straddles the disc and can slide in grooves provided alongside the cylinder body (Figure 27.23).

When the brake is applied, the hydraulic pressure in the fixed cylinder forces the pistons apart, one of which acts directly on the inboard pad and the other indirectly on the outboard pad via the sliding yoke. It should be noted that the latter is also responsible for transmitting the drag forces on the outboard pad to the cylinder body and mounting.

This particular form of floating caliper brake has the following good points:

1 In common with the floating caliper and cylinder type, brake fluid temperature can be reduced.
2 Its frame construction allows an increased diameter of disc within a road wheel of limited size.

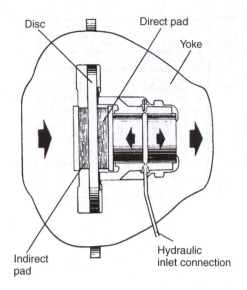

Figure 27.23 Floating caliper and fixed cylinder disc brake (*Girling*)

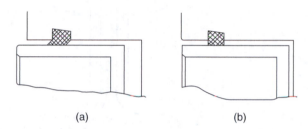

Figure 27.24 Operation of a fluid seal: (a) brake applied (b) brake released

On the other hand it could be at a disadvantage in respect of the following:

1 The overall length of the caliper frame does not make for a compact installation of the outboard friction pad.
2 It does not lend itself to dual hydraulic circuits for safer braking.

Wear indication and replacement of disc pads

Although disc brakes generally lend themselves to simple servicing procedures they can nevertheless be the victims of potentially dangerous neglect. This is because, unlike many drum brakes, they are self-adjusting and can therefore function for long periods without calling attention to themselves.

The automatic adjustment facility is of the one-shot type and is conveniently derived from the elastic recovery of the wheel cylinder seal, through which the sliding movement of the piston takes place (Figure 27.24). That is, the seal retracts the piston into the cylinder bore when the brake is released, but only just sufficient to restore a normal running clearance between the brake pad and the disc. To guarantee a definite running clearance between the pads and disc, a 'low drag' seal arrangement may be used. The seal groove in the caliper wheel cylinder is then provided with a tapered outer edge. Its purpose is to allow more seal flexure and increased retraction of the piston into its cylinder bore when the brakes

are released, thereby eliminating frictional drag between the pads and disc to the benefit of reduced fuel consumption.

If the disc brake pads are allowed to wear beyond the generally recommended limit of 3 mm (1/8 in) lining thickness remaining, the effect may be twofold.

1 The reduced thickness of heat insulation offered by the pad material between the rubbing surfaces and the brake fluid in the caliper cylinder(s) introduces a risk of fluid vaporization and possible loss of braking under conditions of severe usage.
2 In the event of the pad material wearing to the extent that its steel backing plate begins to contact the disc, then apart from scoring the latter the action of the brake will become both inefficient and noisy.

Recognition of when the brake pads have reached the end of their useful life may take two forms: on-board indication to the driver, and off-board examination by the service engineer.

On-board indication
The driver is provided with an early warning system of the pads wearing out by either audible or visual indication, or by change in feel at the brake pedal. An audible indication may be obtained simply by adding a squealer spring to the inboard brake pad. When the latter has worn to a predetermined thickness the spring becomes flush with the pad rubbing surface, so that application of the brake will cause the spring to rub on the disc and emit a high-pitched squeal. A visual indication can be arranged electrically by having a sensor embedded in the brake pad (Figure 27.25). Wearing of the pad eventually exposes the sensor; therefore when the brakes are applied an electrical circuit is completed through contact of the sensor with the disc and illuminates an indicator visible to the driver.

A change in feel at the brake pedal with worn pads can be enforced by their backing plates eventually embracing a cross-leaf spring, which also acts as a retaining device for the brake pads. Since the driver will exert a greater effort on the brake pedal in trying to overcome the blocking effect of the cross-leaf spring, the hard pedal feel indicates that the pads require replacement (Figure 27.42).

Off-board indication
This involves the service engineer in following the general recommendation of brake equipment manufacturers to check visually the state of wear of the brake pads themselves at least every 6000 miles (10 000 km). Each wheel must be removed in turn so that every pad can be viewed for edge thickness, since it is the thinnest one that limits the life of the set. The brake pads are also checked for excessive wear in the DTp vehicle test (still known as the MoT test).

As explained earlier, there are several different forms of disc brake construction, so clearly it is desirable to follow the servicing instructions of the particular manufacturer concerned until complete familiarity with these units is acquired. However, the means by which the disc pads may be changed on the Lockheed type 4×36 MB brake calipers fitted to the front wheels of the Austin Mini Metro 1000 and 1300 are illustrated in Figure 27.25 as a practical example.

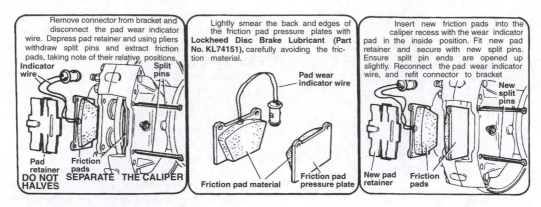

Figure 27.25   Replacing disc brake pads (*Lockheed*)

Brake discs

As in the case of drums, the brake discs rotate with the road wheels and are mounted either outboard or inboard, according to the transmission design requirements for a particular axle. Cast from an alloy iron and machined all over, the discs are shaped like a top hat in order to minimize distortion (Figure 27.18a and b). Instead of directly mounting the disc to the wheel hub, some high-performance cars have the disc attached via an aluminium alloy 'mounting bell'. This saves unsprung weight and minimizes thermally induced distortion of the disc when it is of large diameter. In another installation the disc is slidably mounted on a series of pins, so that it can float and again minimize distortion. Also, it is now fairly common practice for the discs to be ventilated through radial or curved vanes sandwiched between their rubbing surfaces. These vanes therefore act as impellers to improve cooling and reduce disc and pad wear. This cooling effect is enhanced by arranging for the entry for the vanes to be sited outboard of the mounting bell, so that there is least obstruction to the through-flow of air (Figure 27.26). Another cooling feature on some high-performance car disc brakes is the machining of angled grooves across the rubbing surfaces of the disc. These grooves also play a role in scraping clean glazed pads, expelling pad dust and evacuating water splash. Alternatively, the disc may be provided with cross-drillings that communicate with the internal vent spaces of the disc. Any interruption to the rubbing surfaces of the disc will, of course, reduce the effective braking area of the installation, but the cooling benefit is generally regarded as outweighing this consideration.

Similar to a drum brake, the physical dimensions of the disc brake are concerned with the brake torque to be developed and the required thermal capacity. Hence the former is related to the effective radius of the disc friction path, which is necessarily less than that for a brake drum of the same overall diameter, and the latter to the thickness of the disc.

Since both the pads and the disc contribute to the efficient operation of a disc brake, the condition of the disc must receive careful attention in service. In the event of uneven or excessive pad wear, the disc should be examined for significant corrosion or pronounced scoring, the presence of which usually require its replacement. A disc must also run true between its pads, which in practice means that its lateral run-out should not exceed 0.1 mm (0.004 in), otherwise brake judder and a

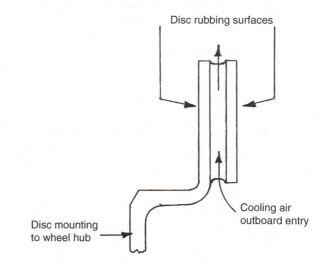

Figure 27.26   Ventilated brake disc

pulsating feel at the brake pedal can be noticed, furthermore the pads can be nudged back beyond their normal seal retraction amount and this leads to increased pedal travel.

Ceramic brake discs

Towards the end of 2000 two German manufacturers, Daimler Chrysler and Porsche, introduced 'ceramic' brake discs for their high-performance cars. Ceramic materials are inorganic and can be used at high temperatures, but are generally thought of as pertaining to pottery and therefore exhibiting brittle characteristics. This implies that such material would simply disintegrate when the brake pads applied pressure to the disc. However, if a carbon fibre reinforced ceramic in combination with silicon is used for the disc, then the ability to resist the effects of heat at elevated temperatures and thereby maintain structural integrity can be exploited to the benefit of brake system performance.

In general terms, the advantages to be gained from using ceramic discs may be summarized as follows:

1 Retain their dimensional stability up to operating temperatures of about 800°C, thereby promoting consistent and judder-free application of the brakes under hard usage.

2 Virtually free from wear with a potential service life several times that of conventional cast iron discs, together with reduced pad wear for minimum servicing.

3 Inherently resistant to corrosion of the disc braking surfaces, which again contributes to a longer life both for the discs and especially the pads, as compared to cast iron discs.

4 Lower density material than cast iron, which not only reduces vehicle overall weight, but also allows a saving in unsprung weight of the suspension system for better ride and handling, especially with modern lighter vehicle structures.

Although ceramic discs retain less heat than heavier cast iron ones, they can still be provided with internal venting channels, so that an outward flow of cooling air can be directed between their rubbing surfaces. Ceramic discs are appreciably more expensive to produce than those made from cast iron, which currently restricts their use to high-performance cars.

27.3 BRAKE FRICTION MATERIALS

General considerations

Reducing the speed of a vehicle requires the addition of a retarding force, applied conventionally through a friction braking system. This converts the kinetic energy of motion into frictional heat energy and dissipates it to the surrounding atmosphere, chiefly by forced conduction, until the vehicle comes to rest, and then by radiation in the absence of air flow over the brakes. Indeed, the brakes may be considered as being a form of heat engine, albeit in the negative sense because they convert mechanical energy into heat rather than vice versa. Life can be much more difficult for the brakes, however, since they are often called upon to decelerate the vehicle in a fraction of the time that the engine is normally expected to take to accelerate it. In other words, the amount of energy handled by the brakes can be the equivalent of many times the power developed by the engine.

Basic requirements

For an acceptable friction brake material these may be summarized, not necessarily in order of priority, as follows:

Resistance to wear To the vehicle user and also the tribologist who studies friction and wear, this requirement is perhaps the most important. A friction material should be hard enough to give a satisfactory wear life, but soft enough to avoid damage to the mating surface of the drum or disc. The most significant factor governing wear is temperature, since the rate of wear increases disproportionately with it.

Correct friction level The friction coefficient of most modern brake lining and pad materials against cast iron normally lies between 0.3 and 0.5. It should be sufficiently high to avoid excessive operating forces in the braking system, but not so high as to cause grabbing of the brakes. In any event the friction level must be high enough for the brakes to provide a minimum retardation sufficient to meet the Department of Transport requirements.

Resistance to heat fade This is a measure of how well the friction level of the material is maintained despite an increase in the braking temperature. Although the basis of the material is little affected by heat, it is the resin used for bonding the material together that can melt and smear itself between the rubbing surfaces and thereby reduce the friction level. For a good quality lining or pad material this will return to its original value on cooling, even after being subjected repeatedly to severe heating, but for a less acceptable material the recovery from fade will be poor.

Minimal friction variations The lining or pad material should be relatively insensitive to variations in operating pressure and rubbing speed. In these respects a less acceptable material would exhibit a reduced friction level at high operating pressures and at high rubbing speeds, neither characteristic being conducive to safe high-speed driving.

Reasonable wet behaviour It is important that linings or pads should recover their original friction level quickly after exposure to water and should not behave erratically during the drying out process. The more porous a material is then the longer it will take to recover.

Quietness of operation Nowadays this is an important comfort factor in vehicle operation. In combination, the friction lining material and the structural features of a brake assembly should contribute to an operational noise level that remains below the audible.

Brake pad and lining materials

As mentioned earlier in connection with clutch centre plate linings (Section 13.3), the growing awareness of the potential health hazards arising from exposure to asbestos fibre has led to the established use of environmentally safer non-asbestos-based friction materials. However, it should be recognized that for technical reasons only chrysotile or white asbestos was normally used in the manufacture of brake friction materials, not the crocidolite or blue asbestos that poses a far more serious health hazard. It is worth recalling that asbestos was an essential ingredient of friction linings and pads primarily because it served as a reinforcing agent, which maintained its strength and stability over a wide range of temperatures and pressures. In this respect it must at least be conceded that asbestos made a significant contribution to the safety of motor vehicles over a period of more than eighty years.

It has taken a great deal of expensive research and development for friction material manufacturers to produce asbestos-free formulations, which not only replace but also improve upon the once conventional asbestos-based materials. Among the alternative materials that have been evaluated are included glass fibre, steel fibre, aluminium wool, mineral wool, aramid fibre and carbon fibre, all of which are appreciably more expensive to use than asbestos.

As examples of modern asbestos-free brake friction materials, it may be of interest to summarize the properties of certain pad and lining materials from the range produced by the Ferodo company. Ferodo 3410F is a disc pad lining material integrally moulded to a steel back plate. Medium grey in appearance, it has a non-asbestos basis of steel fibre and incorporates a blend of selected friction modifying agents, together with a specially developed impregnant. For design purposes its coefficient of friction μ is quoted as 0.4. This material therefore has a high friction level, which is combined with good resistance to fade and wear and makes it suitable for light, medium or heavy duty on passenger cars of all categories. Its operating range in terms of pressure, rubbing speed and temperature is as follows: pressure 0.35–5.17 MN/m^2 (50–750 lbf/in^2); maximum rubbing speed 25 m/s (80 ft/s); maximum temperature 550°C; maximum intermittent temperature 350°C; maximum continuous temperature 225°C.

Ferodo 3651F is a rigid moulded non-asbestos friction material that is made only in the form of radiused linings of either tapered or parallel type for use on the drum brakes of heavy vehicles. Light green in appearance, it has a basis of glass and synthetic fibres in random dispersion and incorporate light alloy particles and selected friction modifiers. A specially developed resin contributes to both the strength and the performance of the material. For design purposes the coefficient of friction μ is quoted as static (cold) 0.4 and dynamic 0.38. This material has a medium/high friction level with excellent frictional stability throughout a wide range of working temperatures, combined with a very good wear resistance, and does not have any ill effect on drum surfaces. Its operating range in terms of pressure, rubbing speed and temperature is as follows: pressure 70–2000 kN/m^2 (10–300 lbf/in^2); maximum rubbing speed 18 m/s (60 ft/s); maximum temperature 400°C; maximum intermittent temperature 300°C; maximum continuous temperature 175°C. For both this and the previously described material, the Ferodo recommended mating surface is good quality, fine grained pearlitic cast iron. This material consists largely of a matrix of what is called pearlite, interspersed with graphite in the form of flakes of various sizes and a small amount of free ferrite and phosphide eutectic (a combination of phosphorus, iron and carbon).

Wear of brake rubbing surfaces

Before discussing the factors that influence the wear life of brake rubbing surfaces, it is perhaps only logical to enquire into the actual nature of wear. In fact this is a topic that has long engaged the activities of research workers, since wear may take several forms dependent on operational and environmental circumstances. However, the form of wear with which we are concerned is often referred to as a ploughing action, and will now be briefly explained.

Although a function of the brake friction material is to distribute load evenly over the metal surface, it can only do this in the general sense of the description, because in reality the rubbing contact between it and the drum or the disc takes place over a whole series of tiny high spots (this would still be true even for precision-ground flat metal surfaces). The increase in pressure at these high spots is such that they momentarily weld themselves together, but as long as there is relative motion between the friction surfaces these tiny welded junctions are continually being broken – thus the ploughing action. It is, of course, this action that produces the friction force and also causes some damage to the rubbing surfaces, or in other words what we recognize as wear.

The factors influencing the wear life of the brake friction material may be summarized as follows:

Temperature attained High rubbing speeds and contact pressures, either singly or in combination (heavy braking),

will increase operating temperatures and have the greatest effect on reducing the wear life of the friction material.

Mating surface condition To reduce wear the rubbing surface of the drum or the disc should have as fine a finish as possible and its material be resistant to scoring from the wear products of the friction material.

Operating environment The ingress of muddy water from wet roads can lead to corrosion of the rubbing surface of the drum or the disc and will increase the rate of wear on the friction material.

Recognizing the various conditions of wear that can arise on the friction material and metal surfaces of brakes better lends itself to illustration rather than words (Figure 27.27). Some examples of serious neglect of drum and disc brakes in service from the Ferodo 'chamber of horrors' collection are also shown in Figures 27.28 and 27.29.

Diagnosing brake wear

Taking for example the drum brake, the various conditions of shoe lining and drum wear may be diagnosed as follows:

Scoring Three possible causes of this are abrasive debris embedded in the lining material; new linings mated with already scored drums; and the drum material too low in hardness for the lining material.

Crazing This is caused by repeated overheating and cooling of the brakes due to heavy usage and appears as randomly oriented cracks on both the friction track of the drum and the lining material.

Heat spotting Either the lining is not sufficiently conformable to the drum, or the latter is distorted so that lining contact occurs only at small heavily loaded areas. Discoloured regions and sometimes cracks appear on the friction track of the drum, this being accompanied by heavy gouging of the lining.

Fading Where this has occurred the lining material degrades or flows at the rubbing surface (an explanation of heat fade has been given earlier). Surface cracks on the brake drum friction track again may be associated with such arduous usage, their presence or otherwise being influenced by the metallurgical specification of the cast iron. Apart from heavy usage of the brakes, fade can also be caused by the brakes dragging on.

Similar considerations apply, of course, in diagnosing disc brake wear.

27.4 HYDRAULIC BRAKE SYSTEMS AND COMPONENTS

General background

An important landmark in the evolution of the motor vehicle braking system was the introduction, on certain American cars of the 1920s, of the Loughead (now known as Lockheed) system of hydraulic footbrake operation. Some thirty years later, the continued development of this principle by various specialist manufacturers of brake system equipment had led

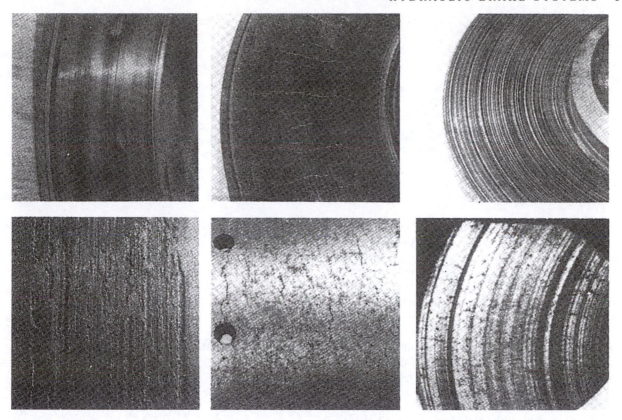

Figure 27.27 Recognizing drum and disc brake wear (*Ferodo*)
Upper line: metal surfaces. Lower line: friction material surfaces. Left: heat spotting. Centre: crazing. Right: scoring

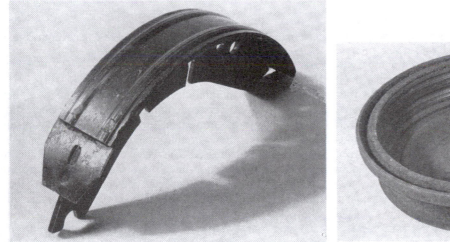

Figure 27.28 Drum brake neglect in service (*Ferodo*)

to the universal adoption of hydraulic footbrakes on cars and light commercial vehicles. The main reasons why hydraulic brakes supplanted the mechanically actuated systems used previously may be summarized as follows:

1 Hydraulic actuation is a more efficient means of transmitting force since, by practically obviating the need for any mechanical linkage, less of the driver's braking effort is wasted in overcoming friction in the system. Furthermore, it retains this higher efficiency by virtue of being self-lubricating and thus provides greater consistency of operation.

2 The hydraulic actuating force supplied to any one axle is automatically balanced between the two brake assemblies,

Figure 27.29 *Disc pad neglect in service (Ferodo)*

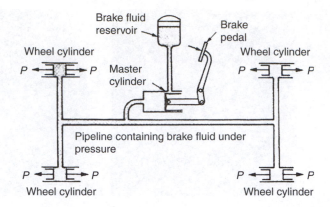

Figure 27.30 Diagrammatic layout of a motor vehicle hydraulic braking system

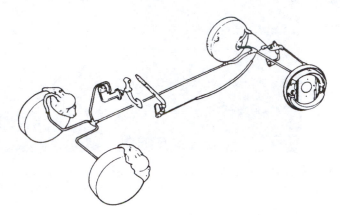

Figure 27.31 A disc front and drum rear, single-line, hydraulic brake installation (*Girling*)

because the pressure applied on a fluid in a closed circuit equalizes itself throughout that circuit. However, it does not correct for variations in brake torque output arising from any differences in friction between the rubbing elements.

3 With the advent of IFS systems in the 1930s the problems of employing geometrically accurate mechanical linkages to the brakes became more difficult, unless enclosed cables were used. This was because the actuating system had to allow not only for independent but also substantially greater movements of the wheels relative to the car structure owing to the softer springing. The flexible hoses of the hydraulic braking system provided the ideal answer to this requirement.

4 In hydraulic actuation the replacement of mechanical linkages by fluid pipelines offers a greater versatility and ease of installation. Such a system also readily lends itself to the incorporation of a brake servo or booster unit to reduce the operating effort required from the driver.

Single-line brake circuit

This represents a very basic hydraulic system in which the emphasis is on the transmission of load, rather than motion, between the driver-operated brake pedal and the friction elements of the wheel brakes. To understand the principle upon which such a system operates, it is necessary to recall the work of the French scientist Blaise Pascal who, in 1658, discovered that pressure on a confined liquid is transmitted equally in all directions and acts with equal force on all equal areas. It thus follows that a perfect pressure balance can be obtained between the actuating units of motor vehicle hydraulic brakes (Figure 27.30), although this does not necessarily mean that the system overall will be perfectly balanced, as will be evident later.

The layout, or hook-up as it is sometimes known, of the now obsolete single-line hydraulic braking system comprises the following components (Figure 27.31).

Master cylinder

For motor cars this unit is usually mounted on the engine side of the body bulkhead structure, through which a push-rod connection is made to a down-hanging or pendant brake pedal that is spring returned to the off or brakes released position. In some light commercial vehicle applications (and earlier car practice) the master cylinder is mounted on the chassis frame where it similarly connects via a push-rod to an up-standing brake pedal. The master cylinder acts as a hydraulic pressure transmitter, because it contains a sealed

piston that slides within a cylindrical bore whenever the brake pedal is depressed.

Wheel cylinders

These act as hydraulic slave cylinders in as much as they are pressure receivers containing sealed pistons which slide in response to the movement of the master cylinder piston resulting from the hydraulic pressure developed when the brake pedal is depressed. The wheel cylinders serve either to expand a pair of friction-lined shoes against their brake drum or to clamp a pair of friction-lined pads vice-like against their brake disc. In the latter case, each pad is of sector shape with the object of providing a uniform rate of both heat generation and wear over its surface area. Either drum or disc, or a combination of both types of brake, may be used according to the layout chosen for any particular vehicle.

Connecting pipes and hoses

These serve as hydraulic pressure conduits and connect the wheel cylinders to the master cylinder. The brake fluid is conveyed mainly by rigid metal tubing suitably formed and sometimes plastic coated to follow the installation path on either the body structure or the chassis frame.

Two important requirements of the piping installation are that the tubing must be supported against vibration and protected against corrosion.

Flexible reinforced rubber hoses are used finally to convey the fluid to the wheel cylinders which, in the brake assemblies of conventional layouts, move relative to the vehicle structure owing to suspension and steering movements of the road wheels.

Brake fluid

In the hydraulic brake system, force is transmitted from the brake pedal to the friction elements of the wheel brakes through the medium of a column of virtually incompressible liquid known as brake fluid.

Although this incompressibility is a characteristic of liquids in general, brake fluid is specially formulated to meet additional and stringent requirements as described later. A reserve of brake fluid is contained in a reservoir connected to the master cylinder.

Divided-line brake circuit

It should be realized that if a serious fluid leakage occurs in the single-line layout the entire hydraulically operated footbrake system can be rendered ineffective. To avoid this potential hazard, the brake system legislative requirements in Europe and the United States now demand a divided-line brake circuit. A failure of either line will still leave two service or wheel brakes in action.

The basic idea of using a divided-line brake circuit is by no means of recent origin. Among some of the early users of the original Lockheed hydraulic brake system there still remained a certain lack of faith in its absolute safety, which could be attributed to the problems of fluid leakage and air ingress that had bedevilled the less successful attempts by other manufacturers to produce a satisfactory hydraulic brake system. This led to a few manufacturers of high-performance cars, commercial vehicles and especially racing cars to adopt a divided-line circuit, so that one line supplied the front brakes and the other one the rear brakes, thereby increasing the safety of the overall system should either line lose its fluid (Figure 27.32a). In practice, this was usually achieved by using a tandem master cylinder, which in effect comprised two master cylinders in line without any fluid communication between them. Less commonly, except on racing cars, two entirely separate single master cylinders were mounted side by side and operated via a balance lever that straddled them. This lever had a restricted movement at its pivot point connection with the brake pedal, so that in the event of one line losing its fluid, the pedal effort could still be transmitted to the other master cylinder.

There were no significant further developments of the divided-line brake circuit until 1963, when the Swedish manufacturer SAAB introduced a diagonally split or X-split brake circuit on their front-wheel-drive SAAB 96. In this system each output from the tandem master cylinder supplies one front brake and its diagonally opposite rear brake, so that in the event of either line failing one-half of the normal maximum braking still remains available from one front and one rear wheel (Figure 27.32b). To maintain car stability under these circumstances it is necessary for the steering geometry to incorporate either centre point steering, or the later developed negative-offset steering, as described in Section 24.2.

Another version of divided-line brake circuit, again from a Swedish manufacturer, was the L-split system introduced on

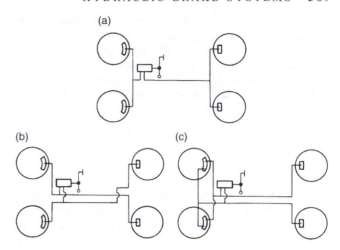

Figure 27.32 Divided-line brake circuits: (a) F/R-split (b) X-split (c) L-split

the rear-wheel-drive 1966 Volvo 144. This more elaborate system is so arranged that each output from the tandem master cylinder supplies one-half of each front brake and a single rear brake (Figure 27.32c). In the event of either line failing there is less imbalance of the brakes than with the X-split circuit.

An important advantage of the X- and L-split circuits is that whichever line fails, the reduction in normal maximum braking remains the same; whereas with the simple front-rear split the reduction that accompanies failure of the front line is much greater than if the rear line fails. The divided-line brake circuit can also be split in other ways with varying degrees of elaboration, each with its advantages and disadvantages, but for modern front-wheel-drive cars the X-split circuit appears to have become established practice, while the simple front-rear split circuit is still generally used on rear-wheel-drive cars.

Single master cylinders

It is now necessary to examine in more detail what is happening in the master cylinder not only when the brakes are being applied, but also when they are being released. For this purpose, the construction and operation of single master cylinders used in residual pressure and non-residual pressure hydraulic brake systems must receive consideration.

Residual pressure type of master cylinder

Master cylinders of the residual pressure type maintain a small pressure in the region of $55\,kN/m^2$ ($8/lbf/in^2$) in the pipelines and wheel cylinders after the brakes have been released.

They were standard equipment in the once conventional all-drum car braking systems, or at least in those where the shoes were positively retracted from the drums when the brakes were released. Master cylinders of this type are now confined mainly to light commercial vehicles employing a similar system of drum brakes. The purpose of maintaining a residual pressure is twofold:

1 To prevent the ingress of air past the wheel cylinder seals when the brakes are released.
2 To ensure a quick response from the brakes during their initial application.

The valve arrangement that maintains that residual pressure also assists in purging air from the hydraulic system when brake bleeding is carried out. This operation becomes necessary when either some part of the system has been temporarily disconnected or after the brake fluid has been renewed.

The construction of a typical residual pressure master cylinder is as follows (Figure 27.33):

Body The main body is bored axially to receive a sliding piston and is drilled radially to provide compensating and supply ports which communicate with either side of the piston head when the brakes are not applied.

Piston This is operated by means of a push-rod connected to the brake pedal and carries two rubber seals, known as the primary and secondary cups. Between these sealing cups the piston skirt is reduced in diameter and the piston head is provided with a series of axial drillings.

Seals The primary cup, against which pressure is generated in the pipelines and wheel cylinders, is acted upon by the piston head but is prevented from adhering to it by a separating washer. The secondary cup is fitted over the flanged push-rod end of the piston to prevent fluid leakage from the open end of the cylinder, which is also protected by a rubber boot dust excluder.

Check valve This is a combination inlet and outlet valve spring loaded against the head of the pressure chamber formed by the master cylinder bore. The same compression spring also serves to return the piston fully against its end retaining ring. Although constructional details vary, a check valve can take the form of an annular rubber cup inlet seal through the centre of which acts a spring-loaded poppet outlet valve.

Fluid reservoir This unit is either formed integrally with, or separately connected to, the master cylinder. It contains a reserve of brake fluid and affords automatic gravity replenishment for the hydraulic system. To meet the demand for increased vehicle safety, the reservoir is generally made from a transparent plastic material, so that the fluid level is readily visible.

The operation of the residual pressure master cylinder during application and release of the brakes is as follows.

Brake application
When the brake pedal is depressed, the initial movement of the push-rod-operated piston causes the sealing lip of the primary cup to pass over the compensating port and henceforth seal off the pressure chamber of the master cylinder. At the same time, the increasing pressure on the primary cup flattens its separating washer, which thus prevents the seal from being extruded through the drillings in the piston head. As pressure builds up it overcomes the spring loading on the outlet check valve, thereby allowing the passage of fluid under pressure into the pipelines and wheel cylinders to actuate the brake assemblies.

Brake release
When the brake pedal is released, the inlet check valve opens to allow the passage of fluid returning under pressure to the master cylinder by the action of the brake shoes pull-off springs. The valve remains open as long as the pressure in the wheel cylinders and pipelines exceeds the required residual pressure of the system: that is, a pressure sufficient to ensure that the sealing cups in the wheel cylinders are kept expanded against their bores. Before the spring-loaded and rapidly returning piston uncovers the compensating port, the pressure drop in the cylinder allows the primary cup to collapse away from the piston head. The separating washer then bows to allow the passage into the pressure chamber of additional fluid drawn from

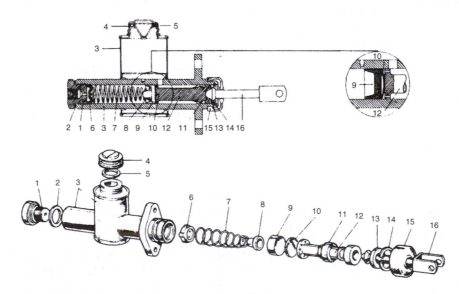

Figure 27.33 Constructional features of a residual pressure master cylinder

1 barrel end plug	6 check valve assembly	11 secondary cup
2 end plug gasket	7 return spring	12 piston
3 barrel and tank assembly	8 return spring retainer	13 and 16 push-rod assembly
4 filler cap	9 main cup	14 circlip
5 filler cap washer	10 piston washer	15 rubber boot

the annular space formed by the reduced skirt of the piston via the holes in the piston head. Fluid is supplied to this source through the supply port of the master cylinder. As the piston finally comes to rest, the primary cup uncovers the compensating port through which the additional fluid drawn into the pressure chamber is released into the reservoir.

Since the quantity of fluid in the wheel cylinders and pipelines is subject to heating and cooling, another function of the compensating port is to allow fluid to flow between the master cylinder and reservoir as it expands or contracts when the brakes are not being used.

Non-residual pressure types of master cylinder

With master cylinders of the non-residual pressure type, the only pressure remaining in the pipelines and wheel cylinders after the brakes are released is a hydrostatic one due to the head of brake fluid in the system. Here it should be appreciated that the *hydrostatic pressure* exerted by a column of liquid is measured in terms of its contained height or *head* and is not related to its cross-section.

The effect of this hydrostatic pressure is such that with drum brakes it can be sufficient to maintain the shoes in light rubbing contact with their drums, while with disc brakes it is insufficient to hinder the elastic recovery of the wheel cylinder seals retaining the pads in close proximity to their discs. To facilitate the bleeding operation, non-residual pressure master cylinders may still incorporate a valve similar to that used in residual pressure versions but modified with a permanent bleed passage to permit final equalization of pressure in the wheel cylinders, pipelines and master cylinder.

Several different constructions of non-residual pressure master cylinders may be encountered. These are the moving seal types with either radial or axial compensating ports, and the stationary seal type with radial compensating ports. A modified check valve with a permanent bleed passage may be fitted in some cases, its sole purpose being to prevent aerated fluid from entering the master cylinder when the brake pedal is pumped during bleeding of the system (Figure 27.34b).

Moving seal with radial compensating port
In both construction and operation this type of master cylinder is similar to its residual pressure counterpart, but differs from it in that the check valve is either omitted or is retained and provided with a permanent bleed passage (Figure 27.34a).

Moving seal with axial compensating port
In this alternative construction, the compensating port is situated at the head of the master cylinder pressure chamber and is controlled by a centre valve with rubber seal (Figure 27.35). A side outlet connection to the wheel cylinders is provided towards the head of the pressure chamber.

When the brake pedal is depressed, the initial movement of the master cylinder piston relieves tension on the centre valve pull-rod and allows the spring-loaded valve to close off the compensating port. Continuing movement of the sealed piston then serves to increase fluid pressure in the master cylinder, pipelines and wheel cylinders. When the brake pedal is released the system pressure falls as the piston retreats and finally intercepts the centre valve pull-rod. This overcomes the closing spring load on the centre valve, which then opens away from the compensating port so that communication is established between the master cylinder and reservoir.

Stationary seal with radial compensating ports
This form of construction is an inversion of the others, because the master cylinder piston moves through a stationary primary or recuperating seal to generate pressure in the hydraulic system (Figure 27.36). The piston carries a secondary sealing cup to prevent fluid leakage from the open end of the cylinder. A series of compensating ports is drilled radially through the hollow front portion of the piston. The primary seal is in fact permitted a very limited end float relative to its seating and is also separated from it by a special shim washer. A series of axially drilled holes is incorporated in the seating immediately behind the shim.

When the brake pedal is depressed, the initial movement of the piston allows its compensating ports to move beyond the primary seal lip. Further movement of the piston generates an increasing pressure in the master cylinder, pipelines and wheel cylinders. The shim washer prevents the pressurized seal from being extruded through the drilled holes in its seating.

When the brake pedal is released, the spring-loaded and rapidly returning piston can produce a pressure drop in the cylinder which allows the primary seal to float slightly

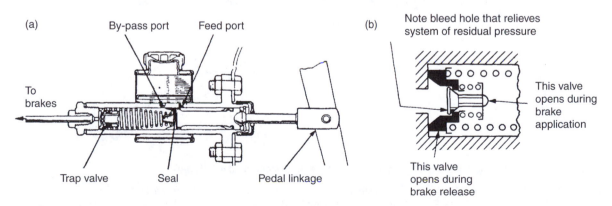

Figure 27.34 (a) Constructional features of a non-residual pressure master cylinder with radial compensating port (b) another form of check or trap valve that facilitates brake bleeding

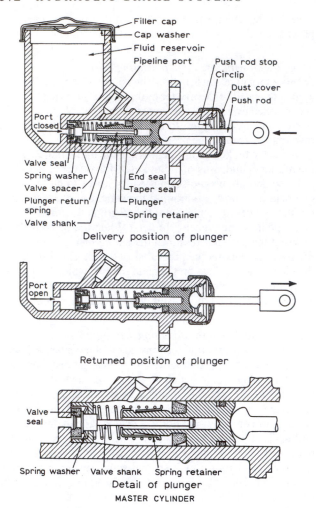

Figure 27.35 Construction and operation of a non-residual pressure master cylinder with an axial compensating port (*Girling*)

forward of its seating so that additional fluid then passes into the pressure chamber. As the piston finally comes to rest, its compensating ports are uncovered by the primary seal and the additional fluid is then released to the reservoir.

An advantage of the second and third forms of master cylinder construction is that the primary seal lip does not have to pass over a compensating port drilled in the wall of the pressure chamber, which action may eventually lead to nibbling of the seal. A further advantage of the third form of master cylinder is that the primary seal acts against a sliding male surface to minimize seal wear, it being found more economical to develop the high standard of surface finish on the piston than in its bore.

Tandem master cylinders

Tandem master cylinders vary in detail construction and especially in respect of their compensating port control. Basically they comprise a common housing that is usually produced from cast iron for optimum service life, in the cylinder bore of which slide two spring-returned pistons with rubber seals. These pistons may either be completely independent in their movements (Figure 27.37) or be linked together such as to have only limited independence (Figure 27.38) both arrangements being able to compensate automatically for any unequal displacement of fluid in operation. The piston that resides near the mouth of the cylinder bore is conventionally referred to as the primary piston and the other one as the secondary piston, neither term intending to imply the normal operating sequence of the pistons. According to design requirements, the return travel of the secondary piston may be limited by a stop pin or screw. The operating chambers of the two pistons are supplied with fluid from either an integral or a separately attached reservoir, the latter being produced from a translucent plastic moulding to provide a ready visual indication of fluid levels. In both reservoir constructions an internal vertical baffle is used to preserve independence of fluid supply for each line of the divided brake circuit.

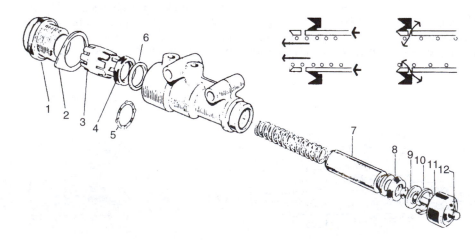

Figure 27.36   Construction and operation of a stationary seal type of master cylinder (*Girling*)

1 end cap	5 steel shim	9 washer
2 gasket	6 nylon backing ring	10 circlip
3 seal support	7 plunger	11 dust cover
4 recuperating seal	8 end seal	12 push-rod

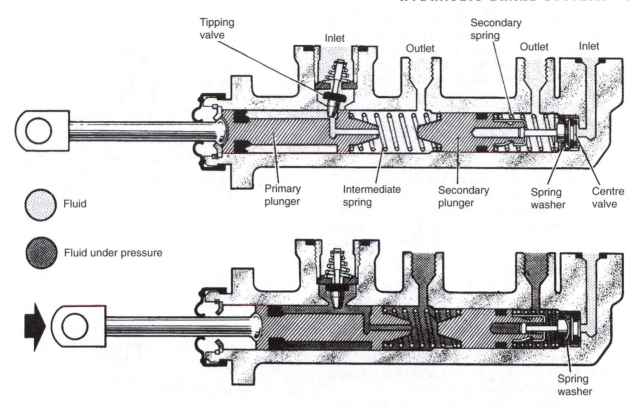

Figure 27.37 Tandem master cylinder with tipping and centre valves (*Girling*)

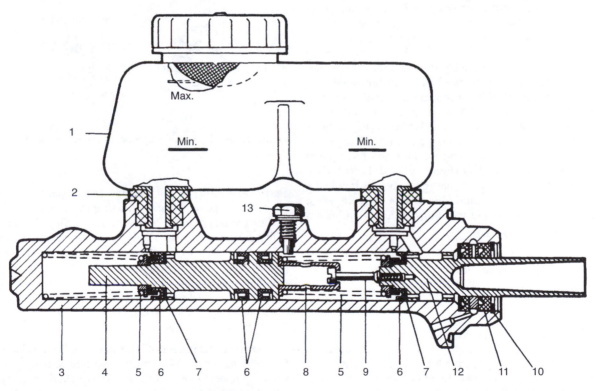

Figure 27.38 Tandem master cylinder with linked pistons (*ATE-SAAB*)

1 container	6 piston seals	10 lock ring
2 rubber seal	7 washer	11 seal ring
3 housing	8 sleeve	12 primary piston
4 secondary piston	9 screw	13 screw
5 spring		

When the brakes are applied in normal operation, the movement of the master cylinder push-rod urges the primary piston forwards. This action closes the compensating port for the primary pressure chamber, either as a direct result of piston seal movement, or by piston movement allowing a tipping valve to reseat. The ensuing rise in pressure ahead of the primary piston causes a similar forwards movement of the secondary piston. This action likewise closes the compensating port for the secondary pressure chamber, either again as a direct result of piston seal movement, or by piston movement in this case allowing a centre valve to reseat. The same pressure is therefore generated ahead of both pistons and therefore in each line of the divided circuit.

In the event of pressure being lost in the line served by the primary piston, this piston will continue to move forwards until it contacts the secondary piston, thereby ensuring that pressure in the other line is maintained. Conversely, if pressure is lost in the line served by the secondary piston, this piston will move forwards until it bottoms against the blind end of the cylinder, so that pressure build-up between the primary and secondary pistons is still available in the other line. Clearly, the travel of the master cylinder push-rod will become greater whichever line of the divided circuit suffers loss of pressure, but in such an emergency it will still be capable of transmitting force to the piston serving the undamaged line and therefore actuate the respective brakes hydraulically.

The earlier reference to linking of the primary and secondary pistons in some tandem master cylinder constructions deserves further explanation. The advantages claimed for this feature are twofold. First, it enables the primary piston spring to be preloaded in position so that initial movement of the primary piston is transmitted without relative compression to the secondary piston, thereby closing both compensating ports simultaneously and allowing the same rate of pressure rise in each line. And second, the linking anchor screw prevents the weaker spring of the secondary piston from remaining compressed when the brakes are released, since otherwise the compensating port for the secondary pressure chamber could remain closed owing to incomplete return of the piston.

A pressure differential warning actuator may be required with tandem master cylinders (Figure 27.39). It comprises a self-centring shuttle valve that becomes hydraulically unbalanced in the event of either line losing pressure, when it moves to complete an electrical circuit and illuminate a facia-mounted warning light.

Wheel cylinders for drum brakes

The construction of a typical single-piston wheel cylinder is as follows (Figure 27.40a):

Body The wheel cylinder body is usually spigot located into the brake back plate assembly and retained by a pair of setscrews. It has a blind-ended cylinder bore to receive the sliding piston and is drilled radially to provide threaded connections for the fluid pipeline and bleeder screw.

Piston The outer end of the piston engages the toe of the adjacent brake shoe either directly or indirectly through the medium of a short push-rod. It is usually spring loaded against the brake shoe at all times.

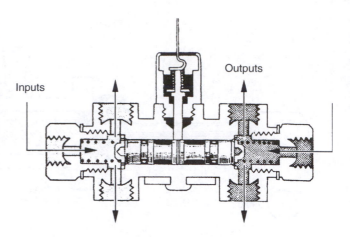

Figure 27.39 Differential pressure warning actuator

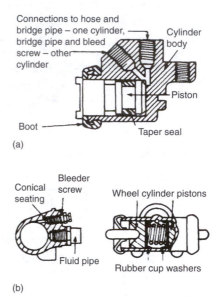

Figure 27.40 Hydraulic wheel cylinders: (a) single (b) double piston

Seals A pressure seal is provided at the inner end of the piston and may take the form of a simple cup washer abutting against the piston head, or an annular sealing ring engaging with a groove in the piston body. In the former case a cup filler, or less commonly a spreader washer, is interposed between the piston seal and return spring to maintain the cup lip in contact with the cylinder wall. The cup filler also reduces the working volume of fluid in the cylinder that is subject to heating. A rubber sealing boot is fitted to exclude dust from the open end of the cylinder.

In operation, depression of the brake pedal displaces fluid from the master cylinder through the connecting pipes to the wheel cylinders. The additional brake fluid entering each cylinder pushes the sealed sliding piston outwards, a movement transmitted simultaneously to the adjacent brake shoe which is thus forced against the drum. The greater the pedal effort exerted by the driver the higher will be the pressure generated in the wheel cylinder, as will be the piston force

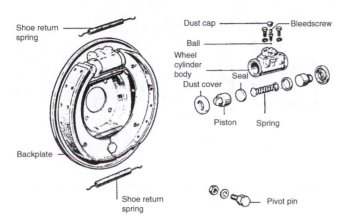

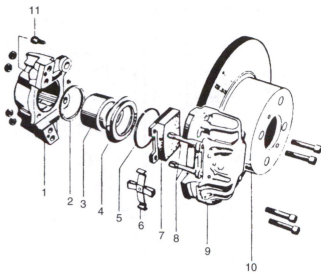

Figure 27.41 Hydraulically operated leading and trailing shoe drum brake (*Girling*)

Figure 27.42 The hydraulically operated disc brake in practice, front-wheel assembly (*ATE-SAAB*)

1 brake housing half
2 piston seal
3 piston
4 gasket
5 gasket retainer
6 retaining spring
7 brake pad
8 locking pins
9 brake housing half
10 brake drum and disc
11 bleeder screw

transmitted to the brake shoe. When the brake pedal is released, the brake shoe pull-off springs return the piston to its initial position in the wheel cylinder and this forces excess brake fluid back to the master cylinder.

Although a single-piston wheel cylinder is usually single acting, in some brakes it is made double acting by slidably mounting its body on the back plate, as described later (Figure 27.54).

The detail construction and mounting of a typical double-piston wheel cylinder (Figure 27.40b) is similar to that of the single-piston unit, apart from the obvious differences of the body being through bored and the duplication of its piston and sealing arrangements. Likewise in operation, both pistons are pushed outwards against their adjacent brake shoes, because the wheel cylinder is double acting. The pistons are returned to their initial position in the usual manner by pull-off springs acting on the brake shoes (Figure 27.41).

Wheel cylinders for disc brakes

We take for example fixed caliper designs. Two opposed hydraulic cylinders are incorporated in either a single- or two-piece caliper construction which straddles the disc and is attached to the axle flange. In the former case, the cylinder bores are machined through from one leg of the caliper to a blind end in the other leg, a screwed plug being used to seal the open-ended cylinder. Two-piece caliper constructions are generally favoured in modern practice so that blind-ended cylinder bores can be machined into each leg of the caliper, thereby dispensing with the need for a screwed sealing plug (Figure 27.19 and 27.42).

A hydraulic pressure balance between the two opposed cylinders is obtained either by an external bridging pipe or internal drillings through the caliper bridge itself. The sliding movement of each hollow piston in its cylinder is sealed by a stationary rubber ring retained within a groove near the rim of the cylinder. By virtue of its enclosed air space, a hollow piston minimizes the amount of heat transferred to the brake fluid from the pad backplate during braking. The operation of the disc brake is similar to that of the drum brake except, of course, that the thrust forces developed at the wheel cylinder pistons arise from pressing the pad liners into rubbing contact

with the brake disc. Thin steel shims with cut-outs are generally used between the piston and pad back plate, so as to secure the best distribution of pressure on the friction lining and avoid brake squeal.

It was earlier inferred in Section 27.2 that the disc brake is virtually non-self-energizing; but why only virtually and not totally, considering that we are simply bringing together two plane surfaces with a clamping load applied normal to them? In fact a mild self-energizing effect inevitably arises from the thickness of the friction pad, which imposes an offset between its rubbing surface and the abutment for its back plate. During braking the frictional drag developed between the pad and its disc therefore creates a turning moment about this offset (Figure 27.43), so that the leading edge of the pad makes a relatively harder contact against the disc than the trailing edge, even though the pad receives a centrally applied clamping load. Of greater significance than the mild self-energizing effect is the unequal wearing over the length of each pad and its influence on pedal feel. In a more recent Japanese disc brake installation this problem has been countered by providing each front caliper with four opposed pistons, the pair of pistons towards the leading edge of the pads being smaller in diameter than the pair towards the trailing edge (Figure 27.44). This arrangement is claimed to confer a better pedal feel under all braking conditions.

Wedge-operated brake expanders

In a wedge-operated brake expander mechanism, the lever and rotary cam method of transferring the actuating force to the brake shoe is replaced by a sliding wedge and tappets (Figure

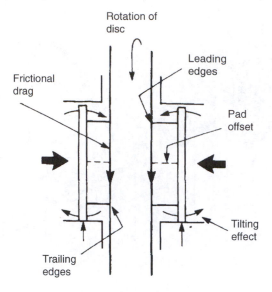

Figure 27.43 Tilting effect on disc brake pads during braking

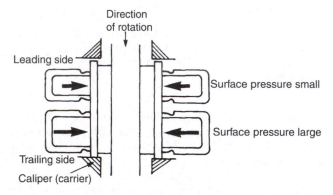

Figure 27.44 Disc pad apply pistons graduated in diameter for better pedal feel (*Colt Mitsubishi*)

27.8). Both the sliding wedge and tappets are of cylindrical form, the difference between them being that the wedge has a conical waisted portion whereas the tappets each have an inclined end face. The action is simply that of drawing the wedge partially through the tappets, which directly engage the toes of the brake shoes and thus bring the friction liners into contact with the drums.

Expander mechanisms of this type offer the following advantages:

1 They possess a high mechanical efficiency, because the friction between the wedge and tappets can be reduced to a minimum by interposing rollers.
2 The wedge expander unit can be easily lubricated and sealed, thus enabling it to retain its efficiency.
3 It can readily provide a floating-cam effect, either by allowing the expander body to slide on the back plate or by incorporating a sliding insert in the wedge.
4 By virtue of acting at right angles to the plane of the brake shoes, wedge expanders are convenient to use with mechanical (especially the handbrake) hydraulic and compressed air braking systems.

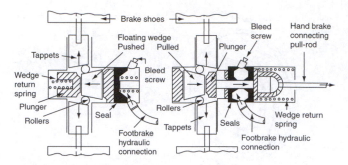

Figure 27.45 Pusher and puller hydraulically operated wedge expanders in diagrammatic form

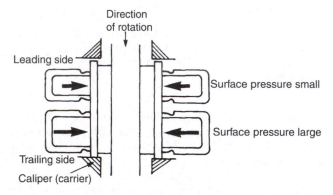

Figure 27.46 Weight transfer during braking

A possible disadvantage of the wedge expander unit has sometimes been its tendency towards sticking, i.e. not returning fully to the off position following release of the brakes.

Although wedge-type expanders were initially developed for mechanical braking system applications on cars, they were eventually supplanted by the purely hydraulic system with which they were originally intended to compete. However, wedge-type expander units continued to find application on commercial vehicles. For medium duty they are hydraulically operated with transversely mounted pusher wheel cylinders on the front brakes and similarly mounted puller wheel cylinders on the rear brakes, the latter arrangement providing convenient lost-motion pull-rod connections to the handbrake system (Figure 27.45). For application to heavy vehicles the wedge-type expander is used in conjunction with compressed air operation as described in Section 28.

Brake apportioning valves

To achieve the shortest emergency stopping distance, all four wheels of a car need to be almost on the point of locking, so that in theory the braking force developed at the road surface level should always be divided between the front and rear wheels in the same ratio as their respective vertical loading. In practice this ideal state of affairs is difficult to attain, because apart from being dependent on the position and amount of load carried, the relative vertical loading is also influenced by the transfer of weight from the rear to the front wheels during actual braking (Figures 19.41 and 27.46). Since this dynamic weight transfer is proportional to the deceleration of the car, it follows that in a simple conventional hydraulic braking system with a fixed ratio of braking to the front and rear wheels, the ratio chosen will be correct

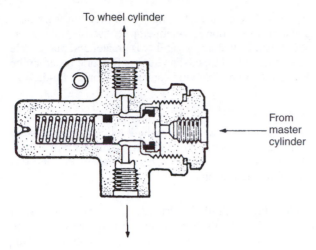

Figure 27.47 Lockheed pressure limiting valve

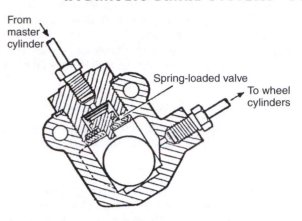

Figure 27.48 Lockheed inertia valve

at one rate of deceleration only. Whereas such a fixed ratio, which usually approximates to the static distribution of weight between the front and rear wheels, is quite satisfactory for most normal braking conditions, under emergency braking the rear wheels could lock prematurely and seriously affect car controllability.

An established approach to reducing this problem, especially on nose-heavy front-wheel-drive cars, has been to impose an additional control over the line pressure supplied to the rear brakes. This may be accomplished by the following types of sensing valve:

Pressure limiting valve Under light braking this type of valve allows the free passage of fluid under pressure to the rear brakes in the normal manner (Figure 27.47). With heavier braking, once the fluid pressure exceeds a predetermined maximum it overcomes the return spring loading on the valve plunger and allows it to move forwards. This seals off the two outlets to the rear brakes, thereby preventing them from locking prematurely, any further increase in system line pressure then being transmitted to the front brakes only.

Pressure reducing valve This represents a refinement of the simple pressure limiting valve, because it maintains a constant braking force at the rear wheels despite thermal expansion of their brake drums during prolonged heavy braking. Under these conditions, a ball valve is unseated by the plunger and allows fluid to pass until the dropping pressure trapped in the rear brake line has been restored to its predetermined maximum, thereby reducing any tendency for the rear wheels to lock prematurely.

Inertia conscious valve A simple version of this type of deceleration sensitive valve consists of a body, a steel ball and a valve that is lightly spring returned. The angular mounting of the valve body is so contrived that under light braking the ball remains at the lower end of its inclined ramp and maintains the valve open, thereby allowing the free passage of fluid under pressure to the rear brakes (Figure 27.48). With heavier braking the inertia force acting on the ball causes it to roll up its bore and close the valve. This seals off the outlet to the rear brakes so that again any further increase in line pressure

will be transmitted to the front brakes only. The inertia conscious valve also provides automatic correction of braking ratio on down or up gradients.

Load conscious valve Valves of this type are typically pressure reducing in construction, but connect to the rear axle via a lever arm and flexible link. Relative changes in ride height are thus gradually transmitted to the valve plunger, which can then reposition itself so that the proportion of line pressure received by the rear brakes will be either increased or decreased to reflect loading conditions on the rear axle. This type of pressure modulating valve is particularly suited to commercial vehicle application, where variations in the distribution of static load are usually more important than weight transfer during braking.

Brake system apportioning valves must be compatible with divided-line brake circuits, so that two rear apportioning valves can be required for an X-split circuit (Figure 27.32). They must also be compatible with anti-lock brake systems (Figure 29.6) and may be electronically operated in conjunction with the system itself.

27.5 HYDRAULIC BRAKE FLUIDS

The properties of brake fluids

Although synthetic fluids in their various formulations are no strangers to the hydraulic equipment engineer, that fluid which is used in the hydraulic braking systems of most motor vehicles has long been regarded with a certain amount of curiosity by automotive service engineers. This is probably because the manufacturers of brake fluids do not normally volunteer the chemical composition of their products, but in any event such information can be meaningful only to a lubrication chemist. However, since we are required to have some knowledge of the constituents and properties of brake fluid, the following basic information should suffice.

Constituents
A conventional polyglycol ether brake fluid is socalled because it includes the following chemical components:

1 Glycols to ensure that the fluid is compatible with the rubber materials in the braking system.

2 Glycol ethers to enable the fluid to flow freely and therefore have the desired viscosity.
3 Polyglycols to confer the required lubricity on the fluid.
4 Corrosion inhibitors to protect the metal parts of the braking system.

Freezing and boiling points

A typical high-quality brake fluid would remain chemically stable over a temperature range of $-65°C$ to $+220°C$.

Viscosity

The maximum acceptable viscosity of a brake fluid is limited by its satisfactory behaviour at $-40°C$. That is, the brakes should not respond sluggishly during either application or release at this low temperature.

Specifications

Since 1945 the Society of Automotive Engineers (SAE) in America has issued specifications relating to the minimum required performance of brake fluids, as also, does the Department of Transportation (DOT) in that country. More recently the European International Standards Organization (ISO) has issued its standard on brake fluid. These specifications are internationally recognized and are therefore generally labelled on the sealed container in which the fluid is sold, being expressed as SAE J1703, DOT 3,4 and 5, and ISO 4925 for conventional brake fluids of the polyglycol ether and the later developed borate ester types. The specifications are similar and differ only in respect of low temperature viscosity, minimum 'dry' boiling point of the pure brake fluid, and minimum 'wet' boiling point of the brake fluid after treatment with moisture. As a matter of background interest, a DOT 3 brake fluid has minimum dry and wet boiling points of $205°C$ and $140°C$ respectively, which compare to $230°C$ and $155°C$ for a DOT 4 fluid. They are periodically revised to take into account the increasing severity of duty imposed upon brake fluids, especially in terms of heat input not only from application of the brakes themselves as car performance has risen, but also from higher underbonnet temperatures where some of the brake equipment is housed. A high quality brake fluid may, of course, be formulated to surpass the requirements of these specifications.

Uncommonly a fluid with a mineral basis may be specified, such as the LHM (Liquide Hydraulique Minérale) fluid used in the central hydraulic system that includes the brakes on certain Cirtroën models. This fluid is identified with a green dye and cannot be mistaken with other coloured dyes for conventional brake fluid. It is vitally important to consult the vehicle manufacturer's recommendations in respect of brake fluid, because only the brake fluid specified for a given system is fully compatible with the rubber components and metals in contact with the fluid. The result of using incorrect fluid could be malfunctioning of the brakes with potentially disastrous consequences.

A brief mention must also be made of silicone based brake fluids, which are sometimes used for specialized applications. Brake fluids of this type are non-hygroscopic (a term explained later) and can therefore meet a DOT 5 specification by virtue of having a higher minimum wet boiling point. They also possess good low-temperature viscosity characteristics, but their bulk compressibility is greater and may create a spongy feel to brake pedal operation. Silicone based brake fluids are not compatible with conventional brake fluids.

The enemies of brake fluid

The satisfactory operation of a hydraulic braking system must ultimately depend upon the integrity of its brake fluid, which can deteriorate in the presence of air, water and dirt for the following reasons:

Air Pure brake fluid is virtually incompressible. If it contains air then the fluid mass becomes compressible to an extent corresponding to the amount of air entrained. This results in the brakes becoming less responsive during application and is indicated by a combination of excessive travel and spongy feel of the brake pedal.

Water Conventional (as opposed to mineral based) brake fluid is by its nature hygroscopic, which simply means that is tends to absorb moisture from the surrounding air. As a result the boiling point of the fluid can be seriously lowered, especially when it is considered that a water content of 5 per cent by weight will virtually halve the boiling point, and this may occur after three years' service life. The important point to appreciate is that if during severe usage of the brakes a fluid reaches its boiling point, gas pockets are formed in the top of the wheel cylinders. At worst the presence of this vapour will, between brake applications, force fluid back into the reservoir until suddenly insufficient fluid can be trapped between the master cylinder and wheel cylinder pistons to operate the brakes at all. It is to avoid such a dangerous situation as this that brake equipment manufacturers normally recommend the system fluid be changed every 18 months to 2 years.

Dirt This is, of course, the enemy of all hydraulic systems, where consistent operation depends upon maintaining fine clearances and effective seals. For this reason scrupulous cleanliness must be exercised when servicing hydraulic brake systems.

Removing air from hydraulic brakes

The purging of air from a hydraulic brake system is commonly referred to as bleeding the brakes (Figure 27.49). Unless the presence of air is indicated by a spongy feel of the brake pedal, bleeding should only be required after some part of the system has been disconnected, or following renewal of the brake fluid.

For the purpose of bleeding the brakes a series of air bleed valves with bleeder screws are employed, which communicate with the interior chambers of the various hydraulic units at their uppermost points. These units include in all cases the wheel cylinders of drum and disc brakes and in some cases the master cylinder as well. The expulsion of air from each unit is achieved by attaching what is termed a bleeder tube to the particular air bleed valve and immersing its other end in a small quantity of new fluid contained in a glass jar. If then the

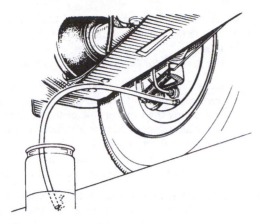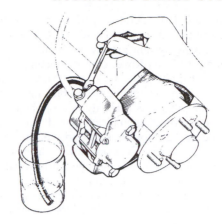

Figure 27.49 Brake bleeding (*Lockheed*)

air bleed valve is opened a fraction of a turn and the brake system pressurized either manually by repeatedly depressing the brake pedal, or automatically by means of a commercial pressure bleeding unit connected to the master cylinder reservoir, any air leaving the system will be made evident by the bubbles rising in the jar of fluid. When the bubbles no longer appear the bleeder screw of the air valve is retightened on to its seating. The procedure is, of course, repeated at the other air bleed valves in the system.

Not so many years ago, in the era of the very basic single-line hydraulic braking system described earlier, it would have been possible to include here a simple set of instructions that would have covered the brake bleeding procedure for practically all vehicle applications. Nowadays matters are a good deal more complicated, because from safety and other considerations the modern hydraulic braking system has reached a high level of sophistication. This in turn has led to many variations in brake bleeding procedures, which can require different pedal actions during manual bleeding, different sequences of bleeding that relate to divided-line safety systems, and with certain of the latter simultaneous bleeding of the front and rear brakes. Furthermore, there are additional factors that usually have to be taken into account when bleeding anti-lock brake systems (ABS) (Section 29), and especially those systems where a pump charged accumulator provides a reserve of pressurized fluid. Also, ABS system modulators must never be drained of fluid, because as with other hydraulic equipment of this nature, there can be difficulty in purging them of air when the system is refilled, and it is for this reason that replacement modulator units are supplied already primed with fluid and temporarily sealed. It is therefore most important that the manufacturer's instructions are faithfully adhered to when carrying out brake-bleeding operations.

Caution Care should be taken in the handling of brake fluids, since they contain agents that damage painted surfaces. They are also toxic to varying degrees, so that contact with sensitive parts of the body must be avoided. Medical advice must be sought immediately if brake fluid is inadvertently swallowed.

27.6 VACUUM SERVO-ASSISTED BRAKING

General background

Until the increasing weight and speed of commercial vehicles dictated the use of power-operated compressed air brakes, as opposed to power-assisted vacuum brakes, the latter proved to be a competent and reliable system of braking. In brief, the installation comprised a master servo, two front wheel servos and a vacuum reservoir. The piston-operated master servo converted vacuum, created by either manifold depression or exhauster pump according to whether the vehicle was petrol or diesel engined, into power for assisting brake application. It was so designed that its output was proportional to the effort applied at the brake pedal. The master servo was incorporated in a simple mechanical linkage from the brake pedal to the rear brakes, and thus assisted the driver's effort. There was no mechanical connection between front and rear brakes, the two front wheel servos being mounted on the stub axles and communicating via flexible vacuum hose pipes directly with the master servo. As the degree of vacuum admitted to the cylinder of the master servo was proportional to the driver's effort on the brake pedal, the same degree of vacuum necessarily existed in the wheel servos to apply the front brakes. Hence, the mechanically operated rear brakes and the vacuum-operated front brakes were servo assisted. The vacuum reservoir had sufficient capacity to provide at least four brake applications with the engine stationary. A non-return valve was used to allow air to be drawn from the vacuum reservoir, but not in a reverse direction. Where used the exhauster pump was typically of the sliding vane rotary type, and was engine driven either via the timing gears or in tandem with the fuel injection pump, or alternatively belt driven from the engine or the gearbox.

A pioneer in the application of vacuum servo assistance to the braking systems of both commercial vehicles and cars was the Belgian engineer Albert Dewandre, who first developed the idea in the mid 1920s.

Servo assistance for car brakes

Sir Henry Royce is credited with having once said that 'the speed at which a car can be driven with safety is determined by its ability to stop.' It is therefore perhaps to have been expected that, since the advent of four-wheel braking in the mid 1920s, the braking system has always been one of the outstanding features of Rolls-Royce cars, combining reliable stopping power with light pedal pressures. Although such attributes are now taken for granted in the braking systems of most modern cars, this was certainly not true before the general adoption of servo-assisted disc brakes. It is therefore relevant to try and account for the superiority of the earlier Rolls-Royce braking systems. The explanation may be found, first, in the use of strictly non-self-energizing drum brakes at the wheels, and second, in the addition of a friction servo motor to multiply the effort exerted on the brake pedal by the driver. In other words, the system used low-factor wheel brakes to provide consistent braking, together with a servo mechanism to reduce what would otherwise be a high pedal effort to operate them.

Since these two basic approaches can now be found in the servo-operated disc brake systems of many modern cars, they likewise account for their much improved braking performance, or at least greatly contribute to it. The reasons for the use of a servo unit in the braking system may therefore be summarized as follows:

1 It permits the use of low-factor wheel brakes with less sensitivity to changes in lining friction as they become heated.
2 The driver effort on the brake pedal can be reduced even further than with some high-factor, non-servo-assisted brakes.
3 A reduction can be made in the total pedal travel required because leverage in the system can be made much less.

Modern vacuum servo units

These may be either of the indirect- or direct-acting variety. The former is actuated by hydraulic pressure from the master cylinder and pressurizes the wheel cylinders via a hydraulic slave cylinder. This type of servo unit may therefore be installed in any convenient space available. For example, twin servo units of this type are housed in a front wing compartment of the high-grade Bristol car. A direct-acting (or integral) servo unit is, as its description suggests, directly actuated by the brake pedal and better lends itself on most cars to the modern requirement for dual hydraulic braking systems.

In either event its purpose is to reduce the driver effort required on the brake pedal. It performs this function by exposing a servo piston to the difference between atmospheric and engine intake manifold pressures. The piston is provided with either a sliding or a rolling seal against its cylindrical housing or chamber.

If used in conjunction with a diesel engine, the servo unit must be connected to either an engine or electrically driven vacuum pump. With earlier designs of vacuum servo units, each side of the piston was exposed to atmospheric pressure when the brakes were released and then manifold depression was allowed to act on the leading side of the piston during their application. This was known as the air suspended type of vacuum servo. In established modern practice, however, the vacuum suspended type of vacuum servo is used in which each side of the piston is exposed to manifold depression when the brakes are released and then atmospheric pressure is admitted to the trailing side of the piston during their application.

The reasons for this reversal in roles are threefold:

1 Less demand on the source of depression during brake application provides a more responsive control.
2 Since only the trailing side of the piston need be exposed to the atmosphere there is a reduced risk of damage by the ingress of foreign matter.
3 A simpler valve arrangement may be used requiring only one port each for the vacuum and atmospheric connections.

A fail-safe feature common to all types of vacuum servo-operated braking systems is that, in the event of failure of the vacuum source, the hydraulic circuit(s) remain unaffected although, of course, the driver pedal effort must be increased.

Operation of vacuum suspended servo unit

Referring first to the general arrangement drawing that names the parts of a typical direct-acting servo unit (Figure 27.50) and then to the simplified diagrams (Figure 27.51), the operating principle may be summarized as follows:

1 With the brakes released the servo piston is fully retracted by its return spring. Similarly the input rod connection to the brake pedal is retracted by the return spring for the latter. In this position of the input rod the vacuum port is open and there is a depression acting on both sides of the servo piston, which can thus be described as being suspended in vacuum.
2 When the brakes are applied the pedal urges the input rod forward, so that first the piston control valve closes the vacuum port and then its continued movement opens the atmospheric port. Air then enters the servo unit with the result that atmospheric pressure acts behind the servo piston and assists the input rod in travelling forwards; the piston in turn actuates the output push-rod connecting to the master cylinder plunger. During release of the brakes the air that entered the servo unit is exhausted via the opening vacuum port, first to the front portion of the chamber and thence past the non-return valve to the intake manifold.

Two additional points that deserve mention are the following:

3 When the brakes are held on, as distinct from being applied by increasing effort on the pedal, both vacuum and atmospheric ports remain closed so that a state of equilibrium exists between the master cylinder and the servo unit.
4 The function of the rubber reaction disc is to impart a certain degree of feel or feedback into the pedal effort being exerted by the driver, because when this is sufficient to open the atmospheric port of the piston control valve, it also means that the valve plunger head has penetrated the reaction disc. Since the latter then acts as a return spring to eject the valve plunger, the driver is made aware of this by an opposing force felt at the brake pedal.

Emergency brake assist

An emergency brake assist system can serve a two-fold purpose. First, it provides a built-in means of overcoming the

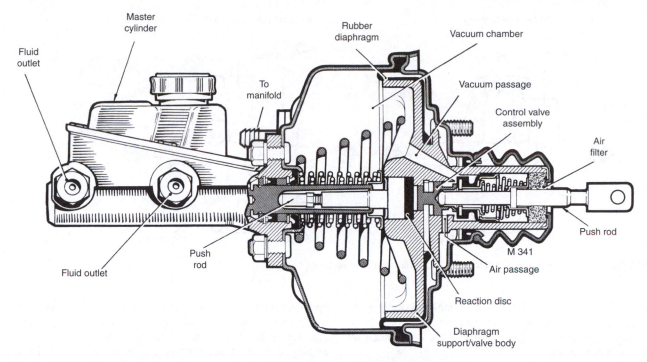

Figure 27.50 Cross-section through a direct-acting suspended vacuum brake servo (*Lockhead*)

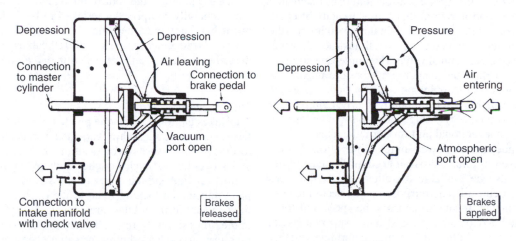

Figure 27.51 Operating principle of a direct-acting suspended vacuum brake servo

psychological barrier that may exist, when a driver is reluctant to brake hard enough in an emergency situation for fear of losing control of the vehicle: and second, to compensate for any physical limitations that a driver may have in exerting sufficient pedal effort under these circumstances. In both cases the system assists the driver to slow the vehicle in a shorter period of time, thereby exploiting the full potential of anti-lock brakes.

Brake assist systems, now sometimes abbreviated to BAS, generally fall into two categories. The simplest arrangement comprises a two-stage vacuum servo or booster unit; while a more comprehensive installation includes an independent power source of pressurized brake fluid, which can be delivered to the main hydraulic operating system when required.

A two-stage suspended vacuum servo unit is essentially a modification of a conventional unit. In one version a pair of depression chambers are mounted in tandem with the hydraulic master-cylinder, the rear chamber being larger in diameter than the front one. Under normal braking the diaphragm in the rear chamber acts on the push-rod of the master-cylinder (Figure 27.52a). This occurs when air admitted by the plunger control valve linked to the brake pedal acts behind the diaphragm and creates a pressure difference across it. During hard braking, such as in an emergency, a pre-set coil spring in the plunger control valve can then be overcome. This allows increased travel of the plunger control valve, which admits air not only behind the diaphragm in the rear chamber, but also behind the smaller diameter diaphragm in the front chamber

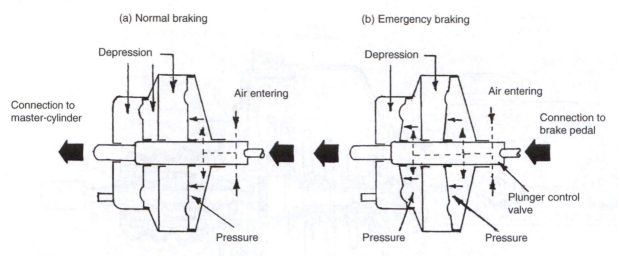

Figure 27.52   Basic principle of emergency brake assist vacuum servo unit

(Figure 27.52b). Hence, there is an additional apply force transmitted to the push-rod of the master-cylinder, which utilizes more effectively the driver effort at the brake pedal for greater stopping power.

In another version of a two-stage vacuum servo unit a single depression chamber and diaphragm is retained, but the plunger control valve incorporates a solenoid. This is actuated by an electronic control system that senses rapidity of brake pedal movement. Under emergency braking conditions, the solenoid is therefore actuated to allow full, instead of partial, opening of the plunger control valve. The immediate build-up of air pressure behind the diaphragm therefore maximizes servo output for greater stopping power, which otherwise may not have been available if related purely to driver effort at the brake pedal. This type of two-stage vacuum servo unit acts in the manner of a conventional unit for normal braking.

Where an independent power source of pressurized brake fluid is employed for brake assist, it is controlled electronically and may also be available for other vehicle functions such as traction control. The purpose of the electronic control is to monitor by means of sensors the speed and force with which the brake pedal is pressed, and compare this data to a stored reference value. It can then determine whether additional line pressure for the brakes is required to give maximum emergency braking. In this event, the brake assist system is commanded to release high-pressure fluid from an accumulator into the main braking system, thereby increasing line pressure above that otherwise related to driver pedal effort. An electrically driven pump is used to charge the hydraulic accumulator, which stores fluid at a pressure in the region of $17.5 \, \text{MN/m}^2$ ($2500 \, \text{lbf/in}^2$).

Brake assist systems do, of course, reduce their contribution to brake application, when the driver eases pedal effort and the emergency has passed.

27.7   THE PARKING BRAKE SYSTEM

General layout

The regulations governing the construction and use of motor vehicles requires their brakes to have a second independent means of operation. This function is performed by a mechanically actuated parking brake system, which in conventional practice is hand operated. Hydraulic connections are not employed for the handbrake system owing to the possibility of the fluid leaking slowly past the seals when the brakes are left applied for any length of time.

In early practice, the control linkage of car handbrake systems generally comprised a series of rods, shafts and levers, but it has long since become established practice to use a simpler arrangement of part-enclosed cables. The following describes the layout and construction of a modern handbrake system.

Driver control

This takes the form of either a pivoted or a sliding handbrake lever incorporating a pawl and ratchet mechanism. As the lever is pulled towards its on position, the spring-loaded pawl slides over the stationary ratchet teeth. When the handbrake is fully applied and the driver releases hold on the lever, the pawl transmits a locking force to the ratchet teeth owing to their special form and thus prevents the lever from returning to the off position. To return the handbrake to its off position, pressure on a release button or catch overcomes the spring load on the pawl to disengage it from the ratchet.

Operating cables

These provide a flexible connection between the handbrake lever mechanism and the wheel brake assemblies. They comprise a cable inner core that is always loaded in tension and connects between the handbrake lever mechanism and the relay levers on the wheel brake assemblies. Partly enclosed within a spirally wound outer casing, the cable is anchored by this casing at one end to the wheel brake assembly and at its other end to the understructure of the car. The purpose of the casing is to serve as a continuous guide and thus prevent the inner cable from straightening. Since it is only necessary to provide control movement in the brake apply direction, the inner cable is not subject to compressive loads that could buckle it and is pulled to its off position by return springs in the brake assemblies.

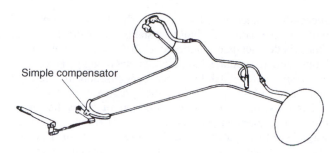

Figure 27.53 Handbrake system with simple compensator

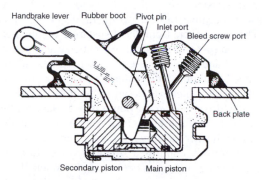

Figure 27.54 Single-piston wheel cylinder with integral handbrake mechanism

Equalizer mechanism

In order to ensure that each of the two brake assemblies operated by the handbrake is applied with the same force, the inner cables are not connected immediately to the handbrake lever mechanism, but through a simple compensating or equalizer device (Figure 27.53). This usually takes the form of a centrally pivoted balance beam with the cables connecting to its outer ends. Alternatively, a continuous inner cable may be used to connect the two brake assemblies, in which case its apex is guided around either a stirrup-mounted pulley or a saddle piece equalizer. To provide an overall adjustment for the handbrake system a screw-thread device is incorporated in series with the handbrake lever mechanism and equalizer device. A silicone grease is typically used to maintain lubrication for the enclosed part of the cable run.

Hydraulic wheel cylinder with integral handbrake mechanism

It was earlier mentioned that a single-piston hydraulic wheel cylinder can be made double acting by allowing its body freedom to slide on the back plate. In one design of leading and trailing shoe rear brake a wheel cylinder of this type is used with added provision for mechanical operation of the shoes by the handbrake (Figure 27.54). Although this makes for a more complicated design of wheel cylinder it does avoid the need for a separate lever-operated connecting linkage between the shoes, where the available space is usually fairly restricted by the bearing housings for the rear hubs. However, the latter type is the more widely used (Figure 27.15).

Mechanical operation of the single-piston hydraulic wheel cylinder is accomplished by making the piston in two halves.

The sealed inner half is acted upon by the footbrake hydraulic system, while the outer half, or buffer piston, is actuated by a cranked lever pivoting in the cylinder body and connected to the handbrake system. A specially shaped slot is cut out of the two-piece piston to accept the inner end of the cranked lever. The length of the slot is such that hydraulic operation of the piston assembly does not disturb the cranked lever.

When the brake pedal is depressed, the build-up of hydraulic pressure in the wheel cylinder produces a two fold action. First, it moves the sealed inner piston which in turn carries the outer buffer piston with it and so applies the leading shoe. Second, it causes an opposite movement of the slidably mounted cylinder body, the effect of which is to apply the trailing shoe.

When the handbrake is operated, the inner end of the cranked lever forces only the buffer piston to move and apply the leading shoe, while the reaction force at the lever pivot causes the cylinder body to slide and apply the trailing shoe.

Handbrake arrangement for disc brakes

The addition of a handbrake mechanism to the disc brake poses certain problems, which partly account for the mixed system of front disc and rear drum brakes being used in conventional practice, although an alternative approach is to build a small drum handbrake mechanism into the disc hub, as mentioned earlier.

Problems arise because the disc brake is to all intents and purposes non-self-energizing in its action and does, in fact, have the lowest brake factor of all automotive brakes. Hence, it follows that the force required to operate it is necessarily the highest. As far as the handbrake system is concerned, this implies that a large leverage must be provided between the driver control and the disc pads. Therefore a small amount of wear at the pads can be magnified into a greatly increased travel at the handbrake lever. It is for this reason that an automatic adjusting device is generally specified for such systems. Also, the operating linkage must be made as rigid as possible, because any unwanted flexing would again increase the travel at the handbrake lever.

In a modern example of automatic handbrake adjuster for a floating caliper disc brake, the input lever actuates a cam and push-rod that transmits an apply force, via the adjusting mechanism (Figure 27.55), to the disc pad piston. When a predetermined amount of pad wear has occurred, and in the course of normal operation of the service brake hydraulic piston, the adjuster pawl clicks over the piston ratchet to establish a constant adjustment for the handbrake. As in other automatic adjusters for brakes, there is a small amount of in-built lost motion to prevent over-adjustment and also permit proper relaxation of the pads after each operation of the service hydraulic brakes (their mode of self-adjustment having been explained earlier).

Electric parking brake

More recently there has been a trend towards electric parking brakes for some high-grade passenger cars. The system basically comprises two electromechanical actuators that are integrated with the floating calipers for the rear disc brakes. For this type of system a clamping force on the brake pads can be derived from a jacking screw and screwed sleeve mechanism.

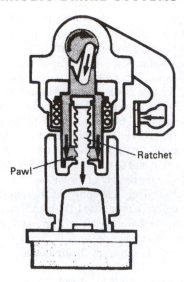

Figure 27.55   Automatic adjuster for disc handbrake (*Girling*)

The jacking screw acts directly upon the brake pad backplate and is prevented from rotating, whilst the screwed sleeve is axially constrained and receives a rotational movement from a concentrically mounted electric motor. Therefore when the handbrake is operated and the motor energized, the rotation of the screwed sleeve compels the jacking screw to transmit an endwise force on the pad backplate to apply the brake. A conventional hydraulic service brake is retained with this system. By virtue of its electrical operation the parking brake can be computer controlled, so that rolling back on hill starts can be avoided and emergency braking can be signalled via a stability control system.

27.8   HYDRAULIC POWER BRAKES

General background

So far we have considered only those hydraulic brake systems where the transmission of apply pressure to the wheel cylinders is either wholly, or partially, dependent upon the effort expended by the driver on the brake pedal. Another category of hydraulic brake system is the fully powered type, the operation of which is analogous to the driver merely depressing a pedal-controlled admission valve which either indirectly, or directly, subjects the wheel cylinders to a build-up of hydraulic pressure from a separate energy source and hence applies the brakes. In a practical system it is however arranged for the depression of the brake pedal to be resisted by a moderate force that is proportional to the forces being applied to the brakes, so that the driver can better feel the amount of braking being used.

Although the idea of hydraulic power braking had received investigation in an earlier period of motor vehicle history, it was not until the mid 1940s that such a system became available, originally on public service vehicles and then some ten years later on passenger cars, albeit on a limited scale. A notable example of hydraulic power braking for a city bus was the Lockheed system, which was later developed in conjunction with London Transport for installation on their successful Routemaster bus in the early 1960s. As compared with an alternative air brake system, the hydraulic power brakes better met their requirements for a lighter-weight installation, a more accurate and quicker responding pedal control, and also greater reliability since condensation troubles and entrainment of foreign matter in the system are avoided.

For passenger cars the introduction of hydraulic power brakes by Citroën on their revolutionary DS19 model in 1955 was a significant step forward in the concept of a central hydraulic system, which additionally provided an energy source for the automatic control of the clutch and gear change, power assistance for the steering, and automatic height control for the suspension. Since the late 1970s there has been an increasing incentive to develop hydraulic power brakes for cars. This has arisen from the legislative requirements for emission control equipment, the associated use of petrol injection systems and the wider use of diesel engines, all of which have tended to diminish the available intake manifold depression for operating a conventional vacuum servo unit. Indeed, a separately driven exhauster pump is necessary for this purpose in the case of diesel-engined cars. More recently, there has been the further incentive of combining the power source with electronically controlled solenoid valves to provide an anti-lock braking system, as described in Section 29.

Types of hydraulic power brake

In terms of being fluid power control systems, there are basically two circuit arrangements that may be used for hydraulic power brakes, which are classified as 'open centre' and 'closed centre'.

Open-centre control system
This represented the earliest type of hydraulic power brake and simply comprised three units: a hydraulic pump of the positive displacement type that was permanently driven from the transmission system, a conventional master cylinder combined with a follow-up servo valve of the piston type that was linked to the brake pedal, and a fluid supply tank.

With the brake released there was no end contact between the master cylinder and servo valve pistons, so that there was a free circulation of fluid from the pump to the supply tank via the central passage of the servo valve (Figure 27.56a). When the driver applied the brakes by depressing the pedal, the forward movement of the servo valve closed its mouth against a seating formed on a rearward extension of the master cylinder piston, thereby stopping what had otherwise been the free return of fluid through its central passage to the supply tank. A build-up of pressure from the pump therefore occurred between the exposed areas of the larger-diameter master cylinder and smaller-diameter servo valve pistons (Figure 27.56b), which in the former case applied the brakes in the manner of a conventional hydraulic brake system and in the latter case imposed a moderate resisting force at the pedal to provide the driver with a sense of feel as to the amount of braking being used. The magnitude of the pressure build-up in the hydraulic power system and therefore the braking effort generated indirectly by the master cylinder system remained proportional to the effort expended by the driver on the brake pedal. In other words, whenever the separating force on the servo valve

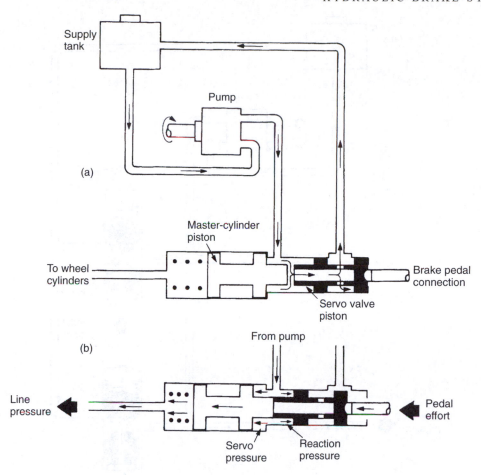

Figure 27.56 Schematic arrangement and operation of early-type open-centre control for hydraulic power brakes

exceeded that due to the opposing pedal force, the servo valve opened just sufficiently to allow a restricted amount of the fluid being pumped to return via its central passage to the supply tank, which prevented any further rise in pressure in both the hydraulic power and master cylinder circuits.

In later applications of the open-centre control system a pedal-operated and spring-returned distribution valve of the spool type is used. This still allows free circulation of fluid between the pump and supply tank when the brakes are released (Figure 27.57a), but establishes a direct connection between the pumped fluid and the wheel cylinders when line pressure is required to apply the brake (Figure 27.57b). A true hydraulic power brake system is therefore obtained, as there is now no need for a separate master cylinder circuit. The driver is again provided with a feel of the amount of braking being used, because the distribution spool valve has a restricted central passage that allows the brake line pressure to act beneath the valve and exert a moderate reaction force against brake pedal effort.

With an open-centre system it will therefore be clear that the working pressure for hydraulic power braking is created by the servo or distribution valve, which provides a modulated restriction to the pumped fluid returning to the supply tank. In this respect it may be compared with the control valve of a power-assisted steering system (Section 25.3). This

type of hydraulic power braking system has the merits of simplicity and, since the system is in a depressurized state with the brake released, it assists in limiting external leaks and extending the life of the components. However, it does have the significant disadvantage that pump delivery is reduced at low vehicle speeds, which then makes it difficult to maintain powerful and responsive application of the brakes. To overcome this problem, later open-centre systems have been used in combination with a closed-centre type, the arrangement of the distribution spool valves for such a system being shown in Figure 27.59.

Closed-centre control system
This represents the modern and more comprehensive type of true hydraulic power brake system. It basically comprises a supply tank from which the fluid is fed to a high-pressure pump driven from the engine. The pump then delivers fluid at a pressure higher than the highest pressure required by the braking system to a gas-charged hydraulic accumulator, where the fluid under pressure is stored until it is released to apply the brakes. This is accomplished by depressing a pedal-operated and spring-returned distribution valve of the spool type, which releases the fluid directly to the wheel cylinders (Figure 27.58b). As in the case of the later type of open-centre control system, the distribution spool valve

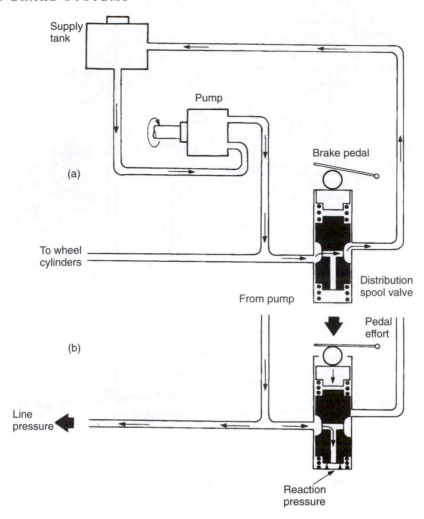

Figure 27.57 Schematic arrangement and operation of later-type open-centre control for hydraulic power brakes

has a restricted central passage that allows the brake line pressure to act beneath the valve and therefore create a reaction force to provide the driver with braking feel. When the brakes are released the fluid is returned via the distribution spool valve to the supply tank (Figure 27.58a). A pressure regulation valve is included in the system to direct the excess output from the pump back to the supply tank; in addition, the accumulator has a non-return valve so that the brakes may be applied a limited number of times after the pump has stopped. Another important advantage that is available from any true hydraulic power braking system is that there is virtually an unlimited displacement of fluid to operate the brakes, which makes the system less sensitive to fluid vaporization, air entrainment and maladjustment of the brakes.

In contrast to an open-centre control system, it will therefore be noticed that when the brakes are released there is no free circulation of fluid through the distribution spool valve. It is only when the latter is depressed by the pedal sufficiently to establish connection between the fluid stored under pressure by the accumulator and the brake lines leading to the wheel cylinders that braking occurs. That is, the braking effort depends upon removing, rather than imposing, a restriction on the fluid being supplied under pressure.

As with other types of brake installation, hydraulic power brake systems have long been duplicated from safety considerations. In commercial vehicle practice this usually takes the form of a tandem brake valve, wherein a primary distribution spool valve moves above a secondary one to operate the front and rear brakes respectively. For passenger car installations two distribution spool valves are commonly mounted side by side with a balance beam linkage connecting them to the brake pedal.

27.9 MAINTENANCE OF HYDRAULIC BRAKES

Preventive maintenance

There is an ancient saying that 'decay is inherent in all material things', and this is no less true for the components of hydraulic brake systems. Servicing for brake safety in terms of preventive and general maintenance has therefore always received careful consideration from brake equipment manufacturers.

Preventive maintenance is concerned with the scheduled replacement, or overhaul, of hydraulic components when in the judgement of the manufacturer a reasonable working life has been achieved. Typically the scheduled periods for

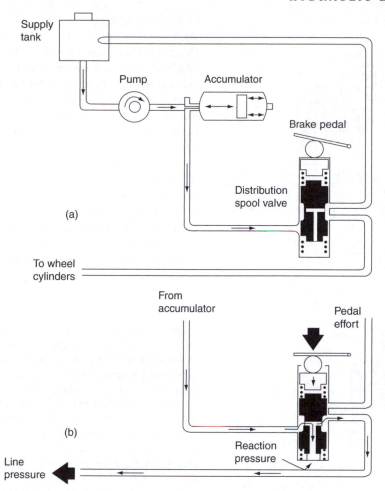

Figure 27.58 Schematic arrangement and operation of a closed-centre control for hydraulic power brakes

passenger cars and light commercial vehicles range from 40 000– 60 000 miles (64 000–96 000 km) or three years, whichever occurs first, and may depend for example on the type of brake calipers used. It is then advised that all hydraulic cylinders, hoses, apportioning valves and servo units are replaced by new, guaranteed components or, if this should not prove possible, the existing cylinders and servo unit may be overhauled using the appropriate service kit.

General maintenance

Weekly   Check the level of the brake fluid in the reservoir and, if necessary, top up to the required level. A sudden fall in fluid level indicates leakage from the system, which must be investigated and rectified without delay.

Every 1000 miles (1600 km)   Adjust drum brakes unless, of course, automatic adjustment is provided for them.

Every 5000 miles (8000 km)   Remove the brake drums and check for wear of the linings, which must not be allowed to wear down to the rivets. In the case of bonded linings these should not be worn below 1.5 mm (1/16 in) from the metal shoe.

Every 5000 miles (8000 km)   Check the disc pads for wear; if they have worn to 3 mm (1/8 in) thickness, replacements should be fitted. This check may not be required where on-board wear indicators are incorporated.

Every 5000 miles (8000 km)   Examine carefully all flexible brake hoses and their unions and also the full length of each metal pipeline for fluid leakage and any signs of general deterioration, such as chafing or corrosion. Renew any unserviceable part without delay.

Every 10 000 miles (16 000 km)   Clean down the back plate and lubricate the tips of brake shoes, the shoe platforms and the adjuster with an approved brake grease.

Every 18 months to 2 years   Renew the fluid in the braking system.

The vehicle manufacturer's own recommended servicing schedule for the hydraulic brake system should always be consulted for any special requirements. For example, the air filter of a vacuum servo unit may need renewing at certain intervals.

Hydraulic brake misbehaviour in service

Assuming that any misbehaviour of the braking system is not being influenced by unequal tyre pressures, or defective operation of the steering or the suspension systems, a purely

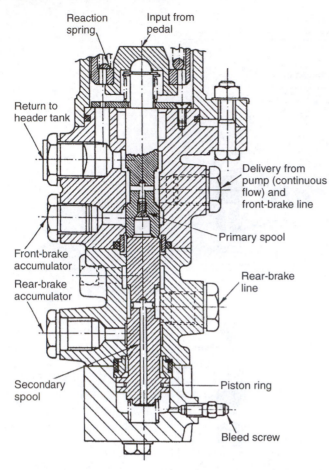

Figure 27.59 Tandem hydraulic power brake valve (*London Transport*)

general guide to the possible faults developed with hydraulic brakes is as follows:

Pedal travel excessive
Drum brakes need manual adjustment
Drum brakes' automatic adjusters seizing
Disc hubs too much end float
Discs running out-of-true
Distorted disc pad damping shims.
Pedal slowly sinks
External leakage from hydraulic system
Internal leakage past master cylinder main seal.
Pedal feels spongy
Air trapped in hydraulic system
Badly relined brake shoes.
Pedal feels springy
Shoe linings not bedded in
Master cylinder mounting defective.
Pedals feels hard
Glazed or contaminated shoe linings
Glazed or contaminated disc pads
Wheel cylinder pistons seizing
Caliper cylinder pistons seizing

Brake servo unit inoperative.
Brakes pulling to one side
Other side shoe linings contaminated
Other side disc pads contaminated
Other side wheel cylinder piston seizing.
Other side caliper cylinder piston seizing
Brakes locking prematurely
Brake system apportioning valve defective.
Brakes binding on release
Drum brakes over-adjusted
Handbrake over-adjusted
Handbrake mechanism seizing
Wheel cylinder piston seizing
Caliper cylinder piston seizing
Master cylinder reservoir vent blocked
Master cylinder compensating port blocked
No clearance at master cylinder push-rod.

Brake squeal

Complaints of brake squeal in service are as old as the motor industry itself. In the early years of the motor vehicle, the explanation and remedies for this phenomenon now seem deceptively simple. It was usually attributed either to dirt clogging the pores in the surface of the brake lining and causing it to become glazed, or to the presence of high spots between the lining and the brake drum. So in service it was the practice to remove the brake shoes and clean their linings, using a stiff wire brush and a flammable agent, which is perhaps best not recalled in the safety conscious climate of today. Alternatively, the linings were hand-bedded using a rasp to remove the high spots and especially to avoid premature heel and toe contact before readjusting the brakes. If these measures did not provide a convincing cure, then it was not unknown for owners to be assured that squealing brakes were at least efficient ones!

In referring to brake squeal in the 1930s a distinguished research engineer of that period, L.H. Dawtrey, wrote 'This is one of the most difficult problems which the designer has to contend with as evidenced by the large variety of vehicles which suffer from the disability.' By then it was realized that the explanation for squeal in drum brakes could involve a number of factors. These included insufficient rigidity of the brake drum and the backplate (or anchor plate as it was then often termed) on which the shoes were mounted, incorrect operating geometry for the shoes, too high friction coefficient for the brake lining and lack of vibration damping measures.

Later research by both vehicle manufacturers and brake equipment suppliers appeared to confirm that brake squeal is caused by a resonant vibration of the rotating and stationary parts of the brake assembly, whether it be of the drum or the later disc type. When the resonant frequency of vibration comes into the audible range, then it is heard as brake squeal. The source of the vibration energy must, of course, originate between the rubbing surfaces of the brake before it is transmitted to other parts of the assembly.

In practice the natural vibration period of the brake drums and, less commonly, the brake discs, has sometimes been suppressed by fitting a frictional damping ring around their peripheries. A classic past example of the former was the

Rolls-Royce spring-loaded damper ring, which encircled the drum near its open end and acted in the manner of a crankshaft torsional vibration damper (Section 1.9). The same principle may also be applied to modern brake discs, one method used being to spring a split steel ring into a deep groove machined in the periphery of the disc.

Again, as a matter of historical interest, Rolls-Royce also incorporated an elegant example of brake shoe damping, where a spring-loaded short auxiliary shoe with an offset pivot was mounted towards the toe of each main brake shoe, so that light application of the brakes was concentrated on the smaller auxiliary shoes. Less exotic ways of damping brake shoe vibration include attaching a lead weight to the shoe web, spring-loading the shoe web into contact with the backplate, and strategic slotting of the shoe web to alter its flexibility and hence the distribution of pressure between the lining and drum.

In the case of disc brakes, a similar redistribution of pressure between the pad and disc is required to dampen vibration. It is generally achieved by offsetting in the circumferential direction the centre of pressure on the pad lining. In effect the apply force from the caliper piston is then no longer concentrated in the middle of the pad area, as would otherwise be expected. To modify the pressure distribution a suitably positioned cutaway portion on the rim of the caliper piston may be used or, as in later practice (Section 27.4), a thin steel shim with an asymmetrical cutaway portion is inserted between the rim of the caliper piston and the pad backplate. Alternatively, the pad may be supplied with the lining already offset on the backplate, or the lining can be strategically tapered.

In all cases of brake squeal it is essential to seek advice from the vehicle manufacturer before departing from the recommended build instructions and friction material specifications.

Personal safety precautions in servicing the braking system

Before jacking up either the front or the rear wheels on to axle stands in order to work on the braking system, the other pair of wheels should be chocked securely against rotation. The inhalation of asbestos dust can be injurious to health, so when cleaning out drum brake assemblies on older vehicles the dust should be removed using a vacuum cleaner or by wiping with a damp rag. Further reference should be made to the precautionary guidance issued by the Health & Safety Executive in their publication HS(G)67 titled 'Health & safety in MOTOR VEHICLE REPAIR'. Brake fluid has already been described as a toxic substance and should be handled with care; in particular, keep it away from the eyes.

27.10 BRAKE EFFICIENCY AND TESTING

Basic considerations and requirements

The total retarding force acting on a vehicle includes braking force, tyre drag, friction losses in the wheel bearings and the transmission system, the force exerted by the vehicle to drive the engine, the effect of road gradient and wind resistance. If all these forces together so happen to equal the weight of the vehicle, then it would slow down or decelerate at the rate of approximately 9.8 metres per second per second ($9.8 \, m/s^2$)

or 32 feet per second per second ($32 \, ft/s^2$). This is the same rate as that at which a body falling freely would accelerate, were it not for air resistance, under the force of gravity. The rate of acceleration or deceleration, due to a body being acted upon by a force equal to its own weight, is known as g. It is therefore convenient to talk in terms of $0.5 \, g$, $0.85 \, g$ and so on, instead of expressing the actual value in either m/s^2 or ft/s^2. Likewise it is convenient to talk of braking force as a percentage of the weight of the vehicle, so that for example a 50 per cent braking force would provide $0.5 \, g$ deceleration on a level road.

In automobile practice, however, it has long been customary to refer to the braking force relative to the weight of the vehicle in terms of braking efficiency, so that a $0.5 \, g$ deceleration is conveniently expressed as a 50 per cent braking efficiency. In theory, this relationship holds good only as long as the coefficient of friction or adhesion between the tyre and road surface is unity or, in other words, the horizontal braking force at the wheel divided by the downward load on the wheel is equal to one. This limitation is nevertheless accepted as being a reasonable basis on which the braking efficiency of a vehicle can be assessed. Summing up, then, if the brakes are producing a retarding force that is equal say to three-quarters of the weight of the vehicle, the braking efficiency is said to be 75 per cent.

In Britain under The Road Vehicles (Construction and Use) Regulations, the basic minimum requirements for braking efficiency that cover most categories of new vehicle are 50 per cent service brakes, 25 per cent secondary brakes and 16 per cent CV parking brakes. The EC minimum requirements for braking efficiency are higher for motor cars and lower for commercial vehicles, but are based on actual stopping distances.

Testing brake efficiency

Several methods of testing brake efficiency have been used in service, past and present, which may be classified under the following headings:

Pendulum decelerometer This type of instrument, which is usually identified as a Tapley meter, is basically a calibrated pendulum device. During braking the pendulum swings forwards because of its inertia, the angle of swing bearing a direct physical relationship to the deceleration of the vehicle (Figure 27.60a). The movement of the pendulum is amplified and held as a scale reading of percentage brake efficiency.
U-tube decelerometer With this type of instrument, deceleration of the vehicle is indicated by movement of a liquid column in a U-tube, which is mounted parallel to the longitudinal axis. Deceleration causes the transfer of liquid from one limb to the other of the U-tube (Figure 27.60b), the limb in which the liquid falls being calibrated and made of smaller bore to amplify the scale reading of percentage brake efficiency.
Platform brake-testing machine This form of a permanently installed equipment was originally introduced in America, where in some states the testing of brakes was early made compulsory. It comprises four spring-returned sliding platforms on to which the vehicle is driven at low speed, the brakes being applied when each wheel is on its appropriate platform (Figure 27.60c). The sliding reaction of the platforms is transmitted as

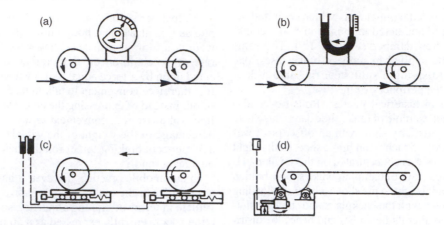

Figure 27.60   Schematic arrangements of brake-testing methods: (a) pendulum decelerometer (b) U-tube decelerometer (c) platform brake-testing machine (d) roller brake-testing machine

movement in liquid columns, which are calibrated to indicate braking force.

Roller brake-testing machine   This is another form of permanently installed equipment, modern versions of which typically comprise a flush-fitting roller bed with two pairs of electrically driven rollers, these being so arranged that both wheels at either end of the vehicle can rest upon them and be tested simultaneously. During testing the braked wheels are forced to rotate by the rollers (Figure 27.60d), the torque reaction developed at the electric motors being sensed to provide an indication of braking force at each wheel, this being displayed on console gauges.

Since the testing of brake efficiency in service is usually now concerned with the DTp (referred to as MoT) brake performance test on a vehicle, it should be mentioned that this test has to be carried out using an approved roller brake-testing machine, except where this may not be appropriate as for example with vehicles having permanently engaged four-wheel drive. In this and other cases where approval has been granted, a decelerometer instrument is used. The former method of brake testing is regarded as being safer and more efficient for the DTp requirement. The DTp (MoT) *Tester's Manual* should be consulted for details of the brake performance test and the equipment used in accordance with the manufacturer's instructions. Since the operation of roller brake-testing machines or 'rolling roads' as they are now known and other brake testing equipment can pose potential safety hazards, unless suitable precautions are observed by personnel, it is advisable to consult the Health & Safety Executive publication HS(G)67 titled 'Health & Safety in MOTOR VEHICLE REPAIR'.

28   Air and endurance brake systems

28.1   PRINCIPLES OF AIR BRAKES

General background

It was as long ago as 1868 that the American engineer George Westinghouse first patented his invention for an automatic compressed air brake for railway trains, which soon proved superior to other types of braking system including his own earlier development of a vacuum-operated brake. Although compressed air brakes for railway trains were soon adopted in America and many other countries, another one hundred years passed before they appeared to any extent on railway trains in Britain. The change from vacuum to compressed air braking in fact coincided with the demise of the steam engine and the introduction of the diesel. With a steam engine the vacuum was created by an ejector, wherein the action of a jet of steam was made to exhaust air from the pipes, cylinders and reservoirs of the system. Therefore with a diesel engine either a vacuum pump was necessary, or it could be just as economical to fit a pressure pump for compressed air braking, with its greater power.

Rather interestingly a somewhat similar pattern of progress occurred with heavy vehicles on the road (although much more rapidly, of course) when the petrol engine was superseded by the diesel. With the petrol-engined vehicle the once popular vacuum-operated system utilized the depression existing in the intake manifold, but since the diesel engine could not offer this facility it similarly became advantageous to mount an engine-driven compressor, rather than an exhauster, and install compressed air brakes, again with their greater power.

Both railway and heavy-vehicle engineers have therefore considered it preferable to utilize a source of compressed air, generally at $700 \, \text{kN/m}^2$ ($100 \, \text{lbf/in}^2$) or more, rather than a source of vacuum that must always be less than atmospheric pressure, or to put it another way a negative pressure no greater than $105 \, \text{kN/m}^2$ ($15 \, \text{lbf/in}^2$). More specifically, the higher operating pressures that are made possible with compressed air brakes allow a reduction in size of the system components, accompanied by quicker application and release characteristics.

More recent years have seen the operation of heavy vehicles where both gross weight and speed capability are much greater than ever before. This in turn has led to an increasing degree of sophisticated engineering in air brake systems, not least to meet the safety directives of the EC (European Community). Indeed, it may be said that air brake technology has virtually become a separate branch of motor vehicle engineering.

Advantages of air brakes

For medium to heavy commercial vehicles, compressed air braking systems, or air brakes as they are generally known, offer the following advantages:

1 As an operating medium, air costs nothing and is always available.
2 The system will tolerate a certain amount of air leakage without failing completely.
3 Large operating forces to expand the brake shoes can readily be generated.
4 A supply of compressed air is a convenient source of energy to operate various ancillary equipment on the vehicle.

Basic installation of air brakes

Commercial vehicle air brake installations constitute a true power as distinct from power-assisted braking systems. This is because there is no direct connection in any mechanical or hydraulic sense between the brake treadle and the wheel brake chambers, although the driver is provided with a certain degree of feel related to system air pressure during braking. The functions of an air brake system are to compress, store, meter and deliver a volume of air under pressure to the wheel brake actuating chambers.

In simple terms, the operation of a basic single-circuit air brake system (Figure 28.1) is such that when the brake treadle is depressed one of two related control valves is opened, so that air under pressure from the reservoir can pass through the control valve and into each wheel brake actuating chamber. Here the compressed air acts against a diaphragm, its resulting movement being transmitted via a push-rod to either the operating lever of the brake camshaft, or the wedge of a brake expander unit, which forces the shoes against the brake drum. As the brake treadle is released, the previously mentioned control valve closes and the other one is opened, thereby allowing the air under pressure in the brake actuating chambers to be exhausted to the atmosphere and the shoe return springs to release the brakes. In the event of the system air pressure falling below a safe working minimum, either a warning light or a buzzer is automatically activated in the driving compartment.

The components of an air brake system are most conveniently considered under the following headings:

Compression and storage
System control
System actuation.

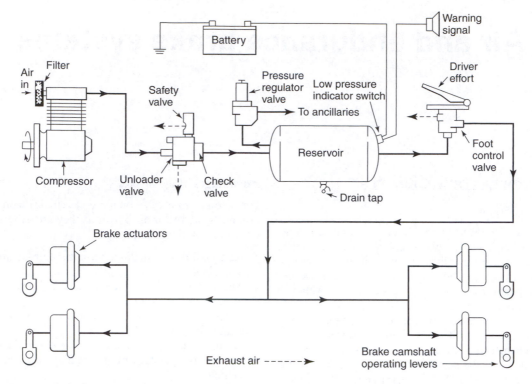

Figure 28.1   Layout of a basic single-circuit air brake system

28.2   COMPRESSION AND STORAGE

Air compressor

This generates a supply of compressed air to operate the braking system and also the vehicle ancillary services. In modern installations it is usually mounted on, and either belt or gear driven from the engine. It takes the form of a reciprocating piston pump that may have a single or twin in-line cylinders. The cylinder head is provided with inlet and delivery valves of the spring-loaded disc type, which are arranged to open in opposite directions. Of rugged construction, the main body of the air compressor is typically cast from high-grade iron. An air-cooled compressor can be recognized by cooling fins integrally cast into the cylinder block and head, while a water-cooled compressor is jacketed and has a smooth exterior with connections to the engine cooling system. The main and big-end bearings of the compressor receive oil under pressure from the engine lubrication system, the small-end bearing and cylinder wall being splash lubricated. Oil return and breathing is into the engine sump.

In operation, the compressor receives a clean supply of air either via the engine intake manifold or from a separate filter mounted on the engine cylinder block. During the down-stroke of the piston a partial vacuum or depression is created in the cylinder space above the piston, so that atmospheric pressure acting above the inlet valve forces it to open against its spring loading and allows air to enter the cylinder (Figure 28.2a). As the piston commences its up-stroke, the increasing pressure of air in the cylinder allows the inlet valve to close under the influence of its spring loading. The continued movement of the piston on its up-stroke then further increases the air pressure in the

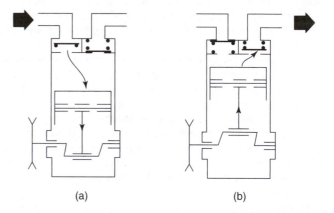

Figure 28.2   Operating principle of air compressor: (a) intake stroke (b) delivery stroke

cylinder, until it becomes sufficient to overcome the spring loading acting on the delivery valve (Figure 28.2b). Air is then discharged under pressure through this valve for delivery to a storage reservoir. As the piston commences its next down-stroke, and simultaneous with the reopening of the inlet valve, the increasing depression in the cylinder allows the delivery valve to close under the influence of its spring loading, thus preventing the compressed air from returning to the cylinder.

Air reservoirs

Several reservoirs are necessary to serve the separate main service, emergency secondary and, where applicable, trailer brake circuits, which comprise the modern comprehensive air brake system. Their purpose is to provide storage capacity of compressed air at the required pressure for immediate effective braking and other ancillary services, and also to maintain the pressure supply after the engine has stopped. The total air storage capacity of the reservoirs is subject to legislative requirements. It is decided on such factors as the air delivery rate of the compressor, the size and number of the wheel brake actuating chambers, the frequency of brake application as related to vehicle type and operation, and the further demands of the ancillary services.

Since they act as pressure vessels, air reservoir tanks are designed with a large factor of safety and constructed from sheet steel with welded seams. Each reservoir is provided with either a plug or a drain valve at its lowest point, so that any condensation or sediment that accumulates may be regularly drained off. This is necessary because, depending on humidity, the air entering the compressor contains a certain amount of water vapour, and as it passes through the compressor it can also acquire oil mist from cylinder wall lubrication. The emulsion of water and oil can have adverse effects on air brake components, while any freezing of the water content in could weather could seriously affect the operation of the braking system.

Two related items of protective equipment that may be included in the air brake compression and storage system are the air dryer and alcohol injector. An air dryer is fitted in the compressor air discharge line to the reservoirs and is mounted external to the engine compartment, such that the motion of the vehicle induces a cooling air flow over the dryer body. Although the main purposes of an air dryer will be self-evident, it is also intended to exclude oil droplets, carbon particles and any other contaminants from the compressed air before it enters the system reservoirs, so that the provision of cool, clean and dry air for the control and actuation systems will increase both service life and operating efficiency of their components. An air dryer is a basically simple unit and operates in three stages that involve cooling, filtering and drying. The heated air entering from the compressor is first cooled by contact with the dryer body interior and deposits water and oil condensate into a lower sump. The cooled air then passes through an oil filter to remove any remaining liquid droplets or solid contaminants, before entering a desiccant chamber. This contains a cartridge of microcrystalline pellets that possess a strong affinity for water and present an exceedingly large surface area to the air flowing over them. By this means the water vapour still in the air is adsorbed by the desiccator, the air being progressively dried before finally entering an upper purge chamber from whence it flows, via an outlet check valve, to the reservoirs. Dried air remaining in the purge chamber is utilized to expel sediment from the sump of the dryer and to remove adsorbed moisture from the desiccant pellets. This process occurs whenever the reservoir air pressure signals the governor to unload the compressor, because air from the unloader line is then allowed to open a purge valve in the base of the dryer, which causes a sudden decompression and reverse flow of air through the unit.

The function of an alcohol injector is to deliver a positive and metered quantity of alcohol into the air being discharged under pressure from the compressor, so that it is carried by the air stream to the rest of the system. Its purpose is to lower the freezing point of any water vapour entrained in the compressed air supply, which reduces the risk of brake misbehaviour through moisture freezing in the system during low-temperature operation of the vehicle. An alcohol injector essentially comprises a simple spring-returned plunger pump with an alcohol reservoir. The injection process is accomplished by depression of the plunger over its working stroke, which occurs whenever the reservoir air pressure signals the governor to unload the compressor, since air from the unloader line is also directed above the enlarged head of the plunger. A non-return valve at the outlet of the injector isolates the entire unit from system air pressure when the pump is not being activated. Provision is made to lock the plunger out of action when the injector is no longer required after winter use, and, of course, the reservoir is then allowed to remain empty. Mineralized methylated spirit or ethyl alcohol are used in the injector unit, the latter generally being preferred. In either case it should be recognized that these substances are highly flammable and give off toxic fumes.

Governor valve

Also known as an air limiter valve, this is mounted in a return line between the compressor and reservoirs to maintain air storage pressure between normal operating limits; typical cut-out and cut-in pressures for a governor valve are 735 kN/m^2 (105 lbf/in^2) and 630 kN/m^2 (90 lbf/in^2) respectively. The governor valve exerts this control by directing air at cut-out pressure to an unloader plunger in the compressor head, which holds the inlet valve off its seat so that no further air compression is possible in the cylinder. Otherwise, the compressor would pump air continuously to the reservoirs regardless of operating pressure requirements. When the compressor is rendered inactive in this way it is said to run light.

The governor valve is intended to be progressive in its operation, which depends upon the opposing forces created by a control or pressure setting spring and air storage pressure acting against either a sliding piston or a flexible diaphragm. We take the latter construction as an example.

The diaphragm is clamped at its outer edges by the valve body and at its centre by the head of a hollow valve plunger, above which acts the control spring. A combination inlet and exhaust disc valve is initially spring loaded into contact with the lower tip of the valve plunger.

In operation, a supply of air at storage pressure is returned to the governor valve and enters a chamber beneath the diaphragm. As the pressure increases against the diaphragm it flexes to lift the valve plunger and compress the control spring. Continued lifting of the valve plunger will first allow the inlet and exhaust valve to cover the exhaust passage seat, following which the tip of the valve plunger will lose contact with the valve altogether. When this happens the compressed air acting beneath the diaphragm passes through the hollow valve plunger and is directed to the unloader plunger in the compressor (Figure 28.3). Simultaneous with this action, the air pressure exerts an additional upward force on the control spring by virtue of acting on the area of the plunger itself. It is the presence of this additional force that allows a substantial

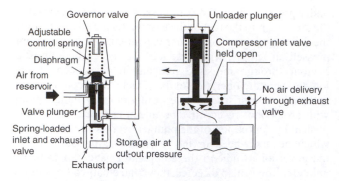

Figure 28.3   Schematic arrangement and operation of governor valve

drop in pressure of the storage air before charging recommences. In other words, it determines the required cut-out and cut-in pressures. When the storage air pressure drops below the cut-in point of the governor valve, the control spring first restores contact between the plunger tip and the inlet and exhaust valve to isolate the air supply to the unloader plunger. With reducing air pressure it then causes the plunger to unseat the inlet and exhaust valve and allow the air supply to the unloader plunger to be vented to atmosphere. The air compressor now resumes its normal operation of charging the reservoirs.

Unloader valve

An unloader valve is installed in the supply line between the compressor and reservoir where it performs a similar function to that of a governor valve. The essential difference between these alternative types of pressure control valve is that, with an unloader valve, the governor valve and unloader plunger are combined into a single unit. Its operation is such that instead of rendering the compressor inactive when cut-out pressure is reached, the compressor continues pumping air that is simply vented to the atmosphere via the unloader valve.

Safety valve

This provides a safeguard against any excessive build-up of air pressure in the storage system in the event of either a governor, or an unloader valve, failing to operate. It may be located at either the compressor, the unloader valve or the reservoir, and simply comprises a ball check valve that is spring loaded on to an orifice seating formed in a brass housing. If the storage air pressure rises above a preset limit, which is typically in the range of $896 \, kN/m^2$ ($130 \, lbf/in^2$) to $1103 \, lbf/m^2$ ($160 \, lbf/in^2$), the ball check valve is lifted against its spring loading and allows excess air to be vented to the atmosphere through the exhaust port. The safety valve remains closed during all normal operation of the air brake system.

Safety tests
A periodic check for any abnormal leakage from the air brake system typically involves the following procedure. First, run the engine until the driver air gauges for the front and rear

brake circuits show a pressure of $735 \, kN/m^2$ ($105 \, lbf/in^2$) or 7 bar and stop the engine, then observe that the pressure does not drop more than $52 \, kN/m^2$ ($7.5 \, lbf/in^2$) or 0.5 bar over a period of 4 minutes. Second, run the engine until the air gauges again show a pressure of $735 \, kN/m^2$ ($105 \, lbf/in^2$) or 7 bar, next fully depress the brake pedal and stop the engine, then keeping the pedal depressed for 2 minutes observe that the pressure does not drop more than $41 \, kN/m^2$ ($6.0 \, lbf/in^2$) or 0.4 bar. If these drops in pressure are exceeded then the system should receive urgent investigation.

The fitting of anti-lock brake systems to heavy vehicles (Section 29.5) is now a requirement of European Union legislation. This has made it necessary for the driver to check the functioning of the system before commencing a journey. For this purpose the satisfactory operation of the system is indicated by a warning lamp on the dashboard, a second warning lamp being provided for a trailer combination. The signal that appears after switching on should extinguish once the vehicle reaches a speed of about 6 mile/h (10 km/h), when the anti-lock facility normally becomes operative.

28.3   SYSTEM CONTROL

Foot-operated brake valves

These are treadle operated. Their function is to provide control of system air pressure, during application and release of the brakes which is precise and proportional to driver effort. Consider first a single foot brake valve. This essentially comprises a treadle-operated telescopic plunger and piston assembly, which acts upon a combination inlet and exhaust valve at its foot. The plunger acts upon the piston through the medium of what is called a graduating spring, this being provided with a retaining collar through which can slide the stem of the piston. Below the piston is a lighter return spring. The lower body of the piston is made both hollow and ported, so that not only does it serve as an exhaust valve seating on brake application, but also it can allow the passage of exhaust air from the actuating chambers on brake release. In the normal position, the combination inlet and exhaust valve is spring loaded against the inlet valve seat, the foot of the piston remaining clear of the upper exhaust valve.

When the driver depresses the brake treadle, a force is transmitted through an interposed roller to the spring-loaded telescopic plunger and piston assembly, the downward motion of which causes the exhaust seat of the piston to close on to the exhaust side of the combination valve. Further depression of the treadle causes the piston to unseat the inlet side of the combination valve, which then allows compressed air from the reservoir to pass through the valve and on to the wheel brake actuators (Figure 28.4a). This compressed air is also allowed access to the underside of the piston, via a bleed hole in the valve body. Therefore when this upward pressure just exceeds the downward one being exerted on the piston through the medium of the plunger and graduating spring, the piston lifts just sufficiently to allow the inlet side of the combination valve to close into its seating. At the same time the piston still maintains a closed exhaust passage against the exhaust side of the combination valve (Figure 28.4b). In this

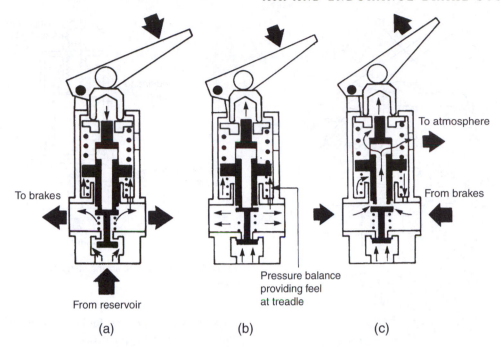

To atmosphere

To brakes

From brakes

From reservoir

Pressure balance
providing feel
at treadle

(a) (b) (c)

Figure 28.4 Schematic arrangement and operation of foot-operated brake valve: (a) brakes applied (b) brakes held (c) brakes released

balance or lapped condition the brakes are held on with a force that is proportional to the effort being applied by the driver, who is therefore provided with a sense of feel as to the amount of braking being used.

Any change in the braking effort exerted by the driver on the treadle will alter the balance point. If the treadle is depressed further, transmitting additional force through the graduating spring, the downward movement of the piston will again unseat the inlet side of the combination valve, thereby increasing the air pressure delivered to the brake actuators. Similarly, this additional pressure will act against the underside of the piston until it exceeds that imposed by the graduating spring force. The piston will then rise to allow the inlet side of the combination valve to close on to its seating with the exhaust side remaining closed. Another balance point is thus established with the brakes being more firmly held on, in accordance with the increased braking effort exerted by the driver.

Conversely, if the driver partially releases the treadle then the reduced force transmitted by the graduating spring will allow the piston to rise, owing to the air pressure beneath it and its return spring. When this occurs the inlet side of the combination valve remains closed and the piston retracts from the exhaust side, which allows some air to be exhausted from the brake actuators until the drop in pressure allows the piston to sink just sufficiently to close off the exhaust side of the combination valve. Once again a new point of balance is reached, but this time at a lower pressure as reflected by the reduced braking effort exerted by the driver. Of course, when the driver fully releases the treadle the piston returns to its highest position, so that the inlet side of the combination valve remains seated to cut off compressed air supply from the reservoirs, and the exhaust side is unseated to exhaust all

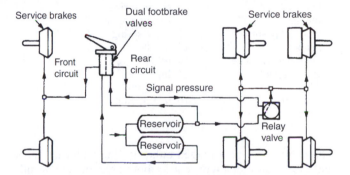

Figure 28.5 Layout of a basic dual-circuit air brake system for a six-wheel rigid vehicle

air pressure from the brake actuators to release the brakes (Figure 28.4c).

The air braking systems installed in modern heavy vehicles in Britain must be designed to meet The Road Vehicles (Construction and Use) Regulations and also EC Directives. One result of this has been the fitting of dual-circuit braking systems for rigid heavy vehicles, which was first encouraged during the mid 1960s and then some ten years later became mandatory. For example, the foot-operated service brake of a six-wheeled vehicle is split into a front service brake which is operative on the front axle brake assemblies, and a rear service brake which is operative on the foremost and rearmost rear axle brake assemblies (Figure 28.5). If a fault occurs in either of the two systems the other one must operate independently, so it becomes necessary to provide dual foot brake valves with each receiving air from its own reservoir.

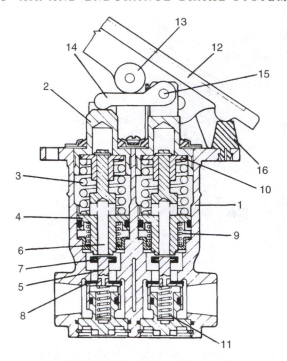

Figure 28.6 Section of Bendix Westinghouse dual foot-operated brake valves (*Seddon Atkinson*)

1 body	9 return spring
2 plunger	10 spring retainer
3 graduating spring	11 valve spring
4 piston	12 treadle
5 inlet/exhaust valve	13 roller
6 exhaust port	14 link
7 supply port	15 pivot pin
8 inlet valve	16 treadle stop

The valves may be arranged either in line with each other and known as a dual-concentric foot valve, or side by side and straddled by a balance beam. The latter is acted upon at its centre by the treadle roller to provide matched air pressures for the two circuits (Figure 28.6).

Pressure regulating valve

This type of valve is used to ensure that adequate air pressure exists in that portion of the system preceding the valve, before any pressure is allowed to build up in the system beyond it. For example, it can ensure that a brake reservoir receives charging priority over a reservoir that supplies the ancillary services. When this type of valve is installed between reservoirs it is also known as a pressure protection valve.

A pressure regulating valve is of simple construction and comprises a valve body, a regulating spring and either a diaphragm- or a piston-controlled delivery valve. In operation, the valve remains closed until reservoir air pressure acting on the underside of either its diaphragm or piston builds up sufficiently to overcome the regulating spring load. Air under pressure can then flow to the ancillary services (Figure 28.7). In the event of a system failure beyond the valve, the air pressure will drop until it reaches the cut-out pressure at which the valve

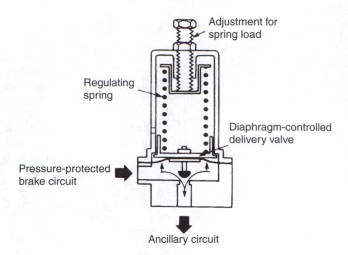

Figure 28.7 Schematic arrangement and operation of pressure regulating valve

closes. This prevents any further loss of air under pressure from the circuit being protected.

Single and double check valves

A single check valve is also known as a non-return valve, and its purpose is to allow the flow of air in one direction only. This type of valve is fitted at the entry to the air reservoirs according to system requirements, and prevents loss of pressure in the event of an air pipe or coupling suffering a fracture. The body of the check valve is screwed directly into the entry port of a reservoir, and its internal components comprise no more than a spring-loaded rubber valve with suitable guidance for the stem and a seating for the valve.

The valve is intended to operate when the pressure of air at the outlet side is equal to, or less than, the pressure at the inlet side. Air is therefore allowed to flow through the unit once the pressure at the inlet port forces the valve off its seat, but it cannot flow in the reverse direction because the combination of pressure and spring load will return the valve to its seating. To meet EEC requirements a more sophisticated version of check valve is now used, this being known as circuit protection valve. It has the characteristic of automatically isolating a defective circuit from the air supply line, so that air pressure in the remaining circuits is kept at a level high enough to maintain acceptable braking efficiency.

A double check valve performs a change-over function in allowing air under pressure from either one of two control valve systems to supply a brake actuator, while isolating the other in the event of failure. This type of valve simply comprises a body with two inlets and one outlet, the inlet passage being sleeved to guide a rubber shuttle valve, which is provided with a seating at each end of the passage.

In operation, the shuttle action of the double check valve ensures that only air at the higher of the two pressures delivered to its inlet connections will pass to the outlet and on to the brake actuator. This is because the air delivered at higher pressure will compel the valve to move over and seal off the opposite inlet connection where the air pressure is lower (Figure 28.8).

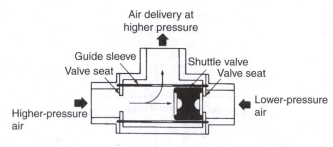

Figure 28.8 Schematic arrangement and operation of double check valve

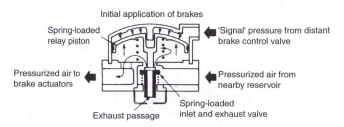

Figure 28.9 Schematic arrangement and operation of relay valve

Relay valves

For the effective operation of an air brake system it is important to obtain a rapid pneumatic balance or equalization of air pressure at all brake actuators. Otherwise, there would be an unacceptable delay in the application of the rear brakes of a rigid long-wheelbase vehicle. To overcome this difficulty it can be arranged for the rear brakes to receive their pressurized air supply more directly from a separate nearby reservoir via a relay valve. This valve can then be rapidly triggered by a signal pressure from a brake control valve, since only a small volume of air need be passed through a narrow-bore pipeline to provide the signal, and the apply pressure of the brakes remains proportional to the signal pressure.

The relay valve comprises an upper air chamber with a signal pressure port and delivery ports in the chamber base. Sliding within the chamber is a spring-loaded relay piston with a valve seat carried at the lower end of its stem. The relay valve lower body contains the by now familiar combination inlet and exhaust valve, which is made hollow and is normally spring loaded against the valve inlet port at the base of the air chamber. Valve inlet or reservoir ports are also provided in the lower body.

In operation, the large area of the relay piston ensures that there is a quick response to signal pressure from a brake control valve. The resulting movement of the piston against its spring first closes the exhaust passage through the combination inlet and exhaust valve, and then depresses this valve so that the inlet seat is uncovered. Storage air from the reservoir can now pass through the inlet port of the valve body into the air chamber beneath the piston, and leave via the delivery ports to supply the brake actuators (Figure 28.9). When the combined force of the piston and valve return springs, together with the air pressure acting beneath the piston, just exceeds the signal pressure acting above the piston it will lift sufficiently to allow reseating of the combination valve. With the inlet seat now covered by the combination valve and the exhaust passage through this valve sealed by contact with the piston stem, the relay valve is in a state of balance. That is, those brakes operated by air supplied via the relay valve are applied just as quickly and held on with the same force as the other brakes, in all cases this force being proportional to the effort exerted by the driver on the control valve.

Other more elaborate versions of relay valve may be used for certain applications. One example is the inverse relay valve that was once used with tractive unit and trailer combinations with a single-line connection. It was so arranged that air passed through it to the trailer reservoir and when storage pressure fell with brake application it triggered the relay valve, thereby admitting air to the trailer brakes from their nearby reservoir. This system had the disadvantage that no air could be supplied to the trailer reservoir during prolonged periods of downhill braking. Another example is the relay emergency valve, which is associated with modern tractive unit and trailer combinations. Apart from speeding up the application and release of the trailer brakes, it also has an emergency valve that allows automatic application of the trailer brakes should either the air pressure in the feed or emergency line to the trailer reservoir fall too low, or in the event of the trailer breaking loose from the tractive unit. For this reason it is known to American engineers as a breakaway valve.

Differential protection valve

Also known as an anti-compounding valve, its purpose is to prevent simultaneous application of both service air and secondary spring brakes, which otherwise could result in mechanical overstressing and damage to the actuating mechanism between the brake chambers and the brake shoes. This type of valve is installed after the relay valve associated with the hand control valve for the spring brake system, which will be described later.

The valve body is provided with inlet connections from the hand control and footbrake valves, and outlet connections to the ports of the spring brake chambers. In a typical construction the valve assembly comprises two concentric pistons, the outer one being spring loaded against a seating that controls the inlet port for service line air, and the inner one being spring loaded away from a seating that controls the inlet port for secondary line air. The latter arrangement is necessary to ensure that during normal driving a pressurized supply of air is made available to hold off the spring brakes (Figure 28.10). This secondary pressurized air also acts in conjunction with the spring loading on both pistons, so that the outer piston covers the service air inlet port.

The differential protection valve will be activated should the driver not only apply the service air brakes from the foot treadle, but also operate the hand control valve to release air from the secondary spring brakes and apply them as well. This unwanted application of the spring brakes is prevented because, as the pressure in the secondary line falls, the pressure in the service line overcomes the spring load on the inner piston so that it moves to cover its seating and prevents

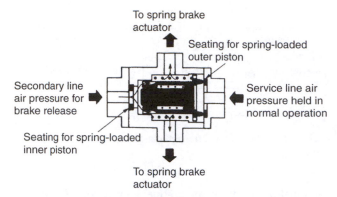

Figure 28.10   Schematic arrangement and operation of differential protection valve

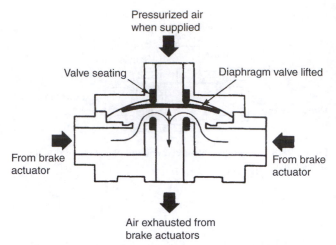

Figure 28.11   Schematic arrangement and operation of quick release valve

further loss of air through the secondary inlet port. Further increase in service line pressure then overcomes the spring load on the outer piston, which moves to uncover its seating so that service line air is admitted to the spring brake chambers. This supply of pressurized air from the service line thus replaces that exhausted from the secondary line, thereby ensuring that the spring brakes remain released during application of the service brakes. When the service brakes are released, the reduction in service line pressure allows first the outer piston to reseat and cover the service inlet port and next the inner piston to return and uncover the secondary inlet port. The secondary spring brakes will then be either applied or released according to driver operation of the hand control valve.

Quick release valves

These valves are installed adjacent to the wheel brake actuators, so that air can be rapidly evacuated from the actuators when the line pressure is released. This avoids the delay in response associated with air from all the actuators being evacuated simultaneously through long runs of air line, before reaching the atmosphere via the exhaust port of the control valve. Quick release valves can therefore perform two functions in an air brake system, since they can allow quicker release of the service air brakes and quicker application of the secondary spring brakes.

This type of valve is of simple construction, and typically comprises a valve body with two ports connecting to a pair of brake actuators and a third port that exhausts to the atmosphere. Another port is included in the cover of the valve body and connects to the brake control valve. Sandwiched between the valve body and cover is a flat rubber diaphragm that can seal against either an upper or a lower seating. When the service air brakes are applied, pressurized air is supplied to the port in the valve cover, which presses the diaphragm on to its lower seating to seal off the exhaust port. The incoming air now flexes the outer edge of the diaphragm and passes out of the ports connecting to the brake actuators. During release of the brakes the air pressure in the actuators will be greater than the reducing line pressure at the valve

cover port, so that the diaphragm is lifted away from its lower seating and allows the air from the actuators to escape rapidly to the atmosphere through the valve body exhaust port (Figure 28.11).

Load sensing valve

The purpose of fitting a load sensing valve is to ensure that the braking force generated at a particular axle is proportional to the load borne by that axle. With an air brake system the valve will therefore control the air pressure in the brake actuators for the regulated axle, according to the load on that axle and brake line pressure. The valve senses the extent of rear axle loading by being made responsive to deflection of the road springs. It is therefore mounted on the chassis frame and connects to the axle through a shock absorbing linkage, so that it can ignore the effects of spring deflection caused merely by road surface irregularities.

In one type of load sensing valve a balance beam is connected at one end to a control piston and at the other end to a balance piston, the fulcrum of the beam being afforded by a pair of slidably mounted rollers. These rollers are automatically repositioned along the beam by a control rod, which is responsive to variations in axle load. The control piston is spring loaded towards the balance beam and incorporates a combination inlet and exhaust valve, while the balance piston is also spring loaded towards the balance beam and engages its end through a lost-motion fork and pin arrangement.

When the driver applies the brakes, the line pressure received by the load sensing valve acts against the control piston to force it downwards. This movement of the control piston allows the combination inlet and exhaust valve to cover the exhaust port in the valve body and uncover the inlet passage in the piston. Pressurized air can then flow through the control piston to the brake actuators and also act against the top of the balance piston. The latter is likewise forced downwards and tends to raise the control piston owing

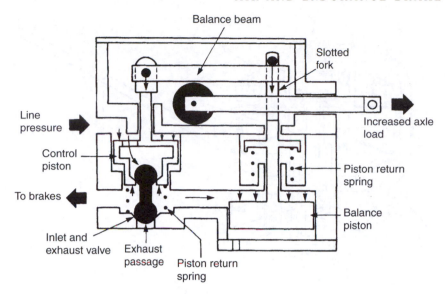

Figure 28.12 Schematic arrangement and operation of load sensing valve

to the see-saw effect of the balance beam. This upwards or return movement of the control piston continues until a point is reached where the combination valve covers both the inlet and exhaust passages (Figure 28.12). This point corresponds to a state of equilibrium between the control and balance pistons, with the air pressure in the brake actuators remaining constant. The brakes are therefore being held on with a force that is proportional not only to the effort being applied by the driver, but also to the load on the axle. Now if we increase the load on the regulated axle, the control rod of the load sensing valve will withdraw to reposition the fulcrum rollers closer to the connecting fork of the balance piston. It thus follows that when the brakes are applied, there will be a greater build-up of pressure in the actuators before it is held constant. This is because more pressure must be applied to the balance piston before a state of equilibrium is reached, owing to the less favourable leverage exerted by the balance piston on the beam.

In the Bendix type of load sensing valve the two-piece housing contains an upper assembly of sliding control piston, diaphragm and the familiar combination inlet and exhaust valve, which operates in conjunction with a lower assembly of a hollow sliding stem valve that is made responsive to suspension deflection and hence vehicle loading (Figure 28.13). There are two ingenious features associated with the operation of this type of load sensing valve. The first of these is an intermeshing arrangement of tapered radial fins for controlling the effective area of the piston diaphragm. One set of fins is fixed in the housing and the other set moves with the control piston, so that the effective area of the diaphragm and therefore the upthrust due to air pressure experienced by the piston depend upon the extent to which the fins of the latter lend support for the flexing diaphragm. Secondly, there is an air-pressure-controlled plunger that can temporarily clamp the ball-pin connection between the sliding stem valve and

its torsionally flexible operating linkage. This isolates the action of the valve against minor suspension disturbances during the process of braking, while avoiding unnecessary wear in the connection by allowing it greater freedom of movement when the brakes are released.

When a brake application is made, the line pressure received by the load sensing valve therefore establishes a positive connection with the operating linkage, via an external pipe and the stem valve plunger, and also forces down the control piston until its spring-loaded combination inlet and exhaust valve is unseated by contact with the mouth of the stem valve. This results in the exhaust passage through the stem valve being closed off and the inlet passage in the control piston being opened up, so that air under pressure is not only directed through the load sensing valve to the brake actuators, but also gains access to the underside of the control piston diaphragm. When the force acting upwards on the diaphragm becomes equal to that acting downwards on the piston, the latter achieves a state of balance and the combined inlet and exhaust valve assumes a lapped condition with the air pressure in the brake actuators remaining constant. The effect of the vehicle being heavily laden will be to increase the height to which the stem valve rises in the housing. This means that the control piston will also be working in a higher position and by the same token giving less support to its diaphragm. Since this will reduce its effective area closer to that of the control piston itself, there needs to be a greater air pressure acting on the diaphragm before the opposing forces on the piston become balanced and the combination inlet and exhaust valve assumes a lapped condition. In other words, the output pressure to the brake actuators is now less reduced in relation to the input line pressure and thus corresponds with the heavier loading of the vehicle. When the brakes are released, the air pressure in the brake actuator line will lift the control piston and unseat the combination inlet and exhaust

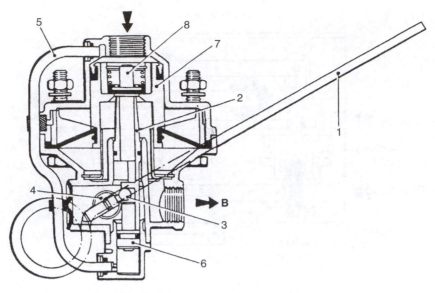

Figure 28.13   Section of Bendix Westinghouse load sensing valve (*Seddon Atkinson*)

1 operating lever
2 stem valve
3 ball-pin
4 operating shaft

5 external pipe
6 plunger
7 piston
8 inlet/exhaust valve

valve from the mouth of the stem valve, which allows the air to escape down through its exhaust passage and thence to the atmosphere, via a rubber check valve in the unit body.

Introduction to electronically-controlled air brakes

A new development in air brake system control for heavy vehicles is that in which a quicker acting electronic control is superimposed on the traditional pneumatic control, so that the latter essentially performs a safety backup function, although there are long-term development aims to make this feature redundant if legislation permits. Systems of this type are now designated as 'electronic braking systems' (EBS) or, in popular jargon, 'braking-by-wire systems'.

In basic principle the foot-operated brake valves assembly is also equipped with a potentiometer device, so that the extent to which the driver applies the brakes can be signalled to an electronic control unit (ECU). This then issues appropriate command signals to electro-pneumatic relay valves, which serve their nearby brake actuators. The function of these relay valves is first to isolate the slower acting pneumatic back-up control, second to admit air more rapidly at a regulated pressure to the brake actuators, and third to release this pressure and restore the back-up control when braking is no longer required.

By virtue of the brakes responding more quickly to the driver pressing the pedal, it will be evident that shorter stopping distances and hence improved vehicle safety is the primary objective of an electronic braking system. This type of control system can also be extended to confer other advantages, because the addition of load sensors can make it responsive to rear axle loading and load transfer during braking, so that overbraking on any particular axle can be avoided to the benefit of reduced liner or pad wear. Imbalance between

tractor and trailer decelerations can likewise be sensed at the fifth-wheel coupling of an articulated vehicle or at the coupling arrangements of a drawbar combination and monitored by the ECU, which then issues command signals to the trailer electro-pneumatic relay valve that regulates the air pressure accordingly. Electronic braking systems are, of course, compatible with established anti-lock brake and traction control systems.

28.4   SYSTEM ACTUATION

Air brake actuators

These are also known as brake chambers. One is mounted externally to each wheel brake. Through the medium of a diaphragm element they convert the energy stored in the compressed air into the mechanical force and movement required to actuate the brake shoes. Owing to their bulky nature they cannot be accommodated within the brake drums, and therefore act upon either lever and cam or wedge and tappet shoe expanders instead of directly on the brake shoes.

Reference has been made in Section 27.1 to the S-cam and rollers type of brake expander and its use on air-braked heavy vehicles. A further development of the fixed-cam brake takes the form of a cam and struts type of expander mechanism. Here the shoe tip rollers associated with S-cam operation (Figure 27.4) are replaced by ball-ended struts and sliding tappets. Each strut or push-rod locates at one end in a spherical recess within the cam and at the other end in a spherical recess inside its sliding tappet (Figure 28.14). Rotation of the camshaft therefore causes the struts to separate the tappets and expand the shoe tips. The advantages claimed for this particular construction are that it provides a sealed and lubricated enclosure for the expander mechanism, and that the shoe tip

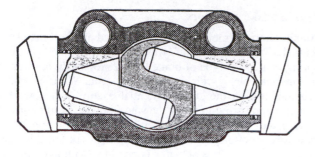

Figure 28.14 Air brake cam and struts expander (*Seddon Atkinson*)

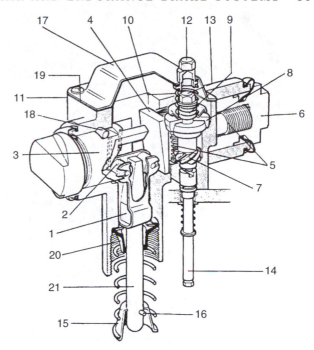

Figure 28.15 Air brake wedge expander (*Lucas Girling*)

1 wedge	12 manual override socket
2 roller	13 manual override pinion head
3 plain tappet	14 manual override stem
4 roller tappet	15 spring retainer
5 adjuster screw	16 pin/circlip
6 input tappet	17 expander cover
7 adjuster pinion	18 tappet stop pin
8 drive cone	19 gasket
9 overload spring	20 seal
10 cone spring	21 wedge push-rod
11 housing	

forces are always constrained to act in-line with the sliding tappets, whereas with an S-cam and rollers the angularity of the shoe tip forces is such that their action is not independent of the direction of drum rotation.

Reference has also been made in Section 27.1 to the continuing use of wedge-type brake expanders for commercial vehicles. Those operated by compressed air for heavy vehicles were originally developed in America during the mid 1960s. The advantages of replacing the lever and cam by a wedge and tappets were perceived as reducing unsprung weight by virtue of eliminating the camshaft and slack adjuster; greater structural rigidity resulting from the brake actuator directly attacking the wedge; shorter actuating stroke with smaller air chamber, thereby minimizing air consumption and promoting quicker application and release; ready accommodation of both air brake and spring brake actuators; and allowing the use of dual-circuit air systems by incorporating twin wedge expanders in each brake. Hardly a disadvantage but more an essential requirement for an air-operated wedge brake is an in-built automatic adjuster, which ensures that the wedge travel between the sliding tappets never becomes excessive or runs out.

Various shoe arrangements may be employed with air brake wedge expanders. A single wedge expander may be used in conjunction with either a simple arrangement of leading and trailing shoes to give a floating-cam characteristic or, via a bell-crank and struts linkage, an arrangement that provides two leading shoes in the forwards direction and leading and trailing shoes in reverse, while the use of twin wedge expanders confers a two leading shoe action in both the forwards and reverse directions. Here it should be noted that in some countries these different shoe arrangements are better known as simplex, duplex and duo-duplex brakes respectively. A modern example of the latter type of air wedge brake is the Girling Twinstop. In this design the air brake actuators are stem mounted from the wedge expanders, their mounting tubes being screwed directly into the wedge expander bodies. Each expander incorporates an integral push-rod and wedge with the usual rollers interposed between the wedge and the inclined faces of its sliding tappets (Figure 28.15). The wedge also provides a positive caged guidance for the rollers, so as properly to control their movements during brake actuation. An automatic adjustment facility with a manual override is embodied in one of the sliding tappets in each expander, the opposing sliding tappet remaining solid.

During normal brake application the wedge is urged inwards, via its push-rod, by the air brake actuator. It therefore imposes a separating force on the sliding tappets through the interposed rollers. Since there are two wedge expanders, each shoe receives movement at both ends and is bodily brought into contact with the rotating drum. Once this occurs both shoes move round slightly, until the trailing end of each is arrested by its adjacent sliding tappet, which has been forced back against an abutment in the expander housing. The opposing tappets then continue to transmit a shoe tip force to the leading ends of the shoes, so that a two leading shoe action is obtained. This same sequence of events does, of course, occur in the opposite sense with reverse rotation of the drum and again confers a two leading shoe action on the brake.

Of simple construction, a single-diaphragm brake actuator comprises a reinforced rubber diaphragm that is sandwiched between a two-piece pressed steel casing, which is held together by a clamping ring and provided with suitable mounting studs. The diaphragm is furnished with a push-rod connection to the shoe expander mechanism and is spring returned to the brakes released position (Figure 28.16).

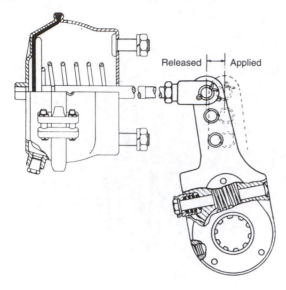

Figure 28.16   Air brake actuator and slack adjuster (*Bendix Westinghouse*)

In some installations where a rear brake actuator operates in conjunction with a mechanical handbrake, the diaphragm utilizes a pull-rod connection to the shoe expander mechanism.

In operation, the pressurized air enters the inlet port of the brake actuator and compels the diaphragm and pushrod assembly to move against the return spring and apply the brakes. The apply force developed is proportional to the effective area of the diaphragm and the pressure of the air admitted to the actuator. As the brakes are released and the air pressure reduces, the diaphragm and push-rod assembly is returned to its original position by the actuator return spring and the pull-off springs for the brake shoes.

During the mid 1960s the legally permitted gross weights of heavy vehicles were increased, and this led to various improvements being incorporated in their air brake systems to meet more stringent regulations. Included among these improvements was the use in some systems of double-diaphragm brake actuators for improved safety. With this arrangement the two diaphragms are separated at their clamping edges by a spacer ring, so that pressurized air can be admitted through a rear inlet port to the secondary diaphragm and a side inlet port to the service diaphragm. The secondary diaphragm has a forward thickening of its central portion to complement the thickness of the spacer ring, and it is also provided with a small-diameter rear lip seal that covers the rear inlet port. In the event of the service diaphragm leaking then clearly the brake can still be operated by using the secondary system, since this part of the chamber remains self-contained. The situation is a little different with a leaking secondary diaphragm, because in this case the return spring or pressurized air acting on the central portion of the diaphragm forces its sealing lip to cover the rear inlet port and prevents the escape of air into the secondary system. Since this then provides the equivalent of a self-contained chamber, normal operation of the service diaphragm is restored.

Spring brake actuators

Another improvement that was introduced into air brake systems during the mid 1960s was the secondary and parking spring brake, this type of brake having previously been used in American heavy-vehicle braking systems. A spring brake actuator utilizes the stored potential energy of a powerful compression spring to apply the wheel brake. During normal driving the spring must therefore be held in a compressed state to ensure that the brake remains released. For this purpose the spring brake actuator is provided with a supply of pressurized air via a hand control valve, the latter being described later. Since air pressure is reduced to apply a spring brake actuator and increased to apply an air brake actuator, the air brake engineer distinguishes between the two forms of air supply by referring to them as inverse air and upright air respectively.

Although air pressure is released from a spring brake for brake application, there still remains sufficient compression in the expanded spring to exert the required force for actuation of the brake. A spring brake may be either applied gradually by the hand control valve for the purpose of secondary or emergency braking, or applied fully to hold the brakes on for parking, thereby replacing the once conventional handbrake that had direct mechanical linkage to the wheel brakes. The spring brake also possesses an important fail-safe feature, because the brakes will be automatically applied should a failure occur in the air pressure circuit for the secondary and parking brake functions.

A spring brake is typically mounted in tandem with a conventional single-diaphragm service air brake, each brake operating independently of the other (Figure 28.17). The service brake chamber is therefore supplied with upright air from the footbrake valve and the spring brake chamber with inverse air from the hand control valve; simultaneous application of service and spring brakes is prevented by the differential protection valve described earlier. In construction the spring brake chamber contains either a diaphragm or a sliding piston with air seal, so that when these are subject to increasing air pressure the powerful spring is compressed to release the brake. When air pressure is reduced, either the diaphragm or the piston moves in the opposite direction under the influence of spring load. Its central stem then bears against the back of the service brake diaphragm, thereby actuating the brake through the usual push-rod connection. Some form of screw release mechanism is provided for manually relieving the brake assemblies of spring brake load, which allows the brake assemblies to be safely serviced, or if necessary the vehicle to be moved in the absence of air pressure.

Lock actuators

Introduced as an alternative to the spring brake, the lock actuator was intended to simplify the parking brake function of heavy vehicles. When signalled to do so from a hand control valve, the lock actuator will arrest the return movement of the guided stem shaft of an air brake diaphragm to hold the brakes applied. Similar to the spring brake, air pressure is released from a lock actuator to hold the brake applied (Figure 28.18).

Compressed air pressure ▨

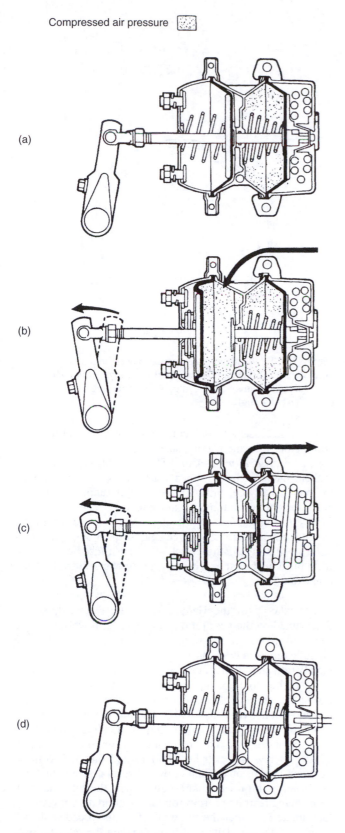

(a)

(b)

(c)

(d)

Figure 28.17 Spring brake actuator operation (*Seddon Atkinson*): (a) normal driving (b) service braking (c) secondary/park braking (d) mechanical release

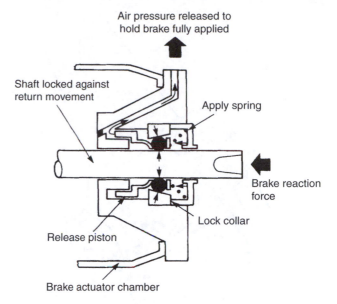

Air pressure released to hold brake fully applied

Shaft locked against return movement

Apply spring

Brake reaction force

Lock collar

Release piston

Brake actuator chamber

Figure 28.18 Schematic arrangement and operation of lock actuator

With a lock actuator the stem shaft of the air brake diaphragm is encircled by hourglass-shaped rollers, which can be either spring loaded against the conical bore of a lock collar so their wedging effect prevents return movement of the stem shaft, or forced away from the conical bore of the lock collar by air pressure acting against a release piston and thereby restoring free movement of the stem shaft. In a typical installation for a double-diaphragm air brake, the lock actuator is released by first applying the secondary brake to allow the rollers to unwedge themselves, following which they remain free on the shaft as long as air pressure is maintained against their release piston. The secondary brake is then, of course, released to drive away the vehicle.

Air brake adjusters

Maintaining correct adjustment in an air brake system is a matter of considerable importance, because it is related to what is sometimes termed brake force build-up time, during which the braking force at the wheels is building up to a maximum value. If adjustment is neglected and the clearance between the shoe linings and the drum becomes excessive, the amount of air that must pass into the actuator chamber to apply the brakes is necessarily greater, which therefore lengthens the brake force buildup time and increases the stopping distance.

The adjustment of cam-operated air brakes is accomplished by what is known as a slack adjuster, the mechanism of which is embodied in the lever arm used between each brake actuator push-rod and expander cam for the shoes. By means of a lockable worm and wheel mechanism the rotational position of the expander cam may be manually adjusted relative to that of the lever arm (Figure 28.16). Hence the most favourable operating geometry for the actuating linkage may be preserved throughout the life of the shoe linings, and by the same token the actuator diaphragm

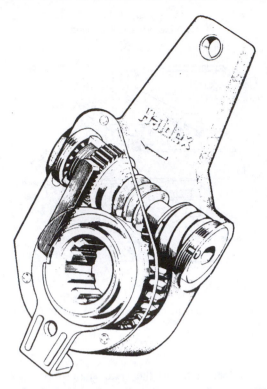

Figure 28.19 Cut-away view showing internal mechanism of Haldex automatic slack adjuster (*Haldex Division, Garphyttan Ltd*)

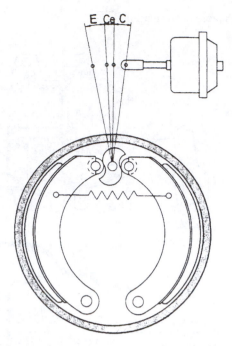

Figure 28.20 Identifying the angular movements of an air brake slack adjuster (*Haldex Division, Garphyttan Ltd*) (C) – Clearance angle (Ce) – Excess clearance angle (E) – Elasticity angle

is never over-extended. During normal brake operation the entire lever arm cum slack adjuster does, of course, rotate bodily with the brake camshaft.

Automatic slack adjusters have increased in popularity since the late 1970s, because they can maintain almost constant the running clearance between the shoe linings and their drums, whilst making allowance for thermal expansion of the drums during heavy braking by not taking up all the perceived excess slack. In more recent years they have become especially relevant to the responsive operation of anti-lock air brake systems, and indeed since late 1994 automatic slack adjusters have been a mandatory fitment under EC legislation for all newly registered heavy vehicles and trailers. Automatic slack adjusters basically incorporate a lost-motion ratchet and pinion drive to a spring-loaded worm shaft and friction clutch arrangement. Since the ratchet reacts against a fixed link that serves as a datum point, any excess clearance due to lining wear is taken up on the return stroke of the brake actuator. A visual indication of lining wear may also be incorporated in the mechanism of these slack adjusters. A long-established and widely used type of automatic slack adjuster for cam-operated air brakes is that known as the Haldex (Figure 28.19), which confers almost constant and correct clearance between the brake linings and the drum. To accomplish this the action of the slack adjuster is able to distinguish between three phases of angular movement (Figure 28.20) as follows:

Clearance angle (C)   This corresponds to the normal running clearance between the brake linings and the drum, which is necessary to prevent any brake drag and to allow cooling between the friction surfaces.

Excess clearance angle (Ce)   This is additional to the normal clearance angle and appears as the brake linings continue to wear in service, prior to the normal running clearance being restored by automatic adjustment.

Elasticity angle (E)   this occurs as the brake shoes are expanded with increasing force against the brake drum temporarily dilating it, the effect of which must be excluded from the adjustment process to prevent over-adjustment and brake drag.

The take-up of the Haldex automatic slack adjuster is determined by the gear ratio of the lever arm worm and the camshaft wheel. It is, of course, most important that the manufacturer's advice be sought in relation to the correct installation and maintenance of an automatic slack adjuster.

As earlier mentioned, automatic shoe adjustment is an essential feature of the air-operated wedge brake. For modern duo-duplex versions it is accomplished by incorporating a jacking screw device within one of the sliding tappets in each expander unit. This mechanism takes the form of an adjuster screw that can be either turned directly within a threaded tappet, or translated by a threaded sleeve that can be turned within a counterbored tappet, the latter in both cases being restrained from rotation. Automatic provision is then made for either the screw itself, or its threaded sleeve, to receive a small degree of turning when the return movement of the tappet increases owing to wear of the shoe lining. This results in the shoe tip being jacked further away from the end of the sliding tappet, so that upon release of the brake the correct shoe to drum clearance is restored. The automatic means of turning either the adjuster screw, or its threaded sleeve, basically relies upon an additional helical connection being established between the sliding tappet and its adjuster, combined with a lost-motion pawl and ratchet or friction clutch control system, so that excess return movement of the sliding tappet causes a turning moment to be exerted on either the screw or its sleeve. An example of the former system is shown in Figure 28.15, where the required helical connection is established between a skew gear that turns the adjuster screw and a meshing adjuster pinion, which also provides a manual override adjustment. When the outward movement of the sliding tappet exceeds the predetermined backlash between the gear and pinion, the reaction from their helical teeth lifts and declutches the pinion against its spring loading, thereby allowing it freedom to turn; but on the return movement of the tappet the pinion drops and is held clutched, so that the helical sliding between their teeth becomes sufficient to turn the gear wheel and hence the adjuster screw. The manual override adjusters serve the three-fold purpose of initially setting the shoes to drum clearance, retracting the shoes to facilitate drum removal and retracting the expanders when fitting new shoes.

28.5 HAND-OPERATED BRAKE VALVES AND OTHER EQUIPMENT

Hand-operated brake valves

There are several types of hand-operated brake valves used in heavy-vehicle air brake systems, their detail design varying in accordance with system requirements, but usually they are associated with the secondary and parking functions of spring brake actuators. Although the secondary brake system is only used if there is a failure in the foot-operated service brake system, it is still desirable that the hand-operated brake valve should exert a progressive controlling action over the spring brakes. In the language of the air brake engineer, this means that such a valve must provide a graduable inverse air pressure, since in applying the spring brakes for secondary braking we are concerned with gradually reducing air pressure. For the parking function all the air pressure is exhausted from the spring brakes, the handle of the valve being made self-locking in the parked position.

In construction a hand-operated brake valve typically comprises a control piston that is acted upon by a lower return spring and an upper control spring. The amount of compression in the latter is varied by a rotary cam arrangement that is positioned by movement of the valve handle, which has a friction-damped motion for sensitive control. When the spring brakes are released the control spring pressure is greatest and it is least when they are applied. Spring loaded against the inlet port of the valve body and towards the exhaust port of the control piston is again the familiar combination inlet and exhaust valve (Figure 28.21). With the hand lever in the off position, the increase in control spring pressure causes the piston to bear down on the combination inlet and exhaust valve, which uncovers the inlet port and covers the exhaust port. This allows air under pressure from the secondary/park reservoir to pass through the valve and on to the spring brake actuators. Their delivery air pressure

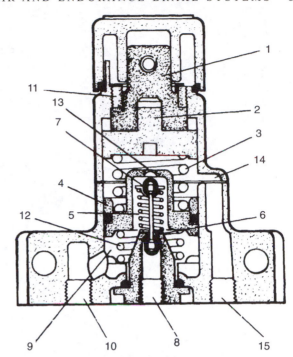

Figure 28.21 Section of Bendix Westinghouse hand control valve (*Seddon Atinson*)

1 cam follower	9 valve chamber
2 cam	10 delivery port
3 spring	11 adjusting ring
4 piston	12 piston return spring
5 inlet/exhaust valve	13 piston vent
6 inlet	14 valve chamber
7 exhaust	15 exhaust port
8 inlet port	

is governed by the preset load exerted by the control spring, this pressure being sufficient to hold off the brakes for normal driving.

When the hand lever is moved from the off position (spring brakes released) and held towards the secondary/park position (spring brakes being applied), the rotary cam allows the control springs to extend and thus reduce the load on the piston. This permits the piston to rise under the influence of the air pressure and return spring force beneath it, until the spring-loaded combination valve covers the body inlet port and uncovers the piston exhaust port. Excess air then escapes through the piston into the chamber above, and from there is released to the atmosphere via the exhaust port of the valve body itself. The reduction in air pressure and therefore application of the spring brakes continues until the control spring moves the piston down sufficiently to cover both inlet and exhaust ports, following which the forces above and below the valve become equal and it assumes a balanced or lapped condition. A constant braking effect is then obtained until there is further movement of the hand lever. The operating characteristics of this type of valve are therefore similar to those of a foot-operated brake valve, but without the same sense of feel, although this feature is present in some designs of hand-operated brake valve.

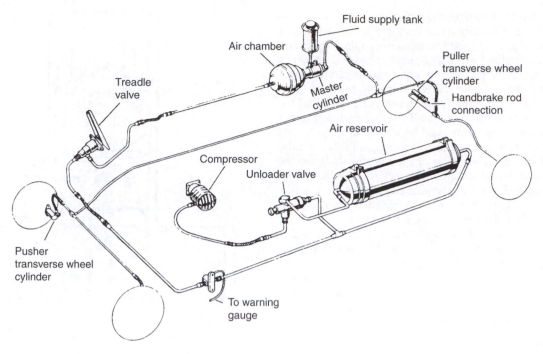

Figure 28.22 Basic layout of an air-over-hydraulic braking system (*Girling*)

Trailer connecting hoses

Since the early 1960s, precoiled nylon air lines, known as 'Susie' hoses, have generally replaced rubber air lines between tractive units and either their semi-trailers or drawbar trailers. The quick release end couplings for the hoses may be of the earlier male and female bayonet type or the later internationally used palm type, these also being known aptly as gladhand couplings in America because of their resemblance to clasped hands. Automatic shut-off valves may be incorporated in both types of coupling, so that air pressure is retained when the feed line to the trailer reservoir is disconnected. The end couplings of the hoses are designed so that they cannot be wrongly connected and the hoses themselves are colour coded for ready identification.

The colours normally used are as follows:

Yellow Service or control line, which supplies a signal air pressure to the relay valve on the trailer for normal operation of the trailer brakes.
Blue Secondary or auxiliary line, which supplies direct air pressure for secondary operation of the trailer brakes in the event of the service brake failing.
Red Feed or emergency line, which supplies air under pressure to the trailer reservoir through a relay emergency valve.
Caution It is very important that the correct procedure for safely connecting and disconnecting these brake line hoses is carefully followed, and this forms part of the driver training for goods vehicles. Reference should be made to The Official Goods Vehicle Driving Manual, published by The Stationery Office for the Driving Standards Agency.

Air-over-hydraulic brakes

For heavy vehicles in the medium weight range, hydraulic actuation of the brakes may be employed in conjunction with a compressed air source of power operation. Systems of this type are therefore known as air-over-hydraulic brakes, and can more readily generate the higher operating forces at the hydraulic wheel cylinders, which otherwise would demand unacceptably large vacuum servo units. Furthermore, the smaller compressed air chamber together with hydraulic master cylinder can be located wherever convenient on the chassis frame.

In a typical air-over-hydraulic brake system (Figure 28.22) an engine-driven compressor supplies air that is stored in the usual reservoir. When the brake pedal or treadle is depressed to operate the control valve, pressure builds up in the air chamber and acts against a diaphragm. Movement of the diaphragm and its associated push-rod then directly forces the piston of the hydraulic master cylinder to displace fluid at equal pressure to the wheel cylinders and thus apply the brakes.

28.6 AIR DISC BRAKES

General background

Following the successful adoption of hydraulically operated disc brakes for passenger cars that began in the mid 1950s, uprated versions of this type of brake were later developed and initially installed on public service vehicles, a notable example of their use being on the motorway coaches of the Midland Red Bus Company in Britain in the early 1960s. Operator experience at the time had tended to show that, so far as public service vehicles were concerned, the disc brake was best suited for long-distance hauls where brake temperatures were lower than those encountered in city running. Hence it was soon recognized that the key to successful operation of disc brakes on commercial vehicles was ensuring adequate dissipation of heat from their discs, otherwise both pad and disc life would become unacceptably short. To meet this requirement in practice meant that the disc had to be of ventilated construction, and designed with the maximum outer and minimum inner diameters together with the maximum thickness which could be installed within the existing wheel space, there being greater difficulties posed by the axle arrangements of commercial vehicles than had previously been the case for passenger cars. It also meant that further research and development had to be undertaken on disc pad materials to provide an increased life. For these reasons it was not until the early 1970s, following a renewed interest in their development in America, that manufacturers elsewhere began to introduce improved designs of disc brake which, together with more suitable friction materials, has led to their gradual acceptance on light to medium commercial vehicles.

The introduction of air-operated disc brakes for heavy vehicles is of more recent origin, because in their case additional design requirements had to be taken into account, not the least of these being the need for a force multiplying mechanism between the disc pads and the air brake actuator chamber. This requirement arises from the relatively low operating pressures of an air brake system, as compared with those of a hydraulic one, which would otherwise demand unrealistic increases in the size of the caliper piston and the amount of air to be moved. That is, the caliper could not accommodate the size of piston required and the brake force build-up time would be unacceptable. Since the presence of a force multiplying mechanism implies a large increase in brake actuator stroke for a small wear reduction in pad thickness, there is an essential requirement for an automatic adjuster that must necessarily be of the mechanical type.

However, the modern air disc brake can offer important advantages over those of an equivalent drum brake, which in general terms are greater structural rigidity, more consistent brake torque, better fade resistance and easier access for maintenance.

Basic construction of air disc brakes

Air disc brakes for heavy vehicles have a basically similar clamping action to later versions of the floating caliper and cylinder, or first-type, hydraulic disc brake used on passenger cars (Section 27.2). However, the construction of the sliding caliper is modified to incorporate a more robust straddle-mounted reaction beam or bridge, instead of a simple claw, to provide an abutment for the outboard pad. The reaction beam is extended chordally across the face of the disc and straddles it at each end, where it unites with the caliper housing (Figure 28.23a). This structural change to the caliper ensures that the outboard disc pad receives a centrally applied clamping load. The backing plates for the outboard and inboard disc pads therefore reside against the

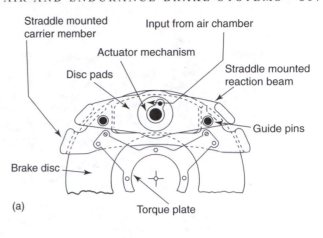

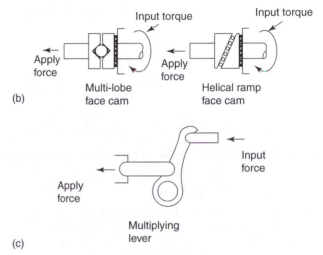

Figure 28.23 Basic arrangement of an air-operated disc brake: (a) straddle-mounted caliper (b) and (c) force multiplying actuators

caliper reaction beam or bridge and an opposite load spreader plate, which is urged towards the disc by the caliper actuating mechanism when the brake is applied. Equality of clamping loads is achieved by virtue of the friction drag forces on both disc pads being reacted against locating lugs, which are formed integral with a straddle-mounted carrier member (Figure 28.24). This is bolted either directly or indirectly to an axle flange, an intermediate torque reaction plate being used in the latter case. The carrier member also incorporates the sealed low-friction guide or slide pins, which support the freely floating caliper assembly (Figure 28.24). Quadruple slide pins are employed in the established Haldex air disc brake, these minimize any tendency for brake drag while improving the durability of the sliding components. The avoidance of brake drag with its adverse effects on both fuel economy and brake wear, is an important consideration to the transport engineer.

Actuators and automatic adjusters for air disc brakes

The actuator housing for the force multiplying mechanism is either integrated with, or separately bolted to, the inboard

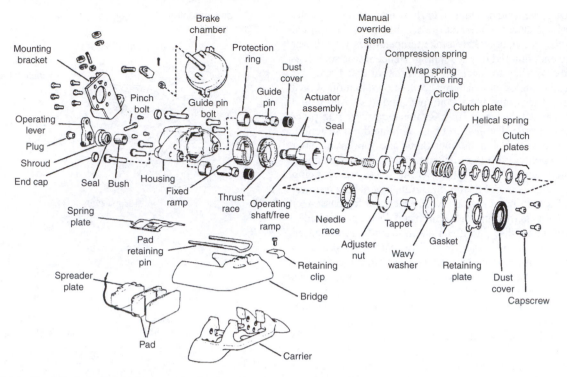

Figure 28.24 Exploded view of Girling reaction beam caliper brake (*Lucas Industries*)

side of the floating caliper and also provides a mounting for the air chamber (Figures 28.24 and 28.25). This is positioned either parallel with, or normal to, the disc face according to the type of force multiplying mechanism used. The former position may be likened to that used for operating rotary cam air drum brakes (Section 28.4), and similarly requires the intervention of a clevis joint where the air chamber push-rod connects to the brake operating lever, the cross-shaft of which transmits a torque input to the force multiplying mechanism (Figure 28.24). This therefore serves not only to convert the partial rotation of the input shaft into a linear motion for the caliper tappet (corresponding to the piston in a hydraulic disc brake), but also to multiply further the actuating force supplied by the air chamber, so that the clamping load imposed on the disc pads generates the required brake torque.

The force multiplying mechanism itself may comprise a pair of opposing face cams with either multi-lobes, or helical ramps, and rolling bearings interposed between their thrust surfaces (Figure 28.23b). An equivalent mechanism used in an American design of brake features a multi-start screw thread on the cross-shaft that engages an externally splined sliding nut. Hence, when the cross-shaft is turned relative to the rotationally fixed cam of the pair, or the sliding nut, the resulting linear movement creates an end-thrust on the load spreader plate for the inboard disc pad. Then by virtue of the caliper being free to slide, this end thrust is also transmitted to the outboard disc pad, so that both pads engage the disc for braking.

The alternative construction for air disc brakes, where the air chamber is mounted normal to the disc with a lever type force multiplying mechanism (Figure 28.23c) may similarly be compared to air drum brakes with wedge instead of cam

expanders for the shoes (Section 28.4). Likewise, the more direct actuation of the disc pads can make for a compact design of brake with a reduction in unsprung weight and a minimum of wearing parts. A modern example of this type of construction is the previously mentioned Haldex DB20 air disc brake (Figure 28.25). Before describing the force multiplying mechanism of this heavy vehicle brake, it may be of interest to quote a few items of dimensional data to put its size into perspective, as follows:

Road wheel size	22.5 in
External diameter of brake disc	430 mm (16.9 in)
Thickness of brake disc (new)	45 mm (1.8 in)
Effective radius of brake disc	172.6 mm (6.8 in)
Swept area of brake disc	1808 cm^2 (280 in^2)
Lining area of each pad (2)	187 cm^2 (29 in^2)

The air chamber is offset above the centre-line of the disc pads, so that its spherically-ended push-rod can engage an upstanding lever. This acts eccentrically upon a wide cross bar, to confer a 15.8:1 mechanical advantage for the apply force at the disc pads. To reduce internal friction partially caged needle roller bearings are used at the pivot points for the straddle-mounted lever and its associated cross bar (Figure 28.25). The force multiplying mechanism is designed to accept a maximum brake chamber force of 13.9 kN (3127 lbf). A spring brake chamber is, of course, used in cases where the disc brake has a parking function.

Brake application (Figure 28.26) The upstanding lever (44) is actuated by the air pressure in the brake chamber (25/26). Since the external and internal radii of the lever (44)

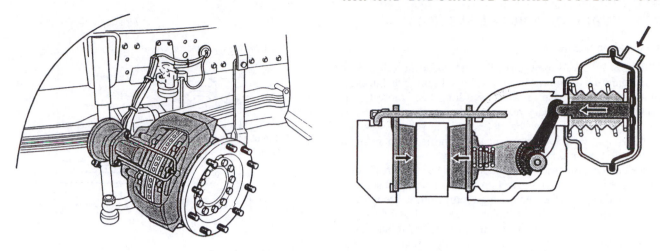

Figure 28.25 Installation and general arrangement of Haldex DB20 air-operated disc brake (*Haldex Brake Products AB, Sweden*)

do not have a common centre, their eccentricity means that the cross bar (41) is forced to move axially in the direction of the brake disc (A). This force is then transmitted from the cross bar (41), which also incorporates twin adjusting screws for setting the predetermined clearance between pads and disc, to the caliper tappets (28) for the inner pad (5). Once this pad (5) comes into contact with the brake disc (A), the caliper (2) is moved on its slide pins so that the outer pad (3) likewise comes into contact with the brake disc.

Brake release (Figure 28.26) The return spring (38) forces the cross bar (41) back into its rest position, so that the design clearance between the pads (5) and the brake disk (A) is restored.

In principle there are two types of automatic adjuster, which may be classified as 'stroke sensing' and 'clearance sensing'. The former type ignores the clearance existing between the friction elements, and adjusts the brake when the air chamber stroke exceeds a predetermined limit. In contrast, the latter type adjusts the brake when the actual clearance between the friction elements exceeds a predetermined limit. Hence, it follows that the possibility of the brakes being over-, or under-, adjusted is less likely to occur where a clearance sensing system is used.

The Haldex DB20 air disc brake operates according to the clearance sensing principle. Similar to identifying the angular movements of their automatic slack adjuster for air drum brakes (Figure 28.20), the braking sequence is divided into three phases (Figure 28.26). Namely, Design clearance 'C', Excess clearance 'Ce' (which is to be adjusted out) and Elasticity 'E'. Basically, the housing of the adjuster (62) is held in position against the internal radius of the upstanding lever (44) by a guide pin (47). This compels the adjuster housing (62) to follow the movement of the lever (44). The rotational motion is transferred from housing (62) to adjustment spring (63), which in turn, after passing the design clearance, transmits the motion to the companion sleeve (65), friction spring (66) and hub (67). If excess clearance is present, the rotation of the hub actuates a synchronized bevel gear drive for the twin adjustment screws incorporated in the

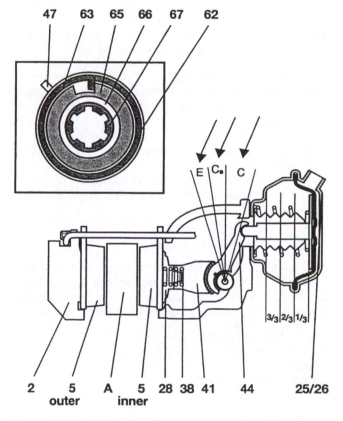

Figure 28.26 Application, release and automatic adjustment of Haldex DB20 disc brake (*Haldex Brake Products AB, Sweden*)

cross bar (41), which then restore the correct clearance between the pads and disc.

Caution It is, vitally important to consult the brake and vehicle manufacturers' instructions when servicing the air disc brakes of heavy vehicles, which should only be performed by trained personnel.

28.7 ENDURANCE BRAKE SYSTEMS

General background

This is a relatively new term that has been introduced by European legislation into heavy vehicle braking technology, and embraces the use of what have long been more familiarly known as 'auxiliary retarders'. An auxiliary retarder as its name suggests may be defined as a vehicle-slowing, rather than a vehicle-stopping, device and is not therefore a substitute for the service braking system.

Reference has earlier been made to the heat fade resistance of brake friction materials (Section 27.3). In general this is of a high standard, but if extremely high temperatures are reached during prolonged spells of braking then the loss of friction can seriously reduce the effectiveness of the brakes and result in brake fade. This can pose very real safety problems in the case of heavy vehicles operating under exceptionally hilly conditions and not least those carrying hazardous goods, since the diesel engine lacks intake manifold depression to create a retarding effect on the overrun. In any event the additional work now required of the service brakes has been heightened still further by changing operating conditions. The ZF Company of Germany have identified these as follows:

Higher vehicle gross weights
Cargo volume maximized by fitting smaller diameter wheels with smaller braking friction areas, thus offering reduced service braking efficiency
Increased driveline efficiency
Improved Cd-value and, therefore, reduced aerodynamic drag (Section 30.8)
Better tyres affording reduced rolling resistance
Greater traffic volumes and irregular traffic flows have led to more frequent braking

Types of auxiliary brake retarder

Various types of auxiliary retarder for augmenting the friction brakes of heavy vehicles have therefore been devised, beginning with the Swiss-designed Oetiker exhaust brake that first appeared in 1925, and they may be classified as follows:

Exhaust brake retarder
Compression brake retarder
Hydraulic retarder
Electromagnetic retarder
Friction retarder

In modern practice auxiliary retarders generally have to be compatible with anti-lock air brake systems, their retarding effect being interrupted when ABS is in operation.

Exhaust brake retarder

The exhaust brake represents a widely used, relatively inexpensive and compact form of retarder. It essentially comprises a shut-off valve in the exhaust system (Figure 28.27) which enables the engine of an overruning vehicle to act as a low-pressure air compressor, so that the work done by the pistons in compressing the air exerts an additional retarding

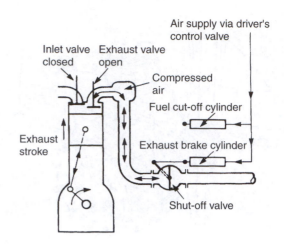

Figure 28.27 Schematic arrangement of an exhaust brake retarder

torque on the crankshaft and hence a braking effect on the transmission and wheels.

Rotary butterfly or sliding gate types of shut-off valve are used in the exhaust brake and must be carefully designed to avoid seizure, and also not to impede the normal flow of exhaust gases when in the open position. They are installed in the exhaust system between the manifold and silencer or, in the case of turbocharged engines, between the manifold and turbocharger. The shut-off valve may be operated either manually or automatically and in both cases provision must be made to cut off the fuel injection before the exhaust brake is applied, so that pure air only is compressed in the exhaust system. This occurs on what would normally be the exhaust strokes of the engine pistons and continues to the point when the inlet valves begin to open. These then allow the air compressed in the exhaust system at about 280–350 kN/m^2 (40–50 lbf/in^2) to escape via the cylinders into the intake manifold and thus equalize pressure before the cycle is repeated. An outstanding advantage of this type of retarder is that changing down to a lower gear will multiply the retarding torque imposed on the road wheels, because this increases with engine speed.

Apart from an increases in the life of the brake linings and drums that is to be expected from the use of any retarder, other advantages usually claimed for exhaust brake operation are those of reducing the danger of engine overspeeding on the overrun, and better maintaining engine temperature under these conditions because in compressing the air the engine is converting mechanical energy into heat. Possible disadvantages are related to any leakages developing in the exhaust brake installation and that maximum retarding torques are limited by engine displacement. A further consideration is that pulsations may be created in the engine intake system, which arise from a combination of cylinder pressures and intake valves opening during exhaust braking and can induce dirt to migrate through the air cleaner element, particularly if this is of the dry type. The use of a suitably tuned wave suppressor installed between the engine and air cleaner may therefore be required to dampen these pulsations,

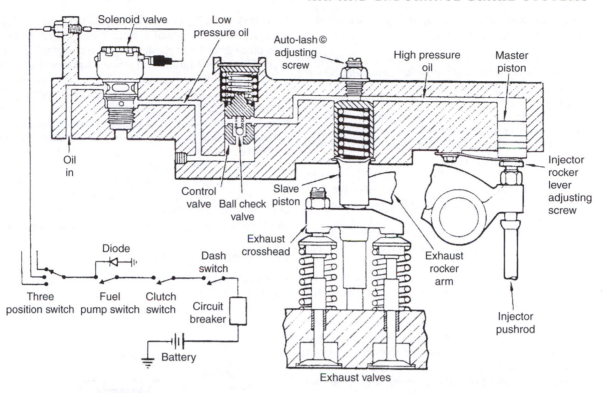

Figure 28.28 Schematic arrangement of Jacobs Engine Brake (*Jacobs Europe*)

this normally being a decision that is taken by the engine manufacturer.

Compression brake retarder

The engine itself has to receive some modification for installing this type of auxiliary retarder, which when operated, temporarily converts the power-producing diesel engine into a power-absorbing air compressor. A widely used retarder of this type is the Jacobs Engine Brake, which was introduced in 1961 and is universally known to transport engineers as the 'Jake Brake' (this being in fact a registered trademark of the Jacobs Vehicle Equipment Company in America). The background of the Jake Brake is worth a brief mention, because it was originally developed by C.L. Cummins, the American automotive diesel engine pioneer referred to earlier (Section 8.8). History records that he narrowly avoided a serious accident at a freight train crossing on a mountainous downgrade in California, when the brakes of an early diesel-engined truck he was demonstrating failed to hold. It was this experience which convinced him of the need for an independent engine brake, so that less reliance had to be placed on the brake linings and prayer under such circumstances.

Considering first the effect of engine compression braking without a Jake Brake, when the vehicle is driving the engine on the overrun, energy is transferred from the driving wheels into performing the engine compression strokes, which thereby hinders rotation of the wheels and creates a retarding force on the vehicle. However, this beneficial braking effect is negated on what would normally be the ensuing engine power strokes (the

fuel injection being cut off), because when the pistons pass their TDC positions and begin their downward strokes, the energy stored in the compressed and heated air in the cylinders is returned to the pistons and hence helps, rather than hinders, the rotation of the driving wheels. In other words we have a swings and roundabouts situation where essentially no energy is absorbed and no net retarding work is done on the vehicle.

Considering next the effect of engine compression braking with a Jake Brake, the ingenuity of this device resides in its ability to release the air pressure in the cylinders in what is termed 'exhaust blowdown', before completion of their compression strokes. Essentially no stored energy is therefore returned to the pistons on what would normally be their power strokes, which otherwise would help rotation of the driving wheels. Stated another way, there is a positive loss of energy from the compressed and heated air to the engine exhaust and cooling systems, which represents the retarding work done on the vehicle.

At the heart of the Jake Brake is an electrically-controlled hydromechanical actuating system (Figure 28.28) which opens the exhaust valves in each cylinder as the piston approaches the end of its compression stroke, thereby releasing the compressed air by exhaust blowdown. A transfer of motion is therefore required between either the injector valve rocker adjusting screw of a common-rail unit injector (Section 8.8 and which betrays the American origins of the Jake Brake) or, less commonly, an exhaust valve rocker adjusting screw of an alternate cylinder. Either arrangement provides the necessary synchronizing link between engine camshaft motion and compression brake timing for the engine cylinders.

When the Jake Brake is operated the all-important exhaust blowdown process occurs as follows:

1 Energizing the solenoid valve allows engine lubricating oil to flow under pressure into the passageway connecting both the master piston and the slave piston.
2 The oil pressure acting on the master piston causes it to move downwards, until it comes into contact with the injector valve rocker adjusting screw.
3 When the injector valve rocker is activated as in a normal injection cycle, its rising adjuster screw forces upwards the master piston. This has the effect of directing oil under increased pressure to the slave piston, because the ball check in the control valve closes and traps the oil for motion transfer.
4 The high-pressure oil therefore compels the spring-returned slave piston to move downwards and intercept the exhaust valves cross-head (Section 2.6). Hence, the exhaust valves are momentarily opened, which releases the compressed air in the cylinder to the exhaust system and the atmosphere.
5 Since this blowdown of compressed air to atmospheric pressure precludes the return of energy to the engine piston on what would normally be its power stroke, the work done in compressing the air to provide auxiliary retardation is not returned during the expansion process and therefore cannot detract from the braking effect.

The manufacturer's safety precautions should be strictly observed in relation to the installation, operation and maintenance of the Jake Brake.

Hydraulic retarder

Of the various retarders used, this type, together with its associated cooling equipment, generally tends to represent the most expensive and often the heaviest installation. The hydraulic retarder operates on the principle of a fluid coupling that is being deliberately kept stalled, so that the vortex motion of the circulating fluid converts into heat some of the mechanical energy being transmitted by the propeller shaft and therefore exerts a braking effect on the road wheels. In construction the hydraulic retarder differs from that of a conventional transmission fluid coupling, as described earlier, in comprising a propeller-shaft-driven double-sided impeller, which is rotated between two inwardly facing and permanently fixed turbine members (Figure 28.29). In hydraulic retarder terminology, the impeller and turbine members are referred to as the rotor and stators respectively. The purpose of this duplex arrangement with a double-sided rotor and twin stators is to increase the maximum braking effect of the retarder.

During normal driving when braking is not required the retarder is empty of oil, but when it is brought into operation, typically by initial depression of the normal brake pedal, the braking effect can be closely controlled according to the amount of oil admitted to the retarder via a pneumatically actuated valve. To dissipate the heat being generated by the retarder, arrangements are made for the fluid in circulation to be passed through either a heat exchanger in the engine cooling system, or a separate oil-to-air heat exchanger in the case of an air-cooled engine. The main advantage of the hydraulic

retarder is that it develops a much greater braking effect than other types, and especially from high speeds. In common with other drive line retarders it still remains effective in the event of an engine failure or a missed downward gear change.

This type of drive line retarder can be installed as either a 'remote' or a 'close-coupled' unit. In the former case it may conveniently replace the intermediate bearing of a divided propeller shaft, while in the latter it can be integrated with the transmission gearbox as in the modern ZF-Intarder and driven through step-up gearing to gain high retarding torques even at low propeller shaft speeds (Figure 28.30). An incidental advantage of this particular arrangement is that there can be a shared oil circulation and heat exchanger system between the transmission gearbox and the Intarder, which reduces oil ageing by virtue of continuous cooling.

Electromagnetic retarder

In principle, the electromagnetic retarder imposes a braking effect on the vehicle drive line and hence the road wheels that is derived from magnetic drag forces. These are created

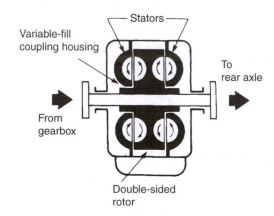

Figure 28.29   Schematic arrangement of a hydraulic retarder

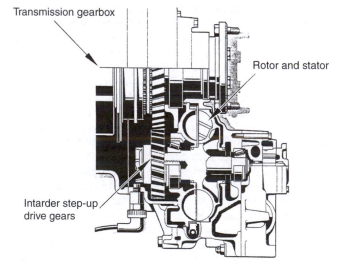

Figure 28.30   Cutaway view of integral ZF-Intarder (*ZF GB*)

by what is called an eddy current effect, the eddy currents being induced in either a single or a pair of rotors by a stator member (Figure 28.31) which comprises a series of electromagnetic pole pieces and is energized from the vehicle electrical system. Since the eddy currents may be regarded as short-circuited electrical currents, they likewise produce a heating effect in the rotor(s), so that some of the mechanical energy being transmitted by the drive line is converted into heat as part of the retardation process. This heat is dissipated to the air by virtue of the propeller-shaft-driven rotor(s) being provided with impeller vanes and therefore acting also as a cooling fan. During normal driving, when braking is not required, no current flows through the electromagnetic pole pieces and the unit remains inactive.

A long-established example of this type of auxiliary retarder is the Telma unit that consists of only two basic components.

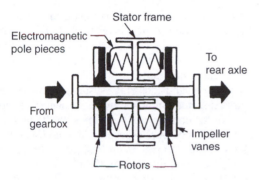

Figure 28.31   Schematic arrangement of an electromagnetic retarder

These are the fixed stator assembly, which can be either mounted directly to the gearbox or to the driving axle, or mid-mounted independently between the chassis frame members (Figure 28.32), and the rotor assembly that is permanently coupled to the propeller shaft. There is, of course, no physical contact between these two assemblies, so that a friction-free operation is ensured. Among the advantages of the Telma retarder are that it is relatively easy to retrofit and by its nature offers optimum integration with ABS systems. It is also self-cooling and self-regulating, its retarding ability being independent of engine speed or gear selected. Telma retarders for buses are automatically controlled through the footbrake and so are used in every brake application.

Friction retarder

Less commonly, a friction-type retarder can be used that takes the form of a wet multiplate friction clutch, which in some cases may be incorporated in the vehicle gearbox (Figure 16.5). For the installation shown the multiplate clutch is conveniently arranged so that it can either be fully engaged to operate reverse gear, or partially engaged via the brake pedal control to act as a friction retarder in the transmission system. This particular arrangement therefore contributes to a compact installation with little increase in weight for the retarder. The stationary plates of the clutch are produced from cast iron and the rotating ones from sintered bronze. A separate lubrication system is used in conjunction with a heat exchanger to ensure adequate cooling for the retarder. Advantages of the friction type of auxiliary brake retarder are those of providing a rapid response and maintaining an effective control until the vehicle is brought to a standstill.

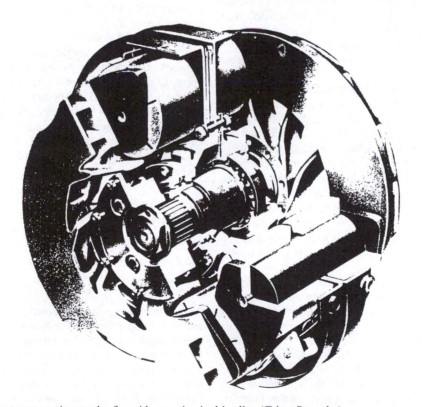

Figure 28.32   Telma electromagnetic retarder for mid-mounting in drive line (*Telma Retarder*)

29 Anti-lock brakes and traction control

29.1 BACKGROUND TO ANTI-LOCK BRAKING

One of the most serious hazards of driving and a significant contributor to the accident toll is the locking of wheels under heavy braking on slippery surfaces. The consequent loss of braking efficiency will greatly increase the braking distance and, coupled with this, will be loss of steering response if the front wheels lock and even worse a loss of directional stability if the rear wheels lock. Furthermore, the vehicle will exhibit an unwelcome sensitivity to road surface irregularities and camber, so that even small disturbing forces can cause it to slew sideways, making control even more difficult. In other words, if a wheel is braked to the point of locking its sideways grip is reduced to zero.

It was mentioned in Chapter 27 that to achieve the shortest emergency distance, all wheels need to be almost on the point of locking. This is because the rolling friction developed by a tyre that is still exerting a rotational grip against the road surface is always greater than the locked wheel friction developed by a tyre that is simply sliding over the road surface, especially when this is wet and slippery. Experienced drivers have long recognized that emergency stopping under these conditions is best accomplished by what is known as cadence braking, the efficacy of which can readily be verified on a skid pan. This technique involves rapid pumping of the brake pedal, so that steering control is retained with each momentary release of the brakes, whilst stopping distance is being shortened as much as possible.

Anti-lock braking systems seek automatically to reproduce a similar cycle of events by means of wheel speed sensors. These detect the point at which a wheel is about to lock and signal the intention to a control mechanism, which momentarily either holds constant or actually reduces the braking force to the wheel concerned, the off-on braking cycle being repeated as long as the wheel is about to lock. Ideally then, an anti-lock braking system should be even more swiftly acting than an experienced driver, not only in minimizing straight-line stopping distances on slippery surfaces, but also and more importantly in allowing the driver to avoid obstacles by retaining steering control where otherwise this would be lost with locked wheels.

It is perhaps of interest to recall that the basic idea of what was then called an anti-skid control was first employed for railway braking systems by the Westinghouse Air Brake Company in America, but it was in the further development of this idea and its application to the disc brakes of aircraft that this same company found the widest demands in the late 1940s. By the early 1950s similar anti-skid controls, such as the Dunlop Maxaret, were also being developed for the aircraft industry. So it was as a result of the experience gained in this field that the possibility of adapting the Maxaret system for use on motor vehicles was investigated by the Dunlop company and an installation tested by the Road Research Labortory. At this juncture it must be appreciated that the objectives sought in applying the wheel anti-lock principle to aircraft and motor vehicle braking systems were somewhat different. With an aircraft the objectives are to prevent at least expensive tyre damage and at worst a potentially dangerous tyre burst, by avoiding wheel locking however briefly it may occur during maximum braking through the landing run. In comparison with an aircraft tyre, the tyre of a motor vehicle is relatively lightly loaded and any damage caused to it by wheel locking is far less significant, but what is really important is any loss of directional control or stability as already mentioned.

Early thinking on anti-lock brake systems for motor vehicles was therefore directed not only towards passenger car applications, but also for commercial vehicles with the jack-knifing problem on articulated ones especially in mind (Section 30.5). Although the Dunlop Maxaret anti-lock principle was eventually adapted to the braking system of the specialist Jensen FF four-wheel-drive car in 1966, it was not until the advent of electronic control systems and electromagnetically sensed wheel speeds, beginning with the introduction of the Anti Blockier System (ABS) in 1978 by Robert Bosch working in conjunction with Mercedes-Benz, that the development of anti-lock braking systems started in earnest and has led to their increasing adoption. Indeed, since 2004 they have become standard equipment on passenger cars in Europe and their operating requirements are defined by regulation.

Finally, it should be mentioned that not all authorities agree with the designation 'anti-lock brake system'. The Society of Automotive Engineers in America originally preferred the term 'wheel slip control system', defining this as 'a system which automatically controls rotational wheel slip during braking'.

29.2 BASIC COMPONENTS OF ANTI-LOCK BRAKE SYSTEMS

The various components used in anti-lock braking systems and their purpose may be summarized as follows.

Wheel speed sensors

As their name suggests, these constantly measure wheel speeds and signal this information either electronically or, less commonly, mechanically to the anti-lock control system.

An electromagnetic sensor comprises a coil winding with a permanent magnet core and pole pin, the latter being accurately aligned in a radial or an axial position with a toothed sensor ring, but separated from its teeth by a small although critical air gap of typically 1 mm (0.040 in). A shrunk-on mounting for the sensor ring can be furnished on the hub flange of a non-driven wheel, or on the neck of a constant-velocity joint housing for a driven wheel, while in some cases it may be mounted from the pinion shaft of the final drive to serve both rear wheels. It cannot be mounted from parts directly subject to brake heating as this would affect the magnetic properties of its metal. The teeth of the sensor ring therefore act as a signal generator, so that when each passing tooth faces the pole pin of the sensor the changing magnetic field induces a voltage across the winding. Above walking pace this produces a frequency of signal that varies in proportion to road wheel speed and thus supplies the required input to the control system.

A mechanical sensor depends on the rotational inertia of a small flywheel to create a signal of imminent wheel lock during braking. Each freely mounted flywheel receives its drive through a spring-loaded ball-and-ramp clutch mechanism, the input shaft for which can be conveniently belt driven from a transmission drive shaft so that wheel speed can be sensed. In operation the flywheel will tend to overrun the shaft driving it when the rotational deceleration of its associated road wheel exceeds a predetermined amount. The inertia of the flywheel will then compel the balls to roll up their ramps, thereby creating an axial component of force within the clutch mechanism. This movement of separation is directly utilized by the control system as a signal of imminent wheel locking.

Electronic control unit

Except in the case of hydromechanical anti-lock braking systems, a microprocessor-based electronic control unit is required to interpret the signals received from the wheel sensors and issue the necessary commands to the system modulator. In simple terms its purpose is to detect any significant slow-down of wheel speed relative to vehicle speed, which could result in the braking force at any wheel exceeding the available tyre to road adhesion force. The rapidity and accuracy with which the electronic control unit can function is such that the braking force can typically be modulated from three to twelve times a second, according to road surface conditions, thereby conferring smooth controlled braking. With anti-lock air brakes for heavy vehicles, the braking force modulation is limited to three or four times a second so as to minimize air consumption.

Apart from computing brake anti-lock requirements, another important function of the electronic control unit is that of monitoring not only its own state of health but also that of other related components, so as to ensure safe operation of the system. Should there be any malfunction it must therefore evaluate whether only part, or the whole, of the system must be switched out of action and then retain the nature of the fault in its memory for service investigation. A warning light informs the driver if a malfunction in the system has resulted in the electronic control unit being automatically switched off. In this event the system is failsafe, because normal braking control without anti-lock intervention remains available to the driver.

System modulator

This component represents the final link in the operating chain of an anti-lock braking system, since in all except hydromechanical versions it receives the commands from the electronic control unit and translates them, independently of driver action, into valve-based sequences that modulate either hydraulic or pneumatic pressure in the individual brake lines. As earlier indicated, the process of modulation can simply mean holding the line pressure constant or, if further correction is required, momentarily relieving and then restoring it. The modulator therefore requires its own source of supply pressure, which in electrohydraulic versions usually takes the form of an electrically driven pump capable of generating high pressure in the region of 14 to $17.5 \, \text{MN/m}^2$ (2000 to $2500 \, \text{lbf/in}^2$). This pump is sometimes arranged to operate in conjunction with a hydraulic accumulator for storage purposes.

An ABS in which hydraulic servo brakes (Section 27.8) and their modulator share a common hydraulic power supply is known as an 'Integrated ABS'. This is distinct from a system with vacuum servo-assisted brakes and an independent hydraulic power supply for their modulator, which is sometimes referred to as an 'Add-on ABS' even though the units may be installed together.

With a hydromechanical anti-lock braking system, the hydraulic pump is mechanically driven from the same source that drives the flywheel wheel speed sensors. In the case of electropneumatic modulators for anti-lock air brakes on heavy vehicles, the continuous supply of compressed air made available by the compression and storage facilities of the braking system proper eliminates the need for an independent source of supply pressure.

It therefore follows that it is really the functional differences between the various types of modulator which determine the particular operating principle of an anti-lock braking system.

29.3   TYPES OF ANTI-LOCK BRAKE SYSTEM

The modulating principles of anti-lock hydraulic braking systems may be classified as follows:
Plunger return
Power recharge
Dynamic inflow

Plunger return

This principle of modulation was early developed to superimpose an anti-lock facility on a conventional hydraulic brake system, and interest in it has recently been revived. An essential feature of this type of system is the introduction of a cylinder and plunger modulator into each brake line between the master cylinder and the wheel cylinders, so that by moving the plunger up and down within its cylinder it becomes possible to vary the volume of a particular brake line and therefore modulate the line pressure and hence braking force generated at that wheel. Except under conditions of imminent wheel locking, the plunger is maintained at its lowest position in the cylinder by fluid at high pressure directed above it from an independently driven pump and storage accumulator, while a downward projecting pin from the plunger holds open an isolating valve that allows the free passage of fluid between the master cylinder and wheel

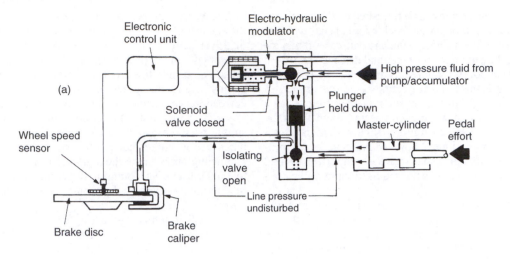

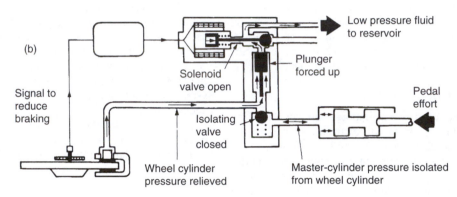

Figure 29.1 Plunger return principle of anti-lock brake modulation

cylinder (Figure 29.1a). Normal braking up to maximum pressure is therefore accomplished with the least volume of brake line, because there is always sufficient pressure acting above the plunger to restrain it from rising against master cylinder pressure acting below it.

However, if a wheel begins to lock the electronic control unit commands the appropriate solenoid valve in the module to shut off the fluid supplied under pressure to the plunger. As a result the plunger rises in its cylinder, the effect of which is twofold: first, it allows the isolating valve for the master cylinder to close, so that no further increase in wheel cylinder pressure is possible no matter how much effort the driver exerts on the pedal; and second, it increases the volume and therefore relieves the pressure in the line to the wheel cylinder, which momentarily releases the braked wheel (Figure 29.1b). Line pressure is of course restored to the wheel cylinder when the electronic control unit commands the solenoid valve to open, so that fluid at a higher pressure can force the plunger down its cylinder to reduce the volume of the brake line and re-establish a connection between the master cylinder and the wheel cylinder, via the now open isolating valve.

A modern application of the plunger return principle of modulation may be found in the Lucas Girling stop control

system (SCS). Each modulator in this hydro-mechanical anti-lock brake system comprises an integrated wheel speed sensor, mechanically driven pump and brake pressure controller (Figure 29.2). Comparing this arrangement with that showing the general principles of a plunger return system of modulation (Figure 29.1), it will be recognized that what are termed the deboost piston, cut-off valve and dump valve correspond to the plunger, isolating and solenoid valves respectively.

The advantages generally claimed for the plunger return principle of modulation are that it avoids the need to open a conventional hydraulic brake system for the release of brake fluid, when line pressure is reduced for anti-lock control; the large open area of the isolating valve offers minimum restriction to the build-up of master cylinder pressure during normal braking; and the system is less costly than others because it does not involve the use of highly developed valve assemblies.

Power recharge

This has been the most widely used principle of modulation in anti-lock brake systems. Here again it is superimposed

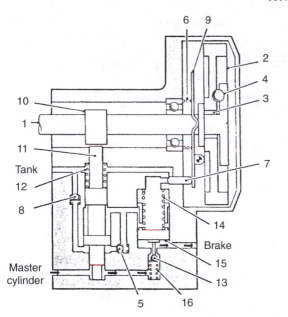

Figure 29.2 Lucas Girling SCS modulator unit (*Lucas Girling*)
1 drive shaft
2 flywheel
3 flywheel bearing
4 ball-and-ramp drive and clutch
5 pump outlet valve
6 flywheel spring
7 dump valve
8 pump inlet valve
9 dump valve lever
10 eccentric cam
11 pump piston
12 piston spring
13 cut-off valve
14 deboost piston spring
15 deboost piston
16 cut-off valve spring

upon a conventional hydraulic brake system, but with the essential difference that a brake line has to be opened in order to reduce pressure for anti-lock control, rather than simply increasing its volume as in plunger return modulation.

In principle a solenoid-operated multifunctional hydraulic shuttle valve serves to modulate the brake pressure between the master cylinder and either a single or a pair of wheel cylinders. During normal braking up to maximum pressure the spring-returned solenoid valve assembly is not energized, so that its upper valve remains open while the lower valve is held closed. In this position the valve assembly provides a free passage for the flow of brake fluid between the master cylinder and wheel cylinder (Figure 29.3a). If a wheel begins to lock during braking, the electronic control unit initially commands the modulator to energize the appropriate solenoid with one-half maximum current, which raises the valve assembly to an intermediate position with both its upper and lower valves held closed. This has the effect of not only isolating the wheel cylinder from any further rise in master cylinder pressure, no matter how much effort the driver exerts on

the pedal, but also holding constant the wheel cylinder pressure (Figure 29.3b).

If this action of simply preventing any further rise in wheel cylinder pressure is not sufficient to correct the imminent locking of the wheel, then the electronic control unit commands the modulator to energize the solenoid with maximum current and therefore raise the valve assembly to its fullest extent. This results in the upper valve remaining closed to isolate master cylinder pressure, but the lower valve being opened to bleed off brake fluid to the return pump via the accumulator, thereby relieving wheel cylinder pressure to free the wheel (Figure 29.3c). At the same time the modulator is commanded to switch on the return pump, so that the fluid bled from the wheel cylinder line can be directed under pressure back to the master cylinder line and the inlet side of the solenoid valve assembly ready to recharge the system for the next cycle. This occurs when the modulator is commanded to switch off the pump and de-energize the solenoid, which allows the valve assembly to drop to the normal braking position. Each time a wheel cylinder line is depressurized, the recharging pressure in the master line cases a slight raising and pulsating at the brake pedal. It should nevertheless be held firmly applied, since any sympathetic pumping of the pedal by the driver will simply increase the time taken to stop.

The power recharge principle of modulation has long been incorporated in certain versions of the Bosch ABS and has been developed to provide a high level of vehicle control under difficult braking conditions commensurate with the higher cost of this type of system.

Dynamic inflow

This principle of modulation was an essential feature of a new concept in anti-lock brake systems introduced by the German company of Alfred Teves (ATE) in the mid 1980s, wherein brake operation, servo amplification and anti-lock modulation were combined in a fully integrated system. An additional low-cost version of this widely used system incorporates a further development of the dynamic inflow principle of modulation and serves to illustrate its application (Figure 29.4).

If the electronic control unit detects a locking tendency at any wheel, it commands modulation of the brake line pressure through solenoid-controlled inlet and outlet valves in the conventional manner. However, when it is necessary to reduce line pressure by opening an outlet valve, the excess volume of brake fluid is conveyed directly to the supply tank. At the same time an electrically driven pump is switched on and supplies fluid under pressure to both brake circuits served by the tandem master cylinder. This has the effect of slowly forcing the master cylinder pistons back to the point at which their central spring-loaded ball valves are unseated by axial stop-pins and limit the pressure (Figure 29.5). The significance of these central valves in the dynamic inflow principle of modulation is that they ensure the pressure generated by the pump remains proportional to pedal effort. They also allow the excess volume of brake fluid not required for modulation replacement to be returned to the supply tank, thereby avoiding the need for a hydraulic accumulator in the system.

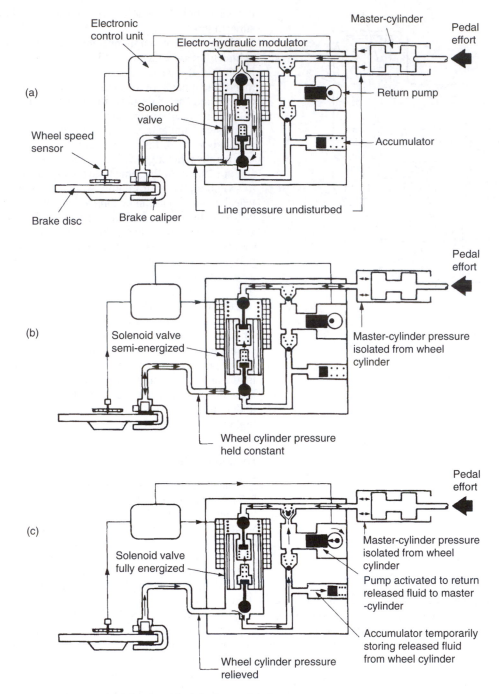

Figure 29.3   Pressure recharge principle of anti-lock brake modulation

29.4   OUTPUT CONTROL CHANNELS FOR ANTI-LOCK BRAKES

The extent to which each wheel is independently controlled with anti-lock brakes determines the number of output channels that need to be provided by the modulator system. Anti-lock brakes were originally introduced for rear-wheel-drive cars having a divided-line brake circuit with a front/rear split (Section 27.4). On these it was found that a four-channel anti-lock control, where each wheel had its own speed sensor and each wheel cylinder could have its apply pressure either held constant or relieved, was not necessarily the ideal arrangement. Although such a system behaved well in terms of stopping distance and steering response when braking on ordinary slippery surfaces, it could be at a disadvantage if the nature of the road surface offered more grip to the wheels on one side of the car than on the other, or what is sometimes grandly referred to as an asymmetrical split grip roadway.

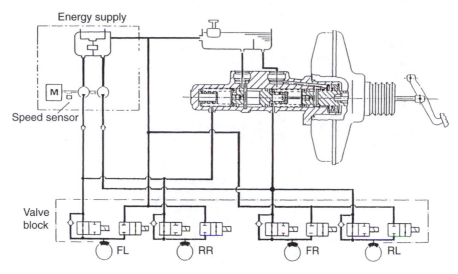

Figure 29.4 Anti-lock brakes with X-split circuit (*Alfred Teves*)

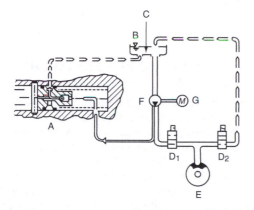

Figure 29.5 Dynamic inflow principle of anti-lock brake modulation (*Alfred Teves*)

A Master cylinder	D$_2$ Solenoid valve (outlet)
B Fluid level sensor	E Disc brake
C Reservoir	F Non-return valve
D$_1$ Solenoid valve (inlet)	G Electrically driven pump

This resulted in the side-to-side braking forces being unequal for the front and rear wheels, which could impose an unstable yawing movement or sideways turning on the car.

To avoid this condition three-channel control systems were developed, so that the front wheels could be speed sensed and controlled individually, while the rear wheels could be speed sensed but controlled as a pair by a floating balance or 'select-low' valve. This means that the maximum braking effort that can be applied to the wheel on the high-grip side of the car is limited to the maximum that can be applied to the other wheel on the low-grip side, so that neither wheel is allowed to lock and compromise directional stability.

For modern front-wheel-drive cars that conventionally have a divided-line brake circuit with an X-split (Section 27.4), a three-channel control with four sensors can similarly

be used (Figure 29.4). Alternatively a less costly two-channel control system with fewer components may be employed. Such a system takes advantage of international legislation that governs the design of passenger vehicle braking systems, which decrees that under all conditions of loading the front wheels must ultimately lock before the rear ones to maintain directional stability. It is, of course, for this reason that the lightly laden rear wheels of front-wheel-drive cars are provided with brake apportioning valves to prevent them from locking (Section 27.4). From this it follows that if a two-channel two-sensor system imposes an anti-lock control on the front wheels only, then through their diagonal hydraulic connections to the rear wheels these too will benefit from a fair degree of control, since in any event they should be prevented from locking by their apportioning valves. A clearly identifiable example of a two-channel control system for front-wheel-drive cars with an X-split brake circuit is the earlier mentioned Lucas Girling SCS, which incorporates a separate anti-lock modulator with integrated speed sensor for each front wheel (Figure 29.6).

In the case of a four-wheel drive vehicle, the anti-lock brake system used on the earlier mentioned Mitsubishi Galant car (Section 21.4) features two diagonal output channels, together with speed sensors at all four wheels that operate in conjunction with a 'G' sensor, which monitors the overall amount of vehicle deceleration (Figure 29.7). A microprocessor-based electronic control unit receives signals from each sensor on independent input channels. It then uses this information to modulate the hydraulic line pressure for the brakes of the right front and left rear wheels on one output channel, and the left front and right rear wheels on the other channel, in the event of incipient wheel locking under braking. A select-low valve is also incorporated in the system, the function of which as earlier explained is to ensure that the maximum braking effort that can be applied to the wheel on the high-grip side of the car is limited to the maximum that can be utilized by the other wheel on the low-grip side. Another point of interest is that the viscous coupling

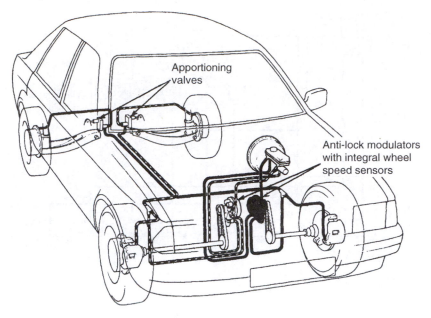

Figure 29.6 Lucas Girling SCS system with two-channel control (*Lucas Girling*)

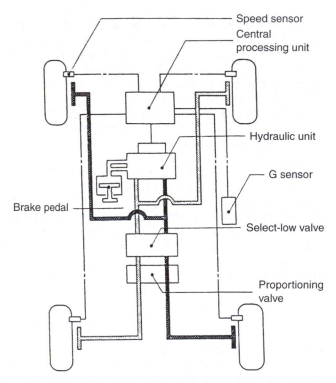

Figure 29.7 Four-wheel drive with anti-lock braking system (*Colt-Mitsubishi*)

unit (VCU) of this 4WD system (Section 21.4) also behaves as an anti-lock brake modulator. By virtue of acting as a limited slip coupling between the front and rear wheels, it automatically inhibits any one pair of wheels from locking up on a sudden application of the brakes.

29.5 ANTI-LOCK AIR BRAKES FOR HEAVY VEHICLES

There is clearly no less a requirement for anti-lock braking on heavy vehicles than there is for passenger cars and, in particular, the hazards of over-braking and skidding with articulated vehicles are widely acknowledged. In fact the gradual adoption of ABS for heavy vehicles, trailers and buses is becoming mandatory under EC braking legislation.

The principles of anti-lock braking as earlier described for hydraulic systems broadly apply to the air brakes of heavy vehicles, but there is a greater multiplicity of components and, as would be expected, basic differences in those that modulate the air pressure in the brake lines.

In a typical installation for a 6 × 4 vehicle the wheel speed sensing equipment is mounted within the drum brake assemblies and comprises six toothed rings, which act in conjunction with axially disposed electro-magnetic sensors. For anti-lock control three separate electronic control and modulator units are mounted from the vehicle chassis frame, their locations being adjacent to the front and tandem rear axles. Each modulator is provided with electropneumatic control valves and intervenes in an appropriate brake line between a system relay valve and a pair of axle brake actuators or chambers. The modulator control valves have three operational modes and can be commanded by the electronic control unit either to hold constant, relieve or restore air pressure directed via the relay valve from the foot-operated brake valve to the wheel brake actuators. Diaphragm controlled 'hold' and 'exhaust' valves are used in the modulator, their respective closing and opening being obtained by admitting system air pressure during braking into their operating chambers through solenoid valves, which are energized from the electronic control unit (Figure 29.8). Unlike an anti-lock hydraulic brake system, where it is necessary to conserve the fluid supply, the

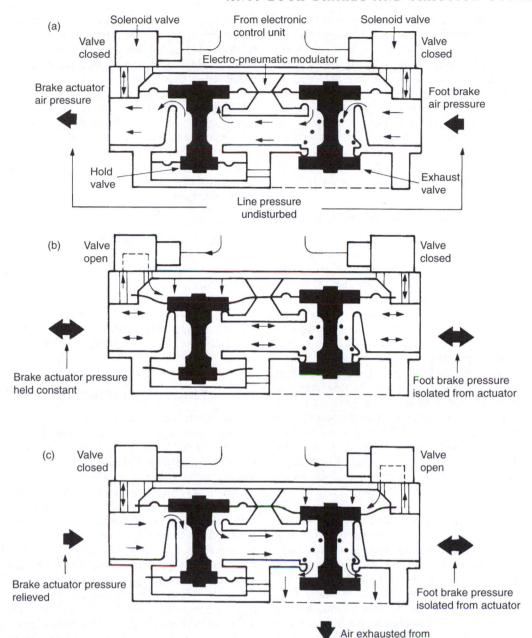

Figure 29.8 Modulation principle for anti-lock air brakes

air exhausted from the modulator is simply dumped to the atmosphere when line pressure is relieved to prevent wheel locking. Similar to an anti-lock hydraulic brake system, however, the electronic control unit possesses a comprehensive self-diagnostic capability and is failsafe in operation.

Anti-lock air brakes can also be installed on multi-axle trailers, a separate electronic control and modulator unit being used with wheel speed sensing being taken from the middle axle of a tri-axle trailer.

Anti-lock brake systems for both light and heavy vehicles are now generally accepted as being reliable in extended service. However, unless any malfunctioning of the system can be attributed to readily identifiable faults, such as defective electrical connections, broken cables, incorrectly gapped wheel speed sensors or damaged sensors; then resort has to be made to specialist diagnostic test equipment, which is used in conjunction with the comprehensive test schedules issued by the vehicle manufacturer.

29.6 TRACTION CONTROL SYSTEMS

General background

During moving off and acceleration, a traction control system performs a similar safety function to anti-lock braking by preventing the driving wheels from slipping, which therefore helps both to maintain directional control and to improve traction under adverse driving conditions, especially on road surfaces that are slippery on one side of the vehicle only. Since a traction control system may be regarded as a logical but inverse development of anti-lock brakes and can therefore utilize some of the same technology, it is perhaps to be expected that this type of system was pioneered by Robert Bosch working in conjunction with Mercedes-Benz, their anti-slip regulation system being introduced in 1987 and known as Antriebs-Schlupf-Regelung (ASR). Other manufacturers have since introduced traction control systems that may similarly be integrated with anti-lock hydraulic and now air brake systems.

Basic operating features of a traction control system

The two essential functions that must be performed by a traction control system are the automatic braking of a single driving wheel that is about to spin and the automatic throttling back of the engine if both driving wheels are about to spin.

Operating the wheel brake independently of the driver in this way not only is the most effective method of rapidly preventing a driving wheel from spinning, but also can arrest the momentum of a driving wheel whose grip suddenly changes to slip and begins to spin as the load on the power train is removed. It will be appreciated that, owing to the effect of the final drive differential, the application of a braking torque on a single driving wheel that is about to slip will result in an equivalent driving torque being applied to the other wheel that is gripping. Traction can therefore be maintained because the system is then acting in the manner of a differential lock. The final link in the system is that of imposing a drive torque control on the power train, so the available torque at the driving wheels is limited to that required for effective traction without slip.

Brake torque control of a traction control system is based on the existing components of an anti-lock braking system. It is therefore managed by an electronic control unit that combines the functions of both systems. The electrohydraulic modulator system used for passenger car anti-lock braking is extended to include additional valves for switching from braking to traction control modes, pressure modulation and pressure limitation. For smooth operation and, in the case of front-wheel-drive cars, minimum steering interference there is a slower build-up of hydraulic pressure for brake torque control as compared with that required for anti-lock braking. As the driving wheels must be braked without any action from the driver, the hydraulic energy is supplied from a high-pressure accumulator that is charged by an electrically driven pump. When signals received from the wheel speed sensors indicate that intervention by the traction control system is required, the electronic control unit commands the modulator switch-over valve to connect the hydraulic accumulator with the appropriate brake cylinder of the wheel beginning to spin. As a modulator valve is provided for each driving wheel, it is thus possible to brake either wheel independently of the other. The modulator valve holds the brake line pressure constant as soon as the wheel speed stops rising and then relieves it when the wheel slip has been brought under control.

Engine torque control is obtained by replacing the conventional mechanical linkage between the accelerator pedal and engine with an electronically based accelerator controller, through which the driver indirectly controls either the throttle valve of a petrol engine or the metering valve of a diesel engine. Based on signals received from the wheel speed sensors, the system therefore allows commands from the traction electronic unit to take precedence over driver action at the accelerator pedal, so that if necessary the engine can be automatically throttled back for traction control even though the driver may unwisely be urging greater acceleration. Such a system involves the sensing of accelerator position by a potentiometer, whose signals are then transmitted to the electronic control unit, which in turn commands an electrically driven actuator to operate the appropriate engine controls. A limited-movement mechanical connection can be retained between the accelerator pedal and engine controls, so that in the event of a fault developing in the electronic control system the vehicle can still be driven. The traction control system is, of course, switched off instantly when the driver applies the brake in the normal manner.

The general layout of the Teves integrated anti-lock braking and traction control system, as applied to a front-wheel-drive car, is shown in Figure 29.9.

29.7   VEHICLE DYNAMICS CONTROL

Whereas anti-lock brake and traction control systems are intended to function only when critical driving conditions are reached in the essentially straight-ahead direction, that is they prevent either a braked wheel from locking or a driven wheel from spinning, a vehicle dynamics control (VDC) or electronic stability programme (ESP) system extends this function to stabilize the vehicle when cornering, or in other words it influences the distribution of wheel lateral forces before critical conditions are reached. Such a system has been developed by Robert Bosch of Germany, again in conjunction with Mercedes-Benz, which constantly compares the direction steered by the driver with the actual direction of the vehicle. If there is a significant deviation from the former, the electronic control unit recognizes this and automatically intervenes to resolve the conflict by either actuating the brakes or adjusting the power from the engine, thereby redistributing the lateral forces acting at the wheels and stabilizing the vehicle.

Since determining the actual course of the vehicle is a relatively complicated business, a vehicle dynamics control system naturally requires components additional to those already associated with ABS/ASR systems (Figure 29.10). In particular a lateral accelerometer or G sensor is required that provides a sensitive response to the forces generated during cornering, and a yaw rate sensor that measures the speed at which the car rotates about its vertical axis. The latter sensor may be regarded as being at the heart of the VDC system, and is in fact a complex instrument that has been derived from aviation practice and adapted to the extreme environmental conditions of the motor vehicle. Two further sensors are also involved, these being a steering angle sensor that signals the intended course steered by the driver, and a braking pressure sensor.

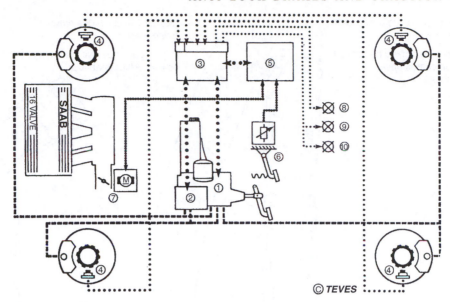

Figure 29.9 Layout of an advanced integrated anti-lock brake and traction control system (*Alfred Teves*)

1 Hydraulic unit with integrated ABS function
2 TCS valve block
3 Electronic ABS/TCS controller
4 ABS sensors (front axle ABS/TCS)
5 Electronic throttle controller } Electronic accelerator
6 Accelerator pedal sensor

7 Throttle actuator
8 ABS warning lamp
9 TCS warning lamp
10 TCS function tell-tale lamp

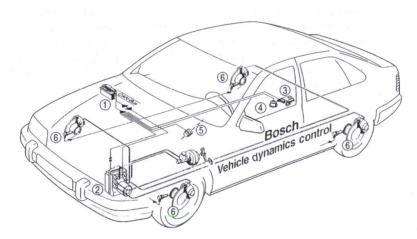

Figure 29.10 Layout of Bosch Vehicle Dynamics Control (VDC) system ('*Photo:Bosch*')

1 Electronic Control Unit (ABS/ASR/VDC) with CAN link to Vehicle Powertrain Control Module
2 Hydraulic Modulator with Pressure Sensor
3 Yaw Rate Sensor
4 Lateral Accelerometer
5 Steering Angle Sensor
6 Wheel Speed Sensors

For an understanding of how the VDC system responds to actual driving situations, it is first necessary to recall how the relative slip angles developed by the tyres as the vehicle is steered round a bend, determine whether or not it will understeer or oversteer (Figures 22.4 and 22.15). Another relevant factor is that the cornering ability of a tyre is reduced and therefore contributes to that end of the vehicle running wide, when either the tractive or braking forces generated between its tread and the road surface are increased. If therefore a car understeers and swerves off course when negotiating a bend, the VDC system compensates for the reduced cornering ability of the front wheels by automatically braking the inner rear wheel to impose a restoring yaw moment (Figure 29.11). Furthermore, the speed of the car is automatically reduced to

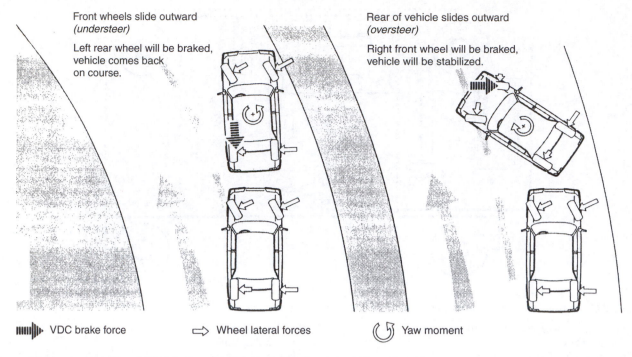

Figure 29.11   Action of Bosch Vehicle Dynamics Control (VDC) system ('*Photo:Bosch*')

an appropriate level, either by throttling the engine or by intervention of the braking system on selected wheels, or both. Conversely, if an oversteering condition is encountered, then the VDC system compensates for the reduced cornering ability of the rear wheels by automatically braking the outer front wheel to impose a restoring yaw moment, but this time in the opposite sense (Figure 29.11). The effectiveness of the VDC system in modifying the lateral stabilizing forces

acting on a vehicle is not confined solely to cornering, but also to achieving optimum driving stability during rapid lane-change manoeuvres. However, it must be stressed that neither anti-lock brakes, traction control, nor vehicle dynamics control systems can defy the basic laws of physics and cater for what is sometimes referred to by development engineers as 'super-lunatic' driving behaviour!

30 Vehicle structure and aerodynamics

30.1 INTEGRAL BODY CONSTRUCTION

General background

The integration of frame and body functions into an all-embracing shell is variously known as integral, unitary, chassisless and originally by the French as monocoque construction. A significant step towards integral construction was the introduction of the all-steel welded body, a concept that was developed in America by the Budd Manufacturing Company for Dodge cars in the late 1920s. Although a separate chassis frame was still used, it no longer functioned as the total load-carrying member.

Integral construction proper was introduced by Citroën in France during the mid 1930s and was evolved in collaboration with the previously mentioned Budd Company. It was based upon the principle of stressing the outer skin of the body shell to perform a load-carrying function, which contributed to the strength and rigidity of the inner structural members, thereby eliminating the need for a separate chassis frame. Perhaps it may be of interest to mention that the monocoque method of providing a more efficient structure was also well known to aircraft engineers and featured in such historic World War II aircraft as Blenheims and Spitfires.

Finally we come to modern practice, where the principle just described of pure integral construction is often modified in various ways, notably in combining it with separate subframes for mounting various parts of the vehicle mechanism. The aim here is chiefly to suppress noise, vibration and harshness from the body interior by the extensive use of rubber isolators and also to simplify the assembly of the mechanical units to the body structure. This particular form of integral or semi-integral construction was pioneered by Daimler-Benz of Germany in the early 1950s. Figure 30.1 illustrates the main parts of a modern integral construction body.

The basic structure

In conventional practice the passenger car body is produced from overlapping sheet metal, fastened by a multiplicity of spot welds often robotically applied. For example, the steel panels that form the body structure of the modern Citroën C5 car are united by no less than 4200 spot welds. The body framework is constructed by joining together thin-walled closed-section members, and where these are subject to particularly large torsional and bending stresses, such as in the case of the body sills, they may be reinforced by the inclusion of longitudinal partitions (Figure 30.4). Similarly the frame joints may require to be locally reinforced by additional flanged members, since much depends on their structural integrity to maintain stiffness of the body and hence preserve correct alignment of the various units it supports.

A further requirement is to avoid resonant vibration of the body structure, which demands a high degree of torsional stiffness or resistance to twisting. It generally follows that a body structure that has the required degree of torsional stiffness also possesses acceptable beam stiffness or resistance to bending. The body is subject to beam loading that arises from its own weight and that of its occupants, together with the weight of the engine and transmission units. It is therefore stressed in compression along its upper body and in tension along its underbody. In modern practice this has led to the selective use of high-strength steel panels in the upper part of the body and ultra high-strength panels in its lower part. As compared to standard steels, these specialized alloy steels confer greater strength so that the thickness of the panels can be reduced to save weight. Furthermore, production methods can now be such that any given panel may be varied in thickness, thereby matching the particular stress levels imposed upon it. That is, an increased wall thickness can be concentrated in areas subject to greatest loading.

The individual members comprising what is termed the 'body less doors' are most conveniently identified in relation to the following assemblies: underframe, side frames, and roof frame. With regard to the all-important underframe assembly, which forms the entire lower portion of the body structure, this is now more often referred to as the 'platform'. Its importance in terms of the considerable research and development costs involved is such that since the mid-1990s platform sharing between different models of high volume manufacturers has become increasingly common to spread these costs. A notable pioneer in this respect was Fiat Auto, who introduced a common floorpan for their Fiat 127 and 128 models in 1970.

Underframe assembly

This may be regarded as the backbone of the modern integral construction body, since it forms the lower portion of the car and provides the major part of its strength (Figure 30.2). Its middle section comprises a floor ribbed for stiffness and flanked on each side by the inner sill members, and usually it has a central tunnel which houses either the propeller shaft, remote gear control linkage or exhaust system, according to the particular layout of the car. The floor panel terminates at its front end in the toe board and dash panel subassembly and at its rear end in the heel board and rear seat pan subassembly. This typically incorporates the boot floor in one large pressing. Box-section cross-members are provided beneath the front seat and at the leading edge of the rear seat pan. A pair of box-section longitudinal members or longerons, running beneath and either partly or wholly the length of the floor panel inboard of the sills, are extended forwards and rearwards to the body extremities. If the

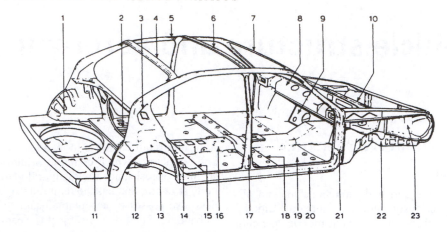

Figure 30.1 Main parts of a modern integral construction body (*Alfa-Romeo*)

1 rear seat squab bulkhead
2 rear seat pan
3 quarter panel
4 rear window (backlight) reinforcement panel
5 cantrails
6 roof bow member
7 windscreen header panel
8 dash panel
9 toe board
10 service compartment bulkhead
11 boot floor
12 wheel arch
13 longerons
14 D post
15 heel board
16 cenral tunnel
17 B-C post
18 cross-member
19 floor panel
20 sills
21 A post
22 longerons
23 wing valances

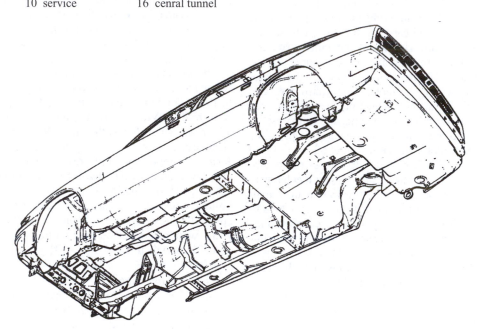

Figure 30.2 Underside view of integral construction body for front engine rear-wheel drive car (*Pressed Steel Fisher*)

longerons are interrupted along their length, the front pair extend inwards as far as the cross-member beneath the front seats, but the rear pair extend no further inwards than the heel board cross-member. Full-length longerons are used for sports utility vehicles (SUV) where integral body construction is employed. The underframe assembly is completed by the front wing valances and wheel arches, which are united with the dash panel and front pair of longerons.

Side frame assemblies
Each of these comprises an outer sill member, door posts, quarter panel including rear wing, and roof cantrails. With a four-door body the front, middle and rear door posts are designated A, B-C and D posts respectively. In all cases they can pose complex joining conditions with their surrounding structure. Their upper ends are united by what is termed a cantrail member, which is also of box-section construction. A side frame is completed by the rear wing valance and wheel arch, this assembly then being integrated with the rear seat pan and boot floor. Each front wing is attached separately both to its underframe valance and wheel arch and side frame door A post.

Roof frame assembly
This serves to close the essentially box-like structure of the conventional car. It comprises, of course, the roof panel

flanked on each side by the cantrails mentioned previously and may be cross-braced by roof bow members. The roof panel joins the windshield header panel at its front end and the reinforcement panel for either a rear window or tailgate at its rear end.

Subframes

A subframe is a rigid, detachable assembly that is connected to the lower body structure through rubber isolation mounts. At the front end of the body it supports an independent suspension system, steering mechanism and the power unit. Acknowledging the latter function it is sometimes referred to as an 'engine cradle'. Since the suspension system and power unit are themselves connected to the subframe through isolation bushings and mounts, it will be evident that there is a double filtering effect against the transmission of vibration and noise to the car body interior. Another requirement to be met by the front subframe in modern practice is that it must safely redistribute the forces of impact in the event of a head-on collision. A four-point mounting is a typical arrangement for the front subframe. At the rear end of the body a subframe supports an independent suspension system, either wholly or partly according to its layout, and the final drive unit. Similar to the application of a front subframe, the suspension system and final drive unit are likewise connected to the rear subframe through isolation bushings and mounts. It necessarily follows that front and rear subframes, which are pre-assembled with the units they support, readily lend themselves to high volume production methods.

Front and rear subframes were traditionally fabricated from several steel stampings welded together and locally reinforced. Since Ford of America introduced a 'hydroformed' front subframe in the mid-1990s, this method of manufacture is now widely used as indeed it is for various other automotive components. Hydroforming or tubular hydroforming is basically a process in which a tube is inserted in the closed cavity of a die, which is formed to correspond to the shape of the finished component. The tube is then partially sealed and filled under high pressure with a hydraulic fluid, so that the increasing pressure forces the tube to assume the internal shape of the die. Manufacturing advantages of hydroforming include more economic assembly arising from fewer parts, the absence of welded flanges and hence a saving in weight, and from a design point of view the ability to accommodate changes in cross-section area. An alternative to the steel subframe is one fabricated from aluminium alloy, using for example extruded sections as introduced in 1999 by General Motors for their Chevrolet Impala, or hydroformed tubing that has recently featured in German practice. An aluminium alloy subframe can be advantageous in terms of high rigidity and low weight.

Advantages of integral body construction

As compared with the once conventional separate frame and body for car build, the following advantages are usually attributed to integral body construction in its various forms:

Stiffer structure for the same weight
Improved passenger accommodation
Safer in a collision

More economical to manufacture.

Stiffer structure for the same weight

This derives mainly from the elimination of the body-to-frame flexible mountings and provides the opportunity for increasing all-round car performance, without incurring a weight penalty. In particular the body is less prone to shake, which helps to improve both ride and handling qualities. The greater stiffness also contributes to improved durability of the structure, because of its better resistance to the weakening effects of vibration. From this it follows that there should be less tendency for body squeaks and rattles to develop. Another important consideration is that the greater stiffness or rigidity of the structure minimizes distortion at the body apertures.

Improved passenger accommodation

The desire to achieve this has always been one of the main reasons for a manufacturer adopting integral construction, and it is gained in two ways. First, the structure is inherently better adapted to the established modern styling requirement for a wide body and hence contributes to greater seating space. Second, it enables the space between minimum ground clearance and the roof to be more efficiently utilized, thereby improving headroom despite a low roof line. This is because the absence of chassis frame side members allows a lower floor relative to the body sills, while the simple cross-members that lend stability to the body floor may be buried under the seats.

Safer in a collision

For some years now increasing attention has been directed by road safety research organizations and manufacturers towards what is called the passive safety of a car (Figure 30.3a). This term refers to the better protection against injury for the car occupants once an accident has occurred. In fact international legislation is now in force which, among many other passive safety requirements, seeks effectively to limit the amount of deformation that the passenger compartment of a car may suffer in relatively low-speed accident situations including front, side and rear impacts.

From our knowledge of basic physics we may recall that the second of Sir Isaac Newton's three laws of motion tells us that when a force acts upon a body, it produces an acceleration which is proportional to the magnitude of the force and inversely proportional to the mass of the body, or in perhaps more familiar terms:

$$\text{force} = \text{mass} \times \text{acceleration (or deceleration)}$$

From this it should be evident that it is the rapidity with which the vehicle is stopped that influences the force imposed upon the occupants. Therefore, the very sudden stopping of a vehicle in a crash can accelerate the occupants, with respect to the vehicle, such the force of impact may either injure or kill them. In simulated crash situations human volunteers (suitably restrained by energy-absorbing harness) have withstood peak acceleration forces of just over $20\,g$ or, stated another way, just over 20 times the gravitational force acting on their own body. On the other hand, it is generally recognized that it would be difficult to sustain life with peak accelerations of $60\,g$ or more.

(a)

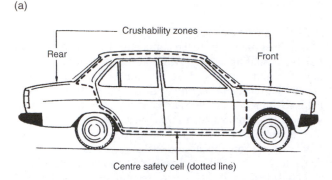

(b)

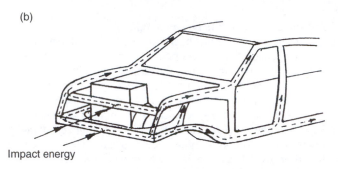

Figure 30.3 Passive safety of integral body construction: (a) crushability zones (b) frontal impact energy dispersion

One method of minimizing the peak deceleration of a crashing vehicle and therefore the acceleration of its occupants is to absorb the energy of impact over a fractionally longer period, by allowing the extremities of the body structure to collapse in a progressive manner. In general terms, this can mean that if a vehicle travelling at a speed in the range of 25 to 45 mile/h (40 to 72 km/h) collides head on with a rigid barrier, then provided that the front end structure of the body can become crushed to a distance of about 0.6 m (2 ft) the peak acceleration suffered by the occupants should not be greater than 40 g.

The greater protection afforded by an integral construction body is achieved mainly by designing the structure so that, unhampered by a separate rigid chassis frame, the front and rear portions of the body may collapse progressively under impact and preserve more nearly intact the middle passenger compartment. Front impact forces are therefore distributed partly through the lower main longerons into the stiffened door sills and floor pan, and partly through the upper rails of the engine compartment into the stiffened A posts and roof panel (Figure 30.3b). Rear impact forces are similarly distributed via the boot compartment into the body structure. A transverse engine installation, especially with in-line cylinders, and a spacious boot compartment both contribute to ample front and rear crush zones. In other words, the front and rear portions of the body are deliberately made more crushable than the occupied middle cell. This particular feature of safety body design, involving controlled crushability zones, was pioneered by Daimler-Benz in the early 1950s. Later research by this company further showed that the most

severe and also most common type of frontal impact occurs when two approaching vehicles collide with an overlap, rather than fully head-on. Since this implies that the forces of impact must be contained mainly by one side of the body structure, it has led to the use of forked longitudinal members, which better distribute the forces of impact into the appropriate load-bearing sections of the rigid passenger compartment and provide increased protection for the footwell (Figure 30.3b).

Other safety features that may be directly applied to the passenger compartment or 'safety cell' as it is sometimes termed, include steel stiffening beams that span the roof cantrails, which in conjunction with reinforced door posts provide better protection in roll-over accidents; a steel stiffening beam that spans the A posts at scuttle height to lend additional support for the steering column, which not only reduces steering-wheel incursion in the event of frontal impact, but also allows effective air bag deployment; and steel stiffening beams installed across the interior of each door, these being termed 'side intrusion bars' and are attached at their ends to reinforced hinges and catches, so that side impacts can be more readily absorbed by the central body structure (Figure 30.4). To ensure a better controlled collapse upon impact for more recent lightened front-end body structures, the forward portions of the longerons and upper rails may be provided with a series of either grooves across their faces or notches along their edges (Figures 30.4 and 30.5). These V-shaped indentations are known as 'fold initiators' and serve to stabilize the axial collapse of the members by encouraging them to deform in a convoluted manner. In low-speed frontal impacts the yielding of these members therefore protects against more serious distortion of the central body structure and has the incidental advantage of reducing accident damage repair costs.

Another established safety feature, which complements the controlled crushability of the modern car, is the collapsible steering column to prevent the driver becoming impaled in a severe accident. To limit the rearward displacement of the steering column into the passenger compartment, some designs of energy-absorbing column incorporate either a convoluted metal section that folds like a Japanese lantern, or plastic pins that shear, when the column is subjected to a telescopic impact load. Other designs of anti-penetration column have deliberately misaligned and universally jointed sections that afford a similar protection for the driver (Figure 23.37).

More economical to manufacture
Although the initial tooling costs of integral body construction can reach staggering proportions, the long-term and necessarily high-volume manufacturing costs are reduced for several reasons. These include the more efficient use of materials (especially the saving that results from the elimination of a separate chassis frame and its associated body mountings), the reduction in the number of parts handled, and the automation of production build techniques.

A more recent design concept for the integral construction body is the pre-assembled modular front end, which can then be assembled to the main body structure as a single unit and therefore reduce assembly time and hence production costs. This system favours an 'open' front end construction, where

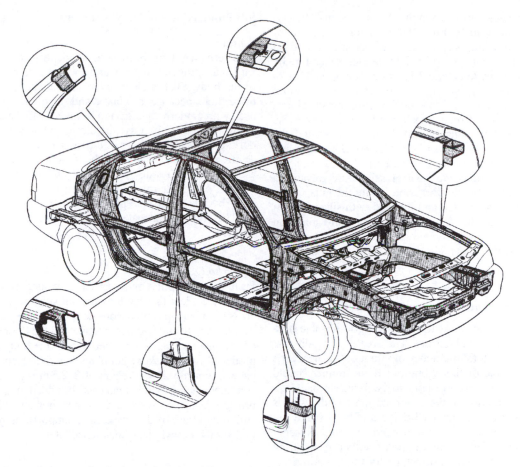

Figure 30.4   Integral construction body sections for stiffness and safety (*Ford*)

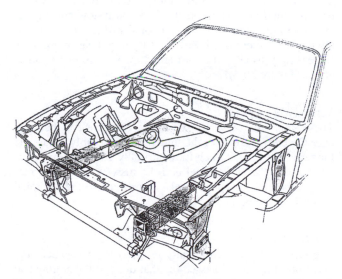

Figure 30.5   Energy absorbing front-end body structure (*Jaguar*)

a transverse impact beam is attached to the lower rails of the engine compartment after the body has been assembled, rather than a 'closed' front end which already has the transverse member welded to it during body assembly.

Less favourable considerations

These are usually concerned with the following aspects of vehicle operation: increase in noise transmission, and earlier deterioration from corrosion.

Increase in noise transmission

The integral construction body tends to be more affected by panel drumming and road rumble noises, as compared with the separate frame and body. Therefore, resort was made early to such devices as the swaging of floor and other body panels, so as to stiffen them and raise their natural frequency of vibration; the application of sound-deadening treatments on panels; and the liberal use of sound-absorbing materials in the offending areas of the body. As mentioned earlier, the semi-integral type of body construction better lends itself to the suppression of noise, vibration and harshness from the body interior.

Earlier deterioration from corrosion

The reason for this is that any corrosive attack on the body panels tends to have a greater weakening effect on the structure as a whole, than would be the case if a separate chassis frame of relatively thicker material was used to support the body. An integral construction body with its panels pressed from sheet metal in thicknesses typically ranging from 1.2 mm (0.048 in) for mild steel down to 0.7 mm (0.028 in) for modern high-strength but also ductile steels, the latter being first used on Japanese cars in the interests of weight saving, is therefore by its nature more prone to deterioration from corrosion. Protection from corrosion involves, in the first instance, the careful design of structural members to avoid wherever possible joints exposed to the entry of road splash and to ensure that those which are exposed are suitably sealed against it. Second, the body structure as a whole must be subjected to selective and thoroughly applied anti-corrosion chemical treatments. In some body constructions the steel pressings are galvanized in areas especially prone to corrosion, while the pillars may, for example, be wax flooded and injected with silicone foam to repel moisture. Another feature protecting against corrosion is the use of anti-chipping screens, which for the Citroën C5 car are used on the underside of the body, sump and rear bumper.

Stiffness requirements of integral construction

An integral construction body must possess considerable stiffness or rigidity for several reasons. The greater its rigidity the higher will be the natural frequency of vibration and hence the lower the amplitude of this vibration. If the former should be too low and unhappily correspond to the frequency of excitation caused by vibration of the unsprung mass of the wheels and suspension, then resonant vibration and body shake can occur. Clearly the body must also be rigid enough to preserve correct alignment of the mounting points either for the front and rear suspension systems direct, or for their intermediate subframes. Similarly, the body shell must be able to resist deflection at its apertures for the doors, so that metal-to-metal clearances and optimum sealing conditions are maintained. The build quality of the modern car is often perceived by what are termed its 'shutlines'. That is, the consistency and size of gap existing between its doors and their surrounding body panels, a gap now in the region of 4 mm (0.15 in) compares to one of about twice this amount acceptable in earlier body build. Figure 30.4 illustrates various sections contributing to stiffness in an integral construction body.

In resisting the various deflections, the integral construction body is subject to a combination of loads that may be summarized as being mainly the following:

Vertical bending loads imposed by the gravitational force acting through the centre of gravity of the car and reacted against at the mounting points for the front and rear suspension systems.

Horizontal bending loads resulting from cornering inertia force acting through the centre of gravity of the car and again reacted against at the mounting points for the front and rear suspension systems.

Torsional loads created by the twisting effect on the body of a single wheel climbing an obstacle and reacted against across the mounting points for the front and rear suspension systems. The maximum torsional load condition would occur if the car were supported on only three of its four wheels.

Stiffening effects of body members

Scuttle

This term has long been used to identify that portion of a car connecting the bonnet and the passenger compartment of the body. Such a clear line of demarcation is now hardly discernible, because the curved and steeply raked windscreen used with modern styling has a lower edge that extends almost to the dash panel. It is, therefore, the latter which really defines the scuttle portion of the modern car. Structurally, the scuttle portion contributes to the torsional stiffness by acting as a forward cross-tie between the underframe and the side frames. A similar role is performed towards the rear of the body by any transverse panel or rear bulkhead that may be used behind the rear seat squab and which spans the rear wing valances.

Underbody assembly

From the description given earlier of the underframe it should be fairly evident that the longitudinal sills, central tunnel and cross-members promote considerable resistance to bending over the length and width respectively of the floor area. For this reason these strengthening members of the body structure are produced from thicker steel plate than that used for the body panels, typically in the region of 2–2.5 mm (0.078–0.097 in). A tunnel bracing member may also be added to the floorpan, to provide increased resistance to side impact collisions. The stiffening effect of the suspension supports, as related to the underbody as a whole, is mentioned later.

Roof bow assembly

In early designs of integral construction the roof bow assembly was generally thicker in section and of greater curvature than is found in modern practice, so that it now contributes rather less to the bending and torsional stiffness of the body, although as mentioned earlier stiffening beams are now increasingly used in the roof structure for safety considerations (Figure 30.4).

Suspension supports

These generally serve to extend the resistance to bending over the entire length of the underbody assembly. Since the longeron members which they normally comprise are integrated with the floor panel and the inswept front and rear wing valances, the supports also contribute to the torsional stiffness of the body.

Bulkhead brace

The term 'bulkhead' originally referred to the vertical flat partition of the scuttle assembly that divided the engine and passenger compartments. In modern practice an often less obvious bulkhead is formed by the complex-shaped pressing that comprises the dash panel. A bulk-head brace serves as a rigid triangulating member, in both side and plan views, between the front longerons and the dash panel where it joins the A posts of the side frames. It may comprise a box-section strut that is disposed diagonally across the sloping inner face of each wing valance, or the bracing effect may be implicit in a sturdy box-like construction for the combination of front wing and valance.

Cantrails

Since the modern styled car has much larger windows or, in the language of the body engineer, greater glass areas, there is less space available for structural members in the upper part of the body than was formerly the case. The cantrails (or roof rails) must therefore be of carefully contrived box sections so that they provide the necessary resistance to bending at the joints between the door pillars and the roof. This bending effect is imposed when, in plan view, the roof tries to rotate relative to the floor as the body is loaded in torsion. That is, the cantrails would attempt to move fore and aft relative to the sills below.

Supporting functions of the body

The attachment of the mechanical units to an integral construction body demands an approach different from that adopted for a separate frame and body, because of the relative difference in metal thickness of the load-bearing members. More specifically, it is necessary to spread the load to be supported over a greater area of the thinner sheet metal used with integral construction. (Figure 30.6). Otherwise problems could arise in service from localized distortion of the sheet metal sections, which at least could introduce misalignment and at worst result in failure of a particular mounting system.

Typical examples of the manner in which the above-mentioned technique is applied are as follows.

Engine and gearbox assembly

With an engine and gearbox unit mounted fore-and-aft, the front engine mounts are conveniently carried by either a simple cross-member, or a more complex subframe, which is primarily used to mount the front suspension mechanism to the body structure (Figures 30.11a and c). In the former case, a shallow partially boxed cross-member can be rigidly attached towards the front ends of the longerons (Figure 30.7a).

Alternatively, a subframe may be flexibly attached to the longerons at more widely spaced locations along them (Figure 30.4). In the latter event, the front extremities of the longerons are integrated with a cross-beam to provide the necessary rigid foundation for mounting the sub-frame. The rear gearbox mount is carried by a short cross-member bolted to the underside of the transmission tunnel, this being locally reinforced for the purpose. Less commonly, the engine and gearbox may be carried in their entirety by a flexibly mounted open subframe that extends over the whole length and breadth of the front longerons.

It has long since become established practice to employ a three-point mounting system for the engine together with gearbox. This superseded the earlier four-point mounting system not only to relieve the engine unit of all strain due to deflections of the chassis frame then used, but also to absorb torque fluctuations more effectively by virtue of greater

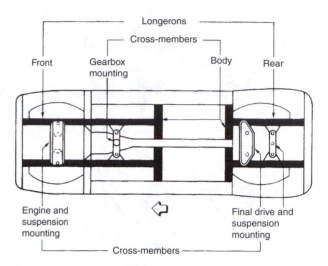

Figure 30.6 Mountings for mechanical units as viewed from underside of integral construction body

mounting flexibility. The torque fluctuations induced by the power impulses from each cylinder are a major source of engine vibration, their effect being most pronounced at idling speeds. They tend to rock the engine unit in opposition to crankshaft rotation, and thus establish the natural axis of oscillation or roll axis of the engine. Hence, it is about this axis that the support mounts must permit the greatest degree of freedom for engine movements. In practice, the roll axis slopes down towards the gearbox end of the engine unit, since the centre of gravity of the engine is higher than that of the gearbox. The actual arrangement of the support mounts may be contrived to produce either a 'centre of gravity' or a 'centre of percussion' system of suspension.

In a typical centre of gravity system, a pair of sandwich mounts supports either the front or, less commonly, the rear of the engine unit. They are arranged in V-formation so that, under the static load of the engine, their projected normals meet on the roll axis (Figure 30.8a). The third mount is of either the cylindrical or the sandwich type, which is arranged to intercept the roll axis at the other end of the engine unit (Figure 30.8b). Hence, the engine is supported about its roll axis which passes through the centres of gravity of both engine and gearbox, so that sideways shake of the engine as it rocks on its mounts is minimized. Furthermore, the main supporting pair of mounts resists the rocking motion of the engine as pure shear loading on their rubbers, thereby providing the most effective isolation of vibrations. To assist in controlling the fore-and-aft movements of the engine unit, either rubber buffer limit stops or a rubber-bushed stay rod may be required, especially where the support mounts are all of the single sandwich type.

With a centre of percussion mounting system, the engine is again supported about its roll axis, but the transverse mounting planes are so disposed that no interaction from disturbing forces occurs between the front and rear mounts. The system is based upon the physical principle that if a

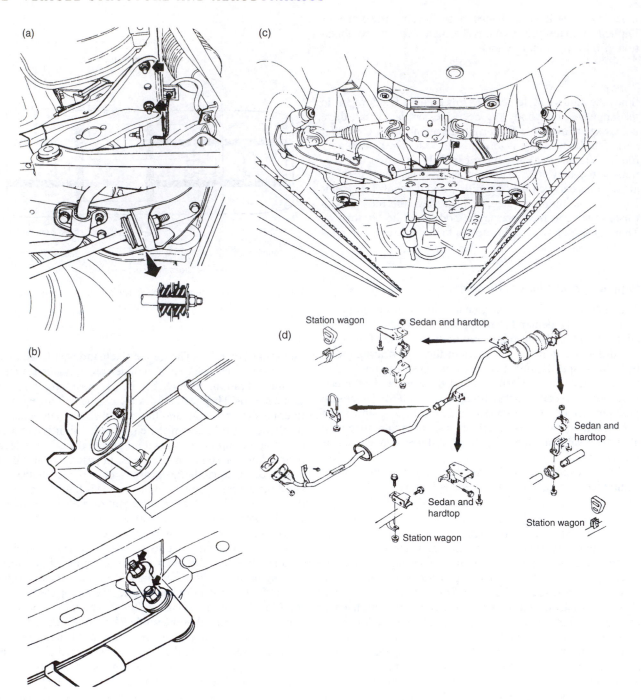

Figure 30.7   Unit mounting methods for integral construction body: (a) cross-member and mounting brackets for engine and front suspension (b) mounting brackets for rear suspension semi-elliptic leaf springs (c) cross-member and mounting brackets for IRS (d) mounting brackets for exhaust system (*Nissan*)

pivoted body is dealt a blow at its centre of percussion, no reaction is produced at the pivot, whereas if the blow is struck at any other point of the body a disturbing force is reacted at the pivot. In other words, the pivot 'doesn't know' when the body is struck at the centre of percussion. As applied to an engine mounting system, the rear mount is thus located at the centre of percussion for the disturbing forces acting upon

the front mounts, and vice versa. This system of mounting is especially applicable to in-line four-cylinder engines, where the vertically acting unbalanced secondary forces may be entirely accommodated by the front support mounts. For this purpose, the latter are located on a horizontal lateral axis that intersects the roll axis in the plane of the unbalanced forces; that is, midway between the cylinders (Figure 30.9). The rear mount is then located at a suitable distance from the centre of percussion for these disturbing forces and is thus not affected by them.

In the case of a transversely mounted engine and gearbox where the support mounts must be able to accommodate the full torque applied to the road wheels, the three mounting points are oriented about the unit as follows. One mount is positioned beneath the unit at the juncture of the bell-housing and engine block and is supported by a detachable third longeron member running across the bottom of the engine compartment to a front cross-beam. A second mount is positioned at cylinder head level to support what would normally be the front of the engine, a pick-up point for this mounting being provided on the reinforced valance structure. The third mount takes the form of either a high-mounted stay rod or, as in later practice, a low-mounted torque arm. To control rocking movements of the engine and gearbox unit, the rubber-bushed ends of the stay rod connect to the rear face of the cylinder head and the upper dash panel structure. The alternative torque arm is rigidly attached at its front end to the bell-housing periphery and at its rear end enters a rubber bushing supported from the lower dash panel structure (Figure 30.10a). With both arrangements the mounting support areas are locally reinforced on the body. For improved vibration control during engine idling, another method of mounting a transverse engine and gearbox is to employ a four-point system. Here the right- and left-hand mounts are placed on the roll axis of the unit and the other two front and rear mounts are inclined to intercept this axis (Figure 30.10b).

Final drive

If the rear wheels are the driven pair and are also independently sprung, it is necessary to provide mounting points for the final drive unit on the car body structure. In conventional practice, however, the final drive unit is supported either wholly, or partly, by the flexibly mounted subframe from which the suspension arms are pivoted (Figure 30.7c). The subframe is attached beneath the forward ends of the rear longerons. When an additional mounting is required for the rear end of the final drive unit, it is usually supported from a stiffening cross-beam integrated with the boot floor and the longerons.

Front suspension

As mentioned previously in connection with engine mounting, the front suspension mechanism is supported from the body structure either directly by means of a rigidly attached and simple cross-member, or indirectly through the medium of a flexibly mounted subframe.

We must now qualify this statement by observing that modern IFS systems of the MacPherson strut and link type require additional supporting features. In particular, a pair of forward anchorage brackets are required where diagonal links are used

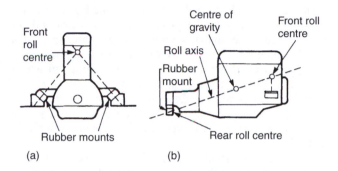

Figure 30.8 Centre of gravity engine mounting system

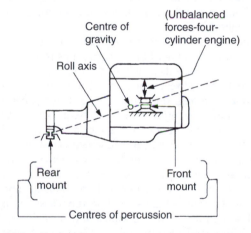

Figure 30.9 Centre of percussion engine mounting system

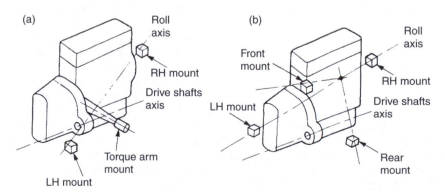

Figure 30.10 Transverse engine mounting systems

to transmit fore-and-aft forces from the track control arms to the body structure (Figure 30.7a), as described in Section 23.2. With this type of system an upper flexible mounting must also be provided for each strut, and, since the suspension spring load is reacted against at the same point, a very rigid and suitably braced MacPherson tower is an essential feature of the wing valance construction (Figure 23.13). Further to increase rigidity and improve car handling by reducing deflections at the strut anchorages, a tower-to-tower cross-brace may be additionally used. The widely spaced attachment points for this type of suspension system do, nevertheless, make a valuable contribution to reducing load concentration on the body front end structure, as indeed was the original design intention.

Rear suspension

In conventional practice the suspension mounting for driven rear wheels, which are independently sprung either by the semi-trailing arm, or by the increasingly popular multilink systems, is effected through the media of a subframe flexibly mounted from the rear longerons (Figures 30.7c and 30.11b). An upper location for each suspension coil spring and shock damper is provided against the upswept portion of the longerons that may be increased in width to accommodate them, or alternatively by the equivalent of MacPherson towers integrated with the rear body structure.

In contrast, the increasingly popular method of suspending non-driven rear wheels by means of a modified strut and link system requires a substantial box-section cross-member from which the transverse links can pivot and against which each suspension coil spring can react if it is separately mounted from the strut (Figure 23.34). Otherwise, MacPherson towers are required in the rear body structure. The cross-member is rigidly attached towards the rear of the body sills, which in this case are extended under the boot floor to furnish a wide base mounting for the rear suspension system (Figure 23.34).

For live axle rear suspension systems the pivot points for flexibly mounting either semi-elliptic leaf springs (Figure 30.7b) or positive locating links are provided by external brackets and internal sleeves integrated with the rear longerons.

Exhaust system

The installation of the flexibly mounted exhaust system to the underframe of an integral construction body varies in detail according to the geographical layout of the system (Figure 30.7d). For a directly routed single system, the front and the rear silencer with tailpipe are supported from mounts that are typically located from rigid body sections towards the centre of the rear seat pan cross-member and the far end of the appropriate rear longeron. With a longer-wheelbase car an additional forward support for the system, taken say from the cross-member under the rear gearbox mount, is also likely to be employed. If any form of dual exhaust system is used then there must, of course, be some duplication of the mounting arrangements. Owing to the high operating temperatures of catalytic converters, their adjacent rubber mountings are silicone based.

Anti-vibration mountings

By the very nature of its rotating, reciprocating and oscillating mechanisms, the running of a car is attended by a certain amount of vibration. Even though such vibration is minimized at source in modern automotive practice, its transmission to the car body interior either simply as vibration or in the form of noise audible to the occupants, or both, is always undesirable. Indeed, its suppression is the declared mission in life of the noise, vibration and harshness (NVH) test engineer, a specialist activity first identified by the Ford Motor Company in the early 1970s when increasing attention began to be given to this problem.

The ideal, but albeit totally unrealistic, method of confining vibration to any particular mechanical unit would be to suspend it in space, so that it could vibrate without causing a disturbance in the car. In reality, the next best thing is to support the unit on flexible anti-vibration mountings in which rubber is generally employed as the spring medium. By using a flexible, as opposed to a rigid, mounting system the vibratory forces of a mechanical unit are reduced to the relatively small spring forces transmitted by the support mountings themselves.

At this point, we should perhaps distinguish between vibration isolation and insulation. For example, the engine rubber mounts not only isolate the car structure from outgoing vibrations of the engine, but also insulate the engine against incoming vibrations from the car structure, so that the overall effect is for vibration of the engine unit to be minimized under all running conditions.

An advantage of using rubber, rather than steel, as the spring medium for anti-vibration mountings is that the sound transmission through them is reduced. Basically this can be attributed to the absence of a metal-to-metal path for the sound to travel along, since rubber in itself is not particularly sound-absorbent owing to its natural frequency being small compared with that of sound. Another advantage is that the internal friction of rubber tends to dampen vibration. However, it is worth reflecting that no matter how much a car may cost, its acceptable functioning depends very much on the few blocks of rubber used to support its mechanical units. The various applications of anti-vibration rubber mountings can be readily identified from the illustrations in Figure 30.11, showing front and rear suspensions systems together with their subframes.

A more recent trend has been towards the use of 'hydroelastic' engine rubber mountings with built-in hydraulic damping, this idea having been first developed by R.E. Rasmussen of General Motors in the early 1960s. In principle, an otherwise conventional conical rubber mountings is internally partitioned to accommodate upper and lower fluid-filled chambers. As the mounting responds to engine movements, the fluid is forced in either direction through an orifice in the partition, thereby exerting a controlled damping force on the spring action of the mounting (Figure 30.12a). This enables a better compromise of high damping at low frequencies and vice versa to be attained, which is especially beneficial in mounting the transverse engines of front-wheel-drive cars.

Since the late 1990s a further refinement in the design of hydraulically damped rubber mountings has been to provide

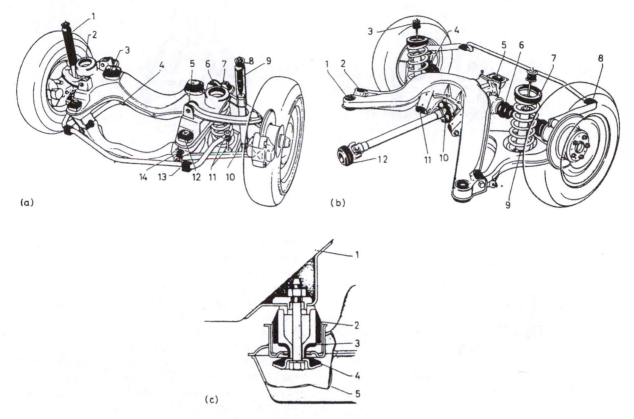

Figure 30.11   Reducing noise, vibration and harshness by the use of rubber parts on axles (*Mercedes-Benz*)

(a) *Front axle*
1 protecting sleeve
2 ring
3 engine mounting
4 bush of the lower lateral wheel fork arm
5/6 bush of the upper transverse wheel fork arm
7 front axle carrier mounting
8 shock damper rings
9 spring washers
10 shock damper mounting
11 pad
12 front axle carrier mounting
13 torsion bar bearing
14 bush of the lower lateral wheel fork arm

(b) *Rear axle*
1 rear axle carrier mounting
2 diagonal wheel fork arm bush
3 shock damper pad
4/6 half-axle gaiters
5 rear axle centre mounting
7 coil spring mounting
8 torsion bar bearing bush
9 shock damper mounting
10 propeller shaft coupling plate
11 diagonal wheel fork arm bush
12 propeller shaft bearing

(c) *Front axle carrier detail*
1 body
2 mounting
3 flange plate
4 bumper pad
5 front axle carrier

them with an adaptive control system. In basic principle this involves the use of a slide valve that is actuated by a linear solenoid, so that an additional orifice between the upper and lower chambers of the mount can either remain covered or be uncovered (Figure 30.12b). Clearly this action will influence the extent of fluid exchange between the two chambers, and hence the degree of damping imposed upon movements of the engine. The solenoid actuator receives its electrical signals via an electronic control system, which recognizes fluctuations in crankshaft rotation as an indication of engine vibration and modifies the damping accordingly.

All the thin metal panels of an integral construction body likely to vibrate or transmit noise are covered with anti-vibration materials of one type or another. They generally take the form of sound-deadening treatments, such as the application under heat of bituminized material direct on to the panels; and sound-absorbing materials in pad form, such as resin-bonded layers of natural and artificial fibres that are cemented to the panels (Figure 30.13). Also, the various box-section stiffening members of the body may be filled with foam products. In all cases the principle is the same: the greater the difference in sound resistance, as defined by the product of sound velocity and material density, between the materials in contact, the more effective is the sound damping.

Checking integral construction body alignment

To verify the accuracy of repair work carried out after accident damage to integral construction car bodies, various

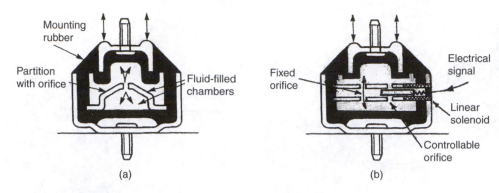

Figure 30.12   Schematic arrangements of hydraulically damped rubber mountings: (a) fixed orifice (b) fixed and controllable orifices

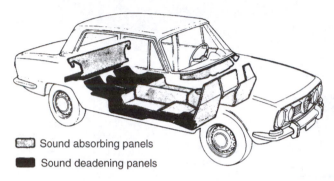

Figure 30.13   Sound-proofing of integral construction body (*Alfa-Romeo*)

forms of alignment checking benches or base panel gauges may be used, which establish the correct location of the mounting points for the suspension and steering assemblies. In the example of body alignment checking bench illustrated (Figure 30.14), all sliding dowels should match freely the respective holes in the body or, at the most, after a slight pressure is applied by hand. The diameter of each dowel is significantly less than the bore of the corresponding hole in the jig, this clearance directly relating to the maximum tolerance allowed for the positioning of the hole in the body.

Body alignment checking equipment must be treated with care, since any loss of accuracy resulting from damage can be reflected in car handling and tyre wear problems.

30.2   ALUMINIUM BODY CONSTRUCTION

General background

The designers of early motor cars soon began to recognize that merely building a more powerful engine was not necessarily the most effective way to improve car performance, if it was to be constrained by excessive weight of the body-work. As early as 1903 we find for example the pioneer French car manufacturer, Panhard et Levassor, offering an optional lighter weight all-aluminium body albeit at an extra cost to the customer. For better quality cars therefore it increasingly became the practice to cover the wooden framework of their bodies with either pressed or hand-beaten,

sheet aluminium panelling. However, this type of lighter weight body construction suffered a decline with the gradual demise of the separate steel chassis frame on which it was built and the widespread adoption of the all-steel integral body construction. There have nevertheless been sporadic attempts to make much more extensive use of aluminium alloy for the structural elements of the car, but with rather less commercial than technical success. This situation may now change with the introduction by Audi in 1994 of the 'Audi Space Frame' (ASF) form of construction for their luxury high-performance A8 model, which is intended to be comparable in terms of both weight and fuel economy to a mid-range saloon of conventional steel construction.

Audi space frame construction

This type of space frame construction should not be confused with that sometimes employed for specialist sports cars, where a web-like structure of tubular steel members is panelled in aluminium, since it more closely resembles a conventional integral body construction. It nevertheless differs from the latter in that instead of utilizing pressed-steel assemblies, an aluminium-silicon alloy is variously deployed in the forms of extruded sections, cast components and sheet panelling. Extruded box-section, straight and curved members comprise the basic structure of the body frame, because their use avoids any loss of rigidity associated with spot-welded seams and they can be produced with variable wall thickness according to the duty they perform, thereby allowing the optimum use of material. Vacuum cast components are employed to confer the required rigidity at the intersecting welded joints for the extruded members, since this method of die-casting in which the die is evacuated before injection of the molten aluminium contributes to a high-strength casting. The sheet panelling for the outer skin of the body is formed from thin-gauge aluminium alloy stampings, these being either riveted or bonded to the frame members.

To enhance passive safety of the ASF body structure in frontal impacts, the longerons beneath the passenger compartment are branched into separate sections, so as to render the entire footwell area extremely resistant to deformation. The forward extending longerons are designed to crumple in specific steps, so that the front end structure can deform

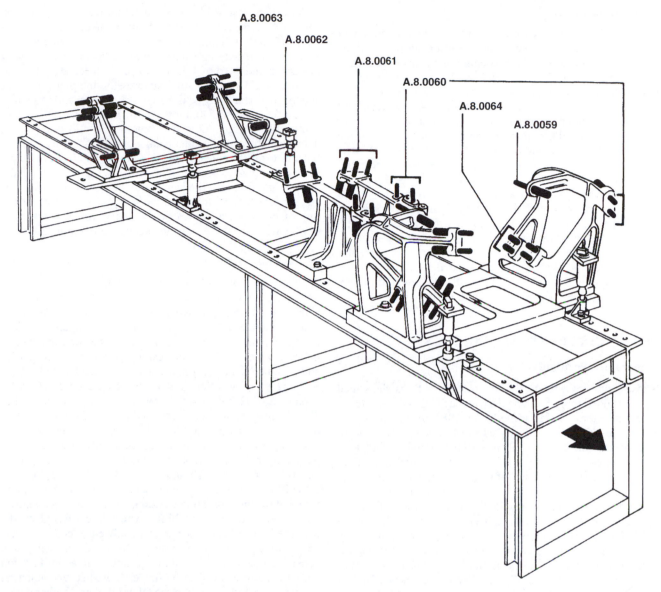

Figure 30.14   An example of a body alignment checking bench: the dowels as fitted are the same on both sides and symmetrical about the longitudinal centreline (*Alfa-Romeo*)

progressively and assist the absorption of energy by the main body structure. Similar considerations have also dictated the design of the rear end structure of the body. To contain sideways impacts the side structures of the body with their broad based pillar-to-sill connections are cross-braced, the sills themselves intersecting generous size crossbars spanning the front and rear of the passenger compartment. Additionally, the doors are provided with reinforcing impact members.

The advantages associated with this new form of aluminium alloy integral body construction may therefore be summarized as follows:

1   It contributes to an overall reduction in the weight of the car, because the engine can be smaller and lighter, the suspension mechanism can be lighter and the fuel tank can be smaller as less fuel is needed to propel a lighter car.

2   The improved power-to-weight ratio of the car allows a better performance for less power output, so that fuel economy and emissions reduction will both benefit.

3   From consideration of 1 and 2 the car will handle better on the road and will also accelerate faster in response to traffic conditions for safer driving.

4   The structural stability of the body helps to reduce vibrations and improves noise control.

5   From consideration of 4 the body also confers better energy absorption thereby enhancing passive safety for the occupants.

6   The greater use of aluminium alloy can be of benefit to the environment, because it has recognized potential in recycling.

Later developments

The 2003 Jaguar luxury car represents a further development in aluminium body construction. It differs from the space frame design in more closely resembling a conventional steel body construction. This is achieved by using pressings of dent-resistant aluminium sheet for the body panels; aluminium extrusions for the roof, doors and safety areas; and aluminium castings where increased strength is required for the door hinge pillars, bolt-on front end, and to mount the engine, transmission and running gear. Other points of interest concern the use of aerospace methods of joining the structural elements. These methods involve using self-piercing rivets that can accommodate up to four layers of material without breaking through the final layer, and structural bonding with epoxy adhesive to provide a stiff connection and a uniform force distribution between the parts to be joined.

From a service point of view any major repairs to an aluminium body are likely to demand specialist treatment, while a generally higher level of skill may also be required for any minor repairs where aluminium welding is involved.

30.3 MULTI-PURPOSE VEHICLES

General background

These represent a relatively new category of vehicle, which has continued to gain in popularity since Chrysler introduced their successful Voyager model in 1983. A multi-purpose vehicle, usually now abbreviated to MPV and earlier known as a 'people carrier', basically gives the impression of a saloon passenger car that has been enlarged both in length and height to accommodate three rows of seats, the seats having a swivelling facility in some designs. There may also be provision to convert seats into tables and for the ready removal of seats to increase cargo space, thus emphasizing the versatility of the MPV. It is typically of box-like form with pronounced sloping of the nose and windscreen, utilizes either a front engine, front-wheel drive or, less commonly, a front engine, rear-wheel drive layout for the mechanical elements, and is intended to ride and handle as nearly as possible like a conventional passenger car. A further development has been the application of a part-time four-wheel drive system to this type of vehicle.

MPV body construction

This generally takes the form of an all-steel integral construction body including sills, which is welded to a ladder-type underframe that extends from the nose to an extreme rear cross-member, so that the whole structure combines strength with lightness and also provides a flat platform for the passenger compartment. The underframe comprises two generously-proportioned longerons which, according to the length of the vehicle, are braced by seven to ten cross-members with their sectional widths increased where appropriate, such as for supporting the engine and transmission units. To complement the front end structure the forward ends of the longerons are supplemented by upper side-members of smaller section, these extending forwards from the A posts at the lower level of the windscreen.

As earlier described for conventional passenger cars, the ends of the longerons and side-members incorporate fold initiators to control deformation in the event of a frontal impact. Front, intermediate and rear stiffening beams are similarly integrated into the roof structure to confer better protection in roll-over accidents. For noise suppression the platform floor and the roof panel may be ribbed to stiffen them and raise their natural frequency of vibration. An MPV may have either conventionally hinged doors or front hinged and rear sliding doors, the object of the latter being not only to facilitate access but also to allow more convenient removal and refitting of the seats according to seating requirements. At the rear of the body a rigid frame is provided to mount the tailgate. The various sections that contribute to the stiffness in the body structure of the modern Nissan Serena MPV, which has a front engine, rear-wheel drive layout, are shown in Figure 30.15.

Sports utility vehicles

Another type of multi-purpose vehicle that has become increasingly popular is the 'sports utility vehicle', usually now abbreviated to SUV, which first appeared on the American market in the early 1990s. It is basically an amalgam of pickup truck and passenger car, but differs from the latter in having a noticeably increased ground clearance, a narrower wheel track and a higher centre of gravity. A further development of the SUV is the Chevrolet Avalanche, introduced by General Motors in 2002, and described by them as an Ultimate Utility Vehicle. In effect it can be converted from an SUV to a pickup truck, simply by removing and stowing the rear window assembly, folding forward the rear seat and lowering what is termed a 'midgate' (similar to a tailgate), thereby creating a generous cargo space. The attractions of an SUV are generally those of providing a more commanding view of the road by virtue of the higher seating position, and possessing a reasonable off-road driving capability, which is conferred by the increased ground clearance and typically enhanced by a four-wheel-drive transmission system. On the debit side safety considerations preclude making abrupt changes in direction and also require speed to be moderated when encountering gusty cross-winds, owing to the taller build and higher centre of gravity of the vehicle.

Structurally, this type of vehicle was originally more closely related to light trucks rather than passenger cars. That is, an often utilitarian body was mounted on a separate ladder chassis frame. This type of construction is now less used than formerly, an integral construction body based on passenger car practice being the preferred option, except where there is a requirement to cater for really arduous off-road driving. An integral body construction generally allows an SUV to be lower in height, while maintaining an acceptable ground clearance, and also to be less heavy. Both of these considerations can prove beneficial to fuel consumption.

30.4 COMMERCIAL VEHICLE CHASSIS FRAMES

General arrangement

The structural foundation of a commercial vehicle lies in its chassis frame, which in long-established practice has taken

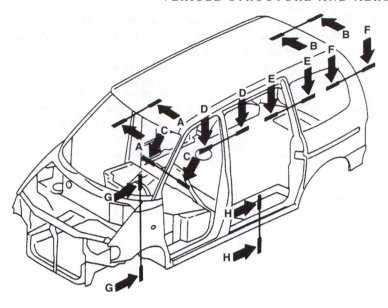

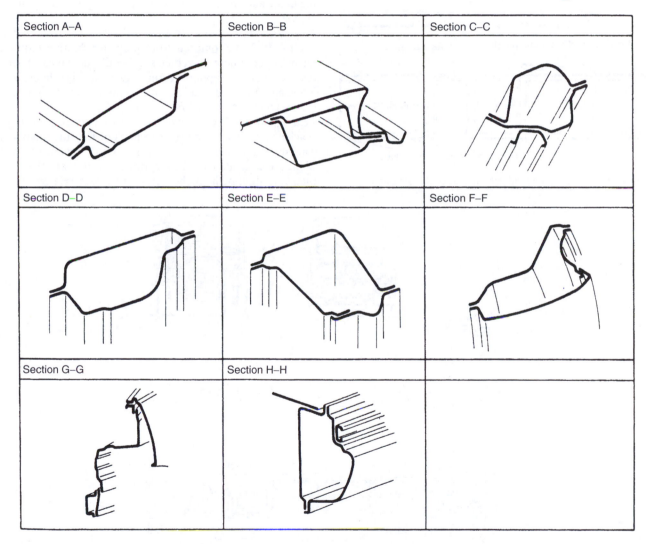

Figure 30.15   Body stiffening sections of Nissan Serena MPV (*Nissan Motor GB*)

an approximately rectangular form resembling a ladder. Indeed, this type of chassis frame is often referred to as a ladder frame, because basically it comprises two side members joined by a series of cross-members (Figure 30.16). Attached to these frame members are numerous mounting brackets for the vehicle mechanism and body (Figure 30.17) their arrangement being typically as follows:

Cab body This is supported from mounting brackets attached to the forward ends of the side members and may in turn incorporate stanchions (upright supports) to secure the radiator. Rubber cushion mountings are generally provided at the cab body mounting points to insulate the cab from vibrations and to minimize any racking strains imposed upon it by the flexibility of the chassis frame. A forward-tilting cab with hand-operated hydraulic tilt mechanism may be used to facilitate maintenance work on the engine unit of some commercial vehicles.

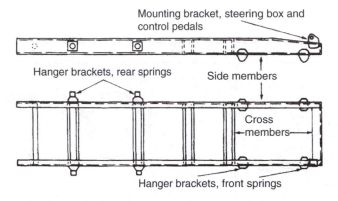

Figure 30.16 A commercial vehicle ladder chassis frame

Steering box An extremely rigid mounting bracket is attached to the front end of the side member which supports the steering box and, in some cases, the control pedals. Rigidity of mounting is usually assisted by the nearby attachment of the front cross-member.

Front springs Extending below the forward portions of the side members are two pairs of hanger brackets from which the front springs and axle are pivoted. To maintain front end alignment and distribute stresses evenly within the frame, the forward cross-members are often positioned to coincide, or nearly so, with the attachment points for the spring hangers. In the case of air sprung front axles, repositioning of the front cross-members together with additional mounting brackets is usually necessary (Figure 23.52).

Power unit This is supported either from the first and second cross-members, or from one of these cross-members and both side members. Suitable attachment brackets are provided at these points for mounting the combined engine, clutch and gearbox unit on rubber cushions. Alternatively, the power unit may be suspended from rubber-bushed links. These various systems of flexible mountings serve to isolate the frame and cab from vibrations caused by the idling diesel engine.

Propeller shaft For reasons that have been explained in Section 19.3, the long length of propeller shafting required on many commercial vehicles generally demands at least one intermediate support bearing. Again, this is rubber cushioned and it is mounted in a bracket underslung from one of the middle cross-members.

Rear springs These pivot from two pairs of outrigger brackets attached to the rear portions of the side members. The rearward cross-members are almost invariably positioned to coincide with these spring mountings and therefore maintain alignment of the sprung rear axle relative to the

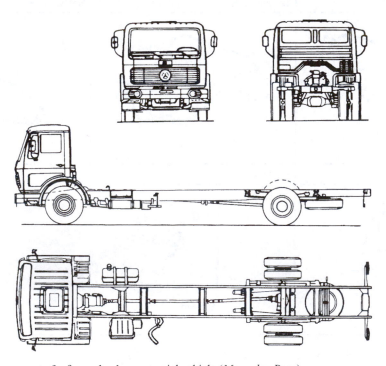

Figure 30.17 General arrangement of a four-wheel commercial vehicle (*Mercedes-Benz*)

chassis frame. In the case of air and rubber springing for rear axles and tandem axle bogies, the repositioning of cross-members together with additional mounting brackets again is usually necessary (Figures 23.40, 23.41 and 23.46).

Spare wheel This may be supported from brackets attached beneath the extreme rear end of the chassis frame, which is usually closed by a final cross-member.

Other services Various other brackets are provided on the frame, especially about its middle portion, chiefly to support the fuel tank on one side and the exhaust system on the other, as well as the reservoirs for the air-operated braking system.

In 1998 Mercedes-Benz introduced a 'two-piece' frame construction in which a common frame-head can be attached to differing lengths of rear ladder frame, according to vehicle application.

Chassis frame construction

So far the commercial vehicle ladder frame has been identified as comprising basically two side members joined by a series of cross-members. It is now necessary to examine this structure in a little more detail, especially with respect to the material sections used, the different methods of joining the frame members, and the means by which they may be reinforced for greater rigidity, as follows.

Material sections
Channel section steel pressings of constant web height are usually employed for the chassis frame side-members. Some examples of their dimensioning for the cold riveted frames used on a Volvo range of trucks are tabulated below:

Although the use of channel section provides adequate resistance to vertical bending loads, it is much less efficient in resisting sideways bending and torsional (twisting) loads (Figure 30.18). However, it does lend itself to the economic production of commercial vehicle frames for the numbers required, and also provides for easy attachment of the cross-members and the various other mounting brackets.

The cross-members must resist torsional deflection of the frame as a whole imposed by rocking movements of the

sprung axles, and also prevent individual twisting of the side members arising from overhung loads such as the fuel tank. Various material sections are used for the cross-members, including I-section beams fabricated from back-to-back channel sections, top-hat and tubular-section beams (Figure 30.19). I-section cross-members are often used towards the front of the frame, because they can better resist bending loads in the horizontal and vertical directions, thereby maintaining front end alignment and supporting the power unit. Top-hat-section cross-members may be used towards the middle of the frame, since they more readily lend themselves to special profiles. These can be required to provide clearance for a component, such as an intermediate bearing for the propeller shaft. Tubular-section cross-members are normally found towards the extreme rear end of the frame, where they offer good resistance against torsional loads.

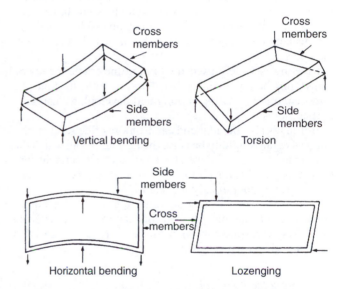

Figure 30.18 Modes of chassis frame deflection (exaggerated for clarity)

Table 30.1 Chassis frame details of Volvo trucks

Model	GVW (kg)	Frame width (mm)	Web height (mm)	Flange width (mm)	Web thickness (mm)	Flange thickness (mm)
FL608 4×2 Rigid	7500	860	220	75	5	5
FL612 4×2 Rigid	12000	862	230	75	6	6
FL618 4×2 Rigid	18000	864	275	90	7	7
FL7 6×4 Rigid	26500	800	275	90	8	10
FL7 8×4 Rigid	34200	800	275	90	8	10
FL10 4×2 Tractor	19500	800	275	85	6	6
FL10 6×4 Tractor	25500	800	275	90	7	7

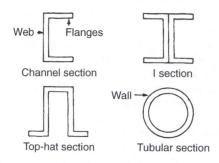

Figure 30.19   Frame member sections

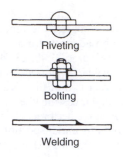

Figure 30.20   Methods of joining the frame members

Joining the frame members

The processes that may be employed to join the side and cross-members are welding, riveting and bolting (Figure 30.20).

Welding gives the most rigid joint, but this method has so far been generally uneconomic for the production numbers involved, especially since machine welding would be required to guarantee the necessary weld uniformity. Welding is therefore usually confined to the fabrication of individual cross-members.

Riveting is widely used for joining the frame members, and it may be done cold rather than hot so as to avoid the rivets contracting during cooling and developing clearances in their holes.

Bolting is also popular and can be more effective than riveting as regards rigidity because the tightness with which the frame members are clamped together can be accurately controlled. A possible disadvantage is that screwed connections may loosen if not properly installed.

A combination of riveting and bolting is sometimes employed; for example, a bolted connection may be confined to the front cross-member so that it can be readily removed to facilitate engine replacement.

Reinforcing the frame joints

To ensure that the inherent rigidity of I- and top-hat-section cross-members is effectively transmitted to the side members, both their mode of attachment and method of reinforcement are given careful attention. Each cross-member is usually attached to the flanges of the side members rather than to the more flexible web. The joints are then reinforced by gusset plates interposed between the top and bottom faces and their mating side member flanges in the case of I-section cross-members, and angled between the front and rear faces and the side member webs where top-hat-section cross-members are used (Figure 30.21).

Apart from reducing stress concentration at the frame joints, these gussets play an important part in resisting sideways bending loads on the frame and also what is known as lozenging of the frame (Figure 30.18). The latter refers to any tendency of the frame side members to move longitudinally relative to each other, which may arise from unequal driving or braking forces being transmitted to the frame.

In long-established practice the chassis frame members of heavy commercial vehicles are produced from carbon-manganese steel pressings.

Chassis cab construction

The cabs mounted on commercial vehicle chassis frames now have to comply with ECE safety standards. In the case of the Volvo trucks mentioned earlier, which also have to meet Swedish safety standards, their cabs are manufactured from sheet steel panels hot-dip galvanized to protect the vehicle from corrosion, whilst the entire cab is treated with an anti-rust compound. Key external panels such as the front grille and step-well area are finished in plastics to protect the cab from road debris. Plastics panels and trim material are labelled and identified to ease recycling.

A cab-over-engine (as opposed to the American cab-behind-engine) chassis layout is generally favoured by European heavy vehicle manufacturers, chiefly because it provides maximum cargo space for a given overall length of vehicle, and also confers better visibility for the driver (Figure 30.17). To facilitate engine maintenance a hydraulically operated tilting mechanism may be fitted to the cab, so that it can be tilted forward about its front anchorage points. According to system design it can accommodate either full over-centre tilting of the cab, or optional partial lifting and full tilting facilities. Such a system basically comprises a pivoting hydraulic ram that is mounted between the chassis frame and the cab, which connects through flexible hoses to a hand-operated, double-acting, pump with integral fluid reservoir. A slow controlled descent of the cab, once it has passed either way over-centre, is imposed by bleed valves in the ram.

Caution   The manufacturers' instructions for its safe operation must be strictly observed.

Checking chassis frame alignment

The time-honoured method of checking chassis frame alignment is by measurement of diagonals, using a plumb-bob and a chalked string. To carry out this check the vehicle should be placed on level ground with a clean surface and the handbrake applied. Initially, the positions of the forward attachment points of the front springs and the rearward attachment points of the rear springs should be transferred to the ground. This is done by chalking an area beneath each position of the plumb-bob and then drawing a cross with a pencil or scriber such that the centre of the cross represents the plumb line. The four crosses on the ground should now accurately represent the extreme corners of the frame, so that if the vehicle is

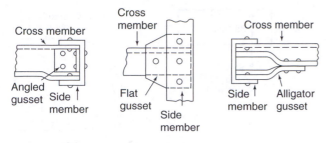

Figure 30.21 Methods of reinforcing the frame joints

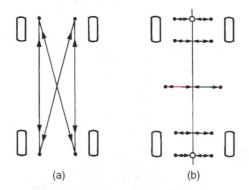

Figure 30.22 Checking chassis frame alignment

moved away the dimensions of the diagonals and lengths of the side members can be determined exactly (Figure 30.22a). If opposite dimensions are within about 6 mm (0.25 in) the frame would generally be regarded as satisfactory, unless the vehicle has received a side impact.

In this case, further points are taken on each side of the frame and plumb lines dropped to the ground. Suitable points are at the rearward attachments of the front springs and the forward attachments of the rear springs, and also two corresponding points on each side of the centre of the frame. There will now be five crosses on the ground on each side of the frame; therefore if the vehicle is moved away again, an exact copy of the essential points of the frame are pictured on the ground. The points at each end of the frame can next be joined by lines and these lines bisected and a line representing the centre of the frame can be marked by stretching a chalked string between them, carefully lifting the centre of the string and allowing it to snap back to the ground so that a white line is marked thereon. From this centreline all the points marked can be checked for correct relative separation (Figure 30.22b), which will indicate whether the frame is bent or not.

Centralized chassis lubrication systems

It is worth recalling that the chassis lubrication requirements of the modern car are exceedingly modest as compared with past practice. For example, in a Ford Popular model of the 1930s one was required to 'lubricate thoroughly with a grease-gun' no fewer than 19 chassis grease nipples every 1000 miles. In the case of larger cars with more elaborate running gear, the demands could be even greater; an example here is the famous Rolls-Royce Silver Ghost of an earlier period, with 25 chassis lubrication points that required attention weekly or every 500 miles. In return for this dedicated attention, one was advised that 'properly lubricated, a Rolls-Royce car will run indefinitely'.

To avoid the chore of hand lubricating the chassis bearings, the original concept of a centralized chassis lubrication system comprising a central pump discharging oil through common supply lines was introduced (perhaps as might be expected) by the American motor industry in the mid 1920s and was first used on Packard cars. Known as the Bijur lubricating system, it was later adopted in Britain, where it continued to find applications on high-grade motor cars, notably Daimler and Rolls-Royce, until the late 1950s. In contrast to these earlier chassis lubrication requirements, the modern car with its rubber-bushed suspension components and 'lubricated-for-life' suspension and steering ball joints generally does not have any chassis lubrication points that require routine attention.

Nowadays, centralized chassis lubrication systems are confined to commercial vehicle applications, which can still feature numerous metal-to-metal bearings throughout their chassis. In a sophisticated modern example of such an installation, up to 60 chassis bearings can be automatically grease lubricated from a central, pneumatically operated, pump assembly. This embodies individual pumping units that feed a multiline distribution system, each bearing being connected to its respective pumping unit by a length of nylon tubing, and receiving precisely metered quantities of grease on an electronically controlled time cycle.

Similar installations are designed to use a transmission grade oil as the chassis lubricant (Figure 30.23).

The automatic chassis lubrication system of a commercial vehicle typically supplies the following points;

Road spring pivot and shackle bearings
Tandem axle balance beam bearings
Front axle king-pins
Track rod ball joints
Drag link ball joints
Power-assisted steering cylinder pivots
Brake camshaft bearings
Brake slack adjusters
Clutch operating shaft bearings
Gear lever linkage
Pedal linkages
Accelerator cross-shaft bearings
Tipping body hinges
Fifth-wheel coupling pivot and jaws.

It must, of course, be added that a centralized chassis lubrication system will only continue to function in an automatic mode, as long as its lubricant reservoir is replenished at intervals specified by the manufacturer. Its pipework and fittings must also be periodically inspected for any accidental damage that could restrict lubricant delivery to any point.

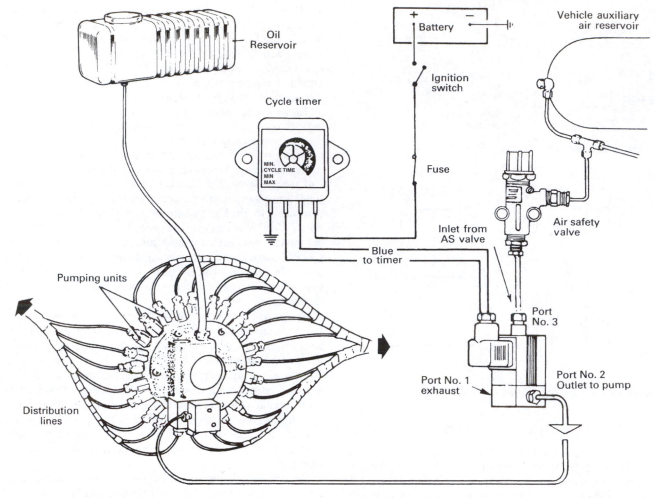

Figure 30.23 Airdromic timed system of automatic chassis lubrication (*Interlube*)

30.5 TRAILER AND CARAVAN COUPLINGS

Trailer couplings

The simplest types of trailer coupling are those used between a four-wheeled, or less commonly six- or eight-wheeled, rigid vehicle towing a drawbar trailer. This type of trailer is identified as having axles at both front and rear, the front axle being able to swivel in the manner of a turntable and connected to the rear of the towing vehicle by means of a cross-pivoted drawbar. The actual forms of coupling used between the rear of the vehicle and the apex of the trailer drawbar are basically the 'pintle hook' and the 'drop pin' (Figure 30.24a and b). Either of these is mounted from the rear cross-member of the vehicle frame and engages the drawbar eye of the trailer. A cushioning rubber sandwich may be incorporated in the coupling mounting, so as to reduce shock transference between the vehicle and trailer. Safe operation of these couplings is provided by swing-down latches for pintle hooks and internal locking for lever retracted drop pins. For more convenient operation of drop pin couplings on heavy vehicles, remote retraction of their pins may be conferred by pneumatic actuators.

More comprehensive forms of coupling are required in the case of articulated vehicles, which comprise a towing vehicle or tractive unit and a semi-trailer. A four- or six-wheeled vehicle may be used as the tractive unit, and the semi-trailer is identified as having from one to three axles at the rear end only. Its front end rests on and is coupled to the tractive unit, which supports 20 per cent or more of the weight of the trailer when it is uniformly loaded. Landing gear legs are provided to support the front end of the trailer when it is uncoupled from the tractive unit.

An early development of the semi-trailer coupling was the ingenious but relatively complicated automatic coupling. This was introduced in the early 1930s by Scammel Lorries Ltd in England for their three-wheeled Mechanical Horse tractive unit and semi-trailer, a combination that remained popular for local haulage over many years. The automatic coupling was later developed for application to more conventional tractive units and semi-trailers, but the limitations placed on its carrying capacity were such that it was generally superseded by the more rugged fifth-wheel type of coupling described later. In brief, the automatic coupling required the rear of the tractive unit to be equipped with a pair of ramps,

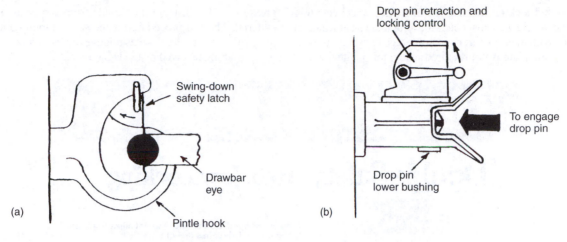

Figure 30.24 Drawbar trailer couplings: (a) pintle hook (b) drop pin

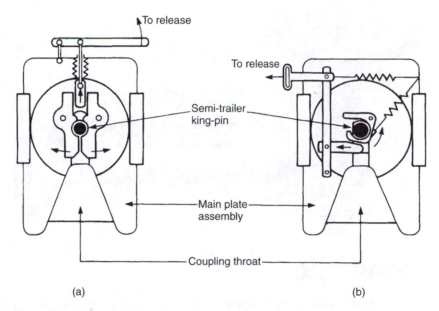

Figure 30.25 Fifth-wheel couplings for semi-trailers: (a) twin jaw (b) hook and wedge

their purpose being to lift the front end of the semi-trailer together with its retractable two-wheeled undercarriage off the ground when the tractive unit was reversed under the trailer. The undercarriage then released and folded away as a pair of flanged rollers mounted on the trailer moved up the ramps until the trailer turntable automatically locked into position on the tractive unit. Weekly greasing of the undercarriage joints was essential to ensure their freedom of movement and safety in operation.

As earlier inferred, the fifth-wheel coupling has long since become established practice for hitching a semi-trailer to its tractive unit. Those parts of the fifth-wheel coupling that are attached to the tractive unit comprise a tilting main plate assembly, which forms the mating lower half of the turntable and is mounted from a support plate that is secured rigidly to the rear of the vehicle frame. The main plate assembly is provided with grease retaining areas on its turntable surface and a V-shaped coupling throat. This terminates in either a twin-jaw or a hook-and-wedge locking mechanism (Figures 30.25a and b). To complete the fifth-wheel coupling a rubbing plate is secured to the underside of the trailer and, together with its central down-pointing king-pin, forms the mating upper half of the turntable.

The coupling process, or locking up as it is sometimes called, is effected when the main plate lock approaches and then grips the king-pin, as the tractive unit is reversed slowly under the front end of the braked semi-trailer. Before this operation is performed the unlocking mechanism of the coupling main plate must be correctly set, which basically involves withdrawing a locking plunger of a twin-jaw lock (Figure 30.26), or the wedge of a hook-and-wedge lock. After coupling, the unlocking mechanism is allowed to spring return to its locked position. In some modern installations of fifth-wheel coupling the unlocking mechanism is pneumatically assisted. It is important that the main plate and locking mechanism of a fifth-wheel coupling are periodically cleaned, inspected for wear and re-lubricated in accordance with manufacturer's instructions. However, lubrication of the main plate is not required where this is fitted with a grease-free, low friction liner made from a special composite material. The full procedure for the safe coupling and uncoupling of a semi-trailer from its tractive unit is, of course, all part of the training received by the HGV driver.

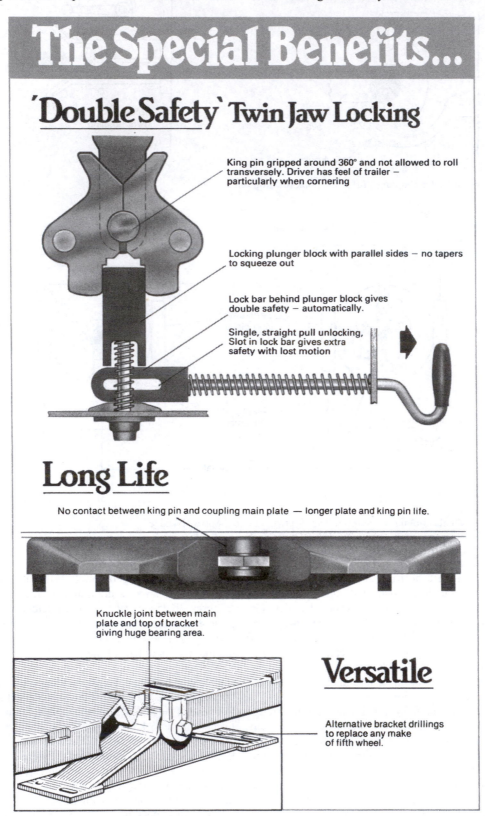

Figure 30.26   Safety features of a modern fifth-wheel coupling (*Davies Magnet*)

Jack-knifing and swing of articulated vehicles

Early opposition to the now widespread adoption of articulated commercial vehicles was based on fears of the tractive unit and semi-trailer jack-knifing, especially when braking on slippery road surfaces. The term *jack-knifing* is used to describe a condition of instability where the semi-trailer pushes the rear of the tractive unit out of line, thereby creating a toggling effect that results in the tractive unit partially rotating and closing on the trailer. Research engineers have long ago established that the most sudden form of jack-knifing occurs with premature locking of the rear brakes on the tractive unit, but that it may similarly occur if the brakes lock prematurely on the trailer and cause it to swing over (Figure 30.27). Jack-knifing can also be provoked by a sudden swerving manoeuvre, even without the brakes being applied.

Apart from ensuring at the design stage that the distribution of braking effort between the axles of articulated vehicles is such as to minimize jack-knifing, a further approach is the fitting of an anti-jack-knife device. A widely accepted type is that manufactured by the British firm of Hope Technical Developments Ltd., which comprises a sealed multidisc clutch unit that is trailer mounted and air pressure operated (Figure 30.28). Its moveable set of discs is splined to a rotatable king-pin, which is also provided with a torque reaction arm that engages with the throat of the fifth-wheel coupling plate on the tractive unit. Therefore when the vehicles brakes are operated, the clutch discs are squeezed together and damp out any dangerous or uncontrolled rotation between the tractive unit and semi-trailer. An important feature of its operation

is that it provides torsional damping, related to the brake application, at the kingpin slightly before braking takes place through the road wheels, so that any would-be tendency to swing is damped out from the start. The early application of this relatively small controlling force does not impair the manoeuvrability of the tractive unit and semi-trailer, so the need to exert a more violent control at later stages is not necessary.

More recently the principles of vehicle dynamics control (Section 29.7) have been applied to tractor and trailer combinations. Such a system is intended to oppose any deviation from correct tracking of the steered vehicle. In simple terms

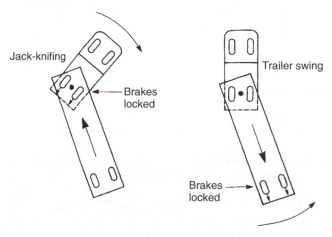

Figure 30.27 Jack-knifing and trailer swing of articulated vehicles

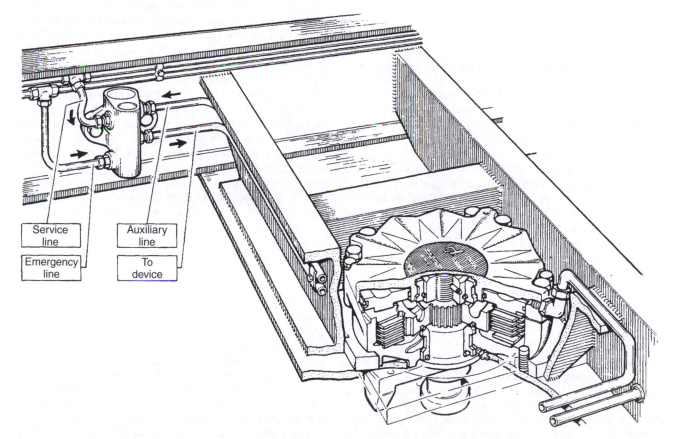

Figure 30.28 Anti-jack-knife device with pneumatic control valve (*Hope Technical Developments*)

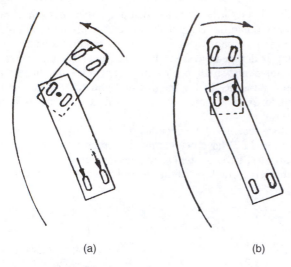

(a)   (b)

Figure 30.29   Correcting oversteer and understeer of articulated vehicles by electronically-controlled selective wheel braking: (a) oversteer – nearside front tractor brake and trailer brakes applied (b) understeer – offside rear tractor brake applied only

if the tractor begins either to oversteer that could lead to a jack-knifing situation, or understeer and run too wide, then the wheel brakes are selectively applied by the electronic control system to correct the deviation. The overall effect is to produce an opposing yawing (or turning) action on the tractor (Figure 30.29) and within reasonable limits of speed restore vehicle stability. This type of system does not interfere with normal braking control.

Caravan couplings

A ball-and-socket type of coupling with a standard towball diameter of 50 mm (2 in) is used for caravan towing by a passenger car. The towball is mounted from a towing bracket that should be specifically designed for the purpose and is typically of T-form, so as to spread the load as widely as possible under the structure of the car. To complete the coupling a socket is formed in what is termed the coupling head, which is usually slidably mounted on the drawbar of the caravan because it also forms part of an overrun braking system. The socket itself comprises an upper cup portion that rests upon the top half of the towball and a lower spring-loaded locking device that engages the underside of the towball (Figure 30.30). A release handle must be lifted before the coupling can be disconnected. Coupling or uncoupling of the caravan is, of course, performed by respectively lowering or raising the coupling head over the towball, care being taken to apply the caravan handbrake before uncoupling on a slope. Maintenance of the coupling head is generally concerned with keeping the inside of the socket housing and its locking mechanism well greased, and applying a grease-gun to nipples that supply the bearings of the slidable drawshaft, typically twice a year.

Various types of stabilizer, employing either friction or hydraulic damping, may also be connected between the car and caravan. Their purpose is to prevent extreme swaying (or snaking as it is known) of the caravan, which could lead to serious instability of both car and caravan. Snaking is usually attributed to excessive towing speed and can be provoked by

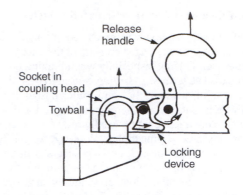

Figure 30.30   Caravan ball-and-socket coupling

cross-wind gusts, wind buffeting from overtaking vehicles, incorrect weight distribution and wrong tyre pressures.

30.6   INTRODUCTION TO VEHICLE AERODYNAMICS

General background

In Section 1.1 it was mentioned that one of the factors opposing the motion of a vehicle was air drag. This assumes increasing importance the faster a vehicle is travelling, because the force opposing motion due to air resistance varies as the square of the speed. Therefore as speeds began to increase, the designers of early cars soon became aware of this energy-consuming factor and gradually tried to avoid sudden changes in body contour, so that the air displaced during the forward movement of the car was disturbed as little as possible. In other words, they began to adopt less box-shaped and more curved body forms that possessed a certain degree of 'streamlining', the purpose of which is to promote a steady airflow with no additional air crossing its path.

The term 'aerodynamic' may be simply defined as the science of air in motion, and as might be expected it first received serious study in connection with early aeroplanes and airships. A notable pioneer in this field was an Austrian engineer, Paul Jaray, who while engaged on airship design in the Germany Zeppelin works during World War I had scientifically developed streamlined forms as a result of both theoretical studies and practical wind tunnel tests on scale models. It was during these investigations that Jaray also conceived the idea of applying the same aerodynamic principles to the design of motor car bodies, whilst recognizing the important difference that a vehicle always moves close to a stationary road surface, which is a complicating factor. Beginning in the early 1920s Jaray designed several fully streamlined car bodies, which sought to deflect the air-flow mainly upwards, as well as rearwards, over the top of the body and then down to the rear end with the minimum of disturbance. Some ten years later a Swiss engineer, Wunibald Kamm, who was researching motor vehicle aerodynamics in Germany found that an extended tail for a streamlined body did not, as previously imagined, maintain a smooth airflow over the entire length of the body, but unhelpfully increased frictional drag of the air because of the larger surface area

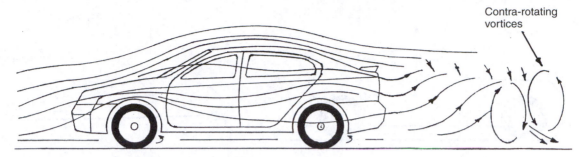

Contra-rotating vortices

Figure 30.31 Schematic airflow around a passenger car

presented by the tail. His requirement was for a streamlined body that tapered moderately towards a relatively blunt end, which coincided with a region where the hitherto smooth airflow would in any event have separated from the body.

However, despite the rigorous application of aerodynamic principles to the motor car body by Jaray, Kamm and other research workers of that era, the seriously streamlined body found only limited acceptance by the motoring public. It did, of course, have its practical disadvantages, which could include restricted interior space, impaired rearward visibility, difficult parking if the rear overhang was increased, reduced accessibility for servicing, and from a purely styling point of view it was rarely pleasing to the eye. Although more restrained forms of body streamlining continued to influence car styling in general, it was not until the energy crisis of the early 1970s that there was a revival of interest in reducing aerodynamic drag to improve fuel consumption rather than car performance. Since then research into vehicle aerodynamics has continued unabated and now constitutes a highly specialized branch of automotive engineering, which has to acknowledge the sometimes conflicting requirements of attractive styling, acceptable handling characteristics in cross-winds, comfortable accommodation and passive safety for the occupants, and economic production.

30.7 BASIC CONSIDERATIONS OF VEHICLE AERODYNAMICS

Reasons for air drag

These are generally considered under the following headings:

Form drag   It will be evident that form drag or profile drag as it is sometimes known, which derives from the basic shape of the vehicle body, represents the main source of aerodynamic drag. A visual impression of the airflow over the streamlined body of a passenger car adapted to modern styling is shown schematically in Figure 30.31. The popularly quoted 'drag coefficient' or 'Cd' value (also known as 'Cx' and 'Cw' values in France and Germany respectively), is used as a measure of how successful a manufacturer has been in producing a low drag body shape. For example, the advanced V6-engined Audi A8 model previously referred to in Section 30.2 is quoted by the manufacturer as having the commendably low Cd of 0.28, which compares to a more usual Cd in excess of 0.30.

The Cd value is non-dimensional and may be expressed as follows:

$$Cd = \frac{D}{0.5\,pAV^2}$$

where D is the aerodynamic drag force (N) measured in a wind tunnel, p is the density of air (kg/m^3), A is the frontal area in terms of the greatest cross-sectional area of the body (m^2) and V is the vehicle speed (km/h).

Hence, it follows that given two cars with the same frontal area and running under identical conditions, the one with the lowest Cd value will generate the least air drag and therefore possesses the aerodynamically more efficient or 'slippery' body shape. This typically displays such features as a gently sloping front end with a pronounced sloping for the rear end in the case of fastback or hatchback bodies, generally well-rounded contours at the junctions between the windscreen, roof panel and side-windows, and a gradual narrowing of the body towards its front and rear ends.

Surface drag   Between the surface of the moving body and the main airflow is a thin boundary layer of air, wherein the air molecules closest to the body surface tend to adhere to it, whilst those closest to the main airflow increase rapidly in speed to catch up with it. The effect is analogous to a pack of playing cards being pushed endwise as described for the lubrication process in Section 4.2. This sets up a shearing action in the air near to the body surface, which creates what is known as 'skin friction'. As a result energy is absorbed from the motive power of the car and dissipated in the form of heat. It will be evident that this type of aerodynamic drag force is chiefly dependent upon the surface area of the body.

Interference drag   Any feature or fitting that interrupts the basic shape of the body, naturally has the effect of disturbing the smooth flow of air over the car. This interference to the airflow arises from superimposed eddy currents and turbulence, the latter being a more irregular swirling of the air, and again this results in energy being absorbed from the motive power of the car. There are perhaps a surprising number of sources that impose interference drag, ranging from such obvious items as number plates, spot lamps and exterior mirrors, to the less obvious ones of door handles, windscreen wipers and window mouldings, which explains of course the modern trend towards flush-fitting of these items. Further to reduce interference drag attempts are also being made to provide a smoother underbody structure for the car,

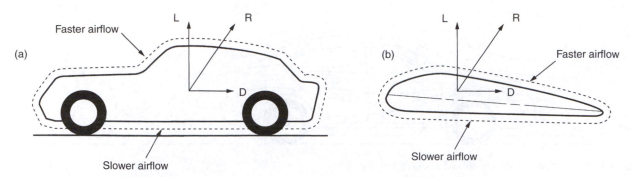

Figure 30.32 Simple comparison between the profile of a car and an aerofoil section for lift induced drag

R resultant aerodynamic force
L lift force
D drag force

together with suitable recessing for the exhaust system and other components wherever this is feasible. In recent practice, smooth plastics panels may be attached to the underside of the body and extend up to the rear suspension arrangements.

Lift induced drag In the streamlined airflow shown schematically for a modern car (Figure 30.31), it will be noticed that the disturbed flow of air downstream or in the 'wake' of the body takes the form of two trailing vortices of swirling air, in which the direction of rotation is upward from the outside of the body and downward to the centre. This pattern indicates that a pressure difference exists between the upper and lower surfaces of the car, which can impose lift induced drag forces at high speeds. It will be evident that, viewed in profile, the air flowing over the top of a car has further to travel than that flowing along its underside and must therefore be travelling faster. This will create a greater reduction in the pressure of the airflow above the car in comparison to that below it, the net result being a lifting effect on the car that resolves into a vertical lift force and a horizontal aerodynamic drag force (Figure 30.32a). In fact we can roughly liken the profile of a car to the aerofoil section of an aircraft wing (Figure 30.32b), but whereas the lifting force on the latter is needless to say very welcome, the opposite is true in the case of the motor car (let alone a Grand Prix car!), because it can cause serious instability at high speeds.

To counteract this aerodynamic lifting effect on modern high-performance cars, laterally mounted air deflectors known as 'airdams' and 'spoilers' may be deployed at the nose and rear deck respectively of the body, so as to reduce the difference in pressure of the air flowing above and below the car, but without significantly adding to the overall aerodynamic drag. The front under-bumper air dam diverts some of the air flow from the slipstream smoothly along the sides of the car, thereby increasing the speed and reducing pressure of the remaining air flow below it, which suppresses lift at the front end of the car. In contrast the rear spoiler slows down a region of fast moving air flow above the car. This not only increases its pressure to inhibit lift at the rear end of the car, but also reduces air turbulence behind it. (Wings of upside down aerofoil section are mounted on Grand Prix cars to achieve a similar but more pronounced effect). A sophisticated application of air dam and spoiler is that found in the Mitsubishi 'Active

Aero System', where an automatically retractable front air dam is used in conjunction with a variable angle rear spoiler. At speeds above 50 mile/h (80 km/h) the air dam is deflected downwards and the rear spoiler is given an increased angle of attack to reduce aerodynamic lift. When the speed falls below 30 mile/h (50 km/h) the air dam is retracted and the rear spoiler returned to its original angle, the overall effect being to maintain constant the Cd value for the car.

Internal flow drag This represents a perhaps less obvious and unwanted contribution to the overall aerodynamic drag imposed on a car. It arises chiefly from the disturbance of the air directed through a stylized intake and passing through the radiator of the engine cooling system, then over the engine and numerous obstructions before escaping via the wing valances and underside of the engine compartment to join the main air flow. Carefully designed encapsulation of the engine compartment has been a recent development, which better manages the air flow emerging below the body. The internal flow drag imposed by the car interior ventilation, heating and air conditioning systems is much less significant. This is because the air passing through them is relatively slow moving, and its entry and exit points have been deliberately sited to take advantage of the air flow over the body (Section 5.6).

30.8 AEROCOUSTICS

It was earlier indicated that a high level of control over noise, vibration and harshness (NVH) has come to be expected of the modern passenger car. This also includes minimizing the effects of wind noise heard in the car interior, especially at the higher speeds of which cars are now capable. The term 'aerocoustics' (borrowed from the aerospace industry) is generally applied to the recognition and the remedies for this particular aspect of NVH. Wind or aerodynamic noise is transmitted as sound waves, which originate from pressure fluctuations in the external air flow. It therefore follows that any features of body design or equipment that interrupt the smooth flow of air over the car are also likely to generate turbulence and wind noise. Some of these sources have already been identified in connection with the previously mentioned interference drag on the body, others include the need for rigid window frames. The importance of maintaining efficient

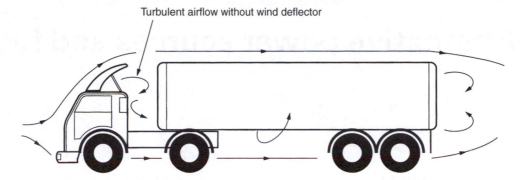

Turbulent airflow without wind deflector

Figure 30.33 Improving the airflow over a heavy vehicle

body sealing cannot, of course, be overemphasized in minimizing the transmission of wind noise to the body interior.

30.9 HEAVY VEHICLE AERODYNAMICS

To conclude this introduction to vehicle aerodynamics mention must be made of its application to heavy vehicles,
because fuel economy is of paramount importance to transport engineers. Clearly the necessarily box-like form of
heavy vehicle bodies and their trailers does not readily lend
itself to aerodynamic treatment, typical Cd values being in
the region of 0.7 or about twice that for a passenger car, but
since the late 1970s research engineers have continued to
investigate means of improving matters. For example, the
leading edges of the cab now tend to have a pronounced
rounding, so that the airflow has less tendency to separate
from the body surface and cause turbulence, which could

otherwise occur with earlier slab-fronted styles of cab. The
body may also have a tapered roof section at the rear.

A familiar bolt-on aid to maintaining a smooth flow of air
over heavy vehicles is the 'wind deflector' mounted above
the cab, which has a curvature designed to transfer with least
disturbance the airflow over the vehicle body or a semi-
trailer (Figure 30.33). In the latter case it helps to prevent
turbulent and therefore energy sapping air currents from
building up and spilling out from the gap between the tractive unit and the semi-trailer, or likewise with a drawbar
trailer outfit. In any event the gap between them should be
no larger than that required for free articulation. This type of
cab-mounted aerodynamic device may form part of what is
termed an 'air management kit', which can include an
under-bumper air dam, side skirts to cover gaps and also
improve stability in cross-winds, and other strategically sited
deflectors according to particular vehicle requirements.

31   Alternative power sources and fuels

31.1   GENERAL BACKGROUND

A student new to the study of modern vehicle technology may be forgiven for assuming that the petrol engine, joined later by the diesel engine, has always reigned supreme during the one hundred years or so history of the motor vehicle. It may therefore be a matter of curiosity to discover that a student taking a City & Guilds of London examination in Motor Car Engineering in say 1909 was expected to answer the following questions: 'Sketch the engine of a steam car with which you are familiar and name the make'; and 'Why are electric cars not more commonly used for country work?'

The present-day, highly refined, petrol-engined passenger car would have been unimaginable to the early motorist, when the future of the petrol engine was by no means assured. It was in fact having to compete with steam and electricity as alternative sources of propulsion. The reciprocating engine for a steam car essentially comprised a boiler, in which steam was generated under pressure by the application of heat to water; together with a steam cylinder and mechanism so that the pressure exerted by the steam was controlled and modified to produce rotary motion at a drive shaft. No clutch or gearbox was required. The latter also applied to the electric car. In this case the principal parts were the batteries, electric motor and a controller.

To put matters into perspective, it is worth recalling what were considered to be the shortcomings of the petrol engine in early cars, as compared to the alternative sources of power. A reporter at the 1899 Automobile Exhibition in Paris listed these shortcomings as uncertain carburation, poor combustion, inefficient cylinder cooling, difficult starting and control of speed, and disagreeable noise and smell. In contrast a motor exhibition held in New York in the same year was entirely devoted to electric vehicles. At that time, the production of electric cars in America was only slightly less than that of steam cars and substantially more than petrol-engined cars. Nor could the performance of record breaking electric and steam cars be underestimated. Again in 1899 the Belgian, Camille Jenatzy, driving an electric car of his own design, was the first man to travel at a speed of over a mile per minute; while in 1906 the American, Fred Marriot, driving a Stanley steam car became the first man to reach a speed in excess of two miles per minute.

At this point we must pause and briefly consider why the petrol-engined car fairly soon superseded its reciprocating steam and battery-electric rivals. In favour of the steam car was its quiet operation, smooth running and useful range of power. However, early steam cars could require some twenty minutes to build up sufficient boiler pressure before the car could move off, and have to stop about every twenty miles or so to take on additional water. The battery-electric car could claim simplicity in construction, smooth running, clean operation and ease of driving. On the debit side, their operating range on fully charged batteries was originally no more than about twenty-five miles at fairly low speeds. Added to this problem was the length of time required to recharge the batteries, which typically involved an over-night charge. Although it would be only fair to add that both steam and electric powered cars were gradually improved, it was the petrol-engined car that showed the greatest potential for further development. In terms of convenience it could travel considerably longer distances on a full tank of petrol, then be quickly refilled. When Henry Ford began to mass-produce his Model T car in 1908, which brought car ownership within the financial reach of millions, the long-term future of the petrol-engined car was assured.

Although there have been many expensive attempts to revive interest in steam driven and battery-electric cars, especially during the 1973 oil crisis, these have not met with any real success. Attaining the required efficiency from a relatively small steam engine poses a considerable problem. In the case of the battery-electric car, the opinion of an early authority on motor vehicles, A. Graham Clark, is still relevant. As long ago as 1914 he wrote: 'The future of the electric car is bound up in the evolution of a light, efficient and cheap battery, and it is on this problem that the keenest intellects have been and are still engaged.' However, the role played by electricity in the more recent developments of hybrid-electric and fuel-cell vehicles will be introduced in Section 32.

31.2   MODERN REQUIREMENT FOR ALTERNATIVE FUELS

Since the oil crisis of the early 1970s research has continued into the feasibility of using alternatives to petroleum based fuels (petrol and diesel). The incentive to do this arises from two main considerations:

1 A long standing recognition that the natural fossil resources from which petroleum based fuels are derived have a finite life.
2 A more recent concern about the atmospheric build-up of carbon dioxide (CO_2) and its contribution to global warming.

The United States Department of Energy definition of an alternative fuel is, therefore, 'Alternative fuels are substantially non-petroleum and yield energy security and environmental benefits.'

31.3 CLASSIFICATION OF ALTERNATIVE FUELS

Alternative fuels for motor vehicles may be listed under the following headings:

Ethanol   Liquefied petroleum gas
Methanol   Hydrogen
Compressed natural gas   Biodiesel
Liquefied natural gas   Electricity

Where an engine has been modified to run on either a conventional or an alternative fuel, it is known as a 'bi-fuel' application. This is distinct from a 'dual-fuel' application, where an engine can run on both conventional and alternative fuels simultaneously.

Ethanol

Also referred to as ethyl alcohol or grain alcohol, ethanol is a liquid fuel derived from corn or other grain, or other agricultural products. It has long been known as an alternative fuel for motor vehicles and possesses a high octane value. It may be of interest to recall that the carburettor of the legendary Ford Model T car could be adjusted so that the engine was able to run on ethanol fuel, which was of benefit to the American farming community who could produce it from fermentation of their crops. In more recent times it has been estimated that more than four million cars run on this fuel in Brazil, where it is fermented from sugar cane. Ethanol is generally well regarded as an automotive fuel, because it promotes greater engine efficiency with lower emissions of atmospheric pollutants. A disadvantage of ethanol as compared to conventional fuels is its lower energy content by volume. In practice a larger size fuel tank is therefore required to compensate for an otherwise reduced mileage range. Ethanol fuel being used in America is a blend of 85 per cent ethanol and 15 per cent unleaded regular petrol (gasoline).

Methanol

This is another alcohol fuel and is also referred to as methyl alcohol or wood alcohol. Although methanol can be derived from a variety of feedstocks, it is now generally produced by a process using natural gas as a feedstock. Its chemical and physical characteristics are similar to those of ethanol, so as might be expected it possesses a high octane value that enhances engine performance. Methanol is also a clean burning fuel that brings benefits in reducing emissions of atmospheric pollutants. In America this property is exploited in 'fuel-flexible' vehicles, which use a blend of 85 per cent methanol and 15 per cent unleaded regular petrol. These vehicles have a sensor in the fuel line that detects the ratio of methanol to petrol in the tank. An on-board electronic control unit then instructs the engine management system to optimize the fuel injection and ignition timing accordingly. As in the case of ethanol, a larger size fuel tank is required to restore vehicle range.

Compressed natural gas

Natural gas is a mixture of hydrocarbons, the main constituent being methane. It is available either from underground or undersea gas wells, or in association with crude oil production. Natural gas will be familiar, of course, as the same fuel used for domestic cooking and heating, and there are abundant supplies in many countries. In fact it has been estimated that over one million vehicles worldwide run on natural gas with Argentina accounting for nearly half of them. The main attraction of using natural gas as a vehicle fuel is its clean burning qualities, which therefore reduce emissions of atmospheric pollutants. It also possesses a high octane value. Natural gas for vehicle applications is generally stored in a compressed state, hence the term 'Compressed Natural Gas' usually abbreviated to CNG. To increase its energy density and therefore save storage space on board vehicles, the gas is compressed to $20 \, MN/m^2$ ($3000 \, lbf/in^2$). For this purpose fuel storage takes the form of purpose designed pressure vessels, which are stronger than conventional fuel tanks and pressure tested to $25 \, MN/m^2$ ($3750 \, lbf/in^2$). In actual operation the natural gas leaving the storage pressure vessels flows, via a master shut-off valve, through a high-pressure fuel regulator. It is then injected at atmospheric pressure into a gas mixer unit, where it is mixed with air ready to enter the engine cylinders and be ignited.

Natural gas as a vehicle fuel may be stored in either a gaseous (CNG) or a liquefied (LNG) state, the basic engine operation being identical in both cases. With LNG the liquid gas is stored cyrogenically at a low temperature of $-162°C$, and requires the use of specially insulated storage vessels. This particular application is therefore better suited to commercial vehicles. An important advantage of running these vehicles on natural gas, whichever system of storage is chosen, is that their engines will be much quieter running than would be the case if they were diesels, thereby making them more compatible with overnight delivery services.

Liquid petroleum gas

Usually referred to simply as 'LPG' liquid petroleum gas mainly consists of propane, which is a product of natural gas processing and petroleum refining. In its natural state propane is a gas, but is turned into a liquid at a moderate pressure of $1120 \, kN/m^2$ ($160 \, lbf/in^2$). When it is drawn from its storage vessel, it changes back into a gas before being burned in the engine cylinders. Propane possesses a high octane value and is more cleanly burning than petrol, although fuel consumption and vehicle performance are generally not quite so good. LPG is used as an alternative fuel in many countries, and especially in bi-fuel conversions for passenger cars and commercial vehicles. However, it is easier and less expensive to convert a petrol-engined vehicle than it is a diesel-engined one. The conversion has to be carried out in accordance with national codes of practice to ensure safe operation. For passenger cars the LPG is stored in a robust doughnut-shaped pressure vessel that is intended to take the place of the spare wheel, which must then be secured in the boot space. In a modern AG Autogas System for petrol engine conversions, the liquid gas flows from the storage vessel and is vaporized in an underbonnet vaporizer/pressure regulator unit. This serves to stabilize the pressure of the gas before it enters a common rail, where it is fed to intake manifold gas injectors in a similar manner to that of the petrol injection system. The whole injection process is controlled by an additional engine management system.

Hydrogen

Discovered in 1766 by the scientist Henry Cavendish, who termed it 'inflammable air', hydrogen is the most abundant, lightest and simplest of the chemical elements. It may be produced either by electrolysis using electrical energy to split water molecules into hydrogen and oxygen, or by synthesis gas production using steam reforming of natural gas. Strictly speaking, hydrogen cannot be considered as a fuel, but rather as a transmitter of energy. For this reason it is playing an important role in the development of fuel cell powered vehicles (Section 32.7), but it can also be utilized as a gaseous fuel by injecting it into the intake manifold of conventional engines. In the latter application it has the immediate attraction that the exhaust gases are free from carbon dioxide, although small amounts of nitrogen oxides, unburned hydrocarbons and carbon monoxide will be emitted from the presence of engine lubrication. In comparison to a petrol fuelled engine, there is some reduction in power and a tendency towards backfiring into the intake manifold. The low density of hydrogen also poses a problem in its on-board storage, because for an equivalent energy content it demands considerably more space than that required for conventional fuels. To counter this problem the hydrogen may be stored either cyrogenically at the extremely low temperature of $-253°C$ when it turns into a liquid, or as a gas at the very high pressure of $30\,MN/m^2$ ($4500\,lbf/in^2$). In both cases there is a weight penalty imposed by the special on-board storage facilities required.

Biodiesel

This fuel is typically blended with standard diesel fuel. It is derived from natural sources such as vegetable oils, which are chemically reacted with an alcohol to produce compounds known as esters. Diesel fuel that results from this process is termed 'biodiesel'. Although this fuel has similar physical properties and offers comparable performance to standard diesel fuel, it demonstrates important advantages in reducing emissions of atmospheric pollutants. These include unburned hydrocarbons, carbon monoxide and particulates.

Another advantage claimed for biodiesel is that it possesses better lubricity than standard diesel fuel. A possible disadvantage of biodiesel is that when used in high-proportion blends or pure, certain types of elastomer that may be used in the vehicle fuel system can be subject to long-term degradation. However, in Britain a modest 5 per cent biodiesel and 95 per cent standard diesel fuel mix is generally favoured.

Electricity

The alternative fuels mentioned so far contain stored chemical energy, which when released through combustion in the engine cylinders provides mechanical power. In contrast the use of electricity as an alternative fuel allows a direct conversion to mechanical power when supplied to an electric motor. Electricity used to power vehicles of the battery-electric type (Section 32.2) is derived from on-board multiple batteries, which act as energy storage devices. These batteries must, of course, be initially charged and subsequently recharged from electricity produced at power stations. This electricity is then transmitted to substations, stepped down to usable lower voltages and distributed for domestic and commercial use. It is therefore from these sources that the batteries of electric vehicles may be recharged, usually overnight when demands on the electrical supply system are least and cost incentives may be offered. The most commendable advantage of operating a battery-electric vehicle is the contribution it makes to cleaner air, because it does not emit any polluting gases. Other advantages relate partly to the lower cost of using electricity as an alternative source of power, and partly to a reduction in maintenance requirements arising from simpler vehicle construction.

To conclude this introduction to vehicle alternative fuels, it must be recognized that accurate comparisons of their relative merits at point of use may not necessarily be all-inclusive. For example, hydrogen is generally regarded as an attractive non-polluting fuel, which can be stored and distributed through pipelines, but it becomes less attractive if the electricity used to produce the hydrogen is generated by burning fossil fuels. In the final analysis the viability of using an alternative fuel relies on it having widespread availability, together with safe and convenient-to-use refuelling or recharging facilities.

32 Battery-electric, hybrid and fuel-cell vehicles

32.1 GENERAL BACKGROUND

The purpose of developing battery-electric, hybrid and fuel-cell sources of power generation is to produce environmentally friendly vehicles. These modern developments are sometimes referred to as 'Clean vehicle technology' and are being vigorously pursued by motor manufacturers. Indeed, at the 10th North American International Auto Show in 1998, John Smith, Chairman of General Motors, was reported as saying 'No car company will be able to survive in the 21st century if it relies solely on internal combustion engines.' However, since clean vehicle technology is in some respects a new and still emerging branch of automotive engineering, the following is intended only as an introduction to the subject.

32.2 BATTERY-ELECTRIC VEHICLES

A battery-electric vehicle or 'BEV' may be defined as one that operates solely on the power provided by on-board rechargeable batteries, which energize an electric motor to drive the wheels. In clean vehicle technology it qualifies as a 'zero emission vehicle', usually abbreviated to 'ZEV', since it emits no pollutants into the atmosphere.

The limitations of early battery-electric passenger cars and the later attempts to revive interest in them, have previously been mentioned (Section 31.1). However, this is not to deny the successful use of battery-electric vehicles for other purposes, such as transporting goods within commercial premises, or urban door-to-door deliveries of milk and bread. In the latter case limitations in their operating range do not represent a serious disadvantage, and for this reason they are sometimes termed 'homing' vehicles.

32.3 LAYOUT OF BATTERY-ELECTRIC VEHICLES

The main units to be accommodated in battery-electric vehicles are a battery pack, electric motor and system controller. Their actual disposition on the vehicle depends on whether it is the front or the rear wheels that are being driven. The former arrangement generally applies to small cars and vans. In this case the electric motor together with a typically 10:1 ratio reduction gear forms a transaxle to drive the wheels (Figure 32.1), similar to the arrangement found with conventional power units. For specialized battery-electric vehicles, such as those used for example in industry, leisure and airports, the electric motor and reduction gear may be integrated with a live rear axle (Figure 32.2). This type of drive is used in the well-known specialist battery-electric vehicles produced by Bradshaw.

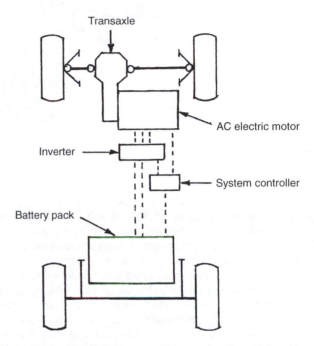

Figure 32.1 Simplified layout of battery-electric vehicle with front-wheel drive

32.4 BASIC UNITS OF BATTERY-ELECTRIC VEHICLES

Battery pack

This unit represents the on-board energy storage system for the vehicle. It comprises an interconnected group of batteries that are treated as a single unit. Since the battery pack is both heavy and bulky it is usually carried low down in the vehicle, typically on underfloor trays.

The three types of battery used in battery-electric vehicles may be categorized as lead-acid, nickel-based and lithium-based. The lead-acid battery will, of course, be most familiar as the ordinary car battery, which is used for starting, ignition, lighting and other duties. For application to the battery-electric vehicle, the lead-acid battery has to be provided with thicker plates to withstand repetitive charge and discharge cycling. These batteries are generally of the maintenance-free type for road vehicles, utilizing a gel rather than a liquid electrolyte that needs periodic topping up. Regarding their performance the lead-acid type yields the least of what is termed 'energy density', which relates battery weight or volume (and hence bulkiness) to the amount of energy stored. Nickel-based

655

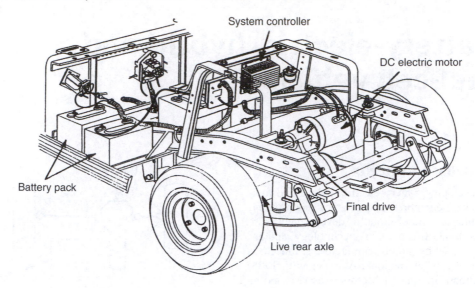

Figure 32.2   Arrangement of battery-electric specialist vehicle (*Bradshaw*)

batteries of the nickel-cadmium and nickel-metal hydride types possess in the region of 50 and 100 per cent greater energy densities respectively, and may also be recharged at a faster rate. Unlike a lead-acid battery that has lead metal and lead dioxide plates, the nickel-cadmium battery has nickel hydroxide and cadmium plates and utilizes an alkali-based electrolyte. Although similar to the nickel-cadmium battery, a special metal alloy replaces the toxic cadmium in the nickel-metal hydride battery. On the debit side, nickel-based batteries are significantly more expensive than lead-acid ones. Finally, lithium-based batteries are being developed for traction applications, following their established use in consumer electrical products such as camcorders. Their chemistry is more complex than that of the previously mentioned types of battery, and a possible disadvantage is the requirement for a battery protection system to prevent overcharging. However, a lithium-based battery can yield about 150 per cent greater energy density than a lead-acid battery and also has a potentially longer life.

Electric motor

The purpose of the electric motor is, of course, to convert electrical energy received from the battery pack into mechanical energy that drives the wheels. Since the current that flows from the battery pack is direct current (DC), it can either be utilized directly to drive a DC motor, or be converted into an alternating current (AC) by what is termed an 'inverter' to drive an AC motor.

Both types of motor, which may have various electrical configurations, can be used to drive electric vehicles. Each has its own advantages and disadvantages, but neither possesses significant advantages over the other. In purely general terms, a DC motor is typically selected for low-speed vehicles, such as specialized carriers; whereas an AC motor now tends to be favoured for on-road vehicles. Technically, a DC motor requires only a simple control system that reduces cost, but it does have a maintenance requirement in respect of its

commutator and brushes, although only after extended service. An AC motor operates with continuous current reversal and therefore does not need a commutator and brushes. This leads to a simpler and lighter construction that only involves a stator and rotor, which accounts for its higher power-to-weight ratio. Although less expensive in itself and requiring virtually no maintenance, an AC motor does demand a more complex and expensive control system.

System controller

Also known as a power converter, the purpose of this unit is to control the flow of electrical energy between the battery pack and the electric motor. It therefore provides the means by which the vehicle speed can be controlled by the driver. Since a battery-electric vehicle must be bi-directional and does not have a reverse gear, the system controller must enable the electric motor to alter its direction of rotation so that the vehicle can be reversed. Another duty performed by the system controller is to provide what is termed 'regenerative braking', which can increase the operating range of the vehicle within the region of 10 to 15 per cent. In effect the electric motor is temporarily made to act as a generator, so that some of the kinetic energy of the slowing vehicle is converted into electrical energy, which can be stored by the battery pack. During this process the motor also imposes a braking effect on the vehicle, independent of its main braking system, thereby reducing brake wear.

Early system controllers for DC motors originally utilized either hand- or foot- operated contactors. In combination with a resistance grid, these selected various circuit arrangements for the battery pack and motor windings, which enabled the speed of the motor and hence that of the vehicle to be controlled in steps. In contrast, a modern system controller is an altogether more complex electronic unit. It is provided with comprehensive feedback signals relating to battery pack and motor operating conditions and can, for instance, limit motor output in the event of overheating. As regards controlling

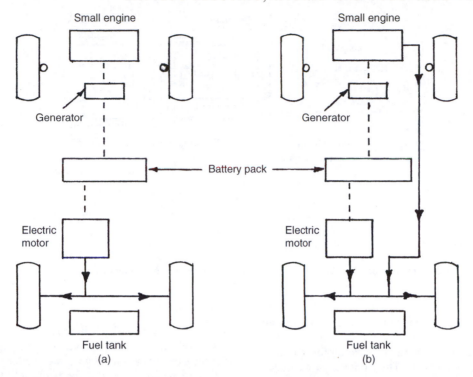

Figure 32.3   Simplified comparison between series and parallel layouts for hybrid-electric vehicles: (a) series (b) parallel

vehicle speed, when the equivalent of a conventional accelerator pedal is depressed, a microprocessor receives an increasing signal voltage, which then commands the control system to increase the available current at the motor, to raise its speed and hence that of the vehicle. Similarly, when a range selector is moved to the reverse position, the control system is commanded to alter the direction of current flow at the motor to reverse the vehicle.

Caution   When servicing battery-electric vehicles it is vitally important to follow the safety precautions recommended by their manufacturers, who normally advise that only specially trained technicians should repair vehicles of this type. Truly it has been said that the battery pack of an electric vehicle should be treated with the same caution and respect as a full fuel tank of a conventionally powered vehicle.

32.5   HYBRID-ELECTRIC VEHICLES

A hybrid-electric vehicle or 'HEV' may be defined as one that combines two sources of power, one of which is electricity. In automotive practice, this means combining the engine and fuel supply of a conventional vehicle with the electric motor and battery pack of an electric vehicle, via a common drive train. The efficiency advantages claimed for hybrid-electric vehicles lie in their ability to offer a better range and performance than battery-electric vehicles, whilst using less fuel and producing fewer emissions than a conventionally powered vehicle.

Although the hybrid-electric vehicle concept received continuous development following the oil crisis of 1973, it was not until the late 1990s that the Japanese companies Toyota and Honda introduced their production Prius and Insight models. The hybrid-electric vehicle actually had much earlier origins, because before World War I the British companies Tilling-Stevens and Daimler both produced what were then known as 'petrol-electric' buses. In the case of the Daimler bus there were two separate petrol engines and electric dynamotors, each pair being coupled together and carried on either side of the bus, providing individual drives to the rear wheels. When the bus was running light the two engines drove the wheels, while the dynamotor acted as a dynamo to charge the batteries. Under heavy load running the batteries automatically supplied electrical energy to the dynamotors, so that they acted as electric motors to increase the power available at the wheels.

32.6   LAYOUT OF HYBRID-ELECTRIC VEHICLES

There are two basic configurations for hybrid-electric vehicles, these being classified as 'series' and 'parallel'.

In a series hybrid layout (Figure 32.3a) a small conventional engine is directly coupled to a generator and cuts in only when required to recharge a battery pack. This supplies energy to an electric motor that drives the wheels, so that when the engine is not running the vehicle operates in zero-emissions mode. When the battery pack state of charge falls below about three-quarters full charge, the engine cuts in so that the generator can recharge the batteries. As the engine is not directly influenced by the performance demands of the vehicle, it can operate within a narrower and more efficient range of speeds, and since it is never required to idle there is a reduction in overall emissions.

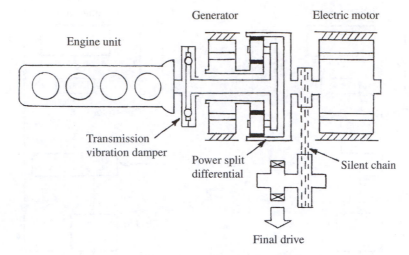

Figure 32.4   Schematic arrangement of a parallel hybrid vehicle with a power split epicyclic differential gearset

In contrast a parallel hybrid layout (Figure 32.3b) provides for either, or both, a small conventional engine and an electric motor to drive the wheels. That is, a mechanical connection is established between not only the electric motor, but also the engine, and the driven wheels. This connection may take the form of an epicyclic differential gear to apportion the power flow from the two energy sources (Figure 32.4). A battery pack is still required, of course, but may be of smaller capacity. Similarly, a medium rather than a heavy-duty electric motor can be used. With a parallel-hybrid the electric motor assists the conventional engine during start-up, acceleration for overtaking and hill climbing. Conversely, at low speed, overrunning down-hill or when standing still, the engine can be switched off altogether. This avoids inefficient operation of the engine and relatively high emissions. As in the case of series hybrid vehicles, the electric motor can also act as a generator when it is being driven by the wheels, thereby providing overrun regenerative braking.

32.7   BASIC UNITS OF HYBRID-ELECTRIC VEHICLES

Engine

The engines used in hybrid-electric vehicles are of either the petrol or the diesel types. For passenger car applications, in-line three- or four-cylinder petrol engines have so far been chosen. These are relatively smaller and lighter than found in conventional cars of equivalent size, because the engine is subject only to average rather than peak loading. The engines are designed for maximum efficiency and embody modern established technical features, such as variable valve timing and direct injection. A transverse engine mounting is adopted as for conventional powered front-wheel drive cars. For commercial vehicle application such as buses, in-line six-cylinder diesel engines are favoured for their high thermal efficiency. These engines feature modern high-pressure, direct-injection technology and are mounted transversely at the rear of the chassis. A fuel tank is, of course, required to supply the engine.

Battery pack

The battery pack, which is smaller than that required for battery-electric vehicles, serves as the energy storage device for the electric motor. High-power battery technologies are therefore being developed for hybrid-electric vehicles. The actual types of battery used are similar to those previously discussed for modern battery-electric vehicles (Section 32.4), nickel-metal hydride and lithium-based batteries currently being favoured. Unlike a battery-electric vehicle, the battery pack of a hybrid-electric vehicle does not require external recharging, because it is maintained in a constant state of charge by output from the engine and by regenerative braking.

Electric motor

The basic units of a hybrid-electric passenger car are necessarily designed to occupy little more underbonnet space than would be the case with a conventionally engined car. A compact and high-torque capacity electric motor is therefore required, which is either wholly or partly responsible for driving the wheels, according to whether a series or a parallel hybrid layout is used. A permanent magnet synchronous type of electric motor is typically employed for this purpose.

System controller

The hybrid-electric vehicle requires an altogether more complex system controller than a battery-electric vehicle. In fact it is the key to the successful operation of this type of vehicle. It is responsible for the electronic control of power flow between the mechanical and electrical elements, so that energy consumption is optimized during all aspects of vehicle operation. The system controller is therefore sometimes referred to as an Integrated Power Module (IPM). More specifically, it exercises control over the power flow into and out of the electric motor, the power output of the engine, the storage of electrical energy and that recovered from regenerative braking, and engine starting procedure. The size, weight and heat dissipation of the module must all be optimized and its reliability demonstrated in extended service.

32.8 FUEL-CELL VEHICLES

Although the purpose of this book has been to concentrate on present-day vehicle technology, it nevertheless seems appropriate at least to introduce the concept of the fuel-cell vehicle. Also known as a fuel-cell electric vehicle or 'FCEV', this type of vehicle is now attracting an increasing amount of public attention.

The basic idea of a fuel cell that directly converts the chemical energy of a fuel into electrical energy was discovered in 1838 by a British lawyer turned scientist called William Grove, who described his invention as a 'gas battery'. Other scientists continued to research the fuel cell, notably Friedrich Otwald, who in 1893 provided the first convincing theoretical explanation of its working. During World War II Thomas Bacon developed a fuel cell for use in Royal Navy submarines, which was later pursued by the Pratt & Whitney company in America to provide capsule power for the Apollo spacecraft. Since the late 1990s major motor manufacturers in America, Europe and Japan have been investigating the fuel cell as a potential power source for vehicles.

32.9 OPERATION AND TYPES OF FUEL CELL

A fuel cell is essentially another form of battery, which, so long as it receives an external supply of fuel, will produce a continuous flow of electrical energy to generate power. Therefore unlike the operation of a storage battery, a fuel cell will never run down or need recharging. In other words, a fuel cell only produces electricity while fuel is being supplied to it.

In practice a fuel cell basically comprises two electrodes with an electrolyte contained between them. One electrode is positive and the other one negative, these being termed the cathode and anode respectively as in any other electrolytic cell.

The function of the electrolyte is to transport the electrically charged particles from one electrode to the other, the reactions at the electrodes being accelerated by catalysts. In simple terms, hydrogen fuel is supplied to the anode of the fuel cell and oxygen from the air enters through the cathode. At the anode the hydrogen atoms are split by the catalyst into protons and electrons, that is positive and negative charged particles, which then pursue different paths. The hydrogen protons travel through the electrolyte to the cathode. Simultaneously, the hydrogen electrons create a separate circuit to provide usable electrical energy, as they too travel to the cathode. Here the oxygen and the hydrogen protons and electrons combine on a catalyst simply to form water (Figure 32.5). To increase the output of electrical energy sufficient to drive a vehicle, fuel cells are combined into groups known as 'stacks'. In a General Motors prototype passenger car a stack of 200 fuel cells is used.

There are several types of fuel cell under development, these being classified according to the kind of electrolyte they use. This in turn determines the chemical reactions that occur in the cell, its catalyst specification, fuel required, and cell operating temperature. Each type of cell naturally has its own advantages and disadvantages, but for motor vehicle applications the 'Polymer Electrolyte Membrane' or 'Proton Exchange Membrane', usually abbreviated to PEM, type of fuel cell would seem to be regarded as the most suitable. This choice is based on its favourable power-to-weight ratio and a relatively low operating temperature of about 80°C, the latter reducing warm-up time for quicker starting. In construction the PEM fuel cell employs a solid polymer as an electrolyte, thereby avoiding the use of corrosive fluids, and its porous carbon electrodes contain a platinum catalyst. The fuel required is pure hydrogen that poses certain onboard storage and supply problems, as previously mentioned in Section 31.4. Alternatively, other fuels may be used but these require the

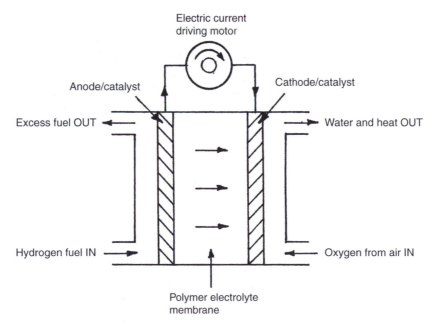

Figure 32.5 Simplified layout of a fuel-cell vehicle with front-wheel drive

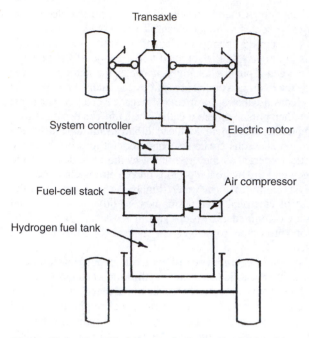

Transaxle

System controller

Electric motor

Fuel-cell stack

Air compressor

Hydrogen fuel tank

Figure 32.6   Simplified layout of a fuel-cell vehicle with front-wheel drive

addition of an onboard reformer to release the hydrogen contained in them.

32.10   LAYOUT OF FUEL-CELL VEHICLES

So far fuel-cell vehicles exist mainly in prototype forms (Figure 32.6). These are being evaluated for performance and reliability in controlled environments selected by their manufacturers, such as for airport transportation and transit bus operation. For passenger car fuel-cell application, it would seem likely that some departure from the present-day conventional layout may ultimately be required, as evidenced by the recent General Motors 'AUTOnomy' prototype vehicle. This features a 'skateboard' chassis construction, which not only allows interchangeable body styles, but also makes provision for the electric motor drive, fuel-cell stack, hydrogen fuel tanks and drive-by-wire electronic controls to be accommodated low down in the chassis framework, thereby lowering the vehicle centre of gravity for improved ride, handling and stability.

Index